# Zebrafish Models of Neurodegenerative Disorders

Advancing our understanding of neurodegenerative diseases demands innovative tools and approaches. *Zebrafish Models in Neurodegenerative Disorders* offers a comprehensive examination of zebrafish as a translational model for studying the molecular and cellular mechanisms of diseases such as Alzheimer's, Parkinson's, and Huntington's diseases. Beginning with the first section informing about the basic molecular mechanisms of neurodegenerative disorders, biomarkers, and therapeutic approaches, the book continues with the detailed exploration of zebrafish physiology, genetics, and behavioural phenotypes as they pertain to neurodegenerative pathologies; insight into cutting-edge technologies, including CRISPR/Cas9 genome editing; and in-depth analysis of zebrafish models, highlighting their strengths in modelling disease progression and therapeutic testing.

# Zebrafish Models of Neurodegenerative Disorders

Edited by
Ebru Emekli-Alturfan

CRC Press
Taylor & Francis Group
Boca Raton London New York

CRC Press is an imprint of the
Taylor & Francis Group, an **informa** business

Designed cover image: Provided by Editor

First edition published 2026
by CRC Press
4 Park Square, Milton Park, Abingdon, Oxon, OX14 4RN

and by CRC Press
2385 NW Executive Center Drive, Suite 320, Boca Raton FL 33431

*CRC Press is an imprint of Taylor & Francis Group, LLC*



*British Library Cataloguing-in-Publication Data*
A catalogue record for this book is available from the British Library

ISBN: 978-1-032-66046-2 (hbk)
ISBN: 978-1-032-67654-8 (pbk)
ISBN: 978-1-032-67655-5 (ebk)

DOI: 10.1201/9781032676555

Typeset in Times
By Deanta Global Publishing Services, Chennai, India

# Contents

# *Preface*

The significant socioeconomic burden and the detrimental effects on quality of life caused by neurodegenerative disorders such as Alzheimer's, Parkinson's, and Huntington's diseases represent one of the most urgent challenges in modern medicine. Despite decades of ongoing research, our understanding of these diseases remains incomplete, and effective treatments remain elusive. Model organisms are essential for the development of innovative research approaches to address these challenges.

Among the various tools available to scientists, the zebrafish (*Danio rerio*) as a model organism has emerged as a powerful and versatile model for studying the complex mechanisms of neurodegeneration. The zebrafish offers unique advantages, such as genetic and physiological similarities to humans, optical transparency for live imaging, and suitability for large-scale genetic and drug screening. These features make the zebrafish an invaluable model for elucidating the mechanisms of diseases, especially neurodegeneration-based diseases, and identifying potential treatments.

This book is designed to fill the gap in the use of zebrafish in modeling neurodegenerative diseases. By compiling the latest developments, methodologies, and studies on the subject, this book aims to serve as a comprehensive resource for both experienced researchers and those new to the field. From understanding the basic molecular mechanisms of neurodegeneration and current therapeutic approaches to exploring translational applications, the chapters here cover a wide range of topics designed to illuminate the utility of zebrafish models in this critical area of research.

I am deeply grateful to all the distinguished authors of this book for contributing their expertise on the subject and to the broader scientific community for their continued efforts to advance our knowledge of the molecular mechanisms of neurodegeneration. I hope that this book will inspire collaboration, innovation, and progress toward understanding and ultimately treating neurodegenerative diseases.

**Ebru Emekli-Alturfan**

## *About the Editor*

**Ebru Emekli-Alturfan** graduated with a Doctor of Veterinary Medicine (DVM) degree from Istanbul University in 2000. She then pursued advanced studies in biochemistry, earning a Ph.D. in the field in 2006, and was appointed as a professor in the Faculty of Dentistry at Marmara University, specifically within the Department of Basic Medical Sciences in 2016. Ebru Emekli-Alturfan leads a research group to investigate molecular mechanisms of neurodegeneration using zebrafish a model organism. Their research encompasses neurodegeneration, neurodevelopment, and its molecular regulation in vertebrate development. To investigate the moleculer foundation of human diseases, they work together with colleagues both within the Medical and Dental School and internationally to develop and implement in *vivo models*. Besides leading a research group, Ebru Emekli-Alturfan teaches medical biochemistry courses at Marmara University's Faculty of Dentistry and Institute of Health Sciences.

# Contributors

**Palak Ahuja**
Department of Pharmacology
ISF College of Pharmacy affiliated to I.K Gujral Punjab Technical University
Jalandhar, Punjab, India

**Efruz İrem Akkuş-Yaylamış**
Department of Medical Biology
Istanbul University-Cerrahpaşa
Istanbul, Türkiye

**Merih Beler**
Institute of Health Sciences
Department of Biochemistry
Marmara University
Istanbul, Türkiye

**Sinem Bireller**
Department of Biochemistry
Acibadem Mehmet Ali Aydinlar University
Istanbul, Türkiye

**Bedia Cakmakoglu**
Department of Molecular Medicine
Aziz Sancar Institute of Experimental Medicine
Istanbul University
Istanbul, Türkiye

**Derya Cansiz**
Department of Medical Biochemistry
Istanbul Medipol University
Istanbul, Türkiye

**Müjgan Cengiz**
Department of Medical Biology
Istanbul University-Cerrahpaşa
Istanbul, Türkiye

**Barış Çerçi**
Medical Biology, Faculty of Medicine
Hacettepe University
Ankara, Türkiye

**Yekta Çulpan**
Department of Medical Pharmacology
Marmara University School of Medicine
Istanbul, Türkiye

**Pervin Rukiye Dinçer**
Medical Biology, Faculty of Medicine
Hacettepe University
Ankara, Türkiye

**Erdinç Dursun**
Department of Neuroscience
Institute of Neurological Sciences
Istanbul University-Cerrahpaşa
Istanbul, Türkiye

**Gizem Eğilmezer**
Institute of Health Sciences
Department of Biochemistry
Marmara University
Istanbul, Türkiye

**Baris Ertugrul**
Department of Molecular Medicine
Aziz Sancar Institute of Experimental Medicine
Istanbul University
Istanbul, Türkiye

**Luis Félix**
Centre for the Research and Technology of Agro-Environmental and Biological Sciences (CITAB)
University of Trás-os-Montes and Alto Douro (UTAD)
Vila Real, Portugal
and
Inov4Agro, Institute for Innovation Capacity Building and Sustainability of Agri-Food Production
Vila Real, Portugal

**Stefan H. Fuss**
Department of Molecular Biology and Genetics
Bogazici University
Istanbul, Turkey

**Duygu Gezen-Ak**
Department of Neuroscience
Institute of Neurological Sciences
Istanbul University-Cerrahpaşa
Istanbul, Turkiye

**Ana Gomes**
Centre for the Research and Technology of Agro-Environmental and Biological Sciences (CITAB)
University of Trás-os-Montes and Alto Douro (UTAD)
Vila Real, Portugal

**Rezzan Gülhan**
Department of Medical Pharmacology
Marmara University School of Medicine
Istanbul, Türkiye

**Samudra Gupta**
S.N. Pradhan Centre for Neurosciences
University of Calcutta
Kolkata, India

**Subhra Prakash Hui**
S.N. Pradhan Centre for Neurosciences
University of Calcutta
Kolkata, India

**Semanur Işıkoğlu**
Institute of Health Sciences
Department of Biochemistry
Marmara University
Istanbul, Türkiye

**Emma Ivantsova**
Department of Physiological Sciences and Center for Environmental and Human Toxicology
College of Veterinary Medicine
University of Florida
Gainesville, Florida, United States

**Atakan Karagöz**
Vocational School of Health Sciences
Feenrbahçe University
Istanbul, Türkiye
and
Institute of Health Sciences
Department of Biochemistry
Marmara University
Istanbul, Türkiye

**Goksu Kasarci Kavsara**
Department of Molecular Medicine
Aziz Sancar Institute of Experimental Medicine
Istanbul University
Istanbul, Türkiye
and
Institute of Graduate Studies in Health Sciences
Istanbul University
Istanbul, Türkiye

**Caghan Kizil**
Department of Neurology
Vagelos College of Physicians and Surgeons
Columbia University Irving Medical Center
New York, United States
and
Taub Institute for Research on Alzheimer's Disease and the Aging Brain
Vagelos College of Physicians and Surgeons
Columbia University Irving Medical Center
New York, United States

**Yigit Kocagöz**
Department of Molecular Biology and Genetics
Kadir Has University
Istanbul, Turkey

**Christopher J. Martyniuk**
Department of Physiological Sciences and Center for Environmental and Human Toxicology
College of Veterinary Medicine
University of Florida
Gainesville, Florida, United States

**Zülal Mızrak**
Institute of Health Sciences
Department of Biochemistry
Marmara University
Istanbul, Türkiye

**Sandra M. Monteiro**
Centre for the Research and Technology of Agro-Environmental and Biological Sciences (CITAB)
University of Trás-os-Montes and Alto Douro (UTAD)
Vila Real, Portugal
and
Inov4Agro, Institute for Innovation Capacity Building and Sustainability of Agri-Food Production
Vila Real, Portugal

**A. Begüm Özel-Korlu**
Faculty of Medicine
Department of Biochemistry
Istanbul Health and Technology University
Istanbul, Türkiye
and
Institute of Health Sciences
Department of Biochemistry
Marmara University
Istanbul, Türkiye

**Elif Ozkok**
Department of Neuroscience
Aziz Sancar Institute of Experimental Medicine
Istanbul University
Istanbul, Türkiye

**Nitasha Rana**
Department of Pharmacology
ISF College of Pharmacy
I.K Gujral Punjab Technical University
Jalandhar, Punjab, India

**Aoishik Roy**
S.N. Pradhan Centre for Neurosciences
University of Calcutta
Kolkata, India

**Arti Singh**
Department of Pharmacology
ISF College of Pharmacy
Jalandhar, Punjab, India
and
Department of Pharmaceutical Sciences
School of Health Sciences & Technology
UPES Dehradun, Uttarakhand

**Charan Singh**
Department of Pharmaceutics
School of Pharmacy
Hemvati Nandan Bahuguna Garhwal University (HNBGU)
Srinagar, Garhwal, Uttarakhand, India

**Mustafa Gani Sürmen**
Department of Molecular Medicine
Hamidiye Institute of Health Sciences
University of Health Sciences
İstanbul, Türkiye

**Saime Sürmen**
Department of Molecular Medicine
Hamidiye Institute of Health Sciences
University of Health Sciences
İstanbul, Türkiye

**Nurdan Tekin**
Department of Medical Pharmacology
Marmara University School of Medicine
Istanbul, Türkiye

**Tuğba Tunalı-Akbay**
Department of Basic Medical Sciences
Faculty of Dentistry
Marmara University
Istanbul, Türkiye

**Fulya Yaylacıoğlu Tuncay**
Medical Biology
Gulhane Medical Faculty
University of Health Sciences
Ankara, Türkiye

**İsmail Ünal**
Department of Medical Biochemistry
School of Medicine
Istanbul Medipol University
Istanbul, Türkiye

**Carlos Venâncio**
Centre for the Research and Technology of Agro-Environmental and Biological Sciences (CITAB)
University of Trás-os-Montes and Alto Douro (UTAD)
Vila Real, Portugal
and
Inov4Agro, Institute for Innovation Capacity Building and Sustainability of Agri-Food Production
Vila Real, Portugal
and
Animal and Veterinary Research Centre (CECAV)
University of Trás-os-Montes and Alto Douro
Vila Real, Portugal
and
Department of Animal Science
School of Agricultural and Veterinary Sciences
University of Trás-os-Montes and Alto Douro
Vila Real, Portugal

**Ayşen Yarat**
Department of Basic Medical Sciences
Faculty of Dentistry
Marmara University
Istanbul, Türkiye

# Section I

# Molecular Mechanisms, Markers, and Pharmacotherapy of Neurodegenerative Disorders

# 1

## *Main Molecular Mechanisms of Neurodegenerative Disorders*

**Erdinç Dursun and Duygu Gezen-Ak**

### Abbreviations

| | |
|---|---|
| **6-OHDA** | 6-Hydroxydopamine |
| **AA** | Arachidonic Acid |
| **AADC** | L-Aromatic Amino Acid Decarboxylase |
| **ABAD** | Alcohol Dehydrogenase Enzyme |
| **AD** | Alzheimer's Disease |
| **AICD** | Beta-Amyloid Precursor Protein Intracellular Domain |
| **ALS** | Amyotrophic Lateral Sclerosis |
| **APP** | Amyloid-Beta Precursor Protein |
| **α-syn** | Alpha-Synuclein |
| **Aβ** | Amyloid Beta |
| **BBB** | Blood-Brain Barrier |
| **CaM** | Calmodulin |
| **CBD** | Corticobasal Degeneration |
| **CDK5** | Cyclin-Dependent Kinase 5 |
| **CJD** | Creutzfeldt-Jakob Disease |
| **CSF** | Cerebrospinal Fluid |
| **DHA** | Docosahexaenoic Acid |
| **DLB** | Dementia with Lewy Bodies |
| **DST** | Disposable Soma Theory |
| **EGFR** | Epidermal Growth Factor Receptor |
| **FAD** | Familial Alzheimer's Disease |
| **FTD** | Frontotemporal Dementia |
| **FTLD** | Frontotemporal Lobar Degeneration |
| **FUS /FET** | Fused-in Sarcoma |
| **GC** | Glial Cytoplasmic Inclusions |
| **GM** | Gangliosides |
| **GPCR** | G-Protein Coupled Receptors |
| **GSK3β** | Glycogen Synthase Kinase 3β |
| **HD** | Huntington's Disease |
| **htt** | Huntingtin |
| **ICAM1** | Intercellular Adhesion Molecule 1 |
| **LB** | Lewy Bodies |
| **L-DOPA** | L-3,4-Dihydroxyphenylalanine |
| **LRRK2** | Leucine-Rich Repeat Kinase 2 |
| **LTD** | Long-Term Depression |
| **MAM** | Mitochondria-Associated ER Membranes |
| **MAP** | Microtubule-Associated Proteins |
| **MPTP** | 1-Methyl-4-Phenyl-1, 2,3,6-Tetrahydropyridine |
| **MSA** | Multiple System Atrophy |
| **mtDNA** | Mitochondrial DNA |
| **NAC** | Non-Aβ Component of AD Amyloid |
| **NFT** | Neurofibrillary Tangles |
| **NHEJ** | Non-Homologous End-Joining |
| **NTR** | N-Terminal Region |
| **OXPHOS** | Oxidative Phosphorylation |
| **p75NTR** | P75 Neurotrophin Receptor |
| **PD** | Parkinson's Disease |
| **PDGFR** | Platelet-Derived Growth Factor Receptor |
| **PINK1** | PTEN-Induced Kinase 1 |
| **PMCA** | Protein Misfolding Cyclic Amplification Assay |
| **PP2A** | Protein Phosphatase 2A |
| **PPI** | Protein-Protein Interaction |
| **PrP** | Prion Protein |
| **$PrP^{C}$** | Cellular Prion Protein |
| **$PrP^{Sc}$** | Pathologic Form of Prion |
| **PSP** | Progressive Supranuclear Palsy |
| **ROS** | Reactive Oxygen Species |
| **RT-QuIC** | Real-Time Quacking Induced Conversion |
| **SAA** | α-syn Seed Amplification Assay |
| **sAPPα** | Soluble Amyloid Precursor Protein α |
| **SNARE** | Soluble N-Ethylmaleimide-Sensitive Protein Binding Protein Receptor |
| **SOD1** | Superoxide Dismutase 1 |
| **TDP-43** | TAR DNA Binding Protein 43 |
| **TFAM** | Mitochondrial Transcription Factor A |
| **ThT** | Thioflavin T |
| **Ub** | Ubiquitin |
| **UPR** | Unfolded Protein Response |
| **UPS** | Ubiquitin-Proteasome System |
| **VPS35** | Vacuolar Protein Sorting-Associated Protein 35 |

DOI: 10.1201/9781032676555-2

# Introduction

## Neurodegeneration and Aging

Neurodegeneration is the progressive loss of neurons in certain areas of the brain and is the main pathological feature of neurodegenerative diseases (Przedborski et al., 2003). Neurodegenerative diseases are still inexorably progressive and lethal. Unlike other cells that can regenerate regularly, neurons are post-mitotic cells, and it is almost impossible to compensate for their loss. The limited neurogenesis in the adult brain cannot cope with the degeneration of the millions of neurons that must function throughout a person's life. As the human lifespan increases, neurons may have difficulty keeping up with this long survival period, which may pave the way for neurodegeneration. It is accepted that the most critical factor for neurodegeneration is age, and it is also suggested that the process has an evolutionary basis in a common ancestor shared by tunicates and vertebrates (Anselmi et al., 2022).

Terminally differentiated neurons that cannot renew themselves may be vulnerable to cellular dysfunction that may occur with age. In fact, the natural aging process covers a wide range of changes such as genomic instability, telomere attrition, epigenetic modifications, loss of proteostasis, mitochondrial dysfunction, cellular aging, stem cell exhaustion, changes in intercellular communication, and immune response (Lopez-Otin et al., 2013). When we look at the neurodegeneration process, we see that it covers many of these topics, and age is the most critical risk factor. Some hypotheses have been put forward to try to explain the relationship between aging and the vulnerability of cells. One is the disposable soma theory (DST) of aging (Kirkwood, 1977). This theory suggests that aging results from a physiological trade-off between investment in somatic maintenance and other biological functions. For example, the growth and differentiation of cells result in the release of free radicals, which over time damage DNA and other molecules (van den Heuvel et al., 2016). Another hypothesis, which defines the capacity of neurons to respond to functional and metabolic stress as "homeostatic reserve," defines normal aging as a functional physiological state characterized by the decline of homeostatic reserves. Accordingly, the decrease in homeostatic reserve with age does not affect the functional resting state of neurons or the satisfaction of basic needs required for neuronal activity. However, when other etiopathogenic factors overlap, such as in conditions requiring excessive activity or neurodegeneration, aged neurons cannot respond to changes as required as a young neuron (Toescu & Verkhratsky, 2003). As a result, according to this hypothesis, while neurons can meet their basic needs with normal aging, when a factor that triggers neurodegeneration occurs, the most sensitive neurons are those with reduced homeostatic reserve.

The "calcium hypothesis of brain aging" suggests that age-related irregularities in calcium balance continuously cause slight changes in intracellular free calcium concentration. Over long periods, as a result of their accumulation, age-related alterations occur in neuron function. By extending this hypothesis to Alzheimer's disease (AD), it has been argued that alterations in the calcium-mediated signaling system in neurodegeneration are the most prominent among the cellular responses that promote neuron dysfunction and death (Khachaturian, 1994; Mattson & Chan, 2003; Thibault et al., 1998; Toescu & Verkhratsky, 2003).

Additionally, cellular senescence represents a state of permanent cell growth arrest and contributes to aging and neurodegeneration. It was reported that astrocytes, microglia, neurons, and endothelial cells in the AD brain express increased levels of senescence-associated proteins, suggesting that cellular senescence may also play a role in AD-type neurodegeneration (Liu, 2022).

These hypotheses can be augmented with many different examples. As a result, age causes the fragility of long-lived neurons to increase in a natural cycle. In fact, if there is no trigger, neurons continue to perform their basic functions. However, when environmental and genetic factors alter the normal conditions, these cells may not be able to provide the necessary responses or cope with the changes.

# Neurodegenerative Diseases

Neurodegenerative diseases represent neurological disorders affecting specific neuronal subsets in specific functional anatomical systems (Przedborski et al., 2003). Neurons communicate through their neurites, axons, and dendrites, and in this way, they govern our senses, movement, speech, memory, and behavior. Disruption of transport and synaptic connections in these extensions leads these cells to death. Thus, different symptoms and, therefore, different neurodegenerative diseases are observed depending on where neurodegeneration begins. Although the term *neurodegeneration* literally describes the degeneration in neurons, we know that glial cells, including astrocytes and oligodendrocytes, are also affected in neurodegenerative disorders (Kovacs, 2017).

Among many different neurodegenerative disorders, the most prominent ones are AD, Parkinson's disease (PD), Huntington's disease (HD), amyotrophic lateral sclerosis (ALS), Frontotemporal dementia (FTD), and prion diseases. According to clinical manifestations, these diseases have been classified into two main groups: 1) dementia and cognitive decline, and 2) movement disorders (Kovacs, 2016). Most of these neurodegenerative diseases are characterized by the intracellular (neurons or glial cells) or extracellular accumulation of specific misfolded proteins with altered physicochemical properties (Kovacs, 2017; Przedborski et al., 2003). For this reason, they are further categorized by their associated pathological protein aggregates. According to their molecular pathological features, neurodegenerative diseases are classified by the name of the major protein that aggregates: tauopathies, alpha-synucleinopathies, TAR DNA binding protein 43 (TDP-43) proteinopathies, fused-in sarcoma (FUS)/FET proteinopathies, prion diseases, trinucleotide repeat diseases, neuroserpinopathy, ferritinopathy, and cerebral amyloidosis. AD is evaluated separately because it involves accumulations of two different major proteins, amyloid beta (Aβ) and tau (Kovacs, 2017) (Table 1.1).

**TABLE 1.1**

Amyloid Fibril Proteins and Related Neurodegenerative Disorders

| Amyloid fibril proteins | Related disease/s | Localization | Additional features | Related genes |
|---|---|---|---|---|
| Aβ | • Alzheimer's disease<br>• Down Syndrome<br>• Cerebral amyloid angiopathy | Extracellular | • Oligomeric intracellular forms<br>• Aβ 1-42 and 1-40 most abundant<br>• Aβ 1-37 and 1-38<br>• Aβ modifications [e.g. phosphorylation (serine 8 or 26); glycosylation; truncated N-terminal and pyroglutamate modification] | *APP*<br>*PSEN1*<br>*PSEN2* |
| Tau | • 3R+4R taupathies (eg. Alzheimer's disease, Niemann-Pick disease)<br>• 3R taupathies (eg. Pick's disease)<br>• 4R taupathies (eg. Corticabasal degeneration, Huntington's disease, Progressive supranuclear palsi) | Intracellular | • Oligomeric forms<br>• 6 isoforms=presence of N terminal (0, 1 or 2 N) and 3 or 4 repeat microtubule binding sites (3R or 4R).<br>• Hyperphosphorylation (e.g. ptau-181, p-tau-217)<br>• Other post-translational modifications (e.g. glycosylation, glycation, nitration, transglutamination) | *MAPT* |
| α-syn | • Parkinson's disease<br>• Dementia with Lewy Bodies<br>• Multiple system atrophy | Intracellular in neurons (Lewy body) or in oligodendrocytes (GCI) | • Oligomeric forms<br>• Most abundant phosphorylation at serine 129<br>• Other post-translational modifications (e.g. glycosylation, nitration, truncated forms) | SNCA<br>PRKN*<br>PINK1<br>DNAJC13<br>LRRK2*<br>PARK7<br>TMEM230<br>LRP10 |
| PrP | • Creutzfeldt Jacob Disease (CJD)<br>• vCJD<br>• Kuru<br>• Fatal insomnia<br>• Gerstmann-Straussler- Scheinker disease<br>• Variably protease-sensitive prionopathy | Extracellular | • Oligomeric forms<br>• Prp 27–30 (most abundant form) | PRNP |

* LB is not seen in all cases

Accumulations of misfolded proteins are seen in the majority of neurodegenerative diseases. However, the proteins that accumulate in different neurodegenerative diseases are different, affect anatomically different brain areas, and may not be in the amyloid structure. Although most neurodegenerative diseases are seen in sporadic form, elucidating their familial forms has contributed significantly to understanding the different pathologies of these diseases (Wolfe, 2018).

Considering the data obtained as a result of genetic studies, the proteins that play a role in pathological changes occurring in sporadic and familial forms overlap. For example, *amyloid-beta precursor protein* (*APP*), *Presenilin 1*, or *Presenilin 2* mutations seen in familial forms of AD play a role in the production of Aβ peptide, which is the core component of extracellular aggregates in AD. Duplications and triplications of alpha-synuclein (α-syn), the main component of Lewy bodies in PD, are responsible for a familial form of PD (Kovacs et al., 2010). These examples can be multiplied by adding microtubule-associated protein tau, prion protein, huntingtin, TDP-43, and fused-in sarcoma (FUS) (Kovacs, 2017).

It is known that no matter which protein accumulation occurs, some common mechanisms are disrupted in neurodegeneration.

## Main Mechanisms in Neurodegenerative Disorders

These are classified under nine sub-headings as previously described (Everse & Coates, 2009; Jellinger, 2010; Wilson et al., 2023).

### Aberrant Proteostasis

As far as we can determine with the current technology level, a cell can produce over 11,000 different proteins (Kulak et al., 2017). Most of these proteins must be folded correctly to perform their cellular functions, and the cell tightly controls these processes.

Protein homeostasis or proteostasis refers to the intracellular mechanisms that ensure balanced protein production/degradation, folding, and traffic. It encompasses a protein quality control system that includes heat shock proteins, unfolded protein response (UPR), ubiquitin-proteasome system (UPS), and autophagy-lysosome system. Under physiological conditions, proteins' folding, lifespan, and functions are strictly controlled by the protein quality control system (Klaips et al., 2018). This control process ensures that proteins function correctly at the right place and time. Proper functioning of protein quality control is essential for highly differentiated post-mitotic neurons to perform their energy-requiring functions for many years (Saibil, 2013).

The synthesis, folding, maintenance of this conformation and eventual reduction of a protein are necessary to ensure proteome balance (Klaips et al., 2018). Heat shock proteins, also called molecular chaperones, take part in the first stage of ensuring this balance. They ensure that proteins are correctly folded in the cell, while their misfolding cannot be corrected, or they ensure that proteins that have completed their half-life are directed towards destruction. On the other hand, chaperones can also intervene and dissolve aggregates that may form due to misfolding (Hartl et al., 2011).

A protein must be removed from the system if it misfolds and/or forms aggregates. Eukaryotic cells generally maintain protein homeostasis through two central systems, UPS and autophagy. Ubiquitin is a tag that marks the target protein to be degraded by lysine, N-terminal methionine, serine, threonine, or cysteine residues (Schmidt et al., 2021; Tanaka & Matsuda, 2014). The process of ubiquitination occurs through a specific enzymatic cascade that works in a hierarchically coordinated manner: E1 ubiquitin-activating enzymes, E2 ubiquitin-conjugating enzymes, E3 ubiquitin ligases, and deubiquitinases that cleave ubiquitin molecules from substrates (Schmidt et al., 2021). Ubiquitin-tagged proteins are usually directed to proteasomes for degradation. After binding to the proteasome, the substrates are deubiquitinated, decompressed, and transferred to the proteolytic 20S core particle for degradation. The autophagy-lysosomal pathway involves the ubiquitin-dependent degradation of the target protein via lysosomes. Autophagy is a pathway that maintains cellular balance by regulating the degradation of cytoplasmic contents. It is critical for the selective capture and degradation of damaged organelles such as mitochondria (mitophagy), invading microbes, and specific cytoplasmic structures such as protein aggregates by autophagosomes before directing them to lysosomes (Corti et al., 2020; Komatsu et al., 2006; Schmidt et al., 2021; Vargas et al., 2023). There are three types of autophagy: bulk autophagy, non-selective autophagy, and selective autophagy. Non-selective autophagy degrades parts of the cytoplasm, some macromolecules, and organelles without discrimination. The targets of selective autophagy are specific and are named according to their target: mitophagy, lysophagy, aggrephagy (protein and RNA aggregates), xenophagy (intracellular pathogens), ER-phagy, pexophagy (peroxisomes), ribophagy (ribosomes), and so on (Vargas et al., 2023). These systems, working together with ubiquitin, are associated with many steps of neurodegeneration. Intracellular inclusions formed by proteins that form aggregates in neurodegenerative diseases, such as alpha-synuclein (α-syn), tau, TDP-43, and even extracellular Aβ plaques, contain abundant ubiquitin (Schmidt et al., 2021; Tanaka & Matsuda, 2014). However, they are also involved in the main pathways disrupted in neurodegenerative diseases such as mitochondrial homeostasis, DNA damage response, endocytosis, neuroinflammation, and neuronal cell death (Schmidt et al., 2021).

## Protein Depositions

### Amyloid Definition

The term *amyloid* was first used by Schleiden and later Virchow to describe iodine-stained deposits seen in the liver. When it was discovered that these structures, which were first thought to be carbohydrates, contained nitrogen, they were understood to be protein structures, but the first name was retained, even though it was incorrect. By staining with Congo red under light microscopy, it was suggested that the amyloid was fibrillar in structure, and transmission electron micrographs confirmed this (Rambaran & Serpell, 2008). To date, 42 amyloid fibril-forming proteins have been identified. Amyloid β (Aβ), alpha-synuclein (Aα-syn), tau (ATau), and prion (APrP) are among these proteins (Buxbaum et al., 2022). Despite the obvious differences in their amino acid sequences and native structures, these amyloidogenic peptides all appear to share a common β-sheet conformation in their polypeptide backbones, and this feature may confer fibrillar, proteolytic resistant, and insoluble properties for all forms of amyloid (Rambaran & Serpell, 2008).

In conclusion, the basic structure of amyloid is fibril. An amyloid fibril consists of twisted protofilaments. An amyloid protofilament is a stack of protein layers in a β-sheet structure that forms an amyloid fibril when twisted around the same stacks. Amyloid fibrils may consist of 1, 2, 3, 4, or many such protofilaments. Protofilaments are connected in parallel via their side chains (Buxbaum et al., 2022).

### Amyloid Beta (Aβ)

One of AD's most distinctive histopathological features is the presence of extracellular amyloid plaques in the brain. Amyloid plaques are formed by the accumulation of small peptides called Aβ. Aβ fragments are produced as a result of proteolytic cleavage of APP, which is a transmembrane protein, and is expressed in many tissues such as nerve cells and also adrenal gland, kidney, heart, spleen, and pancreatic cells. APP is a type 1 membrane glycoprotein involved in several biological activities, including normal brain development, postnatal survival, signaling, intracellular transport, and neuronal homeostasis (Arancio, 2020; Hampel et al., 2021; Wang et al., 2017). It has been reported that APP acts as a mediator for the entry of of Aβ and tau oligomeric forms into the cell and contributes to the spread of oligomeric forms and, thus, the progression of pathology (Puzzo et al., 2017). Considering that Aβ can also increase APP expression (Gezen-Ak et al., 2018), this appears to be a cycle in which they trigger each other.

Proteolytic cleavage of APP occurs in two different pathways, the non-amyloidogenic and amyloidogenic pathways, which trigger the formation of Aβ1-42 peptide. In the non-amyloidogenic pathway, it is cleaved by alpha-secretase (α-secretase) and gamma-secretase (γ-secretase), and a soluble extracellular fragment called soluble amyloid precursor protein α (sAPPα) is produced, which has neurotrophic and neuroprotective functions. Additionally, it can undergo proteolysis to form the intracellular domain (beta-amyloid precursor protein intracellular domain, AICD) and Aβ17-40 and Aβ17-42 peptides (Hunter & Brayne, 2012). In the amyloidogenic pathway, the APP protein is first cleaved by the beta-secretase (β-secretase) enzyme. It forms a large soluble ectodomain (sAPPβ) and a C-terminal C99 peptide. Then, due to the cleavage of the C99 peptide by the γ-secretase enzyme, Aβ variants with a length of 37–43 amino acids, including Aβ1-42, are produced and released into the extracellular space. The remaining part constitutes the intracellular domain (AICD) of APP (Bernabeu-Zornoza et al., 2019).

### *Physiological Roles of Aβ1-42*

Although Aβ1-42 peptides are produced in high amounts in pathological conditions, they are present at low levels in normal brains. For many years, the role of APP and Aβ1-42 in nonpathological conditions was not investigated, and Aβ1-42 was considered a peptide with no physiological role and an incidental product of the catabolism of APP (Duyckaerts et al., 2009; Lesne, 2014). Physiological levels of Aβ1-42 are required for synaptic plasticity. Aβ exerts neurotrophic or neurotoxic effects depending on concentration. These effects are related to the age of individuals as well as factors such as the relative concentration of Aβ1-42 and their cellular environment. At high concentrations, Aβ1-42 causes neurotoxicity and cell death. Aβ was detected at picomolar levels in the interstitial fluid of the normal brain. It has been suggested that low concentrations of Aβ1-42 may modulate trophic signaling and synaptic activity and affect memory and learning. Additionally, physiological functions of soluble Aβ1-40 and Aβ1-42 have been reported to include cholesterol de novo synthesis and control of sphingomyelin levels (Bernabeu-Zornoza et al., 2019; Morley & Farr, 2014). In addition, it has been reported that it plays a role as an antioxidant in stabilizing the blood–brain barrier and regulating gene expression (Gezen-Ak et al., 2022; Soucek et al., 2003).

### *Aβ Fragments and Pathological Features*

Aβ1-40, with a length of 40 amino acids, is the predominant form found *in vivo* and accounts for the highest amount of the Aβ pool in the CSF. The next most abundant alloforms are Aβ1-38, Aβ1-42, and Aβ1-37, respectively. These Aβ variants account for approximately 15%, 10%, and 8% of Aβ forms in cerebrospinal fluid (CSF), respectively (Braun et al., 2022). It has been confirmed that the region with two extra amino acids at the C-terminus of Aβ1-42 is critical for aggregation mechanisms compared to Aβ1-40. Aβ1-42 is the main derivative that is more hydrophobic than other types of Aβ, prone to fibrillation, and accumulates in the brain (Chen et al., 2017; Mehta et al., 2001). Studies have shown that Aβ forms differ in individuals with AD, and although not all of them are seen in all cases, Aβ3-42, Aβ11-42, Aβ4-42, Aβ1-40 and Aβ1-42 are the most dominant forms. It has also been reported that the dominant form detected in all patients without exception is Aβ 1-42 in the first place and Aβ 1-40 in the second place (Noor et al., 2022). Aβ1-42 level and Aβ1-40/1-42 ratio in CSF, together with hyperphosphorylated tau (p-tau), are the most important biomarkers used in the diagnosis of AD (Leuzy et al., 2021; Molinuevo et al., 2018; Randall et al., 2013).

Aβ monomers can come together to form structures ranging in size from low molecular weight oligomers, including dimers, trimers, tetramers, and pentamers, to medium and high molecular weight oligomers, protofibrils, and senile plaques (Chen et al., 2017). All forms of Aβ can coexist in vivo. Aβ forms numerous structures in monomeric and oligomeric states, all resulting in similar fibril structures. Amyloid fibrils of Aβ form a parallel, folded cross-β-sheet structure (Chen et al., 2017; Maji et al., 2009). While fibrils aggregate into larger, insoluble amyloid plaques that form the histological lesions characteristic of AD, Aβ oligomers are soluble structures of heterogeneous sizes and are easy to spread throughout brain regions (Chen et al., 2017). However, binding of Aβ oligomers to neurotransmitter receptors can trigger redistribution of critical synaptic proteins, resulting in hyperactivity of metabotropic and ionotropic glutamate receptors. Because different types of Aβ oligomers have been identified, more than one type of oligomer likely plays a role in disease pathology (Dinamarca et al., 2012; Hector & Brouillette, 2020) (Figure 1.1).

The importance of phosphorylated forms of proteins such as tau or α-syn in the diagnosis or progression of neurodegenerative diseases is known. However, there are very few studies on the effect of Aβ phosphorylation on the molecular properties of amyloid deposits. In these studies, it is thought that phosphorylated forms of Aβ at serine 8 and serine 26 could be associated with AD pathology (Kumar et al., 2020; Kumar & Walter, 2011). Limited studies have shown that Aβ phosphorylation promotes peptide aggregation, increases neurotoxicity, and affects the stability and proteolytic degradation of Aβ (Jamasbi et al., 2017; Kumar & Walter, 2011). It has also been suggested that phosphorylated forms accumulate in neurons in the early stages and are an important component of extracellular plaques during aging (Kumar et al., 2011). The presence of phosphorylated forms has also been demonstrated in human sporadic cases of AD, FAD, CAA, and Down syndrome and in the brains of primates, dogs, and various transgenic mouse models (Kumar et al., 2020). Additionally, pyroglutamate modifications at residues 3 or 11, isomerization/racemization, and glycosylation modifications of Aβ are also reported (Kovacs, 2016).

### *Amyloid Fibrils*

Aβ monomers folded into a β-plated sheet can form low- and high-molecular weight oligomers such as protofibrils and mature fibrils. Fibrils are the primary form of Aβ in plaques and are only soluble in formic acid, unlike detergent-soluble Aβ oligomers. The observation that fibrils can grow rapidly by adding Aβ monomer, but not by adding Aβ oligomers, reduces the possibility of oligomers being intermediate forms (Ono & Watanabe-Nakayama, 2021). Extracellular Aβ aggregates in the parenchyma can be found in different forms such as diffuse, focal, stellate deposits, or, rarely, cotton-wool plaques. Plaques might be seen in the vessel walls of arteries, veins, and capillaries (Duyckaerts et al., 2009). Proteins, lipids, and/or cells contribute to the parenchymal Aβ plaques: ApoE, cholesterol, clusterin, zinc and copper ions, iron, intercellular adhesion molecule 1 (ICAM1), thrombospondin, heparan sulfate proteoglycan, alpha1-antichymotrypsin, Cathepsin D, and so

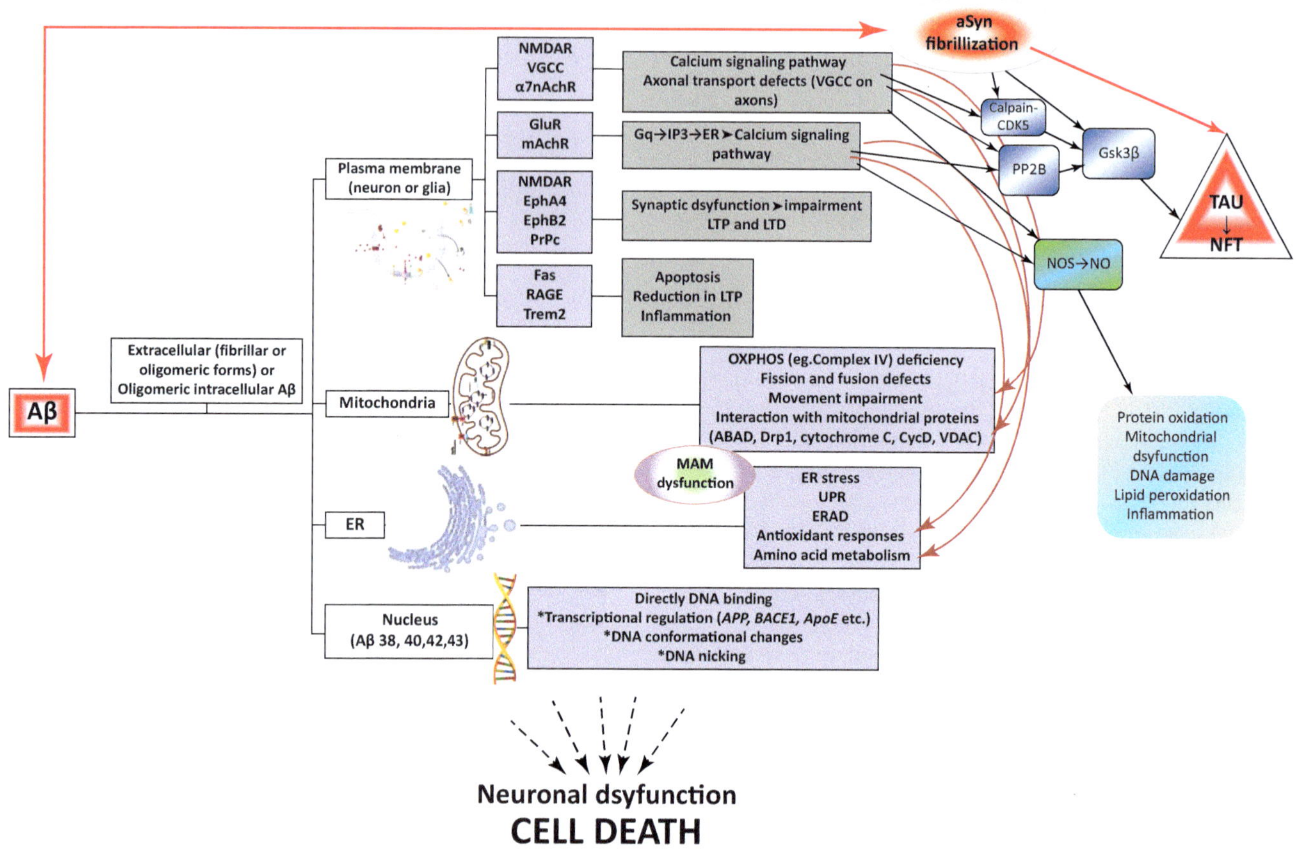


**FIGURE 1.1** Cellular mechanisms associated with Aβ pathology. Amyloid beta (Aβ), endoplasmic reticulum (ER), N-methyl-D-aspartate receptor (NMDAR) voltage gated calcium channel (VGCC), alpha 7 nicotinic acetylcholine receptor (α7nAchR), glutamate receptors (GluR), muscarinic acetylcholine receptors (mAChR), Ephrin (Eph), cellular prion protein ($PrP^c$), receptor for advanced glycation endproducts (RAGE), Triggering receptor expressed on myeloid cells 2 (TREM2), long term potentiation (LTP), Long term depression (LTD), mitochondria-associated ER membranes (MAM), oxidative phosphorylation (OXPHOS), unfolded protein response (UPR), alcohol dehydrogenase enzyme (ABAD), cyclin D (CycD), voltage-dependent anion channel (VDAC), alpha-synuclein (α-syn), protein phosphatase 2B (PP2B), cyclin-dependent kinase 5 (CDK5), glycogen synthase kinase 3β (GSK3β), nitric oxide synthase (NOS), neurofibrillary tangles (NFT).

on. (Duyckaerts et al., 2009). Although a lot of data indicates that the Aβ1-42 peptide is the main component of toxic aggregates, it has been determined that the fibrillary deposits that damage the cerebral vessel walls are formed by Aβ1-40 (Ono & Watanabe-Nakayama, 2021).

It is known that Aβ fibrils formed *in vitro* have a random coil and β-sheet structure. These fibrils, like samples obtained from the human brain, show the cross-β structure and Congo red affinity. However, samples obtained from human brain homogenates have been reported to be polymorphic and structurally quite different from fibrils produced under *in vitro* conditions (Ono & Watanabe-Nakayama, 2021).

## *Alpha-Synuclein (α-syn)*

α-syn was first cloned from synaptic vesicles isolated from Torpedo californica in 1988, and its localization was shown in the nucleus and presynaptic nerve terminals (Maroteaux et al., 1988). Independently, in 1993, α-syn was identified as the non-Aβ component of AD amyloid (NAC) in senile plaques obtained from AD brains (Ueda et al., 1993). In 1995, an RNA whose expression changed during a male zebra finch learning a song was identified and suggested that it may have a critical role in regulating vertebrate neural plasticity. This protein was named synelfin by the researchers and was shown to be a NAC homolog (George et al., 1995). With all this data, we know that α-syn has been identified.

The α-syn protein, a small disordered protein, is encoded by the *SNCA* gene (Bisi et al., 2021). Studies have shown that the protein does not fold without lipids (Davidson et al., 1998; Weinreb et al., 1996; Wu et al., 2008). α-syn protein consists of three regions: a positively charged N-terminal region, a central hydrophobic region (also known as NAC), and an acidic C-terminal region. The N-terminal region of α-syn is a positively charged, containing seven series of 11 amino acid (aa) repeats. Each 11 aa repeat includes a highly conserved hexameric KTKEGV motif. These repeats increase the helical structure formation of α-syn and reduce their tendency to form β-structures. Repeat regions play an important role in the interactions of α-syn with lipids (Emamzadeh, 2016).

α-syn is a high-temperature resistant, 14 kDa cytosolic neuronal protein that is abundantly expressed in the nervous system. α-syn is produced mostly in the neocortex, hippocampus, substantia nigra, thalamus regions, and cerebellum in the brain (Cookson, 2009; Kim et al., 2014). α-syn expression has also been found to be present in various tissues, including muscle, kidney, liver, lung, heart, testicle, colon, blood vessels, lymphocytes, and platelets (Jimenez-Jimenez et al., 2023). Expression of α-syn is stimulated during neuronal development after the neuronal phenotype and synaptic connections are formed (Wang & Hay, 2015). α-syn is normally found in soluble form in the cytoplasm and in regions associated with synaptic vesicle membranes in presynaptic terminals (Maroteaux et al., 1988).

Different post-translational modifications such as nitration, glycosylation, C-terminally truncated species, oligomer forms, and PK-resistant forms are also reported (Kovacs, 2016).

### *Physiological and Pathological Roles of α-syn*

Although its function is not yet fully understood, this protein plays an essential role in cellular events such as synaptic vesicle release, vesicular traffic, fatty acid binding, neuron survival, mitochondrial homeostasis, proteasome function, dopamine metabolism, and chaperone activity (Bras & Outeiro, 2021; McCann et al., 2014). α-syn has been shown to assist in the folding and reopening of synaptic proteins called soluble N-ethylmaleimide-sensitive protein binding protein receptors (SNAREs) and to regulate synaptic vesicle turnover by acting as a molecular chaperone in the formation of this complex (Burre et al., 2010; Martineau et al., 2013; Singh et al., 2014; Sun et al., 2019; Trudler et al., 2021; Wang et al., 2018; Zorec et al., 2016). α-syn involves intracellular vesicular trafficking of proteins (ER to Golgi anterograde transport; Intra-Golgi and retrograde Golgi-to-ER traffic; Trans-Golgi network vesicle budding; Rab regulation of trafficking) (Burre et al., 2010; Lai et al., 2014; Martin-Jimenez et al., 2017; Thayanidhi et al., 2010; Wang et al., 2018; Zambon et al., 2019).

α-syn is involved in the dopamine synthesis pathway, interacting with tyrosine hydroxylase (TH), the rate-limiting enzyme responsible for the conversion of tyrosine to L-3,4-dihydroxyphenylalanine (L-DOPA); and L-aromatic amino acid decarboxylase (AADC), which is the enzyme that catalyzes the conversion of L-DOPA to dopamine (Venda et al., 2010). α-syn participates in dopamine uptake by directly binding to dopamine transporters (Eriksen et al., 2010; Petrelli et al., 2020).

α-syn might influence or is related to different types of signaling mechanisms or molecules (receptor tyrosine kinases, G-protein coupled receptors (GPCR), by Death receptor signaling or signaling by NOTCH/ Hedgehog/ WNT or nuclear receptors) which are responsible for the cell survival (Badawy et al., 2018; Baek et al., 2015; T. Chen et al., 2019; Crews et al., 2008; Kempuraj et al., 2018; Kim et al., 2018; Luo et al., 2019; Sengul et al., 2021; Vargas et al., 2019; Wang et al., 2020).

α-syn could participate in transcriptional regulation by promoting p53 and cdk5 expression (Ho et al., 2019; Kazmierczak et al., 2011; Paiva et al., 2017; Seo & Park, 2020; Turnquist et al., 2016). Furthermore, α-syn and serine 129-phospho-α-syn may help to facilitate the non-homologous end joining (NHEJ) DNA repair in neurons and fully differentiated astroglial cells (Greengard et al., 1994; Kaminsky et al., 2016; Li et al., 2009; Milanese et al., 2018; Paiva et al., 2017; Schaser et al., 2019).

Calcium regulatory protein, also known as calmodulin (CaM), is a messenger protein that gains activity by binding to calcium ions and triggers various mechanisms in short and long-term memory. Both wild-type and mutant α-syn have been shown to interact with CaM in vitro and in vivo (Martinez et al., 2003).

α-syn is related to mitochondrial biogenesis and dynamics (Pantiya et al., 2020; Park et al., 2020; Sarkar et al., 2018; Zambon et al., 2019). It promotes cell death, cell cycle re-entry, S-phase arrest, and increased autophagy in aged post-mitotic cells (Lee et al., 2003; Vasquez et al., 2020). It may be involved in the release of cytochrome C and alteration in BCL2 family proteins. Lewy bodies (LBs) and glial cytoplasmic inclusions (GCIs) are positive for cytochrome c, Apaf-1, and caspase-9 (Danyu et al., 2019; Dolgacheva et al., 2019; Espinosa-Oliva et al., 2019; Kawamoto et al., 2016; Oshima et al., 2016) (Figure 1.2).

### *α-Syn Pathology*

In 1912, Frederick Lewy identified eosinophilic cytoplasmic inclusions, today known as Lewy bodies, in the substantia nigra region of postmortem brain sections of individuals with PD (Lewy F.H., 1912). In 1997, Spillantini et al. showed that the major component of these structures is α-syn (Spillantini et al., 1997). LBs are pathological structures located in the neuron cytoplasm, containing high amounts of α-syn peptide accumulations, and can be seen in neurites and the bodies of cells. These pathological structures in neurites are called "Lewy neurites" (Wakabayashi et al., 2007; Wakabayashi et al., 2013). α-syn-related pathology is seen in dementia with Lewy bodies (DLB), idiopathic Parkinson's disease, and multiple system atrophy (MSA). LBs are found in neurons in the brains of PD and DLB patients, while GCIs are detected in oligodendrocytes in the brains of MSA patients (Estaun-Panzano et al., 2023; Gomperts, 2016).

The definition of LBs as α-syn-dominant pathological structure is due to the fact that these structures are quite complex. The structure of LBs includes α-syn, p62, ubiquitin (Ub), α-syn binding proteins, proteins involved in the Ub-proteasome system, synphilin-1, aggresome- and mitochondria-associated proteins, and cytoskeletal, cytosolic and cellular response proteins (Mahul-Mellier et al., 2020; Wakabayashi et al., 2013). Most of the α-syn accumulated in these structures is phosphorylated at serine 129 (Mahul-Mellier et al., 2020).

α-syn molecules come together to form units with very high molecular weight. The β-sheet-like organization of α-syn can be triggered by changes in aa sequences resulting from pH changes, post-translational modifications, cleavages, or point mutations (Estaun-Panzano et al., 2023). This formation in the amyloid fibril structure is a process based on several stages: primary nucleation, proliferation through growth, elongation, and secondary nucleation (Buell et al., 2014). Primary nucleation forms the first fibrils. Secondary nucleation can lead to an exponential increase in protofibrils and subsequent fibrils. When the combination of monomers begins, the amyloid structure behaves like a "black hole". It attracts more monomers, making the system increasingly energetically stable, which can catalyze the formation of new oligomers (Estaun-Panzano et al., 2023). These pathological forms can be classified as protofibrils (protofilaments), amyloid fibrils, and Lewy pathology, with the smallest being oligomers, increasing in

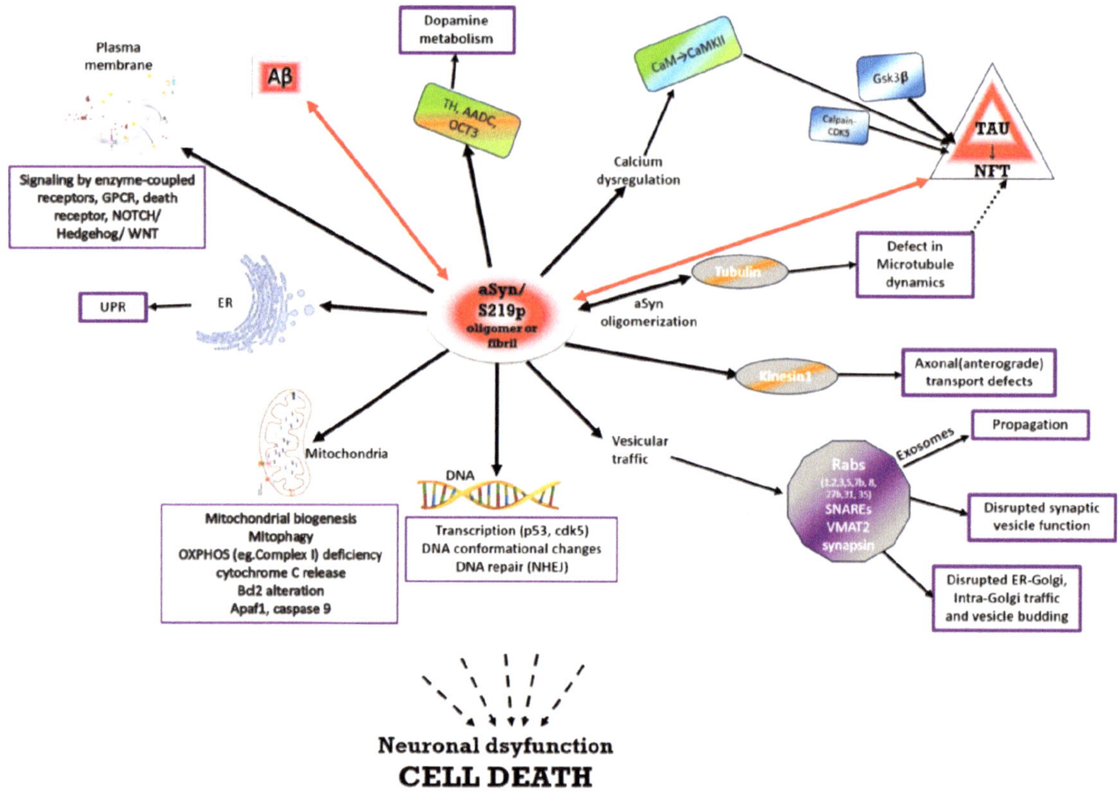


**FIGURE 1.2** Cellular mechanisms associated with α-syn pathology Amyloid beta (Aβ), endoplasmic reticulum (ER), tyrosine hydroxylase (TH), G-protein coupled receptors (GPCR), cyclin-dependent kinase 5 (CDK5), non-homologous end-joining (NHEJ), unfolded protein response (UPR), oxidative phosphorylation (OXPHOS), calmodulin (CaM), calmodulin-dependent protein kinase (CaMK), neurofibrillary tangles (NFT), soluble N-ethylmaleimide-sensitive protein binding protein receptor (SNARE).

size. Oligomers are defined as water-soluble, heterogeneous, and metastable structures (Forloni, 2023; Hijaz & Volpicelli-Daley, 2020). Recent findings indicate that the oligomeric forms may be more toxic and are the main component for the pathology and spread of the disease (Bengoa-Vergniory et al., 2017). The mutations and post-translational modifications of α-syn influence the structure of oligomers. Heterogeneous oligomers can create different morphological structures during assembly, determining their toxicity (Forloni, 2023).

α-syn pathology is not limited to PD. The detection of α-syn accumulations with different morphological features from LBs in the amygdala revealed the existence of co-pathologies involving α-syn (Uchikado et al., 2006). Findings to date indicate that Lewy pathology is present in ~40% of AD cases; in 20% of progressive supranuclear palsy (PSP), corticobasal degeneration (CBD), and sporadic Creutzfeldt-Jakob disease (CJD) cases; in ~15% of frontotemporal lobar degeneration (FTLD); and in ~ 10% of MSA cases. It is thought that αSyn pathology may act as a promoting factor for the accumulation of Aβ, tau, prion, or TDP-43 proteins (Visanji et al., 2021).

### *Tau*

The cytoskeleton maintains the shape and internal organization of cells and provides the mechanical support that allows cells to perform their basic functions, such as division and movement. The cytoskeleton of eukaryotic cells is made of filamentous proteins. All cytoskeletons have three main structures that differ in size and protein composition. One of these is microtubules (the others are actin filaments and intermediate filaments). The morphology and function of a neuron depends on its microtubule cytoskeleton, which shapes its axons and dendrites, performing different tasks of signal reception and transmission (Kelliher et al., 2019).

Another of the most prominent pathological changes seen in AD is neurofibrillary tangles, which are formed by the accumulation of hyperphosphorylated microtubule-associated tau protein in the intracellular space. Tau is a protein expressed predominantly in neurons in the human central and the peripheral nervous system. Tau is physiologically responsible for polymerizing tubulins, microtubule building blocks, and microtubule stability. Expression of tau is developmentally regulated and is encoded from the *MAPT* gene on chromosome 17q21.31 (Goedert & Spillantini, 2019). Six isoforms are found in the adult human brain, depending on the alternative splicing of certain exons (2, 3, and 10), the presence of the N-terminus (0, 1, or 2 N), and the presence of 3 or 4 repeats of microtubule binding sites (3R or 4R). Expression of isoforms varies across developmental stages and even between species. For example, 4R tau is dominant in the adult mouse brain, while the adult chicken brain produces 3R, 4R, and 5R isoforms. In mice, a transition from 3R to 4R tau is observed during the postnatal period, and tau phosphorylation decreases over time (Goedert & Spillantini, 2019). Under physiological conditions, tau exists as a soluble, unfolded protein that assembles and organizes microtubules. In pathological conditions seen in tauopathies, including AD, tau forms insoluble aggregates such as paired helical filaments and straight filaments that form components of neurofibrillary tangles (NFTs) (Chong et al., 2018; Paterno et al., 2022; Wang & Mandelkow, 2016).

The binding of tau protein to microtubules and its separation from microtubules is achieved by the phosphorylation/dephosphorylation mechanism regulated by specific kinases and phosphatases. While tau in dephosphorylated form binds to the microtubule, phosphorylation of tau causes it to detach from the microtubule. While the phosphorylation of tau protein is regulated by glycogen synthase kinase 3β (GSK3β) and cyclin-dependent kinase 5 (CDK5), its dephosphorylation is provided by protein phosphatase 2A (PP2A) (Jouanne et al., 2017). Additional modifications such as N- and C-terminal truncation, glycosylation, glycation, nitration of tyrosine residues, transglutamination, and deamidation; acetylation and oligomeric forms; the banding patterns of C-terminal fragments; and the trypsin-resistant band patterns are also reported (Kovacs, 2016).

In physiological conditions, kinases and phosphatases are controlled and balanced (Tiwari et al., 2019). In normal situations, tau contains 2–3 moles of phosphate; this ratio optimizes the interaction of tau with tubulin and the promotion of microtubule assembly (Holper et al., 2022). However, changes in the expression, activation, or inhibition processes lead to disruption of the kinase/phosphatase balance and tau protein's hyperphosphorylation. Hyperphosphorylation triggers tau to dissociate from microtubules and disorganize the microtubules. Hyperphosphorylated tau proteins come together to form NFTs (Tiwari et al., 2019). Phosphorylated tau seen in AD has been reported to be approximately 2–3 times greater than the phosphate rate in physiological conditions and is found in the cytoplasm of neurons as neuropil threads in dendrites, NFTs in soma and senile plaque neuritic corona in axons (Duyckaerts et al., 2009; Holper et al., 2022; Kovacs, 2016).

Brain tau filaments have a cross-β structure characteristic of amyloids. However, it is still unclear whether phosphorylation is the main factor that triggers tau formation in human diseases. Tau is highly charged and hydrophilic and thus is soluble and stable in aqueous environments over a wide range of pH and temperature. Under pathological conditions, the microtubule-binding domain is densely packed into crossed beta-sheets, forming amyloid fibrils. Assembly of tau into filaments can be triggered by cofactors such as sulfated glycosaminoglycans, RNA, or fatty acids (Fichou et al., 2018). There is evidence that changes in conformation can trigger hyperphosphorylation (Fichou et al., 2018; Goedert & Spillantini, 2019). However, it is still not understood how it forms its unique fibril structure and how this structure maintains its stability (Fichou et al., 2018).

All six tau isoforms are present in Tauopathies: AD, Down's syndrome, Niemann-Pick disease type C, Tangle-only dementia, and so on are 3R + 4R Tauopathies; Pick's disease is 3R Tauopathies; CBD, HD, PSP, etc. are 4R Tauopathies (Goedert & Spillantini, 2019).

### *Prion*

Prion protein (PrP), a proteinaceous infectious particle, is a highly conserved glycosylphosphatidylinositol (GPI)-anchored membrane protein. There are two forms: the normal or cellular isoform, $PrP^{C;}$ and the infectious isoform or scrapie PrP, $PrP^{Sc}$ (Kovac & Curin Serbec, 2022). Many cell types express it under physiological conditions, including cells of the nervous system, circulatory system, and digestive system (Kovac & Curin Serbec, 2022) (Search results < Expression Atlas < EMBL-EBI). The *PRNP* gene in humans encodes PrP, and after its expression, it is transferred to the cell membrane's outer surface and attached to the lipid bilayer with a GPI anchor from the C-terminal region. $PrP^{C}$ exhibits different folding arrangements, including α-helix (42%), β -sheet (3%), and non-folding regions, and can undergo truncation at various sites (Kovac & Curin Serbec, 2022). The most common PrP species is PrP27-30. Further fragments are PrP 11, PrP7-8, PrP14, PrP-CTF12/13, PrP16-17, and PrP17.5-18, and the oligomer form of PrP has also been described (Kovacs, 2016). Although the function of the protein is not yet fully known, its best-defined function is the maintenance of peripheral nerve myelination. In addition, many roles have been stated, such as memory formation, neurite development, iron and copper homeostasis, neuronal development and synaptic plasticity, and neuroprotection in ischemic stroke (Beraldo et al., 2011; Criado et al., 2005; Doeppner et al., 2015; Gasperini et al., 2016; Minamoto et al., 1991; Prestori et al., 2008). As an example of its protective properties, $PrP^{C}$ has been shown to play a role in the response to endoplasmic reticulum stress resulting from the accumulation of unfolded or misfolded proteins in the ER (Wulf et al., 2017). Mice with silenced $PrP^{C}$ have been shown to disrupt hippocampal long-term potentiation (LTP), a type of synaptic plasticity that plays a role in memory formation (Bliss & Collingridge, 1993).

The normal, cellular $PrP^{C}$ is converted into $PrP^{Sc}$ by refolding of a portion of its α-helical and coil structure into β-sheet (Pan et al., 1993). This transition changes the physicochemical properties of the PrP. $PrP^{Sc}$ acts as a template, upon which $PrP^{C}$ is refolded into a nascent $PrP^{Sc}$ molecule (Prusiner, 1998). This causes prion diseases' neurodegenerative and infectious features to appear (Olanow & Prusiner, 2009). In addition to the neurotoxicity known to be caused by misfolded forms, it is thought that loss of normal $PrP^{C}$ function may be an integral part of neurodegenerative processes (Castle & Gill, 2017).

According to their clinical and pathological characteristics, prion diseases are classified as "sporadic," those whose source is unknown; "genetic," those caused by a genetic mutation; and "iatrogenic," those transmitted through organ-tissue transplantation. On the other hand, "variant" types that are transmitted to humans through the consumption of products such as contaminated beef or blood, and "kuru disease" types that are transmitted between humans through cannibalism have also been identified (Johnson, 2005). The sporadic form accounts for 85% of all cases, while the remaining cases are familial (10%–15%) and iatrogenic (less than 1%). Nevertheless, prion diseases are classified as rare diseases and, according to studies, occur at a rate of 1–2 per million (Ironside et al., 2017).

However, recent studies indicate that the incidence of sporadic CJD is increasing worldwide (Uttley et al., 2020).

## The Interaction of Amyloid Structures with Membranes

The plasma membrane separates the cell from the external environment, protects it against external influences, regulates communication with the environment and other cells, and creates a closed compartment for various metabolic reactions. This critical structure undertakes another vital task: nerve cell conduction. It fulfills this task by maintaining intracellular ion balance, primarily through various channels and transporters, and using this balance for nerve transmission. Therefore, neuronal plasma membranes are essential areas where significant changes occur during aging and neurodegeneration processes, and pathology involves various elements.

The cell surface is vital in perceiving and transmitting external stimuli through transmembrane receptors. These transmembrane receptors determine the cell's response through various signal transduction mechanisms. In the cytoplasm, activation of membrane receptors recruits and alters the activity of a network of enzymes, amplifying the incoming stimulus and resulting in a definitive response. In the nucleus, the transcriptional profile is regulated by activating transcription factors and epigenetic mechanisms in response to cellular and environmental changes. Thus, the cell adapts to the new environment (Martin & Dotti, 2022).

The function of integral membrane receptors and the conformational changes necessary for their normal activity depend on lipid–protein or lipid bilayer–protein interactions. For example, these interactions regulate the activity of receptors without directly changing the active site (Andersen & Koeppe, 2007). Besides, lipid content and types determine the localization of membrane receptors and their dimerization ability, movement of integral membrane proteins, endocytosis, and their relation with other adapter proteins (Andersen & Koeppe, 2007). Accordingly, the fundamental components of cell physiology inevitably depend on the correct lipid composition of the plasma membrane (Casares et al., 2019).

On the other hand, aging affects not only the composition and structure of the plasma membrane but also the function of different components presents in the membrane. Changes in the lipid bilayer are fundamental to proteins' distribution, organization, and function in the membrane layers. Therefore, any change in the lipid composition of cell membranes would be expected to affect the function of membrane receptors and how cells perceive the environment (Martin & Dotti, 2022). This proposition formed the basis for the gate theory of aging, which is based on the observation that age-related quantitative and qualitative changes in the plasma membrane-located epidermal growth factor receptor (EGFR) and platelet-derived growth factor receptor (PDGFR) are responsible for the decreased proliferative capacity and general functional impairment of senescent cells. EGFR and PDGFR endocytosis are defective due to reduced functional caveolae (Kawabe et al., 2001; Martin & Dotti, 2022; Park et al., 2000; Reenstra et al., 1996; Yeo & Park, 2002).

In older individuals, lipid changes appear to be more pronounced in the lipid rafts of the plasma membrane. Namely, changes were observed in the composition of lipid rafts isolated from the frontal cortex of individuals without dementia, especially in terms of plasmalogens, arachidonic acid (AA) and docosahexaenoic acid (DHA), phosphatidylinositol, sphingomyelin, sulfatides and cerebrosides, and cholesterol and sterol esters (Diaz et al., 2018). Similar changes have been reported in different regions of aging brains (Colin et al., 2016; Egawa et al., 2016). These data indicate that an aging-related restructuring of lipids in the plasma membrane occurs, and some changes occur early in the aging process.

Various studies show that changes in the plasma membrane in older adults are a normal consequence of metabolic functions that develop throughout life. The free radical theory of aging, in which these events are explained, was reviewed by Zs.-Nagy (Zs-Nagy, 2014). Reactive oxygen species (ROS), by-products of aerobic respiration, interact and damage double bonds, especially those containing fatty acids (Zs-Nagy, 2014). On the other hand, free radicals accumulate during the aging process and cause DHA levels to decrease in membranes (Horrocks & Yeo, 1999). In aged rats, DHA and AA levels have been reported to decrease in the hippocampus, which is associated with lipid peroxidation, altered fatty acid metabolism, and fatty acid desaturase levels (McGahon et al., 1998; McNamara et al., 2008; Terracina et al., 1992).

Another effect of oxidative stress that accumulates during the aging process is to reduce the cholesterol level in the neuron plasma membrane. It has been reported that ROS accumulation increases the expression of the *cyp46A1* gene, which encodes the cholesterol 24-hydroxylase enzyme that reduces cholesterol levels and changes lipid raft composition (Martin et al., 2008; Ohyama et al., 2006; Sodero et al., 2011; Sodero et al., 2012).

An important example of changes in lipid composition is shown in mouse hippocampal neurons. Increased sphingomyelin levels have been detected in aged mouse hippocampal neurons (Trovo et al., 2011). The percentage and content of longer fatty acids esterified to sphingomyelin are significantly increased in different brain regions of aged rats. This has been associated with increased membrane viscosity, or packing density, in the aging brain (Giusto et al., 1992).

Aβ, tau, α-syn, superoxide dismutase 1 (SOD1), huntingtin, prion misfolding, and aggregation are major pathological components of neurodegenerative diseases (Goedert, 2015; Goedert et al., 2010). Information regarding the pathophysiological impact of binding of misfolded protein deposits to the plasma membrane, as well as the interaction of these aggregates with biomolecules under non-acute experimental conditions, is quite limited (Shrivastava et al., 2017).

Pathogenic peptides have been reported to associate with many molecules on the cell surface. While debate continues as to whether they have specific receptors, their different surface features indicate their different affinities to diverse amount of molecules on the cell membrane. These peptides accumulate in the plasma membrane of both neurons and non-neuronal cells. This accumulation leads to alterations in membrane structure, lipid bilayer fluidity, and regular distribution of membrane proteins. Proteins, lipids of the plasma membrane and extracellular matrix influence and/or contribute to the interaction of misfolded protein assemblies, whether naked or in complex with other secreted proteins (Shrivastava et al., 2017).

### Interaction of Misfolded Proteins with Lipids and Proteins

Misfolded proteins may also interact directly with lipids of the plasma membrane. For example, Aβ interacts with cholesterol (Avdulov et al., 1997; Di Scala et al., 2014) and gangliosides (GM), a class of amphiphilic glycolipids (Molander-Melin et al., 2005; Yanagisawa et al., 1995). Further fibrillization of Aβ peptides occurs (Avdulov et al., 1997) through interaction with cholesterol and GM1 (Hong et al., 2014). Similarly, α-syn associates with negatively charged phospholipids (Pieri et al., 2012), cholesterol, and GM (Fantini et al., 2011; Fantini & Yahi, 2011, 2013) in the outer layer of the plasma membrane. It has been shown that α-syn mechanisms have a high diffusion rate, like lipids, by binding to lipids and can travel long distances on the neuron membrane (Renner et al., 2009; Shrivastava et al., 2013). It has also been shown that tau and huntingtin bind with phospholipids (Burke et al., 2013; Jones et al., 2012; Patel et al., 2015; Pieri et al., 2012), and huntingtin has also been shown to be associated with lipid rafts (Valencia et al., 2010).

Misfolded proteins have also been shown to interact directly with membrane proteins. $PrP^C$ interacts with accumulated PrP, for example, protease resistant PrP (Aguzzi & Falsig, 2012). A recent study showed that $PrP^C$ may be a ligand for Schwann cells' Gpr126, a GPCR (Kuffer et al., 2016). On the other hand, protease resistant PrP has also been shown to bind to $PrP^C$, which is a ligand of mGluR5, a neuronal GPCR that physically interacts with $PrP^C$ (Um et al., 2013). Various membrane proteins have been shown to interact with Aβ (Jarosz-Griffiths et al., 2016): α7-nicotinic acetylcholine receptors (Dineley et al., 2001; Dziewczapolski et al., 2009; Snyder et al., 2005; Wang et al., 2000), leukocyte immunoglobulin-like receptor B2, α3-Na+/K+-ATPase, EphB2, and $PrP^C$-mGluR5 complex (Cisse et al., 2011; Haas et al., 2014; Haas et al., 2016; Hu et al., 2014; Kim et al., 2013; Lauren et al., 2009; Ohnishi et al., 2015; Renner et al., 2010; Um et al., 2013). Although knockout of $PrP^C$ does not appear to eliminate Aβ binding to cells and associated neurotoxicity, the knockout of mGluR5s or mGluR5 inhibition has been shown to delay neurodegenerative process significantly (Balducci et al., 2010; Calella et al., 2010; Hamilton et al., 2014; Hamilton et al., 2016; Kessels et al., 2010). These data indicate the interaction between mGluR5s and Aβ peptides located at the plasma membrane. Besides, mGluR5 and Aβ oligomer co-localization in mouse and human astrocytes has also been reported (Casley et al., 2009; Lim et al., 2013; Shrivastava et al., 2013). Similar is the case for α-syn, which interacts with the α3 subunit of the $N^+K^+$ATPase (Shrivastava et al., 2015). The binding of α-syn fibrils to the lymphocyte activation gene 3 membrane protein has been shown to participate in α-syn endocytosis, spreading, and toxicity (Mao et al., 2016). Membrane proteins to which α-syn binds include neurexin subunits, Aβ precursor-like protein 1, and 78 kDa glucose-related protein (Bellani et al., 2014; Mao et al., 2016; Shrivastava et al., 2015). Finally, interaction of extracellular tau aggregates with muscarinic receptors has been reported (Gomez-Ramos et al., 2008). This diversity in the proteins that misfolded proteins associate within the membrane suggests they do not have a specific receptor. Multiple binding proteins in the plasma membrane indicate that these interactions increase pathogenic protein density, thereby promoting diffusion-dependent intermolecular interactions that induce aggregation and toxicity (Shrivastava et al., 2017).

### Plasma Membrane Integrity

Increasing accumulation of misfolded proteins together with the clustering and further assembly into structures ranging from pore-like to raft-like structures create significant consequences on overall membrane topography, geometry, fluidity, and integrity (Antonny, 2011; Danzer et al., 2007; Garten et al., 2015; Kayed et al., 2003; Kayed et al., 2004; Kostka et al., 2008; Li & Gorfe, 2014; Milanesi et al., 2012). This remodeling is due to membrane adhesion of misfolded proteins, lateral diffusion, further accumulation in and on the plasma membrane, binding to proteins in the membrane and altering their localization, removal of lipids from the lipid bilayer, and changes in lipid composition (Hellstrand et al., 2013; Pandey et al., 2009; Reynolds et al., 2011; van Rooijen et al., 2009). All these events continue until the integrity of the plasma membrane is disrupted by hardening to a point where it ruptures (Pieri et al., 2012; Shrivastava et al., 2017; Varkey et al., 2010).

### Ion Homeostasis

$N^+K^+$ATPase is an ion pump that exchanges 3 $Na^+$ out for 2 $K^+$ into the cell for the cost of an ATP. Given that, it restores intracellular ion imbalance and membrane potential after the action potential. α and β subunits are the functional unit of $N^+K^+$ATPase. The β subunit enables the trafficking of $N^+K^+$ATPase to the membrane (Kaplan, 2002). Neurons express catalytic α1 subunit and neuron-specific α3 subunit which has a lower affinity for $Na^+$ (Azarias et al., 2013). The α3 subunit is a crucial part of the mechanism that controls membrane potential. The brain strongly depends on the proper functioning of $N^+K^+$ATPase α3 subunit. Both Aβ and α-syn have been shown to interact with the α3 subunit of $N^+K^+$ATPase and impair the function of this pump (Ohnishi et al., 2015; Shrivastava et al., 2017; Shrivastava et al., 2015). The binding of α-syn to α3-$N^+K^+$ATPase has been reported to reduce the transporter's capacity to rapidly export sodium ions due to its abnormal aggregation around synapses. Such formations have been shown to induce calcium excitotoxicity. The role of α3 subunit in ALS has been reported (Ruegsegger et al., 2016). Mitochondrial and $N^+K^+$ATPase dysfunctions are common features of PD or AD (Johri & Beal, 2012). Given the recent reports that α-syn and Aβ binding to the α3 subunit of $N^+K^+$ATPase is detrimental, it can be predicted that these neurodegenerative disorders are associated with brain oxidative phosphorylation disruptions, which may end up with compromised energy for both the supplier and the processor (Shrivastava et al., 2017).

### *The Consequences of Plasma Membrane-amyloid Protein Interaction*

Disruption of plasma membrane integrity is a common mechanism of AD, PD, and possibly other proteinopathies. Membrane lipids are essential in Aβ and α-syn aggregation and related toxicity (Bae et al., 2013; Bharadwaj et al., 2018; Dias & Nylandsted, 2021; Fecchio et al., 2013). Lipids, lipid-induced conformational changes, and confinement of proteins to small local regions such as a vesicle membrane, increasing the possibility of intermolecular interactions and promoting aggregation, are factors that affect the accumulation process (Bharadwaj et al., 2018; Dikiy & Eliezer, 2012; Jang et al., 2013; Pacheco et al., 2012; Reynolds et al., 2011; Shrivastava et al., 2017). On the other hand, misfolded proteins can also increase plasma membrane permeability. Protein-induced membrane stiffness, membrane thinning, and deformations, as well as detergent-like effects and pore formation, are important molecular mechanisms that produce this result (Bode et al., 2017; Bode et al., 2019; Danzer et al., 2007; Dettmer et al., 2017; Dias & Nylandsted, 2021; Lee et al., 2014; Ouberai et al., 2013; Pacheco et al., 2012; Pieri et al., 2012; Reynolds et al., 2011; Rovere et al., 2019; Sasahara et al., 2013; Shi et al., 2015; Varkey et al., 2010). In addition, neurons must resist increased oxidative stress and associated lipid peroxidation resulting from energy metabolism or protein aggregation (Alecu & Bennett, 2019; Bae et al., 2013; Gaschler & Stockwell, 2017; Yasumoto et al., 2019). To relieve the excessive surface pressure exerted by added proteins during protein aggregation, membrane lipids can be reshaped by repositioning out of the plane of the membrane, resulting in thinning or warping of the membrane (Ouberai et al., 2013; Shi et al., 2015). Excessive membrane curvature has also been reported for α-syn and Aβ oligomers (Bode et al., 2019; Reynolds et al., 2011; Shi et al., 2015; Varkey et al., 2010), where it causes organelle deformations and dysfunctions. During the aggregation process, oligomers cause extensive lipid extraction from the bilayer such as a detergent-like effect. Over time, this may cause the membranes to rupture (Bode et al., 2019; Reynolds et al., 2011). The degree of membrane effects caused by protein accumulation is thought to depend on the concentration of membrane-bound protein (Fecchio et al., 2013; Reynolds et al., 2011; Shi et al., 2015). Membrane disruption results in massive and immediate cytotoxicity. Pore formation associated with amyloidogenic proteins such as α-syn and Aβ, which disrupt membrane integrity, has been described in various articles (Bode et al., 2017; Danzer et al., 2007; Lorenzen et al., 2014; Pieri et al., 2012; Volles et al., 2001). It has been stated that the calcium flux occurring through these structures is comparable to that produced by pore-forming toxins (Danzer et al., 2007; Dias & Nylandsted, 2021).

## DNA Defects

Data shows that many peptides play a role in neurodegenerative diseases associated with DNA and/or RNA under physiological or pathological conditions. In particular, associations with DNA can cause irreversible phenotypic changes in post-mitotic neurons.

Today, we know that Aβ fragments (Aβ38, Aβ40, Aβ42, and Aβ43) can transport to the nucleus and form aggregates (Barucker et al., 2014; Buckig et al., 2002; Gezen-Ak et al., 2018; Ma et al., 2012). Endogenous Aβ1-42 in the cytoplasm can migrate to the nucleus from the cytoplasm in response to antibiotic treatment in primary cortical neurons (Gezen-Ak et al., 2017). Aβ fragments have the ability to affect gene expression by directly binding to the regulator regions of the specific genes (Bailey et al., 2011; Maloney & Lahiri, 2011; Ohyagi et al., 2005). The presence of Aβ fragments in the nucleus not only affects the transcription but may also change DNA structure. Studies reported that Aβ induces double-stranded DNA condensation in vitro (Yu et al., 2007). Aβ triggers the B turn to Z turn helix alteration in DNA topology in AD brains. In vitro studies also demonstrated that Aβ1-42 and Aβ1-16 can induce a conformational change in pUC 18 supercoiled DNA (Hegde et al., 2004). It was suggested that intracellular zinc is essential for Aβ-nucleic acid interactions (Khmeleva et al., 2016). Furthermore, Aβ has DNA-nicking activity, which triggers DNA damage. The oligomer form of Aβ1-42 has a greater impact on DNA-nicking activity than its monomer and fibrillar forms (Suram et al., 2007).

We know from the earliest studies that α-syn is localized in the cell nucleus (Maroteaux et al., 1988). Further studies showed that α-syn can directly bind to nuclear DNA and become involved in the regulation of the expression of genes (Gezen-Ak et al., 2022; Surguchev & Surguchov, 2017). It also binds histones and participates in epigenetic regulatory mechanisms (Surguchov, 2023). α-syn and tau can alter DNA conformation and stabilize Z-DNA conformation, which changes minor and major grooves and therefore affects gene expression patterns in neurodegeneration (Jiang et al., 2018; Standaert & Childers, 2022; Vasudevaraju et al., 2012). Interestingly, it was reported that physiological α-syn binds double-stranded DNA and becomes involved in leading the DNA repair by non-homologous end-joining (NHEJ). The researchers hypothesized that cytoplasmic deposition of α-syn reduces its nuclear levels, increases double-strand DNA breaks in neurons, and may contribute to apoptosis (Schaser et al., 2019). Increased phosphorylated α-syn levels can lead to DNA double helix breakage and can be associated with markers of cellular senescence in α-syn transgenic models (Yoon et al., 2022) (Figure 1.2).

## Mitochondrial Dysfunction

Mitochondrion is an organelle of endosymbiotic origin, surrounded by a double membrane. In addition to being the energy center of the cell, it is the most important trigger of apoptosis, a regulator of calcium homeostasis, and a producer that ensures membrane formation and responsible for the creation of many catalytic centers. While the majority of mitochondrial proteins are encoded in the nuclear genome and are post-translationally delivered to the mitochondria, a small number of mitochondrial proteins are encoded in the mitochondrial genome (Kornmann, 2014; Pinto & Moraes, 2014).

Mitochondrion is a dynamic organelle, and mitochondrial morphology and transport within the cell affect the metabolism of neurons. Normal mitochondrial function is necessary for neurons to maintain their polarity. Mitochondrial dysfunctions are considered one of the early hallmarks of neurodegeneration (Jeong, 2017; Lezi & Swerdlow, 2012; Reddy, 2009).

Free radicals, which cause significant mitochondrial damage, are constantly produced in eukaryotic cells and must be balanced by antioxidants to maintain redox balance. Imbalances between harmful ROS and the antioxidant defence system cause oxidative stress. ROS triggers nucleic acid breaks, enzyme inactivation, polysaccharide depolymerisation, lipid peroxidation, and other cellular events that cause damage, resulting in cell death. Mitochondria, the largest ROS producers, have a key role in the function and survival of neurons in the brain, the human body's most energy-consuming organ. While long-living, energy-dependent neurons perform very limited glycolysis, they obtain their energy largely from aerobic oxidative phosphorylation (OXPHOS) in mitochondria (Yan et al., 2013). The post-mitotic cells, like neurons, contain high amounts of unsaturated fatty acids in their membranes and are, therefore, very sensitive to ROS and mitochondrial dysfunction (Pinto & Moraes, 2014; Yan et al., 2013). Many neurodegenerative disorders show mitochondrial defects. Mitochondrial dysfunction may occur due to abnormal mitochondrial DNA (mtDNA), mutated nuclear proteins interacting directly or indirectly with mitochondria, or unknown causes. In most cases, it is still unclear where mitochondria stand in the pathology of diseases that cause loss of function in neurons and, ultimately, neuron death and in which step of neurodegeneration mitochondrial dysfunction plays a role (Lezi & Swerdlow, 2012; Reddy, 2009).

Mitochondrial dysfunction, mtDNA mutations, and oxidative stress are considered the most prominent features of age-associated neurodegenerative disorders such as AD, PD, ALS, and others (M. Pinto & C. T. Moraes, 2014; Yan et al., 2013).

## Mitochondrial Dysfunction in AD

Communication between neurons, ATP production, and synaptic activity must be proper for normal cognitive activities. However, these activities are interrupted in elderly individuals and AD patients. Studies showed that Aβ fragments can be transported to mitochondria. Those fragments aggregate in mitochondria under certain conditions (Mao & Reddy, 2011) and disrupt neuronal functions before the clinical onset of AD (Jeong, 2017).

Mitochondrial traffic appears to be highly disrupted in the neurons of AD patients. In healthy neurons, mitochondria move anterogradely from the cell body towards axons, dendrites, and synapses, producing ATP. Then, when sufficient ATP supply is provided, they return to the cell body by retrograde transport. This mechanism is abnormal in AD neurons due to inactive or damaged mitochondria. Synaptic mitochondria defects also affect the release of synaptic vesicles and neurotransmitters through disruptions in ATP production (Reddy, 2009). Studies in Aβ1-42 treated cells have shown reduced respiratory capacity and inhibition of some essential enzymes, including pyruvate dehydrogenase, cytochrome oxidase, and α-ketoglutarate, in isolated mitochondria (Lezi & Swerdlow, 2012). Although there was no significant cell death in rat primary neuron culture cells treated with Aβ1-42 at sublethal concentrations for a short time, intracellular mitochondrial movements were rapidly impaired (Rui et al., 2006). Another observation is that the Aβ1-42 treatment increased the mtDNA copy number in cells, but the number of mitochondria with normal morphology decreased (Lezi & Swerdlow, 2012).

It has been reported that OXPHOS complex IV deficient neurons are seen at a higher rate in AD compared to normal brains, and complex IV deficiency may result from increased levels of mutant mtDNA (Lezi & Swerdlow, 2012; Reddy, 2009). Aβ1-42 plaque density was found to be increased in the transgenic animal models (APP/Ld), carrying mutation that inactivates mitochondrial DNA polymerase γ proofreading (Kukreja et al., 2014). The mRNA levels of OXPHOS subunits had changed in the brains or blood of mild cognitive impairment (MCI) or AD patients (Lunnon et al., 2017; Manczak et al., 2004). In AD brains, Aβ was found to interact with alcohol dehydrogenase enzyme (ABAD), which is a mitochondrial matrix protein and causes oxidative stress and weakening of the respiratory chain (Federico et al., 2012). Aβ1-42 is suggested to interact with Drp1, cytochrome-c, cyclophilin D, and VDAC proteins in mitochondria (Pagani & Eckert, 2011). It is also suggested that Aβ affects proteins involved in mitochondria fission and fusion in neurons and astrocytes (Kim et al., 2017; Zysk et al., 2023). Therefore, mitochondrial dysfunction is observed in AD patients through several pathways, including mtDNA replication, transcription, translation, mitochondrial metabolism, movement, and dynamics (Gezen-Ak et al., 2022, 2023).

Amyloid precursor protein (APP) cleavage by γ-secretase activity occurs predominantly in "mitochondria-associated ER membranes (MAM)," where the endoplasmic reticulum physically and biochemically combines with mitochondria. MAM is responsible for calcium homeostasis, lipid transfer, mitochondrial metabolism-dynamics, apoptosis, and mitophagy (Barazzuol et al., 2020). MAM function and ER–mitochondria relationship increase in AD; therefore, it is suggested that AD is also a disease of ER–mitochondrial communication and MAM dysfunction (M. Pinto & C. T. Moraes, 2014). This hypothesis also explains disorders that occur with diseases other than Aβ cascade and depend on mitochondrial function, such as increases in cholesterol levels, changes in fatty acid and phospholipid metabolism, and, most importantly, imbalances in calcium metabolism. However, it is also noted that these complex mitochondrial dysfunctions are not specific to a particular disease, and such an association may not always be a correct proposition (Lezi & Swerdlow, 2012).

According to the mitochondrial cascade hypothesis of sporadic AD, heredity determines the baseline mitochondrial functions, hereditary and environmental factors determine the rate of mitochondrial changes associated with aging, and, finally, the baseline mitochondrial functions of the individual and the rate of change in these functions indicate the potential for AD to occur (Swerdlow et al., 2014). According to this hypothesis, mitochondrial dysfunction promotes the misfolding and accumulation of Aβ. Whether or not mitochondrial dysfunction is the cause of misfolding and accumulation of Aβ as advocated by this hypothesis, the presence of physiological or pathological Aβ in mitochondria may change the development, progression, and severity of the disease with the contribution of the genetic background (Gezen-Ak et al., 2022, 2023).

### Mitochondrial Dysfunction in PD

Similar to AD, the studies showed an association between PD and mitochondrial dysfunction. The first evidence comes from genetics studies which showed the mitochondrial homeostasis- related genes *SNCA*, *PTEN-induced kinase 1 (PINK1)*, *PARKIN*, *Leucine-rich repeat kinase 2 (LRRK2)*, *DJ1(PARK7)*, *ATPase 13A2 (ATP13A2)*, and *vacuolar protein sorting-associated protein 35 (VPS35)* mutations in familial PD (Bonifati et al., 2003; Funayama et al., 2002; Grunewald et al., 2019; Kitada et al., 1998; Valente et al., 2004). The second evidence is a chemical, 1-methyl-4-phenyl-1, 2,3,6-tetrahydropyridine (MPTP), which blocks the OXPHOS complex I function and leads to PD-like symptoms (Langston et al., 1983; Ramsay et al., 1986).

Although the etiology of the disease is still not fully known, it is suggested that mitochondrial dysfunctions, especially those caused by complexes I and IV in the electron transport chain, are crucial in the development of PD. The high susceptibility of substantia nigra neurons to mitochondrial stress and mtDNA damage is thought to result from the anatomical structure of these neurons. It is suggested that their long, unmyelinated axons increase their susceptibility to damage. Studies have shown that α-syn, the main component of LBs, carries an N-terminal mitochondrion targeting sequence (Devi et al., 2008), and it can be located in the inner and outer mitochondrial membrane and also (Cole et al., 2008; Devi et al., 2008; Li et al., 2007) in the mitochondrial matrix of neurons (Liu et al., 2009; Zhang et al., 2008). α-syn in the mitochondrial membrane attenuated the mitochondrial complex I activity in the brains of PD cases (Devi et al., 2008). Additionally, increased expression of α-syn has been shown to increase nigral pathology by weakening mitochondrial function and increasing oxidative stress. It is also suggested that α-syn plays a role in the mitochondrial apoptotic pathway (Federico et al., 2012). The oligomeric and dopamine-modified forms of α-syn can interact with the TOM20 receptor on the mitochondria's outer membrane and block its binding ability to TOM22, which results in impaired mitochondrial protein import. This event can trigger the production of ROS and the loss of mitochondrial membrane potential (Grunewald et al., 2019).

High frequency of mtDNA point mutations, deletions, or changes in mtDNA copy number were reported in PD patients (Bender et al., 2006; Dolle et al., 2016; Grunewald et al., 2016; Pyle et al., 2016; Pyle et al., 2015). The point mutations or deletions in mtDNA lead the defect in mtDNA transcription and also result in the replication errors (Casoli et al., 2014; Coskun et al., 2004; Hoekstra et al., 2016). Additionally, the relationship of mtDNA polymorphisms and haplotype blocks with PD has been shown (Giannoccaro et al., 2017). The mtDNA-encoded genes and nuclear DNA–encoded *PARKIN* and *PINK1* expressions are altered depending on age in sporadic and familial PD cases compared to the healthy controls (Gezen-Ak et al., 2020). Studies showed that Parkin directly interacts with mitochondrial transcription factor A (TFAM) at the mtDNA D-loop and enhanced mitochondrial transcription (Kuroda et al., 2006).

Recently, the localization of DJ-1, α-syn, PINK1, and Parkin in MAM has been reported. It has been suggested that PINK1 and Parkin interact with several MAM proteins and may act as regulator of the ER–mitochondria communication. The hypothesis arises from the fact that defects in PINK1 and/or Parkin in PD may disrupt MAM and its associated functions, resulting in neuronal dysfunction (Barazzuol et al., 2020).

### Mitochondrial Dysfunction in ALS

Mitochondrial dysfunction is one of the earliest hallmarks of ALS. Studies conducted in ALS patients have determined that mitochondria with strange, giant, and spiky morphologies are more numerous than those with normal morphology. Increased mtDNA deletions have also been found in muscle and brain tissue in ALS patients. Abnormal mitochondrial accumulations in the axons of neurons have been reported. Changes in mitochondrial OXPHOS activities have been reported in ALS patients and animal models (Lezi & Swerdlow, 2012). Decreased complex IV activity has been shown in sporadic ALS cases. Moreover, in some patients with mtDNA mutations, a specific ALS phenotype has been described (Lezi & Swerdlow, 2012; Milena Pinto & Carlos T Moraes, 2014). SOD1, TDP-43, FUS, or C9orf72 mutations were suggested to lead to protein accumulation in mitochondria, disrupt mtDNA transcription, change mitochondrial membrane potential, decrease ATP generation, increase cellular ROS, disrupt calcium homeostasis, induce apoptosis, effect mitochondrial morphology and transport, impair mitophagy, and/or disrupt ER–mitochondria contacts (Smith et al., 2019). MAM disruption was reported to be a common pathological feature in ALS (Watanabe et al., 2023). Motor neurons express large numbers of AMPA receptors in the postsynaptic terminal so that they are exposing them to excessive calcium influx upon stimulation. Because of the low expression of calcium-buffering proteins, these neurons are more dependent on mitochondria for calcium homeostasis than other cells and are therefore more vulnerable to calcium dysregulation than other neuronal populations. Miscommunication between ER and mitochondria resulted in calcium dysregulation, and this event was suggested to be a primary cause of motor neuron death in ALS (Smith et al., 2019).

## Cytoskeletal Abnormalities and Disruption of Cellular and Axonal Transport

The functional complexity of the brain is associated with the complex architecture of networks established by millions of neurons and glial cells. The functionality of this complex structure is based on the coordination and rearrangement of the cytoskeleton. Neurons are highly polarized cells with a complex structure. Correct establishment and maintenance of neuronal morphology is essential for the central nervous system to function adequately. Microtubules, part of the cytoskeleton, play essential roles during neuronal development by supporting neuronal migration, neuronal differentiation, axon guidance, and dendrite arborisation and maintaining neuronal connections after they are formed. On the other hand, nerve transmission occurs along axons. Axon diameter, which affects the transmission of electrical signals, may change depending on neuron type, organelle distribution, the activation state of the neuron, or environmental tension. Changes in axon calibre depend on the axonal cytoskeleton. Dendritic spines are small actin-rich dendritic protrusions. Dendritic spine density, morphology, and size are important for learning and memory (Avila et al., 2020; Soliman et al., 2022).

### Cytoskeleton

The structure, function, and plasticity of neurons depend highly on microtubule organization and microtubule regulatory proteins. Thus, alterations in such organization of cytoskeleton and related proteins in aging and neurodegeneration is a common feature. A significant decrease in microtubule density has been observed in neurons of older people. On the other hand, it has been reported that a significant reduction in the number and length of microtubules in pyramidal neurons in the brain of AD cases was observed compared to healthy individuals (Cash et al., 2003). Again, an increase in tubulin glutamylation has been reported in Alzheimer's patients (Vu et al., 2017). Decreased levels of acetylated tubulin, an indicator of a more stable microtubule subpopulation, have been reported in other neurodegenerative disorders such as HD (Dompierre et al., 2007). Significant changes in microtubule-associated proteins (MAP) have been reported during aging and in neurodegenerative diseases such as AD. Examples of these changes include an age-related decrease in MAP2 in the hippocampus of rodents and an abnormal location of MAP5 in AD brains (Chauhan & Siegel, 1997; Di Stefano et al., 2001). One of the most important changes seen in neurodegenerative diseases is tau-related changes. In AD and other tauopathies, tau accumulates in the somato-dendritic compartment of neurons in affected brain regions, forming NFTs (Braak & Braak, 1996). It has been reported that the increase in tau phosphorylation is associated with the redistribution of tau from the axon to the somato-dendritic compartment (Delacourte et al., 1990; Holzer et al., 1994; Liu et al., 1993; Probst et al., 1996). Other post-translational modifications, such as O-glycosylation, ubiquitination, nitration, and truncating, may also participate in pathological changes in tau localization and function (Alquezar et al., 2020). While impairment of the capacity of tau to bind to microtubules is thought to lead to disruptions in microtubule organization in affected neurons, questions have been raised by the observation that degradation of tau, whether chronic or acute, does not have a significant effect on axonal microtubule stability (Harada et al., 1994; Tint et al., 1998). Findings that tau may have a protective effect in various special situations in experiments where tau is knocked down further question the "loss of function" hypothesis (Lopes et al., 2016; Roberson et al., 2007; Tai et al., 2020). Therefore, the "toxic gain of function" hypothesis remains on the agenda, expressed as disease-related post-translational modifications and oligomeric tau species inducing neurotoxicity and participating in disease processes (Selkoe & Hardy, 2016). However, it is not known whether the "gain of toxic function" results from a disorder in microtubules or a change in the relationship of tau with other proteins to which it binds. For example, tau's non-microtubule-binding N-terminal region (NTR) has been reported to interact with many proteins that may contribute to its physiological and pathological role (Trushina et al., 2019). Various analyses have shown that proteins interacting with NTR may play a role in fundamental cellular mechanisms (Brandt et al., 2020). On the other hand, the "tau spread hypothesis," developed to explain the apparent and regular spread of NFTs in AD, has also recently come to the fore (Uemura et al., 2020). Accordingly, pathological forms of tau can be transferred between neurons. Various disease models have shown that extracellular tau proteins are internalized by cells and that internalized tau can induce intracellular tau aggregates (Kfoury et al., 2012; Rauch et al., 2020; Takahashi et al., 2015). It is thought that the propagation properties of tau may depend on its post-translational modifications, aggregation state, and isoforms (Soliman et al., 2022; Vaquer-Alicea et al., 2021).

### Axonal Transport and Motor Proteins

Axonal transport is impaired in AD and other tauopathies (Jellinger, 2010; Schindowski et al., 2008; Stokin & Goldstein, 2006). Proteolytic cleavage of APP occurs before differentiation into axonal transport vesicles (Muresan et al., 2009). APP acts via rapid axonal transport and within vesicular complexes containing presenilin and BACE1 and acts as a receptor for anterograde motor kinesin. APP misregulation can lead to fast axonal transport impairment (Pigino et al., 2009). On the other hand, many factors disrupt Aβ-related neuronal transport pathways (Shah et al., 2009). It is also known that hyperphosphorylation of tau causes its inability to bind to microtubules, destabilizing and deteriorating microtubule structures, thus affecting axonal transport. On the other hand, it has been reported that 3R and 4R tau proteins show different effects on mitochondrial axonal transport (Stoothoff et al., 2009). It may also be the case that changes in axonal transport switch normal tau to a pathogenic state by triggering various stress signals (Falzone et al., 2009). On the other hand, it has been reported that tau deletion or overexpression does not affect axonal transport (Yuan et al., 2008).

α-syn is highly expressed in neurons found at significant levels in presynaptic terminals and is mainly transported slowly. It is conceivable that the significant decline in axonal transport due to age may lead to α-syn accumulation over time and lead to axonal transport disorders seen in synucleinopathies (Jellinger, 2010; Li et al., 2004). In DLB and PD, small α-syn aggregates accumulate in presynaptic terminals and are associated with the loss of dendritic spines (Kramer & Schulz-Schaeffer, 2007).

In SOD-1 transgenic models of ALS, the transport of neurofilament proteins is delayed, and defects in the rapid transport of mitochondria are early defects in neurodegeneration. Mutant SOD1 may affect motor neurons' function and viability in ALS by impairing axonal transport and protein turnover (Rothstein, 2009; Strom et al., 2008).

Striatal degeneration is observed in HD. Huntingtin (htt) protein accumulates in vesicles and moves by rapid axonal transport in this disease. Various animal models have shown that pathological polyQ repeats and the htt protein mediate these changes. Htt is involved in the regulation of transcription, trafficking of certain proteins, vesicle transport, postsynaptic signaling, and cell death. Disturbances in normal protein function and toxic gain of function of mutant htt lead to the disruption of these basic cellular mechanisms (Gil & Rego, 2008; Jellinger, 2010). Some studies reported that the polyQ repeat aggregates disrupt fast axonal transport. Besides, enhanced sensibility of striatal neurons to mutant htt directed axonal transport defects was also reported (Her & Goldstein, 2008).

## Synaptic and Neuronal Network Dysfunction

The synapse is the main information transfer unit in the nervous system. Synapses are computational devices that transmit and transform action-potential encoded information between neurons (Sudhof, 2012). In neuronal networks, synapses work for the formation of learning, memory, and behavior (Taoufik et al., 2018). Synapse stabilization and elimination are essential events for proper synaptic function and neuronal network.

Neurotransmitter synthesis, membrane integrity, endomembrane system, ion balance, mitochondrial function, and axonal transport and also cytoskeletal components, vesicular trafficking (Rab and SNARE proteins), exocytosis, neurotransmitter receptors, protein, lipid, and RNA metabolism must be tightly regulated for synaptic integrity (Tonnies & Trushina, 2017; Wilson et al., 2023). Synaptic failure is one of the earliest hallmarks of neurodegeneration (Arendt, 2009), and when we look at all these cellular events, it is unsurprising that synaptic defects occur in neurodegeneration. Glial cells also play a critical role in synaptic function, and their dysfunctions lead to the alteration of synaptic contacts, resulting in abnormal networking and behavioral abnormalities (Wilton et al., 2019).

α-syn is a presynaptic protein, and the synuclein family members have a role in controlling synapse size and synaptic vesicle distribution (Vargas et al., 2017). In DLB brains, the majority of α-syn aggregates are in presynaptic terminals, resulting in complete loss of dendritic spines at the postsynaptic cell. Overexpression of α-syn or its mutants inhibits neurotransmitter release, alters growth factor secretion, decreases the size of the synaptic vesicle recycling pool, and changes the structure of the nerve terminal, triggering axonal degeneration and autophagy in neurons or glia (Sengul et al., 2021; Taoufik et al., 2018; Vargas et al., 2017). These observations suggest that synaptic loss is a central event in the initiation of neurodegeneration in synucleopathies (Kramer & Schulz-Schaeffer, 2007; Taoufik et al., 2018). Furthermore, some genes in familial PD are encoded synaptic proteins such as SYNJ1, DNAJC6, and DNAJC13 (Wilson et al., 2023).

Proper synaptic function needs to have low levels of Aβ, but pathological Aβ levels induce synaptotoxicity in AD (Palop & Mucke, 2010). During AD progress, before the formation of pathological hallmarks (extracellular amyloid plaques and intracellular NFTs), Aβ deposits begin to be enriched at synapses. Aβ treatment leads to a decreased number of dendritic spines, loss of synaptic proteins, and changes in the level of neurotransmitter receptors on the postsynaptic membrane. These events disrupt synaptic plasticity, LTP, and long-term depression (LTD) (Chen et al., 2019). Low or high molecular weight Aβ oligomers are present in the synaptic regions. Low molecular weight Aβ oligomers were suggested to be more toxic for synapses (Chen et al., 2019). Aβ oligomers can also trigger uncontrolled presynaptic calcium influx and enhance exocytosis of neurotransmitter release, therefore inducing spontaneous postsynaptic activity (Brito-Moreira et al., 2011). The cell surface molecules on postsynaptic membranes can serve as receptors for Aβ such as neurotransmitter receptors, GPCRs, p75 neurotrophin receptor (p75NTR), Ephrin A4, Ephrin B2, $PrP^{C}$, neuroligin, and leukocyte immunoglobulin-like receptor subfamily B member 2 (Chen et al., 2019; Ratan et al., 2023; Yaar et al., 1997). For example, Aβ oligomers bind to NMDA receptors at the postsynaptic plasma membrane and can induce synaptotoxic effects depending on the subunit composition of NMDA receptors (Decker et al., 2010; Tackenberg et al., 2013).

## Neuroinflammation

The blood–brain barrier (BBB) of the cerebrovascular endothelium is a significant factor that grants only restricted access to the immune cells and immune mediators to the central nervous system. The concept of the brain being immunoprivileged is debated in relatively recent studies (Dias-Carvalho et al., 2024). Glial cells, including astrocytes, microglia, oligodendrocytes, and neurons, may all contribute to the immunological responses in the central nervous system (Carson et al., 2006; Tian et al., 2009). Recent studies indicated that T cells could migrate into the central nervous system, although no significant pathological condition exists. They may use CSF and localize on meningeal membranes to be used as a first line of defense (Filiano et al., 2017; Strazielle et al., 2016). Agrawal et al. reported that peripheral immune cells, 90% of T and the rest of B cells, exist in the CSF. This indicates their constant interaction with central nervous system antigens (Agrawal et al., 2022). The brain and the rest of the body can also interact via cytokines, small peptides released in response to injury or infection. These cytokines can be classified as pro- or anti-inflammatory (Dias-Carvalho et al., 2022; Yarlagadda et al., 2009). TNF-α, IL-6, and IL-1β, which are pro-inflammatory cytokines, may damage the BBB by destroying tight junctions of the neurovascular unit and, therefore, altering its permeability (Dias-Carvalho et al., 2024; Disdier et al., 2017). Such infiltration activates resting microglia, which in turn secrete pro-inflammatory cytokines. Such interaction evolves into further interaction between microglia, astrocytes, and the surrounding tissue, a complex process called *neuroinflammation* (Sun et al., 2022). Astrocytes known to support neurons by supplying trophic factors alter this secretion process in pathological conditions (Sengul et al., 2021). Neuroinflammation may lead to the disruption of cellular mechanisms, including synaptic transmission, post-translational modifications, protein cleavage and aggregation, cell death, and neurogenesis (Bi et al., 2021; Chugh et al., 2015; Connolly et al., 2022; Javadpour et al., 2021; Lodeiro et al., 2017; Soto-Faguas et al., 2021).

One significant effect of aging on the brain is the increased permeability of the BBB (Batterman et al., 2021; Chu & Pratico, 2009). The aging brain significantly alters the function of the neuroglial cells, which in turn affects the brain's function (Lopez-Teros et al., 2022; Verkerke et al., 2021). Astrocyte senescence resembles astrogliosis. How astrocytes regulate the BBB is reviewed by Pociute et al. (Pociute et al., 2024). However, astrocytes' pro-inflammatory response is much smaller (Lopez-Teros et al., 2022; Verkerke et al., 2021). Aging also impacts microglia. These changes include morphological alterations like shortened processes, attenuated arborization, or abnormal cytoplasmic structures, which may lead to high inflammatory responses or dysfunction (von Bernhardi et al., 2010).

All neurodegenerative disorders are accompanied by neuroinflammation and a specific immune response as a part of their pathological context (Brosseron et al., 2014). A recent study indicated that Aβ, α-syn, and their combination exert significant but different alterations in the immune response of neuronal cells (Keskin et al., 2023). For a long time, it has been known that the cerebral cortex of AD cases had a significantly higher number of reactive microglia than their age-matched controls (Carpenter et al., 1993). Even the AD brains had higher levels of the IL-1β (Cacabelos et al., 1994). Although many studies have been conducted, there are conflicting results on the peripheral cytokines, whether they are involved in AD or not (Dursun et al., 2015; Gezen-Ak et al., 2013). Brosseron et al. reviewed many controversial or inconclusive reports, even on the frequently studied cytokine levels such as TNFα, IL-1, and IL-6 (Brosseron et al., 2014). A meta-analysis and systematic review by Shen et al. included 170 studies focused on peripheral inflammatory markers and noted that their levels were significantly altered in AD (Shen et al., 2019). We know that peripheral cytokines may disrupt brain processes and induce a local response in the AD brain, where we see reactive astrocytes and microglia (Sakakibara et al., 2019). In AD brains, microglia phagocyte Aβ, yet the internalized Aβ fate is unclear. Although whether Aβ is degraded or results in other processes is unknown, studies indicate that the microglia can retain Aβ for a long time (Paresce et al., 1997). Another study demonstrated that the activated microglia gain the capacity to degrade Aβ in its lysosomes (Majumdar et al., 2007). Given that the most abundant glial cell in central nervous system is the astrocytes, their role in the development and progression of AD should not be ignored. CSF levels of GFAP and S100B, which are thought to be neuroinflammatory markers, increased in AD cases (Fukuyama et al., 2001; Petzold et al., 2003). Astrocytes produce neprilysin, an enzyme involved in Aβ clearance, yet its expression decreases in older mice (Apelt et al., 2003). Besides, APP/DeltaPS1 double-transgenic mice overexpressing neprilysin had reduced Aβ and amyloid load, oxidative stress, and inflammation (El-Amouri et al., 2008). Studies are reporting that cytokines like IL-1β, IL-6, TNF- α, IFN-γ, and TGF-β can induce γ-secretase and increase cleavage of APP to produce Aβ (Liao et al., 2004). It is also known that Aβ induces glial cells and a cytokine response (Gonzalez-Reyes et al., 2017).

Literature indicates that neuroinflammation is not only a significant part of AD but also a part of PD. As in AD, inflammation has two sides in PD. Whether it might be the cause or a contributing factor is debated. For a while, we were aware that non-steroidal anti-inflammatory drug intake in mid-life was associated with attenuated risk of PD development (Wahner et al., 2007). In 2018, Lutters et al. reported that encephalitis lethargica, an influenza virus, showed PD-like symptoms (Lutters et al., 2018). Other studies similarly reported that neuroinflammation caused by a pathogen and reaching the basal ganglia might lead to PD (Pajares et al., 2020). Besides, the active microglia number in the substantia nigra of PD cases was higher than that of age-matched healthy controls (Mirza et al., 2000). Given the fact that dopaminergic neurons are located in the substantia nigra, and that part exhibits more active microglia, one may assume that the PD-related brain region is more prone to neuroinflammation (Yan et al., 2014). However, single-nucleus RNA sequencing studies indicated that microglia, astrocytes, and oligodendrocytes were involved in PD-related neuroinflammation (Smajic et al., 2022). In addition to that, a significant amount of CD8-positive T cell infiltration was observed in substantia nigra pars compacta of PD cases (Galiano-Landeira et al., 2020). A hypothesis also suggests pathogenic infections can manifest PD through the gut or nasal cavity (Braak et al., 2003). Such a hypothesis arises from the potential of pathogens to induce an immune response, which in turn triggers α-syn misfolding and spreading via the olfactory tract or the vagal nerve (Rietdijk et al., 2017; Sharma et al., 2019). On the other hand, the studies noted that either MPTP or 6-hydroxydopamine (6-OHDA)-induced PD models exhibit neurodegeneration related to increased cytokines, especially IL-6 and TNF-α (Barnum et al., 2014; Ciesielska et al., 2007). The molecular action of the cytokines requires a JAK-STAT pathway (Seif et al., 2017). *In vitro* and *in vivo* studies, including microglia, indicated that α-syn induces IL-6, TNF-α, and iNOS via JAK-STAT, and inhibition of the pathway resulted in suppression of microglial response in the substantia nigra (Qin et al., 2016). NF–κB signaling, which regulates inflammatory responses, is also observed in the substantia nigra of PD cases and MPTP-induced PD mice models, and inhibition of this cascade suppressed cytokine response and attenuated dopaminergic neuronal loss (Ghosh et al., 2007). The molecular mechanisms contributing neuroinflammation in PD can be augmented with other examples (Cossu et al., 2023).

## Neuronal Death

Historically, cell death was classified into three distinct categories - apoptosis, autophagy, and necrosis - based primarily on morphological characteristics and the mechanisms of cellular debris clearance. As we know, apoptotic cells exhibit shrinked cytoplasm, condensed chromatin, and fragmented nuclear DNA. They had blebbing plasma membrane forming small vesicles which are phagocytosed by neighboring cells. The cells that demonstrate autophagy manifest extensive cytoplasmic vacuoles and lysosomal degradation. The necrotic cells display no distinctive features of apoptotic or phagocytic cell deaths. They end with the disposal of cell particles, yet no phagocytic and lysosomal involvement occurs.

On the other hand, the last couple of decades have seen discussions about how to define different kinds of cell deaths. At last, the Nomenclature Committee on Cell Death published the terminology that can be used to define different kinds of cell death combining morphological grounds; major cell death modalities on a genetic, biochemical, pharmacological, and functional backgrounds, essential or accessory aspects of the death process and the identification of dead cells with irreversible plasma membrane permeabilization or complete cellular fragmentation (Galluzzi et al., 2018). All mechanisms defined as regulated cell deaths, and all is initiated and advanced by molecular mechanisms that exhibit some significant associations. Each type of regulated cell death may exhibit morphological features. That spectrum may range from a fully necrotic type to a fully apoptotic and an immunomodulatory type. Even the immune profile can range from anti-inflammatory to pro-inflammatory or immunogenic. This perspective enabled the committee to name more than 12 cell death mechanisms (Galluzzi et al., 2018).

The most studied cell death mechanisms in neurodegenerative disorders seem to be apoptosis, autophagy, necrosis, necroptosis, ferroptosis, and pyroptosis. Before discussing the related studies, the basic mechanisms involved in these cell death modes will be explained.

Intrinsic apoptosis is initiated by a variety of microenviromental alterations like growth factor withdrawal, DNA damage, ER stress, increased ROS, replication stress, and microtubular disruptions. Widespread mitochondrial outer membrane permeabilization is a significant step for this type of regulated cell death, and it is irreversible. Executioner molecular pathways include Bax, Bcl2, and caspases, mainly CASP3 (Brumatti et al., 2010; Czabotar et al., 2014; Galluzzi et al., 2016; Galluzzi et al., 2018; Pihan et al., 2017; Roos et al., 2016; Tait & Green, 2010; Vitale et al., 2017). Extrinsic apoptosis can be started with the disruptions and alterations of the extracellular microenvironment. It is usually advanced by two types of plasma membrane receptors. The death receptors require binding of the cognate ligands in order to activate. Death receptors include FAS (CD95 or APO-1), TNFR1, TRAILR1 (or DR4), and TRAILR2 (or DR5). The dependence receptors activate when their specific ligand level becomes less than a certain threshold. The mechanism involves CASP8 and CASP3 as an executioner caspases (Aggarwal et al., 2012; Fleten et al., 2016; Galluzzi et al., 2018; Mehlen & Bredesen, 2011; von Karstedt et al., 2017; Wajant, 2002).

Necrosis usually induces an inflammatory response via damage-associated molecular patterns released from the dying cells. Freezing, burning, and mechanical trauma can induce necrosis (Feoktistova & Leverkus, 2015; Goel et al., 2022; Sachet et al., 2017; Silva, 2010; Vanlangenakker et al., 2012). Generally, no definite cellular signaling pathway is known for necrosis, alhough some forms have well-defined mechanisms such as mitochondrial permeability transition pore activation (Feoktistova & Leverkus, 2015; Kinnally et al., 2011; Vanlangenakker et al., 2012). Necroptosis is also a regulated form of cell death. It shows morphological features of necrosis, and the signaling cascade involves RIP kinases, necrosome formation, and activation of MLKL (Galluzzi et al., 2018; Goel et al., 2022; Vanden Berghe et al., 2010). Necroptosis can be triggered by the alterations or disruptions of the extracellular or intracellular microenvironment. It involves death receptors, including FAS and TNFR1 (Degterev et al., 2008; Degterev et al., 2005; Galluzzi et al., 2014; Vercammen et al., 1998; Vercammen et al., 1997), or pathogen recognition receptors like TLR3, TLR4, and Z-DNA binding protein 1 (Galluzzi et al., 2018; Kaiser et al., 2013; Upton et al., 2010, 2012). Besides, necroptosis was reported to regulate adaptive functions when stress responses failed (Dara et al., 2016; Kaczmarek et al., 2013; Weinlich et al., 2017; Zhang et al., 2016).

Autophagy mediates the removal of damaged organelles, dysfunctional cellular components, and internalized protein aggregates through formation of autophagosomes that subsequently fuse with lysosomes, enabling degradation by lysosomal hydrolases (Cui et al., 2021; Goel et al., 2022; Green & Llambi, 2015; Pereira et al., 2016). Different stress conditions like absence of growth factors, nutrients, or oxygen and ER-stress can regulate autophagy via PKA, AMPK, or mTORC1 (Galluzzi et al., 2017; Parzych & Klionsky, 2014). Beclin1, Vsp34, Vsp15, and LC3 are involved in the fusion of the autophagosomes with lysosomes (Galluzzi et al., 2017; Goel et al., 2022; Green & Llambi, 2015). Chaperone-mediated autophagy is distinct from that involving the removal of specific proteins containing a KFERQ-like pentapeptide motif with the help of hsc70, co-chaperones like hsp90, and hsp40 (Bejarano & Cuervo, 2010; Galluzzi et al., 2017).

Ferroptosis is initiated by oxidative stress. Ferroptotic cell death is iron-dependent. It can be controlled by GPX4 and inhibited by iron chelators, butylated hydroxytoluene, lipophilic antioxidants, vitamin E, or lipid peroxidation inhibitors (i.e: liproxstatin-1, ferrostatin-1) (Galluzzi et al., 2018; Li et al., 2020; Stockwell et al., 2017). Ferroptosis is associated with damaged mitochondria with ruptured outer membranes and cristae loss, condensed chromatin, cytoplasmic swelling, and plasma membrane rupture (Goel et al., 2022; Li et al., 2020).

Pyroptosis, a form of lytic cell death, critically depends on the formation of plasma membrane pores by members of the gasdermin protein family, usually as a response to inflammatory caspase activation (Galluzzi et al., 2018). Inflammasomes, inflammatory caspases activation, cleavage of Gasdermin-D, IL- 1®, and IL-18 response are its characteristics (Cui et al., 2021; Fang et al., 2020; Tan et al., 2014; Yu et al., 2021). Pyroptosis can be either inflammasome dependent or independent (Goel et al., 2022).

Immunogenic cell death activates an adaptive immune response in immunocompetent hosts, lysosome-dependent cell death involves lysosomal membrane permeabilization, cathepsins, mitochondrial outer membrane permeabilization, and caspases. Mitochondrial permeability transition (MPT)–driven necrosis triggered by perturbations of the intracellular microenvironment and cyclophilin D, and autosis, a specific form of autophagy that critically depends on the plasma membrane Na+/K+-ATPase (Galluzzi et al., 2018), should not be ignored.

Postmortem studies indicated that a significantly high number of neurons from AD brains had DNA fragmentation and autophagic vacuoles. A correspondingly high number of neurons had TUNEL staining, which is an indicator of apoptotic cell death (Smale et al., 1995; Troncoso et al.,

1996). Case control studies reported that the entorhinal cortex, frontal cortex, and hippocampi exhibit high levels of caspase-3 (Selznick et al., 1999; Stadelmann et al., 1999; Su et al., 2001). Involvement of caspase-8 and caspase-3/7 in regulating microglia activation in PD and AD has been demonstrated (Burguillos et al., 2011; Matsui et al., 2006; Rohn et al., 2001). Previous studies suggested that intracellular Aβ accumulation is related with p53-mediated transcriptional upregulation of Bax, reduction of Bcl-2 and Bcl-xl levels, and upregulation of Bad activity in frontal cortical and hippocampal neurons (Kitamura et al., 1998; Zhang et al., 2021). On the other hand, cells with swollen morphology, DNA fragmentation, and absence of apoptotic morphology indicated that non-apoptotic cell death mechanisms were also possible in AD (Lassmann et al., 1995; Lucassen et al., 1997).

An activated necroptosis pathway is also observed in AD brains. Necroptosis was positively correlated with Braak staging (a neuropathological classification system used to indicate the progression and severity of neurofibrillary tau pathology in AD and other tauopathies) and inversely correlated cognitive scores. Phosphorylated tau aggregates were co-localized with chaperones. Granulovacuolar degeneration was inversely correlated with the neuronal density in hippocampus and frontal cortex. MLKL and p-MLKL upregulation accompanied by TNF-α signaling was also observed in AD. Studies reported an association between with necroptosis activation via TNF signaling and hippocampal neuron loss. The activation of MEG3 has been found to be related to necroptosis in AD. Blocking MEG3 up-regulation improved neuronal survival and was associated with down-regulated expression of necrosome proteins (Balusu et al., 2023; Caccamo et al., 2017; Jayaraman et al., 2021; Koper et al., 2020; Xu et al., 2021; Yamoah et al., 2020).

Autophagic vacuoles were reported in dystrophic neuritis of AD brains. Double membrane encapsulated autophagosomes around intracellular components were observed. AD progression was correlated with downregulated autophagy kinase complexes, BECN1-PIK3C3, ULK1/2-FIP200, and NRBF2. Autophagy related genes were upregulated in the entorhinal cortex (Lachance et al., 2019; Lipinski et al., 2010; Nixon, 2013; Nixon et al., 2005).

Some evidence also indicated the involvement of parthanatos in AD. These include increased PARylation of nuclear proteins; co-localization of Aβ, tau, and MAP2 with PARP1/PAR; and the differential expression of nucleolar PARP-1 between AD, MCI cases, and healthy individuals (Love et al., 1999; Narne et al., 2017; Regier et al., 2019).

Elevation of cyclophilin D expression in mitochondria was reported in AD. In addition to that, the absence of cyclophilin D reduced mitochondrial and neuronal damage and improved learning and memory in AD. These indicate the presence of mitochondrial permeability transition pore related death pathways in AD (Du et al., 2008).

NLRP3 activation and its contribution to the pathogenesis in APP/PS1 transgenic mice were shown, which indicated the role of pyroptosis in AD (Heneka et al., 2013).

A positive correlation between iron levels in AD brains and the progression of AD, a negative correlation between iron levels and cognitive functions, high levels of cortical iron in MCI due to plaque load, increased lipid peroxidation, and attenuated glutathione were shown, indicating the role of ferroptosis (Ansari & Scheff, 2010; Ayton et al., 2017; Chiang et al., 2017; Hare et al., 2013; Jenkins et al., 2020; van Bergen et al., 2016; Yoo et al., 2010). The types of regulated cell death and experimental models of neurodegenerative disorders can be augmented with other examples (Goel et al., 2022). Rather than the fate of the cell, the type of cell death might indicate certain targets to use for diagnostic or treatment strategies.

## Propagation

It has been shown that proteins with an amyloid fibril structure can pass from cell to cell in the brain like a prion-like infectious particle and spread. The first identified protein with this feature was the prion protein, and today, proteins with this feature are defined as "proteins that acquire alternative conformations that become self-propagating" (Jucker & Walker, 2018). These proteins can self-replicate and spread only by changes in protein conformation, regardless of the DNA sequence.

The pathologic form of prion ($PrP^{Sc}$) can self-multiply by templating the conversion of normally produced α-helical monomer $PrP^{C}$ into additional copies of $PrP^{Sc}$ (Arshad & Watts, 2023; Jucker & Walker, 2018). Misfolded, pathogenic $PrP^{Sc}$ molecules bind together, forming a complex multimeric β-sheet-rich structure. These structures grow, divide, proliferate, and spread to more areas, eventually occupying several regions of the nervous system and causing the death of the cells (Arshad & Watts, 2023; Puoti et al., 2012). The similarity of the amino acid sequences of $PrP^{C}$ and $PrP^{Sc}$ comes to the fore in the spread of prion protein. While the spread occurs very quickly when they belong to the same species, spread is limited when the amino acid sequence changes in proteins belonging to different species. Ultimately, the characteristics of both seeds and host influence the infectivity of $PrP^{Sc}$ as well as the nature of the resulting disease (Arshad & Watts, 2023).

Studies have shown that the mechanisms used by the prion protein may also be valid for proteins in the amyloid structure that play a role in the pathology of other neurodegenerative diseases. It has been shown that Aβ, tau, and α-syn, as well as htt, SOD1, and TDP-43 proteins, can spread via a prion-like mechanism (Bassil et al., 2020; Jucker & Walker, 2018; Olanow & Prusiner, 2009; Pan et al., 1993). However, one of the ways prion diseases occur is through "transmission." To date, no evidence of transmission of non-prion neurodegenerative diseases between species or individuals has been reported. For this reason, the term "prionoid" was used for these proteins that have prion properties but are not prions (Aguzzi & Rajendran, 2009).

Today, one of the most prominent proteins with its prionoid feature is α-syn. *In vivo* and *in vitro*, fibrillar α-syn (f-α-syn) behaves just like $PrP^{Sc}$ and grows by incorporating monomer α-syns in its environment (Dunning et al., 2012; Steiner et al., 2018). *In vitro*, if fibrillar α-syn (f-α-syn) is applied to the solution containing monomer α-syn, the monomers in the environment are largely incorporated into the f-α-syn structure under the necessary conditions. The conversion of this structure can be demonstrated in electron microscopy and with the thioflavin T (ThT) dye, which can be detected by fluorometric measurement when bound to amyloid structures. The conversion properties of f-α-syn also show the basis for its pathogenic properties (Fairfoul et al., 2016; Okuzumi et al., 2021).

α-syn, which is taken into the cell, multiplies within it, and is transported to distant regions by following intercellular extensions, must be able to pass to the next cell. Exosomes, exocytosis, direct passage through the cell membrane, cell injury, and transportation by nanotubes are mechanisms that have been proposed for this transition (Lee et al., 2010). Unlike these, it is possible for f-α-syn to be released out of the cell after the death of cells under stress. Eventually, the oligomeric and fibrillar forms of α-syn are taken up by another neuron cell via endocytosis, and this transition does not require direct contact between cells (Lee et al., 2012; Lee et al., 2010; Steiner et al., 2011).

Recent studies have discussed the cross-talk between Aβ, α-syn, and tau in pathology and propagation (Bassil et al., 2020; Lu et al., 2020; Puangmalai et al., 2022; Sengupta & Kayed, 2022; Shim et al., 2022; Twohig & Nielsen, 2019). For example, Aβ plaques have been shown to accelerate endogenous α-syn seeding. The induction pooling of α-syn within the synaptic terminals by Aβ plaques is "primed" to cross-seed tau. Additionally, Aβ plaques trigger the disassembly of microtubules and cause the release of tau proteins, which facilitates tau–α-syn interactions. Furthermore, α-syn-preformed fibrils enhance tau hyperphosphorylation (Bassil et al., 2020) (Figure 1.3).

### *Self-copy, Spread, and Diagnosis*

The self-replicating property of amyloid proteins has been used in the development of a diagnostic test. The Protein Misfolding Cyclic Amplification assay (PMCA), which was first developed in 2001 based on this principle (Saborio et al., 2001), was later followed by real-time Quacking Induced Conversion (RT-QuIC) assay (Atarashi et al., 2011). RT-QuIC assay is the first ante-mortem test and is considered the gold standard for diagnosing sporadic CJD. It is a technique that tracks the aggregation kinetics of peptides and distinguishes patients based on their aggregation rates (Dong & Satoh, 2021; Fiorini et al., 2020). In this technique, which can directly measure these bioactive protein seeds, the disease-specific "protein seeds" self-replicate, using natural recombinant protein substrates to extend amyloid fibrils (Dong & Satoh, 2021; Standke & Kraus, 2023). This technique, used for ante-mortem diagnosis of prion diseases, is now of great interest for use in the diagnosis of all synucleopathies and is also referred to as the α-syn seed amplification assay (SAA) (Bargar et al., 2021; Concha-Marambio et al., 2023).

## Conclusion

Neurodegenerative diseases are usually sporadic, and a small percentage have a genetic background. They are thought to arise due to a combination of genetic, epigenetic, and environmental factors. High-throughput technologies such as RNA sequencing can potentially contribute to understanding the molecular mechanisms of neurodegenerative diseases, although such large amounts of data create a significant obstacle to interpretation. In recent years, network biology has become a robust platform that integrates multiple omics data to explore molecular networks in the context of health and disease (Santiago et al., 2017). To date, with the genomic technologies used, approximately 3,000 genetic variations associated with AD have been reported (http://www.alzgene.org). Although these studies are valuable for understanding the genetic variation associated with AD, they do not elucidate the multifactorial mechanisms that lead to the disease. Network-based approaches that combine multi-omics

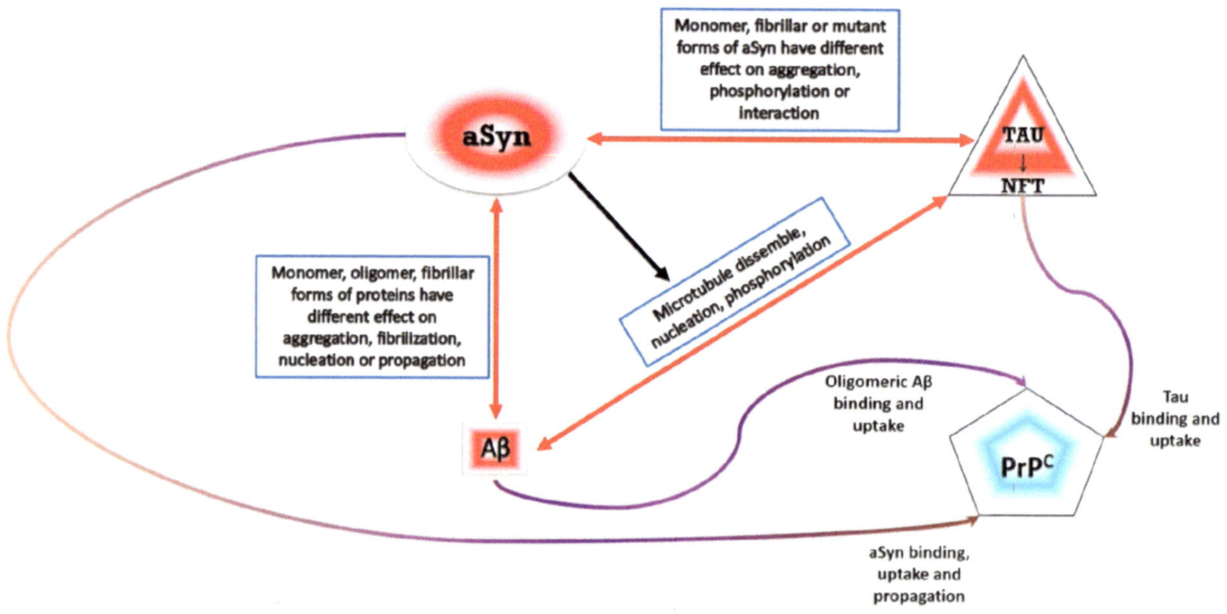


**FIGURE 1.3** The cross-talk between Aβ, α-syn, tau and Prion Alpha-synuclein (α-syn), amyloid beta (Aβ), neurofibrillary tangles (NFT), cellular prion protein (PrP$^c$).

information can be used as a helpful resource in systematically interpreting these results and providing fundamental insight into the mechanisms of disease (Basu et al., 2021; Santiago et al., 2017). In analyses using such approaches, it has been reported that, for example, in AD, important proteins in the MAPK/ERK pathway form a decreased cluster of nodes, while clathrin and various clathrin accessory proteins form an increased cluster of nodes (Hallock & Thomas, 2012). In another study conducted with the integration of genome-wide association, genome-wide expression profiling, and a protein–protein interaction (PPI) network, 108 potential risk factors for AD related to neurogenesis, the MAPK pathway, endocytosis, and calcium metabolism were reported. These studies suggested that among these genes, EGFR and ACTB may be the two most significant risk genes in AD (Talwar et al., 2014). Weighted gene co-expression network analysis showed that microglia and myelination are involved in the pathogenesis of AD and that there are differences in biological pathways between late-onset AD and DLB (Santiago et al., 2017).

Evaluating the data obtained with these new approaches together will enable us to better understand the molecular mechanisms of these diseases in the future and may allow the development of new diagnosis and treatment methods. Such perspectives also require robust modeling of neurodegenerative disorders and their pathologies.

## REFERENCES


Aggarwal, B. B., Gupta, S. C., & Kim, J. H. (2012). Historical perspectives on tumor necrosis factor and its superfamily: 25 years later, a golden journey. *Blood*, *119*(3), 651–665. https://doi.org/10.1182/blood-2011-04-325225

Agrawal, S., Baulch, J. E., Madan, S., Salah, S., Cheeks, S. N., Krattli, R. P. Jr., Subramanian, V. S., Acharya, M. M., & Agrawal, A. (2022). Impact of IL-21-associated peripheral and brain crosstalk on the Alzheimer's disease neuropathology. *Cell Mol Life Sci*, *79*(6), 331. https://doi.org/10.1007/s00018-022-04347-6

Aguzzi, A., & Falsig, J. (2012). Prion propagation, toxicity and degradation. *Nat Neurosci*, *15*(7), 936–939. https://doi.org/10.1038/nn.3120

Aguzzi, A., & Rajendran, L. (2009). The transcellular spread of cytosolic amyloids, prions, and prionoids. *Neuron*, *64*(6), 783–790. https://doi.org/10.1016/j.neuron.2009.12.016

Alecu, I., & Bennett, S. A. L. (2019). Dysregulated lipid metabolism and its role in alpha-synucleinopathy in Parkinson's disease. *Front Neurosci*, *13*, 328. https://doi.org/10.3389/fnins.2019.00328

Alquezar, C., Arya, S., & Kao, A. W. (2020). Tau post-translational modifications: dynamic transformers of tau function, degradation, and aggregation. *Front Neurol*, *11*, 595532. https://doi.org/10.3389/fneur.2020.595532

Andersen, O. S., & Koeppe, R. E., 2nd. (2007). Bilayer thickness and membrane protein function: an energetic perspective. *Annu Rev Biophys Biomol Struct*, *36*, 107–130. https://doi.org/10.1146/annurev.biophys.36.040306.132643

Ansari, M. A., & Scheff, S. W. (2010). Oxidative stress in the progression of Alzheimer disease in the frontal cortex. *J Neuropathol Exp Neurol*, *69*(2), 155–167. https://doi.org/10.1097/NEN.0b013e3181cb5af4

Anselmi, C., Kowarsky, M., Gasparini, F., Caicci, F., Ishizuka, K. J., Palmeri, K. J., Raveh, T., Sinha, R., Neff, N., Quake, S. R., Weissman, I. L., Voskoboynik, A., & Manni, L. (2022). Two distinct evolutionary conserved neural degeneration pathways characterized in a colonial chordate. *Proc Natl Acad Sci U S A*, *119*(29), e2203032119. https://doi.org/10.1073/pnas.2203032119

Antonny, B. (2011). Mechanisms of membrane curvature sensing. *Annu Rev Biochem*, *80*, 101–123. https://doi.org/10.1146/annurev-biochem-052809-155121

Apelt, J., Ach, K., & Schliebs, R. (2003). Aging-related down-regulation of neprilysin, a putative beta-amyloid-degrading enzyme, in transgenic Tg2576 Alzheimer-like mouse brain is accompanied by an astroglial upregulation in the vicinity of beta-amyloid plaques. *Neurosci Lett*, *339*(3), 183–186. https://doi.org/10.1016/s0304-3940(03)00030-2

Arancio, O. (2020). What does the APP family do in the brain? *Neuron*, *108*(4), 583–585. https://doi.org/10.1016/j.neuron.2020.11.003

Arendt, T. (2009). Synaptic degeneration in Alzheimer's disease. *Acta Neuropathol*, *118*(1), 167–179. https://doi.org/10.1007/s00401-009-0536-x

Arshad, H., & Watts, J. C. (2023). Genetically engineered cellular models of prion propagation. *Cell Tissue Res*, *392*(1), 63–80. https://doi.org/10.1007/s00441-022-03630-z

Atarashi, R., Satoh, K., Sano, K., Fuse, T., Yamaguchi, N., Ishibashi, D., Matsubara, T., Nakagaki, T., Yamanaka, H., Shirabe, S., Yamada, M., Mizusawa, H., Kitamoto, T., Klug, G., McGlade, A., Collins, S. J., & Nishida, N. (2011). Ultrasensitive human prion detection in cerebrospinal fluid by real-time quaking-induced conversion. *Nat Med*, *17*(2), 175–178. https://doi.org/10.1038/nm.2294

Avdulov, N. A., Chochina, S. V., Igbavboa, U., Warden, C. S., Vassiliev, A. V., & Wood, W. G. (1997). Lipid binding to amyloid beta-peptide aggregates: preferential binding of cholesterol as compared with phosphatidylcholine and fatty acids. *J Neurochem*, *69*(4), 1746–1752. https://doi.org/10.1046/j.1471-4159.1997.69041746.x

Avila, J., Sousa, M. M., & Sayas, C. L. (2020). Editorial: Shaping the brain by neuronal cytoskeleton: from development to disease and degeneration. *Front Cell Neurosci*, *14*, 12. https://doi.org/10.3389/fncel.2020.00012

Ayton, S., Fazlollahi, A., Bourgeat, P., Raniga, P., Ng, A., Lim, Y. Y., Diouf, I., Farquharson, S., Fripp, J., Ames, D., Doecke, J., Desmond, P., Ordidge, R., Masters, C. L., Rowe, C. C., Maruff, P., Villemagne, V. L., Australian Imaging, B., Lifestyle Research, G., … Bush, A. I. (2017). Cerebral quantitative susceptibility mapping predicts amyloid-beta-related cognitive decline. *Brain*, *140*(8), 2112–2119. https://doi.org/10.1093/brain/awx137

Azarias, G., Kruusmagi, M., Connor, S., Akkuratov, E. E., Liu, X. L., Lyons, D., Brismar, H., Broberger, C., & Aperia, A. (2013). A specific and essential role for Na,K-ATPase alpha3 in neurons co-expressing alpha1 and alpha3. *J Biol Chem*, *288*(4), 2734–2743. https://doi.org/10.1074/jbc.M112.425785

Badawy, S. M. M., Okada, T., Kajimoto, T., Hirase, M., Matovelo, S. A., Nakamura, S., Yoshida, D., Ijuin, T., & Nakamura, S. I. (2018). Extracellular alpha-synuclein drives sphingosine 1-phosphate receptor subtype 1 out of lipid rafts, leading to impaired inhibitory G-protein signaling. *J Biol Chem, 293*(21), 8208–8216. https://doi.org/10.1074/jbc.RA118.001986

Bae, E. J., Ho, D. H., Park, E., Jung, J. W., Cho, K., Hong, J. H., Lee, H. J., Kim, K. P., & Lee, S. J. (2013). Lipid peroxidation product 4-hydroxy-2-nonenal promotes seeding-capable oligomer formation and cell-to-cell transfer of alpha-synuclein. *Antioxid Redox Signal, 18*(7), 770–783. https://doi.org/10.1089/ars.2011.4429

Baek, H. J., Yoon, J. H., Ann, E. J., Kim, M. Y., Ahn, J. S., Kim, S. H., Jo, E. H., Lee, H. J., & Park, H. S. (2015). Alpha-synuclein negatively regulates Notch1 intracellular domain protein stability through promoting interaction with Fbw7. *Neurosci Lett, 600*, 6–11. https://doi.org/10.1016/j.neulet.2015.05.053

Bailey, J. A., Maloney, B., Ge, Y. W., & Lahiri, D. K. (2011). Functional activity of the novel Alzheimer's amyloid beta-peptide interacting domain (AbetaID) in the APP and BACE1 promoter sequences and implications in activating apoptotic genes and in amyloidogenesis. *Gene, 488*(1–2), 13–22. https://doi.org/10.1016/j.gene.2011.06.017

Balducci, C., Beeg, M., Stravalaci, M., Bastone, A., Sclip, A., Biasini, E., Tapella, L., Colombo, L., Manzoni, C., Borsello, T., Chiesa, R., Gobbi, M., Salmona, M., & Forloni, G. (2010). Synthetic amyloid-beta oligomers impair long-term memory independently of cellular prion protein. *Proc Natl Acad Sci U S A, 107*(5), 2295–2300. https://doi.org/10.1073/pnas.0911829107

Balusu, S., Horre, K., Thrupp, N., Craessaerts, K., Snellinx, A., Serneels, L., T'Syen, D., Chrysidou, I., Arranz, A. M., Sierksma, A., Simren, J., Karikari, T. K., Zetterberg, H., Chen, W. T., Thal, D. R., Salta, E., Fiers, M., & De Strooper, B. (2023). MEG3 activates necroptosis in human neuron xenografts modeling Alzheimer's disease. *Science, 381*(6663), 1176–1182. https://doi.org/10.1126/science.abp9556

Barazzuol, L., Giamogante, F., Brini, M., & Cali, T. (2020). PINK1/Parkin Mediated Mitophagy, Ca(2+) Signalling, and ER-mitochondria contacts in Parkinson's disease. *Int J Mol Sci, 21*(5). https://doi.org/10.3390/ijms21051772

Bargar, C., Wang, W., Gunzler, S. A., LeFevre, A., Wang, Z., Lerner, A. J., Singh, N., Tatsuoka, C., Appleby, B., Zhu, X., Xu, R., Haroutunian, V., Zou, W. Q., Ma, J., & Chen, S. G. (2021). Streamlined alpha-synuclein RT-QuIC assay for various biospecimens in Parkinson's disease and dementia with Lewy bodies. *Acta Neuropathol Commun, 9*(1), 62. https://doi.org/10.1186/s40478-021-01175-w

Barnum, C. J., Chen, X., Chung, J., Chang, J., Williams, M., Grigoryan, N., Tesi, R. J., & Tansey, M. G. (2014). Peripheral administration of the selective inhibitor of soluble tumor necrosis factor (TNF) XPro(R)1595 attenuates nigral cell loss and glial activation in 6-OHDA hemiparkinsonian rats. *J Parkinsons Dis, 4*(3), 349–360. https://doi.org/10.3233/JPD-140410

Barucker, C., Harmeier, A., Weiske, J., Fauler, B., Albring, K. F., Prokop, S., Hildebrand, P., Lurz, R., Heppner, F. L., & Huber, O. (2014). Nuclear translocation uncovers the amyloid peptide Aβ42 as a regulator of gene transcription. *Journal of Biological Chemistry, 289*(29), 20182–20191.

Bassil, F., Brown, H. J., Pattabhiraman, S., Iwasyk, J. E., Maghames, C. M., Meymand, E. S., Cox, T. O., Riddle, D. M., Zhang, B., Trojanowski, J. Q., & Lee, V. M. (2020). Amyloid-beta (Abeta) plaques promote seeding and spreading of alpha-synuclein and tau in a mouse model of lewy body disorders with abeta pathology. *Neuron, 105*(2), 260–275 e266. https://doi.org/10.1016/j.neuron.2019.10.010

Basu, A., Ash, P. E., Wolozin, B., & Emili, A. (2021). Protein interaction network biology in neuroscience. *Proteomics, 21*(3–4), e1900311. https://doi.org/10.1002/pmic.201900311

Batterman, K. V., Cabrera, P. E., Moore, T. L., & Rosene, D. L. (2021). T cells actively infiltrate the white matter of the aging monkey brain in relation to increased microglial reactivity and cognitive decline. *Front Immunol, 12*, 607691. https://doi.org/10.3389/fimmu.2021.607691

Bejarano, E., & Cuervo, A. M. (2010). Chaperone-mediated autophagy. *Proc Am Thorac Soc, 7*(1), 29–39. https://doi.org/10.1513/pats.200909-102JS

Bellani, S., Mescola, A., Ronzitti, G., Tsushima, H., Tilve, S., Canale, C., Valtorta, F., & Chieregatti, E. (2014). GRP78 clustering at the cell surface of neurons transduces the action of exogenous alpha-synuclein. *Cell Death Differ, 21*(12), 1971–1983. https://doi.org/10.1038/cdd.2014.111

Bender, A., Krishnan, K. J., Morris, C. M., Taylor, G. A., Reeve, A. K., Perry, R. H., Jaros, E., Hersheson, J. S., Betts, J., Klopstock, T., Taylor, R. W., & Turnbull, D. M. (2006). High levels of mitochondrial DNA deletions in substantia nigra neurons in aging and Parkinson disease. *Nat Genet, 38*(5), 515–517. https://doi.org/10.1038/ng1769

Bengoa-Vergniory, N., Roberts, R. F., Wade-Martins, R., & Alegre-Abarrategui, J. (2017). Alpha-synuclein oligomers: a new hope. *Acta Neuropathol, 134*(6), 819–838. https://doi.org/10.1007/s00401-017-1755-1

Beraldo, F. H., Arantes, C. P., Santos, T. G., Machado, C. F., Roffe, M., Hajj, G. N., Lee, K. S., Magalhaes, A. C., Caetano, F. A., Mancini, G. L., Lopes, M. H., Americo, T. A., Magdesian, M. H., Ferguson, S. S., Linden, R., Prado, M. A., & Martins, V. R. (2011). Metabotropic glutamate receptors transduce signals for neurite outgrowth after binding of the prion protein to laminin gamma1 chain. *FASEB J, 25*(1), 265–279. https://doi.org/10.1096/fj.10-161653

Bernabeu-Zornoza, A., Coronel, R., Palmer, C., Monteagudo, M., Zambrano, A., & Liste, I. (2019). Physiological and pathological effects of amyloid-beta species in neural stem cell biology. *Neural Regen Res, 14*(12), 2035–2042. https://doi.org/10.4103/1673-5374.262571

Bharadwaj, P., Solomon, T., Malajczuk, C. J., Mancera, R. L., Howard, M., Arrigan, D. W. M., Newsholme, P., & Martins, R. N. (2018). Role of the cell membrane interface in modulating production and uptake of Alzheimer's beta amyloid protein. *Biochim Biophys Acta Biomembr, 1860*(9), 1639–1651. https://doi.org/10.1016/j.bbamem.2018.03.015

Bi, H. R., Zhou, C. H., Zhang, Y. Z., Cai, X. D., Ji, M. H., Yang, J. J., Chen, G. Q., & Hu, Y. M. (2021). Neuron-specific deletion of presenilin enhancer2 causes progressive astrogliosis and age-related neurodegeneration in the cortex independent of the Notch signaling. *CNS Neurosci Ther, 27*(2), 174–185. https://doi.org/10.1111/cns.13454

Bisi, N., Feni, L., Peqini, K., Perez-Pena, H., Ongeri, S., Pieraccini, S., & Pellegrino, S. (2021). Alpha-synuclein: an all-inclusive trip around its structure, influencing factors and applied techniques. *Front Chem, 9*, 666585. https://doi.org/10.3389/fchem.2021.666585

Bliss, T. V., & Collingridge, G. L. (1993). A synaptic model of memory: long-term potentiation in the hippocampus. *Nature, 361*(6407), 31–39. https://doi.org/10.1038/361031a0

Bode, D. C., Baker, M. D., & Viles, J. H. (2017). Ion channel formation by amyloid-beta42 oligomers but not amyloid-beta40 in cellular membranes. *J Biol Chem, 292*(4), 1404–1413. https://doi.org/10.1074/jbc.M116.762526

Bode, D. C., Freeley, M., Nield, J., Palma, M., & Viles, J. H. (2019). Amyloid-beta oligomers have a profound detergent-like effect on lipid membrane bilayers, imaged by atomic force and electron microscopy. *J Biol Chem, 294*(19), 7566–7572. https://doi.org/10.1074/jbc.AC118.007195

Bonifati, V., Rizzu, P., van Baren, M. J., Schaap, O., Breedveld, G. J., Krieger, E., Dekker, M. C., Squitieri, F., Ibanez, P., Joosse, M., van Dongen, J. W., Vanacore, N., van Swieten, J. C., Brice, A., Meco, G., van Duijn, C. M., Oostra, B. A., & Heutink, P. (2003). Mutations in the DJ-1 gene associated with autosomal recessive early-onset parkinsonism. *Science, 299*(5604), 256–259. https://doi.org/10.1126/science.1077209

Braak, H., & Braak, E. (1996). Evolution of the neuropathology of Alzheimer's disease. *Acta Neurol Scand Suppl, 165*, 3–12. https://doi.org/10.1111/j.1600-0404.1996.tb05866.x

Braak, H., Del Tredici, K., Rub, U., de Vos, R. A., Jansen Steur, E. N., & Braak, E. (2003). Staging of brain pathology related to sporadic Parkinson's disease. *Neurobiol Aging, 24*(2), 197–211. https://doi.org/10.1016/s0197-4580(02)00065-9

Brandt, R., Trushina, N. I., & Bakota, L. (2020). Much more than a cytoskeletal protein: physiological and pathological functions of the non-microtubule binding region of tau. *Front Neurol, 11*, 590059. https://doi.org/10.3389/fneur.2020.590059

Bras, I. C., & Outeiro, T. F. (2021). Alpha-synuclein: mechanisms of release and pathology progression in synucleinopathies. *Cells, 10*(2). https://doi.org/10.3390/cells10020375

Braun, G. A., Dear, A. J., Sanagavarapu, K., Zetterberg, H., & Linse, S. (2022). Amyloid-beta peptide 37, 38 and 40 individually and cooperatively inhibit amyloid-beta 42 aggregation. *Chem Sci, 13*(8), 2423–2439. https://doi.org/10.1039/d1sc02990h

Brito-Moreira, J., Paula-Lima, A. C., Bomfim, T. R., Oliveira, F. B., Sepulveda, F. J., De Mello, F. G., Aguayo, L. G., Panizzutti, R., & Ferreira, S. T. (2011). Abeta oligomers induce glutamate release from hippocampal neurons. *Curr Alzheimer Res, 8*(5), 552–562. https://doi.org/10.2174/156720511796391917

Brosseron, F., Krauthausen, M., Kummer, M., & Heneka, M. T. (2014). Body fluid cytokine levels in mild cognitive impairment and Alzheimer's disease: a comparative overview. *Mol Neurobiol*. https://doi.org/10.1007/s12035-014-8657-1

Brumatti, G., Salmanidis, M., & Ekert, P. G. (2010). Crossing paths: interactions between the cell death machinery and growth factor survival signals. *Cell Mol Life Sci, 67*(10), 1619–1630. https://doi.org/10.1007/s00018-010-0288-8

Buckig, A., Tikkanen, R., Herzog, V., & Schmitz, A. (2002). Cytosolic and nuclear aggregation of the amyloid beta-peptide following its expression in the endoplasmic reticulum. *Histochem Cell Biol, 118*(5), 353–360. https://doi.org/10.1007/s00418-002-0459-2

Buell, A. K., Dobson, C. M., & Knowles, T. P. (2014). The physical chemistry of the amyloid phenomenon: thermodynamics and kinetics of filamentous protein aggregation. *Essays Biochem, 56*, 11–39. https://doi.org/10.1042/bse0560011

Burguillos, M. A., Deierborg, T., Kavanagh, E., Persson, A., Hajji, N., Garcia-Quintanilla, A., Cano, J., Brundin, P., Englund, E., Venero, J. L., & Joseph, B. (2011). Caspase signalling controls microglia activation and neurotoxicity. *Nature, 472*(7343), 319–324. https://doi.org/10.1038/nature09788

Burke, K. A., Hensal, K. M., Umbaugh, C. S., Chaibva, M., & Legleiter, J. (2013). Huntingtin disrupts lipid bilayers in a polyQ-length dependent manner. *Biochim Biophys Acta, 1828*(8), 1953–1961. https://doi.org/10.1016/j.bbamem.2013.04.025

Burre, J., Sharma, M., Tsetsenis, T., Buchman, V., Etherton, M. R., & Sudhof, T. C. (2010). Alpha-synuclein promotes SNARE-complex assembly in vivo and in vitro. *Science, 329*(5999), 1663–1667. https://doi.org/10.1126/science.1195227

Buxbaum, J. N., Dispenzieri, A., Eisenberg, D. S., Fandrich, M., Merlini, G., Saraiva, M. J. M., Sekijima, Y., & Westermark, P. (2022). Amyloid nomenclature 2022: update, novel proteins, and recommendations by the International Society of Amyloidosis (ISA) Nomenclature Committee. *Amyloid, 29*(4), 213–219. https://doi.org/10.1080/13506129.2022.2147636

Cacabelos, R., Alvarez, X. A., Fernandez-Novoa, L., Franco, A., Mangues, R., Pellicer, A., & Nishimura, T. (1994). Brain interleukin-1 beta in Alzheimer's disease and vascular dementia. *Methods Find Exp Clin Pharmacol, 16*(2), 141–151. https://www.ncbi.nlm.nih.gov/pubmed/8007743

Caccamo, A., Branca, C., Piras, I. S., Ferreira, E., Huentelman, M. J., Liang, W. S., Readhead, B., Dudley, J. T., Spangenberg, E. E., Green, K. N., Belfiore, R., Winslow, W., & Oddo, S. (2017). Necroptosis activation in Alzheimer's disease. *Nat Neurosci, 20*(9), 1236–1246. https://doi.org/10.1038/nn.4608

Calella, A. M., Farinelli, M., Nuvolone, M., Mirante, O., Moos, R., Falsig, J., Mansuy, I. M., & Aguzzi, A. (2010). Prion protein and Abeta-related synaptic toxicity impairment. *EMBO Mol Med, 2*(8), 306–314. https://doi.org/10.1002/emmm.201000082

Carpenter, A. F., Carpenter, P. W., & Markesbery, W. R. (1993). Morphometric analysis of microglia in Alzheimer's disease. *J Neuropathol Exp Neurol, 52*(6), 601–608. https://doi.org/10.1097/00005072-199311000-00007

Carson, M. J., Doose, J. M., Melchior, B., Schmid, C. D., & Ploix, C. C. (2006). CNS immune privilege: hiding in plain sight. *Immunol Rev, 213*, 48–65. https://doi.org/10.1111/j.1600-065X.2006.00441.x

Casares, D., Escriba, P. V., & Rossello, C. A. (2019). Membrane lipid composition: effect on membrane and organelle structure, function and compartmentalization and therapeutic avenues. *Int J Mol Sci, 20*(9), 346. https://doi.org/10.3390/ijms20092167

Cash, A. D., Aliev, G., Siedlak, S. L., Nunomura, A., Fujioka, H., Zhu, X., Raina, A. K., Vinters, H. V., Tabaton, M., Johnson, A. B., Paula-Barbosa, M., Avila, J., Jones, P. K., Castellani, R. J., Smith, M. A., & Perry, G. (2003). Microtubule reduction in Alzheimer's disease and aging is independent of tau filament formation. *Am J Pathol, 162*(5), 1623–1627. https://doi.org/10.1016/s0002-9440(10)64296-4

Casley, C. S., Lakics, V., Lee, H. G., Broad, L. M., Day, T. A., Cluett, T., Smith, M. A., O'Neill, M. J., & Kingston, A. E. (2009). Up-regulation of astrocyte metabotropic glutamate receptor 5 by amyloid-beta peptide. *Brain Res*, *1260*, 65–75. https://doi.org/10.1016/j.brainres.2008.12.082

Casoli, T., Di Stefano, G., Spazzafumo, L., Balietti, M., Giorgetti, B., Giuli, C., Postacchini, D., Fattoretti, P., & Conti, F. (2014). Contribution of non-reference alleles in mtDNA of Alzheimer's disease patients. *Ann Clin Transl Neurol*, *1*(4), 284–289. https://doi.org/10.1002/acn3.42

Castle, A. R., & Gill, A. C. (2017). Physiological functions of the cellular prion protein. *Front Mol Biosci*, *4*, 19. https://doi.org/10.3389/fmolb.2017.00019

Chauhan, N., & Siegel, G. (1997). Age-dependent organotypic expression of microtubule-associated proteins (MAP1, MAP2, and MAP5) in rat brain. *Neurochem Res*, *22*(6), 713–719. https://doi.org/10.1023/a:1027306227402

Chen, G. F., Xu, T. H., Yan, Y., Zhou, Y. R., Jiang, Y., Melcher, K., & Xu, H. E. (2017). Amyloid beta: structure, biology and structure-based therapeutic development. *Acta Pharmacol Sin*, *38*(9), 1205–1235. https://doi.org/10.1038/aps.2017.28

Chen, T., Wang, Q., Chao, D., Xia, T. C., Sheng, S., Li, Z. R., Zhao, J. N., Wen, G. Q., Ding, G., & Xia, Y. (2019). Delta-opioid receptor activation attenuates the oligomer formation induced by hypoxia and/or alpha-synuclein overexpression/mutation through dual signaling pathways. *Mol Neurobiol*, *56*(5), 3463–3475. https://doi.org/10.1007/s12035-018-1316-1

Chen, Y., Fu, A. K. Y., & Ip, N. Y. (2019). Synaptic dysfunction in Alzheimer's disease: mechanisms and therapeutic strategies. *Pharmacol Ther*, *195*, 186–198. https://doi.org/10.1016/j.pharmthera.2018.11.006

Chiang, G. C., Mao, X., Kang, G., Chang, E., Pandya, S., Vallabhajosula, S., Isaacson, R., Ravdin, L. D., Alzheimer's Disease Neuroimaging, I., & Shungu, D. C. (2017). Relationships among cortical glutathione levels, brain amyloidosis, and memory in healthy older adults investigated in Vivo with (1)H-MRS and pittsburgh compound-B PET. *AJNR Am J Neuroradiol*, *38*(6), 1130–1137. https://doi.org/10.3174/ajnr.A5143

Chong, F. P., Ng, K. Y., Koh, R. Y., & Chye, S. M. (2018). Tau proteins and tauopathies in Alzheimer's disease. *Cell Mol Neurobiol*, *38*(5), 965–980. https://doi.org/10.1007/s10571-017-0574-1

Chu, J., & Pratico, D. (2009). The 5-lipoxygenase as a common pathway for pathological brain and vascular aging. *Cardiovasc Psychiatry Neurol*, *2009*, 174657. https://doi.org/10.1155/2009/174657

Chugh, D., Ali, I., Bakochi, A., Bahonjic, E., Etholm, L., & Ekdahl, C. T. (2015). Alterations in brain inflammation, synaptic proteins, and adult hippocampal neurogenesis during epileptogenesis in Mice Lacking Synapsin2. *PLoS One*, *10*(7), e0132366. https://doi.org/10.1371/journal.pone.0132366

Ciesielska, A., Joniec, I., Kurkowska-Jastrzebska, I., Przybylkowski, A., Gromadzka, G., Czlonkowska, A., & Czlonkowski, A. (2007). Influence of age and gender on cytokine expression in a murine model of Parkinson's disease. *Neuroimmunomodulation*, *14*(5), 255–265. https://doi.org/10.1159/000113432

Cisse, M., Halabisky, B., Harris, J., Devidze, N., Dubal, D. B., Sun, B., Orr, A., Lotz, G., Kim, D. H., Hamto, P., Ho, K., Yu, G. Q., & Mucke, L. (2011). Reversing EphB2 depletion rescues cognitive functions in Alzheimer model. *Nature*, *469*(7328), 47–52. https://doi.org/10.1038/nature09635

Cole, N. B., Dieuliis, D., Leo, P., Mitchell, D. C., & Nussbaum, R. L. (2008). Mitochondrial translocation of alpha-synuclein is promoted by intracellular acidification. *Exp Cell Res*, *314*(10), 2076–2089. https://doi.org/10.1016/j.yexcr.2008.03.012

Colin, J., Gregory-Pauron, L., Lanhers, M. C., Claudepierre, T., Corbier, C., Yen, F. T., Malaplate-Armand, C., & Oster, T. (2016). Membrane raft domains and remodeling in aging brain. *Biochimie*, *130*, 178–187. https://doi.org/10.1016/j.biochi.2016.08.014

Concha-Marambio, L., Pritzkow, S., Shahnawaz, M., Farris, C. M., & Soto, C. (2023). Seed amplification assay for the detection of pathologic alpha-synuclein aggregates in cerebrospinal fluid. *Nat Protoc*, *18*(4), 1179–1196. https://doi.org/10.1038/s41596-022-00787-3

Connolly, M. G., Bruce, S. R., & Kohman, R. A. (2022). Exercise duration differentially effects age-related neuroinflammation and hippocampal neurogenesis. *Neuroscience*, *490*, 275–286. https://doi.org/10.1016/j.neuroscience.2022.03.022

Cookson, M. R. (2009). alpha-Synuclein and neuronal cell death. *Mol Neurodegener*, *4*, 9. https://doi.org/10.1186/1750-1326-4-9

Corti, O., Blomgren, K., Poletti, A., & Beart, P. M. (2020). Autophagy in neurodegeneration: new insights underpinning therapy for neurological diseases. *J Neurochem*, *154*(4), 354–371. https://doi.org/10.1111/jnc.15002

Coskun, P. E., Beal, M. F., & Wallace, D. C. (2004). Alzheimer's brains harbor somatic mtDNA control-region mutations that suppress mitochondrial transcription and replication. *Proc Natl Acad Sci U S A*, *101*(29), 10726–10731. https://doi.org/10.1073/pnas.0403649101

Cossu, D., Hatano, T., & Hattori, N. (2023). The role of immune dysfunction in Parkinson's disease development. *Int J Mol Sci*, *24*(23). https://doi.org/10.3390/ijms242316766

Crews, L., Mizuno, H., Desplats, P., Rockenstein, E., Adame, A., Patrick, C., Winner, B., Winkler, J., & Masliah, E. (2008). Alpha-synuclein alters Notch-1 expression and neurogenesis in mouse embryonic stem cells and in the hippocampus of transgenic mice. *J Neurosci*, *28*(16), 4250–4260. https://doi.org/10.1523/JNEUROSCI.0066-08.2008

Criado, J. R., Sanchez-Alavez, M., Conti, B., Giacchino, J. L., Wills, D. N., Henriksen, S. J., Race, R., Manson, J. C., Chesebro, B., & Oldstone, M. B. (2005). Mice devoid of prion protein have cognitive deficits that are rescued by reconstitution of PrP in neurons. *Neurobiol Dis*, *19*(1–2), 255–265. https://doi.org/10.1016/j.nbd.2005.01.001

Cui, J., Zhao, S., Li, Y., Zhang, D., Wang, B., Xie, J., & Wang, J. (2021). Regulated cell death: discovery, features and implications for neurodegenerative diseases. *Cell Commun Signal*, *19*(1), 120. https://doi.org/10.1186/s12964-021-00799-8

Czabotar, P. E., Lessene, G., Strasser, A., & Adams, J. M. (2014). Control of apoptosis by the BCL-2 protein family: implications for physiology and therapy. *Nat Rev Mol Cell Biol*, *15*(1), 49–63. https://doi.org/10.1038/nrm3722

Danyu, L., Yanran, L., Xiuna, J., Ying, C., Sudan, P., Tianen, Z., Zhifen, Z., Dezhi, Z., Kaixun, H., Yingyu, X., & Enxiang, T. (2019). Alpha-Synuclein induced mitochondrial dysfunction via cytochrome c oxidase subunit 2 in SH-SY5Y cells. *Exp Cell Res*, *378*(1), 57–65. https://doi.org/10.1016/j.yexcr.2019.02.006

Danzer, K. M., Haasen, D., Karow, A. R., Moussaud, S., Habeck, M., Giese, A., Kretzschmar, H., Hengerer, B., & Kostka, M. (2007). Different species of alpha-synuclein oligomers induce calcium influx and seeding. *J Neurosci*, *27*(34), 9220–9232. https://doi.org/10.1523/JNEUROSCI.2617-07.2007

Dara, L., Liu, Z. X., & Kaplowitz, N. (2016). Questions and controversies: the role of necroptosis in liver disease. *Cell Death Discov*, *2*, 16089. https://doi.org/10.1038/cddiscovery.2016.89

Davidson, W. S., Jonas, A., Clayton, D. F., & George, J. M. (1998). Stabilization of alpha-synuclein secondary structure upon binding to synthetic membranes. *J Biol Chem*, *273*(16), 9443–9449. https://doi.org/10.1074/jbc.273.16.9443

Decker, H., Jurgensen, S., Adrover, M. F., Brito-Moreira, J., Bomfim, T. R., Klein, W. L., Epstein, A. L., De Felice, F. G., Jerusalinsky, D., & Ferreira, S. T. (2010). N-methyl-D-aspartate receptors are required for synaptic targeting of Alzheimer's toxic amyloid-beta peptide oligomers. *J Neurochem*, *115*(6), 1520–1529. https://doi.org/10.1111/j.1471-4159.2010.07058.x

Degterev, A., Hitomi, J., Germscheid, M., Ch'en, I. L., Korkina, O., Teng, X., Abbott, D., Cuny, G. D., Yuan, C., Wagner, G., Hedrick, S. M., Gerber, S. A., Lugovskoy, A., & Yuan, J. (2008). Identification of RIP1 kinase as a specific cellular target of necrostatins. *Nat Chem Biol*, *4*(5), 313–321. https://doi.org/10.1038/nchembio.83

Degterev, A., Huang, Z., Boyce, M., Li, Y., Jagtap, P., Mizushima, N., Cuny, G. D., Mitchison, T. J., Moskowitz, M. A., & Yuan, J. (2005). Chemical inhibitor of nonapoptotic cell death with therapeutic potential for ischemic brain injury. *Nat Chem Biol*, *1*(2), 112–119. https://doi.org/10.1038/nchembio711

Delacourte, A., Flament, S., Dibe, E. M., Hublau, P., Sablonniere, B., Hemon, B., Sherrer, V., & Defossez, A. (1990). Pathological proteins tau 64 and 69 are specifically expressed in the somatodendritic domain of the degenerating cortical neurons during Alzheimer's disease. Demonstration with a panel of antibodies against tau proteins. *Acta Neuropathol*, *80*(2), 111–117. https://doi.org/10.1007/BF00308912

Dettmer, U., Ramalingam, N., von Saucken, V. E., Kim, T. E., Newman, A. J., Terry-Kantor, E., Nuber, S., Ericsson, M., Fanning, S., Bartels, T., Lindquist, S., Levy, O. A., & Selkoe, D. (2017). Loss of native alpha-synuclein multimerization by strategically mutating its amphipathic helix causes abnormal vesicle interactions in neuronal cells. *Hum Mol Genet*, *26*(18), 3466–3481. https://doi.org/10.1093/hmg/ddx227

Devi, L., Raghavendran, V., Prabhu, B. M., Avadhani, N. G., & Anandatheerthavarada, H. K. (2008). Mitochondrial import and accumulation of alpha-synuclein impair complex I in human dopaminergic neuronal cultures and Parkinson disease brain. *J Biol Chem*, *283*(14), 9089–9100. https://doi.org/10.1074/jbc.M710012200

Di Scala, C., Chahinian, H., Yahi, N., Garmy, N., & Fantini, J. (2014). Interaction of Alzheimer's beta-amyloid peptides with cholesterol: mechanistic insights into amyloid pore formation. *Biochemistry*, *53*(28), 4489–4502. https://doi.org/10.1021/bi500373k

Di Stefano, G., Casoli, T., Fattoretti, P., Gracciotti, N., Solazzi, M., & Bertoni-Freddari, C. (2001). Distribution of map2 in hippocampus and cerebellum of young and old rats by quantitative immunohistochemistry. *J Histochem Cytochem*, *49*(8), 1065–1066. https://doi.org/10.1177/002215540104900818

Dias-Carvalho, A., Ferreira, M., Ferreira, R., Bastos, M. L., Sa, S. I., Capela, J. P., Carvalho, F., & Costa, V. M. (2022). Four decades of chemotherapy-induced cognitive dysfunction: comprehensive review of clinical, animal and in vitro studies, and insights of key initiating events. *Arch Toxicol*, *96*(1), 11–78. https://doi.org/10.1007/s00204-021-03171-4

Dias-Carvalho, A., Sa, S. I., Carvalho, F., Fernandes, E., & Costa, V. M. (2024). Inflammation as common link to progressive neurological diseases. *Arch Toxicol*, *98*(1), 95–119. https://doi.org/10.1007/s00204-023-03628-8

Dias, C., & Nylandsted, J. (2021). Plasma membrane integrity in health and disease: significance and therapeutic potential. *Cell Discov*, *7*(1), 4. https://doi.org/10.1038/s41421-020-00233-2

Diaz, M., Fabelo, N., Ferrer, I., & Marin, R. (2018). "Lipid raft aging" in the human frontal cortex during nonpathological aging: gender influences and potential implications in Alzheimer's disease. *Neurobiol Aging*, *67*, 42–52. https://doi.org/10.1016/j.neurobiolaging.2018.02.022

Dikiy, I., & Eliezer, D. (2012). Folding and misfolding of alpha-synuclein on membranes. *Biochim Biophys Acta*, *1818*(4), 1013–1018. https://doi.org/10.1016/j.bbamem.2011.09.008

Dinamarca, M. C., Rios, J. A., & Inestrosa, N. C. (2012). Postsynaptic receptors for amyloid-beta oligomers as mediators of neuronal damage in Alzheimer's disease. *Front Physiol*, *3*, 464. https://doi.org/10.3389/fphys.2012.00464

Dineley, K. T., Westerman, M., Bui, D., Bell, K., Ashe, K. H., & Sweatt, J. D. (2001). Beta-amyloid activates the mitogen-activated protein kinase cascade via hippocampal alpha7 nicotinic acetylcholine receptors: in vitro and in vivo mechanisms related to Alzheimer's disease. *J Neurosci*, *21*(12), 4125–4133. https://doi.org/10.1523/JNEUROSCI.21-12-04125.2001

Disdier, C., Chalansonnet, M., Gagnaire, F., Gate, L., Cosnier, F., Devoy, J., Saba, W., Lund, A. K., Brun, E., & Mabondzo, A. (2017). Brain inflammation, blood brain barrier dysfunction and neuronal synaptophysin decrease after inhalation exposure to titanium dioxide nano-aerosol in aging rats. *Sci Rep*, *7*(1), 12196. https://doi.org/10.1038/s41598-017-12404-5

Doeppner, T. R., Kaltwasser, B., Schlechter, J., Jaschke, J., Kilic, E., Bahr, M., Hermann, D. M., & Weise, J. (2015). Cellular prion protein promotes post-ischemic neuronal survival, angioneurogenesis and enhances neural progenitor cell homing via proteasome inhibition. *Cell Death Dis*, *6*(12), e2024. https://doi.org/10.1038/cddis.2015.365

Dolgacheva, L. P., Berezhnov, A. V., Fedotova, E. I., Zinchenko, V. P., & Abramov, A. Y. (2019). Role of DJ-1 in the mechanism of pathogenesis of Parkinson's disease. *J Bioenerg Biomembr*, *51*(3), 175–188. https://doi.org/10.1007/s10863-019-09798-4

Dolle, C., Flones, I., Nido, G. S., Miletic, H., Osuagwu, N., Kristoffersen, S., Lilleng, P. K., Larsen, J. P., Tysnes, O. B., Haugarvoll, K., Bindoff, L. A., & Tzoulis, C. (2016). Defective mitochondrial DNA homeostasis in the substantia nigra in Parkinson disease. *Nat Commun*, *7*, 13548. https://doi.org/10.1038/ncomms13548

Dompierre, J. P., Godin, J. D., Charrin, B. C., Cordelieres, F. P., King, S. J., Humbert, S., & Saudou, F. (2007). Histone deacetylase 6 inhibition compensates for the transport deficit in Huntington's disease by increasing tubulin acetylation. *J Neurosci*, *27*(13), 3571–3583. https://doi.org/10.1523/JNEUROSCI.0037-07.2007

Dong, T. T., & Satoh, K. (2021). The latest research on RT-QuIC assays: a literature review. *Pathogens*, *10*(3), 305. https://doi.org/10.3390/pathogens10030305

Du, H., Guo, L., Fang, F., Chen, D., Sosunov, A. A., McKhann, G. M., Yan, Y., Wang, C., Zhang, H., Molkentin, J. D., Gunn-Moore, F. J., Vonsattel, J. P., Arancio, O., Chen, J. X., & Yan, S. D. (2008). Cyclophilin D deficiency attenuates mitochondrial and neuronal perturbation and ameliorates learning and memory in Alzheimer's disease. *Nat Med*, *14*(10), 1097–1105. https://doi.org/10.1038/nm.1868

Dunning, C. J., Reyes, J. F., Steiner, J. A., & Brundin, P. (2012). Can Parkinson's disease pathology be propagated from one neuron to another? *Prog Neurobiol*, *97*(2), 205–219. https://doi.org/10.1016/j.pneurobio.2011.11.003

Dursun, E., Gezen-Ak, D., Hanagasi, H., Bilgic, B., Lohmann, E., Ertan, S., Atasoy, I. L., Alaylioglu, M., Araz, O. S., Onal, B., Gunduz, A., Apaydin, H., Kiziltan, G., Ulutin, T., Gurvit, H., & Yilmazer, S. (2015). The interleukin 1 alpha, interleukin 1 beta, interleukin 6 and alpha-2-macroglobulin serum levels in patients with early or late onset Alzheimer's disease, mild cognitive impairment or Parkinson's disease. *J Neuroimmunol*, *283*, 50–57. https://doi.org/10.1016/j.jneuroim.2015.04.014

Duyckaerts, C., Delatour, B., & Potier, M. C. (2009). Classification and basic pathology of Alzheimer disease. *Acta Neuropathol*, *118*(1), 5–36. https://doi.org/10.1007/s00401-009-0532-1

Dziewczapolski, G., Glogowski, C. M., Masliah, E., & Heinemann, S. F. (2009). Deletion of the alpha 7 nicotinic acetylcholine receptor gene improves cognitive deficits and synaptic pathology in a mouse model of Alzheimer's disease. *J Neurosci*, *29*(27), 8805–8815. https://doi.org/10.1523/JNEUROSCI.6159-08.2009

Egawa, J., Pearn, M. L., Lemkuil, B. P., Patel, P. M., & Head, B. P. (2016). Membrane lipid rafts and neurobiology: age-related changes in membrane lipids and loss of neuronal function. *J Physiol*, *594*(16), 4565–4579. https://doi.org/10.1113/JP270590

El-Amouri, S. S., Zhu, H., Yu, J., Marr, R., Verma, I. M., & Kindy, M. S. (2008). Neprilysin: an enzyme candidate to slow the progression of Alzheimer's disease. *Am J Pathol*, *172*(5), 1342–1354. https://doi.org/10.2353/ajpath.2008.070620

Emamzadeh, F. N. (2016). Alpha-synuclein structure, functions, and interactions. *J Res Med Sci*, *21*, 29. https://doi.org/10.4103/1735-1995.181989

Eriksen, J., Jorgensen, T. N., & Gether, U. (2010). Regulation of dopamine transporter function by protein-protein interactions: new discoveries and methodological challenges. *J Neurochem*, *113*(1), 27–41. https://doi.org/10.1111/j.1471-4159.2010.06599.x

Espinosa-Oliva, A. M., Garcia-Revilla, J., Alonso-Bellido, I. M., & Burguillos, M. A. (2019). Brainiac caspases: beyond the wall of apoptosis. *Front Cell Neurosci*, *13*, 500. https://doi.org/10.3389/fncel.2019.00500

Estaun-Panzano, J., Arotcarena, M. L., & Bezard, E. (2023). Monitoring alpha-synuclein aggregation. *Neurobiol Dis*, *176*, 105966. https://doi.org/10.1016/j.nbd.2022.105966

Everse, J., & Coates, P. W. (2009). Neurodegeneration and peroxidases. *Neurobiol Aging*, *30*(7), 1011–1025. https://doi.org/10.1016/j.neurobiolaging.2007.10.007

Fairfoul, G., McGuire, L. I., Pal, S., Ironside, J. W., Neumann, J., Christie, S., Joachim, C., Esiri, M., Evetts, S. G., Rolinski, M., Baig, F., Ruffmann, C., Wade-Martins, R., Hu, M. T., Parkkinen, L., & Green, A. J. (2016). Alpha-synuclein RT-QuIC in the CSF of patients with alpha-synucleinopathies. *Ann Clin Transl Neurol*, *3*(10), 812–818. https://doi.org/10.1002/acn3.338

Falzone, T. L., Stokin, G. B., Lillo, C., Rodrigues, E. M., Westerman, E. L., Williams, D. S., & Goldstein, L. S. (2009). Axonal stress kinase activation and tau misbehavior induced by kinesin-1 transport defects. *J Neurosci*, *29*(18), 5758–5767. https://doi.org/10.1523/JNEUROSCI.0780-09.2009

Fang, Y., Tian, S., Pan, Y., Li, W., Wang, Q., Tang, Y., Yu, T., Wu, X., Shi, Y., Ma, P., & Shu, Y. (2020). Pyroptosis: a new frontier in cancer. *Biomed Pharmacother*, *121*, 109595. https://doi.org/10.1016/j.biopha.2019.109595

Fantini, J., Carlus, D., & Yahi, N. (2011). The fusogenic tilted peptide (67–78) of alpha-synuclein is a cholesterol binding domain. *Biochim Biophys Acta*, *1808*(10), 2343–2351. https://doi.org/10.1016/j.bbamem.2011.06.017

Fantini, J., & Yahi, N. (2011). Molecular basis for the glycosphingolipid-binding specificity of alpha-synuclein: key role of tyrosine 39 in membrane insertion. *J Mol Biol*, *408*(4), 654–669. https://doi.org/10.1016/j.jmb.2011.03.009

Fantini, J., & Yahi, N. (2013). The driving force of alpha-synuclein insertion and amyloid channel formation in the plasma membrane of neural cells: key role of ganglioside- and cholesterol-binding domains. *Adv Exp Med Biol*, *991*, 15–26. https://doi.org/10.1007/978-94-007-6331-9_2

Fecchio, C., De Franceschi, G., Relini, A., Greggio, E., Dalla Serra, M., Bubacco, L., & Polverino de Laureto, P. (2013). Alpha-Synuclein oligomers induced by docosahexaenoic acid affect membrane integrity. *PLoS One*, *8*(11), e82732. https://doi.org/10.1371/journal.pone.0082732

Federico, A., Cardaioli, E., Da Pozzo, P., Formichi, P., Gallus, G. N., & Radi, E. (2012). Mitochondria, oxidative stress and neurodegeneration. *J Neurol Sci*, *322*(1–2), 254–262. https://doi.org/S0022-510X(12)00258-4 [pii] 10.1016/j.jns.2012.05.030

Feoktistova, M., & Leverkus, M. (2015). Programmed necrosis and necroptosis signalling. *FEBS J*, *282*(1), 19–31. https://doi.org/10.1111/febs.13120

Fichou, Y., Lin, Y., Rauch, J. N., Vigers, M., Zeng, Z., Srivastava, M., Keller, T. J., Freed, J. H., Kosik, K. S., & Han, S. (2018). Cofactors are essential constituents of stable and seeding-active tau fibrils. *Proc Natl Acad Sci U S A*, *115*(52), 13234–13239. https://doi.org/10.1073/pnas.1810058115

Filiano, A. J., Gadani, S. P., & Kipnis, J. (2017). How and why do T cells and their derived cytokines affect the injured and healthy brain? *Nat Rev Neurosci*, *18*(6), 375–384. https://doi.org/10.1038/nrn.2017.39

Fiorini, M., Iselle, G., Perra, D., Bongianni, M., Capaldi, S., Sacchetto, L., Ferrari, S., Mombello, A., Vascellari, S., Testi, S., Monaco, S., & Zanusso, G. (2020). High diagnostic accuracy of RT-QuIC assay in a prospective study of patients with suspected sCJD. *Int J Mol Sci*, *21*(3), 880. https://doi.org/10.3390/ijms21030880

Fleten, K. G., Florenes, V. A., Prasmickaite, L., Hill, O., Sykora, J., Maelandsmo, G. M., & Engesaeter, B. (2016). hvTRA, a novel TRAIL receptor agonist, induces apoptosis and sustained growth retardation in melanoma. *Cell Death Discov*, *2*, 16081. https://doi.org/10.1038/cddiscovery.2016.81

Forloni, G. (2023). Alpha Synuclein: neurodegeneration and inflammation. *Int J Mol Sci*, *24*(6), 5914. https://doi.org/10.3390/ijms24065914

Fukuyama, R., Izumoto, T., & Fushiki, S. (2001). The cerebrospinal fluid level of glial fibrillary acidic protein is increased in cerebrospinal fluid from Alzheimer's disease patients and correlates with severity of dementia. *Eur Neurol*, *46*(1), 35–38. https://doi.org/10.1159/000050753

Funayama, M., Hasegawa, K., Kowa, H., Saito, M., Tsuji, S., & Obata, F. (2002). A new locus for Parkinson's disease (PARK8) maps to chromosome 12p11.2-q13.1. *Ann Neurol*, *51*(3), 296–301. https://doi.org/10.1002/ana.10113

Galiano-Landeira, J., Torra, A., Vila, M., & Bove, J. (2020). CD8 T cell nigral infiltration precedes synucleinopathy in early stages of Parkinson's disease. *Brain*, *143*(12), 3717–3733. https://doi.org/10.1093/brain/awaa269

Galluzzi, L., Baehrecke, E. H., Ballabio, A., Boya, P., Bravo-San Pedro, J. M., Cecconi, F., Choi, A. M., Chu, C. T., Codogno, P., Colombo, M. I., Cuervo, A. M., Debnath, J., Deretic, V., Dikic, I., Eskelinen, E. L., Fimia, G. M., Fulda, S., Gewirtz, D. A., Green, D. R., … Kroemer, G. (2017). Molecular definitions of autophagy and related processes. *EMBO J*, *36*(13), 1811–1836. https://doi.org/10.15252/embj.201796697

Galluzzi, L., Kepp, O., Krautwald, S., Kroemer, G., & Linkermann, A. (2014). Molecular mechanisms of regulated necrosis. *Semin Cell Dev Biol*, *35*, 24–32. https://doi.org/10.1016/j.semcdb.2014.02.006

Galluzzi, L., Kepp, O., & Kroemer, G. (2016). Mitochondrial regulation of cell death: a phylogenetically conserved control. *Microb Cell*, *3*(3), 101–108. https://doi.org/10.15698/mic2016.03.483

Galluzzi, L., Vitale, I., Aaronson, S. A., Abrams, J. M., Adam, D., Agostinis, P., Alnemri, E. S., Altucci, L., Amelio, I., Andrews, D. W., Annicchiarico-Petruzzelli, M., Antonov, A. V., Arama, E., Baehrecke, E. H., Barlev, N. A., Bazan, N. G., Bernassola, F., Bertrand, M. J. M., Bianchi, K., … Kroemer, G. (2018). Molecular mechanisms of cell death: recommendations of the nomenclature committee on cell death 2018. *Cell Death Differ*, *25*(3), 486–541. https://doi.org/10.1038/s41418-017-0012-4

Garten, M., Prevost, C., Cadart, C., Gautier, R., Bousset, L., Melki, R., Bassereau, P., & Vanni, S. (2015). Methyl-branched lipids promote the membrane adsorption of alpha-synuclein by enhancing shallow lipid-packing defects. *Phys Chem Chem Phys*, *17*(24), 15589–15597. https://doi.org/10.1039/c5cp00244c

Gaschler, M. M., & Stockwell, B. R. (2017). Lipid peroxidation in cell death. *Biochem Biophys Res Commun*, *482*(3), 419–425. https://doi.org/10.1016/j.bbrc.2016.10.086

Gasperini, L., Meneghetti, E., Legname, G., & Benetti, F. (2016). In absence of the cellular prion protein, alterations in copper metabolism and copper-dependent oxidase activity affect iron distribution. *Front Neurosci*, *10*, 437. https://doi.org/10.3389/fnins.2016.00437

George, J. M., Jin, H., Woods, W. S., & Clayton, D. F. (1995). Characterization of a novel protein regulated during the critical period for song learning in the zebra finch. *Neuron*, *15*(2), 361–372. https://doi.org/10.1016/0896-6273(95)90040-3

Gezen-Ak, D., Alaylioglu, M., Genc, G., Sengul, B., Keskin, E., Sordu, P., Gulec, Z. E. K., Apaydin, H., Bayram-Gurel, C., Ulutin, T., Yilmazer, S., Ertan, S., & Dursun, E. (2020). Altered transcriptional profile of mitochondrial DNA-encoded OXPHOS subunits, mitochondria quality control genes, and intracellular ATP levels in blood samples of patients with Parkinson's disease. *J Alzheimers Dis*, *74*(1), 287–307. https://doi.org/10.3233/JAD-191164

Gezen-Ak, D., Alaylioglu, M., Yurttas, Z., Camoglu, T., Sengul, B., Isler, C., Yasar Kina, U., Keskin, E., Atasoy, I. L., Kafardar, A. M., Uzan, M., Annweiler, C., & Dursun, E. (2023). Vitamin D receptor regulates transcription of mitochondrial DNA and directly interacts with mitochondrial DNA and TFAM. *J Nutr Biochem*, *116*, 109322. https://doi.org/10.1016/j.jnutbio.2023.109322

Gezen-Ak, D., Atasoy, I. L., Candas, E., Alaylioglu, M., & Dursun, E. (2018). The transcriptional regulatory properties of amyloid beta 1–42 may include regulation of genes related to neurodegeneration. *Neuromolecular Med*, *20*(3), 363–375. https://doi.org/10.1007/s12017-018-8498-6

Gezen-Ak, D., Atasoy, I. L., Candas, E., Alaylioglu, M., Yilmazer, S., & Dursun, E. (2017). Vitamin D receptor regulates amyloid beta 1–42 production with protein disulfide isomerase A3. *ACS Chem Neurosci*, *8*(10), 2335–2346. https://doi.org/10.1021/acschemneuro.7b00245

Gezen-Ak, D., Dursun, E., Hanagasi, H., Bilgic, B., Lohman, E., Araz, O. S., Atasoy, I. L., Alaylioglu, M., Onal, B., Gurvit, H., & Yilmazer, S. (2013). BDNF, TNFalpha, HSP90, CFH, and IL-10 serum levels in patients with early or late onset Alzheimer's disease or mild cognitive impairment. *J Alzheimers Dis*, *37*(1), 185–195. https://doi.org/4032G85782821N23

Gezen-Ak, D., Yurttas, Z., Camoglu, T., & Dursun, E. (2022). Could amyloid-beta 1–42 or alpha-synuclein interact directly with mitochondrial DNA? A hypothesis. *ACS Chem Neurosci*, *13*(19), 2803–2812. https://doi.org/10.1021/acschemneuro.2c00512

Ghosh, A., Roy, A., Liu, X., Kordower, J. H., Mufson, E. J., Hartley, D. M., Ghosh, S., Mosley, R. L., Gendelman, H. E., & Pahan, K. (2007). Selective inhibition of NF-kappaB activation prevents dopaminergic neuronal loss in a mouse model of Parkinson's disease. *Proc Natl Acad Sci U S A*, *104*(47), 18754–18759. https://doi.org/10.1073/pnas.0704908104

Giannoccaro, M. P., La Morgia, C., Rizzo, G., & Carelli, V. (2017). Mitochondrial DNA and primary mitochondrial dysfunction in Parkinson's disease. *Mov Disord*, *32*(3), 346–363. https://doi.org/10.1002/mds.26966

Gil, J. M., & Rego, A. C. (2008). Mechanisms of neurodegeneration in Huntington's disease. *Eur J Neurosci*, *27*(11), 2803–2820. https://doi.org/10.1111/j.1460-9568.2008.06310.x

Giusto, N. M., Roque, M. E., & Ilincheta de Boschero, M. G. (1992). Effects of aging on the content, composition and synthesis of sphingomyelin in the central nervous system. *Lipids*, *27*(11), 835–839. https://doi.org/10.1007/BF02535859

Goedert, M. (2015). NEURODEGENERATION. Alzheimer's and Parkinson's diseases: the prion concept in relation to assembled Abeta, tau, and alpha-synuclein. *Science*, *349*(6248), 1255555. https://doi.org/10.1126/science.1255555

Goedert, M., Clavaguera, F., & Tolnay, M. (2010). The propagation of prion-like protein inclusions in neurodegenerative diseases. *Trends Neurosci*, *33*(7), 317–325. https://doi.org/10.1016/j.tins.2010.04.003

Goedert, M., & Spillantini, M. G. (2019). Ordered assembly of tau protein and neurodegeneration. *Adv Exp Med Biol*, *1184*, 3–21. https://doi.org/10.1007/978-981-32-9358-8_1

Goel, P., Chakrabarti, S., Goel, K., Bhutani, K., Chopra, T., & Bali, S. (2022). Neuronal cell death mechanisms in Alzheimer's disease: an insight. *Front Mol Neurosci*, *15*, 937133. https://doi.org/10.3389/fnmol.2022.937133

Gomez-Ramos, A., Diaz-Hernandez, M., Rubio, A., Miras-Portugal, M. T., & Avila, J. (2008). Extracellular tau promotes intracellular calcium increase through M1 and M3 muscarinic receptors in neuronal cells. *Mol Cell Neurosci*, *37*(4), 673–681. https://doi.org/10.1016/j.mcn.2007.12.010

Gomperts, S. N. (2016). Lewy body dementias: dementia with lewy bodies and parkinson disease dementia. *Continuum (Minneap Minn)*, *22*(2 Dementia), 435–463. https://doi.org/10.1212/CON.0000000000000309

Gonzalez-Reyes, R. E., Nava-Mesa, M. O., Vargas-Sanchez, K., Ariza-Salamanca, D., & Mora-Munoz, L. (2017). Involvement of astrocytes in Alzheimer's disease from a neuroinflammatory and oxidative stress perspective. *Front Mol Neurosci*, *10*, 427. https://doi.org/10.3389/fnmol.2017.00427

Green, D. R., & Llambi, F. (2015). Cell death signaling. *Cold Spring Harb Perspect Biol*, *7*(12), a006080. https://doi.org/10.1101/cshperspect.a006080

Greengard, P., Benfenati, F., & Valtorta, F. (1994). Synapsin I, an actin-binding protein regulating synaptic vesicle traffic in the nerve terminal. *Adv Second Messenger Phosphoprotein Res*, *29*, 31–45. https://doi.org/10.1016/s1040-7952(06)80005-4

Grunewald, A., Kumar, K. R., & Sue, C. M. (2019). New insights into the complex role of mitochondria in Parkinson's disease. *Prog Neurobiol*, *177*, 73–93. https://doi.org/10.1016/j.pneurobio.2018.09.003

Grunewald, A., Rygiel, K. A., Hepplewhite, P. D., Morris, C. M., Picard, M., & Turnbull, D. M. (2016). Mitochondrial DNA depletion in respiratory chain-deficient Parkinson disease neurons. *Ann Neurol*, *79*(3), 366–378. https://doi.org/10.1002/ana.24571

Haas, L. T., Kostylev, M. A., & Strittmatter, S. M. (2014). Therapeutic molecules and endogenous ligands regulate the interaction between brain cellular prion protein (PrPC) and metabotropic glutamate receptor 5 (mGluR5). *J Biol Chem*, *289*(41), 28460–28477. https://doi.org/10.1074/jbc.M114.584342

Haas, L. T., Salazar, S. V., Kostylev, M. A., Um, J. W., Kaufman, A. C., & Strittmatter, S. M. (2016). Metabotropic glutamate receptor 5 couples cellular prion protein to intracellular signalling in Alzheimer's disease. *Brain*, *139*(Pt 2), 526–546. https://doi.org/10.1093/brain/awv356

Hallock, P., & Thomas, M. A. (2012). Integrating the Alzheimer's disease proteome and transcriptome: a comprehensive network model of a complex disease. *OMICS*, *16*(1–2), 37–49. https://doi.org/10.1089/omi.2011.0054

Hamilton, A., Esseltine, J. L., DeVries, R. A., Cregan, S. P., & Ferguson, S. S. (2014). Metabotropic glutamate receptor 5 knockout reduces cognitive impairment and pathogenesis in a mouse model of Alzheimer's disease. *Mol Brain*, *7*, 40. https://doi.org/10.1186/1756-6606-7-40

Hamilton, A., Vasefi, M., Vander Tuin, C., McQuaid, R. J., Anisman, H., & Ferguson, S. S. (2016). Chronic pharmacological mGluR5 inhibition prevents cognitive impairment and reduces pathogenesis in an Alzheimer disease mouse model. *Cell Rep*, *15*(9), 1859–1865. https://doi.org/10.1016/j.celrep.2016.04.077

Hampel, H., Hardy, J., Blennow, K., Chen, C., Perry, G., Kim, S. H., Villemagne, V. L., Aisen, P., Vendruscolo, M., Iwatsubo, T., Masters, C. L., Cho, M., Lannfelt, L., Cummings, J. L., & Vergallo, A. (2021). The amyloid-beta pathway in Alzheimer's disease. *Mol Psychiatry*, *26*(10), 5481–5503. https://doi.org/10.1038/s41380-021-01249-0

Harada, A., Oguchi, K., Okabe, S., Kuno, J., Terada, S., Ohshima, T., Sato-Yoshitake, R., Takei, Y., Noda, T., & Hirokawa, N. (1994). Altered microtubule organization in small-calibre axons of mice lacking tau protein. *Nature*, *369*(6480), 488–491. https://doi.org/10.1038/369488a0

Hare, D., Ayton, S., Bush, A., & Lei, P. (2013). A delicate balance: iron metabolism and diseases of the brain. *Front Aging Neurosci*, *5*, 34. https://doi.org/10.3389/fnagi.2013.00034

Hartl, F. U., Bracher, A., & Hayer-Hartl, M. (2011). Molecular chaperones in protein folding and proteostasis. *Nature*, *475*(7356), 324–332. https://doi.org/10.1038/nature10317

Hector, A., & Brouillette, J. (2020). Hyperactivity induced by soluble amyloid-beta oligomers in the early stages of Alzheimer's disease. *Front Mol Neurosci*, *13*, 600084. https://doi.org/10.3389/fnmol.2020.600084

Hegde, M. L., Anitha, S., Latha, K. S., Mustak, M. S., Stein, R., Ravid, R., & Rao, K. S. (2004). First evidence for helical transitions in supercoiled DNA by amyloid Beta Peptide (1–42) and aluminum: a new insight in understanding Alzheimer's disease. *J Mol Neurosci*, *22*(1–2), 19–31. https://doi.org/10.1385/jmn:22:1-2:19

Hellstrand, E., Nowacka, A., Topgaard, D., Linse, S., & Sparr, E. (2013). Membrane lipid co-aggregation with alpha-synuclein fibrils. *PLoS One*, *8*(10), e77235. https://doi.org/10.1371/journal.pone.0077235

Heneka, M. T., Kummer, M. P., Stutz, A., Delekate, A., Schwartz, S., Vieira-Saecker, A., Griep, A., Axt, D., Remus, A., Tzeng, T. C., Gelpi, E., Halle, A., Korte, M., Latz, E., & Golenbock, D. T. (2013). NLRP3 is activated in Alzheimer's disease and contributes to pathology in APP/PS1 mice. *Nature*, *493*(7434), 674–678. https://doi.org/10.1038/nature11729

Her, L. S., & Goldstein, L. S. (2008). Enhanced sensitivity of striatal neurons to axonal transport defects induced by mutant huntingtin. *J Neurosci*, *28*(50), 13662–13672. https://doi.org/10.1523/JNEUROSCI.4144-08.2008

Hijaz, B. A., & Volpicelli-Daley, L. A. (2020). Initiation and propagation of alpha-synuclein aggregation in the nervous system. *Mol Neurodegener*, *15*(1), 19. https://doi.org/10.1186/s13024-020-00368-6

Ho, D. H., Seol, W., & Son, I. (2019). Upregulation of the p53-p21 pathway by G2019S LRRK2 contributes to the cellular senescence and accumulation of alpha-synuclein. *Cell Cycle*, *18*(4), 467–475. https://doi.org/10.1080/15384101.2019.1577666

Hoekstra, J. G., Hipp, M. J., Montine, T. J., & Kennedy, S. R. (2016). Mitochondrial DNA mutations increase in early stage Alzheimer disease and are inconsistent with oxidative damage. *Ann Neurol*, *80*(2), 301–306. https://doi.org/10.1002/ana.24709

Holper, S., Watson, R., & Yassi, N. (2022). Tau as a Biomarker of Neurodegeneration. *Int J Mol Sci*, *23*(13). https://doi.org/10.3390/ijms23137307

Holzer, M., Holzapfel, H. P., Zedlick, D., Bruckner, M. K., & Arendt, T. (1994). Abnormally phosphorylated tau protein in Alzheimer's disease: heterogeneity of individual regional distribution and relationship to clinical severity. *Neuroscience*, *63*(2), 499–516. https://doi.org/10.1016/0306-4522(94)90546-0

Hong, S., Ostaszewski, B. L., Yang, T., O'Malley, T. T., Jin, M., Yanagisawa, K., Li, S., Bartels, T., & Selkoe, D. J. (2014). Soluble Abeta oligomers are rapidly sequestered from brain ISF in vivo and bind GM1 ganglioside on cellular membranes. *Neuron*, *82*(2), 308–319. https://doi.org/10.1016/j.neuron.2014.02.027

Horrocks, L. A., & Yeo, Y. K. (1999). Health benefits of docosahexaenoic acid (DHA). *Pharmacol Res*, *40*(3), 211–225. https://doi.org/10.1006/phrs.1999.0495

Hu, N. W., Nicoll, A. J., Zhang, D., Mably, A. J., O'Malley, T., Purro, S. A., Terry, C., Collinge, J., Walsh, D. M., & Rowan, M. J. (2014). mGlu5 receptors and cellular prion protein mediate amyloid-beta-facilitated synaptic long-term depression in vivo. *Nat Commun*, *5*, 3374. https://doi.org/10.1038/ncomms4374

Hunter, S., & Brayne, C. (2012). Relationships between the amyloid precursor protein and its various proteolytic fragments and neuronal systems. *Alzheimers Res Ther*, *4*(2), 10. https://doi.org/10.1186/alzrt108

Ironside, J. W., Ritchie, D. L., & Head, M. W. (2017). Prion diseases. *Handb Clin Neurol*, *145*, 393–403. https://doi.org/10.1016/B978-0-12-802395-2.00028-6

Jamasbi, E., Separovic, F., Hossain, M. A., & Ciccotosto, G. D. (2017). Phosphorylation of a full length amyloid-beta peptide modulates its amyloid aggregation, cell binding and neurotoxic properties. *Mol Biosyst*, *13*(8), 1545–1551. https://doi.org/10.1039/c7mb00249a

Jang, H., Connelly, L., Arce, F. T., Ramachandran, S., Kagan, B. L., Lal, R., & Nussinov, R. (2013). Mechanisms for the insertion of toxic, fibril-like beta-amyloid oligomers into the membrane. *J Chem Theory Comput*, *9*(1), 822–833. https://doi.org/10.1021/ct300916f

Jarosz-Griffiths, H. H., Noble, E., Rushworth, J. V., & Hooper, N. M. (2016). Amyloid-beta receptors: the good, the bad, and the prion protein. *J Biol Chem*, *291*(7), 3174–3183. https://doi.org/10.1074/jbc.R115.702704

Javadpour, P., Askari, S., Rashidi, F. S., Dargahi, L., Ahmadiani, A., & Ghasemi, R. (2021). Imipramine alleviates memory impairment and hippocampal apoptosis in STZ-induced sporadic Alzheimer's rat model: possible contribution of MAPKs and insulin signaling. *Behav Brain Res*, *408*, 113260. https://doi.org/10.1016/j.bbr.2021.113260

Jayaraman, A., Htike, T. T., James, R., Picon, C., & Reynolds, R. (2021). TNF-mediated neuroinflammation is linked to neuronal necroptosis in Alzheimer's disease hippocampus. *Acta Neuropathol Commun*, *9*(1), 159. https://doi.org/10.1186/s40478-021-01264-w

Jellinger, K. A. (2010). Basic mechanisms of neurodegeneration: a critical update. *J Cell Mol Med*, *14*(3), 457–487. https://doi.org/10.1111/j.1582-4934.2010.01010.x

Jenkins, N. L., James, S. A., Salim, A., Sumardy, F., Speed, T. P., Conrad, M., Richardson, D. R., Bush, A. I., & McColl, G. (2020). Changes in ferrous iron and glutathione promote ferroptosis and frailty in aging Caenorhabditis elegans. *Elife*, *9*. https://doi.org/10.7554/eLife.56580

Jeong, S. (2017). Molecular and cellular basis of neurodegeneration in Alzheimer's disease. *Mol Cells*, *40*(9), 613–620. https://doi.org/10.14348/molcells.2017.0096

Jiang, K., Rocha, S., Westling, A., Kesarimangalam, S., Dorfman, K. D., Wittung-Stafshede, P., & Westerlund, F. (2018). Alpha-synuclein modulates the physical properties of DNA. *Chemistry*, *24*(58), 15685–15690. https://doi.org/10.1002/chem.201803933

Jimenez-Jimenez, F. J., Alonso-Navarro, H., Garcia-Martin, E., Santos-Garcia, D., Martinez-Valbuena, I., & Agundez, J. A. G. (2023). Alpha-synuclein in peripheral tissues as a possible marker for neurological diseases and other medical conditions. *Biomolecules*, *13*(8). https://doi.org/10.3390/biom13081263

Johnson, R. T. (2005). Prion diseases. *Lancet Neurol*, *4*(10), 635–642. https://doi.org/10.1016/S1474-4422(05)70192-7

Johri, A., & Beal, M. F. (2012). Mitochondrial dysfunction in neurodegenerative diseases. *J Pharmacol Exp Ther*, *342*(3), 619–630. https://doi.org/10.1124/jpet.112.192138

Jones, E. M., Dubey, M., Camp, P. J., Vernon, B. C., Biernat, J., Mandelkow, E., Majewski, J., & Chi, E. Y. (2012). Interaction of tau protein with model lipid membranes induces tau structural compaction and membrane disruption. *Biochemistry*, *51*(12), 2539–2550. https://doi.org/10.1021/bi201857v

Jouanne, M., Rault, S., & Voisin-Chiret, A. S. (2017). Tau protein aggregation in Alzheimer's disease: an attractive target for the development of novel therapeutic agents. *Eur J Med Chem*, *139*, 153–167. https://doi.org/10.1016/j.ejmech.2017.07.070

Jucker, M., & Walker, L. C. (2018). Propagation and spread of pathogenic protein assemblies in neurodegenerative diseases. *Nat Neurosci*, *21*(10), 1341–1349. https://doi.org/10.1038/s41593-018-0238-6

Kaczmarek, A., Vandenabeele, P., & Krysko, D. V. (2013). Necroptosis: the release of damage-associated molecular patterns and its physiological relevance. *Immunity*, *38*(2), 209–223. https://doi.org/10.1016/j.immuni.2013.02.003

Kaiser, W. J., Sridharan, H., Huang, C., Mandal, P., Upton, J. W., Gough, P. J., Sehon, C. A., Marquis, R. W., Bertin, J., & Mocarski, E. S. (2013). Toll-like receptor 3-mediated necrosis via TRIF, RIP3, and MLKL. *J Biol Chem*, *288*(43), 31268–31279. https://doi.org/10.1074/jbc.M113.462341

Kaminsky, N., Bihari, O., Kanner, S., & Barzilai, A. (2016). Connecting malfunctioning glial cells and brain degenerative disorders. *Genomics Proteomics Bioinformatics*, *14*(3), 155–165. https://doi.org/10.1016/j.gpb.2016.04.003

Kaplan, J. H. (2002). Biochemistry of Na,K-ATPase. *Annu Rev Biochem*, *71*, 511–535. https://doi.org/10.1146/annurev.biochem.71.102201.141218

Kawabe, J. I., Grant, B. S., Yamamoto, M., Schwencke, C., Okumura, S., & Ishikawa, Y. (2001). Changes in caveolin subtype protein expression in aging rat organs. *Mol Cell Endocrinol*, *176*(1-2), 91–95. https://doi.org/10.1016/s0303-7207(01)00472-5

Kawamoto, Y., Ayaki, T., Urushitani, M., Ito, H., & Takahashi, R. (2016). Activated caspase-9 immunoreactivity in glial and neuronal cytoplasmic inclusions in multiple system atrophy. *Neurosci Lett*, *628*, 207–212. https://doi.org/10.1016/j.neulet.2016.06.036

Kayed, R., Head, E., Thompson, J. L., McIntire, T. M., Milton, S. C., Cotman, C. W.,& Glabe, C. G. (2003). Common structure of soluble amyloid oligomers implies common mechanism of pathogenesis. *Science*, *300*(5618), 486–489. https://doi.org/10.1126/science.1079469

Kayed, R., Sokolov, Y., Edmonds, B., McIntire, T. M., Milton, S. C., Hall, J. E., & Glabe, C. G. (2004). Permeabilization of lipid bilayers is a common conformation-dependent activity of soluble amyloid oligomers in protein misfolding diseases. *J Biol Chem*, *279*(45), 46363–46366. https://doi.org/10.1074/jbc.C400260200

Kazmierczak, A., Czapski, G. A., Adamczyk, A., Gajkowska, B., & Strosznajder, J. B. (2011). A novel mechanism of non-Abeta component of Alzheimer's disease amyloid (NAC) neurotoxicity. Interplay between p53 protein and cyclin-dependent kinase 5 (Cdk5). *Neurochem Int*, *58*(2), 206–214. https://doi.org/10.1016/j.neuint.2010.11.018

Kelliher, M. T., Saunders, H. A., & Wildonger, J. (2019). Microtubule control of functional architecture in neurons. *Curr Opin Neurobiol*, *57*, 39–45. https://doi.org/10.1016/j.conb.2019.01.003

Kempuraj, D., Selvakumar, G. P., Thangavel, R., Ahmed, M. E., Zaheer, S., Kumar, K. K., Yelam, A., Kaur, H., Dubova, I., Raikwar, S. P., Iyer, S. S., & Zaheer, A. (2018). Glia maturation factor and mast cell-dependent expression of inflammatory mediators and proteinase activated receptor-2 in neuroinflammation. *J Alzheimers Dis*, *66*(3), 1117–1129. https://doi.org/10.3233/JAD-180786

Keskin, E., Gezen-Ak, D., & Dursun, E. (2023). Amyloid beta,alpha-synuclein and amyloid beta-alpha-synuclein combination exert significant but different alterations in inflammatory response profile in differentiated human SH-SY5Y cells. *ACS Omega*, *8*(48), 45519–45534. https://doi.org/10.1021/acsomega.3c05585

Kessels, H. W., Nguyen, L. N., Nabavi, S., & Malinow, R. (2010). The prion protein as a receptor for amyloid-beta. *Nature*, *466*(7308), E3–4; discussion E4–5. https://doi.org/10.1038/nature09217

Kfoury, N., Holmes, B. B., Jiang, H., Holtzman, D. M., & Diamond, M. I. (2012). Trans-cellular propagation of Tau aggregation by fibrillar species. *J Biol Chem*, *287*(23), 19440–19451. https://doi.org/10.1074/jbc.M112.346072

Khachaturian, Z. S. (1994). Calcium hypothesis of Alzheimer's disease and brain aging. *Ann N Y Acad Sci*, *747*, 1–11. https://doi.org/10.1111/j.1749-6632.1994.tb44398.x

Khmeleva, S. A., Radko, S. P., Kozin, S. A., Kiseleva, Y. Y., Mezentsev, Y. V., Mitkevich, V. A., Kurbatov, L. K., Ivanov, A. S., & Makarov, A. A. (2016). Zinc-mediated binding of nucleic acids to amyloid-beta aggregates: role of histidine residues. *J Alzheimers Dis*, *54*(2), 809–819. https://doi.org/10.3233/JAD-160415

Kim, D. I., Lee, K. H., Oh, J. Y., Kim, J. S., & Han, H. J. (2017). Relationship between beta-amyloid and mitochondrial dynamics. *Cell Mol Neurobiol*, *37*(6), 955–968. https://doi.org/10.1007/s10571-016-0434-4

Kim, S., Lim, J., Bang, Y., Moon, J., Kwon, M. S., Hong, J. T., Jeon, J., Seo, H., & Choi, H. J. (2018). Alpha-synuclein suppresses retinoic acid-induced neuronal differentiation by targeting the glycogen synthase kinase-3beta/beta-catenin signaling pathway. *Mol Neurobiol*, *55*(2), 1607–1619. https://doi.org/10.1007/s12035-016-0370-9

Kim, T., Vidal, G. S., Djurisic, M., William, C. M., Birnbaum, M. E., Garcia, K. C., Hyman, B. T., & Shatz, C. J. (2013). Human LilrB2 is a beta-amyloid receptor and its murine homolog PirB regulates synaptic plasticity in an Alzheimer's model. *Science*, *341*(6152), 1399–1404. https://doi.org/10.1126/science.1242077

Kim, W. S., Kagedal, K., & Halliday, G. M. (2014). Alpha-synuclein biology in Lewy body diseases. *Alzheimers Res Ther*, *6*(5), 73. https://doi.org/10.1186/s13195-014-0073-2

Kinnally, K. W., Peixoto, P. M., Ryu, S. Y., & Dejean, L. M. (2011). Is mPTP the gatekeeper for necrosis, apoptosis, or both? *Biochim Biophys Acta*, *1813*(4), 616–622. https://doi.org/10.1016/j.bbamcr.2010.09.013

Kirkwood, T. B. (1977). Evolution of ageing. *Nature*, *270*(5635), 301–304. https://doi.org/10.1038/270301a0

Kitada, T., Asakawa, S., Hattori, N., Matsumine, H., Yamamura, Y., Minoshima, S., Yokochi, M., Mizuno, Y., & Shimizu, N. (1998). Mutations in the parkin gene cause autosomal recessive juvenile parkinsonism. *Nature*, *392*(6676), 605–608. https://doi.org/10.1038/33416

Kitamura, Y., Shimohama, S., Kamoshima, W., Ota, T., Matsuoka, Y., Nomura, Y., Smith, M. A., Perry, G., Whitehouse, P. J., & Taniguchi, T. (1998). Alteration of proteins regulating apoptosis, Bcl-2, Bcl-x, Bax, Bak, Bad, ICH-1 and CPP32, in Alzheimer's disease. *Brain Res*, *780*(2), 260–269. https://doi.org/10.1016/s0006-8993(97)01202-x

Klaips, C. L., Jayaraj, G. G., & Hartl, F. U. (2018). Pathways of cellular proteostasis in aging and disease. *J Cell Biol*, *217*(1), 51–63. https://doi.org/10.1083/jcb.201709072

Komatsu, M., Waguri, S., Chiba, T., Murata, S., Iwata, J., Tanida, I., Ueno, T., Koike, M., Uchiyama, Y., Kominami, E., & Tanaka, K. (2006). Loss of autophagy in the central nervous system causes neurodegeneration in mice. *Nature*, *441*(7095), 880–884. https://doi.org/10.1038/nature04723

Koper, M. J., Van Schoor, E., Ospitalieri, S., Vandenberghe, R., Vandenbulcke, M., von Arnim, C. A. F., Tousseyn, T., Balusu, S., De Strooper, B., & Thal, D. R. (2020). Necrosome complex detected in granulovacuolar degeneration is associated with neuronal loss in Alzheimer's disease. *Acta Neuropathol*, *139*(3), 463–484. https://doi.org/10.1007/s00401-019-02103-y

Kornmann, B. (2014). Quality control in mitochondria: use it, break it, fix it, trash it. *F1000Prime Rep*, *6*, 15. https://doi.org/10.12703/P6-15

Kostka, M., Hogen, T., Danzer, K. M., Levin, J., Habeck, M., Wirth, A., Wagner, R., Glabe, C. G., Finger, S., Heinzelmann, U., Garidel, P., Duan, W., Ross, C. A., Kretzschmar, H., & Giese, A. (2008). Single particle characterization of iron-induced pore-forming alpha-synuclein oligomers. *J Biol Chem*, *283*(16), 10992–11003. https://doi.org/10.1074/jbc.M709634200

Kovac, V., & Curin Serbec, V. (2022). Prion protein: the molecule of many forms and faces. *Int J Mol Sci*, *23*(3), 1232. https://doi.org/10.3390/ijms23031232

Kovacs, G. G. (2016). Molecular pathological classification of neurodegenerative diseases: turning towards precision medicine. *Int J Mol Sci*, *17*(2), 189. https://doi.org/10.3390/ijms17020189

Kovacs, G. G. (2017). Concepts and classification of neurodegenerative diseases. *Handb Clin Neurol*, *145*, 301–307. https://doi.org/10.1016/B978-0-12-802395-2.00021-3

Kovacs, G. G., Botond, G., & Budka, H. (2010). Protein coding of neurodegenerative dementias: the neuropathological basis of biomarker diagnostics. *Acta Neuropathol*, *119*(4), 389–408. https://doi.org/10.1007/s00401-010-0658-1

Kramer, M. L., & Schulz-Schaeffer, W. J. (2007). Presynaptic alpha-synuclein aggregates, not Lewy bodies, cause neurodegeneration in dementia with Lewy bodies. *J Neurosci*, *27*(6), 1405–1410. https://doi.org/10.1523/JNEUROSCI.4564-06.2007

Kuffer, A., Lakkaraju, A. K., Mogha, A., Petersen, S. C., Airich, K., Doucerain, C., Marpakwar, R., Bakirci, P., Senatore, A., Monnard, A., Schiavi, C., Nuvolone, M., Grosshans, B., Hornemann, S., Bassilana, F., Monk, K. R., & Aguzzi, A. (2016). The prion protein is an agonistic ligand of the G protein-coupled receptor Adgrg6. *Nature*, *536*(7617), 464–468. https://doi.org/10.1038/nature19312

Kukreja, L., Kujoth, G. C., Prolla, T. A., Van Leuven, F., & Vassar, R. (2014). Increased mtDNA mutations with aging promotes amyloid accumulation and brain atrophy in the APP/Ld transgenic mouse model of Alzheimer's disease. *Mol Neurodegener*, *9*, 16. https://doi.org/10.1186/1750-1326-9-16

Kulak, N. A., Geyer, P. E., & Mann, M. (2017). Loss-less nano-fractionator for high sensitivity, high coverage proteomics. *Mol Cell Proteomics*, *16*(4), 694–705. https://doi.org/10.1074/mcp.O116.065136

Kumar, S., Lemere, C. A., & Walter, J. (2020). Phosphorylated Abeta peptides in human down syndrome brain and different Alzheimer's-like mouse models. *Acta Neuropathol Commun*, *8*(1), 118. https://doi.org/10.1186/s40478-020-00959-w

Kumar, S., Rezaei-Ghaleh, N., Terwel, D., Thal, D. R., Richard, M., Hoch, M., Mc Donald, J. M., Wullner, U., Glebov, K., Heneka, M. T., Walsh, D. M., Zweckstetter, M., & Walter, J. (2011). Extracellular phosphorylation of the amyloid beta-peptide promotes formation of toxic aggregates during the pathogenesis of Alzheimer's disease. *EMBO J*, *30*(11), 2255–2265. https://doi.org/10.1038/emboj.2011.138

Kumar, S., & Walter, J. (2011). Phosphorylation of amyloid beta (Abeta) peptides - a trigger for formation of toxic aggregates in Alzheimer's disease. *Aging (Albany NY)*, *3*(8), 803–812. https://doi.org/10.18632/aging.100362

Kuroda, Y., Mitsui, T., Kunishige, M., Shono, M., Akaike, M., Azuma, H., & Matsumoto, T. (2006). Parkin enhances mitochondrial biogenesis in proliferating cells. *Hum Mol Genet*, *15*(6), 883–895. https://doi.org/10.1093/hmg/ddl006

Lachance, V., Wang, Q., Sweet, E., Choi, I., Cai, C. Z., Zhuang, X. X., Zhang, Y., Jiang, J. L., Blitzer, R. D., Bozdagi-Gunal, O., Zhang, B., Lu, J. H., & Yue, Z. (2019). Autophagy protein NRBF2 has reduced expression in Alzheimer's brains and modulates memory and amyloid-beta homeostasis in mice. *Mol Neurodegener*, *14*(1), 43. https://doi.org/10.1186/s13024-019-0342-4

Lai, Y., Kim, S., Varkey, J., Lou, X., Song, J. K., Diao, J., Langen, R., & Shin, Y. K. (2014). Nonaggregated alpha-synuclein influences SNARE-dependent vesicle docking via membrane binding. *Biochemistry*, *53*(24), 3889–3896. https://doi.org/10.1021/bi5002536

Langston, J. W., Ballard, P., Tetrud, J. W., & Irwin, I. (1983). Chronic Parkinsonism in humans due to a product of meperidine-analog synthesis. *Science*, *219*(4587), 979–980. https://doi.org/10.1126/science.6823561

Lassmann, H., Bancher, C., Breitschopf, H., Wegiel, J., Bobinski, M., Jellinger, K., & Wisniewski, H. M. (1995). Cell death in Alzheimer's disease evaluated by DNA fragmentation in situ. *Acta Neuropathol*, *89*(1), 35–41. https://doi.org/10.1007/BF00294257

Lauren, J., Gimbel, D. A., Nygaard, H. B., Gilbert, J. W., & Strittmatter, S. M. (2009). Cellular prion protein mediates impairment of synaptic plasticity by amyloid-beta oligomers. *Nature*, *457*(7233), 1128–1132. https://doi.org/10.1038/nature07761

Lee, J., Gillman, A. L., Jang, H., Ramachandran, S., Kagan, B. L., Nussinov, R., & Teran Arce, F. (2014). Role of the fast kinetics of pyroglutamate-modified amyloid-beta oligomers in membrane binding and membrane permeability. *Biochemistry*, *53*(28), 4704–4714. https://doi.org/10.1021/bi500587p

Lee, S. J., Desplats, P., Lee, H. J., Spencer, B., & Masliah, E. (2012). Cell-to-cell transmission of alpha-synuclein aggregates. *Methods Mol Biol*, *849*, 347–359. https://doi.org/10.1007/978-1-61779-551-0_23

Lee, S. J., Desplats, P., Sigurdson, C., Tsigelny, I., & Masliah, E. (2010). Cell-to-cell transmission of non-prion protein aggregates. *Nat Rev Neurol*, *6*(12), 702–706. https://doi.org/10.1038/nrneurol.2010.145

Lee, S. S., Kim, Y. M., Junn, E., Lee, G., Park, K. H., Tanaka, M., Ronchetti, R. D., Quezado, M. M., & Mouradian, M. M. (2003). Cell cycle aberrations by alpha-synuclein overexpression and cyclin B immunoreactivity in Lewy bodies. *Neurobiol Aging*, *24*(5), 687–696. https://doi.org/10.1016/s0197-4580(02)00196-3

Lesne, S. E. (2014). Toxic oligomer species of amyloid-beta in Alzheimer's disease, a timing issue. *Swiss Med Wkly*, *144*, w14021. https://doi.org/10.4414/smw.2014.14021

Leuzy, A., Cullen, N. C., Mattsson-Carlgren, N., & Hansson, O. (2021). Current advances in plasma and cerebrospinal fluid biomarkers in Alzheimer's disease. *Curr Opin Neurol*, *34*(2), 266–274. https://doi.org/10.1097/WCO.0000000000000904

Lewy, F.H. (1912). Paralysis agitans. I. Pathologische Anatomie. In V. M. Lewandowsky (Ed.), *Lewandowsky's Handbuch der Neurologie, 3 Band: Spez Neurologie II* (pp. 920– 933). Springer.

Lezi, E., & Swerdlow, R. H. (2012). Mitochondria in neurodegeneration. *Adv Exp Med Biol*, *942*, 269–286. https://doi.org/10.1007/978-94-007-2869-1_12

Li, H., & Gorfe, A. A. (2014). Membrane remodeling by surface-bound protein aggregates: insights from coarse-grained molecular dynamics simulation. *J Phys Chem Lett*, *5*(8), 1457–1462. https://doi.org/10.1021/jz500451a

Li, J., Cao, F., Yin, H. L., Huang, Z. J., Lin, Z. T., Mao, N., Sun, B., & Wang, G. (2020). Ferroptosis: past, present and future. *Cell Death Dis*, *11*(2), 88. https://doi.org/10.1038/s41419-020-2298-2

Li, J., Han, Y. R., Plummer, M. R., & Herrup, K. (2009). Cytoplasmic ATM in neurons modulates synaptic function. *Curr Biol*, *19*(24), 2091–2096. https://doi.org/10.1016/j.cub.2009.10.039

Li, W., Hoffman, P. N., Stirling, W., Price, D. L., & Lee, M. K. (2004). Axonal transport of human alpha-synuclein slows with aging but is not affected by familial Parkinson's disease-linked mutations. *J Neurochem*, *88*(2), 401–410. https://doi.org/10.1046/j.1471-4159.2003.02166.x

Li, W. W., Yang, R., Guo, J. C., Ren, H. M., Zha, X. L., Cheng, J. S., & Cai, D. F. (2007). Localization of alpha-synuclein to mitochondria within midbrain of mice. *Neuroreport*, *18*(15), 1543–1546. https://doi.org/10.1097/WNR.0b013e3282f03db4

Liao, Y. F., Wang, B. J., Cheng, H. T., Kuo, L. H., & Wolfe, M. S. (2004). Tumor necrosis factor-alpha, interleukin-1beta, and interferon-gamma stimulate gamma-secretase-mediated cleavage of amyloid precursor protein through a JNK-dependent MAPK pathway. *J Biol Chem*, *279*(47), 49523–49532. https://doi.org/10.1074/jbc.M402034200

Lim, D., Iyer, A., Ronco, V., Grolla, A. A., Canonico, P. L., Aronica, E., & Genazzani, A. A. (2013). Amyloid beta deregulates astroglial mGluR5-mediated calcium signaling via calcineurin and Nf-kB. *Glia*, *61*(7), 1134–1145. https://doi.org/10.1002/glia.22502

Lipinski, M. M., Zheng, B., Lu, T., Yan, Z., Py, B. F., Ng, A., Xavier, R. J., Li, C., Yankner, B. A., Scherzer, C. R., & Yuan, J. (2010). Genome-wide analysis reveals mechanisms modulating autophagy in normal brain aging and in Alzheimer's disease. *Proc Natl Acad Sci U S A*, *107*(32), 14164–14169. https://doi.org/10.1073/pnas.1009485107

Liu, G., Zhang, C., Yin, J., Li, X., Cheng, F., Li, Y., Yang, H., Ueda, K., Chan, P., & Yu, S. (2009). Alpha-Synuclein is differentially expressed in mitochondria from different rat brain regions and dose-dependently down-regulates complex I activity. *Neurosci Lett*, *454*(3), 187–192. https://doi.org/10.1016/j.neulet.2009.02.056

Liu, R. M. (2022). Aging, cellular senescence, and Alzheimer's disease. *Int J Mol Sci*, *23*(4), 1989. https://doi.org/10.3390/ijms23041989

Liu, W. K., Dickson, D. W., & Yen, S. H. (1993). Heterogeneity of tau proteins in Alzheimer's disease. Evidence for increased expression of an isoform and preferential distribution of a phosphorylated isoform in neurites. *Am J Pathol*, *142*(2), 387–394. https://www.ncbi.nlm.nih.gov/pubmed/7679548

Lodeiro, M., Puerta, E., Ismail, M. A., Rodriguez-Rodriguez, P., Ronnback, A., Codita, A., Parrado-Fernandez, C., Maioli, S., Gil-Bea, F., Merino-Serrais, P., & Cedazo-Minguez, A. (2017). Aggregation of the inflammatory S100A8 precedes abeta plaque formation in transgenic APP mice: positive feedback for S100A8 and abeta productions. *J Gerontol A Biol Sci Med Sci*, *72*(3), 319–328. https://doi.org/10.1093/gerona/glw073

Lopes, S., Vaz-Silva, J., Pinto, V., Dalla, C., Kokras, N., Bedenk, B., Mack, N., Czisch, M., Almeida, O. F., Sousa, N., & Sotiropoulos, I. (2016). Tau protein is essential for stress-induced brain pathology. *Proc Natl Acad Sci U S A*, *113*(26), E3755–3763. https://doi.org/10.1073/pnas.1600953113

Lopez-Otin, C., Blasco, M. A., Partridge, L., Serrano, M., & Kroemer, G. (2013). The hallmarks of aging. *Cell*, *153*(6), 1194–1217. https://doi.org/10.1016/j.cell.2013.05.039

Lopez-Teros, M., Alarcon-Aguilar, A., Lopez-Diazguerrero, N. E., Luna-Lopez, A., & Konigsberg, M. (2022). Contribution of senescent and reactive astrocytes on central nervous system inflammaging. *Biogerontology*, *23*(1), 21–33. https://doi.org/10.1007/s10522-022-09952-3

Lorenzen, N., Nielsen, S. B., Buell, A. K., Kaspersen, J. D., Arosio, P., Vad, B. S., Paslawski, W., Christiansen, G., Valnickova-Hansen, Z., Andreasen, M., Enghild, J. J., Pedersen, J. S., Dobson, C. M., Knowles, T. P., & Otzen, D. E. (2014). The role of stable alpha-synuclein oligomers in the molecular events underlying amyloid formation. *J Am Chem Soc*, *136*(10), 3859–3868. https://doi.org/10.1021/ja411577t

Love, S., Barber, R., & Wilcock, G. K. (1999). Increased poly(ADP-ribosyl)ation of nuclear proteins in Alzheimer's disease. *Brain*, *122 (Pt 2)*, 247–253. https://doi.org/10.1093/brain/122.2.247

Lu, J., Zhang, S., Ma, X., Jia, C., Liu, Z., Huang, C., Liu, C., & Li, D. (2020). Structural basis of the interplay between alpha-synuclein and Tau in regulating pathological amyloid aggregation. *J Biol Chem*, *295*(21), 7470–7480. https://doi.org/10.1074/jbc.RA119.012284

Lucassen, P. J., Chung, W. C., Kamphorst, W., & Swaab, D. F. (1997). DNA damage distribution in the human brain as shown by in situ end labeling; area-specific differences in aging and Alzheimer disease in the absence of apoptotic morphology. *J Neuropathol Exp Neurol*, *56*(8), 887–900. https://doi.org/10.1097/00005072-199708000-00007

Lunnon, K., Keohane, A., Pidsley, R., Newhouse, S., Riddoch-Contreras, J., Thubron, E. B., Devall, M., Soininen, H., Kloszewska, I., Mecocci, P., Tsolaki, M., Vellas, B., Schalkwyk, L., Dobson, R., Malik, A. N., Powell, J., Lovestone, S., Hodges, A., & AddNeuroMed, C. (2017). Mitochondrial genes are altered in blood early in Alzheimer's disease. *Neurobiol Aging*, *53*, 36–47. https://doi.org/10.1016/j.neurobiolaging.2016.12.029

Luo, Z., Ahlers-Dannen, K. E., Spicer, M. M., Yang, J., Alberico, S., Stevens, H. E., Narayanan, N. S., & Fisher, R. A. (2019). Age-dependent nigral dopaminergic neurodegeneration and alpha-synuclein accumulation in RGS6-deficient mice. *JCI Insight*, *5*(13). https://doi.org/10.1172/jci.insight.126769

Lutters, B., Foley, P., & Koehler, P. J. (2018). The centennial lesson of encephalitis lethargica. *Neurology*, *90*(12), 563–567. https://doi.org/10.1212/WNL.0000000000005176

Ma, J., Goryaynov, A., Sarma, A., & Yang, W. (2012). Self-regulated viscous channel in the nuclear pore complex. *Proc Natl Acad Sci U S A*, *109*(19), 7326–7331. https://doi.org/10.1073/pnas.1201724109

Mahul-Mellier, A. L., Burtscher, J., Maharjan, N., Weerens, L., Croisier, M., Kuttler, F., Leleu, M., Knott, G. W., & Lashuel, H. A. (2020). The process of Lewy body formation, rather than simply alpha-synuclein fibrillization, is one of the major drivers of neurodegeneration. *Proc Natl Acad Sci U S A*, *117*(9), 4971–4982. https://doi.org/10.1073/pnas.1913904117

Maji, S. K., Wang, L., Greenwald, J., & Riek, R. (2009). Structure-activity relationship of amyloid fibrils. *FEBS Lett*, *583*(16), 2610–2617. https://doi.org/10.1016/j.febslet.2009.07.003

Majumdar, A., Cruz, D., Asamoah, N., Buxbaum, A., Sohar, I., Lobel, P., & Maxfield, F. R. (2007). Activation of microglia acidifies lysosomes and leads to degradation of Alzheimer amyloid fibrils. *Mol Biol Cell*, *18*(4), 1490–1496. https://doi.org/10.1091/mbc.e06-10-0975

Maloney, B., & Lahiri, D. K. (2011). The Alzheimer's amyloid beta-peptide (Abeta) binds a specific DNA Abeta-interacting domain (AbetaID) in the APP, BACE1, and APOE promoters in a sequence-specific manner: characterizing a new regulatory motif. *Gene*, *488*(1–2), 1–12. https://doi.org/10.1016/j.gene.2011.06.004

Manczak, M., Park, B. S., Jung, Y., & Reddy, P. H. (2004). Differential expression of oxidative phosphorylation genes in patients with Alzheimer's disease: implications for early mitochondrial dysfunction and oxidative damage. *Neuromolecular Med*, *5*(2),147–162. https://doi.org/10.1385/NMM:5:2:147

Mao, P., & Reddy, P. H. (2011). Aging and amyloid beta-induced oxidative DNA damage and mitochondrial dysfunction in Alzheimer's disease: implications for early intervention and therapeutics. *Biochim Biophys Acta*, *1812*(11), 1359–1370. https://doi.org/10.1016/j.bbadis.2011.08.005

Mao, X., Ou, M. T., Karuppagounder, S. S., Kam, T. I., Yin, X., Xiong, Y., Ge, P., Umanah, G. E., Brahmachari, S., Shin, J. H., Kang, H. C., Zhang, J., Xu, J., Chen, R., Park, H., Andrabi, S. A., Kang, S. U., Goncalves, R. A., Liang, Y., … Dawson, T. M. (2016). Pathological alpha-synuclein transmission initiated by binding lymphocyte-activation gene 3. *Science*, *353*(6307). https://doi.org/10.1126/science.aah3374

Maroteaux, L., Campanelli, J. T., & Scheller, R. H. (1988). Synuclein: a neuron-specific protein localized to the nucleus and presynaptic nerve terminal. *J Neurosci*, *8*(8), 2804–2815. https://doi.org/10.1523/JNEUROSCI.08-08-02804.1988

Martin-Jimenez, C. A., Garcia-Vega, A., Cabezas, R., Aliev, G., Echeverria, V., Gonzalez, J., & Barreto, G. E. (2017). Astrocytes and endoplasmic reticulum stress: a bridge between obesity and neurodegenerative diseases. *Prog Neurobiol*, *158*, 45–68. https://doi.org/10.1016/j.pneurobio.2017.08.001

Martin, M. G., & Dotti, C. G. (2022). Plasma membrane and brain dysfunction of the old: Do we age from our membranes? *Front Cell Dev Biol*, *10*, 1031007. https://doi.org/10.3389/fcell.2022.1031007

Martin, M. G., Perga, S., Trovo, L., Rasola, A., Holm, P., Rantamaki, T., Harkany, T., Castren, E., Chiara, F., & Dotti, C. G. (2008). Cholesterol loss enhances TrkB signaling in hippocampal neurons aging in vitro. *Mol Biol Cell*, *19*(5), 2101–2112. https://doi.org/10.1091/mbc.e07-09-0897

Martineau, M., Shi, T., Puyal, J., Knolhoff, A. M., Dulong, J., Gasnier, B., Klingauf, J., Sweedler, J. V., Jahn, R., & Mothet, J. P. (2013). Storage and uptake of D-serine into astrocytic synaptic-like vesicles specify gliotransmission. *J Neurosci*, *33*(8), 3413–3423. https://doi.org/10.1523/JNEUROSCI.3497-12.2013

Martinez, J., Moeller, I., Erdjument-Bromage, H., Tempst, P., & Lauring, B. (2003). Parkinson's disease-associated alpha-synuclein is a calmodulin substrate. *J Biol Chem*, *278*(19), 17379–17387. https://doi.org/10.1074/jbc.M209020200

Matsui, T., Ramasamy, K., Ingelsson, M., Fukumoto, H., Conrad, C., Frosch, M. P., Irizarry, M. C., Yuan, J., & Hyman, B. T. (2006). Coordinated expression of caspase 8, 3 and 7 mRNA in temporal cortex of Alzheimer disease: relationship to formic acid extractable abeta42 levels. *J Neuropathol Exp Neurol*, *65*(5), 508–515. https://doi.org/10.1097/01.jnen.0000229238.05748.12

Mattson, M. P., & Chan, S. L. (2003). Neuronal and glial calcium signaling in Alzheimer's disease. *Cell Calcium*, *34*(4–5), 385–397. https://doi.org/10.1016/s0143-4160(03)00128-3

McCann, H., Stevens, C. H., Cartwright, H., & Halliday, G. M. (2014). Alpha-synucleinopathy phenotypes. *Parkinsonism Relat Disord*, *20 Suppl 1*, S62–67. https://doi.org/10.1016/S1353-8020(13)70017-8

McGahon, B., Murray, C. A., Clements, M. P., & Lynch, M. A. (1998). Analysis of the effect of membrane arachidonic acid concentration on modulation of glutamate release by interleukin-1: an age-related study. *Exp Gerontol*, *33*(4), 343–354. https://doi.org/10.1016/s0531-5565(97)00130-7

McNamara, R. K., Liu, Y., Jandacek, R., Rider, T., & Tso, P. (2008). The aging human orbitofrontal cortex: decreasing polyunsaturated fatty acid composition and associated increases in lipogenic gene expression and stearoyl-CoA desaturase activity. *Prostaglandins Leukot Essent Fatty Acids*, *78*(4–5), 293–304. https://doi.org/10.1016/j.plefa.2008.04.001

Mehlen, P., & Bredesen, D. E. (2011). Dependence receptors: from basic research to drug development. *Sci Signal*, *4*(157), mr2. https://doi.org/10.1126/scisignal.2001521

Mehta, P. D., Pirttila, T., Patrick, B. A., Barshatzky, M., & Mehta, S. P. (2001). Amyloid beta protein 1–40 and 1–42 levels in matched cerebrospinal fluid and plasma from patients with Alzheimer disease. *Neurosci Lett*, *304*(1–2), 102–106. https://doi.org/10.1016/s0304-3940(01)01754-2

Milanese, C., Cerri, S., Ulusoy, A., Gornati, S. V., Plat, A., Gabriels, S., Blandini, F., Di Monte, D. A., Hoeijmakers, J. H., & Mastroberardino, P. G. (2018). Activation of the DNA damage response in vivo in synucleinopathy models of Parkinson's disease. *Cell Death Dis*, *9*(8), 818. https://doi.org/10.1038/s41419-018-0848-7

Milanesi, L., Sheynis, T., Xue, W. F., Orlova, E. V., Hellewell, A. L., Jelinek, R., Hewitt, E. W., Radford, S. E., & Saibil, H. R. (2012). Direct three-dimensional visualization of membrane disruption by amyloid fibrils. *Proc Natl Acad Sci U S A*, *109*(50), 20455–20460. https://doi.org/10.1073/pnas.1206325109

Minamoto, T., Sawaguchi, K., Ogino, T., & Mai, M. (1991). Anisakiasis of the colon: report of two cases with emphasis on the diagnostic and therapeutic value of colonoscopy. *Endoscopy*, *23*(1), 50–52. https://doi.org/10.1055/s-2007-1010609

Mirza, B., Hadberg, H., Thomsen, P., & Moos, T. (2000). The absence of reactive astrocytosis is indicative of a unique inflammatory process in Parkinson's disease. *Neuroscience*, *95*(2), 425–432. https://doi.org/10.1016/s0306-4522(99)00455-8

Molander-Melin, M., Blennow, K., Bogdanovic, N., Dellheden, B., Mansson, J. E., & Fredman, P. (2005). Structural membrane alterations in Alzheimer brains found to be associated with regional disease development; increased density of gangliosides GM1 and GM2 and loss of cholesterol in detergent-resistant membrane domains. *J Neurochem*, *92*(1), 171–182. https://doi.org/10.1111/j.1471-4159.2004.02849.x

Molinuevo, J. L., Ayton, S., Batrla, R., Bednar, M. M., Bittner, T., Cummings, J., Fagan, A. M., Hampel, H., Mielke, M. M., Mikulskis, A., O'Bryant, S., Scheltens, P., Sevigny,

J., Shaw, L. M., Soares, H. D., Tong, G., Trojanowski, J. Q., Zetterberg, H., & Blennow, K. (2018). Current state of Alzheimer's fluid biomarkers. *Acta Neuropathol*, *136*(6), 821–853. https://doi.org/10.1007/s00401-018-1932-x

Morley, J. E., & Farr, S. A. (2014). The role of amyloid-beta in the regulation of memory. *Biochem Pharmacol*, *88*(4), 479–485. https://doi.org/10.1016/j.bcp.2013.12.018

Muresan, V., Varvel, N. H., Lamb, B. T., & Muresan, Z. (2009). The cleavage products of amyloid-beta precursor protein are sorted to distinct carrier vesicles that are independently transported within neurites. *J Neurosci*, *29*(11), 3565–3578. https://doi.org/10.1523/JNEUROSCI.2558-08.2009

Narne, P., Pandey, V., Simhadri, P. K., & Phanithi, P. B. (2017). Poly(ADP-ribose)polymerase-1 hyperactivation in neurodegenerative diseases: the death knell tolls for neurons. *Semin Cell Dev Biol*, *63*, 154–166. https://doi.org/10.1016/j.semcdb.2016.11.007

Nixon, R. A. (2013). The role of autophagy in neurodegenerative disease. *Nat Med*, *19*(8), 983–997. https://doi.org/10.1038/nm.3232

Nixon, R. A., Wegiel, J., Kumar, A., Yu, W. H., Peterhoff, C., Cataldo, A., & Cuervo,A. M. (2005). Extensive involvement of autophagy in Alzheimer disease: an immuno-electron microscopy study. *J Neuropathol Exp Neurol*, *64*(2), 113–122. https://doi.org/10.1093/jnen/64.2.113

Noor, A., Zafar, S., Shafiq, M., Younas, N., Siegert, A., Mann, F. A., Kruss, S., Schmitz, M., Dihazi, H., Ferrer, I., & Zerr, I. (2022). Molecular profiles of amyloid-beta proteoforms in typical and rapidly progressive Alzheimer's disease. *Mol Neurobiol*, *59*(1), 17–34. https://doi.org/10.1007/s12035-021-02566-9

Ohnishi, T., Yanazawa, M., Sasahara, T., Kitamura, Y., Hiroaki, H., Fukazawa, Y., Kii, I., Nishiyama, T., Kakita, A., Takeda, H., Takeuchi, A., Arai, Y., Ito, A., Komura, H., Hirao, H., Satomura, K., Inoue, M., Muramatsu, S., Matsui, K., … Hoshi, M. (2015). Na, K-ATPase alpha3 is a death target of Alzheimer patient amyloid-beta assembly. *Proc Natl Acad Sci U S A*, *112*(32), E4465–4474. https://doi.org/10.1073/pnas.1421182112

Ohyagi, Y., Asahara, H., Chui, D. H., Tsuruta, Y., Sakae, N., Miyoshi, K., Yamada, T., Kikuchi, H., Taniwaki, T., Murai, H., Ikezoe, K., Furuya, H., Kawarabayashi, T., Shoji, M., Checler, F., Iwaki, T., Makifuchi, T., Takeda, K., Kira, J., & Tabira, T. (2005). Intracellular Abeta42 activates p53 promoter: a pathway to neurodegeneration in Alzheimer's disease. *FASEB J*, *19*(2), 255–257. https://doi.org/10.1096/fj.04-2637fje

Ohyama, Y., Meaney, S., Heverin, M., Ekstrom, L., Brafman, A., Shafir, M., Andersson, U., Olin, M., Eggertsen, G., Diczfalusy, U., Feinstein, E., & Bjorkhem, I. (2006). Studies on the transcriptional regulation of cholesterol 24-hydroxylase (CYP46A1): marked insensitivity toward different regulatory axes. *J Biol Chem*, *281*(7), 3810–3820. https://doi.org/10.1074/jbc.M505179200

Okuzumi, A., Hatano, T., Fukuhara, T., Ueno, S., Nukina, N., Imai, Y., & Hattori, N. (2021). Alpha-synuclein seeding assay using RT-QuIC. *Methods Mol Biol*, *2322*, 3–16. https://doi.org/10.1007/978-1-0716-1495-2_1

Olanow, C. W., & Prusiner, S. B. (2009). Is Parkinson's disease a prion disorder? *Proc Natl Acad Sci U S A*, *106*(31), 12571–12572. https://doi.org/10.1073/pnas.0906759106

Ono, K., & Watanabe-Nakayama, T. (2021). Aggregation and structure of amyloid beta-protein. *Neurochem Int*, *151*, 105208. https://doi.org/10.1016/j.neuint.2021.105208

Oshima, R., Hasegawa, T., Tamai, K., Sugeno, N., Yoshida, S., Kobayashi, J., Kikuchi, A., Baba, T., Futatsugi, A., Sato, I., Satoh, K., Takeda, A., Aoki, M., & Tanaka, N. (2016). ESCRT-0 dysfunction compromises autophagic degradation of protein aggregates and facilitates ER stress-mediated neurodegeneration via apoptotic and necroptotic pathways. *Sci Rep*, *6*, 24997. https://doi.org/10.1038/srep24997

Ouberai, M. M., Wang, J., Swann, M. J., Galvagnion, C., Guilliams, T., Dobson, C. M., & Welland, M. E. (2013). Alpha-Synuclein senses lipid packing defects and induces lateral expansion of lipids leading to membrane remodeling. *J Biol Chem*, *288*(29), 20883–20895. https://doi.org/10.1074/jbc.M113.478297

Pacheco, C., Aguayo, L. G., & Opazo, C. (2012). An extracellular mechanism that can explain the neurotoxic effects of alpha-synuclein aggregates in the brain. *Front Physiol*, *3*, 297. https://doi.org/10.3389/fphys.2012.00297

Pagani, L., & Eckert, A. (2011). Amyloid-Beta interaction with mitochondria. *Int J Alzheimers Dis*, *2011*, 925050. https://doi.org/10.4061/2011/925050

Paiva, I., Pinho, R., Pavlou, M. A., Hennion, M., Wales, P., Schutz, A. L., Rajput, A., Szego, E. M., Kerimoglu, C., Gerhardt, E., Rego, A. C., Fischer, A., Bonn, S., & Outeiro, T. F. (2017). Sodium butyrate rescues dopaminergic cells from alpha-synuclein-induced transcriptional deregulation and DNA damage. *Hum Mol Genet*, *26*(12), 2231–2246. https://doi.org/10.1093/hmg/ddx114

Pajares, M., A, I. R., Manda, G., Bosca, L., & Cuadrado, A. (2020). Inflammation in Parkinson's Disease: Mechanisms and Therapeutic Implications. *Cells*, *9*(7). https://doi.org/10.3390/cells9071687

Palop, J. J., & Mucke, L. (2010). Amyloid-beta-induced neuronal dysfunction in Alzheimer's disease: from synapses toward neural networks. *Nat Neurosci*, *13*(7), 812–818. https://doi.org/10.1038/nn.2583

Pan, K. M., Baldwin, M., Nguyen, J., Gasset, M., Serban, A., Groth, D., Mehlhorn, I., Huang, Z., Fletterick, R. J., Cohen, F. E., & et al. (1993). Conversion of alpha-helices into beta-sheets features in the formation of the scrapie prion proteins. *Proc Natl Acad Sci U S A*, *90*(23), 10962–10966. https://doi.org/10.1073/pnas.90.23.10962

Pandey, A. P., Haque, F., Rochet, J. C., & Hovis, J. S. (2009). Clustering of alpha-synuclein on supported lipid bilayers: role of anionic lipid, protein, and divalent ion concentration. *Biophys J*, *96*(2), 540–551. https://doi.org/10.1016/j.bpj.2008.10.011

Pantiya, P., Thonusin, C., Chattipakorn, N., & Chattipakorn, S. C. (2020). Mitochondrial abnormalities in neurodegenerative models and possible interventions: Focus on Alzheimer's disease, Parkinson's disease, Huntington's disease. *Mitochondrion*, *55*, 14–47. https://doi.org/10.1016/j.mito.2020.08.003

Paresce, D. M., Chung, H., & Maxfield, F. R. (1997). Slow degradation of aggregates of the Alzheimer's disease amyloid beta-protein by microglial cells. *J Biol Chem*, *272*(46), 29390–29397. https://doi.org/10.1074/jbc.272.46.29390

Park, J. H., Burgess, J. D., Faroqi, A. H., DeMeo, N. N., Fiesel, F. C., Springer, W., Delenclos, M., & McLean, P. J. (2020). Alpha-synuclein-induced mitochondrial dysfunction is mediated via a sirtuin 3-dependent pathway. *Mol Neurodegener, 15*(1), 5. https://doi.org/10.1186/s13024-019-0349-x

Park, W. Y., Park, J. S., Cho, K. A., Kim, D. I., Ko, Y. G., Seo, J. S., & Park, S. C. (2000). Up-regulation of caveolin attenuates epidermal growth factor signaling in senescent cells. *J Biol Chem, 275*(27), 20847–20852. https://doi.org/10.1074/jbc.M908162199

Parzych, K. R., & Klionsky, D. J. (2014). An overview of autophagy: morphology, mechanism, and regulation. *Antioxid Redox Signal, 20*(3), 460–473. https://doi.org/10.1089/ars.2013.5371

Patel, N., Ramachandran, S., Azimov, R., Kagan, B. L., & Lal, R. (2015). Ion Channel Formation by Tau Protein: Implications for Alzheimer's Disease and Tauopathies. *Biochemistry, 54*(50), 7320–7325. https://doi.org/10.1021/acs.biochem.5b00988

Paterno, G., Bell, B. M., Gorion, K. M., Prokop, S., & Giasson, B. I. (2022). Reassessment of neuronal tau distribution in adult human brain and implications for tau pathobiology. *Acta Neuropathol Commun, 10*(1), 94. https://doi.org/10.1186/s40478-022-01394-9

Pereira, P. A., Alvim-Soares, A., Bicalho, M. A., Moraes, E. N., Malloy-Diniz, L., Paula, J. J., Romano-Silva, M. A., & Miranda, D. M. (2016). Lack of association between genetic polymorphism of circadian genes (PER2, PER3, CLOCK and OX2R) with late onset depression and Alzheimer's disease in a sample of a Brazilian population (Circadian genes, late-onset depression and Alzheimer's disease). *Curr Alzheimer Res, 13*(12), 1397–1406. https://doi.org/10.2174/1567205013666160603005630

Petrelli, F., Dallerac, G., Pucci, L., Cali, C., Zehnder, T., Sultan, S., Lecca, S., Chicca, A., Ivanov, A., Asensio, C. S., Gundersen, V., Toni, N., Knott, G. W., Magara, F., Gertsch, J., Kirchhoff, F., Deglon, N., Giros, B., Edwards, R. H., … Bezzi, P. (2020). Dysfunction of homeostatic control of dopamine by astrocytes in the developing prefrontal cortex leads to cognitive impairments. *Mol Psychiatry, 25*(4), 732–749. https://doi.org/10.1038/s41380-018-0226-y

Petzold, A., Jenkins, R., Watt, H. C., Green, A. J., Thompson, E. J., Keir, G., Fox, N. C., & Rossor, M. N. (2003). Cerebrospinal fluid S100B correlates with brain atrophy in Alzheimer's disease. *Neurosci Lett, 336*(3), 167–170. https://doi.org/10.1016/s0304-3940(02)01257-0

Pieri, L., Madiona, K., Bousset, L., & Melki, R. (2012). Fibrillar alpha-synuclein and huntingtin exon 1 assemblies are toxic to the cells. *Biophys J, 102*(12), 2894–2905. https://doi.org/10.1016/j.bpj.2012.04.050

Pigino, G., Morfini, G., Atagi, Y., Deshpande, A., Yu, C., Jungbauer, L., LaDu, M., Busciglio, J., & Brady, S. (2009). Disruption of fast axonal transport is a pathogenic mechanism for intraneuronal amyloid beta. *Proc Natl Acad Sci U S A, 106*(14), 5907–5912. https://doi.org/10.1073/pnas.0901229106

Pihan, P., Carreras-Sureda, A., & Hetz, C. (2017). BCL-2 family: integrating stress responses at the ER to control cell demise. *Cell Death Differ, 24*(9), 1478–1487. https://doi.org/10.1038/cdd.2017.82

Pinto, M., & Moraes, C. T. (2014). Mitochondrial genome changes and neurodegenerative diseases. *Biochim Biophys Acta, 1842*(8), 1198–1207. https://doi.org/S0925-4439(13)00337-2

Pociute, A., Pivoriunas, A., & Verkhratsky, A. (2024). Astrocytes dynamically regulate the blood-brain barrier in the healthy brain. *Neural Regen Res, 19*(4), 709–710. https://doi.org/10.4103/1673-5374.382248

Prestori, F., Rossi, P., Bearzatto, B., Laine, J., Necchi, D., Diwakar, S., Schiffmann, S. N., Axelrad, H., & D'Angelo, E. (2008). Altered neuron excitability and synaptic plasticity in the cerebellar granular layer of juvenile prion protein knock-out mice with impaired motor control. *J Neurosci, 28*(28), 7091–7103. https://doi.org/10.1523/JNEUROSCI.0409-08.2008

Probst, A., Tolnay, M., Langui, D., Goedert, M., & Spillantini, M. G. (1996). Pick's disease: hyperphosphorylated tau protein segregates to the somatoaxonal compartment. *Acta Neuropathol, 92*(6), 588–596. https://doi.org/10.1007/s004010050565

Prusiner, S. B. (1998). Prions. *Proc Natl Acad Sci U S A, 95*(23), 13363–13383. https://doi.org/10.1073/pnas.95.23.13363

Przedborski, S., Vila, M., & Jackson-Lewis, V. (2003). Neurodegeneration: what is it and where are we? *J Clin Invest, 111*(1), 3–10. https://doi.org/10.1172/JCI17522

Puangmalai, N., Sengupta, U., Bhatt, N., Gaikwad, S., Montalbano, M., Bhuyan, A., Garcia, S., McAllen, S., Sonawane, M., Jerez, C., Zhao, Y., & Kayed, R. (2022). Lysine 63-linked ubiquitination of tau oligomers contributes to the pathogenesis of Alzheimer's disease. *J Biol Chem, 298*(4), 101766. https://doi.org/10.1016/j.jbc.2022.101766

Puoti, G., Bizzi, A., Forloni, G., Safar, J. G., Tagliavini, F., & Gambetti, P. (2012). Sporadic human prion diseases: molecular insights and diagnosis. *Lancet Neurol, 11*(7), 618–628. https://doi.org/10.1016/S1474-4422(12)70063-7

Puzzo, D., Piacentini, R., Fa, M., Gulisano, W., Li Puma, D. D., Staniszewski, A., Zhang, H., Tropea, M. R., Cocco, S., Palmeri, A., Fraser, P., D'Adamio, L., Grassi, C., & Arancio, O. (2017). LTP and memory impairment caused by extracellular Abeta and Tau oligomers is APP-dependent. *Elife, 6*. https://doi.org/10.7554/eLife.26991

Pyle, A., Anugrha, H., Kurzawa-Akanbi, M., Yarnall, A., Burn, D., & Hudson, G. (2016). Reduced mitochondrial DNA copy number is a biomarker of Parkinson's disease. *Neurobiol Aging, 38*, 216 e217–216 e210. https://doi.org/10.1016/j.neurobiolaging.2015.10.033

Pyle, A., Brennan, R., Kurzawa-Akanbi, M., Yarnall, A., Thouin, A., Mollenhauer, B., Burn, D., Chinnery, P. F., & Hudson, G. (2015). Reduced cerebrospinal fluid mitochondrial DNA is a biomarker for early-stage Parkinson's disease. *Ann Neurol, 78*(6), 1000–1004. https://doi.org/10.1002/ana.24515

Qin, H., Buckley, J. A., Li, X., Liu, Y., Fox, T. H., 3rd, Meares, G. P., Yu, H., Yan, Z., Harms, A. S., Li, Y., Standaert, D. G., & Benveniste, E. N. (2016). Inhibition of the JAK/STAT pathway protects against alpha-synuclein-induced neuroinflammation and dopaminergic neurodegeneration. *J Neurosci, 36*(18), 5144–5159. https://doi.org/10.1523/JNEUROSCI.4658-15.2016

Rambaran, R. N., & Serpell, L. C. (2008). Amyloid fibrils: abnormal protein assembly. *Prion, 2*(3), 112–117. https://doi.org/10.4161/pri.2.3.7488

Ramsay, R. R., Salach, J. I., & Singer, T. P. (1986). Uptake of the neurotoxin 1-methyl-4-phenylpyridine (MPP+) by mitochondria and its relation to the inhibition of the mitochondrial oxidation of NAD+-linked substrates by MPP+. *Biochem Biophys Res Commun*, *134*(2), 743–748. https://doi.org/10.1016/s0006-291x(86)80483-1

Randall, C., Mosconi, L., de Leon, M., & Glodzik, L. (2013). Cerebrospinal fluid biomarkers of Alzheimer's disease in healthy elderly. *Front Biosci (Landmark Ed)*, *18*(3), 1150–1173. https://doi.org/10.2741/4170

Ratan, Y., Rajput, A., Maleysm, S., Pareek, A., Jain, V., Pareek, A., Kaur, R., & Singh, G. (2023). An insight into cellular and molecular mechanisms underlying the pathogenesis of neurodegeneration in Alzheimer's disease. *Biomedicines*, *11*(5), 1398. https://doi.org/10.3390/biomedicines11051398

Rauch, J. N., Luna, G., Guzman, E., Audouard, M., Challis, C., Sibih, Y. E., Leshuk, C., Hernandez, I., Wegmann, S., Hyman, B. T., Gradinaru, V., Kampmann, M., & Kosik, K. S. (2020). LRP1 is a master regulator of tau uptake and spread. *Nature*, *580*(7803), 381–385. https://doi.org/10.1038/s41586-020-2156-5

Reddy, P. H. (2009). Role of mitochondria in neurodegenerative diseases: mitochondria as a therapeutic target in Alzheimer's disease. *CNS spectrums*, *14*(S7), 8–13.

Reenstra, W. R., Yaar, M., & Gilchrest, B. A. (1996). Aging affects epidermal growth factor receptor phosphorylation and traffic kinetics. *Exp Cell Res*, *227*(2), 252–255. https://doi.org/10.1006/excr.1996.0274

Regier, M., Liang, J., Choi, A., Verma, K., Libien, J., & Hernandez, A. I. (2019). Evidence for decreased nucleolar PARP-1 as an early marker of cognitive impairment. *Neural Plast*, *2019*, 4383258. https://doi.org/10.1155/2019/4383258

Renner, M., Choquet, D., & Triller, A. (2009). Control of the postsynaptic membrane viscosity. *J Neurosci*, *29*(9), 2926–2937. https://doi.org/10.1523/JNEUROSCI.4445-08.2009

Renner, M., Lacor, P. N., Velasco, P. T., Xu, J., Contractor, A., Klein, W. L., & Triller, A. (2010). Deleterious effects of amyloid beta oligomers acting as an extracellular scaffold for mGluR5. *Neuron*, *66*(5), 739–754. https://doi.org/10.1016/j.neuron.2010.04.029

Reynolds, N. P., Soragni, A., Rabe, M., Verdes, D., Liverani, E., Handschin, S., Riek, R., & Seeger, S. (2011). Mechanism of membrane interaction and disruption by alpha-synuclein. *J Am Chem Soc*, *133*(48), 19366–19375. https://doi.org/10.1021/ja2029848

Rietdijk, C. D., Perez-Pardo, P., Garssen, J., van Wezel, R. J., & Kraneveld, A. D. (2017). Exploring Braak's hypothesis of Parkinson's disease. *Front Neurol*, *8*, 37. https://doi.org/10.3389/fneur.2017.00037

Roberson, E. D., Scearce-Levie, K., Palop, J. J., Yan, F., Cheng, I. H., Wu, T., Gerstein, H., Yu, G. Q., & Mucke, L. (2007). Reducing endogenous tau ameliorates amyloid beta-induced deficits in an Alzheimer's disease mouse model. *Science*, *316*(5825), 750–754. https://doi.org/10.1126/science.1141736

Rohn, T. T., Head, E., Nesse, W. H., Cotman, C. W., & Cribbs, D. H. (2001). Activation of caspase-8 in the Alzheimer's disease brain. *Neurobiol Dis*, *8*(6), 1006–1016. https://doi.org/10.1006/nbdi.2001.0449

Roos, W. P., Thomas, A. D., & Kaina, B. (2016). DNA damage and the balance between survival and death in cancer biology. *Nat Rev Cancer*, *16*(1), 20–33. https://doi.org/10.1038/nrc.2015.2

Rothstein, J. D. (2009). Current hypotheses for the underlying biology of amyotrophic lateral sclerosis. *Ann Neurol*, *65 Suppl 1*, S3–9. https://doi.org/10.1002/ana.21543

Rovere, M., Powers, A. E., Jiang, H., Pitino, J. C., Fonseca-Ornelas, L., Patel, D. S., Achille, A., Langen, R., Varkey, J., & Bartels, T. (2019). E46K-like alpha-synuclein mutants increase lipid interactions and disrupt membrane selectivity. *J Biol Chem*, *294*(25), 9799–9812. https://doi.org/10.1074/jbc.RA118.006551

Ruegsegger, C., Maharjan, N., Goswami, A., Filezac de L'Etang, A., Weis, J., Troost, D., Heller, M., Gut, H., & Saxena, S. (2016). Aberrant association of misfolded SOD1 with Na(+)/K(+)ATPase-alpha3 impairs its activity and contributes to motor neuron vulnerability in ALS. *Acta Neuropathol*, *131*(3), 427–451. https://doi.org/10.1007/s00401-015-1510-4

Rui, Y., Tiwari, P., Xie, Z., & Zheng, J. Q. (2006). Acute impairment of mitochondrial trafficking by beta-amyloid peptides in hippocampal neurons. *J Neurosci*, *26*(41), 10480–10487. https://doi.org/10.1523/JNEUROSCI.3231-06.2006

Saborio, G. P., Permanne, B., & Soto, C. (2001). Sensitive detection of pathological prion protein by cyclic amplification of protein misfolding. *Nature*, *411*(6839), 810–813. https://doi.org/10.1038/35081095

Sachet, M., Liang, Y. Y., & Oehler, R. (2017). The immune response to secondary necrotic cells. *Apoptosis*, *22*(10), 1189–1204. https://doi.org/10.1007/s10495-017-1413-z

Saibil, H. (2013). Chaperone machines for protein folding, unfolding and disaggregation. *Nat Rev Mol Cell Biol*, *14*(10), 630–642. https://doi.org/10.1038/nrm3658

Sakakibara, Y., Sekiya, M., Saito, T., Saido, T. C., & Iijima, K. M. (2019). Amyloid-beta plaque formation and reactive gliosis are required for induction of cognitive deficits in App knock-in mouse models of Alzheimer's disease. *BMC Neurosci*, *20*(1), 13. https://doi.org/10.1186/s12868-019-0496-6

Santiago, J. A., Bottero, V., & Potashkin, J. A. (2017). Dissecting the molecular mechanisms of neurodegenerative diseases through network biology. *Front Aging Neurosci*, *9*, 166. https://doi.org/10.3389/fnagi.2017.00166

Sarkar, S., Malovic, E., Harischandra, D. S., Ngwa, H. A., Ghosh, A., Hogan, C., Rokad, D., Zenitsky, G., Jin, H., Anantharam, V., Kanthasamy, A. G., & Kanthasamy, A. (2018). Manganese exposure induces neuroinflammation by impairing mitochondrial dynamics in astrocytes. *Neurotoxicology*, *64*, 204–218. https://doi.org/10.1016/j.neuro.2017.05.009

Sasahara, K., Morigaki, K., & Shinya, K. (2013). Effects of membrane interaction and aggregation of amyloid beta-peptide on lipid mobility and membrane domain structure. *Phys Chem Chem Phys*, *15*(23), 8929–8939. https://doi.org/10.1039/c3cp44517h

Schaser, A. J., Osterberg, V. R., Dent, S. E., Stackhouse, T. L., Wakeham, C. M., Boutros, S. W., Weston, L. J., Owen, N., Weissman, T. A., Luna, E., Raber, J., Luk, K. C., McCullough, A. K., Woltjer, R. L., & Unni, V. K. (2019). Alpha-synuclein is a DNA binding protein that modulates DNA repair with implications for Lewy body disorders. *Sci Rep*, *9*(1), 10919. https://doi.org/10.1038/s41598-019-47227-z

Schindowski, K., Belarbi, K., & Buee, L. (2008). Neurotrophic factors in Alzheimer's disease: role of axonal transport. *Genes Brain Behav*, *7 Suppl 1*(1), 43–56. https://doi.org/10.1111/j.1601-183X.2007.00378.x

Schmidt, M. F., Gan, Z. Y., Komander, D., & Dewson, G. (2021). Ubiquitin signalling in neurodegeneration: mechanisms and therapeutic opportunities. *Cell Death Differ*, *28*(2), 570–590. https://doi.org/10.1038/s41418-020-00706-7

Seif F, Khoshmirsafa M, Aazami H, Mohsenzadegan M, Sedighi G, Bahar M. (2017). The role of JAK-STAT signaling pathway and its regulators in the fate of T helper cells. Cell Commun Signal. 2017 Jun 21;15(1):23.

Selkoe, D. J., & Hardy, J. (2016). The amyloid hypothesis of Alzheimer's disease at 25 years. *EMBO Mol Med*, *8*(6), 595–608. https://doi.org/10.15252/emmm.201606210

Selznick, L. A., Holtzman, D. M., Han, B. H., Gokden, M., Srinivasan, A. N., Johnson, E. M., Jr., & Roth, K. A. (1999). In situ immunodetection of neuronal caspase-3 activation in Alzheimer disease. *J Neuropathol Exp Neurol*, *58*(9), 1020–1026. https://doi.org/10.1097/00005072-199909000-00012

Sengul, B., Dursun, E., Verkhratsky, A., & Gezen-Ak, D. (2021). Overexpression of alpha-Synuclein reorganises growth factor profile of human astrocytes. *Mol Neurobiol*, *58*(1), 184–203. https://doi.org/10.1007/s12035-020-02114-x

Sengupta, U., & Kayed, R. (2022). Amyloid beta, Tau, and alpha-Synuclein aggregates in the pathogenesis, prognosis, and therapeutics for neurodegenerative diseases. *Prog Neurobiol*, *214*, 102270. https://doi.org/10.1016/j.pneurobio.2022.102270

Seo, J., & Park, M. (2020). Molecular crosstalk between cancer and neurodegenerative diseases. *Cell Mol Life Sci*, *77*(14), 2659–2680. https://doi.org/10.1007/s00018-019-03428-3

Shah, S. B., Nolan, R., Davis, E., Stokin, G. B., Niesman, I., Canto, I., Glabe, C., & Goldstein, L. S. (2009). Examination of potential mechanisms of amyloid-induced defects in neuronal transport. *Neurobiol Dis*, *36*(1), 11–25. https://doi.org/10.1016/j.nbd.2009.05.016

Sharma, S., Awasthi, A., & Singh, S. (2019). Altered gut microbiota and intestinal permeability in Parkinson's disease: pathological highlight to management. *Neurosci Lett*, *712*, 134516. https://doi.org/10.1016/j.neulet.2019.134516

Shen, X. N., Niu, L. D., Wang, Y. J., Cao, X. P., Liu, Q., Tan, L., Zhang, C., & Yu, J. T. (2019). Inflammatory markers in Alzheimer's disease and mild cognitive impairment: a meta-analysis and systematic review of 170 studies. *J Neurol Neurosurg Psychiatry*, *90*(5), 590–598. https://doi.org/10.1136/jnnp-2018-319148

Shi, Z., Sachs, J. N., Rhoades, E., & Baumgart, T. (2015). Biophysics of alpha-synuclein induced membrane remodelling. *Phys Chem Chem Phys*, *17*(24), 15561–15568. https://doi.org/10.1039/c4cp05883f

Shim, K. H., Kang, M. J., Youn, Y. C., An, S. S. A., & Kim, S. (2022). Alpha-synuclein: a pathological factor with Abeta and tau and biomarker in Alzheimer's disease. *Alzheimers Res Ther*, *14*(1), 201. https://doi.org/10.1186/s13195-022-01150-0

Shrivastava, A. N., Aperia, A., Melki, R., & Triller, A. (2017). Physico-pathologic mechanisms involved in neurodegeneration: misfolded protein-plasma membrane interactions. *Neuron*, *95*(1), 33–50. https://doi.org/10.1016/j.neuron.2017.05.026

Shrivastava, A. N., Kowalewski, J. M., Renner, M., Bousset, L., Koulakoff, A., Melki, R., Giaume, C., & Triller, A. (2013). Beta-amyloid and ATP-induced diffusional trapping of astrocyte and neuronal metabotropic glutamate type-5 receptors. *Glia*, *61*(10), 1673–1686. https://doi.org/10.1002/glia.22548

Shrivastava, A. N., Redeker, V., Fritz, N., Pieri, L., Almeida, L. G., Spolidoro, M., Liebmann, T., Bousset, L., Renner, M., Lena, C., Aperia, A., Melki, R., & Triller, A. (2015). Alpha-synuclein assemblies sequester neuronal alpha3-Na+/K+-ATPase and impair Na+ gradient. *EMBO J*, *34*(19), 2408–2423. https://doi.org/10.15252/embj.201591397

Silva, M. T. (2010). Secondary necrosis: the natural outcome of the complete apoptotic program. *FEBS Lett*, *584*(22), 4491–4499. https://doi.org/10.1016/j.febslet.2010.10.046

Singh, P., Jorgacevski, J., Kreft, M., Grubisic, V., Stout, R. F., Jr., Potokar, M., Parpura, V., & Zorec, R. (2014). Single-vesicle architecture of synaptobrevin2 in astrocytes. *Nat Commun*, *5*, 3780. https://doi.org/10.1038/ncomms4780

Smajic, S., Prada-Medina, C. A., Landoulsi, Z., Ghelfi, J., Delcambre, S., Dietrich, C., Jarazo, J., Henck, J., Balachandran, S., Pachchek, S., Morris, C. M., Antony, P., Timmermann, B., Sauer, S., Pereira, S. L., Schwamborn, J. C., May, P., Grunewald, A., & Spielmann, M. (2022). Single-cell sequencing of human midbrain reveals glial activation and a Parkinson-specific neuronal state. *Brain*, *145*(3), 964–978. https://doi.org/10.1093/brain/awab446

Smale, G., Nichols, N. R., Brady, D. R., Finch, C. E., & Horton, W. E., Jr. (1995). Evidence for apoptotic cell death in Alzheimer's disease. *Exp Neurol*, *133*(2), 225–230. https://doi.org/10.1006/exnr.1995.1025

Smith, E. F., Shaw, P. J., & De Vos, K. J. (2019). The role of mitochondria in amyotrophic lateral sclerosis. *Neurosci Lett*, *710*, 132933. https://doi.org/10.1016/j.neulet.2017.06.052

Snyder, E. M., Nong, Y., Almeida, C. G., Paul, S., Moran, T., Choi, E. Y., Nairn, A. C., Salter, M. W., Lombroso, P. J., Gouras, G. K., & Greengard, P. (2005). Regulation of NMDA receptor trafficking by amyloid-beta. *Nat Neurosci*, *8*(8), 1051–1058. https://doi.org/10.1038/nn1503

Sodero, A. O., Trovo, L., Iannilli, F., Van Veldhoven, P., Dotti, C. G., & Martin, M. G. (2011). Regulation of tyrosine kinase B activity by the Cyp46/cholesterol loss pathway in mature hippocampal neurons: relevance for neuronal survival under stress and in aging. *J Neurochem*, *116*(5), 747–755. https://doi.org/10.1111/j.1471-4159.2010.07079.x

Sodero, A. O., Vriens, J., Ghosh, D., Stegner, D., Brachet, A., Pallotto, M., Sassoe-Pognetto, M., Brouwers, J. F., Helms, J. B., Nieswandt, B., Voets, T., & Dotti, C. G. (2012). Cholesterol loss during glutamate-mediated excitotoxicity. *EMBO J*, *31*(7), 1764–1773. https://doi.org/10.1038/emboj.2012.31

Soliman, A., Bakota, L., & Brandt, R. (2022). Microtubule-modulating agents in the fight against neurodegeneration: will it ever work? *Curr Neuropharmacol*, *20*(4), 782–798. https://doi.org/10.2174/1570159X19666211201101020

Soto-Faguas, C. M., Sanchez-Molina, P., & Saura, C. A. (2021). Loss of presenilin function enhances tau phosphorylation and aggregation in mice. *Acta Neuropathol Commun*, *9*(1), 162. https://doi.org/10.1186/s40478-021-01259-7

Soucek, T., Cumming, R., Dargusch, R., Maher, P., & Schubert, D. (2003). The regulation of glucose metabolism by HIF-1 mediates a neuroprotective response to amyloid beta peptide. *Neuron*, *39*(1), 43–56. https://doi.org/10.1016/s0896-6273(03)00367-2

Spillantini, M. G., Schmidt, M. L., Lee, V. M., Trojanowski, J. Q., Jakes, R., & Goedert, M. (1997). Alpha-synuclein in Lewy bodies. *Nature*, *388*(6645), 839–840. https://doi.org/10.1038/42166

Stadelmann, C., Deckwerth, T. L., Srinivasan, A., Bancher, C., Bruck, W., Jellinger, K., & Lassmann, H. (1999). Activation of caspase-3 in single neurons and autophagic granules of granulovacuolar degeneration in Alzheimer's disease. Evidence for apoptotic cell death. *Am J Pathol*, *155*(5), 1459–1466. https://doi.org/10.1016/S0002-9440(10)65460-0

Standaert, D. G., & Childers, G. M. (2022). Alpha-synuclein-mediated DNA damage, STING activation, and neuroinflammation in Parkinson's disease. *Proc Natl Acad Sci U S A*, *119*(17), e2204058119. https://doi.org/10.1073/pnas.2204058119

Standke, H. G., & Kraus, A. (2023). Seed amplification and RT-QuIC assays to investigate protein seed structures and strains. *Cell Tissue Res*, *392*(1), 323–335. https://doi.org/10.1007/s00441-022-03595-z

Steiner, J. A., Angot, E., & Brundin, P. (2011). A deadly spread: cellular mechanisms of alpha-synuclein transfer. *Cell Death Differ*, *18*(9), 1425–1433. https://doi.org/10.1038/cdd.2011.53

Steiner, J. A., Quansah, E., & Brundin, P. (2018). The concept of alpha-synuclein as a prion-like protein: ten years after. *Cell Tissue Res*, *373*(1), 161–173. https://doi.org/10.1007/s00441-018-2814-1

Stockwell, B. R., Friedmann Angeli, J. P., Bayir, H., Bush, A. I., Conrad, M., Dixon, S. J., Fulda, S., Gascon, S., Hatzios, S. K., Kagan, V. E., Noel, K., Jiang, X., Linkermann, A., Murphy, M. E., Overholtzer, M., Oyagi, A., Pagnussat, G. C., Park, J., Ran, Q., … Zhang, D. D. (2017). Ferroptosis: a regulated cell death nexus linking metabolism, redox biology, and disease. *Cell*, *171*(2), 273–285. https://doi.org/10.1016/j.cell.2017.09.021

Stokin, G. B., & Goldstein, L. S. (2006). Axonal transport and Alzheimer's disease. *Annu Rev Biochem*, *75*, 607–627. https://doi.org/10.1146/annurev.biochem.75.103004.142637

Stoothoff, W., Jones, P. B., Spires-Jones, T. L., Joyner, D., Chhabra, E., Bercury, K., Fan, Z., Xie, H., Bacskai, B., Edd, J., Irimia, D., & Hyman, B. T. (2009). Differential effect of three-repeat and four-repeat tau on mitochondrial axonal transport. *J Neurochem*, *111*(2), 417–427. https://doi.org/10.1111/j.1471-4159.2009.06316.x

Strazielle, N., Creidy, R., Malcus, C., Boucraut, J., & Ghersi-Egea, J. F. (2016). T-lymphocytes traffic into the brain across the Blood-CSF -barrier: evidence using a reconstituted choroid plexus epithelium. *PLoS One*, *11*(3), e0150945. https://doi.org/10.1371/journal.pone.0150945

Strom, A. L., Gal, J., Shi, P., Kasarskis, E. J., Hayward, L. J., & Zhu, H. (2008). Retrograde axonal transport and motor neuron disease. *J Neurochem*, *106*(2), 495–505. https://doi.org/10.1111/j.1471-4159.2008.05393.x

Su, J. H., Zhao, M., Anderson, A. J., Srinivasan, A., & Cotman, C. W. (2001). Activated caspase-3 expression in Alzheimer's and aged control brain: correlation with Alzheimer pathology. *Brain Res*, *898*(2), 350–357. https://doi.org/10.1016/s0006-8993(01)02018-2

Sudhof, T. C. (2012). The presynaptic active zone. *Neuron*, *75*(1), 11-25. https://doi.org/10.1016/j.neuron.2012.06.012

Sun, J., Wang, L., Bao, H., Premi, S., Das, U., Chapman, E. R., & Roy, S. (2019). Functional cooperation of alpha-synuclein and VAMP2 in synaptic vesicle recycling. *Proc Natl Acad Sci U S A*, *116*(23), 11113–11115. https://doi.org/10.1073/pnas.1903049116

Sun, Y., Koyama, Y., & Shimada, S. (2022). Inflammation from peripheral organs to the brain: how does systemic inflammation cause neuroinflammation? *Front Aging Neurosci*, *14*, 903455. https://doi.org/10.3389/fnagi.2022.903455

Suram, A., Hegde, M. L., & Rao, K. S. (2007). A new evidence for DNA nicking property of amyloid beta-peptide (1–42): relevance to Alzheimer's disease. *Arch Biochem Biophys*, *463*(2), 245–252. https://doi.org/10.1016/j.abb.2007.03.015

Surguchev, A. A., & Surguchov, A. (2017). Synucleins and gene expression: ramblers in a crowd or cops regulating traffic? *Front Mol Neurosci*, *10*, 224. https://doi.org/10.3389/fnmol.2017.00224

Surguchov, A. (2023). Alpha-Synuclein and mechanisms of epigenetic regulation. *Brain Sci*, *13*(1), 150. https://doi.org/10.3390/brainsci13010150

Swerdlow, R. H., Burns, J. M., & Khan, S. M. (2014). The Alzheimer's disease mitochondrial cascade hypothesis: progress and perspectives. *Biochim Biophys Acta*, *1842*(8), 1219–1231. https://doi.org/10.1016/j.bbadis.2013.09.010

Tackenberg, C., Grinschgl, S., Trutzel, A., Santuccione, A. C., Frey, M. C., Konietzko, U., Grimm, J., Brandt, R., & Nitsch, R. M. (2013). NMDA receptor subunit composition determines beta-amyloid-induced neurodegeneration and synaptic loss. *Cell Death Dis*, *4*(4), e608. https://doi.org/10.1038/cddis.2013.129

Tai, C., Chang, C. W., Yu, G. Q., Lopez, I., Yu, X., Wang, X., Guo, W., & Mucke, L. (2020). Tau reduction prevents key features of autism in mouse models. *Neuron*, *106*(3), 421–437 e411. https://doi.org/10.1016/j.neuron.2020.01.038

Tait, S. W., & Green, D. R. (2010). Mitochondria and cell death: outer membrane permeabilization and beyond. *Nat Rev Mol Cell Biol*, *11*(9), 621–632. https://doi.org/10.1038/nrm2952

Takahashi, M., Miyata, H., Kametani, F., Nonaka, T., Akiyama, H., Hisanaga, S., & Hasegawa, M. (2015). Extracellular association of APP and tau fibrils induces intracellular aggregate formation of tau. *Acta Neuropathol*, *129*(6), 895–907. https://doi.org/10.1007/s00401-015-1415-2

Talwar, P., Silla, Y., Grover, S., Gupta, M., Agarwal, R., Kushwaha, S., & Kukreti, R. (2014). Genomic convergence and network analysis approach to identify candidate genes in Alzheimer's disease. *BMC Genomics*, *15*(1), 199. https://doi.org/10.1186/1471-2164-15-199

Tan, M. S., Tan, L., Jiang, T., Zhu, X. C., Wang, H. F., Jia, C. D., & Yu, J. T. (2014). Amyloid-beta induces NLRP1-dependent neuronal pyroptosis in models of Alzheimer's disease. *Cell Death Dis*, *5*(8), e1382. https://doi.org/10.1038/cddis.2014.348

Tanaka, K., & Matsuda, N. (2014). Proteostasis and neurodegeneration: the roles of proteasomal degradation and autophagy. *Biochim Biophys Acta*, *1843*(1), 197–204. https://doi.org/10.1016/j.bbamcr.2013.03.012

Taoufik, E., Kouroupi, G., Zygogianni, O., & Matsas, R. (2018). Synaptic dysfunction in neurodegenerative and neurodevelopmental diseases: an overview of induced pluripotent stem-cell-based disease models. *Open Biol*, *8*(9), 180138. https://doi.org/10.1098/rsob.180138

Terracina, L., Brunetti, M., Avellini, L., De Medio, G. E., Trovarelli, G., & Gaiti, A. (1992). Arachidonic and palmitic acid utilization in aged rat brain areas. *Mol Cell Biochem*, *115*(1), 35–42. https://doi.org/10.1007/BF00229093

Thayanidhi, N., Helm, J. R., Nycz, D. C., Bentley, M., Liang, Y., & Hay, J. C. (2010). Alpha-synuclein delays endoplasmic reticulum (ER)-to-Golgi transport in mammalian cells by antagonizing ER/Golgi SNAREs. *Mol Biol Cell*, *21*(11), 1850–1863. https://doi.org/10.1091/mbc.e09-09-0801

Thibault, O., Porter, N. M., Chen, K. C., Blalock, E. M., Kaminker, P. G., Clodfelter, G. V., Brewer, L. D., & Landfield, P. W. (1998). Calcium dysregulation in neuronal aging and Alzheimer's disease: history and new directions. *Cell Calcium*, *24*(5–6), 417–433. https://doi.org/10.1016/s0143-4160(98)90064-1

Tian, L., Rauvala, H., & Gahmberg, C. G. (2009). Neuronal regulation of immune responses in the central nervous system. *Trends Immunol*, *30*(2), 91–99. https://doi.org/10.1016/j.it.2008.11.002

Tint, I., Slaughter, T., Fischer, I., & Black, M. M. (1998). Acute inactivation of tau has no effect on dynamics of microtubules in growing axons of cultured sympathetic neurons. *J Neurosci*, *18*(21), 8660–8673. https://doi.org/10.1523/JNEUROSCI.18-21-08660.1998

Tiwari, S., Atluri, V., Kaushik, A., Yndart, A., & Nair, M. (2019). Alzheimer's disease: pathogenesis, diagnostics, and therapeutics. *Int J Nanomedicine*, *14*, 5541–5554. https://doi.org/10.2147/IJN.S200490

Toescu, E. C., & Verkhratsky, A. (2003). Neuronal ageing from an intraneuronal perspective: roles of endoplasmic reticulum and mitochondria. *Cell Calcium*, *34*(4–5), 311–323. https://doi.org/10.1016/s0143-4160(03)00142-8

Tonnies, E., & Trushina, E. (2017). Oxidative stress, synaptic dysfunction, and Alzheimer's disease. *J Alzheimers Dis*, *57*(4), 1105–1121. https://doi.org/10.3233/JAD-161088

Troncoso, J. C., Sukhov, R. R., Kawas, C. H., & Koliatsos, V. E. (1996). In situ labeling of dying cortical neurons in normal aging and in Alzheimer's disease: correlations with senile plaques and disease progression. *J Neuropathol Exp Neurol*, *55*(11), 1134–1142. https://doi.org/10.1097/00005072-199611000-00004

Trovo, L., Van Veldhoven, P. P., Martin, M. G., & Dotti, C. G. (2011). Sphingomyelin upregulation in mature neurons contributes to TrkB activity by Rac1 endocytosis. *J Cell Sci*, *124*(Pt 8), 1308–1315. https://doi.org/10.1242/jcs.078766

Trudler, D., Sanz-Blasco, S., Eisele, Y. S., Ghatak, S., Bodhinathan, K., Akhtar, M. W., Lynch, W. P., Pina-Crespo, J. C., Talantova, M., Kelly, J. W., & Lipton, S. A. (2021). Alpha-Synuclein oligomers induce glutamate release from astrocytes and excessive extra-synaptic NMDAR activity in neurons, thus contributing to synapse loss. *J Neurosci*, *41*(10), 2264–2273. https://doi.org/10.1523/JNEUROSCI.1871-20.2020

Trushina, N. I., Bakota, L., Mulkidjanian, A. Y., & Brandt, R. (2019). The evolution of tau phosphorylation and interactions. *Front Aging Neurosci*, *11*, 256. https://doi.org/10.3389/fnagi.2019.00256

Turnquist, C., Horikawa, I., Foran, E., Major, E. O., Vojtesek, B., Lane, D. P., Lu, X., Harris, B. T., & Harris, C. C. (2016). p53 isoforms regulate astrocyte-mediated neuroprotection and neurodegeneration. *Cell Death Differ*, *23*(9), 1515–1528. https://doi.org/10.1038/cdd.2016.37

Twohig, D., & Nielsen, H. M. (2019). Alpha-synuclein in the pathophysiology of Alzheimer's disease. *Mol Neurodegener*, *14*(1), 23. https://doi.org/10.1186/s13024-019-0320-x

Uchikado, H., Lin, W. L., DeLucia, M. W., & Dickson, D. W. (2006). Alzheimer disease with amygdala Lewy bodies: a distinct form of alpha-synucleinopathy. *J Neuropathol Exp Neurol*, *65*(7), 685–697. https://doi.org/10.1097/01.jnen.0000225908.90052.07

Ueda, K., Fukushima, H., Masliah, E., Xia, Y., Iwai, A., Yoshimoto, M., Otero, D. A., Kondo, J., Ihara, Y., & Saitoh, T. (1993). Molecular cloning of cDNA encoding an unrecognized component of amyloid in Alzheimer disease. *Proc Natl Acad Sci U S A*, *90*(23), 11282–11286. https://doi.org/10.1073/pnas.90.23.11282

Uemura, N., Uemura, M. T., Luk, K. C., Lee, V. M., & Trojanowski, J. Q. (2020). Cell-to-cell transmission of tau and alpha-synuclein. *Trends Mol Med*, *26*(10), 936–952. https://doi.org/10.1016/j.molmed.2020.03.012

Um, J. W., Kaufman, A. C., Kostylev, M., Heiss, J. K., Stagi, M., Takahashi, H., Kerrisk, M. E., Vortmeyer, A., Wisniewski, T., Koleske, A. J., Gunther, E. C., Nygaard, H. B., & Strittmatter, S. M. (2013). Metabotropic glutamate receptor 5 is a coreceptor for Alzheimer abeta oligomer bound to cellular prion protein. *Neuron*, *79*(5), 887–902. https://doi.org/10.1016/j.neuron.2013.06.036

Upton, J. W., Kaiser, W. J., & Mocarski, E. S. (2010). Virus inhibition of RIP3-dependent necrosis. *Cell Host Microbe*, *7*(4), 302–313. https://doi.org/10.1016/j.chom.2010.03.006

Upton, J. W., Kaiser, W. J., & Mocarski, E. S. (2012). DAI/ZBP1/DLM-1 complexes with RIP3 to mediate virus-induced programmed necrosis that is targeted by murine cytomegalovirus vIRA. *Cell Host Microbe*, *11*(3), 290–297. https://doi.org/10.1016/j.chom.2012.01.016

Uttley, L., Carroll, C., Wong, R., Hilton, D. A., & Stevenson, M. (2020). Creutzfeldt-Jakob disease: a systematic review of global incidence, prevalence, infectivity, and incubation. *Lancet Infect Dis*, *20*(1), e2-e10. https://doi.org/10.1016/S1473-3099(19)30615-2

Valencia, A., Reeves, P. B., Sapp, E., Li, X., Alexander, J., Kegel, K. B., Chase, K., Aronin, N., & DiFiglia, M. (2010). Mutant huntingtin and glycogen synthase kinase 3-beta accumulate in neuronal lipid rafts of a presymptomatic knock-in mouse model of Huntington's disease. *J Neurosci Res*, *88*(1), 179–190. https://doi.org/10.1002/jnr.22184

Valente, E. M., Abou-Sleiman, P. M., Caputo, V., Muqit, M. M., Harvey, K., Gispert, S., Ali, Z., Del Turco, D., Bentivoglio, A. R., Healy, D. G., Albanese, A., Nussbaum, R., Gonzalez-Maldonado, R., Deller, T., Salvi, S., Cortelli, P., Gilks, W. P., Latchman, D. S., Harvey, R. J., … Wood, N. W. (2004). Hereditary early-onset Parkinson's disease caused by mutations in PINK1. *Science*, *304*(5674), 1158–1160. https://doi.org/10.1126/science.1096284

van Bergen, J. M., Li, X., Hua, J., Schreiner, S. J., Steininger, S. C., Quevenco, F. C., Wyss, M., Gietl, A. F., Treyer, V., Leh, S. E., Buck, F., Nitsch, R. M., Pruessmann, K. P., van Zijl, P. C., Hock, C., & Unschuld, P. G. (2016). Colocalization of cerebral iron with Amyloid beta in mild cognitive impairment. *Sci Rep*, *6*, 35514. https://doi.org/10.1038/srep35514

van den Heuvel, J., English, S., & Uller, T. (2016). Disposable soma theory and the evolution of maternal effects on ageing. *PLoS One*, *11*(1), e0145544. https://doi.org/10.1371/journal.pone.0145544

van Rooijen, B. D., Claessens, M. M., & Subramaniam, V. (2009). Lipid bilayer disruption by oligomeric alpha-synuclein depends on bilayer charge and accessibility of the hydrophobic core. *Biochim Biophys Acta*, *1788*(6), 1271–1278. https://doi.org/10.1016/j.bbamem.2009.03.010

Vanden Berghe, T., Vanlangenakker, N., Parthoens, E., Deckers, W., Devos, M., Festjens, N., Guerin, C. J., Brunk, U. T., Declercq, W., & Vandenabeele, P. (2010). Necroptosis, necrosis and secondary necrosis converge on similar cellular disintegration features. *Cell Death Differ*, *17*(6), 922–930. https://doi.org/10.1038/cdd.2009.184

Vanlangenakker, N., Vanden Berghe, T., & Vandenabeele, P. (2012). Many stimuli pull the necrotic trigger, an overview. *Cell Death Differ*, *19*(1), 75–86. https://doi.org/10.1038/cdd.2011.164

Vaquer-Alicea, J., Diamond, M. I., & Joachimiak, L. A. (2021). Tau strains shape disease. *Acta Neuropathol*, *142*(1), 57–71. https://doi.org/10.1007/s00401-021-02301-7

Vargas, J. N. S., Hamasaki, M., Kawabata, T., Youle, R. J., & Yoshimori, T. (2023). The mechanisms and roles of selective autophagy in mammals. *Nat Rev Mol Cell Biol*, *24*(3), 167–185. https://doi.org/10.1038/s41580-022-00542-2

Vargas, J. Y., Loria, F., Wu, Y. J., Cordova, G., Nonaka, T., Bellow, S., Syan, S., Hasegawa, M., van Woerden, G. M., Trollet, C., & Zurzolo, C. (2019). The Wnt/Ca(2+) pathway is involved in interneuronal communication mediated by tunneling nanotubes. *EMBO J*, *38*(23), e101230. https://doi.org/10.15252/embj.2018101230

Vargas, K. J., Schrod, N., Davis, T., Fernandez-Busnadiego, R., Taguchi, Y. V., Laugks, U., Lucic, V., & Chandra, S. S. (2017). Synucleins have multiple effects on presynaptic architecture. *Cell Rep*, *18*(1), 161–173. https://doi.org/10.1016/j.celrep.2016.12.023

Varkey, J., Isas, J. M., Mizuno, N., Jensen, M. B., Bhatia, V. K., Jao, C. C., Petrlova, J., Voss, J. C., Stamou, D. G., Steven, A. C., & Langen, R. (2010). Membrane curvature induction and tubulation are common features of synucleins and apolipoproteins. *J Biol Chem*, *285*(42), 32486–32493. https://doi.org/10.1074/jbc.M110.139576

Vasquez, V., Mitra, J., Wang, H., Hegde, P. M., Rao, K. S., & Hegde, M. L. (2020). A multi-faceted genotoxic network of alpha-synuclein in the nucleus and mitochondria of dopaminergic neurons in Parkinson's disease: emerging concepts and challenges. *Prog Neurobiol*, *185*, 101729. https://doi.org/10.1016/j.pneurobio.2019.101729

Vasudevaraju, P., Guerrero, E., Hegde, M. L., Collen, T. B., Britton, G. B., & Rao, K. S. (2012). New evidence on alpha-synuclein and Tau binding to conformation and sequence specific GC* rich DNA: relevance to neurological disorders. *J Pharm Bioallied Sci*, *4*(2), 112–117. https://doi.org/10.4103/0975-7406.94811

Venda, L. L., Cragg, S. J., Buchman, V. L., & Wade-Martins, R. (2010). Alpha-synuclein and dopamine at the crossroads of Parkinson's disease. *Trends Neurosci*, *33*(12), 559–568. https://doi.org/10.1016/j.tins.2010.09.004

Vercammen, D., Brouckaert, G., Denecker, G., Van de Craen, M., Declercq, W., Fiers, W., & Vandenabeele, P. (1998). Dual signaling of the Fas receptor: initiation of both apoptotic and necrotic cell death pathways. *J Exp Med*, *188*(5), 919–930. https://doi.org/10.1084/jem.188.5.919

Vercammen, D., Vandenabeele, P., Beyaert, R., Declercq, W., & Fiers, W. (1997). Tumour necrosis factor-induced necrosis versus anti-Fas-induced apoptosis in L929 cells. *Cytokine*, *9*(11), 801–808. https://doi.org/10.1006/cyto.1997.0252

Verkerke, M., Hol, E. M., & Middeldorp, J. (2021). Physiological and pathological ageing of astrocytes in the human brain. *Neurochem Res*, *46*(10), 2662–2675. https://doi.org/10.1007/s11064-021-03256-7

Visanji, N. P., Kovacs, G. G., & Lang, A. E. (2021). The discovery of alpha-synuclein in Lewy pathology of Parkinson's disease: the inspiration of a revolution. *Mov Disord Clin Pract*, *8*(8), 1189–1193. https://doi.org/10.1002/mdc3.13312

Vitale, I., Manic, G., De Maria, R., Kroemer, G., & Galluzzi, L. (2017). DNA damage in stem cells. *Mol Cell*, *66*(3), 306–319. https://doi.org/10.1016/j.molcel.2017.04.006

Volles, M. J., Lee, S. J., Rochet, J. C., Shtilerman, M. D., Ding, T. T., Kessler, J. C., & Lansbury, P. T., Jr. (2001). Vesicle permeabilization by protofibrillar alpha-synuclein: implications for the pathogenesis and treatment of Parkinson's disease. *Biochemistry*, *40*(26), 7812–7819. https://doi.org/10.1021/bi0102398

von Bernhardi, R., Tichauer, J. E., & Eugenin, J. (2010). Aging-dependent changes of microglial cells and their relevance for neurodegenerative disorders. *J Neurochem*, *112*(5), 1099–1114. https://doi.org/10.1111/j.1471-4159.2009.06537.x

von Karstedt, S., Montinaro, A., & Walczak, H. (2017). Exploring the TRAILs less travelled: TRAIL in cancer biology and therapy. *Nat Rev Cancer*, *17*(6), 352–366. https://doi.org/10.1038/nrc.2017.28

Vu, H. T., Akatsu, H., Hashizume, Y., Setou, M., & Ikegami, K. (2017). Increase in alpha-tubulin modifications in the neuronal processes of hippocampal neurons in both kainic acid-induced epileptic seizure and Alzheimer's disease. *Sci Rep*, *7*, 40205. https://doi.org/10.1038/srep40205

Wahner, A. D., Bronstein, J. M., Bordelon, Y. M., & Ritz, B. (2007). Nonsteroidal anti-inflammatory drugs may protect against Parkinson disease. *Neurology*, *69*(19), 1836–1842. https://doi.org/10.1212/01.wnl.0000279519.99344.ad

Wajant, H. (2002). The Fas signaling pathway: more than a paradigm. *Science*, *296*(5573), 1635–1636. https://doi.org/10.1126/science.1071553

Wakabayashi, K., Tanji, K., Mori, F., & Takahashi, H. (2007). The Lewy body in Parkinson's disease: molecules implicated in the formation and degradation of alpha-synuclein aggregates. *Neuropathology*, *27*(5), 494–506. https://doi.org/10.1111/j.1440-1789.2007.00803.x

Wakabayashi, K., Tanji, K., Odagiri, S., Miki, Y., Mori, F., & Takahashi, H. (2013). The Lewy body in Parkinson's disease and related neurodegenerative disorders. *Mol Neurobiol*, *47*(2), 495–508. https://doi.org/10.1007/s12035-012-8280-y

Wang, H. Y., Lee, D. H., D'Andrea, M. R., Peterson, P. A., Shank, R. P., & Reitz, A. B. (2000). beta-Amyloid(1-42) binds to alpha7 nicotinic acetylcholine receptor with high affinity. Implications for Alzheimer's disease pathology. *J Biol Chem*, *275*(8), 5626–5632. https://doi.org/10.1074/jbc.275.8.5626

Wang, J., Gu, B. J., Masters, C. L., & Wang, Y. J. (2017). A systemic view of Alzheimer disease - insights from amyloid-beta metabolism beyond the brain. *Nat Rev Neurol*, *13*(10), 612–623. https://doi.org/10.1038/nrneurol.2017.111

Wang, T., & Hay, J. C. (2015). Alpha-synuclein toxicity in the early secretory pathway: how it drives neurodegeneration in Parkinsons disease. *Front Neurosci*, *9*, 433. https://doi.org/10.3389/fnins.2015.00433

Wang, T. Y., Ma, Z., Wang, C., Liu, C., Yan, D. Y., Deng, Y., Liu, W., Xu, Z. F., & Xu, B. (2018). Manganese-induced alpha-synuclein overexpression impairs synaptic vesicle fusion by disrupting the Rab3 cycle in primary cultured neurons. *Toxicol Lett*, *285*, 34–42. https://doi.org/10.1016/j.toxlet.2017.12.024

Wang, X. W., Yuan, L. J., Yang, Y., Zhang, M., & Chen, W. F. (2020). IGF-1 inhibits MPTP/MPP(+)-induced autophagy on dopaminergic neurons through the IGF-1R/PI3K-Akt-mTOR pathway and GPER. *Am J Physiol Endocrinol Metab*, *319*(4), E734-E743. https://doi.org/10.1152/ajpendo.00071.2020

Wang, Y., & Mandelkow, E. (2016). Tau in physiology and pathology. *Nat Rev Neurosci*, *17*(1), 5–21. https://doi.org/10.1038/nrn.2015.1

Watanabe, S., Murata, Y., Oka, Y., Oiwa, K., Horiuchi, M., Iguchi, Y., Komine, O., Sobue, A., Katsuno, M., Ogi, T., & Yamanaka, K. (2023). Mitochondria-associated membrane collapse impairs TBK1-mediated proteostatic stress response in ALS. *Proc Natl Acad Sci U S A*, *120*(47), e2315347120. https://doi.org/10.1073/pnas.2315347120

Weinlich, R., Oberst, A., Beere, H. M., & Green, D. R. (2017). Necroptosis in development, inflammation and disease. *Nat Rev Mol Cell Biol*, *18*(2), 127–136. https://doi.org/10.1038/nrm.2016.149

Weinreb, P. H., Zhen, W., Poon, A. W., Conway, K. A., & Lansbury, P. T., Jr. (1996). NACP, a protein implicated in Alzheimer's disease and learning, is natively unfolded. *Biochemistry*, *35*(43), 13709–13715. https://doi.org/10.1021/bi961799n

Wilson, D. M., 3rd, Cookson, M. R., Van Den Bosch, L., Zetterberg, H., Holtzman, D. M., & Dewachter, I. (2023). Hallmarks of neurodegenerative diseases. *Cell*, *186*(4), 693–714. https://doi.org/10.1016/j.cell.2022.12.032

Wilton, D. K., Dissing-Olesen, L., & Stevens, B. (2019). Neuron-glia signaling in synapse elimination. *Annu Rev Neurosci*, *42*, 107–127. https://doi.org/10.1146/annurev-neuro-070918-050306

Wolfe, M. S. (2018). *The molecular and cellular basis of neurodegenerative diseases: underlying mechanisms*. Elsevier/Academic Press.

Wu, K. P., Kim, S., Fela, D. A., & Baum, J. (2008). Characterization of conformational and dynamic properties of natively unfolded human and mouse alpha-synuclein ensembles by NMR: implication for aggregation. *J Mol Biol*, *378*(5), 1104–1115. https://doi.org/10.1016/j.jmb.2008.03.017

Wulf, M. A., Senatore, A., & Aguzzi, A. (2017). The biological function of the cellular prion protein: an update. *BMC Biol*, *15*(1), 34. https://doi.org/10.1186/s12915-017-0375-5

Xu, C., Wu, J., Wu, Y., Ren, Z., Yao, Y., Chen, G., Fang, E. F., Noh, J. H., Liu, Y. U., Wei, L., Chen, X., & Sima, J. (2021). TNF-alpha-dependent neuronal necroptosis regulated in Alzheimer's disease by coordination of RIPK1-p62 complex with autophagic UVRAG. *Theranostics*, *11*(19), 9452–9469. https://doi.org/10.7150/thno.62376

Yaar, M., Zhai, S., Pilch, P. F., Doyle, S. M., Eisenhauer, P. B., Fine, R. E., & Gilchrest, B. A. (1997). Binding of beta-amyloid to the p75 neurotrophin receptor induces apoptosis: a possible mechanism for Alzheimer's disease. *J Clin Invest*, *100*(9), 2333–2340. https://doi.org/10.1172/JCI119772

Yamoah, A., Tripathi, P., Sechi, A., Kohler, C., Guo, H., Chandrasekar, A., Nolte, K. W., Wruck, C. J., Katona, I., Anink, J., Troost, D., Aronica, E., Steinbusch, H., Weis, J., & Goswami, A. (2020). Aggregates of RNA binding proteins and ER chaperones linked to exosomes in granulovacuolar degeneration of the Alzheimer's disease brain. *J Alzheimers Dis*, *75*(1), 139–156. https://doi.org/10.3233/JAD-190722

Yan, J., Fu, Q., Cheng, L., Zhai, M., Wu, W., Huang, L., & Du, G. (2014). Inflammatory response in Parkinson's disease (review). *Mol Med Rep*, *10*(5), 2223–2233. https://doi.org/10.3892/mmr.2014.2563

Yan, M. H., Wang, X., & Zhu, X. (2013). Mitochondrial defects and oxidative stress in Alzheimer disease and Parkinson disease. *Free Radic Biol Med*, *62*, 90–101. https://doi.org/S0891-5849(12)01823-0

Yanagisawa, K., Odaka, A., Suzuki, N., & Ihara, Y. (1995). GM1 ganglioside-bound amyloid beta-protein (A beta): a possible form of preamyloid in Alzheimer's disease. *Nat Med*, *1*(10), 1062–1066. https://doi.org/10.1038/nm1095-1062

Yarlagadda, A., Alfson, E., & Clayton, A. H. (2009). The blood brain barrier and the role of cytokines in neuropsychiatry. *Psychiatry (Edgmont)*, *6*(11), 18–22. https://www.ncbi.nlm.nih.gov/pubmed/20049146

Yasumoto, T., Takamura, Y., Tsuji, M., Watanabe-Nakayama, T., Imamura, K., Inoue, H., Nakamura, S., Inoue, T., Kimura, A., Yano, S., Nishijo, H., Kiuchi, Y., Teplow, D. B., & Ono, K. (2019). High molecular weight amyloid beta(1-42) oligomers induce neurotoxicity via plasma membrane damage. *FASEB J*, *33*(8), 9220–9234. https://doi.org/10.1096/fj.201900604R

Yeo, E. J., & Park, S. C. (2002). Age-dependent agonist-specific dysregulation of membrane-mediated signal transduction: emergence of the gate theory of aging. *Mech Ageing Dev*, *123*(12), 1563–1578. https://doi.org/10.1016/s0047-6374(02)00092-1

Yoo, M. H., Gu, X., Xu, X. M., Kim, J. Y., Carlson, B. A., Patterson, A. D., Cai, H., Gladyshev, V. N., & Hatfield, D. L. (2010). Delineating the role of glutathione peroxidase 4 in protecting cells against lipid hydroperoxide damage and in Alzheimer's disease. *Antioxid Redox Signal*, *12*(7), 819–827. https://doi.org/10.1089/ars.2009.2891

Yoon, Y. S., You, J. S., Kim, T. K., Ahn, W. J., Kim, M. J., Son, K. H., Ricarte, D., Ortiz, D., Lee, S. J., & Lee, H. J. (2022). Senescence and impaired DNA damage responses in alpha-synucleinopathy models. *Exp Mol Med*, *54*(2), 115–128. https://doi.org/10.1038/s12276-022-00727-x

Yu, H., Ren, J., & Qu, X. (2007). Time-dependent DNA condensation induced by amyloid beta-peptide. *Biophys J*, *92*(1), 185–191. https://doi.org/10.1529/biophysj.106.093559

Yu, P., Zhang, X., Liu, N., Tang, L., Peng, C., & Chen, X. (2021). Pyroptosis: mechanisms and diseases. *Signal Transduct Target Ther*, *6*(1), 128. https://doi.org/10.1038/s41392-021-00507-5

Yuan, A., Kumar, A., Peterhoff, C., Duff, K., & Nixon, R. A. (2008). Axonal transport rates in vivo are unaffected by tau deletion or overexpression in mice. *J Neurosci*, *28*(7), 1682–1687. https://doi.org/10.1523/JNEUROSCI.5242-07.2008

Zambon, F., Cherubini, M., Fernandes, H. J. R., Lang, C., Ryan, B. J., Volpato, V., Bengoa-Vergniory, N., Vingill, S., Attar, M., Booth, H. D. E., Haenseler, W., Vowles, J., Bowden, R., Webber, C., Cowley, S. A., & Wade-Martins, R. (2019).

Cellular alpha-synuclein pathology is associated with bioenergetic dysfunction in Parkinson's iPSC-derived dopamine neurons. *Hum Mol Genet*, *28*(12), 2001–2013. https://doi.org/10.1093/hmg/ddz038

Zhang, L., Qian, Y., Li, J., Zhou, X., Xu, H., Yan, J., Xiang, J., Yuan, X., Sun, B., Sisodia, S. S., Jiang, Y. H., Cao, X., Jing, N., & Lin, A. (2021). BAD-mediated neuronal apoptosis and neuroinflammation contribute to Alzheimer's disease pathology. *iScience*, *24*(9), 102942. https://doi.org/10.1016/j.isci.2021.102942

Zhang, L., Zhang, C., Zhu, Y., Cai, Q., Chan, P., Ueda, K., Yu, S., & Yang, H. (2008). Semi-quantitative analysis of alpha-synuclein in subcellular pools of rat brain neurons: an immunogold electron microscopic study using a C-terminal specific monoclonal antibody. *Brain Res*, *1244*, 40–52. https://doi.org/10.1016/j.brainres.2008.08.067

Zhang, X., Fan, C., Zhang, H., Zhao, Q., Liu, Y., Xu, C., Xie, Q., Wu, X., Yu, X., Zhang, J., & Zhang, H. (2016). MLKL and FADD are critical for suppressing progressive lymphoproliferative disease and activating the NLRP3 inflammasome. *Cell Rep*, *16*(12), 3247–3259. https://doi.org/10.1016/j.celrep.2016.06.103

Zorec, R., Verkhratsky, A., Rodriguez, J. J., & Parpura, V. (2016). Astrocytic vesicles and gliotransmitters: slowness of vesicular release and synaptobrevin2-laden vesicle nanoarchitecture. *Neuroscience*, *323*, 67–75. https://doi.org/10.1016/j.neuroscience.2015.02.033

Zs-Nagy, I. (2014). Aging of cell membranes: facts and theories. *Interdiscip Top Gerontol*, *39*, 62–85. https://doi.org/10.1159/000358900

Zysk, M., Beretta, C., Naia, L., Dakhel, A., Pavenius, L., Brismar, H., Lindskog, M., Ankarcrona, M., & Erlandsson, A. (2023). Amyloid-beta accumulation in human astrocytes induces mitochondrial disruption and changed energy metabolism. *J Neuroinflammation*, *20*(1), 43. https://doi.org/10.1186/s12974-023-02722-z

# 2

## *Molecular Markers of Neurodegenerative Disorders*

**Elif Ozkok**

### Abbreviations

| | |
|---|---|
| **AD** | Alzheimer's Disease |
| **ALS** | Amyotrophic Lateral Sclerosis |
| **ANXA3** | AnnexinA3 |
| **APOE** | Apolipoprotein E |
| **APP** | Amyloid Precursor Protein |
| **AQP9** | Aquaporin 9 |
| **BACE1** | β-Site Amyloid Precursor Protein Cleaving Enzyme 1 |
| **CNS** | Central Nerve System |
| **CSF** | Cerebrospinal Fluid |
| **CYSTM** | Cysteine-Rich, Transmembrane (TM) Module |
| **DLB** | Dementia with Lewy Bodies |
| **DA** | Dopamine |
| **DAT** | Dopamine Transporter |
| **FTD** | Frontotemporal Dementia |
| **GWAS** | Genome-Wide Association Studies |
| **GFAP** | Glial Fibrillary Acidic Protein |
| **HD** | Huntington's Disease |
| **HY** | Hoehn And Yahr Staging |
| **LBD** | Lewy Body Dementia |
| **LRRK2** | Leucine-Rich Repeat Kinase 2 |
| **MCI** | Mild Cognitive Impairment |
| **miRNA** | MicroRNA |
| **MSA** | Multiple System Atrophy |
| **mHTT** | Mutant Huntingtin Protein |
| **NF** | Neurofilament |
| **NfL** | Neurofilament Light Chain |
| **PET** | Positron Emission Tomography |
| **PC** | Phosphatidylcholine |
| **PD** | Parkinson's Disease |
| **PE** | Phosphatidylethanolamine |
| **PINK1** | PTEN Induced Putative Kinase 1 |
| **PLC** | Phospholipase C |
| **PSEN** | Presenilin |
| **SM** | Sphingomyelin |
| **UCH-L1** | Ubiquitin Carboxy-Terminal Hydrolase L1 |
| **WHO** | The World Health Organization |
| **NFTs** | Neurofibrillary Tangles |
| **tNAA** | Total N-Acetyl Aspartate |
| **TREM2** | Triggering Receptor Expressed On Myeloid Cells 2 |

### Introduction

The human brain is the most complex organ, consisting of different cell types, namely neurons and glial cells that work together in coordination. The homeostatic structure that regulates this complex architecture of the brain shows a gradual decrease in its functionality with age, affecting cognitive, social, and motor abilities and leading to the improvement of various neurodegenerative diseases (Campisi et al., 2019; Arora, 2008). Aging is broadly defined as a gradual decline in the living organism's intrinsic ability to defend, maintain, and repair itself to keep working efficiently. It has attracted attention throughout the history of human beings. Along with the whole organism, the functional capabilities of the brain gradually degrade upon aging, manifesting as declines in learning and memory, attention, decision-making capacity, sensory perception, and motor management (Alexander et al., 2012; Dykiert et al., 2012; Levin et al., 2014).

By 2050, there will be 2.1 billion individuals worldwide who are 60 years of age or older, according to the World Health Organization (WHO), with 80% of them residing in low- and middle-income nations (Ageing and Health: World Health Org, 2024). The prediction of this increase in demographic structure is that the incidence of age-related diseases such as Alzheimer's disease (AD), cardiovascular diseases, and cancer will double every 5 years after the age of 60 (Melzer et al., 2020). According to estimates for the United States, 130 million people will have Alzheimer's disease (AD) by 2050 (Klyucherev et al., 2022), and the number of Parkinson's disease (PD) patients is expected to reach 1.6 million by 2037 (Yang et al., 2020).

During the aging process, important indicators such as impaired bioenergetics, weakened adaptive neuroplasticity and resilience, abnormal neuronal network activity, dysfunctional neuronal calcium homeostasis, accumulation of molecules and organelles that have undergone oxidative modification, and inflammation emerge (Mattson and Arumugam, 2018; Bieri et al., 2023; Zia et al., 2021; Keshavarz et al., 2023; Sikora et al., 2021; Tripati, 2012; Wahl et al., 2021). Thus,

DOI: 10.1201/9781032676555-3

age-associated changes render the aging brain susceptible to degenerative disorders including AD, PD, stroke, and various kinds of dementia. The brain is remarkably sensitive to the effects of aging, displaying as changes in structure and cognitive capacity, as well as increased risk for developing certain neurological disorders. Brain health refers to the maintenance of brain functions in several aspects: 1) cognitive health – the ability to adequately think, learn, and remember; 2) motor function – the ability to control movements and balance; 3) emotional health – the ability to interpret and respond to emotions; and 4) tactile function – the ability to feel and respond to sensations of touch, including pressure, pain, and temperature (Cognitive Health and Older Adults, 2023). Aging also sets off a decline in the regenerative capacity of brain cells, such as decreased neurogenesis and oligodendrogenesis. Brain aging is associated also with a decline in the function of essential neurotransmitter systems such as dopamine and acetylcholine, which can lead to cognitive impairment. Brain aging is associated with changes in brain structure, such as the loss of gray matter volume and changes in white matter microstructure (Drayer, 1988). It is also associated with the accumulation of toxic protein aggregates, such as beta-amyloid, tau, and alpha-synuclein which can disrupt neuronal function and contribute to the development of neurodegenerative diseases (Hou et al., 2019; Bourdenx et al., 2017).

The concept of a "biomarker", derived from the combination of "biological" and "marker", encompasses a broad range of medical signs. These signs provide objective evidence of a patient's health condition and can be consistently and accurately quantified. This is distinct from medical symptoms, which are subjective sensations or complaints reported by the patient. The spectrum of biomarkers is broad, from basic measurements such as pulse and blood pressure to complex laboratory evaluations of blood and other biological samples. Historically, medical signals have always formed an integral part of clinical practice, and biomarkers represent the sum of objective and quantifiable indicators that advanced laboratory sciences can consistently measure. Despite the uncertainty surrounding the complexities of biomarker applications in both research and clinical settings, advances in drug discovery and biomedical science have brought the concept of biomarkers to a significant level.

## The Necessity of Biomarker Use

The increasing need for tools to develop accurate and specific biomarkers is a challenge faced in the clinic for neurodegenerative diseases, and the heterogeneity observed between individuals in the progression of the disease (Cedazo-Minguez and Winblad, 2010; Walker, 2016). Moreover, more than one neurodegenerative disease can coexist in the same patient, such as AD, Lewy Body Dementia (LBD), and cerebral amyloid angiopathy leading to terminal intracerebral hemorrhage. Biomarkers, in addition to clinical history and imaging, can help identify concurrent disease processes. Screening for neurodegenerative disease is another indication for biomarkers, as they allow for distinguishing normal aging from pathological cognitive decline, and can appear in cerebrospinal fluid (CSF) or plasma years before the onset of symptoms.

Biomarkers are very important in following the physiological process in the pathogenesis of a disease; they can be useful not only for diagnosis but also for therapeutic management of the treatment of the disease. If it can be measured accurately, it can directly reflect the progression of the disease and the effect of treatment on it (Mohapatra et al., 2020). Molecular biomarkers are the most widely used drug targets and can be investigated using techniques such as proteomics, genetics, and advanced imaging techniques. The pathogenesis process of neurodegeneration such as AD, PD, Huntington's disease (HD), and Amyotrophic Lateral Sclerosis (ALS) is only partially understood. The use of biomarkers has helped to understand the estimating of pathogenesis and early symptoms of the disease and has enabled the determination of drug targets, and it has allowed very important results to be obtained for early diagnosis; however, most of them are still not fully investigated.

## Type of Biomarkers

Clinically available biomarkers can be broadly classified as prognostic, predictive, pharmacodynamic, or surrogate biomarkers based on their utility in clinics. Prognostic biomarkers are used to identify patients with certain risk factors that are not usually found in the normal population; such biomarkers can reflect the development and progression of a particular disorder (Califf, 2018; Henry and Hayes, 2012). Prognostic ones for most central nerve system (CNS) disorders, such as PD or depression, are not fully developed, and the pathogenesis and risk of developing these disorders are largely unknown (Cova and Priori, 2018). Predictive biomarkers are used to predict different outcomes of a particular therapy or treatment (Chen et al., 2015). Pharmacodynamic biomarkers help us understand the effect of a particular therapy or drug treatment, as they help to better determine dose selection in terms of avoiding toxicity to normal tissues/cells and low clinical response rate. Such biomarkers are used to measure whether the drug reaches its target after treatment, and, if it produces a pharmacological response, what is the degree of response. Surrogate markers are the term for a biomarker designed to replace a clinical endpoint. A biomarker is a "candidate" surrogacy marker, whereas a surrogate marker is a test that is used to gauge the results of a certain treatment. This is the primary distinction between the two types of markers. It can allow us to predict clinical benefits by considering factors such as pathophysiology, epidemiology, therapeutics, or other mechanisms.

## Biomarkers of Neurodegenerative Diseases

Measuring biomarkers of neurodegenerative diseases in the blood presents a challenge that requires sensitive and specific tests. As a result of the selectivity of the hematoencephalic barrier, which prevents the free passage of molecules between the CNS and blood compartments and affects the volume ratio of blood to CSF, concentrations of CNS-derived proteins are much lower in blood.

The heterogeneity of neurodegenerative diseases in the clinic affects the development of early diagnosis and effective medical procedures as well as therapeutic interventions. Omics techniques reveal the more complex status of the molecular pathways underlying the development of neurodegenerative diseases and allow for the distinction between individuals based on certain molecular characteristics. These omic techniques consisting of the detection disease-related DNA sequence variants (genomics), transcriptome and non-coding RNA profiling (transcriptomics), genome-wide identification of DNA-protein interactions (epigenomics), analyses of protein–protein interactions (proteomics), and analyses of metabolic systems (metabolomics).

## Fluid Biomarkers

The best validated fluid biomarker in neurodegenerative diseases is CSF, but serum biomarkers are being developed in terms of accuracy and predictive value, the main problem being the blood–brain barrier, which reduces the movement of markers into the systemic circulation. Among nonspecific CSF biomarkers, markers of axonal damage and axonal degeneration, including total tau protein (t-tau) and neurofilament light chains (NfL) are the most studied. Among other nonspecific markers, these are markers of synaptic, lysosomal, and mitochondrial dysfunction; proteins involved in the formation of intracellular proteins or their degradation and clearance; abnormally altered proteins; and markers of glial activation.

Blood proteins and lipids can lead to measurement errors and impact the association of biomarkers with dementia risk. Serum biomarkers are valuable in terms of lower cost, less invasiveness, and increased accessibility of tests in outpatient clinics and can allow screening of the general population. Improvements in accuracy have been achieved for amyloid-beta 42/40 (Aβ 42/40), pTau and NfL ratio, but no biomarker yet allows definitive diagnosis.

### *Amyloid β Biomarkers*

Amyloid Precursor Protein (APP) encodes a type 1 transmembrane glycoprotein that is cleaved either by a non-amyloidogenic pathway (normal state) or by an amyloidogenic pathway (diseased state). APP is composed of 770 amino acids, of which Aβ contains 28 aminoacids and an additional 14 aminoacids from the transmembrane domain of APP. The soluble P3 peptide is released when α-secretase splits and releases the large soluble ectodomain APPsα into the medium at the cleavage site. Meanwhile, the C-terminal fragment C83 remains in the membrane and is further digested by γ-secretase at position 711 (Nhan et al., 2015; Sun et al., 2015).

Alternatively, in a diseased state, the aberrant cleavage is performed by β-secretase, releasing cleaved APPsβ, and the C-terminal segment insoluble Aβ peptides are created when γ-secretase further cleaves C99, which is kept in the membrane. When γ-secretase cleaves both C83 and C99, the APP intracellular domain is released into the cytoplasm. It is soluble and is then carried to the nucleus for additional gene expression activity (Zhang and Song, 2013). In the pathogenesis of amyloid beta, amyloid precursor protein (APP), a protein integrally located in the plasma membrane structure, undergoes differential cleavage by β-site amyloid precursor protein cleaving enzyme 1 (BACE1) and γ-secretase enzymes to form insoluble Aβ fibrils. Aβ then oligomerizes and polymerizes into insoluble amyloid fibrils that aggregate into plaques (Selkoe and Hardy 2016). This polymerization causes the activation of kinase enzymes, leading to hyperphosphorylation of microtubule-associated tau (τ) protein and polymerization into insoluble neurofibrillary tangles (NFTs). The assembly of plaques and neurofibrillary bundles is followed by microglia involvement surrounding the plaques (Gulisano et al., 2018). They induce microglial activation and local inflammatory response and contribute to neurotoxicity.

There are two primary categories of Aβ polymers that directly contribute to the development of plaque and neurotoxicity (Iwatsubo et al., 1994; Mak et al., 1994; Miller et al., 1993).

Plaques impair synaptic signaling and ion transport and increase oxidative stress in neurons (Tönnies and Trushina, 2017; Guo et al., 2020; Long and Holtzman, 2019). The two main Aβ peptides, Aβ40 and Aβ42, differ according to the length of their peptide chains and can be detected in CSF, serum, and PET. They behave differently in the CNS: Aβ42 is much more neurotoxic and disruptive to cellular processes than Aβ40. The CSF Aβ42/Aβ40 ratio is more specific and sensitive than CSF Aβ42 alone. Recently, the focus has been on the Aβ biomarker plasma levels, particularly the Aβ 42/40 ratio. (Zou et al., 2020). Moreover, it was found that the plasma Aβ 42/40 ratio can reliably predict Aβ positron emission tomography (PET) status.

### *Tau Protein Biomarkers*

Intracellular tau protein undertakes an important role in the stabilization of microtubule structure in neurons. Plaques formed by Aβ lead to harmful interactions between tau protein and kinase enzymes and promote the formation of NFTs that cause the tau proteins in the cell to be phosphorylated at high levels and interfere with signal transmission in the cell. Increases in p-Tau, especially tau phosphorylated at threonine amino acids located at positions 181, 217, and 231, can serve as a biomarker for AD. pTau 181, 217 and 231 can be used together with Aβ biomarkers to diagnose and stage the progression of AD (Blennow and Hampel, 2003; Janelidze et al., 2020). While phosphorylation at threonine levels 217 and 181 in pTau protein is unchanged in frontotemporal dementia (FTD), it is significantly higher in AD, in which case it can help us distinguish FTD from AD. It has been determined that the levels of pTau 181, pTau 217, and pTau 231 determined in CSF samples in preclinical AD are increased. pTau 231 examined in CSF in early stage AD is the first tau biomarker that can be detected. With the developments in plasma analyses, there are studies on promising results in predicting AD pathology with the quantification of pTau 217 and 231. There is information that pTau 217 can predict cognitive decline better than other tau biomarkers in people with preclinical AD in plasma samples.

In addition, plasma pTau 231 has been found to be related to PET markers better than pTau 181. pTau has limitations as a biomarker; in short, the presence of pTau does not always indicate the development of symptoms, it cannot predict overlap with other diseases and the threshold for positivity, and it cannot be used in overlapping diseases. Increased pTau has a limited predictive value, and some persons may never develop symptoms (Dubois et al., 2021). Therefore, it is recommended as a best practice to use pTau in conjunction with other biomarkers in the appropriate clinical setting. In addition, further studies are needed to define the appropriate threshold for determining pTau positivity.

### *Neurofilament (NF) Biomarkers*

Neurofilament proteins are structural compounds of axons of neurons. Expression of these proteins is especially abundant in big myelinated axons (Friede and Samorajski, 1970). Accordingly, high NF protein levels are found in disorders with subcortical pathology in CSF, such as vascular dementia and normal pressure hydrocephalus. CSF levels are also high in frontotemporal dementia, whereas they are normal in most cases of AD. In CSF, NF proteins may therefore be valuable in distinguishing between AD, FTD, and subcortical dementia.

## Lipidomic Studies Related to Alzheimer's Disease

Studies in the lipidomics have accelerated the understanding of lipid metabolism and lipid-related diseases and have accelerated biomarkers studies for the diagnosis and prognosis of systemic diseases including diabetes and cancer, neurodegenerative diseases among them. The brain is the second richest lipid-containing tissue in the body after adipose tissue and consists of cholesterol; phospholipids, such as phosphatidylcholine (PC) and phosphatidylethanolamine (PE), and sphingolipids.

The first studies on AD pathogenesis and lipid metabolism were based on studies that detected a dysregulation in cholesterol transport from astrocytes to neurons. Initial studies have demonstrated that a genetic variant of Apolipoprotein E (APOE), encoded by the APOE4 allele, is one of the risk factors for AD; it is associated with changes in cholesterol and sphingolipids (Lane and Farlow, 2005; Bandaru et al., 2009). Although no direct relationship has been determined between cholesterol and amyloidogenesis in Alzheimer's disease, the role of brain cholesterol in APP processing in the membrane lipid structure has been demonstrated (Fabelo et al., 2014).

In neuronal membranes, which are rich in cholesterol and sphingolipids, enzymes such as beta-site APP cleavage enzyme 1 (BACE1) and gamma-secretase can bind to AD-related transmembrane proteins. Studies have shown that cholesterol plays a role in reducing the distance between APP and BACE1 in the lipid membrane structure and that it affects BACE1 and gamma-secretase enzyme activities. It has been shown that sphingolipid metabolism in the brain samples of patients with AD increases the formation of Amyloid beta42, causes increased formation of acid sphingomyelinase and acid ceramidase, and leads to a decrease in sphingomyelin (SM) and ceramide levels (He et al., 2010). Ceramides can stabilize the BACE1 enzyme and prolong its half-life, thus increasing the rate of Amyloid beta. Glycosphingolipids also contribute to the formation of amyloid fibrils, and complexes formed by the binding of GM1 ganglioside to Amyloid beta have been associated with Amyloid beta oligomers seen in CSF (Hong et al., 2014). Sphingolipids, which are the basic components of membrane structure, affect APP processing, Amyloid beta production, and subsequent amyloid aggregation. In the brain samples of patients with AD, changes in various types of phospholipids including PC, PE, phosphatidylinositol, and plasmalogen PE (PE-P) levels, as well as phospholipid metabolizing enzymes such as phospholipase C (PLC) and phospholipase D, have been detected (Han et al., 2001).

In a study conducted in AD, it was reported that changes in plasma phospholipid levels could be used as a new biomolecular diagnostic marker. Another study reported that the decrease in phospholipid levels was expected to accurately predict whether a cognitively normal person would develop mild cognitive impairment (MCI) or AD within two years (Wurtman, 2015). A published work reported results regarding a blood-based biomarker panel with over 90% accuracy in detecting preclinical AD compared to that obtained from a CSF study (Mapstone et al., 2014).

## Genetic Biomarkers in Alzheimer's Disease

### *Rare Mutations*

Individuals with familial AD who present symptoms before the age of 60 years constitute early-onset familial AD cases, accounting for less than 1% of cases, and this is caused by mutations in three genes: β-amyloid precursor protein (APP), presenilin-1 (PSEN1), and presenilin-2 (PSEN2). Most mutations in these genes result in autosomal dominant forms of early-onset AD (Zekanowski et al., 2003; Rachakonda et al., 2004). APP, located on the 21st chromosome, encodes the amyloid-β precursor protein. APP is the focus of the amyloid cascade hypothesis, which suggests that the buildup of β-amyloid peptide is the basis of AD pathology (Berezovska et al., 2003). Alternative splicing sites of the APP produce different APP proteins, and aggregated plaques initiate a variety of kinase reactions that ultimately lead to hyperphosphorylated tau and the formation of neurofibrillary tangles (Hardy and Higgins, 1992). The PSEN1 and PSEN2 genes on the 14th and 1st chromosomes, respectively, encode presenilins, which form the catalytic subunit of the γ-secretase. The γ-secretase is the complex responsible for the degradation of amyloid precursor proteins leading to the formation of β-amyloid peptides (Delabio et al., 2014). These genes' fundamental mutations probably result in increased β-amyloid synthesis overall (APP duplications, for example) or increased aggregation susceptibility of β-amyloid peptides (Lanoiselée et al., 2017).

The triggering receptor expressed in the myeloid cells-2 (TREM2) gene on the 6th chromosome is a transmembrane receptor belonging to the immunoglobulin superfamily and activates strong immune remodeling in case of damage to the tissue. The pathway of TREM2 is important in limiting the spread of tissue damage (Deczkowska et al., 2020), and amyloid and tau diseases are impacted by uncommon TREM2 mutations that increase the risk of developing AD.

### Common Variants

Sporadic AD cases constitute the majority of patients diagnosed and have no discernible heredity, affecting individuals older than 60 years, and have cumulative effects of common, rare, and environmental contributions to disease risk (Tcw and Goate, 2017). The apolipoprotein E (APOE) gene, located on chromosome 19, is a key component of the production, turnover, and clearance of plasma lipoproteins and has three alleles (ε2, ε3, and ε4), and possession of the ε4 allele is the most common genetic risk factor for AD (Riedel et al., 2016). The APOE ε4 allele is the strongest genetic risk factor, accounting for up to 25% of the heritability in AD (Bagyinszky et al., 2020), while the ε2 allele is the strongest genetic protective factor in AD (Serrano-Pozo et al., 2021). Nonetheless, significant pathology in primary tauopathies has been linked to the APOE ε2/ε2 genotype (Reiman et al., 2020).

A genome-wide association studies (GWAS) meta-analysis to research the genetic etiology of AD identified 42 new loci in addition to the 75 known loci. These loci emphasize genes that are known to be involved in AD-associated processes, including innate immunity, endocytosis, and the metabolism of tau and amyloid.

### MicroRNA (miRNA)

The use of miRNAs as biomolecular diagnostic biomarkers provides the detection with high specificity. Furthermore, unlike large RNA molecules such as mRNAs, miRNAs are well protected from deterioration in formalin, paraffin-embedded tissues (Xi et al., 2007; Grasso et al., 2014).

In between AD and normal aging is a state known as mild cognitive impairment (and other dementias) and is generally defined as the first stage at which clinical symptoms become apparent (DeCarli, 2003). miRNA biomarkers have been shown in plasma samples to identify MCI, and the initial miRNA pool was selected from known brain and neuron-enriched miRNAs (Sheinerman et al., 2012).

In later studies, researchers identified two sets of diagnostic marker pairs: the "miR-132 family," consisting of miR-128/miR-491-5p, miR-132/miR-491-5p; and the "miR-134 family," consisting of miR-874/miR-491-5p and miR-134/miR-370, miR-323-3p/miR-370, and miR-382/miR-370, with very high sensitivity and specificity of 79%–100% and 79%–95%, respectively. In another longitudinal study, the identified miRNA diagnostic marker pairs were able to detect MCI in the majority of patients at the symptomatic stage, 1–5 years before clinical diagnosis (Sheinerman et al., 2012; Grasso et al., 2014).

## Alpha-Synuclein Biomarker

Alpha-synuclein plays a role in the formation of vesicles in presynaptic terminals. The misfolding of α-syn and its aggregation caused to the Lewy bodies constitute the main feature of PD. These aggregates negatively affect the signaling between cells. Early detection of disease is of great importance when motor function is impaired in PD; it is understood that the degeneration of nigral neurons is very advanced. α-syn, as a biomarker for PD in CSF and plasma, has been recorded as lower total α-syn levels in CSF in individuals with PD compared to healthy controls. Oligomeric α-syn and phosphorylated α-syn have been shown to be increased in CSF. Oligomeric α-syn in CSF allows better discrimination between PD patients and the healthy controls than CSF α-syn levels or plasma oligomeric α-syn. Moreover, CSF oligomeric α-syn can differentiate between AD and PD or DLB.

α-Synuclein can be reliably found in CSF, and levels are reduced on average in PD, dementia with Lewy bodies (DLB), and multiple system atrophy (MSA). Ultrasensitive cell-free aggregation assays that indirectly detect misfolded prion protein and other prion-like proteins have been developed. Such seeding aggregation assays (e.g., real-time concussion-induced transformation and protein misfolding cyclic amplification) can be applied to determine α-synuclein pathological forms in the CSF. CSF-based assays have demonstrated high diagnostic accuracy for Lewy body disease (including PD and dementia with Lewy bodies [DLB]), with sensitivity >90% and specificity >97% in neuropathologically confirmed parkinsonian cohorts. While CSF analysis remains the gold standard for detecting misfolded α-synuclein, its invasive collection procedure limits widespread clinical utility. The development of reliable detection methods in more accessible biofluids (e.g., blood, saliva) or tissues would represent a significant advancement in Lewy body disorder diagnostics.

Preliminary results indicate that skin biopsies, which include nerve terminals, can be used in seeding aggregation assays to reliably detect misfolded α-synuclein in PD, DLB, and MSA. Other potential peripheral tissues for the detection of misfolded α-synuclein include the olfactory mucosa and the submandibular gland.

Two important biomolecular diagnostic markers are used to realize the onset of PD. These include (1) the loss of the dopamine transporter (DAT) detected by PET imaging and (2) the presence of the α-synuclein protein found in Lewy body lesions. DAT ensures the uptake of dopamine (DA) into dopaminergic neurons by an electrogenic, $Na^+$ and $Cl^-$ transport-related mechanism. DAT also plays a role in the uptake of toxins that cause Parkinson's syndrome. For this reason, striatal localization, especially putamen DAT concentration, is considered the best molecular diagnostic marker with high sensitivity for early stages of PD (Marek et al., 2001; Rachakonda et al., 2004; Shinto et al., 2014).

## Molecular Biomarkers of Parkinson's Disease

In just a small number of families, PD is inherited in an autosomal recessive or autosomal dominant. The genes for α-synuclein (SNCA) and leucine-rich repeat kinase2 (LRRK2) as well as parkin (PARK2), ubiquitin carboxy-terminal hydrolase L1 (UCH-L1), PTEN Induced Putative Kinase 1 (PINK1), and the oncogene DJ1 (DJ1) were discovered to be mutated for late-onset disease and early-onset disease, respectively (Grasso et al., 2014). Numerous gene products from the autosomal dominant versions of the mutant genes have been connected to oxidative stress, mitochondrial malfunction, and improper handling of the gene products' aberrant or defective forms (e.g., oligomeric α-synuclein) (Miller and O'Callaghan, 2015). Mutations in the Parkin gene cause the formation and accumulation of plaques and tangles. The early-onset variant of Parkinsonism has been linked to over 70 mutations in the big Parkin gene. Up to 50% of familial cases of autosomal recessive juvenile Parkinsonism may be caused by mutations in the Parkin gene (Pankratz et al., 2002). A study was conducted to determine whether a gene signature that could support early PD diagnosis could be detected in the blood of patients with early/mild PD, focusing specifically on genes found to be altered in the substantia nigra of sporadic PD (Molochnikov et al., 2012). The study findings suggest that a five-gene panel is capable of diagnosing early/mild PD and may have a possible diagnostic value for the detection of asymptomatic PD before the overt expression of the disorder (Molochnikov et al., 2012).

Another gene ubiquitin carboxy-terminal hydrolase L1 (*UCH-L1*) located on the 4th chromosome encodes a protein that belongs to the deubiquitinating enzymes family. The protein UCH-L1, which makes about 1% of brain protein, is thought to hydrolyze ubiquitinated peptides in order to recycle ubiquitin. This enzyme helps to change damaged proteins that could otherwise build up in the cell to hazardous levels (Leroy et al., 1998). The PINK1 gene product is an active protein in the mitochondria. Its mutations increase sensitivity to cellular stress and have been linked to early forms of PD. Also, two homozygous mutations in the *PINK1* related with PD were detected in Spanish and Italian families. This discovery offered more proof that PARK6 is linked to PINK1 mutations (Valente et al., 2004). Additionally, the DJ-1 gene has mutations linked to PARK7 (Bonifati et al., 2003). LRRK2 mutations were first identified in a few British and Basque families with late-onset PD. Subsequent studies have identified mutations in this gene in other families with PD (such as European Ashkenazi Jewish families in North Africa and the Middle East) and in a small percentage of people with apparently sporadic PD.

### *miRNA*

The discovery of characteristic Lewy bodies and Lewy neuritis in PD is due to a mutation in the α-synuclein gene. As α-synuclein is also detected in other synucleinopathies, it is necessary to use other diagnostic techniques to increase specificity and sensitivity for PD (Davidsson and Sjörgren, 2005). Biological fluids are an excellent source for biomarkers because their proximity to cells reflects their biological state and they are easy and cost-effective to obtain (Shinde et al., 2015). With the increasing importance of miRNAs in biofluids, the development of markers for PD has great potential. A study using qRT-PCR suggested that miR-1, miR-22-5p, and miR-29 expressions in peripheral blood allow distinguishing PD patients from healthy individuals, and also that miR-16-2-3p, miR-26a-2-3p, and miR30a allow distinguishing between treated and untreated patients (Margis et al., 2011). A recent study using next-generation sequencing reported that 16 miRNAs, including miR-16, miR-20a, and miR-320, were altered in PD patients compared to controls (Soreq et al., 2013; Grasso et al., 2014).

### *Glial Fibrillary Acidic Protein (GFAP)*

Glial fibrillary acidic protein (GFAP) is a filamentous protein expressed in astrocytes in the CNS. High levels of GFAP in the CSF or blood are associated with degeneration of astrocytes and have been proposed as a marker for astroglial cell death, astrogliosis, neuroinflammation, and synaptic degeneration. Ultrasensitive immunoassay platforms enable reliable measurements of GFAP levels in the CSF and blood. So far, previous studies have suggested GFAP levels as a potential diagnostic biomarker for differentiating patients with FTLD from healthy controls or patients with AD and indicating disease severity in patients with FTLD. Elevated blood GFAP levels have also been found to be associated especially with GRN mutations and more rapid temporal atrophy in patients with FTLD. Moreover, higher levels of GFAP have been detected in the CSF of patients with FTLD compared with patients with AD, dementia with Lewy bodies, or healthy controls, further substantiating the idea that degeneration of astrocytes is a prominent feature in FTLD (Heller et al., 2020).

## Huntingtin Protein

HD is a neurodegenerative disease that is heritable and can impact mental, cognitive, and motor abilities. HD symptoms include personality changes and a decline in cognitive abilities (Mastrokolias et al., 2015). The pathology is caused by the expansion of CAG repeat in the *HTT* gene, resulting in a mutant huntingtin protein (mHTT). Mutant protein aggregate formation and neuronal cell loss, with transcriptional dysregulation, are important features of HD brain tissue (Runne et al., 2008). mHTT levels were shown by an ultrasensitive single-molecule counting (SMC) mHTT immunoassay for the first time in CSF of individuals carrying HD mutation (Wild et al., 2015). It is important to have a disease progression marker that should be able to identify changes before clinical symptoms. Huntingtin is widely expressed, and blood gene expression changes may reflect mutant huntingtin-specific modifications. Because leukocytes are used to regulate the immune system, blood is a perfect source for detecting HD events like peripheral inflammation. Microarray technology has also been used in research to detect HD blood mRNA alterations, although these findings have been challenging to confirm across studies (Lovrecic et al., 2009). The validation of markers for HD has been always challenging as the disease presents itself through a variety of symptoms and progression rates.

Due to variations in the clinical phenotype of HD and possible confounding effects of pharmacological and environmental factors, a variety of biomarkers are used, some of which may be effective in monitoring the course of HD. The identification of disturbed homoeostasis in HD has led to the identification of numerous possible indicators. Cross-sectional MRS was used in a recent study to separate putaminal metabolites in pre-existing and early HD patients from controls (Sturrock et al., 2015). It was discovered that whereas the gliosis marker myo-inositol was higher in early HD, total N-acetyl aspartate (tNAA) was lower in both pre-manifest HD and early HD compared to healthy individuals. According to a different study, HD gene carriers' putamen and caudate nucleus exhibit metabolite alterations around the onset of the disease (Van Den Bogaard et al., 2011). These associations between the disease burden score and total NAA imply that this metabolite could be helpful in determining the neurochemical reactions to treatment. Vasopressin plays a part in maintaining fluid homeostasis, and elevated serum levels of this hormone have been linked to HD (Wood et al., 2008). They are a potential biomarker in HD due to elevated levels of plasma lipid peroxide, lactic acid, 4-hydroxynoneal, and malondialdehyde, as well as elevated levels of 8-hydroxy-2-deoxyguanosine (8OHdG), a sign of oxidative DNA damage (Weir et al., 2011). Glutathione peroxidase and copper-zinc superoxide dismutase levels in HD patients' erythrocytes were shown to be lower (Chen et al., 2007). Patients with HD have been found to have higher amounts of TNF-a, clusterin, and interleukins 4, 6, 8, 10, and 23 in their postmortem brain and plasma samples (Dalrymple et al., 2007).

Along with the other biomarkers mentioned above, the inflammatory profile discrepancies between gene carriers and control provide possible biochemical indicators for HD. All of these biomarkers would enhance the safety and efficacy of clinical trials and enable precise assessment of the efficacy of novel treatments.

## Molecular Biomarkers of Huntington's Disease

The gene H2A histone family, member Y (H2AFY), was overexpressed in the blood of HD patients (Hu et al., 2011). In particular, the study shows that HD patients overexpress H2AFY by 1.6 times. According to a recent study, gene expression profiling using Fluidigm and next-generation sequencing technology produced a list of genes that are highly expressed in HD blood and may serve as HD markers (Mastrokolias et al., 2015). According to Aziz et al. (2010), there is evidence linking cognitive impairment in HD to changes in circadian rhythm, which may be related to prokineticin2 (PROK2). In HD models, pharmacological sleep imposition restores PROK2 dysregulation and delays cognitive deterioration. A possible biomarker of the development of familial HD is PROK2. Zincfinger protein238 (ZNF238) is a transcriptional repressor involved in brain development and myogenesis, and evidence suggests that gene repression mechanisms are also linked to HD (Zhai et al., 2005). According to a recent study on gene expression, when four other genes' levels rise in tandem with the mRNA levels of Aquaporin 9 (AQP9), the existence of AQP9 and its rise could indicate either central or peripheral inflammatory processes (Mesko et al., 2010).

AnnexinA3 (ANXA3) and Cysteine-rich, transmembrane (TM) module (CYSTM) are two other potential biomarkers for HD (Borovecki et al., 2005; Venancio and Aravind, 2010). ANXA3 is found to be upregulated in neuronal injury models, and CYSTM is involved in stress response, specifically heavy metal tolerance.

### *Exosomes*

Exosomes are vesicles of nanomolar size secreted extracellularly from CNS neurons and glial cells, derived from endosomes in eukaryotic cells. They play critical roles in intercellular communication, angiogenesis, immune regulation, and neurodegenerative processes. They facilitate the transfer of bioactive molecules - including proteins, lipids, mRNAs, and microRNAs - between CNS cells and peripheral targets, thereby modulating cellular metabolism and function. They contribute to neurodegenerative processes in three ways: 1) by targeting mRNA genes involved in regulating or inhibiting protein translation or degradation; 2) by directly affecting neuroinflammation through toll-like receptor binding or by regulating mRNA expression; and 3) by causing a defect in miRNA production itself. These vesicles originating from the CNS can be used to study pathological events occurring in the brain. Based on these findings, they can be considered as potentially effective biomarkers for monitoring disease progression. They have become extremely important in the search for early screening markers. In vitro studies suggest that they have important roles in regulating synaptic activity as they regulate neurotransmitter release and neuronal excitability. Exosomal vesicles (EVs) containing APP, amyloid beta, and tau lead to increased Aβ and tau pathologies in AD pathology. It has also been reported that EVs contain APP, C-terminal fragments of APP, and various isoforms of Aβ (Goate et al., 1991). In addition, ApoE4 causes increased exosome production due to loss of endosome-lysosome function due to heterozygous or homozygous genotype (Raber et al., 1998; Holtzman et al., 2000; Caselli et al., 2009). It has been reported that Aβ is reduced in CSF-derived exosomes from severe AD patients (Simonsen et al., 2017), while in another study, Aβ is shown to be high in plasma-derived exosomes, which may be confounded by other sources (Hornung et al., 2020). Soluble Aβ42 and Aβ42 levels were found to be higher in astrocytic-derived exosomes compared to neuronal exosomes. An increase in the ratio of p-tau to total tau was found in AD patients (Goetzl et al., 2016).

In PD, the exosomal mechanism plays a role in transmitting early molecular changes of the disease to other cells. LRRK2 genetic mutations have been found to be associated with the lysosomal autophagy system (Paisan-Ruiz et al., 2013). The G2019S mutation in LRRK2 has been found to be associated with disruption in the LAS pathway, and the formation of a more aggregated form by disruption of alpha-synuclein homeostasis may cause further depigmentation of dopaminergic neurons. While exosomes facilitate α-synuclein aggregate propagation, their upregulation during impaired lysosomal function appears to provide neuroprotection by enhancing extracellular clearance of pathological proteins, revealing a potential therapeutic avenue for synucleinopathies. (Ben Gedalya et al., 2009; Cabin et al., 2002). The study on astrocytic and oligodendrocyte-derived exosomes obtained from plasma of PD patients showed a high increase in early stage PD patients and was confirmed by the severity of the disease. Protein profiling of exosomes derived from plasma of PD patients from stages one, two, and three, according to Hoehn and Yahr (HY) staging, suggests that apolipoprotein A1 may be a potential biomarker for monitoring PD disease progression (Kitamura et al., 2018).

A study on 239T cells reported the increased presence of polyQGFP and transcript RNA in EVs overexpressing HTT-exon1polyQ-GFP. The uptake of EVs by neurons was increased in the presence of polyQ-GFP RNA, but did not lead to further toxicity. Another study on plasma-derived exosomes from HD patients showed the presence of 13 significantly upregulated miRNAs (Kumar et al., 2017). The therapeutic role of exosomes derived from astrocytic stem cells (ASC-exo) in an in vitro model of HD was shown to lead to a decrease in the production of mHTT aggregates (Lee et al., 2016).

## Conclusion

Biomarkers in neurodegenerative diseases are of great importance in both the diagnosis and the detection and monitoring of the disease's progression. It is expected that more specific and personalized markers will be revealed by taking advantage of today's increasing technological opportunities, and it is expected that research on this subject will continue to be updated.

## REFERENCES

Alexander, G. E., Ryan, L., Bowers, D., Foster, T. C., Bizon, J. L., Geldmacher, D. S., & Glisky, E. L. (2012). Characterizing cognitive aging in humans with links to animal models. *Front Aging Neurosci*, *4*, 21. https://doi.org/10.3389/fnagi.2012.00021

Arora, B. P. (2008). Anti-aging medicine. *Indian J Plast Surg*, *41*(Suppl), S130–S133.

Aziz, N. A., Anguelova, G. V., Marinus, J., Lammers, G. J., & Roos, R.A. (2010). Sleep and circadian rhythm alterations correlate with depression and cognitive impairment in Hungtinton's disease. *Parkinsonism Relat Disord*, *16*(5), 345–350. https://doi.org/10.1016/j.parkreldis.2010.02.009

Bagyinszky, E., Van Giau, V., & An, S. A. (2020). Transcriptomics in Alzheimer's disease: Aspects and challenges. *Int J Mol Sci*, *21*(10), 3517. https://doi.org/10.3390/ijms21103517

Bandaru, V. V. R., Troncoso, J., Wheeler, D., Pletnikova, O., Wang, J., Conant, K., & Haughey, N. J. (2009). ApoE4 disrupts sterol and sphingolipid metabolism in Alzheimer's but not normal brain. *Neurobiol Aging, 30*(4), 591–599. https://doi.org/10.1016/j.neurobiolaging.2007.07.024

Ben Gedalya, T., Loeb, V., Israeli, E., Altschuler, Y., Selkoe, D. J., & Sharon, R. (2009). Alpha-synuclein and polyunsaturated fatty acids promote clathrin-mediated endocytosis and synaptic vesicle recycling. *Traffic*, *10*(2), 218–234. https://doi.org/10.1111/j.1600-0854.2008.00853.x

Berezovska, O., Ramdya, P., Skoch, J., Wolfe, M. S., Bacskai, B. J., & Hyman, B. T. (2003). Amyloid precursor protein associates with a nicastrin-dependent docking site on the presenilin 1-gamma-secretase complex in cells demonstrated by fluorescence lifetime imaging. *J Neurosci, 23*(11), 4560–4566. https://doi.org/10.1523/JNEUROSCI.23-11-04560.2003

Bieri, G., Schroer, A. B., Villeda, S. A. (2023). Blood-to-brain communication in aging and rejuvenation. *Nat Neurosci, 26*(3), 379–393. https://doi.org/10.1038/s41593-022-01238-8

Blennow, K., & Hampel, H. (2003). CSF markers for incipient Alzheimer's disease. *Lancet Neurol, 2*(10), 605–613. https://doi.org/10.1016/s1474-4422(03)00530-1

Bonifati, V., Rizzu, P., van Baren, M. J., Schaap, O., Breedveld, G. J., Krieger, E., Dekker, M. C., Squitieri, F., Ibanez, P., Joosse, M., van Dongen, J. W., Vanacore, N., van Swieten, J. C., Brice, A., Meco, G., van Duijn, C. M., Oostra, B. A., Heutink, P. (2003). Mutations in the DJ-1 gene associated with autosomal recessive early-onset Parkinsonism. *Science*, *299*(5604), 256–269. https://doi:10.1126/science.1077209

Borovecki, F., Lovrecic, L., Zhou, J., Jeong, H., Then, F., Rosas, H. D., Hersch, S. M., Hogarth, P., Bouzou, B., Jensen, R. V., & Krainc, D. (2005). Genome-wide expression profiling of human blood reveals biomarkers for Hungtinton's disease. *Proc Natl Acad Sci USA*, *102*(31), 11023–11028. https://doi.org/10.1073/pnas.0504921102

Bourdenx, M., Koulakiotis, N.S., Sanoudou, D., Bezard, E., Dehay, B., & Tsarbopoulos, A. (2017). Protein aggregation and neurodegeneration in prototypical neurodegenerative diseases: Examples of amyloidopathies, taupathies and synucleinopathies. *Prog Neurobiol, 155*, 171–193. https://doi.org/10.1016/j.pneurobio.2015.07.003

Cabin, D. E., Shimazu, K., Murphy, D., Cole, N. B., Gottschalk, W., McIlwain, K. L., Orrison, B., Chen, A., Ellis, C. E., Paylor, R., Lu, B., & Nussbaum, R. L. (2002). Synaptic vesicle depletion correlates with attenuated synaptic responses to prolonged repetitive stimulation in mice lacking α-synuclein. *J Neurosci*, *22*(20), 8797–8807. https://doi.org/10.1523/JNEUROSCI.22-20-08797.2002

Califf, R. M. (2018). Biomarker definitions and their applications. *Exp Biol Med (Maywood), 243*(3), 213–221. https://doi.org/10.1177/1535370217750088

Campisi, J., Kapahi, P., Lithgow, G. J., Melov, S., Newman, J. C., & Verdin, E. (2019). From discoveries in ageing research to therapeutics for healthy ageing. *Nature, 571*(7764), 183–192. https://doi.org/10.1038/s41586-019-1365-2

Caselli, R. J., Dueck, A. C., Osborne, D., Sabbagh, M. N., Connor, D. J., Ahern, G. L., Baxter, L. C., Rapcsak, S. Z., Shi, J., & Woodruff, B. K. (2009). Longitudinal modeling of age-related memory decline and the APOE ε4 effect. *N Engl J Med*, *361*(3), 255–263. https://doi.org/10.1056/NEJMoa0809437

Cedazo-Minguez, A., & Winblad, B. (2010). Biomarkers for Alzheimer's disease and other forms of dementia: clinical needs, limitations and future aspects. *Exp Gerontol*, *45*(1), 5–14. https://doi.org/10.1016/j.exger.2009.09.008

Chen, C. M., Wu, Y. R., Cheng, M. L., Liu, J. L., Lee, Y. M., Lee, P. W., Soong, B. W., & Chiu, D. T. (2007). Increased oxidative damage and mitochondrial abnormalities in the peripheral blood of Huntington's disease patients. *Biochem Biophys Res Commun*, *359*(2), 335–340. https://doi.org/10.1016/j.bbrc.2007.05.093.

Chen, J. J., Lu, T. P., Chen, Y. C., & Lin, W. J. (2015). Predictive biomarkers for treatment selection: statistical considerations. *Biomark Med, 9*(11), 1121–1135. https://doi.org/10.2217/bmm.15.84

Cognitive Health and Older Adults. (2023). https://www.nia.nih.gov/health/cognitive-health-and-older-adults.

Cova, I., & Priori, A. (2018). Diagnostic biomarkers for Parkinson's disease at a glance: where are we? *J Neural Transm (Vienna), 125*(10), 1417–1432. https://doi.org/10.1007/s00702-018-1910-4

Dalrymple, A., Wild, E. J., Joubert, R., Sathasivam, K., Björkqvist, M., Petersén, A., Jackson, G. S., Isaacs, J. D., Kristiansen, M., Bates, G. P., Leavitt, B. R., Keir, G., Ward, M., & Tabrizi, S. J. (2007). Proteomic profiling of plasma in Huntington's disease reveals neuroinflammatory activation and biomarker candidates .*J Proteome Res, 6*(7), 2833–2840. https://doi: 10.1021/pr0700753

Davidsson, P., & Sjögren, M. (2005). The use of proteomics in biomarker discovery in neurodegenerative diseases. *Disease Markers, 21*(2), 81–92. https://doi.org/10.1155/2005/848676

DeCarli, C. (2003). The role of cerebrovascular disease in dementia. *Neurologist, 9*(3), 123–136. https://doi.10.1097/00127893-200305000-00001

Deczkowska, A., Weiner, A., & Amit, I. (2020). The physiology, pathology, and potential therapeutic applications of the TREM2 signaling pathway. *Cell, 181*(6), 1207–1217. https://doi.org/10.1016/j.cell.2020.05.003

Delabio, R., Rasmussen, L., Mizumoto, I., Viani, G. A., Chen, E., Villares, J., Costa, I. B., Turecki, G., Linde, S. A., Smith, M. C., & Payão, P. L. (2014). PSEN1 and PSEN2 gene expression in alzheimer's disease brain: a new approach. *J Alzheimer's Dis, 42*(3), 757–760. https://doi.org/10.3233/JAD-140033

Drayer, B. P. (1988). Imaging of the aging brain. Part I. Normal findings. *Radiology, 166*(3), 785–796. https://doi.org/10.1148/radiology.166.3.3277247

Dubois, B., Villain N., Frisoni G. B., Rabinovici, G. D., Sabbagh, M., Cappa, S., Bejanin, A., Bombois, S., Epelbaum, S., Teichmann, M., Habert, M. O., Nordberg, A., Blennow, K., Galasko, D., Stern Y., Rowe, C. C., Salloway, S., Schneider, L. S., Cummings, J. L., & Feldman, H. H. (2021). Clinical diagnosis of Alzheimer's disease: recommendations of the International Working Group. *Lancet Neurol, 20*(6), 484–496. https://doi.org/10.1016/S1474-4422(21)00066-1

Dykiert, D., Der, G., Starr, J. M., & Deary, I. J. (2012). Age differences in intra-individual variability in simple and choice reaction time: systematic review and meta-analysis. *PLoS One, 7*(10), e45759. https://doi.org/10.1371/journal.pone.0045759

Fabelo, N., Martin, V., Marin, R., Moreno, D., Ferrer, I., & Diaz, M. (2014). Altered lipid composition in cortical lipid rafts occurs at early stages of sporadic Alzheimer's disease and facilitates APP/BACE1 interactions. *Neurobiol Aging, 35*(8), 1801–1812. https://doi.org/10.1016/j.neurobiolaging.2014.02.005

Friede, R. L., & Samorajski, T. (1970). Axon caliber related to neurofilaments and microtubules in sciatic nerve fibers of rats and mice. *Anat Rec, 167*(4), 379–87. https://doi.org/10.1002/ar.1091670402

Goate, A., Chartier-Harlin, M. C., Mullan, M., Brown, J., Crawford, F., Fidani, L., Giuffra, L., Haynes, A., Irving, N., & James, L., et al. (1991). Segregation of a missense mutation in the amyloid precursor protein gene with familial Alzheimer's disease. *Nature, 349*(6311), 704–706. https://doi.org/10.1038/349704a0

Goetzl, E. J., Mustapic, M., Kapogiannis, D., Eitan, E., Lobach, I. V., Goetzl, L., Schwartz, J. B., & Miller, B. L. (2016). Cargo proteins of plasma astrocyte-derived exosomes in Alzheimer's disease. *FASEB J, 30*(11), 3853–3859. https://doi.org/10.1096/fj.201600756R

Grasso, M., Piscopo P., Confaloni A., & Denti, M. A. (2014). Circulating miRNAs as biomarkers for neurodegenerative Disorders. *Molecules, 19*(5), 6891–6910. https://doi.org/10.3390/molecules19056891

Gulisano, W., Maugeri, D., Baltrons, M. A., Fà, M., Amato, A., Palmeri, A., D'Adamio, L., Grassi, C., Devanand, D. P., Honig, L. S., Puzzo, D., & Arancio, O. (2018). Role of Amyloid-β and tau proteins in Alzheimer's disease: Confuting the Amyloid Cascade. *J Alzheimers Dis, 64*(s1), S611–S631. https://doi.org/10.3233/JAD-179935

Guo, T., Zhang, D., Zeng, Y., Huang, T. Y., Xu, H., & Zhao, Y. (2020). Molecular and cellular mechanisms underlying the pathogenesis of Alzheimer's disease. *Mol Neurodegeneration, 15*(1), 40. https://doi.org/10.1186/s13024-020-00391-7

Han, X., Holtzman, D. M., & McKeel, Jr., D. W. (2001). Plasmalogen deficiency in early Alzheimer's disease subjects and in animal models: molecular characterization using electrospray ionization mass spectrometry. *J Neurochem, 77*(4), 1168–1180. https://doi.org/10.1046/j.1471-4159.2001.00332.x

Hardy, J. A., & Higgins, G. A. (1992). Alzheimer's disease: the amyloid cascade hypothesis. *Science, 256*(5054), 184–185. https://doi.org/10.1126/science.1566067

He, X., Huang, Y., Li, B., Gong, C. X., & Schuchman, E. H. (2010). Deregulation of sphingolipid metabolism in Alzheimer's disease. *Neurobiol Aging, 31*(3), 398–408. https://doi.org/10.1016/j.neurobiolaging.2008.05.010

Heller, C., Foiani, M. S., Moore, K., Convery, R., Bocchetta, M., Neason, M., Cash, D. M., Thomas, D., Greaves, C. V., Woollacott, I. O., Shafei, R., Van Swieten, J. C., Moreno, F., Sanchez-Valle, R., Borroni, B., Laforce, R. Jr., Masellis, M., Tartaglia, M. C., Graff, C., Galimberti, D., Rowe, J. B., Finger, E., Synofzik, M., Vandenberghe, R., de Mendonca, A., Tagliavini, F., Santana, I., Ducharme, S., Butler, C. R., Gerhard, A., Levin, J., Danek, A., Frisoni, G., Sorbi, S., Otto, M., Heslegrave, A. J., Zetterberg, H., Rohrer, J. D., & GENFI. (2020). Plasma glial fibrillary acidic protein is raised in progranulin-associated frontotemporal dementia. *J Neurol Neurosurg Psychiatry, 91*(3), 263–270. https://doi.org/10.1136/jnnp-2019-321954

Henry, N. L., & Hayes, D. F. (2012). Cancer biomarkers. *Mol Oncol, 6*(2), 140–146. https://doi.org/10.1016/j.molonc.2012.01.010

Holtzman, D. M., Bales, K. R., Tenkova, T., Fagan, A. M., Parsadanian, M., Sartorius, L. J., Mackey, B., Olney, J., McKeel, D., Wozniak, D., & Paul, S. M. (2000). Apolipoprotein E isoform-dependent amyloid deposition and neuritic degeneration in a mouse model of Alzheimer's disease. *Proc Natl Acad Sci USA, 97*(6), 2892–2897. https://doi.org/10.1073/pnas.050004797

Hong, S., Ostaszewski, B. L., Yang, T., O'Malley, T. T., Jin, M., Yanagisawa, K., Li, S., Bartels, T., & Selkoe, D. J. (2014). Soluble Aβ oligomers are rapidly sequestered from brain

ISF in vivo and bind GM1 ganglioside on cellular membranes. *Neuron, 82*(2), 308–319. https://doi.org/10.1016/j.neuron.2014.02.027

Hornung, S., Dutta, S., & Bitan, G. (2020). CNS-derived blood exosomes as a promising source of biomarkers: opportunities and challenges. *Front Mol Neurosci, 13*, 38. https://doi.org/10.3389/fnmol.2020.00038

Hou, Y., Dan, X., Babbar, M., Wei, Y., Hasselbalch, S. G., Croteau, D. L., & Bohr, V. A. (2019). Ageing as a risk factor for neurodegenerative disease. *Nat Rev Neurol, 15*(10), 565–581. https://doi.org/10.1038/s41582-019-0244-7

Hu, Y., Chopra, V., Chopra, R., Locascio, J. J., Liao, Z., Ding, H., Zheng, B., Matson, W. R., Ferrante, R. J., Rosas, H. D., Hersch, S. M., & Scherzer, C. R. (2011). Transcriptional modulator H2A histone family, member Y (H2AFY) marks Huntington disease activity in man and mouse. *Proc Natl Acad Sci U S A, 108*(41), 17141–17146. https://doi.org/10.1073/pnas.1104409108

Iwatsubo, T., Odaka, A., Suzuki, N., Mizusawa, H., Nukina, N., & Ihara, Y. (1994). Visualization of A β42(43) and A β40 in senile plaques with end-specific A β monoclonals: evidence that an initially deposited species is A beta 42(43). *Neuron, 13*(1), 45–53. https://doi.org/10.1016/0896-6273(94)90458-8

Janelidze, S., Stomrud, E., Smith, R., Palmqvist, S., Mattsson, N., Airey, D. C., Proctor, N. K., Chai, X., Shcherbinin, S., Sims, J. R., Triana-Baltzer, G., Theunis, C., Slemmon, R., Mercken, M., Kolb, H., Dage, J. L., & Hansson, O. (2020). Cerebrospinal fluid p-tau217 performs better than p-tau181 as a biomarker of Alzheimer's disease. *Nat Commun, 11*(1), 1683. https://doi.org/10.1038/s41467-020-15436-0

Keshavarz, M., Xie, K., Schaaf, K., Bano, D., & Ehninger, D. (2023). Targeting the "hallmarks of aging" to slow aging and treat age-related disease: fact or fiction? *Mol Psychiatry, 28*(1), 242–255. https://doi.org/10.1038/s41380-022-01680-x

Kitamura, Y., Kojima, M., Kurosawa, T., Sasaki, R., Ichihara, S., Hiraku, Y., Tomimoto, H., Murata, M., & Oikawa, S. (2018). Proteomic profiling of exosomal proteins for blood-based biomarkers in Parkinson's disease. *Neuroscience, 392*, 121–128. https://doi.org/10.1016/j.neuroscience.2018.09.017

Klyucherev, T. O., Olszewski, P., Shalimova, A. A., Chubarev, V. N., Tarasov, V. V., Attwood, M. M., Syvänen, S., & Schiöth, H. B. (2022). Advances in the development of new biomarkers for Alzheimer's disease. *Transl Neurodegener, 11*(1), 25. https://doi.org/10.1186/s40035-022-00296-z

Kumar, S., Vijayan, M., Bhatti, J., & Reddy, P. H. (2017). MicroRNAs as peripheral biomarkers in aging and age-related diseases. *Progress in molecular biology and translational science*, 146, 47–94. https://doi.org/10.1016/bs.pmbts.2016.12.013

Lane, R. M., & Farlow, M. R. (2005). Lipid homeostasis and apolipoprotein E in the development and progression of Alzheimer's disease. *J Lipid Res, 46*(5), 949–968. https://doi.org/10.1194/jlr.M400486-JLR200

Lanoiselée, H. M., Nicolas, G., Wallon, D., Rovelet-Lecrux, A., Lacour, M., Rousseau, S., Richard, A. C., Pasquier, F., Rollin-Sillaire, A., Martinaud, O., Quillard-Muraine, M., de la Sayette, V., Boutoleau-Bretonniere, C., Etcharry-Bouyx, F., Chauviré, V., Sarazin, M., le Ber, I., Epelbaum, S., Jonveaux, T., Rouaud, O., Ceccaldi, M., Félician, O., Godefroy, O., Formaglio, M., Croisile, B., Auriacombe, S., Chamard, L., Vincent, J. L., Sauvée, M., Marelli-Tosi, C., Gabelle, A., Ozsancak, C., Pariente, J., Paquet, C., Hannequin, D., Campion, D., collaborators of the CNR-MAJ project. (2017). APP, PSEN1, and PSEN2 mutations in early-onset alzheimer disease: a genetic screening study of familial and sporadic cases. *PLoS Med, 14*, e1002270. https://doi.org/10.1371/journal.pmed.1002270

Lee, M., Liu, T., Im, W., & Kim, M. (2016). Exosomes from adipose-derived stem cells ameliorate phenotype of Huntington's disease in vitro model. *Eur J Neurosci, 44*, 2114–2119. https://doi.org/10.1111/ejn.13275

Leroy, E., Boyer, R., Auburger, G., Leube, B., Ulm, G., Mezey, E., Harta, G., Brownstein, M. J., Jonnalagada, S., Chernova, T., Dehejia, A., Lavedan, C., Gasser, T., Steinbach, P. J., Wilkinson, K. D., & Polymeropoulos, M. H. (1988). The ubiquitin pathway in Parkinson's disease. *Nature, 395*(6701), 451–452. https://doi.org/10.1038/26652

Levin, O., Fujiyama, H., Boisgontier, M. P., Swinnen, S. P., & Summers, J. J. (2014). Aging and motor inhibition: a converging perspective provided by brain stimulation and imaging approaches. *Neurosci Biobehav Rev, 43*, 100–117. https://doi.org/10.1016/j.neubiorev.2014.04.001

Long, J. M., & Holtzman, D. M. (2019). Alzheimer disease: An update on pathobiology and treatment strategies. *Cell, 179*(2), 312–339. https://doi.org/10.1016/j.cell.2019.09.001

Lovrecic, L., Kastrin, A., Kobal, J., Pirtosek, Z., Krainc, D., & Peterlin, B. (2009). Gene expression changes in blood as a putative biomarker for Huntington's disease. *Mov Disord, 24*(15), 2277–2281. https://doi.org/10.1002/mds.22477

Mak, K., Yang, F., Vinters, H. V., Frautschy, S. A., & Cole, G. M. (1994). Polyclonals to β-amyloid (1–42) identify most plaque and vascular deposits in Alzheimer cortex, but not striatum. *Brain Res, 667*(1), 138–142. https://doi.org/10.1016/0006-8993(94)91725-6

Mapstone, M., Cheema, A. K., Fiandaca, M. S., Zhong, X., Mhyre, T. R., MacArthur, L. H., Hall, W. J., Fisher, S. G., Peterson, D. R., Haley, J. M., Nazar, M. D., Rich, S. A., Berlau, D. J., Peltz, C. B., Tan, M. T., Kawas, C. H., & Federoff, H. J. (2014). Plasma phospholipids identify antecedent memory impairment in older adults. *Nat Med, 20*(4), 415–418. https://doi.org/10.1038/nm.3466

Marek, K., Innis, R. B., van Dyck, C., Fussell, B., Early, M., Eberly, S., Oakes, D., & Seibyl, J. (2001). [123I]beta-CIT SPECT imaging assessment of the rate of Parkinson's disease progression. *Neurology, 57*(11), 2089–2094. https://doi.org/10.1212/wnl.57.11.2089

Margis, R., Margis, R., & Rieder, C. R. M. (2011). Identification of blood microRNAs associated to Parkinson's disease. *Journal of Biotechnology, 152*(3), 96–101. https://doi.org/10.1016/j.jbiotec.2011.01.023

Mastrokolias, A., Ariyurek, Y., Goeman, J. J., van Duijn, E., Roos, R. A., van der Mast, R. C., van Ommen, G. B., den Dunnen, J. T., 't Hoen, P. A., & van Roon-Mom, W. M. (2015). Hungtinton's disease biomarker progression profile identified by transcriptome sequencing in peripheral blood. *Eur J Hum Genet, 23*(10), 1349–1356. https://doi.org/10.1038/ejhg.2014.281

Mattson, M. P., & Arumugam, T. V. (2018). Hallmarks of brain aging: adaptive and pathological modification by metabolic states. *Cell Metab, 27*(6), 1176–1199. https://doi.org/10.1016/j.cmet.2018.05.011

Melzer, D., Pilling, L. C., & Ferrucci, L. (2020). The genetics of human ageing. *Nat Rev Genet, 21*(2), 88–101. https://doi.org/10.1038/s41576-019-0183-6

Mesko, B., Poliska, S., Szegedi, A., Szekanecz, Z., Palatka, K., Papp, M., & Nagy, L. (2010). Peripheral blood gene expression patterns discriminate among chronic inflammatory diseases and healthy controls and identify novel targets. *BMC Med Genomics*, *3*, 15. https://doi.org/10.1186/1755-8794-3-15

Miller, D. B., & O'Callaghan, J. P. (2015). Biomarkers of Parkinson's disease: present and future. *Metabolism*, *64*(3 Suppl 1), S40–S46. https://doi.org/10.1016/j.metabol.2014.10.030.

Miller, D. L., Papayannopoulos, I. A., Styles, J., Bobin, S. A., Lin, Y. Y., Biemann, K., & Iqbal, K. (1993). Peptide compositions of the cerebrovascular and senile plaque core amyloid deposits of Alzheimer's disease. *Arch Biochem Biophys, 301*(1), 41–52. https://doi.org/10.1006/abbi.1993.1112

Mohapatra, D., Jena, S., Prusty, S. K., Sahu, P. K. (2020). Biomarkers of Alzheimer's Disease: A Review. *Sys Rev Pharm*, 11(6), 151–158. https://doi.org/10.31838/srp.2020.6.24

Molochnikov, L., Rabey, J. M., Dobronevsky, E. et al. A molecular signature in blood identifies early Parkinson's disease. Mol Neurodegeneration 7, 26 (2012).

Molochnikov, L., Rabey, J. M., Dobronevsky, E., Bonucelli, U., Ceravolo, R., Frosini, D., Grünblatt, E., Riederer, P., Jacob, C., Aharon-Peretz, J., Bashenko, Y., Youdim, M. B., & Mandel, S. A. (2012). A molecular signature in blood identifies early Parkinson's disease. *Mol Neurodegener, 7*, 26. https://doi.org/10.1186/1750-1326-7-26

Nhan, H. S., Chiang, K., & Koo, E. H. (2015). The multifaceted nature of amyloid precursor protein and its proteolytic fragments: friends and foes. *Acta Neuropathol, 129*(1), 1–19. https://doi.org/10.1007/s00401-014-1347-2

Paisan-Ruiz, C., Lewis, P. A., & Singleton, A. B. (2013). LRRK2: cause, risk, and mechanism. *J Parkinson Dis*, *3*(2), 85–103. https://doi.org/10.3233/JPD-130192

Pankratz, N., Nichols, W. C., Uniacke, S. K., Halter, C., Rudolph, A., Shults, C., Conneally, P. M., Foroud, T., & the Parkinson Study Group. (2002). Genome screen to identify susceptibility genes for Parkinson disease in a sample without Parkin mutations. *Am J Hum Genet*, *71*(1), 124–135. https://doi.org/10.1086/341282

Raber, J., Wong, D., Buttini, M., Orth, M., Bellosta, S., Pitas, R. E., Mahley, R. W., & Mucke, L. (1998). Isoform-specific effects of human apolipoprotein E on brain function revealed in ApoE knockout mice: increased susceptibility of females. *Proc Natl Acad Sci USA*, *95*(18), 10914–10919. https://doi.org/10.1073/pnas.95.18.10914

Rachakonda, V., Pan, T., Le, W. T. H., & Le, W. D. (2004). Biomarkers of neurodegenerative disorders: how good are they? *Cell Res 14*(5), 349–358. https://doi.org/10.1038/sj.cr.7290235

Reiman, E. M., Arboleda-Velasquez, J. F., Quiroz, Y. T., Huentelman, M. J., Beach, T. G., Caselli, R. J., Chen, Y., Su, Y., Myers, A. J., Hardy, J., Paul Vonsattel, J., Younkin, S. G., Bennett, D. A., De Jager, P. L., Larson, E. B., Crane, P. K., Keene, C. D., Kamboh, M. I., Kofler, J. K. & Alzheimer's Disease Genetics Consortium. (2020). Exceptionally low likelihood of Alzheimer's dementia in APOE2 homozygotes from a 5,000-person neuropathological study. *Nat Commun*, *11*, 1–11. https://doi.org/10.1038/s41467-019-14279-8

Riedel, B. C., Thompson, P. M., & Brinton, R. D. (2016). Age, APOE and sex: triad of risk of Alzheimer's disease. *J Steroid Biochem Mol Biol*, *160*, 134–147. https://doi.org/10.1016/j.jsbmb.2016.03.012

Runne, H., Regulier, E., Kuhn, A., Zala, D., Gokce, O., Perrin, V., Sick, B., Aebischer, P., Deglon, N., Luthi-Carter, R. (2008). Dysregulation of gene expression in primary neuron models of Hungtinton's Disease shows that polyglutamine-related effects on the striatal transcriptome may not be dependent on brain circuitry. *Journal of Neuroscience*, *28*(39), 9723–9731. https://doi.org/10.1523/JNEUROSCI.3044-08.2008

Selkoe, D. J., & Hardy, J. (2016). The amyloid hypothesis of Alzheimer's disease at 25 years. *EMBO -Mol Med*, *8*(6), 595–608. https://doi.org/10.15252/emmm.201606210

Serrano-Pozo, A., Das, S., & Hyman, B. T. (2021). APOE and Alzheimer's disease: advances in genetics, pathophysiology, and therapeutic approaches. *Lancet Neurology*, *20*(1), 68–80. https://doi.org/10.1016/S1474-4422(20)30412-9. Erratum in: *Lancet Neurol*, 20(2), e2. https://doi.org/10.1016/S1474-4422(21)00004-1

Sheinerman, K. S., Tsivinsky, V. G., Crawford, F., Mullan, M. J., Abdullah, L., & Umansky, S. R. (2012). Plasma microRNA biomarkers for detection of mild cognitive impairment. *Aging (Albany NY)*, *4*(9), 590–605. https://doi.org/10.18632/aging.100486

Shinde, S., Mukhopadhyay, S., Mohsen, G., & Khoo, S. K. (2015). Biofluid-based microRNA biomarkers for Parkinson's disease: an overview and update. *AIMS Medical Science*, *2*(1), 15–25. https://doi.org/10.3934/medsci.2015.1.15

Shinto, L., Quinn, J., Montine, T., Dodge, H. H., Woodward, W., Baldauf-Wagner, S., Waichunas, D., Bumgarner, L., Bourdette, D., Silbert, L., & Kaye, J. (2014). A randomized placebo-controlled pilot trial of omega-3 fatty acids and alpha lipoic acid in Alzheimer's disease. *J Alzheimers Dis*, *38*(1), 111–120. https://doi.org/10.3233/JAD-130722

Sikora, E., Bielak-Zmijewska, A., Dudkowska, M., Krzystyniak, A., Mosieniak, G., Wesierska, M., & Wlodarczyk, J. (2021). Cellular senescence in brain aging. *Front Aging Neurosci*, *13*, 646924. https://doi.org/10.3389/fnagi.2021.646924

Simonsen, A. H., Herukka, S.-K., Andreasen, N., Baldeiras, I., Bjerke, M., Blennow, K., Engelborghs, S., Frisoni, G. B., Gabryelewicz, T., Galluzzi, S., Handels, R., Kramberger, M. G., Kulczyńska, A., Molinuevo, J. L., Mroczko, B., Nordberg, A., Oliveira, C. R., Otto, M., Rinne, J. O., Rot, U., Saka, E., Soininen, H., Struyfs, H., Suardi, S., Visser, P. J., Winblad, B., Zetterberg, H., & Waldemar, G. (2017). Recommendations for CSF AD biomarkers in the diagnostic evaluation of dementia. *Alzheimer's Dement*, *13*(3), 274–284. https://doi.org/10.1016/j.jalz.2016.09.008

Soreq, L., Salomonis, N., Bronstein, M., Greenberg, D. S., Israel, Z., Bergman, H., & Soreq, H. (2013). Small RNA sequencing-microarray analyses in Parkinson leukocytes reveal deep brain stimulation-induced splicing changes that classify brain region transcriptomes. *Front Mol Neurosci*, *6*, 10. https://doi.org/10.3389/fnmol.2013.00010

Steverson, M. (2024). *Ageing and health*. World Health Organization. https://www.who.int/news-room/fact-sheets/detail/ageing-and-health

Sturrock, A., Laule, C., Wyper, K., Milner, R. A., Decolongon, J., Dar Santos, R., Coleman, A. J., Carter, K., Creighton, S., Bechtel, N., Bohlen, S., Reilmann, R., Johnson, H. J., Hayden, M. R., Tabrizi, S. J., Mackay, A. L., & Leavitt, B. R. (2015). A longitudinal study of magnetic resonance spectroscopy Huntington's disease biomarkers. *Mov Disord*, *30*(3), 393–401. https://doi.org/10.1002/mds.26118

Sun, X., Chen, W. D., & Wang, Y. D. (2015). β-Amyloid: the key peptide in the pathogenesis of Alzheimer's disease. *Front Pharmacol*, *6*, 221. https://doi.org/10.3389/fphar.2015.00221

Tcw, J., & Goate, A. M. (2017). Genetics of β-amyloid precursor protein in Alzheimer's disease. *Cold Spring Harb Perspect Med*, 7, a024539. https://doi.org/10.1101/cshperspect.a024539

Tönnies, E., & Trushina, E. (2017). Oxidative, stress. Synaptic dysfunction, and Alzheimer's disease. *J Alzheimers Dis*, *57*(4), 1105–1121. https://doi.org/10.3233/JAD-161088

Tripathi, A. (2012). New cellular and molecular approaches to ageing brain. *Ann Neurosci*, *19*(4), 177–182. https://doi.org/10.5214/ans.0972.7531.190410

Valente, E. M., Abou-Sleiman, P. M., Caputo, V., Muqit, M. M., Harvey, K., Gispert, S., Ali, Z., Del Turco, D., Bentivoglio, A. R., Healy, D. G., Albanese, A., Nussbaum, R., González-Maldonado, R., Deller, T., Salvi, S., Cortelli, P., Gilks, W. P., Latchman, D. S., Harvey, R. J., Dallapiccola, B., Auburger, G., & Wood, N. W. (2004). Hereditary early-onset Parkinson's disease caused by mutations in PINK1. *Science*, *304*(5674), 1158–1160. https://doi.org/10.1126/science.1096284

Van den Bogaard, S. J., Dumas, E. M., Acharya, T. P., Johnson, H., Langbehn, D. R., Scahill, R. I., Tabrizi, S. J., van Buchem, M. A., van der Grond, J., Roos, R. A., & TRACK-HD Investigator Group. (2011). Early atrophy of pallidum and accumbens nucleus in Huntington's disease. *J Neurol*, *258*(3), 412–420. https://doi.org/10.1007/s00415-010-5768-0

Venancio, T. M., & Aravind, L. (2010). CYSTM, a novel cysteine-rich transmembrane module with a role in stress tolerance across eukaryotes. *Bioinformatics*, *26*(2),149–152. https://doi.org/10.1093/bioinformatics/btp647

Wahl, D., Cavalier, A. N., & LaRocca, T. J. (2021). Novel strategies for healthy brain aging. *Exercise Sport Sci Rev*, *49*(2), 115–125. https://doi.org/10.1249/JES.0000000000000242

Walker L. C. (2021). Proteopathic Strains and the Heterogeneity of Neurodegenerative Diseases. *Annu Rev Genet*, *50*, 329–346. https://doi.org/10.1146/annurev-genet-120215-034943

Weir, D. W., Sturrock, A., & Leavitt, B. R. (2011). Development of biomarkers for Huntington's disease. *Lancet Neurol*, *10*(6), 573–590. https://doi.org/10.1016/S1474-4422(11)70070-9

Wild, E. J., Boggio, R., Langbehn, D., Robertson, N., Haider, S., Miller, J. R., Zetterberg, H., Leavitt, B. R., Kuhn, R., Tabrizi, S. J., Macdonald, D., & Weiss, A. (2015). Quantification of mutant huntingtin protein in cerebrospinal fluid from Huntington's disease patients. *J Clin Invest*, *125*(5),1979–1986. https://doi.org/10.1172/JCI80743

Wood, N. I., Goodman, A. O., van der Burg, J. M., Gazeau, V., Brundin, P., Björkqvist, M., Petersén, A., Tabrizi, S. J., Barker, R. A., & Morton, A. J. (2008). Increased thirst and drinking in Huntington's disease and the R6/2 mouse. *Brain Res Bull*, *76*(1-2),70–79. https://doi.org/10.1016/j.brainresbull.2007.12.007

Wurtman, R. (2015). Biomarkers in the diagnosis and management of Alzheimer's disease. *Metabolism Clin and Experimental*, *64*, S47–S50. https://doi.org/10.1016/j.metabol.2014.10.034

Xi, Y., Nakajima, G., Gavin, E., Morris, C. G., Kudo, K., Hayashi, K., & Ju, J. (2007). Systematic analysis of microRNA expression of RNA extracted from fresh frozen and formalin-fixed paraffin-embedded samples. *RNA*, *13*(10), 1668–1674. https://doi.org/10.1261/rna.642907

Yang, W., Hamilton, J. L., Kopil, C., Beck, J. C., Tanner, C. M., Albin, R. L., Ray Dorsey, E., Dahodwala, N., Cintina, I., Hogan, P., & Thompson, T. (2020). Current and projected future economic burden of Parkinson's disease in the U.S. *NPJ Parkinsons Dis*, *6*, 15. https://doi.org/10.1038/s41531-020-0117-1

Zekanowski, C., Styczyńska, M., Pepłońska, B., Gabryelewicz, T., Religa, D., Ilkowski, J., Kijanowska-Haładyna, B., Kotapka-Minc, S., Mikkelsen, S., Pfeffer, A., Barczak, A., Łuczywek, E., Wasiak, B., Chodakowska-Zebrowska, M., Gustaw, K., Łaczkowski, J., Sobów, T., Kuźnicki, J., & Barcikowska, M. (2003). Mutations in presenilin 1, presenilin 2 and amyloid precursor protein genes in patients with early-onset Alzheimer's disease in Poland. *Exp Neurol*, *184*(2), 991–996. https://doi.org/10.1016/S0014-4886(03)00384-4

Zhai, W., Jeong, H., Cui, L,, Krainc, D, & Tjian, R. (2005). In vitro analysis of huntingtin-mediated transcriptional repression reveals multiple transcription factor targets. *Cell*, *123*(7),1241–1253. https://doi.org/10.1016/j.cell.2005.10.030

Zhang, X., & Song, W. (2013). The role of APP and BACE1 trafficking in APP processing and amyloid-β generation. *Alzheimer's Res Ther*, *5*(5), 46. https://doi.org/10.1186/alzrt211

Zia, A., Pourbagher-Shahri, A. M., Farkhondeh, T., & Samarghandian, S. (2021). Molecular and cellular pathways contributing to brain aging. *Behav Brain Funct*, *17*(1), 6. https://doi.org/10.1186/s12993-021-00179-9

Zou, K., Abdullah, M., & Michikawa, M. (2020). Current biomarkers for Alzheimer's disease: from CSF to blood. *J Pers Med*, *10*(3), 85. https://doi.org/10.3390/jpm10030085

# 3

# *Pharmacotherapy of Neurodegenerative Diseases*

**Nurdan Tekin, Yekta Çulpan and Rezzan Gülhan**

## Abbreviations

| | |
|---|---|
| **AD** | Alzheimer's Disease |
| **ALS** | Amyotrophic Lateral Sclerosis |
| **ALSFRS-R** | Amyotrophic Lateral Sclerosis Functional Rating Scale - Revised |
| **ARIA** | Amyloid-Related Imaging Abnormalities |
| **AADC** | Aromatic L-Amino Acid Decarboxylase |
| **COMT** | Catechol-O-Methyltransferase |
| **DBS** | Deep Brain Stimulation |
| **DMD** | Disease-Modifying Drug |
| **FDA** | Food And Drug Administration |
| **GABA** | Gamma-Aminobutyric Acid |
| **HD** | Huntington's Disease |
| **$IgG_1$** | Immunoglobulin Gamma 1 |
| **LRRK-2** | Leucine-Rich Repeat Kinase-2 |
| **MAO-B** | Monoamine Oxidase Type-B |
| **MS** | Multiple Sclerosis |
| **NMDA** | N-Methyl-D-Aspartate |
| **PD** | Parkinson's Disease |
| **PET** | Positron Emission Tomography |
| **SSRIs** | Selective Serotonin Reuptake Inhibitors |
| **SOD-1** | Superoxide Dismutase-1 |
| **VMAT2** | Vesicular Monoamine Transporter Type 2 |

## Introduction

Neurodegenerative diseases are a group of brain and spinal cord disorders characterized by a progressive deterioration of structure and function of neurons and glial cells, which results in cognitive decline, motor dysfunction, and other neurological symptoms. This gradual deterioration, at the end, can lead to death (Gadhave et al., 2024). Neurodegenerative diseases comprise a wide range of debilitating conditions, including Alzheimer's disease (AD), Parkinson's disease (PD), Huntington's disease (HD), multiple sclerosis (MS), and amyotrophic lateral sclerosis (ALS). They affect millions of people worldwide and their incidence is increasing, in part, due to the aging global population. Each of these conditions displays a different clinical presentation, depending upon the types of degenerating neurons and their spatial organization within the nervous system (Wilson et al., 2023).

While age is the most significant risk factor for neurodegenerative diseases, recent findings indicate a complex interplay between genetic predispositions and environmental factors in their development. Several key pathophysiological mechanisms that underlie neurodegeneration in these disorders have been identified, such as protein misfolding and aggregation, synaptic and neuronal network dysfunction, energy homeostasis disruption, cytoskeletal abnormalities, and inflammation (Gadhave et al., 2024). Understanding these mechanisms is crucial for developing pharmacotherapies for symptom management and disease modification.

The current therapeutic landscape for neurodegenerative diseases remains largely focused on symptom management (Table 3.1). For instance, dopamine replacement is still the gold standard treatment for Parkinson's disease, while cholinesterase inhibitors and glutamate regulators are used to address the cognitive symptoms of Alzheimer's disease. However, these therapies do not affect neurodegeneration and are unable to halt or reverse the disease progression. Thus, there is a demanding need for developing disease-modifying therapies that can target the underlying pathophysiological processes (Lamptey et al., 2022).

This chapter provides an overview of existing treatment modalities and emerging therapeutic strategies for the pharmacotherapy of neurodegenerative diseases.

## Alzheimer's Disease

Alzheimer's disease (AD) is the most common neurodegenerative disorder, affecting approximately 10% of individuals aged 65 and older. This disease is characterized by debilitating symptoms, including a gradual decline in cognitive function. The hallmark pathological changes in AD include the formation of hyperphosphorylated tau proteins and the aggregation of beta-amyloid, leading to amyloid plaques, a key diagnostic feature in affected neurons (Kumar et al., 2024). Traditionally, as with other neurodegenerative diseases, treatment for AD has focused on symptomatic relief rather than modifying the disease itself. However, a significant breakthrough was achieved recently with the Food and Drug Administration's (FDA) approval of the first disease-modifying drug (DMD) for patients with advanced AD (Espay et al., 2024).

The primary goals of AD treatment are to alleviate cognitive impairment and manage any accompanying psychiatric and behavioral symptoms. With the introduction of DMDs, there is now the potential not only to slow further neurodegeneration but also to reverse some of the underlying pathophysiological processes of the disease (Table 3.2).

DOI: 10.1201/9781032676555-4

**TABLE 3.1**

Current Therapeutic Drug Classes Used in Neurodegenerative Diseases

| Neurological Disorder | Drug Class | Mechanism | Drugs |
|---|---|---|---|
| **Alzheimer's Disease** | Cholinesterase Inhibitors | Prevent the knockdown of acetylcholine | Donepezil, rivastigmine, galantamine |
| | Glutamate regulators | Antagonize N-methyl-D-aspartate (NMDA) receptor to improve signal-to-noise ratio of glutamatergic transmission | Memantine |
| | Amyloid-directed antibody | Acts by targeting and removing amyloid-beta plaques | Aducanumab |
| **Parkinson's Disease** | Dopamine supplements | Replenish the decreased dopamine levels | Levodopa |
| | Decarboxylase inhibitors | Prevent peripheral breakdown of levodopa | Carbidopa, benserazide |
| | Dopamine agonist | Produces dopamine-like effects | Apomorphine hydrochloride, pergolide, pramipexole dihydrochloride, ropinirole hydrochloride, rotigotine, bromocriptine |
| | MAO-B inhibitors | Inhibit monoamine oxidase B to prevent dopamine breakdown | Selegiline, rasagiline, safinamide |
| | COMT inhibitors | Inhibit catechol-O-methyltransferase to prolong the effect of levodopa | Entacapone, tolcapone, opicapone |
| **Amyotrophic Lateral Sclerosis** | Glutamate-receptor antagonist | Inhibits glutamate receptors | Riluzole |
| | Free-radical scavenger | Scavenges free radicals | Edaravone |
| **Huntington's Disease** | VMAT2 inhibitors | Inhibit vesicular monoamine transporter 2 to reduce dopamine levels | Tetrabenazine, deutetrabenazine |

**TABLE 3.2**

Summary of Alzheimer's Disease Treatment Approaches

| Treatment Type | Mechanism of Action | Indications | Clinical Benefits | Common Adverse Effects |
|---|---|---|---|---|
| **Cholinesterase Inhibitors** Donepezil, Rivastigmine, Galantamine | Increase acetylcholine levels by inhibiting acetylcholinesterase, thus enhancing cholinergic transmission | Mild to moderate Alzheimer's dementia | Modest improvement in cognition, behavior, ADLs | Nausea, diarrhea, bradycardia, insomnia |
| **NMDA Receptor Antagonists** Memantin | Blocks activation of NMDA receptors, which may reduce excitotoxicity and provide neuroprotection | Moderate to severe Alzheimer's dementia | Modest improvements in cognition and function | Dizziness, headache, confusion |
| **Amyloid-Targeted Therapies** Lecanemab, Donanemab | Monoclonal antibodies targeting amyloid plaques, believed to modify disease progression | Mild cognitive impairment or mild dementia with confirmed amyloid pathology | Slows disease progression (modest effect) | Brain edema, hemosiderosis |
| **Antioxidants** Vitamin E | Acts as a free radical scavenger, theoretically providing neuroprotection | Mild to moderate Alzheimer's dementia | Delays functional progression (modest effect) | Risk of increased mortality in high doses |
| **Non-pharmacologic Therapies** Cognitive rehabilitation, exercise | Improve quality of life and delay progression through lifestyle and supportive measures | All stages of Alzheimer's dementia | Enhanced daily function, reduced caregiver burden | Rarely associated with adverse effects |

## Cholinesterase Inhibitors

Since the clinical introduction of tacrine in 1993 for mild-to-moderate AD, cholinesterase inhibitors have been a cornerstone in the treatment of AD. Although the hepatotoxic profile of tacrine limited its widespread use, other cholinesterase inhibitors remain central in treating mild-to-moderate AD (Balázs et al., 2021).

Cholinesterase inhibitors are indirect cholinomimetics that enhance cholinergic transmission by elevating acetylcholine levels. This mechanism addresses a key pathophysiological aspect of AD: reduced acetylcholine availability in specific brain regions. Acetylcholine, released by cholinergic neurons, is rapidly degraded in the synaptic cleft by cholinesterase enzymes. By inhibiting these enzymes, cholinesterase inhibitors elevate acetylcholine concentrations, thereby potentially ameliorating cognitive function. Three cholinesterase inhibitors – donepezil, rivastigmine, and galantamine – are currently approved by the FDA for AD treatment. Although these agents share a similar mechanism of action, but each exhibits unique pharmacokinetic and pharmacodynamic properties.

### *Donepezil*

Donepezil is a rapid, reversible acetylcholinesterase inhibitor. Administered orally or transdermally, it is well-absorbed from gastrointestinal tract and exhibits high protein binding. Metabolized primarily via hepatic pathways (CYP2D6 and CYP3A4), donepezil has an oral elimination half-life of approximately 70 hours, extending to 91 hours for transdermal preparations (Press & Alexander, 2024).

The standard initiation dose is 5 mg once daily, which can be increased to 10 mg if well tolerated. Sustained-release formulations deliver 23 mg daily, while transdermal patches start at 5 mg/24 hours, with titration to 10 mg/24 hours for enhanced efficacy.

Donepezil's adverse effects are predominantly due to increased peripheral cholinergic activity, including nausea, vomiting, and diarrhea, typically observed at treatment initiation and generally resolving with continued therapy. Cardiovascular side effects, particularly those associated with increased vagal tone, include conduction abnormalities (e.g., AV block, QT interval prolongation) hypertension, and syncope, especially in patients with underlying cardiovascular disease.

### *Rivastigmine*

Rivastigmine is a pseudo-irreversible inhibitor of both acetylcholinesterase and butyrylcholinesterase. It binds reversibly to these enzymes and additionally carbamylates the active site, resulting in prolonged action. Although the drug has a short elimination half-life (2 hours), these pharmacological effects extend up to 10 hours. Rivastigmine, being highly lipid-soluble, is rapidly absorbed from the gastrointestinal tract and crosses the blood–brain barrier with ease. The primary route of metabolism is cholinesterase-mediated hydrolysis within the brain, followed by hepatic metabolism, with renal excretion of metabolites (Press & Alexander, 2024).

For moderate-to-severe AD, rivastigmine is initially dosed at 1.5 mg twice daily, which may be titrated to 3 mg as tolerated. Transdermal patches, used for sustained release, are typically initiated at 4.6 to 9.5 mg per 24 hours.

Adverse effects are similar to those observed with other cholinesterase inhibitors; however, rivastigmine has been associated with a slightly higher incidence of fatalities. Dermatological reactions, including allergic contact dermatitis and local skin irritation, are notable adverse effects related to the transdermal formulation.

### *Galantamine*

Galantamine, a naturally derived alkaloid, is a reversible acetylcholinesterase inhibitor primarily enhancing cholinergic transmission. Unlike other acetylcholinesterase inhibitors, galantamine possesses a unique affinity for nicotinic receptors. This dual action – both inhibiting acetylcholinesterase and modulating nicotinic receptors – contributes to its beneficial effects on cognitive function. Additionally, galantamine has been shown to increase levels of glutamate and serotonin. This additional mechanism of action may provide a broader spectrum of neurochemical benefits, potentially augmenting cognitive and behavioral outcomes in patients.

The adverse effect profile of galantamine is generally similar to that of donepezil, with common side effects including gastrointestinal symptoms such as nausea and vomiting. Since most newly diagnosed AD cases are managed with a trial of cholinesterase inhibitors, the choice between donepezil, galantamine, and rivastigmine should be tailored to the individual patient. Key considerations include ease of use, tolerance, and cost, as well as clinician and patient preference. Given the common gastrointestinal side effects associated with cholinesterase inhibitors, strategies to enhance patient tolerance – such as gradual dose titration over several months or administering the medication with meals – may be beneficial.

## NMDA Receptor Antagonists

### *Memantine*

Memantine is an N-methyl-D-aspartate (NMDA) glutamate receptor antagonist. The clinical effects of memantine are not related to cholinergic transmission but rather to its neuroprotective properties. Glutamate is the primary excitatory neurotransmitter in the central nervous system, and NMDA receptor activation leads to ionic flow, especially calcium ($Ca^{2+}$), into neurons. An elevated intracellular calcium level is associated with neuronal cell death; therefore, blocking this receptor with memantine may help protect neurons from further degeneration. Memantine is often used as an adjunct to cholinesterase inhibitors in patients with advanced AD (Press & Buss, 2024).

The initial dose of memantine is 5 mg daily, which can be increased weekly up to a maximum dose of 20 mg, if tolerated by the patient. It is important to note that discontinuation of treatment should be done by gradually tapering the dose to avoid potential withdrawal effects.

Memantine is generally well tolerated, with the most common adverse effects being headache and dizziness. Although rare, NMDA receptor antagonism has been associated with psychotic symptoms; therefore, caution is advised when prescribing memantine to patients with dementia-related psychotic symptoms.

## Amyloid-Targeted Therapies

Until recently, treatments for Alzheimer's disease, as with many other neurodegenerative disorders, were limited to symptomatic relief without targeting the underlying pathology. The introduction of monoclonal antibodies targeting amyloid-beta has not only been a breakthrough in the treatment of AD but has also shifted the therapeutic paradigm for neurodegenerative diseases.

In 2023, two amyloid-targeted therapies, lecanemab and donanemab, were approved by the FDA. Clinical trials demonstrated that these agents are highly effective at reducing amyloid plaques. Although the clinical outcomes were somewhat more modest than anticipated, these drugs do not fully cure the disease but are effective in significantly slowing its progression, thereby reducing disease-related mortality and morbidity (Zhang et al., 2023).

Amyloid-targeted therapies are specifically designed for patients with mild cognitive impairment who exhibit documented amyloid pathology, confirmed through an amyloid PET scan or lumbar puncture. Prior to initiating treatment, it is recommended to assess apolipoprotein E ε4 (ApoE ε4) status, as this genotype is associated with an increased risk of adverse effects.

### *Lecanemab*

Lecanemab is a humanized immunoglobulin gamma 1 ($IgG_1$) monoclonal antibody that targets both the soluble and insoluble forms of amyloid-beta protofibrils. This therapeutic agent functions by reducing the accumulation of amyloid-beta plaques, a pathophysiological hallmark of AD. Lecanemab undergoes degradation by proteolytic enzymes, with an elimination half-life of approximately 7 days. The recommended dose is 10 mg/kg, administered intravenously every two weeks, with the dose calculated based on the patient's actual body weight to ensure optimal therapeutic efficacy.

### *Donanemab*

Donanemab is also a humanized IgG1 monoclonal antibody, directed against the insoluble N-truncated pyroglutamate form of amyloid-beta. Similar to lecanemab, donanemab reduces amyloid-beta plaque formation. It is also metabolized by proteolytic enzymes, with an elimination half-life of approximately 12.1 days. Donanemab is administered intravenously, starting with an initial dose of 700 mg every four weeks for three doses. Maintenance therapy is then continued with 1,400 mg every four weeks until amyloid plaques reach minimal levels, as observed on PET imaging.

While immunotherapy has provided new hope for a wide range of conditions, concerns have arisen about the increased risk of significant adverse effects associated with these treatments, and amyloid-targeted therapies are no exception. Both lecanemab and donanemab carry a black box warning due to the potential risk of amyloid-related imaging abnormalities (ARIA).

### Amyloid-Related Imaging Abnormalities (ARIA)

An immune-mediated response to amyloid-targeted agents can manifest in the cerebral vascular walls, resulting in capillary leakage into the perivascular spaces and hemorrhaging within the cortical and leptomeningeal arteries. ARIA is characterized by vasogenic edema and hemosiderin deposition, often presenting as microhemorrhages and macrohemorrhages. Clinical symptoms of ARIA may include abnormal gait, dizziness, headache, nausea, and seizures (Hampel et al., 2023).

Although the precise mechanisms underlying ARIA are not yet fully understood, the presence of the Apolipoprotein E ε4 allele has been identified as a significant risk factor. Therefore, it is advisable to test for ApoE ε4 status prior to starting treatment with amyloid-targeted therapies.

Management of ARIA is determined by the severity of symptoms; asymptomatic patients with mild ARIA may continue treatment under close monitoring, whereas in more severe cases, treatment should be withheld until symptoms and radiographic abnormalities have resolved.

## Parkinson's Disease

Parkinson's disease (PD) is the second most common neurodegenerative disease after AD (Ou et al., 2021). The prevalence of PD is 1–2 ‰ in the general population and 9.5 ‰ in the population over 65 years of age (Hirtz et al., 2007; von Campenhausen et al., 2005). PD is characterized by degeneration of dopaminergic neurons in the substantia nigra, causing tremor, rigidity, and bradykinesia, which are the classic clinical symptoms of PD (Hirsch et al., 2013). Patients may also experience non-motor symptoms that impair their quality of life, such as intestinal issues, sleep disturbances, and apathy (Sveinbjornsdottir, 2016). In the management of PD, the patient's symptoms and signs, age, stage of disease, and functional status should be considered. Despite the lack of a cure, there are numerous symptomatic pharmacologic, nonpharmacologic, and surgical treatments that can improve motor and nonmotor function as well as quality of life over the course of the disease. This chapter focuses on the pharmacological management of PD. Surgical therapies and nonpharmacologic treatments are not discussed in detail here.

Dopaminergic therapy is the mainstay of pharmacologic treatment for PD. For this purpose, levodopa, dopamine receptor agonists, apomorphine, monoamine oxidase type-B (MAO-B) inhibitors, Catechol-O-methyltransferase (COMT) inhibitors, amantadine, istradefylline, and some muscarinic receptor antagonists are used. They are all regarded as symptomatic treatments; none of them have been proven to be neuroprotective or disease-modifying.

### Levodopa

The FDA approved the levodopa in 1970 and approved the combined treatment with levodopa and carbidopa in 1975. In spite of the decades, levodopa is the initial therapy to control motor symptoms in the treatment of patients with early PD, generally. It has better effects on motor function, everyday activities, and quality of life when compared to other medications and classes (Connolly & Lang, 2014). It is still the gold standard treatment of PD.

Levodopa is a prodrug that is converted to dopamine (DA) via dopa decarboxylase in the central nervous system, compensating for the dopamine deficiency (Haddad et al., 2017). The conversion of levodopa to dopamine by aromatic L-amino acid decarboxylase (AADC) begins in the intestinal mucosa. Dopamine formed in the periphery cannot pass to the brain and causes peripheral side effects. The concomitant use of levodopa with peripheral dopa decarboxylase inhibitors, carbidopa, and benserazide increases antiparkinsonian efficacy and reduces peripheral side effects (Zahoor et al., 2018).

It is used orally and taking it with food delays its absorption. The time taken to reach peak plasma concentration is between 0.5 and 2 hours after an oral dose, and serum elimination half-life is 1–3 hours (Brunton et al., 2018). 25 mg/100 mg carbidopa/levodopa is the most commonly prescribed form, and this form is used 3–4 times daily according to clinical relevancy. The dose is progressively raised after being started low as well. It is more effective for bradykinesia than tremor. Levodopa undergoes first-pass metabolism mainly by two pathways: decarboxylation by AADC and O-methylation by COMT, and it is primarily renally excreted (Aminoff, 2024).

Levodopa's most frequent adverse effects include headache, nausea, dizziness, and somnolence. Serious side effects include dyskinesia, confusion, hallucinations, agitation, psychosis, orthostatic hypotension, and impulse control disorders (compulsive sexual behavior, compulsive shopping, pathological gambling, binge eating disorder, and punding) (Gandhi & Saadabadi, 2023; Voon et al., 2017). After 3–4 years, there are fluctuations in the effect of the drug (on-off syndrome). The response to levodopa gradually decreases and increasing the dose becomes ineffective. Not taking the drug for a certain period may improve the responses. Levodopa should not be stopped suddenly in patients with PD because abruptly stopping the medication has been associated with a syndrome resembling neuroleptic malignant syndrome or akinetic crisis (Brunton et al., 2018).

The FDA approved the use of carbidopa-levodopa enteral suspension in PD patients for reducing clinically problematic fluctuations in dopaminergic drug concentration and thereby decreasing motor fluctuations. Pharmacokinetic problems are being tried to be solved with new technologies.

## Dopamine Receptor Agonists

These act directly on the postsynaptic dopamine receptors and, unlike levodopa, do not need to be converted by enzymes into an active metabolite. The older dopamine agonists, bromocriptine and pergolide, are ergot derivatives and are rarely used to treat parkinsonism nowadays because the newer agents, pramipexole, ropinirole, and rotigotine, have fewer side effects than older dopamine agonists. There is no proof that one agonist is better than another; patients may react to one of these agents but not to another. A meta-analysis shows that dopamine agonists reduce motor complications when compared to levodopa, but it also shows that agonists cause more significant side effects and worse symptom control (Stowe et al., 2008). Dopamine agonists may be preferred in younger patients with a high risk of dyskinesia as an alternative to early levodopa (Pringsheim et al., 2021).

Pramipexole and ropinirole selectively act on $D_2$ and $D_3$ receptors. Rotigotine has activity at $D_1$ class sites in addition to acting at $D_2$ sites (Brunton et al., 2018). Ropinirole and pramipexole are used orally, and rotigotine is used transdermally. Ropinirole and pramipexole are usually used three times a day, and rotigotine is used once daily. Dopamine agonists typically have a longer duration of action (8–24 hours) than levodopa (6–8 hours) (Goldenberg, 2008). Dopamine agonists may produce hallucinosis or confusion like levodopa and cause orthostatic hypotension, so starting with low doses and titrating slowly may be beneficial. Stimulation of the dopaminergic system may also cause patients to become prone to gambling, increase in sexual activity, and excessive spending.

### *Apomorphine*

Apomorphine is also a dopaminergic agonist that interacts mainly with postsynaptic $D_2$ receptors in the caudate nucleus and putamen. The FDA has approved apomorphine as a "rescue therapy" to treat acute intermittent "off" episodes in patients whose response to dopaminergic therapy fluctuates. Subcutaneous injection of apomorphine hydrochloride rapidly reaches the blood and then the brain, leading to clinical benefit that begins very quickly and persists for up to 2 hours (Aminoff, 2024). Nausea is seen in patients up to 30% using apomorphine, and most are started on anti-emetic drugs before treatment. It should not be used with $5\text{-}HT_3$ receptor blockers to relieve the side effect of nausea because severe hypotension may occur. Dyskinesias, drowsiness, chest pain, hypotension, behavioral or mental abnormalities, and injection-site bruising are also common side effects.

## MAO-B Inhibitors

MAO-B inhibitors do not have strong dopaminergic effects, but they are well tolerated and can be administered once daily. It has been demonstrated that MAO-B inhibitors are a moderately effective early PD symptom treatment, reducing disability, the need for levodopa, and the incidence of motor fluctuations (Ives et al., 2004). They can be preferred instead of levodopa in patients who prefer once-daily medication and only need modest benefit. In advanced PD, rasagiline adjunctive therapy dramatically lessens levodopa-related wearing off symptoms (Olanow et al., 2008).

There are three MAO-B inhibitors that are available for use in patients with PD: selegiline, rasagiline, and safinamide. These agents selectively and irreversibly inactivate MAO-B, which is responsible for most of the oxidative metabolism of dopamine in the brain. Selegiline is generally used in patients orally twice daily, while rasagiline is used once daily. Safinamide is usually given orally once daily as adjunctive therapy with levodopa to help with motor fluctuations or monotherapy.

Selegiline may cause anxiety, insomnia, and other adverse symptoms due to amphetamine and methamphetamine metabolites, while rasagiline does not give rise to the undesirable amphetamine metabolites. MAO-B inhibitors' most common side effects are nausea and headache. Drug interactions can be problematic even though selective MAO-B inhibitors are usually well tolerated.

## COMT Inhibitors

COMT inhibitors are not used alone to treat PD; rather, they are used combined with levodopa to potentiate and prolong levodopa's effects by diminishing its peripheral metabolism. Levodopa's relative bioavailability rises as a result of a decrease in its clearance. Also, it reduces "off" time when used as adjunctive therapy with levodopa (Pahwa et al., 2006).

COMT inhibitors in clinical use include entacapone, opicapone, and tolcapone.

Patients using levodopa who have experienced response fluctuations may benefit from these medications, which could result in a smoother response, longer on-time, and the possibility of lowering the daily dose of levodopa. Entacapone is more commonly used than tolcapone because tolcapone has a risk of hepatotoxicity. The pharmacologic effects of the two are similar, and both are rapidly absorbed, bound to plasma proteins, and metabolized before excretion. Entacapone (200 mg) is taken up to six times daily with each dose of levodopa, while tolcapone is started at a standard dosage of 100 mg three times daily (twice that amount may be required). Opicapone is a newer long-acting COMT inhibitor that is taken once daily at bedtime in a 50-mg dose (25 mg daily in moderate liver impairment). When combined with levodopa, it reduces "off" time comparable to entacapone (Ferreira et al., 2016). While entacapone and opicapone are peripherally acting, tolcapone is both peripherally and centrally acting.

The COMT inhibitors' side effects, which include confusion, nausea, and dyskinesias, are partially linked to increased exposure to levodopa. Decreasing the dose of levodopa up to 30% may be required. Diarrhea, orthostatic hypotension, sleep disturbances, and an orange color of the urine are additional side effects. Tolcapone may cause liver damage.

#### *Amantadine*

Amantadine, an antiviral agent, has weak anti-parkinsonism properties. Its mechanism of action remains unclear. It has anticholinergic effects, inhibits N-methyl-D-aspartate (NMDA) glutamate receptors, and seems to change dopamine release in the striatum (Rascol et al., 2021). It has been reported to block adenosine $A_{2A}$ receptors. Patients on levodopa who experience dose-related fluctuations and dyskinesias might benefit from amantadine as an adjuvant. Formulations are available for use once or twice a day. Its side effects (dizziness, lethargy, sleep disturbance, nausea, anticholinergic effects) are mild and reversible. It may also cause livedo reticularis, impulse-control disorders, hallucinations, convulsions, and psychosis.

#### *Istradefylline*

Istradefylline is an analog of caffeine and is an adenosine $A_{2A}$ receptor antagonist like amantadine, and it is approved by the FDA for adjunctive treatment for "off" periods in PD in 2019. It is taken once daily orally to reduce "off" periods and improve motor function and is generally well tolerated (Mizuno & Kondo, 2013). The side effects include sleep disturbances, hallucinations, nausea, constipation, and dizziness.

### Muscarinic Receptor Antagonists

Antimuscarinic drugs used in the treatment of PD include benztropine mesylate, biperiden, and trihexyphenidyl (Aminoff, 2024). They have little effect on bradykinesia, but they may lessen parkinsonian tremor and rigidity. They are used in drug-induced parkinsonism. These drugs are generally used with dopamimetic therapy. Antimuscarinic drugs are poorly tolerated by the elderly and cognitively impaired, and they have several undesirable peripheral and central nervous system effects.

### Surgical or Lesioning Procedures

Deep brain stimulation (DBS) is used as surgical procedure for the treatment of PD. For the treatment of clinical fluctuations or dyskinesias that occur in moderate parkinsonism, stimulation of the globus pallidus or subthalamic nucleus using an implanted electrode and stimulator has produced positive results. DBS is effective in improving 'on' time without troubling dyskinesias, motor function, or quality of life in patients with advanced PD, but it has been associated with an increased risk of serious adverse events related to surgery (Weaver et al., 2009). Conventional thalamotomy and pallidotomy, conventional subthalamotomy, and magnetic resonance imaging-guided focused ultrasound are other lesioning procedures that are studied in advanced PD.

### Emerging and Investigational Therapies

Gene therapy studies are being conducted for the treatment of PD. The viruses as the vector for the genes were infused into some brain regions for glutamic acid decarboxylase, aromatic acid decarboxylase, tyrosine hydroxylase, GTP-cyclohydrolase 1, and neurturin. Although some of the results are promising, more research is required in this area.

Phase II clinical trials are taking place for monoclonal antibodies against α-synuclein. Additionally, a number of substances that block leucine-rich repeat kinase-2 (LRRK-2) activity are being studied in clinical settings. Clinical trials are also looking into compounds that either decrease the substrate of the protein encoded by the risk gene GBA, β-glucocerebrosidase, or increase its activity. Experimental stem cell treatments are under development. Autologous induced pluripotent stem cells (iPSCs) and dopamine neurons from human embryonic stem cells (hESCs) are also being investigated in PD for neural transplantation (Hallett et al., 2015; Kriks et al., 2011).

## Amyotrophic Lateral Sclerosis

Amyotrophic lateral sclerosis (ALS) is an incurable progressive neurodegenerative disorder that has a median survival of 3–5 years and causes muscle weakness, disability, and eventually death (Brown & Al-Chalabi, 2017). The incidence is 2.08 per 100,000 people and the prevalence is 5.40 per 100,000 people in Europe (Chiò et al., 2013). There is no curative treatment for ALS. Current treatments focus on slowing disease progression, alleviating symptoms, and improving quality of life. This section focuses on the disease-modifying treatment for ALS; therefore symptom-based management for ALS is not mentioned.

The FDA has approved three drugs for ALS until now: riluzole (1995), edaravone (2017), and tofersen (accelerated approval in 2023).

### Riluzole

Riluzole inhibits glutamate release; in addition, it blocks postsynaptic NMDA and kainate-type glutamate receptors and inhibits voltage-dependent $Na^+$ channels (Brunton et al., 2018). However, the mechanism of action in ALS is not clear. Clinical studies have shown that riluzole extends median survival by 3 months (Miller et al., 2012). Although the impact of riluzole on ALS is minimal, it represents a critical therapeutic advancement in the management of a disease that has not responded to any prior therapies. The recommended dose is 50 mg twice daily orally, and it is well tolerated. Hepatic transaminases can be elevated with riluzole treatment; therefore, liver function tests are monitored periodically (Bensimon et al., 2002). Nausea, oral hypoesthesia, dizziness, asthenia, neutropenia, and pulmonary disorders have been reported as the side effects of riluzole.

### Edaravone

Edaravone was the first drug for ALS to be approved following the FDA's approval of riluzole. It is a free-radical scavenger that is thought to reduce oxidative stress, although the precise mechanism of action is unknown. The benefits of edaravone have been demonstrated primarily in studies of patients with early-stage ALS (Writing Group; Edaravone (MCI-186) ALS 19 Study Group. 2017). It has intravenous and oral forms. The most common adverse reactions are injection-site contusion, gait disturbance, and headache.

### Tofersen

Pathologic genetic variants in superoxide dismutase-1 (SOD-1) lead to toxic protein misfolding and aggregation and are responsible for 15% of familial ALS cases (Abati et al., 2020). The antisense oligonucleotide tofersen was approved for use in patients with ALS associated with SOD1 pathologic variants. Slower declines in the Amyotrophic Lateral Sclerosis Functional Rating Scale – Revised (ALSFRS-R) score, vital capacity, and grip strength have been observed in patients treated with tofersen (Miller et al., 2022). The route of administration is intrathecal. Common side effects associated with the use of Tofersen include pain at the injection site, headache, fatigue, and arthralgias, whereas serious side effects include myelitis, radiculitis, aseptic meningitis, and elevated intracranial pressure.

### Emerging and Investigational Therapies

A histone deacetylase inhibitor called sodium phenylbutyrate lessens the endoplasmic reticulum's adaptive stress response. Taurursodiol, also called ursodoxicoltaurine, preserves mitochondrial integrity by lowering membrane permeability, which seems to raise the threshold of cellular apoptotic response. Sodium phenylbutyrate-taurursodiol (PB-TURSO), a coformulation of both drugs, is used to lessen the death of neurons. In a short-term trial, the ALSFRS-R score during a 24-week period showed that sodium phenylbutyrate-taurursodiol caused a slower functional decline than a placebo (Paganoni et al., 2020).

The Healey ALS Platform Trial is investigating the safety and efficacy of multiple investigational products, including CNM-Au8, pridopidine, ABBV-CLS-7262, and DNL343. Separately, other potential ALS treatments such as ibudilast (MN166), AKV9 (Nu-9), levosimendan, reldesemtiv, masitinib, and neurotrophic factor (NTF)-secreting mesenchymal stromal cells (MSC-NTF cells) are being studied in independent trials.

## Huntington's Disease

Huntington's disease (HD) is an inherited neurodegenerative disorder characterized by progressive cognitive impairment and motor incoordination, which often manifests as involuntary, jerk-like movements of the extremities (chorea), and personality changes (Ghosh & Tabrizi, 2018). The primary pathophysiological feature of HD is attributed to a trinucleotide expansion in the HTT gene. HD is marked by widespread neuronal loss, especially in the cortex and striatum, with early damage to medium spiny neurons that project to the lateral globus pallidus. Dopamine, glutamate, and gamma-aminobutyric acid (GABA) are the main neurotransmitters affected and the focus of pharmacotherapy.

Current treatments are primarily symptomatic and supportive, as no disease-modifying therapies exist to slow disease progression. Supportive interventions, such as physical therapy, occupational therapy, and home care can help improve the well-being of both patients and caregivers.

Available pharmacological therapies include VMAT2 inhibitors for the management of chorea, second-generation antipsychotics to alleviate both motor and psychiatric symptoms, and selective serotonin reuptake inhibitors (SSRIs) or tricyclic antidepressants to relieve anxiety and depression (Frank, 2014). Additionally, other drugs, such as NMDA receptor antagonists, antiepileptics, benzodiazepines, and cholinesterase inhibitors, may be used to address various symptoms (Table 3.3).

### VMAT2 Inhibitors

Vesicular monoamine transporter type 2 (VMAT2) inhibitors are recommended as first-line treatment for patients with chorea, except in those with depressive or suicidal symptoms, as these medications may exacerbate such conditions. For patients with depression, second-generation antipsychotics are generally the more suitable option. If a patient does not respond to VMAT2 inhibitors or antipsychotics, the American Academy of Neurology guidelines recommend considering alternative agents, such as amantadine and riluzole, alongside VMAT2 inhibitors.

VMAT2 inhibitors primarily function by decreasing dopaminergic neurotransmission. VMAT2 is a vesicular membrane protein essential for the storage and release of monoamines. Once neurotransmitters are synthesized in the presynaptic neuron, they are transported into vesicles, which later fuse with the cell membrane to release the neurotransmitters into the synaptic cleft. Inhibiting VMAT2 disrupts this process, leading to a presynaptic depletion of monoamines, especially dopamine.

TABLE 3.3

Pharmacological Treatments for Symptom Management in Huntington's Disease

| Symptom | Treatment Options | Mechanism of Action | Common Adverse Effects | Additional Considerations |
|---|---|---|---|---|
| **Motor Symptoms** | **VMAT2 Inhibitors** Tetrabenazine Deutetrabenazine | Inhibits vesicular monoamine transporter 2 (VMAT2), decreasing dopamine release | Fatigue, sedation, somnolence, insomnia, depression | May require CYP2D6 genotyping due to interactions affecting drug metabolism<br>Should be avoided in patients with depression due to the risk of worsening depression and suicidality |
| | **NMDA Receptor Antagonists** Amantadine | Blocks NMDA receptors, potentially reducing excitotoxicity | Insomnia, hallucinations, dizziness | Caution with elderly due to risk of confusion and hallucinations |
| | **Antiepileptics** Levetiracetam | Modulates synaptic neurotransmitter release, reducing irritability and seizures | Fatigue, dizziness, mood changes | Monitor for behavioral side effects |
| | **Antiepileptics** Topiramate | Blocks voltage-dependent sodium channels, enhances GABA activity, reduces glutamate | Cognitive dysfunction, weight loss, fatigue | Use caution with hepatic and renal impairment |
| | **Benzodiazepines** Clonazepam Lorazepam | Enhance GABAergic activity | Drowsiness, dependence, memory impairment | Used in the short term only for severe episodes of chorea |
| **Psychiatric Symptoms** | **Antipsychotics*** Olanzapine Risperidone Aripiprazole | Blocks dopamine receptors, managing aggression and psychosis | Weight gain, sedation, metabolic syndrome | May worsen motor symptoms; use the lowest effective dose |
| | **Antidepressants** Fluoxetine Sertraline Venlafaxine | Increases serotonin/norepinephrine levels to stabilize mood | Nausea, insomnia, sexual dysfunction | Monitor for increased risk of suicidality, especially in younger patients |
| **Cognitive Symptoms** | **Cholinesterase Inhibitors** Donepezil Rivastigmine Galantamine | Inhibits acetylcholinesterase to enhance cholinergic function | Nausea, diarrhea, bradycardia | Adverse effects in HD patients can be significant; monitor closely |

Three FDA-approved VMAT2 inhibitors are available: tetrabenazine, deutetrabenazine, and valbenazine. While these drugs exhibit similar efficacy, the choice often depends on availability and cost. If all three are accessible, deutetrabenazine and valbenazine may be preferred over tetrabenazine due to their longer half-lives, which make dosing more convenient (Wu et al., 2024).

### *Tetrabenazine*

The starting dose of tetrabenazine is 12.5 mg once daily. The dose is typically increased until there is an improvement in chorea or until the patient reaches their maximum tolerated dose, with most patients achieving optimal outcomes at a target dose of approximately 50 mg per day. Tetrabenazine has low and variable bioavailability, mainly due to its extensive first-pass metabolism. It is metabolized in the liver via CYP2D6 to its active metabolites. In cases of hepatic impairment, metabolism is significantly reduced, resulting in 7–190-fold higher plasma concentrations.

The manufacturer recommends CYP2D6 genotyping for patients requiring doses over 50 mg/day to determine if they are extensive metabolizers. In this group, the maximum dose may be increased to 100 mg/day. Concomitant use of CYP2D6 inhibitors (e.g., paroxetine, fluoxetine) may increase the drug's plasma concentration, requiring a 50% reduction in tetrabenazine dosage (Suchowersky, 2024). Common adverse effects include fatigue, sedation, somnolence, insomnia, depression, restlessness (akathisia), agitation, and nausea.

### *Deutetrabenazine*

Deutetrabenazine, a deuterated form of tetrabenazine, has a similar mechanism of action, adverse effect profile, and general pharmacokinetics. However, it offers a more favorable pharmacokinetic profile due to attenuated metabolism and an extended half-life, which enhances dosing convenience and may improve patient adherence.

The initial dose of deutetrabenazine is 6 mg twice daily, with weekly titration based on patient response and tolerance, up to a maximum dose of 48 mg/day. In patients taking CYP2D6 inhibitors, the maximum recommended dose is 36 mg/day (Suchowersky, 2024).

### *Valbenazine*

Valbenazine is a selective VMAT2 inhibitor with a prolonged half-life, allowing for once-daily dosing.

### Antipsychotics

Antipsychotics are used to alleviate both chorea and psychiatric symptoms (e.g., psychosis, agitation) by reducing dopaminergic neurotransmission. Antipsychotics are classified into two groups: first-generation antipsychotics (e.g., haloperidol, chlorpromazine), which primarily block dopaminergic receptors; and second-generation antipsychotics, which act on both dopaminergic and other receptors (e.g., serotonergic receptors; (Weickert & Goldberg, 2005). Due to the high incidence of adverse effects with first-generation antipsychotics, second-generation agents are generally preferred. In HD, risperidone, olanzapine, and aripiprazole are the most commonly used antipsychotics. Clozapine and quetiapine are generally not recommended due to their limited efficacy in managing chorea, except when used at very high doses. For severe cases of chorea, haloperidol and chlorpromazine may be considered if their potential therapeutic benefits justify the risk of adverse effects (Unti et al., 2017). Common adverse effects of antipsychotics include sedation, weight gain, parkinsonism, dystonia, akathisia, and hypotension.

## REFERENCES

Abati, E., Bresolin, N., Comi, G., & Corti, S. (2020). Silence superoxide dismutase 1 (SOD1): a promising therapeutic target for amyotrophic lateral sclerosis (ALS). *Expert Opin Ther Targets*, *24*(4), 295–310. https://doi.org/10.1080/14728222.2020.1738390

Aminoff, M. J. (2024). Pharmacologic Management of Parkinsonism & Other Movement Disorders. In T. W. Vanderah (Ed.), *Katzung's Basic & Clinical Pharmacology, 16th edition*. McGraw-Hill. accessmedicine.mhmedical.com/content.aspx?aid=1204140355

Balázs, N., Bereczki, D., & Kovács, T. (2021). Cholinesterase inhibitors and memantine for the treatment of Alzheimer and non-Alzheimer dementias. *Ideggyogy Sz*, *74*(11–12), 379–387. https://doi.org/10.18071/isz.74.0379

Bensimon, G., Lacomblez, L., Delumeau, J. C., Bejuit, R., Truffinet, P., & Meininger, V. (2002). A study of riluzole in the treatment of advanced stage or elderly patients with amyotrophic lateral sclerosis. *J Neurol*, *249*(5), 609–615. https://doi.org/10.1007/s004150200071

Brown, R. H., & Al-Chalabi, A. (2017). Amyotrophic lateral sclerosis. *N Engl J Med*, *377*(2), 162–172. https://doi.org/10.1056/NEJMra1603471

Brunton, L. L., Knollmann, B. C., & Hilal-Dandan, R. (2018). *Goodman & Gilman's the Pharmacological Basis of Therapeutics*. McGraw Hill Medical New York.

Chiò, A., Logroscino, G., Traynor, B. J., Collins, J., Simeone, J. C., Goldstein, L. A., & White, L. A. (2013). Global epidemiology of amyotrophic lateral sclerosis: a systematic review of the published literature. *Neuroepidemiology*, *41*(2), 118–130. https://doi.org/10.1159/000351153

Connolly, B. S., & Lang, A. E. (2014). Pharmacological treatment of Parkinson disease: a review. *Jama*, *311*(16), 1670–1683. https://doi.org/10.1001/jama.2014.3654

Espay, A. J., Kepp, K. P., & Herrup, K. (2024). Lecanemab and donanemab as therapies for Alzheimer's disease: an illustrated perspective on the data. *eNeuro*, *11*(7). https://doi.org/10.1523/eneuro.0319-23.2024

Ferreira, J. J., Lees, A., Rocha, J. F., Poewe, W., Rascol, O., & Soares-da-Silva, P. (2016). Opicapone as an adjunct to levodopa in patients with Parkinson's disease and end-of-dose motor fluctuations: a randomised, double-blind, controlled trial. *Lancet Neurol*, *15*(2), 154–165. https://doi.org/10.1016/s1474-4422(15)00336-1

Frank, S. (2014). Treatment of Huntington's disease. *Neurotherapeutics*, *11*(1), 153–160. https://doi.org/10.1007/s13311-013-0244-z

Gadhave, D. G., Sugandhi, V. V., Jha, S. K., Nangare, S. N., Gupta, G., Singh, S. K., Dua, K., Cho, H., Hansbro, P. M., & Paudel, K. R. (2024). Neurodegenerative disorders: mechanisms of degeneration and therapeutic approaches with their clinical relevance. *Ageing Res Rev*, *99*, 102357. https://doi.org/10.1016/j.arr.2024.102357

Gandhi, K. R., & Saadabadi, A. (2023). Levodopa (l-dopa). In *StatPearls*. StatPearls Publishing.

Ghosh, R., & Tabrizi, S. J. (2018). Clinical features of Huntington's disease. *Adv Exp Med Biol*, *1049*, 1–28. https://doi.org/10.1007/978-3-319-71779-1_1

Goldenberg, M. M. (2008). Medical management of Parkinson's disease. *P t*, *33*(10), 590–606.

Haddad, F., Sawalha, M., Khawaja, Y., Najjar, A., & Karaman, R. (2017). Dopamine and levodopa prodrugs for the treatment of Parkinson's disease. *Molecules*, *23*(1). https://doi.org/10.3390/molecules23010040

Hallett, P. J., Deleidi, M., Astradsson, A., Smith, G. A., Cooper, O., Osborn, T. M., Sundberg, M., Moore, M. A., Perez-Torres, E., Brownell, A. L., Schumacher, J. M., Spealman, R. D., & Isacson, O. (2015). Successful function of autologous iPSC-derived dopamine neurons following transplantation in a non-human primate model of Parkinson's disease. *Cell Stem Cell*, *16*(3), 269–274. https://doi.org/10.1016/j.stem.2015.01.018

Hampel, H., Elhage, A., Cho, M., Apostolova, L. G., Nicoll, J. A. R., & Atri, A. (2023). Amyloid-related imaging abnormalities (ARIA): radiological, biological and clinical characteristics. *Brain*, *146*(11), 4414–4424. https://doi.org/10.1093/brain/awad188

Hirsch, E. C., Jenner, P., & Przedborski, S. (2013). Pathogenesis of Parkinson's disease. *Mov Disord*, *28*(1), 24–30. https://doi.org/10.1002/mds.25032

Hirtz, D., Thurman, D. J., Gwinn-Hardy, K., Mohamed, M., Chaudhuri, A. R., & Zalutsky, R. (2007). How common are the "common" neurologic disorders? *Neurology*, *68*(5), 326–337. https://doi.org/10.1212/01.wnl.0000252807.38124.a3

Ives, N. J., Stowe, R. L., Marro, J., Counsell, C., Macleod, A., Clarke, C. E., Gray, R., & Wheatley, K. (2004). Monoamine oxidase type B inhibitors in early Parkinson's disease: meta-analysis of 17 randomised trials involving 3525 patients. *BMJ*, *329*(7466), 593. https://doi.org/10.1136/bmj.38184.606169.AE

Kriks, S., Shim, J. W., Piao, J., Ganat, Y. M., Wakeman, D. R., Xie, Z., Carrillo-Reid, L., Auyeung, G., Antonacci, C., Buch, A., Yang, L., Beal, M. F., Surmeier, D. J., Kordower, J. H., Tabar, V., & Studer, L. (2011). Dopamine neurons derived from human ES cells efficiently engraft in animal models of Parkinson's disease. *Nature*, *480*(7378), 547–551. https://doi.org/10.1038/nature10648

Kumar, A., Sidhu, J., Lui, F., & Tsao, J. W. (2024). Alzheimer Disease. In: *StatPearls*. StatPearls Publishing. https://www.ncbi.nlm.nih.gov/books/NBK499922/

Lamptey, R. N. L., Chaulagain, B., Trivedi, R., Gothwal, A., Layek, B., & Singh, J. (2022). A review of the common neurodegenerative disorders: current therapeutic approaches and the potential role of nanotherapeutics. *Int J Mol Sci*, *23*(3). https://doi.org/10.3390/ijms23031851

Miller, R. G., Mitchell, J. D., & Moore, D. H. (2012). Riluzole for amyotrophic lateral sclerosis (ALS)/motor neuron disease (MND). *Cochrane Database Syst Rev*, *2012*(3), Cd001447. https://doi.org/10.1002/14651858.CD001447.pub3

Miller, T. M., Cudkowicz, M. E., Genge, A., Shaw, P. J., Sobue, G., Bucelli, R. C., Chiò, A., Van Damme, P., Ludolph, A. C., Glass, J. D., Andrews, J. A., Babu, S., Benatar, M., McDermott, C. J., Cochrane, T., Chary, S., Chew, S., Zhu, H., Wu, F., … Fradette, S. (2022). Trial of antisense oligonucleotide tofersen for SOD1 ALS. *N Engl J Med*, *387*(12), 1099–1110. https://doi.org/10.1056/NEJMoa2204705

Mizuno, Y., & Kondo, T. (2013). Adenosine A2A receptor antagonist istradefylline reduces daily OFF time in Parkinson's disease. *Mov Disord*, *28*(8), 1138–1141. https://doi.org/10.1002/mds.25418

Olanow, C. W., Hauser, R. A., Jankovic, J., Langston, W., Lang, A., Poewe, W., Tolosa, E., Stocchi, F., Melamed, E., Eyal, E., & Rascol, O. (2008). A randomized, double-blind, placebo-controlled, delayed start study to assess rasagiline as a disease modifying therapy in Parkinson's disease (the ADAGIO study): rationale, design, and baseline characteristics. *Mov Disord*, *23*(15), 2194–2201. https://doi.org/10.1002/mds.22218

Ou, Z., Pan, J., Tang, S., Duan, D., Yu, D., Nong, H., & Wang, Z. (2021). Global trends in the incidence, prevalence, and years lived with disability of parkinson's disease in 204 countries/territories from 1990 to 2019. *Front Public Health*, *9*, 776847. https://doi.org/10.3389/fpubh.2021.776847

Paganoni, S., Macklin, E. A., Hendrix, S., Berry, J. D., Elliott, M. A., Maiser, S., Karam, C., Caress, J. B., Owegi, M. A., Quick, A., Wymer, J., Goutman, S. A., Heitzman, D., Heiman-Patterson, T., Jackson, C. E., Quinn, C., Rothstein, J. D., Kasarskis, E. J., Katz, J., … Cudkowicz, M. E. (2020). Trial of sodium phenylbutyrate-taurursodiol for amyotrophic lateral sclerosis. *N Engl J Med*, *383*(10), 919–930. https://doi.org/10.1056/NEJMoa1916945

Pahwa, R., Factor, S. A., Lyons, K. E., Ondo, W. G., Gronseth, G., Bronte-Stewart, H., Hallett, M., Miyasaki, J., Stevens, J., & Weiner, W. J. (2006). Practice parameter: treatment of Parkinson disease with motor fluctuations and dyskinesia (an evidence-based review): report of the Quality Standards Subcommittee of the American Academy of Neurology. *Neurology*, *66*(7), 983–995. https://doi.org/10.1212/01.wnl.0000215250.82576.87

Press, D., & Alexander, M. (2024). *Cholinesterase inhibitors in the treatment of dementia* (DeKosky, S.T, Schmader, K.E., & Wilterdink, J.L.). UpToDate Inc. https://www.uptodate.com

Press, D., & Buss, S. (2024). *Treatment of Alzheimer disease* (DeKosky, S.T, Schmader, K.E., & Wilterdink, J.L, Eds.). UpToDate Inc. https://www.uptodate.com.

Pringsheim, T., Day, G. S., Smith, D. B., Rae-Grant, A., Licking, N., Armstrong, M. J., de Bie, R. M. A., Roze, E., Miyasaki, J. M., Hauser, R. A., Espay, A. J., Martello, J. P., Gurwell, J. A., Billinghurst, L., Sullivan, K., Fitts, M. S., Cothros, N., Hall, D. A., Rafferty, M., … Lang, A. E. (2021). Dopaminergic therapy for motor symptoms in early Parkinson disease practice guideline summary: a report of the AAN guideline subcommittee. *Neurology*, *97*(20), 942–957. https://doi.org/10.1212/wnl.0000000000012868

Rascol, O., Fabbri, M., & Poewe, W. (2021). Amantadine in the treatment of Parkinson's disease and other movement disorders. *Lancet Neurol*, *20*(12), 1048–1056. https://doi.org/10.1016/s1474-4422(21)00249-0

Stowe, R. L., Ives, N. J., Clarke, C., van Hilten, J., Ferreira, J., Hawker, R. J., Shah, L., Wheatley, K., & Gray, R. (2008). Dopamine agonist therapy in early Parkinson's disease. *Cochrane Database Syst Rev,* (2), Cd006564. https://doi.org/10.1002/14651858.CD006564.pub2

Suchowersky, O. (2024). *Huntington disease: management* (Hurtig, H.I., Eichler A.FEds.). UpToDate Inc. https://www.uptodate.com

Sveinbjornsdottir, S. (2016). The clinical symptoms of Parkinson's disease. *J Neurochem*, *139*(Suppl 1), 318–324. https://doi.org/10.1111/jnc.13691

Unti, E., Mazzucchi, S., Palermo, G., Bonuccelli, U., & Ceravolo, R. (2017). Antipsychotic drugs in Huntington's disease. *Expert Rev Neurother*, *17*(3), 227–237. https://doi.org/10.1080/14737175.2016.1226134

von Campenhausen, S., Bornschein, B., Wick, R., Bötzel, K., Sampaio, C., Poewe, W., Oertel, W., Siebert, U., Berger, K., & Dodel, R. (2005). Prevalence and incidence of Parkinson's disease in Europe. *Eur Neuropsychopharmacol*, *15*(4), 473–490. https://doi.org/10.1016/j.euroneuro.2005.04.007

Voon, V., Napier, T. C., Frank, M. J., Sgambato-Faure, V., Grace, A. A., Rodriguez-Oroz, M., Obeso, J., Bezard, E., & Fernagut, P. O. (2017). Impulse control disorders and levodopa-induced dyskinesias in Parkinson's disease: an update. *Lancet Neurol*, *16*(3), 238–250. https://doi.org/10.1016/s1474-4422(17)30004-2

Weaver, F. M., Follett, K., Stern, M., Hur, K., Harris, C., Marks, W. J., Jr., Rothlind, J., Sagher, O., Reda, D., Moy, C. S., Pahwa, R., Burchiel, K., Hogarth, P., Lai, E. C., Duda, J. E., Holloway, K., Samii, A., Horn, S., Bronstein, J., … Huang, G. D. (2009). Bilateral deep brain stimulation vs best medical therapy for patients with advanced Parkinson disease: a randomized controlled trial. *Jama*, *301*(1), 63–73. https://doi.org/10.1001/jama.2008.929

Weickert, T. W., & Goldberg, T. E. (2005). First- and second-generation antipsychotic medication and cognitive processing in schizophrenia. *Curr Psychiatry Rep*, *7*(4), 304–310. https://doi.org/10.1007/s11920-005-0085-5

Wilson, D. M., Cookson, M. R., Van Den Bosch, L., Zetterberg, H., Holtzman, D. M., & Dewachter, I. (2023). Hallmarks of neurodegenerative diseases. *Cell*, *186*(4), 693–714. https://doi.org/10.1016/j.cell.2022.12.032

Writing Group; Edaravone (MCI-186) ALS 19. (2017). Safety and efficacy of edaravone in well defined patients with amyotrophic lateral sclerosis: a randomised, double-blind, placebo-controlled trial. *Lancet Neurol*, *16*(7), 505–512. https://doi.org/10.1016/s1474-4422(17)30115-1

Wu, D., Chen, Q., Yu, Z., Huang, B., Zhao, J., Wang, Y., Su, J., Zhou, F., Yan, R., Li, N., Zhao, Y., & Jiang, D. (2024). Transport and inhibition mechanisms of human VMAT2. *Nature*, *626*(7998), 427–434. https://doi.org/10.1038/s41586-023-06926-4

Zahoor, I., Shafi, A., & Haq, E. (2018). Pharmacological treatment of Parkinson's disease. *Exon Publications*, 129-144.

Zhang, Y., Chen, H., Li, R., Sterling, K., & Song, W. (2023). Amyloid β-based therapy for Alzheimer's disease: challenges, successes and future. *Signal Transduct Target Ther*, *8*(1), 248. https://doi.org/10.1038/s41392-023-01484-7

# Section II

# Zebrafish Model Organism to Study Neurodegenerative Disorders

# 4

# *Zebrafish Model of Neurodegenerative Disorders*

**Nitasha Rana, Palak Ahuja, Charan Singh and Arti Singh**

## Abbreviations

| | |
|---|---|
| **3-NP** | 3-Nitropropionic Acid |
| **6-OHDA** | 6-Hydroxydopamine |
| **AD** | Alzheimer's Disease |
| **ALS** | Amyotrophic Lateral Sclerosis |
| **ALSFRS-R** | Amyotrophic Lateral Sclerosis Functional Rating Scale |
| **AOAA** | Aminooxy Acetic Acid |
| **APP** | Amyloid Precursor Protein |
| **ASO** | Antisense Oligonucleotide |
| **ATP** | Adenosine Triphosphate |
| **BBB** | Blood Brain Barrier |
| **BDNF** | Brain-Derived Neurotrophic Factor |
| **CAG** | Cytosine-Adenine-Guanine |
| **Cas** | CRISPR-Associated System |
| **CNS** | Central Nervous System |
| **COMT** | Catechol O-methyltransferase |
| **CRISPR** | Clustered Regularly Interspaced Short Palindromic Repeats |
| **DA** | Dopamine |
| **DALYs** | Disability Adjusted Life Years |
| **DAT** | Dopamine Transporter |
| **DJ-1** | Parkinsonism Associated Deglycase |
| **DPRs** | Dipeptide Repeat Proteins |
| **DsRed** | Discosoma Red Fluorescent Protein |
| **EAE** | Experimental Autoimmune Encephalomyelitis |
| **EB** | Ethidium Bromide |
| **EBV** | Epstein-Barr Virus |
| **FTDP-17** | Frontotemporal Dementia and Parkinsonism Linked to Chromosome 17 |
| **FUS** | Fused in Sarkoma |
| **GFAP** | Glial Fibrillary Acidic Protein |
| **GFP** | Green Fluorescent Protein |
| **Gln-Glu** | Glutamine-Glutamate |
| **Glu** | Glutamate |
| **GoF** | Gain-of-Function |
| **GSK-3β** | Glycogen Synthase Kinase 3 Beta |
| **HD** | Huntington's Disease |
| **Htt** | Huntingtin Gene |
| **HRE** | Hexanucleotide Repeat Expansion |
| **KA** | Kainic Acid |
| **LMN** | Lower Motor Neurons |
| **LoF** | Loss-of-Function |
| **LRRK2** | Leucine-Rich Repeated Kinase 2 |
| **MAO-A/B** | Monoamine Oxidase A/B |
| **MHC** | Major Histocompatibility Complex |
| **Mhtt** | Mutant Huntingtin Protein |
| **MOG35–55** | Myelin Oligodendrocyte Glycoprotein |
| **MPP+** | 1-Methyl-4-Phenylpyridinium |
| **MPTP** | 1-Methyl-4phenyl-1,2,3,6-Tetrahydropyridine |
| **MS** | Multiple Sclerosis |
| **mTOR** | The Mammalian Target Of Rapamycin |
| **MCI** | Mitochondrial Complex I |
| **MTR** | Metronidazole |
| **NADPH** | Nicotinamide Adenine Dinucleotide Phosphate Hydrogen |
| **NDD** | Neurodegenerative Disease |
| **NFTs** | Neurofibrillary Tangles |
| **NMDA** | N-Methyl-D-Aspartate |
| **NMJ** | Neuromuscular Junction |
| **NTR** | Nitroreductase |
| **OA** | Okadaic Acid |
| **ORF** | Open Reading Frame |
| **OTA** | Ochratoxin A |
| **PARKIN** | E3 Ubiquitin-Protein Ligase |
| **PART** | Primary Age-Related Tauopathy |
| **PD** | Parkinson's Disease |
| **PINK 1** | PTEN Induced Kinase 1 |
| **PNS** | Peripheral Nervous System |
| **PP1** | Protein Phosphatase 1 |
| **PSEN** | Presenilin |
| **QA** | Quinolinic Acid |
| **RA** | Rheumatoid Arthritis |
| **RMS** | Relapsing Multiple Sclerosis |
| **ROS** | Reactive Oxygen Species |
| **RRMS** | Relapsing-Remitting Multiple Sclerosis |
| **SDH** | Succinate Dehydrogenase |
| **SG** | Stress Granule |
| **SMA** | Spinal Muscular Atrophy |
| **SN** | Substantia Nigra |
| **SNc** | Substantia Nigra Pars Compacta |
| **SOD1** | Superoxide Dismutase Type 1 |
| **SPMS** | Secondary Progressive Multiple Sclerosis |
| **TALEN** | Transcription Activator-Like Effector Nuclease |
| **TDP-43** | Transactive Response DNA Binding Protein |
| **UMN** | Upper Motor Neurons |
| **UVB** | Ultraviolet B |
| **vD** | Vitamin D |
| **WT** | Wilms Tumor Protein |
| **ZFNs** | Zinc Finger Nucleases |
| **ZFP** | Zinc Finger Protein |

DOI: 10.1201/9781032676555-6

## Introduction

Neurodegenerative diseases (NDDs) are a diverse set of neurological disorders that have a negative impact on the lives of millions of individuals globally. These diseases involve the gradual deterioration of neurons in either the central nervous system (CNS) or the peripheral nervous system (PNS) (Wilson et al., 2023). Neurodegenerative disorders pose a significant risk to human well-being. The prevalence of these age-related illnesses is rising due to the recent growth in the older population (Wilson et al., 2023). The deterioration of neural networks and the inability of neurons to regenerate effectively due to their permanently specialized biological terms lead to the disruption of the fundamental communication pathways, ultimately resulting in impaired memory, cognition, behavior, and sensory and/or motor function (Wilson et al., 2023). Neurodegenerative disorders include Alzheimer's disease (AD), Parkinson's disease (PD), Huntington's disease (HD), amyotrophic lateral sclerosis (ALS), multiple sclerosis (MS), frontotemporal dementia, and spinocerebellar ataxias. These disorders exhibit a wide range of pathophysiological characteristics, with some leading to memory and cognitive problems, while others impact a person's motor skills, speech, and respiration (Abeliovich & Gitler, 2016; Canter et al., 2016; Taylor et al., 2016; Wyss-Coray, 2016). Neurodegeneration varies based on the specific brain regions (frontotemporal lobes, extrapyramidal system, spinal nerve pathways) where degenerative processes take place, in addition to the clinical manifestations (Dugger & Dickson, 2017). Although neurodegenerative diseases share common pathogenesis features, there remain numerous unresolved intricacies that influence the onset and advancement of neuronal damage. These complexities perplex researchers and doctors, making the search for new therapeutic targets for these diseases even more challenging. A crucial concern involves the alterations that take place at the cellular level and dictate the pathological process, namely the impairment of specific organelles inside the cell (Blagov et al., 2023). The optimal functioning of mitochondria is crucial for neurons, as the transmission of nerve impulses is an energy-intensive process that necessitates the production of substantial amounts of adenosine triphosphate (ATP) (Blagov et al., 2023).

## Prevalence of Neurodegenerative Diseases

Presently, the global population of individuals with dementia exceeds 55 million, with over 60% of them residing in low- and middle-income nations. Annually, there is an approximate total of 10 million newly reported cases (World Health Organization (WHO), 2022). Among the NDD, PD has the highest global prevalence rate at 40%, followed by AD, ALS, HD, and MS (26%, 16%, 10%, and 8%) (Figure 4.1). The incidence of PD has increased twofold during the previous 25 years. In 2019, global estimates indicated that there were more than 8.5 million individuals diagnosed with PD. In 2019, it is estimated that PD caused 5.8 million disability-adjusted life years (DALYs), which is an 81% increase compared to the year 2000. Additionally, PD resulted in 329,000 fatalities, which is almost a 100% increase since 2000 (WHO, 2022). They calculated global HD prevalence as 2.7 per 100,000. An updated systematic review and meta-analysis of 2010–2022 literature was released in 2022. They used 13 of 18 incidence studies and 23 of 27 prevalence studies for the meta-analysis. Global HD incidence grew to 0.48 cases per 100,000 person-years and prevalence to 4.88 per 100,000 (Medina et al., 2022). It is estimated that more than 1.8 million people around the world

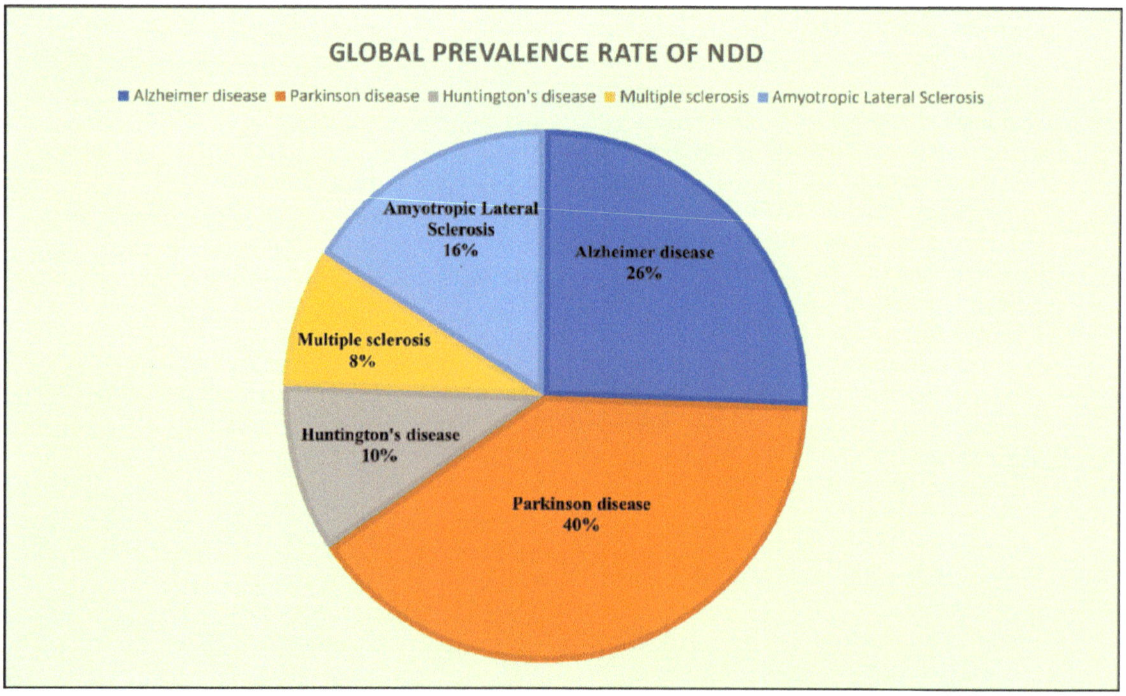


**FIGURE 4.1** Prevalence rate of NDD.

have multiple sclerosis. Although it can affect people of any age, it is more prevalent in young adults and females than in any other age group (WHO, 2022). The prevalence and incidence of ALS were found to be 4.42 (95% confidence interval (CI) 3.92–4.96) per one million people and 1.59 (95% CI 1.39–1.81) per one million person-years, respectively, throughout the entire world using crude data. After the age of 70–79, the prevalence and incidence of ALS climbed (Xu et al., 2020).

## Pathophysiology of Neurodegenerative Diseases

Even though different neurodegenerative illnesses manifest themselves in different parts of the brain and have different causes, it has been found that these diseases may be influenced by the same cellular and molecular functioning (Teleanu et al., 2022). The presence of oxidative stress, which is caused by the uncontrollable generation of reactive oxygen species (ROS), has been observed to be a characteristic that is common to neurodegenerative disorders (Barnham et al., 2004; Onose et al., 2022). A phenomenon known as excitotoxicity describes glutamate's high level of neurotoxicity, even though it plays a widespread role as a neurotransmitter (Rang et al., 2011). It is the mitochondria that are responsible for energy metabolism, and their failure to function properly might result in oxidative stress. Damage to mitochondria, which results in the release of cytochrome c into the cytosol, is also responsible for the beginning of the process of apoptosis. For neurons to survive, mitochondrial integrity is necessary, and mitochondrial dysfunction is considered to be a significant contributor to a wide variety of neurodegenerative illnesses (Rang et al., 2011) (Figure 4.2).

### Oxidative Stress

Because of the brain's high oxygen utilization, poorer antioxidant enzymes, and high amount of readily oxidized polyunsaturated fatty acids, as well as the terminal-differentiation characteristic of neurons, the brain is particularly susceptible to oxidative stress as compared to the other organs in the body for many reasons (Li et al., 2013). The results of endogenous and external processes that involve oxygen and nitrogen are assumed to be the compounds that make up the biological oxidants that are responsible for oxidative damage. Both aerobic respiration and cellular metabolism, as well as the defense against pathogens, lead to the production of reactive species that include oxygen (Simonian & Coyle, 1996). Oxidative stress is defined as the breakdown of balance between pro-oxidant and antioxidant levels. High quantities of ROS can lead to cell death or oxidative stress because of their reactivity (Liu et al., 2017). On the other hand, ionizing radiation and medications that exploit ROS formation as a mode of action are examples of exogenous sources of ROS production. Furthermore, ROS can be formed as by-products as a result of the metabolism of natural substances in the environment. The endogenous sources, on the other hand, are either mitochondrial or non-mitochondrial enzymes that are responsible for manufacturing ROS (Teleanu et al., 2022). In addition, the organism's metalloenzymes utilize the interaction between oxygen and metal ions to activate molecular oxygen as ROS with the resulting free radicals being essential elements of a healthy metabolism. Nevertheless, since ROS are also harmful, cells have evolved intricate mechanisms to control metal ion interactions and the production of ROS. Consequently, when regulatory processes malfunction, the characteristics that cells employ for beneficial functions become harmful (Barnham et al., 2004). Substantial data substantiates a correlation between oxidative stress and the onset of neurodegenerative disorders. The elevated creation of ROS by neurons and the subsequent buildup of oxidative damage that accompanies aging are strongly associated with the degree of neurodegeneration (Niedzielska et al., 2016).

### Excitotoxicity

Glutamatergic neurons are the primary excitatory system in the brain and have a crucial role in various neurophysiological functions. Glutamate is the principal neurotransmitter

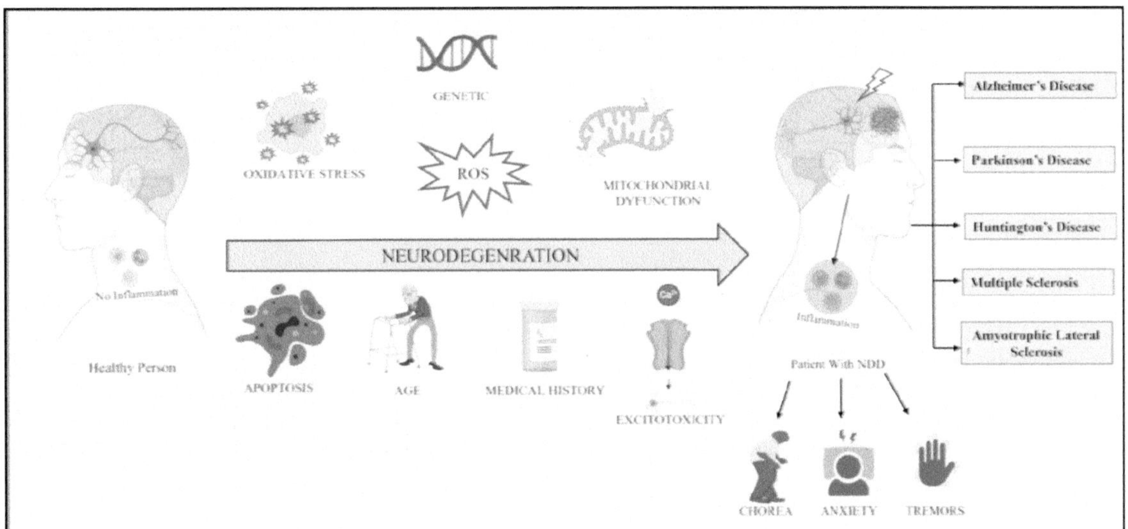


**FIGURE 4.2** Pathophysiology of NDD.

responsible for perception and cognition in the brain, typically inducing an excitatory response. This response is formed through the interaction of glutamate with receptors that make up cation channels. Overstimulation of glutamate receptors can lead to impaired functioning and death of neurons, a phenomenon known as excitotoxicity (Dong et al., 2009). Excessive accumulation of calcium is the primary cause of excitotoxicity. The processes by which this phenomenon takes place and ultimately results in the demise of cells are outlined below (Rang et al., 2011). Excessive activation of both ionotropic and metabotropic glutamate receptors has been definitively linked to the damage to nerve cells seen in various neurodegenerative conditions, such as AD, HD, and ALS (Rang et al., 2011). Glutamate and $Ca^{2+}$ are widely present chemical signals that play a crucial role in brain activity, both outside and inside cells. However, it is concerning that when these signals become unregulated, they can cause significant damage to cells. The energy metabolism of mitochondria serves as a protective mechanism. However, when mitochondrial activity is compromised, it makes neurons susceptible to excitotoxic damage. This can contribute to the development of many neurodegenerative disorders, such as PD (Rang et al., 2011). The entry of calcium ions ($Ca^{2+}$) triggers excitatory processes that involve the production of free radicals, impairment of mitochondrial function, and activation of several enzymes, including those responsible for the production and breakdown of arachidonic acid. The enzymes mentioned consist of several isoforms of phospholipase A2, cyclooxygenase-2, and lipoxygenases (Dong et al., 2009). In addition, dysfunctional mitochondrial activity might result in the release of cytochrome c, which plays a crucial role in initiating apoptosis. The involvement of excitotoxicity in ischemic brain damage is widely recognized, and it is also considered to contribute to the development of other neurodegenerative disorders (Rang et al., 2011).

### Apoptosis

Apoptosis can be triggered by diverse cell surface stimuli. The cell undergoes systematic disassembly, and the smaller remnants are eliminated by macrophages without inducing inflammation. Apoptotic cells can be distinguished by a staining approach that detects the distinctive DNA fractures. Apoptosis can be triggered by many signaling pathways, but in all instances, the ultimate pathway leading to cell death involves the activation of a group of proteases called caspases, which disable different proteins within the cell (Rang et al., 2011). So far, scientists have discovered and designated 10 distinct caspases, which are referred to as caspases 1–10. Caspases are produced as inert precursors. Multiple caspases, specifically caspase 3, have been demonstrated to be activated in neuronal death models (Armstrong et al., 1997; Yansheng Du et al., 1997; Yangshen Du et al., 1997; Hara et al., 1997; McCarthy et al., 1997). Many researchers have observed the presence of oxidative stress in conjunction with neuronal death, which is worth mentioning. While it is true that an increase in ROS generation can trigger apoptosis, there is currently no conclusive data that directly connects these two processes in the neurodegenerative process. In the subsequent sections, we have condensed the information about extensively researched neurodegenerative disorders such as AD, PD, ALS, spinal muscular atrophy (SMA), HD, and diabetic encephalopathy. In these diseases, there is compelling evidence indicating the role of apoptosis in the death of neurons (Okouchi et al., 2007). Neuronal apoptosis is typically inhibited by neuronal growth factors, such as nerve growth factor and brain-derived neurotrophic factor. These proteins are released and necessary for the survival of several groups of neurons in the central nervous system. The growth factors control the expression of two gene products, Bax and Bcl-2. Bax promotes apoptosis, whereas Bcl-2 prevents apoptosis. Inhibiting apoptosis by targeting specific locations along these pathways is a promising approach for the development of neuroprotective medications, although it has not yet yielded positive results (Rang et al., 2011).

## Importance of Animal Models in Research

To study and improve human biology and health, non-human animals have long been used in research (Phillips & Roth, 2019). Animal models mirror numerous human diseases, including systemic autoimmune diseases, rheumatoid arthritis (RA), epilepsy, Alzheimer's disease, cardiovascular diseases, atherosclerosis, diabetes, and others (Mukherjee et al., 2022). In human studies, it is hard to tell how varied life circumstances affect physiological, behavioral, and health results. However, animal models allow researchers to precisely regulate all aspects that the animal is exposed to and assess their effects individually or simultaneously (Phillips & Roth, 2019). Pre-clinical trials with animal models have two stages: first, if the drug shows desirable efficacy, further studies are done; second, if the drug is safe, it is given to small human volunteer groups, while the animal trial evaluates its long-term effects (Badyal & Desai, 2014; Van Dam & De Deyn, 2011). Animal models are traditionally employed in drug discovery to establish and substantiate non-clinical "proof-of-concept" about the safety, effectiveness, and target of interest for particular therapeutic compounds (Singh & Seed, 2021).

## Importance of the Zebrafish Model in Research

The selection process for an animal model in biomedical research is highly complex, as not all models are suitable due to multiple constraints. Several variables must be carefully considered when choosing an optimal animal model for biomedical experiments. The primary criteria involve selecting models that closely resemble both animal species and humans in terms of physiological and/or pathological features (Mukherjee et al., 2022). The rapid progress of clustered regularly interspaced short palindromic repeats (CRISPR) and next-generation sequencing technologies has greatly expedited disease modeling in zebrafish, leading to a deeper comprehension of the molecular pathways behind human genetic illnesses (Choi et al., 2021). Zebrafish are presently regarded as a suitable model organism for studying development, genetics, immunology, behavior, physiology, and nutrition. Scientists have utilized zebrafish to discover novel medicines for the treatment and prevention of several significant human diseases (Teame et al., 2019). Currently, zebrafish are becoming increasingly recognized as a prospective model for PD because of the occurrence of nigrostriatal-like neuronal injury in zebrafish caused by neurotoxins, which leads to PD-like motor dysfunctions comparable to those shown in mammalian models (Saluja et al., 2021).

## Models of Disease

### Parkinson's Disease

PD is a complex, advancing neurological disorder marked by tremors, stiffness, and slow movement (Kouli et al., 2018). The etiology of PD remains uncertain; however, both genetic and environmental variables are deemed significant. Additionally, some potential association locations that could impact the development of PD have been reported (Schapira, 2009). Several studies propose that the beginning and course of PD involve various additional mechanisms, including aberrant protein clearance, mitochondrial failure, and neuroinflammation (Kouli et al., 2023). Treatments that target dopamine are generally effective in alleviating early motor symptoms (Armstrong & Okun, 2020).

#### *Genetic Models of Parkinson's Disease*

One of the initial techniques for generating models was subjecting zebrafish to mutagenic chemicals and subsequently conducting thorough phenotyping to detect fish with desired mutations (Kelsh, 2023; Li et al., 2023). Zinc-finger nucleases (ZFNs) are an alternative approach to gene inactivation, involving the use of restriction enzymes that specifically cleave a targeted segment of DNA. ZFNs exhibit high precision in selectively targeting particular nucleotide sequences. However, occasional mutations may arise at the site of repair. Moreover, this procedure is costly and can pose challenges in terms of its design (Park & Kim, 2023; Saraswat et al., 2023).

##### *E3 Ubiquitin-protein Ligase (PARKIN, PARK2)*

Parkin mutations are responsible for the predominant autosomal recessive variant of PD. Over 100 purported pathogenic mutations have been documented in the parkin gene, the majority of which result in protein dysfunction and typically trigger early-onset PD. Parkin functions as an E3 ubiquitin ligase and plays a role in the ubiquitin-proteasome system. Furthermore, parkin plays a direct role in maintaining the quality of mitochondria when combined with PINK1 (Qi et al., 2023). The Parkin gene encodes an E3 ubiquitin ligase that participates in the proteasome degradation pathway and may have a role in mitochondrial function, alongside the PD-related genes PINK1 and DJ-1 (Yamanaka & Matsui, 2023). The zebrafish parkin gene exhibits a 62% resemblance to its human counterpart and is expressed widely in both larval and adult zebrafish (Yamanaka & Matsui, 2023). The zebrafish parkin gene is active both during the early stages of development and in mature tissues of adult individuals (Paduraru et al., 2023). The zebrafish parkin gene exhibits a conservation rate exceeding 62% with its homolog in humans. It encodes proteins consisting of 458 amino acids, which share functional similarities with human proteins in 92% of their regions (Zhu et al., 2023). A study conducted on zebrafish demonstrated that the number of dopaminergic neurons reduced by 25% within 3 days due to the successful knockdown of parkin morpholino. Additionally, when the zebrafish were exposed to 1-methyl-4-phenylpyridinium ($MPP^+$), a 50% reduction in dopaminergic neurons was observed compared to the control embryos (Cansiz et al., 2023). Conversely, a different method of reducing the expression of parkin in zebrafish did not result in the loss of dopaminergic neurons but did lead to an elevated vulnerability to cell death caused by stress. The morphology of the mitochondria remained unaffected after the temporary suppression of the parkin gene; however, electron density was seen in the t-tubules. T-tubules contain a high concentration of L-type $Ca^{2+}$ channels, which have recently been discovered to contribute to the development of sporadic PD (Bandmann & Burton, 2010).

##### *PTEN-induced Kinase 1 (PINK 1, PARK6)*

PINK1 mutations are the second most prevalent factor contributing to the occurrence of autosomal recessive PD at an early age. The PINK1 gene encodes a protein consisting of 581 amino acids. This protein is typically found with a mitochondrial targeting motif at its N-terminal and a conserved serine-threonine kinase domain (Gandhi et al., 2006; Xi et al., 2011). The zebrafish pink1 gene, which shares 54% similarity with the human PINK1 gene, is expressed throughout the body of zebrafish larvae. However, in adult zebrafish, pink1 is only present in specific areas surrounding the ventricles and in certain dopaminergic neurons in the diencephalon region of the brain (Anichtchik et al., 2008). Neurodegeneration of dopaminergic neurons has been observed in zebrafish with mutant pink1 genes, specifically 5 days after fertilization (Flinn et al., 2013). The fish with reduced expression of the pink1 gene exhibited a notable decrease in swimming activity and a more pronounced reduction in the number of neurons that produce tyrosine hydroxylase (TH), as compared to the control group treated alone to MPTP. Tetrahydrobiopterin is the enzyme that restricts the rate of production in the synthetic pathway for dopamine. It is commonly employed to distinguish dopaminergic cells (Doyle & Croll, 2022). Temporary inhibition of pink1 has been demonstrated to induce mitochondrial impairment, including reduced crista and mitochondria abundance. Studies have indicated that zebrafish with reduced pink1 expression exhibit a strong susceptibility of dopaminergic neuron clusters to MPTP (Sallinen et al., 2010). According to a study, no decrease in the dopaminergic neuron population of the ventral diencephalon was found in zebrafish after transitory silencing of the pink1 gene (Xi et al., 2010).

##### *Parkinsonism-associated Deglycase (DJ-1, PARK7)*

DJ-1 mutations infrequently result in the development of early onset autosomal recessive PD. DJ-1, belonging to the THIJ/PFPL/DJ-1 protein family, serves multiple roles, such as acting as a redox-sensitive chaperone and safeguarding mitochondria from oxidative damage (Xi et al., 2011). The zebrafish ortholog of the dj-1 gene codes for a protein consisting of 189 amino acids, which shares 83% similarity with the human DJ-1 protein. It is predominantly found in the central nervous system and is also found in the intestine and adult tissues throughout early development (Bai et al., 2006). The protein sequences of human (BC008188) and mouse (BC002187) DJ-1 were compared to the zebrafish genome using the Danio rerio Blast Server, which may be accessed at http://www.sanger.ac.uk/cgi-bin/blast/submitblast/d_rerio. The genomic sequence of DJ-1 in Danio rerio was discovered in the contig zkp123Hii.00426. Furthermore, DJ-1 has undergone complete annotation in Ensembl, which may be accessed at http://www.ensembl.org/Danio_rerio/index.html. To validate the in-silico results in vitro, the complete coding region of DJ-1 (570 base pairs) was amplified and sequenced utilizing the specified primers: The forward primer sequence is 5'-TGCAGACACGCACAGG TTAT-3', while the reverse primer sequence is 5'-AGCGGT-GCGATTGTAAATG-3' (Bretaud et al., 2007). Multiple studies have demonstrated that inhibiting the expression of dj-1 has resulted in a decrease in dopaminergic neurons in the striatum. However, no neuronal loss has been detected in the substantia nigra (Ünal & Emekli-Alturfan, 2019).

#### *Leucine-rich Repeat Kinase 2 (LRRK2, PARK8)*

LRRK2 mutations are the primary cause of autosomal dominant PD (Ünal & Emekli-Alturfan, 2019). The LRRK2 gene codes for a substantial, 2,527-amino acid, a 286-kDa protein with several domains, which is classified as a member of the ROCO family (Seegobin et al., 2020). The amino acid sequences and predicted structures of zebrafish and human LRRK2 are largely conserved. The zebrafish gene is very prevalent in the brain throughout embryonic development and is present in numerous tissues and organs of the adult zebrafish (Ünal & Emekli-Alturfan, 2019). Zebrafish, similar to Drosophila and C. elegans, have a single ortholog of the human LRRK2 gene, known as zLRRK2. This zLRRK2 gene shares a 47% sequence identity in terms of amino acids with the human LRRK2 gene (Sheng et al., 2010) (Figure 4.3).

### Chemical Induced Models of Parkinson's Disease

Research has discovered that certain chemical compounds, such as 6-OHDA and MPTP, can specifically damage dopaminergic neurons and cause symptoms similar to PD (Zeng et al., 2018). Neurotoxins previously employed to generate PD in rats are now being applied to zebrafish, yielding favorable outcomes. This section explains the neurotoxins commonly employed to produce PD in zebrafish (Ünal & Emekli-Alturfan, 2019).

#### *1-Methyl-4-phenyl-1,2,3,6-tetrahydropyridine (MPTP)*

The initial observation revealed that MPTP caused severe parkinsonism in four individuals who were intravenously administered an illegal substance containing the neurotoxic (Zeng et al., 2018). MPTP, a derivative resulting from the chemical production of a meperidine analog similar to heroin, is the primary and commonly employed hazardous model in animal investigations (Ünal & Emekli-Alturfan, 2019). According to a study, MPTP hindered mitochondrial complex I (MCI) and triggered the activation of the sncga gene, which is accountable for the buildup of gamma synuclein in 24-hour post-fertilization zebrafish. In addition, MPTP suppressed the expression of parkin/pink1/dj-1, resulting in the impairment of the normal

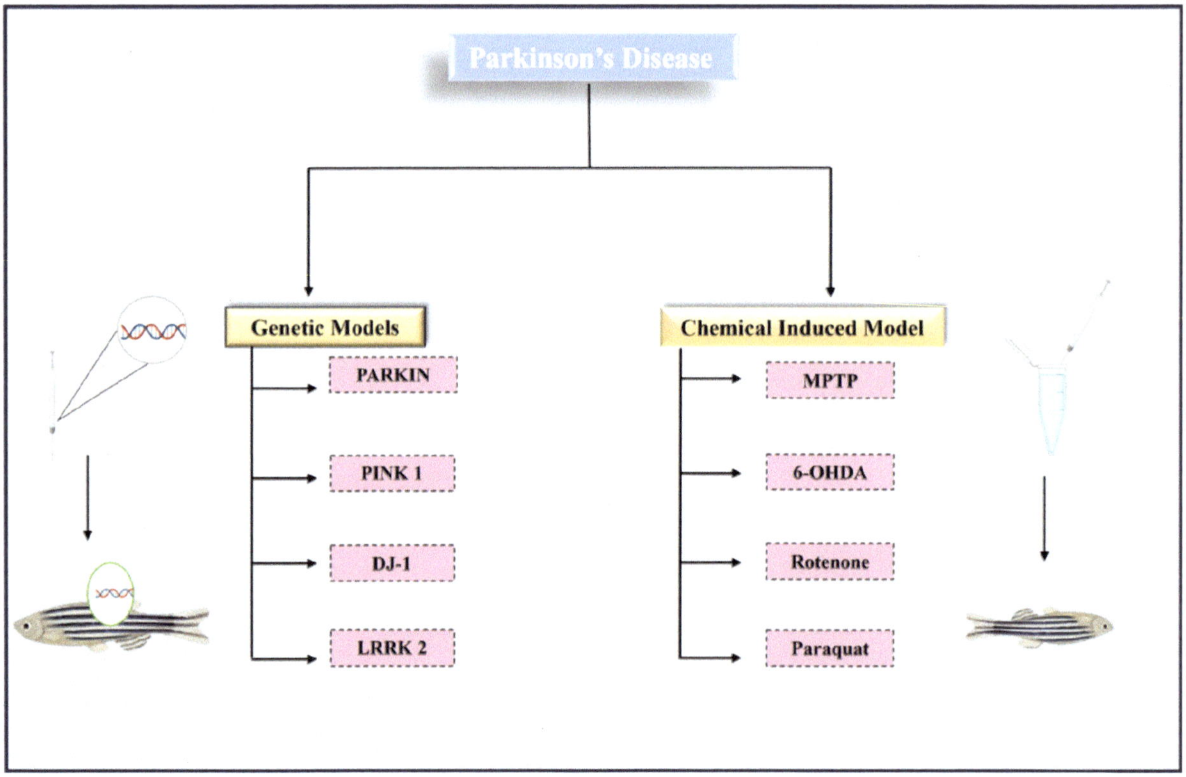


**FIGURE 4.3** Parkinson's models.

function of the parkin/pink1/dj-1 system in eliminating damaged mitochondria (Ünal & Emekli-Alturfan, 2019). MPTP can increase glutamate (Glu) levels and critical enzymes in the glutamine–glutamate (Gln–Glu) cycle in the striatum. These changes are associated with the death of neurons and impairment of motor function (Lu et al., 2018). The neurotoxicity caused by MPTP has been demonstrated to be suppressed by the monoamine oxidase-B inhibitor l-deprenyl, dopamine transporter (DAT) inhibitor nomifensine, or by inactivating DAT using morpholinos in zebrafish larvae. Furthermore, the impact of MPTP on swimming behavior and touch responsiveness in zebrafish larvae was also improved by the administration of l-deprenyl and the suppression of DAT (Ünal & Emekli-Alturfan, 2019). The study found that $MPP^+$ decreased the quantity of posterior tubercular neurons and locomotor activity. However, paenolum was able to prevent neurodegeneration and locomotor dysfunction in this particular paradigm (Lu et al., 2015).

#### *6-Hydroxydopamine (6-OHDA)*

An animal model of PD can be created by injecting 6-OHDA (Ungerstedt et al., 1974) directly into the brain. This injection causes mitochondrial dysfunction in the dopaminergic neurons located in the substantia nigra pars compacta (SNc) (Zeng et al., 2018). The enhancement of complex I inhibition by 6-OHDA was observed when monoamine oxidase A/B (MAO-A/B) was inhibited, indicating that the neurotoxicity is caused by 6-OHDA itself rather than its oxidation products (Glinka & Youdim, 1995). Once inside cells, 6-OHDA undergoes auto-oxidation or metabolic degradation and produces hydrogen peroxide, superoxide, and hydroxyl radicals. This process causes lipid peroxidation, protein oxidation, and DNA oxidation and finally results in oxidative stress and mitochondrial dysfunction (Zeng et al., 2018). The administration of 6-OHDA may induce an elevation in glutamate and d-aminobutyric acid levels while causing a reduction in glutamine levels in the striatum. This indicates a probable alteration in the balance of the Gln–Glu cycle between astrocytes and neurons (Lei & Powers, 2013). Disruption of the Gln–Glu cycle would disturb the equilibrium between excitatory and inhibitory brain processes, potentially resulting in lasting anomalies in glutamatergic and GABAergic activity (Zeng et al., 2018).

#### *Rotenone*

Rotenone, a member of the rotenoid family, is a commonly employed chemical for creating experimental models of PD. Rotenoids are naturally occurring poisonous compounds found in the stem, seed, and root of several plants (Bové et al., 2005; Xiong et al., 2012). Due to its high hydrophobicity, rotenone can readily traverse the blood–brain barrier (BBB) and bio-membranes, independent of DAT (Zeng et al., 2018). Unlike MPTP, rotenone triggers the development of α-syn fibrils (Ünal & Emekli-Alturfan, 2019). For instance, the application of rotenone in zebrafish could result in reduced motor performance, impaired sense of smell, and lower dopamine (DA) levels (Khotimah et al., 2015). A recent study demonstrated that exposing adult zebrafish to a concentration of 2 μg l-1 rotenone over 4 weeks resulted in a significant drop of 40%–50% in TH expression. Furthermore, the decline in DA levels was found to be associated with a reduced ability to move and exhibit altered locomotor activity, which is a characteristic motor symptom (Wang et al., 2017). As previously mentioned, rotenone is a powerful agent that can effectively trigger both motor and nonmotor signs of PD in a zebrafish model (Robea et al., 2020).

#### *Paraquat*

Paraquat is mostly utilized as a herbicide, and its mechanism of action closely resembles that of MPTP (Robea et al., 2020). Paraquat is capable of traversing BBB. Pre-treating L-valine, a substrate with a strong attraction to the neutral amino acid transporter, significantly decreased the ability of paraquat to cross the BBB. This indicates that the process of penetration is facilitated by the neutral amino acid transporter (Zeng et al., 2018). Paraquat reduces the activity of MCI in the substantia nigra (SN), and the suppression of MCI leads to the generation of ROS (Ünal & Emekli-Alturfan, 2019). Paraquat manipulates the pentose phosphate pathway to enhance the production of nicotinamide adenine dinucleotide phosphate hydrogen (NADPH)-reducing equivalents and promoting paraquat redox cycling (R. Powers et al., 2017a). Paraquat further induces an elevation in glucose absorption, the relocation of glucose transporters to the plasma membrane, and the activation of AMP-activated protein kinase (Zeng et al., 2018). The apoptotic pathway mediated by mitochondria is responsible for the pro-apoptotic effects of paraquat. This mechanism leads to the increased expression of pro-apoptotic members of the Bcl-2 family, followed by the release of cytochrome c and activation of caspase-3 (Fei et al., 2008). A solitary administration of paraquat is adequate to stimulate microglia and make dopaminergic neurons more susceptible to degeneration with successive injections (Zeng et al., 2018) (Figure 4.3).

### Huntington's Disease

HD is the predominant hereditary neurodegenerative disorder, marked by uncontrollable motor movements and impairments in cognitive and emotional functioning (Finkbeiner, 2011). Specifically, it has been proposed that the condition is caused by an expansion of cytosine-adenine-guanine (CAG) trinucleotides in exon 1 of the huntingtin gene (HTT), which encodes for polyglutamine (polyQ) at the N-terminal region (MacDonald et al., 1993). The CAG sequence often exhibits a repetition of 9–35 times, with a median average ranging from 17 to 20 repeats (Saudou & Humbert, 2016). Fortunately, numerous novel HD therapies are presently undergoing clinical trials specifically designed to address the disease at its source by reducing the quantities of mutant huntingtin protein (mHTT). Presently, there is a significant focus on antisense oligonucleotide (ASO) therapeutics. These therapies involve the binding of ASOs to pre-RNA or mRNA molecules, resulting in the modification of protein expression by processes such as RNA degradation, translation inhibition, or splicing manipulation. Potential medicines currently being developed in the pre-clinical stage include zinc-finger protein (ZFP) therapies, transcription activator-like effector nuclease (TALEN) therapies, and CRISPR/CRISPR-associated system (Cas) therapies (Ferguson et al., 2022). The clinical manifestations of HD are traditionally categorized into three domains: motor, cognitive, and mental. The motor symptoms exhibit a progressive nature and mostly manifest as hyperkinetic movements characterized by involuntary chorea (Nopoulos, 2016). An in-vitro investigation demonstrated that verapamil, clonidine, calpastatin, and 2'5'-dideoxyadenosine are drugs that hinder the aggregation of HTT in cell culture models. Additionally, these compounds limit the development of HTT aggregates in vertebrate models of HD (Kumar et al., 2021).

## Genetic Models of Huntington's Disease

### *Deletion of N17 Coupled with 97Q Expansion (mHTT-ΔN17-exon1)*

HD is a result of an elongated polyQ sequence located at the N-terminus of the HTT (MacDonald et al., 1993). Protein–protein interactions are responsible for the typical distribution of HTT inside cells, excluding the polyQ areas (Truant et al., 2007). The N-terminal 17 amino acids of HTT (N17) have been proposed as an essential component in controlling the localization and stability of the protein (Steffan et al., 2004). After extensive recombination, the presence of EGFP$^+$ protein aggregates may be observed in both mHTT-exon1 and mHTT-ΔN17-exon1 lines. Remarkably, these fish have normal development for up to 5 weeks, after which the mHTT-ΔN17-exon1 lines start displaying aberrant movement and swimming behavior that gradually deteriorate until the fish become incapable of swimming by approximately 12 weeks of age (Veldman et al., 2015). The mHTT-exon1 lines exhibit a less severe decline in swimming ability, which becomes noticeable only at 4 months of age and advances at a slower rate. In addition, they introduced the mHTT-ΔN17-exon1 line into transgenic Cre driver lines specific to neurons, glia, muscle, or vasculature. Only fish that exhibited mHTT-ΔN17-exon1 expression exclusively in neurons displayed a gradual movement problem (Veldman et al., 2015). The newly developed zebrafish model of HD will be extremely beneficial for studying the underlying processes of mutant polyQ toxicity and for identifying prospective treatments or genetic factors that can improve disease-like symptoms in living organisms (Stone & Perkins, 1981).

### *Htt-Loss of Function Model*

HTT is a substantial protein consisting of about 3,100 amino acids. It is mostly found in the brain and testes. The crucial factor here is the existence of numerous HEAT repeats, which frequently facilitate protein–protein interactions, alongside polyglutamine and nearby polyproline (Li et al., 2006). This characteristic is present in all vertebrates, including zebrafish Htt (Takano & Gusella, 2002). HTT interacts with several effector proteins and also contributes to transcription and trafficking mechanisms that can alter the processing and localization of various other proteins. Hence, the lack of typical HTT function may have broader implications on cellular physiology than now acknowledged (Schulte & Littleton, 2011). The presence of CAG repeats in exon 1 of the huntingtin gene (Htt) leads to the formation of extended PolyQ repeats in the huntingtin protein (Rana et al., 2023). This ultimately causes HD, a neurodegenerative disorder that manifests as an autosomal dominant subtype of disease (Chia et al., 2022). The mHTT modifies cellular processes both in vitro and in vivo by enlisting wild-type HTT. Mos-mediated knockdown of Htt in zebrafish can lead to a partial drop in Htt levels. This could provide fresh insights into the various neuronal responses to Htt loss (Diekmann et al., 2009). Nevertheless, the impairment of HTT function is dependent on the elongation of polyQ and is expedited as individuals age, while it also safeguards certain aspects of Htt's developmental functions (Kumar et al., 2021). The decreased expression of genes typically seen in the brain of zebrafish (six1, dlx3b, and emx3) indicates that the suppression of Htt mRNA translation disrupted the development of the neural plate in the frontmost area (Diekmann et al., 2009). The expression and distribution of the apical marker ZO-1 were shown to be reduced 24 hours after fertilization. This impairment led to erroneous cells in the diencephalic neural tube and cellular clusters in the brain ventricles. It was also discovered that the expression of ZO-1 is dependent on HTT (Chia et al., 2022). In addition, mutant Htt forms aggregates with other proteins, such as CREB binding protein, leading to a substantial decrease in the availability of proteins in the cell (Diekmann et al., 2009). Crucially, zebrafish can survive without huntingtin, unlike mice, which could aid in gaining a more comprehensive understanding of the neurodevelopmental processes involved in HD. Studies conducted on zebrafish indicate that the malfunctioning of early Htt may only have a restricted impact on neural precursors that are influenced by the absence of normal Htt at later stages of development (Chia et al., 2022) (Figure 4.4).

### *Huntington's Disease Model with BDNF Loss*

Htt engages in interactions with several proteins and exhibits diverse functions in cellular processes, including vesicle trafficking, apoptosis inhibition, and transcription regulation. Htt plays multiple roles that are essential for the functioning of brain-derived neurotrophic factor (BDNF) (Diekmann et al., 2009). BDNF is a member of the neurotrophin family. Additionally, this group includes neurotrophin 3 (NT3), nerve growth factor (NGF), and neurotrophin 4/5 (Kumar et al., 2021). PreproBDNF, proBDNF, and mature BDNF are distinct variants of BDNF derived from the bdnf gene. Nevertheless, the formation of mature BDNF requires the involvement of multiple metabolic pathways. The signal peptide regulates the proteolytic processing of preproBDNF, resulting in the generation of proBDNF (Anastasia et al., 2013). There is a hypothesis suggesting that invertebrates also utilize the same neurotrophin signaling molecules as vertebrates (van Kesteren et al., 1998). Zebrafish have been demonstrated to possess three distinct proBDNF isoforms, namely NP_001295578.1, NP_571670.2, and NP_001295577.1 (Lucini et al., 2018). The zebrafish with reduced Htt expression exhibited decreased levels of BDNF and demonstrated behaviours that closely resembled those observed in the fish with reduced BDNF expression (Diekmann et al., 2009). The addition of recombinant BDNF protein to the fish growth medium can help alleviate the consequences of the phenotype resulting from the downregulation of BDNF expression in zebrafish. This downregulation of BDNF expression in zebrafish produces results that are very comparable to those created by Htt-MO injection (Diekmann et al., 2009). To counteract the consequences of reduced BDNF expression in zebrafish, which closely resemble the effects of Htt-MO injection, one possible approach is to supplement the fish growth medium with recombinant BDNF protein (Kumar et al., 2021). Zebrafish that were administered BDNF-MO injections at a dosage that caused a 50% reduction in BDNF protein (11 ng) had a strong resemblance to Htt knockdown fish. The elevated levels of BDNF-MO resulted in premature mortality, as they were intended to eliminate BDNF expression, hence hindering the investigation of developmental traits (Diekmann et al., 2009). Normalizing BDNF signaling could serve as a unique therapeutic target for treating HD (Smith-Dijak et al., 2019).

#### *Huntington's Disease Model with Q102-GFP Aggregation*

Scientists created HD zebrafish models using transgenic technology. Schiffer et al. (2007) introduced 102 polyglutamine repeats into the N-terminal fragment of the HTT protein fused with GFP (Q102-GFP) in zebrafish. They observed that this mutant protein accumulated in large SDS-insoluble inclusions in the zebrafish embryos, which mimics a significant characteristic of HD pathology (Wang & Cao, 2021). The presence of the mutant HTT protein led to an elevation in atypical structure and the occurrence of programmed cell death, known as apoptosis, in zebrafish embryos. Subsequent research revealed that the presence of soluble mutant HTT protein forms is the cause of toxicity and abnormal polyglutamine formations in zebrafish (Schiffer et al., 2007). Additionally, the study revealed that the toxicity of the substance can be mitigated by the heat-shock proteins Hsp40 and Hsp70. Significantly, this HD transgenic model allowed the identification of two inhibitors, 293G02 and 306H03, that effectively prevent the aggregation of Q102-GFP in vivo. These inhibitors belong to the N'-benzylidene-benzohydrazide class. A separate investigation involved the creation of a genetically modified zebrafish line that consistently produced a Q71-GFP fusion protein. This protein was regulated by the rhodopsin promoter and was used to examine FDA-approved medications in order to discover new mechanisms that induce autophagy (Wang & Cao, 2021). Three medications (L-type $Ca^{2+}$ channel antagonists, the $K^{+}$ATP channel opener minoxidil, and the Gi signaling activator clonidine) were identified as participants in a cyclical mTOR-independent circuit that controls autophagy. This pathway contains numerous potential targets for triggering autophagy and decreasing the accumulation of aggregates (Williams et al., 2008) (Figure 4.4).

### Chemical Induced Models of Huntington's Disease

#### *3-Nitropropionic Acid (3-NP)*

3-NP, aminooxy acetic acid (AOAA), malonate, and succinate are among the mitochondrial inhibitors used to mimic HD (Wiprich et al., 2020). 3-NP is a naturally existing plant and mycotoxin that has been associated with neurological diseases in both grazing animals and humans (Mitchell, 2006). The 3-NP model is a dependable tool for investigating HD because of its ability to replicate a sequence of cell death that is observed in the HD brain, particularly the impairment of mitochondrial activity (Ramaswamy et al., 2007). 3-NP induces irreversible inhibition of succinate dehydrogenase (SDH), an enzyme found in mitochondrial complex II of the respiratory chain. Moreover, 3-NP can traverse the BBB, which allows it to harm the CNS even when administered systemically (Brouillet et al., 1999; Kaur et al., 2017; Túnez & Santamaría, 2009; Túnez et al., 2010). Like 3-NP, ochratoxin A (OTA) is also a food-origin mycotoxin, and exposure of zebrafish to OTA leads to reduced growth, as well as the development of

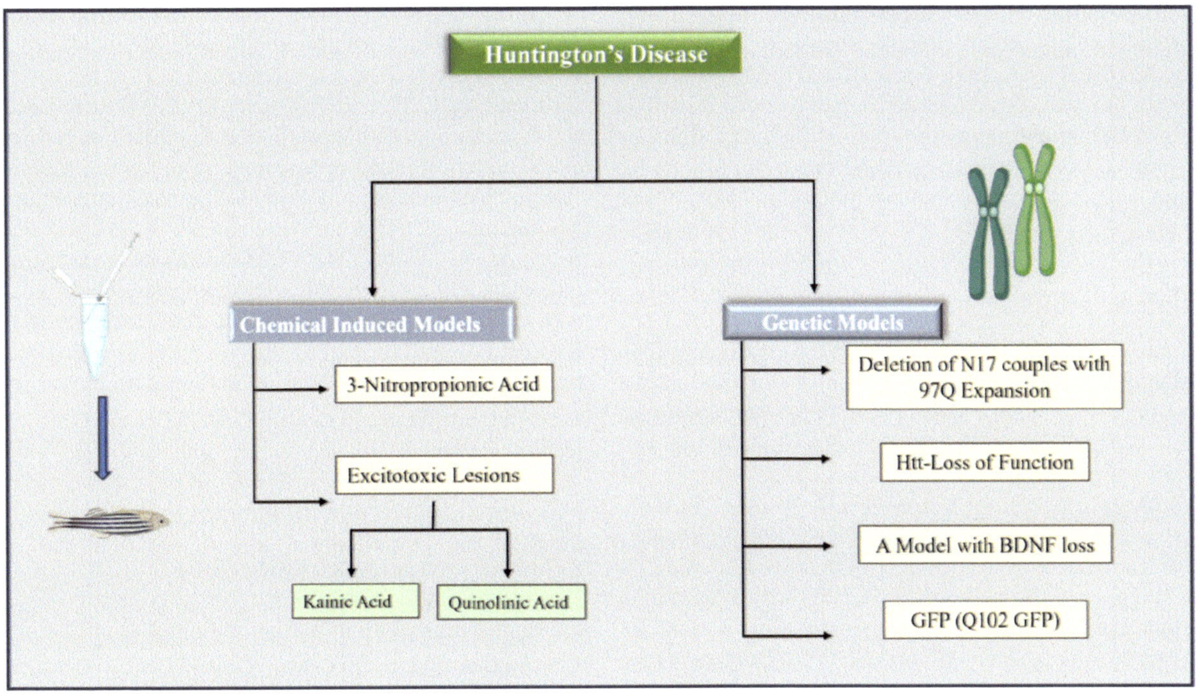


**FIGURE 4.4** Huntington models.

craniofacial deformities and body curvatures (Haq et al., 2016; Khezri et al., 2018). 3-NP is a toxic substance that impacts the process of energy metabolism. As a result, it is anticipated to have an impact on the movement patterns of zebrafish larvae. The absence of an effect of 3-NP on locomotor measures may be attributed to adaptive responses to the decrease in energy that occurs during this period of development (Wiprich et al., 2020). While HD is known for its movement abnormalities, it is essential for models to also accurately simulate the disease's more severe cognitive impairments, including difficulties in working memory and attention (Palfi et al., 2000). Similar to the neuronal death observed in the brains of individuals with HD, 3-NP induces toxicity in the striatum by initiating the degeneration of GABAergic medium spiny neurons in that region (Beal et al., 1993).

*Excitotoxic Models of Huntington's Disease*

Kainic acid (KA) served as an excitotoxin in initial excitotoxic models of neurodegeneration (Coyle & Schwarcz, 1976). KA induced neuropathological lesions and, at greater concentration, destroyed nerve fibers. This led to the exploration of alternate excitotoxins, such as ibotenic acid and quinolinic acid (QA) (Henderson & Hirsch, 1949). Excitotoxins are a group of chemical compounds that imitate glutamate or excitatory amino acids (Young et al., 1988). QA is a compound that activates the N-methyl-aspartate (NMDA) receptors. Its excitotoxicity leads to the death of neurons, which closely resembles the neuronal loss observed in HD (Lelos & Dunnett, 2018). Due to the inability of QA to pass through the BBB, it is directly delivered into the striatum (Foster et al., 1984). Furthermore, infusions of QA result in more localized damage in areas like the striatum and hippocampus compared to kainic acid. Kainic acid, on the other hand, tends to induce epileptic foci and cause less specific and more widespread cell loss (Lelos & Dunnett, 2018). QA's ability to induce cell death and perhaps mimic the mechanism of neuronal death seen in HD brains is a significant reason for its extensive utilization in HD studies (Beal et al., 1991; Ferrante et al., 1993; Roberts et al., 1993). QA induces neurodegenerative features in animal models of HD by enhancing the entry of $Ca^{2+}$ ions, reducing ATP production, and triggering excitotoxic cell death in the striatum, a brain region abundant in NMDA receptors (Stone & Perkins, 1981) (Figure 4.4).

## Multiple Sclerosis

A chronic inflammatory disease of the CNS, multiple sclerosis (MS) causes demyelination, which in turn causes axonal and neuronal degeneration and, ultimately, greater impairment (Reich, 2018). MS is a complicated illness; in addition to several well-known environmental factors, such as lower vitamin D (vD) levels or ultraviolet B light (UVB) exposure, Epstein-Barr virus (EBV) infection, obesity, smoking, and mutations of several genes may also raise the risk of developing the condition (Fragoso, 2014). Strongly associated with UVB radiation, which increases the generation of cutaneous vitamin D, is the latitudinal gradient in MS prevalence. Some factors have linked vD to the causative pathway of MS, including low vD levels, decreased vD consumption, decreased outdoor exercise, and greater vulnerability to the disease linked to genetic variants causing low vD levels (Sintzel et al., 2018). Demyelinating plaques are caused by perivenular inflammatory lesions, which are the histological hallmark of MS (Karussis, 2014). T-lymphocytes, predominantly MHC class I restricted $CD8^+$ T-cells, are found in the inflammatory infiltrates; B-cells and plasma cells are also present, although in somewhat smaller quantities (Lassmann, 2013). Astroglial proliferation (gliosis) and neurodegeneration, as well as inflammation with demyelination, are the two pathological features that define MS. The peripheral nerve system (PNS) is unaffected by tissue injury in MS (Hauser & Cree, 2020). There are two possible clinical outcomes for MS: progressive or relapsing. Relapsing multiple sclerosis (RMS) is the most frequent kind of disease, with discrete bouts of neurological impairment followed by partial, total, or no remission. Relapses typically become less frequent with time, but a steady worsening frequently takes over, leading to continuous progression referred to as secondary progressive MS (SPMS) (Lublin et al., 2014). According to one idea, MS initially manifests as an inflammatory illness (during RMS), but later on, neurodegeneration that is unaffected by inflammatory reactions takes over as the primary cause of the disease's progression (during SPMS) (Correale et al., 2017). A hallmark pathologic feature of patients with progressive disease is the steady enlargement of preexisting white matter lesions, and focal white matter damage is also common in SPMS patients (Frischer et al., 2015; Luchetti et al., 2018). Disease-modifying treatments alter how MS progresses by suppressing or changing immune activity. They mainly have an anti-inflammatory effect during the relapsing phase of MS; they lower the relapse rate, lessen the buildup of magnetic resonance imaging lesions, and stabilize, postpone, and, in certain situations, slightly improve disability (Hauser & Cree, 2020). Ocrelizumab, Opicinumab, Ibudilast, and Cladribine are the medications used in MS. Ocrelizumab is a B-cell-depleting antibody that targets CD20. There were 732 patients with a primary progressive illness course in the key ORATORIO study. Ocrelizumab lowered the rate of brain shrinkage by 17.5% and the chance of disability advancement at 12 weeks by 24% (Montalban et al., 2017). A lytic monoclonal antibody called ofatumumab is an immunoglobulin 1κ (IgG1κ) that attaches to CD20. When taken subcutaneously, ofatumumab differs from ocrelizumab. A new paradigm in B-cell targeted therapy may be achieved with the MS drug ofatumumab, which is presently undergoing development. Its goal is to reduce B cell levels rather than eradicate them (Faissner & Gold, 2019). Cladribine is a purine nucleoside that accumulates nuclear material and causes the target cell to die when it is phosphorylated in cells with high levels of deoxycytidine kinase (Beutler, 1992).

### Genetic Models of Multiple Sclerosis

*Tg (mbpa:gal4-vp16;uas:NTRmCherry) Transgenic Zebrafish*

The mbpa promoter controls the conditional ablation of mature oligodendrocytes in *Tg (mbpa:gal4-vp16;uas:NTRmCherry)* transgenic zebrafish. The bacterial cytotoxin gene nsfB encodes nitroreductase (NTR), which, while expressed, is not cytotoxic. However, when treated with the prodrug metronidazole (MTZ), this enzyme changes MTZ into cytotoxins, allowing for the targeted ablation of Schwann and oligodendrocyte cells. Therefore, when these transgenic zebrafish are exposed to a media containing MTZ, oligodendrocytes, and Schwann cells rapidly ablate, which is followed by demyelination of the nervous system (Kim et al., 2021) (Figure 4.5).

### Chemical Induced Models of Multiple Sclerosis

#### *Experimental Autoimmune Encephalomyelitis (EAE) Model*

More recently, an EAE zebrafish model was developed for in vivo rapid screen for MS development (Kulkarni et al., 2017). Subcutane injection of myelin oligodendrocyte glycoprotein –35–55 (MOG35–55) close to the spinal vertebrae led to decreased survival, decreased body weight, paralysis, and disability. When used as an EAE model, zebrafish have the benefit of having clinical symptoms that manifest quickly, with a noticeable disease phenotype 3 days after vaccination and a significant impairment 7 days after fertilization. The viability of employing this model as a screening tool was reported, demonstrating that fingolimod, dexamethasone, and dimethyl fumarate therapy decreased the effects of EAE in a dose-dependent way. Before screening in rodent EAE models, which is ultimately far more time-consuming and costly, this model has the potential to be a very helpful as an in vivo screening method for medicines (Getts et al., 2012; Merrill, 2009).

#### *Ethidium Bromide (EB) Model*

Research on the processes of demyelination and remyelination in MS can be assisted by using EB, a straightforward method for inducing neural cell death. Zhu et al. (2016) determined the ideal concentration and duration of treatment for the demyelination agent EB using the motility assay in order to create a zebrafish demyelination model. The application of 0.1, 1, 10, 25, 50, 75, and 100 μM EB to zebrafish at 2 dpf for 24–48 hours, or 0.1 μM EB for 72 hours, did not cause a decrease in motility distance or demyelination. All zebrafish treated with 0.1, 1, 10, 25, 50, 75, and 100 μM of EB for 72 hours showed varying degrees of demyelination and motility distance reduction. There was a dose-dependent decrease in motility distance and demyelination; 100% of the zebrafish treated with 75 μM of EB for 72 hours showed a reduction in motility distance and demyelination without any other tissue or organ toxic phenotypes that could be seen (Chen et al., 2014) (Figure 4.5).

### Alzheimer's Disease

Alzheimer's disease (AD) is a long-term neurodegenerative condition linked to a steady deterioration in cognitive function. Based on research, amyloid plaques and neurofibrillary tangles (NFTs) are the primary neuropathological features associated with AD (Koehler et al., 2018). The limbic brain regions in the AD brain frequently exhibit at least mild atrophy. In most AD patients, the major motor and somatosensory cortices remain intact, but the frontal and temporal cortices frequently exhibit ventricular enlargement and atrophy of the gyri. Studies on

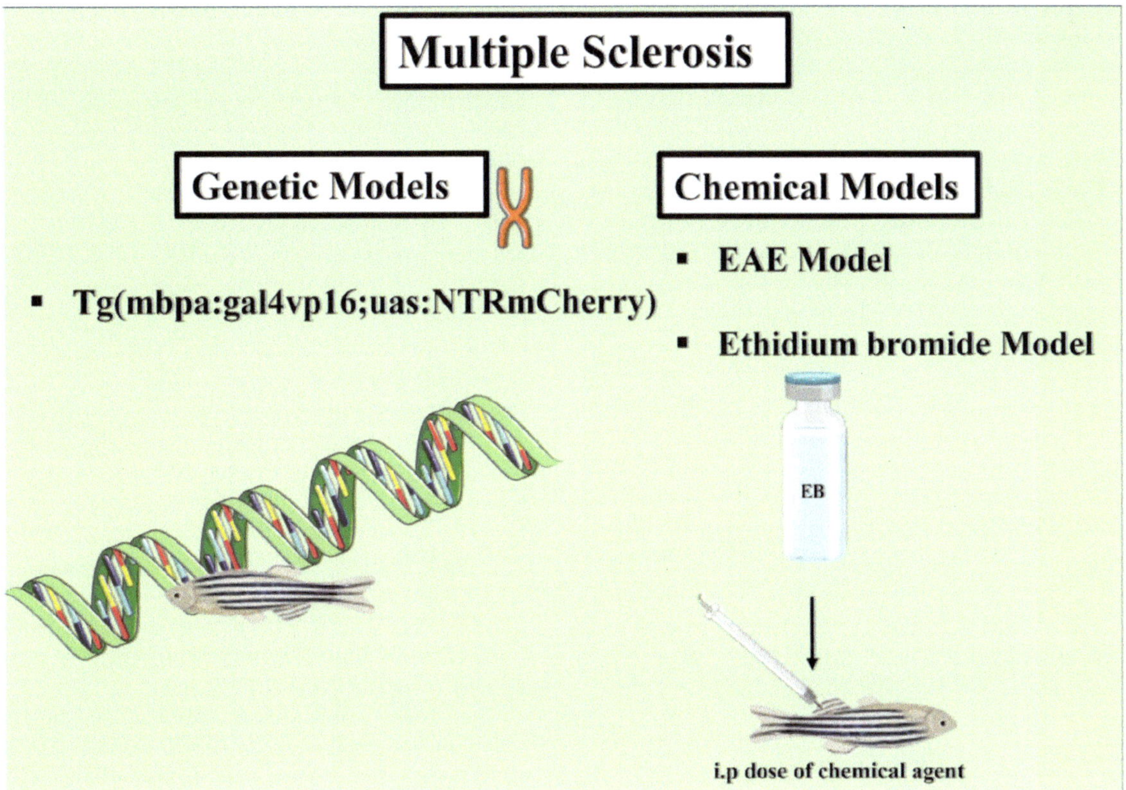


**FIGURE 4.5** Multiple sclerosis models.

functional imaging have emphasized the importance of the precuneus and posterior cingulate gyrus (Koehler et al., 2018). Recent years have seen some novel developments in the field of dementia research, including the identification of transactivation response DNA binding protein 43 kDa (TDP-43) and the introduction of the neuropathological notion of primary age-related tauopathy (PART) (Rami et al., 2012). The clinical manifestation of AD begins with recent (short-term) memory deficiencies, wordfinding issues, and language impairments, and it progressively advances to global cognitive impairment (Becker et al., 1992). Eventually, AD spreads to encompass diffuse cortical functions, which results in the incapacity to oversee everyday tasks. AD also results in a variety of behavioral and psychological changes that can be quite upsetting for carers as well as patients (Rocca et al., 2010; Zhao et al., 2016). The characteristic of AD's cognitive manifestation is impaired memory. The first stage is marked by mild memory impairments (such as losing things, having trouble recalling names and remembering conversations, and skipping appointments) (López & DeKosky, 2008). Nonetheless, myoclonic movements and extrapyramidal signs are the most common neurological symptoms linked to AD, and their existence influences the disease's clinical course (Heyman et al., 1999; Lopez et al., 1997; Stern et al., 1994). The N-methyl-D-aspartate receptor antagonist and cholinesterase inhibitors are two examples of the current treatments for AD that can temporarily lessen psychiatric symptoms but cannot reverse or stop the disease's progression. Many multinational pharmaceutical companies have tried many clinical trials based on amyloid cleansing as a cure, but none of them have been successful from an amyloid perspective. Numerous treatments, such as neurotransmitter mutations, anti-inflammatory and neuroprotective interventions, behavioral submission, cognitive augmentation, and anti-tau and anti-amyloid interventions, have been created and studied by researchers (Huang et al., 2017). Numerous cutting-edge treatments are being tested in different stages of clinical trials. The therapeutic efficacy of elenbecestat, β-secretase inhibitor, was examined in a 24-month research to assess its safety and efficacy in AD in two Phase III clinical trials MISSION AD1 and MISSION AD2. These trials are anticipated to continue through December 2023. The recently developed agent sodium oligomannate (GV-971), an oligosaccharide produced from marine sources, may enhance cognitive function in AD. The primary constituent of senile plaques, β-amyloid (Aβ) aggregates are linked to the pathophysiology of AD. Sodium oligomannate attaches to several amyloid sites, further destabilizes, stops the synthesis and accelerates the removal of Aβ (Wang et al., 2019). Apiprazole, a dopamine system stabilizer, restores dopamine activity in the cortical regions known to be associated with cognitive impairment and reduces the hyperactivity of dopaminergic neurons that cause psychosis. It attenuates neuropathy and may be able to reverse the neurotoxicity caused by Aβ in AD-related psychosis (Coleman et al., 2017).

## Genetic Models of Alzheimer's Disease

### *Amyloid-β42 (Aβ42)*

There are three types of toxic Aβ plaques found in AD patients: senile, diffuse, and neuritic (Hey et al., 2018). The β- and γ-secretases of amyloid precursor protein (APP) catalyze the proteolytic cleavage of Aβ, a process known as the amyloidogenic pathway. The two main forms of Aβ oligomers are Aβ1-42, which is a byproduct of the amyloidogenic process; and Aβ1-40, which is a physiologically important molecule in the brain (Kiper & Freeman, 2022). It builds up in neurons and causes apoptosis, activation of microglia, loss of synapses, and impairments in memory in the adult zebrafish brain (Bhattarai et al., 2016). Zebrafish APP models have been developed using several approaches to tackle research questions related to AD. Cameron and colleagues, for instance, discovered that exposing growing zebrafish to monomeric Aβ peptides might cause angiogenesis. They did this by utilizing a transgenic zebrafish that expresses green fluorescent protein (GFP) in vascular endothelial cells. Tg(kdr:eGPF)s843, a transgenic zebrafish model, was created by microinjecting a linearized construct made up of a promoter, eGFP tag, and zebrafish gene. This is significant because it can be utilised to ascertain the consequences of hypervascularization in adult transgenic zebrafish and the long-term effects of Aβ exposure in the cerebral vasculature (Cameron et al., 2012). Studies utilising this paradigm revealed that APP and Aβ have a conserved physiological role in zebrafish (Lee & Cole, 2007).

### *Tau-induced Alzheimer's Disease Model*

In AD and similar tauopathies, tau, the microtubule-associated protein, produces insoluble filaments that build up as NFTs. Zebrafish tauopathy comes in a variety of forms. Two of these tauopathy models have human tau coupled to GFP expressed transiently under the control of either the neuronal HuC/D promoter or the gata2 promoter (Bai et al., 2007). By expressing TAU0N4R under the direction of the enolase promoter, a stable model was created (Paquet et al., 2009). The cytoskeleton was disrupted in the transgenic zebrafish produced with FTDP-17 mutation, a pathology seen in human NFTs. Using this model, scientists were able to examine the functional effects of human tau mutations at specific post-translational modification sites as well as those linked to hereditary dementias (such as G272V, K280, P301S, S305N, and R406W) (Kiper & Freeman, 2022). The zebrafish genome was modified to incorporate both human TAUP301L and discosoma red fluorescent protein (DsRed), enabling the co-expression of DsRed fluorescence and the real-time detection of tau-expressing cells. It is significant to remember that this mutant zebrafish line showed NFTs at 5 weeks. Characterizing tau hyperphosphorylation in adult zebrafish brains was made possible by the effective and efficient generation of zebrafish expressing human TAUP301L. Research revealed that this expression of human TAUP301L produced hyperphosphorylated tau but no NFT (Takashima et al., 1998). This tauopathy model may be used in future research to assess the role of tau phosphorylation in neurodegeneration and to screen treatments for reducing tau phosphorylation in response to environmental perturbations (Kiper & Freeman, 2022). Human 4R/0N Tau was expressed in stable transgenic zebrafish, the most prevalent Tau species seen in progressive supranuclear palsy (Yamada et al., 1996; Yokota et al., 2018). Under transcriptional regulation of the eno2 promoter, the transgene was expressed. Human 4R/0N Tau is widely expressed in the CNS, and this expression continues into adulthood. In adult brain regions of pathogenic significance to progressive supranuclear palsy, such as the optic tectum, refractile Tau accumulations that resembled neurofibrillary tangles were seen within neuronal cell bodies and proximal axons. A potentially crucial step in validating the model system will be made possible by analyses intended to determine whether age-related pathology accumulates in these transgenic lines, similar to the progressive changes observed in the human disease and recapitulated in certain murine models. This is made possible by the Tau transgene's stable expression into adulthood in this model. Moreover, extensive transgenic expression in neurons may facilitate the investigation of elements contributing to cellular susceptibility to tauopathy (Sengoku, 2020).

### *Presenilin Mutation as a Model of Alzheimer's Disease*

The aspartyl protease complexes that cause APP's γ-secretase cleavage contain a substantial amount of presenilin (PSEN) activity (Nornes et al., 2003). The main part of γ-secretase, which is in charge of sequentially cleaving amyloid precursor proteins by proteolysis and producing amyloid-β peptides, is encoded by the PSEN1 and PSEN2 genes. According to earlier research, tau and GSK-3β bind to the same PSEN1 areas (Ochalek et al., 2017), and mutations in PSEN1 can lessen tau phosphorylation by reducing cytoskeletal connections (Campbell et al., 2006). Two of the four loci targeted for researching the pathogenic etiology of AD are PSEN1 and PSEN2. Patients with early-onset AD may benefit from treatments that target PSENs. 200+ PSEN1 mutations and about 20 PSEN2 mutations are thought to be directly related to the onset of familial AD (Hin et al., 2018). A frameshift mutation in PSEN2 that causes a "premature" stop codon was modeled by Hin and colleagues using zebrafish (Babin et al., 1997). This early-onset familial AD mutation of PSEN2 causes a shortened open reading frame (ORF) due to a premature stop in the sequence. Upon examination, several pathological changes in the mutant zebrafish brain were discovered to be similar to those in the brains of AD patients (Zon & Peterson, 2005) (Figure 4.6).

## Chemical Induced Models of Alzheimer's Disease

### *Okadaic acid (OA)*

Within the aquatic environment, one of the most prevalent poisons is okadaic acid. Numerous species of dinofagellates, a group of phytoplankton, generate it. It was first isolated from the Halichondria okadai black sponge (Yang et al., 2014). The enzymes that dephosphorylate tau protein are called protein phosphatases 1 (PP1) and 2A (PP2A). When tau is not dephosphorylated, microtubule dissociation and the subsequent production of hazardous NFTs are prevented (Goedert et al., 1995; Koehler et al., 2018). Okadaic acid (OA) inhibits PP1 and PP2A, the two primary cytosolic broad-specificity protein phosphatases (Bialojan & Takai, 1988). Adult zebrafish experience a cognitive impairment due to OA; different groups are tested on their memory and learning using spatial

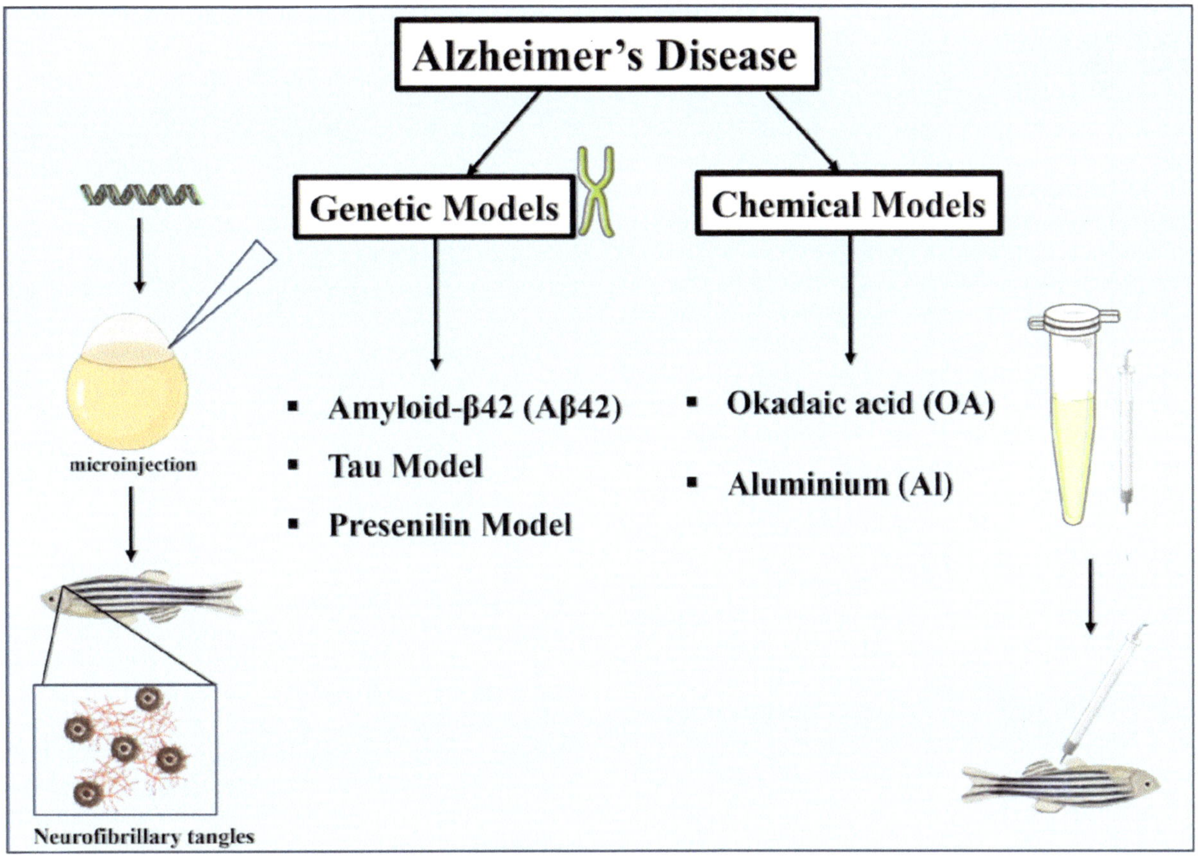


**FIGURE 4.6** Alzhemier's disease models.

alternation. The expression of tau, phosphorylated tau (ptau), Aβ fragments, and glycogen synthase kinase α/β (pGSK3α/β) is altered by OA, leading to pathophysiology similar to AD. Toxic oligomer formation brought on by aberrant Aβ protein accumulation is the mechanism behind synaptic dysfunction. Damage of this kind may have promoted changes in glutamate receptors, hyperexcitability of circuits, and signaling pathways linked to synaptic plasticity (Crews & Masliah, 2010).

*Aluminium (Al)*

Aluminium at 100 μM reduced the amount of glial fibrillary acidic protein (GFAP) in zebrafish larvae, which resulted in neurological damage at the level of glial cells (Monaco et al., 2017). In adult zebrafish, aluminum chloride ($AlCl_3$) at 50 μg/L (pH 5.8) exhibited AD-like behavior in locomotor activity tests and activity-avoidance conditioned paradigms. This finding demonstrates how aluminum exposure in an acidic environment affects zebrafish locomotor activity as well as their capacity for learning and memory (He et al., 2012). In an additional investigation, 50 μg/L of $AlCl_3$ treatment for 96 hours at pH 5.8 significantly raised (36%) AChE activity and reduced locomotor activity; however, at pH 6.8, it did not affect locomotor activity of zebrafish (Senger et al., 2011) (Figure 4.6).

## Amyotrophic Lateral Sclerosis

The multisystem neurodegenerative disease amyotrophic lateral sclerosis (ALS), commonly referred to as Lou Gehrig's or Charcot's disease, is heterogeneous in terms of its etiology, neuropathology, and clinical manifestations (Bonifacino et al., 2021). Gene mutations (SOD1, TDP-43, FUS, and C90RF72), excitotoxicity, protein aggregation, mitochondrial dysfunction, and the reactivity of glial cells, including astrocytes and microglia, are the most frequent causes (Bonifacino et al., 2021). Variable degeneration of the upper motor neurons (UMN) and lower motor neurons (LMN) characterizes the unique clinical presentations associated with ALS. The kind of motor neuron affected, the age and location of the disease's onset, family history, the extent of extramotor involvement, and the length of the illness can all influence the phenotypic variability of ALS (Swinnen et al., 2018). Focal muscle weakness is a sign of spinal ALS patients, and it can begin proximally or distally in the upper and lower extremities. As the illness worsens, patients may have clonus; weakness; fasciculation; loss of muscular mass; hyporeflexia, which is linked to LMN injury; hyperreflexia, which includes the Hoffmann and Babinski responses; and, in rare instances, modest hypertonia, which suggests UMN involvement. Patients may experience limb spasticity, which impairs manual dexterity, balance, and gait as the condition progresses. Respiratory problems can occasionally appear. Dysarthria, dysphagia, and tongue fasciculations are the first symptoms of bulbar muscular weakening in patients with bulbar ALS type (Audag et al., 2019). Riluzole, 50 mg twice daily, provides anti-glutamatergic actions by preventing glutamate from being released from nerve terminals to settle voltage-gated sodium channels in the inactivated state, and prolongs the mean patient survival by 3–6 months; it is still the sole licensed disease-modifying medication in most European countries (Masrori & Van Damme, 2020). Edaravone, a free radical scavenger, has been researched in ALS more recently. After 6 months of treatment, a Phase III randomized double-blind research of ALS patients who were chosen for the study found that the deterioration in ALSFRS-R scores was considerably less after receiving intravenous edaravone 60 mg/day for two weeks each month (Abe et al., 2017). A beneficial effect on the decline of ALSFRS-R was suggested by a randomized controlled trial that used masitinib at a dose of 4.5 mg/kg/day in addition to riluzole, at least in patients exhibiting normal disease course (Mora et al., 2020). Nutrition can be improved by making dietary adjustments, and if calorie intake is inadequate or swallowing becomes dangerous, a gastrostomy tube may be a possibility. In addition to aided communication – which uses customized software – speech therapy is frequently required. The preferred course of action for prolonging life in cases of respiratory insufficiency is noninvasive ventilation (Masrori & Van Damme, 2020).

## Genetic Models of Amyotrophic Lateral Sclerosis

*Cu/Zn SOD1 Mutations*

Human $SOD1^{G93A}$, $SOD1^{G37R}$, or $SOD1^{A4V}$ mRNAs were introduced into embryos to create an ALS model in zebrafish. This resulted in aberrant motor neuron axon branching and decreased axon length. In a dose-dependent way, the latter mutation produced the most impacted phenotype (Lemmens et al., 2007). These SOD1 mutations all caused the major symptoms of ALS, including adult-onset muscle atrophy and paralysis, early neuromuscular junction (NMJ) alterations, oxidative stress vulnerability, and untimely death (Benedetti et al., 2016; Powers et al., 2017b).

*C9orf72 Knockdown Model*

The most common cause of ALS is GGGGCC hexanucleotide repeat expansion (HRE) within chromosome 9 open reading frame 72 known as the C9orf72 gene. According to available data, HREs cause neurodegeneration by causing RNA foci and/or dipeptide repeat proteins (DPR) to accumulate. In the zebrafish model, both a gain-of-function (GoF) and a loss-of-function (LoF) of C9orf72 have been studied (Fortier et al., 2020). The deletion of the C9orf72 sequence resulted in defects in spontaneous and evoked swimming, motor axonopathy, axonal degeneration, disrupted arborization and shorter axons at early developmental stages, impaired neuronal development, and significant decrease in lifespan, which summarises many of the clinical features of ALS. The expression of the human WT C9orf72 mRNA restored these impairments, demonstrating the phenotypic specificity of the induction (Ciura et al., 2013). This model was defined by an accumulation of RNA foci and DPRs in the central nervous system and muscle. To replicate the elevated DPR levels noted in the central nervous system of C9orf72 patients, transgenic zebrafish models have been generated (Bonifacino et al., 2021). It was possible to create lines with 80 repeats of the GGGGCC sequence (ggggcc80) without ATG, and lines with ATG as the translation initiater codon replaced in front of the 80xGGGGCC repeat sequence that forces the the translation of the poly-GA protein connected to GFP (GA80-GFP) (Ohki et al., 2017). RNA foci in spinal cord neurons were seen in both lines. While the expression of GA80-GFP showed high toxicity, the expression of ggggcc80-GFP was only slightly toxic. This difference in toxicity was mitigated, though, by the ability of morpholino antisense to interfere with the translation of GA80-GFP, indicating that DPR reduction may be a useful therapeutic strategy for ALS patients with C9orf72 mutations (Green & Planchart, 2018).

*FUS Mutations*

Both sporadic and familial types of ALS are caused by mutations in the FUS gene, and more than 50 distinct FUS gene mutations have been linked to ALS so far. ALS mutations lead to the mislocalization of the nuclear protein to the cytoplasm, resultign in the accumulation of the cytoplasmic protein inclusions. Either gain of harmful function in the cytoplasm or loss of FUS activity in the nucleus lead to the disease. Accordingly, both LoF and GoF of FUS function lead to impaired presynaptic function at the neuromuscular junction (NMJ) in the zebrafish model (Armstrong & Drapeau, 2013). In the zebrafish spinal cord, the expression of the human R495X mutation in FUS resulted in the abrogation of a putative nuclear localization signal and a notable cytoplasmic accumulation of the protein, which was in some ways distinct from the observations made for recessive (H517Q) and dominant (R521G) FUS mutants. In addition, under circumstances of heat shock or oxidative stress, the ALS-linked FUS mutants aggregated into perinuclear stress granules (SGs), but not the WT protein (Bosco et al., 2010). Furthermore, mutant FUS mislocalized from the nucleus to the cytoplasm in cells other than motor neurons in zebrafish expressing GFP-tagged WT or mutant R521C human FUS. It was shown that human FUS has an innate tendency to aggregate, regardless of disease-associated mutations or particular cell types, as both WT and FUSR521C are localized at SGs. The R521C mutation did, however, cause an increase in the relative cytosolic to nuclear FUS, which significantly increased the SG assembly and persistence inside susceptible cells – though these cells were not necessarily motor neurons (Acosta et al., 2014). As was previously observed, homozygous mutants have shorter lifespans, and motor impairments due to loss of the FUS orthologue in zebrafish were also linked to specific cellular defects such as NMJ fragmentation and decreased motor neuron length. Furthermore, short Tau isoform expression is favored by FUS LoF modification of Tau transcripts (Bourefis et al., 2020).

*TDP-43-A315T Model*

When compared to controls expressing the human WT protein, zebrafish with the TAR DNA-binding protein 43 (TDP-43)-A315T mutation exhibited a significant decrease in motor axon length and an increase in aberrant branching. Furthermore, it was found that the WT and mutant TPD-43 proteins were localized in the nucleus, suggesting that pathogenic abnormalities can be caused without cytoplasmic translocation. Additionally, there was paralysis, neurodegeneration, oxidative stress, and a limited life span (Bonifacino et al., 2021). It's interesting to note that expressing human wild-type TDP-43 somewhat prevented identical motor impairments and axonopathy produced by endogenous TDP-43 knockdown. This shows that pathogenic mutations may produce both LoF and GoF variants, and it also highlights the significance of TDP-43 functionality (Kabashi et al., 2010) (Figure 4.7).

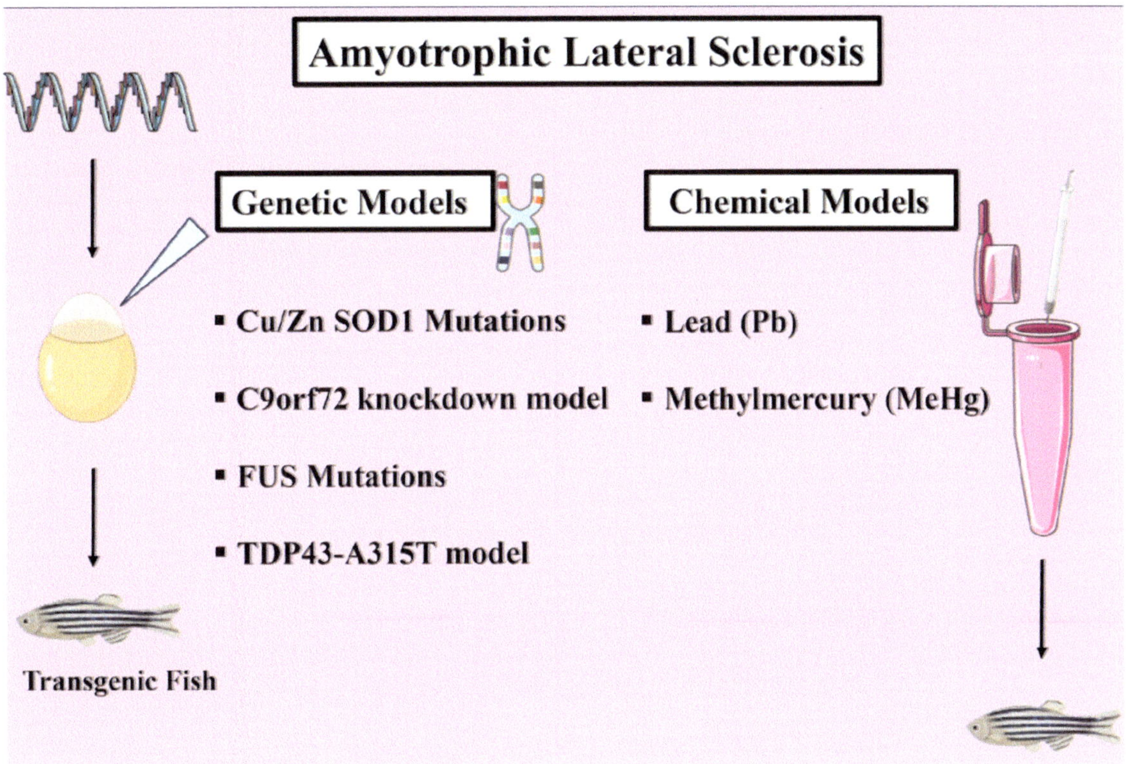


**FIGURE 4.7** ALS models.

## *Chemical Induced Models of Amyotrophic Lateral Sclerosis*

### *Lead (Pb)*

One of the main environmental variables thought to cause ALS is heavy metal exposure, including Pb. Even at modest dosages, Pb's multiple binding activity disrupts numerous molecules and processes, causing harm to most systems in the body, including the nervous system. Exposure to Pb at developmental stages has been studied at concentrations between 10 ppb and ~2 ppm. Studies reveal that Pb modifies several genes linked to the development of the nervous system at low concentrations (less than 100 ppb) (Lee et al., 2014). The GABAergic system's gene expression varied during embryonic development between 24 and 72 hpf, and the amount of GABA varied depending on the developmental stage and Pb dosage between 10 and 100 ppb (Wirbisky et al., 2014). These alterations may be linked to shorter neuronal axons and lower activity (hypoactivity) (Zhu et al., 2016). Zebrafish have changed color preferences and impaired adult learning at levels above 100 ppb (Bault et al., 2015).

### *Methylmercury (MeHg)*

Methylmercury (MeHg) is a neurotoxic agent that decreases myelin basic protein, damages oligodendrocytes, degenerates white matter, produces apoptosis and excitotoxicity, and causes axonal neurodegeneration and motor neuron degeneration. Research on adult zebrafish has indicated that mercury inhibited the activity and gene expressions of adenosine deaminase in the brain at levels < 250 μM (Senger et al., 2010). On the other hand, the expression of genes that function in the antioxidant responses, apoptosis, mitochondrial homeostasis, and metal chelation did not change by MeHg consumption in zebrafish brain samples (Gonzalez et al., 2005). In zebrafish brain, MeHg as well as inorganic mercury exposure through diet activated the production of metallothionein at high concentrations. Increased metallothionein by mercury was attributed to the *mt3*, the brain isoform of MT in zebrafish (Gentès et al., 2015).

Mid-range levels of inorganic mercury exposure (between 720 ppb and 6 ppm) demonstrated significant Hg accumulation in the brain, delayed hatching, increased mortality, and the induction of hyperactive behavior coupled with decreased levels of the neurotransmitters serotonin and dopamine. On the other hand, low levels of inorganic mercury exposures caused altered swim behavior, decreased foraging efficiency, and inhibition of membrane adenosine deaminase (Liu et al., 2013). Studies at high dosages (10–13 ppm) revealed changes in proteins linked to oxidative phosphorylation and gap junctions, significant increases in mt2, and mitochondrial dysfunction. Remarkably, alterations in these pathways seemed connected to the mTOR pathway, which MeHg may have triggered by causing the generation of reactive oxygen species (Liu et al., 2013). Humans are most commonly exposed to mercury in their diets, particularly when they eat tainted fish. It is interesting to note that zebrafish progeny raised on diets containing mercury-contaminated fish meant for human consumption showed increased morphological abnormalities, neurobehavioral defects, dysregulated gene expression profiles, and increased mortality (Green & Planchart, 2018). These findings suggest that similar exposure routes to mercury, independent of phylogeny, result in similar physiological outcomes (Liu et al., 2013) (Figure 4.7).

## Conclusion

Finding a successful treatment for NDD is critically needed, as the disorder is affecting an increasing number of people. Nevertheless, the only available treatment at this time is symptomatic. By modulating glutamatergic transmission by blockers of N-methyl-D-aspartate (NMDA) receptors or maintaining the levels of ACh and cholinergic transmission by AChE inhibitors, current AD medications can only momentarily improve learning and memory. It has been demonstrated that administering L-DOPA, dopamine agonists, MAO inhibitors, inhibitors of catechol-O-methyltransferase (COMT), and anticholinergics can momentarily alleviate motor and nonmotor PD symptoms without stopping the disease's progression. The FDA-approved medications daravone and riluzole for ALS only slightly improve survival. Due to the clinical variability of NDD, several pharmaceutical treatments intended for the condition have failed in clinical trials during the past few decades. Additionally, because the majority of NDD pathologies are irreversible, disease-modifying therapies are only beneficial when applied early in the course of the illness. To replicate the modification in various clinical subtypes and at various phases of the disease, we must thus produce more pre-clinical NDD animal models. A unique technique for evaluating possible treatments and learning more about disease pathways has been made possible by the recent development of novel zebrafish models. The current research on the zebrafish model as a helpful tool for studying several neurological disorders, including AD, PD, HD, ALS, and MS, is highlighted in this chapter. Zebrafish offer several benefits and serve as a crucial basis for determining the pathophysiology of NDD and for medication discovery. Current understanding of many chemicals and genetic models is provided by the production of cognitive impairment and dementia in zebrafish as a model. Utilizing zebrafish – more especially, genetically engineered zebrafish – is an alternate, cost-effective, and efficient method that researchers can employ to look into many aspects of neurodegenerative diseases. Compared to in vivo chemical screening, in vitro chemical screening offers a far higher throughput, while in vivo models can replicate intricate bodily systems such as the BBB. Both in vitro and in vivo drug screening have been conducted using zebrafish, and many novel compounds are beneficial against NDD. According to reports based on zebrafish study results, the aforementioned genetic, physiological, metabolic, and pharmacologic conservation enable us to view the zebrafish model as a useful tool for finding new drug targets, developing novel drugs, and simulating human diseases. As evidence that the zebrafish model is a transferable model system, several chemicals found through screening experiments in Danio rerio have entered clinical trials. Additionally, it can be applied in the future to the throughput screening of any disease to find innovative therapeutic medications. The benefits of adopting the zebrafish model over alternative models, such as its small size and the transparency of its embryos and larvae, are also significant. The model's minimal maintenance costs make it an affordable option for studies with little funding. Furthermore, the use of zebrafish instead of rodents in the drug development process has opened our eyes to new possibilities for treating acute illnesses like ALS, AD, PD, HD, and MS. In addition, to discover new medications other than current inhibitors or proteins that target NDD, further molecular processes need to be found. Additionally, drugs with low or no specificity must be researched and developed. Finally, it is important to ascertain the compounds' pharmacological medication safety. Drugs will be more easily transferred from the lab to the clinic with the use of zebrafish.

## REFERENCES

Abe, K., Aoki, M., Tsuji, S., Itoyama, Y., Sobue, G., Togo, M., Hamada, C., Tanaka, M., Akimoto, M., & Nakamura, K. (2017). Safety and efficacy of edaravone in well defined patients with amyotrophic lateral sclerosis: a randomised, double-blind, placebo-controlled trial. *The Lancet Neurology*, *16*(7), 505–512.

Abeliovich, A., & Gitler, A. D. (2016). Defects in trafficking bridge Parkinson's disease pathology and genetics. *Nature*, *539*(7628), 207–216.

Acosta, J. R., Goldsbury, C., Winnick, C., Badrock, A. P., Fraser, S. T., Laird, A. S., Hall, T. E., Don, E. K., Fifita, J. A., & Blair, I. P. (2014). Mutant human FUS Is ubiquitously mislocalized and generates persistent stress granules in primary cultured transgenic zebrafish cells. *PLoS One*, *9*(6), e90572.

Anastasia, A., Deinhardt, K., Chao, M. V., Will, N. E., Irmady, K., Lee, F. S., Hempstead, B. L., & Bracken, C. (2013). Val66Met polymorphism of BDNF alters prodomain structure to induce neuronal growth cone retraction. *Nature communications*, *4*(1), 2490.

Anichtchik, O., Diekmann, H., Fleming, A., Roach, A., Goldsmith, P., & Rubinsztein, D. C. (2008). Loss of PINK1 function affects development and results in neurodegeneration in zebrafish. *Journal of Neuroscience*, *28*(33), 8199–8207.

Armstrong, G. A., & Drapeau, P. (2013). Loss and gain of FUS function impair neuromuscular synaptic transmission in a genetic model of ALS. *Human Molecular Genetics*, *22*(21), 4282–4292.

Armstrong, M. J., & Okun, M. S. (2020). Diagnosis and treatment of Parkinson disease: a review. *Jama*, *323*(6), 548–560.

Armstrong, R. C., Aja, T. J., Hoang, K. D., Gaur, S., Bai, X., Alnemri, E. S., Litwack, G., Karanewsky, D. S., Fritz, L. C., & Tomaselli, K. J. (1997). Activation of the CED3/ICE-related protease CPP32 in cerebellar granule neurons undergoing apoptosis but not necrosis. *Journal of Neuroscience*, *17*(2), 553–562.

Audag, N., Goubau, C., Toussaint, M., & Reychler, G. (2019). Screening and evaluation tools of dysphagia in adults with neuromuscular diseases: A systematic review. *Therapeutic Advances in Chronic Disease*, *10*, 2040622318821622.

Babin, P. J., Thisse, C., Durliat, M., Andre, M., Akimenko, M.-A., & Thisse, B. (1997). Both apolipoprotein E and AI genes are present in a nonmammalian vertebrate and are highly expressed during embryonic development. *Proceedings of the National Academy of Sciences*, *94*(16), 8622–8627.

Badyal, D. K., & Desai, C. (2014). Animal use in pharmacology education and research: The changing scenario. *Indian Journal of Pharmacology*, *46*(3), 257.

Bai, Q., Garver, J. A., Hukriede, N. A., & Burton, E. A. (2007). Generation of a transgenic zebrafish model of Tauopathy using a novel promoter element derived from the zebrafish eno2 gene. *Nucleic Acids Research*, *35*(19), 6501–6516.

Bai, Q., Mullett, S. J., Garver, J. A., Hinkle, D. A., & Burton, E. A. (2006). Zebrafish DJ-1 is evolutionarily conserved and expressed in dopaminergic neurons. *Brain Research*, *1113*(1), 33–44.

Bandmann, O., & Burton, E. A. (2010). Genetic zebrafish models of neurodegenerative diseases. *Neurobiology of Disease*, *40*(1), 58–65.

Barnham, K. J., Masters, C. L., & Bush, A. I. (2004). Neurodegenerative diseases and oxidative stress. *Nature reviews Drug Discovery*, *3*(3), 205–214.

Bault, Z. A., Peterson, S. M., & Freeman, J. L. (2015). Directional and color preference in adult zebrafish: Implications in behavioral and learning assays in neurotoxicology studies. *Journal of Applied Toxicology*, *35*(12), 1502–1510.

Beal, M. F., Brouillet, E., Jenkins, B. G., Ferrante, R., Kowall, N., Miller, J., Storey, E., Srivastava, R., Rosen, B., & Hyman, B. (1993). Neurochemical and histologic characterization of striatal excitotoxic lesions produced by the mitochondrial toxin 3-nitropropionic acid. *Journal of Neuroscience*, *13*(10), 4181–4192.

Beal, M. F., Ferrante, R., Swartz, K., & Kowall, N. (1991). Chronic quinolinic acid lesions in rats closely resemble Huntington's disease. *Journal of Neuroscience*, *11*(6), 1649–1659.

Becker, J. T., Lopez, O. L., & Wess, J. (1992). Material-specific memory loss in probable Alzheimer's disease. *Journal of Neurology, Neurosurgery & Psychiatry*, *55*(12), 1177–1181.

Benedetti, L., Ghilardi, A., Rottoli, E., De Maglie, M., Prosperi, L., Perego, C., Baruscotti, M., Bucchi, A., Del Giacco, L., & Francolini, M. (2016). INaP selective inhibition reverts precocious inter-and motorneurons hyperexcitability in the Sod1-G93R zebrafish ALS model. *Scientific Reports*, *6*(1), 24515.

Beutler, E. (1992). Cladribine (2-chlorodeoxyadenosine). *The Lancet*, *340*(8825), 952–956.

Bhattarai, P., Thomas, A. K., Cosacak, M. I., Papadimitriou, C., Mashkaryan, V., Froc, C., Reinhardt, S., Kurth, T., Dahl, A., & Zhang, Y. (2016). IL4/STAT6 signaling activates neural stem cell proliferation and neurogenesis upon Amyloid-β42 aggregation in adult zebrafish brain. *Cell Reports*, *17*(4), 941–948.

Bialojan, C., & Takai, A. (1988). Inhibitory effect of a marine-sponge toxin, okadaic acid, on protein phosphatases. Specificity and kinetics. *Biochemical Journal*, *256*(1), 283–290.

Blagov, A., Borisov, E., Grechko, A., Popov, M., Sukhorukov, V., & Orekhov, A. (2023). The role of impaired mitochondrial transport in the development of neurodegenerative diseases. *Journal of Integrative Neuroscience*, *22*(4), 86.

Bonifacino, T., Zerbo, R. A., Balbi, M., Torazza, C., Frumento, G., Fedele, E., Bonanno, G., & Milanese, M. (2021). Nearly 30 years of animal models to study amyotrophic lateral sclerosis: A historical overview and future perspectives. *International journal of molecular sciences*, *22*(22), 12236.

Bosco, D. A., Lemay, N., Ko, H. K., Zhou, H., Burke, C., Kwiatkowski Jr, T. J., Sapp, P., McKenna-Yasek, D., Brown Jr, R. H., & Hayward, L. J. (2010). Mutant FUS proteins that cause amyotrophic lateral sclerosis incorporate into stress granules. *Human Molecular Genetics*, *19*(21), 4160–4175.

Bourefis, A.-R., Campanari, M.-L., Buee-Scherrer, V., & Kabashi, E. (2020). Functional characterization of a FUS mutant zebrafish line as a novel genetic model for ALS. *Neurobiology of Disease*, *142*, 104935.

Bové, J., Prou, D., Perier, C., & Przedborski, S. (2005). Toxin-induced models of Parkinson's disease. *NeuroRx*, *2*(3), 484–494.

Bretaud, S., Allen, C., Ingham, P. W., & Bandmann, O. (2007). p53-dependent neuronal cell death in a DJ-1-deficient zebrafish model of Parkinson's disease. *Journal of Neurochemistry*, *100*(6), 1626–1635.

Brouillet, E., Conde, F., Beal, M., & Hantraye, P. (1999). Replicating Huntington's disease phenotype in experimental animals. *Progress in Neurobiology*, *59*(5), 427–468.

Cameron, D. J., Galvin, C., Alkam, T., Sidhu, H., Ellison, J., Luna, S., & Ethell, D. W. (2012). Alzheimer's-related peptide amyloid-β plays a conserved role in angiogenesis. *PLoS One*, *7*(7), e39598.

Campbell, W. A., Yang, H., Zetterberg, H., Baulac, S., Sears, J. A., Liu, T., Wong, S. T., Zhong, T. P., & Xia, W. (2006). Zebrafish lacking Alzheimer presenilin enhancer 2 (Pen-2) demonstrate excessive p53-dependent apoptosis and neuronal loss. *Journal of Neurochemistry*, *96*(5), 1423–1440.

Cansiz, D., Unal, I., Beler, M., Ustundag, U. V., Ak, E., Emekli-Alturfan, E., & Alturfan, A. A. (2023). The effect of acetic acid-induced pain in Parkinson's disease model in zebrafish. *NeuroToxicology*, *99*, 14–23.

Canter, R. G., Penney, J., & Tsai, L.-H. (2016). The road to restoring neural circuits for the treatment of Alzheimer's disease. *Nature*, *539*(7628), 187–196.

Chen, J.-B., Gao, H.-W., Zhang, Y.-L., Zhang, Y., Zhou, X.-F., Li, C.-Q., & Gao, H.-P. (2014). Developmental toxicity of diclofenac and elucidation of gene regulation in zebrafish (Danio rerio). *Scientific Reports*, *4*(1), 4841.

Chia, K., Klingseisen, A., Sieger, D., & Priller, J. (2022). Zebrafish as a model organism for human neurodegenerative diseases. *Frontiers in Molecular Neuroscience*, *376*, 15. https://doi.org/10.3389/fnmol.2022.940484

Choi, T.-Y., Choi, T.-I., Lee, Y.-R., Choe, S.-K., & Kim, C.-H. (2021). Zebrafish as an animal model for biomedical research. *Experimental & Molecular Medicine*, *53*(3), 310–317.

Ciura, S., Lattante, S., Le Ber, I., Latouche, M., Tostivint, H., Brice, A., & Kabashi, E. (2013). Loss of function of C9orf72 causes motor deficits in a zebrafish model of amyotrophic lateral sclerosis. *Annals of Neurology*, *74*(2), 180–187.

Coleman, P. J., Gotter, A. L., Herring, W. J., Winrow, C. J., & Renger, J. J. (2017). The discovery of suvorexant, the first orexin receptor drug for insomnia. *Annual Review of Pharmacology and Toxicology*, *57*, 509–533.

Correale, J., Gaitán, M. I., Ysrraelit, M. C., & Fiol, M. P. (2017). Progressive multiple sclerosis: From pathogenic mechanisms to treatment. *Brain*, *140*(3), 527–546.

Coyle, J. T., & Schwarcz, R. (1976). Lesion of striatal neurons with kainic acid provides a model for Huntington's chorea. *Nature*, *263*(5574), 244–246.

Crews, L., & Masliah, E. (2010). Molecular mechanisms of neurodegeneration in Alzheimer's disease. *Human Molecular Genetics*, *19*(R1), R12–R20.

Diekmann, H., Anichtchik, O., Fleming, A., Futter, M., Goldsmith, P., Roach, A., & Rubinsztein, D. C. (2009). Decreased BDNF levels are a major contributor to the embryonic phenotype of huntingtin knockdown zebrafish. *Journal of Neuroscience*, *29*(5), 1343–1349.

Dong, X.-x., Wang, Y., & Qin, Z.-h. (2009). Molecular mechanisms of excitotoxicity and their relevance to pathogenesis of neurodegenerative diseases. *Acta Pharmacologica Sinica*, *30*(4), 379–387.

Doyle, J. M., & Croll, R. P. (2022). A Critical review of zebrafish models of Parkinson's disease. *Frontiers in Pharmacology*, *13*, 835827.

Du, Y., Bales, K. R., Dodel, R. C., Hamilton-Byrd, E., Horn, J. W., Czilli, D. L., Simmons, L. K., Ni, B., & Paul, S. M. (1997). Activation of a caspase 3-related cysteine protease is required for glutamate-mediated apoptosis of cultured cerebellar granule neurons. *Proceedings of the National Academy of Sciences*, *94*(21), 11657–11662.

Du, Y., Dodel, R. C., Bales, K. R., Jemmerson, R., Hamilton-Byrd, E., & Paul, S. M. (1997). Involvement of a caspase-3-like cysteine protease in 1-methyl-4-phenylpyridinium-mediated apoptosis of cultured cerebellar granule neurons. *Journal of -Neurochemistry*, *69*(4), 1382–1388.

Dugger, B. N., & Dickson, D. W. (2017). Pathology of neurodegenerative diseases. *Cold Spring Harbor perspectives in Biology*, *9*(7), a028035.

Faissner, S., & Gold, R. (2019). Progressive multiple sclerosis: Latest therapeutic developments and future directions. *Therapeutic Advances in Neurological Disorders*, *12*, 1756286419878323.

Fei, Q., McCormack, A. L., Di Monte, D. A., & Ethell, D. W. (2008). Paraquat neurotoxicity is mediated by a Bak-dependent mechanism. *Journal of Biological Chemistry*, *283*(6), 3357–3364.

Ferguson, M. W., Kennedy, C. J., Palpagama, T. H., Waldvogel, H. J., Faull, R. L., & Kwakowsky, A. (2022). Current and possible future therapeutic options for Huntington's disease. *Journal of Central Nervous System Disease*, *14*, 11795735221092517.

Ferrante, R. J., Kowall, N. W., Cipolloni, P., Storey, E., & Beal, M. F. (1993). Excitotoxin lesions in primates as a model for Huntington's disease: histopathologic and neurochemical characterization. *Experimental Neurology*, *119*(1), 46–71.

Finkbeiner, S. (2011). Huntington's disease. *Cold Spring Harbor Perspectives in Biology*, *3*(6), a007476.

Flinn, L. J., Keatinge, M., Bretaud, S., Mortiboys, H., Matsui, H., De Felice, E., Woodroof, H. I., Brown, L., McTighe, A., & Soellner, R. (2013). TigarB causes mitochondrial dysfunction and neuronal loss in PINK1 deficiency. *Annals of Neurology*, *74*(6), 837–847.

Fortier, G., Butti, Z., & Patten, S. A. (2020). Modelling C9orf72-related amyotrophic lateral sclerosis in zebrafish. *Biomedicines*, *8*(10), 440.

Foster, A. C., Miller, L. P., Oldendorf, W. H., & Schwarcz, R. (1984). Studies on the disposition of quinolinic acid after intracerebral or systemic administration in the rat. *Experimental Neurology*, *84*(2), 428–440.

Fragoso, Y. D. (2014). Modifiable environmental factors in multiple sclerosis. *Arquivos de neuro-psiquiatria*, *72*, 889–894.

Frischer, J. M., Weigand, S. D., Guo, Y., Kale, N., Parisi, J. E., Pirko, I., Mandrekar, J., Bramow, S., Metz, I., & Brück, W. (2015). Clinical and pathological insights into the dynamic nature of the white matter multiple sclerosis plaque. *Annals of Neurology*, *78*(5), 710–721.

Gandhi, S., Muqit, M., Stanyer, L., Healy, D., Abou-Sleiman, P., Hargreaves, I., Heales, S., Ganguly, M., Parsons, L., & Lees, A. (2006). PINK1 protein in normal human brain and Parkinson's disease. *Brain*, *129*(7), 1720–1731.

Getts, D. R., Martin, A. J., McCarthy, D. P., Terry, R. L., Hunter, Z. N., Yap, W. T., Getts, M. T., Pleiss, M., Luo, X., & King, N. J. (2012). Microparticles bearing encephalitogenic peptides induce T-cell tolerance and ameliorate experimental autoimmune encephalomyelitis. *Nature Biotechnology*, *30*(12), 1217–1224.

Glinka, Y. Y., & Youdim, M. B. (1995). Inhibition of mitochondrial complexes I and IV by 6-hydroxydopamine. *European Journal of Pharmacology: Environmental Toxicology and Pharmacology*, *292*(3–4), 329–332.

Goedert, M., Jakes, R., Qi, Z., Wang, J., & Cohen, P. (1995). Protein phosphatase 2A Is the major enzyme in brain that dephosphorylates τ protein phosphorylated by proline-directed protein kinases or cyclic AMP-dependent protein kinase. *Journal of Neurochemistry*, *65*(6), 2804–2807.

Green, A. J., & Planchart, A. (2018). The neurological toxicity of heavy metals: A fish perspective. *Comparative Biochemistry and Physiology Part C: Toxicology & Pharmacology*, *208*, 12–19.

Haq, M., Gonzalez, N., Mintz, K., Jaja-Chimedza, A., De Jesus, C. L., Lydon, C., Welch, A. Z., & Berry, J. P. (2016). Teratogenicity of ochratoxin A and the degradation product, ochratoxin α, in the zebrafish (Danio rerio) embryo model of vertebrate development. *Toxins*, *8*(2), 40.

Hara, H., Friedlander, R. M., Gagliardini, V., Ayata, C., Fink, K., Huang, Z., Shimizu-Sasamata, M., Yuan, J., & Moskowitz, M. A. (1997). Inhibition of interleukin 1β converting enzyme family proteases reduces ischemic and excitotoxic neuronal damage. *Proceedings of the National Academy of Sciences*, *94*(5), 2007–2012.

Hauser, S. L., & Cree, B. A. (2020). Treatment of multiple sclerosis: A review. *The American Journal of Medicine*, *133*(12), 1380–1390.e1382.

He, X., Zhong, Z.-M., & Che, Y. (2012). Locomotor activity and learning and memory abilities in Alzheimer's disease induced by aluminum in an acid environment in zebrafish. https://doi.org/10.3724/SP.J.1141.2012.02231

Henderson, L., & Hirsch, H. M. (1949). Quinolinic acid metabolism. 1. Urinary excretion by the rat following tryptophan and 3-hydroxyanthranilic acid administration. *Journal of Biological Chemistry*, *181*, 667–675.

Hey, J. A., Yu, J. Y., Versavel, M., Abushakra, S., Kocis, P., Power, A., Kaplan, P. L., Amedio, J., & Tolar, M. (2018). Clinical pharmacokinetics and safety of ALZ-801, a novel prodrug of tramiprosate in development for the treatment of Alzheimer's disease. *Clinical Pharmacokinetics*, *57*, 315–333.

Heyman, A., Fillenbaum, G., Gearing, M., Mirra, S., Welsh-Bohmer, K., Peterson, B., & Pieper, C. (1999). Comparison of Lewy body variant of Alzheimer's disease with pure Alzheimer's disease: Consortium to Establish a registry for Alzheimer's disease, Part XIX. *Neurology*, *52*(9), 1839–1839.

Hin, N., Newman, M., Kaslin, J., Douek, A. M., Lumsden, A., Zhou, X.-F., Mañucat-Tan, N. B., Ludington, A., Adelson, D. L., & Pederson, S. (2018). Accelerated brain aging towards transcriptional inversion in a zebrafish model of familial Alzheimer's disease. *BioRxiv*, 262162.

Huang, M., Jiang, X., Liang, Y., Liu, Q., Chen, S., & Guo, Y. (2017). Berberine improves cognitive impairment by promoting autophagic clearance and inhibiting production of β-amyloid in APP/tau/PS1 mouse model of Alzheimer's disease. *Experimental Gerontology*, *91*, 25–33.

Kabashi, E., Lin, L., Tradewell, M. L., Dion, P. A., Bercier, V., Bourgouin, P., Rochefort, D., Bel Hadj, S., Durham, H. D., & Velde, C. V. (2010). Gain and loss of function of ALS-related mutations of TARDBP (TDP-43) cause motor deficits in vivo. *Human Molecular Genetics*, *19*(4), 671–683.

Karussis, D. (2014). The diagnosis of multiple sclerosis and the various related demyelinating syndromes: a critical review. *Journal of Autoimmunity*, *48*, 134–142.

Kaur, N., Jamwal, S., Kaur Gill, H., & Bansal, P. K. (2017). Animal models of Huntington's disease. *Animal Models of Neurological Disorders: Principle and Working Procedure for Animal Models of Neurological Disorders*, 43–57. https://doi.org/10.1007/978-981-10-5981-0_4

Kelsh, R. N. (2023). Myron *Gordon Award Lecture 2023: Painting the neural crest: How studying pigment cells illuminates neural crest cell biology*. Pigment Cell & Melanoma Research.

Khezri, A., Herranz-Jusdado, J. G., Ropstad, E., & Fraser, T. W. (2018). Mycotoxins induce developmental toxicity and behavioural aberrations in zebrafish larvae. *Environmental Pollution*, *242*, 500–506.

Khotimah, H., Sumitro, S. B., & Widodo, M. A. (2015). Zebrafish Parkinson's model: Rotenone decrease motility, dopamine, and increase α-synuclein aggregation and apoptosis of zebrafish brain. *International Journal of PharmTech Research*, *4*, 614–621.

Kim, S., Lee, D.-W., Schachner, M., & Park, H.-C. (2021). Small compounds mimicking the adhesion molecule L1 improve recovery in a zebrafish demyelination model. *Scientific Reports*, *11*(1), 5878.

Kiper, K., & Freeman, J. L. (2022). Use of zebrafish genetic models to study etiology of the amyloid-beta and neurofibrillary tangle pathways in Alzheimer's disease. *Current Neuropharmacology*, *20*(3), 524.

Koehler, D., Shah, Z. A., Hensley, K., & Williams, F. E. (2018). Lanthionine ketimine-5-ethyl ester provides neuroprotection in a zebrafish model of okadaic acid-induced Alzheimer's disease. *Neurochemistry International*, *115*, 61–68.

Kouli, A., Spindler, L. R., Fryer, T. D., Hong, Y. T., Malpetti, M., Aigbirhio, F. I., White, S. R., Camacho, M., O'Brien, J. T., & Williams-Gray, C. H. (2023). Neuroinflammation is linked to dementia risk in Parkinson's disease. *Brain*, *147*(3), 923–935. https://doi.org/10.1093/brain/awad322

Kouli, A., Torsney, K. M., & Kuan, W.-L. (2018). Parkinson's disease: etiology, neuropathology, and pathogenesis. *Exon Publications*, *8*, 3–26.

Kulkarni, P., Yellanki, S., Medishetti, R., Sriram, D., Saxena, U., & Yogeeswari, P. (2017). Novel zebrafish EAE model: A quick in vivo screen for multiple sclerosis. *Multiple Sclerosis and Related Disorders*, *11*, 32–39.

Kumar, V., Singh, C., & Singh, A. (2021). Zebrafish an experimental model of Huntington's disease: Molecular aspects, therapeutic targets and current challenges. *Molecular Biology Reports*, *48*, 8181–8194.

Lassmann, H. (2013). Pathology and disease mechanisms in different stages of multiple sclerosis. *Journal of the Neurological Sciences*, *333*(1–2), 1–4.

Lee, J.-A., & Cole, G. J. (2007). Generation of transgenic zebrafish expressing green fluorescent protein under control of zebrafish amyloid precursor protein gene regulatory elements. *Zebrafish*, *4*(4), 277–286.

Lee, S., Page-McCaw, P., & Gamse, J. T. (2014). Kctd12 and Ulk2 partner to regulate dendritogenesis and behavior in the habenular nuclei. *PLoS One*, *9*(10), e110280.

Lei, S., & Powers, R. (2013). NMR metabolomics analysis of Parkinson's disease. *Current Metabolomics*, *1*(3), 191–209.

Lelos, M. J., & Dunnett, S. B. (2018). Generating excitotoxic lesion models of Huntington's disease. *Huntington's Disease*, 209–220. https://doi.org/10.1007/978-1-4939-7825-0_11

Lemmens, R., Van Hoecke, A., Hersmus, N., Geelen, V., D'Hollander, I., Thijs, V., Van Den Bosch, L., Carmeliet, P., & Robberecht, W. (2007). Overexpression of mutant superoxide dismutase 1 causes a motor axonopathy in the zebrafish. *Human Molecular Genetics*, *16*(19), 2359–2365.

Li, C. Y., Boldt, H., Parent, E., Ficklin, J., James, A., Anlage, T. J., Boyer, L. M., Pierce, B. R., Siegfried, K. R., & Harris, M. P. (2023). Genetic tools for the study of the mangrove killifish, Kryptolebias marmoratus, an emerging vertebrate model for phenotypic plasticity. *Journal of Experimental Zoology Part B: Molecular and Developmental Evolution*, *342*(3), 164–177.

Li, J., O, W., Li, W., Jiang, Z.-G., & Ghanbari, H. A. (2013). Oxidative stress and neurodegenerative disorders. *International Journal of Molecular Sciences*, *14*(12), 24438–24475.

Li, W., Serpell, L. C., Carter, W. J., Rubinsztein, D. C., & Huntington, J. A. (2006). Expression and characterization of full-length human huntingtin, an elongated HEAT repeat protein. *Journal of Biological Chemistry*, *281*(23), 15916–15922.

Liu, F., Pogoda, H.-M., Pearson, C. A., Ohyama, K., Löhr, H., Hammerschmidt, M., & Placzek, M. (2013). Direct and indirect roles of Fgf3 and Fgf10 in innervation and vascularisation of the vertebrate hypothalamic neurohypophysis. *Development*, *140*(5), 1111–1122.

Liu, Z., Zhou, T., Ziegler, A. C., Dimitrion, P., & Zuo, L. (2017). Oxidative stress in neurodegenerative diseases: from molecular mechanisms to clinical applications. *Oxidative Medicine and Cellular Longevity*, *2017*, 2525967.

López, O. L., & DeKosky, S. T. (2008). Clinical symptoms in Alzheimer's disease. *Handbook of Clinical Neurology*, *89*, 207–216.

Lopez, O. L., Wisnieski, S. R., Becker, J. T., Boiler, F., & DeKosky, S. T. (1997). Extrapyramidal signs in patients with probable Alzheimer disease. *Archives of Neurology*, *54*(8), 969–975.

Lu, X.-L., Lin, Y.-H., Wu, Q., Su, F.-J., Ye, C.-H., Shi, L., He, B.-X., Huang, F.-W., Pei, Z., & Yao, X.-L. (2015). Paeonolum protects against MPP+-induced neurotoxicity in zebrafish and PC12 cells. *BMC Complementary and Alternative Medicine*, *15*(1), 1–10.

Lu, Y., Zhang, X., Zhao, L., Yang, C., Pan, L., Li, C., Liu, K., Bai, G., Gao, H., & Yan, Z. (2018). Metabolic disturbances in the striatum and substantia nigra in the onset and progression of MPTP-induced Parkinsonism model. *Frontiers in Neuroscience*, *12*, 90.

Lublin, F. D., Reingold, S. C., Cohen, J. A., Cutter, G. R., Sørensen, P. S., Thompson, A. J., Wolinsky, J. S., Balcer, L. J., Banwell, B., & Barkhof, F. (2014). Defining the clinical course of multiple sclerosis: The 2013 revisions. *Neurology*, *83*(3), 278–286.

Luchetti, S., Fransen, N. L., van Eden, C. G., Ramaglia, V., Mason, M., & Huitinga, I. (2018). Progressive multiple sclerosis patients show substantial lesion activity that correlates with clinical disease severity and sex: a retrospective autopsy cohort analysis. *Acta neuropathologica*, *135*, 511–528.

Lucini, C., D'Angelo, L., Cacialli, P., Palladino, A., & De Girolamo, P. (2018). BDNF, brain, and regeneration: Insights from zebrafish. *International Journal of Molecular Sciences*, *19*(10), 3155.

MacDonald, M. E., Ambrose, C. M., Duyao, M. P., Myers, R. H., Lin, C., Srinidhi, L., Barnes, G., Taylor, S. A., James, M., & Groot, N. (1993). A novel gene containing a trinucleotide repeat that is expanded and unstable on Huntington's disease chromosomes. *Cell*, *72*(6), 971–983.

Masrori, P., & Van Damme, P. (2020). Amyotrophic lateral sclerosis: A clinical review. *European Journal of Neurology*, *27*(10), 1918–1929.

McCarthy, M. J., Rubin, L. L., & Philpott, K. L. (1997). Involvement of caspases in sympathetic neuron apoptosis. *Journal of Cell Science*, *110*(18), 2165–2173.

Medina, A., Mahjoub, Y., Shaver, L., & Pringsheim, T. (2022). Prevalence and incidence of Huntington's disease: An updated systematic review and meta-analysis. *Movement Disorders*, *37*(12), 2327–2335.

Merrill, J. E. (2009). In vitro and in vivo pharmacological models to assess demyelination and remyelination. *Neuropsychopharmacology*, *34*(1), 55–73.

Mitchell, I. J. (2006). Animal Models of Huntington's Disease. World Scientific. In *Animal Models Of Neuropsychiatric Diseases,* 119–158. https://doi.org/10.1142/9781860948022_0003

Monaco, A., Grimaldi, M., & Ferrandino, I. (2017). Aluminium chloride-induced toxicity in zebrafish larvae. *Journal of Fish Diseases*, *40*(5), 629–635.

Montalban, X., Hauser, S. L., Kappos, L., Arnold, D. L., Bar-Or, A., Comi, G., De Seze, J., Giovannoni, G., Hartung, H.-P., & Hemmer, B. (2017). Ocrelizumab versus placebo in primary progressive multiple sclerosis. *New England Journal of Medicine*, *376*(3), 209–220.

Mora, J. S., Genge, A., Chio, A., Estol, C. J., Chaverri, D., Hernández, M., Marín, S., Mascias, J., Rodriguez, G. E., & Povedano, M. (2020). Masitinib as an add-on therapy to riluzole in patients with amyotrophic lateral sclerosis: a randomized clinical trial. *Amyotrophic Lateral Sclerosis and Frontotemporal Degeneration*, *21*(1–2), 5–14.

Mukherjee, P., Roy, S., Ghosh, D., & Nandi, S. (2022). Role of animal models in biomedical research: a review. *Laboratory Animal Research, 38*(1), 18.

Niedzielska, E., Smaga, I., Gawlik, M., Moniczewski, A., Stankowicz, P., Pera, J., & Filip, M. (2016). Oxidative stress in neurodegenerative diseases. *Molecular Neurobiology, 53*, 4094–4125.

Nopoulos, P. C. (2016). Huntington disease: a single-gene degenerative disorder of the striatum. *Dialogues in Clinical Neuroscience, 18*(1), 91–98.

Nornes, S., Groth, C., Camp, E., Ey, P., & Lardelli, M. (2003). Developmental control of Presenilin1 expression, endoproteolysis, and interaction in zebrafish embryos. *Experimental Cell Research, 289*(1), 124–132.

Ochalek, A., Mihalik, B., Avci, H. X., Chandrasekaran, A., Téglási, A., Bock, I., Giudice, M. L., Táncos, Z., Molnár, K., & László, L. (2017). Neurons derived from sporadic Alzheimer's disease iPSCs reveal elevated TAU hyperphosphorylation, increased amyloid levels, and GSK3B activation. *Alzheimer's Research & Therapy, 9*(1), 1–19.

Ohki, Y., Wenninger-Weinzierl, A., Hruscha, A., Asakawa, K., Kawakami, K., Haass, C., Edbauer, D., & Schmid, B. (2017). Glycine-alanine dipeptide repeat protein contributes to toxicity in a zebrafish model of C9orf72 associated neurodegeneration. *Molecular Neurodegeneration, 12*(1), 1–11.

Okouchi, M., Ekshyyan, O., Maracine, M., & Aw, T. Y. (2007). Neuronal apoptosis in neurodegeneration. *Antioxidants & Redox Signaling, 9*(8), 1059–1096.

Onose, G., Anghelescu, A., Blendea, D., Ciobanu, V., Daia, C., Firan, F. C., Oprea, M., Spinu, A., Popescu, C., & Ionescu, A. (2022). Cellular and Molecular Targets for Non-Invasive, Non-Pharmacological Therapeutic/Rehabilitative Interventions in Acute Ischemic Stroke. *International Journal of Molecular Sciences, 23*(2), 907.

Paduraru, E., Iacob, D., Rarinca, V., Plavan, G., Ureche, D., Jijie, R., & Nicoara, M. (2023). Zebrafish as a potential model for neurodegenerative diseases: A focus on toxic metals implications. *International Journal of Molecular Sciences, 24*(4), 3428.

Palfi, S., Leventhal, L., Goetz, C. G., Hantraye, P., Roitberg, B. Z., Sramek, J., Emborg, M., & Kordower, J. H. (2000). Delayed onset of progressive dystonia following subacute 3-nitropropionic acid treatment in Cebus apella monkeys. *Movement disorders: official journal of the Movement Disorder Society, 15*(3), 524–530.

Paquet, D., Bhat, R., Sydow, A., Mandelkow, E.-M., Berg, S., Hellberg, S., Fälting, J., Distel, M., Köster, R. W., & Schmid, B. (2009). A zebrafish model of tauopathy allows in vivo imaging of neuronal cell death and drug evaluation. *The Journal of Clinical Investigation, 119*(5), 1382–1395.

Park, J.-H., & Kim, H. (2023). Harnessing CRISPR/Cas9 for Enhanced Disease Resistance in Hot Peppers: A Comparative Study on CaMLO2-Gene-Editing Efficiency across Six Cultivars. *International Journal of Molecular Sciences, 24*(23), 16775.

Phillips, N. L. H., & Roth, T. L. (2019). Animal models and their contribution to our understanding of the relationship between environments, epigenetic modifications, and behavior. *Genes, 10*(1), 47.

Powers, R., Lei, S., Anandhan, A., Marshall, D. D., Worley, B., Cerny, R. L., Dodds, E. D., Huang, Y., Panayiotidis, M. I., & Pappa, A. (2017a). Metabolic investigations of the molecular mechanisms associated with Parkinson's disease. *Metabolites, 7*(2), 22.

Powers, S., Kwok, S., Lovejoy, E., Lavin, T., & Sher, R. B. (2017b). Editor's highlight: embryonic exposure to the environmental neurotoxin BMAA negatively impacts early neuronal development and progression of neurodegeneration in the Sod1-G93R Zebrafish model of amyotrophic lateral sclerosis. *Toxicological Sciences, 157*(1), 129–140.

Qi, R., Sammler, E., Gonzalez-Hunt, C. P., Barraza, I., Pena, N., Rouanet, J. P., Naaldijk, Y., Goodson, S., Fuzzati, M., & Blandini, F. (2023). A blood-based marker of mitochondrial DNA damage in Parkinson's disease. *Science Translational Medicine, 15*(711), eabo1557.

Ramaswamy, S., McBride, J. L., & Kordower, J. H. (2007). Animal models of Huntington's disease. *Ilar Journal, 48*(4), 356–373.

Rami, L., Sala-Llonch, R., Solé-Padullés, C., Fortea, J., Olives, J., Lladó, A., Pena-Gómez, C., Balasa, M., Bosch, B., & Antonell, A. (2012). Distinct functional activity of the precuneus and posterior cingulate cortex during encoding in the preclinical stage of Alzheimer's disease. *Journal of Alzheimer's Disease, 31*(3), 517–526.

Rana, N., Singh, C., & Singh, A. (2023). Modeling Huntington's disease: An insight on in-vitro and in-vivo models. *Behavioural Brain Research, 459*, 114757.

Rang, H. P., Dale, M. M., Ritter, J. M., Flower, R. J., & Henderson, G. (2011). *Rang & Dale's pharmacology.* Elsevier Health Sciences.

Reich, D. (2018). Lucchinetti CF and Calabresi PA. *Multiple Sclerosis. The New England Journal of Medicine, 378*, 169–180.

Robea, M.-A., Balmus, I.-M., Ciobica, A., Strungaru, S., Plavan, G., Gorgan, L. D., Savuca, A., & Nicoara, M. (2020). Parkinson's disease-induced zebrafish models: Focussing on oxidative stress implications and sleep processes. *Oxidative Medicine and Cellular Longevity, 2020*, 1–15.

Roberts, R. C., Ahn, A., Swartz, K. J., Beal, M. F., & DiFiglia, M. (1993). Intrastriatal injections of quinolinic acid or kainic acid: Differential patterns of cell survival and the effects of data analysis on outcome. *Experimental Neurology, 124*(2), 274–282.

Rocca, P., Leotta, D., Liffredo, C., Mingrone, C., Sigaudo, M., Capellero, B., Rocca, G., Simoncini, M., Pirfo, E., & Bogetto, F. (2010). Neuropsychiatric symptoms underlying caregiver stress and insight in Alzheimer's disease. *Dementia and Geriatric Cognitive Disorders, 30*(1), 57–63.

Sallinen, V., Kolehmainen, J., Priyadarshini, M., Toleikyte, G., Chen, Y.-C., & Panula, P. (2010). Dopaminergic cell damage and vulnerability to MPTP in Pink1 knockdown zebrafish. *Neurobiology of Disease, 40*(1), 93–101.

Saluja, D., Jhanji, R., Kaushal, S., Verma, B., Sharma, N., Singh, R., Agrawal, S., Yadav, M., Kumar, A., & Singh, C. (2021). Importance of zebrafish as an efficient research model for the screening of novel therapeutics in neurological disorders. *CNS & Neurological Disorders-Drug Targets (Formerly Current Drug Targets-CNS & Neurological Disorders), 20*(2), 145–157.

Saraswat, P., Chaturvedi, A., & Ranjan, R. (2023). Zinc Finger Nuclease (ZFNs) and Transcription Activator-like Effector Nucleases (TALENs) based genome editing in enhancement of anticancer activity of plants. In *Plant-derived anticancer drugs in the OMICS era* (pp. 281–293). Apple Academic Press.

Saudou, F., & Humbert, S. (2016). The biology of huntingtin. *Neuron, 89*(5), 910–926.

Schapira, A. H. (2009). Etiology and pathogenesis of Parkinson disease. *Neurologic Clinics, 27*(3), 583–603.

Schiffer, N. W., Broadley, S. A., Hirschberger, T., Tavan, P., Kretzschmar, H. A., Giese, A., Haass, C., Hartl, F. U., & Schmid, B. (2007). Identification of anti-prion compounds as efficient inhibitors of polyglutamine protein aggregation in a zebrafish model. *Journal of Biological Chemistry, 282*(12), 9195–9203.

Schulte, J., & Littleton, J. T. (2011). The biological function of the Huntingtin protein and its relevance to Huntington's disease pathology. *Current Trends in Neurology, 5*, 65.

Seegobin, S. P., Heaton, G. R., Liang, D., Choi, I., Blanca Ramirez, M., Tang, B., & Yue, Z. (2020). Progress in LRRK2-associated Parkinson's disease animal models. *Frontiers in Neuroscience, 14*, 674.

Senger, M. R., Rosemberg, D. B., Seibt, K. J., Dias, R. D., Bogo, M. R., & Bonan, C. D. (2010). Influence of mercury chloride on adenosine deaminase activity and gene expression in zebrafish (Danio rerio) brain. *Neurotoxicology, 31*(3), 291–296.

Senger, M. R., Seibt, K. J., Ghisleni, G. C., Dias, R. D., Bogo, M. R., & Bonan, C. D. (2011). Aluminum exposure alters behavioral parameters and increases acetylcholinesterase activity in zebrafish (Danio rerio) brain. *Cell Biology and Toxicology, 27*, 199–205.

Sengoku, R. (2020). Aging and Alzheimer's disease pathology. *Neuropathology, 40*(1), 22–29.

Sheng, D., Qu, D., Kwok, K. H. H., Ng, S. S., Lim, A. Y. M., Aw, S. S., Lee, C. W. H., Sung, W. K., Tan, E. K., & Lufkin, T. (2010). Deletion of the WD40 domain of LRRK2 in Zebrafish causes Parkinsonism-like loss of neurons and locomotive defect. *PLoS Genetics, 6*(4), e1000914.

Simonian, N., & Coyle, J. (1996). Oxidative stress in neurodegenerative diseases. *Annual Review of Pharmacology and Toxicology, 36*(1), 83–106.

Singh, V. K., & Seed, T. M. (2021). How necessary are animal models for modern drug discovery? *Expert Opinion on Drug Discovery, 16*(12), 1391–1397.

Sintzel, M. B., Rametta, M., & Reder, A. T. (2018). Vitamin D and multiple sclerosis: A comprehensive review. *Neurology and Therapy, 7*, 59–85.

Smith-Dijak, A. I., Sepers, M. D., & Raymond, L. A. (2019). Alterations in synaptic function and plasticity in Huntington disease. *Journal of Neurochemistry, 150*(4), 346–365.

Steffan, J. S., Agrawal, N., Pallos, J., Rockabrand, E., Trotman, L. C., Slepko, N., Illes, K., Lukacsovich, T., Zhu, Y.-Z., & Cattaneo, E. (2004). SUMO modification of Huntingtin and Huntington's disease pathology. *Science, 304*(5667), 100–104.

Stern, Y., Albert, M., Brandt, J., Jacobs, D., Tang, M., Marder, K., Bell, K., Sano, M., Devanand, D. P., & Bylsma, F. (1994). Utility of extrapyramidal signs and psychosis as predictors of cognitive and functional decline, nursing home admission, and death in Alzheimer's disease: Prospective analyses from the Predictors study. *Neurology, 44*(12), 2300–2300.

Stone, T. W., & Perkins, M. N. (1981). Quinolinic acid: a potent endogenous excitant at amino acid receptors in CNS. *European journal of pharmacology, 72*(4), 411–412.

Swinnen, B., Bento-Abreu, A., Gendron, T. F., Boeynaems, S., Bogaert, E., Nuyts, R., Timmers, M., Scheveneels, W., Hersmus, N., & Wang, J. (2018). A zebrafish model for C9orf72 ALS reveals RNA toxicity as a pathogenic mechanism. *Acta Neuropathologica, 135*, 427–443.

Takano, H., & Gusella, J. F. (2002). The predominantly HEAT-like motif structure of huntingtin and its association and coincident nuclear entry with dorsal, an NF-kB/Rel/dorsal family transcription factor. *BMC Neuroscience, 3*, 1–13.

Takashima, A., Murayama, M., Murayama, O., Kohno, T., Honda, T., Yasutake, K., Nihonmatsu, N., Mercken, M., Yamaguchi, H., & Sugihara, S. (1998). Presenilin 1 associates with glycogen synthase kinase-3β and its substrate tau. *Proceedings of the National Academy of Sciences, 95*(16), 9637–9641.

Taylor, J. P., Brown Jr, R. H., & Cleveland, D. W. (2016). Decoding ALS: from genes to mechanism. *Nature, 539*(7628), 197–206.

Teame, T., Zhang, Z., Ran, C., Zhang, H., Yang, Y., Ding, Q., Xie, M., Gao, C., Ye, Y., & Duan, M. (2019). The use of zebrafish (Danio rerio) as biomedical models. *Animal Frontiers, 9*(3), 68–77.

Teleanu, D. M., Niculescu, A.-G., Lungu, I. I., Radu, C. I., Vladâcenco, O., Roza, E., Costăchescu, B., Grumezescu, A. M., & Teleanu, R. I. (2022). An overview of oxidative stress, neuroinflammation, and neurodegenerative diseases. *International Journal of Molecular Sciences, 23*(11), 5938.

Truant, R., Atwal, R. S., & Burtnik, A. (2007). Nucleocytoplasmic trafficking and transcription effects of huntingtin in Huntington's disease. *Progress in Neurobiology, 83*(4), 211–227.

Túnez, I., & Santamaría, A. (2009). Modelo de enfermedad de Huntington inducido con ácido 3-nitropropiónico. *Rev. neurol.(Ed. impr.)*, 430–434.

Túnez, I., Tasset, I., Pérez-De La Cruz, V., & Santamaría, A. (2010). 3-Nitropropionic acid as a tool to study the mechanisms involved in Huntington's disease: past, present and future. *Molecules, 15*(2), 878–916.

Ünal, İ., & Emekli-Alturfan, E. (2019). Fishing for Parkinson's Disease: A review of the literature. *Journal of Clinical Neuroscience, 62*, 1–6.

Ungerstedt, U., Ljungberg, T., & Steg, G. (1974). Behavioral, physiological, and neurochemical changes after 6-hydroxydopamine-induced degeneration of the nigro-striatal dopamine neurons. *Advances in Neurology, 5*, 421–426.

Van Dam, D., & De Deyn, P. P. (2011). Animal models in the drug discovery pipeline for Alzheimer's disease. *British Journal of Pharmacology, 164*(4), 1285–1300.

van Kesteren, R. E., Fainzilber, M., Hauser, G., van Minnen, J., Vreugdenhil, E., Smit, A. B., Ibáñez, C. F., Geraerts, W. P., & Bulloch, A. G. (1998). Early evolutionary origin of the neurotrophin receptor family. *The EMBO Journal, 17*(9), 2534–2542.

Veldman, M. B., Rios-Galdamez, Y., Lu, X.-H., Gu, X., Qin, W., Li, S., Yang, X. W., & Lin, S. (2015). The N17 domain mitigates nuclear toxicity in a novel zebrafish Huntington's disease model. *Molecular Neurodegeneration, 10*(1), 1–16.

Wang, J., & Cao, H. (2021). Zebrafish and medaka: Important animal models for human neurodegenerative diseases. *International Journal of Molecular Sciences*, *22*(19), 10766.

Wang, X., Sun, G., Feng, T., Zhang, J., Huang, X., Wang, T., Xie, Z., Chu, X., Yang, J., & Wang, H. (2019). Sodium oligomannate therapeutically remodels gut microbiota and suppresses gut bacterial amino acids-shaped neuroinflammation to inhibit Alzheimer's disease progression. *Cell Research*, *29*(10), 787–803.

Wang, Y., Liu, W., Yang, J., Wang, F., Sima, Y., Zhong, Z.-m., Wang, H., Hu, L.-F., & Liu, C.-F. (2017). Parkinson's disease-like motor and non-motor symptoms in rotenone-treated zebrafish. *NeuroToxicology*, *58*, 103–109.

Williams, A., Sarkar, S., Cuddon, P., Ttofi, E. K., Saiki, S., Siddiqi, F. H., Jahreiss, L., Fleming, A., Pask, D., & Goldsmith, P. (2008). Novel targets for Huntington's disease in an mTOR-independent autophagy pathway. *Nature Chemical Biology*, *4*(5), 295–305.

Wilson, D. M., Cookson, M. R., Van Den Bosch, L., Zetterberg, H., Holtzman, D. M., & Dewachter, I. (2023). Hallmarks of neurodegenerative diseases. *Cell*, *186*(4), 693–714.

Wiprich, M. T., Zanandrea, R., Altenhofen, S., & Bonan, C. D. (2020). Influence of 3-nitropropionic acid on physiological and behavioral responses in zebrafish larvae and adults. *Comparative Biochemistry and Physiology Part C: Toxicology & Pharmacology*, *234*, 108772.

Wirbisky, S. E., Weber, G. J., Lee, J.-W., Cannon, J. R., & Freeman, J. L. (2014). Novel dose-dependent alterations in excitatory GABA during embryonic development associated with lead (Pb) neurotoxicity. *Toxicology Letters*, *229*(1), 1–8.

WHO (2022). Parkinson disease: a public health approach. Technical brief. Geneva, 2022. Licence: CC BY-NC-SA 3.0 IGO.

Wyss-Coray, T. (2016). Ageing, neurodegeneration and brain rejuvenation. *Nature*, *539*(7628), 180–186.

Xi, Y., Noble, S., & Ekker, M. (2011). Modeling neurodegeneration in zebrafish. *Current Neurology and Neuroscience Reports*, *11*, 274–282.

Xi, Y., Ryan, J., Noble, S., Yu, M., Yilbas, A. E., & Ekker, M. (2010). Impaired dopaminergic neuron development and locomotor function in zebrafish with loss of pink1 function. *European Journal of Neuroscience*, *31*(4), 623–633.

Xiong, N., Long, X., Xiong, J., Jia, M., Chen, C., Huang, J., Ghoorah, D., Kong, X., Lin, Z., & Wang, T. (2012). Mitochondrial complex I inhibitor rotenone-induced toxicity and its potential mechanisms in Parkinson's disease models. *Critical Reviews in Toxicology*, *42*(7), 613–632.

Xu, L., Liu, T., Liu, L., Yao, X., Chen, L., Fan, D., Zhan, S., & Wang, S. (2020). Global variation in prevalence and incidence of amyotrophic lateral sclerosis: a systematic review and meta-analysis. *Journal of Neurology*, *267*, 944–953.

Yamada, M., Itoh, Y., Otomo, E., Suematsu, N., & Matsushita, M. (1996). Dementia of the Alzheimer type and related dementias in the aged: DAT subgroups and senile dementia of the neurofibrillary tangle type. *Neuropathology*, *16*(2), 89–98.

Yamanaka, T., & Matsui, H. (2023). Modeling familial and sporadic Parkinson's diseases in small fishes. *Development, Growth & Differentiation*, *66*(1), 4–20.

Yang, N., Sheridan, A., & Wexler, P. (2014). *Encyclopedia of toxicology*. Elsevier.

Yokota, O., Miki, T., Ikeda, C., Nagao, S., Takenoshita, S., Ishizu, H., Haraguchi, T., Kuroda, S., Terada, S., & Yamada, N. (2018). Neuropathological comorbidity associated with argyrophilic grain disease. *Neuropathology*, *38*(1), 82–97.

Young, A. B., Greenamyre, J. T., Hollingsworth, Z., Albin, R., D'Amato, C., Shoulson, I., & Penney, J. B. (1988). NMDA receptor losses in putamen from patients with Huntington's disease. *Science*, *241*(4868), 981–983.

Zeng, X.-S., Geng, W.-S., & Jia, J.-J. (2018). Neurotoxin-induced animal models of Parkinson disease: pathogenic mechanism and assessment. *ASN Neuro*, *10*, 1759091418777438.

Zhao, Q.-F., Tan, L., Wang, H.-F., Jiang, T., Tan, M.-S., Tan, L., Xu, W., Li, J.-Q., Wang, J., & Lai, T.-J. (2016). The prevalence of neuropsychiatric symptoms in Alzheimer's disease: systematic review and meta-analysis. *Journal of Affective Disorders*, *190*, 264–271.

Zhu, B., Wang, Q., Shi, X., Guo, Y., Xu, T., & Zhou, B. (2016). Effect of combined exposure to lead and decabromodiphenyl ether on neurodevelopment of zebrafish larvae. *Chemosphere*, *144*, 1646–1654.

Zhu, Y., Auer, F., Gelnaw, H., Davis, S. N., Hamling, K. R., May, C. E., Ahamed, H., Ringstad, N., Nagel, K. I., & Schoppik, D. (2023). Scalable Apparatus to Measure Posture and Locomotion (SAMPL): A high-throughput solution to study unconstrained vertical behavior in small animals. *bioRxiv*, 2023.2001. 2007.523102.

Zon, L. I., & Peterson, R. T. (2005). In vivo drug discovery in the zebrafish. *Nature Reviews Drug Discovery*, *4*(1), 35–44.

# 5

## *Zebrafish Model to Investigate Neurodegenerative Aspects of Autism Spectrum Disorders*

**Gizem Eğilmezer and Ebru Emekli-Alturfan**

### Abbreviations

| | |
|---|---|
| **Aβ** | Amyloid β Peptide |
| **AD** | Alzheimer's Disease |
| **α-syn** | α-Synuclein |
| **ALS** | Amyotrophic Lateral Sclerosis |
| **APP** | Amyloid Precursor Protein |
| **APLP1** | APP-Like Protein 1 |
| **APLP2** | APP-Like Protein 2 |
| **ASD** | Autism Spectrum Disorder |
| **CHOP** | C/EBP Homologous Protein |
| **DIA1R** | Deleted in Autism-1 Related |
| **ENU** | N-Ethyl-N-Nitrosourea |
| **EEG** | Electroencephalography |
| **FAD** | Familial Alzheimer's Diseases |
| **GABA** | Gamma-Aminobutyric Acid |
| **MAPs** | Microtubule-Associated Proteins |
| **MS** | Multiple Sclerosis |
| **PC** | Purkinje Cells |
| **PD** | Parkinson's Disease |
| **TSC** | Tuberous Sclerosis Complex |
| **TSC2** | Tuberous Sclerosis Complex 2 |
| **ZFN** | Zinc Finger Nuclease |

### Introduction

Despite the differences in terms of their developmental processes and molecular mechanisms, the discovery of common points between neurodegenerative and neurodevelopmental disorders in recent years is remarkable. Although the neurodegeneration process has a chronic course, symptoms in neurodegenerative diseases appear only when the damage and damage-related loss in neuron populations is large (Ten Kate et al., 2018; Leuzy et al., 2018; Salomone et al., 2012; Schapira et al., 2017). For this reason, late intervention in neurodegenerative diseases also hinders the efficiency of symptomatic treatments, which already have limited effectiveness (Jęśko et al., 2020).

Accumulation of proteins with misfolding properties occurs as a characteristic feature of Parkinson's disease (PD) and Alzheimer's disease (AD), two major neurodegenerative diseases. Microtubule-associated proteins (MAPs) interact with the microtubules of the cytoskeleton, ensuring the stability of the cell and its internal structures and the transport of components within the cell. Of these, tau is known as type II MAP and is associated with AD because neurodegeneration in AD is accompanied by massive accumulation of tau along with amyloid β (Aβ) peptide. α-Synuclein (α-syn) is a presynaptic neuronal protein, and its abnormal soluble oligomeric structures, called protofibrils, are associated with genetic and neuropathological disorders that occur in PD.

Autism spectrum disorders (ASD), on the other hand, are neurodevelopmental disorders with symptoms that begin in early childhood and are characterized by significant deficits in social communication and interaction and limited, repetitive behaviors and interests (Hodges et al., 2020). Its incidence in society has increased over time, and currently this prevalence is around 1% (Maenner et al., 2020).

ASD is an important disorder with a chronic course that negatively affects the daily lives and social environments of both the affected individuals and their families. It is thought that genetic and environmental factors play a role together in the formation of this disease, but the etiology of ASD has not been fully elucidated (Pendergrass et al., 2014). Although the etiology of the disease has not yet been fully elucidated, neuroanatomical and functional changes are found in different brain regions of ASD patients, and abnormalities have been reported most frequently in the amygdala, frontal cortex, temporal lobe, and cerebellum related to language and speech functions (Lombarda et al., 2011). In addition, it has been shown that there are limbic system anomalies in ASD, smaller neurons in the hippocampus, and dendritic branches decrease over time (Raymond et al., 1995).

In studies, it has been reported that there are many thin and short-range axons in the anterior cingulate cortex of the brains of people with ASD and a decrease in long-range axons and abnormal dendritic branching. Also, it has been determined that the myelin thickness of axons in the orbitofrontal cortex decreases and thus the axon diameter decreases (Zikopoulos et al., 2010). These regions where structural abnormalities


DOI: 10.1201/9781032676555-7

have been reported are brain regions associated with attention, social interactions, and understanding emotions. Irregularities in some skeletal proteins found in the neuron and synapse structure are thought to be associated with ASD (Guang et al., 2018).

In addition, connections have been found between ASD and conditions that negatively affect the neurodevelopmental process, such as disruptions in neurogenesis mechanisms, synaptic plasticity, and pathologies in synaptic functions (Gilbert et al., 2017).

It has been reported that some factors cause damage to neurons due to the production of neurotoxic substances directly or indirectly through immune system activation, and this situation is linked to the etiopathogenesis of ASD (Kern et al., 2013).

In various studies, it has been found that people suffering from ASD have smaller microcolumns in the frontal and temporal cortex, fewer neurons in the amygdala, and degeneration of Purkinje cells in the cerebellum (Schumann et al., 2006). Contrary to all these, there are also thoughts that there is no data related to neurodegeneration in ASD (World Health Organization, 2005).

In recent years, possible common mechanisms between ASD and neurodegenerative diseases have been suggested (Alexiou et al., 2018; Khan et al., 2016; Lahiri et al., 2013). Changes in the metabolism of key proteins involved in neurodegeneration, such as Aβ, tau, and possibly α-syn, have been suggested as possible mechanisms. In this section, current evidence on the relation between the neurodegenerative pathways and the molecular mechanisms in ASD development will be presented in zebrafish as a model organism.

## Autism Spectrum Disorder with Controversial Neurodegenerative Aspects

### History of Autism Spectrum Disorders and Differential Diagnosis

ASD is a neurodevelopmental disorder characterized by marked deficits in social and communication domains and characteristic behaviors and/or limited interests. The term autism, derived from the ancient Greek word "authos" meaning "self," was first used by Bleuler in 1911 to describe the detachment of schizophrenia patients from the outside world (Donvan & Zucker, 2018). The use of the term "autism" differently from how it is expressed in schizophrenia was made by Leo Kanner in 1943; autism, previously known as "early onset schizophrenia", "childhood psychosis," or "childhood schizophrenia", was evaluated for the first time as a disorder different from schizophrenia by Kanner. In these cases, common features such as difficulty in communicating with the environment, echolalia, reaction to differences, insistence on sameness, late onset of speech, wrong choice of pronouns, and repetitive behaviors were observed (Kanner, 1943). Autism was first defined as a "spectrum disorder" by Wing and Gould in 1979, and it was noted that it could occur in different severities and types in individuals with ASD (Wing & Gould,1979).

### Differential Diagnosis of Autism Spectrum Disorders

As the name suggests, there are a number of clinical diseases that have some characteristics with ASD, therefore a comprehensive diagnostic assessment is essential including the evaluation of the situations specified below.

Intellectual disability: ASD should be considered when the social communication level is below what is expected based on the general development level as the disease is often seen together with intellectual disability (Fuentes et al., 2014).

Reactive attachment disorder: Deficiencies in social and communication areas may be observed in a way that resembles ASD symptoms due to attachment problems, inadequate care, and loveless growth in childhood, when basic conditions are not fully met (Volkmar et al., 2014).

Visual and hearing impairment: Children with ASD reacting late or poorly to auditory stimuli and having speech problems may lead to suspicions of a hearing impairment. In order to understand this precisely, hearing tests must be performed (Fuentes et al., 2014).

Language disorders: Language problems experienced by children affect their interaction with the outside world, and this can be confused with ASD symptoms. In order to resolve this confusion, it is necessary to look at the way children with this problem express their interests and desires, their interest in their environment, and their ability to use gestures and facial expressions, and then decide whether there are symptoms of ASD (Volkmar et al., 2014).

Very early onset schizophrenia: The reason ASD has similarities with this condition is the decrease in emotional expression and the symptoms of negativism, in which individuals resist the commands given. Very early onset schizophrenia begins to show symptoms after a normal developmental period. In this case, the general symptoms of ASD are not seen (Fuentes et al., 2014).

Selective mutism: Anxiety and certain communication problems are common problems in this group; children in this group do not communicate in unfamiliar environments. They communicate normally in the environment they know and there is no problem with their social skills. Thus, their differences from ASD can be understood.

Landau-Kleffner syndrome: Individuals suffering from Landau-Kleffner syndrome develop normally until the age of three or four. After this period, seizures or sleep electroencephalogram (sleep EEG) abnormalities begin and are characterized by a major decline in language development over time. The best method for differential diagnosis is a neurological examination and EEG (Fuentes et al., 2014).

### Etiology of Autism Spectrum Disorders

ASD and the factors that cause this disorder have often been the focus of research since ASD was first described. While it was previously thought that incorrect and inappropriate parental behavior caused autism, it has become increasingly accepted that autism is a brain disease with a biological aspect. Studies conducted so far have focused on investigating the structural and functional changes in the brain, the age at which these changes begin, and the factors that lead to these changes, but the etiology of ASD has not been fully elucidated. According to the current literature, ASD is thought to be a multifactorial, complex, and heterogeneous neurodevelopmental disorder in which many genetic and environmental factors play a role. Factors that play a role in the etiology of ASD are listed below.

## Genetic Factors

Studies to date indicate that ASD is mostly caused by genetic inheritance or environmental conditions in utero that can lead to epigenetic modifications in the zygote genome.

Considering that there are more than 600 known ASD-associated genes, it is important to understand the cellular mechanisms that enable ASD behaviors to occur. Hundreds of genes have been identified as ASD risk factors or ASD-related genes. This means that mutations in these genes are associated with the presence of ASD in the individual. Polymorphisms in DNA can be observed through point mutations, copy number variants, deletions, insertions, epigenetic changes, and different genes on different chromosomes.

It has been determined that most of these genes play a role in cellular processes that are of great importance for the general brain structure in neuron development and regulation. This is consistent with the notion that ASD is a neurodevelopmental disorder, as the phenotypic expression of many ASD-related genes results in changes in neuron structure and regulation that impact social and behavioral development. Considering that genetic change can cause a change in the structure and function of the encoded molecules, polymorphisms in neurogenesis and synaptogenesis genes can affect the concentration of neurons and synapses in normal functioning and therefore lead to atypical neuron signaling. In fact, these changes are related to ASD, which may result from synaptogenesis factor polymorphisms. Therefore, one possible implication is that changes in genes leading to neurogenesis or synaptogenesis factors are associated with ASD, given that the neurodevelopmental disorder is also associated with atypical sensory processing and excitation (Kenyon, 2024).

Recent studies have found that variations in the oxytocin and vasopressin (OXT-AVP) pathway, as well as AGTR2 gene variants, may negatively affect the physiological processes responsible for learning, social behavior, and memory observed in individuals with ASD (Rylaarsdam et al., 2019). Current studies show that there is a relationship between a disorder in the secretory function of the Golgi apparatus, which negatively affects neural structure and function, and the development of ASD symptoms.

In addition to many genes responsible for the secretory function of the Golgi, abnormal post-translational modification of various proteins involved in this function has occurred in children diagnosed with ASD. The human c3orf58 gene, found at chromosome position 3q24, has recently been renamed DIA1 because of its deletion in ASD. Accordingly, human genes C3orf58 (Deleted in Autism- DIA-1) and cXorf36 or Deleted in Autism-1 Related (DIA1R) are among the genes related with ASD (Doan et al., 2019). In parallel, some studies have proven the vital role of DIA-I in X-linked ASD and mental retardation (Sajdel-Sulkowska et al., 2019).

Additionally, studies in rodents have shown the increasing importance of serotonin and serotonin receptors in neuropsychiatric disorders, brain development, and ASD (Fatemi, 2002). The amounts of serotonin in the blood are significantly reduced in mothers and children with ASD, but in other family members this level is expectedly normal. On the contrary, other studies have shown that circulating serotonin levels increase in individuals diagnosed with ASD, and this is explained by the decrease in the expression of the serotonin transporter 5HT2A, which ensures the removal of free serotonin (Muller et al., 2016 ).

## Epigenetic Factors

Epigenetic modifications indicate differences in gene expression and, accordingly, in the organism, through mechanisms such as DNA methylation and histone modification, without external conditions cause alterations in the DNA sequence. It has been stated in various studies that situations such as environmental toxins, infection, heavy metals, and drug ingredients used by the mother with natural or unnatural contact during pregnancy can lead to various DNA modifications and problems in early fetal brain development (Siu and Weksberg, 2017).

Although connections between autism and epigenetics have been shown, some studies suggested that epigenetics had no effect on autism. In a study, it was observed that there was no difference in the global methylation rate between the brain of a healthy individual and the brain of an individual with autism (Ginsberg et al., 2012). In this study Ginsberg et al., reported no difference in the methylation of candidate genes. It was concluded that the differences in gene expressions in autism may result from regulatory mechanisms other than gene methylation. Conflicting studies on the subject create the need to further investigate the role of epigenetics in autism (Eshraghi et al., 2018).

## Environmental Factors

### *Maternal Exposure*

Various factors such as pollutants like pesticides that the mother is in contact with in the external environment during pregnancy, medications used by the mother, infections, poor quality and inadequate nutrition, and nutrition deficient in essential vitamins and minerals may negatively affect the fetal brain development and nervous system formation process and increase the likelihood of ASD (Qin et al., 2024).

### *Early Developmental Stages*

In the early developmental stages, the brain develops rapidly and neural networks form. A problem that occurs during this critical period can increase the risk of ASD by interfering with the normal process of brain structure and functioning. For example, lack of early social communication and delayed language development may be among the first symptoms of ASD.

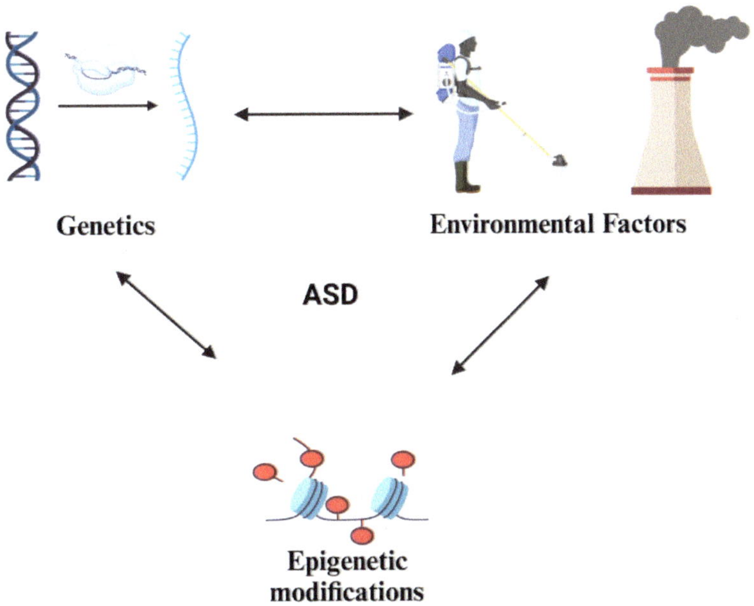


**FIGURE 5.1** Genetic-environmental interactions in ASD.

These abnormalities cause difficulties in the brain's ability to process information in the nervous system, establish connections, and adapt to the external environment (Pinto-Martin et al., 2005; Qin et al., 2024).

*Environmental and Genetic Interactions*

As seen in Figure 5.1, ASD occurs when the environmental and genetic factors interact with each other. Certain genetic sensitivities may occur in response to various environmental stimuli, which may lead to the formation of ASD (Wender et al., 2017; Qin et al., 2024).

## Autism Spectrum Disorder and Neuron Damage

The existence of neuronal damage that occurs in the development of ASD continues to be a matter of debate. The loss of function at the neurological level that occurs in individuals with ASD can also be referred to as neurodegeneration or a type of encephalopathy. When neurodegeneration is mentioned, we are talking about various losses of neurons at the structural or functional level. In neurodegenerative diseases such as PD, AD, or Huntington's disease, early clinical symptoms occur as neurological loss, and this is also observed in children diagnosed with ASD and whose symptoms progress (Kern et al., 2013).

In a study, neuronal damage that increased with age was detected in individuals with ASD (Courchesne et al., 2007). Additionally, evidence shows that the neuronal structure of children with ASD differs from the neuronal structure of healthy individuals. It has been stated that in ASD cases, the number of dendritic extensions and the cell body size of the neuron decrease compared to the expected neuron structure (Figure 5.2). This also indicates that neurogenesis and synaptogenesis factors are important for genetic screening (Kenyon, 2024).

### Role of Neurodegeneration and Neuronal Cell Loss in Autism Spectrum Disorders

There are many studies on neuron loss in individuals diagnosed with ASD. In a study, the increase in the level of C/EBP homologous protein (CHOP), which is involved in the regulation of genes encoding proteins that play a role in proliferation, differentiation, expression and energy metabolism, suggests that apoptosis will be increased in the brains of individuals with autism because CHOP first causes cell death caused by endoplasmic reticulum stress. In the same study, in order to obtain more precise data on apoptosis, the activation of Caspase 8, the initiator type of the caspase required for the extrinsic apoptotic pathway, was examined, and it was shown that the activation of Caspase 8 increased. Accordingly, it is possible to talk about neuron loss due to increased apoptotic activities in the autistic brain compared to the control brain (Dong et al., 2018).

In addition, the decrease in realin, which enables the migration of neurons during early brain development; a decrease in the amount of Bcl-2, which has inhibitory properties on apoptosis; and a significant increase in the amount of p53, which causes apoptosis, were detected. All these changes cause mental retardation (Palmen et al., 2004).

Likewise, research on the amygdala region shows that abnormal growth results in a decrease in the number of neurons (Schumann et al., 2006). Kemper and Bauman (1993) were the first to see the abnormality in the microscopic integrity of the amygdala. According to their research on six autism samples, including individuals between the ages of 9 and 29 (some of the samples had epilepsy), the neurons in the nuclei of the amygdala in autism cases were much smaller and denser than normal compared to the age-matched control group. In a period following this study, Schumann and Amaral (2006) studied nine autism cases aged 10–44 without epilepsy and a control group consisting of the same number of men of the same age to investigate the number and size of neurons in the

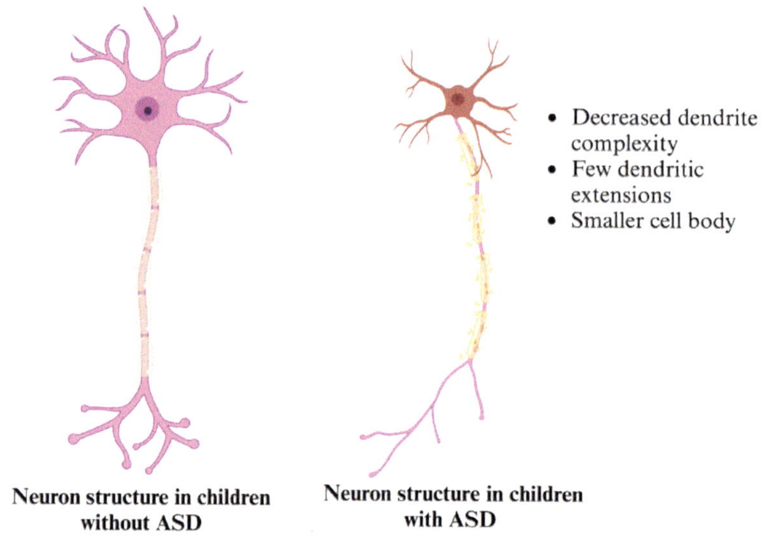


**FIGURE 5.2** This image shows the differences in neuron structure in children with and without a diagnosis of ASD. Compared to neurons from children without an ASD diagnosis, children with ASD have smaller cell body and decreased dendritic extensions and dendrite complexity. Created with Biorender.

amygdala. Significantly fewer neurons were detected in the amygdala and lateral nuclei of individuals belonging to the autism group compared to the control group. According to the data suggested and obtained by Kemper and Bauman (1993), an increase in the number of neurons and a decrease in the size of neurons were not obtained.

The number of Purkinje cells (PC) in the cerebellum vermis in the ASD group was found to be lower than the neurotypical controls (Whitney et al., 2008; Ritvo et al., 1986). In other studies, neuronal pathology in the brains of individuals with autism provides an opportunity to explain neurodegeneration to some extent. In a study investigating the brains of individuals diagnosed with ASD, degeneration and microglial activation were observed in PCs (Vargas et al., 2005). Another study revealed a connection between PC loss and gliosis, which refers to neuroglial cell proliferation that develops after neurodegeneration. This result suggests that these differences are not neurodevelopmental but acquired (Casanova, 2007).

It was also observed that individuals with ASD contained fewer neurons in the amygdala and in certain regions of the brain compared to the control group. In another study, it was found that the neurons in the fusiform gyrus in the brain of individuals with ASD were fewer and smaller in size compared to the control group (van Kooten et al., 2008). Moreover, pyramidal cells, a type of multipolar nerve cells found in the cerebral cortex, hippocampus, and amygdala regions, were reported to be less abundant in the brains of individuals with ASD, compared to controls (Camacho et al., 2013).

## Aggregation of Common Proteins in Neurodegenerative Diseases and Autism Spectrum Disorders

Both AD and PD target specific groups of neurons and regions in the brain, causing unique symptoms and progressions. The clinical symptoms observed in AD are related to the loss of neurons located mainly in the hippocampus and cerebral neocortex in the brain and the degree of this loss. While AD mostly affects cognitive functions, PD primarily causes impairments in motor functions. Aggregation of amyloid peptide (Aβ). Aβ and tau proteins are associated with neurodegeneration resulting in neuron loss observed in AD (Jęśko et al., 2020).

Degeneration in PD affects a diverse population of CNS neurons in the substantia nigra and is associated with abnormal accumulation of α-syn protein. On the other hand, research has shown that in ASD, abnormalities are observed in some specific signaling pathways in the cerebral cortex and cerebellum that are associated with PD. In silico studies also support this situation, and it has been suggested that APP, presenilin-1, or tau may play a role in these pathways (Abraham et al., 2019).

## Amyloid Precursor Protein

Transmembrane amyloid precursor protein (APP) is expressed by microglia, astrocytes, oligodendrocytes, and brain neurons. The physiological function of APP is not fully understood. In addition to regulating neurite outgrowth and assistance, synaptogenesis, and neurotransmission, APP (and hydrolysis products of APP) have been linked to neural protection regarding cytotoxicity or ischemia, cell proliferation, neurotrophic signaling, positive control of cell migration, and suppression of adhesion. APP may participate in antimicrobial signaling and function as a pro-inflammatory factor (Snitz et al., 2015). Since Aβ is frequently found in the healthy brain, its important physiological roles and early significance in cognitive decline have been argued against, as well as Aβ's critical role in the pathological cascade.

APP can be processed by a number of controlled intramembrane proteolysis processes using enzymes known as secretases. The extracellular domain of APP is first removed from each of the two most extensively studied proteolytic sequences, and the residual membrane-bound portion is subsequently cleaved. The release or elimination of Aβ might result from the proteolytic cleavage of APP, which is known as the amyloidogenic or non-amyloidogenic route, respectively. Under healthy settings, APP is processed by α-cleavage, which involves the disintegrin and metalloprotease α-secretase ADAM10. Membrane-anchored zinc-dependent proteases called ADAM proteins are involved in ectodomain shedding, which is the process of removing extracellular domains from transmembrane proteins such as adhesion molecules, Notch, and APP that are crucial for brain development and homeostasis.

APP is cut by α-Secretase inside the Aβ sequence, which stops the synthesis of amyloid. sAPPα, a sizable soluble fragment that includes the extracellular (N-terminal) portion of APP and preserves a portion of the Aβ sequence, is released by α-cleavage. sAPPα has been linked to synaptic formation, neurogenesis, neuroprotection, and glial function modulation. The transmembrane region of the remaining C-terminal APP C83 (or CTF83) stub is cut inside by γ-Secretase, a transmembrane protein complex that contains presenilins and interacts with α-secretase both physically and functionally. This cleavage releases the C-terminal intracellular domain (AICD) and the p3 peptide, which is effectively the shortened Aβ17-40 or Aβ17-42, including its hydrophobic transmembrane portion (Galzitskaya et al., 2018).

## Amyloid Precursor Protein in Neurodegenerative Diseases

Extracellular amyloid accumulation in the brain observed in AD is one of the characteristic features of the disease. The deposits mainly contain Aβ, which is derived from APP, a transmembrane protein 40–42 amino acids long. There are two different antagonist mechanisms that metabolize APP. The soluble N-terminal ectodomain (sAPPα) is released during alpha secretase cleavage, which drives the primary route. This route takes place in the Aβ sequence, preventing the formation of Aβ. Early in the development of the brain, sAPPα has already been identified (Löffler and Huber 1992). The APP family also includes APP-like proteins 1 and 2 (APLP1 and APLP2), which share significant sequence similarity with APP. While APP and APLP2 are expressed in many tissues, APLP1 is only found in the brain. Alpha secretase is cleaved by the metalloproteases ADAM 10 and ADAM 17.

There is an alternative pathway for the degradation pf APP, in which beta secretase is involved and Aβ is generated. In this pathway, Aβ is formed as a result of the degradation of APP by beta secretase followed by gamma secretase. Presenilin serves as the catalytic part of γ-Secretase, a complex consisting of four distinct integral membrane proteins, including a distinctive membrane-embedded aspartyl protease (De Strooper et al., 2012). The soluble APP beta protein (sAPPβ), which is produced by β secretase cleavage, has the identical sequence as sAPPα except for the last C-terminal amino acids.

The APP intracellular domain separates from the cell membrane and penetrates the nucleus to control gene expression following the splitting of the Aβ peptide.

The Aβ peptide in AD pathogenesis first causes the synthesis of soluble neurotoxic oligomers, which in turn causes fibrils to develop in amyloid deposits. Presenilin or APP hereditary mutations lead to a rise in Aβ breakdown in familial Alzheimer's diseases (FAD).

Some individuals with sporadic disorders have an increase in cleavage due to a deficit of the miRNAs that regulate the production of beta secretase, resulting in a high level of the enzyme. A decrease in Aβ-degrading enzymes causes an increase in Aβ peptide in sporadic illnesses. Various Aβ peptide species have been found in the brain and existence of isoforms with shortened N and C termini have been reported. Aβ1–40 is the main isoform among them. Conversely, more aggressive and neurotoxic are the less common species Aβ1-42 and its N-terminally shortened isoforms (Portelius et al., 2011).

Alterations have been shown in APP metabolism in neurodegenerative diseases. In particular, increased β-cleavage has been shown in AD through the action of β-Secretase known as the β-site APP cleaving enzyme 1, BACE1, cleaving the N-terminus of the Aβ sequence. The primary β-secretase is BACE1, and BACE-2 cleaves APP in the β-site and within the Aβ sequence to produce several shorter peptides. Certain APP mutations have the ability to prevent this intra-Aβ splitting, which raises BACE2's proportion as a β-secretase (Farzan et al., 2000). Following β-cleavage, Aβ is separated from the intracellular AICD by γ-secretase, which can coexist in a multiprotein complex with BACE1. Many Aβ isoforms with varying lengths (39–42 aa) and levels of neurotoxicity are produced by this process. Aβ is found in human CSF fluid and plasma, and in laboratory settings, it emerges after stimuli from nerve terminals; therefore β-cleavage must also occur in a healthy brain at a low intensity (Puzzo and Arancio, 2013). Amyloidogenic cleavage of APP results in the most common products being Aβ40 and the more toxic Aβ42. One major genetic factor contributing to AD is polymorphism in the apolipoprotein E (APOE) gene, which interacts with Aβ and controls its aggregation and elimination (Liu et al., 2013; Jęśko et al., 2020).

Mutations in the Aβ sequence of the APP promote the formation of fibrils and β-sheet-rich Aβ oligomers. Aggregation rate can also be impacted by post-translational changes of Aβ peptide, including phosphorylation, glycosylation, and oxidation/nitration. Free radicals are released during Aβ oligomerization, which breaks down membranes and surface proteins and eventually damages DNA (Galzitskaya et al., 2018)

The co-expression of APP, secretases, and their post-translational regulators leads to the diversity of Aβ accumulation in different brain areas. At an early stage in the presymptomatic phase, Aβ deposition begins. Individual differences exist in the pattern of Aβ deposition as the disease advances. Aβ accumulates primarily in the neocortex's basal regions, then moves across the entire structure and impacts the hippocampus. In the end, this results in a comparatively stable pattern of wide neocortical deposition (covering cortical areas for motor and sensory functions) toward the end, where Aβ is also detected in subcortical structures. Aβ is distributed fairly widely throughout the depth of the cortex, and some labeling happens in the white matter; this contrasts with the hyperphosphorylated tau accumulation, which is more layer-specific (Braak and Braak, 1991).

### Amyloid Precursor Protein in Autism Spectrum Disorders

The fact that amyloid precursor protein is effective in ASD development as well as neurodegeneration is explained by the effects of high APP/sAPPa on brain development, cognitive functions, and cytoarchitecture. The highest levels of APP expression observed during synaptogenesis, which occurs intensively in the first stages of life, together with the functions of APP in cell proliferation, migration, and neuronal connection, suggest the effects of APP in disorders of ASD. APP-related pathways mediate molecular mechanisms that lead the cell to either growth via sAPPα or death via Aβ. The balance here is striking, and its disruption may be associated with the development of both ASD and AD. Accordingly, dysregulation of APP and its metabolites in ASD have been shown to potentially contribute to brain overgrowth, possibly caused by abnormal proteolysis of APP (Sokol and Lahiri 2023). Redundant sAPPα and its downstream signal pathway is associated with ASD and the accumulation of Aβ is a characteristic feature of AD (Lahiri et al., 2013). Increased sAPP and sAPP α and decreased Aβ peptides have been reported in the plasma samples and brain tissues of children with ASD (Sokol et al., 2006; Bailey et al., 2008; Sokol and Lahiri 2023). ASD has been associated with enlarged brains, or megalencephaly, and white matter expansion in ASD may be caused by neuroinflammation instead of an increased number of neurons and axons (Lee et al., 2017).

G-protein-coupled (metabotropic) glutamate receptor (mGluR)-mediated neuronal activity enhances APP translation and causes sAPPα release at synapses. Accordingly, the upregulation of sAPPα has been suggested to be associated with increased glutamatergic neurotransmission found in ASD (Lahiri et al., 2013).

ASD is characterized by obsessive behaviors and defective social communication skills in behavior. Furthermore, there is a strong correlation between ASD and intellectual disability, indicating that ASD significantly impairs cognitive function. Normal CNS development, particularly in the cerebral cortex, is necessary for cognitive function. Intellectual impairment can result from any of these mechanisms being disturbed. Studies conducted in recent years have revealed the relationship between the mitogen activated protein kinase (MAP) pathway, which is involved in many intracellular cell pathways, and ASD. Extracellular signal–regulated kinase (ERK)/MAPK signaling has critical roles in behavior and cognitive functions, and ASD has been linked to the ERK/MAPK cascade (Vithayathil et al., 2018). Modified metabolism of APP/Aβ may impact brain development via alternative signaling pathways. APP and other cleavage products including sAPPβ and CTFs, phosphorylate ERK and initiate Ras/MAPK pathway, which is one of the key regulators of cell proliferation and results in increased neurogenesis (Rohe et al., 2008; Venezia et al., 2006).

#### *α-Synuclein*

One of the most prevalent proteins in the brain is α-synuclein. As a presynaptic soluble or membrane-binding protein, α-synuclein may play a part in plasticity based on its pattern of expression. It is found in the brain, primarily attached to synaptic vesicle membranes in neuronal presynaptic terminals. It is also found in the Golgi apparatus, endolysosomal system, nucleus, and endoplasmic reticulum. The olfactory bulb, cerebral cortex, substantia nigra, striatum, and substantia nigra exhibit the highest levels of α-Synuclein.

The main physiological function of α-synuclein includes neurotransmission through its involvement in exocytosis, endocytosis, and neurotransmitter release. α-synuclein interacts with proteins associated with membrane traffic and is required for synaptic plasticity and axon guidance, as well as nerve regeneration (Liu et al., 2004).

The macromolecular complex known as the soluble N-ethylmaleimide-sensitive-factor attachment receptor (SNARE) is primarily engaged in membrane fusion activities, namely the fusing of the pre-synaptic plasma membrane and the synaptic vesicle (SV) membrane for the release of neurotransmitters (Südhof and Rothman, 2009). The establishment of a stable four-helix arrangement between VAMP2 (also called synaptobrevin2), syntaxin-1, and SNAP25 triggers the formation of the SNARE complex in neurons. This complex is influenced by a variety of substances, such as proteins, lipids, and metabolic ions like calcium. Of the three main SNARE proteins, two are found on the plasma membrane while VAMP2 is found on the SV membrane (Wang et al., 2020).

Membrane-bound α-synuclein regulates neurotransmitter release by maintaining the balance of synaptic vesicle pools by regulating processes such as vesicle clustering and docking by interacting with VAMP2, the synaptic vesicle membrane, and synaptic vesicle–bound synapsins (Bendor et al.,2013). Specifically, α-Syn has been involved in controlling neuronal cholesterol in lipid rafts, which are essential for modulating synaptic integrity, vesicle cycling, and synaptic localization. Conversely, cholesterol promotes interactions among oligomers of α-syn (García-Sanz et al., 2021).

Phospholipase D2 (PLD2) is an enzyme involved in membrane signaling and vesicle trafficking and breaks down phosphatidylcholine to release choline and phosphatidic acid (PA). The overexpression of PLD2 in the rat substantia nigra pars compacta (SNc) resulted in the neurodegeneration of dopaminergic neurons (Gorbatyuk et al., 2010). α-Synuclein interacts and inhibits PLD2 and consequently suppresses PLD2-induced nigral degeneration (Gorbatyuk et al., 2010).

Recent research has examined the relationship between α-synuclein and acidic phosphoinositides, revealing that α-synuclein colocalizes with the phosphorylated active form of the clathrin adaptor protein 2 and phosphatidylinositol 4,5-bisphosphate. It is noteworthy that clathrin-regulated endocytosis in central neurons is altered by α-synuclein mutations. Accordingly, mutant and aberrant α-synuclein affects lipid homeostasis and impairs vesicle trafficking in accordance with its physiological roles (Navarro-Paya et al., 2022).

#### *α-Synuclein in Neurodegenerative Diseases*

The clinical characteristic of Parkinson's disease (PD) is intracellular protein aggregates, primarily composed of α-Syn seen in Lewy bodies (LBs) within dopaminergic neurons in the substantia nigra. α-syn disruptions and aggregates have typically been associated with PD, but more lately, AD has also been connected to them (Calabresi et al., 2023). The neurodegenerative disorders linked to α-Syn aggregates are collectively referred to as "synucleinopathies" (Jasutkar and Mouradian, 2022).

As α-synuclein interacts with membrane trafficking-related proteins, it can cause SNARE dysregulation and disruptions to synaptic homeostasis when it is overexpressed or downregulated. Alterations in the levels of synaptic proteins are caused by the build-up or depletion of α-synuclein.

α-syn forms oligomers and aggregates and specifically affects the survival of neuronal or glial populations due to mutations in the SNCA gene that codes for α-syn. Abnormalities in the post-translational modifications of α-syn protein (like phosphorylation or modification by free radicals), or its overexpression primarily occurs in PD (Jesko et al., 2020).

The biochemical analysis of senile plaques in AD led to the discovery of synuclein. A segment from the center of α-synuclein (61–95), now known as the non-Aβ component (NAC), accumulates at elevated concentrations in plaques despite being less common than the Aβ peptide. Synuclein was shown to have a role in the pathophysiology of both AD and dementia with Lewy bodies (DLB) (Bendor et al., 2013).

Disorders in the α-syn mechanism cause the formation of oligomers and aggregates of this protein, and these aggregates cause the death of neurons and glial cells. Defects in the α-Syn mechanism can be listed as mutations in the SNCA gene coding for α-syn, disorders in the post-translational modification of the α-syn protein due to phosphorylation or the effect of free radicals, or excessive increases in α-syn expression (Vidović et al., 2022). The α-helical form is considered the non-pathogenic form, and the transition from the monomeric α-helix structure to the β-sheet structure is held responsible for the misfolding of α-Syn and its aggregation in fibrils (Meade et al., 2019). The toxicity of α-syn is linked to its structure, quantity, and tendency to aggregate; however the exact connection between the two is still up for discussion (Alam et al., 2019).

The tendency for α-Syn aggregation is dependent on certain cellular microenvironment circumstances (temperature, ionic strength, pH), macromolecular overpopulation, and the presence or absence of other ligands/partners (specific lipid molecules, membranes, amyloidogenic proteins, metal ions, and intermediate toxic species) (Jasutkar and Mouradian, 2022). Furthermore, Post-Translational Modifications (PTMs) alter α-syn's size, charge, structure, and/or conformation, affecting its toxicity, aggregation, and membrane binding. There are more than 300 distinct PTMs known for α-syn, including lipidation, oxidation, acylation, ubiquitination, glycosylation, and phosphorylation (Schmidt et al.,2022; Vidović and Rikalovic, 2022).

## Microglial Activation and Neurodegeneration in Autism Spectrum Disorders

Microglia, which has a major impact on brain development, neuronal networks, and injury repair, and microglial cells are the primary cells that maintain the line of defense throughout the brain parenchyma. Recent studies indicate that microglial cells are one of the phagocytic cell types that can destroy all cells or cellular infrastructures in the brain, especially synapses in humans and mice (Bar et al., 2019).

During fetal development, microglia progenitor cells migrate and cluster in the brain before cerebral vascular branching occurs and the blood–brain barrier (BBB) is fully formed (Theoharides et al., 2011). During early brain neurogenesis processes, excess synaptic connections must be eliminated in order to maintain appropriate connections through synaptic shaping, which is important for maintaining nerve function under normal conditions. Microglial cells are of great importance in destroying immature synapses after neurogenesis by eliminating these synaptic connections (Cornell et al., 2022).

In parallel with this idea, in a study conducted on mice, the spine thickness and the frequency of excitatory postsynaptic connections at minimal levels were higher in mice with defective microglia function compared to the control group. Therefore, it is conceivable that the abnormal function of microglia, affected by genetic and environmental risk factors, may lead to neurodevelopmental impairment in ASD (Lee et al., 2015).

Research shows that microglia in the brains of individuals with ASD have distinct abnormalities in both structural features and brain functions, which can lead to disruptions in social contacts and social communication (Tang et al., 2014). In brain tissues obtained from 11 individuals diagnosed with ASD, microglia were found to be significantly activated in all brain regions, especially in the cerebellum (Vargas et al., 2005). The cerebral cortex of individuals diagnosed with ASD shows morphological changes such as increased microglial neuron cell body (soma) size and elongation of neuronal filopodia, resulting in differences in microglial activity in the brain of individuals with ASD (Tetreault et al., 2012).

There are different views and studies on the role of microglia in neurodegeneration. Another study suggested that microglia can temporarily transform into an ameboid form in which neurons can phagocytose (Brown et al., 2012). In addition, in ongoing neuroimmune activities, microglia produce cytokines with high toxicity for neurons, and this neurotoxicity has a preliminary importance in neurodegeneration (Rock et al., 2004).

Neurodegeneration may result from damage to presynaptic terminals prior to loss of the cell body. There are also some opposing views regarding the involvement of microglia in synaptic stripping and synapse degeneration (Perry et al., 2010).

According to research, it has been found that there is an ongoing neuroinflammatory process involving microglial activation in different parts of the brain in individuals with ASD (Morgan et al., 2010; Suzuki et al., 2013). Another study found that the process of neuronal degeneration and neuroglial activation commonly occurred in the Purkinje cell layer and the granular cell layer of the cerebellum, and that these findings were compatible and consistent with the ongoing postnatal neurodegeneration and neuroinflammation process (Vargas et al., 2005).

### *Increase in Pro-Inflammatory Cytokines and Neurodegeneration in Autism Spectrum Disorders*

The pathophysiology of ASD, which is not yet clear and has many subtexts, and the lack of a fully effective treatment today have led to further consideration and investigation of different etiological factors. The observation of different inflammatory responses and neuroimmunity abnormalities throughout the lives of individuals diagnosed with autism suggests that inflammation plays a role in the etiology of ASD, and studies continue to be conducted in this direction.

Some postmortem studies have found significant neuroinflammatory changes in various brain regions of patients with ASD, confirming these ideas (Vargas et al., 2005). Increased synthesis and increased density of microglial cells were detected in different brain areas, including the dorsolateral prefrontal cortex and cerebellum (Vargas et al., 2005; Morgan et al., 2010, Tetreault et al., 2012). These findings were supported by positron emission tomography (PET) studies conducted in ASD patients (Suzuki et al., 2013).

During the 2–8 weeks after fertilization, genetic and environmental risk factors may affect the inflammatory response of newborns, causing differences in the normal development of the brain after birth (Adams-Chapman et al., 2006). These genetic and environmental factors can have an impact on neuronal functioning and neuroimmunity by causing microglia activation, or on chronic neuroinflammation by directly regulating neuronal function (Depino, 2013).

Research on the role of microglial activation and neuroinflammation in ASD has been established, and microglial function abnormalities in ASD are still an important research topic (Carson et al., 2006; Frick et al., 2016). After microglial activation, proliferation, migration, phagocytosis, presentation of antigens to T cells, release of reactive oxygen species, activation of several genes and proteins such as inducible nitric oxide synthase (iNOS), COX1, COX2, and various pro-inflammatory cytokines including IL-1ß and TNF-α are released, leading to chronic and long-term inflammation in ASD (Monnet-Tschudi et al., 2011).

These changes that occur in ASD are linked to some extent with the behavioral features of the disease, because this ongoing neuroinflammation and unexpected inflammatory response in the brain causes cognitive dysfunction (Lucchiana et al., 2014; Agusti et al., 2018).

The hippocampus (Donvan & Zucker, 2018) and cerebellum () are the two brain regions related to limited social communication and repetitive behaviors, which are symptoms of ASD. In a study, it was found that valproic acid, used as an environmental risk factor, caused increased activation in the hippocampus and cerebellum through long-term glial activation (Lucchiana et al., 2014).

Studies show that the activation of astrocytes and microglia in the cortex and cerebellum is associated with the pro-inflammatory markers such as IL-6, TNFα (Chez et al., 2007), TGF-ß1, IFN-1, IFN-γ, and IL-8 (Garbett et al., 2008). Moreover, increased expression of these inflammatory cytokines and genes related to immunological reactions have been reported in different types of brain regions of individuals diagnosed with ASD (Chez et al., 2010).

All these results show thats neuroinflammation may contribute to the behavioral differences observed in autism, and accordingly, keeping microglia activation under control and suppressing the production of cytokines and free radicals can be used therapeutically in the diagnosis and treatment of ASD (Eissa et al., 2020).

### *Neurodegeneration and Oxidative Stress in Autism Spectrum Disorders*

One of the underlying causes of neurodegenerative diseases is the excessive production of reactive oxygen species (ROS) and the resulting oxidative stress (Klein et al., 2003).

Free radicals and other reactive oxygen species are produced in all living things. Any free radical that contains oxygen can be called ROS. Free radicals and other ROS are unstable atoms, molecules, or various ions consisting of unpaired electrons. ROS are undesirable compounds because the unpaired electron reacts with other ions and molecules by oxidation or takes an electron from other molecules and pairs that electron. This leads to degradation of other molecules and cell damage (Gutman, 2002). In fact, one of the most important problems that arises after this event is that ROS steals electrons from lipid membranes. As a result, the lipid membrane is oxidized, and lipid peroxidation occurs. Lipid peroxidation damages the integrity and fluidity of the cell membrane structure. This series of events leads to cell death (Efe et al., 1999).

In the usual process, it is possible to talk about a certain balance between ROS production and the antioxidant amount of the cell (Anderson, 2004). However, when oxidative stress occurs, an imbalance between free radicals and antioxidants is noted. The brain is highly vulnerable to oxidative stress, with limited antioxidant capacity, higher energy requirements, and higher lipid and iron content (Juurlink et al., 1998).

It has been observed that ROS levels increase in the relevant brain regions in the postmortem brain tissues of individuals with neurodegenerative disorders such as PD, AD, and ALS (Anderson, 2004). Additionally, neuronal damage due to oxidative stress has been emphasized as the basis of psychiatric diseases such as schizophrenia, bipolar disorder, depression, and attention deficit and hyperactivity disorder (Popa-Wagner et al., 2013).

It has been stated that some children diagnosed with ASD may show symptoms of ASD due to brain damage or neuronal cell death due to trauma after birth. According to this finding, it has been stated that neurodegeneration in autism may be affected by oxidative stress and toxicity (Kern et al., 2006).

In addition to all these data, genetic and environmental factors can differentiate the immune response by increasing the production of free radicals, leading to glial activation, increasing the inflammatory response and ultimately mediating the clinical symptoms of autism through its effect on neurons (Huuskonen et al., 2005).

Postmortem studies have shown that oxidative stress markers are increased in the brains of individuals diagnosed with ASD. A study reported that DNA oxidation in individuals with autism increased twofold in the frontal cortex, temporal cortex, and cerebellum compared to the control group (Chauhan et al., 2012).

In another study, it was observed that the lipofuscin density in the brains of individuals with autism increased compared to the control group due to oxidative stress, and it was emphasized that the affected areas were cortical regions related to communication (Lopez-Hurtado&Prieto, 2008).

8-oxo-guanosine, an RNA oxidative damage marker, can be used to address various aspects of damage and degeneration in the brain caused by oxidative stress in neurodegenerative diseases. It has been determined that the cerebellum levels of this substance in individuals with autism show a significant increase compared to the control group (Sajdel-Sulkowska et al., 2011).

## Experimental Models of Autism Spectrum Disorders

### *In Vitro Models of Autism Spectrum Disorders*

Cellular models provide great benefits in the diagnosis and treatment of diseases. The use of advantageous cellular models, which are easy to produce and easy to maintain, and especially under laboratory conditions, is widespread.

The effects of ASD are first observed in central nervous system cells associated with higher cognitive functions. The inability to obtain these cell types from biopsies poses a problem in researching autism using cell cultures as a model. And the fact that neurons have a low proliferation rate leads to their inability to be cultured in the long term and, accordingly, to the difficulty of creating model cell lines (Shi et al., 2017). In fact, this situation was reversed in a study by finding mechanisms that allow adult somatic cells to be reprogrammed. Thus, it enables the conversion of differentiated cell lines into induced pluripotent stem cells (hiPSCs) by expressing four genes known as Yamanaka factors (Oct3/4, Sox2, Klf4, c-Myc), and this provides a great advantage in stem cell studies (Takahashi et al., 2006).

HiPSCs renew themselves quickly and it is possible to talk about their high differentiation potential. Since HiPSCs can be differentiated into various cell types from the nervous system, they enable the investigation of the basis of neurodegenerative diseases. Thus, specialized in vitro models can be created (Juopperi et al., 2011).

In a study, a TSC cerebellar Purkinje cell (PC) model with patient-derived human-induced pluripotent stem cells (hiPSCs) was developed to understand the molecular mechanisms that may lead to cerebellar abnormalities in ASD and TSC (tuberous sclerosis complex), resulting in pathogenic TSC2 mutations. hiPSC-derived PCs from patients with AD were found to exhibit mTORC1 pathway hyperactivation, defects in neuronal differentiation and RNA regulation, hypoexcitability, and reduced synaptic activity compared to those derived from controls. It is possible to talk about abnormal PC function during the development process in individuals suffering from these diseases (Sundberg et al., 2018).

In another in vitro study, human pluripotent stem cells carrying a rare RELN variant (RELN-del) were produced using targeted genome editing. These cells differentiated into homogeneous dopaminergic neurons. In this study, RELN-del was found to induce impaired reelin signaling and reduce the expression levels of genes associated with cell migration in neural cells (Arioka et al., 2018).

### *Animal Models in Autism Spectrum Disorders Research*

The complex history of ASD has historically been studied using animal models since it was impossible to create human neural cell cultures with infinite capacity for multiplication. Because they establish the role of certain genes in the disorder's etiology, animal models are particularly helpful for researching diseases of the central nervous system.

There are primary methods that have been employed to find animal models of ASD. The first method is forward genetics, wherein phenotypes like ASD are found in the chosen animal model, and subsequently the molecular causes of the observed changes are clarified. Reverse genetics is the second strategy, whereby specific mutations are inserted into the animal model's genome, after which the phenotype is identified (Sakai et al., 2018). The most common animal model utilized in neuroscience research is the rodent *Mus musculus*. This has less to do with mice being superior models than other animals and more to do with the fact that the mouse genome had been sequenced initially and that manipulation tools were created more quickly (Sakai et al., 2018).

#### *Rodent Models in Autism Spectrum Disorders Research*

Rodents such as *Mus musculus* are widely used in neuroscience studies. First of all, sequencing the mouse genome and faster development of tools to manipulate it means that rodents are used more frequently than other models in practice. Of course, in addition to these rodent models, there are also a wide variety of model organisms available today; some of these are models of great importance and usefulness in ASD modeling, such as *Rattus norvegicus* (Ellenbroek et al., 2016) or *Danio rerio* (Sakai et al., 2018). This diversity suggests that over time, research on ASD can be developed with new animal models (Meshalkina et al., 2018).

Brain structure and function are more complex and detailed in humans compared to rodents. This difference means that there is a wide variety of brain regions, cell numbers, and intercellular connections. All of these distinguish these two living species from each other mentally and behaviorally. In this case, it will be very difficult for a rodent model to fully express the processes occurring in individuals with ASD as expected. The mechanisms affecting ASD symptoms offer the opportunity to be investigated in rodent models accompanied by gene editing technologies that allow the development of animals with phenotypes similar to those observed in humans. Although *Mus musculus* and *Rattus norvegicus* are rodents belonging to the same class and sharing many of the same characteristics, there are very different data between them in terms of physiology, behavior, and pharmacological foundations that lead to results that can be obtained from both living species (Pensado-López et al., 2020).

When comparing mouse and rat brains in terms of neurophysiology, it is possible to mention some distinctive features. Examples of these differences include the fact that some neurotransmitters and their receptors have different distributions in both species and the distinctive features in the neurogenesis processes of both species. These also appear to affect brain regions such as the hippocampus or cortical regions that play a role in ASD development (Wöhr et al., 2013).

When mice and rats are examined in terms of behavior, both species live in hierarchical groups where it is possible to talk about complex social interaction networks, but the communication established between individuals is different in both species. Mice are more possessive and aggressive in terms of protecting their territory than rats but are still less impulsive. Mice are easier to train, show more consistent development during studies, and are not as affected by human presence as rats. In connection with all these features, it is possible to mention some distinctive features in their cognitive capacities (Ellenbroek et al., 2016).

Rodents have been a preclinical tool to elucidate many aspects of the causes of ASD and to test new potential treatments. The reason for such success is that they can summarize the core symptoms of ASD, such as impairments in social interaction, communication, and the presence of repetitive behaviors (Patterson, 2011).

Certain mouse strains with ASD-like phenotypes, such as the BTBR T+tf/J strain, have been generated through inbreeding procedures. This type includes social behavior disorders (decreased interaction between individuals, dislike of prior interaction), communication disorders (altered patterns or responses to both USV and scent cue), and repetitive behaviors (self-grooming, burying behaviors, and preferences for certain objects or spaces). They are frequently preferred because they allow the ASD symptoms observed in humans to be obtained and examined. Additionally, BTBR mice have difficulty receiving learning-related commands and become very anxious when threatened.

This frequently used type is also preferred because it molecularly reveals the differences observed in brain structure and development in individuals with ASD. Disruption of several ASD-related genes such as kynurenine 3-hydroxylase, Disc1, a protein involved in neuroprotection and dopamine signaling, and Ext1, a protein involved in the synthesis of precursor molecules, has also been detected in this species (Meyza et al., 2013).

Rats, on the other hand, are used as an efficient model in neurodevelopmental diseases, including ASD, in terms of social skills and behavioral aspects that are more complex than in mice. The first available rat knockout models to study ASD were generated in 2010 using ZFN and ENU-induced mutagenesis (Wöhr et al., 2013). Studies on rat models used today have increasingly grown in number, including genetic models created with specific ASD-related genes and some pharmacological experimental models.

Although they are guiding and explanatory models in ASD research, more manageable animal models such as zebrafish have begun to be preferred due to some undeniable disadvantages.

### *Zebrafish Models in Autism Spectrum Disorders Research*

Zebrafish have begun to be used frequently to examine and study risk genes associated with neurodevelopmental diseases. Some of the reasons why researchers chose zebrafish for this advantageous choice are rapid growth and development of transparent embryos in the external environment, which allows direct observation of the developing nervous system in the early stages; and high-throughput pharmacological screens to identify small molecule suppressors of simple behavioral phenotypes, which is not possible in rodent models. They have remarkable traceability to perform, are produced quickly and at low cost, and allow the genetic manipulation technique, which has been facilitated by the continuation of CRISPR/Cas9 gene editing technologies that have become popular recently (Sakai et al., 2018).

Studies in zebrafish models indicate a translational link between zebrafish and humans. Structurally, the zebrafish brain has the same major subdivisions as the mammalian brain: forebrain, midbrain, hindbrain, and spinal cord (Guo, 2009). Many brain regions in zebrafish and mammals show structural homology, including the thalamus, optic tectum, and cerebellum. Morphogenesis in the zebrafish brain lasts for three days, and the mechanisms behind the formation of different brain structures such as the neural tube or telencephalon differ from mammals (Kimmel et al., 1995). However, zebrafish and mammals show homology in terms of cell types, intercellular connections, differentiation, signaling pathways, and gene expression patterns, as well as the most important brain regions and main subdivisions (Mueller et al., 2009).

Zebrafish have neuroanatomical regions that are almost equivalent to the human brain at some functional and structural levels, such as the lateral, dorsal, and medial pallium, which contain many of the same features as the human hippocampus, neocortex, and amygdala. In addition to all these, genes associated with early development are expressed in similar ways in the brains of zebrafish and mammals, and the main neurotransmitters in the mammalian brain such as GABA, glutamate, dopamine, norepinephrine, serotonin, histamine, and acetylcholine are also located in zebrafish (Guo, 2009).

In terms of autism, this structural homology with zebrafish is of serious importance for critical processes such as the cerebellar structure and the differentiation of this structure. Approximately 22 hours after fertilization, the cerebellar becomes evident (Kimmel et al., 1995), and 3 days later, the differentiation of glutamatergic and GABAergic excitatory or inhibitory neurons begins, respectively; the layers are formed 5 days after fertilization (Bae et al., 2009). This entire developmental process is equivalent to the pregnancy process in mammals, but although they are located in different expression domains, expression of atoh1 genes by cerebellar progenitors is synthesized in stimulatory cells and ptf1a expression leads to the formation of inhibitory cells (Kani et al., 2010).

In the human brain, glutamatergic neurons consist of granule cells, and GABAergic neurons consist of Purkinje cells. In the zebrafish brain, such cells are developed in three parts: molecular, Purkinje cells, and granule layers (Bae et al., 2009). Purkinje cells are effective on cerebellar nerve connections and function. Purkinje cell recognizes synaptic signals, processes these signals, and transmits this information to the cerebellar nuclei through efferent projections. These nuclei connect the cerebellum to the brain and spinal cord. In connection with this whole process, cognitive functions such as motor, language, emotional, and sensory functions are regulated (Sudarov, 2013). As mentioned under the heading of neuronal cell loss in ASD, it should be underlined that purkinje cell has a major role in the normal functioning of the nervous system and that the decrease in the amount and size of purkinje cells, which is one of the most reproducible and clearly observed differences in most ASD cases, is inevitable (Jeong et al., 2014). Despite all these studies, the specific role of purkinje cells in the development of ASD-like phenotypes has not been fully proven due to its complexity.

A study identified certain mutations in a nuclear receptor (RORa) required for cerebellar development in families with variable neurodevelopmental delay and intellectual disability, including cognitive, motor, and behavioral phenotypes. A zebrafish model was created using CRISPR/Cas9. At the end of the experiment, a decrease in the number of Purkinje and granule cells was detected, and in the light of these results, the neuroanatomical aspects of the brain with ASD were clarified (Guissart et al., 2018). These results once again show that the use of zebrafish is the right choice to investigate the basis of ASD and other neurodevelopmental diseases and to develop treatment methods.

In genetic studies, the zebrafish genome sequencing project was first initiated at the Wellcome Trust Sanger Institute in 2001, and a high-quality sequence assembly of the zebrafish genome was published in 2013. According to this review, approximately 70% of the human genome has a zebrafish ortholog, and more than 80% of these genes are associated with human diseases (Howe et al., 2013).

Regarding developmental studies, it is possible to say the following, as stated in the previous sections: Expression patterns in early developmental genes in zebrafish and humans are homologous to each other, and between both groups, main neurotransmitter systems such as GABA, glutamate, norepinephrine, cholinergic and dopaminergic pathways, as well as glial cells were also identified in a common manner. In a parallel study, a method was developed to understand the cellular elements associated with behavioral patterns, and evolutionarily conserved cellular and molecular systems involved in basic neuromodulatory circuits were demonstrated (Lovett-Barron et al., 2017).

From a behavioral perspective, results have also been obtained that zebrafish share many behavioral patterns with humans, including physiological, emotional, and social responses (Meshalkina et al., 2018).

## Conclusion

In conclusion, all these data prove once again that the zebrafish is a suitable research model in terms of neural connections, related genes, and physiopathology underlying the etiology of all other neurodevelopmental diseases, including ASD.

## REFERENCES


Abraham, J.R., Szoko, N., Barnard, J., et al., 2019. Proteomic investigations of autism brain identify known and novel pathogenetic processes. Sci. Rep. 9 (1), 13118.

Adams-Chapman, I., Stoll, B. J. (2006). Neonatal infection and long-term neurodevelopmental outcome in the preterm infant. Curr. Opin. Infect. Dis. 19 (3), 290–297. doi: 10.1097/01.qco.0000224825.57976.87

Agustí, A., García-Pardo, M. P., López-Almela, I., Campillo, I., Maes, M., Romaní-Pérez, M., & Sanz, Y. (2018). Interplay between the gut-brain axis, obesity and cognitive function. *Frontiers in neuroscience*, *12*, 155.

Alam P, Bousset L, Melki R, Otzen DE. α-synuclein oligomers and fibrils: a spectrum of species, a spectrum of toxicities. J Neurochem. 2019 Sep;150(5):522-534. doi: 10.1111/jnc.14808.

Alexiou, A., Soursou, G., Yarla, N.S., & Md Ashraf, G. (2018). Proteins Commonly Linked to Autism Spectrum Disorder and Alzheimer's Disease. *Curr Protein Pept Sci.* 19(9),876-880.

Almeida, C., Oliveira, R., Soares, R., & Barata, P. (2020). Influence of gut microbiota dysbiosis on brain function: a systematic review. *Porto biomedical journal*, *5*(2).

Anderson, J.K. (2004). Oxidative stress in neurodegeneration: cause or consequence? Nat Neurosci Rev. 5, S18-S25.

Appleton, J. (2018). The gut-brain axis: Influence of microbiota on mood and mental health. *Integrative Medicine: A Clinician's Journal*, *17*(4), 28.

Arioka, Y., Shishido, E., Kubo, H. et al. (2018). Single-cell trajectory analysis of human homogenous neurons carrying a rare RELN variant. Transl Psychiatry 8, 129.

Bae, Y.K. Kani, S. Shimizu, T. Tanabe, K. Nojima, H. Kimura, Y. Shin-ichi, H. Hibi, M. (2009). Anatomy of zebrafish cerebellum and screen for mutations affecting its development. Dev. Biol. 330, 406–426

Bailey A. R., Giunta B. N., Obregon D., Nikolic W. V., Tian J., Sanberg C. D., et al. (2008). Peripheral biomarkers in autism: secreted amyloid precursor protein-alpha as a probable key player in early diagnosis. Int. J. Clin. Exp. Med. 1, 338–344.

Bar E, Barak B. Microglia roles in synaptic plasticity and myelination in homeostatic conditions and neurodevelopmental disorders. Glia. (2019) 67:2125–41. doi: 10.1002/glia.23637.

Basnet, R. M., Zizioli, D., Taweedet, S., Finazzi, D., & Memo, M. (2019). Zebrafish larvae as a behavioral model in neuropharmacology. *Biomedicines*, *7*(1), 23.

Bendor, J.T., Logan, T.P. & Edwards, R.H. (2013). The function of α-synuclein. Neuron. 18;79(6):1044-66. doi: 10.1016/j.neuron.2013.09.004

Bhuiyan, P., Chen, Y., Karim, M., Dong, H., & Qian, Y. (2021). Bidirectional communication between mast cells and the gut-brain axis in neurodegenerative diseases: avenues for therapeutic intervention. *Brain research bulletin*, *172*, 61-78.

Borrelli, L., Aceto, S., Agnisola, C., De Paolo, S., Dipineto, L., Stilling, R. M., Dinan, T. G., Cryan, J. F., Menna, L. F., & Fioretti, A. (2016). Probiotic modulation of the microbiota-gut-brain axis and behaviour in zebrafish. *Scientific reports*, *6*(1), 30046.

Braak, H. & Braak, E., (1991). Neuropathological stageing of Alzheimer-related changes. Acta Neuropathol. 82 (4), 239–259.

Brown, G.C. & Neher, J.J. (2012). Eaten alive! Cell death by primary phagocytosis: 'phagoptosis'. Trends Biochem Sci. 7, 325–332.

Butt, R. L., & Volkoff, H. (2019). Gut microbiota and energy homeostasis in fish. *Frontiers in endocrinology*, *10*, 9.

Calabresi, P., Mechelli, A., Natale, G., Volpicelli-Daley, L., Di Lazzaro, G., Ghiglieri, V. (2023). Alpha-synuclein in Parkinson's disease and other synucleinopathies: from overt neurodegeneration back to early synaptic dysfunction. Cell Death Dis. 1,14(3),176.

Camacho, J., Combs, Z., Schumman, C., Amaral, D., Martinez-Cerdeno, V. (2013). Postmortem analysis of cell density in autistic temporal cortex. Soc Neurosci Annu Meet http://sfari.org/news-and-opinion/conference-news/2010/society-for-neuroscience-2010/autism-brains-characterized-by-fewer-excitatory-neurons#sthash.OzOcSsEo.dpuf

Carabotti, M., Scirocco, A., Maselli, M. A., & Severi, C. (2015). The gut-brain axis: interactions between enteric microbiota, central and enteric nervous systems. *Annals of gastroenterology: quarterly publication of the Hellenic Society of Gastroenterology*, *28*(2), 203.

Carson, M. J., Doose, J. M., Melchior, B., Schmid, C. D., Ploix, C. C. (2006). CNS immune privilege: hiding in plain sight. Immunol. Rev. 213, 48–65.

Casanova, M.F. (2007). The neuropathology of autism. Brain Pathol. 417, 422-433.

Casas, A., Parra, F., & Blancas, J. (2015). Evolution of humans and by humans. *Evolutionary ethnobiology*, 21-36.

Catron, T. R., Gaballah, S., & Tal, T. (2019). Using zebrafish to investigate interactions between xenobiotics and microbiota. *Current Pharmacology Reports*, *5*, 468-480.

Chauhan, A., Audhya, T., Chauhan, V. (2012). Brain region-specific glutathione redox imbalance in autism, Neurochem Res. 37(8),1681-1689.

Chen, Z., Zhijie, C., Yuting, Z., Chan, L., Shilin, X., Qichun, Z., Jing, L., & Zhixian, M. (2021). The ameliorative effects of isorhynchophylline on morphine dependence are mediated through the microbiota-gut-brain Axis. *Frontiers in Pharmacology*, *12*, 526923.

Cheng, Y., Chen, C., & Zhang, F. (2023). Immunity orchestrates a bridge in gut-brain axis of neurodegenerative diseases. *Ageing Research Reviews*, *85*, 101857.

Chez, M. G., Dowling, T., Patel, P. B., Khanna, P., and Kominsky, M. (2007). Elevation of tumor necrosis factor-alpha in cerebrospinal fluid of autistic children. Pediatr. Neurol. 36 (6), 361–365.

Chez, M. G., & Guido-Estrada, N. (2010). Immune therapy in autism: historical experience and future directions with immunomodulatory therapy. Neurotherapeutics 7 (3), 293–301

Chia, K., Klingseisen, A., Sieger, D., & Priller, J. (2022). Zebrafish as a model organism for neurodegenerative disease. *Frontiers in Molecular Neuroscience*, *15*, 940484.

Cornell, J., Salinas, S., Huang, H.Y., Zhou, M. (2022). Microglia regulation of synaptic plasticity and learning and memory. Neural Regen Res. 17,705–16. doi: 10.4103/1673-5374.322423.

Courchesne, E., Pierce, K., Schumann, C.M., Redcay, E., Buckwalter, J.A., Kennedy, D.P., Morgan, J. (2007). Mapping early brain development in autism. Neuron. 25,56(2),399-413.

Davenport, E. R., Sanders, J. G., Song, S. J., Amato, K. R., Clark, A. G., & Knight, R. (2017). The human microbiome in evolution. *BMC biology*, *15*(1), 1-12.

Davis, D. J., Bryda, E. C., Gillespie, C. H., & Ericsson, A. C. (2016a). Microbial modulation of behavior and stress responses in zebrafish larvae. *Behavioural brain research*, *311*, 219-227.

Davis, D. J., Doerr, H. M., Grzelak, A. K., Busi, S. B., Jasarevic, E., Ericsson, A. C., & Bryda, E. C. (2016b). Lactobacillus plantarum attenuates anxiety-related behavior and protects against stress-induced dysbiosis in adult zebrafish. *Scientific reports*, *6*(1), 33726.

Depino, A. M. (2013). Peripheral and central inflammation in autism spectrum disorders. Mol. Cell Neurosci. 53, 69–76. doi: 10.1016/j.mcn.2012.10.003

De Strooper, B., Iwatsubo, T., Wolfe, M.S. (2012). Presenilins and γ-secretase: structure, function, and role in Alzheimer Disease. Cold Spring Harb Perspect Med. 2(1),a006304.

Dethlefsen, L., McFall-Ngai, M., & Relman, D. A. (2007). An ecological and evolutionary perspective on human–microbe mutualism and disease. *Nature*, *449*(7164), 811-818.

Dicks, L. M., Dreyer, L., Smith, C., & Van Staden, A. D. (2018). A review: the fate of bacteriocins in the human gastro-intestinal tract: do they cross the gut–blood barrier? *Frontiers in Microbiology*, *9*, 2297.

Doan, R.N., Lim, E.T., De Rubeis, S. et al. (2019). Recessive gene disruptions in autism spectrum disorder. *Nat Genet* 51, 1092–1098.

Dong, D., Zielke, H.R., Yeh, D., Yang, P. (2018). Cellular stress and apoptosis contribute to the pathogenesis of autism spectrum disorder. Autism Res. 11(7),1076-1090.

Donvan, J., & Zucker, C. (2018). In a different key: The story of autism: Book Reviews. J Hist Behav Sci. 54(1), 66-7.

Efe, H., Deger, O., Kirci, D., Karahan, S.C., Orem, A., Calapoglu, M. (1999). Decreased neutrolphil antioxidative enzyme activity and increased lipid peroxidation in hyperlipoproteinemic human subjects. Clin Chem Acta. 279, 155-165.

Eissa, N., Sadeq, A., Sasse, A., & Sadek, B. (2020). Role of neuroinflammation in autism spectrum disorder and the emergence of brain histaminergic system. Lessons also for BPSD?. Frontiers in Pharmacology, 11, 886.

Ellenbroek, B. Youn, J. (2016). Rodent models in neuroscience research: Is it a rat race? Dis. Model. Mech. 9, 1079–1087.

Eshraghi, A.A., Liu, G., Kay, S.S., Eshraghi, R.S., Mittal, J., Moshiree, B., Mittal, R. (2018). Epigenetics and Autism Spectrum Disorder: Is There a Correlation? Front Cell Neurosci. 27,12:78.

Farzan, M., Schnitzler, C.E., Vasilieva, N., et al., (2000). BACE2, a beta -secretase homolog, cleaves at the beta site and within the amyloid-beta region of the amyloid-beta precursor protein. *Proc. Natl. Acad. Sci. U.S.A.* 97 (17), 9712–9717.

Fatemi, S. (2002). The role of Reelin in pathology of autism. Mol Psychiatry 7, 919–920.

Frick, L., Rapanelli, M., Abbasi, E., Ohtsu, H., Pittenger, C. (2016). Histamine regulation of microglia: Gene-environment interaction in the regulation of central nervous system inflammation. Brain Behav. Immun. 57, 326–337.

Fuentes, J., Bakare, M., Munir, K., et al. (2014). Autism spectrum disorder, IACAPAP Etextb child Adolesc Ment Heal Geneva Int Assoc Child Adolesc Psychiatry Allied Prof., 2,1-35.

García-Sanz, P., Aerts, J.M.F.G., Moratalla, R. (2021). The role of cholesterol in α-synuclein and Lewy body pathology in GBA1 Parkinson's disease. Mov Disord. 36,1070–1085.

Galzitskaya, O.V., Galushko, E.I., Selivanova, O.M. (2018). Studies of the process of amyloid formation by Aβ peptide. Biochemistry (Mosc.) 83 (Suppl. 1), S62–S80.

Garbett, K., Ebert, P. J., Mitchell, A., Lintas, C., Manzi, B., Mirnics, K., et al. (2008). Immune transcriptome alterations in the temporal cortex of subjects with autism. Neurobiol. Dis. 30 (3), 303–311.

Ghaisas, S., Maher, J., & Kanthasamy, A. (2016). Gut microbiome in health and disease: Linking the microbiome–gut–brain axis and environmental factors in the pathogenesis of systemic and neurodegenerative diseases. *Pharmacology & therapeutics*, *158*, 52-62.

Ghosh, S., Whitley, C. S., Haribabu, B., & Jala, V. R. (2021). Regulation of intestinal barrier function by microbial metabolites. *Cellular and molecular gastroenterology and hepatology*, *11*(5), 1463-1482.

Gilbert, P. (2017). Definitions and controversies. *Compassion: Concepts, research and applications*, 3-15.

Ginsberg, M. R., Rubin, R. A., Falcone, T., Ting, A. H., Natowicz, M. R. (2012). Brain transcriptional and epigenetic associations with autism. PLoS ONE 7:e44736. 10.1371/journal .pone.0044736

Gorbatyuk, O.S., Li, S., Nguyen, F.N., Manfredsson, F.P., Kondrikova, G., Sullivan, L.F., Meyers, C., Chen, W., Mandel, R.J., Muzyczka, N. (2010). α-Synuclein Expression in Rat Substantia Nigra Suppresses Phospholipase D2 Toxicity and Nigral Neurodegeneration. Mol Ther. 18,1758–1768.

Guang, S., Pang, N., Deng, X., Yang, L., He, F., Wu, L., Chen, C., Yin, F., & Peng, J. (2018). Synaptopathology Involved in Autism Spectrum Disorder. *Front Cell Neurosci.* 21,12:470.

Gubert, C., Kong, G., Renoir, T., & Hannan, A. J. (2020). Exercise, diet and stress as modulators of gut microbiota: Implications for neurodegenerative diseases. *Neurobiology of disease*, *134*, 104621.

Guissart, C. Latypova, X. Rollier, P. Khan, T.N. Stamberger, H. McWalter, K. Cho, M.T. Kjaergaard, S. Weckhuysen, S. Lesca, G. et al. (2018). Dual Molecular Effects of Dominant RORA Mutations Cause Two Variants of Syndromic Intellectual Disability with Either Autism or Cerebellar Ataxia. Am. J. Hum. Genet. 102, 744–759.

Guo, S. (2009). Using zebrafish to assess the impact of drugs on neural development and function. Expert Opin. Drug Discov. 4, 715–726.

Gutman, J. (2002). Glutathione- Your Bodies Most Powerful Protector. 3rd edition. Montreal: Communications Kudoca Inc.

Hang, Z., Zhou, L., Xing, C., Wen, Y., & Du, H. (2023). The Blood-brain Barrier, a Key Bridge to Treat Neurodegenerative Diseases. *Ageing Research Reviews*, 102070.

Hodges, H., Fealko, C., & Soares, N. (2020). Autism spectrum disorder: definition, epidemiology, causes, and clinical evaluation. *Transl Pediatr*, 9(Suppl 1):S55-S65.

Holmqvist, S., Chutna, O., Bousset, L., Aldrin-Kirk, P., Li, W., Björklund, T., Wang, Z.-Y., Roybon, L., Melki, R., & Li, J.-Y. (2014). Direct evidence of Parkinson pathology spread from the gastrointestinal tract to the brain in rats. *Acta neuropathologica*, *128*, 805-820.

Hou, K., Wu, Z.-X., Chen, X.-Y., Wang, J.-Q., Zhang, D., Xiao, C., Zhu, D., Koya, J. B., Wei, L., & Li, J. (2022). Microbiota in health and diseases. *Signal transduction and targeted therapy*, *7*(1), 135.

Howe, K. Clark, M.D. Torroja, C.F; Torrance, J. Berthelot, C. Muffato, M. Collins, J.E. Humphray, S. McLaren, K. Matthews, L. et al. (2013). The zebrafish reference genome sequence and its relationship to the human genome. Nature , 496, 498–503.

Huuskonen, J., Suuronen, T., Miettinen, R., van Groen, T., Salminen, A. (2005). A refined in vitro model to study inflammatory responses in organotypic membrane culture of postnatal rat hippocampal slices. J Neuroinflammation, 2,25.

Jasutkar G.H. & Mouradian, M.M. (2022). Therapeutics in the pipeline targeting α-synuclein for Parkinson's disease. Pharmacol. Rev. 74,207–237.

Jeong, J.W. Tiwari, V.N. Behen, M.E. Chugani, H.T. Chugani, D.C. (2014). In vivo detection of reduced purkinje cell fibers with diffusion MRI tractography in children with autistic spectrum disorders. Front. Hum. Neurosci. 8, 110.

Jęśko, H., Cieślik, M., Gromadzka, G., & Adamczyk, A. (2020). Dysfunctional proteins in neuropsychiatric disorders: From neurodegeneration to autism spectrum disorders. *Neurochem Int.* 141,104853.

Juopperi, T.A. Song, H. Ming, G.-L. (2011). Modeling neurological diseases using patient-derived induced pluripotent stem cells. Future Neurol. 6, 363–373.

Juurlink, B.H. & Paterson, P.G. (1998). Review of oxidative stress in brain and spinal cord injury: suggestions for pharmacological and nutritional management strategies. J Spinal Cord Med. 21, 309-334.

Kani, S. Bae, Y.K. Shimizu, T. Tanabe, K. Satou, C. Parsons, M.J. Scott, E., Higashijima, S.I. Hibi, M. (2010). Proneural gene-linked neurogenesis in zebrafish cerebellum. Dev. Biol. 343, 1–17.

Kemper, T.L. & Bauman, M.L. (1993) The contribution of neuropathologic studies to the understanding of autism. Neurol. Clin. 11, 175–187.

Kenyon, Júlia Braga. (2024). The Complexity Behind the Genetics of Autism Spectrum Disorder. Honors Theses. 145

Kern, J.K., Geier, D.A., Sykes, L.K., &Geier, M.R. (2013). Evidence of neurodegeneration in autism spectrum disorder. Transl Neurodegener. 8;2(1):17.

Kern, J.K. & Jones, A.M. (2006) Evidence of toxicity, oxidative stress, and neuronal insult in autism, J Toxicol Environ Heal - Part B Crit Rev. 9(6),485-499.

Kesika, P., Suganthy, N., Sivamaruthi, B. S., & Chaiyasut, C. (2021). Role of gut-brain axis, gut microbial composition, and probiotic intervention in Alzheimer's disease. *Life sciences*, *264*, 118627.

Kimmel, C.B. Ballard, W.W. Kimmel, S.R. Ullmann, B. Schilling, T.F. (1995). Stages of embryonic development of the zebrafish. Dev. Dyn. 203, 253–310.

Klein, J.A. & Ackerman, S.L. (2003). Oxidative stress, cell cycle, and neurodegeneration. J Clin Invest. 111, 785-793.

Jia, P.P., Junaid, M., Wen, P.P., Yang, Y.F., Li, W.G., Yang, X.G., & Pei, D.S. (2021). Role of germ-free animal models in understanding interactions of gut microbiota to host and environmental health: A special reference to zebrafish. *Environmental pollution*, *279*, 116925.

Kanner, L. (1943). Autistic disturbances of affective contact. Nerv Child J Psychopathol Psychother Ment Hyg Guid Child 2, 217–50.

Kern, J.K., Geier, D.A., Sykes, L.K., &Geier, M.R. (2013). Evidence of neurodegeneration in autism spectrum disorder. *Transl Neurodegener.* 8;2(1):17.

Khan, S.A., Khan, S.A., Narendra, A.R., Mushtaq, G., Zahran, S.A., Khan, S., & Kamal, M.A. (2016). Alzheimer's Disease and Autistic Spectrum Disorder: Is there any Association? *CNS Neurol Disord Drug Targets.*15(4),390-402.

Lahiri, D.K., Sokol, D.K., Erickson, C., Ray, B., Ho, C.Y., & Maloney, B. (2013). Autism as early neurodevelopmental disorder: evidence for an sAPPα-mediated anabolic pathway. *Front Cell Neurosci.* 21,7,94.

Lee, A. S., Azmitia, E. C., Whitaker-Azmitia, P. M. (2017). Developmental microglial priming in postmortem autism spectrum disorder temporal cortex. Brain Behav. Immun. 62, 193–202.

Lee, E.J., Choi, S.Y., Kim, E. (2015). NMDA receptor dysfunction in autism spectrum disorders. Curr Opin Pharmacol. 20:8–13.

Lee, J.G., Cho, H.J., Jeong, Y.M., & Lee, J.S. (2021). Genetic approaches using zebrafish to study the microbiota–gut–brain axis in neurological disorders. *Cells*, *10*(3), 566.

Leuzy, A., Rodriguez-Vieitez, E., Saint-Aubert, L., Chiotis, K., Almkvist, O., Savitcheva, I., Jonasson, M., Lubberink, M., Wall, A., Antoni, G., & Nordberg, A. (2018). Longitudinal uncoupling of cerebral perfusion, glucose metabolism, and tau deposition in Alzheimer's disease. *Alzheimers Dement,* 14(5):652-663.

Li, P., Zhang, J., Liu, X., Gan, L., Xie, Y., Zhang, H., & Si, J. (2022). The function and the affecting factors of the zebrafish gut microbiota. *Frontiers in Microbiology*, *13*, 903471.

Liu, C.C., Kanekiyo, T., Xu, H., Bu, G. (2013). Apolipoprotein E and Alzheimer disease: risk, mechanisms and therapy. Nat. Rev. Neurol. 9, 106–118.

Liu, S., Ninan, I., Antonova, I., et al. (2004). alpha-Synuclein produces a long-lasting increase in neurotransmitter release. EMBO J. 23 (22), 4506–4516.

Lombardo, M.V., Chakrabarti, B., Bullmore, E.T., MRC AIMS Consortium, &Baron-Cohen, S. (2011). Specialization of right temporo-parietal junction for mentalizing and its relation to social impairments in autism. *Neuroimage.* 1,56(3):1832-8.

Lopez-Hurtado, E. & Prieto, J.J. (2008). A microscopic study of language-related cortex in autism. Am J Biochem Biotechnol. 4(2),130-145.

López Nadal, A., Ikeda-Ohtsubo, W., Sipkema, D., Peggs, D., McGurk, C., Forlenza, M., Wiegertjes, G. F., & Brugman, S. (2020). Feed, microbiota, and gut immunity: using the zebrafish model to understand fish health. *Frontiers in immunology*, *11*, 512428.

Lovett-Barron, M., Andalman, A.S., Allen, W.E., Vesuna, S., Kauvar, I., Burns, V.M., Deisseroth, K. (2017). Ancestral Circuits for the Coordinated Modulation of Brain State. Cell, 171, 1411–1423.e17.

Löffler, J. & Huber, G. (1992). Beta-amyloid precursor protein isoforms in various rat brain regions and during brain development. J. Neurochem. 59, 1316–1324.

Lu, H., Li, P., Huang, X., Wang, C. H., Li, M., & Xu, Z. Z. (2021). Zebrafish model for human gut microbiome-related studies: Advantages and limitations. *Medicine in Microecology*, *8*, 100042.

Lucchina, L. & Depino, A. M. (2014). Altered peripheral and central inflammatory responses in a mouse model of autism. Autism Res. 7 (2), 273–289.

Maenner, M.J., Warren, Z., Williams, A.R., Amoakohene, E., Bakian, A.V., Bilder, D.A., ....... & Shaw, K.A. (2020). Prevalence and Characteristics of Autism Spectrum Disorder Among Children Aged 8 Years - Autism and Developmental Disabilities Monitoring Network, 11 Sites, United States, *MMWR Surveill Summ.* 24,72(2),1-14.

Meshalkina, D.A., Kizlyk, M.N., Kysil, E.V., Collier, A.D., Echevarria, D.J., Abreu, M.S., Barcellos, L.J.G., Song, C., Warnick, J.E., Kyzar, E.J., et al. (2018). Zebrafish models of Autism Spectrum Disorder. Exp. Neurol. 299, 207–216.

Martinez, J. E., Kahana, D. D., Ghuman, S., Wilson, H. P., Wilson, J., Kim, S. C., Lagishetty, V., Jacobs, J. P., Sinha-Hikim, A. P., & Friedman, T. C. (2021). Unhealthy lifestyle and gut dysbiosis: a better understanding of the effects of poor diet and nicotine on the intestinal microbiome. *Frontiers in endocrinology*, *12*, 667066.

Meade, R.M., Fairlie, D.P., Mason, J.M. (2019). Alpha-synuclein structure and Parkinson's disease—Lessons and emerging principles. Mol. Neurodegener. 14,29.

Meijer, A. H., van der Vaart, M., & Spaink, H. P. (2014). Real-time imaging and genetic dissection of host–microbe interactions in zebrafish. *Cellular microbiology*, *16*(1), 39-49.


Meshalkina, D.A., Kizlyk, M.N., Kysil, E.V., Collier, A.D., Echevarria, D.J., Abreu, M.S., Barcellos, L.J.G., Song, C., Warnick, J.E., Kyzar, E.J., et al. (2018). Zebrafish models of Autism Spectrum Disorder. Exp. Neurol. 299, 207–216.


Meyza, K.Z., Defensor, E.B., Jensen, A.L., Corley, M.J., Pearson, B.L., Pobbe, R.L.H., Bolivar, V.J., Blanchard, D.C., Blanchard, R.J. (2013). The BTBR T+tf/J mouse model for Autism Spectrum Disorders–in search of biomarkers. Behav. Brain Res. 251, 25–34.

Monnet-Tschudi, F., Defaux, A., Braissant, O., Cagnon, L., Zurich, M. G. (2011). Methods to assess neuroinflammation. Curr. Protoc. Toxicol. 50,12.19.1-12.19.20.

Morgan, J.T., Chana, G., Pardo, C.A., Achim, C., Semendeferi, K., Buckwalter, J., Courchesne, E., Everall, I.P. (2010). Microglial activation and increased microglial density observed in the dorsolateral prefrontal cortex in autism. Biol Psychiatry, 68(4),368-376.

Muller, C.L., Anacker, A.M.J., Veenstra-VanderWeele, J. (2016). The serotonin system in autism spectrum disorder: From biomarker to animal models. Neuroscience. 3;321:24-41.

Mueller, T., Wullimann, M.F. (2009). An evolutionary interpretation of teleostean forebrain anatomy. Brain Behav. Evol. 74, 30–42.

Navarro-Paya, C., Sanz-Hernandez, M., De Simone, A. (2022). Plasticity of membrane binding by the central region of α-synuclein. Front Mol Biosci. 9,857217.

Ng, A. N., de Jong-Curtain, T. A., Mawdsley, D. J., White, S. J., Shin, J., Appel, B., Dong, P. D. S., Stainier, D. Y., & Heath, J. K. (2005). Formation of the digestive system in zebrafish: III. Intestinal epithelium morphogenesis. *Developmental biology*, *286*(1), 114-135.

Okubo, R., Koga, M., Katsumata, N., Odamaki, T., Matsuyama, S., Oka, M., Narita, H., Hashimoto, N., Kusumi, I., & Xiao, J. (2019). Effect of bifidobacterium breve A-1 on anxiety and depressive symptoms in schizophrenia: A proof-of-concept study. *Journal of affective disorders*, *245*, 377-385.

Paduraru, E., Iacob, D., Rarinca, V., Plavan, G., Ureche, D., Jijie, R., & Nicoara, M. (2023). Zebrafish as a potential model for neurodegenerative diseases: a focus on toxic metals implications. *International Journal of Molecular Sciences*, *24*(4), 3428.

Palmen, S.J., van Engeland, H., Hof, P.R., Schmitz, C. (2004). Neuropathological findings in autism. Brain : a journal of neurology. 127(Pt 12),2S72-83.

Patterson, P.H. (2011). Modeling autistic features in animals. Pediatr. Res. 69, 34R–40R.

Pendergrass, S., Girirajan, S., Selleck, S. (2014). Uncovering the etiology of autism spectrum disorders: genomics, bioinformatics, environment, data collection and exploration, and future possibilities. *Pac Symp Biocomput*. 422-6.

Perry, V.H. & O'Connor, V. (2010). The role of microglia in synaptic stripping and synaptic degeneration: a revised perspective. ASN Neuro 2: e00047.

Peterson, C. T. (2020). Dysfunction of the microbiota-gut-brain axis in neurodegenerative disease: the promise of therapeutic modulation with prebiotics, medicinal herbs, probiotics, and synbiotics. *Journal of evidence-based integrative medicine*, *25*, 2515690X20957225.

Phelps, D., Brinkman, N. E., Keely, S. P., Anneken, E. M., Catron, T. R., Betancourt, D., Wood, C. E., Espenschied, S. T., Rawls, J. F., & Tal, T. (2017). Microbial colonization is required for normal neurobehavioral development in zebrafish. *Scientific reports*, *7*(1), 11244.

Plaza-Díaz, J., Solís-Urra, P., Rodríguez-Rodríguez, F., Olivares-Arancibia, J., Navarro-Oliveros, M., Abadía-Molina, F., & Álvarez-Mercado, A. I. (2020). The gut barrier, intestinal microbiota, and liver disease: molecular mechanisms and strategies to manage. *International Journal of Molecular Sciences*, *21*(21), 8351.

Popa-Wagner, A., Mitran, S., Sivanesan, S., Chang, E., Buga, A-M. (2013). ROS and brain diseases: the good, the bad, and the ugly. Oxid Med Cell Longev., 963520.

Portelius, E., Mattsson, N., Andreasson, U., Blennow, K., Zetterberg H. (2011) Novel Abeta Isoforms in Alzheimer s Disease -Their Role in Diagnosis and Treatment. Curr. Pharm. Des. 17, 2594–2602.

Puzzo, D., Arancio, O. (2013). Amyloid-β peptide: dr. Jekyll or mr. Hyde? J Alzheimers Dis 33, S111–S120.

Qin, L., Wang, H., Ning, W. et al. (2024). New advances in the diagnosis and treatment of autism spectrum disorders. Eur J Med Res 29, 322.

Raymond, G. V., Bauman, M. L., &Kemper, T. L. (1995). Hippocampus in autism: a golgi analysis. *Acta Neuropathol.* 91, 117–119.

Ritvo, E.R., Freeman, B.J., Scheibel, A.B., et al. (1986). Lower Purkinje cell counts in the cerebella of four autistic subjects: initial findings of the UCLA-NSAC Autopsy Research Report. The American Journal of Psychiatry. 143(7),862-866.

Rock, R.B., Gekker, G., Hu, S., Sheng, W.S., Cheeran, M., Lokensgard, J.R., Peterson, P.K. (2004). Role of microglia in central nervous system infections. Clin Microbiol Rev 17, 942-964.

Rohe, M., Carlo, A.S., Breyhan, H., Sporbert, A., Militz, D., Schmidt, V., Wozny, C., Harmeier, A., Erdmann, B., Bales, K.R., Wolf, S., Kempermann, G., Paul, S.M., Schmitz, D., Bayer, T.A., Willnow, T.E., Andersen, O.M. (2008). Sortilin-related receptor with A-type repeats (SORLA) affects the amyloid precursor protein-dependent stimulation of ERK signaling and adult neurogenesis. J Biol Chem. 23,283(21),14826-34.

Rook, G., Bäckhed, F., Levin, B. R., McFall-Ngai, M. J., & McLean, A. R. (2017). Evolution, human-microbe interactions, and life history plasticity. *The Lancet*, *390*(10093), 521-530.

Rutsch, A., Kantsjö, J. B., & Ronchi, F. (2020). The gut-brain axis: how microbiota and host inflammasome influence brain physiology and pathology. *Frontiers in immunology*, *11*, 604179.

Rylaarsdam, L., Guemez-Gamboa, A. (2019). Genetic causes and modifiers of autism spectrum disorder. Front. Cell. Neurosci, 13, 385.

Sajdel-Sulkowska, E.M., Xu, M., McGinnis, W., Koibuchi, N. (2011). Brain regionspecific changes in oxidative stress and neurotrophin levels in autism spectrum disorders (ASD), Cerebellum, 10(1),43-48.

Sakai C., Ijaz S., Hoffman E.J. (2018). Zebrafish models of neurodevelopmental disorders: Past, present, and future. Front. Mol. Neurosci. 11,294.

Salomone, S., Caraci, F., Leggio, G.M., Fedotova, J., & Drago, F. (2012). New pharmacological strategies for treatment of Alzheimer's disease: focus on disease modifying drugs. *Br J Clin Pharmacol,* 73(4),504-17.

Sampson, T. R., Challis, C., Jain, N., Moiseyenko, A., Ladinsky, M. S., Shastri, G. G., Thron, T., Needham, B. D., Horvath, I., & Debelius, J. W. (2020). A gut bacterial amyloid promotes α-synuclein aggregation and motor impairment in mice. *Elife*, *9*, e53111.

Schapira, A.H.V., Chaudhuri, K.R., & Jenner, P. (2017) Non-motor features of Parkinson disease. *Nat Rev Neurosci.* 18(7),435-450.

Schmidt S., Vogt Waisenhorn, D.M., Wurs, W. (2022). Chapter 5—"Parkinson's disease—A role of non-enzymatic post-translational modifications in disease onset and progression?" Mol. Aspect Med. 101096.

Schumann, C.M. & Amaral, D.G. (2006). Stereological analysis of amygdala neuron number in autism. J Neurosci. 26(29):7674-7679.

Shepherd, I., & Eisen, J. (2011). Development of the zebrafish enteric nervous system. *Methods in cell biology*, *101*, 143-160.

Shi, Y., Inoue, H., Wu, J.C., Yamanaka, S. (2017). Induced pluripotent stem cell technology: A decade of progress. Nat. Rev. Drug Discov. 16, 115–130.

Singh, A., Dawson, T. M., & Kulkarni, S. (2021). Neurodegenerative disorders and gut-brain interactions. *The Journal of clinical investigation*, *131*(13).

Siu, M.T. & Weksberg, R. (2017). Epigenetics of autism spectrum disorder. Adv Exp Med Biol. 2017;978:63–90.

Snitz, B.E., Weissfeld, L.A., Cohen, A.D., Lopez, O.L., Nebes, R.D., Aizenstein, H.J., McDade, E., Price, J.C., Mathis, C.A., Klunk, W.E. (2015). Subjective Cognitive Complaints, Personality and Brain Amyloid-beta in Cognitively Normal Older Adults. Am J Geriatr Psychiatry. 23(9),985-93.

Sokol, D. K., Demao, C., Farlow, M. R., Dunn, D. W., Maloney, B., Zimmer J. A., et al. (2006). High levels of Alzheimer beta-amyloid precursor protein (APP) in children with severely autistic behavior and aggression. J. Child Neurol. 21, 444–449.

Sokol, D.K. & Lahiri, D.K. (2023). APPlications of amyloid-β precursor protein metabolites in macrocephaly and autism spectrum disorder. Front Mol Neurosci. 20,16,1201744.

Sommer, F., & Bäckhed, F. (2013). The gut microbiota—masters of host development and physiology. *Nature reviews microbiology*, *11*(4), 227-238.

Stilling, R. M., Bordenstein, S. R., Dinan, T. G., & Cryan, J. F. (2014). Friends with social benefits: host-microbe interactions as a driver of brain evolution and development? *Frontiers in cellular and infection microbiology*, *4*, 147.

Sudarov, A. (2013). Defining the role of cerebellar purkinje cells in autism spectrum disorders. Cerebellum 12, 950–955.

Südhof, T. C. & Rothman, J. E. (2009). Membrane fusion: grappling with SNARE and SM proteins. Science 323, 474–477.

Sundberg, M., Tochitsky, I., Buchholz, D.E. et al. (2018). Purkinje cells derived from TSC patients display hypoexcitability and synaptic deficits associated with reduced FMRP levels and reversed by rapamycin. Mol Psychiatry 23, 2167–2183.

Suzuki K, Sugihara G, Ouchi Y, et al. (2013). Microglial activation in young adults with autism spectrum disorder. Arch Gen Psychiatry, 70(1),49-58.

Takahashi, K. & Yamanaka, S. (2006). Induction of pluripotent stem cells from mouse embryonic and adult fibroblast cultures by defined factors. Cell, 126, 663–676.

Tan, F., Limbu, S. M., Qian, Y., Du, Z.-Y., & Zhang, M. (2019). The responses of germ-free zebrafish (Danio rerio) to varying bacterial concentrations, colonization time points, and exposure duration. *Frontiers in Microbiology*, *10*, 477775.

Tang, G., Gudsnuk, K., Kuo, S-H., Cotrina, M.L., Rosoklija, G., Sosunov, A., et al. (2014). Loss of mTOR-dependent macroautophagy causes autistic-like synaptic pruning deficits. Neuron. 83:1131–43.

Ten Kate, M., Ingala, S., Schwarz, A.J., Fox, N.C., Chételat, G., van Berckel, B.N.M., Ewers, M., Foley, C., Gispert, J.D., Hill, D., Irizarry, M.C., Lammertsma, A.A., Molinuevo, J.L., Ritchie, C., Scheltens, P., Schmidt, M.E., Visser, P.J., Waldman, A., Wardlaw, J., Haller, S., & Barkhof, F. (2018). Secondary prevention of Alzheimer's dementia: neuroimaging contributions. *Alzheimers Res Ther. 30*;10(1):112.

Tetreault, N.A., Hakeem, A.Y., Jiang, S., Williams, B.A., Allman, E., Wold, B.J. (2012). Microglia in the cerebral cortex in autism. J Autism Dev Disord. 42:2569–84.

Theoharides, T.C. & Zhang, B. (2011). Neuro-inflammation, blood-brain barrier, seizures and autism. J Neuroinflammation. 8,16

Ufnal, M., & Pham, K. (2017). The gut-blood barrier permeability–a new marker in cardiovascular and metabolic diseases? *Medical hypotheses*, *98*, 35-37.

Valdes, A. M., Walter, J., Segal, E., & Spector, T. D. (2018). Role of the gut microbiota in nutrition and health. *Bmj*, *361*.

van Kooten, I.A., Palmen, S.J., von Cappeln, P., Steinbusch, H.W., Korr, H., Heinsen, H., Hof, P.R., van Engeland, H., Schmitz, C. (2008). Neurons in the fusiform gyrus are fewer and smaller in autism. Brain, 131(Pt 4):987-999.

Vargas, D.L., Nascimbene, C., Krishnan, C., Zimmerman, A.W., Pardo, C.A. (2005). Neuroglial activation and neuroinflammation in the brain of patients with autism. Ann Neurol, 57, 304.

Venezia, V., Nizzari, M., Repetto, E., Violani, E., Corsaro, A., Thellung, S., Villa, V., Carlo, P., Schettini, G., Florio, T., Russo, C. (2006). Amyloid precursor protein modulates ERK-1 and -2 signaling. Ann N Y Acad Sci. 1090:455-65.

Ventura, M., Turroni, F., Motherway, M. O. C., MacSharry, J., & van Sinderen, D. (2012). Host–microbe interactions that facilitate gut colonization by commensal bifidobacteria. *Trends in microbiology*, *20*(10), 467-476.

Vidović, M. & Rikalovic, M.G. (2022) Alpha-Synuclein Aggregation Pathway in Parkinson's Disease: Current Status and Novel Therapeutic Approaches. Cells. 24,11(11),1732.

Vithayathil, J., Pucilowska, J., Landreth, G.E. (2018). ERK/MAPK signaling and autism spectrum disorders. Prog Brain Res. 241,63-112.

Volkmar, F., Siegel, M., Woodbury-Smith, M., King, B., McCracken, J., State, M. (2014). Practice parameter for the assessment and treatment of children and adolescents with autism spectrum disorder. J Am Acad Child Adolesc Psychiatry, 53(2),237-257.

Wallace, K. N., Akhter, S., Smith, E. M., Lorent, K., & Pack, M. (2005). Intestinal growth and differentiation in zebrafish. *Mechanisms of development*, *122*(2), 157-173.

Walter, J., & Ley, R. (2011). The human gut microbiome: ecology and recent evolutionary changes. *Annual review of microbiology*, *65*, 411-429.

Wang, C., Lau, C. Y., Ma, F., & Zheng, C. (2021). Genome-wide screen identifies curli amyloid fibril as a bacterial component promoting host neurodegeneration. *Proceedings of the National Academy of Sciences*, *118*(34), e2106504118.

Wang, C., Tu, J., Zhang, S. et al. (2020). Different regions of synaptic vesicle membrane regulate VAMP2 conformation for the SNARE assembly. Nat Commun 11, 1531.

Wang, X., Zhang, J.-B., He, K.-J., Wang, F., & Liu, C.-F. (2021). Advances of zebrafish in neurodegenerative disease: from models to drug discovery. *Frontiers in Pharmacology*, *12*, 713963.

Wei, W., Wang, S., Xu, C., Zhou, X., Lian, X., He, L., & Li, K. (2022). Gut microbiota, pathogenic proteins and neurodegenerative diseases. *Frontiers in Microbiology*, *13*, 959856.

Wender, C.L. & Veenstra-VanderWeele, J. (2017). Challenge and potential for research on gene-environment interactions in autism spectrum disorder. In: Tolan P, Leventhal B, editors. Gene-environment transactions in developmental psychopathology. Berlin: Springer; 2. pp. 157–176.

Whitney, E.R., Kemper, T.L., Bauman, M.L., Rosene, D.L., Blatt, G.J. (2008). Cerebellar Purkinje cells are reduced in a subpopulation of autistic brains: a stereological

Wing L.& Gould, J. (1979). Severe impairments of social interaction and associated abnormalities in children: Epidemiology and classification. Journal of autism and developmental disorders. 9(1),11-29.

Wöhr, M.& Scattoni, M.L. (2013).Behavioural methods used in rodent models of Autism Spectrum Disorders: Current standards and new developments. Behav. Brain Res. 251, 5–17.

World Health Organization (2013). The Global Advisory Committee on Vaccine Safety Statement on Thiomersal. Weekly Epidemiological Record on 7 January 2005.WHO; 2013.

Xia, H., Chen, H., Cheng, X., Yin, M., Yao, X., Ma, J., Huang, M., Chen, G., & Liu, H. (2022). Zebrafish: an efficient vertebrate model for understanding role of gut microbiota. *Molecular Medicine*, *28*(1), 1-20.

Zheng, D., Liwinski, T., & Elinav, E. (2020). Interaction between microbiota and immunity in health and disease. *Cell research*, *30*(6), 492-506.

Zheng, Y., Bonfili, L., Wei, T., & Eleuteri, A. M. (2023). Understanding the Gut–Brain Axis and Its Therapeutic Implications for Neurodegenerative Disorders. *Nutrients*, *15*(21), 4631.

Zhong, X., Li, J., Lu, F., Zhang, J., & Guo, L. (2022). Application of zebrafish in the study of the gut microbiome. *Animal Models and Experimental Medicine*, *5*(4), 323-336.

Zikopoulos, B. & Barbas, H. (2010). Changes in prefrontal axons may disrupt the network in autism, J Neurosci., 30(44),14595-14609.

# 6

# *Neurogenesis in the Zebrafish Olfactory Epithelium in Response to Neurotoxicity and Injury*

**Yigit Kocagöz and Stefan H. Fuss**

## Abbreviations

| | |
|---|---|
| **ADAM** | A Disintegrin and Metalloprotease |
| **AKT** | Protein Kinase B (AKT Kinase) |
| **ATP** | Adenosine Triphosphate |
| **BrdU** | 5-Bromo-2'-Deoxyuridine |
| **celsr1a** | Cadherin EGF LAG Seven-Pass G-Type Receptor 1a |
| **CldU** | 5-Chloro-2'-Deoxyuridine |
| **CNS** | Central Nervous System |
| **DAMP** | Damage-Associated Molecular Pattern |
| **DPM** | Diesel Particulate Matter |
| **D-NALT** | Diffuse Nasal-Associated Lymphoid Tissue |
| **dpl** | Days Post Lesion |
| **EdU** | Ethynyl-2'-Deoxyuridine |
| **EGF** | Epidermal Growth Factor |
| **EGFR** | Epidermal Growth Factor Receptor |
| **GBC** | Globose Basal Cell |
| **GFP** | Green Fluorescent Protein |
| **HBC** | Horizontal Basal Cell |
| **HB-EGF** | Heparin-Binding Epidermal Growth Factor-Like Growth Factor |
| **hpl** | Hours Post Lesion |
| **IdU** | 5-Iodo-2'-Deoxyuridine |
| **IL-1b** | Interleukin 1 Beta |
| **IL-6** | Interleukin 6 |
| **ILC** | Interlamellar Curve |
| **JAK** | Janus Kinase |
| **MAP kinase** | Mitogen-Activated Protein Kinase |
| **NALT** | Nasal-Associated Lymphoid Tissue |
| **OE** | Olfactory Epithelium |
| **O-NALT** | Organized Nasal-Associated Lymphoid Tissue |
| **OMP/ompb** | Olfactory Marker Protein / b |
| **OR** | Olfactory Receptor |
| **OSN** | Olfactory Sensory Neuron |
| **PCNA** | Proliferating Cell Nuclear Antigen |
| **pH3** | Phosphorylated Histone 3 |
| **PI3K** | Phosphatidylinositol 3-Kinase |
| **PM2.5** | Particulate Matter with A Size $\leq 2.5$ µm |
| **PNS** | Peripheral Nervous System |
| **SNS** | Sensory/Non-Sensory Border |
| **STAT** | Signal Transducer and Activator of Transcription |
| **TGF-b** | Tumor Growth Factor Beta |
| **TNF-a** | Tumor Necrosis Factor Alpha |
| **trpc2** | Transient Receptor Potential C2 |
| **TUNEL** | Terminal Deoxynucleotidyl Transferase Dutp Nick End Labeling |

## Introduction

Neurodegenerative disorders typically are the result of compromised nerve cell function or disturbances in tissue homeostasis in the nervous system (Wilson et al., 2023). As detailed in the remaining chapters of this book, dysregulation of gene expression, formation of intracellular or pericellular protein and RNA aggregates, exposure to cytotoxic concentrations of metabolites and neurotransmitters, inflammation, or disturbances in energy metabolism are the primary causes of neuronal cell death in the central (CNS) and peripheral nervous system (PNS) (Jellinger, 2010; Wilson et al., 2023). In many cases, genetic predispositions have been identified which result in degenerative phenotypes that affect all or specific subsets of nerve cells, depending on their differential sensitivity to the defect (Pihlstrøm et al., 2017; Pandya et al., 2021; Ciurea et al., 2023). Thus, for the most part, neurodegeneration is an endogenous process that develops and intensifies within the limits of the nervous system (Lanigan et al., 2021). However, exposure to environmental toxins, pollutants, and pathogens have also been shown to either damage nerve cells directly, or to exaggerate ongoing neurodegenerative processes by sensitizing cells (Cannon and Greenamyre, 2011; Nabi and Tabassum, 2022).

Typically, nerve cells are embedded deep inside the body where they are protected from environmental influences by surrounding tissues and complex organ structures. For instance, the brain is shielded by the cranial bones, cushioned by cerebrospinal fluid, and largely isolated from the somatic



DOI: 10.1201/9781032676555-8

metabolism by the blood–brain barrier, which restricts the intake of metabolites, pathogens, and toxic agents to the brain parenchyma (Sweeney et al., 2019; Czarniak et al., 2023). Related barrier functions also exist in the PNS that isolate nerve cells from metabolic influences of their surrounding tissue environment (Reinhold and Rittner, 2017). Peripheral sensory structures, such as touch- and heat-sensitive nerve endings in the skin, are further protected by the integument (Abraira and Ginty, 2013), while retinal neurons and auditory hair cells are embedded within the larger complexes of the eye and inner ear, respectively (Bossy, 1970). Despite the presence of these protective physical barriers, central, peripheral, and sensory neurons can be severely affected by exposure to environmental toxicants and pathogens that are able to penetrate the nervous system (Genc et al., 2012; Dando et al., 2014). Often, complex and mutual relationships exist between the progression of neurodegenerative disorders, ageing, and the breakdown of protective barriers that exacerbate the disease phenotype (Knox et al., 2022).

Different from other nerve cells, however, the specific function of chemosensory neurons requires their direct and unprotected contact with the external chemical environment (Genter and Doty, 2019). Taste cells of the tongue and palate sample chemicals that pass through the oral cavity with the ingested food and drink, and which function as tastants when interacting with appropriate molecular taste receptors (Roper, 2022). In addition to relevant chemical stimuli, taste cells are also exposed to toxic food contaminants or other potentially harmful substances that occur naturally in the diet. Olfactory sensory neurons (OSNs) of the nose occupy the olfactory epithelium (OE) that lines the nasal cavity and recognize a wide range of environmental chemicals (Mombaerts 2004a). Smell sensation requires that the surrounding medium – air for terrestrial organisms; water for limnic and marine species – is actively presented to OSNs by sniffing or through guided water currents, respectively (Wachowiak, 2011; Døving et al., 1977). OSNs extend cytoplasmic processes, non-motile cilia or microvilli, into the nasal cavity to detect environmental chemicals through different classes of olfactory receptor (OR) proteins (Mombaerts, 2004a). As a consequence, in addition to sampling relevant odorant molecules, OSNs are constantly bombarded with a wide range of chemicals and pathogens, some of which can have cytotoxic effects (Figure 6.1A) (Genter and Doty, 2019).

Environmental pollution, occupational exposure to hazardous chemicals, and viral or bacterial infections are the main cause of olfactory dysfunction in humans, either by affecting OSNs directly or by disrupting the structural integrity of the OE (Genter and Doty, 2019). Therefore, OSN degeneration and renewal are continuous processes that start from early organismal age and persist lifelong (Brann and Firestein, 2014). Moreover, OSNs project axons to glomeruli in the olfactory bulb (OB) of the forebrain, where they make synaptic contacts onto second order projection neurons, the mitral and tufted cells, that connect divergently to multiple higher brain structures (Mombaerts, 2006; Imamura et al., 2020). Therefore, the OE is also a potential route for the unimpeded entry of chemicals and pathogens into the brain (Figure 6.1B), which may have deleterious consequences for the nervous system beyond the peripheral OE (Dando et al., 2014).

Robust structural, cellular, and molecular mechanisms are implemented in the OE that anticipate and counter the continuous loss of OSNs but also the consequences of physical tissue injury (Schwob, 2005). At the most basic level, the olfactory mucosa is endowed with anatomical and chemical barriers to limit the breach of toxicants and pathogens into the tissue and to prevent them from damaging OE cells (Zhang et al., 2023). The apical surface of the OE is covered by a thin layer of mucus that traps pathogens but also contains antioxidants and enzymes that break down toxic substances to limit harmful exposure (Yoshikawa et al., 2018). Sustentacular cells are the resident glia cells of the OE and establish a closely sealed tissue border by forming tight junctions between themselves and OSN dendrites to separate OE cell bodies from the external environment (Liang, 2020). Breakdown of the apical barrier by chemical or physical damage results in disturbances of the internal ionic and metabolic milieu, which has detrimental effects on OSNs (Zhang et al., 2023). The OE is also a tissue that, similar to epithelial barriers in the lung, skin, and intestine, is endowed with a rich complement of immune cells,

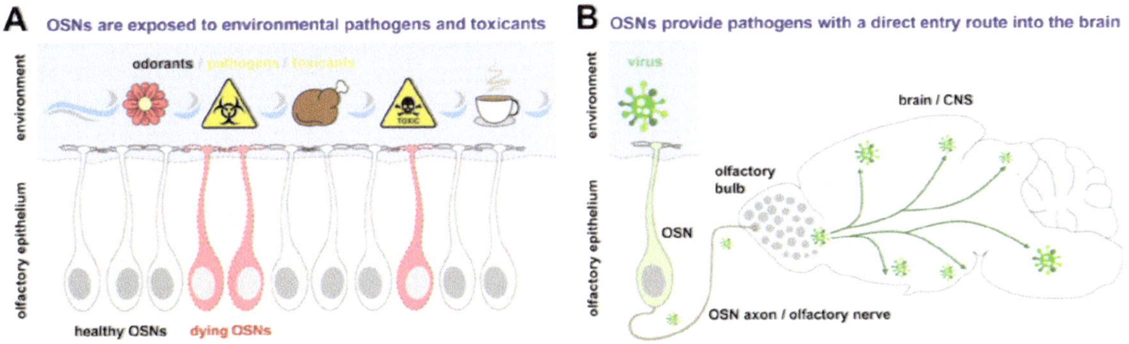


**FIGURE 6.1** The olfactory epithelium is in direct contact with the external environment **A.** Olfactory sensory neurons (OSNs) extend cellular processes (sensory cilia or microvilli) into the nasal cavity through which they detect odorant molecules that reach the surface of the olfactory epithelium. Their unprotected contact with the external environment makes them vulnerable to toxicants and pathogens that enter the nasal cavity, and which may have cytotoxic effects.**B.** The olfactory epithelium also provides a direct entry route for pathogens and toxicants into the brain. OSNs project axons to olfactory bulb glomeruli through the olfactory nerve, along which viruses and other pathogens/toxicants can enter the brain by bypassing the blood–brain barrier.

which eliminate pathogens and prevent them from entering the OE and the brain (Tacchi et al., 2014; Sepahi and Salinas, 2016; Palominos and Whitlock, 2021; Zhang et al., 2023; Wellford and Moseman, 2024).

Even though these physical, chemical, and immune barriers exist, extensive OSN cell death occurs regularly and at a considerable rate (Mackay-Sim and Kittel, 1991; Carr and Farbman, 1993; Gaun et al., 2022). To replace dying or damaged OSNs, the OE is equipped with two distinct types of neurogenic stem/progenitor cells that are capable of regenerating lost OSNs under physiological and acute injury conditions (Schwob et al., 2017; Yu and Wu, 2017). This is different from most other regions of the CNS and PNS, where postembryonic neurogenesis and nerve cell regeneration is sparse, except for the forebrain ventricles and the hippocampus (Iismaa et al., 2018; Kemperman et al., 2015; Lim and Alvarez-Buylla, 2016). Thus, the peripheral OE constitutes a unique site of high neurogenic activity, which, however, shows characteristic differences to the classical proliferation centers in the brain. Especially the division of labor between constitutively active and injury-induced stem cells more closely resembles epithelial tissues, such as skin and intestine (Gonzales and Fuchs, 2017; Hageman et al., 2020), than the adult neurogenic stem cell niches in the subventricular zone and hippocampus.

Understanding how cellular composition, niche factors, signaling pathways, and sensitivity to environmental stressors differ between the OE and other regions of the nervous system with reduced or absent neuroregenerative capacity is, therefore, an important and active avenue of research that bears potential for the development of therapeutic approaches to counter the consequences of neurodegenerative disorders and traumatic brain injury (Shohayeb et al., 2018; Jurkowski et al., 2020). Because the OE comprises a system that effectively deals with sporadic nerve cell degeneration but also acute episodes of traumatic injury that affect large numbers of OSNs simultaneously (Schwob et al., 2017; Yu and Wu, 2017), valuable insight about the molecular regulation of neurogenic stem cell niches may be gained (Llorente et al., 2022). In addition, comparing highly regenerative organisms, such as zebrafish, with mammals that have reduced regenerative capacity may shed light on the function of molecular pathways and regulators lost over ~450 million years of divergent evolution (Massoz et al., 2021).

About 70% of human genes have an orthologue in the zebrafish genome (Howe et al., 2013) and the zebrafish OE bears significant cellular and molecular similarities to the OE of higher vertebrates (Korsching et al., 1997; Calvo-Ochoa et al., 2021). This makes the zebrafish OE an attractive and experimentally accessible model to elucidate cellular responses and signaling pathways that are active during nerve cell degeneration and restoration of olfactory function. Even though the events that take place in the OE are not typically considered to be a neurodegenerative disorder, the OE is a tissue that is prone to environmental degeneration, which may have relevance for understanding nerve cell toxicity in general or the etiology of degenerative disorders, which can be influenced by the presence of toxicants (Cannon and Greenamyre, 2011; Nabi and Tabassum, 2022). Here we will review the anatomical structure of the zebrafish OE, its sensitivity to toxicants and pathogens, and the cellular mechanisms that contribute to OSN maintenance under physiological conditions and tissue regeneration in response to acute injury.

## Structural Organization and Cellular Architecture of the Zebrafish OE

The zebrafish OE has a morphology that is typical for most teleost species, although deviations from the basic plan exist (Hansen and Zielinski, 2005; Olivares and Schmachtenberg, 2019). The olfactory pathway of the zebrafish originates from OSNs that are contained within the paired olfactory organs that are situated in the nasal cavities on the dorsal anterior side of the head (Figure 6.2A, B). OSNs connect to the olfactory bulb of the forebrain through axons that form the olfactory nerve (Hansen and Zeiske, 1998). The dorsal surface of the nasal cavity is covered by skin tissue that forms funnel-shaped nostrils at the anterior end (Hansen and Zielinski, 2005). Swimming movement of the fish pushes water through the nostril into the olfactory cavity, which passes across the OE and exits from a posterior outflow opening (Døving et al., 1977). Each olfactory organ has an overall flower-like appearance (Figure 6.2B), which is formed by bilaterally symmetric stacks of 6–11 lamellae, depending on the age of the fish, that project radially from the central structure of the midline raphe (Hansen and Zeiske, 1998). The convoluted pseudostratified OE lines the surfaces on both sides of each lamella, facing water-filled spaces through which odorants are delivered to OSNs (Figure 6.2B, C). A repeating structural unit of the OE is formed by U-shaped "epithelial folds" that comprise of a continuous stretch of epithelial tissue along the surface of two neighboring lamellae and their junction at the central interlamellar curves (ILCs).

The zebrafish OE can be divided into a central sensory region that contains mature OSNs and a surrounding ring of structural cells that further support OE function (Figure 6.2C, D, E) (Hansen and Zeiske, 1998; Celik et al., 2002; Sato et al., 2005). The sensory OE comprises four primary cell types, which include OSNs and sustentacular glial cells in addition to two types of stem/progenitor cell populations, the globose basal cells (GBCs) and horizontal basal cells (HBCs) (Figure 6.2E) (Hansen and Zeiske, 1998; Bayramli et al., 2017; Demirler et al., 2020). The peripheral non-sensory OE largely contains mucus-producing goblet cells and cells with beating cilia, which generate a laminar flow of water across the sensory OE surface to support the delivery of odorant molecules to OSNs (Hansen and Zeiske, 1998; Reiten et al., 2017). Mutants of cilia function show impaired odorant delivery and lack of odorant responses in the OE and olfactory bulb (Reiten et al., 2017). On horizontal cross-sections through the OE, mature OSNs occupy the inner two-thirds of each epithelial fold and extend from the ILC to the sharp border between sensory and non-sensory cells (SNS), as indicated by immunohistochemistry or RNA *in situ*-hybridization against various neuronal markers (Figure 6.2D) (Celik et al. 2002; Sato et al. 2005; Iqbal and Byrd-Jacobs 2010; Bayramli et al., 2017). Sustentacular glial cells are regularly interspersed between OSNs (Figure 6.2E) but appear to extend at least somewhat into the non-sensory

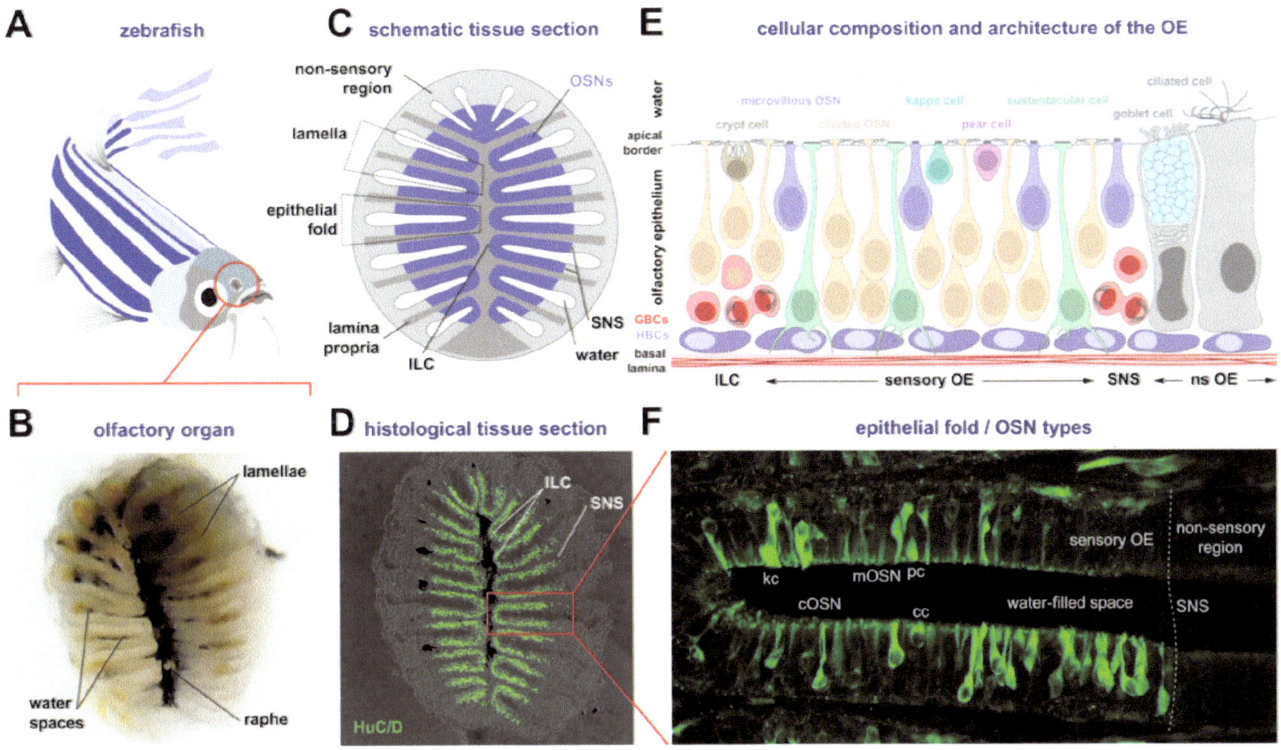


**FIGURE 6.2** Structural organization of the zebrafish olfactory epithelium **A.** The paired olfactory organs of the zebrafish are located on the dorsal-anterior side of the head, just anterior to the eye. **B.** The zebrafish olfactory organ has a flower-like appearance and consists of multiple stacks of bilaterally symmetric lamellae that originate from the central structure of the midline raphe. The spaces between lamellae are accessible by water entering the nasal cavity through the nostril. **C.** Schematic representation of a tissue section through the zebrafish olfactory organ. The convoluted olfactory epithelium (OE) lines the surface of the lamellae, which enclose water-filled spaces (white) through which odorants are delivered to OSNs. The epithelium is continuous between the surfaces of neighboring lamellae, giving rise to U-shaped epithelial folds. The lamina propria (dark grey) runs through the center of each lamella and is composed of blood vasculature and the basal lamina of the epithelium. OSNs (blue) occupy the inner two-thirds of the tissue from the interlamellar curves (ILCs) to the sensory/non-sensory border (SNS) and are surrounded by a non-sensory region (light grey), which contains structural and supportive cells. **D.** Histological tissue section through the OE counterstained by immunohistochemistry against the pan-neuronal marker HuC/D (green), which marks all nerve cells. OSNs occupy the central two-thirds of the tissue and their presence defines the sensory OE. **E.** Cellular architecture of the zebrafish OE, which is composed of five morphologically distinct OSN types, stem cells, glia cells, and supportive cells. Crypt (brown), kappe (dark green), and pear (magenta) cells possess large spherical cell bodies that are located directly underneath the apical surface of the OE. They extend microvilli (kappe, pear cells) or microvilli and cilia (crypt cells) into the water space above the epithelium to detect odorants. Microvillous (blue) and ciliated (tan) OSNs are situated in successively deeper tissue layers. Their apical dendrites are decorated with microvilli or non-motile cilia that function as odorant sensors. Sustentacular glial cells (light green) are regularly spaced across the tissue and span the entire thickness of the OE. They extend apical projections that form tight junctions with OSN dendrites, creating a diffusion barrier, while their endfeet processes envelop horizontal basal cells (HBCs, dark blue) and contact the basal lamina. HBCs are injury-responsive stem cells that form a band of cells in the most basal stratum of the tissue. Continuously active globose basal cells (GBCs, red) are localized to the ILC and the SNS. The non-sensory OE contains mucus-producing goblet cells and cells with motile cilia that generate a laminar water flow across the OE surface, which supports odorant delivery to OSNs. **F.** Morphologically distinct OSN subtypes revealed by biocytin-backtracing through the olfactory nerve. Large spherical crypt (cc) and kappe (kc) cells are located in apical tissue layers. Microvillous OSNs (mOSNs) possess short and thick dendrites and are located in more apical strata than the cell bodies of ciliated OSNs (cOSNs). A sharp transition between the region containing neurons (sensory OE) and the non-sensory region can be seen at the SNS (yellow line).

periphery (Demirler et al., 2020). GBC progenitors are sparse and are localized to the inner and peripheral margins of the sensory OE at the ILC and SNS, respectively (Bayramli et al., 2017). In contrast, HBCs are uniformly distributed throughout the basal layers of the sensory region (Demirler et al., 2020). GBCs and HBCs have distinct functions and are active under different tissue conditions. In zebrafish and mammals, GBCs generate a steady stream of new OSNs to replace sporadically dying cells under physiological conditions, while HBCs remain largely dormant in the intact tissue (Schwob et al., 2017; Yu and Wu, 2017; Bayramli et al., 2017; Demirler et al., 2020). HBCs become selectively activated by acute tissue injury and transition through a lineage that includes GBC-like cells to restore larger numbers of OSNs, eventually other cell types, that have been lost because of the injury (Leung et al., 2007; Iwai et al., 2008; Kocagöz et al., 2022).

The zebrafish OE contains five morphologically and functionally distinct OSN subtypes (Figure 6.2E, F) that are characteristic for the different chemosensory subsystems that exist in other vertebrates (Munger et al., 2009). The main morphotype are the ciliated OSNs, which are equivalent to OSNs of the main OE of mammals (Hansen and Zeiske, 1998). These cells can be recognized by expression of specific marker genes, such as the *olfactory marker protein b* (*ompb*) gene, and occupy basal and intermediate positions along the vertical dimension of the tissue (Celik et al., 2002; Sato et al., 2005). Ciliated

OSNs extend long dendrites, which, as their name implies, are decorated with 3–7 long, non-motile sensory cilia that reach beyond the apical diffusion barrier that is formed by sustentacular cells (Hansen and Zeiske, 1998). They express either one of the 176 conventional OR genes (Alioto and Ngai, 2005; Niimura and Nei, 2005) or of the 112 trace amine-associated receptors (TAARs) (Hussain et al., 2013) and signal through the G-protein subunit $G\alpha_{olf}$ (Hansen et al., 2003). These receptors detect fish-relevant odorants, including nucleotides, bile salts, and amine-containing compounds (Korsching, 2020), but also the courtship pheromone prostaglandin F2α (Yabuki et al., 2016). Ciliated cells extend axons to the olfactory bulb, where they segregate into regional subsets of glomeruli along the dorsal, ventral, and medial surfaces (Sato et al., 2005).

The second major OSN population are the microvillous cells that predominantly locate to intermediate and apical regions of the OE (Sato et al., 2005). Their dendrites are shorter than those of ciliated cells and bear multiple microvilli that interact with chemicals in the water (Hansen and Zeiske, 1998). These cells more closely resemble the characteristic sensory neurons of the vomeronasal organ of other species (Torres et al., 2023). In zebrafish, they can be recognized by expression of the *transient receptor potential type C2* gene (*trpc2*) (Sato et al., 2005) and the signal transduction component $G\alpha_o$ (Hansen et al., 2003). They also express vomeronasal receptor-related genes of the V2R/OlfC family (Korsching, 2020) and have been shown to be involved in the detection of amino acids, which are relevant food cues in fish (Lipschitz and Michel, 2002; Koide et al., 2009). Their axons give rise to a dense cluster of glomeruli in the lateral olfactory bulb that resemble the smaller and more diffuse glomeruli of the accessory olfactory bulb of rodents (Sato et al., 2005; Koide et al., 2009).

In addition, three unique and fish-specific OSN classes have been identified with the crypt, kappe, and pear cells (Figure 6.2E, F) (Hansen and Finger, 2000; Ahuja et al., 2014; Wakisaka et al., 2017). These cells have in common that they possess large spherical cell bodies, which are situated directly underneath the apical surface. Pear and kappe cells extend microvilli, while crypt cells bear microvilli and cilia that function as sensory appendages. For crypt and kappe cells, specific single glomeruli have been described in the mediolateral olfactory bulb by virtue of their immunoreactivity to TrkA/S100 and $G\alpha_o$, respectively (Ahuja et al., 2013; Ahuja et al., 2014). A specific glomerulus in the lateral olfactory bulb has been identified for pear cells, which respond to adenosine, a breakdown product of nucleotides, which are also food cues in fish (Wakaisaka et al., 2017). The specific function of crypt cells is less clear, but their involvement in social interaction and kin recognition has been suggested (Biechl et al., 2016), while the function of kappe cells and their occurrence in other fish species remains currently unknown (Ahuja et al., 2014).

Thus, the zebrafish OE is characterized by the presence of multiple OSN cell types that segregate along different epithelial strata. Beginning from the apical, water-exposed surface, crypt, kappe, and pear cells are the predominant type (Hansen an Zeiske, 1998; Ahuja et al., 2014; Wakisaka et al., 2017). Between and below these cells follow several layers of microvillous OSNs and, even further inside, the cell bodies of ciliated OSNs (Hansen and Zeiske, 1998; Sato et al., 2005). HBC progenitor cells form the most basal cell population and are directly attached to the basal lamina (Demirler et al., 2020). Sustentacular cells possess large cell bodies that are situated immediately above HBCs but extend protrusions that span the entire thickness of the OE (Hansen and Zeiske, 1998; Demirler et al., 2020). At their apical tips, they show widenings that occupy the space between apical chemosensory cells and dendrites of ciliated and microvillous OSNs with which they form tight junctions (Hansen and Zeiske, 1998; Liang, 2020). At their basal side, sustentacular cells also extend fingerlike protrusions that wrap around HBCs to contact the underlying basal lamina (Demirler et al., 2020).

## Exposure of the Human OE to Environmental Toxicants and Pathogens

The main disturbances that result in olfactory dysfunction in humans are exposure to toxic chemicals and infections by viruses and parasites (Genter and Doty, 2019). For humans, environmental and occupational exposure to higher-than-normal concentrations of chemicals and dust particles has been shown to adversely affect smell sensation and to cause tissue damage (Genter and Doty, 2019; Werner and Nies, 2018). While the prevalence of olfactory dysfunction is high, ranging between 2.5% and 25% in the general population, work-related deficits may account for a substantial fraction with approximately 5% of all cases (Werner and Nies, 2018). The exact prevalence, however, is difficult to determine and depends on multiple confounding circumstances, such as duration of exposure, toxicant concentration, and age of the individual. In addition, allergic rhinitis is highly prevalent and ranges between 12% to 46% across different countries (Bousquet et al., 2020).

A major symptom of patients during the recent COVID-19 pandemic was anosmia, the complete or partial loss of smell sensation, illustrating vividly how airborne pathogens can affect the function of the olfactory system (Anastassopoulou et al., 2024; Kim et al., 2024). The cause of the loss of function after SARS-CoV-2 infection are still not completely understood but, at least in part, involved degeneration of OSNs (Bryche et al., 2020; Tsioumpekou et al., 2023)). Postmortem samples taken from 85 patients revealed that sustentacular glial cells, but not OSNs, were the primary target of SARS-CoV-2 infection (Khan et al., 2022). Only sustentacular cells express the entry receptor angiotensin-converting enzyme 2, which renders them sensitive to virus infection (Fodoulian et al., 2020). Nevertheless, massive structural damage and loss of olfactory cilia from OSNs have been observed in the SARS-CoV2-infected hamster (Bryche et al., 2020) and zebrafish OE (Kraus et al., 2022). Thus, disruption of OE integrity, breakdown of the apical border, disturbances in internal ionic conditions, and release of inflammatory cytokines triggered by the virus infection may have resulted in subsequent OSN death or loss of olfactory signal transduction and are the most likely cause of the anosmia (Karki and Kanneganti, 2022; Tsukahara et al., 2023).

Other respiratory pathogens that have been shown to affect the olfactory mucosa by causing the common cold, rhinitis, and sinusitis are influenza, respiratory syncytial, and rhinoviruses, but also the bacteria *Staphylococcus aureus* and *Streptococcus pneumoniae* (Scadding, 2008; Casadei and Salinas, 2019). Besides blockage of airflow, damage to the OE structure often occurs, most likely because of the inflammatory immune response that accompanies these conditions (Imamura and Hasegawa-Ishii, 2016). The OE is endowed with a vast nasal-associated lymphoid tissue (NALT), which includes both the specialized tonsil-like lymphoid tissues of the organized NALT (O-NALT) and the diffuse distribution of immune cells around the olfactory mucosa (D-NALT, Sepahi and Salinas, 2016). While O-NALT is common in higher vertebrates, teleosts, including zebrafish, only contain the diffuse system (Tacchi et al., 2014). In addition to biological pathogens, other environmental substances can also stimulate the immune system and trigger allergic reactions. Risk factors for allergic rhinitis in humans include pollen, dust, dust mites, pets, and molds (Bousquet et al., 2020). Again, the ensuing immune response and cytokine storm seem to be the main cause of tissue damage (Karki and Kanneganti, 2022). Therefore, OSNs, do not have to be the direct target of the viral infection or toxicant activity to undergo immune system-induced cell death. In fact, a genetic mouse model of chronic rhinosinusitis, in which the inflammatory cytokine TNF-a was overexpressed in sustentacular cells in the mouse, showed loss of OSNs and severely reduced odorant responses (Lane et al., 2010). TNF-a recruits and activates cells of the innate immune system, neutrophils and macrophages, which most likely play an important role in the observed cell death (Muendlein et al., 2022; Tsioumpeko et al., 2023). Thus, OE inflammation is not only a reactive response to tissue disturbances, such as the presence of pathogens or damaged cells, but can be itself the underlying cause of tissue damage.

In addition to airborne pathogens, a main source of cytotoxic substances that have an effect on the nervous system are environmental pollutants (Genc et al., 2012). Ambient particulate matter is a major contributor to air pollution and is composed of a large variety of chemical components with different particle sizes and properties (Potter et al., 2021). Being a product of fuel combustion and industrial processes, organic carbon compounds, polycyclic aromatic hydrocarbons, sulfates, and nitrates are major components of hazardous particulate matter (Lin et al., 2022). Metals, such as manganese (Mn), zinc (Zn), lead (Pb), copper (Cu), nickel (Ni), and barium (Ba) often adsorb to particulate matter and contribute further to its toxic effect (Potter et al., 2021). While larger particles above 10 µm can be trapped by the mucus layers of the conducting airways and olfactory mucosa, components with smaller sizes (< 2,5 µm; Particulate matter (PM) 2.5) can leak into the CNS via the olfactory nerve and cause serious neurotoxic effects (Lin et al., 2022). Cell culture studies demonstrate that direct contact with PM2.5 can cause mitochondrial dysfunction, decrease cell viability, and downregulate neuroprotective genes in brain cells.

As mentioned above, the OE also provides a direct entry route for pathogens and chemical agents into the CNS (Figure 6.1B). Because OSNs extend axons directly to the olfactory bulb, invasion along the olfactory nerve bypasses the blood–brain barrier (Dando et al., 2014). While this connection is an important pathway to deliver therapeutic drugs to the brain (Jeong et al., 2023), it also poses considerable risks. Common pathogens that can follow a nasal route are viruses, such as herpes simplex type 1, vesicular stomatitis, and rabies virus, in addition to bacteria such as *Streptococcus pneumoniae*, *Burkholderia pseudomallei*, and *Listeria monocytogenes*. A particularly dramatic example can be seen with the brain-eating-amoeba *Naegleria fowleri*. Exposure to the protozoan can occur during swimming or contact with contaminated water, and the pathogen penetrates the OE to enter the brain, where it causes meningencephalitis (Siddiqui et al., 2016). Brain damage can be observed already 96 h after pathogen infection in rodents, resulting in the overall high mortality rate of 90%. Air pollutants, such as particulate matter and other airborne toxicants, can also use the nasal pathway to enter the brain parenchyma (Genc et al., 2012). In the adult and developing brain, PM2.5 has been shown to cause cell cycle arrest, neuronal cell death, inflammation, and the induction of oxidative stress, in addition to epigenetic changes and direct damage of genetic material (Potter et al., 2021; You et al., 2022; Shang et al., 2023).

## Zebrafish as a Model to Test Environmental Toxicity

Zebrafish has long been used as a simple vertebrate model to study the effect of toxins on development and tissue function. By now, zebrafish has become a widely used living biosensor to monitor the presence of common and industrial pollutants in the environment and to assess their effect on embryo/cell viability and teratogenesis (Bambino and Chu, 2017; Li et al., 2023). The most straightforward assay is to assess embryo mortality upon exposure to candidate toxicants (Figure 6.3A) (Ali et al., 2011). More comprehensive scales have been developed to score for defined teratogenic effects using gross morphological aberrations (DarT), metabolic criteria (mDarT), and changes in gene expression (Gene-DarT) (Nagel, 2002; Busquet et al., 2008; Weil et al., 2009). Several transgenic reporter lines have been developed that respond to the presence or harmful effects of certain environmental pollutants in different zebrafish tissues (Figure 6.3B) (Lai et al., 2021; Silva Brito et al., 2022). Their relevance includes ecotoxicological assessment of water pollution, reproductive toxicology, endocrine changes, and developmental aberrations, including neurodevelopmental defects. The usage of a whole organism model bears severe advantages over cell culture systems because it allows for the examination of behavioral effects, endocrine and metabolic function, and whole tissue responses, including intra- and intercellular signaling changes (Choi et al., 2021). Zebrafish assays, therefore, provide more comprehensive insight into organismal effects, rather than isolated studies on "naked cells" in the absence of surrounding tissue structures that could diminish or exacerbate the toxic effect on individual cells.

The rapid development of the zebrafish nervous system, which occurs within hours rather than weeks or months, enables the analysis of developmental defects over a short experimental period or in real time, which makes it particularly suitable for toxicological studies but also for drug screening (d'Amora

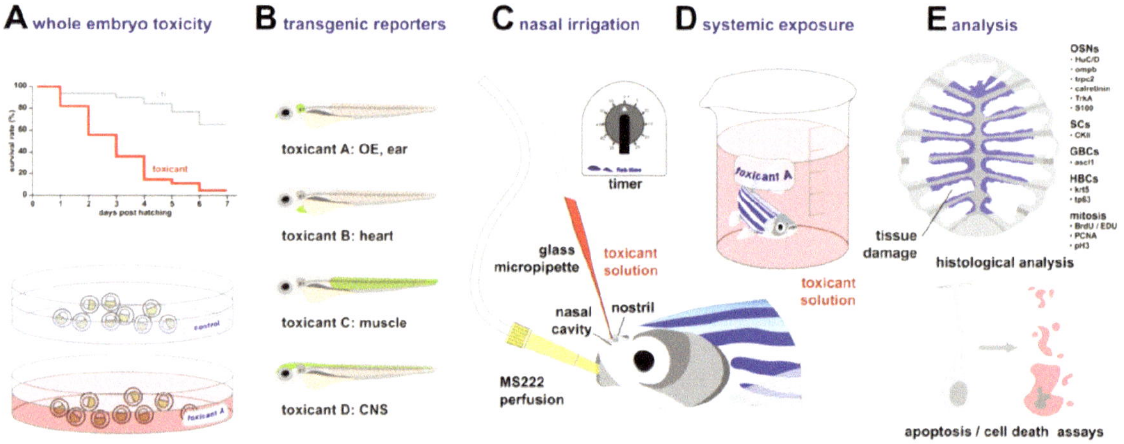


**FIGURE 6.3** Approaches to study cytotoxicity in zebrafish and the olfactory epithelium **A.** A simple toxicity test is to compare embryo viability in the presence of potential toxicants relative to a control group. The toxicant is delivered systemically, but embryos can also be scored for specific morphological defects, such as heart malformation, scoliosis, edema formation and others, in addition to overall survivability. **B.** Different transgenic reporter lines are available, which express reporter genes (e.g. green fluorescent protein; green) under the control of different promoters that are sensitive to cell/tissue stress. Tissues eventually show differential sensitivity to different toxicants. **C.** A specific method to test the effect of toxicants onto the olfactory epithelium (OE) is to irrigate the nasal cavity with solutions containing the potentially cytotoxic substance (red). The solution can be delivered through a glass micropipette into the nostril and fills the entire nasal cavity. The exposure time can be controlled but is limited to several minutes, unless the fish are continuously perfused with a low concentration of the anesthetics MS-222. **D.** Systemic exposure can also be applied to adult zebrafish, similar to the larvae shown in A, but larger quantities of the cytotoxic agent are needed because of the larger water volume that is required to maintain the fish. **E.** The effect of toxicants onto the OE can be analyzed by histological methods, either scoring for gross morphological tissue damage, the loss of specific OE cell populations, for which reliable molecular markers are available (OSNs: olfactory sensory neurons, SCs: sustentacular glial cells, GBCs: globose basal cells, HBCs: horizontal basal cells), or the occurrence of apoptotic cells.

and Giordani, 2018). Different read-outs can be employed to assess toxicity, such as changes in gene expression, nerve cell morphogenesis, and behavioral studies. Sophisticated transgenic models are available that respond to specific molecules, such as the Tg(cyp1a:GFP) transgenic line that reports dioxin and other aromatic hydrocarbons, which are a common byproducts of industrial processes (Xu et al., 2015). More general reporters that respond to a broad range of diverse stressors have also been developed. For instance, the huORFZ zebrafish line, in which green fluorescent protein (GFP) expression is controlled by the micro-ORF of the upstream open reading frame (uORF) of human DNA-damage-inducible transcript 3, responds with reporter gene expression to several heavy metal ions but also to chemicals that disrupt the endocrine system in a tissue specific manner (Lee et al., 2014). Among those tested, the OE responded to cadmium (Cd) and nickel (Ni) ions, suggesting that cellular or tissue stress was caused by their presence. Another useful approach that can be employed in neurotoxicity studies is to compare morphological features in the presence and absence of suspected toxicants (Lai et al., 2021). For example, when Tg(nkx2.2a:mEGFP) embryos, expressing GFP in the CNS, were exposed to various neurotoxic compounds, such as acetaminophen, atenolol, atrazine, ethanol, and lindane, changes in the length of ventral axons could be observed, suggesting that these substances have detrimental effects on nervous system development (Zhang and Gong, 2013). Importantly, the axon length assay was about 10 times more sensitive than standard teratogenesis tests based on the DarT scale (Nagel, 2002). The examples in which ventral axon shortening has been used for neurotoxicity assessment are diverse, include other genetically labeled cell populations, and provide the opportunity to observe morphological alterations *in vivo* and real time (Kanungo et al., 2011; Yang et al., 2011).

## The Zebrafish OE as a Sensitive Biosensor for Environmental Neurotoxicity

More relevant in the current context, the effect of different classes of chemicals has also been tested on zebrafish OE cells in a large number of studies (Tierney et al., 2010; Calvo-Ochoa and Byrd-Jacobs, 2019). As mentioned above, OSNs are particularly suited to report toxic effects because they are more directly exposed to the environment than other neuron populations. Nevertheless, olfactory neurotoxicity in zebrafish is a sensitive assay that may have a much broader relevance for human health beyond the OE (Legradi et al., 2018; Fitzgerald et al., 2021). Environmental disturbances that have been shown to affect olfactory function in zebrafish, and in part the structural integrity of OE cells, comprise a wide range of conditions and substances that include but are not limited to changes in pH and salinity but also environmental pollutants, such as metal ions, pesticides, herbicides, detergents, particulate matter, and microplastics (Tierney et al., 2010; Calvo-Ochoa and Byrd-Jacobs, 2019). However, direct histological observations of tissue injury and/or OSN cell death are not available for all substances. Instead, the toxicant effects have been tested in behavioral assays in some studies, suggesting that some pollutants may affect olfactory signaling, while others have a direct cytotoxic effect on OE cells (Tierney et al., 2010).

The stratified distribution of different OSN morphotypes (Figure 6.2E) suggests that apical cells, including crypt, kappe, pear, and microvillous cells, are more prone to chemical insult and toxicant exposure. Especially, corrosive agents would first be encountered by apical cells before reaching deeper tissue layers. Surprisingly, this is not the case for all substances that have been tested for their cytotoxic effects on zebrafish OSNs. In fact, the heavy metal ions Cu and Zn have been demonstrated to selectively affect ciliated OSNs, particularly at low concentrations (Lazzari et al., 2017; Hentig and Byrd-Jacobs, 2016; Ma et al., 2018). Ciliated and microvillous cells were equally sensitive to Ni exposure, while crypt cells were not affected (Lazzari et al., 2019). Crypt cells, however, have been shown to be damaged by high concentrations of urea (Bettini et al., 2016), and along with microvillous but not ciliated cells in the presence of mercury (Hg; Lazzari et al., 2022). Even more surprisingly, the detergent Triton X-100, which results in abrasion of the OE from the apical surface (Cancalon, 1983), also showed a more pronounced effect on ciliated cells despite their localization to deeper tissue layers (Paskin et al., 2011). This suggests that cytotoxicity in the OE is not simply determined by the accessibility of the toxicant to cells but may also depend on intrinsic molecular factors and/or morphological features that render specific cell types more vulnerable to different toxicants (Spurgeon et al., 2020). The higher sensitivity of the deeper located ciliated cells may in part be explained by their protruding cilia, which are about 2–3 μm long and essentially provide a larger membrane area for toxicant attack than the short (0.5 μm) microvilli (Hansen and Zeiske, 1998). It has been demonstrated in zebrafish but also in rodents that cilia can be shed by detergent exposure or the $Ca^{2+}$ shock method, which may be responsible for the cell death seen in ciliated cells (Friedrich and Korsching, 1997; Mayer et al., 2008). Yet, cilia can also regrow after ablation without obvious harm to the cell (Adamek et al., 1984; Friedrich and Korsching, 1997). Therefore, intrinsic, but unidentified, properties of ciliated OSNs may also provide additional determinants for their susceptibility to toxic substances. In contrast, the overall resilience of crypt cells to many cytotoxic agents may be due to their narrow surface opening, the crypt, that limits direct water flux and accessibility by pollutants (Lazzari et al., 2022).

The exact molecular signatures that render neurons selectively susceptible to different stressors are of great interest, especially in the brain, where different neuronal subtypes are more prone to undergo degeneration in the context of degenerative disorders (Nabi and Tabassum, 2022). For instance, dopaminergic neurons selectively die in Parkinson's disease (Surmeier, 2018), whereas amyotrophic lateral sclerosis affects upper motor neurons (Ovsepian et al., 2024). Selective vulnerability of different nerve cell classes to Huntington's disease has also been shown (Mitchel and Griffith, 2003). As the most common form of dementia, Alzheimer's disease also selectively and severely affects central cholinergic neurons (Davies and Maloney, 1976), while frontotemporal dementia preferentially impacts von Economo neurons (Seeley et al., 2006). The molecular underpinnings of cell type–specific susceptibility are not always known; however, it is plausible to assume that substantial diversity exists between different neuron types in their energy demand, sensitivity to reactive oxygen species, molecular receptor expression, and niche context, which includes neighboring cells and extracellular matrix composition (Calabresi et al., 2000; Wang and Michaelis, 2010; Pandya and Patani, 2021). In Alzheimer's disease, which is often characterized by irreversible aggregation of amyloid beta peptide and Tau-protein, cortical and subcortical neurons become less susceptible to pathological adverse effects when the cell bodies are surrounded by chondroitin sulfate proteoglycan–containing extracellular matrix of the perineuronal net (Suttkus et al., 2016). A neuroprotective role of the perineuronal net has also been described for $FeCl_3$-induced neurodegeneration (Suttkus et al., 2014; Ueno et al., 2018). Neurodegenerative disorders can also cause metabolic changes in other cells of neural tissues, such as glia cells, that regulate metabolites and many other processes throughout the brain parenchyma (Stoklund Dittlau and Freude, 2024). Not surprisingly, astrocytes, as the predominant glia cells, are also differentially susceptible to disturbances (Zimmer et al., 2024), and alterations in astrocyte physiology may contribute to the progression of disorders by generating secondary vulnerability of specific neuronal types (Cooper et al., 2020). Understanding the molecular and structural underpinnings of the selective vulnerability of the five different OSN subtypes of the zebrafish OE to different toxicants may, thus, contribute further to this important area of research.

## Experimental Approaches to Study OSN Neurotoxicity and OE Regeneration in Zebrafish

The small size of adult zebrafish of 2–3 cm and their aquatic lifestyle make it easy to expose fish to xenobiotics by simply supplying them with the tank water (Figure 6.3D). This is even more straightforward and cost efficient in early embryos, which only measure several millimeters ($\leq$ 19 mm) and can be kept in large quantities in moderate water volumes (Figure 6.3A) (Singleman and Holtzman, 2014). For instance, exposure of zebrafish larvae to a 10 mM solution of $CuSO_4$ for 4 h resulted in an increased number of dying OE cells, as measured by TUNEL staining (Palominos and Whitlock, 2021). However, in this setup, systemic effects will occur that often result in detrimental side effects, such as unwanted death of the entire animal. Thus, a more specific approach to test the effect of toxicants onto OE cells, or to induce OSN death for regeneration studies, is the direct irrigation of the olfactory cavity with solutions through the nostril (Figure 6.3C). Olfactory regeneration studies in fish were spearheaded by Paul Cancalon in the early 1980s using $ZnSO_4$ and the detergent Triton X-100 in the channel catfish (Cancalon, 1982, 1983). Irrigation with these compounds triggered a concentration-dependent de- and regeneration of OE cells and sensitivity to odorants. The Triton X-100 method has been adapted to zebrafish, where it has the same effect (Iqbal and Byrd-Jacobs, 2010). Zebrafish can be easily anesthetized by immersion in Tricaine (Methanesulfonate, MS-222) at 160 mg/ml concentration and are capable of surviving for a few minutes outside the water with little support. The fish can be wrapped with wet paper towel or placed into an incision made in a simple sponge. We routinely employ irrigation of the OE with $\leq$ 1% Triton X-100, dissolved in PBS, for up to 90 s with two consecutive applications through a small micropipette at 45 s intervals directly through the nostril. As explained in more detail below, Triton X-100 treatment results in rapid degeneration of nearly all OSNs over the first 24 h and a remarkably fast recovery of the nerve cell population within 5–7 days post lesion (dpl) (Iqbal and Byrd-Jacobs, 2010; Kocagöz et al., 2022).

The same approach can be used to test other substances for their candidate cytotoxic effect but is limited to short exposure times. Extended application of candidate toxicants or drugs, such as small molecule inhibitors, agonists, and growth factors, requires modifications to keep the fish alive during the procedure (Figure 6.3C). We have had good success and high recovery rates by continuously perfusing fish with a half concentration of MS-222 (80 mg/ml) in freshly aerated tank water through the mouth and gills. Using a simple gravity-fed perfusion system, fish can be kept anesthetized for up to 30 minutes without detectable harm to the animal (Sireci et al., 2024). It is advisable to replace the agent several times throughout the procedure to prevent dilution and active outflow from the olfactory cavity powered by the beating cilia of the nonsensory OE. However, as described above, the apical border that is formed by processes of sustentacular glia cells constitutes a significant barrier to tissue penetration when drugs and inhibitors are applied directly to the OE. Mild and short co-application of Triton X-100 has been used to deliver dye molecules directly to OSNs (Friedrich and Korsching, 1997). However, the method includes the transient loss of olfactory cilia, which must regenerate over 24–48 h before the OE adopts pre-manipulation conditions.

The irrigation method is not limited to chemicals but can also be used to test biological agents. Recently, it has been demonstrated that intranasal delivery of SARS-Covid 2 spike protein causes severe structural and functional alterations in the organ accompanied by a strong neuroimmune response and olfactory dysfunction (Kraus et al., 2022). Similar to the Triton X-100 injury model, intranasal administration was chosen to investigate the pathological effects of viral proteins specifically on the OE structure and to prevent systemic effects. Likewise, intranasally delivered growth factors and small molecule inhibitors that affect intra- and extracellular signaling in OE cells can be applied without causing toxicity (Sireci et al., 2024), demonstrating that the use of intranasal administration is not restricted to toxicity testing but can also be used to study signaling events that are related to nerve cell de- and regeneration or OE development.

Mechanical injury approaches, such as stab lesions with fine needles through the OE, are also possible but technically more challenging because of the small size of the olfactory organ of only ~500 μm diameter (Bayramli et al., 2017). Even though the first 4–6 pairs of OE lamellae are visible through the nostril from the outside, it is difficult to target specific positions, which would be necessary for reidentification of the injury site during later analysis. Coating the injury needles with membrane dyes, such as DiI or DiO, may help, but these dyes do not easily sustain tissue staining by immunohistochemistry or RNA *in situ*-hybridization that may be necessary to analyze the tissue effect (Elberger and Honig, 1990) unless fixable modifications of the dyes are used (Jensen and Berg, 2016).

A physical injury method that can be used to selectively kill all OSNs is bulbectomy, the surgical removal of the olfactory bulb. This method, which is frequently applied in rodent models (Hendriksen et al., 2015), is not generally preferred in zebrafish due to its small size and the inaccessible anatomical location of the olfactory bulb directly underneath the forebrain. However, successful application of the procedure has been reported (Maaswinkel and Li, 2003). An alternative and easier method to selectively kill OSNs is to cut the axon bundle between the olfactory bulb and the OE, which also triggers retrograde degeneration (Kocagöz et al., 2022). Successful nerve transection displays certain phenotypes in the adult OE, such as the increase in apoptotic body formation, a decrease in the number of cells that stain positive for the pan-neuronal marker HuC/D, epithelial thinning, and a robust compensatory proliferation response during the days following the procedure. Recently, removable wax plug insertions into the olfactory cavity have been shown to induce OE degeneration, most likely by mechanical stress rather than sensory deprivation (Scheib et al., 2019). Continuous presence of the plugs for 7 d results in near complete loss not only of sensory neurons but essentially the entire organ structure. Surprisingly, however, fish recover quickly and can regrow the entire OE within only 1–3 weeks and reestablish axonal connections to the olfactory bulb.

Genetic methods may provide additional specificity to target different OE cell types. A widely used inducible system to selectively ablate specific cell populations in zebrafish has been developed with the nitroreductase system (Curado et al., 2008; Sharrock et al., 2022). Expression of the bacterial enzyme under control of tissue-/cell type-specific promoters renders cells vulnerable to the pro-drug metronidazole. The system has been successfully used in cardiac and liver tissues (Curado et al., 2008; Chlebowski et al., 2017) but has not yet been employed to selectively destroy any of the different OE cell types in zebrafish. Suitable promoters to target OSNs are available with the HuC promoter (Park et al., 2000), which in the OE drives expression in all OSNs or the *ompb* and *trpc2* promoters, which are selectively active in ciliated and microvillous cells, respectively (Celik et al., 2002; Sato et al., 2005). Crypt, kappe, and pear cells can be identified by the expression of specific markers, such as TrkA/S100, $Ga_o$, and the A2c adenosine receptor, respectively (Ahuja et al., 2013; Ahuja et al., 2014; Wakisaka et al, 2017), but specific transgenic lines for these genes have so far not been employed. In contrast, selective expression in sustentacular cells may be achieved through usage of the *six4b* gene promoter (Palominos et al., 2022). The genetic methods developed to target OE cells are not limited to nitroreductase, and other genetically encoded agents to drive cell death, such as diphtheria toxin, could also be used (Jimenez et al., 2021). More complicated approaches that allow for temporal control of the transgene, such as the CreERT2/loxP or tetracyclin transactivator systems, are also available (Stieger et al., 2009; Chekuru et al., 2017).

Because the OE is not isolated from the blood circulation through the blood–brain barrier or a related system, drugs and small molecule inhibitors can be systemically delivered to reach the OE through an intraperitoneal route. The method works well for inhibitors of signaling pathways, most likely because the agents are given at high concentration and reach the tissue through the rich vascular system of the lamina propria in the center of olfactory lamellae (Hansen and Zeiske, 1998; Palominos and Whitlock, 2021). The general application, however, increases the chance of exerting systemic effects, which may be unwanted. As a note of caution, we were less successful in directing proteins and smaller peptides, such as cytokines and chemokines, to the OE using intraperitoneal injection, probably because they were sequestered along the dispersion route and drop down to low concentrations before encountering the OE tissue (Demirler, Sireci, and Fuss, unpublished). Direct injection into the blood circulation either through the cardinal vein (Cianciolo Cosentino et al., 2010) or the retro-orbital plexus (Pugach et al., 2009) may work for some reagents in addition to extended nasal irrigation (Sireci et al., 2024).

The effect of OSN degeneration that is induced by the various methods and agents described above can be assessed using a broad palette of behavioral and morphological assays (Figure 6.3E). In addition to transgenic lines that express reporter genes in specific cell populations of the OE, immunohistochemistry against the pan-neuronal marker HuC/D is the most convenient because it labels the entire OSN population (Demirler et al., 2020). Anti-calretinin staining can also be used, but calretinin is expressed preferentially in microvillous cells during larval stages, whereas it is expressed by ciliated and microvillous cells in the adult, sometimes confusing interpretation of results if specific cell types are of interest (Koide et al., 2009; Kress et al., 2015). RNA *in situ*-hybridization against *ompb* and *trpc2* allows for the selective assessment of ciliated and microvillous cells, respectively (Sato et al., 2005), while riboprobes directed against specific OR types reveal specific OSN subpopulations (Bayramli et al., 2017). If OSN subtypes are not of interest, reduction in OSN density and thinning of the OE can be observed and quantified using optical tools and basic imaging software (Iqbal and Byrd-Jacobs, 2019; Bettini et al., 2016; Kocagöz et al., 2022). For instance, application of 1% Triton X-100 results in the loss of 88% of HuC/D-positive cells over the first 24 hpl when measured as changes in optical density of the antibody signal (Kocagöz et al., 2022). Surprisingly, apoptotic cell death is rare in the OE, and only few if any cells can be observed in the intact tissue, while alternative cell death assays, such as TUNEL staining, reveal many dying OSNs (Bayramli et al., 2017). Both Triton X-100 irrigation and olfactory nerve transection result in an increased number of caspase 3-positive cells, suggesting that naturally occurring OSN cell death in the intact OE is the result of different molecular programs (Bayramli et al., 2017; Kocagöz et al., 2022).

Specific antibodies to reveal other OE cell populations also exist. Crypt cells can be visualized with antibodies against the calcium-binding protein S100 and TrkA (Sandulescu et al., 2011; Ahuja et al., 2013), while kappe cells are labeled by $G\alpha_o$ immunoreactivity. However, the specificity may not necessarily be due to expression of the proteins but because of cross-reactivity of the (manufacturer-/batch-specific) antibodies with unknown antigens (Korsching, 2020). Nevertheless, S100 transcript expression has been used in many studies to assess crypt cell susceptibility (Tilton et al., 2008; Volz et al., 2020a; Song et al., 2021). Sustentacular cells can be visualized by anti-cytokeratin immunoreactivity, which labels their intracellular filaments, sometimes obscuring cell boundaries (Demirler et al., 2020). Alternative markers, such as Sox2, can also be employed, but overlap with expression in HBCs and GBCs. Unfortunately, zebrafish sustentacular cells are not reactive to many of the available antibodies that label sustentacular cells in rodents, including Pax6, Six2, or GFAP (Guo et al., 2010; Sattler et al., 2011). Zebrafish HBCs express the markers tumor protein 63 (tp63) and Keratin5 (Demirler et al., 2020) similar to their mammalian counterparts (Schwob et al., 2017), and good antibodies are available. GBCs, on the other hand, are characterized by expression of the proneural factor Ascl1, while later stages of the OSN lineage can be identified by expression of neurod1/4, neuregulin, and gap43 (Bayramli et al., 2017; Aguillon et al., 2018).

Degeneration of OE cells by cytotoxic agents triggers a proliferative injury response in the OE (Iqbal and Byrd-Jacobs, 2010; Demirler et al., 2020; Palominos and Whitlock 2021; Kocagöz et al., 2022), which can be visualized by a variety of methods. Anti-PCNA or anti-pH3 staining by immunohistochemistry is robust and faithfully labels acutely dividing cells (Bayramli et al., 2017; Kocagöz et al., 2022; Sireci et al., 2024). Alternative methods can be used, which integrate over longer intervals and can be used in different combinations and at different time points to set up sophisticated analysis regimes (Bayramli et al., 2017). Conjugated thymidine analogs, such as the halogenated derivatives BrdU, CldU, and IdU or the ethynylated variant EdU, can be applied directly to the tank water or injected intraperitoneally (Lindsey et al., 2018). Our experience is that intraperitoneally injected thymidine analogs are active for 12–24 h after injection and that results are comparable to 24 h incubations in tank water.

Another useful approach to follow the effect of cytotoxic agents is to test for odor-induced neuronal activity. OSN cell death goes along with a decreased sensitivity to odorants, and the specific defects can be informative about which OSN cell types have been affected (Lipschitz and Michel, 2002; Døving et al., 2011). Electro-olfactogram recordings measure bulk electrical responses from the OE, the olfactory nerve, or the olfactory bulb surface (Baldwin and Scholz, 2005). Diminished or absent responses to amino acids are usually interpreted as degeneration of microvillous cells, while bile acids or nucleotides are preferred to assess ciliated cells (Heffern et al., 2018). The same logic applies to behavioral tests, which either score innate or learned attraction and avoidance to specific odorants (Braubach et al., 2009; Kermen et al., 2020). A more involved method is to use $Ca^{2+}$ imaging in the OE or olfactory bulb (Friedrich and Korsching, 1997; Fuss and Korsching, 2001; Dieris et al., 2024). The technique relies on the binding of fluorescent dyes, such as Fluo4-AM, to released $Ca^{2+}$ ions when voltage-gated $Ca^{2+}$-channels open in responsive cells upon depolarization of OSNs (Schild and Restrepo, 1998). $Ca^{2+}$ imaging can be used both to assess the responsiveness of the olfactory system to different odorants and to reveal olfactory and non-olfactory functions of other relevant biomolecules (Demirler et al., 2020; Dieris et al., 2024).

## The Effect of Specific Toxicants on the Zebrafish OE

Olfactory neurotoxicity studies in zebrafish cover a wide range of chemically diverse substances that are emitted into the environment from a multitude of different sources (Tierney et al., 2010; Calvo-Ochoa and Byrd-Jacobs, 2019). Of specific environmental and occupational interest are metal ions but also pesticides, solvents, and particulate matter, such as remnants of organic combustion, and microplastics. Zebrafish has also been used to study the effect of viruses that cause rhinitis in humans (Casadei and Salinas, 2019; Fajar et al., 2024). However, so far, most viruses were not tested for their specific effects in the olfactory tissue, but zebrafish larvae were utilized as a whole animal model of infection. A particularly interesting study, however, assessed the effect of SARS-CoV-2 infection on OE integrity and function (Kraus et al., 2022). Curiously, the spike surface protein of the virus alone was sufficient to cause edema, widespread OSN degeneration, and loss of sensitivity to odorants. Similar to the OE of mammals (Bryche et al., 2020), SARS-CoV-2 affected sustentacular cells and induced a strong immune response (Kraus et al., 2022), which may be causative for the observed damage. These elegant studies clearly demonstrate that observations from the zebrafish OE can be relayed to higher vertebrates, including humans. Infection of the zebrafish OE with a member of the rhabdovirus family, which includes the rabies virus that can infect nerve cells in humans (Bastos et al., 2023), induced apoptotic cell death selectively in crypt cells because they express the required entry protein TrkA (Sepahi et al., 2019). Surprisingly, crypt cells also seem to "smell" the presence of the virus to elicit an immune response. The authors showed that virus exposure evokes electrical signals in OSNs and subsequent activation of the immune system, which could be blocked by the TrkA receptor inhibitor AG879 (Sepahi et al., 2019; Kraus et al., 2023).

More than any other substance class, however, metal ions have been studied for their effect on olfactory toxicity in zebrafish. One possible explanation for the strong detrimental effects of metal ions on sensory organs may be the competition of divalent cations with $Ca^{2+}$ and their selective blockage of ion channels, resulting in sensory dysfunction and subsequent cell death (Manetti, 2023). A comparative study assessed the effect of the four metal ions Cd, Zn, arsenic (As), and chromium (Cr), both individually and in combination (Heffern et al., 2018). While no changes in olfactory sensitivity were observed in response to Cr and As, Cd and Zn diminished odorant responses. Environmental sources of Cd include mining, refineries, and waste disposal of batteries, pigments, and plastics (Mendez-Amenta and Rios, 2007). Cd exposure reduces olfactory sensitivity to alarm substance, which is a complex odorant released by injured conspecifics (Jesuthasan and Mathuru, 2008; Masuda et al., 2024). Cd also disrupts olfactory-mediated aversive behavior of zebrafish larvae to taurocholic acid and L-cysteine, which are mediated by ciliated and microvillous cells, respectively (Heffern et al., 2018). At the gene expression level, Cd resulted in reduction of *ompb* and *trpc2* expression but not of *s100*, suggesting that microvillous and ciliated cells have been affected but not crypt cells (Volz et al., 2020a). Although the full effect of Cd on OSNs is not known, oxidative stress appears to play a critical role during Cd-induced olfactory dysfunction. Cd exposure causes upregulation of the *nuclear factor-erythroid 2 related factor 2* (*nrf2*) gene, a transcription factor that regulates the expression of antioxidant and other cytoprotective genes (Wang and Gallagher, 2013). Direct upregulation of the antioxidant genes *peroxiredoxin 1* (*prdx1*), *heme oxygenase 1* (*hmox1*), and *metallothionein 2* (*mt2*) could be demonstrated by expression profiling (Volz et al., 2020a), suggesting that Cd causes severe oxidative stress, which could be the main cause of OSN cell death that was observed by apoptotic markers in the larval OE (Matz and Krone, 2007).

Although Zn is an essential component of several proteins and enzymes, it can be toxic in excessive concentrations (Hussain et al., 2022). Mining, smelting metals, steel production, or burning coal and waste can result in environmental pollution with Zn. Exposure of the OE to $ZnSO_4$ results in severe morphological changes of the OE, which include the fusion of lamellae, neuroinflammation, loss of neurons, and anosmia (Cancalon 1982; Hentig and Byrd-Jacobs, 2016; Abreu et al., 2017). Zn appears to preferentially affect ciliated cells as was revealed by electron microscopy and behavioral studies. The apical OE surface was devoid of cilia, while microvilli were still present (Hentig and Byrd-Jacobs, 2016). At the behavioral level, Zn treatment reduced the sensitivity to bile salts, which are odorants for ciliated cells, while amino acid detection by ciliated cells was less severely affected (Heffern et al., 2018).

Zinc oxide nanoparticles are an engineered nanomaterial that is commonly found in pharmaceuticals, biomedical products, cosmetics, sunscreen, food additives, and industrial products because of their anti-bacterial, anti-fungal, and catalytic function, in addition to their strong absorption of UV light (Jiang and Cai, 2018). Using the Tg(EpRE:mCherry) transgenic reporter fish line, which reports oxidative stress (Mourabit et al., 2019), it was shown that exposure to Zinc oxide nanoparticles or $ZnSO_4$ elicited strong and dose-dependent responses in the OE (Takesono et al., 2023). The particles also induced neutrophil recruitment and local inflammation, which was followed by reduced responsivity to the decay-associated odorant cadaverin, which is detected by TAAR expressing ciliated OSNs and typically triggers strong behavioral avoidance responses (Hussain et al., 2013). Accordingly, expression of the Tg(EpRE:mCherry) reporter was found primarily in calretinin-negative ciliated OSNs (Takesono et al., 2023).

Ni is introduced into the environment from natural and anthropogenic sources, such as combustion of organic energy sources (coal, diesel, fuel), the incineration of waste, and sewage, but also tobacco smoke, dental implants, stainless steel kitchen utensils, and inexpensive jewelry (Cempel and Nikel, 2006; Begum et al., 2022). Ni causes behavioral deficits in fish and displays adverse effects on the OE by reducing the overall thickness of lamellae (Dew et al., 2014; Lazzari et al., 2019). However, the adult OE architecture is not uniformly disturbed by Ni, and the ventral half was more affected than the dorsal half, possibly due to differences either in cellular composition or water flow dynamics between the two halves of the OE (Lazzari et al., 2019). The authors also suggested a preferential sensitivity of ciliated OSNs, while microvillous OSNs were affected less, and crypt cells were not affected at all. Different diverse mechanisms have been proposed that may contribute to Ni neurotoxicity, in which the metal ion may interfere with energy production, induce oxidative stress, change neurotransmitter release and sensitivity, generally disturb metal ion homeostasis, and generate epigenetic changes, in addition to directly stimulating apoptotic pathways (Ijomone, 2021).

Another heavy metal that poses a risk to natural life is uranium (U). Because of its high density, depleted U is used in aircraft counterweights, radiation shielding, and armor plating (Ran et al., 2020). Free U can be lethal in zebrafish within 4 days at a concentration of 3.05 mg/L when exposed acutely (Labrot et al., 1999). In natural ecosystems close to U-rich sources, water levels can reach up to 12 mg/L (Salonen, 1994). Even below lethal concentrations, it has been observed that modest U levels cause transcriptional changes in multiple organs of zebrafish (Lerebours et al., 2009). In a study focusing on the effect of chronic U exposure on sensory organs, it was observed that the highest bioaccumulation in fish occurred in the OE, followed by the olfactory bulb (Faucher et al., 2012). U caused severe morphological alteration in both organs and necrosis of ciliated OSNs, possibly due to its high affinity to ion channels on the sensory cilia.

Similar to U, cobalt (Co) is an environmental contaminant whose effects on aquatic life have been observed but less studied than other heavy metals (Stubblefield et al., 2020). Sources of Co contamination include the metal industry, ceramics, cement, recycling of electric and electronic devices, and pigment production for the paint industry (Leyssens et al., 2017). One study conducted on cichlid fish, cavefish, and goldfish showed that fish spending time in water with sublethal doses of $CoCl_2$ showed a significant decrease in the number of viable cells in the OE after 18 h of treatment (Butler et al., 2016). Co induces noticeable morphological changes already after 3 h of exposure, which persist even after the fish are returned to Co-free water. High-dose exposure to Co results in olfaction-related behavioral changes in adult zebrafish, such as reduced schooling behavior and increased surfacing (Hussainzada et al., 2014). In addition to behavioral changes, the OE is severely damaged upon acute exposure to $CoCl_2$, and tissue abnormalities emerge, such as fusion of olfactory lamellae and increased necrosis.

Mn is another metal that can cause severe neurotoxicity upon overexposure. Unlike the other metals listed above, Mn can be found in almost every redox state and is involved in various natural metabolic functions such as biomolecule synthesis and enzyme activation (Li and Yang, 2018). A tight regulation of Mn ions in the body is therefore essential for healthy physiology. When overabundant, Mn accumulates in the dopamine-rich areas of the CNS, oxidizes dopamine, and eventually disrupts synaptic neurotransmission (Soares et al., 2020). Prolonged exposure causes locomotion deficits and a decrease in dopamine levels, making the metal a useful molecule to model Parkinson's disease in zebrafish (Nadig et al., 2022). In addition to disruption of synaptic signaling, Mn exposure severely affects sensory organ development in zebrafish, including the eye and OE (Alba-Gonzalez et al., 2024). OE size is reduced and apical processes of OSNs are enlarged in the presence of Mn (Pignatelli and Belluzzi, 2017). Surprisingly, the olfactory bulb, where dopaminergic interneurons are common, was not affected. The observed alterations in the OE morphology also affect olfactory sensation, as adult animals experience impairments in their sensitivity to amino acids (Nadig et al., 2022). Yet, the morphological changes in the OE and cell death await more detailed analysis.

The effects of Cu toxicity on the fish olfactory system have been examined in more detail than any of the other metal ions that were discussed above. In various teleost species, such as rainbow trout and coho salmon, Cu exposure has been observed to cause changes in olfactory-mediated behaviors (Baldwin et al., 2003; Razmara and Pyle, 2021). Electro-olfactogram measurements, which record electrophysiological properties of OSNs in response to odorant stimulation, revealed a decrease in overall electrical activity upon Cu exposure (Sandahl et al., 2007). In zebrafish, Cu damage is time- and dose-dependent (Ma et al., 2018). Microvillous OSNs are more resistant to short-term (< 3hr) exposure and low-concentrations ($\leq$ 0.25 uM), while both ciliated and microvillous OSNs are equally and severely damaged after long-term (24 h) contact, even at low concentration. OSN damage is accompanied by a decrease in aversive responsiveness to taurocholic acid in olfactory-mediated behavioral tests, implying that ciliated OSNs were damaged. An independent study supported the distinct susceptibility of ciliated OSNs to Cu but also pointed out that crypt neurons were not harmed by the exposure (Lazzari et al., 2017). In contrast, however, a transcriptome analysis performed after Cu contact also reported downregulation of the crypt cell marker *s100* (Tilton et al., 2008), suggesting that Cu exerts selective and diverse impairments on different cell types. It causes cell death in ciliated OSNs but alteration of gene expression in crypt cells.

Finally, Hg, as one of the most dangerous heavy metals, can accumulate in internal (liver, muscles) and surface organs (gills, skin) at detectable levels in zebrafish (Vieira et al., 2021). In zebrafish larvae, it has been observed that Hg accumulates in OSNs of the olfactory pits and interferes with tubulin stability (MacDonald et al., 2016). OE exposure during adulthood results in epithelial thinning and reduction in the optical density of both ciliated and microvillous OSNs (Lazzari et al., 2022). One interesting finding of this study is that Hg also reduces the number of TrkA-positive crypt cells, which are largely resistant to most of the other heavy metals. Hg-induced injury appears to exert different biochemical effects from other heavy metals, as it affects microvillous and crypt OSNs more than ciliated cells, which overall constitute the most susceptible cell group for other metal ions. Because the exact mechanisms of Hg toxicity are not well understood (MacDonald et al., 2015), the unique selectiveness among OSN subtypes may provide interesting insight into its neuron-specific toxicity.

Beyond heavy metals, a study on diesel particulate matter (DPM) may be of particular interest because DPM is a common air pollutant in the human environment, and the effect on the zebrafish OE can be easily related to the OE of terrestrial organisms (Song et al., 2021). DPM constitutes a heterogenous substance that contains many diverse organic chemicals but also elements and heavy metals (Potter et al., 2021). Severe damage to sensory cilia can be seen already after 1 d of exposure, which progresses further with continued DPM presence (Song et al., 2021). These changes go along with changes in olfactory sensitivity at the behavioral level. DPM-exposed fish were unable to detect mixtures of amino acids in a behavioral test. At the histological level, a reduction in the number of calretinin-positive OSNs could be observed. The results were confirmed by qRT-PCR experiments, demonstrating a severe reduction in *s100*-positive crypt neurons but also *omp*-positive ciliated cells, along with an increase in markers of inflammation and oxidative stress. Absence of odor responses to amino acids, however, suggest that microvillous cells were also damaged. Inflammation of the olfactory mucosa, oxidative stress, and mitochondrial dysfunction were also observed in a human study after exposure to PM2.5, further underlining the validity and relevance of the zebrafish model (Calderón-Garcidueñas, 2010; Chew et al., 2020).

A recent but similarly sensitive issue is the contamination with plastic compounds, in particular the ubiquitous occurrence of microplastics in the environment and food sources (Ziani et al., 2023). Microplastics are geographically widespread, especially in aquatic ecosystems, and are present in large quantities. Even though the authors did not perform a histological examination of the OE, both altered behavioral responses to odorants and expression changes of signal transduction and receptor genes could be observed in the goldfish, implying a toxic effect on OSNs (Shi et al., 2021). Interestingly, microplastics also had synergistic effects with Cu ions and increased zebrafish embryo lethality when both agents were administered together (Santos et al., 2021).

Urea and other uremic toxins that damage the urinary system, such as phenol or indole derivatives, attract attention because, in addition to external contact, they can also be produced by the organism itself (Xu et al., 2023). Elevated levels of uremic toxins due to chronic kidney disease and impaired renal clearance can have a direct toxic effect on neurons and amplify cerebrovascular damage, eventually leading to cognitive impairment (Assem et al., 2018; Watanabe et al., 2021). Chronic kidney diseases have effects on olfactory sensitivity, and treatments like hemodialysis or renal transplantation have been shown to improve olfactory function in patients by reducing uremic toxin load (Griep et al., 1997; Landis et al., 2011). In zebrafish embryo experiments, it was observed that urea exposure caused an increase in $G\alpha_{olf}$ expression, a marker for ciliated OSNs (Ferrando et al., 2014). To increase the systemic concentration of the toxin in adult zebrafish experiments, Bettini and colleagues added sublethal doses of urea to the tank water and then measured epithelial thickness and the occupation of the OE by OSNs (Bettini et al., 2016). While short-term exposure (48 and 96 h) had no effect on epithelial thickness, prolonged exposure (30 d) caused a 29% reduction. The decrease in HuC/D-positive mature OSNs was also moderate, but crypt cells appear to be the most sensitive OSN. While it is not known through which mechanism urea affects neurons in fish, a recent study revealed that nitric oxide synthesis can be affected by urea exposure, which may disrupt homeostasis of some neurons and leads to subsequent death (Cacialli et al., 2024).

Lastly, pesticides constitute another critical type of toxicants that are most frequently released into nature, and which pose a considerable risk for human health. In the context of olfaction, pesticides have been investigated little, and the available studies assess olfaction largely from a behavioral perspective (Tierney et al., 2006, 2007; Volz et al., 2020b), while the impact on molecular mechanisms lag behind. For instance, the effects of the organophosphate chlorpyrifos, a frequently preferred pesticide affecting the insect nervous system, were examined at the transcriptional level (Tilton et al., 2011). Exposure to the pesticide downregulates genes related to metabolic processes and olfaction, such as *ompb*, *calbindin*, *s100* and *olfactory receptor 2.1*, but shows little effect on genes related to signal transduction. Another pesticide, rotenone, which is frequently used as insecticide and piscicide to eradicate poisonous fish species, has been observed to cause olfactory dysfunction and Parkinson's disease–like neurological symptoms in zebrafish (Wang et al., 2017). Originally obtained from plants, rotenone is known for its inhibitory impact on mitochondrial complex I and cellular respiration (Palmer, 1968). Mechanistic effects on the fish olfactory system have not been investigated; however, in mice, the compound has been shown to trigger mitochondrial stress in the olfactory bulb and apoptosis of dopaminergic neurons (Sasajima et al., 2015). Rotenone causes structural alterations in the glomerular layer and reduces the density of both dopaminergic neurons and astrocyte processes in rats through oxidative stress (Voronkov et al., 2017).

## OSNs Have a Limited Lifespan and Are Short Lived in Comparison to Other Neurons

OSNs have a finite lifetime even in the absence of damaging agents, which ranges between one month in zebrafish (Bayramli et al., 2017) to several months in mammals (Mackay-Sim and Kittel, 1991; Holl, 2018; Gaun et al., 2022). Thus, the OSN population naturally undergoes lifelong continuous turnover under physiological conditions, which makes the OE an interesting system to study adult neurogenesis and the factors that influence neurogenic cell proliferation and tissue maintenance. The full range of factors determining OSN lifespan have not been resolved, nor have the cellular and molecular mechanisms by which ageing OSNs are eliminated from the OE. In the mouse OE, neuronal activity, positional effects, and maturity conditions have been shown to influence OSN survival times (Holcomb et al., 1995; Watt et al., 2004; Savya et al., 2019; Gaun et al., 2022). Yet, determining the exact life span of individual OSNs is difficult and likely multifactorial. A major complication is that not all newborn OSNs reach maturity and are eliminated shortly after their birth (Savya et al., 2019). Accordingly, elevated numbers of dying cells can be observed in the zebrafish OE at sites where new OSNs are added to the OE (Bayramli et al., 2017). The most likely reasons for the death of newborn OSNs is that they either fail to express appropriate chemosensory receptors or cannot establish functional connections to glomeruli in the OB (Schwob et al., 1992; Tian and Ma, 2008). With few exceptions, ciliated OSNs only express a single OE or TAAR gene

(Mombaerts, 2004b), a condition that is known as the "one neuron – one receptor" rule and which most likely applies to the zebrafish OE as well (Dang et al., 2018). OSNs expressing the same chemoreceptor gene converge onto the same glomerulus in the OB (Mombaerts, 2006), which is critical for encoding odor information in the nervous system (Fulton et al., 2024). Neurons that do not obey the "one neuron – one receptor" rule may not be able to extend axons to appropriate glomeruli and may be eliminated by activity-dependent competition (Wang et al., 2017). Low-level expression of multiple OR genes during early ontogenetic OSN development has been demonstrated (Hanchate et al., 2015; Tan et al., 2015). Trophic input from the OB is essential for OSN survival, explaining why a certain fraction of immature OSNs may die if they fail to establish synaptic contacts (Schwob et al., 1992). Thus, scoring OSNs from their time of birth may underestimate their average lifetime because a large number of early dying cells are included (Gaun et al., 2022). Another complication in determining lifespan is that label-retaining stem cells will keep producing label-positive cells, even in the absence of the proliferation marker when thymidine analogs are used (Taupin, 2007). To circumvent these issues, different genetic lineage tracing approaches have been taken in the mouse OE to determine the lifetime of OSNs after they reach functional maturity (Holl, 2018; Gaun et al., 2022). The OSN lineage progresses from HBC/GBC stem cells through transit amplifying progenitors and immature OSNs in zebrafish and rodents (Bayramli et al., 2017; Schwob et al., 2017). The lineage can be followed by the successive expression of the different molecular markers detailed above. The ultimate mature OSN marker is OMP, expression of which shortly precedes OR gene expression (Rodriguez-Gil et al., 2015). Tracking the lifespan of OSNs from the onset of OMP expression reports longer survival times than birth dating approaches using incorporation of thymidine analogs (Mackay-Sim and Kittel, 1991; Holl, 2018; Gaun et al., 2022). A very recent report demonstrated that mouse OSNs may have a half-life typically exceeding 6 months and that survival depends on organismal age and positional factors that correlate with nasal airflow, neuronal activity, and the expression of protective enzymes that function as antioxidants (Gaun et al., 2022).

Using BrdU incorporation assays, combined with expression of the mature marker HuC/D, that are subject to the cautions described above, we have measured a half-life of zebrafish OSNs of 19 days but also detected around 8% of OSNs that persisted beyond the 3-month analysis (Bayramli et al., 2017). Regional influences may be less severe in the zebrafish OSN, which measures only 500 µm across and is most likely not subject to regional differences in local water flow or odorant exposure. Nevertheless, positional differences between the dorsal and ventral OE have been reported for Ni toxicity (Lazzari et al., 2019). More importantly, we observed a hotspot of OSN cell death in mid-lamellar regions that correlate with the presence of ontogenetically older OSNs (Bayramli et al., 2017). New neurons are generated at the ILC and SNS in the zebrafish OE and drift away from their site of neurogenesis to accumulate in mid-lamellar positions where cells originating from both proliferation zones meet. OSN generation can be quite high, and we estimated that around 5,500 new OSNs are added to the zebrafish OE over a 24 h period. These numbers suggest that the entire OE population would undergo near-complete turnover every month and several rounds over the 3–5 year lifespan of the animal in captivity (Gerhard et al., 2002).

Aged OSNs in the rodent OE are, at least in part, eliminated by apoptosis (Holcomb et al., 1995; Cowan and Roskams, 2002). In as much OSN lifetime is also determined by intrinsic factors or if OSN cell death depends solely on external influences is currently unknown (Gaun et al., 2022). The persistence of a certain fraction of long-lived cells in rodents and zebrafish somehow argues against a fixed lifespan, and external factors appear more dominant in determining OSN cell death. In the zebrafish OE, increased cell death in mid-lamellar regions can be observed by TUNEL staining but not by direct apoptotic markers, such as activated caspase, suggesting that aged OSNs are eliminated by non-apoptotic cell death pathways (Bayramli et al., 2017). Intriguingly, the mean lifetime of zebrafish OSNs of 28 days correlates well with the time that elapses before newborn OSNs that were generated from the opposing proliferation zones meet in mid-lamellar positions, suggesting at least the partial contribution of an active elimination mechanism. This again is similar to the situation in the intestinal crypt, where enterocytes generated in the base of the crypt meet at the tip of intestinal villi, where they are actively shed from the tissue (Patterson and Watson, 2017).

## Two Modes of Neurogenesis in the Zebrafish Olfactory System

Regardless of the exact cause and underlying cellular mechanisms, a large fraction of functionally mature OSNs is eliminated from the OE on a daily basis. To maintain sensory perception, the continuous loss of sporadically dying OSNs needs to be compensated for by an active cell proliferation mechanism that resupplies new neurons (Brann and Firestein, 2014). Moreover, direct mechanical or chemical injury can result in the sudden death of many OSNs that die simultaneously (Figure 6.4A). The vertebrate OE is endowed with two distinct stem/progenitor cell populations that specifically contribute to OSN neurogenesis in the intact and the acutely injured OE (Figure 6.4D) (Schwob et al., 2017; Kocagöz et al., 2022). GBCs are continuously active neuronal progenitor cells that are largely responsible for the maintenance of the OE and the replacement of sporadically dying neurons (Figure 6.4B, top panel). Activated GBCs generate daughter cells that progress through a fixed neuronal lineage to generate terminally differentiated OSNs (Cau et al., 1997; Schwob et al., 2017). HBCs, on the other hand, are dormant stem cells that are selectively activated by acute injury conditions and contribute to rapid restoration of the OE when large numbers of OSNs need to be generated (Figure 6.4B, bottom panel) (Leung et al., 2007; Iwai et al., 2008; Kocagöz et al., 2022). The first cell type that is generated by HBCs are cells that appear molecularly similar, if not identical, to GBCs, which then ultimately differentiate to become OSNs (Gadye et al., 2017; Kocagöz et al., 2022). In this regard, the OE stem cell niche again is more similar to epithelial tissues than to the two dominant stem cell niches in the brain. The skin and intestine also include

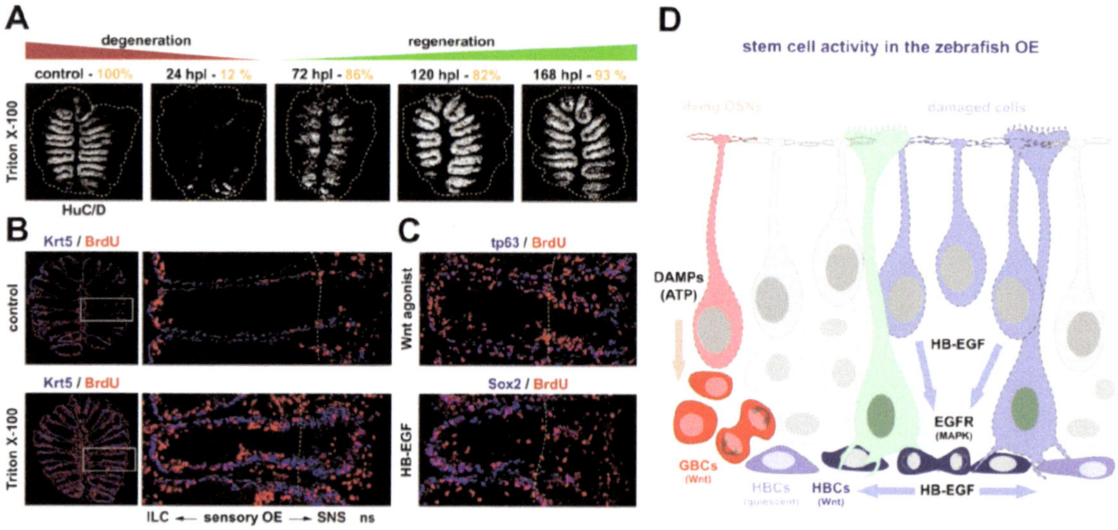


**FIGURE 6.4** The regenerative response of the adult zebrafish olfactory epithelium **A.** Irrigation of the adult zebrafish olfactory epithelium (OE; Figure 3C) with a 1% solution of the detergent Triton X-100 results in rapid degeneration of HuC/D-positive olfactory sensory neurons (OSNs; white) by 24 h post lesion (hpl). Subsequently, the tissue regenerates until the OSN population is restored between 120 and 168 hpl. The yellow dotted line denotes the outline of the tissue section. **B.** Immunohistochemistry against the horizontal basal cell (HBC) marker Keratin 5 (blue) and the mitotic activity marker BrdU (red) on fish that have been incubated with the thymidine analog for 24 h prior to analysis. A full tissue section is shown on the left, whereas the higher power view on the right illustrates a single epithelial fold, as indicated by the grey box. In the intact OE (control; top panel) mitotic activity occurs at the interlamellar curves (ILC) and the sensory/non-sensory border (SNS; yellow dotted line) but not in the center of the sensory OE. The activity originates from continuously dividing GBCs, which are positioned to the ILC and SNS and contribute to OSN maintenance under physiological conditions. No mitotic activity can be seen in HBCs. In the Triton X-100-lesioned OE (bottom panel), mitotic activity originating from HBCs (magenta) is induced across the entire OE, including the sensory region. **C.** Activation of key molecular pathways is required for HBC proliferation. Stimulation of the OE with Wnt pathway agonists (top) induces mitotic activity (BrdU; red) across the sensory OE, including tp63-positive HBCs (blue). The signaling factor HB-EGF is strongly and transiently upregulated in the injured tissue. Exposure of the OE to recombinant HB-EGF (bottom) activates Sox2-positive cells (blue), which include both HBCs and HBC-derived globose basal cells (GBCs). **D.** Model of stem cell activity in the intact and injured zebrafish OE. Sporadically dying OSNs (light red) release damage-associated molecular patterns (DAMPs), such as the intracellular energy carrier ATP. ATP selectively stimulates GBCs that are positioned at the ILC and SNS. GBCs produce a continuous stream of new OSNs that contribute to tissue maintenance under physiological conditions. Their rate of OSN production is matched to the number of dying cells through the available tissue concentration of the DAMP. Large-scale damage of OE tissue structure and cells by physical or chemical disturbances results in upregulation of the signaling factor HB-EGF. HB-EGF signals through the EGF receptor (EGFR) that is specifically expressed by HBCs. HB-EGF stimulation of HBC results in their mitotic expansion and generation of a transient population of GBC-like cells in the sensory OE. Binding of HB-EGF to EGFR activates intracellular MAP kinase signaling, which promotes cell proliferation, growth, and differentiation. Activated HBCs recruit neighboring cells into the regenerative response through paracrine stimulation with HB-EGF. Wnt signaling is generally required for mitotic activity in HBCs and GBCs.

different stem cell types that selectively contribute to tissue maintenance and regeneration following injury (Gonzales and Fuchs, 2017; Hageman et al., 2020). Crypt base columnar cells of the intestine are, like GBCs, constitutively active stem cells that undergo rapid cycling and provide new progenitors to the transit amplifying zone (Capdevila et al., 2024). New enterocytes migrate out of the crypt towards the tip of intestinal villi, where they are shed from the tissue (Patterson and Watson, 2017). The intestinal crypt also contains the so called +4 cells, named after their characteristic position about 4–6 cells from the base (Hageman et al., 2020; Rees et al., 2020). The +4 cells are quiescent reserve stem cells that respond to injury conditions, much like HBCs in the OE. Activated +4 cells generate Lgr5-positive cells that closely resemble crypt base columnar cells, while activated HBCs generate Lgr5-positive GBC-like cells in the mouse OE (Chen et al., 2014). Lgr5 expression has been helpful in identifying actively cycling stem cells in multiple organs and organisms. Lgr5 is a component of the Wnt signaling pathway, which promotes mitotic activity (Leung et al., 2018). The zebrafish genome, however, does not contain a gene coding for Lgr5, but recently, *celsr1a* has been identified in zebrafish, which may be a functional equivalent (Li et al., 2020). Transcriptome analysis of the injured zebrafish OE shows that *celsr3* but not *celsr1a* is upregulated 3-fold at 3 dpl when HBC-derived *ascl1* cells occur (Kocagöz et al., 2022; Demirler, Kocagöz, unpublished), but tissue expression of *celsr3* by OE cell types has not yet been investigated. Despite the gap in knowledge about the molecular definition of the two OE stem cell niches in zebrafish, *wnt2*, *wnt4b*, *wnt5b*, and *wnt10* are highly expressed in the intact OE, and *wnt7aa* and *wnt7bb* appear to be upregulated following injury, suggesting that Wnt signaling plays a similarly important role (Kocagöz et al., 2022).

GBCs are a heterogenous cell population that comprises active progenitor cells but also transit-amplifying cells. Cells of the early lineage can be identified by expression of the proneural factor *ascl1* and the stem cell marker Sox2 in zebrafish and mouse (Schwob et al., 2017; Demirler et al., 2020). Ascl1 is a downstream target of the Notch signaling cascade, which is expressed in the Notch low daughter cell during asymmetric cell divisions, and which drives neuronal differentiation (Sueda and Kageyam, 2020). In contrast, the Ascl1-negative/ Notch-high cell retains its stem cell identity. Intriguingly, in the intact zebrafish OE, cells with *ascl1* expression are regionalized and can only be observed in the center of the ILC and at the SNS flanking the region that is occupied by OSNs from both ends of the lamella. This is different from the rodent OE, in which GBCs are evenly distributed throughout the entire sensory OE and are located in a layer that is slightly more apical than HBCs (Schwob et al., 2017). The mechanisms that localize active GBCs to the ILC and SNS in the zebrafish OE are not known and curious. Lineage tracing in the mouse OE revealed conflicting results regarding the role of HBCs during OE maintenance. In one study, HBC activity contributed prominently to regular OSN turnover (Iwai et al., 2008), whereas in another, HBCs remained mitotically silent and did not generate a significant number of GBCs and other OE cells over a 5-month period (Leung et al, 2007). Thus, the possibility exists that all GBCs are HBC derived, at least in the postembryonic OE (Packard et al., 2011). In zebrafish, localized signaling factors could provide input to HBCs at the ILC and SNS to constantly generate active GBCs. In line with this assumption, we observed increased HBC activity at the ILC and SNS in comparison to mid-lamellar regions of the intact zebrafish OE (Kocagöz et al., 2022). Tissue injury by Triton X-100, on the other hand, results in the transient induction of evenly distributed Ascl1-positive GBCs across the entire zebrafish OE.

HBCs are considered to be a reserve stem cell population that remains mitotically quiescent unless severe disturbances to OE integrity occur (Leung et al., 2007; Herrick et al., 2017). These cells are attached to the basal lamina of the OE and are evenly distributed across the OE in rodents but also zebrafish (Carter et al., 2004; Demirler et al., 2020). Thus, GBCs and HBCs occupy similar positions in the rodent OE but are spatially segregated in zebrafish. HBCs of both taxa can be identified by their expression of the transcription factor tp63 and Keratin 5, another similarity that they share with epithelial stem cells but not the neurogenic niches of the brain (Yang et al., 1999; Romano et al., 2012). Activated HBCs generate GBC-like cells, which differentiate along the neuronal lineage to generate OSNs (Gadye et al., 2017; Kocagöz et al., 2022). The lineage potential in zebrafish has not been investigated, but in the mouse OE, HBCs can also give rise to other non-neuronal cell types, such as sustentacular cells and Bowman's gland cells, which do not exist in zebrafish (Leung et al., 2007; Iwai et al., 2008). It remains equally unexplored whether different types of chemosensory neurons, such as ciliated, microvillous, crypt, kappe, and pear cells, can arise from the same GBC/HBC or if different stem/progenitor pools with restricted lineage potential exist. The HBC response is transient and ceases when a sufficient number of GBCs has been generated to restore the neuronal population of the zebrafish OE, approximately around 5 dpl (Iqbal and Byrd-Jacobs, 2010; Kocagöz et al., 2022). The signals that activate injury-responsive HBCs are an area of high interest and will be discussed in detail below. Because of their uniform distribution across the OE tissue, HBC activation results in neurogenesis from within the sensory area of the zebrafish OE (Figure 6.4B, bottom panel), thus, the region between the constitutively active proliferation sites at the ILC and SNS (Kocagöz et al., 2022). OSN maturation progresses along the basal to apical axis in rodents, while apical and mid-basal layers are occupied by microvillous and ciliated OSNs in the zebrafish OE, respectively (Sato et al., 2005). Therefore, stratification in the zebrafish OE is largely a consequence of morphological OSN subtype rather than functional maturity.

## Maintenance Neurogenesis in the Intact OE

For the conditions discussed above, some OSNs are continuously lost from the OE. If these cells were not replaced at an equal rate, the organism would lose its ability to perceive odor information over time because of neuronal depletion (Calvo-Ochoa et al., 2021). Therefore, OSN loss and neurogenesis must be carefully balanced to avoid sensory deficits but also to prevent hyperproliferation of the OE. Since OSN loss may result, at least in part, from environmental influences, the number of dying cells may fluctuate depending on the moment-by-moment exposure to toxic substances. In the mouse OE, a positive correlation between the frequency of dying and dividing cells has been shown (Gaun et al., 2022). Thus, specific signaling mechanisms must be in place that communicate between the OSN population and GBCs, which are the main contributors to maintenance neurogenesis in the intact OE.

GBC activity could be restrained by inhibitory signals arising from mature OSNs. Once OSN density reaches sufficient levels, GBC activity would slow down or cease to prevent hyperproliferation. Evidence for negative feedback by OSNs stems from early cell culture experiments in which neurogenesis from olfactory progenitors was suppressed when they were co-cultured with mature OSN (Mumm et al., 1996). Factors controlling the rate of OSN neurogenesis have subsequently been identified with the TGF-b family member GDF11 (also known as BMP11) and ActivinbB, which cause cell cycle arrest through p21Kip (Wu et al., 2003; Gokoffski et al., 2011). Interestingly, GDF11 and ActivinbB intercept the olfactory lineage at different developmental stages (Gokoffski et al., 2011). While Activin slows down expansion of Sox2- and Ascl1-positive cells of the early GBC lineage, GDF11 inhibits immediate neuronal progenitors of the late lineage. The same study also identified follistatin as a negative regulator of this inhibition; however, the tissue source of this signal remains obscure. Regardless, the rate of GBC proliferation and lineage progression may be under the tight influence of positive and negative regulatory factors emanating from the overlying OSN population.

The list of factors and signaling molecules that activate OSN neurogenesis, however, is much longer and includes a collection of structurally and functionally diverse molecules that range from growth factors to neurotransmitters (Buckland and Cunningham, 1998; Feron et al., 1999; Plendl et al., 1999; Simpson et al., 2003; Beites et al., 2005; Jia et al., 2009; Jia and Hegg, 2012; Chen et al., 2020; Sireci et al., 2024). Some of these factors are relevant to drive lineage progression forward, while others may have a direct effect on GBC mitotic activity. An interesting class of molecules that could make GBCs sensitive to the moment-by-moment rate of OSN death are commonly known as damage-associated molecular patterns (DAMPS) (Venereau et al., 2015). Disintegrating cells release large quantities of molecularly diverse components that are usually restricted to their intracellular compartment. Thus, an increase in the extracellular concentration of these molecules could be a sensitive indicator of the rate of cell death as their concentration would be proportional to the number of dying cells and could further modulate GBC activity to establish a match between cell proliferation and death. A common DAMP that has been shown to be relevant in various injury contexts is the intracellular energy carrier ATP and related purine compounds (Kang et al., 2024), which are present in millimolar intracellular concentrations (Traut, 1994). Because it is rapidly degraded by enzymes, it can act as a time-resolved DAMP that communicates fluctuating tissue states (Venereau et al., 2015). Elevated OSN cell death would result in an increased tissue concentration of available ATP, while ATP levels are reduced under conditions when fewer OSNs die spontaneously.

We have experimentally tested the potential function of ATP as a relevant DAMP in the zebrafish OE (Demirler et al., 2020). Increasing the systemic concentration of ATP resulted in elevated cell proliferation and OSN neurogenesis. The effect could be blocked by simultaneous application of the inhibitor Suramin, suggesting that the effect is mediated by P2Y receptors (Lambrecht et al., 2002). Interestingly, ATP stimulation selectively activated OSN neurogenesis from the ILC and SNS, thus, the positions where GBCs are located, but it did not have any influence on mid regions of the sensory OE that are occupied by HBCs (Demirler et al., 2020). Sensitivity of GBCs to purine compounds could be confirmed by $Ca^{2+}$ activity recordings from acute slices through the zebrafish OE. Interestingly, despite the observation that purine compounds also activate $Ca^{2+}$ signals in HBCs, HBC proliferation did not occur (Demirler et al., 2020). This paradox may be a consequence of the normally closed OE structure that is created by tight junctions of the apical border (Liang, 2020). ATP released during physical damage of the OE may leak out through the injury site and may not be a reliable signal to trigger robust HBC responses. In the intact tissue, however, the molecule is contained within the OE and may reach sufficient levels to selectively stimulate GBCs. This appears meaningful because precocious HBC activation would result in hyperproliferation, which is inappropriate if only a small number of OSNs undergo cell death under physiological conditions.

## The OE Response to Acute Injury

The OE also has an exceptionally high capacity for acute structural and functional repair in comparison to other neuronal tissues (Iismaa et al., 2018). Damaged neurons are replaced in rapid time and functional connections are reestablished, even when severe structural injury occurs (Schwob et al., 2017; Yu and Wu, 2017). The regenerative capacity of the zebrafish OE, however, is unprecedented among taxa when it comes to the speed with which OE structure and function are restored. The loss of nearly the entire OSN population is repaired within only 5–7 days following degeneration (Figure 6.4A) (Iqbal and Byrd-Jacobs, 2010; Kocagöz et al., 2022), while it takes up to several months in rodents (Schwob et al., 2005).

Tissue injury activates HBCs (Figure 6.4B, bottom panel), which are mitotically dormant in the intact rodent OE (Herrick et al., 2017), while slow-cycling HBCs can be found in the intact zebrafish OE (Kocagöz et al., 2022). However, the absence of Ascl1-positive GBC-like cells in mid-lamellar positions suggests that HBCs predominantly undergo self-renewing symmetrical cell divisions, rather than asymmetric divisions that generate GBCs. Different injury-related events could provide input into HBC activity. Tissue damage results in structural changes that include the disruption of cell–cell contacts or the upregulation of injury markers by resident cells and immune cells that are recruited to the injury site (Soliman and Barreda, 2022). Elegant experiments using genetic cell ablation in the mouse OE showed that selective elimination of OSNs is not sufficient to trigger an injury response in HBCs (Herrick et al., 2017). This principle seems to be conserved in the zebrafish OE (Kocagöz et al., 2022). Transection of the olfactory nerve, which results in OSN degeneration without damaging other OE cell types, activated maintenance neurogenesis by GBCs at the ILC and SNS, further supporting the concept that GBC activity is modulated by OSN-derived DAMPS but only if the tissue structure remains intact to contain the signaling molecules.

In the mouse OE, the loss of sustentacular cells has been shown to drive HBC activation (Herrick et al., 2017). Sustentacular cells span the entire OE from the apical surface to the basal lamina, where they make close contacts with HBCs through endfeet processes. Sustentacular cells survey the entire OE and, therefore, are predestined to function as damage sensors. They express the Notch ligand Jagged 1, which stimulates Notch receptors on HBCs. Thus, under physiological conditions, HBCs show high Notch activity, which upregulates tp63 expression to induce mitotic dormancy (Herrick et al., 2017). Both downregulation of tp63 and Notch signaling result in HBC expansion and OSN neurogenesis (Fletcher et al., 2011; Schnittke et al., 2015). Preliminary results from the zebrafish OE, however, suggest that Notch pathway function is not equivalent. Inhibition of gamma secretase, the main enzyme that cleaves the actively signaling Notch intracellular domain (Olsauskas-Kuprys et al., 2013), does not promote HBC activity as would be expected if Notch signaling had the same function in zebrafish (Efe and Fuss, unpublished observation). Rather, Notch inhibition results in reduced mitotic activity and OSN production in the context of tissue injury. This mode of action is more in line with the function of the Notch signaling pathway in other tissues and organisms, where it regulates lineage commitment rather than mitotic activity directly (Tatsumi et al., 2010; Basak et al., 2012; Sueda and Kageyama, 2020). However, the relationship between Notch signaling, mitotic activity, and cell differentiation is complex and depends on cell type, tissue context, developmental stage, and lineage progression.

We have performed transcriptome profiling by RNA-sequencing of the injured zebrafish OE over relevant time points of de- and regeneration (Kocagöz et al., 2022). The first 24 hpl are characterized by a progressive reduction in HuC/D-positive OSNs. Accordingly, we observe a strong reduction in the expression of genes that are expressed by OSNs and are involved in olfactory signal transduction. Among the signaling pathways that are activated at relevant time points after the injury, TGF-b, Wnt, and epidermal growth factor (EGF) signaling stand out. Canonical Wnt signaling involves the transcriptional activator b-catenin, which is released from degradation in actively signaling cells (Clevers and Nusse., 2012). Accordingly, b-catenin-positive cells can be observed in GBCs, the ILC and SNS in the intact OE, while b-catenin expression is strongly upregulated by basal cells when injury occurs (Kocagöz et al., 2022). Loss- and gain-of-function experiments showed that Wnt is both necessary and sufficient to drive mitotic activity in HBCs. Stimulation with synthetic Wnt agonists or inhibition of the destruction complex result in increased mitotic activity in tp63-positive HBCs (Figure 6.4C, top panel). The injury-induced proliferation response is severely diminished when Wnt signaling is inhibited following injury. However, active Wnt signaling is required in nearly all dividing cells (Niehrs and Acebron, 2012). While Wnt components are upregulated in the injured zebrafish OE (Kocagöz et al., 2022) tissue, b-catenin levels are also sensitive to converging inputs, such as TGF-b, fibroblast growth factor, and Notch signaling (Katoh and Katoh, 2007).

EGF signaling is another molecular pathway that appears to be activated early after experimental OE lesions (Kocagöz et al., 2022). The zebrafish genome contains six genes that code for members of the ErbB family of receptor tyrosine kinases that form functional EGF receptors. Of those, *erbb1/egfra* is selectively upregulated 2.3-fold at 4 hpl (Sireci et al., 2024). EGFR can form homodimeric receptors, which interact with ligands of the EGF family, including EGF, Tgf-α, HB-EGF, betacellulin, amphiregulin, epigen, and epiregulin (Linggi and Carpenter, 2006). Of those, *hbegfa* was highly and transiently upregulated at the 4 hpl time point. *Hbegfa* reaches 14-fold higher expression levels at 4 hpl before declining to only 4-fold higher levels at 12 and reverts to pre-injury levels between 24 and 72 hpl (Sireci et al., 2024). Tissue analysis of *egfra* expression revealed that the receptor is expressed by HBCs. The ligand *hbegfa* is expressed in a few sporadic cells in the intact OE, which resembled HBCs. Upon injury, all major OE cell types, OSNs, sustentacular cells, and HBCs, upregulated *hbegfa* expression. The trigger of elevated *hbegfa* expression is currently not known but most likely is downstream of cytokine expression by injured cells or recruited immune cells (Takenobu et al., 2003).

HB-EGF is produced as a single-pass transmembrane protein that is activated by cleavage of the extracellular signaling domain by specific sheddases, which belong to the family of ADAM metalloproteases (Higashiyama and Nanba, 2005; Blobel, 2005). The soluble extracellular domain binds to and activates EGFR, which has been shown to signal through multiple different intracellular signaling routes, such as MAP kinase, PI3K/AKT, and JAK/STAT signaling in different cell types (Wee and Wang, 2017). Stimulation of the OE with recombinant HB-EGF mimicked the injury response by activating HBC proliferation and OSN neurogenesis (Figure 6.4C, bottom panel) (Sireci et al., 2024), while inhibition of the HB-EGF/EGFR signaling route at different levels severely diminished OE regeneration and OSN production. Inhibition of metalloproteases, sequestration of the soluble ectodomain of Hbegfa, and inhibition of receptor kinase activity all had the same effect. At the intracellular level, inhibition of MAP kinase and AKT signaling also prevented HBC proliferation and OSN regeneration. Consistently, both inhibition of EGFR and reducing the concentration of soluble Hbegfa in the context of tissue injury prevented the formation of pERK-positive HBCs. Thus, HB-EGF is an important component of the response of the zebrafish OE to injury. In the injured mouse, activation of EGFR and HBC activation by the non-canonical ligand NrCam has been suggested (Chen et al., 2020). In zebrafish, however, *nrcama* is downregulated, not upregulated, following tissue damage (Sireci et al., 2024). Since Notch signaling does not play the same role in the zebrafish OE, HB-EGF signaling may be the main activator. Interestingly, the signaling pathway has been shown to play similar roles in the injured zebrafish retina and spinal cord (Wan et al., 2012; Cigliola et al., 2023), where it also promotes progenitor cell expansion and neurogenesis, suggesting that HB-EGF may be a conserved signal that has important relevance for nerve cell regeneration in zebrafish. Whether these findings can be generalized to the human nervous system, however, awaits further investigation.

## The Next Frontier: Immune Cell Contribution to OE Regeneration

Even though they are not considered to be OE cells in a strict sense, the OE is densely occupied by immune cells (Tacchi et al., 2014; Sepahi et al., 2016; Palominas and Whitlock, 2021; Palominos et al., 2022). Neutrophils and macrophages of the innate immune system are abundant in the mammalian (Winkley et al., 2021; Wellford et al., 2024) and zebrafish OE (Palominas and Whitlock, 2021; Palominos et al., 2022). As already detailed above, part of their function is to prevent the breach of foreign pathogens into the OE and the CNS (Wellford and Moseman, 2024). The downside of the protective immune barrier, however, is the observation that acute and chronic immune activation can result in tissue damage, OSN cell death, and loss of smell sensation (Lane et al., 2010; LaFever et al., 2022).

A different aspect of the rich immune environment in OE tissue might be that innate and adaptive immune cells may also have supportive effects on OE regeneration. Because the OE is not shielded by the blood–brain barrier and is in contact with the systemic circulation, immune cells can easily be recruited (Getchell et al., 2002). In the zebrafish OE, blood vessels run in the lamina propria of each lamella, and larger blood lacunae form in the lamellar tips and the midline raphe (Palominos and Whitlock, 2021). There are two major consequences of this open organization that distinguishes the OE from the CNS. The main immune cell type in the brain are microglia, which in zebrafish are derived from macrophages that are recruited to and trapped in the brain before the blood–brain barrier closes (Casano et al., 2016; Xu et al., 2016). The OE, on the other hand, also contains neutrophils that are not found in the brain and very likely contains other immune cells of the erythroid and lymphoid lineages (Winkley et al., 2021; Palominos et al., 2022; Wellford et al., 2024). Focal brain lesions in zebrafish result in attraction of resident microglia to the site of injury (Sieger et al., 2012), but the extent and role of other immune cells has not been investigated in detail. In principle, traumatic injury and damage to blood vessels could enable infiltration of other cell types (Alam et al., 2020). In adult and larval models of OE injury, it has been shown that both neutrophils and macrophages are recruited (Palominos and Whitlock, 2021; Palominos et al., 2022). Preliminary data from our research suggests that neutrophil depletion impairs successful OE regeneration (Mollamustafaoglu and Fuss, unpublished). The mechanism of action, how neutrophils contribute to OE regeneration, however, has not been worked out. In the zebrafish brain and other tissues, transient acute inflammation is a prerequisite for successful regeneration (Kyritsis et al., 2012; Iribarne et al., 2021). Microglia/macrophages are recruited to injured cells through the release of chemokines and release immunomodulatory signaling molecules into the surrounding tissue. In particular, the pro-inflammatory cytokines IL-1b and IL-6 have been shown to be released by activated macrophages/microglia, and the cytokines CXCL1-3, CXCL8 are strongly involved in immune cell recruitment (Bouchery and Harris, 2019). In the mouse OE, regeneration depends on the CXCR4 receptor and its ligand CXCL12 (Senf et al., 2021). Comparative studies between zebrafish and mammals have revealed that a main difference appears to be that inflammation only occurs during the early phase after injury and is quickly resolved in zebrafish, while it persists longer in mammals (Diotel et al., 2020). In line with this assumption, it has been demonstrated that extending the inflammation phase in zebrafish experimentally results in reduced regeneration after injury (Palsamy et al., 2023). Thus, a current concept is that the high regenerative capacity of the zebrafish CNS can be attributed to the transient nature of the immune response. In as much this principle is also true for the OE awaits further investigation.

## Conclusions

Zebrafish represent a simple and easy-to-maintain research model that allows for the assessment of toxic effects of environmental pollutants and other chemicals that humans can be exposed to in an occupational context. The effect of these substances on development and function of the nervous system can be studied in larval and adult zebrafish. Thus, information derived from the zebrafish model allows for relevant insight into the molecular and cellular mechanisms that are influenced by everyday toxicants. The olfactory system, because of its exposed nature and direct contact with the environment, is a particularly insightful system to study the effect of toxicants on nerve cells – their degenerative, but also their regenerative response. The stem cell niches in the zebrafish OE are reasonably well understood and bear many similarities to the mammalian OE but also show unique differences that are specific to zebrafish. Relevant molecular factors have been described in the zebrafish OE (Figure 6.4D) that regulate nerve cell regeneration, which may have relevance for human health and therapeutic approaches.

## Acknowledgements

The authors acknowledge support from Bogazici University BAP (project numbers: 17P01P8, 21BM1 to SHF) and the Scientific and Technological Research Council of Turkey (TÜBITAK; grant numbers: 113T038 and 119Z081 to SHF; TÜBITAK-BIDEB 2211-A and 2218 fellowships to YK).

## REFERENCES

Abraira, V. E., & Ginty, D. D. (2013). The sensory neurons of touch. *Neuron*, *79*(4), 618–639. https://doi.org/10.1016/j.neuron.2013.07.051

Abreu, M. S., Giacomini, A. C., Rodriguez, R., Kalueff, A. V., & Barcelloeirs, L. J. (2017). Effects of $ZnSO_4$-induced peripheral anosmia on zebrafish behavior and physiology. *Behavioral Brain Research*, *320*, 275–281. https://doi.org/10.1016/j.bbr.2016.12.014

Adamek, G. D., Gesteland, R. C., Mair, R. G., & Oakley, B. (1984). Transduction physiology of olfactory receptor cilia. *Brain Research*, *310*(1), 87–97. https://doi.org/10.1016/0006-8993(84)90012-x

Aguillon, R., Batut, J., Subramanian, A., Madelaine, R., Dufourcq, P., Schilling, T. F., & Blader, P. (2018). Cell-type heterogeneity in the early zebrafish olfactory epithelium is generated from progenitors within preplacodal ectoderm. *eLife*, *7*, e32041. https://doi.org/10.7554/eLife.32041

Ahuja, G., Bozorg Nia, S., Zapilko, V., Shiriagin, V., Kowatschew, D., Oka, Y., & Korsching, S. I. (2014). Kappe neurons, a novel population of olfactory sensory neurons. *Scientific Reports*, *4*, 4037. https://doi.org/10.1038/srep04037

Ahuja, G., Ivandic, I., Saltürk, M., Oka, Y., Nadler, W., & Korsching, S. I. (2013). Zebrafish crypt neurons project to a single, identified mediodorsal glomerulus. *Scientific Reports*, *3*, 2063. https://doi.org/10.1038/srep02063

Alam, A., Thelin, E. P., Tajsic, T., Khan, D. Z., Khellaf, A., Patani, R., & Helmy, A. (2020). Cellular infiltration in traumatic brain injury. *Journal of Neuroinflammation*, *17*(1), 328. https://doi.org/10.1186/s12974-020-02005-x

Alba-González, A., Dragomir, E. I., Haghdousti, G., Yáñez, J., Dadswell, C., González-Méndez, R., Wilson, S. W., Tuschl, K., & Folgueira, M. (2024). Manganese overexposure alters neurogranin expression and causes behavioral deficits in larval zebrafish. *International Journal of Molecular Sciences*, *25*(9), 4933. https://doi.org/10.3390/ijms25094933

Ali, S., van Mil, H. G., & Richardson, M. K. (2011). Large-scale assessment of the zebrafish embryo as a possible predictive model in toxicity testing. *PLoS One*, *6*(6), e21076. https://doi.org/10.1371/journal.pone.0021076

Alioto, T. S., & Ngai, J. (2005). The odorant receptor repertoire of teleost fish. *BMC Genomics*, *6*, 173. https://doi.org/10.1186/1471-2164-6-173

Anastassopoulou, C., Davaris, N., Ferous, S., Siafakas, N., Boufidou, F., Anagnostopoulos, K., & Tsakris, A. (2024). The molecular basis of olfactory dysfunction in COVID-19 and long COVID. *Lifestyle Genomics*, *17*(1), 42–56. https://doi.org/10.1159/000539292

Assem, M., Lando, M., Grissi, M., Kamel, S., Massy, Z. A., Chillon, J. M., & Hénaut, L. (2018). The impact of uremic toxins on cerebrovascular and cognitive disorders. *Toxins (Basel)*, *10*(7), 303. https://doi.org/10.3390/toxins10070303

Baldwin, D. H., Sandahl, J. F., Labenia, J. S., & Scholz, N. L. (2003). Sublethal effects of copper on coho salmon: Impacts on nonoverlapping receptor pathways in the peripheral olfactory nervous system. *Environmental Toxicology and Chemistry*, *22*(10), 2266–2274. https://doi.org/10.1897/02-428

Baldwin, D. H., & Scholz, N. L. (2005). The electro-olfactogram: An in vivo measure of peripheral olfactory function and sublethal neurotoxicity in fish. In G. K. Ostrander (Ed.), *Techniques in aquatic toxicology* (pp. 257–276). CRC Press. https://doi.org/10.1201/9780203501597.ch14

Bambino, K., & Chu, J. (2017). Zebrafish in toxicology and environmental health. *Current Topics in Developmental Biology*, *124*, 331–367. https://doi.org/10.1016/bs.ctdb.2016.10.007

Basak, O., Giachino, C., Fiorini, E., Macdonald, H. R., & Taylor, V. (2012). Neurogenic subventricular zone stem/progenitor cells are Notch1-dependent in their active but not quiescent state. *Journal of Neuroscience*, *32*(16), 5654–5666. https://doi.org/10.1523/JNEUROSCI.0455-12.2012

Bastos, V., Pacheco, V., Rodrigues, É. D. L., Moraes, C. N. S., Nóbile, A. L., Fonseca, D. L. M., Souza, K. B. S., do Vale, F. Y. N., Filgueiras, I. S., Schimke, L. F., Giil, L. M., Moll, G., Cabral-Miranda, G., Ochs, H. D., Vasconcelos, P. F. D. C., de Melo, G. D., Bourhy, H., Casseb, L. M. N., & Cabral-Marques, O. (2023). Neuroimmunology of rabies: New insights into an ancient disease. *Journal of Medical Virology*, *95*(10), e29042. https://doi.org/10.1002/jmv.29042

Bayramli, X., Kocagöz, Y., Sakizli, U., & Fuss, S. H. (2017). Patterned arrangements of olfactory receptor gene expression in zebrafish are established by radial movement of specified olfactory sensory neurons. *Scientific Reports*, *7*(1), 5572. https://doi.org/10.1038/s41598-017-06041-1

Begum, W., Rai, S., Banerjee, S., Bhattacharjee, S., Mondal, M. H., Bhattarai, A., & Saha, B. (2022). A comprehensive review on the sources, essentiality and toxicological profile of nickel. *RSC Advances*, *12*(15), 9139–9153. https://doi.org/10.1039/d2ra00378c

Beites, C. L., Kawauchi, S., Crocker, C. E., & Calof, A. L. (2005). Identification and molecular regulation of neural stem cells in the olfactory epithelium. *Experimental Cell Research*, *306*(2), 309–316. https://doi.org/10.1016/j.yexcr.2005.03.027

Bettini, S., Lazzari, M., Ferrando, S., Gallus, L., & Franceschini, V. (2016). Histopathological analysis of the olfactory epithelium of zebrafish (Danio rerio) exposed to sublethal doses of urea. *Journal of Anatomy*, *228*(1), 59–69. https://doi.org/10.1111/joa.12397

Biechl, D., Tietje, K., Gerlach, G., & Wullimann, M. F. (2016). Crypt cells are involved in kin recognition in larval zebrafish. *Scientific Reports*, *6*, 24590. https://doi.org/10.1038/srep24590

Blobel, C. P. (2005). ADAMs: Key components in EGFR signalling and development. *Nature Reviews Molecular Cell Biology*, *6*(1), 32–43. https://doi.org/10.1038/nrm1548

Bossy, J. (1970). *Atlas of neuroanatomy and special sense organs*. Saunders.

Bouchery, T., & Harris, N. (2019). Neutrophil-macrophage cooperation and its impact on tissue repair. *Immunology and Cell Biology*, *97*(3), 289–298. https://doi.org/10.1111/imcb.12241

Bousquet, J., Anto, J. M., Bachert, C., Baiardini, I., Bosnic-Anticevich, S., Walter Canonica, G., Melén, E., Palomares, O., Scadding, G. K., Togias, A., & Toppila-Salmi, S. (2020). Allergic rhinitis. *Nature Reviews Disease Primers*, *6*(1), 95. https://doi.org/10.1038/s41572-020-00227-0

Brann, J. H., & Firestein, S. J. (2014). A lifetime of neurogenesis in the olfactory system. *Frontiers in Neuroscience*, *8*, 182. https://doi.org/10.3389/fnins.2014.00182

Braubach, O. R., Wood, H. D., Gadbois, S., Fine, A., & Croll, R. P. (2009). Olfactory conditioning in the zebrafish (Danio rerio). *Behavioral Brain Research*, *198*(1), 190–198. https://doi.org/10.1016/j.bbr.2008.10.044

Bryche, B., St Albin, A., Murri, S., Lacôte, S., Pulido, C., Ar Gouilh, M., Lesellier, S., Servat, A., Wasniewski, M., Picard-Meyer, E., Monchatre-Leroy, E., Volmer, R., Rampin, O., Le Goffic, R., Marianneau, P., Meunier, N. (2020). Massive transient damage of the olfactory epithelium associated with infection of sustentacular cells by SARS-CoV-2 in golden Syrian hamsters. *Brain, Behavior, and Immunity*, *89*, 579–586. https://doi.org/10.1016/j.bbi.2020.06.032

Buckland, M. E., & Cunningham, A. M. (1998). Alterations in the neurotrophic factors BDNF, GDNF and CNTF in the regenerating olfactory system. *Annals of the New York Academy of Sciences*, *855*, 260–265. https://doi.org/10.1111/j.1749-6632.1998.tb10579.x

Busquet, F., Nagel, R., von Landenberg, F., Mueller, S. O., Huebler, N., & Broschard, T. H. (2008). Development of a new screening assay to identify proteratogenic substances using zebrafish Danio rerio embryo combined with an exogenous mammalian metabolic activation system (mDarT). *Toxicological Sciences*, *104*(1), 177–188. https://doi.org/10.1093/toxsci/kfn065

Butler, J. M., Field, K. E., & Maruska, K. P. (2016). Cobalt chloride treatment used to ablate the lateral line system also impairs the olfactory system in three freshwater fishes. *PLoS One*, *11*(7), e0159521. https://doi.org/10.1371/journal.pone.0159521

Cacialli, P., Ricci, S., Frabetti, F., Ferrando, S., & Franceschini, V. (2024). Exposure of zebrafish embryos to urea affects NOS1 gene expression in neuronal cells. *Environments*, *11*(3), 41. https://doi.org/10.3390/environments11030041

Calabresi, P., Centonze, D., & Bernardi, G. (2000). Cellular factors controlling neuronal vulnerability in the brain: A lesson from the striatum. *Neurology*, *55*(9), 1249–1255. https://doi.org/10.1212/wnl.55.9.1249

Calderón-Garcidueñas, L., Franco-Lira, M., Henríquez-Roldán, C., Osnaya, N., González-Maciel, A., Reynoso-Robles, R., Villarreal-Calderon, R., Herritt, L., Brooks, D., Keefe, S., Palacios-Moreno, J., Villarreal-Calderon, R., Torres-Jardón, R., Medina-Cortina, H., Delgado-Chávez, R., Aiello-Mora, M., Maronpot, R. R. (2010). Urban air pollution: Influences on olfactory function and pathology in exposed children and young adults. *Experimental Toxicology Pathology*, *62*(1), 91–102. https://doi.org/10.1016/j.etp.2009.02.117

Calvo-Ochoa, E., & Byrd-Jacobs, C. A. (2019). The olfactory system of zebrafish as a model for the study of neurotoxicity and injury: Implications for neuroplasticity and disease. *International Journal of Molecular Sciences*, *20*(7), 1639. https://doi.org/10.3390/ijms20071639

Calvo-Ochoa, E., Byrd-Jacobs, C. A., & Fuss, S. H. (2021). Diving into the streams and waves of constitutive and regenerative olfactory neurogenesis: Insights from zebrafish. *Cell and Tissue Research*, *383*(1), 227–253. https://doi.org/10.1007/s00441-020-03334-2

Cancalon, P. (1982). Degeneration and regeneration of olfactory cells induced by ZnSO4 and other chemicals. *Tissue and Cell*, *14*(4), 717–733. https://doi.org/10.1016/0040-8166(82)90061-1

Cancalon, P. (1983). Influence of a detergent on the catfish olfactory mucosa. *Tissue and Cell*, *15*(2), 245–258. https://doi.org/10.1016/0040-8166(83)90020-4

Cannon, J. R., & Greenamyre, J. T. (2011). The role of environmental exposures in neurodegeneration and neurodegenerative diseases. *Toxicological Sciences*, *124*(2), 225–250. https://doi.org/10.1093/toxsci/kfr239

Capdevila, C., Miller, J., Cheng, L., Kornberg, A., George, J. J., Lee, H., Botella, T., Moon, C. S., Murray, J. W., Lam, S., Calderon, I., Malagola, E., Whelan, G., Lin, C. S., Han, A., Wang, T. C., Sims, P. A., & Yan, K. S. (2024). Time-resolved fate mapping identifies the intestinal upper crypt zone as an origin of Lgr5+ crypt base columnar cells. *Cell*, *187*(12), 3039–3055.e14. https://doi.org/10.1016/j.cell.2024.05.001

Carr, V. M., & Farbman, A. I. (1993). The dynamics of cell death in the olfactory epithelium. *Experimental Neurology*, *124*(2), 308–314. https://doi.org/10.1006/exnr.1993.1201

Carter, L. A., MacDonald, J. L., & Roskams, A. J. (2004). Olfactory horizontal basal cells demonstrate a conserved multipotent progenitor phenotype. *Journal of Neuroscience*, *24*(25), 5670–5683. https://doi.org/10.1523/JNEUROSCI.0330-04.2004

Casadei, E., & Salinas, I. (2019). Comparative models for human nasal infections and immunity. *Developmental and Comparative Immunology*, *92*, 212–222. https://doi.org/10.1016/j.dci.2018.11.022

Casano, A. M., Albert, M., & Peri, F. (2016). Developmental apoptosis mediates entry and positioning of microglia in the zebrafish brain. *Cell Reports*, *16*(4), 897–906. https://doi.org/10.1016/j.celrep.2016.06.033

Cau, E., Gradwohl, G., Fode, C., & Guillemot, F. (1997). Mash1 activates a cascade of bHLH regulators in olfactory neuron progenitors. *Development*, *124*(8), 1611–1621. https://doi.org/10.1242/dev.124.8.1611

Celik, A., Fuss, S. H., & Korsching, S. I. (2002). Selective targeting of zebrafish olfactory receptor neurons by the endogenous OMP promoter. *European Journal of Neuroscience*, *15*(5), 798–806. https://doi.org/10.1046/j.1460-9568.2002.01913.x

Cempel, M., & Nikel, G. (2006). Nickel: A review of its sources and environmental toxicology. *Polish Journal of Environmental Studies*, *15*(3), 375–382.

Chekuru, A., Kuscha, V., Hans, S., & Brand, M. (2017). Ligand-controlled site-specific recombination in zebrafish. *Methods in Molecular Biology*, *1642*, 87–97. https://doi.org/10.1007/978-1-4939-7169-5_6

Chen, M., Tian, S., Yang, X., Lane, A. P., Reed, R. R., & Liu, H. (2014). Wnt-responsive $Lgr5^+$ globose basal cells function as multipotent olfactory epithelium progenitor cells. *The Journal of Neuroscience*, *34*(24), 8268–8276. https://doi.org/10.1523/JNEUROSCI.0240-14.2014

Chen, Z. H., Luo, X. C., Yu, C. R., & Huang, L. (2020). Matrix metalloprotease-mediated cleavage of neural glial-related cell adhesion molecules activates quiescent olfactory stem cells via EGFR. *Molecular Cellular Neuroscience*, *108*, 103552. https://doi.org/10.1016/j.mcn.2020.103552

Chew, S., Lampinen, R., Saveleva, L., Korhonen, P., Mikhailov, N., Grubman, A., Polo, J. M., Wilson, T., Komppula, M., Rönkkö, T., Gu, C., Mackay-Sim, A., Malm, T., White, A. R., Jalava, P., & Kanninen, K. M. (2020). Urban air particulate matter induces mitochondrial dysfunction in human olfactory mucosal cells. *Particle and Fibre Toxicology*, *17*(1), 18. https://doi.org/10.1186/s12989-020-00352-4

Chlebowski, A. C., La Du, J. K., Truong, L., Massey Simonich, S. L., & Tanguay, R. L. (2017). Investigating the application of a nitroreductase-expressing transgenic zebrafish line for high-throughput toxicity testing. *Toxicology Reports*, *4*, 202–210. https://doi.org/10.1016/j.toxrep.2017.04.005

Choi, T. Y., Choi, T. I., Lee, Y. R., Choe, S. K., & Kim, C. H. (2021). Zebrafish as an animal model for biomedical research. *Experimental & Molecular Medicine*, *53*(3), 310–317. https://doi.org/10.1038/s12276-021-00571-5

Cianciolo Cosentino, C., Roman, B. L., Drummond, I. A., & Hukriede, N. A. (2010). Intravenous microinjections of zebrafish larvae to study acute kidney injury. *Journal of Visualized Experiments*, *42*, 2079. https://doi.org/10.3791/2079

Cigliola, V., Shoffner, A., Lee, N., Ou, J., Gonzalez, T. J., Hoque, J., Becker, C. J., Han, Y., Shen, G., Faw, T. D., Abd-El-Barr, M. M., Varghese, S., Asokan, A., & Poss, K. D. (2023). Spinal cord repair is modulated by the neurogenic factor Hb-egf under direction of a regeneration-associated enhancer. *Nature Communications*, *14*(1), 4857. https://doi.org/10.1038/s41467-023-40486-5

Ciurea, A. V., Mohan, A. G., Covache-Busuioc, R. A., Costin, H. P., Glavan, L. A., Corlatescu, A. D., & Saceleanu, V. M. (2023). Unraveling molecular and genetic insights into neurodegenerative diseases: Advances in understanding Alzheimer's, Parkinson's, and Huntington's diseases and amyotrophic lateral sclerosis. *International Journal of Molecular Sciences*, *24*(13), 10809. https://doi.org/10.3390/ijms241310809

Clevers, H., & Nusse, R. (2012). Wnt/β-catenin signaling and disease. *Cell*, *149*(6), 1192–1205. https://doi.org/10.1016/j.cell.2012.05.012

Cooper, M. L., Pasini, S., Lambert, W. S., D'Alessandro, K. B., Yao, V., Risner, M. L., & Calkins, D. J. (2020). Redistribution of metabolic resources through astrocyte networks mitigates neurodegenerative stress. *Proceedings of the National Academy of Sciences*, *117*(31), 18810–18821. https://doi.org/10.1073/pnas.2009425117

Cowan, C. M., & Roskams, A. J. (2002). Apoptosis in the mature and developing olfactory neuroepithelium. *Microscopy Research and Technique*, *58*(3), 204–215. https://doi.org/10.1002/jemt.10150

Curado, S., Stainier, D. Y., & Anderson, R. M. (2008). Nitroreductase-mediated cell/tissue ablation in zebrafish: A spatially and temporally controlled ablation method with applications in developmental and regeneration studies. *Nature Protocols*, *3*(6), 948–954. https://doi.org/10.1038/nprot.2008.58

Czarniak, N., Kamińska, J., Matowicka-Karna, J., & Koper-Lenkiewicz, O. M. (2023). Cerebrospinal fluid-basic concepts review. *Biomedicines*, *11*(5), 1461. https://doi.org/10.3390/biomedicines11051461

d'Amora, M., & Giordani, S. (2018). The utility of zebrafish as a model for screening developmental neurotoxicity. *Frontiers in Neuroscience*, *12*, 976. https://doi.org/10.3389/fnins.2018.00976

Dando, S. J., Mackay-Sim, A., Norton, R., Currie, B. J., St John, J. A., Ekberg, J. A., Batzloff, M., Ulett, G. C., & Beacham, I. R. (2014). Pathogens penetrating the central nervous system: Infection pathways and the cellular and molecular mechanisms of invasion. *Clinical Microbiology Reviews*, *27*(4), 691–726. https://doi.org/10.1128/CMR.00118-13

Dang, P., Fisher, S. A., Stefanik, D. J., Kim, J., & Raper, J. A. (2018). Coordination of olfactory receptor choice with guidance receptor expression and function in olfactory sensory neurons. *PLoS Genetics*, *14*(1), e1007164. https://doi.org/10.1371/journal.pgen.1007164

Davies, P., & Maloney, A. J. (1976). Selective loss of central cholinergic neurons in Alzheimer's disease. *Lancet*, *2*(8000), 1403. https://doi.org/10.1016/s0140-6736(76)91936-x

Demirler, M. C., Sakizli, U., Bali, B., Kocagöz, Y., Eski, S. E., Ergönen, A., Alkiraz, A. S., Bayramli, X., Hassenklöver, T., Manzini, I., & Fuss, S. H. (2020). Purinergic signalling selectively modulates maintenance but not repair neurogenesis in the zebrafish olfactory epithelium. *FEBS Journal*, *287*(13), 2699–2722. https://doi.org/10.1111/febs.15170

Dew, W. A., Azizishirazi, A., & Pyle, G. G. (2014). Contaminant-specific targeting of olfactory sensory neuron classes: Connecting neuron class impairment with behavioural deficits. *Chemosphere*, *112*, 519–525. https://doi.org/10.1016/j.chemosphere.2014.02.047

Dieris, M., Kowatschew, D., Hassenklöver, T., Manzini, I., & Korsching, S. I. (2024). Calcium imaging of adult olfactory epithelium reveals amines as important odor class in fish. *Cell and Tissue Research*, *396*(1), 95–102. https://doi.org/10.1007/s00441-024-03859-w

Diotel, N., Lübke, L., Strähle, U., & Rastegar, S. (2020). Common and distinct features of adult neurogenesis and regeneration in the telencephalon of zebrafish and mammals. *Frontiers in Neuroscience*, *14*, 568930. https://doi.org/10.3389/fnins.2020.568930

Døving, K. B., Dubois-Dauphin, M., Holley, A., & Jourdan, F. (1977). Functional anatomy of the olfactory organ of fish and the ciliary mechanism of water transport. *Acta Zoologica*, *58*(4), 245–255. https://doi.org/10.1111/j.1463-6395.1977.tb00260.x

Døving, K. B., Hansson, K. A., Backström, T., & Hamdani, E. L. H. (2011). Visualizing a set of olfactory sensory neurons responding to a bile salt. *Journal of Experimental Biology*, *214*(Pt 1), 80–87. https://doi.org/10.1242/jeb.046607

Elberger, A. J., & Honig, M. G. (1990). Double-labeling of tissue containing the carbocyanine dye DiI for immunocytochemistry. *Journal of Histochemistry & Cytochemistry*, *38*(5), 735–739. https://doi.org/10.1177/38.5.2110209

Fajar, S., Dwi, S. P., Nur, I. S., Wahyu, A. P., Sukamto, S. M., Winda, A. R., Nastiti, W., Andri, F., & Firzan, N. (2024). Zebrafish as a model organism for virus disease research: Current status and future directions. *Heliyon*, *10*(13), e33865. https://doi.org/10.1016/j.heliyon.2024.e33865

Faucher, K., Floriani, M., Gilbin, R., & Adam-Guillermin, C. (2012). Uranium-induced sensory alterations in the zebrafish Danio rerio. *Aquatic Toxicology*, 124–125, 94–105. https://doi.org/10.1016/j.aquatox.2012.08.004

Féron, F., Vincent, A., & Mackay-Sim, A. (1999). Dopamine promotes differentiation of olfactory neuron in vitro. *Brain Research*, *845*(2), 252–259. https://doi.org/10.1016/s0006-8993(99)01959-9

Ferrando, S., Gallus, L., Gambardella, C., Marchesotti, E., Ravera, S., Franceschini, V., & Masini, M. A. (2014). Effects of urea on the molecules involved in the olfactory signal transduction: A preliminary study on Danio rerio. *Fish Physiology and Biochemistry*, *40*(6), 1793–1800. https://doi.org/10.1007/s10695-014-9968-x

Fitzgerald, J. A., Könemann, S., Krümpelmann, L., Županič, A., & Vom Berg, C. (2021). Approaches to test the neurotoxicity of environmental contaminants in the zebrafish model: From behavior to molecular mechanisms. *Environmental Toxicology and Chemistry*, *40*(4), 989–1006. https://doi.org/10.1002/etc.4951

Fletcher, R. B., Prasol, M. S., Estrada, J., Baudhuin, A., Vranizan, K., Choi, Y. G., & Ngai, J. (2011). p63 regulates olfactory stem cell self-renewal and differentiation. *Neuron*, *72*(5), 748–759. https://doi.org/10.1016/j.neuron.2011.09.009

Fodoulian, L., Tuberosa, J., Rossier, D., Boillat, M., Kan, C., Pauli, V., Egervari, K., Lobrinus, J. A., Landis, B. N., Carleton, A., & Rodriguez, I. (2020). SARS-CoV-2 receptors and entry genes are expressed in the human olfactory neuroepithelium and brain. *iScience*, *23*(12), 101839. https://doi.org/10.1016/j.isci.2020.101839

Friedrich, R. W., & Korsching, S. I. (1997). Combinatorial and chemotopic odorant coding in the zebrafish olfactory bulb visualized by optical imaging. *Neuron*, *18*(5), 737–752. https://doi.org/10.1016/s0896-6273(00)80314-1

Fulton, K. A., Zimmerman, D., Samuel, A., Vogt, K., & Datta, S. R. (2024). Common principles for odor coding across vertebrates and invertebrates. *Nature Reviews Neuroscience*, *25*(7), 453–472. https://doi.org/10.1038/s41583-024-00822-0

Fuss, S. H., & Korsching, S. I. (2001). Odorant feature detection: activity mapping of structure response relationships in the zebrafish olfactory bulb. *Journal of Neuroscience*, *21*(21), 8396–407. https://doi.org/10.1523/JNEUROSCI.21-21-08396.2001

Gadye, L., Das, D., Sanchez, M. A., Street, K., Baudhuin, A., Wagner, A., Cole, M. B., Choi, Y. G., Yosef, N., Purdom, E., Dudoit, S., Risso, D., Ngai, J., & Fletcher, R. B. (2017). Injury activates transient olfactory stem cell states with diverse lineage capacities. *Cell Stem Cell*, *21*(6), 775–790. e9. https://doi.org/10.1016/j.stem.2017.10.014

Gaun, V., Martens, J. R., & Schwob, J. E. (2022). Lifespan of mature olfactory sensory neurons varies with location in the mouse olfactory epithelium and age of the animal. *Journal of Comparative Neurology*, *530*(12), 2238–2251. https://doi.org/10.1002/cne.25330

Genc, S., Zadeoglulari, Z., Fuss, S. H., & Genc, K. (2012). The adverse effects of air pollution on the nervous system. *Journal of Toxicology*, *2012*, 782462. https://doi.org/10.1155/2012/782462

Genter, M. B., & Doty, R. L. (2019). Toxic exposures and the senses of taste and smell. *Handbook of Clinical Neurology*, *164*, 389–408. https://doi.org/10.1016/B978-0-444-63855-7.00022-8

Gerhard, G. S., Kauffman, E. J., Wang, X., Stewart, R., Moore, J. L., Kasales, C. J., Demidenko, E., & Cheng, K. C. (2002). Life spans and senescent phenotypes in two strains of zebrafish (Danio rerio). *Experimental Gerontology*, *37*(8–9), 1055–1068. https://doi.org/10.1016/s0531-5565(02)00088-8

Getchell, T. V., Subhedar, N. K., Shah, D. S., Hackley, G., Partin, J. V., Sen, G., & Getchell, M. L. (2002). Chemokine regulation of macrophage recruitment into the olfactory epithelium following target ablation: Involvement of macrophage inflammatory protein-1alpha and monocyte chemoattractant protein-1. *Journal of Neuroscience Research*, *70*(6), 784–793. https://doi.org/10.1002/jnr.10432

Gokoffski, K. K., Wu, H. H., Beites, C. L., Kim, J., Kim, E. J., Matzuk, M. M., Johnson, J. E., Lander, A. D., & Calof, A. L. (2011). Activin and GDF11 collaborate in feedback control of neuroepithelial stem cell proliferation and fate. *Development*, *138*(19), 4131–4142. https://doi.org/10.1242/dev.065870

Gonzales, K. A. U., & Fuchs, E. (2017). Skin and its regenerative powers: An alliance between stem cells and their niche. *Developmental Cell*, *43*(4), 387–401. https://doi.org/10.1016/j.devcel.2017.10.001

Griep, M. I., Van der Niepen, P., Sennesael, J. J., Mets, T. F., Massart, D. L., & Verbeelen, D. L. (1997). Odour perception in chronic renal disease. *Nephrology Dialysis Transplantation*, *12*(10), 2093–2098. https://doi.org/10.1093/ndt/12.10.2093

Guo, Z., Packard, A., Krolewski, R. C., Harris, M. T., Manglapus, G. L., & Schwob, J. E. (2010). Expression of pax6 and sox2 in adult olfactory epithelium. *Journal of Comparative Neurology*, *518*(21), 4395–4418. https://doi.org/10.1002/cne.22463

Hageman, J. H., Heinz, M. C., Kretzschmar, K., van der Vaart, J., Clevers, H., & Snippert, H. J. G. (2020). Intestinal regeneration: Regulation by the microenvironment. *Developmental Cell*, *54*(4), 435–446. https://doi.org/10.1016/j.devcel.2020.07.009

Hanchate, N. K., Kondoh, K., Lu, Z., Kuang, D., Ye, X., Qiu, X., Pachter, L., Trapnell, C., & Buck, L. B. (2015). Single-cell transcriptomics reveals receptor transformations during olfactory neurogenesis. *Science*, *350*(6265),1251–5. https://doi.org/10.1126/science.aad2456

Hansen, A., & Finger, T. E. (2000). Phyletic distribution of crypt-type olfactory receptor neurons in fishes. *Brain Behavior and Evolution*, *55*(2), 100–110. https://doi.org/10.1159/000006645

Hansen, A., Rolen, S. H., Anderson, K., Morita, Y., Caprio, J., & Finger, T. E. (2003). Correlation between olfactory receptor cell type and function in the channel catfish. *Journal of Neuroscience*, *23*(28), 9328–9339. https://doi.org/10.1523/JNEUROSCI.23-28-09328.2003

Hansen, A., & Zeiske, E. (1998). The peripheral olfactory organ of the zebrafish, Danio rerio: An ultrastructural study. *Chemical Senses*, *23*(1), 39–48. https://doi.org/10.1093/chemse/23.1.39

Hansen, A., & Zielinski, B. S. (2005). Diversity in the olfactory epithelium of bony fishes: Development, lamellar arrangement, sensory neuron cell types and transduction components. *Journal of Neurocytology*, *34*(3–5), 183–208. https://doi.org/10.1007/s11068-005-8353-1

Heffern, K., Tierney, K., & Gallagher, E. P. (2018). Comparative effects of cadmium, zinc, arsenic, and chromium on olfactory-mediated neurobehavior and gene expression in larval zebrafish (Danio rerio). *Aquatic Toxicology*, *201*, 83–90. https://doi.org/10.1016/j.aquatox.2018.05.016

Hendriksen, H., Korte, S. M., Olivier, B., & Oosting, R. S. (2015). The olfactory bulbectomy model in mice and rats: One story or two tails? *European Journal of Pharmacology*, *753*, 105–113. https://doi.org/10.1016/j.ejphar.2014.10.033

Hentig, J. T., & Byrd-Jacobs, C. A. (2016). Exposure to zinc sulfate results in differential effects on olfactory sensory neuron subtypes in adult zebrafish. *International Journal of Molecular Sciences*, *17*(9), 1445. https://doi.org/10.3390/ijms17091445

Herrick, D. B., Lin, B., Peterson, J., Schnittke, N., & Schwob, J. E. (2017). Notch1 maintains dormancy of olfactory horizontal basal cells, a reserve neural stem cell. *Proceedings of the National Academy of Sciences*, *114*(28), E5589–E5598. https://doi.org/10.1073/pnas.1701333114

Higashiyama, S., & Nanba, D. (2005). ADAM-mediated ectodomain shedding of HB-EGF in receptor cross-talk. *Biochimica et Biophysica Acta*, *1751*(1), 110–117. https://doi.org/10.1016/j.bbapap.2004.11.009

Holcomb, J. D., Mumm, J. S., & Calof, A. L. (1995). Apoptosis in the neuronal lineage of the mouse olfactory epithelium: Regulation in vivo and in vitro. *Developmental Biology*, *172*(1), 307–323. https://doi.org/10.1006/dbio.1995.0025

Holl, A. M. (2018). Survival of mature mouse olfactory sensory neurons labeled genetically perinatally. *Molecular and Cellular Neuroscience*, *88*, 258–269. https://doi.org/10.1016/j.mcn.2018.02.005

Howe, K., Clark, M. D., Torroja, C. F., Torrance, J., Berthelot, C., Muffato, M., et al. (2013). The zebrafish reference genome sequence and its relationship to the human genome. *Nature*, *496*, 498–503. https://doi.org/10.1038/nature11992

Hussain, A., Saraiva, L. R., Ferrero, D. M., Ahuja, G., Krishna, V. S., Liberles, S. D., & Korsching, S. I. (2013). High-affinity olfactory receptor for the death-associated odor cadaverine. *Proceedings of the National Academy of Sciences*, *110*(48), 19579–19584. https://doi.org/10.1073/pnas.1318596110

Hussain, S., Khan, M., Sheikh, T. M. M., Mumtaz, M. Z., Chohan, T. A., Shamim, S., & Liu, Y. (2022). Zinc essentiality, toxicity, and its bacterial bioremediation: A comprehensive insight. *Frontiers in Microbiology*, *13*, 900740. https://doi.org/10.3389/fmicb.2022.900740

Hussainzada, N., Lewis, J. A., Baer, C. E., Ippolito, D. L., Jackson, D. A., & Stallings, J. D. (2014). Whole adult organism transcriptional profiling of acute metal exposures in male zebrafish. *BMC Pharmacology and Toxicology*, *15*, 15. https://doi.org/10.1186/2050-6511-15-15

Iismaa, S. E., Kaidonis, X., Nicks, A. M., Bogush, N., Kikuchi, K., Naqvi, N., Harvey, R. P., Husain, A., & Graham, R. M. (2018). Comparative regenerative mechanisms across different mammalian tissues. *NPJ Regenerative Medicine*, *3*, 6. https://doi.org/10.1038/s41536-018-0044-5

Ijomone, O. M. (2021). Neurotoxicity of nickel. In M. Aschner & L. G. Costa (Eds.), *Advances in Neurotoxicology*, *5*, 263–284. Academic Press. https://doi.org/10.1016/bs.ant.2020.11.004

Imamura, F., & Hasegawa-Ishii, S. (2016). Environmental toxicants-induced immune responses in the olfactory mucosa. *Frontiers in Immunology*, *7*, 475. https://doi.org/10.3389/fimmu.2016.00475

Imamura, F., Ito, A., & LaFever, B. J. (2020). Subpopulations of projection neurons in the olfactory bulb. *Frontiers in Neural Circuits*, *14*, 561822. https://doi.org/10.3389/fncir.2020.561822

Iqbal, T., & Byrd-Jacobs, C. (2010). Rapid degeneration and regeneration of the zebrafish olfactory epithelium after triton X-100 application. *Chemical Senses*, *35*(5), 351–61. https://doi.org/10.1093/chemse/bjq019

Iribarne, M. (2021). Inflammation induces zebrafish regeneration. *Neural Regeneration Research*, *16*(9), 1693–1701. https://doi.org/10.4103/1673-5374.306059

Iwai, N., Zhou, Z., Roop, D. R., & Behringer, R. R. (2008). Horizontal basal cells are multipotent progenitors in normal and injured adult olfactory epithelium. *Stem Cells*, *26*(5), 1298–1306. https://doi.org/10.1634/stemcells.2007-0891

Jellinger, K. A. (2010). Basic mechanisms of neurodegeneration: A critical update. *Journal of Cellular and Molecular Medicine*, *14*(3), 457–487. https://doi.org/10.1111/j.1582-4934.2010.01010.x

Jensen, K. H., & Berg, R. W. (2016). CLARITY-compatible lipophilic dyes for electrode marking and neuronal tracing. *Scientific Reports*, *6*, 32674. https://doi.org/10.1038/srep32674

Jeong, S. H., Jang, J. H., & Lee, Y. B. (2023). Drug delivery to the brain via the nasal route of administration: Exploration of key targets and major consideration factors. *Journal of Pharmaceutical Investigation*, *53*(1), 119–152. https://doi.org/10.1007/s40005-022-00589-5

Jesuthasan, S. J., & Mathuru, A. S. (2008). The alarm response in zebrafish: Innate fear in a vertebrate genetic model. *Journal of Neurogenetics*, *22*(3), 211–228. https://doi.org/10.1080/01677060802298475

Jia, C., Doherty, J. P., Crudgington, S., & Hegg, C. C. (2009). Activation of purinergic receptors induces proliferation and neuronal differentiation in Swiss Webster mouse olfactory epithelium. *Neuroscience*, *163*(1), 120–128. https://doi.org/10.1016/j.neuroscience.2009.06.040

Jia, C., & Hegg, C. C. (2012). Neuropeptide Y and extracellular signal-regulated kinase mediate injury-induced neuroregeneration in mouse olfactory epithelium. *Molecular and Cellular Neurosciences*, *49*(2), 158–170. https://doi.org/10.1016/j.mcn.2011.11.004

Jiang, J., Pi, J., & Cai, J. (2018). The advancing of zinc oxide nanoparticles for biomedical applications. *Bioinorganic Chemistry and Applications*, *2018*, 1062562. https://doi.org/10.1155/2018/1062562

Jimenez, E., Slevin, C. C., Colón-Cruz, L., & Burgess, S. M. (2021). Vestibular and auditory hair cell regeneration following targeted ablation of hair cells with diphtheria toxin in zebrafish. *Frontiers in Cellular Neuroscience*, *15*, 721950. https://doi.org/10.3389/fncel.2021.721950

Jurkowski, M. P., Bettio, L., Woo, E. K., Patten, A., Yau, S. Y., & Gil-Mohapel, J. (2020). Beyond the hippocampus and the SVZ: Adult neurogenesis throughout the brain. *Frontiers in Cellular Neuroscience*, *14*, 576444. https://doi.org/10.3389/fncel.2020.576444

Kang, N., Ji, Z., Li, Y., Gao, J., Wu, X., Zhang, X., Duan, Q., Zhu, C., Xu, Y., Wen, L., Shi, X., & Liu, W. (2024). Metabolite-derived damage-associated molecular patterns in immunological diseases. *FEBS Journal*, *291*(10), 2051–2067. https://doi.org/10.1111/febs.16902

Kanungo, J., Lantz, S., & Paule, M. G. (2011). In vivo imaging and quantitative analysis of changes in axon length using transgenic zebrafish embryos. *Neurotoxicology and Teratology*, *33*(6), 618–623. https://doi.org/10.1016/j.ntt.2011.08.013

Karki, R., & Kanneganti, T. D. (2022). Innate immunity, cytokine storm, and inflammatory cell death in COVID-19. *Journal of Translational Medicine*, *20*(1), 542. https://doi.org/10.1186/s12967-022-03767-z

Katoh, M., & Katoh, M. (2007). WNT signaling pathway and stem cell signaling network. *Clinical Cancer Research*, *13*(14), 4042–4045. https://doi.org/10.1158/1078-0432.CCR-06-2316

Kempermann, G., Song, H., & Gage, F. H. (2015). Neurogenesis in the adult hippocampus. *Cold Spring Harbor Perspectives in Biology*, *7*(9), a018812. https://doi.org/10.1101/cshperspect.a018812

Kermen, F., Darnet, L., Wiest, C., Palumbo, F., Bechert, J., Uslu, O., & Yaksi, E. (2020). Stimulus-specific behavioral responses of zebrafish to a large range of odors exhibit individual variability. *BMC Biology*, *18*(1), 66. https://doi.org/10.1186/s12915-020-00801-8

Khan, M., Clijsters, M., Choi, S., Backaert, W., Claerhout, M., Couvreur, F., Van Breda, L., Bourgeois, F., Speleman, K., Klein, S., Van Laethem, J., Verstappen, G., Dereli, A. S., Yoo, S. J., Zhou, H., Dan Do, T. N., Jochmans, D., Laenen, L., Debaveye, Y., De Munter, P., Gunst, J., Jorissen, M., Lagrou, K., Meersseman, P., Neyts, J., Thal, D. R., Topsakal, V., Vandenbriele, C., Wauters, J., Mombaerts, P., & Van Gerven, L. (2022). Anatomical barriers against SARS-CoV-2 neuroinvasion at vulnerable interfaces visualized in deceased COVID-19 patients. *Neuron*, *110*(23), 3919–3935.e6. https://doi.org/10.1016/j.neuron.2022.11.007

Kim, S., Finlay, J. B., Ko, T., & Goldstein, B. J. (2024). Long-term olfactory loss post-COVID-19: Pathobiology and potential therapeutic strategies. *World Journal of Otorhinolaryngology Head and Neck Surgery*, *10*(2), 148–155. https://doi.org/10.1002/wjo2.165

Knox, E. G., Aburto, M. R., Clarke, G., Cryan, J. F., & O'Driscoll, C. M. (2022). The blood-brain barrier in aging and neurodegeneration. *Molecular Psychiatry*, *27*(6), 2659–2673. https://doi.org/10.1038/s41380-022-01511-z

Kocagöz, Y., Demirler, M. C., Eski, S. E., Güler, K., Dokuzluoglu, Z., & Fuss, S. H. (2022). Disparate progenitor cell populations contribute to maintenance and repair neurogenesis in the zebrafish olfactory epithelium. *Cell and Tissue Research*, *388*(2), 331–358. https://doi.org/10.1007/s00441-022-03597-x

Koide, T., Miyasaka, N., Morimoto, K., Asakawa, K., Urasaki, A., Kawakami, K., & Yoshihara, Y. (2009). Olfactory neural circuitry for attraction to amino acids revealed by transposon-mediated gene trap approach in zebrafish. *Proceedings of the National Academy of Sciences*, *106*(24), 9884–9889. https://doi.org/10.1073/pnas.0900470106

Korsching, S. I. (2020). Taste and smell in zebrafish. In B. Fritzsch (Ed.), *The senses: A comprehensive reference 2* (pp. 466–492). Academic Press. https://doi.org/10.1016/B978-0-12-809324-5.24155-2

Korsching, S. I., Argo, S., Campenhausen, H., Friedrich, R.W., Rummrich, A., & Weth, F. (1997). Olfaction in zebrafish: What does a tiny teleost tell us? *Seminars in Cell & Developmental Biology*, *8*(2), 181–187. https://doi.org/10.1006/scdb.1996.0136

Kraus, A., Garcia, B., Ma, J., Herrera, K. J., Zwaka, H., Harpaz, R., Wong, R. Y., Engert, F., & Salinas, I. (2023). Olfactory detection of viruses shapes brain immunity and behavior in zebrafish. *eLife*, *12*, e88271. https://doi.org/10.7554/eLife.88271.1

Kraus, A., Huertas, M., Ellis, L., Boudinot, P., Levraud, J. P., & Salinas, I. (2022). Intranasal delivery of SARS-CoV-2 spike protein is sufficient to cause olfactory damage, inflammation and olfactory dysfunction in zebrafish. *Brain, Behavior, and Immunity*, *102*, 341–359. https://doi.org/10.1016/j.bbi.2022.03.006

Kress, S., Biechl, D., & Wullimann, M. F. (2015). Combinatorial analysis of calcium-binding proteins in larval and adult zebrafish primary olfactory system identifies differential olfactory bulb glomerular projection fields. *Brain Structure and Function*, *220*(4), 1951–1970. https://doi.org/10.1007/s00429-014-0765-1

Kyritsis, N., Kizil, C., Zocher, S., Kroehne, V., Kaslin, J., Freudenreich, D., Iltzsche, A., & Brand, M. (2012). Acute inflammation initiates the regenerative response in the adult zebrafish brain. *Science*, *338*(6112), 1353–1356. https://doi.org/10.1126/science.1228773

Labrot, F., Narbonne, J. F., Ville, P., Saint Denis, M., & Ribera, D. (1999). Acute toxicity, toxicokinetics, and tissue target of lead and uranium in the clam *Corbicula fluminea* and the worm *Eisenia fetida*: Comparison with the fish *Brachydanio rerio*. *Archives of Environmental Contamination and Toxicology*, *36*(2), 167–178. https://doi.org/10.1007/s002449900457

LaFever, B. J., Kawasawa, Y. I., Ito, A., & Imamura, F. (2022). Pathological consequences of chronic olfactory inflammation on neurite morphology of olfactory bulb projection neurons. *Brain, Behavior, and Immunity Health*, *21*, 100451. https://doi.org/10.1016/j.bbih.2022.100451

Lai, K. P., Gong, Z., & Tse, W. K. F. (2021). Zebrafish as the toxicant screening model: Transgenic and omics approaches. *Aquatic Toxicology*, *234*, 105813. https://doi.org/10.1016/j.aquatox.2021.105813

Lambrecht, G., Braun, K., Damer, M., Ganso, M., Hildebrandt, C., Ullmann, H., Kassack, M. U., & Nickel, P. (2002). Structure-activity relationships of suramin and pyridoxal-5'-phosphate derivatives as P2 receptor antagonists. *Current Pharmaceutical Design*, *8*(26), 2371–2399. https://doi.org/10.2174/1381612023392973

Landis, B. N., Marangon, N., Saudan, P., Hugentobler, M., Giger, R., Martin, P. Y., & Lacroix, J. S. (2011). Olfactory function improves following hemodialysis. *Kidney International*, *80*(8), 886–893. https://doi.org/10.1038/ki.2011.189

Lane, A. P., Turner, J., May, L., & Reed, R. (2010). A genetic model of chronic rhinosinusitis-associated olfactory inflammation reveals reversible functional impairment and dramatic neuroepithelial reorganization. *The Journal of Neuroscience*, *30*(6), 2324–2329. https://doi.org/10.1523/JNEUROSCI.4507-09.2010

Lanigan, L. G., Russell, D. S., Woolard, K. D., Pardo, I. D., Godfrey, V., Jortner, B. S., Butt, M. T., & Bolon, B. (2021). Comparative pathology of the peripheral nervous system. *Veterinary Pathology*, *58*(1), 10–33. https://doi.org/10.1177/0300985820959231

Lazzari, M., Bettini, S., Milani, L., Maurizii, M. G., & Franceschini, V. (2017). Differential response of olfactory sensory neuron populations to copper ion exposure in zebrafish. *Aquatic Toxicology*, *183*, 54–62. https://doi.org/10.1016/j.aquatox.2016.12.012

Lazzari, M., Bettini, S., Milani, L., Maurizii, M. G., & Franceschini, V. (2019). Differential nickel-induced responses of olfactory sensory neuron populations in zebrafish. *Aquatic Toxicology*, *206*, 14–23. https://doi.org/10.1016/j.aquatox.2018.10.011

Lazzari, M., Bettini, S., Milani, L., Maurizii, M. G., & Franceschini, V. (2022). Response of olfactory sensory neurons to mercury ions in zebrafish: An immunohistochemical study. *Microscopy and Microanalysis*, *28*(1), 227–242. https://doi.org/10.1017/S1431927621013763

Lee, H. C., Lu, P. N., Huang, H. L., Chu, C., Li, H. P., & Tsai, H. J. (2014). Zebrafish transgenic line huORFZ is an effective living bioindicator for detecting environmental toxicants. *PLoS ONE*, *9*(3), e90160. https://doi.org/10.1371/journal.pone.0090160

Legradi, J. B., Di Paolo, C., Kraak, M. H. S., van der Geest, H. G., Schymanski, E. L., Williams, A. J., Dingemans, M. M. L., Massei, R., Brack, W., Cousin, X., Begout, M. L., van der Oost, R., Carion, A., Suarez-Ulloa, V., Silvestre, F., Escher, B. I., Engwall, M., Nilén, G., Keiter, S. H., Pollet, D., Waldmann, P., Kienle, C., Werner, I., Haigis, A. C., Knapen, D., Vergauwen, L., Spehr, M., Schulz, W., Busch, W., Leuthold, D., Scholz, S., Vom Berg, C. M., Basu, N., Murphy, C. A., Lampert, A., Kuckelkorn, J., Grummt, T., & Hollert, H. (2018). An ecotoxicological view on neurotoxicity assessment. *Environmental Sciences Europe*, *30*(1), 46. https://doi.org/10.1186/s12302-018-0173-x

Lerebours, A., Gonzalez, P., Adam, C., Camilleri, V., Bourdineaud, J. P., & Garnier-Laplace, J. (2009). Comparative analysis of gene expression in brain, liver, skeletal muscles, and gills of zebrafish (Danio rerio) exposed to environmentally relevant waterborne uranium concentrations. *Environmental Toxicology and Chemistry*, *28*(6), 1271–1278. https://doi.org/10.1897/08-357.1

Leung, C. T., Coulombe, P. A., & Reed, R. R. (2007). Contribution of olfactory neural stem cells to tissue maintenance and regeneration. *Nature Neuroscience*, *10*(6), 720–726. https://doi.org/10.1038/nn1882

Leung, C. T., Tan, S. H., & Barker, N. (2018). Recent advances in Lgr5+ stem cell research. *Trends in Cell Biology*, *28*(5), 380–391. https://doi.org/10.1016/j.tcb.2018.01.010

Leyssens, L., Vinck, B., Van Der Straeten, C., Wuyts, F., & Maes, L. (2017). Cobalt toxicity in humans—a review of the potential sources and systemic health effects. *Toxicology*, *387*, 43–56. https://doi.org/10.1016/j.tox.2017.05.015

Li, C., Barton, C., Henke, K., Daane, J., Treaster, S., Caetano-Lopes, J., Tanguay, R. L., & Harris, M. P. (2020). celsr1a is essential for tissue homeostasis and onset of aging phenotypes in the zebrafish. *eLife*, *9*, e50523. https://doi.org/10.7554/eLife.50523

Li, H., Liu, Y., Chen, Q., Jin, L., & Peng, R. (2023). Research progress of zebrafish model in aquatic ecotoxicology. *Water*, *15*(9), 1735. https://doi.org/10.3390/w15091735

Li, L., & Yang, X. (2018). The essential element manganese, oxidative stress, and metabolic diseases: Links and interactions. *Oxidative Medicine and Cellular Longevity*, *2018*, 7580707. https://doi.org/10.1155/2018/7580707

Liang, F. (2020). Sustentacular cell enwrapment of olfactory receptor neuronal dendrites: An update. *Genes*, *11*(5), 493. https://doi.org/10.3390/genes11050493

Lim, D. A., & Alvarez-Buylla, A. (2016). The adult ventricular-subventricular zone (V-SVZ) and olfactory bulb (OB) neurogenesis. *Cold Spring Harbor Perspectives in Biology*, *8*(5), a018820. https://doi.org/10.1101/cshperspect.a018820

Lin, C. H., Nicol, C. J. B., Wan, C., Chen, S. J., Huang, R. N., & Chiang, M. C. (2022). Exposure to PM2.5 induces neurotoxicity, mitochondrial dysfunction, oxidative stress, and inflammation in human SH-SY5Y neuronal cells. *Neurotoxicology*, *88*, 25–35. https://doi.org/10.1016/j.neuro.2021.10.009

Lindsey, B. W., Douek, A. M., Loosli, F., & Kaslin, J. (2018). A whole brain staining, embedding, and clearing pipeline for adult zebrafish to visualize cell proliferation and morphology in 3-dimensions. *Frontiers in Neuroscience*, *11*, 750. https://doi.org/10.3389/fnins.2017.00750

Linggi, B., & Carpenter, G. (2006). ErbB receptors: New insights on mechanisms and biology. *Trends in Cell Biology*, *16*(12), 649–656. https://doi.org/10.1016/j.tcb.2006.10.008

Lipschitz, D. L., & Michel, W. C. (2002). Amino acid odorants stimulate microvillar sensory neurons. *Chemical Senses*, *27*, 277–286. https://doi.org/10.1093/chemse/27.3.277

Llorente, V., Velarde, P., Desco, M., & Gómez-Gaviro, M.V. (2022). Current understanding of the neural stem cell niches. *Cells*, *11*(19), 3002. https://doi.org/10.3390/cells11193002.

Ma, E. Y., Heffern, K., Cheresh, J., & Gallagher, E. P. (2018). Differential copper-induced death and regeneration of olfactory sensory neuron populations and neurobehavioral function in larval zebrafish. *Neurotoxicology*, *69*, 141–151. https://doi.org/10.1016/j.neuro.2018.10.002

Maaswinkel, H., & Li, L. (2003). Olfactory input increases visual sensitivity in zebrafish: A possible function for the terminal nerve and dopaminergic interplexiform cells. *Journal of Experimental Biology*, *206*(Pt 13), 2201–2209. https://doi.org/10.1242/jeb.00397

MacDonald, T. C., Korbas, M., James, A. K., Sylvain, N. J., Hackett, M. J., Nehzati, S., Krone, P. H., George, G. N., & Pickering, I. J. (2015). Interaction of mercury and selenium in the larval stage zebrafish vertebrate model. *Metallomics*, *7*(8), 1247–1255. https://doi.org/10.1039/c5mt00145e

MacDonald, T. C., Sylvain, N. J., James, A. K., Pickering, I. J., Krone, P. H., & George, G. N. (2016). Effects of inorganic mercury on the olfactory pits of zebrafish larvae. *Metallomics*, *8*(5), 514–517. https://doi.org/10.1039/c6mt00031b

Mackay-Sim, A., & Kittel, P. W. (1991). On the life span of olfactory receptor neurons. *European Journal of Neuroscience*, *3*(3), 209–215. https://doi.org/10.1111/j.1460-9568.1991.tb00081.x

Manetti, P. (2023). Heavy metal-induced neurotoxicity and calcium channel involvement. *Toxicology*, *9*, 236. https://doi.org/10.4172/2476-2067.1000236

Massoz, L., Dupont, M. A., & Manfroid, I. (2021). Zebra-fishing for regenerative awakening in mammals. *Biomedicines*, *9*(1), 65. https://doi.org/10.3390/biomedicines9010065

Masuda, M., Ihara, S., Mori, N., Koide, T., Miyasaka, N., Wakisaka, N., Yoshikawa, K., Watanabe, H., Touhara, K., & Yoshihara, Y. (2024). Identification of olfactory alarm substances in zebrafish. *Current Biology*, *34*(7), 1377–1389.e7. https://doi.org/10.1016/j.cub.2024.02.003

Matz, C. J., & Krone, P. H. (2007). Cell death, stress-responsive transgene activation, and deficits in the olfactory system of larval zebrafish following cadmium exposure. *Environmental Science & Technology*, *41*(14), 5143–5148. https://doi.org/10.1021/es070452c

Mayer, U., Ungerer, N., Klimmeck, D., Warnken, U., Schnölzer, M., Frings, S., & Möhrlen, F. (2008). Proteomic analysis of a membrane preparation from rat olfactory sensory cilia. *Chemical Senses*, *33*(2), 145–162. https://doi.org/10.1093/chemse/bjm073

Méndez-Armenta, M., & Ríos, C. (2007). Cadmium neurotoxicity. *Environmental Toxicology and Pharmacology*, *23*(3), 350–358. https://doi.org/10.1016/j.etap.2006.11.009

Mitchell, I. J., & Griffiths, M. R. (2003). The differential susceptibility of specific neuronal populations: Insights from Huntington's disease. *IUBMB Life*, *55*(6), 293–298. https://doi.org/10.1080/1521654031000153012

Mombaerts, P. (2004a). Genes and ligands for odorant, vomeronasal, and taste receptors. *Nature Reviews Neuroscience*, *5*(4), 263–278. https://doi.org/10.1038/nrn1365

Mombaerts, P. (2004b). Odorant receptor gene choice in olfactory sensory neurons: The one receptor-one neuron hypothesis revisited. *Current Opinion in Neurobiology*, *14*(1), 31–36. https://doi.org/10.1016/j.conb.2004.01.014

Mombaerts, P. (2006). Axonal wiring in the mouse olfactory system. *Annual Review of Cell and Developmental Biology*, *22*, 713–737. https://doi.org/10.1146/annurev.cellbio.21.012804.093915

Mourabit, S., Fitzgerald, J. A., Ellis, R. P., Takesono, A., Porteus, C. S., Trznadel, M., Metz, J., Winter, M. J., Kudoh, T., & Tyler, C. R. (2019). New insights into organ-specific oxidative stress mechanisms using a novel biosensor zebrafish. *Environmental International*, *133*(Pt A), 105138. https://doi.org/10.1016/j.envint.2019.105138

Muendlein, H. I., Connolly, W. M., Cameron, J., Jetton, D., Magri, Z., Smirnova, I., Vannier, E., Li, X., Martinot, A. J., Batorsky, R., & Poltorak, A. (2022). Neutrophils and macrophages drive TNF-induced lethality via TRIF/CD14-mediated responses. *Science Immunology*, *7*(78), eadd0665. https://doi.org/10.1126/sciimmunol.add0665

Mumm, J. S., Shou, J., & Calof, A. L. (1996). Colony-forming progenitors from mouse olfactory epithelium: Evidence for feedback regulation of neuron production. *Proceedings of the National Academy of Sciences*, *93*(20), 11167–11172. https://doi.org/10.1073/pnas.93.20.11167

Munger, S. D., Leinders-Zufall, T., & Zufall, F. (2009). Subsystem organization of the mammalian sense of smell. *Annual Review of Physiology*, *71*, 115–140. https://doi.org/10.1146/annurev.physiol.70.113006.100608

Nabi, M., & Tabassum, N. (2022). Role of environmental toxicants on neurodegenerative disorders. *Frontiers in Toxicology*, *4*, 837579. https://doi.org/10.3389/ftox.2022.837579

Nadig, A. P. R., Huwaimel, B., Alobaida, A., Khafagy, E. S., Alotaibi, H. F., Moin, A., Lila, A. S. A., Suman, M. S., & Krishna, K. L. (2022). Manganese chloride ($MnCl_2$) induced novel model of Parkinson's disease in adult zebrafish; Involvement of oxidative stress, neuroinflammation, and apoptosis pathway. *Biomedical Pharmacotherapy*, *155*, 113697. https://doi.org/10.1016/j.biopha.2022.113697

Nagel, R. (2002). DarT: The embryo test with the zebrafish *Danio rerio*—a general model in ecotoxicology and toxicology. *ALTEX*, *19*(Suppl 1), 38–48.

Niehrs, C., & Acebron, S. P. (2012). Mitotic and mitogenic Wnt signalling. *EMBO Journal*, *31*(12), 2705–2713. https://doi.org/10.1038/emboj.2012.124

Niimura, Y., & Nei, M. (2005). Evolutionary dynamics of olfactory receptor genes in fishes and tetrapods. *Proceedings of the National Academy of Sciences*, *102*(17), 6039–6044. https://doi.org/10.1073/pnas.0501922102

Olivares, J., & Schmachtenberg, O. (2019). An update on anatomy and function of the teleost olfactory system. *PeerJ*, *7*, e7808. https://doi.org/10.7717/peerj.7808

Olsauskas-Kuprys, R., Zlobin, A., & Osipo, C. (2013). Gamma secretase inhibitors of Notch signaling. *OncoTargets and Therapy*, *6*, 943–955. https://doi.org/10.2147/OTT.S33766

Ovsepian, S. V., O'Leary, V. B., & Martinez, S. (2024). Selective vulnerability of motor neuron types and functional groups to degeneration in amyotrophic lateral sclerosis: Review of the neurobiological mechanisms and functional correlates. *Brain Structure and Function*, *229*(1), 1–14. https://doi.org/10.1007/s00429-023-02728-6

Packard, A., Schnittke, N., Romano, R. A., Sinha, S., & Schwob, J. E. (2011). DeltaNp63 regulates stem cell dynamics in the mammalian olfactory epithelium. *Journal of Neuroscience*, *31*(24), 8748–8759. https://doi.org/10.1523/JNEUROSCI.0681-11.2011

Palmer, G., Horgan, D. J., Tisdale, H., Singer, T. P., & Beinert, H. (1968). Studies on the respiratory chain-linked reduced nicotinamide adenine dinucleotide dehydrogenase. XIV. Location of the sites of inhibition of rotenone, barbiturates, and piericidin by means of electron paramagnetic resonance spectroscopy. *Journal of Biological Chemistry*, *243*(4), 844–847.

Palominos, M. F., & Whitlock, K. E. (2021). The olfactory organ is populated by neutrophils and macrophages during early development. *Frontiers in Cell and Developmental Biology*, *8*, 604030. https://doi.org/10.3389/fcell.2020.604030

Palominos, M. F., Calfún, C., Nardocci, G., Candia, D., Torres-Paz, J., & Whitlock, K. E. (2022). The olfactory organ is a unique site for neutrophils in the brain. *Frontiers in Immunology*, *13*, 881702. https://doi.org/10.3389/fimmu.2022.881702

Palsamy, K., Chen, J. Y., Skaggs, K., Qadeer, Y., Connors, M., Cutler, N., Richmond, J., Kommidi, V., Poles, A., Affrunti, D., Powell, C., Goldman, D., & Parent, J. M. (2023). Microglial depletion after brain injury prolongs inflammation and impairs brain repair, adult neurogenesis and pro-regenerative signaling. *Glia*, *71*(11), 2642–2663. https://doi.org/10.1002/glia.24444

Pandya, V. A., & Patani, R. (2021). Region-specific vulnerability in neurodegeneration: Lessons from normal ageing. *Ageing Research Reviews*, *67*, 101311. https://doi.org/10.1016/j.arr.2021.101311

Park, H. C., Kim, C. H., Bae, Y. K., Yeo, S. Y., Kim, S. H., Hong, S. K., Shin, J., Yoo, K. W., Hibi, M., Hirano, T., Miki, N., Chitnis, A. B., & Huh, T. L. (2000). Analysis of upstream elements in the HuC promoter leads to the establishment of transgenic zebrafish with fluorescent neurons. *Developmental Biology*, *227*(2), 279–293. https://doi.org/10.1006/dbio.2000.9898

Paskin, T. R., Iqbal, T. R., & Byrd-Jacobs, C. A. (2011). Olfactory bulb recovery following reversible deafferentation with repeated detergent application in the adult zebrafish. *Neuroscience*, *196*, 276–284. https://doi.org/10.1016/j.neuroscience.2011.09.005

Patterson, A. M., & Watson, A. J. M. (2017). Deciphering the complex signaling systems that regulate intestinal epithelial cell death processes and shedding. *Frontiers in Immunology*, *8*, 841. https://doi.org/10.3389/fimmu.2017.00841

Pignatelli, A., & Belluzzi, O. (2017). Dopaminergic neurones in the main olfactory bulb: An overview from an electrophysiological perspective. *Frontiers in Neuroanatomy*, *11*, 7. https://doi.org/10.3389/fnana.2017.00007

Pihlstrøm, L., Wiethoff, S., & Houlden, H. (2017). Genetics of neurodegenerative diseases: An overview. *Handbook of Clinical Neurology*, *145*, 309–323. https://doi.org/10.1016/B978-0-12-802395-2.00022-5

Plendl, J., Stierstorfer, B., & Sinowatz, F. (1999). Growth factors and their receptors in the olfactory system. *Anatomy and Histology of Embryology*, *28*(2), 73–79. https://doi.org/10.1046/j.1439-0264.1999.00165.x

Potter, N. A., Meltzer, G. Y., Avenbuan, O. N., Raja, A., & Zelikoff, J. T. (2021). Particulate matter and associated metals: A link with neurotoxicity and mental health. *Atmosphere*, *12*(4), 425. https://doi.org/10.3390/atmos12040425

Pugach, E. K., Li, P., White, R., & Zon, L. (2009). Retro-orbital injection in adult zebrafish. *Journal of Visualized Experiments*, *34*, 1645. https://doi.org/10.3791/1645

Ran, Y., Wang, S., Zhao, Y., Li, J., Ran, X., & Hao, Y. (2020). A review of biological effects and treatments of inhaled depleted uranium aerosol. *Journal of Environmental Radioactivity*, *222*, 106357. https://doi.org/10.1016/j.jenvrad.2020.106357

Razmara, P., & Pyle, G. G. (2021). Effect of copper nanoparticles and copper ions on the architecture of rainbow trout olfactory mucosa. *Ecotoxicology and Environmental Safety*, *227*, 112876. https://doi.org/10.1016/j.ecoenv.2021.112876

Rees, W. D., Tandun, R., Yau, E., Zachos, N. C., & Steiner, T. S. (2020). Regenerative intestinal stem cells induced by acute and chronic injury: The saving grace of the epithelium? *Frontiers in Cell and Developmental Biology*, *8*, 583919. https://doi.org/10.3389/fcell.2020.583919

Reinhold, A. K., & Rittner, H. L. (2017). Barrier function in the peripheral and central nervous system—a review. *Pflugers Archiv: European Journal of Physiology*, *469*(1), 123–134. https://doi.org/10.1007/s00424-016-1920-8

Reiten, I., Uslu, F. E., Fore, S., Pelgrims, R., Ringers, C., Diaz Verdugo, C., Hoffman, M., Lal, P., Kawakami, K., Pekkan, K., Yaksi, E., & Jurisch-Yaksi, N. (2017). Motile-cilia-mediated flow improves sensitivity and temporal resolution of olfactory computations. *Current Biology*, *27*(2), 166–174. https://doi.org/10.1016/j.cub.2016.11.036

Rodriguez-Gil, D. J., Bartel, D. L., Jaspers, A. W., Mobley, A. S., Imamura, F., & Greer, C. A. (2015). Odorant receptors regulate the final glomerular coalescence of olfactory sensory neuron axons. *Proceedings of the National Academy of Sciences*, *112*(18), 5821–5826. https://doi.org/10.1073/pnas.1417955112

Romano, R. A., Smalley, K., Magraw, C., Serna, V. A., Kurita, T., Raghavan, S., & Sinha, S. (2012). ΔNp63 knockout mice reveal its indispensable role as a master regulator of epithelial development and differentiation. *Development*, *139*(4), 772–782. https://doi.org/10.1242/dev.071191

Roper, S. D. (2022). Encoding taste: From receptors to perception. *Handbook of Experimental Pharmacology*, *275*, 53–90. https://doi.org/10.1007/164_2021_559

Salonen, L. (1994). 238U series radionuclides as a source of increased radioactivity in groundwater originating from Finnish bedrock. In *Future groundwater resources at risk* (pp. 71–84). IAHS Publication No. 222.

Sandahl, J. F., Baldwin, D. H., Jenkins, J. J., & Scholz, N. L. (2007). A sensory system at the interface between urban stormwater runoff and salmon survival. *Environmental Science & Technology*, *41*(8), 2998–3004. https://doi.org/10.1021/es062287r

Sandulescu, C. M., Teow, R. Y., Hale, M. E., & Zhang, C. (2011). Onset and dynamic expression of S100 proteins in the olfactory organ and the lateral line system in zebrafish development. *Brain Research*, *1383*, 120–127. https://doi.org/10.1016/j.brainres.2011.01.087

Santos, D., Félix, L., Luzio, A., Parra, S., Bellas, J., & Monteiro, S. M. (2021). Single and combined acute and subchronic toxic effects of microplastics and copper in zebrafish (Danio rerio) early life stages. *Chemosphere*, *277*, 130262. https://doi.org/10.1016/j.chemosphere.2021.130262

Sasajima, H., Miyazono, S., Noguchi, T., & Kashiwayanagi, M. (2015). Intranasal administration of rotenone in mice attenuated olfactory functions through the lesion of dopaminergic neurons in the olfactory bulb. *Neurotoxicology*, *51*, 106–115. https://doi.org/10.1016/j.neuro.2015.10.006

Sato, Y., Miyasaka, N., & Yoshihara, Y. (2005). Mutually exclusive glomerular innervation by two distinct types of olfactory sensory neurons revealed in transgenic zebrafish. *Journal of Neuroscience*, *25*(20), 4889–4897. https://doi.org/10.1523/JNEUROSCI.0679-05.2005

Sattler, R., Ayukawa, Y., Coddington, L., Sawa, A., Block, D., Chipkin, R., & Rothstein, J. D. (2011). Human nasal olfactory epithelium as a dynamic marker for CNS therapy development. *Experimental Neurology*, *232*(2), 203–211. https://doi.org/10.1016/j.expneurol.2011.09.002

Savya, S. P., Kunkhyen, T., & Cheetham, C. E. J. (2019). Low survival rate of young adult-born olfactory sensory neurons in the undamaged mouse olfactory epithelium. *Journal of Bioenergetics and Biomembranes*, *51*(1), 41–51. https://doi.org/10.1007/s10863-018-9774-8

Scadding, G. K. (2008). Rhinitis and sinusitis. *Clinical Respiratory Medicine*, *2008*, 409–423. https://doi.org/10.1016/B978-032304825-5.10032-7

Scheib, J. J., Pozzuto, J. M., & Byrd-Jacobs, C. A. (2019). Reversible deafferentation of the zebrafish olfactory bulb with wax plug insertion. *Journal of Neuroscience Methods*, *311*, 47–56. https://doi.org/10.1016/j.jneumeth.2018.10.014

Schild, D., & Restrepo, D. (1998). Transduction mechanisms in vertebrate olfactory receptor cells. *Physiological Reviews*, *78*(2), 429–466. https://doi.org/10.1152/physrev.1998.78.2.429

Schnittke, N., Herrick, D. B., Lin, B., Peterson, J., Coleman, J. H., Packard, A. I., Jang, W., & Schwob, J. E. (2015). Transcription factor p63 controls the reserve status but not the stemness of horizontal basal cells in the olfactory epithelium. *Proceedings of the National Academy of Sciences*, *112*(36), E5068–E5077. https://doi.org/10.1073/pnas.1512272112

Schwob, J. E. (2005). Restoring olfaction: A view from the olfactory epithelium. *Chemical Senses*, *30*(Suppl 1), i131–i132. https://doi.org/10.1093/chemse/bjh149

Schwob, J. E., Jang, W., Holbrook, E. H., Lin, B., Herrick, D. B., Peterson, J. N., & Hewitt Coleman, J. (2017). Stem and progenitor cells of the mammalian olfactory epithelium: Taking poietic license. *Journal of Comparative Neurology*, *525*(4), 1034–1054. https://doi.org/10.1002/cne.24105

Schwob, J. E., Szumowski, K. E., & Stasky, A. A. (1992). Olfactory sensory neurons are trophically dependent on the olfactory bulb for their prolonged survival. *Journal of Neuroscience*, *12*(10), 3896–3919. https://doi.org/10.1523/JNEUROSCI.12-10-03896.1992

Seeley, W. W., Carlin, D. A., Allman, J. M., Macedo, M. N., Bush, C., Miller, B. L., & DeArmond, S. J. (2006). Early frontotemporal dementia targets neurons unique to apes and humans. *Annals of Neurology*, *60*(6), 660–667. https://doi.org/10.1002/ana.21055

Senf, K., Karius, J., Stumm, R., & Neuhaus, E. M. (2021). Chemokine signaling is required for homeostatic and injury-induced neurogenesis in the olfactory epithelium. *Stem Cells*, *39*(5), 617–635. https://doi.org/10.1002/stem.3338

Sepahi, A., & Salinas, I. (2016). The evolution of nasal immune systems in vertebrates. *Molecular Immunology*, *69*, 131–138. https://doi.org/10.1016/j.molimm.2015.09.008

Sepahi, A., Kraus, A., Casadei, E., Johnston, C. A., Galindo-Villegas, J., Kelly, C., García-Moreno, D., Muñoz, P., Mulero, V., Huertas, M., & Salinas, I. (2019). Olfactory sensory neurons mediate ultrarapid antiviral immune responses in a TrkA-dependent manner. *Proceedings of the National Academy of Sciences*, *116*(25), 12428–12436. https://doi.org/10.1073/pnas.1900083116

Shang, M., Tang, M., & Xue, Y. (2023). Neurodevelopmental toxicity induced by airborne particulate matter. *Journal of Applied Toxicology*, *43*(1), 167–185. https://doi.org/10.1002/jat.4382

Sharrock, A. V., Mulligan, T. S., Hall, K. R., Williams, E. M., White, D. T., Zhang, L., Emmerich, K., Matthews, F., Nimmagadda, S., Washington, S., Le, K. D., Meir-Levi, D., Cox, O. L., Saxena, M. T., Calof, A. L., Lopez-Burks, M. E., Lander, A. D., Ding, D., Ji, H., & Ackerley, D. F. (2022). NTR 2.0: A rationally engineered prodrug-converting enzyme with substantially enhanced efficacy for targeted cell ablation. *Nature Methods*, *19*(2), 205–215. https://doi.org/10.1038/s41592-021-01364-4

Shi, W., Sun, S., Han, Y., Tang, Y., Zhou, W., Du, X., & Liu, G. (2021). Microplastics impair olfactory-mediated behaviors of goldfish Carassius auratus. *Journal of Hazardous Materials*, *409*, 125016. https://doi.org/10.1016/j.jhazmat.2020.125016

Shohayeb, B., Diab, M., Ahmed, M., & Ng, D. C. H. (2018). Factors that influence adult neurogenesis as potential therapy. *Translational Neurodegeneration*, *7*, 4. https://doi.org/10.1186/s40035-018-0109-9

Siddiqui, R., Ali, I. K. M., Cope, J. R., & Khan, N. A. (2016). Biology and pathogenesis of *Naegleria fowleri*. *Acta Tropica*, *164*, 375–394. https://doi.org/10.1016/j.actatropica.2016.09.009

Sieger, D., Moritz, C., Ziegenhals, T., Prykhozhij, S., & Peri, F. (2012). Long-range Ca2+ waves transmit brain-damage signals to microglia. *Developmental Cell*, *22*(6), 1138–1148. https://doi.org/10.1016/j.devcel.2012.04.012

Silva Brito, R., Canedo, A., Farias, D., & Rocha, T. L. (2022). Transgenic zebrafish (Danio rerio) as an emerging model system in ecotoxicology and toxicology: Historical review, recent advances, and trends. *Science of The Total Environment*, *848*, 157665. https://doi.org/10.1016/j.scitotenv.2022.157665

Simpson, P. J., Wang, E., Moon, C., Matarazzo, V., Cohen, D. R., Liebl, D. J., & Ronnett, G. V. (2003). Neurotrophin-3 signaling maintains maturational homeostasis between neuronal populations in the olfactory epithelium. *Molecular and Cellular Neuroscience*, *24*(4), 858–874. https://doi.org/10.1016/j.mcn.2003.08.001

Singleman, C., & Holtzman, N. G. (2014). Growth and maturation in the zebrafish, *Danio rerio*: A staging tool for teaching and research. *Zebrafish*, *11*(4), 396–406. https://doi.org/10.1089/zeb.2014.0976

Sireci, S., Kocagöz, Y., Alkiraz, A. S., Güler, K., Dokuzluoglu, Z., Balcioglu, E., Meydanli, S., Demirler, M. C., Erdogan, N. S., & Fuss, S. H. (2024). HB-EGF promotes progenitor cell proliferation and sensory neuron regeneration in the zebrafish olfactory epithelium. *FEBS Journal*, *291*(10), 2098–2133. https://doi.org/10.1111/febs.17033.

Soares, A. T. G., Silva, A. C., Tinkov, A. A., Khan, H., Santamaría, A., Skalnaya, M. G., Skalny, A. V., Tsatsakis, A., Bowman, A. B., Aschner, M., & Ávila, D. S. (2020). The impact of manganese on neurotransmitter systems. *Journal of Trace Elements in Medicine and Biology*, *61*, 126554. https://doi.org/10.1016/j.jtemb.2020.126554

Soliman, A. M., & Barreda, D. R. (2022). Acute inflammation in tissue healing. *International Journal of Molecular Sciences*, *24*(1), 641. https://doi.org/10.3390/ijms24010641

Song, S. J., Park, B., Jo, K., & Kim, C. S. (2021). Damage to olfactory organs of adult zebrafish induced by diesel particulate matter. *International Journal of Molecular Sciences*, *23*(1), 407. https://doi.org/10.3390/ijms23010407

Spurgeon, D., Lahive, E., Robinson, A., Short, S., & Kille, P. (2020). Species sensitivity to toxic substances: Evolution, ecology and applications. *Frontiers in Environmental Science*, *8*, 588380. https://doi.org/10.3389/fenvs.2020.588380

Stieger, K., Belbellaa, B., Le Guiner, C., Moullier, P., & Rolling, F. (2009). In vivo gene regulation using tetracycline-regulatable systems. *Advanced Drug Delivery Reviews*, *61*(7–8), 527–541. https://doi.org/10.1016/j.addr.2008.12.016

Stoklund Dittlau, K., & Freude, K. (2024). Astrocytes: The stars in neurodegeneration? *Biomolecules*, *14*(3), 289. https://doi.org/10.3390/biom14030289

Stubblefield, W. A., Van Genderen, E., Cardwell, A. S., Heijerick, D. G., Janssen, C. R., & De Schamphelaere, K. A. C. (2020). Acute and chronic toxicity of cobalt to freshwater organisms: Using a species sensitivity distribution approach to establish international water quality standards. *Environmental Toxicology and Chemistry*, *39*(4), 799–811. https://doi.org/10.1002/etc.4662

Sueda, R., & Kageyama, R. (2020). Regulation of active and quiescent somatic stem cells by Notch signaling. *Development, Growth & Differentiation*, *62*(1), 59–66. https://doi.org/10.1111/dgd.12626

Surmeier, D. J. (2018). Determinants of dopaminergic neuron loss in Parkinson's disease. *FEBS Journal*, *285*(19), 3657–3668. https://doi.org/10.1111/febs.14607

Suttkus, A., Morawski, M., & Arendt, T. (2016). Protective properties of neural extracellular matrix. *Molecular Neurobiology*, *53*(1), 73–82. https://doi.org/10.1007/s12035-014-8990-4

Suttkus, A., Rohn, S., Weigel, S., Glöckner, P., Arendt, T., & Morawski, M. (2014). Aggrecan, link protein and tenascin-R are essential components of the perineuronal net to protect neurons against iron-induced oxidative stress. *Cell Death & Disease*, *5*(3), e1119. https://doi.org/10.1038/cddis.2014.25

Sweeney, M. D., Zhao, Z., Montagne, A., Nelson, A. R., & Zlokovic, B. V. (2019). Blood-brain barrier: From physiology to disease and back. *Physiological Reviews*, *99*(1), 21–78. https://doi.org/10.1152/physrev.00050.2017

Tacchi, L., Musharrafieh, R., Larragoite, E. T., Crossey, K., Erhardt, E. B., Martin, S. A. M., LaPatra, S. E., & Salinas, I. (2014). Nasal immunity is an ancient arm of the mucosal immune system of vertebrates. *Nature Communications*, *5*, 5205. https://doi.org/10.1038/ncomms6205

Takenobu, H., Yamazaki, A., Hirata, M., Umata, T., & Mekada, E. (2003). The stress- and inflammatory cytokine-induced ectodomain shedding of heparin-binding epidermal growth factor-like growth factor is mediated by p38 MAPK, distinct from the 12-O-tetradecanoylphorbol-13-acetate- and lysophosphatidic acid-induced signaling cascades. *Journal of Biological Chemistry*, *278*(19), 17255–17262. https://doi.org/10.1074/jbc.M211835200

Takesono, A., Dimitriadou, S., Clark, N. J., Handy, R. D., Mourabit, S., Winter, M. J., Kudoh, T., & Tyler, C. R. (2023). Zinc oxide nanoparticles disrupt development and function of the olfactory sensory system impairing olfaction-mediated behaviour in zebrafish. *Environmental International*, *180*, 108227. https://doi.org/10.1016/j.envint.2023.108227

Tan, L., Li, Q., & Xie, X. S. (2015). Olfactory sensory neurons transiently express multiple olfactory receptors during development. *Molecular Systems Biology*, *11*(12), 844. https://doi.org/10.15252/msb.20156639

Tatsumi, K., Okuda, H., Makinodan, M., Yamauchi, T., Makinodan, E., Matsuyoshi, H., Manabe, T., & Wanaka, A. (2010). Transient activation of Notch signaling in the injured adult brain. *Journal of Chemical Neuroanatomy*, *39*(1), 15–19. https://doi.org/10.1016/j.jchemneu.2009.09.003

Taupin, P. (2007). BrdU immunohistochemistry for studying adult neurogenesis: Paradigms, pitfalls, limitations, and validation. *Brain Research Reviews*, *53*(1), 198–214. https://doi.org/10.1016/j.brainresrev.2006.08.002

Tian, H., & Ma, M. (2008). Activity plays a role in eliminating olfactory sensory neurons expressing multiple odorant receptors in the mouse septal organ. *Molecular Cellular Neuroscience*, *38*(4), 484–488. https://doi.org/10.1016/j.mcn.2008.04.006

Tierney, K. B., Taylor, A. L., Ross, P. S., & Kennedy, C. J. (2006). The alarm reaction of coho salmon parr is impaired by the carbamate fungicide IPBC. *Aquatic Toxicology*, *79*(2), 149–157. https://doi.org/10.1016/j.aquatox.2006.06.003

Tierney, K. B., Baldwin, D. H., Hara, T. J., Ross, P. S., Scholz, N. L., & Kennedy, C. J. (2010). Olfactory toxicity in fishes. *Aquatic Toxicology*, *96*(1), 2–26. https://doi.org/10.1016/j.aquatox.2009.09.019

Tierney, K. B., Singh, C. R., Ross, P. S., & Kennedy, C. J. (2007). Relating olfactory neurotoxicity to altered olfactory-mediated behaviors in rainbow trout exposed to three currently-used pesticides. *Aquatic Toxicology*, *81*(1), 55–64. https://doi.org/10.1016/j.aquatox.2006.11.006

Tilton, F. A., Tilton, S. C., Bammler, T. K., Beyer, R., Farin, F., Stapleton, P. L., & Gallagher, E. P. (2008). Transcriptional biomarkers and mechanisms of copper-induced olfactory injury in zebrafish. *Environmental Science & Technology*, *42*(24), 9404–9411. https://doi.org/10.1021/es801636v

Tilton, F. A., Tilton, S. C., Bammler, T. K., Beyer, R. P., Stapleton, P. L., Scholz, N. L., & Gallagher, E. P. (2011). Transcriptional impact of organophosphate and metal mixtures on olfaction: Copper dominates the chlorpyrifos-induced response in adult zebrafish. *Aquatic Toxicology*, *102*(3–4), 205–215. https://doi.org/10.1016/j.aquatox.2011.01.012

Torres, M. V., Ortiz-Leal, I., & Sanchez-Quinteiro, P. (2023). Pheromone sensing in mammals: A review of the vomeronasal system. *Anatomia*, *2*(4), 346–413. https://doi.org/10.3390/anatomia2040031

Traut, T. W. (1994). Physiological concentrations of purines and pyrimidines. *Molecular and Cellular Biochemistry*, *140*(1), 1–22. https://doi.org/10.1007/BF00928361

Tsioumpekou, M., Krijgsman, D., Leusen, J. H. W., & Olofsen, P. A. (2023). The role of cytokines in neutrophil development, tissue homing, function and plasticity in health and disease. *Cells*, *12*(15), 1981. https://doi.org/10.3390/cells12151981

Tsukahara, T., Brann, D. H., & Datta, S. R. (2023). Mechanisms of SARS-CoV-2-associated anosmia. *Physiological Reviews*, *103*(4), 2759–2766. https://doi.org/10.1152/physrev.00012.2023

Ueno, H., Fujii, K., Suemitsu, S., Murakami, S., Kitamura, N., Wani, K., Aoki, S., Okamoto, M., Ishihara, T., & Takao, K. (2018). Expression of aggrecan components in perineuronal nets in the mouse cerebral cortex. *IBRO Reports*, *4*, 22–37. https://doi.org/10.1016/j.ibror.2018.01.002

Vénéreau, E., Ceriotti, C., & Bianchi, M. E. (2015). DAMPs from cell death to new life. *Frontiers in Immunology*, *6*, 422. https://doi.org/10.3389/fimmu.2015.00422

Vieira, H. C., Rodrigues, A. C. M., Soares, A. M. V. M., Abreu, S., & Morgado, F. (2021). Mercury accumulation and elimination in different tissues of zebrafish (Danio rerio) exposed to a mercury-supplemented diet. *Journal of Marine Science and Engineering*, *9*(8), 882. https://doi.org/10.3390/jmse9080882

Volz, S. N., Hausen, J., Nachev, M., Ottermanns, R., Schiwy, S., & Hollert, H. (2020a). Short exposure to cadmium disrupts the olfactory system of zebrafish (Danio rerio) - Relating altered gene expression in the olfactory organ to behavioral deficits. *Aquatic Toxicology*, *226*, 105555. https://doi.org/10.1016/j.aquatox.2020.105555

Volz, S. N., Hausen, J., Smith, K., Ottermanns, R., Schaeffer, A., Schiwy, S., & Hollert, H. (2020b). Do you smell the danger? Effects of three commonly used pesticides on the olfactory-mediated antipredator response of zebrafish (Danio rerio). *Chemosphere*, *241*, 124963. https://doi.org/10.1016/j.chemosphere.2019.124963

Voronkov, D. N., Kutukova, K. A., Ivanov, M. V., & Khudoerkov, R. M. (2017). Immunomorphological changes in the olfactory bulbs of rats after intranasal administration of rotenone. *Bulletin of Experimental Biology and Medicine*, *164*(2), 203–206. https://doi.org/10.1007/s10517-017-3958-9

Wachowiak, M. (2011). All in a sniff: Olfaction as a model for active sensing. *Neuron*, *71*(6), 962–973. https://doi.org/10.1016/j.neuron.2011.08.030

Wakaisaka, N., Miyasaka, N., Koide, T., Masuda, M., Hiraki-Kajiyama, T., & Yoshihara, Y. (2017). An adenosine receptor for olfaction in fish. *Current Biology*, *27*(10), 1437–1447.e4. https://doi.org/10.1016/j.cub.2017.04.014

Wan, J., Ramachandran, R., & Goldman, D. (2012). HB-EGF is necessary and sufficient for Müller glia dedifferentiation and retina regeneration. *Developmental Cell*, *22*(2), 334–347. https://doi.org/10.1016/j.devcel.2011.11.020

Wang, L., & Gallagher, E. P. (2013). Role of Nrf2 antioxidant defense in mitigating cadmium-induced oxidative stress in the olfactory system of zebrafish. *Toxicology and Applied Pharmacology*, *266*(2), 177–186. https://doi.org/10.1016/j.taap.2012.11.010

Wang, Q., Titlow, W. B., McClintock, D. A., Stromberg, A. J., & McClintock, T. S. (2017). Activity-dependent gene expression in the mammalian olfactory epithelium. *Chemical Senses*, *42*(8), 611–624. https://doi.org/10.1093/chemse/bjx028

Wang, X., & Michaelis, E. K. (2010). Selective neuronal vulnerability to oxidative stress in the brain. *Frontiers in Aging Neuroscience*, *2*, 12. https://doi.org/10.3389/fnagi.2010.00012

Wang, Y., Liu, W., Yang, J., Wang, F., Sima, Y., Zhong, Z. M., Wang, H., Hu, L. F., & Liu, C. F. (2017). Parkinson's disease-like motor and non-motor symptoms in rotenone-treated zebrafish. *Neurotoxicology*, *58*, 103–109. https://doi.org/10.1016/j.neuro.2016.11.006

Watanabe, K., Sato, E., Mishima, E., Watanabe, M., Abe, T., Takahashi, N., & Nakayama, M. (2021). Effect of uremic toxins on hippocampal cell damage: Analysis in vitro and in a rat model of chronic kidney disease. *Heliyon*, *7*(2), e06221. https://doi.org/10.1016/j.heliyon.2021.e06221

Watt, W. C., Sakano, H., Lee, Z. Y., Reusch, J. E., Trinh, K., & Storm, D. R. (2004). Odorant stimulation enhances survival of olfactory sensory neurons via MAPK and CREB. *Neuron*, *41*(6), 955–967. https://doi.org/10.1016/s0896-6273(04)00075-3

Wee, P., & Wang, Z. (2017). Epidermal growth factor receptor cell proliferation signaling pathways. *Cancers*, *9*(5), 52. https://doi.org/10.3390/cancers9050052

Wellford, S. A., Chen, C. W., Vukovic, M., Batich, K. A., Lin, E., Shalek, A. K., Ordovas-Montanes, J., Park Moseman, A., & Ashley Moseman, E. (2024). Distinct olfactory mucosal macrophage populations mediate neuronal maintenance and pathogen defense. *Mucosal Immunology*, S1933-0219(24)00075-8. https://doi.org/10.1016/j.mucimm.2024.07.009

Wellford, S. A., & Moseman, E. A. (2024). Olfactory immunology: The missing piece in airway and CNS defense. *Nature Reviews Immunology*, *24*(6), 381–398. https://doi.org/10.1038/s41577-023-00972-9

Werner, S., & Nies, E. (2018). Olfactory dysfunction revisited: A reappraisal of work-related olfactory dysfunction caused by chemicals. *Journal of Occupational Medicine and Toxicology*, *13*, 28. https://doi.org/10.1186/s12995-018-0209-6

Weil, M., Scholz, S., Zimmer, M., Sacher, F., & Duis, K. (2009). Gene expression analysis in zebrafish embryos: A potential approach to predict effect concentrations in the fish early life stage test. *Environmental Toxicology and Chemistry*, *28*(9), 1970–1978. https://doi.org/10.1897/08-627.1

Wilson, D. M. III, Cookson, M. R., Van Den Bosch, L., Zetterberg, H., Holtzman, D. M., & Dewachter, I. (2023). Hallmarks of neurodegenerative diseases. *Cell*, *186*(4), 693–714. https://doi.org/10.1016/j.cell.2022.12.032

Winkley, K., Banerjee, D., Bradley, T., Koseva, B., Cheung, W. A., Selvarangan, R., Pastinen, T., & Grundberg, E. (2021). Immune cell residency in the nasal mucosa may partially explain respiratory disease severity across the age range. *Scientific Reports*, *11*(1), 15927. https://doi.org/10.1038/s41598-021-95532-3

Wu, H. H., Ivkovic, S., Murray, R. C., Jaramillo, S., Lyons, K. M., Johnson, J. E., & Calof, A. L. (2003). Autoregulation of neurogenesis by GDF11. *Neuron*, *37*(2), 197–207. https://doi.org/10.1016/s0896-6273(02)01172-8

Xu, H., Li, C., Li, Y., Ng, G. H., Liu, C., Zhang, X., & Gong, Z. (2015). Generation of Tg(cyp1a) transgenic zebrafish for development of a convenient and sensitive in vivo assay for aryl hydrocarbon receptor activity. *Marine Biotechnology (New York)*, *17*(6), 831–840. https://doi.org/10.1007/s10126-015-9669-1

Xu, J., Wang, T., Wu, Y., Jin, W., & Wen, Z. (2016). Microglia colonization of developing zebrafish midbrain is promoted by apoptotic neuron and lysophosphatidylcholine. *Developmental Cell*, *38*(2), 214–222. https://doi.org/10.1016/j.devcel.2016.06.018

Xu, Y., Bi, W. D., Shi, Y. X., Liang, X. R., Wang, H. Y., Lai, X. L., Bian, X. L., & Guo, Z. Y. (2023). Derivation and elimination of uremic toxins from kidney-gut axis. *Frontiers in Physiology*, *14*, 1123182. https://doi.org/10.3389/fphys.2023.1123182

Yabuki, Y., Koide, T., Miyasaka, N., Wakisaka, N., Masuda, M., Ohkura, M., Nakai, J., Tsuge, K., Tsuchiya, S., Sugimoto, Y., & Yoshihara, Y. (2016). Olfactory receptor for prostaglandin F2α mediates male fish courtship behavior. *Nature Neuroscience*, *19*(7), 897–904. https://doi.org/10.1038/nn.4314

Yang, A., Schweitzer, R., Sun, D., Kaghad, M., Walker, N., Bronson, R. T., Tabin, C., Sharpe, A., Caput, D., Crum, C., & McKeon, F. (1999). p63 is essential for regenerative proliferation in limb, craniofacial and epithelial development. *Nature*, *398*(6729), 714–718. https://doi.org/10.1038/19539

Yang, D., Lauridsen, H., Buels, K., Chi, L. H., La Du, J., Bruun, D. A., Olson, J. R., Tanguay, R. L., & Lein, P. J. (2011). Chlorpyrifos-oxon disrupts zebrafish axonal growth and motor behavior. *Toxicological Sciences*, *121*(1), 146–159. https://doi.org/10.1093/toxsci/kfr028

Yoshikawa, K., Wang, H., Jaen, C., Haneoka, M., Saito, N., Nakamura, J., Adappa, N. D., Cohen, N. A., & Dalton, P. (2018). The human olfactory cleft mucus proteome and its age-related changes. *Scientific Reports*, *8*(1), 17170. https://doi.org/10.1038/s41598-018-35102-2

You, R., Ho, Y. S., & Chang, R. C. (2022). The pathogenic effects of particulate matter on neurodegeneration: A review. *Journal of Biomedical Science*, *29*(1), 15. https://doi.org/10.1186/s12929-022-00799-x

Yu, C. R., & Wu, Y. (2017). Regeneration and rewiring of rodent olfactory sensory neurons. *Experimental Neurology*, 287(Pt 3), 395–408. https://doi.org/10.1016/j.expneurol.2016.06.001

Zhang, R., Zhang, L., Li, P., Pang, K., Liu, H., & Tian, L. (2023). Epithelial barrier in the nasal mucosa, related risk factors and diseases. *International Archives of Allergy and Immunology*, *184*(5), 481–501. https://doi.org/10.1159/000528969

Zhang, X., & Gong, Z. (2013). Fluorescent transgenic zebrafish Tg(nkx2.2a) provides a highly sensitive monitoring tool for neurotoxins. *PLoS One*, *8*(2), e55474. https://doi.org/10.1371/journal.pone.0055474

Ziani, K., Ioniţă-Mîndrican, C. B., Mititelu, M., Neacşu, S. M., Negrei, C., Moroşan, E., Drăgănescu, D., & Preda, O. T. (2023). Microplastics: A real global threat for environment and food safety: A state of the art review. *Nutrients*, *15*(3), 617. https://doi.org/10.3390/nu15030617

Zimmer, T. S., Orr, A. L., & Orr, A. G. (2024). Astrocytes in selective vulnerability to neurodegenerative disease. *Trends in Neurosciences*, *47*(4), 289–302. https://doi.org/10.1016/j.tins.2024.02.008

# 7

# *Epigenetic Regulations During Neurogenesis in Zebrafish*

**Samudra Gupta, Aoishik Roy and Subhra Prakash Hui**

## Abbreviations

| | |
|---|---|
| **EdU** | 5-Ethynyl-2′-Deoxyuridine |
| **BBB** | Blood–Brain Barrier |
| **BMP** | Bone Morphogenetic Protein |
| **BLBP** | Brain Lipid-Binding Protein |
| **BrdU** | Bromodeoxyuridine |
| **CNS** | Central Nervous System |
| **Dpf** | Days Post-Fertilization |
| **Dpi** | Days Post-Injury |
| **DG** | Dentate Gyrus |
| **DMB** | Differentially Methylated Base |
| **DNMT** | DNA Methyltransferase |
| **DV** | Dorso-Ventral |
| **ERG** | Ependymo-Radial Glia |
| **EV** | Extracellular Vesicles |
| **FGF** | Fibroblast Growth Factor |
| **GFAP** | Glial Fibrillary Acidic Protein |
| **HAT** | Histone Acetyltransferase |
| **HDAC** | Histone Deacetylase |
| **hpf** | Hours Post Fertilization |
| **hpi** | Hours Post Injury |
| **MBD** | Methyl-CpG Binding Domain |
| **miRNAs** | microRNAs |
| **MG** | Müller Glia |
| **MGPC** | Müller Glial Progenitor Cells |
| **NPC** | Neural Progenitor Cells |
| **NSC** | Neural Stem Cells |
| **NEC** | Neuro-Epithelial Cells |
| **NMDA** | N-Methyl-D-Aspartate |
| **ncRNA** | Non-Coding RNA |
| **OB** | Olfactory Bulb |
| **OPC** | Oligodendrocyte Precursor Cells |
| **PCNA** | Proliferating Cell Nuclear Antigen |
| **RGC** | Radial Glial Cells |
| **RPC** | Retinal Precursor Cells |
| **RL** | Rhombic Lip |
| **Shh** | Sonic Hedgehog |
| **SCI** | Spinal Cord Injury |
| **SGZ** | Sub Granular Zone |
| **SVZ** | Sub Ventricular Zone |
| **TNF α** | Tumour Necrosis Factor A |
| **UV** | Ultraviolet |
| **VZ** | Ventricular Zone |

## Introduction

The progression whereby undifferentiated neural progenitor cells undergo proliferation and maturation to become fully developed and functional neurons is identified as neurogenesis. Neural progenitors are activated during the initial stages of neurogenesis, and a phase of cell division ensues to augment the progenitor pool. Subsequently, a sequence of events unfolds, whereby committed progenitors are designated and subsequently undergo differentiation into post-mitotic neurons. These processes are synchronized both spatially and temporally to yield diverse types of neural and glial cells, ultimately forming the mature central nervous system (CNS) (Schmidt et al., 2013).

Neurogenesis is a complex process that is essential to many aspects of brain function, which include motor control, memory, learning, and sensory processing. Thus, understanding the mechanisms and regulation of neurogenesis is essential for understanding the complications and the functioning of the nervous system. For proper progression of the neurogenic events, a complex interaction of genetic and epigenetic mechanisms is involved, which is further controlled by various signalling molecules that transmit extracellular cues to the neural progenitors. Zebrafish, distinguished by external fertilization and optically transparent embryos, coupled with the remarkable regenerative capacity of their CNS following injury in both embryonic and adult stages, stands as an exclusive model organism. This distinctive characteristic makes it a vital model system for comprehensive investigations aimed at unravelling the complexities of neurogenesis. Consequently, it provides an exceptional opportunity to explore the cellular and molecular mechanisms governing the dynamic process of neurogenesis.

Epigenetic regulation, the heritable modification of gene expression without alterations in the underlying DNA sequence, has emerged as an important player in influencing the fate of neural progenitors during neurogenesis. In zebrafish, as in other vertebrates, the epigenetic landscape is governed by modifications at the level of DNA and histone, along

DOI: 10.1201/9781032676555-9

with non-coding RNAs, contributing to the precise modulation of gene expression. Understanding the epigenetic landscape in the context of neurogenesis in zebrafish has the potential to reveal conserved mechanisms that can be extrapolated to higher vertebrate groups, including humans.

The enzymatic process of adding methyl groups to cytosine residues within the DNA molecule, facilitated by DNA methyltransferases, serves as a mechanism to suppress gene expression. Investigations in zebrafish have unveiled distinct DNA methylation patterns linked to neurogenic events, underscoring the crucial involvement of this epigenetic modification in influencing the fate of neuronal progenitors during neurogenesis (Powell et al., 2014). Understanding the interplay between DNA methylation patterns and the complex network of extracellular and intracellular molecules regulating the neurogenesis process is crucial for conducting future studies during zebrafish neurogenesis.

Histone modifications, involving post-translational alterations to histone proteins, majorly through methylation, acetylation, and phosphorylation, represent another facet of epigenetic regulation governing chromatin structure and gene accessibility. Histone methylation can induce either transcriptional activation or repression, and this outcome depends on the specific position of the lysine residue in the histone protein and the number of methyl groups added to it (Zhu et al., 2018). Histone acetylation and deacetylation are the two major histone modifications that regulate gene expression patterns among other major histone modifications, which can either enhance or suppress gene expression respectively depending on the methylation of the specific amino acid (Zhu et al., 2018). Histone acetylation involves the addition of acetyl groups to the lysine residues of histone proteins, making the chromatin structure more relaxed and accessible for increased transcriptional activities. Histone deacetylation, on the other hand, removes acetyl group from the histone lysine residues, making the chromatin structure more compact, resulting in transcriptional repression (Feng, Fouse, & Fan, 2007). In zebrafish neurogenesis, numerous researchers have observed specific histone modifications associated with the regulation of key genes that govern neural progenitor activity following injury (Mitra et al., 2018; Kiyooka et al., 2020 ; Shimizu & Kawasaki, 2021; Cavone et al., 2021). Incorporating histone modifications into the broader context of signalling pathways offers a comprehensive perspective on the molecular dynamics orchestrating the activities of neural progenitor cells in zebrafish.

As a significant post-transcriptional modifier, non-coding RNAs, encompassing microRNAs and long non-coding RNAs, play a significant role in contributing to the epigenetic functioning. Specific non-coding RNAs have been identified as crucial regulators of neurogenic processes in zebrafish, influencing the delicate balance between the proliferation and differentiation of neural progenitors, particularly in the context of regenerative neurogenesis (Rajaram et al.2014; Sharma et al., 2019).

Key signalling pathways, including Notch, BMP, and Wnt, among others, communicate spatial and temporal information to developing neural cells. In zebrafish, the observed spatiotemporal regulation of these signalling pathways plays a crucial role in precisely controlling neurogenesis both in developmental and regenerative contexts throughout the CNS (Ramachandran et al., 2011; Alunni et al., 2013; Cardozo et al., 2017). The proliferation, maturation, and differentiation of neuronal progenitors during neurogenesis in the zebrafish CNS are governed by the temporal activation and coordinated functioning of signalling pathways. These pathways involve a synergy of genetic and epigenetic modulators (Zhu et al., 2018).

Although studies on the epigenetic regulation of neurogenesis in mammals have been well-documented (Covic et al., 2010; Sun et al., 2011; Murao et al., 2016), a comprehensive investigation of this process in the zebrafish CNS is still lacking. This chapter aims to provide an overview of the interplay between epigenetic regulation and signalling molecules governing neurogenesis in zebrafish through the integration of insights derived from recent studies. The subsequent sections of this chapter delve into the process of developmental and regenerative neurogenesis occurring in the various regions of zebrafish CNS involving specific epigenetic mechanisms and their cross-talks with different signalling pathways. By deciphering the reported epigenetic mechanisms that govern neurogenesis in zebrafish, we can further explore key regulatory mechanisms involved in other organisms, including humans, a promising endeavour for developing therapeutic strategies in humans to address various neurological conditions and advance the regenerative medicine field.

## Neurogenesis During Development and Regeneration

A set of intricately synchronized processes known as neurogenesis involves the formation, maturation, and integration of neural progenitor cells which ultimately aid in the development of a fully operational nervous system. Neurogenesis occurs at many locations and throughout discrete developmental stages in the vertebrate nervous system. Even though it is most active during the embryonic stages, neurogenesis is nevertheless present in several higher vertebrate species with reduced intensity and in many lower vertebrates with higher intensity into adulthood. Early in the development of vertebrate embryos, the ectoderm initiates neural destiny in response to signals from the underlying notochord. During the initial stage of neurogenesis, neuro-epithelial cells (NEC) within the neural tube wall undergo a transformational process that results in their transformation into radial glial cells (RGC). Following this, the RGCs continue to mature, becoming glial cells that give birth to all different kinds of neurons by direct or indirect pathways (Paridaen & Huttner, 2014).

All vertebrates undergo adult neurogenesis, representing an evolutionarily conserved process. However, there are notable differences in neurogenic niches, neural stem cell identity, and neural plasticity in both normal and injury-induced scenarios (Ghaddar et al., 2021). The CNS generates new neurons continuously within active neurogenic zones. These newly generated neurons can be stimulated for post-injury neuronal repair, although there are differences in repair efficiency across species (Alunni & Bally-Cuif, 2016).

Regenerative neurogenesis is a process that facilitates the restoration of neuronal production in various vertebrate groups following neuronal damage. Higher mammals typically exhibit limited regenerative potential, whereas lower vertebrates, such as fish and amphibians like axolotl and xenopus, undergo extensive regenerative neurogenesis (Kizil et al., 2012; Hui et al., 2013; Diaz Quiroz & Echeverri, 2013; Ghosh & Hui, 2018; Diotel et al., 2020; Rubenstein et al., 2020; Becker & Becker, 2022).

Regenerative neurogenesis in mammals is confined to specific centres within the CNS. In the case of zebrafish, they possess the capability for extensive regenerative neurogenesis, primarily attributed to the presence of a pluripotent non-neuronal cell type known as radial glia. This characteristic is observed in their adult CNS during both embryonic and adult life stages (Figure 7.1) (Ghosh & Hui, 2016).

## Brain

### *Developmental Neurogenesis in Brain*

Neurogenesis in the vertebrate brain encompasses several stages, starting from progenitor proliferation and progressing through neuronal differentiation and maturity. This process results in the generation of a diverse array of neuronal and glial cell types, achieved through expansion, divergence, and terminal division (Raj et al., 2020). NECs are the earliest neurogenic cells in the growing nervous system of mammals. During the development of brain ventricles, NECs mature into RGCs. Previously, RGCs were recognized as a diverse population that aids in the migration of forming neurons (Pinto & Götz, 2007), but recent studies showed that RGCs can also act like neural stem cells (NSC) to yield nearly all the neurons of brain. Apart from that, RGCs produce glial descendants, such as oligodendrocytes and ependymal cells (Diotel et al., 2020). The majority of the cells are not neurogenic in nature, while only RGC-astrocyte-like cells, which reside in specific telencephalic regions – primarily the sub-ventricular zone (SVZ) of the lateral ventricles and the sub-granular zone (SGZ) of the dentate gyrus (DG) of the hippocampus – maintain their neurogenic qualities into adulthood (Dennis et al., 2016; Fares et al., 2019).

Compared to mammals, the zebrafish brain exhibits heightened neurogenic activity in both embryonic and adult conditions. This enhanced neurogenic capacity is attributed to the presence of numerous proliferative niches distributed throughout almost every brain subdivision (Diotel et al. 2020). These proliferative sites correspond essentially to neurogenic niches and are mainly located along the ventricular layers of telencephalon, diencephalon, and rhombencephalon, which also comprise the equivalent regions of two neurogenic niches of mammalian brain, SGZ and SVZ (Grandel et al., 2006; Zupanc & Zupanc, 2006; März et al., 2010). These regions are designated as progenitor pools, which will, in a developmental time period, give rise to the main neurogenic sites including the olfactory bulb, pre-optic region, and optic tectum, which harbour neural progenitors (Hans et al., 2004; Geling et al., 2004; Bae, 2006; Scholpp et al., 2009).

Primary neurogenesis occurs at early developmental stages, while secondary neurogenesis establishes the principal neural framework during late embryonic and larval stages (Manto et al., 2012). It requires strict regulation of self-renewal and differentiation rates of neural progenitors. Numerous signalling pathways and morphogens also aid in the proper proliferation and differentiation of neuronal progenitors and maturation of newly born neurons (Rubenstein et al., 2020).

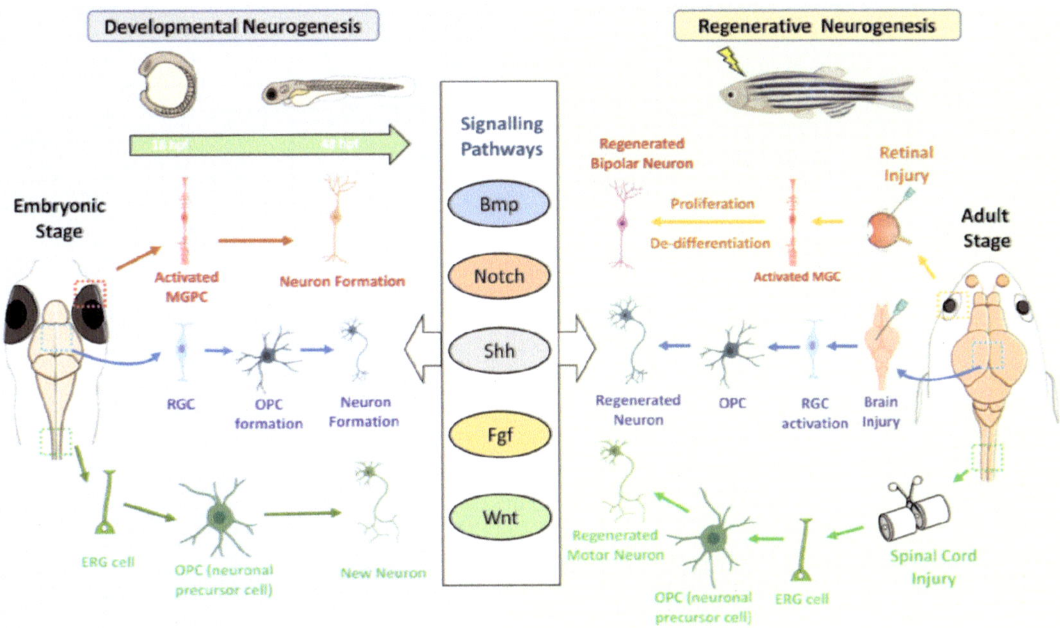


**FIGURE 7.1** Summary of organ-specific key progenitor cells and their development during developmental and regenerative neurogenesis in zebrafish central nervous system and associated major signaling pathways regulating the neurogenesis in both developmental and adult stage of zebrafish. hpf, hour post-injury; RGC, Radial glial cell; MGC, Muller glia cell; MGPC, Muller glia precursor cell; OPC, Oligodendrocyte progenitor cell; ERG, Ependymo-radial glial cell.

### *Regenerative Neurogenesis in Brain*

Progenitor zones with varying levels of occurrence and neurogenic ability have been found in the adult brains of all vertebrates that have been investigated to date; however, the degree of variability of regenerative neurogenesis within the vertebrate group is still enigmatic (Chapouton et al., 2007; Kaslin et al., 2008). Adult mammalian brains have limited neurogenic capacity due to the presence of reduced neural stem cell niches and lineage restriction, while zebrafish exhibit exceptional adult neurogenic potential due to their numerous NSC niches in their brain (Kizil et al., 2012).

The telencephalon, a highly studied and prolific neurogenic niche in the zebrafish brain, comprises ventral and dorsal areas, sub pallium, and olfactory bulbs, with diverse telencephalic progenitors found in one ventral and one dorsal ventricular region (Grandel et al., 2006; März et al., 2010; Kizil et al., 2012 ; Diotel et al., 2020). The ventral progenitor zone is the significant proliferating area which is located dorsoventrally between the dorsal and ventral nuclei of the ventral telencephalon and extends longitudinally from the olfactory bulb (OB) to the anterior commissure (Kizil et al., 2012). Since the cells residing at these locations, which express the glial cell-specific gene markers, are seemed radial and have lengthy processes emanating from their cell bodies, it is likely that they are stem cells similar to radial glia (Grandel & Brand, 2013). The proliferating cells are also located in the periventricular grey region surrounding the peripheral rim of the zebrafish optic tectum (OT) (Grandel et al., 2006; Ito et al., 2010). Progenitor activity is also documented in zebrafish cerebellum (Kizil et al., 2012). Research shows RGCs which originate from the ventricular region undergo proliferation to promote neurogenesis in zebrafish post-injury (Barbosa et al., 2015; Lindsey et al., 2019).

März et al.(2010) identified three primary forms of neurogenic progenitor cells in adult zebrafish telencephalon: type 1, type 2, and type 3. Type 1 cells express RGC markers like GFAP, S100β, BLBP, Vimentin, AroB, nestin, and Sox2 but are non-proliferating in nature, while type 2 cells express the same markers along with PCNA, indicating their proliferative nature. Type 3 are neuroblasts showing progenitor markers but faintly expressing RGC markers and being in proliferating state (März et al., 2010). Brain damage triggers symmetric division of NSCs to generate new neurons, but over time, the number of NSCs decreases, indicating a specific and dynamic response to brain damage (Barbosa et al., 2015).

Injury-induced immune response triggers neurogenesis, which involves microglia activation and infiltration of leukocytes into the injury site. Microglia and other immune cells secrete several proinflammatory molecules like interleukins and other growth factors that are essential for the activation and proliferation of neural progenitor cells (NPC), which subsequently promotes formation of new neurons (Kyritsis et al., 2012; Ghaddar et al., 2021).

## Spinal Cord

### *Developmental Neurogenesis in Spinal Cord*

The vertebrate spinal cord undergoes a complex process of developmental neurogenesis, generating neurons from progenitors during embryonic growth. In the zebrafish spinal cord, there is a distinct spatial organization among different neuronal types. Primary neurons, born early and characterized by their larger size, encompass a variety of cell types, including motor, sensory, and interneurons. These neurons undergo axonogenesis as part of their developmental process. Late-born secondary neurons exhibit characteristics such as smaller size and a higher numerical abundance. These neurons exclusively consist of motor and interneurons. Secondary neurons have finer axons than primary neurons. Motor neurons are primarily located in the ventral position, interneurons in the intermediate position, and sensory neurons in the dorsal position (Ghosh & Hui, 2016).

The spinal cord harbours many proliferating neural precursors. The proliferation rate of the progenitor is uniform across the dorsoventral length of the neural tube and is unaltered by the identity of the precursors (Kicheva et al., 2014). Single-cell RNA sequencing data provides insights into distinct progenitor populations dedicated to each cell type, highlighting the specialized nature of the cellular lineage originating from Olig2-expressing ventral spinal cord progenitor cells (Scott et al., 2021). Progenitor cells in the ventral spinal cord expressing Olig2 exhibit the capacity to generate both motor neurons and oligodendrocyte precursor cells (OPC). Notably, some OPCs maintain their identity as precursor cells, while others undergo further differentiation into myelinating oligodendrocytes. Research indicates that within 1 day post fertilization (dpf), nearly all motor neurons in the spinal cord of zebrafish embryos are generated by progenitors of motor neuron (pMN). This observation highlights the early and predominant role of pMN progenitors in specifying the motor neuron lineage during the initial stages of zebrafish embryonic development (Myers et al., 1986).

### *Regenerative Neurogenesis in Spinal Cord*

The neurons in the main motor cortex of the central nervous system, including the spinal cord, are prone to cell death following injury. Unfortunately, the replacement of these nerve cells is not easily achieved in mammals, representing a challenge in regenerative processes within the CNS of higher vertebrates (Hellenbrand et al., 2021). As compared to mammals, the regenerative ability after spinal cord injury (SCI) in zebrafish is impressively high, and the damage response is characterized by efficient axonal growth, adult neurogenesis, and lack of glial scarring (Hui et al., 2010; Becker & Becker, 2015).

Zebrafish retain radial glia, a type of non-neuronal cell, in their spinal cord, which are often referred to as ependymo-radial glial cells (ERG) and undergo injury-induced proliferation to generate new neurons to accomplish regenerative neurogenesis (Becker & Becker, 2015; Ghosh & Hui, 2016). These cells exhibit diversity; in the undamaged spinal cord, they do not give rise to new neurons. However, they maintain their identity as dorsoventral precursor-region cells established during development (Reimer et al., 2008).

After injury, ventrolateral and surrounding areas express Olig2, leading to the generation of various cell types at different dorsoventral positions in the spinal cord. This suggests that fate-restricted precursor regions have the capability to regenerate different neuronal cell types (Kuscha et al., 2012).

A few ERGs perform a switch from epithelial to mesenchymal in addition to producing neurons, which results in the generation of extended astrocyte-like cells which repair the damaged spinal cord (Klatt Shaw et al., 2021).

The regenerative process is orchestrated by collaborative efforts involving neuronal cells, radial glial cells, and H3P+ cells. Importantly, there is a heightened incorporation of BrdU in close vicinity to the ependyma, indicating increased cellular activity in this region during spinal cord regeneration. These temporal and cellular dynamics contribute to a comprehensive understanding of the events involved in the post-injury regeneration of the spinal cord (Hui et al., 2010).

Two independent studies came up with the same inference that the ventricular zone, or the cells right around the ependymal canal, has the highest rate of cell proliferation (Reimer et al., 2008; Hui et al., 2010).

Within the ventricular area of the ependyma, a subset of cells expressing Hu/NeuroD was observed, suggesting their potential role in migration to the regenerated cords. Remarkably, the damaged spinal cord of the zebrafish demonstrated the capacity to regenerate nearly all damaged axons within a 30-day period (Ghosh & Hui, 2018).

## Retina

### *Developmental Neurogenesis in Retina*

The process of neurogenesis in zebrafish retina begins with retinal ganglionic cells and culminates with Müller glial cells, which together form the laminated structure that is specific to the retina of vertebrates (Jeon, 2007). It has been seen that ganglionic cells are the first cell type to be generated which undergoes their last mitotic divisions at about 24–36 hours post-fertilization (hpf), as shown by the inclusion of the BrdU (a S-phase marker) incorporation. The cells of the inner nuclear layer are produced next; they are born between 36 and 48 hpf (Jeon, 2007). Cone photoreceptor cells emerge later in the developmental timeline, specifically between 48 and 60 hpf (Jeon, 2007). Studies conducted in zebrafish retina at 1 and 2 dpf with BrdU incorporation showed that BrdU labelling was extensively seen in the central retina, especially in the innermost layer, inner sublamina of the inner nuclear layer along with ganglionic layer where the amacrine cells are located. The fact that the cells in the outer and peripheral retinas were unlabelled suggests that they divided after the cells in the inner and centre layers had stopped dividing (Marcus et al., 1999). It is seen that within 3 dpf, neurogenesis in the zebrafish eye's central retina is almost complete and almost within 1 dpf, all the six major kinds of neurons, namely photoreceptor cells, ganglion, bipolar, amacrine, horizontal, and inter-plexiform cells, including the Müller glial cells, are formed (Detrich, 2010). Inner nuclear layer proliferating cells, later described as Müller glia, generate neural progenitors during the course of regenerative neurogenesis and rod precursors during sustained neurogenesis (Lahne et al., 2020). Interkinetic nuclear migration is a stage of the cell cycle progression during which the majority of Müller glia nuclei migrate to the outer nuclear layer. They undergo a single unequal cell division, resulting in the generation of a postmitotic Müller glia and Müller glia–derived progenitors (Lahne et al., 2020). Müller glia in zebrafish serve as multifaceted retinal neural stem cells that produce retinal neurons through restorative and homeostatic developmental processes (Bernardos et al., 2007).

### *Regenerative Neurogenesis in Retina*

Müller glial cells play a crucial role as key retinal neural precursor cells. After injury, they undergo proliferation followed by dedifferentiation to give rise to multipotent retinal progenitors. These cells then migrate to the site of injury, contributing to regenerative neurogenesis in the retina (Bernardos et al., 2007).

Numerous studies on injury-induced retinal regeneration in zebrafish have proposed that Müller glia undergo proliferation and dedifferentiation and express neuronal markers, which is indicative of the fact that Müller glia possess neurogenic abilities (Fausett & Goldman, 2006; Fimbel et al., 2007). Bernardos et al. (2007) through lineage tracing using glial-specific markers showed that the progenitors resulting from the division of Müller glia indeed differentiate into retinal neurons (Bernardos et al., 2007). Their study also showed that Müller glia in zebrafish serve as multipotent retinal stem cells with the capacity to generate various types of retinal neurons. Another study by Powell et al. (2016) showed that following cell death in the retinal nuclear layers due to the application of intravitreal injection of toxin in zebrafish retina, Müller glial cells responded by producing multipotent neural progenitors that migrated to all the nuclear layers and differentiated into specific cell types, suggesting that the Müller Glial progenitors in the injured retina are inherently multipotent (Powell et al., 2016). In a recent study by Lyu et al. (2023), following light-induced and N-methyl D-aspartate (NMDA)–induced toxicity damage to the zebrafish retina using EdU incorporation, it has been seen that neurons derived from Müller glia–derived progenitor cells (MGPC) in these two injury models do not strictly adhere to a cell type–specific pattern. Although they exhibit a consistent inclination toward neurons that are preferentially lost after injury, there is also notable generation of considerable numbers of inner retinal neurons following light damage, and similarly, many photoreceptors are generated following NMDA injury (Lyu et al., 2023).

Subsequent investigations carried out in the larval zebrafish retina have demonstrated the engagement of pro-inflammatory responses, including microglial activation. This activation, in turn, results in the proliferation of Müller glia and subsequent regeneration, particularly in the context of acute and chronic retinal damage (Iribarne & Hyde, 2022). These evidences suggest that zebrafish Müller glia serve as a source of neural progenitor cells that aid in regenerative neurogenesis in injured zebrafish retina.

## Epigenetic Regulation of Neurogenesis

Epigenetic modifications are crucial mechanisms that govern gene silencing and activation at both the transcriptional and translational levels. These modifications play a fundamental role in regulating various biological processes and maintaining cellular identity and function (Zhu et al., 2018). The field of epigenetics studies phenotypic alterations that are self-sustaining, adaptable, and transmissible without effecting DNA's core sequence (Goldberg et al., 2007). Alterations at the level of DNA and histone structure, along with the involvement of non-coding RNAs, play a significant role in regulating gene functioning. These modifications influence transcriptional responses and contribute to our understanding of cell restorative mechanisms in both normal and injured contexts (York et al., 2013; Zhu et al., 2018).

DNA methylation is brought about by the enzyme called DNA methyltransferases (DNMTs) either directly interacting with the transcription factor which binds to DNA promoter, or by attaching with MBD proteins which form interactions with HDAC to transform chromatin into a repressive state. These are the two ways in which DNA methyltransferases function to inhibit the expression of genes (Fan & Hutnick, 2005; Watt & Molloy, 1988).

The lysine residues of the histone protein are acetylated or deacetylated, a reversible process catalyzed by either histone acetyltransferases (HAT) or histone deacetylases (HDAC). While histone acetylation promotes transcription of genes, histone deacetylation suppresses gene transcription (Feng et al., 2007). miRNAs are another class of epigenetic modifier controlling gene activities either directly or through the interaction of DNA and histone modifiers.

Zebrafish and mammals share many critical epigenetic mechanisms and modulators that are evolutionarily conserved (Cavalieri & Spinelli, 2017). In zebrafish, the roles of these epigenetic modifiers have been studied during neurogenesis, mostly in injury-induced contexts. Here in this section, we will try to emphasize the epigenetic modulators affecting neurogenesis in the parts of the zebrafish CNS.

## Epigenetic Regulation of Neurogenesis in Brain

The regulation of regenerative neurogenesis in the zebrafish brain involves not only cues from genetic mechanisms but also significant involvement of epigenetic regulation. Epigenetic processes, such as DNA methylation, histone modifications, and non-coding RNA regulation, play crucial roles in shaping the regenerative capacity of neural progenitor cells and influencing the overall regenerative response in the brain (Kiyooka et al., 2020; Shimizu & Kawasaki, 2021). Studying these modulators and their interactions with signalling molecules is instrumental in regulating the process of neurogenesis in the zebrafish brain.

Studies conducted in the zebrafish optic tectum post injury have demonstrated the involvement of epigenetic modifiers, specifically histone modifiers such as Hdac and Hat, in the process of regenerative neurogenesis in the adult zebrafish brain following injury (Figure 7.2) (Kiyooka et al., 2020; Shimizu & Kawasaki, 2021).

Following a stab injury in the adult zebrafish optic tectum, the injured optic tectum showed a substantial decrease in the expression of both *hdac1* and *hdac3* (Kiyooka et al., 2020). Their study further documented that Hdac inhibition inhibited RGC proliferation while facilitating neuronal differentiation. It was seen that the epigenetic modifier Hdac elucidated its role via the Notch signaling pathway (Kiyooka et al., 2020).

On the contrary, a study was seen conducted using HAT inhibitors in the adult optic tectum which showed a decrease in

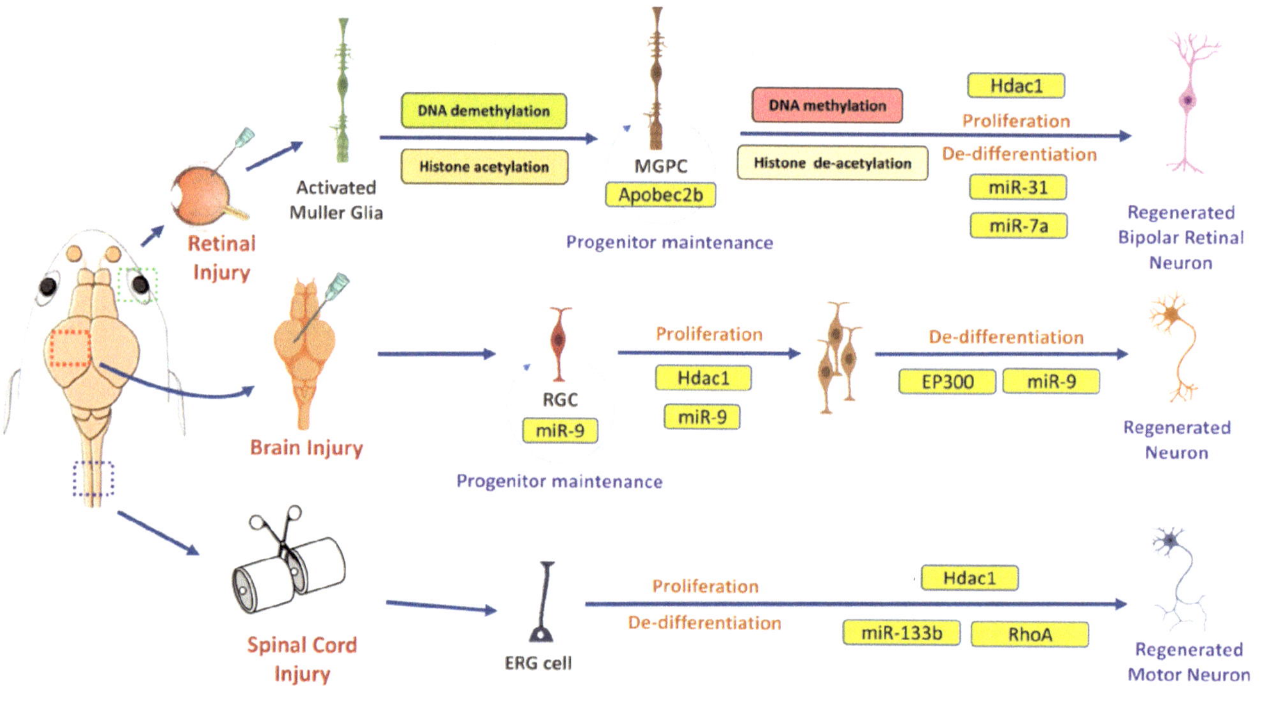


**FIGURE 7.2** Major epigenetic regulators associated with the proliferation and dedifferentiation stages of injury-induced key progenitor cells in different parts of central nervous system in zebrafish. RGC, Radial glial cell; ERG, Ependymo-radial glial cells; MGPC, Müller glial precursor cell; hdac, histone deacetylases; miR, microRNA.

the acetylation of H3K9. It was seen that while RGC proliferation was enhanced by HAT EP300 inhibition, the production of new neurons following injury was reduced. This suggests that EP300 regulates NPC proliferation and differentiation by suppressing histone acetylation and the expression of Notch target genes (Shimizu & Kawasaki, 2021).

microRNAs (miR) were observed as another class of epigenetic modifiers regulating brain neurogenesis. It was seen that in zebrafish brain, *miR-9* is necessary for the maintenance of neural progenitor cells and neurogenesis (Radhakrishnan et al., 2016). *miR-9* governs hindbrain neurogenesis through several mechanisms. It maintains neural progenitor cells by targeting genes such as *zic5* and *her6*, which are involved in progenitor maintenance. Additionally, *miR-9* targets *elavl3/HuC* for neuronal development, thus regulating the differentiation of neural progenitors into neurons in the hindbrain. Overall, *miR-9* plays a crucial role in orchestrating the neurogenic process in the hindbrain by regulating the balance between progenitor maintenance and neuronal differentiation (Radhakrishnan et al., 2016) .

The role of DNA methyltransferases (DNMTs) in neurogenesis of zebrafish brain have been seen through studies. Morpholino-induced knockdown of *dnmt1* and *dnmt3* showed that the presence of Dnmt3 is essential for proper neurogenesis, and its absence leads to significant defects in the zebrafish brain (Rai et al., 2010).

## Epigenetic Regulation of Neurogenesis in Spinal Cord

While extensive research has been conducted to unravel the epigenetic mechanisms governing spinal cord injury (SCI) repair in mammals, the exploration of similar processes in zebrafish, particularly regarding regenerative neurogenesis after SCI, remains limited. Gupta et al. (2023) in their review highlighted the epigenetic mechanism involved in the regenerative aspect of zebrafish spinal cord postinjury (Gupta et al., 2023).

The function of Hdac1 in the ERG during zebrafish spinal cord regeneration was examined in a recent study (McCann, 2019). The quantity of newborn motor neurons, as well as all newly formed neurons, was reduced in larval zebrafish post spinal cord transection due to the potential Hdac1 inhibitor. This inhibitor is presumed to lower the expression of Hdac1 in ERGs (McCann, 2019). It was seen that the *Her4.1* activator line can induce Hdac1 expression in spinal cord cells, primarily ERGs, which are positive for radial glia marker GFAP and negative for neuronal marker HuC. They observed that a decline in the quantity of the newly generated neurons was due to the reduction in the injury-induced ERG proliferation. Thus, in the zebrafish spinal cord, Hdac1 is a beneficial regulator of regeneration and enhances the development of new motor neurons and stimulates the growth of ERGs. Overall Hat blockage in the ERG or higher deacetylation due to Hdac1 upregulation had no impact on the quantity of newly formed motor neurons following spinal cord damage (McCann, 2019). In larval zebrafish spinal cord, in uninjured as well in conditions post injury, it was seen that neurogenesis is inhibited by dominant-negative, cell type–specific inhibition of Hdac1 (Cavone et al., 2021). Cigliola et al. (2023), through epigenetic profiling, reported a tissue regeneration enhancer element (TREE) associated with *hb-egfa* that controls gene expression in spinal cord injury (Cigliola et al., 2023).

Another epigenetic modulator of spinal cord neurogenesis and axonogenesis post injury is microRNAs. Studies on zebrafish adults have demonstrated that exogenous overexpression of *miR-133b* regulates the RhoA signaling system, which not only enhances motor function recovery but also aids in axon regrowth following SCI (Yu et al., 2011). In zebrafish embryos, a model employing two-photon axotomy, it has been observed that the production of *miR-133b* in only a single Mauthner cell prevented axon regeneration. Its suppression increased axon outgrowth through the manipulation of *tppp3*, which may enhance axon regeneration (Huang et al., 2017). These two studies clearly suggest that miR-133b has contrasting roles, highlighting the need for deeper investigation in future research.

## Epigenetic Regulation of Neurogenesis in Retina

Zebrafish Müller glia possess epigenetic plasticity, which allows the glia to undergo a process of reprogramming in response to injury. These retinal progenitor cells (RPC) then develop into various types of retinal cells, therefore repairing the injured retina and regaining eyesight, and the reprogramming of Müller glia is mostly dependent on epigenetic mechanisms, which mainly include histone modification, DNA methylation, and non-coding RNA (Si et al., 2023).

Post retinal damage, gene expressions linked to DNA methylation and demethylation exhibit highly or moderately elevated expression; however, the timing of maximal gene activity varies (Powell, Elsaeidi, & Goldman, 2012; Powell, Grant, Cornblath, & Goldman, 2013). It was seen that *apobec2a* and *apobec2b* contributed to the demethylation of DNA by deaminating cytosine residues of DNA molecules, and knocking down of *apobec2a* and *apobec2b* reduced Müller glia proliferation and regeneration in the optic nerve and retina (Powell et al., 2012). A subsequent investigation, however, found that the changes in DNA methylation during regeneration occur primarily independently of the expression of *apobec2* (Powell et al., 2013). Thus, it is still illusive whether the Apobec2a and Apobec2b proteins control retinal regenerative neurogenesis by signals other than DNA methylation. It is seen that the pluripotency factor promoters and a few genes linked to regeneration remain hypomethylated in Müller glia and their progenitors. Conversely, the less methylated promoter and intragenic regions of the significantly elevated genes in MGPCs at 4 days post-injury (dpi) indicate that the induction of gene expression is unlikely to be caused by, or inversely related to, DNA demethylation (Zhu et al., 2018). It is thought that while DNA demethylation could be necessary, it is insufficient to trigger MG reprogramming. The methylation status of genomic DNA varies dynamically throughout the shift from MGs to MGPCs; it is mostly seen between 0 and 2 dpi, when MGPCs are developing. It has been seen that between 2 and 4 dpi, when MGPCs are proliferating, there is an increase in de novo methylation levels (Powell et al., 2013).

In developing retinal tissue of zebrafish, proliferating cells express genes belonging to the *dnmt1* and *dnmt3-dnmt8* families (Seritrakul & Gross, 2014). The *tet2* and *tet3* combination is necessary for retinal neuron formation in a cell-non-autonomous manner, according to a loss-of-function study in zebrafish, possibly by controlling Notch and Wnt signaling (Seritrakul & Gross, 2017). It is suggested through following studies that DNA demethylation may be the primary mechanism for reprogramming Müller glial cells to start the regenerative neurogenesis response.

In a study by Yamaguchi et al. (2005) it was shown that the transition of zebrafish retina from proliferative state to differentiating state depends on the activity of Hdac1. It was also noted that the ratio of differentiating to proliferating cells rises in direct proportion to Hdac activity, which is indicative of the fact that Hdac proteins influence an essential phase of zebrafish retinal neurogenesis (Yamaguchi et al., 2005).

Hdac additionally regulate genes related to regeneration at the phase of retinal regeneration in zebrafish (Mitra et al., 2018). At the injury site, Hdac play a crucial role in controlling genes linked to regeneration, such as *lin28a* and *ascl1a*, as well as repressors like *her4.1* (Mitra et al., 2018). Her4.1, which likewise suppresses *lin28a* and vital cytokines to limit the growth of Müller glia progenitor cells, is negatively regulated by Hdac activity. Interestingly, whole-retina RNA sequencing analysis also reveals that inhibition of Hdac reversibly limits restoration via its inhibition of key cytokines and other genes crucial for regeneration (Mitra et al., 2018).

Early after retinal injury, Hdac1 expression is downregulated, which encourages Müller glia to dedifferentiate into MGPCs by enhancing the expression of multiple genes and cytokines related to regeneration. Late after retinal injury, Hdac1 expression is upregulated, which inhibits the extent of MGPC proliferation. Hdac therefore play a vital part in controlling retinal regeneration post injury (Si et al., 2023).

Over the last few years, a large number of studies have changed our understanding of ncRNAs from "junk" transcription items to useful regulatory molecules that control a variety of biological processes, such as gene expression, post-transcriptional modification, the transmission of signals, and remodeling of the chromatin (Si et al., 2023). From post-transcriptional control of mRNA integrity to epigenetic silencing, ncRNAs are engaged in regulation of genes at several levels. Consequently, ncRNAs are the primary regulators of a number of physiological and pathological processes (Guttman & Rinn, 2012). Rajaram et al.( 2014) in their study emphasized the role of miRNAs in zebrafish retinal regeneration post injury. They showed that the dicer-dependent miRNA synthesis pathway is necessary for proper retinal healing in zebrafish, as proven by using a model of damaged zebrafish with dicer knockdown. The scientists have clarified the function of differently expressed microRNAs in controlling the quantity of proliferating Müller glia–derived neural progenitors during retinal restoration in mature zebrafish by analyzing the absence of regenerative function in the developing retina. In this model, morpholino-mediated reduction of increased *miR-7a*, *miR-31*, *miR-27c*, *miR-142b*, and *miR-146a* reduced the number of neural precursors propagating in the zebrafish retina during light exposure. According to this study, these miRNAs may accelerate the Müller glial dedifferentiation-mediated process of retinal regeneration (Rajaram et al., 2014). Their study also emphasized on how the expression titers of 23 miRNAs, including *miR-143*, *miR-23a*, and *miR-145*, declined during the period of proliferation. An additional investigation revealed that Oct4 can contribute to Müller glia remodeling by suppressing *miR-143/145* (Sharma et al., 2019).

These findings suggest that the neurogenic capacity of Müller glia and RPC differs based on their different expressions of various epigenetic modifiers. Furthermore, manipulating these modifiers offers a new method for reprogramming Müller glia to stimulate retinal regeneration.

## Epigenetic Regulations Targeting Signaling Pathways During Neurogenesis

Epigenetic mechanisms play a pivotal role in modulating gene expression patterns during neurogenesis, influencing the activation or repression of various signalling pathways that govern various cellular processes. During zebrafish neurogenesis, the intricate interplay between epigenetic regulations and signalling pathways have shed light on the molecular mechanisms that orchestrate neural cell fate determination and differentiation.

Studies conducted on zebrafish brain post-injury reveal an interaction between Hdac and the Notch signalling pathways. Inhibition of Hdac resulted in the inhibition of the expression of Notch pathway genes *her4.1* and *her6*. This observation suggests that injury-induced alteration of gene expression of Notch pathway genes occurs through histone deacetylation (Kiyooka et al., 2020). In another study, the inhibition of HAT EP300 was found to downregulate Notch genes *her4* and *her6*, consequently promoting RGC proliferation in the optic tectum. This suggests that the inhibition of EP300 regulates the proliferation and differentiation of NSCs by suppressing histone acetylation and the expression of Notch target genes and that HAT and HDAC act in a contradictory way to regulate neurogenesis (Shimizu & Kawasaki, 2021).

It has been observed that *miR-9* plays a regulatory role in neurogenesis in the zebrafish brain by interacting with Notch signalling molecules and its downregulatory genes in a complex manner (Radhakrishnan et al., 2016).

A study during zebrafish embryonic neurogenesis showed that *hdac1* plays a crucial role in repressing the expression of Notch target genes. Additionally, *hdac1* is essential for maintaining the production of motoneurons in response to hedgehog signalling. (Cunliffe, 2004).

Numerous studies have been conducted demonstrating the interaction between various epigenetic modifiers and signalling molecules during retinal neurogenesis in zebrafish. In a study by Yamaguchi et al., (2005), they showed Hdac1 modulates neurogenesis in embryonic zebrafish retina by antagonizing Notch and Wnt signaling pathways (Yamaguchi et al., 2005). In the injured zebrafish retina, *let7* miRNA controls the synthesis of *shha*, *shhb*, *ptch1*, *smo*, and *zic2b* mRNAs. The stimulation of pro-regenerative transcription cascades during retinal regeneration, such as *lin28a*, *ascl1a*, *foxn4*, *mmp9*, and *zic2b*, requires the Shh signaling pathway (Kaur et al.,

2018). Subsequent research revealed that glial cell remodeling throughout Müller glial retinal restoration is negatively regulated by the *miR-216a*-Dotl1 axis via the conventional Wnt signaling cascade (Kara et al., 2019). Studies showed that inhibiting transforming growth factor-β signaling, the titer of *miR-200a/b* that mediates Müller glia reprogramming, is increased (Sharma et al., 2020).

## Conclusion

The diverse epigenetic activity seen throughout the different phases of neurogenesis highlights the significance of the regulatory mechanism of epigenetic mediators in orchestrating cell fate decisions. This chapter provides an insight into the epigenetic regulation of the process of neurogenesis in zebrafish along with the integration of epigenetic regulation and signaling cascades in regulating the fate of neural progenitor cells, especially during regenerative neurogenesis. Though our study provides a comprehensive perspective on the epigenetic regulation of zebrafish neurogenesis, future studies and applications in this field have a lot of intriguing opportunities. Firstly, by figuring out the intricate cross-talk between various epigenetic modifiers and their temporal dynamics, we may refine our knowledge of the hierarchical regulatory networks that control neurogenesis. Advanced methodologies such as single-cell epigenomics will be pivotal in capturing the cellular heterogeneity inherent in the neural progenitor population. Additionally, exploring the functional consequences of individual epigenetic modifications on synaptic plasticity and behavioral outcomes will offer a much more thorough representation of the molecular landscape during neurogenesis. Gaining further perception into the mechanism of how environmental stimuli and genetic factors cross-talk with epigenetic regulators will enhance our current knowledge of the process of epigenetic regulation of neurogenesis in the zebrafish nervous system. Understanding the interdependence of epigenetic modifiers and signalling pathways in zebrafish neurogenesis carries significant translational implications. The conserved nature of these regulatory networks across species, including mammals, positions zebrafish as a valuable model for investigating similar interactions in higher vertebrates, including humans. Targeting specific modulators within the epigenetic-signalling axis may provide a novel therapeutic approach for modulating neurogenesis in the context of neurological disorders and neurodegeneration.

It is seen that miRNAs exhibit substantial potential as a novel class of therapeutic agents. However, certain challenges must be considered, including potential side effects and toxicity associated with high doses during in vivo administration. Additionally, the possibility of unexpected off-target effects of specific miRNAs poses another consideration in their therapeutic application (Gupta et al., 2023). In recent times, extracellular vesicles (EVs) have garnered considerable interest as promising therapeutic carriers for delivering cargo to target cells. This attention primarily stems from their capability to traverse the blood–brain barrier, facilitating the delivery of therapeutic agents to the specified target organs (Alvarez-Erviti et al., 2011); (Haney et al., 2015). microRNAs transported by EVs have demonstrated the capability to modulate critical pathways associated with neuroplasticity, axon guidance, and remyelination, suggesting a promising avenue for facilitating regenerative neurogenesis in spinal cord. Moreover, EVs can also transport DNA methylation modifiers, facilitating the regulation of DNA methylation patterns that govern gene expression (Gupta et al., 2023).

EV-mediated epigenetic regulation has a lot of promise, but standardizing its efficiency and safety would require addressing a number of issues. Further important concerns include optimizing processes for the effective loading of particular epigenetic cargo into EVs and strengthening their stability during storage and transit. Thorough preclinical investigation utilizing suitable model organisms and pertinent outcome indicators will be essential in assessing the therapeutic possibilities of EV-driven epigenetic regulation (Gupta et al., 2023).

## REFERENCES

Alunni, A., & Bally-Cuif, L. (2016). A comparative view of regenerative neurogenesis in vertebrates. *Development*, *143*(5), 741–753. https://doi.org/10.1242/dev.122796

Alunni, A., Krecsmarik, M., Bosco, A., Galant, S., Pan, L., Moens, C. B., & Bally-Cuif, L. (2013). Notch3 signaling gates cell cycle entry and limits neural stem cell amplification in the adult pallium. *Development*, *140*(16), 3335–3347. https://doi.org/10.1242/dev.095018

Alvarez-Erviti, L., Seow, Y., Yin, H., Betts, C., Lakhal, S., & Wood, M. J. A. (2011). Delivery of siRNA to the mouse brain by systemic injection of targeted exosomes. *Nature Biotechnology*, *29*(4), 341–345. https://doi.org/10.1038/nbt.1807

Bae, Y. K. (2006). Patterning of proneuronal and inter-proneuronal domains by hairy- and enhancer of split-related genes in zebrafish neuroectoderm. *Development*, *133*(8), 1609–1609. https://doi.org/10.1242/dev.02360

Barbosa, J. S., Sanchez-Gonzalez, R., Di Giaimo, R., Baumgart, E. V., Theis, F. J., Götz, M., & Ninkovic, J. (2015). Neurodevelopment. Live imaging of adult neural stem cell behavior in the intact and injured zebrafish brain. *Science*, *348*(6236), 789–793. https://doi.org/10.1126/science.aaa2729

Becker, C. G., & Becker, T. (2015). Neuronal regeneration from ependymo-radial glial cells: Cook, little pot, cook! *Developmental Cell*, *32*(4), 516–527. https://doi.org/10.1016/j.devcel.2015.01.001

Becker, T., & Becker, C. G. (2022). Regenerative neurogenesis: The integration of developmental, physiological and immune signals. *Development*, *149*(8). https://doi.org/10.1242/dev.199907

Bernardos, R. L., Barthel, L. K., Meyers, J. R., & Raymond, P. A. (2007). Late-stage neuronal progenitors in the retina are radial Müller glia that function as retinal stem cells. *The Journal of Neuroscience*, *27*(26), 7028–7040. https://doi.org/10.1523/JNEUROSCI.1624-07.2007

Cardozo, M. J., Mysiak, K. S., Becker, T., & Becker, C. G. (2017). Reduce, reuse, recycle - Developmental signals in spinal cord regeneration. *Developmental Biology*, *432*(1), 53–62. https://doi.org/10.1016/j.ydbio.2017.05.011

Cavalieri, V., & Spinelli, G. (2017). Environmental epigenetics in zebrafish. *Epigenetics & Chromatin*, *10*(1), 46. https://doi.org/10.1186/s13072-017-0154-0

Cavone, L., McCann, T., Drake, L. K., Aguzzi, E. A., Oprişoreanu, A.-M., Pedersen, E., … Becker, C. G. (2021). A unique macrophage subpopulation signals directly to progenitor cells to promote regenerative neurogenesis in the zebrafish spinal cord. *Developmental Cell*, *56*(11), 1617–1630.e6. https://doi.org/10.1016/j.devcel.2021.04.031

Chapouton, P., Jagasia, R., Bally-Cuif, L. (2007). Adult neurogenesis in non-mammalian vertebrates. *Bioessays*. 29(8):745-57.

Cigliola, V., Shoffner, A., Lee, N., Ou, J., Gonzalez, T. J., Hoque, J., … Poss, K. D. (2023). Spinal cord repair is modulated by the neurogenic factor Hb-egf under direction of a regeneration-associated enhancer. *Nature Communications*, *14*(1), 4857. https://doi.org/10.1038/s41467-023-40486-5

Covic, M., Karaca, E., & Lie, D. C. (2010). Epigenetic regulation of neurogenesis in the adult hippocampus. *Heredity*, *105*(1), 122–134. https://doi.org/10.1038/hdy.2010.27

Cunliffe, V. T. (2004). Histone deacetylase 1 is required to repress Notch target gene expression during zebrafish neurogenesis and to maintain the production of motoneurones in response to hedgehog signalling. *Development*, *131*(12), 2983–2995. https://doi.org/10.1242/dev.01166

Dennis, C. V., Suh, L. S., Rodriguez, M. L., Kril, J. J., & Sutherland, G. T. (2016). Human adult neurogenesis across the ages: An immunohistochemical study. *Neuropathology and Applied Neurobiology*, *42*(7), 621–638. https://doi.org/10.1111/nan.12337

Detrich, I., & William, H. (2010). The Zebrafish: Cellular and Developmental Biology, Part A. Methods in Cell Biology. *Academic Press*, *133*(3), 366.

Diaz Quiroz, J. F., & Echeverri, K. (2013). Spinal cord regeneration: Where fish, frogs and salamanders lead the way, can we follow? *The Biochemical Journal*, *451*(3), 353–364. https://doi.org/10.1042/BJ20121807

Diotel, N., Lübke, L., Strähle, U., & Rastegar, S. (2020). Common and distinct features of adult neurogenesis and regeneration in the telencephalon of zebrafish and mammals. *Frontiers in Neuroscience*, *14*, 568930. https://doi.org/10.3389/fnins.2020.568930

Fan, G., & Hutnick, L. (2005). Methyl-CpG binding proteins in the nervous system. *Cell Research*, *15*(4), 255–261. https://doi.org/10.1038/sj.cr.7290294

Fares, J., Bou Diab, Z., Nabha, S., & Fares, Y. (2019). Neurogenesis in the adult hippocampus: History, regulation, and prospective roles. *The International journal of neuroscience*, *129*(6), 598–611. https://doi.org/10.1080/00207454.2018.1545771

Fausett, B. V., & Goldman, D. (2006). A role for alpha1 tubulin-expressing Müller glia in regeneration of the injured zebrafish retina. *The Journal of Neuroscience*, *26*(23), 6303–6313. https://doi.org/10.1523/JNEUROSCI.0332-06.2006

Feng, J., Fouse, S., & Fan, G. (2007). Epigenetic regulation of neural gene expression and neuronal function. *Pediatric Research*, *61*(5 Pt 2), 58R–63R. https://doi.org/10.1203/pdr.0b013e3180457635

Fimbel, S. M., Montgomery, J. E., Burket, C. T., & Hyde, D. R. (2007). Regeneration of inner retinal neurons after intravitreal injection of ouabain in zebrafish. *The Journal of Neuroscience*, *27*(7), 1712–1724. https://doi.org/10.1523/JNEUROSCI.5317-06.2007

Geling, A., Plessy, C., Rastegar, S., Strähle, U., & Bally-Cuif, L. (2004). Her5 acts as a prepattern factor that blocks neurogenin1 and coe2 expression upstream of Notch to inhibit neurogenesis at the midbrain-hindbrain boundary. *Development*, *131*(9), 1993–2006. https://doi.org/10.1242/dev.01093

Ghaddar, B., Lübke, L., Couret, D., Rastegar, S., & Diotel, N. (2021). Cellular mechanisms participating in brain repair of adult zebrafish and mammals after injury. *Cells*, *10*(2), 391. https://doi.org/10.3390/cells10020391

Ghosh, S., & Hui, S. P. (2016). Regeneration of zebrafish CNS: adult neurogenesis. *Neural Plasticity*, *2016*, 5815439. https://doi.org/10.1155/2016/5815439

Ghosh, S., & Hui, S. P. (2018). Axonal regeneration in zebrafish spinal cord. *Regeneration*, *5*(1), 43–60. https://doi.org/10.1002/reg2.99

Goldberg, A. D., Allis, C. D., & Bernstein, E. (2007). Epigenetics: A landscape takes shape. *Cell*, *128*(4), 635–638. https://doi.org/10.1016/j.cell.2007.02.006

Grandel, H., & Brand, M. (2013). Comparative aspects of adult neural stem cell activity in vertebrates. *Development Genes and Evolution*, *223*(1–2), 131–147. https://doi.org/10.1007/s00427-012-0425-5

Grandel, H., Kaslin, J., Ganz, J., Wenzel, I., & Brand, M. (2006). Neural stem cells and neurogenesis in the adult zebrafish brain: origin, proliferation dynamics, migration and cell fate. *Developmental Biology*, *295*(1), 263–277. https://doi.org/10.1016/j.ydbio.2006.03.040

Gupta, S., Dutta, S., & Hui, S. P. (2023). Regenerative potential of injured spinal cord in the light of epigenetic regulation and modulation. *Cells*, *12*(13). https://doi.org/10.3390/cells12131694

Guttman, M., & Rinn, J. L. (2012). Modular regulatory principles of large non-coding RNAs. *Nature*, *482*(7385), 339–346. https://doi.org/10.1038/nature10887

Haney, M. J., Klyachko, N. L., Zhao, Y., Gupta, R., Plotnikova, E. G., He, Z., … Batrakova, E. V. (2015). Exosomes as drug delivery vehicles for Parkinson's disease therapy. *Journal of Controlled Release*, *207*, 18–30. https://doi.org/10.1016/j.jconrel.2015.03.033

Hans, S., Scheer, N., Riedl, I., Weizsäcker, E., Blader, P., & Campos-Ortega, J. A. (2004). her3, a zebrafish member of the hairy-E(spl) family, is repressed by Notch signalling. *Development*, *131*(12), 2957–2969. https://doi.org/10.1242/dev.01167

Hellenbrand, D. J., Quinn, C. M., Piper, Z. J., Morehouse, C. N., Fixel, J. A., & Hanna, A. S. (2021). Inflammation after spinal cord injury: A review of the critical timeline of signaling cues and cellular infiltration. *Journal of Neuroinflammation*, *18*(1), 284. https://doi.org/10.1186/s12974-021-02337-2

Huang, R., Chen, M., Yang, L., Wagle, M., Guo, S., & Hu, B. (2017). MicroRNA-133b Negatively Regulates Zebrafish Single Mauthner-Cell Axon Regeneration through Targeting tppp3 in Vivo. *Frontiers in Molecular Neuroscience*, *10*, 375. https://doi.org/10.3389/fnmol.2017.00375

Hui, S. P., Dutta, A., & Ghosh, S. (2010). Cellular response after crush injury in adult zebrafish spinal cord. *Developmental Dynamics*, *239*(11), 2962–2979. https://doi.org/10.1002/dvdy.22438

Hui, S. P., Monaghan, J. R., Voss, S. R., & Ghosh, S. (2013). Expression pattern of Nogo-A, MAG, and NgR in regenerating urodele spinal cord. *Developmental Dynamics*, *242*(7), 847–860. https://doi.org/10.1002/dvdy.23976

Iribarne, M., & Hyde, D. R. (2022). Different inflammation responses modulate Müller glia proliferation in the acute or chronically damaged zebrafish retina. *Frontiers in cell and developmental biology*, *10*, 892271. https://doi.org/10.3389/fcell.2022.892271

Ito, Y., Tanaka, H., Okamoto, H., & Ohshima, T. (2010). Characterization of neural stem cells and their progeny in the adult zebrafish optic tectum. *Developmental Biology*, *342*(1), 26–38. https://doi.org/10.1016/j.ydbio.2010.03.008

Jeon, K. W. (2007). *A survey of cell biology, international review of cytology*. Academic Press, 259 (1st ed.,), 248.

Kara, N., Kent, M. R., Didiano, D., Rajaram, K., Zhao, A., Summerbell, E. R., & Patton, J. G. (2019). The miR-216a-Dot1l regulatory axis is necessary and sufficient for müller glia reprogramming during retina regeneration. *Cell Reports*, *28*(8), 2037–2047.e4. https://doi.org/10.1016/j.celrep.2019.07.061

Kaslin, J., Ganz, J., Brand, M. (2008). Proliferation, neurogenesis and regeneration in the non-mammalian vertebrate brain. *Philos Trans R Soc Lond B Biol Sci*. 12;363(1489):101-22.

Kaur, S., Gupta, S., Chaudhary, M., Khursheed, M. A., Mitra, S., Kurup, A. J., & Ramachandran, R. (2018). let-7 microRNA-mediated regulation of shh signaling and the gene regulatory network is essential for retina regeneration. *Cell Reports*, *23*(5), 1409–1423. https://doi.org/10.1016/j.celrep.2018.04.002

Kicheva, A., Bollenbach, T., Ribeiro, A., Valle, H. P., Lovell-Badge, R., Episkopou, V., & Briscoe, J. (2014). Coordination of progenitor specification and growth in mouse and chick spinal cord. *Science*, *345*(6204), 1254927. https://doi.org/10.1126/science.1254927

Kiyooka, M., Shimizu, Y., & Ohshima, T. (2020). Histone deacetylase inhibition promotes regenerative neurogenesis after stab wound injury in the adult zebrafish optic tectum. *Biochemical and Biophysical Research Communications*, *529*(2), 366–371. https://doi.org/10.1016/j.bbrc.2020.06.025

Kizil, C., Kaslin, J., Kroehne, V., & Brand, M. (2012). Adult neurogenesis and brain regeneration in zebrafish. *Developmental Neurobiology*, *72*(3), 429–461. https://doi.org/10.1002/dneu.20918

Klatt Shaw, D., Saraswathy, V. M., Zhou, L., McAdow, A. R., Burris, B., Butka, E., … Mokalled, M. H. (2021). Localized EMT reprograms glial progenitors to promote spinal cord repair. *Developmental Cell*, *56*(5), 613–626.e7. https://doi.org/10.1016/j.devcel.2021.01.017

Kuscha, V., Frazer, S. L., Dias, T. B., Hibi, M., Becker, T., & Becker, C. G. (2012). Lesion-induced generation of interneuron cell types in specific dorsoventral domains in the spinal cord of adult zebrafish. *The Journal of Comparative Neurology*, *520*(16), 3604–3616. https://doi.org/10.1002/cne.23115

Kyritsis, N., Kizil, C., Zocher, S., Kroehne, V., Kaslin, J., Freudenreich, D., … Brand, M. (2012). Acute inflammation initiates the regenerative response in the adult zebrafish brain. *Science*, *338*(6112), 1353–1356. https://doi.org/10.1126/science.1228773

Lahne, M., Nagashima, M., Hyde, D. R., & Hitchcock, P. F. (2020). Reprogramming müller glia to regenerate retinal neurons. *Annual Review of Vision Science*, *6*, 171–193. https://doi.org/10.1146/annurev-vision-121219-081808

Lindsey, B. W., Aitken, G. E., Tang, J. K., Khabooshan, M., Douek, A. M., Vandestadt, C., & Kaslin, J. (2019). Midbrain tectal stem cells display diverse regenerative capacities in zebrafish. *Scientific Reports*, *9*(1), 4420. https://doi.org/10.1038/s41598-019-40734-z

Lyu, P., Iribarne, M., Serjanov, D., Zhai, Y., Hoang, T., Campbell, L. J., … Blackshaw, S. (2023). Common and divergent gene regulatory networks control injury-induced and developmental neurogenesis in zebrafish retina. *Nature Communications*, *14*(1), 8477. https://doi.org/10.1038/s41467-023-44142-w

Manto, M., Gruol, D. L., Schmahmann, J. D., Koibuchi, N., & Rossi, F. (2012). *Handbook of the cerebellum and cerebellar disorders, 4,* (2013th ed., p. 2458). Springer.

Marcus, R. C., Delaney, C. L., & Easter, S. S. (1999). Neurogenesis in the visual system of embryonic and adult zebrafish (Danio rerio). off. *Visual Neuroscience*, *16*(3), 417–424. https://doi.org/10.1017/s095252389916303x

März, M., Chapouton, P., Diotel, N., Vaillant, C., Hesl, B., Takamiya, M., … Strähle, U. (2010). Heterogeneity in progenitor cell subtypes in the ventricular zone of the zebrafish adult telencephalon. *Glia*, *58*(7), 870–888. https://doi.org/10.1002/glia.20971

McCann, T. (2019). The role of histone deacetylase 1 in neuroregeneration in the zebrafish spinal cord. Edinburgh Medical School thesis and dissertation collection. http://hdl.handle.net/1842/35781

Mitra, S., Sharma, P., Kaur, S., Khursheed, M. A., Gupta, S., Ahuja, R., … Ramachandran, R. (2018). Histone deacetylase-mediated müller glia reprogramming through Her4.1-Lin28a axis is essential for retina regeneration in zebrafish. *iScience*, *7*, 68–84. https://doi.org/10.1016/j.isci.2018.08.008

Murao, N., Noguchi, H., & Nakashima, K. (2016). Epigenetic regulation of neural stem cell property from embryo to adult. *Neuroepigenetics*, *5*, 1–10. https://doi.org/10.1016/j.nepig.2016.01.001

Myers, P. Z., Eisen, J. S., & Westerfield, M. (1986). Development and axonal outgrowth of identified motoneurons in the zebrafish. *The Journal of Neuroscience*, *6*(8), 2278–2289. https://doi.org/10.1523/JNEUROSCI.06-08-02278.1986

Paridaen, J. T. M. L., & Huttner, W. B. (2014). Neurogenesis during development of the vertebrate central nervous system. *EMBO Reports*, *15*(4), 351–364. https://doi.org/10.1002/embr.201438447

Pinto, L., & Götz, M. (2007). Radial glial cell heterogeneity--the source of diverse progeny in the CNS. *Progress in Neurobiology*, *83*(1), 2–23. https://doi.org/10.1016/j.pneurobio.2007.02.010

Powell, C., Cornblath, E., Elsaeidi, F., Wan, J., & Goldman, D. (2016). Zebrafish Müller glia-derived progenitors are multipotent, exhibit proliferative biases and regenerate excess neurons. *Scientific Reports*, *6*(1), 24851. https://doi.org/10.1038/srep24851

Powell, C., Cornblath, E., & Goldman, D. (2014). Zinc-binding domain-dependent, deaminase-independent actions of apolipoprotein B mRNA-editing enzyme, catalytic polypeptide

2 (Apobec2), mediate its effect on zebrafish retina regeneration. *The Journal of Biological Chemistry*, *289*(42), 28924–28941. https://doi.org/10.1074/jbc.M114.603043

Powell, C., Elsaeidi, F., & Goldman, D. (2012). Injury-dependent Müller glia and ganglion cell reprogramming during tissue regeneration requires Apobec2a and Apobec2b. *The Journal of Neuroscience*, *32*(3), 1096–1109. https://doi.org/10.1523/JNEUROSCI.5603-11.2012

Powell, C., Grant, A. R., Cornblath, E., & Goldman, D. (2013). Analysis of DNA methylation reveals a partial reprogramming of the Müller glia genome during retina regeneration. *Proceedings of the National Academy of Sciences of the United States of America*, *110*(49), 19814–19819. https://doi.org/10.1073/pnas.1312009110

Radhakrishnan, B., & Alwin Prem Anand, A. (2016). Role of miRNA-9 in brain development. *Journal of Experimental Neuroscience*, *10*, 101–120. https://doi.org/10.4137/JEN.S32843

Rai, K., Jafri, I. F., Chidester, S., James, S. R., Karpf, A. R., Cairns, B. R., & Jones, D. A. (2010). Dnmt3 and G9a cooperate for tissue-specific development in zebrafish. *The Journal of Biological Chemistry*, *285*(6), 4110–4121. https://doi.org/10.1074/jbc.M109.073676

Raj, B., Farrell, J. A., Liu, J., El Kholtei, J., Carte, A. N., Navajas Acedo, J., … Schier, A. F. (2020). Emergence of neuronal diversity during vertebrate brain development. *Neuron*, *108*(6), 1058–1074.e6. https://doi.org/10.1016/j.neuron.2020.09.023

Rajaram, K., Harding, R. L., Bailey, T., Patton, J. G., & Hyde, D. R. (2014). Dynamic miRNA expression patterns during retinal regeneration in zebrafish: Reduced dicer or miRNA expression suppresses proliferation of Müller glia-derived neuronal progenitor cells. *Developmental Dynamics*, *243*(12), 1591–1605. https://doi.org/10.1002/dvdy.24188

Ramachandran, R., Zhao, X.-F., & Goldman, D. (2011). Ascl1a/Dkk/beta-catenin signaling pathway is necessary and glycogen synthase kinase-3beta inhibition is sufficient for zebrafish retina regeneration. *Proceedings of the National Academy of Sciences of the United States of America*, *108*(38), 15858–15863. https://doi.org/10.1073/pnas.1107220108

Reimer, M. M., Sörensen, I., Kuscha, V., Frank, R. E., Liu, C., Becker, C. G., & Becker, T. (2008). Motor neuron regeneration in adult zebrafish. *The Journal of Neuroscience*, *28*(34), 8510–8516. https://doi.org/10.1523/JNEUROSCI.1189-08.2008

Rubenstein, J. L. R., Rakic, P., Chen, B., & Kwan, K. Y. (2020). *Patterning and cell type specification in the developing CNS and PNS: Comprehensive developmental neuroscience* (2nd ed., p. 1122). Academic Press.

Schmidt, R., Strähle, U., & Scholpp, S. (2013). Neurogenesis in zebrafish - from embryo to adult. *Neural Development*, *8*, 3. https://doi.org/10.1186/1749-8104-8-3

Scholpp, S., Delogu, A., Gilthorpe, J., Peukert, D., Schindler, S., & Lumsden, A. (2009). Her6 regulates the neurogenetic gradient and neuronal identity in the thalamus. *Proceedings of the National Academy of Sciences of the United States of America*, *106*(47), 19895–19900. https://doi.org/10.1073/pnas.0910894106

Scott, K., O'Rourke, R., Winkler, C. C., Kearns, C. A., & Appel, B. (2021). Temporal single-cell transcriptomes of zebrafish spinal cord pMN progenitors reveal distinct neuronal and glial progenitor populations. *Developmental Biology*, *479*, 37–50. https://doi.org/10.1016/j.ydbio.2021.07.010

Seritrakul, P., & Gross, J. M. (2014). Expression of the de novo DNA methyltransferases (dnmt3 - dnmt8) during zebrafish lens development. *Developmental Dynamics*, *243*(2), 350–356. https://doi.org/10.1002/dvdy.24077

Seritrakul, P., & Gross, J. M. (2017). Tet-mediated DNA hydroxymethylation regulates retinal neurogenesis by modulating cell-extrinsic signaling pathways. *PLoS Genetics*, *13*(9), e1006987. https://doi.org/10.1371/journal.pgen.1006987

Sharma, P., Gupta, S., Chaudhary, M., Mitra, S., Chawla, B., Khursheed, M. A., & Ramachandran, R. (2019). Oct4 mediates Müller glia reprogramming and cell cycle exit during retina regeneration in zebrafish. *Life Science Alliance*, *2*(5). https://doi.org/10.26508/lsa.201900548

Sharma, P., Gupta, S., Chaudhary, M., Mitra, S., Chawla, B., Khursheed, M. A., … Ramachandran, R. (2020). Biphasic role of Tgf-β signaling during Müller Glia reprogramming and retinal regeneration in zebrafish. *iScience*, *23*(2), 100817. https://doi.org/10.1016/j.isci.2019.100817

Shimizu, Y., & Kawasaki, T. (2021). Histone acetyltransferase EP300 regulates the proliferation and differentiation of neural stem cells during adult neurogenesis and regenerative neurogenesis in the zebrafish optic tectum. *Neuroscience Letters*, *756*, 135978. https://doi.org/10.1016/j.neulet.2021.135978

Si, T.-E., Li, Z., Zhang, J., Su, S., Liu, Y., Chen, S., … Zang, W. (2023). Epigenetic mechanisms of Müller glial reprogramming mediating retinal regeneration. *Frontiers in Cell and Developmental Biology*, *11*, 1157893. https://doi.org/10.3389/fcell.2023.1157893

Sun, J., Sun, J., Ming, G., & Song, H. (2011). Epigenetic regulation of neurogenesis in the adult mammalian brain. *The European Journal of Neuroscience*, *33*(6), 1087–1093. https://doi.org/10.1111/j.1460-9568.2011.07607.x

Watt, F., & Molloy, P. L. (1988). Cytosine methylation prevents binding to DNA of a HeLa cell transcription factor required for optimal expression of the adenovirus major late promoter. *Genes & Development*, *2*(9), 1136–1143. https://doi.org/10.1101/gad.2.9.1136

Yamaguchi, M., Tonou-Fujimori, N., Komori, A., Maeda, R., Nojima, Y., Li, H., … Masai, I. (2005). Histone deacetylase 1 regulates retinal neurogenesis in zebrafish by suppressing Wnt and Notch signaling pathways. *Development*, *132*(13), 3027–3043. https://doi.org/10.1242/dev.01881

York, E. M., Petit, A., & Roskams, A. J. (2013). Epigenetics of neural repair following spinal cord injury. *Neurotherapeutics*, *10*(4), 757–770. https://doi.org/10.1007/s13311-013-0228-z

Yu, Y.-M., Gibbs, K. M., Davila, J., Campbell, N., Sung, S., Todorova, T. I., … Schachner, M. (2011). MicroRNA miR-133b is essential for functional recovery after spinal cord injury in adult zebrafish. *The European Journal of Neuroscience*, *33*(9), 1587–1597. https://doi.org/10.1111/j.1460-9568.2011.07643.x

Zhu, X., Xiao, C., & Xiong, J.-W. (2018). Epigenetic regulation of organ regeneration in zebrafish. *Journal of cardiovascular development and disease*, *5*(4), 57. https://doi.org/10.3390/jcdd5040057

Zupanc, G. K. H., & Zupanc, M. M. (2006). New neurons for the injured brain: Mechanisms of neuronal regeneration in adult teleost fish. *Regenerative Medicine*, *1*(2), 207–216. https://doi.org/10.2217/17460751.1.2.207

# 8

# *Proteomic Approaches on Neurodegenerative Disease Models in Zebrafish*

**Mustafa Gani Sürmen and Saime Sürmen**

## Abbreviations

| | |
|---|---|
| **AD** | Alzheimer's Disease |
| **APP** | Amyloid Precursor Protein |
| **BMAA** | ß-Methylamino-L-Alanine |
| **CSF** | Cerebrospinal Fluid |
| **FACs** | Fluorescent-Activated Cell Sorting |
| **FANs** | Fluorescence Activated Nuclei Sorting |
| **LCM** | Laser Capture Microdissection |
| **MACs** | Magnetically-Activated Cell Sorting |
| **MPTP** | 1-Methyl-4-Phenyl-1,2,3,6-Tetrahydropyridine |
| **MS** | Mass Spectrometry |
| **OKA** | Okadaic Acid |
| **PD** | Parkinson's Disease |
| **rAD** | Rapid Alzheimer's Disease |
| **sAD** | Sporadic Alzheimer's Disease |

## Introduction

Proteins, which have a wide variety of roles within the cell, from enzymatic activities to cytoskeletal organization, enable the execution and regulation of cellular processes by exhibiting a dynamic profile in terms of time and location. However, proteins undergo post-translational modifications in order to perform their functions in the cell, resulting in an enormous variety of proteoforms. Evaluation of cellular activities at the proteome level is critical, as changes in the expression, activity, and interactions of proteins communicating with each other and with other molecules can be an indicator of various diseases (Martinez-Val et al., 2022; Forgrave et al., 2021). On the other hand, the human genome project was completed in the 2000s and there have been remarkable advances in omics technologies that enable the production of large amounts of data. Thus, scientists who previously used gel systems for proteomic studies have, over time, turned to mass spectrometers that enable high-efficiency qualitative and quantitative detection of proteins. In other words, significant advances have been made in proteomics research thanks to mass spectrometry (MS) systems that allow the depth and coverage of the identifiable proteome to be increased (Abdallah et al., 2012). Nowadays, MS-based proteomics, which allows the characterization of thousands of proteins in large numbers of complex samples, stands out as a powerful approach in the study of neurological diseases. Neuroproteomic studies provide detailed information about the organs, tissues, and cells that make up the nervous system in order to better understand these diseases. In order to find markers of neurological diseases and reveal molecular events, proteomic analyses are performed on a wide variety of biological samples, from human brain tissue, which has a very heterogeneous structure due to containing different cell types, to organelle proteome (Fingleton et al., 2021; Hosp & Mann, 2017). In studies conducted in this context, various proteins related to neuroinflammation, energy metabolism, mitochondrial and synaptic dysfunction, and the ubiquitin–proteasome system have been highlighted (Liang et al., 2024; Di Domenico et al., 2017). However, studying the formation mechanisms of damage due to both neurotoxicity and other factors with clinical samples such as human brain tissue and serum involves certain limits and technical difficulties, as will be mentioned in the following sections (Li et al., 2003; Baker et al., 2012; Lee et al., 2019). On the other hand, animal models of Alzheimer's and Parkinson's diseases are valuable preclinical tools for studying toxicity-induced neurodegenerative diseases. In some neurotoxicological studies, molecular changes in animal models have been examined using proteomic approaches to understand the etiology of diseases and provide more specific insights into biological processes (Sowell et al., 2009; Moon, 2022). The proteome of zebrafish has recently been the focus of attention of researchers, especially due to its similarities with the human nervous system. Different developmental stages of zebrafish, especially embryonic stages, are used in ecotoxicology studies examining the effect of different biological conditions on the proteome (Li et al., 2023; Kwon et al., 2023). In this context, this chapter discusses MS-based proteomic studies using the zebrafish model to examine the molecular pathogenesis of neurodegenerative diseases.

## An Overview of Proteomics and Neurodegenerative Diseases

Since the introduction of the concepts of "proteome and proteomics" by Marc Wilkins in the mid-1990s, the scientific world has witnessed numerous technological and methodological advances in protein research. Technological developments,

DOI: 10.1201/9781032676555-10

especially in MS systems, have enabled high-resolution analysis and provided large volumes of data to researchers (Wilkins et al., 1996; Hosp & Mann, 2017).

Moreover, thanks to new software and informatics tools, processing of this large volume of data and interpretation of proteome findings has become easier (Chen et al., 2020). Today's proteomics technology has the ability to perform fast, robust, and reproducible analyses in which more than 10,000 proteins can be examined in human brain tissue (Higginbotham et al., 2020; Ping et al., 2020). In this way, the importance of proteomic research in understanding the molecular pathophysiology of neurodegenerative diseases and improving treatment strategies for the diseases is increasing. Although various neurological diseases have been examined using MS-based proteomic approaches to date, a significant part of the studies have focused on Alzheimer's and Parkinson's diseases. In particular, proteomic strategies specific to cell types and structures belonging to certain regions of the brain have been successfully applied. Thanks to the laser capture microdissection (LCM) method, which can be applied to both fresh frozen and formalin-fixed paraffin-embedded tissues, the proteomes of the hippocampus, temporal cortex, ventral tegmental area, and substantia nigra regions of the brain, including amyloid plaques, were examined (Hondius et al., 2016; Drummond et al., 2015; Drummond et al., 2018; Griesser et al., 2020; Miller et al., 2014). One of these studies, using a localized proteomic approach, revealed proteome differences in amyloid plaques of patients with rapid (rAD) and sporadic (sAD) forms of Alzheimer's disease (AD). The results showed higher levels of neuronal protein and lower levels of astrocyte protein in rAD plaques (Drummond et al., 2017).

In order to investigate proteomic changes during the progression of AD, asymptomatic AD patients (AsymAD: AD pathology without cognitive symptoms) who have Amyloid Beta accumulation but do not show symptoms of cognitive decline have also been included in some studies. While there was an increase in astrocyte and microglial proteins (MDK, COL25A1, MAPT, NTN1, SMOC1, GFAP, LTF, FABP7, CD44, PRDX1, and DDAH2) in the dorsolateral prefrontal cortex tissues of both AsymAD and AD patients, it was observed that synaptic transmission proteins (NPTX2, VGF, RPH3A, and GSTM1) decreased in AD compared to AsymAD (Higginbotham et al., 2020; Johnson et al., 2020, 2018).

It is predicted that especially the increase in oxidative stress and mitochondrial dysfunction may have a strong effect on synapses in the progression of AD pathology. Accumulating proteomic data has also provided important information about the crosstalk between synaptic plasticity, redox, and energy metabolism that is impaired in AD (Butterfield & Boyd-Kimball, 2019; Di Domenico et al., 2017). Another important sample in which neurodegenerative diseases are investigated is cerebrospinal fluid (CSF). In one of these studies, it was reported that approximately 70% of the CSF proteome, a significant part of which consists of synaptic, vascular, myelination, glial immunity, and metabolic proteins, overlaps with brain proteome (Higginbotham et al., 2020). Deeper analyses of the CSF proteome showed that this rate may be higher. Almost 85% of the nearly 6,000 proteins identified in the CSF proteome were also found in the human cerebral cortex (Wang et al., 2020). In these studies, several promising protein biomarker candidates for AD that can be identified in both brain and CSF have also been reported, including BASP1, YWHAZ, C9ORF72, SOD1, MDH1 (Higginbotham et al., 2020), SMOC1, TAU, GFAP, SUCLG2, PRDX3, and NTN1 (Wang et al., 2020). Additionally, a meta-analysis of six studies revealed that 65 of 311 upregulated proteins and 44 of 165 downregulated proteins in CSF samples of AD patients compared with controls overlapped with the brain proteome (Bai et al., 2021).

Lewy body formation and progressive dopaminergic neuron losses in the substantia nigra region of the brain, which are two characteristic pathological findings of Parkinson's disease (PD), make the protein differences in this region interesting. In-depth proteomic analyses revealed downregulation of mitoribosome proteins as the most dysregulated proteins in the substantia nigra (Jang et al., 2023). Substantia nigra synaptosome proteomics also showed decreased expression of 14 proteins indicating mitochondrial dysfunction, including mitochondrial proteins such as Thymidine kinase 2, 39S ribosomal protein L37, neurolysin, and Methionine-tRNA ligase (Plum et al., 2020). A study of protein changes in substantia nigra tissue associated with Lewy body pathology also showed a strong negative correlation of poly(A) RNA binding proteins with Lewy bodies. With the new findings, 5 main mechanisms have been revealed – Arp23 complex-mediated actin nucleation, poly(A) RNA binding, basement membrane and endothelium, synapse, and metabolic process of reactive oxygen species – associated with Lewy body and dopaminergic neuron losses (Petyuk et al., 2021).

On the other hand, various candidate protein markers for Parkinson's have been suggested by proteomic analyses of CSF (e.g. granin family proteins, ubiquitin, β2-microglobulin, and two secretogranin 1 fragments); plasma (e.g. TMSB4X- thymosin beta-4, CHI3L1- Chitinase-3-like protein 1, CCDC154- coiled-coil domain-containing protein 154, TRIM3- tripartite motif-containing protein 3, DHH- desert hedgehog protein, NRP2- neuropilin-2 and CLIC1- chloride intracellular channel protein 1); serum (e.g. PEDF- pigment epithelium-derived factor, gelsolin, apolipoprotein J); saliva (e.g. S100-A16 and ARPC1A- actin-related protein 2/3 complex subunit 1A); tears and urine (e.g. VGF, ENPEP) conducted by different researchers (Rotunno et al., 2020; Magdalinou et al., 2017; Zhao et al., 2022; Dong et al., 2019; Jiang et al., 2019; Figura et al., 2021; Boerger et al., 2019; Virreira Winter et al., 2021). In several studies, proteomic approaches have been applied to examine, in addition to Alzheimer's and Parkinson's diseases, Huntington's disease, multiple sclerosis, prion, spinal muscular atrophy, and amyotrophic lateral sclerosis (Fang et al., 2009; Guldbrandsen et al., 2020; Shi et al., 2015; Kessler et al., 2019; Thompson et al., 2020). Proteomic analysis of clinical samples from patients generally reveals protein differences in the late stages of the disease. However, some findings suggest that the onset of neurodegenerative diseases, especially Alzheimer's and Parkinson's, dates back much earlier than the symptoms (Villemagne et al., 2013; Trifonova et al., 2020). Experimental animals have various advantages in monitoring the developmental process of such diseases and detecting protein changes at an early stage for diagnosis. Mouse, rat, and zebrafish models are widely used to explain the neurodegenerative effects and pathological mechanisms of neurotoxin agents, especially those that interfere with normal brain physiology (Sowell et al., 2009; Moon, 2022; d'Amora & Giordani, 2018). In this chapter, proteomic applications of zebrafish animal models, which have become increasingly indispensable in the research of neurodegenerative diseases, have been extensively examined, including their advantages and disadvantages.

## An Overview of Proteomic Methods

High-resolution mass spectrometers have an important place in the field of proteomics thanks to their ability to identify thousands of proteins with very low sample amounts. Two basic approaches, bottom-up and top-down, are frequently used in MS-based proteomics. These two approaches differ from each other according to whether proteins are analysed intact or not. Bottum-up proteomics, also often referred to as shotgun proteomics, has broad application in the study of neurodegenerative diseases (Carbonara et al., 2021; Filiou et al., 2012). When it comes to shotgun proteomics, there are many methods, techniques, and tools with strengths and weaknesses for each step of the workflow (Duong & Lee, 2023) (Figure 8.1). In this approach, where proteins are digested into peptides using an endopeptidase, peptides are analysed in two ways, labelled or unlabelled, for quantitative protein measurement.

Spectral counting and MS ion density measurement are the two best-known methods for label-free quantification of proteins. While label-free strategies provide advantages in comparing large numbers and types of clinical samples with minimal manipulation, each sample is analysed individually and reproducible measurements can be difficult to ensure due to risks from preparation and device performance. In order to overcome this, it is important to normalize the raw data according to the amount of protein or total peptide during the processing stage (Filiou et al., 2012; Xie et al., 2011).

Labelled methods, on the other hand, are based on the principle of incorporating specific mass labels into peptides chemically, enzymatically, or via metabolic conversion and evaluating the mass difference between labelled and unlabelled ions. Performing labelling at the first stage provides a significant advantage in minimizing errors in the preparation

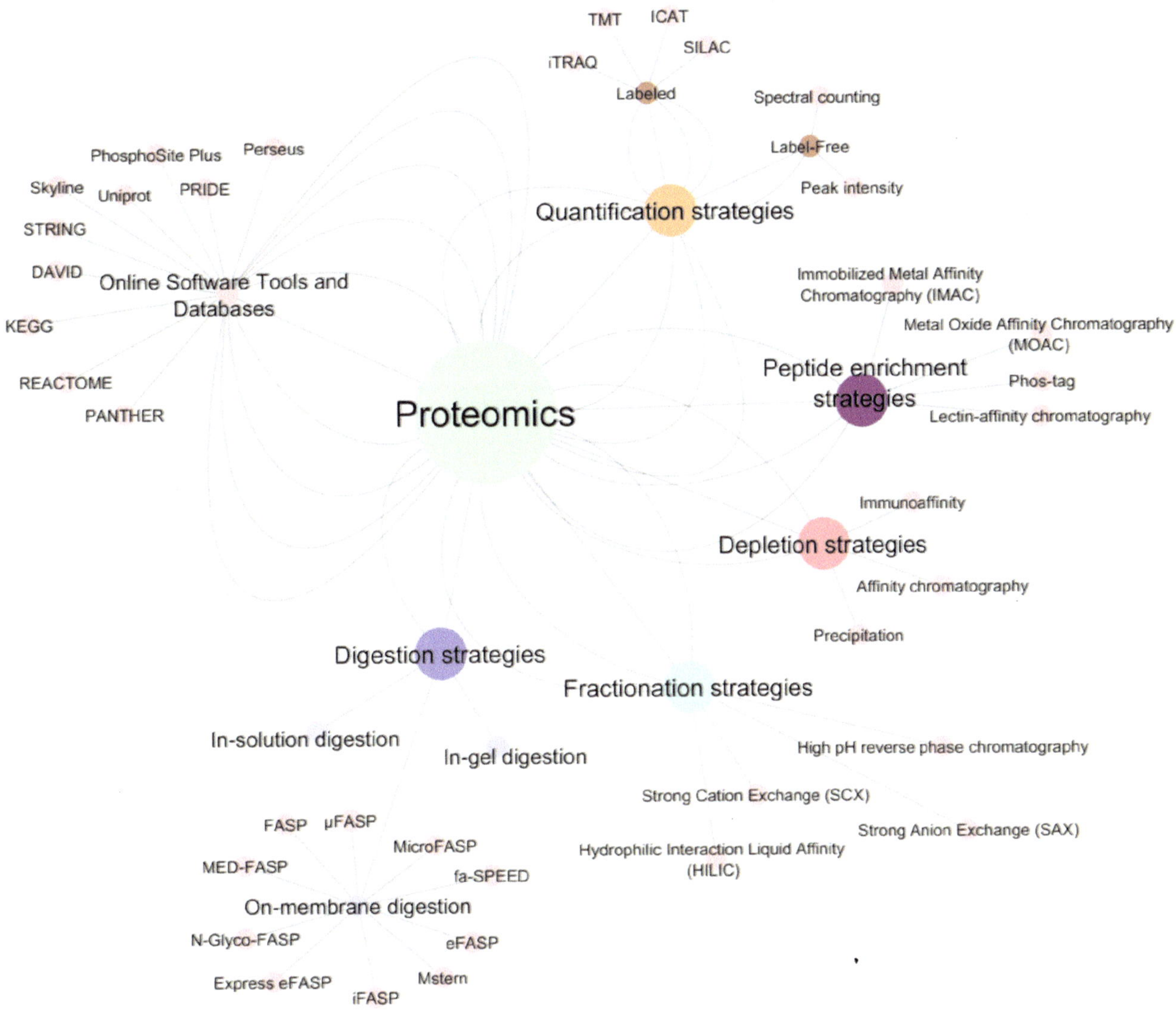


**FIGURE 8.1** Overview of proteomic methods, tools and databases.

stage. The most commonly used labelled methods include SILAC, ITRAQ, and TMT. In the basic principle of SILAC, the best known of metabolic labelling, proteins obtained from cells grown in two different media containing isotope-labelled and unlabelled amino acids are digested, and the resulting light and heavy chain peptides are mixed and analysed. SILAC is a frequently preferred method to reveal quantitative changes in protein expressions in complex samples. While the use of classical SILAC is limited only to cell culture, the development of the super-SILAC method using internal standards has offered the opportunity to measure the proteomes of tissues and model organisms. The developed SILAC methods have been applied to organisms such as yeast, bacteria, roundworms, fruit fly,mice, and zebrafish (Chen et al., 2015).

The high rate of tissue regeneration in zebrafish provides an advantage over metabolic labelling in other model animals. With this method, protein profiles of 9 organs of zebrafish (brain, heart, liver, spleen, muscle, skin, testicles, swim bladder, and gills) were obtained. Replicate analyses showed over 70% overlap (Nolte et al., 2014).

In addition, SILAC is also used to measure protein dynamics and follow organ development in zebrafish. SILAC-based molecular studies in zebrafish, which attempt to explain the vital activities of organisms by revealing protein cycles, provide important insights into how cellular regeneration can be manipulated (Konzer et al., 2013; Nolte et al., 2015; Westman-Brinkmalm et al., 2011). In this context, an improved SILAC fish diet containing Lys6 was used to measure changes during embryonic heart development (Konzer et al., 2013). A similar diet allowed monitoring of proteins directly involved in regeneration in zebrafish fins (Nolte et al., 2015).

On the other hand, many methods and techniques have been developed in the last 20 years aiming to reduce the complexity of biological samples and increase analysis power and reproducibility. The shotgun proteomics approach has a wide range of methods and techniques (Figure 8.1), and a basic workflow is shown in Figure 8.2. In this context, studies based on the shotgun proteomic approach are generally included in this chapter.

## Pros and Cons of Zebrafish Models for Proteomic Studies

Brain tissue, serum, plasma, and CSF are among the most commonly used samples to study neurodegenerative diseases (Raghunathan et al., 2022; Jain & Sathe, 2021; Cilento et al., 2019). However, the brain has an extremely complex organization consisting of billions of cells, and the enormous cell number and diversity here must be taken into account when examining changes associated with neurodegenerative diseases with proteomic methods. Accordingly, it is very important to obtain protein of sufficient purity during the sample preparation process in order to obtain a complete picture of the molecular functioning specific to the cell type. To overcome this problem and obtain homogeneous cell populations, it is necessary to use various techniques such as laser capture microdissection (LCM), fluorescent-activated cell

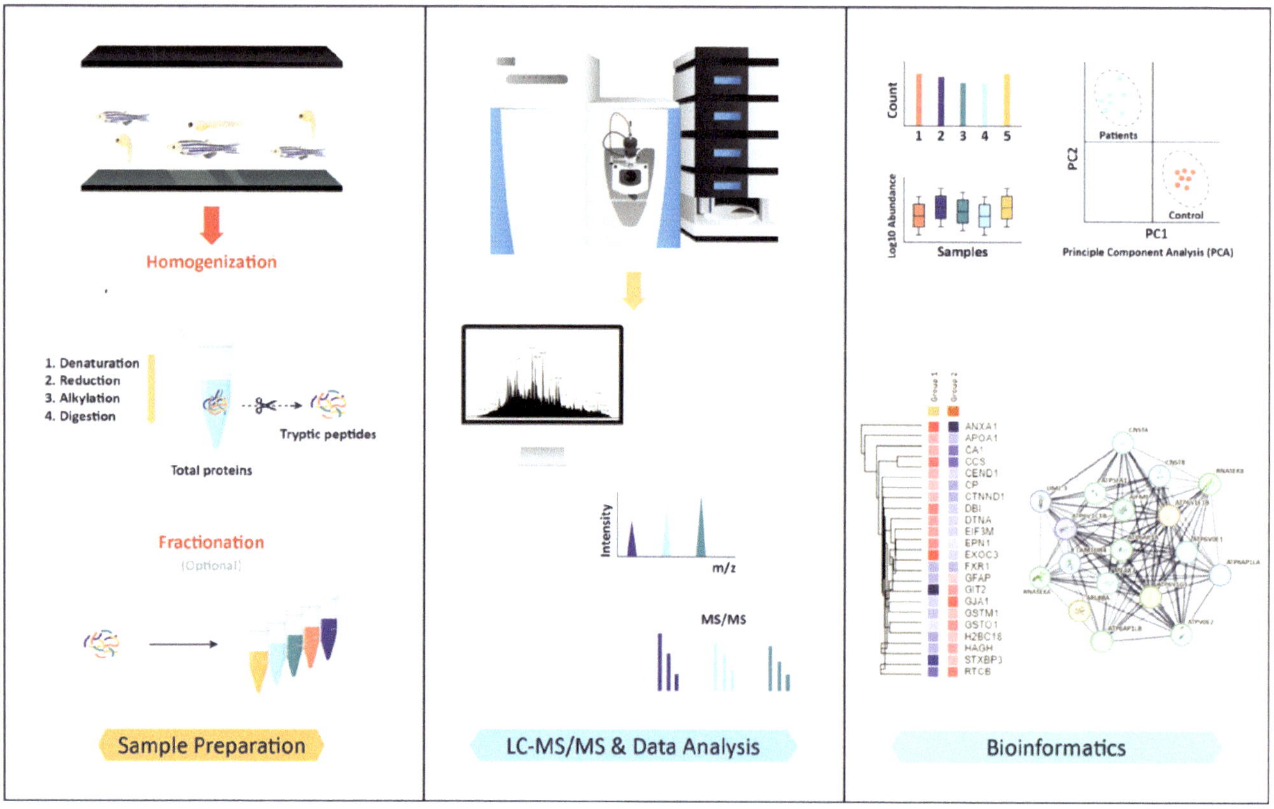


**FIGURE 8.2** General workflow in MS-based proteomics.

sorting (FACs), magnetically activated cell sorting (MACs), and fluorescence activated nuclei sorting (FANs) (Rayaprolu et al., 2021; Dammer et al., 2013). Another issue that should be taken into consideration when studying brain tissue is the time of collection and storage of the samples. Various studies have highlighted that post-mortem interval and storage conditions can have a significant impact on tissue quality and subsequent analysis, especially if quantification or post-translational modification studies are performed (Li et al., 2003; Ferrer et al., 2007).

Although brain tissue provides important clues about disease status and neuropathological processes, its usability is limited. On the other hand, the fact that CSF, which is vital for brain functions, is in contact with the brain surface and contains brain-derived proteins makes it a valuable biological fluid. However, there are a few points that should be taken into consideration in the proteomic analysis of CSF. A small amount of CSF material is obtained using a relatively less invasive method. Since individual differences may affect protein content, factors such as age, gender, and collection time need to be taken into consideration. Moreover, the protein concentration is much lower than in plasma, and since most of the protein content of CSF comes from blood, it is an important problem to remove abundant proteins such as albumin, transthyretin, transferrin, α1-antitrypsin, apolipoprotein A, and immunoglobulins from the sample (Haggmark et al., 2016; Bastos et al., 2017; Carlyle et al., 2018).

Collecting serum and plasma samples is minimally invasive and much easier than CSF. However, blood is in contact with many tissues and organs. Therefore, as a reflection of events in the central nervous system, the diversity and amount of protein biomarkers passing into the blood are likely to be much lower than in the CSF (Carlyle et al., 2018). For example, it has been reported that the concentration of tau protein in plasma is approximately 100 times lower than in CSF (Budelier & Bateman, 2020). Additionally, researchers suffer from high-abundance proteins, such as albumin and globulin in plasma and serum, obscuring the MS analysis. For this reason, in many studies, various immune-depletion, fractionation, and enrichment techniques are applied to detect changing expression differences of low concentration proteins that will reflect changes in the brain. However, removal procedures can cause the loss of low-abundance proteins as well as high-abundance proteins (Lee et al., 2019).

In summary, the heterogeneity of brain tissue and its inability to obtain it easily and the presence of some high abundance proteins in biological fluids such as CSF, serum, and plasma lead to various disadvantages. On the other hand, experimental animals are preferred to avoid such disadvantages or to more clearly reveal the mechanisms that represent the initial stages of neurodegenerative diseases through certain chemical applications and to test pre-clinical drug applications (Sanfeliu et al., 2024; Moon, 2022).

Compared to other vertebrate animal models, zebrafish have become a prominent animal model for obtaining information about metabolic pathways, thanks to their advantages that facilitate mass production, such as their low cost, small size, ease of care, rapid embryonic development, short sexual maturity cycle, and high fertility rate. These desirable characteristics have also made zebrafish attractive for proteomic studies (d'Amora & Giordani, 2018; Chia et al., 2022).

In particular, the production of high numbers of embryos and their rapid development provides a large number of copies for analysis, thus minimizing protein changes due to individual differences (Bai & Tang, 2023). Having its genome sequenced also facilitates protein identification (Howe et al., 2013; Kelkar et al., 2014). To date, the proteome of different developmental stages and various organs of zebrafish has been examined (Purushothaman et al., 2019; Ünal et al., 2023; Konzer et al., 2013; Groh et al., 2011; Nolte et al., 2015; Surmen, 2022; Langebeck-Jensen et al., 2019; De Souza et al., 2009; Ayobahan et al., 2020; Eastlake et al., 2017) (Figure 8.3).

Although there is little evidence about the zebrafish blood tissue, it is known to be similar to the human plasma proteome. Some of these proteins in the plasma proteome are associated with immunity and lipid transport. Additionally, 598 of the 882 known zebrafish plasma proteins have human homologs. Notable findings of the zebrafish plasma proteome include that it has high abundance proteins similar to human plasma; female plasma in particular carries vitellogenins and contains many homologues of human proteins (Li et al., 2016). Serum and plasma samples are valuable clinical samples for the discovery of biomarkers of neurodegenerative diseases. However, the 10 most abundant proteins in the composition of human blood tissue have a ratio of 90%, and the 22 most abundant proteins have a ratio of approximately 99% (Baker et al., 2012; Wang et al., 2020) (Figure 8.4). The obligation to search for biomarkers in the remaining 1% requires deep proteomic analyses to obtain a comprehensive human serum proteome. Although the nervous system of zebrafish is similar to that of humans, since it has a simpler metabolism, the probability of detecting protein markers of the nervous system is higher than in human plasma tissue. However, since the amount of plasma that can be collected is quite limited, a large number of fish are needed. Additionally, unlike males, female plasma proteome contains abundant vitellogenins and may negatively affect the yield of MS analyses. In this context, if a gender-related study will not be carried out, the use of male zebrafish may provide an advantage in proteomic studies (Li et al., 2016). Additionally, the high abundance of vitellogenins in larvae and embryos can also pose problems that limit shotgun proteomic analyses. Although there is a high sample requirement, protein loss, and low reproducibility risks, efforts are being made to increase analysis efficiency by deyolking and fractionation to overcome this. A new deyolking procedure has recently been published that provides a higher number of protein identifications (compared to non-deyolking) from the beginning of embryonic development. Although this procedure significantly increased the number of non-vitogenin proteins in 5 different embryonic stages (1-, 16-, and 32-cell stages, oblong, bud), parallel processing of both yolk and non-yolk samples was recommended to obtain a comprehensive proteome analysis in studies (Purushothaman et al., 2019; Rahlouni et al., 2015). On the other hand, it is possible to reveal more specific proteomes such as synaptosome profile, myelin proteome, and brain plasma membrane proteome in zebrafish by using various purification and enrichment techniques (Bayes et al., 2017; Siems et al., 2021; Smidak et al., 2016).

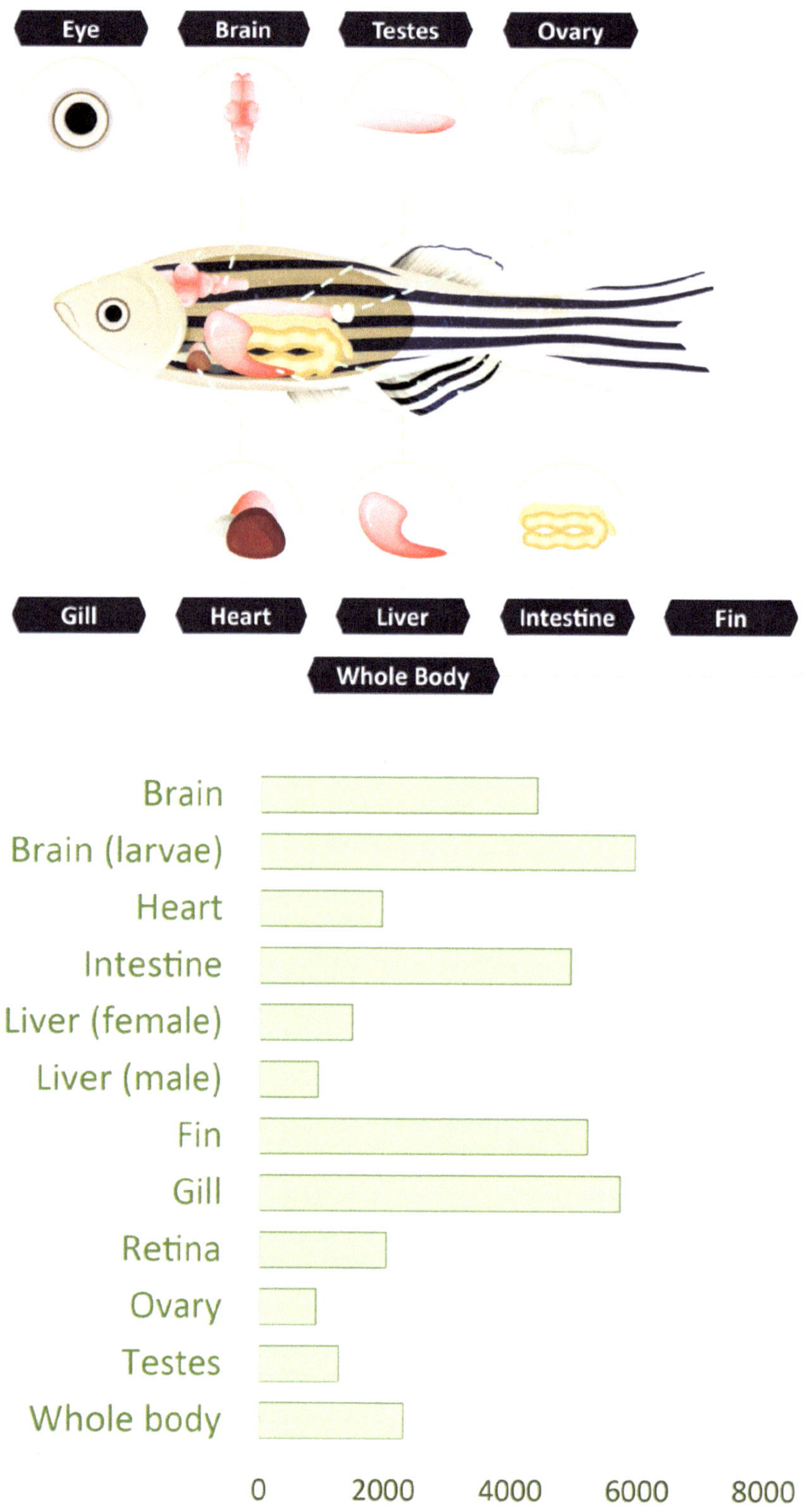


**FIGURE 8.3** Zebrafish organs and identified proteins by bottom-up proteomic approaches.

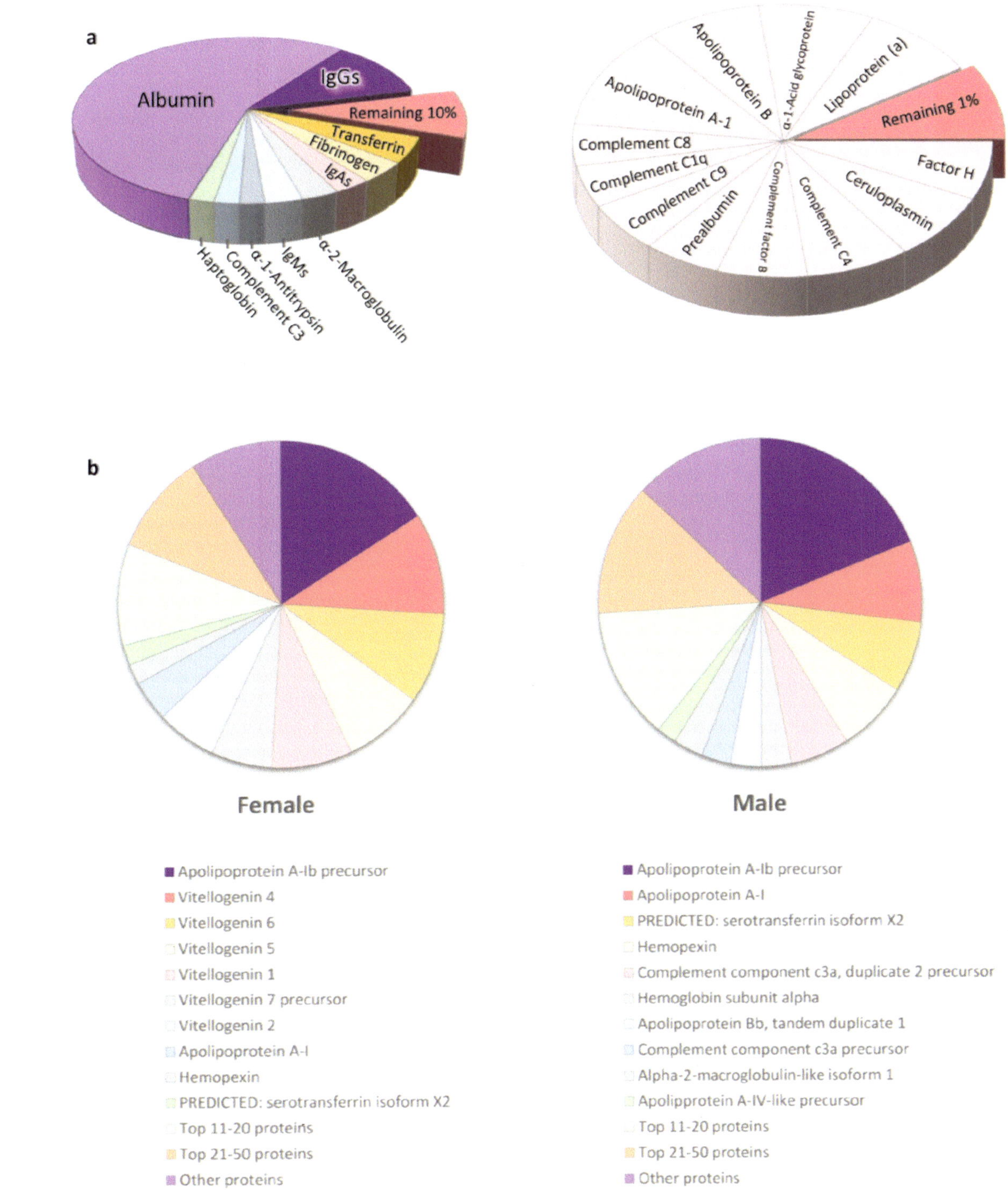


**FIGURE 8.4** Abundance distribution of human and zebrafish plasma proteins. a. human b. zebrafish.

## Proteomic Studies on Zebrafish Models for Neurodegenerative Diseases

Large-scale studies such as genomics and transcriptomics have revealed that approximately 70% of zebrafish genes have at least one human ortholog, and more than 26,000 of them code for proteins (Howe et al., 2013). In addition to its high genetic similarity, the proteome of zebrafish may also bear significant similarities with the human proteome (Li et al., 2016; Groh et al., 2015). However, comparative analysis of transcriptomic and proteomic data has highlighted that the correlation between mRNA and protein levels may be low and that proteomic research is important in examining cellular changes associated with diseases (Groh et al., 2011; Alli Shaik et al., 2014; Hin et al., 2020; Kelkar et al., 2014). Although there are relatively few studies, the number of proteomic studies investigating the pathological effects of various chemical agents and environmental factors on the embryogenic stages and organ systems of zebrafish is increasing every year. While some of these studies focus on the changes in the zebrafish proteome caused by neurotoxic agents, some of them provide important information about the neurodegenerative damage in the Alzheimer's and Parkinson's diseases model caused by these agents (Gebriel et al., 2014; Toni et al., 2019; Frøyset et al., 2016; Kwon et al., 2023; Wang et al., 2023). In this section, research on the zebrafish proteome is examined under two main headings: studies on neurodegenerative diseases and neurotoxic agents.

## Parkinson's Disease

1-Methyl-4-phenyl-1,2,3,6-tetrahydropyridine (MPTP) and rotenone are the major neurotoxins known to induce PD-like motor and non-motor changes (Chia et al., 2022). Proteomic study carried out in the brain tissues of adult zebrafish exposed to MPTP to understand the etiology of PD revealed expression changes in 78 proteins as well as downregulated NEFL (cytoskeleton remodelling of neurofilaments), MUNC13-1 (regulation of synaptic functions), NAV2 (cellular growth and migration), and GAPVD1 (endocytosis) proteins related to neurological pathways (Sarath Babu et al., 2016). Additionally, global proteomic analyses provided insights into the mechanisms of pesticide-induced neurotoxicity and the effects of therapeutic agents in the zebrafish Parkinson's model. It has been suggested that Suvarna Bhasma may have a neuroprotective effect on zebrafish where a decrease in rotenone-induced motor behavior and dopamine levels was observed. According to preliminary findings, PD-related proteins (sncgb, ywhael, ywhah, uchl1, ywhaba, psma6a, ywhabl, ywhaqb, and ywhabb) are differentially expressed with rotenone exposure, while Suvarna Bhasma has been reported to be protective against rotenone (Biswas et al., 2024).

On the other hand, there are various studies showing that the intestinal microbiome can be improved through diet and that some nutritional habits can have a neuroprotective effect. Disturbances between the gut–brain axis, which may occur before motor symptoms, may indicate pathological processes that cause PD. It is thought that the intestinal microbiome and gastrointestinal dysfunctions play a role in the disruptions in the two-way communication between the gut and the brain. The findings in these studies that polyunsaturated fatty acids reduce the risk of PD also support the idea that diet may have important effects in reducing the severity of the disease (Dogra et al., 2022; Alfonsetti et al., 2022).

A study identifying the molecular network of the gut–brain axis demonstrated beneficial effects of erucic acid, a monounsaturated ω9-fatty acid, in a rotenone-induced PD model. Accordingly, erucic acid, as a PPARγ-ligand, was effective in the expression of proteins related to cytoskeletal organization and lipid metabolism, which were significantly dysregulated in both the intestines and brain tissues of zebrafish (Ünal et al., 2023). Additionally, the healing effect of the ketogenic diet against rotenone toxicity was investigated using proteomic methods. Global proteomic and phosphoproteomic analyses of zebrafish have shown that octanoic acid can reverse rotenone-induced expression changes in MAP family members, energy and redox system, stress response, and mitochondrial and cytoskeletal proteins (Sürmen et al., 2021, 2022).

Although dieldrin is another pesticide associated with PD, there is still insufficient information on whether it increases the risk of neurodegeneration. When the molecular networks regulated by dieldrin were evaluated in terms of transcriptomics and proteomics, it was seen that exposure to dieldrin caused significant changes in the proteome of whole-body zebrafish. It has been observed that some proteins related to oxidative respiration and indicating mitochondrial dysfunction, such as rotenone, are affected by dieldrin (Cowie et al., 2017).

In addition, differentiations in proteins known to play a role in mitochondrial metabolism, mitophagy, stress response, redox regulation, and inflammation were also detected in the brain tissues of early adult zebrafish before PD-related phenotypic changes appear. Another advantageous feature of zebrafish for researchers who want to work with experimental animal models is that they are accessible for manipulation and monitoring at all developmental stages. Recently, the DJ-1/park 7 knockout zebrafish model was produced using CRISPR-Cas9 technology to induce PD. It is known from previous studies that DJ-1 deficiency causes changes in both motor and dopaminergic functions in different animal models. It was found that dopaminergic rate-limiting enzyme tyrosine hydroxylase levels remained normal in the young adult and larval stages of the knockout DJ-1 zebrafish model, but decreased in the late adult stage (Edson et al., 2019).

In summary, the accumulated evidence on proteomic changes in zebrafish models as a result of studies on PD sheds light on the mechanisms underlying PD pathogenesis, providing strong insights into the identification of target molecules and the development of therapeutic agents.

## Alzheimer's Disease

Although Alzheimer's is the most common neurodegenerative disease and many molecular studies have been conducted on it, interestingly, much fewer proteomic studies have been conducted on zebrafish models of this disease. Other systems biology approaches were also used in these studies, which addressed the neuropathological mechanisms and treatment methods in Alzheimer's disease from different perspectives (Hin et al., 2020; Siddiqui et al., 2023; Hernández et al., 2020). Although it is estimated that mutations cause some molecular changes at very early ages in individuals carrying some mutations in familial AD, the mechanisms in the brain tissue are not fully known. Recently, modelling of the K115fs mutation of the PSEN2 gene by gene editing has been used to evaluate the oxidative stress-induced PS2V-like expression pattern. The fact that both gene expression regulation and protein abundances in the heterozygous mutant young adult and old adult zebrafish brain are different from normal brain aging indicated that pathological changes occur much before AD symptoms (Hin et al., 2020).

Another study is the proteomic examination of the brain tissue of BACE1 knockout zebrafish. BACE1 is a secretase involved in the production of toxic amyloid β (Aβ) by degrading amyloid precursor protein (APP) and is among the important drug targets for AD. However, in addition to APP,

BACE1 is predicted to have multiple substrates. Identification of substrates of BACE1, which is considered the rate-limiting enzyme of Aβ production, can provide important information in the evaluation of drug response and the development of different enzyme inhibitors (Dislich et al., 2015; Hogl et al., 2013). In this context, in a comprehensive proteomic study in which 5,614 proteins were identified through at least two peptides, a distinctive accumulation of 24 intrinsic membrane proteins was reported in the BACE1 knockout 6-month-old adult zebrafish brain. Other notable findings are that 5 of these proteins (Appa, Appb, Aplp2, Pges2, and Nadl1.2) representing candidate Bace1 substrates are previously validated murine or human orthologs, and 4 (Leucine-rich repeat and fibronectin type III domain-containing protein 1, Novel protocadherin protein, Novel protein similar to vertebrate glypican 1 and Glypican-1 like) are homologues of proteins known as candidate BACE1 substrates (Hogl et al., 2013).

In addition, observed changes in the zebrafish brain proteome associated with factors that induce neurotoxicity have provided new information about the pathogenesis of AD (Jami et al., 2023; Siddiqui et al., 2023). Although manganese is a trace element that may be important for the nervous system, excessive exposure to it is known to play an important role in the development of various neurodegenerative diseases, including AD (Martins et al., 2020). The toxic effect of manganese has been demonstrated in cell culture and mouse and rat models, as well as on zebrafish embryos (Exil et al., 2014; Krishna et al., 2014; Zhao et al., 2009; Hernández et al., 2015). However, the mechanism of manganese-induced neurotoxicity in zebrafish has not been studied sufficiently and requires further examination with proteomic analyses. On the other hand, the integration of proteomic and metallomic data obtained from cell culture samples indicated impaired protein metabolism, including mechanisms such as energy, immune response, apoptosis, protein accumulation, and cell cycle.

It has been reported that MnCl2 type causes expression changes in various AD-related proteins, as well as an excessive increase in ALDOA, ATP5K, COX5A, NDUFA4, and DLD proteins involved in metabolic pathways (Hernández et al., 2020).

Additionally, okadaic acid (OKA) is known as a powerful agent for examining AD-related neurotoxicity mechanisms in vivo and in vitro (Kamat et al., 2013; Chia et al., 2022). Exposure to OKA causes distinct AD pathologies, including tau hyperphosphorylation, plaque formation, and glial activation, as well as learning and memory disorders (Koehler et al., 2018; Kamat et al., 2013; Thawkar et al., 2021). Proteomic analyses in zebrafish brain tissue also support the toxic effects of OKA that inhibit neuronal development. Therefore, the OKA-induced AD zebrafish model is considered promising for drug discovery for the treatment of the disease (Siddiqui et al., 2023; Koehler et al., 2018; Nada et al., 2016).

## Neurotoxicity Studies

Today, a wide variety of chemical agents such as industrial wastes, heavy metals, pesticides, microplastics, and polluting gases that can cause environmental pollution can have damaging effects on body systems (Costa et al., 2020; Prüst et al., 2020; Wang et al., 2021). There are many studies reporting the neurotoxic effects of especially environmental air pollution on the central nervous system. Some of these studies have shown that traffic-related air pollution may play a role in neurodevelopmental and neurodegenerative disorders. Additionally, epidemiological and neurotoxicological studies also support the relationship between air pollution toxicity and the risk of neurodegenerative diseases (Wang et al., 2021; Ha et al., 2022). According to animal studies, the two main mechanisms known to occur in the central nervous system associated with exposure to air pollution are oxidative stress and neuroinflammation (Costa et al., 2020). Diesel exhaust is a major cause of traffic-related air pollution, and several studies investigating neurotoxicity mechanisms have found significant findings that diesel exhaust triggers neuroinflammation. In a study conducted on rats, it was reported that the midbrain region, where exposure to diesel exhaust for 6 months increased the levels of TNF alpha, α Synuclein, and Aβ42, may be more sensitive to neuroinflammation (Levesque et al., 2011).

Studies conducted on zebrafish embryos and larvae have shown that diesel exhaust can cause an increase in IL-1ß, neuron loss, and microglial activation, and a decrease in the expression of autophagy-related genes, in addition to the accumulation of synuclein gamma 1 (Sncg1) equivalent to mammalian α-synuclein (Ha et al., 2022; Barnhill et al., 2020).

Additionally, transcriptomic and proteomic analyses have revealed various biological processes and signalling pathways involved in neurodegenerative events. Proteomic data indicating activation of Phase I and Phase II xenobiotic metabolisms associated with increased protein expressions (Phase I proteins: CYP1A1, CYP3A7, HMOX1 (heme oxygenase 1), CAT (catalase) and CES1 (carboxylesterase 1) and Phase II proteins: UGT1A1 (UDP glucuronosyltransferase family 1 member A1) and GSTP1 (glutathione S-transferase pi 1)) in larvae exposed to diesel exhaust were also supported by transcriptomic analyses (Jami et al., 2023).

On the other hand, the widespread use of plastic, another polluting petroleum product, in daily life makes exposure to nanoplastics inevitable. There is evidence that nanoparticles that pass through the blood–brain barrier and accumulate in the brain associated with this strong exposure have a neurotoxic effect and cause PD-like neurodegeneration (Shan et al., 2022; Liang et al., 2022; Prüst et al., 2020). Additionally, a source of nanoplastics, the effects of polystyrene particles on the proteome of zebrafish larvae have also been associated with PD-like symptoms. Proteomic results revealed significant changes in neurotransmitter disorder–associated (PHGDH, NENF, COX5AB, and ABAT) proteins, in addition to PARK 7 and PRDX2 proteins closely related to PD-pathogenesis (Wang et al., 2023).

In addition to environmental air pollution, neurotoxins that can be found in food and drinking water, even in low doses, can also cause neurodegenerative damage through different mechanisms. Some types of cyanotoxins produced by cyanobacteria have been shown to be neurotoxic in various in vivo and in vitro experiments.

There is a lot of evidence revealing the neurotoxic effect of ß-methylamino-L-alanine (BMAA) formation (Nugumanova et al., 2023; Sini et al., 2021; Yu et al., 2021). When evaluated holistically, behavioural and proteomic analysis results in zebrafish larvae exposed to BMAA alone or together with different cyanotoxins provided important information for new treatment methods focusing on amyotrophic lateral sclerosis–like neurodegeneration (Frøyset et al., 2016; Martin et al., 2021).

Generally speaking, neurodegenerative damage and associated proteomic changes caused by neurotoxic agents in the central nervous system have been reported in numerous neurotoxicological studies (Jami et al., 2023; Wang et al., 2023; Frøyset 2016; Martin et al., 2021). (Table 8.1)

At the same time, the issue of how long-term exposure to these agents affects cognitive development and memory has been of great interest, and it has been shown that the brain is more sensitive to various neurotoxic agents during the critical period of development. In these studies, larvae and embryos of zebrafish are widely used to evaluate developmental neurotoxicity due to their advantages in allowing developmental differences in the nervous system to be easily monitored in vivo (d'Amora & Giordani, 2018). Atrazine, Perfluorooctanesulfonic acid, perfluorobutanesulfonic acid, and Tris(1,3-dichloropropyl) phosphate are among the agents whose neurodevelopmental effects have been investigated using proteomic approaches (Horzmann et al., 2018; Ahkin Chin Tai et al., 2023; Lee et al., 2021; Min et al., 2023; Xu et al., 2023). In addition, changes in various proteins of neurological pathways have been revealed through zebrafish applications to understand the role of different compounds such as

**TABLE 8.1**

Proteomic Investigations of Various Neurotoxic Agents on Different Developmental Stages of Zebrafish

| Stage | Agent | Exposure time/period | Exposure dose | Quantification Method | Highlighted results | Ref |
|---|---|---|---|---|---|---|
| 50 larvae in each replicate (transgenic zebrafish line Tg (atoh1a:dTomato) | PS nanoplastics | for 144 h | 1.0 mg/L | TMT labelling | **122 DEPs**<br>-5 proteins were relevant to neurotransmitter disorders.<br>-Upregulation of PARK7, PHGDH, NENF, COX5AB, and downregulation of ABAT | Wang et al., 2023 |
| 80–100 larvae heads | Diesel exhaust | until 5 dpf | 20 μg/ml | 10-plex TMT labelling | **748 DEPs**<br>-The activation of xenobiotic (Phase I and II) metabolism.<br>-Upregulation of CYP1A1, CYP3A7, HMOX1, UGT1A1, GSTP1, SULT1A1, SULT1C2, SULT2B1 | Jami et al., 2023 |
| 60 larvae (triplicates of 20 each) | PFBS | for 120 hpf | 7, 70, 700, 3500, and 7000 μM | 4-plex TMT labelling | **386 DEPs**<br>Dysregulation of mitochondria and cardiac muscle function, glycolysis/glucogenesis, biosynthesis of amino acid, cytochrome P450 and glutathione-related antioxidant mechanisms | Min et al., 2023 |
| Larvae | BMAA | until 4 dpf | 65mM | Label-free | Dysregulation of proteins associated with glutamate receptor signalling, protein homeostasis, ROS response, and neuronal cell death | Frøyset et al., 2016 |
| 50 embryos (per treatment group) | Atrazine | until 120 hpf | 0, 0.3, 3, and 30 μg/L | Label-free | **21 DEPs**<br>Top upstream regulators TGFB1, peoniflorin, GnRH analog and L-Dopa | Horzmann et al., 2018 |
| 30 larvae (per treatment group) | Atrazine | until 120 hpf | 0, 0.3, 3, and 30 μg/L | Label-free | **62 DEPs**<br>Differentially expressed ATP synthase proteins, collagen proteins, heterogeneous nuclear ribonucleoproteins, and ribosomal proteins | Ahkin Chin Tai et al., 2023 |
| 60 embryos (20 larvae per replicate) | PFOS | at 4 hpf, until 120 hpf | 0.1, 1, 5, and 10 μM | 4-plex TMT labelling | **143 DEPs**<br>Perturbations in the pathways of neurological function, oxidative stress, and energy metabolism. | Lee et al., 2021 |
| Embryos | TDCPP | at 48 hpf, until 144 h | 400 and 600 μg/L | TMT labelling | **1059 DEPs**<br>Dysregulation of focal adhesion, ECM–receptor interactions, signalling transduction, calcium signalling, and MAPK signalling pathways | Xu et al., 2023 |

(*Continued*)

**TABLE 8.1 (CONTINUED)**

Proteomic Investigations of Various Neurotoxic Agents on Different Developmental Stages of Zebrafish

| Stage | Agent | Exposure time/period | Exposure dose | Quantification Method | Highlighted results | Ref |
|---|---|---|---|---|---|---|
| Pools of 8 Larvae | ACR | at 5 dpf, for 72h | 0.5–3.0mM | Label-free | **88 DEPs**<br>-Alteration of transcriptional markers of proteins involved in synaptic vesicle cycle<br>-Formation of ACR adducts on cysteine residues of some synaptic proteins | Prats et al., 2017 |
| Adult zebrafish brain | ACR | for 72 h | 0.75 mM | Label-free | **80 DEPs**<br>-14 of the DEPs were rescued by NAC<br>-141 modified proteins, some of them with catalytic activity, amino acid metabolism, neurotransmission, and cell redox mechanisms | Faria et al., 2019 |
| 5 Adult zebrafish brain for each 3 replicates | MeP | at 4 months old, for 28 days | 10 µg/L | - | **64 DEPs** in male brain<br>**177 DEPs** in female brain<br>11 significantly enriched biological pathways, including response to stress and LPS endotoxin pathways | Hu et al., 2023 |
| Embryos of 10 adult zebrafish | Benzyl benzoate | at 6–8 months old, for 7 days | 1 µg/ mL | Label-free | **83 DEPs**<br>Dysregulation of proteins in structural molecule, lipid transporter, RNA binding, and oxidoreductase activity | Kwon et al., 2023 |
| A total of 300 adult zebrafish whole brain (male and female, 6 months old, 20 fish per batch in duplicates) | MPTP | - | 24-hour intervals (1-dose: 10 µg/µL and 2-dose: 2x10 µg/µL) | 4-plex iTRAQ labelling | **78 DEPs**<br>Dysregulation of proteins associated with cytoskeleton remodelling of neurofilaments, neuronal development, neurophysiological synaptic vesicle fusion, and GABA-B receptor signalling in presynaptic nerve terminals | Sarath Babu et al., 2016 |
| Adult zebrafish brain | M. aeruginosa | at 5 months old, for 96h | - | Label-free | **108 DEPs**<br>- Severe impairment of neurotransmitter transmission and brain information storage and transmission<br>- Susceptibility to neurological diseases<br>- Dysregulation of proteins in synaptic vesicle cycle and terpenoid skeleton biosynthesis pathway | Yu et al., 2021 |
| Embryos | Cyanotoxin BMAA (2,4-DAB isomer) | at 6 hpf, until 6 dpf (time point: 2 dpf and 4 dpf) | 500 µM | Label-free | A total of **102** and **398 DEPs** in the 2 dpf<br>and 4 dpf groups, respectively<br>-BMAA cause neurotoxic effect<br>-Pathways affected by 2,4-DAP include cellular damage and inflammatory response.<br>- Dysregulation of proteins in apoptosis and metabolic function such as glycolysis and gluconeogenesis, calcium signalling (at 4 dpf) | Martin et al., 2021 |

ACR: Acrylamide, BMAA: β-methyl-amino-L-alanine, DEPs: Differentially expressed proteins, dpf: days post-fertilization, hpf: hour post-fertilization, MeP: Methylparaben, MPTP: 1-methyl-4-phenyl-1,2,3,6-tetrahydropyridine, NAC: N-acetylcysteine, PFBS: perfluorobutanesulfonic acid, PFOS: Perfluorooctanesulfonic acid, PS: Polystyrene, TDCPP: Tri(1,3-dichloropropyl) phosphate, TMT: Tandem mass tag, 2,4-DAB: 2,4- diaminobutyric acid acrylamide, methylparaben, and benzyl benzoate (Prats et al., 2017; Faria et al., 2019; Hu et al., 2023; Kwon et al., 2023). In particular, the number of studies on acrylamide neurotoxicity has increased significantly in the last 10 years. The neurotoxic effect of acrylamide, a compound commonly encountered in everyday foods, have been linked to various synaptopathies (Prats et al., 2017). In the zebrafish model developed for acute acrylamide toxicity, acrylamide adducts were detected in the cysteine residues of some proteins related to the synaptic vesicle cycle (Faria et al., 2019).

### Ecotoxicity Studies

Zebrafish have also been an ideal model organism to investigate neurodevelopmental toxicity caused by pollutants frequently released into the aquatic ecosystem. Bioaccumulation of heavy metals, pesticides, and microplastics are the main research topics of aquatic ecotoxicology. It is known that acute or chronic exposure to these types of pollutants, especially during the embryogenic stage, causes serious malformations, neurological damage, and behavioural disorders (Maharaj et al., 2020; Li et al., 2023). On the other hand, studies have shown that the eco-neurotoxic effect is not only caused by chemical pollution and that temperature changes in water also affect movement, learning, and memory behaviour (Toni et al., 2019). According to experimental observations, the optimum water temperature for survival of zebrafish varies between 22 and 31 °C (Sfakianakis et al., 2011). However, various studies have reported that acute or chronic exposure to lower or higher temperatures can cause significant effects on zebrafish behaviour and brain proteome (Toni et al., 2019).

In these studies, compared to the control group, zebrafish kept at 18 °C exhibited an anxiety-like state, while they exhibited proactive behaviour in 21-day treatments at 34 °C. In addition, proteomic analyses in zebrafish brain tissue exposed to thermal stress at 18 °C and 34 °C revealed altered expression of various proteins, indicating that neurons, axons, and dendrites were affected. Long-term changes in proteins related to energy metabolism, neurotransmitter release, synaptic plasticity, cytoskeletal organization, and lipid transport that persist after 21 days may contain clues to explain the mechanisms of cognitive impairments (Nonnis et al., 2021; Angiulli et al., 2020; Toni et al., 2019). Protein biomarkers, which stand out with the combined analysis results of these studies, strengthened the relationship between temperature fluctuation and neurotoxicity (Maffioli et al., 2023).

## Conclusion

To date, MS-based proteomics has made significant progress in providing a qualitative and quantitative view of the proteome of diverse biological structures, from single types of cells or subcellular structures to an organism. In parallel with these developments, interest in neuroproteomic research aimed at understanding the molecular pathophysiology of diseases has also increased. The use of proteomic technologies has revealed various proteins that will contribute to the development of better diagnostic and treatment strategies for neurodegenerative diseases. Findings in these studies included changes in the proteomes of clinical samples such as human brain, serum, plasma, CSF, urine, saliva, and tears, as well as animal models. However, proteomic studies conducted with clinical samples have some limitations. First of all, heterogeneity is very high in extremely organized organs such as the brain, and for more specific cell-based results, additional steps such as LCM, FACs, and MACs are required in the preparation phase before MS analysis. There is abundant protein problem in serum, plasma, and CSF samples. On the other hand, brain and CSF are obtained using invasive methods. Moreover, the disease is advanced by the time it is diagnosed, whereas the developmental stage of the disease predates the symptoms. Therefore, information on proteomic changes regarding the initial stage of the disease is insufficient. In this respect, experimental animals seem to provide several advantages. The number of proteomic studies in the zebrafish model has increased significantly in the last 10 years. This chapter also focused on proteomic studies using the zebrafish model to study neurodegenerative diseases. It was revealed that zebrafish, whose genome was completely sequenced, had more than 26,000 protein-coding genes. It was also found that more than half of the zebrafish genes are human orthologs and share high levels of similarities with the human central nervous system.

In this regard, the importance of zebrafish in researching neurodegenerative diseases, especially Parkinson's, has become more clearly understood.

Since the mid-2000s, many studies have been carried out to reveal the developmental stages of zebrafish, the protein profiles of various organs, and the pathological effects of gene regulation or neurotoxic agents, and as a result, comprehensive knowledge has been provided. In addition, comprehensive profiles of cellular structures such as zebrafish brain plasma membrane proteome, myelin proteome, extracellular vesicle, and synapse proteome, which can be a valuable source of data to study neural processes, were also obtained. This chapter focused on studies using proteomic approaches in zebrafish neurodegenerative disease models. If the highlights of proteomic studies in zebrafish are to be mentioned, the findings obtained in modelling neurodegenerative diseases and neurotoxicology studies have been interesting. It has been noted that short-chain fatty acids may have positive effects in improving Parkinson's symptoms caused by rotenone, a pesticide. Another toxicology study tried to explain the neurological effects of diesel exhaust, one of the important causes of today's air pollution, with proteomic changes in the larval stage of zebrafish. Another threatening situation for human health is neuroplastic exposure and climate change. It has been revealed that nano-plastics cause PD-like symptoms and protein expression differences in this process. On the other hand, it has been shown that not only chemical agents but also fluctuations in ambient temperature can affect movement, learning, and memory behaviour in zebrafish. It has been shown that zebrafish exposed to thermal stress can exhibit anxiety or proactive behaviour, and brain proteome analysis strengthened the findings regarding cognitive impairments. Contributions to revealing disease mechanisms using proteomic approaches continue at an increasing pace. However, the number of these studies is quite low, especially for Alzheimer's, which ranks first among the most common neurodegenerative diseases. Today's proteomics technology has the potential to reveal comprehensive protein profiling of subcellular structures. Even in the zebrafish synapse proteome, nearly 10,000 proteins have been identified. As of 2024, information on a total of 47,577 zebrafish proteins is stored in the UniProt Knowledge Base, of which only 3,320 have been manually annotated and reviewed (access date: 30.03.2024). It seems that more proteomic studies are needed to establish a more robust data source. Thanks to developing instruments, methods, and bioinformatics tools, the field of proteomics has a rich infrastructure. In this respect, the role of proteomics, which aims to holistically examine proteins that play a major role in the pathophysiology of neurodegenerative diseases, is very important.

## Acknowledgments

The authors would like to thank Demet Turan for her support of the illustrations in this chapter.

## REFERENCES

Abdallah, C., Dumas-Gaudot, E., Renaut, J., & Sergeant, K. (2012). Gel-based and gel-free quantitative proteomics approaches at a glance. *International Journal of Plant Genomics*, 494572. https://doi.org/10.1155/2012/494572

Ahkin Chin Tai, J. K., Horzmann, K. A., Jenkins, T. L., Akoro, I. N., Stradtman, S., Aryal, U. K., & Freeman, J. L. (2023). Adverse developmental impacts in progeny of zebrafish exposed to the agricultural herbicide atrazine during embryogenesis. *Environment International*, *180*, 108213. https://doi.org/10.1016/j.envint.2023.108213

Alfonsetti, M., Castelli, V., & d'Angelo, M. (2022). Are we what we eat? Impact of diet on the gut-brain axis in Parkinson's disease. *Nutrients*, *14*(2), 380. https://doi.org/10.3390/nu14020380

Alli Shaik, A., Wee, S., Li, R. H., Li, Z., Carney, T. J., Mathavan, S., & Gunaratne, J. (2014). Functional mapping of the zebrafish early embryo proteome and transcriptome. *Journal of Proteome Research*, *13*(12), 5536–5550. https://doi.org/10.1021/pr5005136

Angiulli, E., Pagliara, V., Cioni, C., Frabetti, F., Pizzetti, F., Alleva, E., & Toni, M. (2020). Increase in environmental temperature affects exploratory behaviour, anxiety and social preference in Danio rerio. *Scientific Reports*, *10*(1), 5385. https://doi.org/10.1038/s41598-020-62331-1

Ayobahan, S. U., Eilebrecht, S., Baumann, L., Teigeler, M., Hollert, H., Kalkhof, S., Eilebrecht, E., & Schäfers, C. (2020). Detection of biomarkers to differentiate endocrine disruption from hepatotoxicity in zebrafish (Danio rerio) using proteomics. *Chemosphere*, *240*, 124970. https://doi.org/10.1016/j.chemosphere.2019.124970

Bai, B., Vanderwall, D., Li, Y., Wang, X., Poudel, S., Wang, H., Dey, K. K., Chen, P. C., Yang, K., & Peng, J. (2021). Correction to: Proteomic landscape of Alzheimer's disease: Novel insights into pathogenesis and biomarker discovery. *Molecular Neurodegeneration*, *16*(1), 72. https://doi.org/10.1186/s13024-021-00493-w

Bai, C., & Tang, M. (2023). Progress on the toxicity of quantum dots to model organism-zebrafish. *Journal of Applied Toxicology, JAT*, *43*(1), 89–106. https://doi.org/10.1002/jat.4333

Baker, E. S., Liu, T., Petyuk, V. A., Burnum-Johnson, K. E., Ibrahim, Y. M., Anderson, G. A., & Smith, R. D. (2012). Mass spectrometry for translational proteomics: Progress and clinical implications. *Genome Medicine*, *4*(8), 63. https://doi.org/10.1186/gm364

Barnhill, L. M., Khuansuwan, S., Juarez, D., Murata, H., Araujo, J. A., & Bronstein, J. M. (2020). Diesel exhaust extract exposure induces neuronal toxicity by disrupting autophagy. *Toxicological Sciences: An Official Journal of the Society of Toxicology*, *176*(1), 193–202. https://doi.org/10.1093/toxsci/kfaa055

Bastos, P., Ferreira, R., Manadas, B., Moreira, P. I., & Vitorino, R. (2017). Insights into the human brain proteome: Disclosing the biological meaning of protein networks in cerebrospinal fluid. *Critical Reviews in Clinical Laboratory Sciences*, *54*(3), 185–204. https://doi.org/10.1080/10408363.2017.1299682

Bayés, À., Collins, M. O., Reig-Viader, R., Gou, G., Goulding, D., Izquierdo, A., Choudhary, J. S., Emes, R. D., & Grant, S. G. (2017). Evolution of complexity in the zebrafish synapse proteome. *Nature Communications*, *8*, 14613. https://doi.org/10.1038/ncomms14613

Biswas, S., Chawda, M., Gudi, R., & Bellare, J. (2024). Neuroprotective effects of nanogold-based Ayurveda medicine Suvarna Bhasma against rotenone-induced Parkinson's-like model. *Journal of Ayurveda and Integrative Medicine*, *15*(1), 100854. https://doi.org/10.1016/j.jaim.2023.100854

Boerger, M., Funke, S., Leha, A., Roser, A. E., Wuestemann, A. K., Maass, F., Bähr, M., Grus, F., & Lingor, P. (2019). Proteomic analysis of tear fluid reveals disease-specific patterns in patients with Parkinson's disease - A pilot study. *Parkinsonism & Related Disorders*, *63*, 3–9. https://doi.org/10.1016/j.parkreldis.2019.03.001

Budelier, M. M., & Bateman, R. J. (2020). Biomarkers of Alzheimer disease. *The Journal of Applied Laboratory Medicine*, *5*(1), 194–208. https://doi.org/10.1373/jalm.2019.030080

Butterfield, D. A., & Boyd-Kimball, D. (2019). Redox proteomics and amyloid β-peptide: Insights into Alzheimer disease. *Journal of Neurochemistry*, *151*(4), 459–487. https://doi.org/10.1111/jnc.14589

Carbonara, K., Andonovski, M., & Coorssen, J. R. (2021). Proteomes are of proteoforms: Embracing the complexity. *Proteomes*, *9*(3), 38. https://doi.org/10.3390/proteomes9030038

Carlyle, B. C., Trombetta, B. A., & Arnold, S. E. (2018). Proteomic approaches for the discovery of biofluid biomarkers of neurodegenerative dementias. *Proteomes*, *6*(3), 32. https://doi.org/10.3390/proteomes6030032

Chen, C., Hou, J., Tanner, J. J., & Cheng, J. (2020). Bioinformatics methods for mass spectrometry-based proteomics data analysis. *International Journal of Molecular Sciences*, *21*(8), 2873. https://doi.org/10.3390/ijms21082873

Chen, X., Wei, S., Ji, Y., Guo, X., & Yang, F. (2015). Quantitative proteomics using SILAC: Principles, applications, and developments. *Proteomics*, *15*(18), 3175–3192. https://doi.org/10.1002/pmic.201500108

Chia, K., Klingseisen, A., Sieger, D., & Priller, J. (2022). Zebrafish as a model organism for neurodegenerative disease. *Frontiers in Molecular Neuroscience*, *15*, 940484. https://doi.org/10.3389/fnmol.2022.940484

Cilento, E. M., Jin, L., Stewart, T., Shi, M., Sheng, L., & Zhang, J. (2019). Mass spectrometry: A platform for biomarker discovery and validation for Alzheimer's and Parkinson's diseases. *Journal of Neurochemistry*, *151*(4), 397–416. https://doi.org/10.1111/jnc.14635

Costa, L. G., Cole, T. B., Dao, K., Chang, Y. C., Coburn, J., & Garrick, J. M. (2020). Effects of air pollution on the nervous system and its possible role in neurodevelopmental and neurodegenerative disorders. *Pharmacology & Therapeutics*, *210*, 107523. https://doi.org/10.1016/j.pharmthera.2020.107523

Cowie, A. M., Sarty, K. I., Mercer, A., Koh, J., Kidd, K. A., & Martyniuk, C. J. (2017). Molecular networks related to the immune system and mitochondria are targets for the pesticide dieldrin in the zebrafish (Danio rerio) central nervous system. *Journal of Proteomics*, *157*, 71–82. https://doi.org/10.1016/j.jprot.2017.02.003

Dammer, E. B., Duong, D. M., Diner, I., Gearing, M., Feng, Y., Lah, J. J., Levey, A. I., & Seyfried, N. T. (2013). Neuron enriched nuclear proteome isolated from human brain. *Journal of Proteome Research*, *12*(7), 3193–3206. https://doi.org/10.1021/pr400246t

d'Amora, M., & Giordani, S. (2018). The utility of zebrafish as a model for screening developmental neurotoxicity. *Frontiers in Neuroscience*, *12*, 976. https://doi.org/10.3389/fnins.2018.00976

De Souza, A. G., MacCormack, T. J., Wang, N., Li, L., & Goss, G. G. (2009). Large-scale proteome profile of the zebrafish (Danio rerio) gill for physiological and biomarker discovery studies. *Zebrafish*, *6*(3), 229–238. https://doi.org/10.1089/zeb.2009.0591

Di Domenico, F., Barone, E., Perluigi, M., & Butterfield, D. A. (2017). The triangle of death in Alzheimer's disease brain: The aberrant cross-talk among energy metabolism, mammalian target of rapamycin signaling, and protein homeostasis revealed by redox proteomics. *Antioxidants & Redox Signaling*, *26*(8), 364–387. https://doi.org/10.1089/ars.2016.6759

Dislich, B., Wohlrab, F., Bachhuber, T., Müller, S. A., Kuhn, P. H., Hogl, S., Meyer-Luehmann, M., & Lichtenthaler, S. F. (2015). Label-free quantitative proteomics of mouse cerebrospinal fluid detects β-site APP Cleaving Enzyme (BACE1) protease substrates in vivo. *Molecular & Cellular Proteomics: MCP*, *14*(10), 2550–2563. https://doi.org/10.1074/mcp.M114.041533

Dogra, N., Mani, R. J., & Katare, D. P. (2022). The gut-brain axis: Two ways signaling in Parkinson's disease. *Cellular and Molecular Neurobiology*, *42*(2), 315–332. https://doi.org/10.1007/s10571-021-01066-7

Dong, W., Qiu, C., Gong, D., Jiang, X., Liu, W., Liu, W., Zhang, L., & Zhang, W. (2019). Proteomics and bioinformatics approaches for the identification of plasma biomarkers to detect Parkinson's disease. *Experimental and Therapeutic Medicine*, *18*(4), 2833–2842. https://doi.org/10.3892/etm.2019.7888

Drummond, E. S., Nayak, S., Faustin, A., Pires, G., Hickman, R. A., Askenazi, M., Cohen, M., Haldiman, T., Kim, C., Han, X., Shao, Y., Safar, J. G., Ueberheide, B., & Wisniewski, T. (2017). Proteomic differences in amyloid plaques in rapidly progressive and sporadic Alzheimer's disease. *Acta Neuropathologica*, *133*(6), 933–954. https://doi.org/10.1007/s00401-017-1691-0

Drummond, E. S., Nayak, S., Pires, G., Ueberheide, B., & Wisniewski, T. (2018). Isolation of amyloid plaques and neurofibrillary tangles from archived Alzheimer's disease tissue using laser-capture microdissection for downstream proteomics. *Methods in Molecular Biology (Clifton, N.J.)*, *1723*, 319–334. https://doi.org/10.1007/978-1-4939-7558-7_18

Drummond, E. S., Nayak, S., Ueberheide, B., & Wisniewski, T. (2015). Proteomic analysis of neurons microdissected from formalin-fixed, paraffin-embedded Alzheimer's disease brain tissue. *Scientific Reports*, *5*, 15456. https://doi.org/10.1038/srep15456

Duong, V. A., & Lee, H. (2023). Bottom-up proteomics: Advancements in sample preparation. *International Journal of Molecular Sciences*, *24*(6), 5350. https://doi.org/10.3390/ijms24065350

Eastlake, K., Heywood, W. E., Tracey-White, D., Aquino, E., Bliss, E., Vasta, G. R., Mills, K., Khaw, P. T., Moosajee, M., & Limb, G. A. (2017). Comparison of proteomic profiles in the zebrafish retina during experimental degeneration and regeneration. *Scientific Reports*, *7*, 44601. https://doi.org/10.1038/srep44601

Edson, A. J., Hushagen, H. A., Frøyset, A. K., Elda, I., Khan, E. A., Di Stefano, A., & Fladmark, K. E. (2019). Dysregulation in the brain protein profile of zebrafish lacking the Parkinson's disease-related protein DJ-1. *Molecular Neurobiology*, *56*(12), 8306–8322. https://doi.org/10.1007/s12035-019-01667-w

Exil, V., Ping, L., Yu, Y., Chakraborty, S., Caito, S. W., Wells, K. S., Karki, P., Lee, E., & Aschner, M. (2014). Activation of MAPK and FoxO by manganese (Mn) in rat neonatal primary astrocyte cultures. *PLoS One*, *9*(5), e94753. https://doi.org/10.1371/journal.pone.0094753

Fang, Q., Strand, A., Law, W., Faca, V. M., Fitzgibbon, M. P., Hamel, N., Houle, B., Liu, X., May, D. H., Poschmann, G., Roy, L., Stühler, K., Ying, W., Zhang, J., Zheng, Z., Bergeron, J. J., Hanash, S., He, F., Leavitt, B. R., Meyer, H. E., … McIntosh, M. W. (2009). Brain-specific proteins decline in the cerebrospinal fluid of humans with Huntington disease. *Molecular & Cellular Proteomics: MCP*, *8*(3), 451–466. https://doi.org/10.1074/mcp.M800231-MCP200

Faria, M., Prats, E., Gómez-Canela, C., Hsu, C. Y., Arick, M. A., 2nd, Bedrossiantz, J., Orozco, M., Garcia-Reyero, N., Ziv, T., Ben-Lulu, S., Admon, A., Gómez-Oliván, L. M., & Raldúa, D. (2019). Therapeutic potential of N-acetylcysteine in acrylamide acute neurotoxicity in adult zebrafish. *Scientific Reports*, *9*(1), 16467. https://doi.org/10.1038/s41598-019-53154-w

Ferrer, I., Santpere, G., Arzberger, T., Bell, J., Blanco, R., Boluda, S., Budka, H., Carmona, M., Giaccone, G., Krebs, B., Limido, L., Parchi, P., Puig, B., Strammiello, R., Ströbel, T., & Kretzschmar, H. (2007). Brain protein preservation largely depends on the postmortem storage temperature: Implications for study of proteins in human neurologic diseases and management of brain banks: A BrainNet Europe study. *Journal of Neuropathology and Experimental Neurology*, *66*(1), 35–46. https://doi.org/10.1097/nen.0b013e31802c3e7d

Figura, M., Sitkiewicz, E., Świderska, B., Milanowski, Ł., Szlufik, S., Koziorowski, D., & Friedman, A. (2021). Proteomic profile of Saliva in Parkinson's disease patients: A proof of concept study. *Brain Sciences*, *11*(5), 661. https://doi.org/10.3390/brainsci11050661

Filiou, M. D., Martins-de-Souza, D., Guest, P. C., Bahn, S., & Turck, C. W. (2012). To label or not to label: Applications of quantitative proteomics in neuroscience research. *Proteomics*, *12*(4–5), 736–747. https://doi.org/10.1002/pmic.201100350

Fingleton, E., Li, Y., & Roche, K. W. (2021). Advances in proteomics allow insights into neuronal proteomes. *Frontiers in Molecular Neuroscience*, *14*, 647451. https://doi.org/10.3389/fnmol.2021.647451

Forgrave, L. M., Wang, M., Yang, D., & DeMarco, M. L. (2021). Proteoforms and their expanding role in laboratory medicine. *Practical Laboratory Medicine*, *28*, e00260. https://doi.org/10.1016/j.plabm.2021.e00260

Frøyset, A. K., Khan, E. A., & Fladmark, K. E. (2016). Quantitative proteomics analysis of zebrafish exposed to sub-lethal dosages of β-Methyl-Amino-L-Alanine (BMAA). *Scientific Reports*, *6*, 29631. https://doi.org/10.1038/srep29631

Gebriel, M., Prabhudesai, S., Uleberg, K. E., Larssen, E., Piston, D., Bjørnstad, A. H., & Møller, S. G. (2014). Zebrafish brain proteomics reveals central proteins involved in neurodegeneration. *Journal of Neuroscience Research*, *92*(1), 104–115. https://doi.org/10.1002/jnr.23297

Griesser, E., Wyatt, H., Ten Have, S., Stierstorfer, B., Lenter, M., & Lamond, A. I. (2020). Quantitative profiling of the human substantia nigra proteome from laser-capture microdissected FFPE tissue. *Molecular & Cellular Proteomics: MCP*, *19*(5), 839–851. https://doi.org/10.1074/mcp.RA119.001889

Groh, K. J., Nesatyy, V. J., Segner, H., Eggen, R. I., & Suter, M. J. (2011). Global proteomics analysis of testis and ovary in adult zebrafish (Danio rerio). *Fish Physiology and Biochemistry*, *37*(3), 619–647. https://doi.org/10.1007/s10695-010-9464-x

Groh, K. J., & Suter, M. J. (2015). Stressor-induced proteome alterations in zebrafish: A meta-analysis of response patterns. *Aquatic Toxicology (Amsterdam, Netherlands)*, *159*, 1–12. https://doi.org/10.1016/j.aquatox.2014.11.013

Guldbrandsen, A., Lereim, R. R., Jacobsen, M., Garberg, H., Kroksveen, A. C., Barsnes, H., & Berven, F. S. (2020). Development of robust targeted proteomics assays for cerebrospinal fluid biomarkers in multiple sclerosis. *Clinical Proteomics*, *17*, 33. https://doi.org/10.1186/s12014-020-09296-5

Ha, S. M., Barnhill, L. M., Li, S., & Bronstein, J. M. (2022). Neurotoxicity of diesel exhaust extracts in zebrafish and its implications for neurodegenerative disease. *Scientific Reports*, *12*(1), 19371. https://doi.org/10.1038/s41598-022-23485-2

Häggmark, A., Schwenk, J. M., & Nilsson, P. (2016). Neuroproteomic profiling of human body fluids. *Proteomics. Clinical Applications*, *10*(4), 485–502. https://doi.org/10.1002/prca.201500065

Hernández, R. B., Carrascal, M., Abian, J., Michalke, B., Farina, M., Gonzalez, Y. R., Iyirhiaro, G. O., Moteshareie, H., Burnside, D., Golshani, A., & Suñol, C. (2020). Manganese-induced neurotoxicity in cerebellar granule neurons due to perturbation of cell network pathways with potential implications for neurodegenerative disorders. *Metallomics: Integrated Biometal Science*, *12*(11), 1656–1678. https://doi.org/10.1039/d0mt00085j

Hernández, R. B., Nishita, M. I., Espósito, B. P., Scholz, S., & Michalke, B. (2015). The role of chemical speciation, chemical fractionation and calcium disruption in manganese-induced developmental toxicity in zebrafish (Danio rerio) embryos. *Journal of Trace Elements in Medicine and Biology: Organ of the Society for Minerals and Trace Elements (GMS)*, *32*, 209–217. https://doi.org/10.1016/j.jtemb.2015.07.004

Higginbotham, L., Ping, L., Dammer, E. B., Duong, D. M., Zhou, M., Gearing, M., Hurst, C., Glass, J. D., Factor, S. A., Johnson, E. C. B., Hajjar, I., Lah, J. J., Levey, A. I., & Seyfried, N. T. (2020). Integrated proteomics reveals brain-based cerebrospinal fluid biomarkers in asymptomatic and symptomatic Alzheimer's disease. *Science Advances*, *6*(43), eaaz9360. https://doi.org/10.1126/sciadv.aaz9360

Hin, N., Newman, M., Kaslin, J., Douek, A. M., Lumsden, A., Nik, S. H. M., Dong, Y., Zhou, X. F., Mañucat-Tan, N. B., Ludington, A., Adelson, D. L., Pederson, S., & Lardelli, M. (2020). Accelerated brain aging towards transcriptional inversion in a zebrafish model of the K115fs mutation of human PSEN2. *PLoS One*, *15*(1), e0227258. https://doi.org/10.1371/journal.pone.0227258

Hogl, S., van Bebber, F., Dislich, B., Kuhn, P. H., Haass, C., Schmid, B., & Lichtenthaler, S. F. (2013). Label-free quantitative analysis of the membrane proteome of Bace1 protease knock-out zebrafish brains. *Proteomics*, *13*(9), 1519–1527. https://doi.org/10.1002/pmic.201200582

Hondius, D. C., van Nierop, P., Li, K. W., Hoozemans, J. J., van der Schors, R. C., van Haastert, E. S., van der Vies, S. M., Rozemuller, A. J., & Smit, A. B. (2016). Profiling the human hippocampal proteome at all pathologic stages of Alzheimer's disease. *Alzheimer's & Dementia: The Journal of the Alzheimer's Association*, *12*(6), 654–668. https://doi.org/10.1016/j.jalz.2015.11.002

Horzmann, K. A., Reidenbach, L. S., Thanki, D. H., Winchester, A. E., Qualizza, B. A., Ryan, G. A., Egan, K. E., Hedrick, V. E., Sobreira, T. J. P., Peterson, S. M., Weber, G. J., Wirbisky-Hershberger, S. E., Sepúlveda, M. S., & Freeman, J. L. (2018). Embryonic atrazine exposure elicits proteomic, behavioral, and brain abnormalities with developmental time specific gene expression signatures. *Journal of Proteomics*, *186*, 71–82. https://doi.org/10.1016/j.jprot.2018.07.006

Hosp, F., & Mann, M. (2017). A primer on concepts and applications of proteomics in neuroscience. *Neuron*, *96*(3), 558–571. https://doi.org/10.1016/j.neuron.2017.09.025

Howe, K., Clark, M. D., Torroja, C. F., Torrance, J., Berthelot, C., Muffato, M., Collins, J. E., Humphray, S., McLaren, K., Matthews, L., McLaren, S., Sealy, I., Caccamo, M., Churcher, C., Scott, C., Barrett, J. C., Koch, R., Rauch, G. J., White, S., Chow, W., … Stemple, D. L. (2013). The zebrafish reference genome sequence and its relationship to the human genome. *Nature*, *496*(7446), 498–503. https://doi.org/10.1038/nature12111

Hu, C., Bai, Y., Sun, B., Zhou, X., & Chen, L. (2023). Exposure to methylparaben at environmentally realistic concentrations significantly impairs neuronal health in adult zebrafish. *Journal of Environmental Sciences (China)*, *132*, 134–144. https://doi.org/10.1016/j.jes.2022.07.012

Jain, A. P., & Sathe, G. (2021). Proteomics landscape of Alzheimer's disease. *Proteomes*, *9*(1), 13. https://doi.org/10.3390/proteomes9010013

Jami, M. S., Murata, H., Barnhill, L. M., Li, S., & Bronstein, J. M. (2023). Diesel exhaust exposure alters the expression of networks implicated in neurodegeneration in zebrafish brains. *Cell Biology and Toxicology*, *39*(3), 641–655. https://doi.org/10.1007/s10565-021-09618-9

Jang, Y., Pletnikova, O., Troncoso, J. C., Pantelyat, A. Y., Dawson, T. M., Rosenthal, L. S., & Na, C. H. (2023). Mass spectrometry-based proteomics analysis of human substantia nigra from Parkinson's disease patients identifies multiple pathways potentially involved in the disease. *Molecular & Cellular Proteomics: MCP*, *22*(1), 100452. https://doi.org/10.1016/j.mcpro.2022.100452

Jiang, R., Rong, C., Ke, R., Meng, S., Yan, X., Ke, H., & Wu, S. (2019). Differential proteomic analysis of serum exosomes reveals alterations in progression of Parkinson disease. *Medicine*, *98*(41), e17478. https://doi.org/10.1097/MD.0000000000017478

Johnson, E. C. B., Dammer, E. B., Duong, D. M., Ping, L., Zhou, M., Yin, L., Higginbotham, L. A., Guajardo, A., White, B., Troncoso, J. C., Thambisetty, M., Montine, T. J., Lee, E. B., Trojanowski, J. Q., Beach, T. G., Reiman, E. M., Haroutunian, V., Wang, M., Schadt, E., Zhang, B., … Seyfried, N. T. (2020). Large-scale proteomic analysis of Alzheimer's disease brain and cerebrospinal fluid reveals early changes in energy metabolism associated with microglia and astrocyte activation. *Nature Medicine*, *26*(5), 769–780. https://doi.org/10.1038/s41591-020-0815-6

Johnson, E. C. B., Dammer, E. B., Duong, D. M., Yin, L., Thambisetty, M., Troncoso, J. C., Lah, J. J., Levey, A. I., & Seyfried, N. T. (2018). Deep proteomic network analysis of Alzheimer's disease brain reveals alterations in RNA binding proteins and RNA splicing associated with disease. *Molecular Neurodegeneration*, *13*(1), 52. https://doi.org/10.1186/s13024-018-0282-4

Kamat, P. K., Rai, S., & Nath, C. (2013). Okadaic acid induced neurotoxicity: An emerging tool to study Alzheimer's disease pathology. *Neurotoxicology*, *37*, 163–172. https://doi.org/10.1016/j.neuro.2013.05.002

Kelkar, D. S., Provost, E., Chaerkady, R., Muthusamy, B., Manda, S. S., Subbannayya, T., Selvan, L. D., Wang, C. H., Datta, K. K., Woo, S., Dwivedi, S. B., Renuse, S., Getnet, D., Huang, T. C., Kim, M. S., Pinto, S. M., Mitchell, C. J., Madugundu, A. K., Kumar, P., Sharma, J., … Pandey, A. (2014). Annotation of the zebrafish genome through an integrated transcriptomic and proteomic analysis. *Molecular & Cellular Proteomics: MCP*, *13*(11), 3184–3198. https://doi.org/10.1074/mcp.M114.038299

Kessler, T., Latzer, P., Schmid, D., Warnken, U., Saffari, A., Ziegler, A., Kollmer, J., Möhlenbruch, M., Ulfert, C., Herweh, C., Wildemann, B., Wick, W., & Weiler, M. (2020). Cerebrospinal fluid proteomic profiling in nusinersen-treated patients with spinal muscular atrophy. *Journal of Neurochemistry*, *153*(5), 650–661. https://doi.org/10.1111/jnc.14953

Koehler, D., & Williams, F. E. (2018). Utilizing zebrafish and okadaic acid to study Alzheimer's disease. *Neural Regeneration Research*, *13*(9), 1538–1541. https://doi.org/10.4103/1673-5374.237111

Konzer, A., Ruhs, A., Braun, H., Jungblut, B., Braun, T., & Krüger, M. (2013). Stable isotope labeling in zebrafish allows in vivo monitoring of cardiac morphogenesis. *Molecular & Cellular Proteomics: MCP*, *12*(6), 1502–1512. https://doi.org/10.1074/mcp.M111.015594

Krishna, S., Dodd, C. A., Hekmatyar, S. K., & Filipov, N. M. (2014). Brain deposition and neurotoxicity of manganese in adult mice exposed via the drinking water. *Archives of Toxicology*, *88*(1), 47–64. https://doi.org/10.1007/s00204-013-1088-3

Kwon, Y. S., Park, C. B., Lee, S. M., Zee, S., Kim, G. E., Kim, Y. J., Sim, H. J., Kim, J. H., & Seo, J. S. (2023). Proteomic analysis of zebrafish (Danio rerio) embryos exposed to benzyl benzoate. *Environmental Science and Pollution Research International*, *30*(10), 26375–26386. https://doi.org/10.1007/s11356-022-24081-7

Langebeck-Jensen, K., Shahar, O. D., Schuman, E. M., Langer, J. D., & Ryu, S. (2019). Larval zebrafish proteome regulation in response to an environmental challenge. *Proteomics*, *19*(14), e1900028. https://doi.org/10.1002/pmic.201900028

Lee, H., Sung, E. J., Seo, S., Min, E. K., Lee, J. Y., Shim, I., Kim, P., Kim, T. Y., Lee, S., & Kim, K. T. (2021). Integrated multi-omics analysis reveals the underlying molecular mechanism for developmental neurotoxicity of perfluorooctanesulfonic acid in zebrafish. *Environment International*, *157*, 106802. https://doi.org/10.1016/j.envint.2021.106802

Lee, P. Y., Osman, J., Low, T. Y., & Jamal, R. (2019). Plasma/serum proteomics: Depletion strategies for reducing high-abundance proteins for biomarker discovery. *Bioanalysis*, *11*(19), 1799–1812. https://doi.org/10.4155/bio-2019-0145

Levesque, S., Surace, M. J., McDonald, J., & Block, M. L. (2011). Air pollution & the brain: Subchronic diesel exhaust exposure causes neuroinflammation and elevates early markers of neurodegenerative disease. *Journal of Neuroinflammation*, *8*, 105. https://doi.org/10.1186/1742-2094-8-105

Li, C., Tan, X. F., Lim, T. K., Lin, Q., & Gong, Z. (2016). Comprehensive and quantitative proteomic analyses of zebrafish plasma reveals conserved protein profiles between genders and between zebrafish and human. *Scientific Reports*, *6*, 24329. https://doi.org/10.1038/srep24329

Li, H., Liu, Y., Chen, Q., Jin, L., & Peng, R. (2023). Research progress of zebrafish model in aquatic ecotoxicology. *Water*, *15*(9), 1735. https://doi.org/10.3390/w15091735

Li, J., Gould, T. D., Yuan, P., Manji, H. K., & Chen, G. (2003). Postmortem interval effects on the phosphorylation of signaling proteins. *Neuropsychopharmacology: Official Publication of the American College of Neuropsychopharmacology*, *28*(6), 1017–1025. https://doi.org/10.1038/sj.npp.1300112

Liang, B., Huang, Y., Zhong, Y., Li, Z., Ye, R., Wang, B., Zhang, B., Meng, H., Lin, X., Du, J., Hu, M., Wu, Q., Sui, H., Yang, X., & Huang, Z. (2022). Brain single-nucleus transcriptomics highlights that polystyrene nanoplastics potentially induce Parkinson's disease-like neurodegeneration by causing energy metabolism disorders in mice. *Journal of Hazardous Materials, 430*, 128459. https://doi.org/10.1016/j.jhazmat.2022.128459

Liang, Z., Zhuang, H., Cao, X., Ma, G., & Shen, L. (2024). Subcellular proteomics insights into Alzheimer's disease development. *Proteomics. Clinical Applications*, *18*(3), e2200112. https://doi.org/10.1002/prca.202200112

Maffioli, E., Nonnis, S., Grassi Scalvini, F., Negri, A., Tedeschi, G., & Toni, M. (2023). The neurotoxic effect of environmental temperature variation in adult zebrafish (Danio rerio). *International Journal of Molecular Sciences, 24*(21), 15735. https://doi.org/10.3390/ijms242115735

Magdalinou, N. K., Noyce, A. J., Pinto, R., Lindstrom, E., Holmén-Larsson, J., Holtta, M., Blennow, K., Morris, H. R., Skillbäck, T., Warner, T. T., Lees, A. J., Pike, I., Ward, M., Zetterberg, H., & Gobom, J. (2017). Identification of candidate cerebrospinal fluid biomarkers in parkinsonism using quantitative proteomics. *Parkinsonism & Related Disorders, 37*, 65–71. https://doi.org/10.1016/j.parkreldis.2017.01.016

Maharaj, S., El Ahmadie, N., Rheingold, S., El Chehouri, J., Yang, L., Souders, C. L., 2nd, & Martyniuk, C. J. (2020). Sub-lethal toxicity assessment of the phenylurea herbicide linuron in developing zebrafish (Danio rerio) embryo/larvae. *Neurotoxicology and Teratology, 81*, 106917. https://doi.org/10.1016/j.ntt.2020.106917

Martin, R. M., Bereman, M. S., & Marsden, K. C. (2021). BMAA and MCLR interact to modulate behavior and exacerbate molecular changes related to neurodegeneration in larval zebrafish. *Toxicological Sciences: An Official Journal of the Society of Toxicology, 179*(2), 251–261. https://doi.org/10.1093/toxsci/kfaa178

Martinez-Val, A., Guzmán, U. H., & Olsen, J. V. (2022). Obtaining complete human proteomes. *Annual Review of Genomics and Human Genetics, 23*, 99–121. https://doi.org/10.1146/annurev-genom-112921-024948

Martins, A. C., Jr, Gubert, P., Villas Boas, G. R., Meirelles Paes, M., Santamaría, A., Lee, E., Tinkov, A. A., Bowman, A. B., & Aschner, M. (2020). Manganese-induced neurodegenerative diseases and possible therapeutic approaches. *Expert Review of Neurotherapeutics, 20*(11), 1109–1121. https://doi.org/10.1080/14737175.2020.1807330

Miller, R. A., Winrow, C. J., Spellman, D. S., Song, Q., Reiss, D. R., Conway, J. P., Taylor, R. R., Coleman, P. J., Hendrickson, R. C., & Renger, J. J. (2014). Quantitative proteomics in laser capture microdissected sleep nuclei from rat brain. *Journal of Neurogenetics, 28*(1–2), 136–145. https://doi.org/10.3109/01677063.2014.883389

Min, E. K., Lee, H., Sung, E. J., Seo, S. W., Song, M., Wang, S., Kim, S. S., Bae, M. A., Kim, T. Y., Lee, S., & Kim, K. T. (2023). Integrative multi-omics reveals analogous developmental neurotoxicity mechanisms between perfluorobutanesulfonic acid and perfluorooctanesulfonic acid in zebrafish. *Journal of Hazardous Materials, 457*, 131714. https://doi.org/10.1016/j.jhazmat.2023.131714

Moon C. (2022). New insights into and emerging roles of animal models for neurological disorders. *International Journal of Molecular Sciences, 23*(9), 4957. https://doi.org/10.3390/ijms23094957

Nada, S. E., Williams, F. E., & Shah, Z. A. (2016). Development of a novel and robust pharmacological model of okadaic acid-induced Alzheimer's disease in Zebrafish. *CNS & Neurological Disorders Drug Targets, 15*(1), 86–94. https://doi.org/10.2174/1871527314666150821105602

Nolte, H., Hölper, S., Housley, M. P., Islam, S., Piller, T., Konzer, A., Stainier, D. Y., Braun, T., & Krüger, M. (2015). Dynamics of zebrafish fin regeneration using a pulsed SILAC approach. *Proteomics, 15*(4), 739–751. https://doi.org/10.1002/pmic.201400316

Nolte, H., Konzer, A., Ruhs, A., Jungblut, B., Braun, T., & Krüger, M. (2014). Global protein expression profiling of zebrafish organs based on in vivo incorporation of stable isotopes. *Journal of Proteome Research, 13*(4), 2162–2174. https://doi.org/10.1021/pr5000335

Nonnis, S., Angiulli, E., Maffioli, E., Frabetti, F., Negri, A., Cioni, C., Alleva, E., Romeo, V., Tedeschi, G., & Toni, M. (2021). Acute environmental temperature variation affects brain protein expression, anxiety and explorative behaviour in adult zebrafish. *Scientific Reports, 11*(1), 2521. https://doi.org/10.1038/s41598-021-81804-5

Nugumanova, G., Ponomarev, E. D., Askarova, S., Fasler-Kan, E., & Barteneva, N. S. (2023). Freshwater cyanobacterial toxins, cyanopeptides and neurodegenerative diseases. *Toxins, 15*(3), 233. https://doi.org/10.3390/toxins15030233

Petyuk, V. A., Yu, L., Olson, H. M., Yu, F., Clair, G., Qian, W. J., Shulman, J. M., & Bennett, D. A. (2021). Proteomic profiling of the substantia nigra to identify determinants of lewy body pathology and dopaminergic neuronal loss. *Journal of Proteome Research, 20*(5), 2266–2282. https://doi.org/10.1021/acs.jproteome.0c00747

Ping, L., Kundinger, S. R., Duong, D. M., Yin, L., Gearing, M., Lah, J. J., Levey, A. I., & Seyfried, N. T. (2020). Global quantitative analysis of the human brain proteome and phosphoproteome in Alzheimer's disease. *Scientific Data, 7*(1), 315. https://doi.org/10.1038/s41597-020-00650-8

Plum, S., Eggers, B., Helling, S., Stepath, M., Theiss, C., Leite, R. E. P., Molina, M., Grinberg, L. T., Riederer, P., Gerlach, M., May, C., & Marcus, K. (2020). Proteomic characterization of synaptosomes from human substantia nigra indicates altered mitochondrial translation in Parkinson's disease. *Cells, 9*(12), 2580. https://doi.org/10.3390/cells9122580

Prats, E., Gómez-Canela, C., Ben-Lulu, S., Ziv, T., Padrós, F., Tornero, D., Garcia-Reyero, N., Tauler, R., Admon, A., & Raldúa, D. (2017). Modelling acrylamide acute neurotoxicity in zebrafish larvae. *Scientific Reports, 7*(1), 13952. https://doi.org/10.1038/s41598-017-14460-3

Prüst, M., Meijer, J., & Westerink, R. H. S. (2020). The plastic brain: Neurotoxicity of micro- and nanoplastics. *Particle and Fibre Toxicology, 17*(1), 24. https://doi.org/10.1186/s12989-020-00358-y

Purushothaman, K., Das, P. P., Presslauer, C., Lim, T. K., Johansen, S. D., Lin, Q., & Babiak, I. (2019). Proteomics analysis of early developmental stages of zebrafish embryos. *International Journal of Molecular Sciences, 20*(24), 6359. https://doi.org/10.3390/ijms20246359

Raghunathan, R., Turajane, K., & Wong, L. C. (2022). Biomarkers in Neurodegenerative diseases: Proteomics spotlight on ALS and Parkinson's disease. *International Journal of Molecular Sciences, 23*(16), 9299. https://doi.org/10.3390/ijms23169299

Rahlouni, F., Szarka, S., Shulaev, V., & Prokai, L. (2015). A survey of the impact of deyolking on biological processes covered by shotgun proteomic analyses of zebrafish embryos. *Zebrafish, 12*(6), 398–407. https://doi.org/10.1089/zeb.2015.1121

Rayaprolu, S., Higginbotham, L., Bagchi, P., Watson, C. M., Zhang, T., Levey, A. I., Rangaraju, S., & Seyfried, N. T. (2021). Systems-based proteomics to resolve the biology of Alzheimer's disease beyond amyloid and tau.

*Neuropsychopharmacology: Official Publication of the American College of Neuropsychopharmacology*, *46*(1), 98–115. https://doi.org/10.1038/s41386-020-00840-3

Rotunno, M. S., Lane, M., Zhang, W., Wolf, P., Oliva, P., Viel, C., Wills, A. M., Alcalay, R. N., Scherzer, C. R., Shihabuddin, L. S., Zhang, K., & Sardi, S. P. (2020). Cerebrospinal fluid proteomics implicates the granin family in Parkinson's disease. *Scientific Reports*, *10*(1), 2479. https://doi.org/10.1038/s41598-020-59414-4

Sanfeliu, C., Bartra, C., Suñol, C., & Rodríguez-Farré, E. (2024). New insights in animal models of neurotoxicity-induced neurodegeneration. *Frontiers in Neuroscience*, *17*, 1248727. https://doi.org/10.3389/fnins.2023.1248727

Sarath Babu, N., Murthy, C.hL., Kakara, S., Sharma, R., Brahmendra Swamy, C. V., & Idris, M. M. (2016). 1-Methyl-4-phenyl-1,2,3,6-tetrahydropyridine induced Parkinson's disease in zebrafish. *Proteomics*, *16*(9), 1407–1420. https://doi.org/10.1002/pmic.201500291

Sfakianakis, D. G., Leris, I., & Kentouri, M. (2011). Effect of developmental temperature on swimming performance of zebrafish (Danio rerio) juveniles. *Environmental Biology Fishes*, *90*, 421–427. https://doi.org/10.1007/s10641-010-9751-5

Shan, S., Zhang, Y., Zhao, H., Zeng, T., & Zhao, X. (2022). Polystyrene nanoplastics penetrate across the blood-brain barrier and induce activation of microglia in the brain of mice. *Chemosphere*, *298*, 134261. https://doi.org/10.1016/j.chemosphere.2022.134261

Shi, Q., Chen, L. N., Zhang, B. Y., Xiao, K., Zhou, W., Chen, C., Zhang, X. M., Tian, C., Gao, C., Wang, J., Han, J., & Dong, X. P. (2015). Proteomics analyses for the global proteins in the brain tissues of different human prion diseases. *Molecular & Cellular Proteomics: MCP*, *14*(4), 854–869. https://doi.org/10.1074/mcp.M114.038018

Siddiqui, A., Abidin, S. A. Z., Shah, Z. A., Othman, I., & Kumari, Y. (2023). Behavioural, genomics and proteomic approach to examine Alzheimer's disease in zebrafish. *Comparative Biochemistry and Physiology. Toxicology & Pharmacology: CBP*, *271*, 109636. https://doi.org/10.1016/j.cbpc.2023.109636

Siddiqui, T., Cosacak, M. I., Popova, S., Bhattarai, P., Yilmaz, E., Lee, A. J., Min, Y., Wang, X., Allen, M., İş, Ö., Atasavum, Z. T., Rodriguez-Muela, N., Vardarajan, B. N., Flaherty, D., Teich, A. F., Santa-Maria, I., Freudenberg, U., Werner, C., Tosto, G., Mayeux, R., … Kizil, C. (2023). Nerve growth factor receptor (Ngfr) induces neurogenic plasticity by suppressing reactive astroglial Lcn2/Slc22a17 signaling in Alzheimer's disease. *NPJ Regenerative Medicine*, *8*(1), 33. https://doi.org/10.1038/s41536-023-00311-5

Siems, S. B., Jahn, O., Hoodless, L. J., Jung, R. B., Hesse, D., Möbius, W., Czopka, T., & Werner, H. B. (2021). Proteome profile of myelin in the zebrafish brain. *Frontiers in Cell and Developmental Biology*, *9*, 640169. https://doi.org/10.3389/fcell.2021.640169

Sini, P., Dang, T. B. C., Fais, M., Galioto, M., Padedda, B. M., Lugliè, A., Iaccarino, C., & Crosio, C. (2021). Cyanobacteria, cyanotoxins, and neurodegenerative diseases: Dangerous Liaisons. *International journal of molecular sciences*, *22*(16), 8726. https://doi.org/10.3390/ijms22168726

Smidak, R., Aradska, J., Kirchberger, S., Distel, M., Sialana, F. J., Wackerlig, J., Mechtcheriakova, D., & Lubec, G. (2016). A detailed proteomic profiling of plasma membrane from zebrafish brain. *Proteomics. Clinical Applications*, *10*(12), 1264–1268. https://doi.org/10.1002/prca.201600081

Sowell, R. A., Owen, J. B., & Butterfield, D. A. (2009). Proteomics in animal models of Alzheimer's and Parkinson's diseases. *Ageing Research Reviews*, *8*(1), 1–17. https://doi.org/10.1016/j.arr.2008.07.003

Sürmen, M. G., Sürmen, S., Cansız, D., Ünal, İ., Üstündağ, Ü. V., Alturfan, A. A., Büyükkayhan, D., & Emekli-Alturfan, E. (2022). Amelioration of rotenone-induced alterations in energy/redox system, stress response and cytoskeleton proteins by octanoic acid in zebrafish: A proteomic study. *Journal of Biochemical and Molecular Toxicology*, *36*(5), e23024. https://doi.org/10.1002/jbt.23024

Sürmen, M. G., Sürmen, S., Cansız, D., Ünal, İ., Üstündağ, Ü. V., Alturfan, A. A., & Emekli-Alturfan, E. (2021). Quantitative phosphoproteomics to resolve the cellular responses to octanoic acid in rotenone exposed zebrafish. *Journal of Food Biochemistry*, *45*(10), e13923. https://doi.org/10.1111/jfbc.13923

Thawkar, B. S., & Kaur, G. (2021). Zebrafish as a promising tool for modeling neurotoxin-induced Alzheimer's disease. *Neurotoxicity Research*, *39*(3), 949–965. https://doi.org/10.1007/s12640-021-00343-z

Thompson, A. G., Gray, E., Mäger, I., Thézénas, M. L., Charles, P. D., Talbot, K., Fischer, R., Kessler, B. M., Wood, M., & Turner, M. R. (2020). CSF extracellular vesicle proteomics demonstrates altered protein homeostasis in amyotrophic lateral sclerosis. *Clinical Proteomics*, *17*, 31. https://doi.org/10.1186/s12014-020-09294-7

Toni, M., Angiulli, E., Miccoli, G., Cioni, C., Alleva, E., Frabetti, F., Pizzetti, F., Grassi Scalvini, F., Nonnis, S., Negri, A., Tedeschi, G., & Maffioli, E. (2019). Environmental temperature variation affects brain protein expression and cognitive abilities in adult zebrafish (Danio rerio): A proteomic and behavioural study. *Journal of Proteomics*, *204*, 103396. https://doi.org/10.1016/j.jprot.2019.103396

Trifonova, O. P., Maslov, D. L., Balashova, E. E., Urazgildeeva, G. R., Abaimov, D. A., Fedotova, E. Y., Poleschuk, V. V., Illarioshkin, S. N., & Lokhov, P. G. (2020). Parkinson's disease: Available clinical and promising omics tests for diagnostics, disease risk assessment, and pharmacotherapy personalization. *Diagnostics (Basel, Switzerland)*, *10*(5), 339. https://doi.org/10.3390/diagnostics10050339

Ünal, İ., Cansız, D., Sürmen, M. G., Sürmen, S., Sezer, Z., Beler, M., Üstündağ, Ü. V., Güzel, E., Alturfan, A. A., & Emekli-Alturfan, E. (2023). Identification of molecular network of gut-brain axis associated with neuroprotective effects of PPARδ-ligand erucic acid in rotenone-induced Parkinson's disease model in zebrafish. *The European journal of neuroscience*, *57*(4), 585–606. https://doi.org/10.1111/ejn.15904

Villemagne, V. L., Burnham, S., Bourgeat, P., Brown, B., Ellis, K. A., Salvado, O., Szoeke, C., Macaulay, S. L., Martins, R., Maruff, P., Ames, D., Rowe, C. C., Masters, C. L., & Australian Imaging Biomarkers and Lifestyle (AIBL) Research Group (2013). Amyloid β deposition, neurodegeneration, and cognitive decline in sporadic Alzheimer's disease: A prospective cohort study. *The Lancet. Neurology*, *12*(4), 357–367. https://doi.org/10.1016/S1474-4422(13)70044-9

Virreira Winter, S., Karayel, O., Strauss, M. T., Padmanabhan, S., Surface, M., Merchant, K., Alcalay, R. N., & Mann, M. (2021). Urinary proteome profiling for stratifying patients with familial Parkinson's disease. *EMBO Molecular Medicine*, *13*(3), e13257. https://doi.org/10.15252/emmm.202013257

Wang, H., Dey, K. K., Chen, P. C., Li, Y., Niu, M., Cho, J. H., Wang, X., Bai, B., Jiao, Y., Chepyala, S. R., Haroutunian, V., Zhang, B., Beach, T. G., & Peng, J. (2020). Integrated analysis of ultra-deep proteomes in cortex, cerebrospinal fluid and serum reveals a mitochondrial signature in Alzheimer's disease. *Molecular Neurodegeneration*, *15*(1), 43. https://doi.org/10.1186/s13024-020-00384-6

Wang, J., Ma, T., Ma, D., Li, H., Hua, L., He, Q., & Deng, X. (2021). The impact of air pollution on neurodegenerative diseases. *Therapeutic Drug Monitoring*, *43*(1), 69–78. https://doi.org/10.1097/FTD.0000000000000818

Wang, Y., Wang, J., Cong, J., Zhang, H., Gong, Z., Sun, H., Wang, L., & Duan, Z. (2023). Nanoplastics induce neuroexcitatory symptoms in zebrafish (Danio rerio) larvae through a manner contrary to Parkinsonian's way in proteomics. *The Science of the Total Environment*, *905*, 166898. https://doi.org/10.1016/j.scitotenv.2023.166898

Westman-Brinkmalm, A., Abramsson, A., Pannee, J., Gang, C., Gustavsson, M. K., von Otter, M., Blennow, K., Brinkmalm, G., Heumann, H., & Zetterberg, H. (2011). SILAC zebrafish for quantitative analysis of protein turnover and tissue regeneration. *Journal of Proteomics*, *75*(2), 425–434. https://doi.org/10.1016/j.jprot.2011.08.008

Wilkins, M. R., Sanchez, J. C., Gooley, A. A., Appel, R. D., Humphery-Smith, I., Hochstrasser, D. F., & Williams, K. L. (1996). Progress with proteome projects: why all proteins expressed by a genome should be identified and how to do it. *Biotechnology & Genetic Engineering Reviews*, *13*, 19–50. https://doi.org/10.1080/02648725.1996.10647923

Xie, F., Liu, T., Qian, W. J., Petyuk, V. A., & Smith, R. D. (2011). Liquid chromatography-mass spectrometry-based quantitative proteomics. *The Journal of Biological Chemistry*, *286*(29), 25443–25449. https://doi.org/10.1074/jbc.R110.199703

Xu, Y., Yang, L., Teng, Y., Li, J., & Li, N. (2023). Exploring the underlying molecular mechanism of tri(1,3-dichloropropyl) phosphate-induced neurodevelopmental toxicity via thyroid hormone disruption in zebrafish by multi-omics analysis. *Aquatic Toxicology (Amsterdam, Netherlands)*, *258*, 106510. https://doi.org/10.1016/j.aquatox.2023.106510

Yu, Y., Zhang, Q., Liu, G., Deng, Y., Kang, J., Zhang, F., Lu, T., Sun, L., & Qian, H. (2021). Proteomic analysis of zebrafish brain damage induced by Microcystis aeruginosa bloom. *The Science of the Total Environment*, *795*, 148865. https://doi.org/10.1016/j.scitotenv.2021.148865

Zhao, F., Cai, T., Liu, M., Zheng, G., Luo, W., & Chen, J. (2009). Manganese induces dopaminergic neurodegeneration via microglial activation in a rat model of manganism. *Toxicological Sciences: An Official Journal of the Society of Toxicology*, *107*(1), 156–164. https://doi.org/10.1093/toxsci/kfn213

Zhao, Y., Zhang, Y., Zhang, J., & Yang, G. (2022). Plasma proteome profiling using tandem mass tag labeling technology reveals potential biomarkers for Parkinson's disease: A preliminary study. *Proteomics. Clinical Applications*, *16*(2), e2100010. https://doi.org/10.1002/prca.202100010

# 9

# *Zebrafish as a Model for Functional Analyses of Alzheimer's Disease–Related Genes and Drug Development*

**Caghan Kizil**

## Abbreviations

| | |
|---|---|
| **Aβ42** | Amyloid Beta 42 |
| **AD** | Alzheimer's Disease |
| **ABCA7** | ATP-Binding Cassette Subfamily A Member 7 |
| **Ahr** | Aryl Hydrocarbon Receptor |
| **APOE** | Apolipoprotein E |
| **APP** | Amyloid Precursor Protein |
| **ATP** | Adenosine triphosphate |
| **BDNF** | Brain-Derived Neurotrophic Factor |
| **BBB** | Blood–Brain Barrier |
| **CRISPR** | Clustered Regularly Interspaced Short Palindromic Repeats |
| **EOAD** | Early-Onset Alzheimer's Disease |
| **ER** | Endoplasmic Reticulum |
| **FMNL2** | Formin-Like Protein 2 |
| **FN1** | Fibronectin 1 |
| **GVU** | Gliovascular Unit |
| **GWAS** | Genome-Wide Association Studies |
| **MAPT** | Microtubule-Associated Protein Tau |
| **mRNA** | Messenger RNA |
| **MTD** | Multitarget Drugs |
| **NFTs** | Neurofibrillary Tangles |
| **NPY** | Neuropeptide Y |
| **PSEN1** | Presenilin 1 |
| **PSEN2** | Presenilin 2 |
| **RNA-seq** | RNA Sequencing |
| **SMAD3** | Mothers Against Decapentaplegic Homolog 3 |
| **snRNAseq** | Single-Nucleus RNA Sequencing |
| **SORL1** | Sortilin-Related Receptor |
| **VEGFA** | Vascular endothelial growth factor A |

## Introduction

Alzheimer's disease (AD), a debilitating neurodegenerative disorder, is characterized by hallmark pathologies, including Aβ42 plaques, neurofibrillary tangles, co-manifesting vascular pathology, and neuroinflammation (Reitz et al., 2023; Mayeux and Stern, 2012; Reitz and Mayeux, 2009; Selkoe and Hardy, 2016; Haass and Selkoe, 2007; Selkoe, 2003; Selkoe, 2001; Bellenguez et al., 2022; Kunkle et al., 2019; Carrasquillo et al., 2014; McGowan et al., 2005). The multifactorial etiology of AD encompasses genetic, environmental, and cellular factors that collectively cause synaptic dysfunction, neuronal degeneration, and cognitive decline. Insights from genome-wide association studies (GWAS) and single-cell transcriptomic analyses have delineated the intricate genetic architecture of AD, implicating numerous loci in disease susceptibility and progression (Gunasekaran et al., 2024; Gunasekaran et al., 2023; Yang et al., 2022; Dumitrescu et al., 2020; N'Songo et al., 2017; Carrasquillo et al., 2015; Genin et al., 2011; Felsky et al., 2023; Belloy et al., 2023; Sariya et al., 2021; Cruchaga et al., 2018; Reitz et al., 2010; Farrer et al., 1997). For instance, GWAS meta-analyses have identified over 70 AD-associated loci, implicating pathways such as microglial activity, amyloid processing, and vascular homeostasis as pivotal in AD pathogenesis (Bellenguez et al., 2022). Single-cell transcriptomics has further illuminated the cellular and pathological heterogeneity within AD-affected brains, revealing critical interactions between neuronal, glial, and vascular components that shape disease trajectories (Green et al., 2024; Serrano-Pozo et al., 2024; Sun et al., 2023; Yang et al., 2022; Su et al., 2022; Leng et al., 2021; Olah et al., 2020; Wang et al., 2020; Lau et al., 2020; Mathys et al., 2023; Blanchard et al., 2022; Mathys et al., 2019; Tsartsalis et al., 2024; Is et al., 2022).

Zebrafish has become an important vertebrate model in biological research because of its distinctive combination of genetic and experimental advantages (Chia et al., 2022; Barbereau et al., 2021; Cully, 2019; Caramillo and Echevarria, 2017; Bayes et al., 2017; MacRae and Peterson, 2015; Varshney et al., 2015; Malaga-Trillo et al., 2011; Xi et al., 2011; Newman et al., 2010; Xia, 2010; Lieschke and Currie, 2007; Nüsslein-Volhard and Dahm, 2002; Zon, 1999). Its optical transparency, rapid embryonic development, presence of diverse reporters and gene-edited lines, and ease of genetic manipulation render it an invaluable tool for probing fundamental biological processes. Approximately 75% of human genes have at least one zebrafish ortholog, underscoring its genetic congruence with humans (Howe et al., 2013). This homology extends to genes implicated in neurological and neurodegenerative disorders (Bashirzade et al., 2022; Ding et al., 2019; Bhattarai et al., 2017b; Lopez et al., 2017; Bhattarai et al., 2017a; Tincer



DOI: 10.1201/9781032676555-11

et al., 2016; Newman et al., 2014; Schmid and Haass, 2013; van Tijn et al., 2011; Xi et al., 2011; Fleisch et al., 2010; Xia, 2010; Tomasiewicz et al., 2002; Tayran et al., 2024; Bhattarai et al., 2024; Is et al., 2024b; Siddiqui et al., 2023; Kim et al., 2023; Felsky et al., 2023; Kizil et al., 2022; Lee et al., 2022; Cosacak et al., 2022; Siddiqui et al., 2022; Bhattarai et al., 2022; Siddiqui et al., 2021; Cosacak et al., 2020a; Bhattarai et al., 2020; Cosacak et al., 2017; Bhattarai et al., 2016), making zebrafish a useful model for understanding complex disease mechanisms. In neuroscience, zebrafish offer unparalleled opportunities to investigate brain development, neuroplasticity, and behavioral phenotypes, supported by their well-characterized central nervous system (Rubinstein, 2003; Tomasiewicz et al., 2002; Kelsh and Raible, 2002; Chitnis and Dawid, 1999; Schier, 1997; Haffter et al., 1996; Driever et al., 1996; Granato and Nusslein-Volhard, 1996; Neuhauss et al., 1996; Kimmel et al., 1995; Strahle and Blader, 1994; Eisen, 1991; Fulwiler and Gilbert, 1991; Kimmel et al., 1988). Its capacity for in vivo imaging of neuronal and vascular dynamics, alongside genetic manipulability via CRISPR-Cas9 and transgenic technologies (Li et al., 2016; Varshney et al., 2015; Hruscha et al., 2013; Jao et al., 2013; Julich and Holley, 2024; Than-Trong et al., 2020; Dray et al., 2015; Vacaru et al., 2014; Li et al., 2012; Peri and Nusslein-Volhard, 2008; Redd et al., 2006; Lawson and Weinstein, 2002), have made it a key tool in neurodegenerative disease research, and its uses are still growing.

This chapter will explore the use of zebrafish as a model organism in AD research, highlighting its potential to connect basic science with translational therapeutic strategies, the integrative applications of zebrafish in understanding AD's multifactorial nature, and its value in both preclinical and clinical contexts.

## Zebrafish as a Model for Alzheimer's Disease Pathology

Recent advancements in zebrafish models of AD have yielded insights into the molecular and cellular mechanisms underlying AD pathogenesis. Zebrafish engineered to express modified human amyloid precursor protein (APP) forms, or Aβ42, have modeled critical features of the disease, such as Aβ42 accumulation, oxidative stress, and behavioral deficits (Ozcan et al., 2024; Banote et al., 2020; Abramsson et al., 2013; Cosacak et al., 2019; Bhattarai et al., 2017b; Bhattarai et al., 2017a; Bhattarai et al., 2016). In familial AD, PSEN1 and PSEN2, which are important components of the γ-secretase complex that cleaves APP, are often mutated (Wen et al., 2004, Handler et al., 2000; Holcomb et al., 1998; Scheuner et al., 1996). Zebrafish harboring PSEN1 mutations with altered γ-secretase activity lead to aberrant Aβ42 production, calcium dysregulation, and synaptic defects, and drug treatment for secretase inhibition showed versatile phenotypic effects (Barazzuol et al., 2023; Barthelson et al., 2021; Newman et al., 2020; van Tijn et al., 2011; Xia, 2010; Leimer et al., 1999; Arslanova et al., 2010; Yang et al., 2010). These findings highlight the conserved role of PSEN mutations in modulating key AD-related pathways. Additional to amyloid, zebrafish transgenic lines expressing human tau mutations, including hTauP301L and hTauA152T, have advanced our understanding of the molecular dynamics of tau aggregation, phosphorylation, and cytoskeletal disruption (Barbereau et al., 2021; Lopez et al., 2017; Cosacak et al., 2017; Bai et al., 2007). These models show tau hyperphosphorylation and neurofibrillary tangle formation, a hallmark of human tauopathies, and provide valuable insights into the mechanisms driving tau-induced neuronal damage.

Single-cell/nucleus transcriptomics has revolutionized the understanding of cell-type-specific responses to AD pathology, providing unparalleled insights into molecular dynamics at the cellular level (Green et al., 2024; Tsartsalis et al., 2024; Serrano-Pozo et al., 2024; Sun et al., 2023; Felsky et al., 2023; Cosacak et al., 2022; Yang et al., 2022; Su et al., 2022; Leng et al., 2021; Olah et al., 2020; Wang et al., 2020; Lau et al., 2020; Mathys et al., 2019; Is et al., 2024a; Is et al., 2024b). Single-cell transcriptomic analyses in zebrafish have provided insights into cell-type-specific responses to AD pathology, highlighting common pathological mechanisms in human brains and unique neuroregenerative mechanisms absent in humans (Tayran et al., 2024; Siddiqui et al., 2023; Cosacak et al., 2022; Siddiqui et al., 2021; Cosacak et al., 2020b; Cosacak et al., 2019; Bhattarai et al., 2016). Cross-species transcriptomic integration has shown that approximately 95% of human and zebrafish cell types co-cluster, with shared genetic alterations in neurons, astroglia, vasculature, oligodendrocytes, and immune cells (Cosacak et al., 2022). These findings highlight the value of zebrafish as a model for uncovering both conserved and unique molecular responses, providing a powerful tool for identifying mechanisms that could inform regenerative therapies for human AD.

Single-cell studies have extensively characterized the immune landscape of AD. In humans, microglial states transition from homeostatic to disease-associated phenotypes (Keren-Shaul et al., 2017; Green et al., 2024; Olah et al., 2020; Mathys et al., 2017; Haage and De Jager, 2022). Zebrafish, as well, display a remarkable capacity for modulating microglial polarization, leveraging anti-inflammatory phenotypes to mitigate neurotoxicity and enhance synaptic repair in neurodegeneration and injury (Kizil and Bhattarai, 2018; Kizil et al., 2015; Kizil et al., 2012; Kyritsis et al., 2012; Redd et al., 2006; Siddiqui et al., 2023; Siddiqui et al., 2021; Jurisch-Yaksi et al., 2020; Cosacak et al., 2019; Kizil, 2018; Bhattarai et al., 2016). Recent work by Haage et al. further exemplified the potential of zebrafish for elucidating the molecular pathways underlying microglial dynamics and specific polarization states in AD (Haage et al., 2024).

## Investigation of Genetic Variants and Gene Function in Zebrafish

Presenilin 1 (*PSEN1*) and Presenilin 2 (*PSEN2*) are central components of the γ-secretase complex, crucial for cleaving amyloid precursor protein (APP) to produce amyloid peptides, a hallmark of AD pathology. Zebrafish models (Barthelson et al., 2021; Newman et al., 2020; Newman et al., 2009; Barthelson et al., 2022; Newman et al., 2014) have been useful to investigate the molecular and cellular consequences of PSEN1/PSEN2 mutations. Models of PSEN1 variations reveal disruptions in mitochondrial function, particularly in oxidative phosphorylation, which leads to altered ATP synthesis and lysosomal acidification. These findings suggest that the early impacts of these mutations may initiate a cascade of metabolic stress, contributing to neurodegenerative processes. Moreover, zebrafish harboring different PSEN1 mutations exhibit age-dependent hypoxic stress responses, characterized by accelerated brain aging and disrupted regulation of hypoxia-responsive genes, including those associated with glycolytic metabolism (Newman et al., 2020; van Tijn et al., 2011; Newman et al., 2009). Transcriptomic analyses have also highlighted shared pathways between PSEN1 mutant zebrafish and human brains, particularly in terms of mitochondrial dysregulation and impaired neurogenesis. Several studies investigated the role of presenilin 2 (*psen2*) in zebrafish. In zebrafish, *psen2* knockout line showed the relevance of PSEN2 to ER-mitochondria contacts, locomotor activity, and autophagy (Barazzuol et al., 2023). Hypoxia-induced PS2V isoform showed relevance to oxidative stress and inflammation pathways in AD (Ebrahimie et al., 2016). In another study, altered *psen2* splicing impacted on developmental processes and gene regulation, notably within the Notch signaling pathway (Newman et al., 2009).

Tau protein, encoded by the *MAPT* gene, plays a central role in neurofibrillary tangle (NFT) formation, a hallmark of AD (Jia et al., 2024; Heneka et al., 2024; Iqbal et al., 2005; Goedert et al., 2006; Arnaud et al., 2006; Spillantini et al., 1998; Matsuo et al., 1994). Tau pathologies are characterized by hyperphosphorylation, conformational changes, and aggregation into NFTs, leading to cytoskeletal destabilization and synaptic deficits. Zebrafish tauopathy models expressing human tau variants replicate important aspects of human tauopathies, such as phosphorylation at sites relevant to AD, behavioral deficits, and neuronal degeneration (Barbereau et al., 2021; Lopez et al., 2017; Cosacak et al., 2017; Paquet et al., 2010). They provide a strong platform for studying tau pathophysiology and drug screening, as they showed that zebrafish expressing human tau mutations show early synaptic loss and axonal transport defects, shedding light on how tau dysregulation affects neuronal function (Bhattarai et al., 2022; Barbereau et al., 2021; Ding et al., 2019; Lopez et al., 2017; Cosacak et al., 2017; Santana et al., 2012; Xi et al., 2011; Xia, 2010; Tomasiewicz et al., 2002). Zebrafish models were also utilized to identify compounds that modulate tau phosphorylation and aggregation, highlighting their translational potential (Barbereau et al., 2021; Paquet et al., 2010; Paquet et al., 2009). A transgenic zebrafish line expressing the human hTauP301L variant, which rapidly develops hallmark features of tauopathies, including tau phosphorylation, conformational alterations, and neuronal cell death, allowed for in vivo imaging of tau-induced neuronal toxicity using techniques such as time-lapse microscopy (Paquet et al., 2010; Paquet et al., 2009). Studies in zebrafish expressing hTauP301L have revealed early synaptic loss mirroring early AD phenotypes in humans. However, despite chronic tau hyperphosphorylation, zebrafish brains exhibit a striking resistance to NFT formation (Cosacak et al., 2017). This suggests the presence of intrinsic protective mechanisms that prevent tau aggregation into tangles, which are not manifesting in mammals, proposing therapeutic routes.

The *SORL1* gene, encoding the sortilin-related receptor, plays a critical role in endosomal sorting and trafficking of amyloid precursor protein (APP), influencing Aβ42 production and clearance (Knupp et al., 2020; Holstege et al., 2017; Schmidt et al., 2016; Cuccaro et al., 2016; Vardarajan et al., 2015; Dodson et al., 2008; Lane et al., 2010). SORL1 mutations or variants have been primarily linked to late-onset AD (Reitz et al., 2023; Bellenguez et al., 2022; Cuccaro et al., 2016). Zebrafish models have proven useful for exploring SORL1 function and its involvement in AD-related molecular pathways. Heterozygous mutations in zebrafish *sorl1* orthologues caused alterations in oxidative phosphorylation and mitochondrial bioenergetics, as identified through transcriptomic analyses of zebrafish brains (Barthelson et al., 2022). These changes indicate that SORL1 mutations may drive early cellular stress by impairing mitochondrial function. Moreover, zebrafish models highlight that SORL1 mutations lead to disruptions in endosomal and lysosomal pathways, which are central to APP processing and Aβ42 regulation. Transheterozygous *sorl1* mutant zebrafish exhibit significant upregulation of genes involved in oxidative phosphorylation, suggesting potential compensatory responses to mitochondrial dysfunction (Barthelson et al., 2022). Interestingly, SORL1-related transcriptomic signatures in zebrafish align with human data, further validating zebrafish as a model for studying SORL1-associated AD pathology (Barthelson et al., 2022; Newman et al., 2014). These insights emphasize the utility of zebrafish in dissecting the molecular consequences of SORL1 dysfunction and identifying early intervention targets.

The *ABCA7* gene, a critical player in lipid metabolism and Aβ42 regulation, is a significant genetic risk factor for AD (Tayran et al., 2024; Kawatani et al., 2023; Lyssenko and Pratico, 2021; Aikawa et al., 2018; Cukier et al., 2016; Sakae et al., 2016; Reitz et al., 2013; Ray et al., 2023). Zebrafish models have provided valuable insights into ABCA7's function and its role in modulating AD-related pathology. Recent research using CRISPR-Cas9 to delete the *abca7* gene in zebrafish has shown that it is involved in maintaining synapses, the growth of astrocytes, and neuropeptide signalling (Tayran et al., 2024). In zebrafish, the loss of ABCA7 function leads to decreased astroglial proliferation and reduced synaptic density in the presence of Aβ42 (Aβ) toxicity. Heterozygous ABCA7-deficient zebrafish fail to mount the typical regenerative astroglial response to Aβ, which is critical for neuronal repair. Immunostaining and transcriptomic analyses indicate that astroglial proliferation and neurogenic potential are compromised in these knockout models, highlighting the pivotal role of ABCA7 in maintaining neurogenesis under AD-like conditions (Tayran et al., 2024). A key finding from zebrafish studies is the ABCA7-dependent induction of neuropeptide Y (NPY), a molecule that plays a crucial role in neuroprotection and synaptic integrity. Under Aβ toxicity, ABCA7 knockout zebrafish fail to upregulate NPY expression, resulting in diminished synaptic density and astroglial activity. Furthermore, NPY interacts with brain-derived neurotrophic factor (BDNF) and nerve growth factor receptor (NGFR), forming a signaling axis that governs synaptic resilience and neurogenic responses (Tayran et al., 2024). Neurons in postmortem AD brains significantly downregulate NPY and BDNF levels, which correlates with reduced astroglial proliferation and increased AD pathology. These parallels underscore zebrafish's utility in modeling human ABCA7-related AD mechanisms and identifying therapeutic targets. Zebrafish, therefore, provides an additional model to explore how ABCA7 modulates neural repair mechanisms, and provides implications for developing strategies to enhance brain resilience in human AD.

Vascular dysfunction is increasingly recognized as a critical contributor to AD pathology (Is et al., 2024b; Zhang et al., 2024; Yang et al., 2022; Kim et al., 2020; Zenaro et al., 2017; Thomsen et al., 2017; Halliday et al., 2016; Lathe et al., 2014; Mayeux and Stern, 2012; Greenberg et al., 2004; Zarow et al., 1997; Glenner and Wong, 1984). The blood–brain barrier (BBB), a specialized structure, is essential for maintaining brain homeostasis by regulating the exchange of nutrients, immune cells, and neurotoxic substances between the brain and systemic circulation (Knox et al., 2022; De Silva and Faraci, 2020; Xu et al., 2019; Liebner et al., 2018; Yamazaki and Kanekiyo, 2017; Zenaro et al., 2017; Daneman and Prat, 2015). Dysfunction of the BBB in AD is marked by increased permeability, reduced cerebral blood flow, and the infiltration of neurotoxic substances, which exacerbate amyloid deposition, tau hyperphosphorylation, and neuroinflammation. Gliovascular dysfunction, where astrocytes, pericytes, and endothelial cells fail to maintain the structural and functional integrity of the BBB, intertwines these pathological processes. Recent studies utilizing zebrafish models have illuminated the molecular and cellular mechanisms underlying vascular contributions to AD, with implications for novel therapeutic strategies.

In genome-wide association studies (GWAS) of AD cohorts from different ethnic groups, Formin-like protein 2 (FMNL2), a cytoskeletal regulator from the formin family of genes (Breitsprecher and Goode, 2013; Faix and Grosse, 2006; Otomo et al., 2005), was found to be a key gene connected to vascular risk factors and AD pathology. Amyloid toxicity changes FMNL2's expression (Lee et al., 2022). Zebrafish models have been instrumental in uncovering the dual role of FMNL2 in maintaining BBB integrity and regulating gliovascular dynamics. In zebrafish, Aβ42 exposure triggers upregulation of *fmnl2a* in astrocytes, leading to the detachment of astroglial endfeet from blood vessels – a hallmark of impaired gliovascular remodeling (Lee et al., 2022). This disorganization compromises BBB integrity, facilitating the influx of neurotoxic substances and accelerating neuroinflammation. FMNL2 knockdown in zebrafish alleviates gliovascular disorganization, highlighting its protective role in astrocyte-endothelial interactions. This reduction in FMNL2 expression leads to an increase in extracellular amyloidosis, as FMNL2 is also involved in facilitating microglial clearance of Aβ42 aggregates. This dual role underscores the complexity of FMNL2's function in AD and highlights the need for therapeutic strategies that balance its impact on gliovascular remodeling and amyloid clearance. Furthermore, zebrafish studies have revealed FMNL2's involvement in cytoskeletal remodeling within astrocytes, which is crucial for maintaining endfeet attachment to blood vessels (Lee et al., 2022). By controlling the polymerization and movement of actin filaments, FMNL2 may help regulate the BBB function, making it a possible target for treatments that aim to keep the integrity of the blood vessels in AD.

Single-nucleus RNA sequencing (snRNAseq) studies have shown that the gliovascular unit (GVU) of AD brains is significantly changed, including the physiological and molecular alterations in endothelia, pericytes, and astrocytes, all of which are important for keeping the BBB intact (Tsartsalis et al., 2024; Walchli et al., 2024; Sun et al., 2023; Green et al., 2023; Is et al., 2022; Yang et al., 2022; Oveisgharan et al., 2022). The VEGFA-SMAD3 axis is a key regulatory pathway that is altered in AD, with SMAD3 being upregulated in pericytes and VEGFA being downregulated in astrocytes (Is et al., 2024a; Is et al., 2024b). These transcriptional changes impair the crosstalk between vascular cells and astrocytes, exacerbating BBB dysfunction. In zebrafish models, loss of VEGFA signaling in astrocytes leads to a reduction in vascular repair and angiogenesis, while SMAD3 upregulation in pericytes contributes to abnormal endothelial interactions and increased BBB permeability (Is et al., 2024b). Functional studies in zebrafish demonstrated that altered VEGFA expression in astrocytes regulates SMAD3-mediated vascular dysfunction, BBB integrity, and Aβ42-induced vascular damage (Is et al., 2024b).

Fibronectin 1 (*FN1*) was identified as a key extracellular matrix (ECM) component that modulates vascular pathology in AD (Bhattarai et al., 2024; Noda et al., 2018; Lepelletier et al., 2017; Sun et al., 2021). A rare protective variant in FN1, *rs140926439*, has been linked to a lower risk of AD and a delay in the onset of symptoms in people who carry *APOE4* allele (Bhattarai et al., 2024), a well-established and the strongest genetic risk factor for late-onset AD (Jackson et al., 2024; Belloy et al., 2023; Chai, 2007; Corder et al., 1993). Zebrafish models have provided crucial insights into the mechanisms by which FN1 influences vascular health and its interplay with APOE-related pathways. In zebrafish, loss-of-function mutations in *fn1b*, the ortholog of FN1, revealed a reduction in gliosis and enhanced gliovascular remodeling (Bhattarai et al., 2024). These findings suggest that pathological accumulation of FN1 at the BBB contributes to impaired clearance of toxic Aβ42 aggregates. The zebrafish model further demonstrated that FN1 modulates ECM composition and regulates interactions between astrocytes, microglia, and endothelial cells, which are essential for maintaining BBB integrity. Notably, FN1 knockdown in zebrafish mitigated the vascular deposition of Aβ42 and reduced neuroinflammation, highlighting its potential as a therapeutic target (Bhattarai et al., 2024). The interplay between FN1 and APOE4 is particularly compelling, with its effects mediated in part through alterations in lipid metabolism and vascular function (Kloske et al., 2024; Montagne et al., 2021; Montagne et al., 2020; Koizumi et al., 2018). FN1 alters the clearance of amyloid aggregations and regulates BBB stability and neuroinflammation, offering a promising avenue for therapeutic interventions aimed at mitigating AD pathology.

Considering the examples of *FMNL2*, *SMAD3*, and *FN1*, zebrafish proves to be a powerful model for studying vascular contributions to AD pathology by advancing our understanding of BBB breakdown and the signaling mechanisms between gliovascular cell types. These insights provide a foundation for developing targeted therapies aimed at preserving vascular integrity and mitigating AD progression.

## Zebrafish as a Preclinical Drug Development and Screening Model in Alzheimer's Disease

Zebrafish offer an unparalleled platform for high-throughput drug screening and pharmacological testing aimed at mitigating the effects of AD risk factors. Due to their optical transparency, small size, and genetic tractability, zebrafish enable rapid evaluation of candidate drugs under in vivo conditions, bridging the gap between in vitro assays and mammalian studies. Several compounds targeting Aβ42 aggregation and tau hyperphosphorylation have been extensively tested in zebrafish models. Zebrafish have been widely utilized to screen inhibitors of amyloid fibrillogenesis, identifying compounds that effectively reduce plaque burden and restore behavioral phenotypes, γ-secretase activity, kinase inhibition, and multitarget modulation of AD pathologies (Paquet et al., 2009; Reinhardt et al., 2019; Turgutalp and Kizil, 2024; Turgutalp et al., 2022; Bhattarai et al., 2022; Charbord et al., 2021; Wang et al., 2021; MacRae and Peterson, 2015; Arslanova et al., 2010; Rubinstein, 2003; Haage et al., 2024). Given the multifactorial nature of AD, the traditional one-drug-one-target approach has largely failed to provide effective therapeutic solutions. AD pathogenesis involves interrelated mechanisms, including Aβ42 aggregation, tau hyperphosphorylation, neuroinflammation, oxidative stress, and vascular dysfunction. As such, multitarget-directed drugs (MTDs), designed to simultaneously modulate multiple pathological pathways, offer a promising therapeutic paradigm (Turgutalp and Kizil, 2024; Turgutalp et al., 2022; Bhattarai et al., 2022). Recent studies have demonstrated the utility of zebrafish as a preclinical platform for evaluating MTDs, enabling high-throughput in vivo screening, dynamic imaging, and real-time monitoring of drug effects on neural and vascular systems. Another promising MTD strategy involves dual targeting of Aβ aggregation and tau hyperphosphorylation. By inhibiting key kinases and amyloid-processing enzymes, these compounds modulate core AD pathologies (Reinhardt et al., 2019).

## Conclusion

Zebrafish models have been instrumental in bridging the translational gap between in vitro assays and mammalian models. Their scalability and speed enable rapid validation of pharmacological interventions targeting AD-relevant genes. Improvements in imaging technologies and high-throughput platforms help zebrafish-based drug discovery. This makes it easier to find multitarget drugs that modulate the complex pathological underpinnings of AD. By using the special qualities of zebrafish, we can speed up the process of developing new medicines. Zebrafish present an exceptional opportunity for the integration of multi-omics approaches, enabling a systems biology perspective in AD research and providing a unique level of molecular and cellular resolution in vertebrate brains. This multidisciplinary approach is important for investigating how genetic, environmental, and cellular factors interact to cause AD. Furthermore, zebrafish models hold promise for personalized medicine, as patient-specific genetic variants can be modeled rapidly, providing insights into individual disease mechanisms and potential therapeutic responses. The versatility of zebrafish also extends to the exploration of emerging therapeutic strategies, such as gene editing and regenerative medicine. By harnessing the regenerative capacity of zebrafish neural tissues, researchers can probe the mechanisms underlying neurogenesis and synaptic repair, paving the way for innovative treatments aimed at restoring brain function. Zebrafish are uniquely suited for evaluating the pharmacokinetic and pharmacodynamic properties of pharmacological agents as well as early in vivo efficacy studies. In conclusion, zebrafish can transform the landscape of Alzheimer's disease research, offering a dynamic and versatile platform for both fundamental and translational studies. As we continue to refine and expand the use of zebrafish models, they are poised to play an even greater role in the development of innovative therapies and the advancement of our understanding of this devastating disease.

### Acknowledgements/Funding

I extend my gratitude to both past and present members of the Kizil Lab for their contributions to the studies mentioned in this chapter. This work was conducted without any external funding. Due to space constraints, it was not possible to reference all the relevant literature from our colleagues and the extensive body of work related to the focus of this review.

## REFERENCES

Abramsson, A., Kettunen, P., Banote, R. K., Lott, E., Li, M., Arner, A. & Zetterberg, H. (2013). The Zebrafish Amyloid Precursor Protein-B Is Required for Motor Neuron Guidance and Synapse Formation. *Dev Biol, 381*, 377–388.

Aikawa, T., Holm, M. L. & Kanekiyo, T. (2018). ABCA7 and Pathogenic Pathways of Alzheimer's Disease. *Brain Sci, 8*, Article 27.

Arnaud, L., Robakis, N. K. & Figueiredo-Pereira, M. E. (2006). It May Take Inflammation, Phosphorylation and Ubiquitination to 'Tangle' in Alzheimer's Disease. *Neurodegener Dis, 3*, 313–319.

Arslanova, D., Yang, T., Xu, X., Wong, S. T., Augelli-Szafran, C. E. & Xia, W. (2010). Phenotypic Analysis of Images of Zebrafish Treated with Alzheimer's Gamma-Secretase Inhibitors. *BMC Biotechnol, 10*, 24.

Bai, Q., Garver, J. A., Hukriede, N. A. & Burton, E. A. (2007). Generation of A Transgenic Zebrafish Model Of Tauopathy Using A Novel Promoter Element Derived From The Zebrafish Eno2 Gene. *Nucleic Acids Res, 35*, 6501–6516.

Banote, R. K., Chebli, J., Satir, T. M., Varshney, G. K., Camacho, R., Ledin, J., Burgess, S. M., Abramsson, A. & Zetterberg, H. (2020). Amyloid Precursor Protein-B Facilitates Cell Adhesion During Early Development in Zebrafish. *Sci Rep, 10*, 10127.

Barazzuol, L., Cieri, D., Facchinello, N., Cali, T., Washbourne, P., Argenton, F. & Pizzo, P. (2023). Unraveling Presenilin 2 Functions in A Knockout Zebrafish Line to Shed Light into Alzheimer's Disease Pathogenesis. *Cells, 12*, 376.

Barbereau, C., Cubedo, N., Maurice, T. & Rossel, M. (2021). Zebrafish Models To Study New Pathways In Tauopathies. *Int J Mol Sci, 22*, 4626.

Barthelson, K., Newman, M. & Lardelli, M. (2022). Brain Transcriptomes of Zebrafish and Mouse Alzheimer's Disease Knock-In Models Imply Early Disrupted Energy Metabolism. *Dis Model Mech, 15*. https://doi.org/10.1242/dmm.049187

Barthelson, K., Newman, M., Nowell, C. J. & Lardelli, M. (2021). No Observed Effect on Brain Vasculature of Alzheimer's Disease-Related Mutations in the Zebrafish Presenilin 1 Gene. *Mol Brain, 14*, 22.

Bashirzade, A. A., Zabegalov, K. N., Volgin, A. D., Belova, A. S., Demin, K. A., De Abreu, M. S., Babchenko, V. Y., Bashirzade, K. A., Yenkoyan, K. B., Tikhonova, M. A., Amstislavskaya, T. G. & Kalueff, A. V. (2022). Modeling Neurodegenerative Disorders in Zebrafish. *Neurosci Biobehav Rev, 138*, 104679.

Bayes, A., Collins, M. O., Reig-Viader, R., Gou, G., Goulding, D., Izquierdo, A., Choudhary, J. S., Emes, R. D. & Grant, S. G. (2017). Evolution of Complexity in the Zebrafish Synapse Proteome. *Nat Commun, 8*, 14613.

Bellenguez, C., Kucukali, F., Jansen, I. E., Kleineidam, L., Moreno-Grau, S., Amin, N., Naj, A. C., Campos-Martin, R., Grenier-Boley, B., Andrade, V., Holmans, P. A., Boland, A., Damotte, V., Van Der Lee, S. J., Costa, M. R., Kuulasmaa, T., Yang, Q., De Rojas, I., Bis, J. C., … Del Ser, T. (2022). New Insights into The Genetic Etiology of Alzheimer's Disease and Related Dementias. *Nat Genet*. https://doi.org/10.1002/alz.037973

Belloy, M. E., Andrews, S. J., Le Guen, Y., Cuccaro, M., Farrer, L. A., Napolioni, V. & Greicius, M. D. (2023). APOE Genotype and Alzheimer Disease Risk Across Age, Sex, aAnd Population Ancestry. *Jama Neurol, 80*, 1284–1294.

Bhattarai, P., Cosacak, M. I., Mashkaryan, V., Demir, S., Popova, S., Govindarajan, N., Brandt, K., Zhang, Y., Chang, W., Ampatzis, K. & Kizil, C. (2020). Neuron-Glia Interaction Through Serotonin-BDNF-NGFR Axis Enables Regenerative Neurogenesis in Alzheimer's Model of Adult Zebrafish Brain. *Plos Biol, 18*, E3000585.

Bhattarai, P., Gunasekaran, T. I., Belloy, M. E., Reyes-Dumeyer, D., Julich, D., Tayran, H., Yilmaz, E., Flaherty, D., Turgutalp, B., Sukumar, G., Alba, C., Mcgrath, E. M., Hupalo, D. N., Bacikova, D., Le Guen, Y., Lantigua, R., Medrano, M., Rivera, D., Recio, P., Nuriel, T., Kizil, C. & Vardarajan, B. N. (2024). Rare Genetic Variation In Fibronectin 1 (FN1) Protects Against Apoeepsilon4 in Alzheimer's Disease. *Acta Neuropathol, 147*, 70.

Bhattarai, P., Thomas, A. K., Cosacak, M. I., Papadimitriou, C., Mashkaryan, V., Zhang, Y. & Kizil, C. (2017a). Modeling Amyloid-B42 Toxicity And Neurodegeneration In Adult Zebrafish Brain. *Journal of Visualized Experiments, 128*.

Bhattarai, P., Thomas, A. K., Papadimitriou, C., Cosacak, M. I., Mashkaryan, V., Froc, C., Kurth, T., Dahl, A., Zhang, Y. & Kizil, C. (2016). IL4/STAT6 Signaling Activates Neural Stem Cell Proliferation And Neurogenesis Upon Amyloid-B42 Aggregation In Adult Zebrafish Brain. *Cell Reports, 17*, 941–948.

Bhattarai, P., Thomas, A. K., Zhang, Y. & Kizil, C. (2017b). The Effects of Aging on Amyloid-B42-Induced Neurodegeneration And Regeneration in Adult Zebrafish Brain. *Neurogenesis, 4*, E1322666.

Bhattarai, P., Turgutalp, B. & Kizil, C. (2022). Zebrafish As An Experimental and Preclinical Model for Alzheimer's Disease. *ACS Chem Neurosci, 13*, 2939–2941.

Blanchard, J. W., Akay, L. A., Davila-Velderrain, J., Von Maydell, D., Mathys, H., Davidson, S. M., Effenberger, A., Chen, C. Y., Maner-Smith, K., Hajjar, I., Ortlund, E. A., Bula, M., Agbas, E., Ng, A., Jiang, X., Kahn, M., Blanco-Duque, C., Lavoie, N., Liu, L., Reyes, R., … Tsai, L. H. (2022). APOE4 Impairs Myelination Via Cholesterol Dysregulation In Oligodendrocytes. *Nature, 611*, 769–779.

Breitsprecher, D. & Goode, B. L. (2013). Formins at a Glance. *J Cell Sci, 126*, 1–7.

Caramillo, E. M. & Echevarria, D. J. (2017). Alzheimer's Disease in the Zebrafish: Where Can We Take It? *Behav Pharmacol, 28*, 179–186.

Carrasquillo, M. M., Crook, J. E., Pedraza, O., Thomas, C. S., Pankratz, V. S., Allen, M., Nguyen, T., Malphrus, K. G., Ma, L., Bisceglio, G. D., Roberts, R. O., Lucas, J. A., Smith, G. E., Ivnik, R. J., Machulda, M. M., Graff-Radford, N. R., Petersen, R. C., Younkin, S. G. & Ertekin-Taner, N. (2015). Late-Onset Alzheimer's Risk Variants in Memory Decline, Incident Mild Cognitive Impairment, And Alzheimer's Disease. *Neurobiol Aging, 36*, 60–67.

Carrasquillo, M. M., Khan, Q., Murray, M. E., Krishnan, S., Aakre, J., Pankratz, V. S., Nguyen, T., Ma, L., Bisceglio, G., Petersen, R. C., Younkin, S. G., Dickson, D. W., Boeve, B. F., Graff-Radford, N. R. & Ertekin-Taner, N. (2014). Late-Onset Alzheimer Disease Genetic Variants in Posterior Cortical Atrophy And Posterior AD. *Neurology, 82*, 1455–1462.

Chai, C. K. (2007). The Genetics of Alzheimer's Disease. *Am J Alzheimers Dis Other Demen, 22*, 37–41.

Charbord, J., Ren, L., Sharma, R. B., Johansson, A., Agren, R., Chu, L., Tworus, D., Schulz, N., Charbord, P., Stewart, A. F., Wang, P., Alonso, L. C. & Andersson, O. (2021). In Vivo Screen Identifies A SIK Inhibitor That Induces Beta Cell Proliferation Through A Transient UPR. *Nat Metab, 3*, 682–700.

Chia, K., Klingseisen, A., Sieger, D. & Priller, J. (2022). Zebrafish as a Model Organism for Neurodegenerative Disease. *Front Mol Neurosci, 15*, 940484.

Chitnis, A. B. & Dawid, I. B. (1999). Neurogenesis in Zebrafish Embryos. *Methods Cell Biol, 59*, 367–386.

Corder, E. H., Saunders, A. M., Strittmatter, W. J., Schmechel, D. E., Gaskell, P. C., Small, G. W., Roses, A. D., Haines, J. L. & Pericak-Vance, M. A. (1993). Gene Dose of Apolipoprotein E Type 4 Allele and the Risk of Alzheimer's Disease in Late Onset Families. *Science, 261*, 921–923.

Cosacak, M. I., Bhattarai, P., Bocova, L., Dzewas, T., Mashkaryan, V., Papadimitriou, C., Brandt, K., Hollak, H., Antos, C. L. & Kizil, C. (2017). Human TAUP301L Overexpression Results In TAU Hyperphosphorylation Without Neurofibrillary Tangles in Adult Zebrafish Brain *Scientific Reports, 7*, 12959.

Cosacak, M. I., Bhattarai, P., De Jager, P. L., Menon, V., Tosto, G. & Kizil, C. (2022). Single Cell/Nucleus Transcriptomics Comparison in Zebrafish and Humans Reveals Common and Distinct Molecular Responses to Alzheimer's Disease. *Cells, 11*, 1807.

Cosacak, M. I., Bhattarai, P. & Kizil, C. (2020a). Alzheimer's Disease, Neural Stem Cells and Neurogenesis: Cellular Phase at Single-Cell Level. *Neural Reg Res, 15*, 824–827.

Cosacak, M. I., Bhattarai, P. & Kizil, C. (2020b). Protocol for Dissection and Dissociation of Zebrafish Telencephalon for Single-Cell Sequencing. *STAR Protoc, 1*, 100042.

Cosacak, M. I., Bhattarai, P., Reinhardt, S., Petzold, A., Dahl, A., Zhang, Y. & Kizil, C. (2019). Single-Cell Transcriptomics Analyses of Neural Stem Cell Heterogeneity and Contextual Plasticity in a Zebrafish Brain Model Of Amyloid Toxicity. *Cell Rep, 27*, 1307–1318.E3.

Cruchaga, C., Del-Aguila, J. L., Saef, B., Black, K., Fernandez, M. V., Budde, J., Ibanez, L., Deming, Y., Kapoor, M., Tosto, G., Mayeux, R. P., Holtzman, D. M., Fagan, A. M., Morris, J. C., Bateman, R. J., Goate, A. M., Dominantly Inherited Alzheimer, N., Disease Neuroimaging, I., Study, N.-L. F. & Harari, O. (2018). Polygenic Risk Score of Sporadic Late-Onset Alzheimer's Disease Reveals a Shared Architecture with the Familial and Early-Onset Forms. *Alzheimers Dement, 14*, 205–214.

Cuccaro, M. L., Carney, R. M., Zhang, Y., Bohm, C., Kunkle, B. W., Vardarajan, B. N., Whitehead, P. L., Cukier, H. N., Mayeux, R., St George-Hyslop, P. & Pericak-Vance, M. A. (2016). SORL1 Mutations in Early- and Late-Onset Alzheimer Disease. *Neurol Genet, 2*, E116.

Cukier, H. N., Kunkle, B. W., Vardarajan, B. N., Rolati, S., Hamilton-Nelson, K. L., Kohli, M. A., Whitehead, P. L., Dombroski, B. A., Van Booven, D., Lang, R., Dykxhoorn, D. M., Farrer, L. A., Cuccaro, M. L., Vance, J. M., Gilbert, J. R., Beecham, G. W., Martin, E. R., Carney, R. M., Mayeux, R., Schellenberg, G. D., Byrd, G. S., Haines, J. L., Pericak-Vance, M. A. & Alzheimer's Disease Genetics Consortium. (2016). ABCA7 Frameshift Deletion Associated with Alzheimer Disease in African Americans. *Neurol Genet, 2*, E79.

Cully, M. (2019). Zebrafish Earn Their Drug Discovery Stripes. *Nat Rev Drug Discov, 18*, 811–813.

Daneman, R. & Prat, A. (2015). The Blood-Brain Barrier. *Cold Spring Harb Perspect Biol, 7*, A020412.

De Silva, T. M. & Faraci, F. M. (2020). Contributions of Aging to Cerebral Small Vessel Disease. *Annu Rev Physiol, 82*, 275–295.

Ding, Y., Lei, L., Lai, C. & Tang, Z. (2019). Tau Protein and Zebrafish Models for Tau-Induced Neurodegeneration. *J Alzheimers Dis, 69*, 339–353.

Dodson, S. E., Andersen, O. M., Karmali, V., Fritz, J. J., Cheng, D., Peng, J., Levey, A. I., Willnow, T. E. & Lah, J. J. (2008). Loss Of LR11/SORLA Enhances Early Pathology in a Mouse Model of Amyloidosis: Evidence for a Proximal Role in Alzheimer's Disease. *J Neurosci, 28*, 12877–12886.

Dray, N., Bedu, S., Vuillemin, N., Alunni, A., Coolen, M., Krecsmarik, M., Supatto, W., Beaurepaire, E. & Bally-Cuif, L. (2015). Large-Scale Live Imaging of Adult Neural Stem Cells in Their Endogenous Niche. *Development, 142*, 3592–3600.

Driever, W., Solnica-Krezel, L., Schier, A. F., Neuhauss, S. C., Malicki, J., Stemple, D. L., Stainier, D. Y., Zwartkruis, F., Abdelilah, S., Rangini, Z., Belak, J. & Boggs, C. (1996). A Genetic Screen for Mutations Affecting Embryogenesis in Zebrafish. *Development, 123*, 37–46.

Dumitrescu, L., Mahoney, E. R., Mukherjee, S., Lee, M. L., Bush, W. S., Engelman, C. D., Lu, Q., Fardo, D. W., Trittschuh, E. H., Mez, J., Kaczorowski, C., Hernandez Saucedo, H., Widaman, K. F., Buckley, R., Properzi, M., Mormino, E., Yang, H. S., Harrison, T., Hedden, T., Nho, K.,…., Hohman, T. J. (2020). Genetic Variants and Functional Pathways Associated with Resilience to Alzheimer's Disease. *Brain, 143*, 2561–2575.

Ebrahimie, E., Moussavi Nik, S. H., Newman, M., Van Der Hoek, M. & Lardelli, M. (2016). The Zebrafish Equivalent of Alzheimer's Disease-Associated PRESENILIN Isoform PS2V Regulates Inflammatory and Other Responses to Hypoxic Stress. *J Alzheimers Dis, 52*, 581–608.

Eisen, J. S. (1991). Developmental Neurobiology of the Zebrafish. *J Neurosci, 11*, 311–317.

Faix, J. & Grosse, R. (2006). Staying in Shape with Formins. *Dev Cell, 10*, 693–706.

Farrer, L. A., Cupples, L. A., Haines, J. L., Hyman, B., Kukull, W. A., Mayeux, R., Myers, R. H., Pericak-Vance, M. A., Risch, N. & Van Duijn, C. M. (1997). Effects of Age, Sex, and Ethnicity on the Association Between Apolipoprotein E Genotype and Alzheimer Disease. A Meta-Analysis. APOE and Alzheimer Disease Meta Analysis Consortium. *JAMA, 278*, 1349–1356.

Felsky, D., Santa-Maria, I., Cosacak, M. I., French, L., Schneider, J. A., Bennett, D. A., De Jager, P. L., Kizil, C. & Tosto, G. (2023). The Caribbean-Hispanic Alzheimer's Disease Brain Transcriptome Reveals Ancestry-Specific Disease Mechanisms. *Neurobiol Dis, 176*, 105938.

Fleisch, V. C., Fraser, B. & Allison, W. T. (2010). Investigating Regeneration and Functional Integration of CNS Neurons: Lessons from Zebrafish Genetics and Other Fish Species. *Biochim Biophys Acta, 1812*, 364–380.

Fulwiler, C. & Gilbert, W. (1991). Zebrafish Embryology and Neural Development. *Curr Opin Cell Biol, 3*, 988–991.

Genin, E., Hannequin, D., Wallon, D., Sleegers, K., Hiltunen, M., Combarros, O., Bullido, M. J., Engelborghs, S., De Deyn, P., Berr, C., Pasquier, F., Dubois, B., Tognoni, G., Fievet, N., Brouwers, N., Bettens, K., Arosio, B., Coto, E., Del Zompo, M., Mateo, I.,…. Campion, D. (2011). APOE and Alzheimer Disease: A Major Gene with Semi-Dominant Inheritance. *Mol Psychiatry, 16*, 903–907.

Glenner, G. G. & Wong, C. W. (1984). Alzheimer's Disease: Initial Report of the Purification and Characterization of A Novel Cerebrovascular Amyloid Protein. *Biochem Biophys Res Commun, 120*, 885–890.

Goedert, M., Klug, A. & Crowther, R. A. (2006). Tau Protein, The Paired Helical Filament and Alzheimer's Disease. *J Alzheimers Dis, 9*, 195–207.

Granato, M. & Nusslein-Volhard, C. (1996). Fishing for Genes Controlling Development. *Curr Opin Genet Dev, 6*, 461–468.

Green, G. S., Fujita, M., Yang, H. S., Taga, M., Cain, A., Mccabe, C., Comandante-Lou, N., White, C. C., Schmidtner, A. K., Zeng, L., Sigalov, A., Wang, Y., Regev, A., Klein, H. U., Menon, V., Bennett, D. A., Habib, N. & De Jager, P. L. (2024). Cellular Communities Reveal Trajectories of Brain Ageing and Alzheimer's Disease. *Nature, 633*, 634–645.

Green, G. S., Fujita, M., Yang, H. S., Taga, M., Mccabe, C., Cain, A., White, C. C., Schmidtner, A. K., Zeng, L., Wang, Y., Regev, A., Menon, V., Bennett, D. A., Habib, N. & De Jager, P. L. (2023). Cellular Dynamics Across Aged Human Brains Uncover a Multicellular Cascade Leading to Alzheimer's Disease. *Biorxiv*. https://doi.org/10.1101/2023.03.07.531493.

Greenberg, S. M., Gurol, M. E., Rosand, J. & Smith, E. E. (2004). Amyloid Angiopathy-Related Vascular Cognitive Impairment. *Stroke, 35*, 2616–2619.

Gunasekaran, T. I., Reyes-Dumeyer, D., Faber, K. M., Goate, A., Boeve, B., Cruchaga, C., Pericak-Vance, M., Haines, J. L., Rosenberg, R., Tsuang, D., Mejia, D. R., Medrano, M., Lantigua, R. A., Sweet, R. A., Bennett, D. A., Wilson, R. S., Alba, C., Dalgard, C., Foroud, T., Vardarajan, B. N. & Mayeux, R. (2024). Missense and Loss-of-Function Variants at GWAS Loci in Familial Alzheimer's Disease. *Alzheimers Dement.*

Gunasekaran, T. I., Reyes-Dumeyer, D., Faber, K. M., Goate, A., Boeve, B., Cruchaga, C., Pericak-Vance, M., Haines, J. L., Rosenberg, R., Tsuang, D., Mejia, D. R., Medrano, M., Lantigua, R. A., Sweet, R. A., Bennett, D. A., Wilson, R. S., Foroud, T., Vardarajan, B. N. & Mayeux, R. (2023). Identification of Rare Damaging Missense and Loss of Function Variants in GWAS Loci Using Genome Sequencing Data from Two Cohorts of Familial Late-Onset Alzheimer's Disease. *Medrxiv.* https://psycnet.apa.org/doi/10.1002/alz.14221

Haage, V. & De Jager, P. L. (2022). Neuroimmune Contributions to Alzheimer's Disease: A Focus on Human Data. *Mol Psychiatry, 27*, 3164–3181.

Haage, V., Tuddenham, J. F., Comandante-Lou, N., Bautista, A., Monzel, A., Chiu, R., Fujita, M., Garcia, F. G., Bhattarai, P., Patel, R., Buonfiglioli, A., Idiarte, J., Herman, M., Rinderspacher, A., Mela, A., Zhao, W., Argenziano, M. G., Furnari, J. L., Banu, M. A., Landry, D. W., ... De Jager, P. L. (2024). A Pharmacological Toolkit for Human Microglia Identifies Topoisomerase I Inhibitors as Immunomodulators for Alzheimer's Disease. *Biorxiv.* https://doi.org/10.1101/2024.02.06.579103

Haass, C. & Selkoe, D. J. (2007). Soluble Protein Oligomers in Neurodegeneration: Lessons from the Alzheimer's Amyloid Beta-Peptide. *Nat Rev Mol Cell Biol, 8*, 101–112.

Haffter, P., Granato, M., Brand, M., Mullins, M. C., Hammerschmidt, M., Kane, D. A., Odenthal, J., Van Eeden, F. J., Jiang, Y. J., Heisenberg, C. P., Kelsh, R. N., Furutani-Seiki, M., Vogelsang, E., Beuchle, D., Schach, U., Fabian, C. & Nusslein-Volhard, C. (1996). The Identification of Genes with Unique and Essential Functions in the Development of The Zebrafish, Danio Rerio. *Development, 123*, 1–36.

Halliday, M. R., Rege, S. V., Ma, Q., Zhao, Z., Miller, C. A., Winkler, E. A. & Zlokovic, B. V. (2016). Accelerated Pericyte Degeneration and Blood-Brain Barrier Breakdown in Apolipoprotein E4 Carriers with Alzheimer's Disease. *J Cereb Blood Flow Metab, 36*, 216–227.

Handler, M., Yang, X. & Shen, J. (2000). Presenilin-1 Regulates Neuronal Differentiation During Neurogenesis. *Development, 127*, 2593–2606.

Heneka, M. T., Van Der Flier, W. M., Jessen, F., Hoozemanns, J., Thal, D. R., Boche, D., Brosseron, F., Teunissen, C., Zetterberg, H., Jacobs, A. H., Edison, P., Ramirez, A., Cruchaga, C., Lambert, J. C., Laza, A. R., Sanchez-Mut, J. V., Fischer, A., Castro-Gomez, S., Stein, T. D., Kleineidam, L., .... Riechers, S. P. (2024). Neuroinflammation in Alzheimer Disease. *Nat Rev Immunol.* https://doi.org/10.1038/s41577-024-01104-7

Holcomb, L., Gordon, M. N., Mcgowan, E., Yu, X., Benkovic, S., Jantzen, P., Wright, K., Saad, I., Mueller, R., Morgan, D., Sanders, S., Zehr, C., O'campo, K., Hardy, J., Prada, C. M., Eckman, C., Younkin, S., Hsiao, K. & Duff, K. (1998). Accelerated Alzheimer-Type Phenotype in Transgenic Mice Carrying Both Mutant Amyloid Precursor Protein and Presenilin 1 Transgenes. *Nat Med, 4*, 97–100.

Holstege, H., Van Der Lee, S. J., Hulsman, M., Wong, T. H., Van Rooij, J. G., Weiss, M., Louwersheimer, E., Wolters, F. J., Amin, N., Uitterlinden, A. G., Hofman, A., Ikram, M. A., Van Swieten, J. C., Meijers-Heijboer, H., Van Der Flier, W. M., Reinders, M. J., Van Duijn, C. M. & Scheltens, P. (2017). Characterization of Pathogenic SORL1 Genetic Variants for Association with Alzheimer's Disease: A Clinical Interpretation Strategy. *Eur J Hum Genet, 25*, 973–981.

Howe, K., Clark, M. D., Torroja, C. F., Torrance, J., Berthelot, C., Muffato, M., Collins, J. E., Humphray, S., Mclaren, K., Matthews, L., Mclaren, S., Sealy, I., Caccamo, M., Churcher, C., Scott, C., Barrett, J. C., Koch, R., Rauch, G. J., White, S., Chow, W., Kilian, B., ... Stemple, D. L. (2013). The Zebrafish Reference Genome Sequence and Its Relationship to the Human Genome. *Nature, 496*, 498–503.

Hruscha, A., Krawitz, P., Rechenberg, A., Heinrich, V., Hecht, J., Haass, C. & Schmid, B. (2013). Efficient CRISPR/Cas9 Genome Editing with Low Off-Target Effects in Zebrafish. *Development, 140*, 4982–4987.

Iqbal, K., Alonso Adel, C., Chen, S., Chohan, M. O., El-Akkad, E., Gong, C. X., Khatoon, S., Li, B., Liu, F., Rahman, A., Tanimukai, H. & Grundke-Iqbal, I. (2005). Tau Pathology in Alzheimer Disease and Other Tauopathies. *Biochim Biophys Acta, 1739*, 198–210.

Is, O., Min, Y., Wang, X., Oatman, S. R., Abraham Daniel, A. & Ertekin-Taner, N. (2024a). Multi Layered Omics Approaches Reveal Glia Specific Alterations in Alzheimer's Disease: A Systematic Review and Future Prospects. *Glia.* https://doi.org/10.1002/glia.24652

Is, O., Wang, X., Patel, T., Quicksall, Z., Heckman, M., White, L., Lewis-Tuffin, L., Deniz, K., Tutor-New, F., Carnwath, T., Min, Y., Oatman, S., Reddy, J., Carrasquillo, M., Nguyen, T., Ho, C., Malphrus, K., Nho, K., Saykin, A., Murray, M., Dickson, D., Allen, M. & Ertekin-Taner, N. (2022). Single Nuclei Transcriptome Reveals Perturbed Brain Vascular Molecules in Alzheimer's Disease. *Biorxiv.* https://doi.org/10.1101/2021.12.28.474255

Is, O., Wang, X., Reddy, J., Min, Y., Yilmaz, E., Bhattarai, P., Patel, T., Bergman, J., Quicksall, Z., Heckman, M., Tutor-New, F., Demirdöğen, B. C., White, L., Koga, S., Krause, V., Inoue, Y., Kanekiyo, T., Cosacak, M. I., Nelson, N., Lee, A., ... Ertekin-Taner, N. (2024b). Gliovascular Transcriptional Perturbations in Alzheimer's Disease Reveal Molecular Mechanisms of Blood Brain Barrier Dysfunction. *Nat Commun, 15*, 4758.

Jackson, R. J., Hyman, B. T. & Serrano-Pozo, A. (2024). Multifaceted Roles of APOE in Alzheimer Disease. *Nat Rev Neurol*, 20, 457–474.

Jao, L. E., Wente, S. R. & Chen, W. (2013). Efficient Multiplex Biallelic Zebrafish Genome Editing Using a CRISPR Nuclease System. *Proc Natl Acad Sci U S A, 110*, 13904–13909.

Jia, J., Ning, Y., Chen, M., Wang, S., Yang, H., Li, F., Ding, J., Li, Y., Zhao, B., Lyu, J., Yang, S., Yan, X., Wang, Y., Qin, W., Wang, Q., Li, Y., Zhang, J., Liang, F., Liao, Z. & Wang, S. (2024). Biomarker Changes During 20 Years Preceding Alzheimer's Disease. *N Engl J Med, 390*, 712–722.

Julich, D. & Holley, S. A. (2024). Live Imaging of Fibronectin 1a-Mneongreen and Fibronectin 1b-Mcherry Knock-In Alleles During Early Zebrafish Development. *Cells Dev, 177*, 203900.

Jurisch-Yaksi, N., Yaksi, E. & Kizil, C. 2(020). Radial Glia in the Zebrafish Brain: Functional, Structural, and Physiological Comparison with the Mammalian Glia. *Glia*, 68, 2451–2470.

Kawatani, K., Holm, M. L., Starling, S. C., Martens, Y. A., Zhao, J., Lu, W., Ren, Y., Li, Z., Jiang, P., Jiang, Y., Baker, S. K., Wang, N., Roy, B., Parsons, T. M., Perkerson, R. B., 3rd, Bao, H., Han, X., Bu, G. & Kanekiyo, T. (2023). ABCA7 Deficiency Causes Neuronal Dysregulation by Altering Mitochondrial Lipid Metabolism. *Mol Psychiatry*.

Kelsh, R. N. & Raible, D. W. (2002). Specification of Zebrafish Neural Crest. *Results Probl Cell Differ, 40*, 216–236.

Keren-Shaul, H., Spinrad, A., Weiner, A., Matcovitch-Natan, O., Dvir-Szternfeld, R., Ulland, T. K., David, E., Baruch, K., Lara-Astaiso, D., Toth, B., Itzkovitz, S., Colonna, M., Schwartz, M. & Amit, I. (2017). A Unique Microglia Type Associated with Restricting Development of Alzheimer's Disease. *Cell, 169*, 1276–1290.E17.

Kim, H. W., Hong, J. & Jeon, J. C. (2020). Cerebral Small Vessel Disease and Alzheimer's Disease: A Review. *Front Neurol, 11*, 927.

Kim, Y. A., Siddiqui, T., Blaze, J., Cosacak, M. I., Winters, T., Kumar, A., Tein, E., Sproul, A. A., Teich, A. F., Bartolini, F., Akbarian, S., Kizil, C., Hargus, G. & Santa-Maria, I. (2023). RNA Methyltransferase Nsun2 Deficiency Promotes Neurodegeneration Through Epitranscriptomic Regulation of Tau Phosphorylation. *Acta Neuropathol, 145*, 29–48.

Kimmel, C. B., Ballard, W. W., Kimmel, S. R., Ullmann, B. & Schilling, T. F. (1995). Stages of Embryonic Development of the Zebrafish. *Dev Dyn, 203*, 253–310.

Kimmel, C. B., Sepich, D. S. & Trevarrow, B. (1988). Development Of Segmentation in Zebrafish. *Development, 104*, 197–207.

Kizil, C. (2018). Mechanisms of Pathology-Induced Neural Stem Cell Plasticity and Neural Regeneration in Adult Zebrafish Brain. *Current Pathobiology Reports, 6*, 71–77.

Kizil, C. & Bhattarai, P. (2018). Is Alzheimer's Also A Stem Cell Disease? - The Zebrafish Perspective. *Front Cell Dev Biol, 6*, 159.

Kizil, C., Kyritsis, N. & Brand, M. (2015). Effects of Inflammation on Stem Cells: Together They Strive? *EMBO Rep, 16*, 416–426.

Kizil, C., Kyritsis, N., Dudczig, S., Kroehne, V., Freudenreich, D., Kaslin, J. & Brand, M. (2012). Regenerative Neurogenesis from Neural Progenitor Cells Requires Injury-Induced Expression of Gata3. *Dev Cell, 23*, 1230–1237.

Kizil, C., Sariya, S., Kim, Y. A., Rajabli, F., Martin, E., Reyes-Dumeyer, D., Vardarajan, B., Maldonado, A., Haines, J. L., Mayeux, R., Jimenez-Velazquez, I. Z., Santa-Maria, I. & Tosto, G. (2022). Admixture Mapping of Alzheimer's Disease in Caribbean Hispanics Identifies A New Locus on 22q13.1. *Mol Psychiatry*. https://doi.org/10.1038/s41380-022-01526-6

Kloske, C. M., Belloy, M. E., Blue, E. E., Bowman, G. R., Carrillo, M. C., Chen, X., Chiba-Falek, O., Davis, A. A., Paolo, G. D., Garretti, F., Gate, D., Golden, L. R., Heinecke, J. W., Herz, J., Huang, Y., Iadecola, C., Johnson, L. A., Kanekiyo, T., Karch, C. M., Khvorova, A.,..., Holtzman, D. M. (2024). Advancements in APOE and Dementia Research: Highlights from the 2023 AAIC Advancements: APOE Conference. *Alzheimers Dement, 20*, 6590–6605.

Knox, E. G., Aburto, M. R., Clarke, G., Cryan, J. F. & O'driscoll, C. M. (2022). The Blood-Brain Barrier In Aging and Neurodegeneration. *Mol Psychiatry, 27*, 2659–2673.

Knupp, A., Mishra, S., Martinez, R., Braggin, J. E., Szabo, M., Kinoshita, C., Hailey, D. W., Small, S. A., Jayadev, S. & Young, J. E. (2020). Depletion of the AD Risk Gene SORL1 Selectively Impairs Neuronal Endosomal Traffic Independent of Amyloidogenic APP Processing. *Cell Rep, 31*, 107719.

Koizumi, K., Hattori, Y., Ahn, S. J., Buendia, I., Ciacciarelli, A., Uekawa, K., Wang, G., Hiller, A., Zhao, L., Voss, H. U., Paul, S. M., Schaffer, C., Park, L. & Iadecola, C. (2018). Apoepsilon4 Disrupts Neurovascular Regulation and Undermines White Matter Integrity and Cognitive Function. *Nat Commun, 9*, 3816.

Kunkle, B. W., Grenier-Boley, B., Sims, R., Bis, J. C., Damotte, V., Naj, A. C., Boland, A., Vronskaya, M., Van Der Lee, S. J., Amlie-Wolf, A., Bellenguez, C., Frizatti, A., Chouraki, V., Martin, E. R., Sleegers, K., Badarinarayan, N., Jakobsdottir, J., Hamilton-Nelson, K. L., Moreno-Grau, S., Olaso, R., … Genetic and Environmental Risk in AD/Defining Genetic, Polygenic and Environmental Risk for Alzheimer's Disease Consortium (GERAD/PERADES). (2019). Genetic Meta-Analysis of Diagnosed Alzheimer's Disease Identifies New Risk Loci and Implicates Abeta, Tau, Immunity and Lipid Processing. *Nat Genet, 51*, 414–430.

Kyritsis, N., Kizil, C., Zocher, S., Kroehne, V., Kaslin, J., Freudenreich, D., Iltzsche, A. & Brand, M. (2012). Acute Inflammation Initiates the Regenerative Response in the Adult Zebrafish Brain. *Science, 338*, 1353–1356.

Lane, R. F., Raines, S. M., Steele, J. W., Ehrlich, M. E., Lah, J. A., Small, S. A., Tanzi, R. E., Attie, A. D. & Gandy, S. (2010). Diabetes-Associated Sorcs1 Regulates Alzheimer's Amyloid-Beta Metabolism: Evidence for Involvement of Sorl1 and the Retromer Complex. *J Neurosci, 30*, 13110–13115.

Lathe, R., Sapronova, A. & Kotelevtsev, Y. (2014). Atherosclerosis and Alzheimer–Diseases with a Common Cause? Inflammation, Oxysterols, Vasculature. *BMC Geriatr, 14*, 36.

Lau, S. F., Cao, H., Fu, A. K. Y. & Ip, N. Y. (2020). Single-Nucleus Transcriptome Analysis Reveals Dysregulation of Angiogenic Endothelial Cells and Neuroprotective Glia in Alzheimer's Disease. *Proc Natl Acad Sci U S A, 117*, 25800–25809.

Lawson, N. D. & Weinstein, B. M. (2002). In Vivo Imaging of Embryonic Vascular Development Using Transgenic Zebrafish. *Dev Biol, 248*, 307–318.

Lee, A. J., Raghavan, N. S., Bhattarai, P., Siddiqui, T., Sariya, S., Reyes-Dumeyer, D., Flowers, X. E., Cardoso, S. A. L., De Jager, P. L., Bennett, D. A., Schneider, J. A., Menon, V., Wang, Y., Lantigua, R., Medrano, M., Rivera, D., Jimenez-Velazquez, I. Z., Kukull, W. A., Brickman, A. M., Manly,

J. J., Tosto, G., Kizil, C., Vardarajan, B. & Mayeux, R. (2022). FMNL2 Regulates Gliovascular Interactions and Is Associated with Vascular Risk Factors and Cerebrovascular Pathology in Alzheimer's Disease. *Acta Neuropathol.* https://doi.org/10.1007/s00401-022-02431-6

Leimer, U., Lun, K., Romig, H., Walter, J., Grunberg, J., Brand, M. & Haass, C. (1999). Zebrafish (Danio Rerio) Presenilin Promotes Aberrant Amyloid Beta-Peptide Production and Requires A Critical Aspartate Residue For Its Function In Amyloidogenesis. *Biochemistry, 38*, 13602–13609.

Leng, K., Li, E., Eser, R., Piergies, A., Sit, R., Tan, M., Neff, N., Li, S. H., Rodriguez, R. D., Suemoto, C. K., Leite, R. E. P., Ehrenberg, A. J., Pasqualucci, C. A., Seeley, W. W., Spina, S., Heinsen, H., Grinberg, L. T. & Kampmann, M. (2021). Molecular Characterization of Selectively Vulnerable Neurons in Alzheimer's Disease. *Nat Neurosci, 24*, 276–287.

Lepelletier, F. X., Mann, D. M., Robinson, A. C., Pinteaux, E. & Boutin, H. (2017). Early Changes in Extracellular Matrix in Alzheimer's Disease. *Neuropathol Appl Neurobiol, 43*, 167–182.

Li, L., Yan, B., Shi, Y. Q., Zhang, W. Q. & Wen, Z. L. (2012). Live Imaging Reveals Differing Roles of Macrophages and Neutrophils During Zebrafish Tail Fin Regeneration. *J Biol Chem, 287*, 25353–25360.

Li, M., Zhao, L., Page-Mccaw, P. S. & Chen, W. (2016). Zebrafish Genome Engineering Using the CRISPR-Cas9 System. *Trends Genet, 32*, 815–827.

Liebner, S., Dijkhuizen, R. M., Reiss, Y., Plate, K. H., Agalliu, D. & Constantin, G. (2018). Functional Morphology of the Blood-Brain Barrier in Health and Disease. *Acta Neuropathol, 135*, 311–336.

Lieschke, G. J. & Currie, P. D. (2007). Animal Models of Human Disease: Zebrafish Swim into View. *Nat Rev Genet, 8*, 353–367.

Lopez, A., Lee, S. E., Wojta, K., Ramos, E. M., Klein, E., Chen, J., Boxer, A. L., Gorno-Tempini, M. L., Geschwind, D. H., Schlotawa, L., Ogryzko, N. V., Bigio, E. H., Rogalski, E., Weintraub, S., Mesulam, M. M., Tauopathy Genetics, C., Fleming, A., Coppola, G., Miller, B. L. & Rubinsztein, D. C. (2017). A152T Tau Allele Causes Neurodegeneration That Can Be Ameliorated in A Zebrafish Model By Autophagy Induction. *Brain, 140*, 1128–1146.

Lyssenko, N. N. & Pratico, D. (2021). ABCA7 and the Altered Lipidostasis Hypothesis of Alzheimer's Disease. *Alzheimers Dement, 17*, 164–174.

Macrae, C. A. & Peterson, R. T. (2015). Zebrafish as Tools for Drug Discovery. *Nat Rev Drug Discov, 14*, 721–731.

Malaga-Trillo, E., Salta, E., Figueras, A., Panagiotidis, C. & Sklaviadis, T. (2011). Fish Models in Prion Biology: Underwater Issues. *Biochim Biophys Acta, 1812*, 402–414.

Mathys, H., Adaikkan, C., Gao, F., Young, J. Z., Manet, E., Hemberg, M., De Jager, P. L., Ransohoff, R. M., Regev, A. & Tsai, L. H. (2017). Temporal Tracking of Microglia Activation in Neurodegeneration at Single-Cell Resolution. *Cell Rep, 21*, 366–380.

Mathys, H., Davila-Velderrain, J., Peng, Z., Gao, F., Mohammadi, S., Young, J. Z., Menon, M., He, L., Abdurrob, F., Jiang, X., Martorell, A. J., Ransohoff, R. M., Hafler, B. P., Bennett, D. A., Kellis, M. & Tsai, L. H. (2019). Single-Cell Transcriptomic Analysis of Alzheimer's Disease. *Nature, 570*, 332–337.

Mathys, H., Peng, Z., Boix, C., Victor, M., Leary, N., Babu, S., Abdelhady, G., Jiang, X., Ng, A., Ghafari, K., Kunisky, A., Mantero, J., Galani, K., Lohia, V., Fortier, G., Lotfi, Y., Ivey, J., Brown, H., Patel, P., Chakraborty, N., ... Tsai, L. H. (2023). Single-Cell Atlas Reveals Correlates of High Cognitive Function, Dementia, and Resilience to Alzheimer's Disease Pathology. *Cell.* https://doi.org/10.1016/j.cell.2023.08.039

Matsuo, E. S., Shin, R. W., Billingsley, M. L., Van Devoorde, A., O'connor, M., Trojanowski, J. Q. & Lee, V. M. (1994). Biopsy-Derived Adult Human Brain Tau Is Phosphorylated at Many of the Same Sites as Alzheimer's Disease Paired Helical Filament Tau. *Neuron, 13*, 989–1002.

Mayeux, R. & Stern, Y. (2012). Epidemiology of Alzheimer Disease. *Cold Spring Harb Perspect Med, 2*, a006239.

Mcgowan, E., Pickford, F., Kim, J., Onstead, L., Eriksen, J., Yu, C., Skipper, L., Murphy, M. P., Beard, J., Das, P., Jansen, K., Delucia, M., Lin, W. L., Dolios, G., Wang, R., Eckman, C. B., Dickson, D. W., Hutton, M., Hardy, J. & Golde, T. (2005). Abeta42 Is Essential for Parenchymal and Vascular Amyloid Deposition in Mice. *Neuron, 47*, 191–199.

Montagne, A., Nation, D. A., Sagare, A. P., Barisano, G., Sweeney, M. D., Chakhoyan, A., Pachicano, M., Joe, E., Nelson, A. R., D'orazio, L. M., Buennagel, D. P., Harrington, M. G., Benzinger, T. L. S., Fagan, A. M., Ringman, J. M., Schneider, L. S., Morris, J. C., Reiman, E. M., Caselli, R. J., Chui, H. C., ... Zlokovic, B. V. (2020). APOE4 Leads to Blood-Brain Barrier Dysfunction Predicting Cognitive Decline. *Nature, 581*, 71–76.

Montagne, A., Nikolakopoulou, A. M., Huuskonen, M. T., Sagare, A. P., Lawson, E. J., Lazic, D., Rege, S. V., Grond, A., Zuniga, E., Barnes, S. R., Prince, J., Sagare, M., Hsu, C. J., Ladu, M. J., Jacobs, R. E. & Zlokovic, B. V. (2021). APOE4 Accelerates Advanced-Stage Vascular and Neurodegenerative Disorder in Old Alzheimer's Mice Via Cyclophilin A Independently of Amyloid-Beta. *Nat Aging, 1*, 506–520.

N'songo, A., Carrasquillo, M. M., Wang, X., Burgess, J. D., Nguyen, T., Asmann, Y. W., Serie, D. J., Younkin, S. G., Allen, M., Pedraza, O., Duara, R., Greig Custo, M. T., Graff-Radford, N. R. & Ertekin-Taner, N. (2017). African American Exome Sequencing Identifies Potential Risk Variants at Alzheimer Disease Loci. *Neurol Genet, 3*, E141.

Neuhauss, S. C., Solnica-Krezel, L., Schier, A. F., Zwartkruis, F., Stemple, D. L., Malicki, J., Abdelilah, S., Stainier, D. Y. & Driever, W. (1996). Mutations Affecting Craniofacial Development in Zebrafish. *Development, 123*, 357–367.

Newman, M., Ebrahimie, E. & Lardelli, M. (2014). Using The Zebrafish Model For Alzheimer's Disease Research. *Front Genet, 5*, 189.

Newman, M., Nik, H. M., Sutherland, G. T., Hin, N., Kim, W. S., Halliday, G. M., Jayadev, S., Smith, C., Laird, A. S., Lucas, C. W., Kittipassorn, T., Peet, D. J. & Lardelli, M. (2020). Accelerated Loss of Hypoxia Response in Zebrafish with Familial Alzheimer's Disease-Like Mutation of Presenilin 1. *Hum Mol Genet, 29*, 2379–2394.

Newman, M., Tucker, B., Nornes, S., Ward, A. & Lardelli, M. (2009). Altering Presenilin Gene Activity in Zebrafish Embryos Causes Changes in Expression of Genes with Potential Involvement in Alzheimer's Disease Pathogenesis. *J Alzheimers Dis, 16*, 133–147.

Newman, M., Verdile, G., Martins, R. N. & Lardelli, M. (2010). Zebrafish as a Tool in Alzheimer's Disease Research. *Biochim Biophys Acta, 1812*, 346–352.

Noda, Y., Kuzuya, A., Tanigawa, K., Araki, M., Kawai, R., Ma, B., Sasakura, Y., Maesako, M., Tashiro, Y., Miyamoto, M., Uemura, K., Okuno, Y. & Kinoshita, A. (2018). Fibronectin Type III Domain-Containing Protein 5 Interacts with APP and Decreases Amyloid Beta Production in Alzheimer's Disease. *Mol Brain, 11*, 61.

Nüsslein-Volhard, C. & Dahm, R. (2002). *Zebrafish: A Practical Approach*. Oxford University Press.

Olah, M., Menon, V., Habib, N., Taga, M. F., Ma, Y., Yung, C. J., Cimpean, M., Khairallah, A., Coronas-Samano, G., Sankowski, R., Grun, D., Kroshilina, A. A., Dionne, D., Sarkis, R. A., Cosgrove, G. R., Helgager, J., Golden, J. A., Pennell, P. B., Prinz, M., Vonsattel, J. P. G., ... De Jager, P. L. (2020). Single Cell RNA Sequencing of Human Microglia Uncovers a Subset Associated With Alzheimer's Disease. *Nat Commun, 11*, 6129.

Otomo, T., Otomo, C., Tomchick, D. R., Machius, M. & Rosen, M. K. (2005). Structural Basis of Rho Gtpase-Mediated Activation of the Formin Mdia1. *Mol Cell, 18*, 273–281.

Oveisgharan, S., Dawe, R. J., Yu, L., Kapasi, A., Arfanakis, K., Hachinski, V., Schneider, J. A. & Bennett, D. A. (2022). Frequency And Underlying Pathology of Pure Vascular Cognitive Impairment. *JAMA Neurol*. https://doi.org/10.1001/jamaneurol.2022.3472

Ozcan, G. G., Lim, S., Canning, T., Tirathdas, L., Donnelly, J., Kundu, T. & Rihel, J. (2024). Genetic And Chemical Disruption of Amyloid Precursor Protein Processing Impairs Zebrafish Sleep Maintenance. *Iscience, 27*, 108870.

Paquet, D., Bhat, R., Sydow, A., Mandelkow, E. M., Berg, S., Hellberg, S., Falting, J., Distel, M., Koster, R. W., Schmid, B. & Haass, C. (2009). A Zebrafish Model of Tauopathy Allows in Vivo Imaging of Neuronal Cell Death and Drug Evaluation. *J Clin Invest, 119*, 1382–1395.

Paquet, D., Schmid, B. & Haass, C. (2010). Transgenic Zebrafish as a Novel Animal Model to Study Tauopathies and Other Neurodegenerative Disorders in Vivo. *Neurodegener Dis, 7*, 99–102.

Peri, F. & Nusslein-Volhard, C. (2008). Live Imaging of Neuronal Degradation by Microglia Reveals a Role for V0-Atpase A1 in Phagosomal Fusion in Vivo. *Cell, 133*, 916–927.

Ray, N., Kunkle Br, Hamilton-Nelson, K., Kurup, J., Rajabli, F., Cosacak, M., Kizil, C., Jean-Francois, M., Cuccaro, M., Reyes-Dumeyer, D., Cantwell, L., Kuzma, A., Vance, J. M., Gao, S., Hendrie, H., Baiyewu, O., Ogunniyi, O., Akinyemi, R., Alzheimer's Disease Genetics Consortium, Lee, W.-P., ... Reitz, C. (2023). Extended Genome-Wide Association Study Employing the African Genome Resources Panel Identifies Novel Susceptibility Loci for Alzheimer's Disease in Individuals of African Ancestry. *Alzheimers Dement*. https://doi.org/10.1002/alz.078904

Redd, M. J., Kelly, G., Dunn, G., Way, M. & Martin, P. (2006). Imaging Macrophage Chemotaxis in Vivo: Studies of Microtubule Function in Zebrafish Wound Inflammation. *Cell Motil Cytoskeleton, 63*, 415–422.

Reinhardt, L., Kordes, S., Reinhardt, P., Glatza, M., Baumann, M., Drexler, H. C. A., Menninger, S., Zischinsky, G., Eickhoff, J., Frob, C., Bhattarai, P., Arulmozhivarman, G., Marrone, L., Janosch, A., Adachi, K., Stehling, M., Anderson, E. N., Abo-Rady, M., Bickle, M., Pandey, U. B., ... Sterneckert, J. L. (2019). Dual Inhibition of GSK3beta and CDK5 Protects the Cytoskeleton of Neurons from Neuroinflammatory-Mediated Degeneration in Vitro and in Vivo. *Stem Cell Reports, 12*, 502–517.

Reitz, C., Jun, G., Naj, A., Rajbhandary, R., Vardarajan, B. N., Wang, L. S., Valladares, O., Lin, C. F., Larson, E. B., Graff-Radford, N. R., Evans, D., De Jager, P. L., Crane, P. K., Buxbaum, J. D., Murrell, J. R., Raj, T., Ertekin-Taner, N., Logue, M., Baldwin, C. T., Green, R. C., ... Alzheimer Disease Genetics, C. (2013). Variants In The ATP-Binding Cassette Transporter (ABCA7), Apolipoprotein E ϵ4, and the Risk of Late-Onset Alzheimer Disease in African Americans. *JAMA, 309*, 1483–1492.

Reitz, C. & Mayeux, R. (2009). Use of Genetic Variation as Biomarkers For Alzheimer's Disease. *Ann N Y Acad Sci, 1180*, 75–96.

Reitz, C., Pericak-Vance, M. A., Foroud, T. & Mayeux, R. (2023). A Global View of the Genetic Basis of Alzheimer Disease. *Nat Rev Neurol*, 19, 261–277.

Reitz, C., Tang, M. X., Schupf, N., Manly, J. J., Mayeux, R. & Luchsinger, J. A. (2010). A Summary Risk Score for the Prediction of Alzheimer Disease in Elderly Persons. *Arch Neurol, 67*, 835–841.

Rubinstein, A. L. (2003). Zebrafish: From Disease Modeling to Drug Discovery. *Curr Opin Drug Discov Devel, 6*, 218–223.

Sakae, N., Liu, C. C., Shinohara, M., Frisch-Daiello, J., Ma, L., Yamazaki, Y., Tachibana, M., Younkin, L., Kurti, A., Carrasquillo, M. M., Zou, F., Sevlever, D., Bisceglio, G., Gan, M., Fol, R., Knight, P., Wang, M., Han, X., Fryer, J. D., Fitzgerald, M. L., ... Kanekiyo, T. (2016). ABCA7 Deficiency Accelerates Amyloid-Beta Generation And Alzheimer's Neuronal Pathology. *J Neurosci, 36*, 3848–3859.

Santana, S., Rico, E. P. & Burgos, J. S. (2012). Can Zebrafish Be Used as Animal Model to Study Alzheimer's Disease? *Am J Neurodegener Dis, 1*, 32–48.

Sariya, S., Felsky, D., Reyes-Dumeyer, D., Lali, R., Lantigua, R. A., Vardarajan, B., Jimenez-Velazquez, I. Z., Haines, J. L., Shellenberg, G. D., Pericak-Vance, M. A., Pare, G., Mayeux, R. & Tosto, G. (2021). Polygenic Risk Score for Alzheimer's Disease in Caribbean Hispanics. *Ann Neurol, 90*, 366–376.

Scheuner, D., Eckman, C., Jensen, M., Song, X., Citron, M., Suzuki, N., Bird, T. D., Hardy, J., Hutton, M., Kukull, W., Larson, E., Levy-Lahad, E., Viitanen, M., Peskind, E., Poorkaj, P., Schellenberg, G., Tanzi, R., Wasco, W., Lannfelt, L., Selkoe, D. & Younkin, S. (1996). Secreted Amyloid Beta-Protein Similar to That in the Senile Plaques of Alzheimer's Disease Is Increased in Vivo by the Presenilin 1 and 2 and APP Mutations Linked to Familial Alzheimer's Disease. *Nat Med, 2*, 864–870.

Schier, A. F. (1997). Genetics of Neural Development in Zebrafish. *Curr Opin Neurobiol, 7*, 119–126.

Schmid, B. & Haass, C. (2013). Genomic Editing Opens New Avenues for Zebrafish as a Model for Neurodegeneration. *J Neurochem, 127*, 461–470.

Schmidt, V., Schulz, N., Yan, X., Schurmann, A., Kempa, S., Kern, M., Bluher, M., Poy, M. N., Olivecrona, G. & Willnow, T. E. (2016). SORLA Facilitates Insulin Receptor Signaling in Adipocytes and Exacerbates Obesity. *J Clin Invest, 126*, 2706–2720.

Selkoe, D. J. (2001). Alzheimer's Disease: Genes, Proteins, and Therapy. *Physiol Rev, 81*, 741–766.

Selkoe, D. J. (2003). Folding Proteins in Fatal Ways. *Nature, 426*, 900–904.

Selkoe, D. J. & Hardy, J. (2016). The Amyloid Hypothesis of Alzheimer's Disease at 25 Years. *EMBO Mol Med, 8*, 595–608.

Serrano-Pozo, A., Li, H., Li, Z., Munoz-Castro, C., Jaisa-Aad, M., Healey, M. A., Welikovitch, L. A., Jayakumar, R., Bryant, A. G., Noori, A., Connors, T. R., Hu, M., Zhao, K., Liao, F., Lin, G., Pastika, T., Tamm, J., Abdourahman, A., Kwon, T., Bennett, R. E., ... Das, S. (2024). Astrocyte Transcriptomic Changes Along The Spatiotemporal Progression of Alzheimer's Disease. *Nat Neurosci, 27*, 2384–2400.

Siddiqui, T., Bhattarai, P., Popova, S., Cosacak, M. I., Sariya, S., Zhang, Y., Mayeux, R., Tosto, G. & Kizil, C. (2021). Kyna/Ahr Signaling Suppresses Neural Stem Cell Plasticity And Neurogenesis in Adult Zebrafish Model Of Alzheimer's Disease. *Cells, 10*, 2748.

Siddiqui, T., Cosacak, M. I., Popova, S., Bhattarai, P., Santa-Maria, I., Tosto, G. & Kizil, C. (2022). Neurogenic Induction by Ngfr Reduces Reactive Astrocytic Lcn2/Slc22a17 Signalling, Amyloid Deposition and Tau Phosphorylation. *Biorxiv*, 2022.08.20.504608.

Siddiqui, T., Cosacak, M. I., Popova, S., Bhattarai, P., Yilmaz, E., Lee, A. J., Min, Y., Wang, X., Allen, M., Is, O., Atasavum, Z. T., Rodriguez-Muela, N., Vardarajan, B. N., Flaherty, D., Teich, A. F., Santa-Maria, I., Freudenberg, U., Werner, C., Tosto, G., Mayeux, R., Ertekin-Taner, N. & Kizil, C. (2023). Nerve Growth Factor Receptor (Ngfr) Induces Neurogenic Plasticity by Suppressing Reactive Astroglial Lcn2/Slc22a17 Signaling in Alzheimer's Disease. *Npj Regen Med, 8*, 33.

Spillantini, M. G., Murrell, J. R., Goedert, M., Farlow, M. R., Klug, A. & Ghetti, B. (1998). Mutation in the Tau Gene in Familial Multiple System Tauopathy with Presenile Dementia. *Proc Natl Acad Sci USA, 95*, 7737–7741.

Strahle, U. & Blader, P. (1994). Early Neurogenesis in The Zebrafish Embryo. *Faseb J, 8*, 692–698.

Su, Y., Zhou, Y., Bennett, M. L., Li, S., Carceles-Cordon, M., Lu, L., Huh, S., Jimenez-Cyrus, D., Kennedy, B. C., Kessler, S. K., Viaene, A. N., Helbig, I., Gu, X., Kleinman, J. E., Hyde, T. M., Weinberger, D. R., Nauen, D. W., Song, H. & Ming, G. L. (2022). A Single-Cell Transcriptome Atlas of Glial Diversity in the Human Hippocampus Across the Postnatal Lifespan. *Cell Stem Cell, 29*, 1594–1610.E8.

Sun, N., Akay, L. A., Murdock, M. H., Park, Y., Galiana-Melendez, F., Bubnys, A., Galani, K., Mathys, H., Jiang, X., Ng, A. P., Bennett, D. A., Tsai, L. H. & Kellis, M. (2023). Single-Nucleus Multiregion Transcriptomic Analysis of Brain Vasculature in Alzheimer's Disease. *Nat Neurosci, 26*, 970–982.

Sun, Y., Xu, S., Jiang, M., Liu, X., Yang, L., Bai, Z. & Yang, Q. (2021). Role of the Extracellular Matrix in Alzheimer's Disease. *Front Aging Neurosci, 13*, 707466.

Tayran, H., Yilmaz, E., Bhattarai, P., Min, Y., Wang, X., Ma, Y., Wang, N., Jeong, I., Nelson, N., Kassara, N., Cosacak, M. I., Dogru, R. M., Reyes-Dumeyer, D., Stenersen, J., Reddy, J., Qiao, M., Flaherty, D., Gunasekaran, T. I., Yang, Z., Jurish-Yaksi, N., ... Kizil, C. (2024). ABCA7-Dependent Neuropeptide-Y Signalling Is a Resilience Mechanism Required for Synaptic Integrity in Alzheimer's Disease. *Cell Genomics, 4*, 100642.

Than-Trong, E., Kiani, B., Dray, N., Ortica, S., Simons, B., Rulands, S., Alunni, A. & Bally-Cuif, L. (2020). Lineage Hierarchies and Stochasticity Ensure the Long-Term Maintenance of Adult Neural Stem Cells. *Sci Adv, 6*, Eaaz5424.

Thomsen, M. S., Routhe, L. J. & Moos, T. (2017). The Vascular Basement Membrane in The Healthy and Pathological Brain. *J Cereb Blood Flow Metab, 37*, 3300–3317.

Tincer, G., Mashkaryan, V., Bhattarai, P. & Kizil, C. (2016). Neural Stem/Progenitor Cells in Alzheimer's Disease. *Yale J Biol Med, 89*, 23–35.

Tomasiewicz, H. G., Flaherty, D. B., Soria, J. P. & Wood, J. G. (2002). Transgenic Zebrafish Model of Neurodegeneration. *J Neurosci Res, 70*, 734–745.

Tsartsalis, S., Sleven, H., Fancy, N., Wessely, F., Smith, A. M., Willumsen, N., Cheung, T. K. D., Rokicki, M. J., Chau, V., Ifie, E., Khozoie, C., Ansorge, O., Yang, X., Jenkyns, M. H., Davey, K., Mcgarry, A., Muirhead, R. C. J., Debette, S., Jackson, J. S., Montagne, A., ... Matthews, P. M. (2024). A Single Nuclear Transcriptomic Characterisation of Mechanisms Responsible for Impaired Angiogenesis and Blood-Brain Barrier Function in Alzheimer's Disease. *Nat Commun, 15*, 2243.

Turgutalp, B., Bhattarai, P., Ercetin, T., Luise, C., Reis, R., Gurdal, E., Isaak, A., Biriken, D., Dinter, E., Sipahi, H., Schepmann, D., Junker, A., Wünsch, B., Sippl, W., Gulcan, H., Yarim, M. & Kizil, C. (2022). Discovery of Potent Cholinesterase Inhibition-Based Multi-Target-Directed Lead Compounds For Synaptoprotection in Alzheimer's Disease. *J Med Chem*. https://doi.org/10.1021/acs.jmedchem.2c01003

Turgutalp, B. & Kizil, C. (2024). Multi-Target Drugs for Alzheimer's Disease. *Trends Pharmacol Sci, 45*, 628–638.

Vacaru, A. M., Unlu, G., Spitzner, M., Mione, M., Knapik, E. W. & Sadler, K. C. (2014). In Vivo Cell Biology in Zebrafish - Providing Insights into Vertebrate Development and Disease. *J Cell Sci, 127*, 485–495.

Van Tijn, P., Kamphuis, W., Marlatt, M. W., Hol, E. M. & Lucassen, P. J. (2011). Presenilin Mouse and Zebrafish Models for Dementia: Focus on Neurogenesis. *Prog Neurobiol, 93*, 149–164.

Vardarajan, B. N., Zhang, Y., Lee, J. H., Cheng, R., Bohm, C., Ghani, M., Reitz, C., Reyes-Dumeyer, D., Shen, Y., Rogaeva, E., St George-Hyslop, P. & Mayeux, R. (2015). Coding Mutations in Sorl1 and Alzheimer Disease. *Ann Neurol, 77*, 215–227.

Varshney, G. K., Sood, R. & Burgess, S. M. (2015). Understanding and Editing the Zebrafish Genome. *Adv Genet, 92*, 1–52.

Walchli, T., Ghobrial, M., Schwab, M., Takada, S., Zhong, H., Suntharalingham, S., Vetiska, S., Gonzalez, D. R., Wu, R., Rehrauer, H., Dinesh, A., Yu, K., Chen, E. L. Y., Bisschop, J., Farnhammer, F., Mansur, A., Kalucka, J., Tirosh, I., Regli, L., ... Radovanovic, I. (2024). Single-Cell Atlas of The Human Brain Vasculature Across Development, Adulthood and Disease. *Nature, 632*, 603–613.

Wang, X., Allen, M., Li, S., Quicksall, Z. S., Patel, T. A., Carnwath, T. P., Reddy, J. S., Carrasquillo, M. M., Lincoln, S. J., Nguyen, T. T., Malphrus, K. G., Dickson, D. W., Crook, J. E., Asmann, Y. W. & Ertekin-Taner, N. (2020). Deciphering Cellular Transcriptional Alterations in Alzheimer's Disease Brains. *Mol Neurodegener, 15*, 38.

Wang, X., Zhang, J. B., He, K. J., Wang, F. & Liu, C. F. (2021). Advances of Zebrafish in Neurodegenerative Disease: From Models to Drug Discovery. *Front Pharmacol, 12*, 713963.

Wen, P. H., Hof, P. R., Chen, X., Gluck, K., Austin, G., Younkin, S. G., Younkin, L. H., Degasperi, R., Gama Sosa, M. A., Robakis, N. K., Haroutunian, V. & Elder, G. A. (2004). The Presenilin-1 Familial Alzheimer Disease Mutant P117L Impairs Neurogenesis in the Hippocampus of Adult Mice. *Experimental Neurology, 188*, 224–237.

Xi, Y., Noble, S. & Ekker, M. (2011). Modeling Neurodegeneration in Zebrafish. *Curr Neurol Neurosci Rep, 11*, 274–282.

Xia, W. (2010). Exploring Alzheimer's Disease in Zebrafish. *J Alzheimers Dis, 20*, 981–990.

Xu, L., Nirwane, A. & Yao, Y. (2019). Basement Membrane and Blood-Brain Barrier. *Stroke Vasc Neurol, 4*, 78–82.

Yamazaki, Y. & Kanekiyo, T. (2017). Blood-Brain Barrier Dysfunction and The Pathogenesis of Alzheimer's Disease. *Int J Mol Sci, 18*, 1965.

Yang, A. C., Vest, R. T., Kern, F., Lee, D. P., Agam, M., Maat, C. A., Losada, P. M., Chen, M. B., Schaum, N., Khoury, N., Toland, A., Calcuttawala, K., Shin, H., Palovics, R., Shin, A., Wang, E. Y., Luo, J., Gate, D., Schulz-Schaeffer, W. J., Chu, P., … Wyss-Coray, T. (2022). A Human Brain Vascular Atlas Reveals Diverse Mediators of Alzheimer's Risk. *Nature, 603*, 885–892.

Yang, T., Arslanova, D., Xu, X., Li, Y. M. & Xia, W. (2010). In Vivo Manifestation of Notch Related Phenotypes in Zebrafish Treated with Alzheimer's Amyloid Reducing Gamma-Secretase Inhibitors. *J Neurochem, 113*, 1200–1209.

Zarow, C., Barron, E., Chui, H. C. & Perlmutter, L. S. (1997). Vascular Basement Membrane Pathology and Alzheimer's Disease. *Ann N Y Acad Sci, 826*, 147–160.

Zenaro, E., Piacentino, G. & Constantin, G. (2017). The Blood-Brain Barrier in Alzheimer's Disease. *Neurobiol Dis, 107*, 41–56.

Zhang, Z., Lin, W., Gan, Q., Lei, M., Gong, B., Zhang, C., Henrique, J. S., Han, J., Tian, H., Tao, Q., Potempa, L. A., Stein, T. D., Emili, A. & Qiu, W. Q. (2024). The Influences of Apoe Isoforms on Endothelial Adherens Junctions and Actin Cytoskeleton Responding to Mcrp. *Angiogenesis, 27*, 861–881.

Zon, L. I. (1999). Zebrafish: A New Model for Human Disease. *Genome Res, 9*, 99–100.

# 10

# *Perfluorinated Chemicals, Zebrafish, and Neurodegenerative Disorders*

**Emma Ivantsova and Christopher J. Martyniuk**

## Abbreviations

| | |
|---|---|
| **ACHE** | Acetylcholinesterase |
| **AD** | Alzheimer's Disease |
| **ADHD** | Attention-Deficit/Hyperactivity Disorder |
| **ALS** | Amyotrophic Lateral Sclerosis |
| **APP** | Amyloid Beta A4 Precursor Protein |
| **ASD** | Autism Spectrum Disorder |
| **CNS** | Central Nervous System |
| **DDT** | Dichlorodiphenyltrichloroethane |
| **DNA** | Deoxyribonucleic Acid |
| **GABA** | Gamma Aminobutyric Acid |
| **MANF** | Mesencephalic Astrocyte Derived Neurotrophic Factor |
| **MPP+** | 1-Methyl-4-Phenylpyridinium |
| **MPTP** | 1-Methyl-4-Phenyl-1,2,3,6-Tetrahydropyridine |
| **PD** | Parkinson's Disease |
| **PFDoA** | Perfluorododecanoic Acid |
| **PFAS** | Perfluoroalkyl and Polyfluoroalkyl Substances |
| **PFBA** | Perfluorobutanoic Acid |
| **PFBS** | Perfluorobutane Sulfonic Acid |
| **PFNA** | Perfluorononanoic Acid |
| **PFOA** | Perfluorooctanoic Acid |
| **PFOS** | Perfluorooctane Sulfonic Acid |
| **PRX2** | Peroxiredoxin 2 |
| **PSEN1** | Presenilin 1 |
| **PSEN2** | Presenilin 2 |
| **P53** | Tumor Protein 53 |
| **ROS** | Reactive Oxygen Species |

## Environmental Chemicals and Neurodegenerative Diseases

Acute and chronic exposure to environmental chemicals plays a significant role in promoting toxic mechanisms that underlie neurodegenerative diseases and neurological disorders. While most neurodegenerative diseases are associated with defective genes and specific alleles, the percentage contribution of genetics to the risk of neurodegenerative disease can be relatively low. For instance, while early onset of Parkinson's disease (PD) is thought to be more dependent upon genetic factors, most reported cases of PD are sporadic in nature where a positive familial history has been reported in less than 10% of patients (Thomas & Beal, 2007); however, around 5%–10% of individuals with the disease have a monogenic mutation (Cherian & Divya, 2020). This differs significantly from Alzheimer's disease (AD) as genetic factors are estimated to be associated with 80% of reported cases (Gatz et al., 2006). These genetic effects are often the result of several gene products exerting small effects on the underlying disease. Layered on top of these complex genetic interactions are external factors like diet, exercise, and exposure to environmental chemicals, each of which play a key role in disease onset and progression. In terms of environmental exposures, several chemicals have been associated with an increased risk of neurological and neurodegenerative diseases. Evidence for a link between environmental chemical exposure and neurodegenerative diseases comes from experimental research using animal models of neurodegeneration in epidemiological studies. For example, high levels of dieldrin in blood were associated with increased risk of AD in adults (Singh et al., 2014), and the risk of PD in adults increased by 75% from proximity to agricultural applications of paraquat (Costello et al., 2009).

Several pesticides to date have been associated with an increased risk of neurodegeneration, perhaps some of the most well studied being those associated with PD. PD is characterized by the degeneration of dopamine neurons in the substantia nigra, leading to motor dysfunction. Hallmark mechanisms underlying neuronal degeneration include oxidative stress and mitochondrial dysfunction (Borsche et al., 2021; Cao et al., 2018; Yan et al., 2013). The organochlorine pesticide dieldrin, a chemical related to dichlorodiphenyltrichloroethane (DDT), has long been implicated in neurotoxicity in wildlife, animal research models, and humans (Cao et al., 2018; Kanthasamy et al., 2005; Martyniuk et al., 2010). Paraquat or methyl viologen (N, N′-dimethyl-4,4′-bipyridinium dichloride) is a herbicide that has a very similar structure to that of 1-methyl-4-phenyl-1,2,3,6-tetrahydropyridine (MPTP), which is a precursor to the neurotoxin $MPP^+$. This neurotoxin, like paraquat, is transported into dopamine neurons through dopamine active transporters to induce toxicity and cell death (Rappold et al., 2011). Occupational health studies have demonstrated an increased

DOI: 10.1201/9781032676555-12

risk to PD through the manufacturing of paraquat (reviewed in (Cao et al., 2018). Related to AD, herbicides like glyphosate and organophosphate pesticides (Mostafalou & Abdollahi, 2023) are positively associated with an increased risk of developing the condition. Several other neurological disorders like depression and anxiety are also linked to environmental chemical exposure (Hsiao et al., 2023; Wiersielis et al., 2020). Further complicating our understanding of the environmental influence of chemicals on neurodegenerative diseases are gene environment interaction and the recognition that several chemicals can be present as mixtures in environments where susceptible individuals may live. For example, in an agricultural setting, individuals may be exposed to herbicides as well as chemicals used in personal care products, food packaging, cookware, and other chemicals used in residential areas. These complex chemical mixtures can exert unique properties that further damage neurons and glia over time, causing prolonged stress in the central nervous system (CNS). Abiotic factors are also considered to be important variables in chemical exposure scenarios, as temperature can modulate toxicity of chemicals in both water and air (Kazmi et al., 2022; Noyes & Lema, 2015). Taken together, evidence suggests that exposure to several types of chemicals is related to neurological diseases in genetically susceptible individuals.

## Zebrafish as a Developmental Toxicity Model

Given that there are thousands of chemicals in the environment, it becomes more pressing to quantify their effects in the CNS. Identifying which chemicals have high potential for neurotoxicity is the first step in mitigating risks of neurodegeneration and neurological disorders. To test such a vast number of potential neurotoxins, high-throughput bioassays are needed to prioritize chemicals for further study. While several neuronal cell lines are used for high throughput screening of chemicals, they are often not representative of an integrated CNS. In vivo experiments with rodents are highly valuable for toxicity studies and are often used as experimental models for human neurological diseases like PD and AD (Antony et al., 2011; Braidy et al., 2012). A disadvantage of using mice and rats, however, is the expense and logistics of conducting multiple neurotoxicity assays in many individuals and achieving high experimental power with numbers and replication. Given deficiencies in cell lines and mammalian models, alternative animal models like zebrafish (*Danio rerio*) have quickly positioned themselves as an important intermediate step in neurotoxicity studies.

Zebrafish are utilized as a model species in environmental studies for a variety of reasons. Large-scale experimental analysis can be completed using zebrafish as they have high fecundity, in which several hundred embryos can be laid weekly. These embryos are small, transparent, and develop rapidly externally; therefore, alterations in their structure may easily be observed. Additional advantages in utilizing zebrafish include their low maintenance cost and easy handling. They are also valuable in modelling human diseases as they have been found to contain genes involved in human neurodegenerative diseases and share the same major neurotransmitters and basic central nervous system structure as humans (Horzmann & Freeman, 2018; Panula et al., 2010). These beneficial characteristics of the zebrafish model facilitate the rapid testing of chemicals. Chemicals can be prioritized for further study using zebrafish developmental toxicity assays, and technological applications (i.e., genome editing) can be further advanced to associate screening results to adverse mechanisms of actions (Horzmann & Freeman, 2018). Zebrafish can also be used for in vivo micro-fraction techniques to discover natural drug-like products, which are small particles produced by living organisms (Pitchai et al., 2019), facilitating the development of therapeutic interventions. Another significant advantage of zebrafish in the study of human disease is well characterized behavioral assays that correspond to some rodent behavioral tests (Champagne et al., 2010). Locomotor activity, anti-anxiolytic/anxiolytic behavior, and social behavior are all key behaviors often studied within the context of neurological disease. There are several behavioral tests optimized in zebrafish that complement mammalian behavioral assays (Benvenutti et al., 2021; Champagne et al., 2010; Langova et al., 2020; Stewart et al., 2010). This feature is important because many neurological disorders like PD have significant consequences for motor circuitry (McGregor & Nelson, 2019) while others, like AD, can negatively affect cognition (Corey-Bloom, 2002). Figure 10.1 outlines different behavioral assays that can be conducted at different life stages of zebrafish development; many of the assays have been utilized for studying human diseases. Developmental neurotoxicity has become a major focus of regulatory agencies, and several programs are now in place to study adverse effects on the developing nervous system from chemical agents.

Figure 10.1 shows the developmental profile of zebrafish from early fertilization to adulthood. At each life stage, there are behavioral assays that can be used to uncover neurotoxicity and neurological disorders as zebrafish can mirror human symptoms caused by neurologically related changes (Benvenutti et al., 2021; de Abreu et al., 2020; Farrell et al., 2011; Hamilton et al., 2021; Ogi et al., 2021). For instance, zebrafish exhibiting decreased social behavior and interaction suggest schizophrenic-related withdrawal behavior, while fish exhibiting impaired cognitive abilities and hyperactivity suggest more ADHD-related behaviors.

The zebrafish model has been used extensively to study chemical exposures in the context of neurological disorders. There have been several recent studies on the topic of zebrafish, pesticides, and PD (Kalyn et al., 2023; S.-T. Liu et al., 2023; Min et al., 2023) and AD (H. Liu et al., 2023; Marinho et al., 2019; Wang et al., 2019). We direct the reader to these reviews for more information on the use of zebrafish to understand the mechanisms underlying pesticide exposure and human neurological disease. In the following sections of this chapter, we focus on an emerging chemical class of concern, perfluoroalkyl and polyfluoroalkyl substances (PFAS), providing an overview of evidence for neurotoxicity based on zebrafish studies, highlighting potential links with neurological disorders in humans.

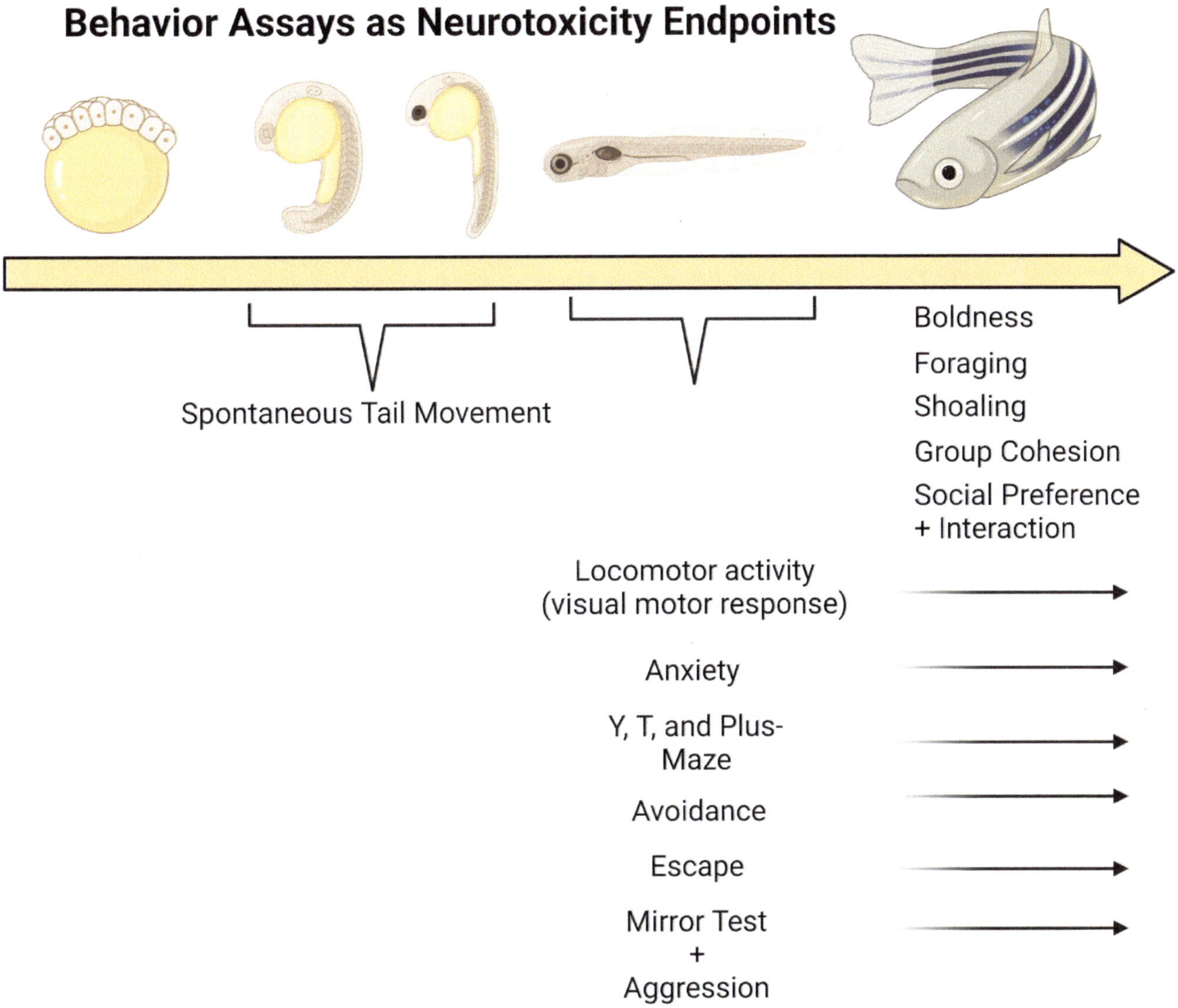


**FIGURE 10.1** Developmental profile of zebrafish from early fertilization to adulthood. At each life stage, there are behavioral assays that can be used to uncover neurotoxicity and neurological disorders (Benvenutti et al., 2021; de Abreu et al., 2020; Farrell et al., 2011; Hamilton et al., 2021; Ogi et al., 2021).

## Perfluorinated Compounds

### Introduction

PFAS are man-made compounds that have been heavily utilized worldwide in industrial and consumer applications (i.e., cleaning products, food packaging, lubricants, non-stick kitchenware, personal care products) since the early 1930s. These compounds share carbon chains with attached fluorine atoms and can be categorized by the presence of functional groups or side chains. This chemical structure contributes to biological, chemical, and thermal stability (Meng et al., 2019); therefore, PFAS can resist biodegradation, hydrolysis, and photolysis. Consequently, PFAS exhibit environmental persistence and, due to their high utilization rates, they are found in waters, sediments, and waste worldwide, in which many studies to date have reported varying concentrations from the low ng/L to high μg/L (An et al., 2023; Gómez-Canela et al., 2012; Llorca et al., 2012). Additionally, PFAS have the potential to bioaccumulate in various organisms, including fish (Stahl et al., 2011), which has been found to be responsible for various developmental and physiological alterations (Hagenaars et al., 2011; Ivantsova et al., 2023; Vogs et al., 2019; Wasel et al., 2022).

Toxicity of PFAS has been found to be positively correlated with increasing carbon chain lengths. Consequently, long-chain PFAS, such as perfluorooctanoic acid (PFOA, $C_8HF_{15}O_2$) and perfluorooctane sulfonic acid (PFOS, $C_8HF_{17}O_3S$), began to be phased out of production in the early 2000s and replaced with short-chain alternatives, including perfluorobutanoic acid (PFBA, $C_4HF_7O_2$) and perfluorobutane sulfonic acid (PFBS, $C_4HF_9O_3S$). Though these alternatives are deemed as "safer" due to their decreased carbon chain length, they still raise concern due to their environmental persistence and toxicity concerns as they can exhibit other unique physiochemical features contributing to these factors.

## Neurotoxicity Mechanisms of PFAS: Insights from Zebrafish

Neurotoxins are responsible for the disruption or death of essential neurons part of the nervous system. Consequently, disrupted neural pathways are associated with neurodegenerative diseases and other neurodevelopmental disorders, including attention-deficit/hyperactivity disorder (ADHD), amyotrophic lateral sclerosis (ALS), AD, Huntington's disease, PD, and schizophrenia. The neurotoxicity of such environmental chemicals can be analyzed through various omics techniques, including metabolomics, proteomics, and transcriptomics, which can reveal disrupted mechanism pathways. Several studies have utilized omics approaches to understand the neurotoxicity of PFAS. For example, zebrafish exposed to PFOS showed transcriptome alterations related to neuromuscular system disruption (Wang et al., 2022), and zebrafish exposed to perfluorononanoic acid (PFNA, $C_9HF_{17}O_2$) have been shown to have altered neuro-related pathways, including neuroinflammatory pathways and calcium signaling (S. Liu et al., 2023). Pathways that can provide insight into neurodegeneration/neurotoxicity include those associated with iron homeostasis, mitochondrial dysfunction, motor dysfunction, and neurotransmitter synthesis (Cao et al., 2018). These pathways are not only studied following individual exposure to chemical compounds, but also in chemical mixtures to analyze the cumulative effect(s) on developmental neurotoxicity (Atzei et al., 2021). Different mechanisms underlying the toxicity of perfluorinated compounds in zebrafish are presented in Table 10.1.

Mitochondria are essential for carrying out biochemical reactions within cells. Their proper function can be disrupted by a variety of factors, including DNA mutations and environmental factors, which can lead to cell death. Additionally, mitochondrial dysfunction has been thought to contribute to the progression of neurodegenerative disorders, including PD (Kalyn et al., 2023). The presence of reactive oxygen species (ROS) is a common sign of oxidative stress within a cell, and accumulated ROS can cause molecular damage. To prevent oxidative damage, mitochondria implement their antioxidant systems. For instance, peroxiredoxin 2 (PRX2) is an antioxidant enzyme containing peroxidase activity to aid in the prevention of oxidative stress in neurons; thus, it is expected to be elevated in a stressful environment. Zhang et al. (2011) observed an elevated expression of PRX2 in zebrafish exposed to PFOS, implying that their detoxification pathways were engaged in the presence of induced oxidative stress. Mitochondrial ROS can also decrease neuronal quantities by inducing pro-apoptotic protein release and mediating

**TABLE 10.1**

Perfluorinated Chemicals and Markers of Neurodegenerative Disease in Zebrafish

| Chemical | Dose | Duration | Outcome of exposure | Reference |
|---|---|---|---|---|
| PFDoA | 0.24, 1.2, 6 mg/L | 5 days | Levels of dopamine increased; *manf* upregulated (0.76-fold); *ache*, *a1-tubulin*, *gfap*, *gap43*, *mbp*, *manf*, *syn2a* downregulated; decrease in swimming speed | Guo et al. (2018) |
| PFNA | 0.01, 0.1, 1, 10, and 100 µg/L | 5 days | Nervous system genes were reduced (*drd1a*, *drd1b*, *drd2a*, *drd2b*, *drd4-rs*, *gria1a*, *gria1b*, *gria2b*, *gria3b*, *grin1b*, *grm1b*, *grm5a*, *camk2a*, *gabbr1b*, *gabra1*, *gabra6b*, *gabrg1*, *cacna1aa*); Number of GABAergic neurons decreased; Neurotransmitters (dopamine, glutamate, serotonin, acetylcholine, noradrenaline) decreased; Swimming behavior inhibited | S. Liu et al. (2023) |
| PFOA | 10, 100, 1000 µg/L | 7 days | Downregulated *manf;* Upregulated *dat*, *nr4a2*, *pax2*, *pax5*, *th1*, and *th2*; spontaneous tail movement, total distance, average movement speed, and moving cumulative duration time decreased | Yu et al. (2021) |
| PFOS | 0.032, 0.32, 3.2 mg/L | 5 days | *manf*, *th1*, *th2*, and *drd1* upregulated and *cdnf* downregulated; Increased dopamine content (1.23-fold); Increased swimming speed; Increased distance moved | Wu et al. (2022) |
| | 0.1–20 µM | 5 days | Altered pathways include the hemoglobin complex, glutamate metabolism, and glycerophospholipid metabolism | Lee et al. (2021) |
| | 1, 3, 5 mg/L | 5 days | Alpha-tubulin dysfunction; Increased PRX2 and CDK5 expression | Zhang et al. (2011) |
| | 0.1 and 1 mg/L | 5 days | Reduced dopaminergic neuron number; distance moved, and total distance travelled reduced | Kalyn et al. (2023) |
| | 100, 500, 1000 µg/L | 5 days | Dopamine content increased; *dat* expression decreased; *drd2a*, *drd3b*, *drd4* expression increased; increased apoptotic brain cells; reduced activity; spontaneous tail curling inhibited; neuroinflammatory and apoptosis gene expression increased | Mahapatra et al. (2023) |
| | 7, 14, 28, 64 µM | 5 days | Less ramified microglial cells; upregulated microglia activation gene (*p2ry12*); anxiety-like thigmotaxis | Paquette et al. (2023) |
| | 0.04, 0.1, 0.4 and 1 µM | 7 days | Embryonic spontaneous tail coiling increased; increased swimming velocity; altered neuromuscular-related genes (*ache*, *chata*, *ngfb*, *ryr3*, *tcf3a*, *acte1*) | Wang et al. (2022) |
| | 4, 20, and 100 µg/L | 21 days | Decreased dopamine levels in the brain; decreased number of dopaminergic neurons; concentration-dependent relationship observed with hyperactivity and cognitive impairment | Zhang et al. (2023) |

inflammation, which triggers apoptosis. Studies have reported that zebrafish exposed to PFOS had upregulated neuroinflammation-related (interleukin-1 beta (*il-1β* ), interleukin 6 (*il-6*), nuclear factor kappa beta (*nf-kβ* ), and tumor necrosis factor alpha (*tnf-α* )) and apoptosis-related (B-cell lymphoma 2 (*bcl-2*), *caspase-3*, tumor protein 53 (*p53*)) gene expression with significantly increased apoptotic cells in the brain (Mahapatra et al., 2023).

Heme-binding results in signal transduction and, within the pathway, specific genes are required for the conversion of compounds. For instance, cytochrome P450 family 27 subfamily B member 1 (*cyp27b1*) is responsible for the production of 1-alpha-hydroxylase, which promotes the conversion of inactive vitamin D into its active form. As a result, neurotransmitter synthesis (e.g., acetylcholine and gamma aminobutyric acid (GABA)) and neural differentiation are supported (Lee et al., 2021). Disruption of *cyp27b1* can thus contribute to vitamin D deficiency, which can lead to neurotoxicity as inadequate GABA production is associated with AD (Jiménez-Balado & Eich, 2021) and seizure occurrence. Iron homeostasis processes are also implicated in human neurodegenerative diseases. For example, ferroptosis is an iron-dependent form of cell death, and altered iron homeostasis, lipid peroxidation, and oxidative injury contribute to ferroptosis and are implicated in AD pathology (Zhang et al., 2021).

Other mechanisms elucidated in the zebrafish model in response to PFAS exposure are those related to cell structure. Microtubule dysfunction can cause the development or progression of neurodegenerative disorders as microtubules are responsible for cell movement. Alpha-tubulin can polymerize into microtubules, which contributes to motor neuron function (Wilson et al., 2002). Zhang et al. (2011) found that alpha-tubulin was downregulated in the brain and spinal cord region in zebrafish following exposure to PFOS. The overall disruption of cells localized to the nervous system may also contribute to neurodegeneration. For instance, microglia are cells located throughout the central nervous system, which have dynamic functions involved with immune and homeostatic reactions (i.e., metabolite and cellular debris removal), and their dysfunction has been noted in various disease models, including AD (Mancuso et al., 2019); however, studies examining the relationship between microglia interactions with neurons and environmental exposures are overall not abundant. One study found that PFOS modulates the homeostatic morphology and function of microglia in zebrafish as cells were less ramified (Paquette et al., 2023). Other studies have also examined the relationship between neurotransmitter dysfunction and neurotoxicity following exposure to PFAS (Guo et al., 2018; S. Liu et al., 2023; Wu et al., 2022). Synaptogenesis establishes neural networks and is also associated with forming neurotransmitter release sites. One study found that PFNA disrupts the synaptogenesis signaling pathway in zebrafish, which decreased GABAergic neurons, neurotransmitters (i.e., dopamine, glutamate, serotonin, acetylcholine, noradrenaline), and their respective gene receptor expressions, which can ultimately induce neurobehavioral effects as swimming behavior in the fish was significantly decreased (Liu et al., 2023).

## Zebrafish as a Model for Neurological Disease and PFAS Exposure

### Parkinson's Disease, Perfluorinated Compounds, and Zebrafish

PD is a progressive neurodegenerative disease with an obscure etiology; however, it is thought to arise due to the loss of dopaminergic cells in the substantia nigra. The absence of midbrain dopaminergic neurons in zebrafish, such as those in the substantia nigra and ventral tegmentum, is the primary distinction between zebrafish and mammalian systems (Nishimura et al., 2015). Though debated, the zebrafish brain region considered analogous to the mammalian midbrain dopaminergic neuron population is the posterior tuberculum in the diencephalon. Studies report that the mechanism of action of toxicity of neurotoxins, such as MPTP, in zebrafish is comparable to humans (Barnhill et al., 2020); thus, zebrafish are intended to exploit toxicity as PD models. Today, perfluorinated compound exposure in developing zebrafish have been implicated in PD through neurobehavioral changes and altered dopamine levels. Dopamine is a neurotransmitter involved in the reward system, which motivates mammals to engage in specific behaviors that promote health and reproductive fitness. Various studies report on the relationship between the dopamine pathway and motor dysfunction. For example, disrupted dopamine levels and altered expression of dopamine-related genes (dopamine neurotrophic factor, *cdnf*; dopamine transporter, *dat*; dopamine receptor D1, D2a, 3b, and 4, *drd1, drd2a, drd3b, drd4;* mesencephalic astrocyte derived neurotrophic factor, *manf*) are correlated with changes in distance travelled and swimming speeds of zebrafish (Wu et al., 2022). Thus, these transcripts and behaviors may be useful for elucidating the effects of chemicals on neurological disorders like PD. In the case or PFAS, studies report that reduced dopaminergic neuron numbers and decreased gene expression in fish exposed to PFOS can reduce motor activity (Kalyn et al., 2023), whereas increased dopaminergic neuron numbers and gene expression in fish exposed to PFOS increases motor activity (Mahapatra et al., 2023; Wu et al., 2022). Other studies report that an inverse relationship between dopamine levels and expression can also relate to motor activity. Yu et al. (2021) observed that upregulated dopamine-related gene expression was correlated to decreased motor activity in zebrafish exposed to PFOA. Additionally, Guo et al. (2018) found that increased dopamine levels and expression of acetylcholinesterase (*ache*), *a1-tubulin*, glial fibrillary acidic protein (*gfap*), growth associated protein 43 (*gap43*), myelin basic protein (*mbp*), *manf*, and synapsin IIa (*syn2a*) decreased motor activity in zebrafish exposed to perfluorododecanoic acid (PFDoA, $C_{12}HF_{23}O_2$). Thus, there is some evidence from zebrafish studies that the dopamine system is targeted by PFAS, which has high relevance to understanding the effect of chemical exposures on PD.

### Alzheimer's Disease, Perfluorinated Compounds, and Zebrafish

AD is a progressive neurodegenerative disease that leads to impaired judgment, memory, and awareness from the irreversible loss of cerebral cortex and hippocampus-related neurons (Murray & Murray, 2007). AD is characterized by extracellular plaques derived from intracellular neurofibrillary tangles containing hyperphosphorylated tau, a microtubule protein (Murray & Murray, 2007). Like PD, microtubule and neurotransmitter disruption can lead to the progression of neurodegenerative diseases. For instance, serotonin is a monoamine neurotransmitter responsible for mood control, and low serotonin levels have been observed in individuals with AD. One hypothesis for low serotonin levels is low tryptophan levels, as tryptophan is an essential amino acid responsible for producing various neurotransmitters, including serotonin. Additionally, metabolism of glutamine has been observed to be compromised in individuals with AD. Zebrafish can be used to investigate not only neurotransmitter disruption, but also genes involved in the pathogenesis of AD, as they contain orthologous genes to humans. Familial AD, the chance of offspring inheriting a mutated gene, is associated with mutated forms of presenilin 1 (PSEN1), presenilin 2 (PSEN2), and amyloid beta A4 precursor protein (APP) (Newman et al., 2014). In zebrafish, psen1 and psen2 are orthologs of PSEN1 and PSEN2 in humans, respectively, and appa and appb are co-orthologs of human APP (Newman et al., 2014). Insufficient data are available on the impact PFAS have on these specific genes; however, some studies do report neurotransmitter precursor disruption. Lee et al. (2021) found that PFOS decreased tryptophan and L-glutamine in zebrafish; thus, the authors attributed these decreased neurotransmitter levels to the behavioral alterations observed, including altered total distance moved and other swimming movements (i.e., bursting, freezing). Future studies should explore more the relationship between PFAS exposure and genes associated with AD to further develop molecular biomarkers that can be rapidly screened in the zebrafish model following exposure to different PFAS.

## Other Neurological Diseases and PFAS Exposure

Schizophrenia is a debilitating disorder characterized by cognitive impairment (e.g., memory deficits), positive symptoms (e.g., hallucinations, hyperactivity, aggression), and negative symptoms (apathy, social withdrawal). Though genetic factors play a role in diagnosis, neurotransmitter dysregulation, including GABA, glutamate, and dopamine, is associated with its pathogenesis. Zebrafish are useful vertebrate models in demonstrating Parkinson's-like phenotypes that mirror human symptoms through behavioral studies, as well as cell determination, neurogenesis, and neuronal migration (Morris, 2009). For instance, behavioral tests analyzing shoaling can characterize social deficits and, if fish are less interactive, this would be suggestive of withdrawal behavior typically examined in schizophrenics (Demin et al., 2019). Studies have also suggested the role that environmental factors play in the development of schizophrenia; recent studies indicate molecular and cellular pathways associated with schizophrenia in zebrafish exposed to PFBA (Wasel et al., 2022).

Autism, or autism spectrum disorder (ASD), is another neurological disorder associated with cognitive, motor, and social deficits. The direct cause of ASD is unknown; however, like other diseases, biological and environmental factors are thought to influence its pathogenesis. Consequently, the zebrafish model has been implemented to not only understand its pathogenesis, but also uncover potential treatments (Meshalkina et al., 2018).

Lastly, another neurological disorder that can be potentially associated following environmental chemical exposure is ADHD, which is characterized by difficulty paying attention and hyperactivity. Previous studies have shown positive correlations between ADHD symptoms in children and the presence of PFNA (Lien et al., 2016) and PFOA (Hoffman et al., 2010). Studies conducted with zebrafish also support a relationship between perfluorinated compounds and ADHD where PFOS exposure has been found to induce hyperactivity and impair cognitive ability in a concentration-dependent manner (Zhang et al., 2023). The field is still in its infancy, and other PFAS may be linked to these disorders; however such relationships require rigorous evaluation. In this regard, the repertoire of diverse behaviors in zebrafish may be useful for testing hypothesis related to PFAS and neurotoxicity.

## Conclusion

Several environmental chemicals have now been implicated in neurodegenerative diseases and disorders. Perfluorinated compounds represent some of the most pressing chemicals currently under study due to new discoveries related to their breakdown products and their widespread environmental presence. Toxicity data on this extensive chemical family for both wildlife and humans is lacking, as studies, until most recently, have focused primarily on PFOS and PFOA. With the understanding that environmental chemical exposures can exacerbate neurodegenerative processes, it becomes increasingly important to elucidate the neurotoxicity mechanisms of different PFAS. Here, we provide examples of how the zebrafish model can be used to evaluate PFAS-induced neurotoxicity, shedding light on mechanisms underlying neurological disorders. The zebrafish model has an extensive repertoire of behavioral assays that are relevant for neurological disorders like PD and ASD, among others. Furthermore, high-throughput behavioral assays using zebrafish facilitate stronger linkages between early molecular events following exposures to chemicals and downstream behavioral responses, which can be useful for isolating mechanisms of neurodegeneration, such as mitochondrial dysfunction and oxidative stress. Integrated omics approaches with phenotypic and behavioral data are also expected to shed new light on potential neurotoxicity of PFAS and to reveal subtle effects within the CNS. Such continued efforts will act to develop safer guidelines for food and water consumption regarding PFAS, further protecting human health.

**Funding:** This research was funded by the University of Florida and the College of Veterinary Medicine (CJM).

**Credit author statement: EI:** investigation, writing original draft, review, and editing; **CJM,** investigation, writing original draft, review, editing, and supervision.

## Declaration of Competing Interest

The authors declare that they have no known competing financial interests or personal relationships that could have appeared to influence the work reported in this paper.

## REFERENCES

An, X., Lei, H., Lu, Y., Xie, X., Wang, P., Liao, J., Liang, Z., Sun, B., & Wu, Z. (2023). Per and polyfluoroalkyl substances (PFASs) in water and sediment from a temperate watershed in China: Occurrence, sources, and ecological risks. *Science of The Total Environment*, *890*, 164207.

Antony, P. M., Diederich, N. J., & Balling, R. (2011). Parkinson's disease mouse models in translational research. *Mammalian Genome*, *22*, 401–419.

Atzei, A., Jense, I., Zwart, E. P., Legradi, J., Venhuis, B. J., van der Ven, L. T., Heusinkveld, H. J., & Hessel, E. V. (2021). Developmental neurotoxicity of environmentally relevant pharmaceuticals and mixtures thereof in a zebrafish embryo behavioural test. *International Journal of Environmental Research and Public Health*, *18*(13), 6717.

Barnhill, L. M., Murata, H., & Bronstein, J. M. (2020). Studying the pathophysiology of Parkinson's disease using zebrafish. *Biomedicines*, *8*(7), 197.

Benvenutti, R., Marcon, M., Gallas-Lopes, M., de Mello, A. J., Herrmann, A. P., & Piato, A. (2021). Swimming in the maze: An overview of maze apparatuses and protocols to assess zebrafish behavior. *Neuroscience & Biobehavioral Reviews*, *127*, 761–778.

Borsche, M., Pereira, S. L., Klein, C., & Grünewald, A. (2021). Mitochondria and Parkinson's Disease: Clinical, Molecular, and Translational Aspects. *Journal of Parkinson's Disease*, *11*(1), 45–60. https://doi.org/10.3233/jpd-201981

Braidy, N., Munoz, P., Palacios, A. G., Castellano-Gonzalez, G., Inestrosa, N. C., Chung, R. S., Sachdev, P., & Guillemin, G. J. (2012). Recent rodent models for Alzheimer's disease: Clinical implications and basic research. *Journal of Neural Transmission*, *119*, 173–195.

Cao, F., Souders Ii, C. L., Perez-Rodriguez, V., & Martyniuk, C. J. (2018). Elucidating conserved transcriptional networks underlying pesticide exposure and Parkinson's disease: A focus on chemicals of epidemiological relevance. *Frontiers in Genetics*, *9*, 701. https://doi.org/10.3389/fgene.2018.00701

Champagne, D. L., Hoefnagels, C. C., De Kloet, R. E., & Richardson, M. K. (2010). Translating rodent behavioral repertoire to zebrafish (Danio rerio): Relevance for stress research. *Behavioural Brain Research*, *214*(2), 332–342.

Cherian, A., & Divya, K. (2020). Genetics of Parkinson's disease. *Acta Neurologica Belgica*, 1–9.

Corey-Bloom, J. (2002). The ABC of Alzheimer's disease: Cognitive changes and their management in Alzheimer's disease and related dementias. *International psychogeriatrics*, *14*(S1), 51–75.

Costello, S., Cockburn, M., Bronstein, J., Zhang, X., & Ritz, B. (2009). Parkinson's disease and residential exposure to maneb and paraquat from agricultural applications in the central valley of California. *American Journal of Epidemiology*, *169*(8), 919–926.

de Abreu, M. S., Maximino, C., Banha, F., Anastácio, P. M., Demin, K. A., Kalueff, A. V., & Soares, M. C. (2020). Emotional behavior in aquatic organisms? Lessons from crayfish and zebrafish. *Journal of Neuroscience Research*, *98*(5), 764–779.

Demin, K. A., Meshalkina, D. A., Volgin, A. D., Yakovlev, O. V., de Abreu, M. S., Alekseeva, P. A., Friend, A. J., Lakstygal, A. M., Zabegalov, K., & Amstislavskaya, T. G. (2019). Developing zebrafish experimental animal models relevant to schizophrenia. *Neuroscience & Biobehavioral Reviews*, *105*, 126–133.

Farrell, T. C., Cario, C. L., Milanese, C., Vogt, A., Jeong, J.-H., & Burton, E. A. (2011). Evaluation of spontaneous propulsive movement as a screening tool to detect rescue of Parkinsonism phenotypes in zebrafish models. *Neurobiology of Disease*, *44*(1), 9–18.

Gatz, M., Reynolds, C. A., Fratiglioni, L., Johansson, B., Mortimer, J. A., Berg, S., Fiske, A., & Pedersen, N. L. (2006). Role of genes and environments for explaining Alzheimer disease. *Archives of General Psychiatry*, *63*(2), 168–174.

Gómez-Canela, C., Barth, J. A., & Lacorte, S. (2012). Occurrence and fate of perfluorinated compounds in sewage sludge from Spain and Germany. *Environmental Science and Pollution Research*, *19*, 4109–4119.

Guo, X., Zhang, S., Lu, S., Zheng, B., Xie, P., Chen, J., Li, G., Liu, C., Wu, Q., & Cheng, H. (2018). Perfluorododecanoic acid exposure induced developmental neurotoxicity in zebrafish embryos. *Environmental Pollution*, *241*, 1018–1026.

Hagenaars, A., Vergauwen, L., De Coen, W., & Knapen, D. (2011). Structure–activity relationship assessment of four perfluorinated chemicals using a prolonged zebrafish early life stage test. *Chemosphere*, *82*(5), 764–772.

Hamilton, T. J., Krook, J., Szaszkiewicz, J., & Burggren, W. (2021). Shoaling, boldness, anxiety-like behavior and locomotion in zebrafish (Danio rerio) are altered by acute benzo [a] pyrene exposure. *Science of The Total Environment*, *774*, 145702.

Hoffman, K., Webster, T. F., Weisskopf, M. G., Weinberg, J., & Vieira, V. M. (2010). Exposure to polyfluoroalkyl chemicals and attention deficit/hyperactivity disorder in US children 12–15 years of age. *Environmental Health Perspectives*, *118*(12), 1762–1767.

Horzmann, K. A., & Freeman, J. L. (2018). Making Waves: New Developments in Toxicology with the Zebrafish. *Toxicological Sciences*, *163*(1), 5–12. https://doi.org/10.1093/toxsci/kfy044

Hsiao, C. C., Yang, A.-M., Wang, C., & Lin, C.-Y. (2023). Association between glyphosate exposure and cognitive function, depression, and neurological diseases in a representative sample of US adults: NHANES 2013–2014 analysis. *Environmental Research*, *237*, 116860.

Ivantsova, E., Lu, A., & Martyniuk, C. J. (2023). Occurrence and toxicity mechanisms of perfluorobutanoic acid (PFBA) and perfluorobutane sulfonic acid (PFBS) in fish. *Chemosphere*, *349*, 140815.

Jiménez-Balado, J., & Eich, T. S. (2021). GABAergic dysfunction, neural network hyperactivity and memory impairments in human aging and Alzheimer's disease. *Seminars in Cell & Developmental Biology*, *116*, 146–159.

Kalyn, M., Lee, H., Curry, J., Tu, W., Ekker, M., & Mennigen, J. A. (2023). Effects of PFOS, F-53B and OBS on locomotor behaviour, the dopaminergic system and mitochondrial function in developing zebrafish (Danio rerio). *Environmental Pollution*, *326*, 121479.

Kanthasamy, A. G., Kitazawa, M., Kanthasamy, A., & Anantharam, V. (2005). Dieldrin-induced neurotoxicity: Relevance to Parkinson's disease pathogenesis. *NeuroToxicology*, *26*(4), 701–719.

Kazmi, S. S. U. H., Wang, Y. Y. L., Cai, Y.-E., & Wang, Z. (2022). Temperature effects in single or combined with chemicals to the aquatic organisms: An overview of thermo-chemical stress. *Ecological Indicators*, *143*, 109354.

Langova, V., Vales, K., Horka, P., & Horacek, J. (2020). The role of zebrafish and laboratory rodents in schizophrenia research. *Frontiers in Psychiatry*, *11*, 703.

Lee, H., Sung, E. J., Seo, S., Min, E. K., Lee, J.-Y., Shim, I., Kim, P., Kim, T.-Y., Lee, S., & Kim, K.-T. (2021). Integrated multi-omics analysis reveals the underlying molecular mechanism for developmental neurotoxicity of perfluorooctanesulfonic acid in zebrafish. *Environment International*, *157*, 106802.

Lien, G.-W., Huang, C.-C., Shiu, J.-S., Chen, M.-H., Hsieh, W.-S., Guo, Y.-L., & Chen, P.-C. (2016). Perfluoroalkyl substances in cord blood and attention deficit/hyperactivity disorder symptoms in seven-year-old children. *Chemosphere*, *156*, 118–127.

Liu, H., Fu, R., Zhang, Y., Mao, L., Zhu, L., Zhang, L., Liu, X., & Jiang, H. (2023). Integrate transcriptomic and metabolomic analysis reveals the underlying mechanisms of behavioral disorders in zebrafish (Danio rerio) induced by imidacloprid. *Science of The Total Environment*, *870*, 161541.

Liu, S., Qiu, W., Li, R., Chen, B., Wu, X., Magnuson, J. T., Xu, B., Luo, S., Xu, E. G., & Zheng, C. (2023). Perfluorononanoic Acid Induces Neurotoxicity via Synaptogenesis Signaling in Zebrafish. *Environmental Science & Technology*, *57*(9), 3783–3793.

Liu, S.-T., Horng, J.-L., Lin, L.-Y., & Chou, M.-Y. (2023). Fenpropathrin causes alterations in locomotion and social behaviors in zebrafish (Danio rerio). *Aquatic Toxicology*, *265*, 106756.

Llorca, M., Farré, M., Picó, Y., Müller, J., Knepper, T. P., & Barceló, D. (2012). Analysis of perfluoroalkyl substances in waters from Germany and Spain. *Science of The Total Environment*, *431*, 139–150.

Mahapatra, A., Gupta, P., Suman, A., Ray, S. S., Malafaia, G., & Singh, R. K. (2023). Unraveling the mechanisms of perfluorooctanesulfonic acid-induced dopaminergic neurotoxicity and microglial activation in developing zebrafish. *Science of The Total Environment*, *887*, 164030.

Mancuso, R., Fryatt, G., Cleal, M., Obst, J., Pipi, E., Monzón-Sandoval, J., Ribe, E., Winchester, L., Webber, C., & Nevado, A. (2019). CSF1R inhibitor JNJ-40346527 attenuates microglial proliferation and neurodegeneration in P301S mice. *Brain*, *142*(10), 3243–3264.

Marinho, C. S., Matias, M. V., Brandão, I. G., Santos, E. L., Machado, S. S., & Zanta, C. L. (2019). Characterization and kinetic study of the brain and muscle acetylcholinesterase from Danio rerio. *Comparative Biochemistry and Physiology Part C: Toxicology & Pharmacology*, *222*, 11–18.

Martyniuk, C. J., Kroll, K. J., Doperalski, N. J., Barber, D. S., & Denslow, N. D. (2010). Genomic and proteomic responses to environmentally relevant exposures to dieldrin: indicators of neurodegeneration? *Toxicological Sciences*, *117*(1), 190–199.

McGregor, M. M., & Nelson, A. B. (2019). Circuit mechanisms of Parkinson's disease. *Neuron*, *101*(6), 1042–1056.

Meng, J., Liu, S., Zhou, Y., & Wang, T. (2019). Are perfluoroalkyl substances in water and fish from drinking water source the major pathways towards human health risk? *Ecotoxicology and Environmental Safety*, *181*, 194–201. https://doi.org/10.1016/j.ecoenv.2019.06.010

Meshalkina, D. A., Kizlyk, M. N., Kysil, E. V., Collier, A. D., Echevarria, D. J., Abreu, M. S., Barcellos, L. J., Song, C., Warnick, J. E., & Kyzar, E. J. (2018). Zebrafish models of autism spectrum disorder. *Experimental Neurology*, *299*, 207–216.

Min, E. K., Lee, H., Sung, E. J., Seo, S. W., Song, M., Wang, S., Kim, S. S., Bae, M. A., Kim, T.-Y., & Lee, S. (2023). Integrative multi-omics reveals analogous developmental neurotoxicity mechanisms between perfluorobutanesulfonic acid and perfluorooctanesulfonic acid in zebrafish. *Journal of Hazardous Materials*, *457*, 131714.

Morris, J. A. (2009). Zebrafish: A model system to examine the neurodevelopmental basis of schizophrenia. *Progress in Brain Research*, *179*, 97–106.

Mostafalou, S., & Abdollahi, M. (2023). The susceptibility of humans to neurodegenerative and neurodevelopmental toxicities caused by organophosphorus pesticides. *Archives of Toxicology*, *97*(12), 3037–3060.

Murray, M. J., & Murray, C. F. (2007). Alzheimer's disease. In *Complications in anesthesia* Ed: John L. Atlee; Elsevier Inc. (pp. 493–495).

Newman, M., Ebrahimie, E., & Lardelli, M. (2014). Using the zebrafish model for Alzheimer's disease research. *Frontiers in Genetics*, *5*, 189.

Nishimura, Y., Murakami, S., Ashikawa, Y., Sasagawa, S., Umemoto, N., Shimada, Y., & Tanaka, T. (2015). Zebrafish as a systems toxicology model for developmental neurotoxicity testing. *Congenital anomalies*, *55*(1), 1–16.

Noyes, P. D., & Lema, S. C. (2015). Forecasting the impacts of chemical pollution and climate change interactions on the health of wildlife. *Current Zoology*, *61*(4), 669–689.

Ogi, A., Licitra, R., Naef, V., Marchese, M., Fronte, B., Gazzano, A., & Santorelli, F. M. (2021). Social preference tests in zebrafish: A systematic review. *Frontiers in Veterinary Science*, *7*, 590057.

Panula, P., Chen, Y.-C., Priyadarshini, M., Kudo, H., Semenova, S., Sundvik, M., & Sallinen, V. (2010). The comparative neuroanatomy and neurochemistry of zebrafish CNS systems of relevance to human neuropsychiatric diseases. *Neurobiology of Disease*, *40*(1), 46–57.

Paquette, S. E., Martin, N. R., Rodd, A., Manz, K. E., Allen, E., Camarillo, M., Weller, H. I., Pennell, K., & Plavicki, J. S. (2023). Evaluation of neural regulation and microglial responses to brain injury in larval zebrafish exposed to perfluorooctane sulfonate. *Environmental Health Perspectives*, *131*(11), 117008.

Pitchai, A., Rajaretinam, R. K., & Freeman, J. L. (2019). Zebrafish as an emerging model for bioassay-guided natural product drug discovery for neurological disorders. *Medicines (Basel)*, *6*(2), 61. https://doi.org/10.3390/medicines6020061

Rappold, P. M., Cui, M., Chesser, A. S., Tibbett, J., Grima, J. C., Duan, L., Sen, N., Javitch, J. A., & Tieu, K. (2011). Paraquat neurotoxicity is mediated by the dopamine transporter and organic cation transporter-3. *Proceedings of the National Academy of Sciences*, *108*(51), 20766–20771.

Singh, N. K., Banerjee, B., Bala, K., Basu, M., & Chhillar, N. (2014). Polymorphism in cytochrome P450 2D6, glutathione S-transferases Pi 1 genes, and organochlorine pesticides in Alzheimer disease: A case–control study in north indian population. *Journal of Geriatric Psychiatry and Neurology*, *27*(2), 119–127.

Stahl, T., Mattern, D., & Brunn, H. (2011). Toxicology of perfluorinated compounds. *Environmental Sciences Europe*, *23*(1), 38.

Stewart, A., Kadri, F., DiLeo, J., Min Chung, K., Cachat, J., Goodspeed, J., Suciu, C., Roy, S., Gaikwad, S., & Wong, K. (2010). The developing utility of zebrafish in modeling neurobehavioral disorders. *International Journal of Comparative Psychology*, *23*(1), 130.

Thomas, B., & Beal, M. F. (2007). Parkinson's disease. *Human molecular genetics*, *16*(R2), R183–R194.

Vogs, C., Johanson, G., Na̍slund, M., Wulff, S., Sjö̍din, M., Hellstrandh, M., Lindberg, J., & Wincent, E. (2019). Toxicokinetics of perfluorinated alkyl acids influences their toxic potency in the zebrafish embryo (Danio rerio). *Environmental Science & Technology*, *53*(7), 3898–3907.

Wang, X., Shi, X., Zheng, S., Zhang, Q., Peng, J., Tan, W., & Wu, K. (2022). Perfluorooctane sulfonic acid (PFOS) exposures interfere with behaviors and transcription of genes on nervous and muscle system in zebrafish embryos. *Science of The Total Environment*, *848*, 157816.

Wang, Y., Wang, L., Chang, W., Zhang, Y., Zhang, Y., & Liu, W. (2019). Neurotoxic effects of perfluoroalkyl acids: Neurobehavioral deficit and its molecular mechanism. *Toxicology Letters*, *305*, 65–72.

Wasel, O., Thompson, K. M., & Freeman, J. L. (2022). Assessment of unique behavioral, morphological, and molecular alterations in the comparative developmental toxicity profiles of PFOA, PFHxA, and PFBA using the zebrafish model system. *Environment International*, *170*, 107642.

Wiersielis, K. R., Samuels, B. A., & Roepke, T. A. (2020). Perinatal exposure to bisphenol A at the intersection of stress, anxiety, and depression. *Neurotoxicology and teratology*, *79*, 106884.

Wilson, S. W., Brand, M., & Eisen, J. S. (2002). Patterning the zebrafish central nervous system. In L. Solnica-Krezel (Ed.), *Pattern formation in zebrafish* (pp. 181–215). Springer.

Wu, L., Dang, Y., Liang, L.X., Gong, Y.C., Zeeshan, M., Qian, Z., Geiger, S. D., Vaughn, M. G., Zhou, Y., & Li, Q.Q. (2022). Perfluorooctane sulfonates induces neurobehavioral changes and increases dopamine neurotransmitter levels in zebrafish larvae. *Chemosphere*, *297*, 134234.

Yan, M. H., Wang, X., & Zhu, X. (2013). Mitochondrial defects and oxidative stress in Alzheimer disease and Parkinson disease. *Free Radical Biology and Medicine*, *62*, 90–101. https://doi.org/10.1016/j.freeradbiomed.2012.11.014

Yu, T., Zhou, G., Cai, Z., Liang, W., Du, Y., & Wang, W. (2021). Behavioral effects of early-life exposure to perfluorooctanoic acid might synthetically link to multiple aspects of dopaminergic neuron development and dopamine functions in zebrafish larvae. *Aquatic Toxicology*, *238*, 105926.

Zhang, G., Zhang, Y., Shen, Y., Wang, Y., Zhao, M., & Sun, L. (2021). The potential role of ferroptosis in Alzheimer's disease. *Journal of Alzheimer's Disease*, *80*(3), 907–925.

Zhang, L., Li, Y.Y., Chen, T., Xia, W., Zhou, Y., Wan, Y.J., Lv, Z.Q., Li, G.Q., & Xu, S.Q. (2011). Abnormal development of motor neurons in perfluorooctane sulphonate exposed zebrafish embryos. *Ecotoxicology*, *20*, 643–652.

Zhang, M., Gu, X., Wu, L., Wan, N., Liu, Y., Xin, Z., Chen, T., Liu, S., Li, M., & Deng, M. (2023). A new mechanistic insight into the association between environmental perfluorooctane sulfonic acid (PFOS) exposure and attention deficit and hyperactivity disorder (ADHD)-like behavior. *NeuroToxicology*, *99*, 254–263.

# 11

# *Effect of Myelination on Neurodegeneration and Neurodegenerative Diseases in the Zebrafish Model*

Efruz İrem Akkuş-Yaylamış and Müjgan Cengiz

## Abbreviations

| | |
|---|---|
| **AD** | Alzheimer's Disease |
| **ALS** | Amyotrophic Lateral Sclerosis |
| **APP** | Amyloid Precursor Protein |
| **Aβ** | Amyloid β Peptide |
| **BACE1** | Beta-Secretase 1 |
| **CNP** | 2',3'-Cyclic Nucleotide 3'-Phosphodiesterase |
| **CNS** | Central Nervous System |
| **dpf** | Days After Fertilization |
| **Krox20/EGR2** | Early Growth Response Protein 2 |
| **ErbB2/ErbB3** | Erb-B2 RECEPTOR TYROSINE KINASE 2/3 |
| **GFP** | Green Fluorescent Protein |
| **hpf** | Hours After Fertilization |
| **IL-7R** | Interleukin 7 Receptor |
| **KO** | Knockout |
| **LB** | Lewy Bodies |
| **LINGO-1** | Leucine-Rich Repeat and Ig Domain Containing 1 |
| **MCT1** | Monocarboxylate Transporter 1 |
| **MRI** | Magnetic Resonance Imaging |
| **MPTP** | 1-Methyl-4-Phenyl-1,2,3,6-Tetrahydropyridine |
| **MTZ** | Metronidazole |
| **MS** | Multiple Sclerosis |
| **MBP** | Myelin Basic Protein |
| **nm** | Nanometer |
| **NRG1** | Neuregulin-1 |
| **NTR** | Nitroreductase |
| **NMDA** | Ionotropic N Methyl-D-Aspartate |
| **Oct6** | Organic Cation/Carnitine Transporter 6 |
| **Olig1** | Oligodendrocyte Transcription Factor 1 |
| **Olig2** | Oligodendrocyte Transcription Factor 2 |
| **Opcs** | Oligodendrocyte Progenitor Cells |
| **PD** | Parkinson's Disease |
| **PNS** | Peripheral Nervous System |
| **RORγt** | RAR-Related Orphan Receptor-γt |
| **ROS** | Reactive Oxygen Species |
| **SOD** | Superoxide Dismutase |
| **Snc** | Substantia Nigra Compacta |
| **SOD** | Superoxide Dismutase |
| **Sox10** | SRY-Box Transcription Factor 10 |
| **T4** | Thyroxine |

## Introduction

### Biology of Axonal Myelination

Almost all axons in the vertebrate nervous system are surrounded by an insulating sheath called myelin, which has a lipoprotein structure in both the central nervous system (CNS) and peripheral nervous system (PNS) (Stadelmann et al., 2019). Myelin protects the axons, enhancing the speed and reliability of nerve impulse conduction (Simons & Nave, 2016). Myelin is a multilayered lipid membrane structure consisting of lamellae with alternating concentric electron-dense and light layers surrounding the axon (Baumann & Pham-Dinh, 2001). The lamellae repeat radially around the axon with a period of 12 nm (Boullerne, 2016). In 1854, Rudolf Virchow derived the term myelin from the Greek word *myelos*, which means bone marrow, in order to define the matter he had observed in large quantities in the brain (Virchow, 1860). His first observations were that myelin was secreted by neurons and was an insulating substance. In 1928, histological silver carbonate staining by Pio Del Rio Hortega showed that myelin was not of neuronal origin but was actually produced by oligodendrocytes (Pérez-Cerdá et al., 2015). Glial cells surround axon segments, and oligodendrocytes in the CNS and Schwan cells in the PNS form myelin, ensuring fast conduction of action potentials and protecting the axon from damage (Kandel, 2000). In the CNS, oligodendrocytes have the capacity to myelinate several axon segments, whereas in the PNS, a single Schwann cell forms a single myelin sheath. Schwann cells, a type of glial cell, play a pivotal role in myelination in the peripheral nervous system (Kandel, 2000). Named after the German anatomist Theodor Schwann, these cells wrap around individual nerve fibers, creating the myelin sheath (Boullerne, 2016). The process begins during embryonic development and continues into



DOI: 10.1201/9781032676555-13

early adulthood. As Schwann cells envelop the axon, they lay down layers of myelin, cover it, and have a protective role (Boullerne, 2016).

The myelination of axons is intended to increase the speed of signal transduction and to separate voltage-sensitive ion channels (such as the K+ channel) into distinct axonal segments: nodes of Ranvier, paranodes, juxtaparanodes, and internodes (Kandel, 2000). Each myelin sheath segment or internode is 150–200 μm in length, and internodes are separated by cavities called nodes of Ranvier, where myelin is lacking (Baumann & Pham-Dinh, 2001). The nodes of Ranvier separating adjacent internodes along myelinated fibers are the sites where the axolemma is exposed to the extracellular environment (Zalc, 2006). The nodes of Ranvier play an important role in nerve impulse transmission (Kandel, 2000). The myelin sheath acts as an electrical insulator in axons, allowing nerve impulses to jump from one node of Ranvier to another. This "saltatory conduction" significantly accelerates the speed of signal transmission along the nerve fiber (Kandel, 2000).

Electron microscopy studies have allowed us to understand the structure of myelin with its characteristic periodic structure of alternating electron dense and light layers (Peters, 1961; Salzer & Zalc, 2016; Snaidero et al., 2014). The major dense line in the myelin structure represents condensed cytoplasmic surfaces. The intraperiod line is composed of tightly packed outer membranes (Peters, 1961). Myelin lamellae are tightly connected by tight junctions between the outer face membranes composed of Claudin-11 (Morita et al., 1999). Compacted myelin provides the high electrical resistance and low capacitance required for rapid conduction of impulses (Peters, 1961; Snaidero et al., 2014). Composition of the myelin structure with opposing lamellae and tight myelin layers is essential for the insulating properties of myelin (Baumann & Pham-Dinh, 2001). Thereby, ion leakage is minimized, and efficient and fast nerve conduction is ensured (Salzer & Zalc, 2016). Such a tight arrangement is an obstacle to the maintenance of molecular metabolism between the myelin sheath and the axon (Stadelmann et al., 2019). Molecular transport pathways are required between compacted myelin layers (Stadelmann et al., 2019). Myelin is less compacted at the inner and outer ends of the spiral and forms inner and outer loops with a small amount of oligodendrocyte cytoplasm (Jean Harry & Toews, 1998). Myelin lamellae terminate near the node of Ranvier in small, enlarged loops containing cytoplasm (Peters, 1961). The loops are arranged in a roughly regular and symmetrical pattern on either side of the node. These ordered loops are called the paranodal region (or paranod) (Baumann & Pham-Dinh, 2001). Paranodal loops are in contact with axons. The area of 10–15 μm adjacent to the paranode is called the juxtaparanode (Stadelmann et al., 2019). Freeze-fracture studies have observed abundant intramembrane cytoplasmic canals on the outer surface of the juxtaparanodal membrane of the axon and between myelin lamellae in paranodal regions (Miller & Pinto Da Silva, 1977; Rash et al., 1997). These cytoplasmic channels represent gap junctions that provide pathways for the distribution of organelles such as peroxisomes and lysosomes among nodes (Stadelmann et al., 2019).

The initiation and progression of myelin formation during the embryonic period occurs through the recognition and differentiation of myelin-forming cells by the axon (Baumann & Pham-Dinh, 2001; Simons & Nave, 2016). Through interactions between axons and myelin-forming cells, the inner myelin membrane initiates growth with lateral extensions; thus myelin surrounds the axon (Jean Harry & Toews, 1998). The myelin sheath is formed in glial cells by the fusion of adjacent inner surfaces of the plasma membrane with no intervening cytoplasm (Jean Harry & Toews, 1998). The cell types that form the myelin sheath are derived from different precursor populations and genetic programs that direct myelin formation (Baumann & Pham-Dinh, 2001). Although there are several molecular or morphological differences between nerve fibers in the PNS and CNS, the basic myelin sheath arrangement and electrophysiological properties are essentially the same (Stadelmann et al., 2019). During myelination of a peripheral nerve fiber, the outer membrane of the Schwann cell surrounds the axon to form a double-membrane structure called the mesaxon, which extends and spirals in layers from a single center around the axon (Boullerne, 2016). As the axon is surrounded, the cytoplasm of the Schwann cell is compressed to form a compact lamellar structure (Boullerne, 2016).

The biological structure of myelin consists of a lipid bilayer interspersed with protein layers. Its composition is similar to that of the plasma membrane, with 70% lipid and 30% protein content, including high concentrations of cholesterol and phospholipids (Müller et al., 2013). In the central nervous system, myelin has two main proteins: myelin basic protein (MBP), a small, positively charged protein located on the cytoplasmic surface of myelin; and proteolipid protein (PLP), a hydrophobic membrane protein (Nave & Trapp, 2008). MBP is required for plasma membrane compaction and cytoplasm extrusion (compacted myelin contains no cytoplasm) (Nave & Trapp, 2008). Both of these proteins in the myelin membrane structure are known to provide structural stability for the sheath. Once formed, myelin membranes are very stable, and the lipid and membrane components are long-lived (Stadelmann et al., 2019). Another myelin-associated protein, 2',3'-cyclic nucleotide 3'-phosphodiesterase (CNP), interacts with the actin cytoskeleton and counteracts the polymerizing forces of MBP, thus forming cytoplasmic canals within the myelin sheath (Figure 11.1) (Nave & Trapp, 2008).

Myelin has a remarkably stable molecular structure, probably due to its lipid composition, which is rich in glycosphingolipids and cholesterol and contains high levels of saturated long-chain fatty acids (Schmitt et al., 2015). It also contains plasmalogens (ether lipids) in abundance (Schmitt et al., 2015). Myelin membrane presents structural stability as well as metabolic stability (Stadelmann et al., 2019). The half-life of cholesterol in the myelin membrane is extended to several months (Poitelon et al., 2020). In a rat study, myelin proteins were found to be among the longest-lived proteins along with histones, nucleoporins, and lamins by mass spectrometry (Toyama et al., 2013). With all these properties, myelin is a stable system, unlike most cell membranes, which are highly dynamic and out of equilibrium (Stadelmann et al., 2019).

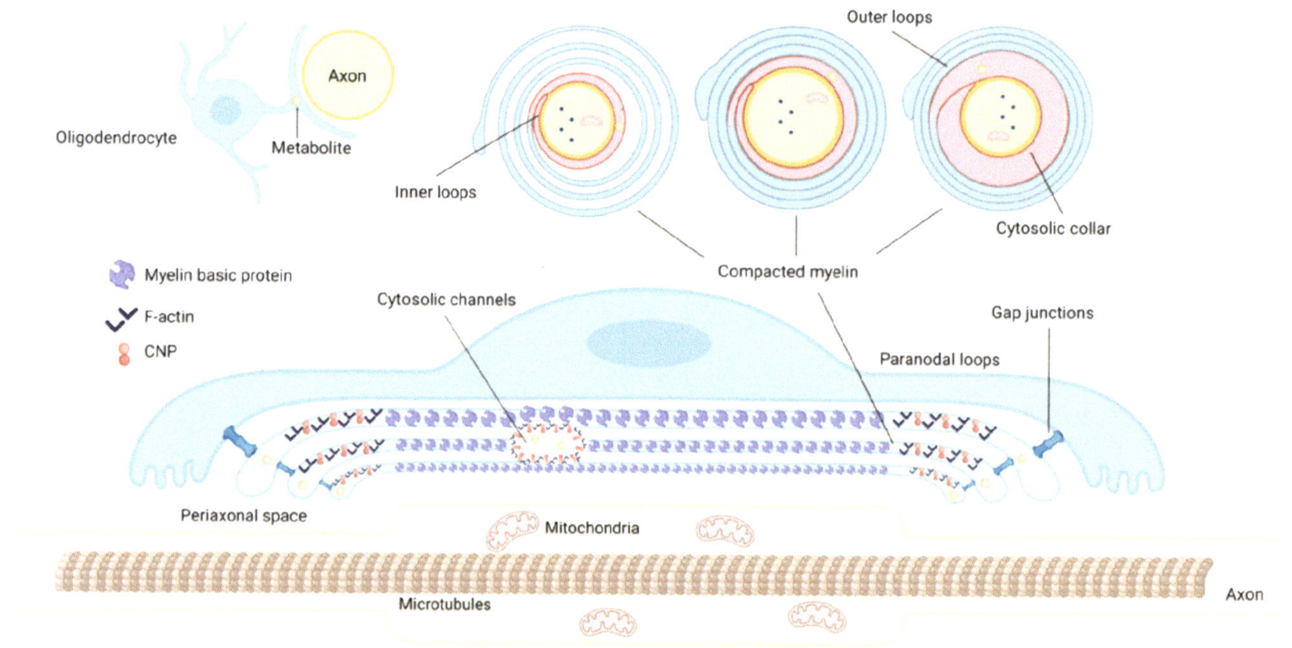


**FIGURE 11.1** Model of myelin formation by oligodendrocytes surrounding an axonal segment.

Structural molecules and metabolites required for membrane production and myelin growth are transported through a cytosolic channel system towards the growing tip, the inner tongue of the myelin layer. The force pushing the inner tongue forward relies on the polymerization–depolymerization cycles of actin filaments. The overlapping plasma membranes are then compressed in a zipper-like manner, forming the myelin sheath. Myelin basic protein (MBP) is important in the formation of compressed myelin membrane layers by squeezing cytoplasmic surfaces tightly packed together. Membrane-associated 2',3'-cyclic nucleotide 3'-phosphodiesterase (CNP) interacts with the actin cytoskeleton and counteracts the compressive force of MBP. This helps maintain uncompressed cytosolic channels in mature sheaths that are critical for metabolite diffusion and continuous delivery of adaxonal membrane proteins (such as MCT1 and NMDA receptors) and thus axonal metabolic support. Several cell adhesion molecules in paranodal regions keep the myelin borders stable. Gap junctions connect myelin paranodal loops at the myelin lateral edges. Created with Biorender.

## Myelination in Zebrafish

Myelination plays crucial roles in the vertebrate nervous system (Jean Harry & Toews, 1998). Myelin protects and insulates axons from damage, allowing rapid and constant transmission of action potentials (Salzer & Zalc, 2016). It also provides trophic and metabolic support to axons and glial cells and is involved in synaptic plasticity, allowing higher cognitive and intellectual functions. Invertebrates lack myelinated nerve tubes, and therefore the surrounding of axons with myelin sheaths is a crucial novelty in vertebrates (Zalc et al., 2008). Since myelin first evolved in vertebrates, zebrafish is one of the simplest vertebrate model systems to study the genetics and development of myelination (Preston & Macklin, 2015). Zebrafish have become a frequently used vertebrate model in developmental studies due to their short life cycle, external development, rapid sexual maturation, optically transparent embryo structure and sharing a gene pool largely similar to humans (Monk & Talbot, 2009; Preston & Macklin, 2015). The translucent structure of zebrafish embryos and larvae enables noninvasive visualisation of internal body structures (D'Rozario et al., 2017).

Zebrafish, with all these advantages, are a robust, rapid, and cost-effective model for studying vertebrate myelination in vivo (D'Rozario et al., 2017). While myelination barely begins in the first month in rodents, in zebrafish embryos a non-compact myelin sheath begins to form 3 days after fertilization (dpf), and by approximately 5 dpf, a compact myelin structure forms (Figure 11.2). This significantly shortens the time required to investigate the molecular mechanisms of myelination (D'Rozario et al., 2017).

The use of zebrafish as a suitable model for the exploration of molecular mechanisms of vertebrate myelination is based on the fact that myelin structure and function are essentially comparable between fish and mammals (Ghosh & Hui, 2018). In mammals and zebrafish, myelination occurs through similar processes, such as the formation of myelin by oligodendrocytes and Schwann cells in the CNS and PNS (Ghosh & Hui, 2018). The physiology of myelination has been well characterized in mammals, but zebrafish have relatively recently become a tool for in vivo analysis of oligodendrocytes and Schwann cells (D'Rozario et al., 2017). Transgenic zebrafish have been developed that express green fluorescent protein (GFP) in the membrane of oligodendrocytes and Schwann cells, making these cells visible. Such transgenic models are very beneficial for studying myelination in vivo (Jung et al., 2010; Kirby et al., 2006).

Biochemical studies have revealed similarities and differences in the myelination process between zebrafish and mammals. The morphogenetic cellular movements and genetic program directing myelination are conserved in zebrafish and mammals (Zalc et al., 2008). The genes involved in the myelination process in zebrafish are homologous to those in mammals (Dubois-Dalcq et al., 2008). Specifically, the evolutionary

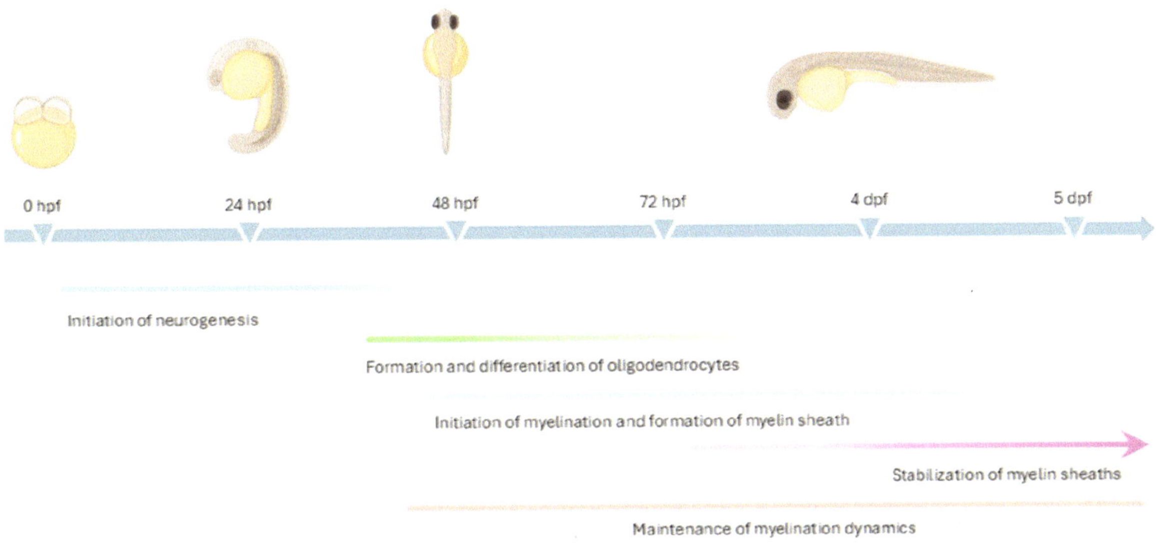


**FIGURE 11.2** Neurogenesis and myelination during development in zebrafish.

Neurogenesis begins in the first hours of development and continues with the differentiation and migration of OPCs around 36 hpf. OPCs soon differentiate into mature myelinating oligodendrocytes that will form myelin sheaths, and myelin sheath formation begins at 72 hpf. Myelin sheath dynamics continue throughout embryogenesis in zebrafish. Created with Biorender.

histories of humans and zebrafish show considerable conservation (Dubois-Dalcq et al., 2008). Homology between zebrafish and humans is also seen in the signaling molecules that play a role in myelination (Baumann & Pham-Dinh, 2001). Zebrafish and mammals share the same neurotransmitters (glutamate, acetylcholine, dopamine, etc.) and their biological functions in the myelination process (D'Rozario et al., 2017). Myelin basic protein, myelin proteolipid protein, and myelin protein zero are components of myelin in zebrafish and mammals, but there are also teleost-specific myelin proteins such as Zwilling-A and Zwilling-B in the CNS (Masson & Nait-Oumesmar, 2023). Transcription factors known to control the formation of oligodendrocyte precursor cells in mammals, including olig1, olig2, and sox10, were found to function similarly in zebrafish (Preston & Macklin, 2015). Signaling pathways including Hedgehog, Notch, Wnt, and ErbB/EGF drive oligodendrocyte formation and are involved in the myelination process in a comparable manner between zebrafish and mammals (Masson & Nait-Oumesmar, 2023; Preston & Macklin, 2015). Similarly, Sox10, ErbB2/ErbB3, Oct6 and Krox20/EGR2 are expressed in Schwann cells of both species (Raphael & Talbot, 2011). Together, these similarities suggest that the genes identified in zebrafish may also have essential functions in the myelination process and oligodendrocyte and glial cell development in mammals.

## Neurodegeneration and Myelin Dynamics in the Zebrafish

Neurodegeneration refers to the progressive and irreversible loss of the structure or function of neurons, leading to a decline in cognitive and motor functions (Ohno & Ikenaka, 2019). Factors such as neuronal activation, energy metabolism, signal transduction, and ion homeostasis support axonal and neuronal survival, and their disruption can also lead to neurodegeneration (Ohno & Ikenaka, 2019). The causes of neurodegeneration are multifactorial and include genetic, environmental, and age-related factors (Verkhratsky et al., 2019). Neurodegenerative diseases, especially those that occur with aging, are progressive disorders that lead to motor or sensory dysfunction through abnormal protein accumulation and selective neuronal and nerve myelin loss (Lee et al., 2012; Verkhratsky et al., 2019).

Proper myelination is essential for normal neurological function, and disruptions in this process can lead to various disorders, notably multiple sclerosis, in which the immune system mistakenly attacks the myelin sheath, disrupting signal transmission (Jean Harry & Toews, 1998). Neurodegeneration represents the pathological process that leads to neuronal loss and cognitive decline, while myelination is a vital physiological process that enables efficient communication within the nervous system (Ettle et al., 2016; Osorio-Querejeta et al., 2017). The relationship between neurodegeneration and myelination is complex and multifaceted, involving intricate interactions between various cellular and molecular processes. While myelination is typically considered a protective and supportive mechanism for neuronal function, disturbances in this process can contribute to or exacerbate neurodegenerative conditions (Saab & Nave, 2017). Furthermore, neurodegeneration adversely affects the integrity and structure of the myelin sheath (Ohno & Ikenaka, 2019; Saab & Nave, 2017). Both areas of research are critical to the understanding and treatment of neurological disorders. Understanding this relationship is crucial for unraveling the mechanisms underlying specific neurological disorders and developing potential therapeutic interventions.

Axonal myelination is important for the proper functioning of many cognitive and behavioral processes in the brain

(Kandel, 2000). In aging and in neurodegenerative diseases, healthy myelin begins to lose its structural integrity (Ettle et al., 2016). Genetic and pathological defects of myelin lead to myelin's absence or structural and morphological disruptions, causing degeneration and death of axons and neurons, resulting in impaired motor activity, cognitive deficits, and neurological disorders (Nave, 2010). The pathological loss of myelin is termed demyelination and occurs in neurodegenerative disorders such as multiple sclerosis, Alzheimer's disease, and Parkinson's disease (Ferrer, 2018; Liu, 2023; Nave, 2010; Osorio-Querejeta et al., 2017). Demyelination is a pathological condition in which the myelin sheath vanishes from the periphery of the axon (Jean Harry & Toews, 1998; Osorio-Querejeta et al., 2017). When a demyelinating damage occurs, CNS has the capability to repair demyelinated axons and regenerate myelin through a process called remyelination (Osorio-Querejeta et al., 2017). Thus, demyelinated axons are remyelinated and new myelin is synthesized (Figure 11.3) (Dubois-Dalcq et al., 2008; Osorio-Querejeta et al., 2017). Oligodendrocyte progenitor cells (OPCs) are responsible for initiating the remyelination process (Lee et al., 2012; Osorio-Querejeta et al., 2017; Simons & Nave, 2016). First, OPCs migrate towards demyelinating lesions and proliferate there, then differentiate into oligodendrocytes that expand their membranes to resurround axons (Verkhratsky et al., 2019). However, disruption of myelin structure and failure of remyelination in the CNS lead to various neurodegenerative diseases (Verkhratsky et al., 2019).

Zebrafish are particularly useful for investigating various aspects of neurodegenerative diseases, including neurodegeneration studies, due to their translucent embryos, rapid development, and genetic similarities to humans (D'Rozario et al., 2017; Ghosh & Hui, 2018). Several genes associated with neurodegenerative diseases in humans are similar in zebrafish (Dubois-Dalcq et al., 2008). This makes them a suitable model for studying the genetic factors and molecular mechanisms that contribute to neurodegeneration. The transparency of their embryos allows the development of the nervous system to be visualized and monitored in real time (Ghosh & Hui, 2018). This transparency facilitates the observation of neuronal structure and function, making it easier to study the progression of neurodegenerative disorders. Zebrafish have a remarkable ability to regenerate neurons, and studying the mechanisms behind this regenerative capacity could provide insights into potential strategies to promote neuronal regeneration in humans, a critical aspect in neurodegenerative diseases (Ghosh & Hui, 2018).

Transgenic zebrafish, in which both oligodendrocytes and Schwann cells express a membrane-targeted GFP, making these cells visible, make it possible to study myelination in vivo (Jung et al., 2010; Kirby et al., 2006). Both myelination and remyelination processes can be monitored in vivo by time-lapse confocal microscopy without the need to sacrifice the animal, provided that oligodendrocytes express a fluorescent protein. However, for the analysis of remyelination, a demyelination model must first be established (Kim et al., 2015; Osorio-Querejeta et al., 2017). In order to create a demyelination model, damage to myelin by laser microsurgery or transgenic organisms in which oligodendrocytes are exposed to metronidazole (MTZ) can be produced (Chung et al., 2013; Curado et al., 2007). To investigate remyelination in vivo, the researchers developed a model of demyelination using a transgenic zebrafish model in which nfsB, a bacterial cytotoxin gene encoding the enzyme nitroreductase (NTR), which can convert prodrugs such as MTZ into cytotoxins, is expressed in oligodendrocytes (Chung et al., 2013). In such models, the remyelination process starts immediately after the manipulation or exposure is terminated (Kim et al., 2015). Thanks to these models, it is possible to evaluate candidate agents for the treatment of demyelinating diseases.

## Multiple Sclerosis and Myelination in Zebrafish Model

Myelin sheaths have important roles in the regulation of the speed and amplitude of action potentials in the nervous system, protective effects with axons, neuronal plasticity, and development of the nervous system (Jean Harry & Toews, 1998). Loss or damage to the sheath results in functional impairments as well as a number of serious neurodegenerative diseases (Ettle et al., 2016; Verkhratsky et al., 2019). Demyelinating diseases are typically characterised by lesions associated with the loss of myelin as it protects axons in the nervous system (Jean Harry & Toews, 1998). Demyelination can affect the entire CNS and is involved in the pathophysiology of various neurodegenerative diseases, including demyelinating disorders such as multiple sclerosis. The demyelination associated with this crippling disease involves autoimmune-based oligodendrocyte damage (Compston & Coles, 2002). Multiple sclerosis (MS) is a chronic, multifocal, relapsing-remitting disease characterised by oligodendrocyte loss and axon demyelination (Confavreux & Compston, 2006). The specific mechanisms of demyelination remain unclear but involve autoimmune and inflammatory responses involving T and B lymphocytes (Kim et al., 2015). Various pro- and anti-inflammatory cytokines affect demyelination in MS patients (Kim et al., 2015). The pathological manifestations of MS are focal inflammatory demyelinated lesions in white matter (Zhu et al., 2019). Neurological symptoms in MS are mainly caused by focal inflammatory demyelinated lesions 'mediated mainly by immune cells such as lymphocytes and macrophages, which impair saltatory conduction (Compston & Coles, 2002; Paz Soldan & Rodriguez, 2002). These impair the transmission of signals in the affected nerves. Subsequently, the reduced transmission causes deficits in sensation, movement, cognition, and various neural functions (Compston & Coles, 2002; Scherer & Wrabetz, 2008). While current MS therapies are intended to modulate the inflammatory component, the latest studies have focused on potential therapies that enhance neuroprotection and remyelination.

Demyelination in the early stages of MS is followed by spontaneous regeneration of myelin sheaths (remyelination) (Paz Soldan & Rodriguez, 2002). However, remyelination unfortunately results in failure in most cases (Dubois-Dalcq et al., 2005). Failure of remyelination is associated with axonal degeneration and axonal loss in white matter lesions in patients with progressive forms of MS (Paz Soldan & Rodriguez, 2002). Chronic and persistent neurological impairment has

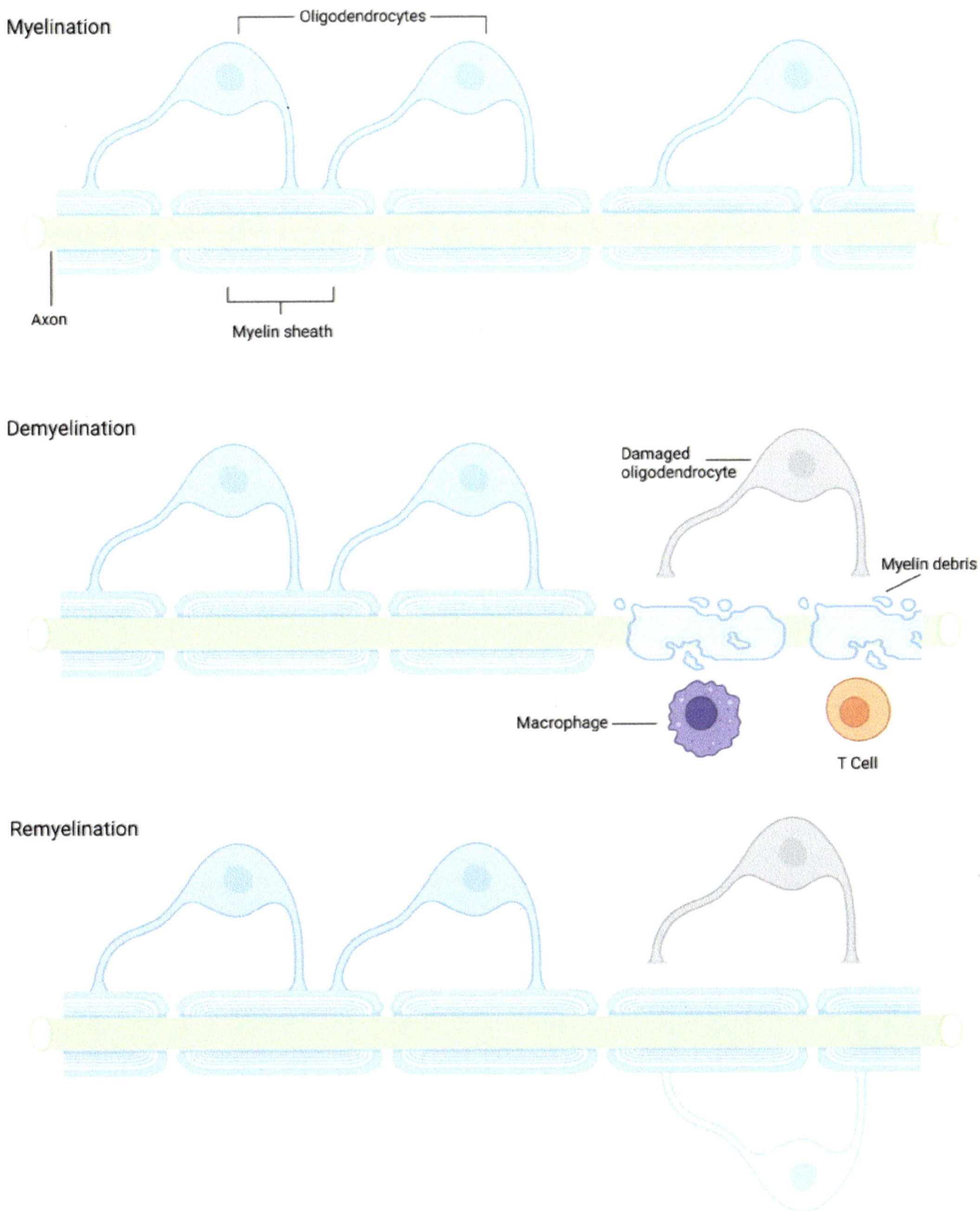


**FIGURE 11.3** Myelination dynamics.

Demyelination is a pathological condition in which the myelin sheath is lost from the periphery of the axon. The CNS has the capacity to repair demyelinated axons and regenerate myelin through a process called remyelination. Following demyelination, which is the result of pathological activation of T cells and macrophages, the myelin sheath is lost, but the underlying axon remains intact. This triggers the naturally occurring regenerative response of remyelination. Oligodendrocyte progenitor cells (OPCs) are responsible for initiating this process. To do this, OPCs migrate towards the lesions, proliferate, and differentiate into oligodendrocytes that expand their membranes to remyelinate axons. Created with Biorender.

been shown to be associated with inflammatory demyelinated axonal loss (Scherer & Wrabetz, 2008). Following demyelination, which in MS is the result of pathological activation of T cells and macrophages, the myelin sheath has disappeared, but the underlying axon remains intact (Dubois-Dalcq et al., 2005). The lack of damage to the axon triggers the naturally occurring regenerative response of remyelination (Karttunen et al., 2017). However, when remyelination is not achieved for certain reasons, such as insufficiency of OPCs or aging, efficient conduction cannot be restored and the supportive role of myelin is lost (Franklin et al., 2002; Goldschmidt et al., 2009). Even if remyelination occurs, abnormally thin and short myelin sheath formation is observed (Goldschmidt et al., 2009). These conditions eventually lead to energy deficiency, reduced axonal metabolic support, impaired axonal transport, and ultimately axonal degeneration (Karttunen et al., 2017).

The most promising way to treat MS is to restore the endogenous remyelination process. Studies focus on the discovery of novel agents to promote remyelination in demyelinating diseases (Karttunen et al., 2017; Kim et al., 2015; Zhu et al., 2019). The researchers investigated the possibility that sulfasalazine, an anti-inflammatory and immunomodulatory drug used in rheumatology and inflammatory bowel disease, may promote remyelination in a transgenic zebrafish model of NTR/MTZ-induced demyelination (Kim et al., 2015). In this reported study, sulfasalazine treatment was shown to promote remyelination and reduce the number of macrophages/microglia in the CNS of demyelinated zebrafish larvae (Kim et al., 2015). These findings suggest that temporal modulation of sulfasalazine can be used to overcome MS by promoting or enhancing remyelination (Kim et al., 2015). In another metronidazole-induced demyelination study, the interleukin-7 receptor (IL-7R), which has been implicated in MS pathogenesis, was examined in a transgenic zebrafish line expressing GFP (Lei et al., 2017). In the aforementioned study, it was shown for the first time by magnetic resonance imaging (MRI) that adult transgenic zebrafish exposed to metronidazole induced demyelination resembling morphological changes associated with early stages of MS (Lei et al., 2017). Furthermore, targeted knockdown of IL-7R showed that IL-7R is required for myelination in embryonic and larval zebrafish. (Lei et al., 2017).

Another research group treated zebrafish embryos with ethidium bromide for 72 hours after fertilization to create a model of demyelination (Zhu et al., 2019). The embryos were then treated with thyroxine (T4), a remyelination agent. RAR-related orphan receptor-γt (RORγt) inhibitors, a potential therapeutic target for chronic autoimmune and inflammatory diseases such as MS, were evaluated for their therapeutic effects on remyelination, axon regeneration, motor neuron promotion, and anti-inflammation (Zhu et al., 2019). T4 was reported to significantly ameliorate ethidium bromide–induced motility dysfunction and myelin damage and to promote myelin basic protein regeneration in demyelinated zebrafish (Zhu et al., 2019).

LINGO-1 (leucine-rich repeat and Ig domain containing 1) has been identified as a negative regulator of oligodendrocyte differentiation (Mi et al., 2005). Its inhibition has been shown to promote neuron survival, axon regeneration, oligodendrocyte differentiation, and remyelination in most mammalian animal models (Lee et al., 2007). One study investigated the inhibition of lingo1b, the homolog of human LINGO-1, in a zebrafish model, and found that lingo1b deficiency enhances myelination and oligodendrocyte differentiation during embryogenesis (Yin & Hu, 2014). In conclusion, lingo1b protein was shown to function as a negative regulator of myelination and oligodendrocyte differentiation during zebrafish development (Yin & Hu, 2014).

In summary, studying myelination in the context of MS using the zebrafish model is of great importance in unraveling the intricacies of this complex neurological disorder. Thanks to the improved zebrafish demyelination models, the development of therapeutic drug research for MS may be accelerated in the future. Moreover, the ability to manipulate genetic and environmental factors in zebrafish provides a powerful platform to investigate the genetic basis of MS and test novel therapeutic interventions. As we attempt to better understand the role of myelination in MS pathology, the zebrafish model is emerging as a valuable ally in the search for more effective therapies and ultimately a cure for this debilitating condition.

## Alzheimer's Disease and Myelination in Zebrafish Model

Alzheimer's disease (AD) is the most common neurodegenerative disease and is characterized by progressive memory loss, abnormalities in daily living, deterioration in cognition, incapacity to be self-sufficient, and late-onset intelligence loss (Van Bebber et al., 2013). Pathologically, AD is caused by the accumulation of extracellular plaques composed of amyloid β peptide (Aβ) and intracellular neurofibrillary tangles composed of hyperphosphorylated tau protein (Haass & Selkoe, 2007). Aβ, which is the major component of plaques in the brains of Alzheimer's patients, is produced from β-amyloid precursor protein by a stepwise proteolytic process (Haass & Selkoe, 2007). Tau is a microtubule-associated protein whose high phosphorylation is a pathological marker in AD (Van Bebber et al., 2013).

The zebrafish model is a suitable organism to study the pathogenesis of AD due to its many advantages in the study of neurodegenerative diseases (Liu, 2023). Mutations in amyloid precursor protein (APP) have been shown to cause AD (Lanoiselée et al., 2017; Zhang et al., 2017). Zebrafish carry two APP genes (appa and appb) that are homologous to human APP. In one study, amyloid precursor protein-b was shown to be involved in early zebrafish development (Banote et al., 2020). The proteases β- and γ-secretases proteolytically produce Aβ from β-amyloid precursor protein (Haass & Selkoe, 2007). β-secretase, also known as Beta-site APP Cleaving Enzyme (BACE)1, cleaves APP and induces Aβ release (Selkoe, 2001; Vassar & Citron, 2000). BACE1, an aspartyl protease, cleaves neuregulin 1 and has been shown to play a role in myelination. Inhibition of BACE1 or β -secretase is an area under investigation for the reduction of amyloid plaques. Previous studies have reported that BACE1 knockout (KO) mice have mild cognitive deficits, hypomyelination, and impaired axon guidance (Hitt et al., 2012; Hu et al., 2006; Ohno et al., 2004). Researchers cloned the BACE1 ortholog (bace1) in zebrafish to investigate its function (Van Bebber et al., 2013). They examined inhibition of Bace1 in a transgenic line expressing membrane-bound GFP under the control of the myelin-specific claudin k promoter (*bace1-/-; claudin k: GFP*). In the referred study, taking advantage of the transparent structure of zebrafish larvae, the *claudin k: GFP* transgene emphasised oligodendrocytes and Schwann cells, and their differentiation was followed in live *claudin k:GFP* transgenic animals from 2 dpf. In conclusion, *bace1-/-; claudin k: GFP* larvae were reported to have a severe reduction in myelination of Schwann cells of the PNS, first observed at 3 dpf. Conversely, oligodendrocytes, the glial cells of the CNS, did not show a defect in myelination in bace1 KO zebrafish larvae. BACE1 is implicated in myelination through degradation of neuregulin-1 (NRG1), a protein with two transmembrane domains. The same research group investigated whether NRG1 type III, which has been identified as a physiological substrate of mouse BACE1 and is required for normal myelination, is also required for myelination in zebrafish. In zebrafish, knock-down of nrg1 type III was shown to result in hypomyelination of the PNS. The researchers also examined if the hypomyelination observed in *bace1-/-* was a secondary effect caused by abnormal Schwann cell development. For this, they analysed Schwann cells using a transgenic line expressing GFP under the control of the *foxd3* promoter (*foxd3:GFP*), which has been established to label glial cells. They detected the presence of Schwann cells but commented that they were not myelinated. In conclusion, it was reported that the hypomyelination defect observed in *bace1-/-* larvae is not due to the absence or mis-migration of PNS glial cells but to a decrease in BACE1-dependent nrg1 type III signaling (Van Bebber et al., 2013).

In conclusion, the investigation of myelination in AD studies using the zebrafish model represents a crucial path forward in neuroscience research. The complex interplay between myelination and neurodegenerative processes sheds light on new therapeutic targets and diagnostic approaches. As the role of myelination in Alzheimer's pathology continues to be investigated, the zebrafish model stands as a valuable tool, offering unique insights that hold promise for future advances in the field of neurodegeneration.

## Zebrafish Models in Parkinson's Disease and Myelination

Parkinson's disease (PD) is a progressive neurodegenerative disorder that occurs mostly in elderly people and results in the loss of dopaminergic neurons in the substantia nigra pars compacta (Bains & Shaw, 1997). This leads to inhibition of dopamine production and accumulation of Lewy bodies (LB), which are aggregates of α-synuclein (Spillantini et al., 1997). These accumulated neuromolecules cause different clinical symptoms like tremor of limbs, radykinesia, rigidity of the muscles, and posture impairment (DeMaagd & Philip, 2015; Venda et al., 2010). Various different symptoms such as depression, hyposmia, fatigue, sleep disorders, and cognitive disorders such as dementia can also be observed in the disease (DeMaagd & Philip, 2015).

Dopamine is synthesized from a tyrosine amino acid with the help of two enzymes (tyrosine hydrolase and amino acid decarboxylase) and is degraded by monoamine oxidase B, catechol-O methyltransferase, and dopamine β-hydroxylase (Venda et al., 2010). If dopamine increases in the neuronal cell, it causes neuronal damage through ROS production (Segura-Aguilar et al., 2014). Also, ROS accumulation in the brain may lead to DNA mutations and afterwards the loss of dopaminergic neurons in the midbrain, called the substantia nigra region (Segura-Aguilar et al., 2014). There are some important studies about increased oxidative stress in the substantia nigra compacta (SNc) in the normal brain, but more dramatic increases in oxidative stress are observed in PD patients (Dias et al., 2013). However, other factors involving inflammation, excitotoxic mechanisms, toxic action of nitric oxide, and mitochondrial dysfunction play roles in the etiology of PD (Andersen, 2005).

Given that ROS can be produced by different biological structures, many ROS functions have been defined mainly according to their reactivity (Villalpando-Rodriguez & Gibson, 2021). Mitochondria and oxidative metabolism are some sources of ROS, so enzymes such as nitric oxide synthase, monoamine oxidase, and xanthine oxidase daily produce huge amounts of the reactive oxygen and nitrosative species superoxide ($O^{\cdot-}$), hydroxyl (OH•), peroxyl (ROO•), and nitric oxide in the brain (Chaudhary et al., 2023). There are some studies about oxidative stress and Parkinson formation. In a postmortem study, a strong relationship between oxidative stress and PD was shown (Valko et al., 2007). The first important change related to oxidative stress in the brain of PD patients is a significant reduction of glutathione in the substantia nigra (Pearce et al., 1997; Sian et al., 1994). Glutathione depletion in dopaminergic cells in culture has been shown to lead to a decrease in mitochondrial complex I activity and mitochondrial function (Hsu et al., 2005; Jha et al., 2000). It has been established that reduced activation of mitochondrial complex I may underlie sporadic Parkinson's disease and that defects in complex I cause α-synuclein aggregation (Moon & Paek, 2015). Complex I inhibition appears to result from nitric oxide production following glutathione depletion in dopaminergic neurons (Hsu et al., 2005). Treatment of glutathione-depleted cultured dopaminergic cells with inhibitors of nitric oxide synthetase (NOS) recovers mitochondrial complex I inhibition (Chinta & Andersen, 2006). It has been shown that iron levels increase in the affected brain regions of Parkinson's disease patients (Chinta & Andersen, 2006; Muñoz et al., 2016).

The most well-known of the PD-related genes that cause Parkinson's disease in humans is the SNCA gene encoding α-synuclein (Spillantini et al., 1997). Synuclein proteins collect in clusters which are called Lewy bodies. α-synuclein mutations are linked to early onset familial PD, and α-synuclein aggregation in Lewy bodies is associated with sporadic cases of PD (Marino et al., 2022; Srinivasan et al., 2021). The physiological function of α-synuclein is still under investigation and is thought to be responsible for regulating the synaptic transmission (Marino et al., 2022).

The neurotoxin 1-methyl-4-phenyl-1,2,3,6-tetrahydropyri dine (MPTP) causes Parkinson's symptoms in animals (Cansız et al., 2022). It is widely used to induce PD as it causes selective degeneration of dopamine neurons in the substantia nigra (Cansız et al., 2022; Ren et al., 2022). In a study conducted with zebrafish, it was reported that the application of MPTP caused locomotor impairment in zebrafish (Bretaud et al., 2004; Zhang et al., 2020). It is characterized by a decrease in the total distance traveled and swimming speed (Cansız et al., 2022; Ren et al., 2022). The loss of neuron cells, or dopamine degeneration, is an important symptom of Parkinson's disease (Samii et al., 2004). For the treatment to relieve these symptoms, anti-PD therapy may act by preventing the progressive destruction of dopamine neurons. YIAEDAER peptides were used to correct this situation (Ren et al., 2022). YIAEDAER (Tyr-Ile-Ala-Glu-Asp-Ala-Glu-Arg) peptide is a multi-purpose peptide isolated from *Neptunea arthritica cumingii* plants (Yu et al., 2020). The peptide was purified from the extract from the inner part of this plant (Ren et al., 2022). This peptide caused an increase in total distance traveled and swimming speed and corrected the locomotor impairment caused by MPTP (Ren et al., 2022). Additionally, this study reported that YIAEDAER peptides would correct MPTP-induced DA destruction, retreat the blood–brain barrier, and ensure the normal functions of cerebral blood flow (Ren et al., 2022).

Aluminum is considered an etiological factor in neurodegenerative diseases, where it has toxic effects on various organs (Inan-Eroglu & Ayaz, 2018). In a study with zebrafish, fish were exposed to Al for 10, 15, and 20 days, and then the neurotoxic effects of aluminum were studied (Capriello et al., 2022). Histological staining showed alterations in the morphology of the brain parenchyma and myelin and activation of neurodegenerative processes (Capriello et al., 2022). The results obtained from this study showed that aluminum significantly affected the histology of brain tissue during the acute periods of exposure to this toxic compound (Inan-Eroglu & Ayaz, 2018). It has also been stated that following this exposure, aluminum subsequently causes changes in the histology of the brain tissue, causing neurodegeneration, and it can cause demyelination (Capriello et al., 2022). There was an improvement/stabilization of overall neurological conditions as confirmed by the lowest level in PARP activity after long exposure. This effect was also supported by gene expression analysis. Aluminum-induced change in the expression of genes closely associated with parkinsonism also indicates increased expression of neuroprotective genes (Capriello et al., 2022). This situation was also followed by the elevation of the expression GFAP.

## Zebrafish Models in Amyotrophic Lateral Sclerosis and Myelination

Amyotrophic lateral sclerosis (ALS), also known as Lou Gehrig's disease, is a progressive neurodegenerative disease affecting neurons in the brain and spinal cord (Brotman et al., 2024). Oligodendrocytes have received little attention in ALS research, although some studies have suggested that oligodendrocytes play a role in the pathogenesis of ALS (Nonneman et al., 2014). Monocarboxylate transporter 1 (MCT1), which is the most prominent MCT, is produced in oligodendrocytes and is important for axonal survival, but its expression is reduced in ALS patients and in the mutant superoxide dismutase 1 ($SOD1^{G93A}$) mouse model of ALS (Lee et al., 2012). While the exact cause of ALS is not fully known, there are several factors that researchers think may contribute to the development of the disease.

In a small percentage of cases, ALS is inherited through genetic mutations. Many of the cases are sporadic (Ghasemi & Brown, 2018). However, ~10 % of ALS patients are familial, and they have mutations in several genes (Kirby et al., 2016). Mutations in genes such as C9orf72, SOD1, and Fused in Sarcoma (FUS) have been associated with familial ALS, where multiple family members are affected by the disease (Bülbül et al., 2018). These genetic mutations can lead to the malfunctioning of motor neurons, which are essential for controlling muscle movement (Wang et al., 2023). While most cases of ALS are sporadic and not directly linked to genetics, environmental factors may play a role in increasing the risk of developing the disease (Oskarsson et al., 2015). Exposure to certain toxins and chemicals, such as heavy metals and pesticides, has been studied as potential risk factors for ALS (Trojsi et al., 2013). However, the exact relationship between environmental exposures and ALS risk is still being investigated (Stipa et al., 2023).

Inflammation in the nervous system is believed to contribute to the progression of ALS (Liu & Wang, 2017). Immune system cells of the CNS, called microglia and astrocytes, become activated and release inflammatory molecules, which can damage motor neurons and accelerate the disease process (You et al., 2023). Understanding how neuroinflammation contributes to ALS could lead to new therapeutic approaches for slowing or halting the progression of the disease (Liu et al., 2021).

Glutamate is a neurotransmitter involved in signal transmission between nerve cells (Zhou & Danbolt, 2014). In ALS, there is evidence of excessive glutamate release and impaired glutamate uptake by surrounding cells (Rosenblum & Trotti, 2017). This leads to an overstimulation of nerve cells, a phenomenon known as excitotoxicity, which can ultimately result in the death of motor neurons (Armada-Moreira et al., 2020). Abnormal accumulation of protein aggregates, such as TDP-43 and FUS, is a hallmark of ALS pathology (Scotter et al., 2015). These misfolded proteins can disrupt cellular functions and lead to the degeneration of motor neurons (Maguire, 2017). Understanding the mechanisms underlying protein aggregation in ALS could provide insights into potential therapeutic targets for the disease (Blokhuis et al., 2013).

A study in the zebrafish animal model showed that myelin degeneration induced by the accumulation of mutant SOD1 (mtSOD1) promotes ALS (Kim et al., 2019). Researchers have previously used the $SOD1^{G93A}$ ALS mouse model, which shows anxiety-like behavioural abnormalities and early learning effects before the onset of motor symptoms (Quarta et al., 2015). In the mentioned study, they used P16-UAS transgenic zebrafish that selectively express mtSOD1 in mature oligodendrocytes (Kim et al., 2019; Yoshida & Macklin, 2005). The toxicity of mtSOD1 is thought to contribute to the degeneration of mature oligodendrocytes by at least two different mechanisms: disruption of the myelin sheath and downregulation of MCT1 (Kim et al., 2019; Lee et al., 2012). They showed that enormous mtSOD1 accumulation inside the myelin sheath caused myelin disruption, resulting in oligodendrocyte degeneration (Kim et al., 2019). They observed transcriptional downregulation of MCT1 in *MBP:mtSOD1zebrafish* (Kim et al., 2019). *MBP:mtSOD1zebrafish* also represented similar patterns of behavioral changes, indicating that myelin defects exist responsible for the psychiatric symptoms seen in patients with ALS and the $SOD1^{G93A}$ mouse model (Lee et al., 2012).

The zebrafish model has emerged as a promising organism for investigating the complex mechanisms underlying ALS, particularly in understanding the role of myelination. As research continues to unravel the mysteries surrounding myelination in ALS using this model system, there is a hopeful expectation that these findings will ultimately translate into effective treatments for this devastating neurodegenerative disease.

## Conclusion

Zebrafish are an important model organism for myelination research due to their ability to regenerate injured axons and rapidly remyelinate them. These properties make them particularly valuable in the study of diseases that develop as a result of injury and cause neurodegeneration. Building upon the zebrafish model's regenerative properties, future research should focus on elucidating the molecular mechanisms that govern axonal injury response and remyelination. Exploring the genetic pathways involved in zebrafish myelination will provide key insights into how these processes might be manipulated to promote repair in humans. Further investigation into the role of glial cells, particularly oligodendrocytes, in zebrafish models will also be critical for understanding their contribution to myelination and regeneration. Additionally, advances in imaging and genetic manipulation tools offer exciting opportunities to track myelination dynamics in real-time, allowing researchers to uncover the spatiotemporal patterns of remyelination. Investigating how environmental factors, pharmacological agents, or stem cell therapies might enhance remyelination could pave the way for novel therapeutic approaches in treating neurodegenerative diseases such as multiple sclerosis and spinal cord injuries.

In conclusion, zebrafish represent a powerful model organism for studying myelination and its regenerative potential, providing valuable insights into the mechanisms of axonal repair and remyelination. Their unique ability to regenerate damaged axons and re-establish myelin makes them an essential tool in the exploration of neurodegenerative diseases and injuries that disrupt myelination. As our understanding of zebrafish neurobiology deepens, it holds great promise for advancing therapeutic strategies aimed at promoting myelination and restoring neuronal function in human neurodegenerative conditions. Future research in this area has the potential to significantly impact the development of novel interventions for diseases like multiple sclerosis and other myelin-related disorders.

## REFERENCES

Andersen, J. K. (2005). Iron dysregulation and Parkinson's disease. *Journal of Alzheimer's Disease*, *6*(s6), S47–S52. https://doi.org/10.3233/JAD-2004-6S602

Armada-Moreira, A., Gomes, J. I., Pina, C. C., Savchak, O. K., Gonçalves-Ribeiro, J., Rei, N., Pinto, S., Morais, T. P., Martins, R. S., Ribeiro, F. F., Sebastião, A. M., Crunelli, V., & Vaz, S. H. (2020). Going the extra (Synaptic) mile: Excitotoxicity as the road toward neurodegenerative diseases. *Frontiers in Cellular Neuroscience*, *14*, 1–27. https://doi.org/10.3389/fncel.2020.00090

Bains, J. S., & Shaw, C. A. (1997). Neurodegenerative disorders in humans: The role of glutathione in oxidative stress-mediated neuronal death. *Brain research. Brain Research Reviews*, *25*(3), 335–358. https://doi.org/10.1016/s0165-0173(97)00045-3

Banote, R. K., Chebli, J., Şatır, T. M., Varshney, G. K., Camacho, R., Ledin, J., Burgess, S. M., Abramsson, A., & Zetterberg, H. (2020). Amyloid precursor protein-b facilitates cell adhesion during early development in zebrafish. *Scientific Reports*, *10*(1), 10127. https://doi.org/10.1038/s41598-020-66584-8

Baumann, N., & Pham-Dinh, D. (2001). *Biology of Oligodendrocyte and Myelin in the Mammalian Central Nervous System*. http://physrev.physiology.org

Blokhuis, A. M., Groen, E. J. N., Koppers, M., van den Berg, L. H., & Pasterkamp, R. J. (2013). Protein aggregation in amyotrophic lateral sclerosis. *Acta Neuropathologica*, *125*(6), 777–794. https://doi.org/10.1007/s00401-013-1125-6

Boullerne, A. I. (2016). The history of myelin. *Experimental Neurology*, *283*(Pt B), 431–445. https://doi.org/10.1016/j.expneurol.2016.06.005

Bretaud, S., Lee, S., & Guo, S. (2004). Sensitivity of zebrafish to environmental toxins implicated in Parkinson's disease. *Neurotoxicology and Teratology*, *26*(6), 857–864. https://doi.org/10.1016/j.ntt.2004.06.014

Brotman, R. G., Moreno-Escobar, M. C., Joseph, J., & Pawar, G. (2024). Amyotrophic Lateral Sclerosis. In: *StatPearls*. Treasure Island, FL: StatPearls Publishing. https://pubmed.ncbi.nlm.nih.gov/32310611/

Bülbül, N. G., Seçil, Y., Başak, N., Beckmann, Y., Türe, H. S., Tunca, C., & Özoğuz, A. (2018). Two families with SOD1 (L144F) and C9orf72 gene mutations and an overview of amyotrophic lateral sclerosis. *Turkish Journal of Neurology*, *24*(2), 159–164. https://doi.org/10.4274/tnd.26214

Cansız, D., Ustundag, U. V., Unal, I., Alturfan, A. A., & Emekli-Alturfan, E. (2022). Morphine attenuates neurotoxic effects of MPTP in zebrafish embryos by regulating oxidant/antioxidant balance and acetylcholinesterase activity. *Drug and Chemical Toxicology*, *45*(6), 2439–2447. https://doi.org/10.1080/01480545.2021.1957558

Capriello, T., Di Meglio, G., De Maio, A., Scudiero, R., Bianchi, A. R., Trifuoggi, M., Toscanesi, M., Giarra, A., & Ferrandino, I. (2022). Aluminium exposure leads to neurodegeneration and alters the expression of marker genes involved to parkinsonism in zebrafish brain. *Chemosphere*, *307*(Pt 1), 135752. https://doi.org/10.1016/j.chemosphere.2022.135752

Chaudhary, P., Janmeda, P., Docea, A. O., Yeskaliyeva, B., Abdull Razis, A. F., Modu, B., Calina, D., & Sharifi-Rad, J. (2023). Oxidative stress, free radicals and antioxidants: potential crosstalk in the pathophysiology of human diseases. *Frontiers in Chemistry*, *11*, 1158198. https://doi.org/10.3389/fchem.2023.1158198

Chinta, S. J., & Andersen, J. K. (2006). Reversible inhibition of mitochondrial complex I activity following chronic dopaminergic glutathione depletion in vitro: Implications for Parkinson's disease. *Free Radical Biology and Medicine*, *41*(9), 1442–1448. https://doi.org/10.1016/j.freeradbiomed.2006.08.002

Chung, A.-Y., Kim, S., Kim, E., Kim, D., Jeong, I., Cha, Y. R., Bae, Y., Park, S. W., Lee, J., & Park, H.-C. (2013). Indian hedgehog b function is required for the specification of oligodendrocyte progenitor cells in the zebrafish CNS. *The Journal of Neuroscience*, *33*(4), 1728–1733. https://doi.org/10.1523/JNEUROSCI.3369-12.2013

Compston, A., & Coles, A. (2002). Multiple sclerosis. *The Lancet*, *359*(9313), 1221–1231. https://doi.org/10.1016/S0140-6736(02)08220-X

Confavreux, C., & Compston, A. (2006). The natural history of multiple sclerosis. *McAlpine's Multiple Sclerosis*, 183–272. https://doi.org/10.1016/B978-0-443-07271-0.50006-9

Curado, S., Anderson, R. M., Jungblut, B., Mumm, J., Schroeter, E., & Stainier, D. Y. R. (2007). Conditional targeted cell ablation in zebrafish: A new tool for regeneration studies. *Developmental Dynamics*, *236*(4), 1025–1035. https://doi.org/10.1002/dvdy.21100

DeMaagd, G., & Philip, A. (2015). Parkinson's disease and its management: Part 1: Disease entity, risk factors, pathophysiology, clinical presentation, and diagnosis. *P & T: A Peer-Reviewed Journal for Formulary Management*, *40*(8), 504–532.

Dias, V., Junn, E., & Mouradian, M. M. (2013). The role of oxidative stress in Parkinson's disease. *Journal of Parkinson's Disease*, *3*(4), 461–491. https://doi.org/10.3233/JPD-130230

D'Rozario, M., Monk, K. R., & Petersen, S. C. (2017). Analysis of myelinated axon formation in zebrafish. *Methods in Cell Biology*, *138*, 383–414. https://doi.org/10.1016/bs.mcb.2016.08.001

Dubois-Dalcq, M., ffrench-Constant, C., & Franklin, R. J. M. (2005). Enhancing central nervous system remyelination in multiple sclerosis. *Neuron*, *48*(1), 9–12. https://doi.org/10.1016/j.neuron.2005.09.004

Dubois-Dalcq, M., Williams, A., Stadelmann, C., Stankoff, B., Zalc, B., & Lubetzki, C. (2008). From fish to man: understanding endogenous remyelination in central nervous system demyelinating diseases. *Brain*, *131*(7), 1686–1700. https://doi.org/10.1093/brain/awn076

Ettle, B., Schlachetzki, J. C. M., & Winkler, J. (2016). Oligodendroglia and myelin in neurodegenerative diseases: More than just bystanders? *Molecular Neurobiology*, *53*(5), 3046–3062. https://doi.org/10.1007/s12035-015-9205-3

Ferrer, I. (2018). Oligodendrogliopathy in neurodegenerative diseases with abnormal protein aggregates: The forgotten partner. *Progress in Neurobiology*, *169*, 24–54. https://doi.org/10.1016/j.pneurobio.2018.07.004

Franklin, R. J. M., Zhao, C., & Sim, F. J. (2002). Ageing and CNS remyelination. *Neuroreport*, *13*(7), 923–928. https://doi.org/10.1097/00001756-200205240-00001

Ghasemi, M., & Brown, R. H. (2018). Genetics of amyotrophic lateral sclerosis. *Cold Spring Harbor Perspectives in Medicine*, *8*(5), a024125. https://doi.org/10.1101/cshperspect.a024125

Ghosh, S., & Hui, S. P. (2018). Axonal regeneration in zebrafish spinal cord. *Regeneration*, *5*(1), 43–60. https://doi.org/10.1002/reg2.99

Goldschmidt, T., Antel, J., König, F. B., Brück, W., & Kuhlmann, T. (2009). Remyelination capacity of the MS brain decreases with disease chronicity. *Neurology*, *72*(22), 1914–1921. https://doi.org/10.1212/WNL.0b013e3181a8260a

Haass, C., & Selkoe, D. J. (2007). Soluble protein oligomers in neurodegeneration: Lessons from the Alzheimer's amyloid β-peptide. *Nature Reviews Molecular Cell Biology*, *8*(2), 101–112. https://doi.org/10.1038/nrm2101

Hitt, B., Riordan, S. M., Kukreja, L., Eimer, W. A., Rajapaksha, T. W., & Vassar, R. (2012). β-Site Amyloid Precursor Protein (APP)-cleaving Enzyme 1 (BACE1)-deficient mice exhibit a Close Homolog of L1 (CHL1) loss-of-function phenotype involving axon guidance defects. *Journal of Biological Chemistry*, *287*(46), 38408–38425. https://doi.org/10.1074/jbc.M112.415505

Hsu, M., Srinivas, B., Kumar, J., Subramanian, R., & Andersen, J. (2005). Glutathione depletion resulting in selective mitochondrial complex I inhibition in dopaminergic cells is via an NO-mediated pathway not involving peroxynitrite: Implications for Parkinson's disease. *Journal of Neurochemistry*, *92*(5), 1091–1103. https://doi.org/10.1111/j.1471-4159.2004.02929.x

Hu, X., Hicks, C. W., He, W., Wong, P., Macklin, W. B., Trapp, B. D., & Yan, R. (2006). Bace1 modulates myelination in the central and peripheral nervous system. *Nature Neuroscience*, *9*(12), 1520–1525. https://doi.org/10.1038/nn1797

Inan-Eroglu, E., & Ayaz, A. (2018). Is aluminum exposure a risk factor for neurological disorders? *Journal of Research in Medical Sciences: The Official Journal of Isfahan University of Medical Sciences*, *23*, 51. https://doi.org/10.4103/jrms.JRMS_921_17

Jean Harry, G., & Toews, A. D. (1998). Myelination, Dysmyelination, and Demyelination. In *Handbook of Developmental Neurotoxicology* (pp. 87–115). https://doi.org/10.1016/B978-012648860-9.50007-8

Jha, N., Jurma, O., Lalli, G., Liu, Y., Pettus, E. H., Greenamyre, J. T., Liu, R. M., Forman, H. J., & Andersen, J. K. (2000). Glutathione depletion in PC12 results in selective inhibition of mitochondrial complex I activity. Implications for Parkinson's disease. *The Journal of Biological Chemistry*, *275*(34), 26096–26101. https://doi.org/10.1074/jbc.M000120200

Jung, S., Kim, S., Chung, A., Kim, H., So, J., Ryu, J., Park, H., & Kim, C. (2010). Visualization of myelination in GFP-transgenic zebrafish. *Developmental Dynamics*, *239*(2), 592–597. https://doi.org/10.1002/dvdy.22166

Kandel, E. R., S. J. H., & J. T. M. (2000). *Principles of Neural Science* (4th ed.). McGraw-Hill, Health Professions Division.

Karttunen, M. J., Czopka, T., Goedhart, M., Early, J. J., Lyons, D. A., & De Castro, F. (2017). Regeneration of myelin sheaths of normal length and thickness in the zebrafish CNS correlates with growth of axons in caliber. *PLoS One*, *12*(5), e0178058. https://doi.org/10.1371/journal.pone.0178058

Kim, S., Chung, A., Na, J. E., Lee, S. J., Jeong, S. H., Kim, E., Sun, W., Rhyu, I. J., & Park, H. (2019). Myelin degeneration induced by mutant superoxide dismutase 1 accumulation promotes amyotrophic lateral sclerosis. *Glia*, *67*(10), 1910–1921. https://doi.org/10.1002/glia.23669

Kim, S., Lee, Y. Il, Chang, K. Y., Lee, D. W., Cho, S. C., Ha, Y. W., Na, J. E., Rhyu, I. J., Park, S. C., & Park, H. C. (2015). Promotion of remyelination by sulfasalazine in a transgenic zebrafish model of demyelination. *Molecules and Cells*, *38*(11), 1013–1021. https://doi.org/10.14348/molcells.2015.0246

Kirby, B. B., Takada, N., Latimer, A. J., Shin, J., Carney, T. J., Kelsh, R. N., & Appel, B. (2006). In vivo time-lapse imaging shows dynamic oligodendrocyte progenitor behavior during zebrafish development. *Nature Neuroscience*, *9*(12), 1506–1511. https://doi.org/10.1038/nn1803

Kirby, J., Al Sultan, A., Waller, R., & Heath, P. (2016). The genetics of amyotrophic lateral sclerosis: Current insights. *Degenerative Neurological and Neuromuscular Disease*, *6*, 49–64. https://doi.org/10.2147/DNND.S84956

Lanoiselée, H.-M., Nicolas, G., Wallon, D., Rovelet-Lecrux, A., Lacour, M., Rousseau, S., Richard, A.-C., Pasquier, F., Rollin-Sillaire, A., Martinaud, O., Quillard-Muraine, M., de la Sayette, V., Boutoleau-Bretonniere, C., Etcharry-Bouyx, F., Chauviré, V., Sarazin, M., le Ber, I., Epelbaum, S., Jonveaux, T., … Campion, D. (2017). APP, PSEN1, and PSEN2 mutations in early-onset Alzheimer disease: A genetic screening study of familial and sporadic cases. *PLoS Medicine*, *14*(3), e1002270. https://doi.org/10.1371/journal.pmed.1002270

Lee, X., Yang, Z., Shao, Z., Rosenberg, S. S., Levesque, M., Pepinsky, R. B., Qiu, M., Miller, R. H., Chan, J. R., & Mi, S. (2007). NGF regulates the expression of axonal LINGO-1 to inhibit oligodendrocyte differentiation and myelination. *The Journal of Neuroscience*, *27*(1), 220–225. https://doi.org/10.1523/JNEUROSCI.4175-06.2007

Lee, Y., Morrison, B. M., Li, Y., Lengacher, S., Farah, M. H., Hoffman, P. N., Liu, Y., Tsingalia, A., Jin, L., Zhang, P.-W., Pellerin, L., Magistretti, P. J., & Rothstein, J. D. (2012). Oligodendroglia metabolically support axons and contribute to neurodegeneration. *Nature*, *487*(7408), 443–448. https://doi.org/10.1038/nature11314

Lei, X., Cai, S., Chen, Y., Cui, J., Wang, Y., Li, Z., & Li, Y. (2017). Down-regulation of interleukin 7 receptor (IL-7R) contributes to central nervous system demyelination. *Oncotarget*, *8*(17), 28395–28407. www.impactjournals.com/oncotarget/

Liu, E., Karpf, L., & Bohl, D. (2021). Neuroinflammation in amyotrophic lateral sclerosis and frontotemporal dementia and the interest of induced pluripotent stem cells to study immune cells interactions with neurons. *Frontiers in Molecular Neuroscience*, *14*. https://doi.org/10.3389/fnmol.2021.767041

Liu, J., & Wang, F. (2017). Role of neuroinflammation in amyotrophic lateral sclerosis: Cellular mechanisms and therapeutic implications. *Frontiers in Immunology*, *8*. https://doi.org/10.3389/fimmu.2017.01005

Liu, Y. (2023). Zebrafish as a model organism for studying pathologic mechanisms of neurodegenerative diseases and other neural disorders. *Cellular and Molecular Neurobiology*, *43*(6), 2603–2620. https://doi.org/10.1007/s10571-023-01340-w

Maguire, G. (2017). Amyotrophic lateral sclerosis as a protein level, non-genomic disease: Therapy with S2RM exosome released molecules. *World Journal of Stem Cells*, *9*(11), 187–202. https://doi.org/10.4252/wjsc.v9.i11.187

Marino, G., Calabresi, P., & Ghiglieri, V. (2022). Alpha-synuclein and cortico-striatal plasticity in animal models of Parkinson disease. *Handbook of Clinical Neurology*, *184*, 153–166. https://doi.org/10.1016/B978-0-12-819410-2.00008-4

Masson, M. A., & Nait-Oumesmar, B. (2023). Emerging concepts in oligodendrocyte and myelin formation, inputs from the zebrafish model. *GLIA*, *71*(5), 1147–1163. https://doi.org/10.1002/glia.24336

Mi, S., Miller, R. H., Lee, X., Scott, M. L., Shulag-Morskaya, S., Shao, Z., Chang, J., Thill, G., Levesque, M., Zhang, M., Hession, C., Sah, D., Trapp, B., He, Z., Jung, V., McCoy, J. M., & Pepinsky, R. B. (2005). LINGO-1 negatively regulates myelination by oligodendrocytes. *Nature Neuroscience*, *8*(6), 745–751. https://doi.org/10.1038/nn1460

Miller, R. G., & Pinto Da Silva, P. (1977). Particle rosettes in the periaxonal Schwann cell membrane and particle clusters in the axolemma of rat sciatic nerve. *Brain Research*, *130*(1), 135–141. https://doi.org/10.1016/0006-8993(77)90848-4

Monk, K. R., & Talbot, W. S. (2009). Genetic dissection of myelinated axons in zebrafish. *Current Opinion in Neurobiology*, *19*(5), 486–490. https://doi.org/10.1016/j.conb.2009.08.006

Moon, H. E., & Paek, S. H. (2015). Mitochondrial dysfunction in Parkinson's disease. *Experimental Neurobiology*, *24*(2), 103–116. https://doi.org/10.5607/en.2015.24.2.103

Morita, K., Sasaki, H., Fujimoto, K., Furuse, M., & Tsukita, S. (1999). Claudin-11/OSP-based tight junctions of myelin sheaths in brain and sertoli cells in Testis. *The Journal of Cell Biology*, *145*(3), 579–588. https://doi.org/10.1083/jcb.145.3.579

Müller, C., Bauer, N. M., Schäfer, I., & White, R. (2013). Making myelin basic protein -from mRNA transport to localized translation. *Frontiers in Cellular Neuroscience*, *7*. https://doi.org/10.3389/fncel.2013.00169

Muñoz, Y., Carrasco, C. M., Campos, J. D., Aguirre, P., & Núñez, M. T. (2016). Parkinson's disease: The mitochondria-iron link. *Parkinson's Disease*, *2016*, 1–21. https://doi.org/10.1155/2016/7049108

Nave, K.-A. (2010). Myelination and support of axonal integrity by glia. *Nature*, *468*(7321), 244–252. https://doi.org/10.1038/nature09614

Nave, K.-A., & Trapp, B. D. (2008). Axon-glial signaling and the glial support of axon function. *Annual Review of Neuroscience*, *31*(1), 535–561. https://doi.org/10.1146/annurev.neuro.30.051606.094309

Nonneman, A., Robberecht, W., & Den Bosch, L. Van. (2014). The role of oligodendroglial dysfunction in amyotrophic lateral sclerosis. *Neurodegenerative Disease Management*, *4*(3), 223–239. https://doi.org/10.2217/nmt.14.21

Ohno, M., Sametsky, E. A., Younkin, L. H., Oakley, H., Younkin, S. G., Citron, M., Vassar, R., & Disterhoft, J. F. (2004). BACE1 deficiency rescues memory deficits and cholinergic dysfunction in a mouse model of Alzheimer's disease. *Neuron*, *41*(1), 27–33. https://doi.org/10.1016/S0896-6273(03)00810-9

Ohno, N., & Ikenaka, K. (2019). Axonal and neuronal degeneration in myelin diseases. *Neuroscience Research*, *139*, 48–57. https://doi.org/10.1016/j.neures.2018.08.013

Oskarsson, B., Horton, D. K., & Mitsumoto, H. (2015). Potential environmental factors in amyotrophic lateral sclerosis. *Neurologic Clinics*, *33*(4), 877–888. https://doi.org/10.1016/j.ncl.2015.07.009

Osorio-Querejeta, I., Sáenz-Cuesta, M., Muñoz-Culla, M., & Otaegui, D. (2017). Models for studying myelination, demyelination and remyelination. *NeuroMolecular Medicine* *19*(2–3), 181–192). https://doi.org/10.1007/s12017-017-8442-1

Paz Soldan, M. M., & Rodriguez, M. (2002). Heterogeneity of pathogenesis in multiple sclerosis: Implications for promotion of remyelination. *The Journal of Infectious Diseases*, *186*(s2), S248–S253. https://doi.org/10.1086/344283

Pearce, R. K., Owen, A., Daniel, S., Jenner, P., & Marsden, C. D. (1997). Alterations in the distribution of glutathione in the substantia nigra in Parkinson's disease. *Journal of Neural Transmission*, *104*(6–7), 661–677. https://doi.org/10.1007/BF01291884

Pérez-Cerdá, F., Sánchez-Gómez, M. V., & Matute, C. (2015). Pío del Río Hortega and the discovery of the oligodendrocytes. *Frontiers in Neuroanatomy*, *9*, 92. https://doi.org/10.3389/fnana.2015.00092

Peters, A. (1961). A radial component of central myelin sheaths. *The Journal of Cell Biology*, *11*(3), 733–735. https://doi.org/10.1083/jcb.11.3.733

Poitelon, Y., Kopec, A. M., & Belin, S. (2020). Myelin fat facts: An overview of lipids and fatty acid metabolism. *Cells*, *9*(4). https://doi.org/10.3390/cells9040812

Preston, M. A., & Macklin, W. B. (2015). Zebrafish as a model to investigate CNS myelination. John Wiley and Sons Inc. *GLIA*, *63*(2), 177–193. https://doi.org/10.1002/glia.22755

Quarta, E., Bravi, R., Scambi, I., Mariotti, R., & Minciacchi, D. (2015). Increased anxiety-like behavior and selective learning impairments are concomitant to loss of hippocampal interneurons in the presymptomatic SOD1(G93A) ALS mouse model. *Journal of Comparative Neurology*, *523*(11), 1622–1638. https://doi.org/10.1002/cne.23759

Raphael, A. R., & Talbot, W. S. (2011). New insights into signaling during myelination in zebrafish. Academic Press Inc. *Current Topics in Developmental Biology*, *97*, 1–19. https://doi.org/10.1016/B978-0-12-385975-4.00007-3

Rash, J. E., Duffy, H. S., Dudek, F. E., Bilhartz, B. L., Whalen, L. R., & Yasumura, T. (1997). Grid-mapped freeze-fracture analysis of gap junctions in gray and white matter of adult rat central nervous system, with evidence for a "panglial syncytium" that is not coupled to neurons. *The Journal of Comparative Neurology*, *388*(2), 265–292. https://doi.org/10.1002/(sici)1096-9861(19971117)388:2<265::aid-cne6>3.0.co;2-#

Ren, Q., Jiang, X., Zhang, S., Gao, X., Paudel, Y. N., Zhang, P., Wang, R., Liu, K., & Jin, M. (2022). Neuroprotective effect of YIAEDAER peptide against Parkinson's disease like pathology in zebrafish. *Biomedicine & Pharmacotherapy = Biomedecine & Pharmacotherapie*, *147*, 112629. https://doi.org/10.1016/j.biopha.2022.112629

Rosenblum, L. T., & Trotti, D. (2017). *EAAT2 and the Molecular Signature of Amyotrophic Lateral Sclerosis* (pp. 117–136). https://doi.org/10.1007/978-3-319-55769-4_6

Saab, A. S., & Nave, K. A. (2017). Myelin dynamics: Protecting and shaping neuronal functions. *Current Opinion in Neurobiology*, *47*, 104–112. https://doi.org/10.1016/j.conb.2017.09.013

Salzer, J. L., & Zalc, B. (2016). Myelination. *Current Biology*, *26*(20), R971–R975. https://doi.org/10.1016/j.cub.2016.07.074

Samii, A., Nutt, J. G., & Ransom, B. R. (2004). Parkinson's disease. *The Lancet*, *363*(9423), 1783–1793. https://doi.org/10.1016/S0140-6736(04)16305-8

Scherer, S. S., & Wrabetz, L. (2008). Molecular mechanisms of inherited demyelinating neuropathies. *Glia*, *56*(14), 1578–1589. https://doi.org/10.1002/glia.20751

Schmitt, S., Cantuti Castelvetri, L., & Simons, M. (2015). Metabolism and functions of lipids in myelin. *Biochimica et Biophysica Acta (BBA) - Molecular and Cell Biology of Lipids*, *1851*(8), 999–1005. https://doi.org/10.1016/j.bbalip.2014.12.016

Scotter, E. L., Chen, H.-J., & Shaw, C. E. (2015). TDP-43 proteinopathy and ALS: Insights into disease mechanisms and therapeutic targets. *Neurotherapeutics*, *12*(2), 352–363. https://doi.org/10.1007/s13311-015-0338-x

Segura-Aguilar, J., Paris, I., Muñoz, P., Ferrari, E., Zecca, L., & Zucca, F. A. (2014). Protective and toxic roles of dopamine in Parkinson's disease. *Journal of Neurochemistry*, *129*(6), 898–915. https://doi.org/10.1111/jnc.12686

Selkoe, D. J. (2001). Alzheimer's disease: Genes, proteins, and therapy. *Physiological Reviews*, *81*(2), 741–766. https://doi.org/10.1152/physrev.2001.81.2.741

Sian, J., Dexter, D. T., Lees, A. J., Daniel, S., Agid, Y., Javoy-Agid, F., Jenner, P., & Marsden, C. D. (1994). Alterations in glutathione levels in Parkinson's disease and other neurodegenerative disorders affecting basal ganglia. *Annals of Neurology*, *36*(3), 348–355. https://doi.org/10.1002/ana.410360305

Simons, M., & Nave, K. A. (2016). Oligodendrocytes: Myelination and axonal support. *Cold Spring Harbor Perspectives in Biology*, *8*(1). https://doi.org/10.1101/cshperspect.a020479

Snaidero, N., Möbius, W., Czopka, T., Hekking, L. H. P., Mathisen, C., Verkleij, D., Goebbels, S., Edgar, J., Merkler, D., Lyons, D. A., Nave, K.-A., & Simons, M. (2014). Myelin membrane wrapping of CNS axons by PI(3,4,5)P3-Dependent polarized growth at the inner tongue. *Cell*, *156*(1–2), 277–290. https://doi.org/10.1016/j.cell.2013.11.044

Spillantini, M. G., Schmidt, M. L., Lee, V. M.-Y., Trojanowski, J. Q., Jakes, R., & Goedert, M. (1997). α-synuclein in Lewy bodies. *Nature*, *388*(6645), 839–840. https://doi.org/10.1038/42166

Srinivasan, E., Chandrasekhar, G., Chandrasekar, P., Anbarasu, K., Vickram, A. S., Karunakaran, R., Rajasekaran, R., & Srikumar, P. S. (2021). Alpha-synuclein aggregation in Parkinson's disease. *Frontiers in Medicine*, *8*, 736978. https://doi.org/10.3389/fmed.2021.736978

Stadelmann, C., Timmler, S., Barrantes-Freer, A., & Simons, M. (2019). Myelin in the central nervous system: Structure, function, and pathology. *Physiological Reviews*, *99*, 1381–1431. https://doi.org/10.1152/physrev.00031.2018.-Oligodendro

Stipa, G., Ancidoni, A., Vanacore, N., & Bellomo, G. (2023). Raw water and ALS: A unifying hypothesis for the environmental agents involved in ALS. *Annals of Neurosciences*, *30*(2), 124–132. https://doi.org/10.1177/09727531221120358

Toyama, B. H., Savas, J. N., Park, S. K., Harris, M. S., Ingolia, N. T., Yates, J. R., & Hetzer, M. W. (2013). Identification of long-lived proteins reveals exceptional stability of essential cellular structures. *Cell*, *154*(5), 971–982. https://doi.org/10.1016/j.cell.2013.07.037

Trojsi, F., Monsurrò, M., & Tedeschi, G. (2013). Exposure to environmental toxicants and pathogenesis of amyotrophic lateral sclerosis: State of the art and research perspectives. *International Journal of Molecular Sciences*, *14*(8), 15286–15311. https://doi.org/10.3390/ijms140815286

Valko, M., Leibfritz, D., Moncol, J., Cronin, M. T. D., Mazur, M., & Telser, J. (2007). Free radicals and antioxidants in normal physiological functions and human disease. *The International Journal of Biochemistry & Cell Biology*, *39*(1), 44–84. https://doi.org/10.1016/j.biocel.2006.07.001

Van Bebber, F., Hruscha, A., Willem, M., Schmid, B., & Haass, C. (2013). Loss of Bace2 in zebrafish affects melanocyte migration and is distinct from Bace1 knock out phenotypes. *Journal of Neurochemistry*, *127*(4), 471–481. https://doi.org/10.1111/jnc.12198

Vassar, R., & Citron, M. (2000). Aβ-generating Enzymes. *Neuron*, *27*(3), 419–422. https://doi.org/10.1016/S0896-6273(00)00051-9

Venda, L. L., Cragg, S. J., Buchman, V. L., & Wade-Martins, R. (2010). α-synuclein and dopamine at the crossroads of Parkinson's disease. *Trends in Neurosciences*, *33*(12), 559–568. https://doi.org/10.1016/j.tins.2010.09.004

Verkhratsky, A., Ho, M. S., Zorec, R., & Parpura, V. (Eds.). (2019). *Neuroglia in Neurodegenerative Diseases, 1175*. https://doi.org/10.1007/978-981-13-9913-8

Villalpando-Rodriguez, G. E., & Gibson, S. B. (2021). Reactive Oxygen Species (ROS) regulates different types of cell death by acting as a rheostat. *Oxidative Medicine and Cellular Longevity, 2021*, 9912436. https://doi.org/10.1155/2021/9912436

Virchow, R. (1860). *Cellular Pathology as Based Upon Physiological and Pathological Histology, 47*(1), 23–25. https://doi.org/10.1111/j.1753-4887.1989.tb02747.x

Wang, H., Guan, L., & Deng, M. (2023). Recent progress of the genetics of amyotrophic lateral sclerosis and challenges of gene therapy. *Frontiers in Neuroscience, 17*. https://doi.org/10.3389/fnins.2023.1170996

Yin, W., & Hu, B. (2014). Knockdown of Lingo1b protein promotes myelination and oligodendrocyte differentiation in zebrafish. *Experimental Neurology, 251*, 72–83. https://doi.org/10.1016/j.expneurol.2013.11.012

Yoshida, M., & Macklin, W. B. (2005). Oligodendrocyte development and myelination in GFP-transgenic zebrafish. *Journal of Neuroscience Research, 81*(1), 1–8. https://doi.org/10.1002/jnr.20516

You, J., Youssef, M. M. M., Santos, J. R., Lee, J., & Park, J. (2023). Microglia and astrocytes in amyotrophic lateral sclerosis: Disease-associated states, pathological roles, and therapeutic potential. *Biology, 12*(10), 1307. https://doi.org/10.3390/biology12101307

Yu, Z., Hu, Z., Song, H., Xu, T., Yang, M., Zhou, C., & Zhang, T. (2020). Aggregation behavior of juvenile Neptunea cumingii and effects on seed production. *Journal of Oceanology and Limnology, 38*(5), 1590–1598. https://doi.org/10.1007/s00343-020-0042-5

Zalc, B. (2006). The acquisition of myelin: A success story. *Novartis Foundation Symposium, 276*, 15–21; discussion 21-5, 54–57, 275–281. https://doi.org/10.1002/9780470032244.ch3

Zalc, B., Goujet, D., & Colman, D. (2008). The origin of the myelination program in vertebrates. *Current Biology, 18*(12), R511–R512. https://doi.org/10.1016/j.cub.2008.04.010

Zhang, S., Wang, Z., Cai, F., Zhang, M., Wu, Y., Zhang, J., & Song, W. (2017). BACE1 cleavage site selection critical for amyloidogenesis and Alzheimer's pathogenesis. *The Journal of Neuroscience, 37*(29), 6915–6925. https://doi.org/10.1523/JNEUROSCI.0340-17.2017

Zhang, S., Yu, Z., Xia, J., Zhang, X., Liu, K., Sik, A., & Jin, M. (2020). Anti-Parkinson's disease activity of phenolic acids from Eucommia ulmoides Oliver leaf extracts and their autophagy activation mechanism. *Food & Function, 11*(2), 1425–1440. https://doi.org/10.1039/c9fo02288k

Zhou, Y., & Danbolt, N. C. (2014). Glutamate as a neurotransmitter in the healthy brain. *Journal of Neural Transmission, 121*(8), 799–817. https://doi.org/10.1007/s00702-014-1180-8

Zhu, X. Y., Guo, S. Y., Xia, B., Li, C. Q., Wang, L., & Wang, Y. H. (2019). Development of zebrafish demyelination model for evaluation of remyelination compounds and RORγt inhibitors. *Journal of Pharmacological and Toxicological Methods, 98*, 106585. https://doi.org/10.1016/j.vascn.2019.106585

# 12

## *Zebrafish as a Model Organism to Evaluate the Neurotoxic Effects of Endocrine Disruptive Chemicals*

**Merih Beler and Ebru Emekli-Alturfan**

### Abbreviations

| | |
|---|---|
| **5-HT** | Serotonin |
| **Aβ** | β-Amyloid |
| **AChE** | Acetylcholinesterase |
| **AD** | Alzheimer's Disease |
| **AhR** | Aryl Hydrocarbon Receptor |
| **ALS** | Amyotrophic Lateral Sclerosis |
| **ASD** | Autism Spectrum Disorder |
| **BCRP** | Breast Cancer Resistance Protein |
| **BP- Bisphenol** | (BPS, BPF, etc.) |
| **BPA** | Bisphenol A |
| **CNS** | Central Nervous System |
| **DMP** | Dimethyl Phthalate |
| **EAE** | Experimental Autoimmune Encephalomyelitis |
| **EDC** | Endocrine-Disrupting Chemicals |
| **ER** | Estrogen Receptor |
| **Htt** | Huntingtin |
| **MeP** | Methylparaben |
| **NMDA** | N-Methyl-D-Aspartate |
| **NO** | Nitric Oxide |
| **PCBs** | Polychlorinated Biphenyls |
| **PD** | Parkinson's Disease |
| **RNS** | Reactive Nitrogen Species |
| **ROS** | Reactive Oxygen Species |
| **SNCA** | α-Synuclein |
| **TBBPA** | Tetrabromobisphenol A |
| **UPR** | Unfolded Protein Response |

### Introduction

Chemicals are an indispensable part of our everyday existence. However, certain substances, referred to as endocrine disruptors, can negatively impact the body's endocrine system, which is responsible for regulating hormones (Monneret, 2017). Endocrine-disrupting chemicals (EDCs) are compounds that can be found in the environment, as well as in food, cosmetics, manufactured goods, and other preserved items. Exogenous substances, which are known to disrupt the regular functioning of the body's neuroendocrine system, are being explored more and more. Certain substances imitate the actions of hormones, while others interfere with the normal levels of hormones in the body, thereby disturbing the regulation of bodily activities (Thakkar et al., 2022b). Endocrine disruptors refer to both naturally occurring and artificially produced compounds that interfere with the regular functioning of the body by imitating or obstructing the actions of hormones (Beler et al., 2022). Being exposed to endocrine disruptors can result in lasting consequences and could also affect the generations that follow (Monneret, 2017). More than 1,000 compounds are either confirmed or strongly suspected to be EDCs among the numerous substances listed in the European Union. On the other hand, the majority of compounds have not undergone endocrine activity testing, which raises the possibility that there are many undiscovered EDCs. The endocrine system is known to be disrupted by a wide range of substances (Bertram et al., 2022). Identifying a compound as an EDC requires proof of adverse effects, evidence of endocrine-disrupting activity, and evidence of a plausible connection between the observed adverse effect and the EDC (Pietro, Forcucci and Chiarelli, 2023). Pharmaceuticals, organochlorine insecticides, polychlorinated biphenyls, dioxin and dioxin-like compounds, plasticizers, and a broad variety of other manufactured and naturally occurring agents are believed to disturb the endocrine system (Yilmaz et al., 2019).

These agents can engage with the many components of the endocrine system and, as a result, impact several biological processes and the overall balance of organisms. Moreover, EDCs are widely present in the environment, leading to continuous human exposure via several paths, including ingestion, cutaneous absorption, and inhalation, as well as parenteral, sublingual, and maternal–fetal routes. EDCs may adversely affect human health across several domains, including cardiovascular, reproductive, neurological, immunological, and developmental systems (Abrantes-Soares et al., 2022). Exposure to certain factors during early stages of life leads to the development of illnesses later in life, including reproductive/endocrine disorders, immunological diseases, cardiovascular diseases, brain/nervous system diseases, and cancer (Kim et al., 2022).



DOI: 10.1201/9781032676555-14

**TABLE 12.1**

Common Endocrine Disrupting Chemicals (Gore et al., 2015; Wuttke et al., 2010)

| EDC | General Chemical Formula | Group | Sources | Effects |
|---|---|---|---|---|
| Bisphenol A<br>Bisphenol AF<br>Bisphenol F<br>Bisphenol S | $C_{15}H_{16}O_2$<br>$(CF_3)2C(C_6H_4OH)_2$<br>$C_{13}H_{12}O_2$<br>$(HOC_6H_4)\ 2SO_2$. | Bisphenols | Epoxy resins, plastic bottles and toys, food cans | Neurodevelopment effect, estrogenic effect, obesogen, reproductive effects, adverse thyroid hormone action |
| DDT | $C_{14}H_9Cl_5$ | Pesticides, organochloride | Fish, contaminated water and soil | Hepatotoxicity, neurotoxicity, carcinogenity |
| Phthalates | $C_8H_4O_4$ | Plasticizers | Soaps, shampoos, nail polishes, film coatings | Negative effects on CNS, obesogen, carcinogen, liver damage |
| PFOA | $C_8HF_{15}O_2$ | Fluorosurfactant | Contaminated food and water, food wrappers, floor waxes | Carcinogen, immune system toxicity, hepatotoxicity |
| Ethylparaben | $C_9H_{10}O_3$ | Parabens | Personal care products | Impaired gonad development, adverse nervous system impact |
| Methylparaben | $C_8H_8O_3$ | Parabens | Personal care products | Steroidogenesis function damage, developmental toxicity, thyroid endocrine impacts |
| TCDD (2,3,7,8-Tetrachlorodi-benzodioxin) | $C_{12}H_4Cl_4O_2$ | Dioxins | Chlorinated herbicide production, chlorine bleaching of paper, smelting | Immunity damage, reproductive defects, atrophy of thymus gland |

Out of the many compounds produced, it is estimated that over 1,000 of them could possess endocrine-acting characteristics. Endocrine-disrupting compounds (EDCs) include pesticides, fungicides, industrial chemicals, plasticizers, nonylphenols, metals, medicinal agents, and phytoestrogens (Yilmaz et al., 2019). The names, chemical formulas, and effects of commonly known endocrine-disrupting chemicals are shown in the table below (Table 12.1)

## Endocrine-Disrupting Chemicals: Mechanisms of Action and Their Effects

It is widely accepted that the activities of EDCs are primarily influenced by the traditional nuclear hormone receptors, including estrogen receptor (ER), androgen receptor (AR), ThR, and aryl hydrocarbon receptor (AhR). There is a growing body of data, both in laboratory experiments and in living organisms, which supports the effects of EDCs on progesterone receptors (PR), glucocorticoid receptors (GR), and peroxisome proliferator–activated receptors (PPARs). EDCs may hijack intracellular signaling pathways mediated by receptors and their coregulatory components, leading to consequences that may not align with the expected effects of the endogenous hormone (Gore et al., 2015).

### Bisphenol A and Bisphenol Analogues

Bisphenol A (BPA; 4,4′-isopropylidenediphenol) is a molecule of organic compounds consisting of two phenol rings linked by a methyl bridge, with two methyl functional groups bonded to the bridge. BPA is used as a substance for manufacturing phenol resins, polyacrylates, and polyesters, primarily for the creation of epoxy resins and polycarbonate plastics. Occasionally, the compounds of BPA exhibit more toxicity or estrogenic potency than BPA itself. The chlorinated derivative, 3,3′-dichlorobisphenol A (3,3′-diClBPA), has a greater level of estrogenic activity compared to BPA. Tetrabromobisphenol A (TBBPA) and 3,3′,5,5′-tetrachlorobisphenol A (3,3′,5,5′-tetraClBPA) have higher toxicity levels compared to BPA (Murata and Kang, 2018). The metabolism of BPA in mammary tissue occurs through two pathways: glucuronidation and sulfation. BPA undergoes glucuronidation by liver microsomes. Yokota et al. (2002) found increased UDP-glucuronosyltransferase (UGT) activities for BPA in microsomes from carp (Cyprinus carpino) intestines. In carp intestines, BPA is mostly metabolized into BPA glucuronide. Metabolism rates of BPA may vary by fish species. The zebrafish liver has a more rapid metabolism of BPA than the rainbow trout liver. A faster metabolism of BPA results in reduced estrogenic sensitivity (Kang et al., 2006).

In mammals, BPA undergoes glucuronidation in microsomes. Studies conducted in rats have demonstrated that cytochrome P450 enzymes play a role in the glucuronidation of BPA in microsomes. BPA can be metabolized by this enzyme while also inhibiting it in the liver. Upon examination of its metabolites, it is observed that through bisphenol-o-quinone, 5-hydroxy BPA, and a bisphenol semiquinone are formed. It is known that these are produced by microsomal P450 enzymes and may lead to increased estrogenic activity or induce toxic effects by interacting with DNA (Kang et al., 2006). After glucuronidation and sulfation, these metabolites combine with bile. The purpose of this metabolism on BPA is to render it soluble in water. When compared, the highest concentration of bisphenols found in adolescent urine was BPA. In addition, children who have been diagnosed with autism spectrum disorder (ASD) have higher amounts of BPA metabolites in their urine due to their exposure to BPA. This exposure to BPA has been linked to the development of neurodevelopmental issues, including ASD (Scaramella et al., 2022). In human tissues, it was predominantly measured in adipose tissue, followed by the liver and the brain. It is known to accumulate in the white matter of the brain. Pathological changes accompanying neurological disorders begin to emerge with the disruption of the blood–brain barrier. Breast cancer resistance protein (BCRP) plays a significant role in protecting the brain from environmental factors. It has been observed that BPA significantly inhibits the permeability function of BCRP, subsequently leading to neurotoxicity (Xu et al., 2024; Kang et al., 2006).

In humans, BPA interacts with ER-α and ER-β to act as an estrogen regulator, thereby affecting metabolism. Following estrogen binding to the ERs' ligand binding domain (LBD), translocation of the ER-ligand complexes into the nucleus, recruitment of co-regulator complexes, and docking to the estrogen response element (ERE) of target gene promoters, the canonical estrogen signaling pathway concludes. When BPA activates ERs and then undergoes signal transduction, it passes its estrogenic effects on to the target cell. Within the ERs, there are subtypes such as the nuclear estrogen receptor α/β (ERα/β), the estrogen-related receptor α/β/γ (ERRα/β/γ), and the membrane estrogen receptor (transmembrane G protein-coupled estrogen receptor, GPER) (Pinto et al., 2019; Yuan et al., 2023). In vitro investigations have shown that many bisphenol analogues may attach and trigger the human ERs, similar to bisphenol A (BPA) (Pinto et al., 2019).

In humans, nuclear hormone receptors represent a large family of 48 transcription factors that play important roles in cell growth and differentiation, development, homeostasis, and other metabolic events by regulating gene networks. They are activated by ligands and are modular proteins consisting of several domains, including a central DNA-binding domain along with ligand-independent and dependent regions. Estrogen receptors, to which BPA and its derivatives bind, are nuclear receptors. In addition to ER-α and ER- β, BPA can also bind to other nuclear receptors such as the androgen receptor (AR), estrogen-related receptor gamma (ERRγ), thyroid hormone receptors TR-α and TR-β, pregnane X receptor (PXR), and peroxisome proliferator–activated receptors PPAR-α, γ and δ (Delfosse et al., 2014).

The zebrafish (*Danio rerio*) serves as an in vivo model for investigating environmental contaminants. Zebrafish possess three estrogen receptors (zfERα, zfERβ2, and zfERβ1) that exhibit around 50% similarity in their amino acid sequences compared to the corresponding estrogen receptors in humans. The presence of zfERα is observed in the heart valves of growing zebrafish larvae, but the zfERβ1 subtype is strongly expressed in the liver. Zebrafish embryos have measurable amounts of zfERβ2 and zfERβ1, but not zfERα, in distinct brain areas during brain development. The temporal correlation between the expression of the cyp19a1b gene and zfERβ2 and zfERβ1 indicates that these receptors are responsible for driving the expression of cyp19a1b (brain aromatase) (Pinto et al., 2019).

17β-estradiol (E2) regulates ER receptors' function and plays crucial roles in the formation and maintenance of several tissues including the mammary gland, uterus, bone, cardiovascular system, and central nervous system. The binding of E2 to ERs triggers a cascade of molecular processes, including the enlistment of members from the steroid receptor coactivator (SRC) family, which ultimately leads to the transcription of certain genes. Bisphenols competitively inhibit binding to ER receptors with E2 and exhibit xenoestrogenic effects. Studies have shown that BPAF and BPC bind more strongly to both ER receptors than BPA (Delfosse et al., 2012). The ranking order for ERa and ERb affinity is as follows: BPC>BPAF>BPB>BPA>BPE>BPF>BPS. The overall order for activity is roughly comparable for ERa and ERb. When the activity ranking is made for ER receptors, the order is as follows: BPF=BPE=BPS>BPA>BPC=BPAF=BPB, among which BPC has the least activity compared to ERα when compared with Erβ (Delfosse et al., 2014).

Thyroid hormones T3 (triiodothyronine) and T4 (thyroxine) play essential functions in growth, development, and metabolism, particularly in vertebral metabolism. Bisphenols are also among thyroid-disrupting chemicals, affecting the hypothalamic-pituitary-thyroid (HPT) axis, which regulates thyroid hormone homeostasis. Studies indicate that BPA acts as an antagonist by binding to TRα and TRβ. Transthyretin is a transport protein involved in the transportation of thyroid hormones across the blood–brain barrier in fish. EDCs such as bisphenols competitively bind to the thyroxine binding sites on transthyretin, inhibiting its transport function. Additionally, EDCs/TDCs can disrupt motor functions and damage the nervous system in aquatic animals by crossing the blood–brain barrier and binding to transthyretin (Li et al., 2023).

BPA-mediated activation of estrogen receptor alpha (ERα) has a role in several aspects of cancer progression, including cell growth, survival, proliferation, migration, invasion, apoptosis, and resistance to anticancer drugs. ERβ, on the other hand, reduces the activity of ERα. BPA can promote the development and multiplication of cancer cells. The inclusion of BPA in ER-positive ovarian and breast cancer cells triggers the stimulation of mitogen-activated protein kinase/extracellular signal-regulated kinase 1/2 (MARK/ERK1/2) and phosphoinositide 3-kinase/Akt (also known as protein kinase B)/ mammalian target of rapamycin (PI3K/Akt/mTOR) signaling pathways, which play crucial roles in cellular growth and proliferation (Murata and Kang, 2018).

It has been observed that BPA triggers adipogenic markers such as leptin, PPARγ, and adipocyte fatty acid–binding protein 4/adipocyte protein 2 (AP2), increasing their expression. Mature adipocytes undergo a significant decrease in insulin-stimulated glucose consumption and ERK1/2 and Akt activation in the presence of BPA. Collectively, exposure to BPA triggers immune cell–mediated inflammatory reactions and enhances the generation of pro-inflammatory substances, such as TNF-α and IL-6, mostly through the activation of MAPK signaling pathways. The inflammatory reactions caused by BPA are associated with a higher susceptibility to inflammatory disorders, such as respiratory diseases, as well as metabolic diseases including obesity, insulin resistance, and metabolic syndrome (Murata and Kang, 2018).

The aryl hydrocarbon receptor (AhR) serves as a chemical effector activated by these substances. BPA compounds can activate AhR. Following activation, aromatic hydrocarbon receptors enter the nucleus within the cell and trigger a cascade of reactions. These processes induce enzymes involved in phase I and phase II xenobiotic metabolism, ultimately leading to toxic effects. In this context, it can be suggested that AhR plays a role in neurotoxicity triggered by bisphenol compounds (Shan, Ma et al., 2023).

## Parabens

Parabens are synthetic preservatives used in a wide range of products, from personal care items to food storage, cosmetics, and pharmaceuticals. Methylparaben, ethylparaben, propylparaben, and butylparaben are common preservatives across a wide spectrum of products (Wei et al., 2024). The ubiquity of parabens in daily life suggests that daily exposure could be common. Through routes such as dermal contact, inhalation, or their high lipophilic properties facilitating metabolism via the gastrointestinal tract through the food chain, parabens can easily enter the metabolism (Ding et al., 2017; Wei et al., 2021). The reason parabens can easily dissolve in solvents such as alcohol, ether, and acetone are because they are chemical homologs of p-hydroxybenzoic acid, either alkyl or aryl. The increase in the antimicrobial property of parabens depends on the length of the alkyl chain; as the alkyl chain elongates, its hydrophobic property increases (Wei et al., 2021). The antibacterial and antifungal activities of parabens are believed to stem from their ability to disrupt intracellular proteins and cell membranes of pathogens. This disruption leads to a change in the enzyme function of microbial cells. In addition, when the amount of chlorination of parabens grew, the adverse effects decreased (Kiran et al., 2024).

The majority of toxicological studies, including in vitro, in silico, and in vivo experiments, focus primarily on the endocrine disrupting effects of methylparaben (MeP). MeP has been found to interfere with the nuclear signaling pathways mediated by estrogen and androgen receptors, resulting in the excessive production of vitellogenin (Hu et al., 2022).

It is known that paraben exposure, similar to bisphenols, creates xenoestrogenic effects, and ethyl and butylparaben have a significant impact on steroidogenesis through cholesterol metabolism. Assessing the disruption of the AR is therefore crucial to improve the risk assessment of different paraben analogues (Ding et al., 2017). From studies, we can infer that parabens can interact with androgen, estrogen, progesterone, glucocorticoid, and PPARs receptors (Wei et al., 2021). According to a study conducted by Wang et al. in 2024, it was observed that methylparaben and butyl paraben activate voltage-gated calcium channels, affect extracellular $Ca^{++}$ influx, and consequently stimulate gonadotropin synthesis. Extracellular signal–regulated kinase 1 (ERK1) and ERK2 is an isoenzyme belonging to the MAPK family, and their activation mediates gonadotropin synthesis. According to the data obtained in the study of Wang et al., methylparaben and butylparaben increase LH and FSH production by activating the ERK1 and ERK2 pathways (Wang et al., 2024).

Within the human body, carboxylesterases metabolize parabens into p-hydroxybenzoic acid. Subsequently, these compounds are conjugated with their respective sulfate and glucuronide derivatives. Finally, together with low molecular-mass parabens (MeP, EtP, and PrP), they are eliminated through urine (Wei et al., 2021). Paraben has been shown to exhibit estrogenic effects via vitellogenin activity in fish studies, while in rat studies, it has been demonstrated to enhance estrogenic activity through competitive binding to estrogen receptors (Karpuzoğlu et al., 2013).

Various alkyl parabens have demonstrated strong affinity for estrogen receptors, whereas MP has been found to impact FSH levels and folliculogenesis and induce histopathological consequences (Ateş et al., 2018). The androgenic effects of parabens vary depending on the characteristics of their side chains. For instance, benzyl and phenyl parabens, which contain a phenyl group in their side chain, exhibit potent androgenic effects by binding to the AR receptors, whereas other parabens are known to have no effect. The recognition and endocrine functioning of parabens through different receptors vary due to physicochemical modifications in their side chains. Studies have indicated that MeP, EtP, PrP, and BuP can bind to thyroid receptors (TR) and activate signaling pathways. Further research is needed to elucidate how different side chains of parabens bind to ERs, and to understand the distinctions between linear and branched chains versus aromatic chains (Wei et al., 2022).

## Phthalates

Di (2-ethylhexyl) phthalate (DEHP) is a commonly used compound derived from phthalic acid, which is a member of the ester family known as phthalates or PAEs (Li et al, 2024). Phthalates are chemical compounds widely used in PET plastics to enhance elasticity and flexibility, as they do not form a chemical bond with plastic products, allowing for easy dispersion into the environment (Chen et al., 2024; Üstündağ et al., 2017). They can infiltrate the body through environmental contamination via air, water, and other pathways owing to their extensive usage in construction materials, the clothing sector, children's toys, medical equipment, and food and beverage packaging, among other areas (Li et al., 2024; Lyu et al., 2024; Chen et al., 2024). The primary reason for their prevalent use as PVC materials lies in their possession of a highly polar lipid chain and a conjugated pi bond attached to the benzene ring. The bonds facilitating polymer formation are van der Waals forces and hydrogen bonds, thereby allowing DEHP to readily leach from plastic products into the environment (Li et al., 2024).

Previous studies on animals and humans have shown that phthalate exposure can lead to several health problems in humans. These include developmental and reproductive toxicity, DNA damage, neuro toxicity, and cardiovascular toxicity (Singh et al., 2024). The gills of fish serve as a connection between the organism and the water environment. They are the primary pathway for chemicals to enter the fish's body, making them a crucial site for the absorption and processing of xenobiotics. Exposure to phthalates has been shown in studies to lead to an increase in free radicals and induction of xenobiotic metabolism, resulting in elevated levels of reactive oxygen species (ROS) and subsequent oxidative stress. The liver plays a significant role in the process of transforming foreign substances (xenobiotics), and specific enzymes such as ALT, AST, and GGT are commonly used as indicators of liver damage caused by exposure to environmental toxins like phthalate and bisphenol A (Sun et al., 2021).

In a study conducted by Lyu et al., examining the regulation of the ROS mechanism through oxidant-antioxidant parameters, it was observed that the defense enzyme superoxide dismutase (SOD) increased in response to oxidative stress, while catalase (CAT) activity showed similarity to SOD. Glutathione peroxidase (GSH-Px), another detoxification enzyme responsible for maintaining redox balance, was found to be inhibited. In the same study, changes in cellular apoptosis due to phthalate exposure were investigated, revealing alterations in the *p53*, *bax*, and *bcl2* genes. *p53* can control several cellular death processes through distinct signaling pathways, including necrotic cell death, iron death, and autophagic death. Bax is a protein that promotes cell death (apoptosis) and has a crucial role in apoptosis that is dependent on mitochondria. The *bcl-2* gene can control the programmed death of cells and enhance the survival of cells as an anti-apoptotic gene. Following phthalate exposure, an increase in *p53* and *bax* expression and a decrease in *bcl2* were observed in the spleens of zebrafish (Lyu et al., 2024). The induction of blood immunotoxicity by DMP occurs through oxidative damage and apoptosis that is reliant on caspase activation. Phthalates-induced oxidative damage has been documented to trigger the liberation of hemoglobin iron from red blood cells, leading to detrimental consequences on the oxygen-carrying capacity of red blood cells (Cao et al., 2024).

The exposure to DMP also leads in the dysregulation of MAPK and calcium pathway signaling. The membrane-associated protein kinase (MAPK) pathway is known to be involved in the buildup of ROS and the process of apoptosis. The calcium transduction pathway is essential in the process of cardiogenesis. Imbalance in the calcium signaling system results in irregular contraction and relaxation of the heart. Higher levels of intracellular $Ca^{2+}$ have been identified as significant catalysts for the generation of ROS, which subsequently disturb the balance of $Ca^{2+}$ inside cells and mitochondria. Exposure to DMP can lead to cardiotoxicity via several metabolic pathways (Cao et al., 2024).

DEHP not only disrupts $Ca^{++}$ homeostasis in mitochondria, but also stimulates the endoplasmic reticulum stress pathway and PPAR-related pathways. The endoplasmic reticulum has a vital function in the synthesis and transportation of proteins. Research has shown that diethyl phthalate triggers endoplasmic reticulum stress, which leads to the activation of the unfolded protein response (UPR). Reactive oxygen species, when they accumulate, cause oxidative stress. This stress leads to the activation of endoplasmic reticulum stress through the increased production of signaling molecules by radicals. The activation of both endoplasmic reticulum stress and UPR results in the initiation of apoptosis and inflammation (Singh et al., 2024).

DEHP disrupts the permeability of the blood–brain barrier, leading to an induction of astrocytes. Neuronal cells have been found to experience neurotoxicity caused by DEHP through many routes, including oxidative stress, Nrf2 activation, mitochondrial failure, and programmed cell death; exposure to DEHP results in adverse effects on the development and functioning of the central nervous system. Lots of studies have documented the neurotoxic impact of DEHP-induced endoplasmic reticulum stress, which depends on the dosage. DEHP has an impact on the development of the hippocampus region of the brain. It has the potential to reduce the number of dopaminergic neurons in the midbrain (Singh et al., 2024).

## Zebrafish as a Model Organism to Investigate the Neurotoxic Effects of Endocrine Disrupting Chemicals

The zebrafish (*Danio rerio*) is extensively utilized for assessing environmental contaminants. Neurotoxicity is primarily caused by excitotoxicity, oxidative stress, and neuroinflammation, which are essential molecular pathways involved in this process (Gu et al., 2022). Zebrafish serve as an economical alternative to a disease model due to both neurobehavioral similarities (e.g., reduced locomotor activity) and the potential to investigate specific markers of neurodegenerative diseases at the molecular level (Kalueff et al., 2014). In addition, zebrafish also have a functioning blood–brain barrier (BBB), which is comparable to that of humans. A wide range of crucial neurotransmitters were identified in the neurotransmitter profile of zebrafish, making it very significant for neuroscientific investigations (Wang and Cao, 2021).

It is widely recognized that a large number of EDCs are extensively used in a variety of industrial processes. These compounds have long half-lives in the environment and are absorbed into the organism through the air, water, and food chain, bioaccumulating (Laws et al., 2021; Kiran et al., 2024). Cosmetic pollutants also include some of these aforementioned endocrine disruptors (e.g., parabens, phthalates). It is observed that these chemicals induce gastrointestinal dysbiosis and affect the brain–gut axis. The brain–gut axis is significant in neurodegenerative diseases such as Alzheimer's and Parkinson's (Kiran et al., 2024).

The brain is a dynamic organ that is always changing, and because its developmental process continues throughout life, it is influenced by environmental factors. The dynamic development of the brain and its relationship with environmental factors also affect an individual's behavior towards the outside world. The connections between neurons involve modulation in response to environmental signals and stimulation. The reason for these connections is the existence of cellular plasticity. Cellular plasticity is an acquired characteristic dependent on stem cells capable of replacing damaged brain cells, including oligodendrocytes, astrocytes, neurons, and synaptic activity (Mesnil et al., 2020).

Gap junctions (GJs) are specialized regions of the cell membrane involved in signal transmission, facilitating the transfer of ions and small molecules between adjacent cells, as well as myelination (Jabeen and Thirumalai, 2013; Mesnil et al., 2020). GJs are composed of proteins called connexins and function as gap junction proteins. It is hypothesized that the integrity of GJ proteins may be implicated in the modulation of BBB permeability, which can be influenced by exposure to chemical pollutants. Disruption of GJ proteins due to conditions affecting intracellular connectivity may impact brain physiology, leading to issues such as memory deficits, behavioral abnormalities, and synaptic transmission impairments (Mesnil et al., 2020).

In vertebrates, connexins are protein subunits that assemble into gap junction plaques. The ortholog of connexin 36 abundant in the brains and retinas of mammals is connexin 35 (Cx35) in zebrafish. Cx35 has first observed one day post fertilization (1 dpf). It is crucial for observing the development of the nervous system due to its expression in the CNS (Jabeen and Thirumalai, 2013).

It is believed that EDCs disrupt hormonal homeostasis, leading to behavioral effects by affecting the function of the nervous system at the molecular level. These chemicals are thought to play a role in neurodegenerative diseases such as AD and PD by influencing the nervous system's function in an interdependent manner (Costa and Cairro, 2024). These EDCs are also known as anthropogenic chemicals, which cause adverse effects on the neuroendocrine system (Yirun et al., 2021).

The neurotransmitter system is preserved in zebrafish from early stages of development, exhibiting distinct resemblances to mammalian systems. This includes the presence of dopaminergic cell clusters in the olfactory bulb and hypothalamus, as well as the production of neurotransmitters such as dopamine (DA), serotonin (5-HT), acetylcholine (ACh), histamine (HA), glutamate, and GABA. The zebrafish has a significant degree of genetic similarity to humans. A thorough sequencing investigation has shown that zebrafish have at least one orthologous gene that is shared with over 70% of all human genes, including several risk genes that have been found in different human neurodegenerative disorders (Chia et al., 2022).

The genes SNCA, PINK1, LRRK2, and Parkin are associated with the development of PD. The genes PSEN1 and PSEN2 are involved in familial AD. The genes SOD1, TARDBP, C9orf72, and FUS are implicated in amyotrophic lateral sclerosis (ALS). Lastly, the gene HD is associated with Huntington's disease (Chia et al., 2022; Rana et al., 2024).

## Use of Zebrafish to Evaluate the Effects of Endocrine Disrupting Chemicals on Parkinson's Disease

The activities of the neuroendocrine system are mostly associated with the actions of neurotransmitters. Research indicates that brain neurotransmitter systems, such as the dopaminergic, adrenergic, serotonergic, and cholinergic systems, have significant effects on the activation and/or inhibition of the neuroendocrine system. The vulnerability of both DA and 5-HT synthesis to the detrimental effects of EDCs is due to the regulation of both systems by estrogen-dependent mechanisms (Yirun et al., 2021).

PD is considered to be one of the most common neurodegenerative diseases worldwide. PD is characterized by the loss of dopaminergic neurons in the substantia nigra and the concurrent accumulation of α-synuclein between cells in various parts of the brain. The major molecular pathways observed in PD include α-synuclein proteostasis, mitochondrial dysfunction, oxidative stress, calcium homeostasis, axonal transport, and neuroinflammation (Poewe et al., 2017; Ünal and Emekli-Alturfan, 2019; Cansiz et al., 2023). The primary motor symptoms include resting tremors, slowed movement (bradykinesia), muscle stiffness (rigidity), and impaired balance (loss of postural reflexes). The disorder has been associated with mitochondrial dysfunction. The presence of BPA leads to mitochondrial toxicity, characterized by a reduction in the activity of mitochondrial enzymes and an increase in the organization of ROS (Landolfi et al., 2017). Zebrafish dopaminergic neurons exhibit sensitivity to oxidative stress (Doyle and Croll, 2022).

Currently, three specific genes have been discovered that commonly lead to the development of PD at early stages. These genes are PARK2, which encodes Parkin; PINK1, which encodes PTEN-induced putative kinase 1 (Pink1); and PARK7, which encodes DJ-1. A zebrafish-specific homologue of the human PARK2 gene, called *park2*, is located on chromosome 13. It produces a protein consisting of 458 amino acids (465 in humans). The sequence of this protein is 62% identical and 75% comparable to the sequence of the human protein. Within functionally significant areas of Parkin, the level of identity climbs to 78% and the level of similarity increases to 93%. The zebrafish *park2* gene is expressed widely during development and in adult tissues, comparable to the human gene. The *pink1* gene, with a similarity of 57.8%, and the *park7* gene, with 83% similarity, are orthologous genes that are crucial in the early stages of zebrafish development. Park7 also plays an essential role in the antioxidant stress response and the survival of neuron existence (Bandmann and Burton, 2010; Wang and Cao, 2021). Observations in zebrafish with temporary gene suppression of *pink1* indicated a reduction in the number of dopaminergic neurons, impaired mitochondrial function, enhanced caspase-3 activity, and elevated levels of ROS. Conversely, treatment with antioxidants reduced the number of ROS and improved the phenotype (Ünal and Emekli-Alturfan, 2019). *α-synuclein* (SNCA) is the primary protein of Lewy bodies, which are commonly seen in PD. Zebrafish have three synuclein genes in addition to snca, namely *sncb, sncg1, and sncg2*, which produce *β-synuclein*, *γ1-synuclein*, and *γ2-synuclein*, respectively. Suppression of *γ1-synuclein* and *β-synuclein* in zebrafish resulted in decreased movement and lower levels of dopamine (Liu, 2023).

Exposure to BPA is known to affect neurotransmitters in the brain. This compound leads to a decrease in dopamine levels by acting through associated pathways in synaptic clefts. Tyrosine hydroxylase is known as the rate-limiting enzyme in dopamine synthesis, and BPA acts to reduce tyrosine hydroxylase in the substantia nigra. It also triggers the inhibition of dopamine receptors by stimulating apoptotic mechanisms in neurons (Figure 12.1) (Qi et al., 2024).

The accumulation of PCBs (polychlorinated biphenyls) seems to result in a decrease in extracellular dopamine levels, most likely due to the inhibition of dopamine production. Estrogens influence dopamine production via regulating the transcription of tyrosine hydroxylase (TH). Studies have indicated that estradiol acts as a transcriptional regulator of TH by increasing ERα activation and reducing ERβ activation. It is believed that EDCs, which have xenoestrogenic functions, also affect the dopaminergic system through this mechanism (Jones and Miller, 2008).

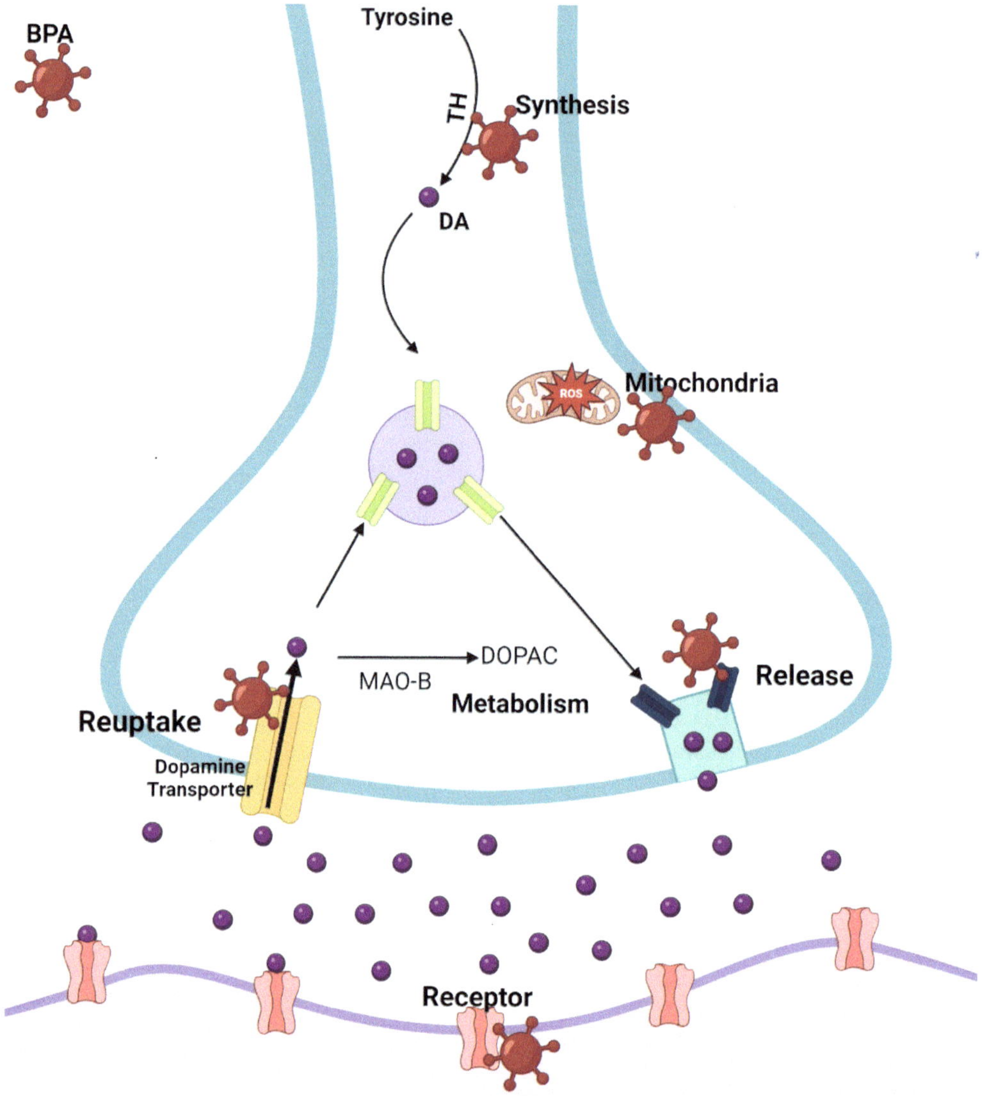


**FIGURE 12.1** Deleterious effects of BPA on dopamine metabolism. Created with BioRender. BPA: Bisphenol A, DA: Dopamine, TH: Tyrosine hydroxylase, DOPAC: 3,4-dihydroxyphenylacetic acid, MAO-B: Monoamine Oxidase B.

Mitochondrial dysfunction is a crucial component of the pathogenesis of PD and plays a significant role in the death of dopaminergic cells. The genes PINK1 and Parkin have been associated with the process of mitophagy, while current research indicates that mutations in the genes LRRK2 and SNCA also have a role in causing mitochondrial malfunction. Multiple chemical compounds that are recognized as environmental risk factors for developing PD primarily affect Complex I of the mitochondrial electron transport chain (ETC) and thus lead to neuronal death (Doyle and Croll, 2022). The impact of mitochondrial dysfunction on programmed cell death, specifically autophagy and apoptosis, was assessed following exposure to dibutyl phthalate (DBP). In addition, the components responsible for mitochondrial oxidative phosphorylation consist of five enzymatic complexes (I, II, III, IV, and V) that are located in the inner membrane or cristae edge. These complexes are formed by subunits encoded by both nuclear DNA and mtDNA, and they work together to maintain the electron transport chain and facilitate ATP production. However, the strong and consistent ability of DBP or its metabolite MBP to interact with the core subunits of complexes I, II, III, and IV suggests that it has the potential to cause diseases that affect metabolic status. Undoubtedly, the presence of DBP exposure led to an increased risk of neurodegeneration in PD, as evidenced by transcriptome studies that showed enrichment of the mitochondrion and metabolism-related pathways (Fan et al., 2023).

Exposure to MeP induces damage to the sympathetic nervous system, which precedes neural degeneration leading to neurodegenerative diseases. In zebrafish, neurobehavioral changes can be monitored at the molecular level through the measurement of acetylcholinesterase (AChE) activity. AChE is a highly effective enzyme in the nervous system that facilitates the breakdown of the neurotransmitter acetylcholine into choline and acetate by hydrolysis. Inhibition of AChE causes an accumulation of acetylcholine in synapses, resulting in a dysfunctional nervous system. This is linked to the use of irreversible AChE activity modulators. Thakkar et al. (2022b) investigated the effect of MeP exposure on AChE activity in the brains of zebrafish. The decreased activity of AChE in the brain tissue of the exposed fish is likely a result of oxidative stress induced by exposure to MeP. Nevertheless, chronic oxidative stress causes harm to the cells, proteins, and DNA of the body, resulting in various neurological disorders including PD (Thakkar et al., 2022a).

## Use of Zebrafish to Evaluate the Effects of Endocrine Disrupting Chemicals on Alzheimer's Disease

Alzheimer's disease (AD) is a continuously degenerative illness of the nervous system, the most common and widespread kind of dementia (Wang and Cao, 2021; Chia et al., 2022). AD is a condition that cannot be reversed and is responsible for around 70% of all instances of dementia (Wang and Cao, 2021). AD is commonly identified by the existence of extracellular β-amyloid (Ab) deposits formed from cleaved amyloid precursor protein (APP), together with intracellular neurofibrillary tangles (NFTs) composed of aggregated hyperphosphorylated tau proteins. The condition causes a progressive reduction in the size of the hippocampus and parietal region of the brain (Chia et al., 2022).

The clearance of Aβ via the vascular route is a crucial molecular mechanism, and damage to this mechanism leads to the development of vascular abnormalities, thereby inducing the formation of AD. Aβ can be degraded by the action of some proteases and insulin-degrading enzymes in the extracellular space. Additionally, neurovascular cell units also lead to lysosomal degradation via endocytosis. The BBB serves as a conduit for the transportation of amyloids, assisting in the reduction of intercellular accumulation, and thus facilitating the removal of amyloids from the brain. Furthermore, it is widely observed that the cerebrospinal fluid (CSF) exhibits variable dynamics in AD (Wang and Cao, 2021).

Zebrafish (*Danio rerio*) models of AD, known as ZAMs, have been set up either the injection of Aβ into the brain or through the over-expression of Aβ and tau isoforms in transgenic constructs. Zebrafish has been chosen as a viable model for studying the microbiota–gut–brain axis in AD (Cacabelos et al., 2021). Adult zebrafish brains can serve as an experimental model for investigating the regenerative capacity of AD through the induction of amyloidosis. In a study conducted by Siddiqui et al. (2021), transcriptomic analysis was performed comparing zebrafish brains injected with Aβ42 to a control group, aiming to elucidate changes in gene expression in neuronal stem cells and neurons. It was observed that IL4 was overexpressed in zebrafish injected with Aβ42, and it was noted that IL4 alone was sufficient to trigger regenerative neurogenesis and stimulate the proliferation of neuronal stem cells (NSCs). One of the functional roles of IL4 in the adult zebrafish brain tissue is its modulation of the pathway involved in the synthesis of serotonin from tryptophan, through which neurogenesis is regulated. IL4 also leads to a decrease in the enzymes involved in converting tryptophan to kynurenic acid. Aryl hydrocarbon receptor 2 serves as the receptor for kynurenic acid, and through this receptor, it functions as a regulator of neurogenesis and neuronal stem cell proliferation (Siddiqui et al., 2021).

Ach is an important neurotransmitter that is widely distributed throughout the human CNS. Understanding the mechanisms and functions of memory is crucial, since it plays a vital role in controlling the release of other neurotransmitters, including GABA. The symptoms of AD occur when there is a decrease in the transmission of cholinergic signals in the hippocampus region and the basal portions of the forebrain. The existence of acetylcholinesterase neurons in the telencephalon indicates the dominance of a cholinergic system in the teleost. Telencephalon plays a crucial role in the zebrafish brain by facilitating the storage of memory. The glutamatergic pathway, which includes the NMDA receptor, is a crucial excitatory path in vertebrates. It plays a vital role in processes such as learning, memory, and synaptic plasticity. Nevertheless, elevated levels of glutamate in the synaptic cleft might have a neurotoxic effect. Glutamate serves as a ligand for several classes of metabotropic and ionotropic receptors. Of the glutamate receptors found in humans, NMDA (N-methyl-D-aspartate) has also been characterized in zebrafish. Research has shown that telencephalon may be influenced by NMDA receptor antagonists and that its activation over an extended period impacts learning and memory (Shenoy et al., 2022).

Neurodegenerative disorders can arise due to malfunctioning brain cells that, in turn, cause alterations in signaling networks. GABA is a neurotransmitter that has the ability to reduce or inhibit the activity of postsynaptic cells, hence regulating brain activities. Dysfunction in GABA neurotransmission is a significant factor in central nervous system diseases. Malfunction of the GABAergic system leads to an imbalance in the transmission of excitatory and inhibitory impulses. This disparity serves as a possible indicator for the early phases of AD (Xu et al., 2020).

AD development is influenced by environmental toxins, including toxic metals, pesticides, industrial chemicals, and air pollution. Prolonged exposure and bioaccumulation over a prolonged period pose a danger of inducing neuroinflammation and increasing susceptibility to developing AD (Thawkar and Kaur, 2021).

Stress hormone cortisol levels in fish tend to rise in response to both short-term and long-term stresses. The hypothalamic–pituitary–adrenal axis controls the secretion of the stress hormone cortisol. When the hypothalamus detects a threat, it secretes corticotropin-releasing factor (CRF), which triggers the pituitary gland to secrete the adrenocorticotropic hormone, which in turn stimulates the adrenal glands to secrete cortisol. Several chemicals, such as zinc chloride (ZnCl2), acrylamide, bisphenol, venlafaxine, and mercury (II) chloride (HgCl2), have been found to cause elevated cortisol levels (Toni et al., 2023).

Both in vitro and in vivo studies have shown that BPA induces brain damage by disrupting protein and lipid structures through a free radical–mediated mechanism (Khan et al.,

2018). Oxidative stress is a critical factor in the pathology of AD, particularly in lipid peroxidation. Exposure to BPA leads to increased production of ROS in the CNS as observed in living organisms. Concurrently, there is a decline in antioxidant presence in various brain regions, resulting in inflammation and accelerated development of diseases like Alzheimer's. This signifies the disruption of the redox system due to BPA exposure (Haridevamuthu et al., 2022). Besides cholinergic dysfunction and neuroinflammation, oxidative stress undoubtedly contributes to AD (Nadiga et al., 2024). Mitochondrial dysfunction plays a significant role in AD. Deficiencies in mitochondrial markers such as phosphofructokinase (PFK) and pyruvate dehydrogenase lead to chronic hypoperfusion. This condition results in the production of reactive oxygen species, leading to mitochondrial dysfunction and subsequent oxidative stress (Armstrong, 2019).

The endocrine system and steroid hormones are linked to various cerebral activities. Specifically, interference with thyroid hormone can result in the decline of cognitive functions in the CNS. Exposure to bisphenols is known to have both neurobehavioral and cognitive negative effects. These disorders encompass memory loss, abnormalities in behavior, and neurodegeneration. In the pathogenesis of AD, the accumulation of amyloid beta in neurons is exacerbated by exposure to BPA. Exposure to BPA leads to an increase in the expression of beta-site amyloid precursor protein cleaving enzyme 1 (*BACE1*), which is an enzyme that plays a crucial role in the degradation of amyloid beta protein and amyloid precursor protein (*APP*) levels in the hippocampus during pregnancy in rats and is a target for drug therapy in AD. It is known that BPA exposure induces sex-dependent upregulation of the *Bace1* gene during the fertilization period, thus increasing the risk of AD. The increase in *BACE1* enzyme along with exposure to BPA is correlated with elevated levels of $A\beta$ . BPA exposure leading to $A\beta$ accumulation in the cerebral cortex and hippocampus also increases phosphorylated tau proteins. Increased ROS and disruption of the insulin signaling pathway, which are neurotoxic effects, are also factors contributing to the accumulation of amyloid beta (Xu et al., 2024).

It is known that BPF induces neurotoxicity and neuroinflammation, while BPS, like other bisphenols, triggers the estrogenic pathway. Studies have demonstrated that this condition leads to cognitive impairment via hippocampal oxidative stress. Similarly, BPA exerts xenoestrogenic effects via the estrogen pathway and affects the hippocampus. Insulin signaling in AD occurs through the activation of insulin receptors via tyrosine kinase. BPA has a disruptive effect on insulin signaling, leading to AD-like neurotoxicity through the impairment of this pathway. It causes an increase in pathological proteins associated with AD such as tau, APP, Ab, and BACE-1 (Costa and Cairrao, 2024).

It is known that exposure to diethyl phthalate leads to mitochondrial damage and increased oxidation in the brains of zebrafish (Lyu et al., 2024). DEHP has been found to have an exacerbating effect on amyloid toxicity in an Alzheimer's model created with Caenorhabditis elegans. In rats, a relationship has been identified between perinatal short-term DEHP exposure and induced AD and insulin resistance (Yen et al., 2021). Exposure to dibutyl phthalate in zebrafish is thought to affect the development of neurodegenerative diseases such as Alzheimer's and Parkinson's through mechanisms involving mitochondrial dysfunction, apoptosis, and autophagy (Fan et al., 2023).

General findings regarding parabens indicate that exposure to parabens in zebrafish leads to increased repetitive swimming behaviors, developmental delays, decreased locomotor activity, and increased anxiety-like behaviors, along with an increase in thigmotaxis. It can be said that parabens affect neurodevelopmental processes rather than neurodegenerative diseases (Merola et al., 2024). In a study conducted by Santiago-Castaneda et al. (2021) using rats, it was observed that propyl paraben induced hippocampal neuroprotection in rats with a traumatic brain injury–induced Alzheimer's model. Considering the estrogenic effect of propylparaben, it has been suggested that it may be protective in neurodegenerative diseases by inducing antiapoptotic mechanisms, inducing proinflammatory cytokines, and reducing oxidative stress. Additionally, sodium channel blockade has been proposed as a protective mechanism against neurodegeneration, and propylparaben has been stated to be a voltage-dependent sodium channel blocker (Santiago-Castaneda et al., 2021). In a study conducted by Kim et al. (2022), the effects of chronic exposure to butylparaben on photosensitivity and memory in zebrafish were investigated. According to this study, BuP can directly penetrate the blood–brain barrier and lead to neurobehavioral disorders. Abnormal development related to learning and memory was observed (Kim et al., 2022).

## Use of Zebrafish to Evaluate the Effects of Endocrine Disrupting Chemicals on Huntington's Disease

Huntington's disease (HD) is an autosomal dominant neurodegenerative disorder linked to significant motor, cognitive, and psychological impairments, located on chromosome 4. The disease results from a mutation in one of the two copies of the gene known as huntingtin (HTT) or interesting transcript 15 (IT15), which encode the protein huntingtin (Htt). Htt is a large protein of 350 kDa that is expressed throughout the brain. The HD gene product does not exhibit sequence similarity to other proteins in the existing database. It is one of the nine neurological diseases resulting from the expansion of a CAG trinucleotide repeat, which encodes an elongated polyglutamine tract within the corresponding disease proteins. Generally, the presence of over 35 CAG repeats increases the likelihood of HD, while exceeding 40 repeats definitively results in HD (Das and Rajanikant, 2014). An expansion to 36–39 CAG repeats increases susceptibility to Huntington's disease. Mutant huntingtin (mHtt) induces a negative gain-of-function impact on neurons by several pathways, including as mitochondrial malfunction, excitotoxicity, neurological inflammation, oxidative stress, neurochemical imbalance, and the depletion of brain-derived neurotrophic factor (BDNF). The striatum's GABAergic medium spiny neurons (MSNs) begin to degenerate in HD patients. Involuntary choreatic movements, behavioral and mental abnormalities, and other symptoms may be seen in HD patients. Further symptoms may include tremors, bradykinesia, impaired cognition, dementia, anxiety, aggressiveness, and changes in sexual behavior (Kumar et al., 2021).

HD is a debilitating disease, and dyshomeostasis in several pathways has been shown to play a critical role in the development of incapacitating diseases like this, including calcium dysregulation, glutamate excitotoxicity, protein misfolding and aggregation, mitochondrial dysregulation, impairment of protein degradation mechanisms, induction of inflammatory responses, and endoplasmic reticulum stress. Oxidative stress in HD characterized by mutant huntingtin accumulation presents protein oxidation, dysfunction in metabolic energetics, metal ion dyshomeostasis, DNA damage, and mitochondrial dysfunction (Bajpai et al., 2024).

The developing neurological system and the blood–brain barrier exhibit heightened vulnerability and sensitivity to various chemicals, including substances from the environment and drugs. Developmental interaction to these chemicals is associated with various neurological disorders (d'Amora and Giordani, 2018).

EDCs are known for their hormone-like signaling effects, and among the potential consequences of disrupted signaling is neurodegeneration, which has been discussed in this section in relation to certain compounds and their general association with neurodegeneration. Additionally, exogenous substances, including these chemicals (like bisphenols, phtalates, pesticides, heavy metals, and other organic pollutants) and radiation, can contribute to the generation of ROS. Antioxidant enzymes such as glutathione peroxidase (GPX), superoxide dismutase (SOD), and mitochondrial catalase play a crucial role in reducing reactive oxygen species (ROS) in mitochondria. Increased ROS production in neurons influences calcium signaling, leading to excitotoxicity and the activation of cell death pathways. Metal ion dyshomeostasis, closely linked to redox reactions within the cellular environment, is frequently observed in neurodegenerative diseases (Ozkemahli et al., 2023; Bajpai et al., 2024).

Increased exposure to these chemicals is suggested to significantly elevate the risk of neurodegenerative diseases, such as HD. Investigating the relationship between EDCs and the HD model in future studies is of great importance for elucidating the underlying mechanisms of this association.

## Use of Zebrafish to Evaluate the Effects of Endocrine Disrupting Chemicals on Amyotrophic Lateral Sclerosis (ALS) and Multiple Sclerosis (MS)

Multiple sclerosis (MS) is an ongoing, immune-related demyelinating disorder affecting the central nervous system. Experimental autoimmune encephalomyelitis (EAE) serves as the most prevalent animal model in multiple sclerosis (MS) research and has significantly contributed to the advancement of several first-line therapies aimed at the inflammatory phase of the disease. EAE is an induced inflammatory disorder of the central nervous system (CNS) that occurs subsequent to the activation of the immune response against CNS-specific antigens. Experimental autoimmune encephalomyelitis (EAE) serves as a model to investigate neuroinflammatory mechanisms and toxin-induced demyelination, enhancing our comprehension of oligodendrocyte biology, demyelination, and remyelination processes (Burrows et al., 2019). The nervous system of zebrafish is quite basic. During the larval stage, imaging of the central nervous system is made possible by their semi-transparent bodies. Several methods exist for inducing MS in zebrafish, one of which involves analyzing neurocircuitry at the synapse level. The demyelination process starts when the zebrafish immune system attacks its own myelin sheath, which is induced. Scientists may study multiple sclerosis (MS) and test novel chemicals for therapeutic medication development by watching demyelination in real time using zebrafish as subjects (Maktabi et al., 2024). Nevertheless, it can be considered that studies predominantly focus on the discovery of therapeutic compounds. However, recent discussions have shifted toward the potential repair of pre-existing myelin and oligodendrocyte damage. In this context, the zebrafish model, particularly known for its remarkable regenerative capacity during the larval stage, has been proposed as a promising tool for investigating repair mechanisms (Rice, 2012).

Amyotrophic lateral sclerosis (ALS) is a progressive and devastating neurodegenerative disorder, representing the most prevalent motor neuron disease and placing significant demands on patients, society, and healthcare systems. In patients with ALS, the degeneration of upper and lower motor neurons results in the loss of motor function, muscle atrophy, and mortality.

Furthermore, frontotemporal neurons and various other neuronal populations are often impacted, resulting in cognitive, behavioral, or sleep impairments in roughly 50% of patients. The pathophysiology of ALS is complex, involving loss of protein and RNA homeostasis, mitochondrial dysfunction, excitotoxicity, oxidative stress, neuroinflammation, glial dysfunction, and various other pathological mechanisms. The biological processes that result in the loss of axonal projections, neuronal death, and muscle denervation are the characteristic of ALS (Oliveira et al., 2024). SOD1 was the first gene identified as associated with ALS. It encodes a ubiquitous Cu/Zn superoxide dismutase that catalyzes the dismutation of superoxide radicals into hydrogen peroxide and dioxygen. However, the mechanism through which mutant SOD1 induces motor neuron damage remains unclear. Current evidence strongly indicates that SOD1 mutations are associated with both loss and gain of toxic functions, resulting in disruptions to mitochondrial dynamics, protein folding, axonal transport, and cellular metabolism (Liguori et al., 2021).

The zebrafish, a small vertebrate, serves as a prominent model organism for studying neurodegenerative diseases. The pathophysiology of amyotrophic lateral sclerosis (ALS) is related to BPA's neurotoxicity mechanisms, which include mitochondrial dysfunction and oxidative stress. Prior research has suggested that exposure to bisphenol A (BPA) or β-methylamino-L-alanine (BMAA) can effectively model sporadic ALS in zebrafish, resulting in motor neuron damage and changes in motor responses. The mechanisms of BMAA neurotoxicity encompass excitotoxicity, potential disturbances in proteostasis, and a reduction in the synthesis of melatonin, a neuroprotective and sleep-regulating hormone (Oliveira et al., 2024).

## Conclusion

EDCs play a direct role in neurodegenerative processes, both by mimicking hormones and affecting signaling pathways, and by participating in oxidative reactions through xenobiotic metabolism, leading to oxidative stress. Zebrafish, due to their ease of breeding and manipulation, are highly suitable model organisms for various disease studies and have gained significant attention in research on neurodegenerative diseases. The specific mechanisms through which EDCs contribute to the development of neurodegenerative diseases can be explored using the zebrafish model. These mechanisms are linked to several pathways, including mitochondrial dysfunction and the activation of neuronal death cascades.

Future studies can shed light on the development and potential treatment strategies for leading neurodegenerative disorders such as ALS, HD, PD, MS, and AD. These may include exploring the effects of antioxidants, repairing oxidative damage, and other approaches using zebrafish. Furthermore, the role of widely encountered EDCs, such as BPA and its derivatives, pesticides, and plastics, should be thoroughly investigated in these processes to understand their contributions to neurodegeneration at a deeper level.

## REFERENCES


Abrantes-Soares, F., Lorigo, M., & Cairrao, E. (2022). Effects of BPA substitutes on the prenatal and cardiovascular systems. *Critical Reviews in Toxicology, 52*(6), 469–498. https://doi.org/10.1080/10408444.2022.2142514

Armstrong, R.A. (2019). Risk factors for Alzheimer's disease. *Folia Neuropathol* 2019; 57 (2): 87-105

Ateş, P. S., Ünal, İ., Üstündağ, Ü. V., Alturfan, A. A., Yiğitbaşı, T., & Emekli-Alturfan, E. (2018). Methylparaben induces malformations and alterations on apoptosis, oxidant-antioxidant status, ccnd1 and myca expressions in zebrafish embryos. *Journal of Biochemical and Molecular Toxicology, 32*(3), e22036. https://doi.org/10.1002/jbt.22036

Bajpai, A., Bharathi, V., & Patel, B. K. (2024). Therapeutic targeting of the oxidative stress generated by pathological molecular pathways in the neurodegenerative diseases, ALS and Huntington's. *European Journal of Pharmacology, 987*, 177187. https://doi.org/10.1016/j.ejphar.2024.177187

Bandmann, O., & Burton, E. A. (2010). Genetic zebrafish models of neurodegenerative diseases. *Neurobiology of Disease, 40*(1), 58–65. https://doi.org/10.1016/j.nbd.2010.05.017

Beler, M., Cansız, D., Ünal, İ, et al. (2022). Bisphenol A reveals its obesogenic effects through disrupting glucose tolerance, oxidant-antioxidant balance, and modulating inflammatory cytokines and fibroblast growth factor in zebrafish. *Toxicology and Industrial Health, 38*(1), 19–28. https://doi.org/10.1177/07482337211054372

Bertram, M. G., Gore, A. C., Tyler C. R., & Brodin, T. (2022). Endocrine-disrupting chemicals. *Current Biology, 32*(13), R727–R730. https://doi.org/10.1016/j.cub.2022.05.063

Burrows, D. J., McGown, A., Jain, S. A., De Felice, M., Ramesh, T. M., Sharrack, B., & Majid, A. (2019). "Animal models of multiple sclerosis: From rodents to zebrafish. *Multiple Sclerosis Journal, 25*(3), 306–324.

Cacabelos, R., Carrera, I., Martínez-Iglesias, O., Cacabelos, N., Naidoo, V. (2021). What is the gold standard model for Alzheimer's disease drug discovery and development? *Expert Opin Drug Discov.* 16(12):1415-1440. doi: 10.1080/17460441.2021.1960502

Cansiz, D., Unal, I., Beler, M., Ustundag, U. V., Ak, E., Emekli-Alturfan, E., & Alturfan, A. A. (2023). The effect of acetic acid-induced pain in Parkinson's disease model in zebrafish. *Neurotoxicology, 99*, 14–23. https://doi.org/10.1016/j.neuro.2023.09.004

Cao, B., Kong, H., Shen, C., She, G., Tian, S., Liu, H., Cui, L., Zhang, Y., He, Q., Xia, Q., & Liu, K. (2024). Dimethyl phthalate induced cardiovascular developmental toxicity in zebrafish embryos by regulating MAPK and calcium signaling pathways. *The Science of the Total Environment, 926*, 171902. https://doi.org/10.1016/j.scitotenv.2024.171902

Chen, H., Wang, Y., & Liang, H. (2024). The combined neurotoxicity of DBP and nano-TiO2 in embryonic zebrafish (Danio rerio) revealed by oxidative activity, neuro-development genes expression and metabolomics changes. *Aquatic Toxicology, 269*, 106881. https://doi.org/10.1016/j.aquatox.2024.106881

Chia, K., Klingseisen, A., Sieger, D., & Priller, J. (2022). Zebrafish as a model organism for neurodegenerative disease. *Frontiers in Molecular Neuroscience, 15*, 940484. https://doi.org/10.3389/fnmol.2022.940484

Costa, H.E., Cairrao, E. (2024). Effect of bisphenol A on the neurological system: a review update. *Arch Toxicol.* 98(1):1-73. doi: 10.1007/s00204-023-03614-0.

d'Amora, M., & Giordani, S. (2018). The utility of zebrafish as a model for screening developmental neurotoxicity. *Frontiers in Neuroscience, 12*, 976.

Das, S., & Rajanikant, G. K. (2014). Huntington disease: Can a zebrafish trail leave more than a ripple? *Neuroscience & Biobehavioral Reviews, 45*, 258–261.

Delfosse, V., Grimaldi, M., le Maire, A., Bourguet, W., & Balaguer, P. (2014). Nuclear receptor profiling of bisphenol-A and its halogenated analogues. *Vitamins and Hormones, 94*, 229–251. https://doi.org/10.1016/B978-0-12-800095-3.00009-2

Delfosse, V., Grimaldi, M., Pons, J. L., Boulahtouf, A., le Maire, A., Cavailles, V., Labesse, G., Bourguet, W., & Balaguer, P. (2012). Structural and mechanistic insights into bisphenols action provide guidelines for risk assessment and discovery of bisphenol A substitutes. *Proceedings of the National Academy of Sciences of the United States of America, 109*(37), 14930–14935. https://doi.org/10.1073/pnas.1203574109

Ding, K., Kong, X., Wang, J., Lu, L., Zhou, W., Zhan, T., Zhang, C., & Zhuang, S. (2017). Side chains of parabens modulate antiandrogenic activity: In vitro and molecular docking studies. *Environmental Science & Technology, 51*(11), 6452–6460. https://doi.org/10.1021/acs.est.7b00951

Di Pietro, G., Forcucci, F., Chiarelli, F. (2023). Endocrine disruptor chemicals and children's health. *International Journal of Molecular Sciences, 24*(3), 2671. https://doi.org/10.3390/ijms24032671

Doyle, J. M., & Croll, R. P. (2022). A critical review of zebrafish models of Parkinson's disease. *Frontiers in Pharmacology, 13*, 835827. https://doi.org/10.3389/fphar.2022.835827

Fan, X., Zhang, D., Hou, T., Zhang, Q., & Wang, Z. (2023). Insight into the health risk implicated in mitochondrial toxicity of dibutyl phthalate exposure on zebrafish (Danio rerio) cells. *Chemosphere*, *326*, 138510. https://doi.org/10.1016/j.chemosphere.2023.138510

Gore, A. C., Chappell, V. A., Fenton, S. E., et al. (2015). EDC-2: The endocrine society's second scientific statement on endocrine-disrupting chemicals. *Endocrine Reviews*, *36*(6), E1–E150. https://doi.org/10.1210/er.2015-1010

Gu, J., Zhu, Y., Guo, M., Yin, X., Liang, M., Lou, X., Chen, J., Zhou, L., Fan, D., Shi, L., Hu, G., & Ji, G. (2022). The potential mechanism of BPF-induced neurotoxicity in adult zebrafish: Correlation between untargeted metabolomics and gut microbiota. *The Science of the Total Environment*, *839*, 156221. https://doi.org/10.1016/j.scitotenv.2022.156221

Haridevamuthu, B., Guru, A., Murugan, R., Sudhakaran, G., Pachaiappan, R., Almutairi, M.H., Almutairi, B.O., Juliet, A., Arockiaraj, J. (2022). Neuroprotective effect of Biochanin a against Bisphenol A-induced prenatal neurotoxicity in zebrafish by modulating oxidative stress and locomotory defects. *Neurosci Lett.* 1;790:136889. doi: 10.1016/j.neulet.2022.136889.

Hu, C., Sun, B., Tang, L., Liu, M., Huang, Z., Zhou, X., & Chen, L. (2022). Hepatotoxicity caused by methylparaben in adult zebrafish. *Aquatic Toxicology*, *250*, 106255. https://doi.org/10.1016/j.aquatox.2022.106255

Jabeen, S., & Thirumalai, V. (2013). Distribution of the gap junction protein connexin 35 in the central nervous system of developing zebrafish larvae. *Frontiers in Neural Circuits*, *7*, 91. https://doi.org/10.3389/fncir.2013.00091

Jones, D. C., & Miller, G. W. (2008). The effects of environmental neurotoxicants on the dopaminergic system: A possible role in drug addiction. *Biochemical Pharmacology*, *76*(5), 569–581. https://doi.org/10.1016/j.bcp.2008.05.010

Kalueff, A. V., Stewart, A. M., & Gerlai, R. (2014). Zebrafish as an emerging model for studying complex brain disorders. *Trends in Pharmacological Sciences*, *35*(2), 63–75. https://doi.org/10.1016/j.tips.2013.12.002

Kang, J. H., Katayama, Y., & Kondo, F. (2006). Biodegradation or metabolism of bisphenol A: From microorganisms to mammals. *Toxicology*, *217*(2–3), 81–90. https://doi.org/10.1016/j.tox.2005.10.001

Karpuzoglu, E., Holladay, S. D., & Gogal, R. M., Jr (2013). Parabens: Potential impact of low-affinity estrogen receptor binding chemicals on human health. *Journal of Toxicology and Environmental Health. Part B, Critical Reviews*, *16*(5), 321–335. https://doi.org/10.1080/10937404.2013.809252

Khan, J., Salhotra, S., Ahmad, S., Sharma, S., Abdi, S.A.H., Banerjee, B.D., Parvez, S., Gupta, S., Raisuddin, S. (2018). The protective effect of α-lipoic acid against bisphenol A-induced neurobehavioral toxicity. *Neurochem Int.* 118:166-175.

Kim, K., Kwon, J. S., Ahn, C., & Jeung, E. B. (2022). Endocrine-disrupting chemicals and their adverse effects on the endoplasmic reticulum. *International Journal of Molecular Sciences*, *23*(3), 1581. https://doi.org/10.3390/ijms23031581

Kiran, N. S., Yashaswini, C., & Chatterjee, A. (2024). Noxious ramifications of cosmetic pollutants on gastrointestinal microbiome: A pathway to neurological disorders. *Life Sciences*, *336*, 122311. https://doi.org/10.1016/j.lfs.2023.122311

Kumar, V., Singh, C., & Singh, A. (2021). Zebrafish an experimental model of Huntington's disease: Molecular aspects, therapeutic targets and current challenges. *Molecular Biology Reports*, *48*, 8181–8194. https://doi.org/10.1007/s11033-021-06787-y

Landolfi, A., Troisi, J., Savanelli, M. C., Vitale, C., Barone, P., & Amboni, M. (2017). Bisphenol A glucuronidation in patients with Parkinson's disease. *Neurotoxicology*, *63*, 90–96. https://doi.org/10.1016/j.neuro.2017.09.008

Laws, M. J., Neff, A. M., Brehm, E., Warner, G. R., & Flaws, J. A. (2021). Endocrine disrupting chemicals and reproductive disorders in women, men, and animal models. *Advances in Pharmacology*, *92*, 151–190. https://doi.org/10.1016/bs.apha.2021.03.008

Li, X., Hu, S., Jiang, N., Yao, X., Wang, C., Wang, Q., Yang, Z., & Wang, J. (2024). Biotoxicity responses of Zebrafish in environmentally relevant concentration of Di (2-ethylhexyl) phthalate. *Environmental Toxicology and Pharmacology*, *107*, 104423. https://doi.org/10.1016/j.etap.2024.104423

Li, Y., Zhang, Z., Wang, J., Shan, Y., Tian, H., Cui, P., & Ru, S. (2023). Zebrafish (Danio rerio) TRβ- and TTR-based electrochemical biosensors: Construction and application for the evaluation of thyroid-disrupting activity of bisphenols. *Environmental Pollution*, *330*, 121745. https://doi.org/10.1016/j.envpol.2023.121745

Liguori, F., Amadio, S., & Volonte, C. (2021). Where and why modeling amyotrophic lateral sclerosis. *International Journal of Molecular Sciences*, *22*(8), 3977.

Liu, Y. (2023). Zebrafish as a model organism for studying pathologic mechanisms of neurodegenerative diseases and other neural disorders. *Cellular and Molecular Neurobiology*, *43*(6), 2603–2620. https://doi.org/10.1007/s10571-023-01340-w

Lyu, L., Tao, Y., Abaakil, K., Gu, Y., Zhong, G., Hu, Y., & Zhang, Y. (2024). Novel insights into DEHP-induced zebrafish spleen damage: Cellular apoptosis, mitochondrial dysfunction, and innate immunity. *The Science of the Total Environment*, *912*, 169324. https://doi.org/10.1016/j.scitotenv.2023.169324

Maktabi, B., Collins, A., Safee, R., Bouyer, J., Wisner, A. S., Williams, F. E., & Schiefer, I. T. (2024). Zebrafish as a model for multiple sclerosis. *Biomedicines*, *12*(10), 2354. https://doi.org/10.3390/biomedicines12102354

Merola, C., Caioni, G., Bertolucci, C., Lucon-Xiccato, T., Savaşçı, B.B., Tait, S., Casella, M., Camerini, S., Benedetti, E., Perugini, M. (2024). Embryonic and larval exposure to propylparaben induces developmental and long-term neurotoxicity in zebrafish model. *Sci Total Environ.* 20;912:168925. doi: 10.1016/j.scitotenv.2023.168925.

Mesnil, M., Defamie, N., Naus, C., & Sarrouilhe, D. (2020). Brain disorders and chemical pollutants: A gap junction link?. *Biomolecules*, *11*(1), 51. https://doi.org/10.3390/biom11010051

Monneret, C. (2017). What is an endocrine disruptor? *Comptes Rendus Biologies*, *340*(9–10), 403–405. https://doi.org/10.1016/j.crvi.2017.07.004

Murata, M., & Kang, J. H. (2018). Bisphenol A (BPA) and cell signaling pathways. *Biotechnology Advances*, *36*(1), 311–327. https://doi.org/10.1016/j.biotechadv.2017.12.002

Nadiga, A.P.R., Suman, Krishna, K.L. (2024). A novel Zebrafish model of Alzheimer's disease by Aluminium chloride; involving nitro-oxidative stress, neuroinflammation and cholinergic pathway. *Eur J Pharmacol.* 15;965:176332. doi: 10.1016/j.ejphar.2024.176332.

Oliveira, N. A. S., Pinho, B. R., Pinto, J., Guedes de Pinho, P., & Oliveira, J. M. A. (2024). Edaravone counteracts redox and metabolic disruptions in an emerging zebrafish model of sporadic ALS. *Free Radical Biology and Medicine, 217*, 126–140.

Ozkemahli, G., P. Erkekoglu, A. Ercan, N. D. Zeybek, N. Yersal and B. Kocer-Gumusel (2023). Effects of single or combined exposure to bisphenol A and mono(2-ethylhexyl)phthalate on oxidant/antioxidant status, endoplasmic reticulum stress, and apoptosis in HepG2 cell line. *Environmental Science and Pollution Research International, 30*(5), 12189–12206.

Pinto, C., Hao, R., Grimaldi, M., Thrikawala, S., Boulahtouf, A., Aït-Aïssa, S., Brion, F., Gustafsson, J. Å., Balaguer, P., & Bondesson, M. (2019). Differential activity of BPA, BPAF and BPC on zebrafish estrogen receptors in vitro and in vivo. *Toxicology and Applied Pharmacology, 380*, 114709. https://doi.org/10.1016/j.taap.2019.114709

Poewe, W., Seppi, K., Tanner, C. M., Halliday, G. M., Brundin, P., Volkmann, J., Schrag, A. E., & Lang, A. E. (2017). Parkinson disease. *Nature Reviews. Disease Primers, 3*, 17013. https://doi.org/10.1038/nrdp.2017.13

Santiago-Castañeda, C., Segovia-Oropeza, M., Concha, L., Orozco-Suárez, S.A., Rocha, L. (2021) Propylparaben Reduces the Long-Term Consequences in Hippocampus Induced by Traumatic Brain Injury in Rats: Its Implications as Therapeutic Strategy to Prevent Neurodegenerative Diseases. *J Alzheimers Dis.* 2021;82(s1):S215-S226. doi: 10.3233/JAD-200914.

Qi, T., Jing, D., Zhang, K., Shi, J., Qiu, H., Kan, C., Han, F., Wu, C., & Sun, X. (2024). Environmental toxicology of bisphenol A: Mechanistic insights and clinical implications on the neuroendocrine system. *Behavioural Brain Research, 460*, 114840. https://doi.org/10.1016/j.bbr.2023.114840

Rana, N., Kapil, L., Singh, C., & Singh, A. (2024). Modeling Huntington's disease: An insight on in-vitro and in-vivo models. *Behavioural Brain Research, 459*, 114757. https://doi.org/10.1016/j.bbr.2023.114757

Rice, J. (2012). Animal models: Not close enough. *Nature, 484*, S9. https://doi.org/10.1038/nature11102

Scaramella., C., Alzagatiti, J.B., Creighton, C., Mankatala, S., Licea, F., Winter, G.M., et al. (2022). 17;9(3):*ENEURO*.0020-22.2022. doi: 10.1523/ENEURO.0020-22.2022.

Shan, J., Ma, X. F., Wu, M. Y., Lin, Y. J., Wang, Y., Wang, R., Li, H. M., Wu, Z. L., & Xu, H. M. (2023). Preliminary study on the role of aryl hydrocarbon receptor in the neurotoxicity of three typical bisphenol compounds (BPA, BPS and TBBPA) at environmentally relevant concentrations to adult zebrafish (Danio rerio). *Heliyon, 9*(6), e16649.

Siddiqui, T., Bhattarai, P., Popova, S., Cosacak, M.I., Sariya S, Zhang Y, Mayeux R, Tosto G, Kizil C. (2021). KYNA/Ahr Signaling Suppresses Neural Stem Cell Plasticity and Neurogenesis in Adult Zebrafish Model of Alzheimer's Disease. *Cells.* 14;10(10):2748. doi: 10.3390/cells10102748.

Singh, J., Jangra, A., & Kumar, D. (2024). Recent advances in toxicological research of di-(2-ethylhexyl)-phthalate: Focus on endoplasmic reticulum stress pathway. *Chemosphere, 356*, 141922. https://doi.org/10.1016/j.chemosphere.2024.141922

Sun, Y., Wang, X., Zhou, Y., Zhang, J., Cui, W., Wang, E., Du, J., Wei, B., Xu, X. (2021). Protective effect of metformin on BPA-induced liver toxicity in rats through upregulation of cystathionine β synthase and cystathionine γ lyase expression. *Sci Total Environ.* 1;750:141685. doi: 10.1016/j.scitotenv.2020.141685.

Thakkar, S., Seetharaman, B., & Ramasamy, V. (2022a). Impact of chronic sub-lethal methylparaben exposure on cardiac hypoxia and alterations in neuroendocrine factors in zebrafish model. *Molecular Biology Reports, 49*(1), 331–340. https://doi.org/10.1007/s11033-021-06878-w

Thakkar, S., Seetharaman, B., Kumar, H. et al. (2022b). Endocrine-disrupting chemicals exposure alter neuroendocrine factors, disrupt cardiac functions and provokes hypoxia conditions in zebrafish model. *Archives of Environmental Contamination and Toxicology, 83*, 201–213. https://doi.org/10.1007/s00244-022-00955-2

Thawkar, B.S., Kaur, G. (2021). Zebrafish as a Promising Tool for Modeling Neurotoxin-Induced Alzheimer's Disease. Neurotox Res. 39(3):949-965. doi: 10.1007/s12640-021-00343-z.

Toni, M., Arena, C., Cioni, C., Tedeschi, G.(2023). Temperature- and chemical-induced neurotoxicity in zebrafish. *Front Physiol.* 3;14:1276941. doi: 10.3389/fphys.2023.1276941.

Ustundag, U. V., Unal, I., Ates, P. S., Alturfan, A. A., Yigitbasi, T., & Emekli-Alturfan, E. (2017) Oxidant-antioxidant status and c-myc expression in BPA and DEHP-exposed zebrafish embryos. *European Journal of Cell Biology, 76*(1), 26–30.

Ünal, İ., & Emekli-Alturfan, E. (2019). Fishing for Parkinson's disease: A review of the literature. *Journal of Clinical Neuroscience: Official Journal of the Neurosurgical Society of Australasia, 62*, 1–6. https://doi.org/10.1016/j.jocn.2019.01.015

Wang, L., Liu, X., Zhao, M., Li, F., & Liu, J. (2024). Disruption of gonadotropin hormone biosynthesis by parabens: A potential development and reproduction-associated adverse outcome pathway. *Environmental Pollution, 347*, 123716. https://doi.org/10.1016/j.envpol.2024.123716

Wang, J., & Cao, H. (2021). Zebrafish and Medaka: Important animal models for human neurodegenerative diseases. *International Journal of Molecular Sciences, 22*(19), 10766. https://doi.org/10.3390/ijms221910766

Wei, F., Cheng, H., & Sang, N. (2022). Comprehensive assessment of estrogenic activities of parabens by in silico approach and in vitro assays. *The Science of the Total Environment, 845*, 157194. https://doi.org/10.1016/j.scitotenv.2022.157194

Wei, F., Mortimer, M., Cheng, H., Sang, N., & Guo, L. H. (2021). Parabens as chemicals of emerging concern in the environment and humans: A review. *The Science of the Total Environment, 778*, 146150. https://doi.org/10.1016/j.scitotenv.2021.146150

Wei, F., Yue, H., Wang, B., Cheng, H., & Sang, N. (2024). Endocrine disrupting effects of parabens in zebrafish (Danio rerio): New insights from transcriptomics, metabolomics, and molecular dynamics simulation. *Chemosphere, 354*, 141682. https://doi.org/10.1016/j.chemosphere.2024.141682

Wuttke, W., Jarry, H., Seidlova-Wuttke, D. (2010). Definition, classification and mechanism of action of endocrine disrupting chemicals. *Hormones, 9*(1),9–15. https://doi.org/10.1007/BF03401276

Xu, Y., Nie, J., Lu, C., Hu, C., Chen, Y., Ma, Y., Huang, Y., & Lu, L. (2024). Effects and mechanisms of bisphenols exposure on neurodegenerative diseases risk: A systemic review. *The Science of the Total Environment*, *919*, 170670. https://doi.org/10.1016/j.scitotenv.2024.170670

Xu, Y., Zhao, M., Han, Y., Zhang, H. (2020) GABAergic Inhibitory Interneuron Deficits in Alzheimer's Disease: Implications for Treatment. *Front Neurosci.* 30;14:660. doi: 10.3389/fnins.2020.00660

Yen, P.L., How, C.M., Hsiu-Chuan, Liao, V. (2021). Early-life and chronic exposure to di(2-ethylhexyl) phthalate enhances amyloid-β toxicity associated with an autophagy-related gene in Caenorhabditis elegans Alzheimer's disease models. *Chemosphere*. 273:128594. doi: 10.1016/j.chemosphere.2020.128594.

Yilmaz, B., Terekeci, H., Sandal, S., et al. (2019). Endocrine disrupting chemicals: Exposure, effects on human health, mechanism of action, models for testing and strategies for prevention. *Reviews in Endocrine and Metabolic Disorders*, *21*, 127–147. https://doi.org/10.1007/s11154-019-09521-z

Yokota, H., Miyashita, N., Yuasa, A. (2002). High glucuronidation activity of environmental estrogens in the carp (Cyprinus carpino) intestine. *Life Sci.* 12;71(8):887-98. doi: 10.1016/s0024-3205(02)01767-8.

Yirun, A., Ozkemahli, G., Balci, A., et al. (2021). Neuroendocrine disruption by bisphenol A and/or di(2-ethylhexyl) phthalate after prenatal, early postnatal and lactational exposure. *Environmental Science and Pollution Research*, *28*, 26961–26974. https://doi.org/10.1007/s11356-021-12408-9

Yuan, M., Chen, S., Zeng, C., Fan, Y., Ge, W., & Chen, W. (2023). Estrogenic and non-estrogenic effects of bisphenol A and its action mechanism in the zebrafish model: An overview of the past two decades of work. *Environment International*, *176*, 107976. https://doi.org/10.1016/j.envint.2023.107976

# 13

## *Using Zebrafish to Explore the Gut–Brain Axis in Neurodegenerative Disorders*

**İsmail Ünal**

### Abbreviations

| | |
|---|---|
| **α-Syn** | α-Synuclein |
| **AD** | Alzheimer's Disease |
| **ALS** | Amyotrophic Lateral Sclerosis |
| **ANS** | Autonomic Nervous System |
| **CNS** | The Central Nervous System |
| **CRISPR/Cas9** | The Clustered Regularly Interspaced Short Palindromic Repeats |
| **ENS** | The Enteric Nervous System |
| **GABA** | Gamma-Aminobutyric Acid |
| **GF** | Germ-Free |
| **GS** | Gastrointestinal System |
| **HD** | Huntington's Disease |
| **HPA** | The Hypothalamic–Pituitary–Adrenal Axis |
| **hpf** | Hours Post-Fertilization |
| **HTT** | Huntingtin Gene |
| **MS** | Multiple Sclerosis |
| **PD** | Parkinson's Disease |
| **SCFA** | Short-Chain Fatty Acids |
| **TLR** | Toll-Like Receptor |

### Introduction

Bacteria, archaea, and eukaryotic microorganisms have dominated the biosphere since the advent of cellular life, nearly 3.8 billion years ago. The eukaryotic cell then began to evolve around 1.5 billion years ago. According to research, an alphaproteobacteria–mediated endosymbiotic event that produced mitochondria is thought to have given origin to this new form of cell. The majority of the approximately 38 animal phyla emerged as a result of the Cambrian explosion, which occurred between 540 and 520 million years ago. Vertebrates emerged early in the development of animals, approximately 20–30 million years after the Cambrian explosion, despite the fact that humans tend to conceive of them as much more recent (Rook et al., 2017). On the other hand, Homo sapiens first appeared 200,000 years ago (Casas et al., 2015). The distribution of microbes throughout and across the human body is indicative of adaptations made to terrestrial life some 400 million years ago (Dethlefsen et al., 2007). The development of agriculture and animal domestication have had a significant role in the recent history of human evolution. Primitive hominids were predominantly omnivorous, surviving on starchy roots and bulbs in times of famine, particularly in savannas with limited edible plant life. The transition to a Neolithic diet resulted in an increase in simple carbohydrates, which the small intestine could then further break down and absorb. An organism can quickly and effectively pick up new skills by obtaining the required genes from another organism. By obtaining gene sequences from other bacteria through horizontal gene transfer, microbes can frequently expand ecologically into new habitats. Multicellular eukaryotes, on the other hand, do not absorb exogenous DNA as readily as bacteria do. Rather, they form symbiotic associations with microorganisms possessing the requisite genes, facilitating the fast and flexible growth of their phenotypic capacities (Walter & Ley, 2011).

According to Root et al. (2017), humans inhabit a microbiological environment that influences every facet of our biology. A biological dogma states that a mammal comes into touch with microbes for the first time during birth through the birth canal. Nonetheless, an increasing amount of data indicates that some bacteria found in the uterus are passed on from the mother, rendering the sterile uterus idea out of date (Stilling et al., 2014). According to Ventura et al. (2012), the colonisation of the human gut microbiota by bacteria from the mother's faecal and vaginal microbiota and other ambient microorganisms starts at birth. The microbiota grows in complexity and abundance after delivery via the birth canal. Breastfeeding and absorbing new microorganisms from the surroundings both contribute to these changes at the community level (Stilling et al., 2014). Before reaching a stable composition, it develops for approximately 2–3 years (Rutsch et al., 2020). Breast milk has the potential to function as a vaccination for nursing babies and is thought to be crucial in forming a bifidobacteria-rich gut microbiota in newborns prior to the weaning phase. Therefore, the microbial communities that infants are exposed to determine the first colonisation of bacteria. Coevolution of bacteria and their hosts in the human intestinal environment can result in the formation of symbiotic partnerships, in which partners get special metabolic activities in addition to other advantages; or commensal relationships, in which neither side suffers (Ventura et al., 2012).



DOI: 10.1201/9781032676555-15

## The Community Within Us: Microbiota

The term "microbiota" refers to live microorganisms that are present in a certain habitat, such as the intestinal and oral microbiota (Hou et al., 2022). It describes the collection of genomes from all microorganisms found in the environment. This includes not only the microbial community (which consists of bacteria, viruses, and other microorganisms), but also metabolites, structural components, and environmental factors. According to this, the microbiome is more diverse than the microbiota (Hou et al., 2022; Kesika et al., 2021). High microbial gene richness, stable core microbiota, and high taxonomic diversity are often signs of a healthy microbiota community (Hou et al., 2022). Anaerobic bacteria make up the majority of the normal gut microbiota, outnumbering facultative anaerobic and aerobic bacteria by a factor of 100 to 1000. The adult gut has $10^{14}$ bacterial cells in total, which is ten times more than the entire number of human cells in the body. Their total genomes have almost 5 million genes, which is twice as much as the genetic potential of the host. A broad spectrum of biochemical and metabolic activities are provided by this vast array of gene products, which enhance host physiology. In actuality, the gut microbiota is regarded as an extra organ since its metabolic capability is comparable to that of the liver (Sommer and Bäckhed, 2013).

As evidenced by recent studies, interactions between hosts and microbes are critical for several aspects of typical "mammalian" physiology, including immunological homeostasis and metabolic activity. For almost a century, research on beneficial host–microbe interactions has been conducted. Recently, methods based on gene sequences have made it possible to characterise complex microbial communities in more detail (Dethlefsen et al., 2007). From birth, the human body coexists with a diverse population of symbiotic and pathogenic microorganisms that colonize the skin, mucosal surfaces, and gastrointestinal tract. Over time, these microbial communities mature into a stable, resilient ecosystem that plays a critical role in defending against external pathogens (Hou et al., 2022). Although the composition of the human microbiota is still being studied, it is known that the skin, mucosal surfaces, and intestinal lumen all contain distinct populations of bacteria. An individual's microbiota can have fingerprint-like uniqueness at the species and strain levels (Dethlefsen et al., 2007). Microbiota can also be found in the mouth cavity, lung, vagina, and skin, although it is less well-established than in the intestines. Among human microbiomes, the oral microbiota is thought to be the second biggest (Hou et al., 2022). In the large intestine, there are around $10^{11}$ microbial cells per millilitre, as opposed to $10^{8}$ in the small intestine. Based on both culture-dependent and culture-independent methods, a detailed examination of a single person's microbiome reveals 150–400 species in their gut. The majority of these species are often found in the phyla and include Actinobacteria, Proteobacteria, Firmicutes, and Bacteroidetes. A number of trends become apparent when we look at the microbiomes of populations all around the world, despite the fact that every individual's microbiome is unique. Age and environmental variables (such as medication usage) can have an impact on human gut microbiota. The diversity of the microbiota generally rises from birth to maturity and falls as people age (beyond 70) (Hou et al., 2022). Cultural and environmental variables are the likely causes of these demographic differences. Across cultures and regions, diets in particular differ greatly. The stomach, small intestine, and large intestine make up the three organs that make up the 20-foot-long human digestive system, although the majority of studies on the human microbiome concentrate on the microbial population in the large intestine as revealed by excrement. Of all the human body's surfaces or organs, this community has the most microbial biomass (Davenport et al., 2017).

When the microbiota is in a healthy state, it affects a host's ability to fight off infections, digest and absorb nutrients, and develop and train various organs as well as the immune system (Rutsch et al., 2020). The organ in the human body with the highest level of immunological activity is the gut (Ventura et al., 2012). The immune system initially interacts with the microbiota in mucosal regions, where it shapes immunological tolerance against commensal microorganisms and maintains mucosal integrity (Rutsch et al., 2020). The immune system of the host is strictly divided to the mucosal surface in a healthy condition in response to the gut microbiota. The intestinal lumen is isolated from the surrounding tissues by a single epithelial layer. The intestinal epithelium is isolated from the residing microorganisms by a thick layer of mucus (Zheng et al., 2020). Distant organs are also impacted by the developing occurrences (Rutsch et al., 2020). Endogenous intestinal mucus and dietary fibres are examples of non-digestible substrates that can be fermented thanks to the assistance of the intestinal microbiota. The development of specialised microorganisms that generate gases and short-chain fatty acids (SCFAs) is aided by this fermentation. Acetate, propionate, and butyrate are the primary SCFAs that are generated. The primary fuel source for human colonocytes is butyrate, which also triggers intestinal gluconeogenesis, improving glucose and energy balance. In order to produce unconjugated and secondary bile acids, which act as metabolic regulators and signalling molecules to affect crucial host pathways, gut microbial enzymes play a role in bile acid metabolism (Valdes et al., 2018). Nutrient utilisation is changed by the gut microbiota and it therefore affects the enteroendocrine cells' secretion of physiologically active peptides, which might have an impact on the gut–brain axis. As the primary humoral effectors, bacterial metabolites are important. The key humoral effectors are bacterial metabolites, especially short-chain fatty acids. These compounds, which are most recognised for their impact on enterocyte feeding, also exhibit strong hormone-like action, possess immunomodulatory qualities, and interact with nerve cells via inducing sympathetic branch stimulation of the autonomic nervous system (ANS) (Appleton, 2018). Through a variety of processes (vitamin synthesis, medication digestion, oligo- and polysaccharide digestion, etc.), gut microorganisms contribute to the metabolism of nutrients and xenobiotics and regulate immunological and neurodendocrine functioning in humans (Agustí et al., 2018).

## Communication Between the Gut and the Brain: The Gut–Brain Axis

The gut–brain axis, a two-way communication channel between the brain and the gastrointestinal system (GS), is influenced by the gut microbiota, which also has an impact on the brain. This connection helps to maintain host homeostasis (Butt & Volkoff, 2019). Hormonal, immunological, and neurological interactions between the stomach and the brain mediate the intricate, two-way communication system known as the gut–brain axis. This is just one more way that the gut microbiota may affect how the brain develops and functions. The central nervous system (CNS), the hypothalamic–pituitary–adrenal (HPA) axis, the enteric nervous system (ENS), the extrinsic fibres of the ANS, and the gut microbiota combine to form the gut–brain axis (Figure 13.1). Vagal and spinal fibres from the GS's external innervations link the stomach to the brain, while efferent sympathetic and parasympathetic fibres from the brain reach the GS (Agustí et al., 2018).

The ANS, which has both sympathetic and parasympathetic limbs, controls efferent signals that go from the CNS to the intestinal wall and afferent signals that originate in the lumen and are sent there via the vagal, spinal, and enteric pathways. The primary axis for transmitting stress in an organism is considered to be the HPA axis, which synchronises its adaptive reactions to various stressors. It is a component of the limbic system, a significant brain area primarily responsible for memory and affective reactions. The brain can affect the activity of intestinal functional effector cells, including immune cells, epithelial cells, smooth muscle cells, interstitial cells of Cajal, and enterochromaffin cells, through the combination of neurological and hormonal communication pathways. However, the gut microbiota, whose part in gut–brain crosstalk has recently been assessed, influences these same cells (Carabotti et al., 2015). The ENS is situated in the GS tract and is a crucial part of the ANS. In order to regulate intestinal processes such as motor activity, secretion, absorption, and immunological defence, the ENS occurs as a complex network of 200–600 million neurons. This network is crucial for preserving intestinal homeostasis and interacting with the microbiota and host systems. The ENS is sometimes referred to as the "second brain" because of its comparable structure and neurochemistry to the CNS. As a result, any mechanism that causes dysfunction in the CNS can also cause dysfunction in the ENS, and vice versa. Through the activation of pattern recognition receptors like Toll-Like Receptors (TLR)-2 and TLR-4, which recognise microbial lipopolysaccharide, peptidoglycan, and viral RNA, the gut microbiota affects ENS growth and function dramatically (Zheng et al., 2023).

## Microbes and Defence Layer: The Intestinal Barrier

The primary barrier that keeps the gut fidelity intact and shields the host from the environment is the intestinal barrier (Plaza-Díaz et al., 2020). Maintaining the organism's homeostasis is another crucial function of the intestinal barrier. It provides that nutrients are removed from the intestinal lumen and limits the entry of dangerous chemicals (Ufnal & Pham, 2017). There are several layers in the intestinal barrier. Mucus covers the inner layer of epithelial cells, keeping infections from sticking to them. Its outer film creates a favourable habitat for commensal bacteria, which are crucial to the intestinal

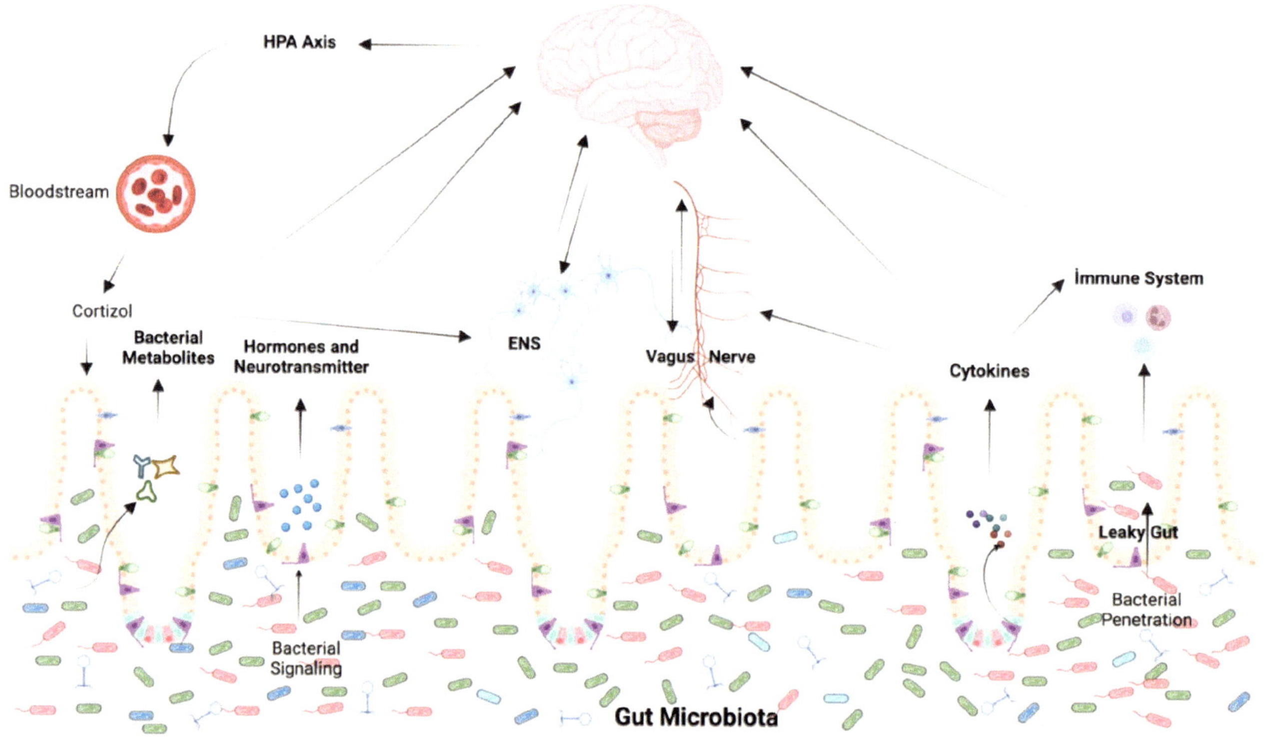


**FIGURE 13.1** Pathways of action, metabolites, and bidirectional communication of the gut–brain axis. ENS: The enteric nervous system, HPA: The hypothalamic–pituitary–adrenal axis (Created with Biorender).

barrier's effective operation. A single layer of enterocytes joined by tight junctions forms the physical barrier, which is essential to the intestinal barrier's selectivity. Additionally, this layer has immune-stimulating Paneth cells and goblet cells that produce mucus. Lastly, proper blood perfusion is ensured by the endothelium and underlying connective tissue of the intestinal capillaries, which has a significant impact on barrier functions (Ufnal & Pham, 2017). Dysbiosis may result from the microbiota's interactions with the intestinal barrier and its impact on its permeability, which can have profound consequences. To reduce the risk of disease, it is critical to maintain a healthy gut microbiota (Dicks et al., 2018). Maintaining the homeostasis inside the GS is critical since disruptions to the intestinal barrier and natural microbiota have been linked to a number of diseases, including diabetes, Alzheimer's disease (AD), irritable bowel syndrome, and stress-related behaviours (Dicks et al., 2018).

### Dysbiosis of the Intestines in Neurodegenerative Diseases

The composition and diversity of the gut microbiota are important for several reasons. In addition to controlling nutrient and metabolite extraction, synthesis, and absorption, gut bacteria also control basic functions like digestion throughout the gastrointestinal system. Additionally, commensal bacteria support the early immune response to pathogenic bacteria by generating bacteriocins, competing for resources, and preserving the integrity of the intestinal epithelium (Gubert et al., 2020). Maintaining a healthy gut microbiota is essential for preserving regular gut physiology and helps ensure that signals are correctly sent along the gut–brain axis. This is a cross-talk including numerous overlapping pathways between the stomach and the brain that enables the person to sustain their health (Almeida et al., 2020). An imbalance of the gut microbiota might be referred to as "gut dysbiosis," a general phrase. In dysbiosis, pathogenic microorganisms (pathobionts) proliferate and beneficial microbial input or signalling is lost (Martinez et al., 2021). As a consequence of improper signal transmission along the gut–brain axis brought on by intestinal dysbiosis, intestinal physiology may be adversely affected, which may break intestinal permeability (Almeida et al., 2020). The brain's aberrant buildup of pathogenic proteins may be directly or indirectly influenced by the gut microbiota. On the one hand, the gut microbiota can directly cause pathogenic protein synthesis by acting on the intestine, and it can also enter the brain through the vagus nerve. The gut microbiota can ignite a cytokine storm that leads to an overabundance of inflammation in the body, which in turn leads to an aberrant build-up of pathogenic proteins in the brain. Furthermore, a more severe inflammatory and disease response is triggered by this aberrant accumulation of harmful proteins from many origins (Wei et al., 2022). According to Ghosh et al. (2021), intestinal barrier failure is linked to the systemic inflammatory response, which exacerbates the pathophysiological condition of the underlying diseases. According to Almeida et al. (2020), it may result in an inflammatory condition that allows cytokines to enter the circulation and go to the brain. According to Ghosh et al. (2021), dysregulation of epithelial apoptosis/enterocyte death, disruption of mucus, and increased paracellular permeability as a result of tight junction disruption all contribute to increased intestinal barrier permeability. Thus, the aberrant process of pathogenic protein accumulation in neurodegenerative diseases, which is either directly or indirectly mediated by the gut microbiota in the brain, is anticipated to offer a theoretical foundation for future intestinal microorganism suppression of pathogenic protein accumulation in neurodegenerative diseases including AD, Parkinson's Disease (PD), and Huntington's Disease (HD) (Wei et al., 2022).

## As a Model Organism, Zebrafish

Because of their tiny size, high fecundity, early optical transparency, quick external growth, reasonable laboratory rearing, and availability of genetic resources, zebrafish are valued in life sciences research. As of right now, zebrafish are among the most often used vertebrate models in biological study domains, including toxicity, nutrition, development, reproduction, and genetics (Xia et al., 2022). The zebrafish model organism has several benefits. Firstly, its small size, short reproductive cycle, high egg-laying rate, quick development, and conveniently accessible rearing conditions allow for the provision of a large number of zebrafish samples for experimental investigations. Second, there is a significant degree of physiological and genetic similarity between humans and zebrafish, with 18,029 (71.4%) human genes having at least one zebrafish homolog and 2,601 (82%) human disease–associated genes. Third, zebrafish larvae are frequently employed in germ-free (GF) research. Zebrafish larvae's transparency makes it possible to sample the gut microbiota thoroughly and image it in real time, which simplifies the research of microbiota communities and host–microbiota interactions. Fourth, the applicability of zebrafish in studies of the intestinal microbiota has been expanded through the use of fluorescent protein-expressing bacteria and/or zebrafish transgenic models and mutant strains, microbiota transfer techniques, and construction of knockout zebrafish. Zebrafish are now often employed as research models by scientists to examine the gut microbiota because of these benefits (Li et al., 2022).

### Zebrafish Intestinal Structure and Development

Three major phases of intestinal development are experienced by larval zebrafish. From 26 hours post-fertilization (hpf) to 52 hpf, Stage I is characterised by the creation of lumens and the tubularization of a thin endodermal strip. In zebrafish, this is the period of development from hatching till mouth opening. Before 76 hpf, the embryonic stage from mouth opening to anus opening, intestinal epithelial cells proliferate and polarise in Stage II. Stage III is the developmental stage that lasts from the anus opening to the yolk's full absorption (76–126 hpf). The intestinal epithelium undergoes remodelling and differentiation as its defining characteristics. This indicates that in Stage III larval zebrafish, most of the digestive system is operational (Ng et al., 2005; Wallace et al., 2005). Compared to the larval intestine, the adult intestine is more complexly structured, including an S-shaped ring that folds back on itself. In terms of development, structural composition, and biological roles, the intestines of zebrafish and mammals are rather similar. Based on the transcriptome data, the adult zebrafish intestines can be separated into seven segments (S1 to S7) that are equivalent in length and extend from the anterior to posterior area. There are common architectural distinctions between the digestive systems of zebrafish and mammals, despite the fact that their guts are rather comparable in terms of development, structural composition, and biological roles. The oesophagus, stomach, small intestine (duodenum, jejunum, and ileum), and colon are the four distinct portions that make up the incredibly lengthy and intricate gastrointestinal system in mammals. Xia et al. (2022) report that adult zebrafish lack a stomach and have a simple conical intestine that is divided into three sections: anterior intestinal bulb, middle intestine, and posterior intestine. Without a stomach, the oesophagus in zebrafish joins straight to the intestine (Wallace et al., 2005). The gut of zebrafish carries out fundamental physiological processes such as nutrition absorption, digesting, and metabolic homeostasis. There are three layers to the intestinal wall of zebrafish; the mucosa (consisting of epithelium and lamina propria), muscularis externa (circular and longitudinal smooth muscle), and serosa layer. Zebrafish have irregularly folded intestinal epithelium. In cross-section, these folds resemble the villi that are seen in the small intestine of mammals (Xia et al., 2022).

## The Development of Zebrafish Microbiota

Zebrafish have a complex ecosystem of commensal bacteria in their digestive systems. Their intestines are colonised by at least 500–1,000 distinct species of bacteria. Following hatching, bacterial colonisation of the zebrafish gut occurs concurrently with the digestive tract's differentiation (Li et al., 2022). The microbial composition of mice is closer to humans than to zebrafish. Fish gut contents are predicted to have $10^7$–$10^{11}$ bacteria, but mouse and human gut contents are expected to contain $10^8$ and $10^{14}$ bacteria, respectively (Lu et al., 2021). Developmental stage (age), sex, and the host's immune system are a few host variables. The characteristics of the zebrafish gut microbiota can be influenced by exogenous variables such as water chemistry, feeding procedures, pathogen host infrastructure, and meal composition. The intestinal microbiome of zebrafish was remarkably unaffected by gender. The bacterial composition of the gut microbiota of male and female zebrafish shares similarities (Li et al., 2022). Fish have considerably simpler intestinal microbiomes, intestinal structure and function, and intestinal gene expression than humans (Lu et al., 2021). Zebrafish go through three stages of development: larva, fry, and adult. Whereas bacteroidetes and firmicutes are the predominant phyla in the human intestine, proteobacteria members dominate the gut microbiota of zebrafish at every stage of their life cycle (Li et al., 2022; Lu et al., 2021). Both γ- and α-proteobacteria are classified as proteobacteria; however, γ-proteobacteria are more common in the gut of young fish, and α-proteobacteria continue to increase until 21 days post-fertilization. In the larval stage, individual differences in the gut microbiota are negligible. By 4 dpf, there are less germs in the mouth, pharynx, oesophagus, and proximal intestine on the day the mouth opens. The amount of gut bacteria grows between 4 and 8 dpf. Age-related differences in individuals increase throughout time. In the embryos and larvae of zebrafish, aeromonas and pseudomonas species predominate. According to Li et al. (2022), the Fusobacteria phylum is more common in adult zebrafish. Proteobacteria, fusobacteria, firmicutes, bacteroidetes, actinobacteria, and verrucomicrobia form the majority of the gut microbiome of zebrafish (Lu et al., 2021).

## Vagus Nerve and Immunity in Zebrafish

Zebrafish vagus nerve develops at around 3–4 dpf. Tryptophan metabolites and *Edwardsiella tarda* infection have been shown to activate the vagus' afferent sensory circuit (Lee et al., 2021). Enteroendocrine cells can be stimulated by the gut microbiota to release gut hormones. These hormones primarily affect the hypothalamus and affect the CNS through the bloodstream and vagal afferent fibres. Additionally, enteroendocrine cells relay signals from the gut microbiota to control the vagal and intestinal pathways. According to Zhong et al. (2022), the gut microbiota has the ability to synthesise tryptophan catabolite, which stimulates enteroendocrine cells and swiftly activates intestinal and vagus nerve cells.

Similar to all other vertebrates, the zebrafish ENS originates from the neural crest (Shepherd & Eisen, 2011). Composed of muscle layers, submucosal/myelinenteric plexus, related glia, and enteric neurons, the zebrafish ENS is a peripheral nervous system generated from the neural crest. The zebrafish ENS mediates the communication between the gut and the CNS. As with mammals, it has been demonstrated to control intestinal motility via neurons that release neurotransmitters like gamma-aminobutyric acid (GABA), acetylcholine, histamine, dopamine, and serotonin (Lee et al., 2021).

The innate and adaptive immune systems of zebrafish and humans are quite similar. Due to a lack of adaptive immunity in their early stages of life (up to 4–6 weeks after fertilization), larval zebrafish are completely dependent on their innate immune system to fend against microbial invasion (Xia et al., 2022). The gut microbiota and the immune system interact, changing the microbiome's diversity and composition in ways that affect host health. The variety and composition of the intestinal microbiota in zebrafish are significantly influenced by the immune system. At 5dpf, lymphocytes in the gut start to participate in the adaptive immune process, causing zebrafish to exhibit an adaptive immune response (Li et al., 2022). The zebrafish immune system also demonstrated a significant degree of similarities with mammal immune systems. According to Lee et al. (2021), zebrafish have numerous similarities with mammals in terms of immune cell lineages, immunological receptors, and inflammatory mediators such complement, cytokines, and interleukins. Zebrafish are an excellent animal model to clarify the effects of immune systems on the gut microbiota community because of their embryonic transparency (Xia et al., 2022). Variations in the gut microbiota have the potential to impact the host's epigenetic structure, which might have long-term effects on certain gene areas (Zhong et al., 2022).

## Zebrafish as a Model Organism to Investigate the Gut–Brain Axis in Neurodegenerative Diseases

Research models, in their most basic form, are representations of intricate issues found in a variety of disciplines, including chemistry, physics, and biology. Within the field of science, the term "model" can also refer to other experimental processes or non-human participants. Rats and mice have long been the animals utilised in neuroscience studies. In addition to mammalian models, zebrafish are thought to be a useful substitute (Paduraru et al., 2023). Initially, teratogenesis, hazardous chemicals, and water pollution were monitored using zebrafish. Zebrafish were first presented by Streisinger et al. in 1981 as a useful animal model that resembles the mammalian system (X. Wang et al., 2021). Numerous scientific investigations, including those in toxicology, pharmacology, neuroscience, and genetics, employ this organism (Paduraru et al., 2023). It has been established that zebrafish are a useful tool for studying neurodegenerative diseases including ALS, PD, and AD (X. Wang et al., 2021). The neuroanatomical and neurochemical pathways of the zebrafish brain bear essential similarities to those of humans, and the fish CNS may mimic features of human brain diseases, including protein aggregation, neuronal degeneration, and glial cell activation (Chia et al., 2022). Zebrafish brains differ from human brains in several ways, despite the fact that their general cerebral cell and brain architecture is quite similar in both types. For instance, the brains of zebrafish lack a cortex or hippocampus but instead have analogous regions that work similarly to those of mammals. This allows for behaviour that is similar to that of mammals in an experimental context (Paduraru et al., 2023). Zebrafish are fast-developing animals. Zebrafish may create hundreds of embryos every week after they reach sexual maturity, which occurs at 12 weeks of age. By 72 hpf, the CNS and all main organs are fully developed. In terms of its structure and related neural circuits, such as the majority of significant neurochemical signal transduction pathways, the zebrafish brain has anatomical and functional similarities to the human brain (Chia et al., 2022). All the essential elements needed for neurotransmission, including transporters, synthesis and metabolism enzymes, and neurotransmitter receptors, are present in the zebrafish brain (Basnet et al., 2019). Zebrafish have a neurotransmitter system that is conserved from an early developmental stage. It is similar to mammalian systems in that it can produce neurotransmitters like glutamate, GABA, acetylcholine, histamine, dopamine, and serotonin, as well as dopaminergic cell clusters in the olfactory bulb and hypothalamus (Chia et al., 2022). Regarding intestinal gene function, cell structure, and molecular mechanism, zebrafish and humans are quite similar. Gut microorganisms may also be linked to highly conserved profiles in the domains of lipid metabolism, the immune system, the nervous system, and the molecular pathways of cancer-related cell proliferation and death. Additionally, because their transparent embryos and larvae are easy to manipulate, zebrafish are a good model organism for studying the role of microbiota in controlling host development. Researchers can use this and the benefit of body transparency (Figure 13.2) to create fluorescence-based assays that tag particular genes or microbes in order to carry out large-scale, high-throughput screening of in vivo interactions between microbes and xenobiotics or prebiotics, as well as combinatorial gut microbe cocktails (Lu et al., 2021).

While it is generally accepted that zebrafish microbiome colonisation happens during hatching, there is also evidence to show that microbiome components are delivered immediately during spawning. According to Lopez Nadal et al. (2020), the colonisation cycle may be broken down into four stages: (1) organisms migrate from the environment to fish; (2) these bacteria adapt in the fish's gut; (3) microorganisms migrate from the host to the environment; and (4) microbes adapt to their environment. Zebrafish are a useful vertebrate model for studying the microbiome because they have several benefits over mammalian systems. In contrast to mice and rats, which need time-consuming axenic breeding colonies and isolation chambers, zebrafish embryos are simply turned germ-free, making experimental setup simpler while preserving physiological relevance. Axenic zebrafish can also be colonised by immersion in water that contains a single microbial strain, a regulated combination of microorganisms, or a heterogeneous mixture of bacteria (Catron et al., 2019). Zebrafish that had been traditionally colonised and subjected to broad-spectrum antibiotics exhibited hyperactivity, mimicking the locomotor hyperactivity seen in axenic larvae. Moreover, colonisation with a single species of bacteria was shown to resemble the stimulation of normal locomotor activity caused by a variety of different microorganisms, but neither heat-killed bacteria nor microbiota-associated models did. Results corroborate the idea that aberrant neurobehavioural development results from disruptions in the microbiota during specific developmental stages in the early years (Phelps et al., 2017). In the study involving zebrafish, administration of *Lactobacillus plantarum* changed the β-diversity of the gut microbiota, but it did not modify the gut microbiota's fundamental architecture. According to this report, several routes showed enrichment following this modification. Administration of *Lactobacillus plantarum* was found to modify GABAergic and serotonergic signalling, protect against stress-induced dysbiosis of the gut microbiota, and dramatically decrease anxiety-related behaviours in zebrafish (Davis et al., 2016b).

The zebrafish model has gained popularity in genetic investigations because of its small body size, large brood size, robust forward and reverse genetics, physiological and genomic parallels to humans, ideal body transparency for in vivo imaging, and potential for in vivo chemical screening. Because of these benefits, zebrafish are now the primary model organism used to investigate physiological, pathological, and developmental processes. In order to comprehend the underlying mechanisms, it has been noted that the model, which consists of a sizable collection of mutant and transgenic animals with the ability to visualise pertinent organs and signalling pathways, may imitate a range of human disease states (Lee et al., 2021). Zebrafish represent a great vertebrate model for forward genetic screening, and their usefulness is further increased by efficient methods for targeted mutagenesis and gene knockdown (Meijer et al., 2014). In loss-of-function investigations, morpholino oligonucleotides have been frequently utilised in zebrafish to temporarily reduce the endogenous expression of genes of interest due to their

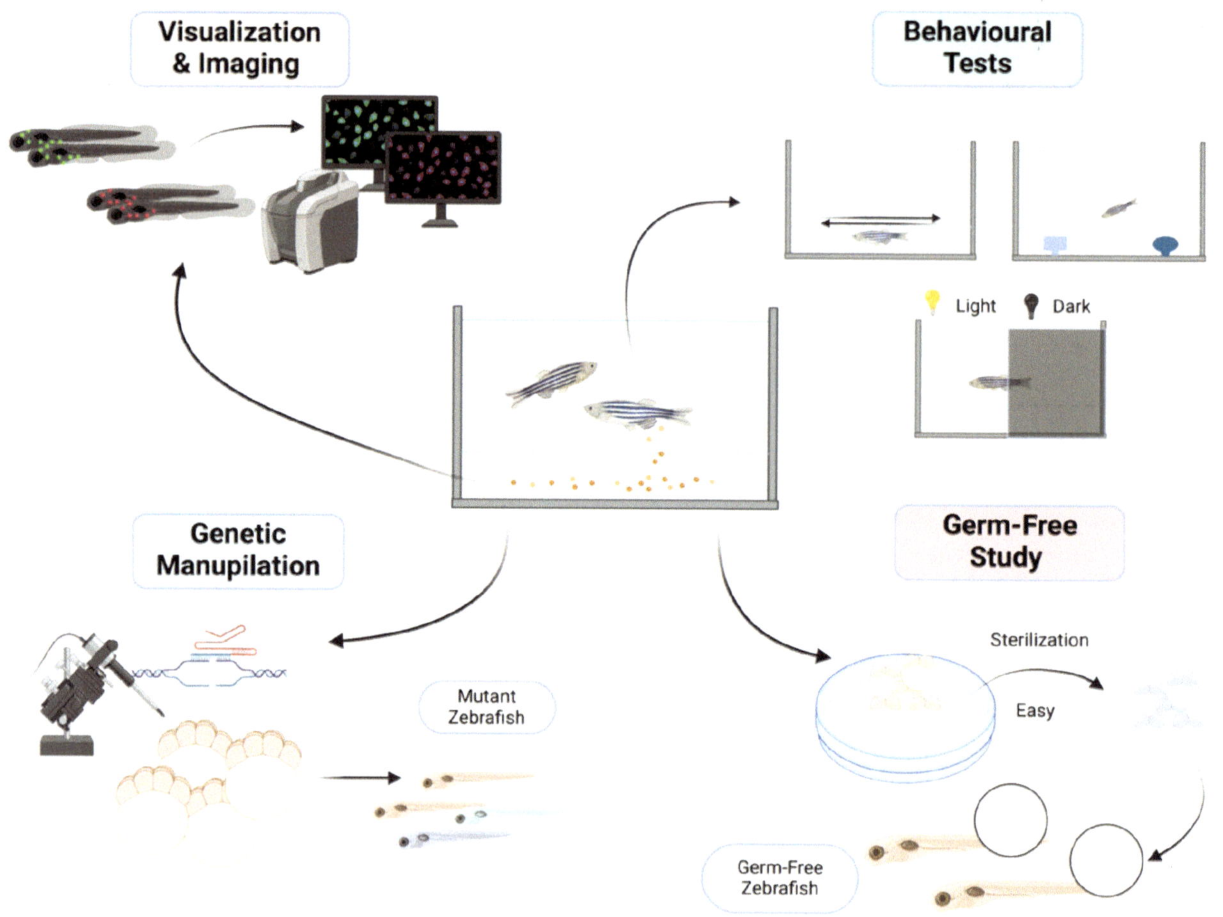


**FIGURE 13.2** Significant advantages of zebrafish for investigating the effect of the gut–brain axis on neurodegenerative diseases (Created with Biorender).

effectiveness and simplicity. These morpholinos can be engineered to impede translation by attaching to translation initiation sites, or to impede RNA splicing by attaching to splicing sites that impact the expression of the mRNA during zygotic or maternal zygotic processes, respectively. Even though the administration of morpholinos has the potential to be toxic and cause misinterpreted gene functions and unexpected phenotypes, if proper usage guidelines are carefully followed, this technique can be a useful loss-of-function genetic tool to quickly test gene function. Mutant knockout zebrafish lines are necessary for permanent loss-of-function research, and the zebrafish model is a perfect system to use state-of-the-art genome editing techniques. The use of genome editing technologies, such as zinc finger nucleases, transcription activator–like effector nucleases, and clustered regularly interspaced short palindromic repeats (CRISPR/Cas9) editing, led to the availability and widespread use of gene-targeted zebrafish knockouts created through reverse genetic methods. In vivo, it has been adjusted to enable quick and precise target gene editing. Zebrafish reverse genetics methods have been modified to enable the creation of mutants on demand since the CRISPR/Cas9 system was successfully validated to create zebrafish knockouts in vivo. Additionally, it has been used for target-specific transcriptional control, precise base pair editing, and the induction of knockin mutations (Lee et al., 2021).

Throughout development, the gut microbiota's establishment is influenced by both host genetics and environmental influences. Zebrafish are a great model organism for researching the changes in the gut microbiota during development and the relationship between host genes and gut microorganisms because of their short life cycle and ability to be genetically modified (Lu et al., 2021). Since zebrafish are transparent during the embryonic and larval stages of development, high-resolution imaging of the interactions between microbes and hosts in vertebrates is best achieved during these life stages. Zebrafish transparency provides excellent opportunities for in vivo research of the interaction between microorganisms and host cells as well as high-resolution imaging of cellular migratory processes in response to infections. Fluorescent reporter lines that identify several zebrafish immune cell types or subcellular spaces where pathogens are found have been created to aid in vivo imaging. In vivo imaging is a major advantage of the zebrafish model for cellular interaction research. The ability to combine imaging with sophisticated forward and reverse genetics methods, as well as the complete accessibility of state-of-the-art genomic tools like whole genome epigenetic analyses and RNA deep sequencing, have recently significantly increased this advantage (Meijer et al., 2014).

In 2004, the GF zebrafish model was effectively developed. GF zebrafish offer several benefits that facilitate the cultivation of gnotobiotic organisms and exploration of the relationship between the microbiota and host. First, from the moment of fertilization until early maturity, zebrafish larvae are translucent, making them ideal for in vivo investigations of intestinal microbes. Second, after they reach sexual maturity (about 2–4 months), zebrafish have high fecundity (each female lays 200–300 eggs), meaning that over the year, eggs are normally deposited every 4–7 days under standard experimental circumstances. Third, raising zebrafish larvae in a lab setting is a simple process. These characteristics, along with the many genomic, anatomical, and physiological similarities between zebrafish and mammals, as well as the cell types they share with mammals, such as enterocytes, secretory goblet cells, and enteroendocrine cells, make GF zebrafish useful models for learning about the roles of the gut microbiota (Tan et al., 2019). Because the chorion's surface has been sterilised, it is simple to create GF zebrafish embryos, which permits suitable regulation of their microbial environment (Xia et al., 2022). Acquiring adult GF zebrafish remains challenging, nonetheless, given their 30-day maximum life duration (Zhong et al., 2022). Of these well-established animal models, the GF zebrafish offers a practical, affordable, and dependable model due to its close resemblance to the human genome and disease, standard large-scale embryo production, and visible growth through the early stages. The GF zebrafish model has been used to investigate immune systems by regulating the microbial population, which allows researchers to examine the innate immune system dependently. This is because the zebrafish lacks a functional adaptive immune system in their early life, up to 7 dpf (Jia et al., 2021). Research has demonstrated that when GF mice are compared to certain pathogen-free mice, they have reduced anxiety-like behaviours and enhanced motor activity (Chen et al., 2021). In a research on GF zebrafish larvae, it was discovered that the larvae lacked a microbiota, which resulted in substantial changes to their locomotor and anxiety-related behaviours. Additionally, it was noted that specific reactions to acute stressors vanished in the larvae that were not exposed to microbes (Davis et al., 2016a).

Zebrafish models are valid for many more reasons than only their physical and genetic similarity to humans. Their similar behaviours also have a part in it. Zebrafish can be used to simulate neurological and mental disorders because of their vast variety of complex behaviours, which include social, anxiety, learning and memory, and defence behaviours. The behavioural paradigms of zebrafish and mammals are quite similar, suggesting that many behaviours are species-specific and have been preserved via evolution. Zebrafish behavioural research has been further aided by the recent development of high-throughput monitoring of zebrafish embryos and larvae, which has improved the objectivity, reproducibility, and efficiency of the field. In terms of important behaviours employed in neuropharmacology, such as stress and anxiety, zebrafish larvae are identical to adult zebrafish and other vertebrates. Zebrafish larvae's neural activity can also be tracked or observed when the larvae perform a behaviour. Both structurally and neurochemically, the zebrafish brain is comparable to that of humans and rats. Zebrafish have molecular and structural homology in several brain areas linked to human diseases. For instance, much as in humans and rodents, the habenula and amygdala in zebrafish are involved in regulating emotional behaviour. In both humans and zebrafish, hyperactivation of the habenula results in comparable situations including depression and stress-related behaviours. This illustrates how the brains of zebrafish and mammals operate similarly (Basnet et al., 2019). An average 25% reduction in symptoms associated with anxiety and sadness has been observed in human cases when individuals with schizophrenia received the probiotic *Bifidobacterium breve* A1 strain (Okubo et al., 2019). The novel tank diving test, wading test, touch test, and predator avoidance test are among the experiments conducted on juvenile and adult zebrafish. These tests may be used to gauge avoidance behaviours, social interactions, and anxiety-like behaviours (Catron et al., 2019). Because of their shoaling characteristic and active social behaviour, which can be statistically analysed, zebrafish are often utilised in CNS-regulated behavioural investigations (Lu et al., 2021). Similarly, studies on rats and mice have long measured a range of anxiety-like behaviours as well as other learning and memory behaviours. But compared to the zebrafish model – where more than 3,000 embryos may be screened in a week for some larval behavioural experiments – these are far less effective and more expensive (Catron et al., 2019).

Neurodegeneration diseases, which mostly comprise AD, PD, amyotrophic lateral sclerosis (ALS), HD, and multiple sclerosis (MS), are classified as irreversible and severe diseases marked by substantial neuronal loss (Cheng et al., 2023; Hang et al., 2023). The main characteristics of AD, a common degenerative neurological disease, are the development of neurofibrillary tangles from intracellular hyperphosphorylation of tau proteins and amyloid plaques outside of neurons, which lead to memory loss and cognitive dysfunction. Clinical symptoms of PD are similar to those of AD in that the majority of symptoms are motor (myotonia, bradykinesia, resting tremor), with non-motor symptoms including sadness, anxiety, insomnia, constipation, and exhaustion. The pathogenesis of PD is characterised by the selective absence of nigrostriatal dopaminergic neurons, the development of Lewy bodies (mostly composed of α-synuclein (α-Syn)), and a reduction in striatal dopamine content (Cheng et al., 2023). ALS is another neurodegenerative disease. The hallmark of the disease is the loss of voluntary muscular movement brought on by the degeneration of both lower motor neurons (found in the brainstem and spinal cord) and higher motor neurons (found in the brain's motor cortex) (Singh et al., 2021). The pathophysiological mechanisms driving the degeneration of motor neurons in ALS are intricate and multifaceted, involving the interplay of hereditary and environmental variables (Bhuiyan et al., 2021). TDP-43+ cytoplasmic inclusion bodies in motor neurons are the pathological hallmarks of ALS (Singh et al., 2021). Another prevalent neurodegenerative disease is MS, which is defined as a CNS degenerative disease with localised inflammation-related impairment of motor and cognitive function. The predominant aetiology of multiple sclerosis is abnormal astrocyte proliferation accompanied by axonal or neuronal loss and demyelination (Cheng et al., 2023). The huntingtin gene (HTT) mutation is the cause of HD, a genetic neurodegenerative disease. Mutations in HTT involve a CAG trinucleotide repeat, which is translated into tiny fragmented proteins during elongation, which then aggregate into disease-causing cellular debris. While striatal GABAergic medium-sized spiny neurons projecting to the substantia nigra and globus pallidus are extremely selectively affected by neurodegeneration in early stages of HD, later stages exhibit considerable atrophy over a wide range of brain areas, leading to severe symptoms across the CNS (Singh et al., 2021).

Studies indicate that dysbiosis of the microbiota may have a role in neurological conditions including neurodegeneration and mood disorders. A portion of this disease might be caused by the gut–brain–microbiota axis. Through neuroendocrine, neuroimmune, and direct neural pathways like the vagus nerve, gut microorganisms affect this two-way communication mechanism. There is mounting evidence that microbes can synthesise neurotransmitters, encourage intestinal epithelial cells to produce serotonin, create bioactive compounds, change epigenetic regulation through fermentation byproducts, and release metabolites that can penetrate the blood–brain barrier and enter the bloodstream. Furthermore, both the aged and those suffering from neurodegenerative diseases exhibit a decline in gut microbial richness. Disrupted gut–brain axis signalling during gut dysbiosis results in inflammation and oxidative stress, as well as abnormal metabolism and immunological response (Peterson, 2020). There is growing evidence on the significance of the gut microbiota in the pathophysiology of long-term neurodegenerative diseases including AD and PD. In PD, disruption of the GS frequently happens years before the disease is identified. According to a hypothesis put out by Braak and associates, the disease starts in the intestines and travels up the gut–brain axis, which includes the spinal cord and vagus nerve, to the brain. In postmortem cases of early PD, Lewy bodies in the ENS, a characteristic of the disease, have been discovered (Ghaisas et al., 2016). In a zebrafish study, it was shown that the administration of antibiotics altered the morphine-dependent behaviour of the fish, the features of the microbes in the intestines, and the genes connected to the microbiota–gut–brain axis in the intestine and brain. Furthermore, it has been shown that the use of isorhynchophylline has a healing effect that vanishes when antibiotics are administered. Thus, it has been demonstrated that the brain and intestinal microbiota interact, and dysbiosis in the microbes in the gut may impact the body's ability to use isorhynchophylline (Chen et al., 2021). After 28 days of dietary supplementation with the probiotic *Lactobacillus rhamnosus*, differences were observed in the behaviour of the treated and control groups in terms of flocking, brain expression levels of brain-derived neurotrophic factor, and genes related to serotonin signalling and metabolism in zebrafish. Furthermore, the microbiota showed a notable rise in firmicutes and a declining trend in proteobacteria. This work showed that endogenous neuroactive compounds in zebrafish might be modulated by certain microorganisms (Borrelli et al., 2016). Most research on the connection between the gut–brain axis and neurodegenerative diseases has been done using mice and rats as model species. A research using monocolonization mice demonstrated that colonisation with curli-producing *Escherichia coli (E. coli)* induced α-Syn disease in the brain and gut of mice overexpressing human amyloid α-Syn. The results of the same investigation showed that *E. coli* curli expression exacerbated α-Syn-induced behavioural impairments. Curli-producing *E. coli* that monocolonizes α-Syn-overexpressing mice worsens motor impairment and gastrointestinal dysfunction and causes αSyn aggregation and inflammation in the brain and gut (Sampson et al., 2020). α-Syn and other α-Syn variants found in brain lysates from PD patients were injected into the intestinal wall of rats and demonstrated to be delivered in a time-dependent way via the vagus nerve to the dorsal motor nucleus of the vagus in the brainstem (Holmqvist et al., 2014). This work provided evidence that various forms of α-Syn may diffuse from the stomach to the brain and that aggregated α-Syn is transported by microtubule-associated transport in neurons. C.elegans that expressed human α-Syn were given E. coli mutants for consumption. It has been discovered that in E. coli, genetic deletion or pharmacological inhibition of the curli main subunit CsgA reduces α-syn-induced neuron mortality, restores mitochondrial health, and enhances neuron functioning. It was shown that bacterially produced CsgA co-localized with α-Syn in neurons and that this promoted α-Syn aggregation by cross-seeding (C. Wang et al., 2021).

## Conclusion

As a model organism, zebrafish have become more popular recently, and it appears that this popularity will continue in the near future. The presence of genetic homologues of human genes, its availability to genetic manipulation and screening techniques, the ease of use in the GF model, and its high reproducibility thanks to high egg production have made zebrafish a valuable model organism for research on neurodegenerative diseases. In recent years, a new window has been opened on microbiota and the gut–brain axis.

Any selected bacteria or the examined bacterial group can colonize zebrafish embryos generated as GF, simplifying the research on the effects on embryos. Therefore, utilising this fantastic benefit of zebrafish embryos can make the investigation of the impacts of microbiota much more convenient.

Studying cellular migrations and interactions in the relationship between the microbiota and host is made easier by the zebrafish model. We may anticipate that studying genes and their functions in the gut–brain axis will get simpler with the aid of evolving gene-editing technology and the zebrafish model. Because of their behavioural similarities to humans, zebrafish provide an invaluable model in behavioural studies. Neurological and psychological symptoms are seen in neurodegenerative diseases before, during, and following diagnosis.

The brain's ability to operate depends on the neurotransmitters that are primarily responsible for signal transmission between various neuronal and glial cell types. Serotonin, dopamine, and GABA are among the neurotransmitters that the gut microbiota may either produce itself or stimulate the production thereof. As a result, the gut microbiota has come to be known as the second brain. Neurodegenerative diseases have been associated with intestinal dysbiosis. In this regard, abnormalities occurring in the gut–brain axis, in particular in the intestinal production of neurotransmitters under the influence of the intestinal microbiota, may have neurological and psychological consequences. Zebrafish are a suitable model organism to investigate this mechanism due to the features summarized in this chapter. Because of their many behavioural similarities to humans, zebrafish are therefore frequently employed in behavioural investigations allowing researchers studying neurodegenerative diseases to monitor neurological and psychological processes.

In conclusion, zebrafish is a useful model that may be used to study the influence of the gut–brain axis in neurodegenerative illnesses. We suggest that the popularity of zebrafish in this area will increase in the upcoming years.

## REFERENCES


Agustí, A., García-Pardo, M. P., López-Almela, I., Campillo, I., Maes, M., Romaní-Pérez, M., & Sanz, Y. (2018). Interplay between the gut-brain axis, obesity and cognitive function. *Frontiers in Neuroscience*, *12*, 155.

Almeida, C., Oliveira, R., Soares, R., & Barata, P. (2020). Influence of gut microbiota dysbiosis on brain function: A systematic review. *Porto Biomedical Journal*, *5*(2), 1–8.

Appleton, J. (2018). The gut-brain axis: Influence of microbiota on mood and mental health. *Integrative Medicine: A Clinician's Journal*, *17*(4), 28.

Basnet, R. M., Zizioli, D., Taweedet, S., Finazzi, D., & Memo, M. (2019). Zebrafish larvae as a behavioral model in neuropharmacology. *Biomedicines*, *7*(1), 23.

Bhuiyan, P., Chen, Y., Karim, M., Dong, H., & Qian, Y. (2021). Bidirectional communication between mast cells and the gut-brain axis in neurodegenerative diseases: Avenues for therapeutic intervention. *Brain Research Bulletin*, *172*, 61–78.

Borrelli, L., Aceto, S., Agnisola, C., De Paolo, S., Dipineto, L., Stilling, R. M., Dinan, T. G., Cryan, J. F., Menna, L. F., & Fioretti, A. (2016). Probiotic modulation of the microbiota-gut-brain axis and behaviour in zebrafish. *Scientific Reports*, *6*(1), 30046.

Butt, R. L., & Volkoff, H. (2019). Gut microbiota and energy homeostasis in fish. *Frontiers in Endocrinology*, *10*, 9.

Carabotti, M., Scirocco, A., Maselli, M. A., & Severi, C. (2015). The gut-brain axis: Interactions between enteric microbiota, central and enteric nervous systems. *Annals of Gastroenterology: Quarterly Publication of the Hellenic Society of Gastroenterology*, *28*(2), 203.

Casas, A., Parra, F., & Blancas, J. (2015). Evolution of humans and by humans. In U. P. Albuquerque, P. M. de Medeiros, & A. Casas (Eds.), *Evolutionary ethnobiology* (pp. 21–36). Springer.

Catron, T. R., Gaballah, S., & Tal, T. (2019). Using zebrafish to investigate interactions between xenobiotics and microbiota. *Current Pharmacology Reports*, *5*, 468–480.

Chen, Z., Zhijie, C., Yuting, Z., Chan, L., Shilin, X., Qichun, Z., Jing, L., & Zhixian, M. (2021). The ameliorative effects of isorhynchophylline on morphine dependence are mediated through the microbiota-gut-brain Axis. *Frontiers in Pharmacology*, *12*, 526923.

Cheng, Y., Chen, C., & Zhang, F. (2023). Immunity orchestrates a bridge in gut-brain axis of neurodegenerative diseases. *Ageing Research Reviews*, *85*, 101857.

Chia, K., Klingseisen, A., Sieger, D., & Priller, J. (2022). Zebrafish as a model organism for neurodegenerative disease. *Frontiers in Molecular Neuroscience*, *15*, 940484.

Davenport, E. R., Sanders, J. G., Song, S. J., Amato, K. R., Clark, A. G., & Knight, R. (2017). The human microbiome in evolution. *BMC Biology*, *15*(1), 1–12.

Davis, D. J., Bryda, E. C., Gillespie, C. H., & Ericsson, A. C. (2016a). Microbial modulation of behavior and stress responses in zebrafish larvae. *Behavioural Brain Research*, *311*, 219–227.

Davis, D. J., Doerr, H. M., Grzelak, A. K., Busi, S. B., Jasarevic, E., Ericsson, A. C., & Bryda, E. C. (2016b). Lactobacillus plantarum attenuates anxiety-related behavior and protects against stress-induced dysbiosis in adult zebrafish. *Scientific Reports*, *6*(1), 33726.

Dethlefsen, L., McFall-Ngai, M., & Relman, D. A. (2007). An ecological and evolutionary perspective on human–microbe mutualism and disease. *Nature*, *449*(7164), 811–818.

Dicks, L. M., Dreyer, L., Smith, C., & Van Staden, A. D. (2018). A review: The fate of bacteriocins in the human gastro-intestinal tract: Do they cross the gut–blood barrier? *Frontiers in Microbiology*, *9*, 2297.

Ghaisas, S., Maher, J., & Kanthasamy, A. (2016). Gut microbiome in health and disease: Linking the microbiome–gut–brain axis and environmental factors in the pathogenesis of systemic and neurodegenerative diseases. *Pharmacology & Therapeutics*, *158*, 52–62.

Ghosh, S., Whitley, C. S., Haribabu, B., & Jala, V. R. (2021). Regulation of intestinal barrier function by microbial metabolites. *Cellular and Molecular Gastroenterology and Hepatology*, *11*(5), 1463–1482.

Gubert, C., Kong, G., Renoir, T., & Hannan, A. J. (2020). Exercise, diet and stress as modulators of gut microbiota: Implications for neurodegenerative diseases. *Neurobiology of Disease*, *134*, 104621.

Hang, Z., Zhou, L., Xing, C., Wen, Y., & Du, H. (2023). The blood-brain barrier, a key bridge to treat neurodegenerative diseases. *Ageing Research Reviews*, *91*, 102070.

Holmqvist, S., Chutna, O., Bousset, L., Aldrin-Kirk, P., Li, W., Björklund, T., Wang, Z.-Y., Roybon, L., Melki, R., & Li, J.-Y. (2014). Direct evidence of Parkinson pathology spread from the gastrointestinal tract to the brain in rats. *Acta Neuropathologica*, *128*, 805–820.

Hou, K., Wu, Z.-X., Chen, X.-Y., Wang, J.-Q., Zhang, D., Xiao, C., Zhu, D., Koya, J. B., Wei, L., & Li, J. (2022). Microbiota in health and diseases. *Signal Transduction and Targeted Therapy*, *7*(1), 135.

Kesika, P., Suganthy, N., Sivamaruthi, B. S., & Chaiyasut, C. (2021). Role of gut-brain axis, gut microbial composition, and probiotic intervention in Alzheimer's disease. *Life Sciences*, *264*, 118627.

Jia, P.-P., Junaid, M., Wen, P.-P., Yang, Y.-F., Li, W.-G., Yang, X.-G., & Pei, D.-S. (2021). Role of germ-free animal models in understanding interactions of gut microbiota to host and environmental health: A special reference to zebrafish. *Environmental Pollution*, *279*, 116925.

Lee, J.-G., Cho, H.-J., Jeong, Y.-M., & Lee, J.-S. (2021). Genetic approaches using zebrafish to study the microbiota–gut–brain axis in neurological disorders. *Cells*, *10*(3), 566.

Li, P., Zhang, J., Liu, X., Gan, L., Xie, Y., Zhang, H., & Si, J. (2022). The function and the affecting factors of the zebrafish gut microbiota. *Frontiers in Microbiology*, *13*, 903471.

López Nadal, A., Ikeda-Ohtsubo, W., Sipkema, D., Peggs, D., McGurk, C., Forlenza, M., Wiegertjes, G. F., & Brugman, S. (2020). Feed, microbiota, and gut immunity: Using the zebrafish model to understand fish health. *Frontiers in Immunology*, *11*, 512428.

Lu, H., Li, P., Huang, X., Wang, C. H., Li, M., & Xu, Z. Z. (2021). Zebrafish model for human gut microbiome-related studies: Advantages and limitations. *Medicine in Microecology*, *8*, 100042.

Martinez, J. E., Kahana, D. D., Ghuman, S., Wilson, H. P., Wilson, J., Kim, S. C., Lagishetty, V., Jacobs, J. P., Sinha-Hikim, A. P., & Friedman, T. C. (2021). Unhealthy lifestyle and gut dysbiosis: A better understanding of the effects of poor diet and nicotine on the intestinal microbiome. *Frontiers in Endocrinology*, *12*, 667066.

Meijer, A. H., van der Vaart, M., & Spaink, H. P. (2014). Real-time imaging and genetic dissection of host–microbe interactions in zebrafish. *Cellular Microbiology*, *16*(1), 39–49.

Okubo, R., Koga, M., Katsumata, N., Odamaki, T., Matsuyama, S., Oka, M., Narita, H., Hashimoto, N., Kusumi, I., & Xiao, J. (2019). Effect of bifidobacterium breve A-1 on anxiety and depressive symptoms in schizophrenia: A proof-of-concept study. *Journal of Affective Disorders*, *245*, 377–385.

Ng, A. N., de Jong-Curtain, T. A., Mawdsley, D. J., White, S. J., Shin, J., Appel, B., Dong, P. D. S., Stainier, D. Y., & Heath, J. K. (2005). Formation of the digestive system in zebrafish: III. Intestinal epithelium morphogenesis. *Developmental Biology*, *286*(1), 114–135.

Paduraru, E., Iacob, D., Rarinca, V., Plavan, G., Ureche, D., Jijie, R., & Nicoara, M. (2023). Zebrafish as a potential model for neurodegenerative diseases: A focus on toxic metals implications. *International Journal of Molecular Sciences*, *24*(4), 3428.

Peterson, C. T. (2020). Dysfunction of the microbiota-gut-brain axis in neurodegenerative disease: The promise of therapeutic modulation with prebiotics, medicinal herbs, probiotics, and synbiotics. *Journal of Evidence-Based Integrative Medicine*, *25*, 2515690X20957225.

Phelps, D., Brinkman, N. E., Keely, S. P., Anneken, E. M., Catron, T. R., Betancourt, D., Wood, C. E., Espenschied, S. T., Rawls, J. F., & Tal, T. (2017). Microbial colonization is required for normal neurobehavioral development in zebrafish. *Scientific Reports*, *7*(1), 11244.

Plaza-Díaz, J., Solís-Urra, P., Rodríguez-Rodríguez, F., Olivares-Arancibia, J., Navarro-Oliveros, M., Abadía-Molina, F., & Álvarez-Mercado, A. I. (2020). The gut barrier, intestinal microbiota, and liver disease: Molecular mechanisms and strategies to manage. *International Journal of Molecular Sciences*, *21*(21), 8351.

Rook, G., Bäckhed, F., Levin, B. R., McFall-Ngai, M. J., & McLean, A. R. (2017). Evolution, human-microbe interactions, and life history plasticity. *The Lancet*, *390*(10093), 521–530.

Rutsch, A., Kantsjö, J. B., & Ronchi, F. (2020). The gut-brain axis: how microbiota and host inflammasome influence brain physiology and pathology. *Frontiers in Immunology*, *11*, 604179.

Sampson, T. R., Challis, C., Jain, N., Moiseyenko, A., Ladinsky, M. S., Shastri, G. G., Thron, T., Needham, B. D., Horvath, I., & Debelius, J. W. (2020). A gut bacterial amyloid promotes α-synuclein aggregation and motor impairment in mice. *Elife*, *9*, e53111.

Shepherd, I., & Eisen, J. (2011). Development of the zebrafish enteric nervous system. *Methods in cell Biology*, *101*, 143–160.

Singh, A., Dawson, T. M., & Kulkarni, S. (2021). Neurodegenerative disorders and gut-brain interactions. *The Journal of Clinical Investigation*, *131*(13). https://doi.org/10.1172/jci143775

Sommer, F., & Bäckhed, F. (2013). The gut microbiota—masters of host development and physiology. *Nature Reviews Microbiology*, *11*(4), 227–238.

Stilling, R. M., Bordenstein, S. R., Dinan, T. G., & Cryan, J. F. (2014). Friends with social benefits: Host-microbe interactions as a driver of brain evolution and development? *Frontiers in Cellular and Infection Microbiology*, *4*, 147.

Tan, F., Limbu, S. M., Qian, Y., Du, Z.-Y., & Zhang, M. (2019). The responses of germ-free zebrafish (Danio rerio) to varying bacterial concentrations, colonization time points, and exposure duration. *Frontiers in Microbiology*, *10*, 477775.

Ufnal, M., & Pham, K. (2017). The gut-blood barrier permeability–a new marker in cardiovascular and metabolic diseases? *Medical Hypotheses*, *98*, 35–37.

Valdes, A. M., Walter, J., Segal, E., & Spector, T. D. (2018). Role of the gut microbiota in nutrition and health. *BMJ*, *361*, 36–44.

Ventura, M., Turroni, F., Motherway, M. O. C., MacSharry, J., & van Sinderen, D. (2012). Host–microbe interactions that facilitate gut colonization by commensal bifidobacteria. *Trends in Microbiology*, *20*(10), 467–476.

Wallace, K. N., Akhter, S., Smith, E. M., Lorent, K., & Pack, M. (2005). Intestinal growth and differentiation in zebrafish. *Mechanisms of Development*, *122*(2), 157–173.

Walter, J., & Ley, R. (2011). The human gut microbiome: ecology and recent evolutionary changes. *Annual Review of Microbiology*, *65*, 411–429.

Wang, C., Lau, C. Y., Ma, F., & Zheng, C. (2021). Genome-wide screen identifies curli amyloid fibril as a bacterial component promoting host neurodegeneration. *Proceedings of the National Academy of Sciences*, *118*(34), e2106504118.

Wang, X., Zhang, J.-B., He, K.-J., Wang, F., & Liu, C.-F. (2021). Advances of zebrafish in neurodegenerative disease: From models to drug discovery. *Frontiers in Pharmacology*, *12*, 713963.

Wei, W., Wang, S., Xu, C., Zhou, X., Lian, X., He, L., & Li, K. (2022). Gut microbiota, pathogenic proteins and neurodegenerative diseases. *Frontiers in Microbiology*, *13*, 959856.

Xia, H., Chen, H., Cheng, X., Yin, M., Yao, X., Ma, J., Huang, M., Chen, G., & Liu, H. (2022). Zebrafish: An efficient vertebrate model for understanding role of gut microbiota. *Molecular Medicine*, *28*(1), 1–20.

Zheng, D., Liwinski, T., & Elinav, E. (2020). Interaction between microbiota and immunity in health and disease. *Cell Research*, *30*(6), 492–506.

Zheng, Y., Bonfili, L., Wei, T., & Eleuteri, A. M. (2023). Understanding the gut–brain axis and its therapeutic implications for neurodegenerative disorders. *Nutrients*, *15*(21), 4631.

Zhong, X., Li, J., Lu, F., Zhang, J., & Guo, L. (2022). Application of zebrafish in the study of the gut microbiome. *Animal Models and Experimental Medicine*, *5*(4), 323–336.

# 14

## *Glyphosate Exposure and Neurological Outcomes in Zebrafish*

**Ana Gomes, Sandra M. Monteiro, Carlos Venâncio and Luís Félix**

### Abbreviations

| | |
|---|---|
| **EPSPS** | 5-Enolpyruvylshikimate-3-Phosphate Synthase |
| **PEP** | Phosphoenolpyruvate |
| **AMPA** | Aminomethylphosphonic Acid |
| **AChE** | Acetylcholinesterase |

### Introduction

Glyphosate (N-(phosphonomethyl) glycine) is the main active compound in several herbicide formulations widely used in agriculture (Serra et al., 2021). It was synthesised in 1950 by Henri Martin, showing no pharmaceutical application. After two decades, its herbicidal functions were described and introduced to the market under the trade name Roundup® (Székács and Darvas, 2012). This herbicide reduces the synthesis of aromatic amino acid compounds (e.g., phenylalanine, tyrosine, and tryptophan) in plants (Uren Webster et al., 2014). The enzyme 5-enolpyruvylshikimate-3-phosphate synthase (EPSPS) is inhibited due to the competition of glyphosate with the enolpyruvyl moiety of phosphoenolpyruvate (PEP) for binding to the active site of EPSPS (Dill et al., 2010). Therefore, glyphosate inhibits the synthesis of necessary proteins and secondary metabolites through the Shikimate pathway (Dill, 2005).

In animals, glyphosate is considered safer than other herbicides. Nevertheless, the widespread and improper use of glyphosate leads to its persistence in the environment, raising some issues regarding its potential environmental impacts on non-target species (Costas-Ferreira et al., 2022). In fact, its presence is commonly registered in rivers (Tu Vu et al., 2023; Geng et al., 2021; Masiol et al., 2018), lakes (Kujawski and Nowakowski, 2023; Berman et al., 2018), groundwater (Van Stempvoort et al., 2016; Lutri et al., 2020) and sediments (Ronco et al., 2016; Feltracco et al., 2022) in concentrations ranging from 0.1 to 1489 μg/L (Alza-Camacho et al., 2016; Brovini et al., 2021; Montiel-Leon et al., 2019). The highest glyphosate concentrations were registered in regions with intense use of herbicides (Brovini et al., 2021). In addition to its environmental risks, the presence of glyphosate residues in human urine samples has raised concerns about its increased exposure in the human population and its potential side effects (Nova et al., 2020; Mills et al., 2017). Aminomethylphosphonic acid (AMPA) is the main metabolite of glyphosate degradation (Grandcoin et al., 2017). The presence of this metabolite has also been determined in soils (Pelosi et al., 2022; Silva et al., 2018) and water bodies (Desmet et al., 2016; Kujawski and Nowakowski, 2023) in concentrations between 0.02 and 234 μg/L (de Matos et al., 2023; Carretta et al., 2019; Ronco et al., 2016). The exposure of non-mammalians to AMPA may provoke oxidative and genetic damage (Figure 14.1) (de Brito Rodrigues et al., 2019; Ivantsova et al., 2022).

According to the World Health Organization, glyphosate residues are not harmful to human health (WHO, 2022). Nonetheless, some studies have reported that glyphosate may induce neurotoxic effects through the inhibition of acetylcholinesterase (AChE) in non-mammalian species (Cattaneo et al., 2011; Lajmanovich et al., 2011; Gholami-Seyedkolaei et al., 2013; Pala, 2019). In zebrafish (*Danio rerio*), an attractive model for ecotoxicological studies (Scholz et al., 2008; Choi et al., 2021), glyphosate has shown to induce malformations (Gaur and Bhargava, 2019; Sulukan et al., 2017), oxidative damage (Lanzarin et al., 2021; Liu et al., 2022), behavioural impairments (Lanzarin et al., 2020; Bridi et al., 2017) and mortality (Panetto et al., 2019; Gaur and Bhargava, 2019) (Figure 14.1).

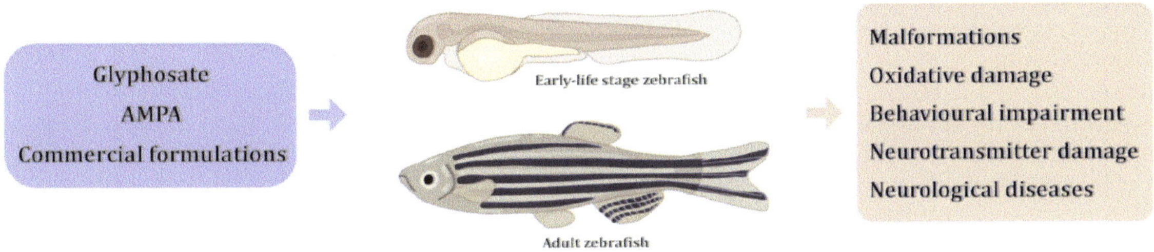


**FIGURE 14.1** Adverse effects due to glyphosate or glyphosate-based commercial formulations and aminomethylphosphonic acid (AMPA) in zebrafish.

DOI: 10.1201/9781032676555-16

**TABLE 14.1**

Studies on the Neurotoxic Effects of Glyphosate or Its Commercial Formulations in Zebrafish

| Compound | Concentration (µg/mL) | Start of exposure | Time of exposure | Neurotoxic effects | Reference |
|---|---|---|---|---|---|
| **Early life stage** | | | | | |
| Glyphosate | 0.01, 0.1 and 1 | 4 hpf | 96 h | Alteration of 18 genes expression related to neurological development at 1 µg/mL. | (Terrazas-Salgado *et al.*, 2022) |
| Roundup® | 0.0048 | 3 hpf | 120 h | Increase in the AChE enzymatic activity in the first generation exposed and in their offspring. | (Pompermaier *et al.*, 2022a; Pompermaier *et al.*, 2022b) |
| Glyphosate | 1, 5, 10, and 50 | 2 hpf | 96 h | Defects in craniofacial development at 50 µg/mL. | (Diaz-Martin *et al.*, 2021a) |
| Glyphosate | 5, 10 and 50 | 2 hpf | 96 h | Reduction of acetylated α-tubulin levels and alteration of microtubule stability (10 and 50 µg/mL). | (Diaz-Martin *et al.*, 2021b) |
| Glyphosate | $5.0\times10^{-5}$, 0.0001, 0.0005, 0.001, 0.01, 0.1, 1, and 10 | 1.5 hpf | 120 h | Induction of midbrain microglia morphological reactivity (1 µg/mL) and gene families' alterations involved in neuronal functions and synaptic transmission (0.001 and 1 µg/mL). | (Forner-Piquer *et al.*, 2021) |
| Glyphosate | 0.01, 0.1, 0.5, 1, 5, and 10 | 0.75 hpf | 96 h | Damage of motoneurons expression at all concentrations. | (Zhang *et al.*, 2017) |
| Glyphosate or Roundup® | 50 | 5 hpf | 24 h | Abnormal brain development, loss of delineated brain ventricles, and inhibition of genes expressed in fore and midbrain regions of the brain. | (Roy *et al.*, 2016) |
| **Adult stage** | | | | | |
| Glyphosate | 1 and 5 | 6 months | 96 h | Induction of the expression of 5HT4R and GNAT2 and disruption of circadian rhythm (5 µg/mL). In the offspring, induce necrosis and degeneration in the brain, activation of anxiety, light signal transduction, and circadian rhythm pathways. | (Sulukan *et al.*, 2023b; Sulukan *et al.*, 2023a) |
| Roundup® | 0.015 and 0.5 | 8–12 months | 14 d | At 0.015 µg/mL the activity of AChE increased and, at 0.5 µg/mL, decreased. | (Falfushynska *et al.*, 2022) |
| Glyphosate | 0.0003 and 0.003 | Unspecified | 14 d | Alteration of the monoaminergic neurotransmitters profile and expression of genes involved in the dopaminergic system in all concentrations of glyphosate at all concentrations. | (Faria *et al.*, 2021) |
| Scout® | 0.065, 1, and 10 | Unspecified | 7 d | Inhibition of cell viability and mitochondrial complex enzymatic activity (0.065 and 1 µg/mL). Modulation of gene expression related to mitochondrial complexes, induction of reactive species production, and hyperpolarization of the mitochondrial membrane in the brain at all concentrations of Scout®. | (Pereira *et al.*, 2018) |
| Glyphosate | 5 and 10 | Unspecified | 24 and 96 h | It did not cause alterations in AChE activity but altered the gene expression of *ache* (10 µg/mL). | (Lopes *et al.*, 2017) |

hpf – hours post-fertilization.

There is controversy about the potential neurotoxicity of glyphosate. Although the European Food Safety Agency (EFSA) classified glyphosate as non-neurotoxic (EFSA, 2023) and the European Commission renewed its use for 10 years (2023–2033) (EUR-Lex, 2023), recent studies have suggested a neurotoxic potential for glyphosate (Barbosa et al., 2001; Terrazas-Salgado et al., 2022; Faria et al., 2021). Zebrafish exhibit less development in several parts, particularly the telencephalon, but a comparable organization of the nervous system and neurotransmitter systems to that observed in humans (de Esch et al., 2012; Tropepe and Sive, 2003). Therefore, it is an appropriate experimental model for neurotoxicology studies (de Esch et al., 2012). Table 14.1 shows the studies on the neurotoxic effects of glyphosate or its commercial formulations in zebrafish.

In the present study, a systematic review was carried out of the neurotoxic effects of glyphosate and its commercial formulations in zebrafish. This chapter aims to understand the impact of glyphosate exposure on development and behavioural responses in zebrafish. Furthermore, the neurotransmission effects and associated neurological diseases were also discussed.

## Effects of Glyphosate and AMPA in Zebrafish

### Developmental Effects

Glyphosate exposure alters nervous system development (Table 14.2), inducing craniofacial malformations (Diaz-Martin et al., 2021a; Bridi et al., 2017), total brain ventricle

**TABLE 14.2**
Studies on the Embryonic Developmental Effects of Glyphosate or its Commercial Formulations in Zebrafish

| Compound | Concentration (μg/mL) | Start of exposure | Time of exposure | Development effects | References |
|---|---|---|---|---|---|
| Glyphosate or Roundup® | 0.25, 0.5, and 1 | 3 hpf | 96 h | Pericardiac and yolk sac oedema, head and tail malformations, and spine curvature at 0.5 and 1 μg/mL of Roundup®. | (Ames *et al.*, 2023) |
| Glyphosate | 0.7, 7, and 35 | 1–1.5 hpf | 120 h | Head, yolk sac, and tail malformations, curvature in the spine and developmental retardation at 7 and 35 μg/mL. Inhibition of the expression of growth receptor hormones (*ghrα* and *ghrβ*) at 0.7 and 7 μg/mL. | (Liu *et al.*, 2022) |
| Glyphosate | $5\times10^{-5}$, 0.0001, 0.0005, 0.001, 0.01, 0.1, 1, and 10 | 1.5 hpf | 120 h | Decrease in the number of rest microglia at 1 μg/mL. | (Forner-Piquer *et al.*, 2021) |
| Glyphosate | 1, 5, 10, and 50 | 2 hpf | 24 h | Cranium malformations and decrease in the ceratohyal cartilage length at 50 μg/mL. The exposed embryos presented bone deficits and spine curvature as adults (10 and 50 μg/mL). | (Diaz-Martin *et al.*, 2021a) |
| Roundup® Ultra Max | 2, 5, and 8.5 | 2.5 hpf | 72 h | Increase in larvae length and eye diameter (8.5 μg/mL). Decrease in the forebrain, midbrain, and hindbrain vesicle length (8.5 μg/mL). | (Lanzarin *et al.*, 2019) |
| Glyphosate | 1, 5, 10, and 100 | 4 hpf | 96 h | Yolk sac malformations, spine curvature, pericardiac oedema, and body malformations (head and tail) at all concentrations used. | (Sulukan *et al.*, 2017) |
| Glyphosate | 0.1, 1, 10, 100, 200, 400, and 600 | 0.75 hpf | 96 h | Developmental retardation (≥ 100 μg/mL). Inhibition of the *ntl* gene expression (notochord formation and cell movement during gastrulation) and increase of the *krox20* gene expression (hindbrain segmentation) at 1, 10, and 100 μg/mL. | (Zhang *et al.*, 2017) |
| Glyphosate or Roundup® | 0.01, 0.065, and 0.5 | 3 dpf | 96 h | Decrease in larvae length (0.01–0.5 μg/mL of Roundup®) and ocular distance (0.5 μg/mL of glyphosate). | (Bridi *et al.*, 2017) |
| Glyphosate or Roundup® | 50 | 5 hpf | 24 h | Partial or total loss of brain ventricles. Inhibition of the expression genes involved in the nervous system (*pax2*, *otx2* and *ephA4*) and eye (*pax6*) development. | (Roy *et al.*, 2016) |

hpf – hours post-fertilization; dpf – days post-fertilization.

loss, flattened brain (Roy et al., 2016), and head malformations (Ames et al., 2023) in zebrafish. The developmental malformations in glyphosate-exposed zebrafish may be mediated through estrogenic-dependent signalling pathways (Sulukan et al., 2017; Diaz-Martin et al., 2021a). In zebrafish embryos, exposure to glyphosate inhibited the expression of carbonic anhydrase (Sulukan et al., 2017). A zinc-containing metalloenzyme which participates in several biological processes, such as estrogenic receptor activity (Bonfanti et al., 2018; Sulukan et al., 2017). The dysregulation of the estrogenic receptor alpha (Erα) has also been observed in zebrafish embryos exposed to glyphosate (Diaz-Martin et al., 2021a).

The changes in several genes' expression seem to indicate that glyphosate exposure may influence nervous system development. In fact, the alteration of the expression of genes involved in neurogenesis (*nav2a*), synaptic transmission (*aqba1a*), and brain development (*pax2*, *otx2*, and *epha4a*) has been observed after glyphosate exposure in zebrafish (Terrazas-Salgado et al., 2022; Roy et al., 2016). Genes coding for glutamate receptor (*grin2*, *gria3*, *grm4*, and *grik5*), gamma-aminobutyric acid (GABA) receptor (*gabra* and *gabrb*), and calcium channels (*cacna*) were upregulated in glyphosate-exposed zebrafish (Forner-Piquer et al., 2021). The mechanism associated with glutamate excitotoxicity and calcium overload involves calcium influx through the activation of N-methyl-D-aspartate (NMDA) glutamate receptors and voltage-dependent calcium channels (Forner-Piquer et al., 2021; Cattani et al., 2017). However, while this mechanism has been described in rats, further elucidation is necessary in zebrafish.

In zebrafish, glyphosate exposure also induces variations in genes coding for growth receptors hormones (*ghrα* and *ghrβ*), bone development (*bsp*), notochord and tail regions formation (*nav2* and *ntl*), and eye development (*pax6*) (Liu et al., 2022; Terrazas-Salgado et al., 2022; Zhang et al., 2017; Bridi et al., 2017; Diaz-Martin et al., 2021a). Indeed, several studies have reported developmental retardation (Zhang et al., 2017; Liu et al., 2022) and malformations in bones (Diaz-Martin et al., 2021a), tail (Liu et al., 2022; Sulukan et al., 2017; Lanzarin et al., 2019; Diaz-Martin et al., 2021a; Ames et al., 2023) and eyes (Bridi et al., 2017; Roy et al., 2016; Lanzarin et al., 2019). The registration of other malformations such as pericardial and yolk sac oedemas and swim bladder deficiency have also been observed in zebrafish exposed to glyphosate (Liu et al., 2022; Sulukan et al., 2017; Ames et al., 2023).

The malformations induced during embryonic development are commonly accompanied by increases in reactive oxygen species production and cellular apoptosis (Sulukan et al., 2017; Lanzarin et al., 2021; Liu et al., 2022; Roy et al., 2016). Glyphosate's characteristics related to polarity and solubility allow it to penetrate the zebrafish chorion (Ames et al., 2022). Once absorbed by the embryos, this herbicide accumulates and induces structural damage (Zhang et al., 2017). Furthermore, glyphosate can disrupt protective cellular mechanisms such as antioxidant and biotransformation metabolisms (Lanzarin et al., 2021; Ames et al., 2023; Falfushynska et al., 2022).

### Behavioural Responses

Glyphosate exposure alters behavioural responses in zebrafish (Table 14.3) (Sulukan et al., 2023a; Lopes et al., 2022). In this species, the neurulation, the process of brain and spinal cord formation, is established by 24 hours post-fertilization (hpf) when the first neurons are interconnected by axons (de Esch et al., 2012). The embryos begin to exhibit vigorous spontaneous movements and touch-evoked responses characterized by disorganized movements (Legradi et al., 2015). The exposure to glyphosate has been shown to impair these movements in zebrafish embryos (Lopes et al., 2022; Durante et al., 2023; Zhang et al., 2017). Thence, the alterations in spontaneous movements might influence the hatching rate of embryos (Jin et al., 2009). The literature reported that the damage to the primary motoneurons due to glyphosate exposure lead to spontaneous movement alterations and, consequently, hatching rate variations (Zhang et al., 2017).

The zebrafish nervous system is fully functional by 120 hpf and it is possible to access well-established behavioural patterns (Legradi et al., 2015). Glyphosate exposure impairs the behavioural activity of zebrafish, including velocity, rest time, and motor coordination (Forner-Piquer et al., 2021; Bridi et al., 2017; Pompermaier et al., 2022a). The incorrect functioning of the cholinergic system associated with AChE variations may be a mechanistic hypothesis (Falfushynska et al., 2022). Furthermore, glyphosate is an acid which causes pain and aversion, and may alter the activity responses in zebrafish (da Costa Chaulet et al., 2019).

The memory impairment has been described as a consequence of glyphosate exposure in zebrafish (Bridi et al., 2017). These alterations are also related to the cholinergic system impairment and could impact social interactions (Talyn et al., 2023).

Exposure to glyphosate may also alter the anxiety-related behavioural responses in zebrafish (Faria et al., 2021; da Costa Chaulet et al., 2019; Pompermaier et al., 2022b; Sulukan et al., 2023a). This herbicide has been described as an endocrine disruptor of the hypothalamus-pituitary-internal axis (HPI), altering the levels of the hormone cortisol (Koakoski et al., 2014; Cericato et al., 2008). The influence of glyphosate in the development of the central nervous system of zebrafish may also alter the response-ability to an aversive stimulus (Bridi et al., 2017; Cericato et al., 2008; Pompermaier et al., 2022b; Pompermaier et al., 2022a; Pompermaier et al., 2020). These behavioural responses may make fish more vulnerable to predation (Lanzarin et al., 2020).

## Neurotransmission Effects and Neurological Diseases

The acetylcholinesterase (AChE) is an essential regulatory enzyme responsible for hydrolysing the neurotransmitter acetylcholine (Gholivand et al., 2009). This neurotransmitter is mainly present in the central and peripheral areas of the nervous system (Woolf and Butcher, 2011). The AChE suppression leads to acetylcholine accumulation in the cholinergic synapses and, subsequently, the stimulation of muscarinic and nicotinic cholinergic receptors (Gholivand et al., 2009). On the other hand, the AChE stimulation results in a rapid acetylcholine degradation and an inhibition of the cholinergic receptors, inducing effects on cognitive functions (Soreq and Seidman, 2001). In fact, the alteration of AChE in zebrafish exposed to glyphosate is strongly associated with behavioural variations as described above (Table 14.3).

Glyphosate exposure may induce glutamate and GABA imbalance (Forner-Piquer et al., 2021). The authors related an upregulation of genes coding for glutamate receptor (*grin2*, *gria3*, *grm4* and *grik5*) and GABA receptor (*gabra* and *gabrb*) to glyphosate exposure in zebrafish (Forner-Piquer et al., 2021). Glutamate is an excitatory neurotransmitter important for the nervous system's flexibility, while GABA is important in regulating brain activity (Zhou and Danbolt, 2014; Ge et al., 2007). These amino-acid neurotransmitters are responsible for maintaining homeostasis, and its dysregulation is associated with Parkinson's and Huntington's diseases in humans (Figure 14.2) (Zhang et al., 2019; Andre et al., 2010). However, it is necessary to clarify whether neurotransmitter dysregulation is associated with neurological diseases in zebrafish.

The dysregulation of the neurotransmitters dopamine, serotonin, and norepinephrine and its acidic metabolites (3,4-dihydroxyphenylacetic acid (DOPAC) and homovanilic acid (HVA)) due to glyphosate exposure, in zebrafish, is documented in the literature (Faria et al., 2021). Furthermore, the upregulation of serotonin receptor gene (5-*ht4r*) and the downregulation of genes coding for synthesis (*th1* and *th2*), degradation (*comtb*) and transport (*scl6a3*) of dopamine was also observed (Sulukan et al., 2023a; Faria et al., 2021). Dopamine is a neurotransmitter that participates in motor coordination, while norepinephrine functions as a neurotransmitter in the autonomic nervous system, including the sympathetic and parasympathetic nervous systems (Bjorklund and Dunnett, 2007; McCorry, 2007). Serotonin controls functions related to the sleep cycle, ingestion, aggression, and depression (Jenkins et al., 2016). The alteration of these amino neurotransmitters is associated with anxiety and exploratory and social behaviour disorders. Indeed, previous studies demonstrated a dysregulation of these neurotransmitters and anxiety-related behavioural responses in zebrafish exposed to glyphosate (Sulukan et al., 2023a; Sulukan et al., 2023b; Faria et al., 2021). Furthermore, glyphosate exposure may induce short-chain fatty acid production in the gut microbiome of zebrafish, leading to an increase in biosynthesis and liberation of serotonin by the intestinal enteroendocrine cells (Bellot et al., 2023). Multiple sclerosis,

**TABLE 14.3**

Studies on the Behavioural Effects of Glyphosate or its Commercial Formulations and Aminomethylphosphonic Acid (Ampa) in Zebrafish.

| Compound | Concentration (μg/mL) | Start of exposure | Time of exposure | Behavioural effects | Reference |
|---|---|---|---|---|---|
| **Early life stage** | | | | | |
| Glyphosate | 0.0169, 0.169, and 1.69 | 5–6 hpf | 7 d | Glyphosate (0.0169 and 1.69 μg/mL) increased the behavioural activity. | (Ivantsova *et al.*, 2022) |
| Roundup® | 0.0048 | 3 hpf | 120 h | Inhibition of distance travelled, rotations number, absolute turn angle, line crossings, and entries in the central area. The number of individuals decreases in the aversive stimulus area. | (Pompermaier *et al.*, 2022a) |
| Roundup® | 0.0048 | 3 hpf | 120 h | Increase in the number of rotations, line crossings, time in the central zone, entries into the central zone, and distance travelled. Increase in the number of individuals in the central zone in response to an aversive stimulus. | (Pompermaier *et al.*, 2022b) |
| Glyphosate | $5\times10^{-5}$, 0.0001, 0.0005, 0.001, 0.01, 0.1, 1, and 10 | 1.5 hpf | 120 h | Decrease in the distance travelled, mean speed, rotations, number and mobile time (≥0.1 μg/mL). | (Forner-Piquer *et al.*, 2021) |
| Glyphosate | 1, 5, 10, and 50 | 2 hpf | 24 h | Increase in tail movements at 50 μg/mL. | (Diaz-Martin *et al.*, 2021a) |
| Roundup® | 1, 2, and 5 | 2.5 hpf | 72 h | Altered locomotion in response to an aversive stimulus at 5 μg/mL. | (Lanzarin *et al.*, 2020) |
| Glyphosate or Roundup® | 0.01, 0.065, and 0.5 | 3 dpf | 96 h | Inhibition of distance travelled (0.01 and 0.5 μg/mL of glyphosate or Roundup®) and absolute turn angle (0.01 μg/mL of glyphosate or Roundup®). Increase in the mobile time (0.065 and 0.5 μg/mL of Roundup®) and the number of individuals in non-stimuli areas (≥0.01 μg/mL of glyphosate and 0.065 and 0.5 μg/mL of Roundup®). | (Bridi *et al.*, 2017) |
| Glyphosate | 0.01, 0.1, 0.5, 1, 5, and 10 | 0.75 hpf | 48 h | Increase in locomotion and inhibition of rest time at all concentrations. | (Zhang *et al.*, 2017) |
| **Adult stage** | | | | | |
| Glyphosate | 0.0003 and 0.003 μg/L | Unspecified | 14 d | Increase in the time and distance travelled in the bottom area of the aquarium (0.003 μg/mL). Decrease in the time spent near the conspecifics in a social preference test (0.003 μg/mL). | (Bellot *et al.*, 2023) |
| Glyphosate | 1 and 5 | 6 months | 96 h | Decrease in the travelled distance, mean speed, and angular speed (5 μg/mL). Increase in the time immobile in the bottom area of the aquarium (5 μg/mL). The offspring showed an increase in locomotion and thigmotaxis. | (Sulukan *et al.*, 2023b; Sulukan *et al.*, 2023a) |
| Roundup® | 0.0048 | 180 dpf | 120 h | Increase in the distance travelled and absolute turn angle. Inhibition of the latency to enter the top zone of the aquarium. Ability impairment to react to an aversive stimulus. | (Pompermaier *et al.*, 2022a) |
| Glyphosate | 0.0003 and 0.003 | Unspecified | 14 d | Inhibition of the time in the upper area of the aquarium and capacity to shoal (0.003 μg/mL). | (Faria *et al.*, 2021) |
| Roundup® | 0.52 | Unspecified | 30 min | Locomotion decreased, entries and time in the central area increased in response to an aversive stimulus. | (Pompermaier *et al.*, 2020) |
| Roundup® | 1, 3 and 5 | Unspecified | 96 h | Increase in the number of rotations (3 μg/mL), crossings between zones, and time in the upper area of the aquarium (3 and 5 μg/mL). | (da Costa Chaulet *et al.*, 2019) |
| Glyphosate or Roundup® | 0.01, 0.065, and 0.5 | 6–8 months | 96 h | Decrease in the distance travelled, mean speed, and line crossings (0.5 μg/mL of glyphosate and 0.065 and 0.5 μg/mL of Roundup®). Impairment of memory at 0.5 μg/mL of Roundup®. Decrease in ocular distance in response to an aggressive-like stimulus at all concentrations of glyphosate and Roundup®. | (Bridi *et al.*, 2017) |
| Roundup® | 0.00659 and 5.2 | Unspecified | 150 sec | Decrease in the number of rotations and mean speed (0.00659 μg/mL). Aversion to glyphosate at 5.2 μg/mL. | (da Rosa *et al.*, 2016) |

hpf – hours post-fertilization; dpf – days post-fertilization

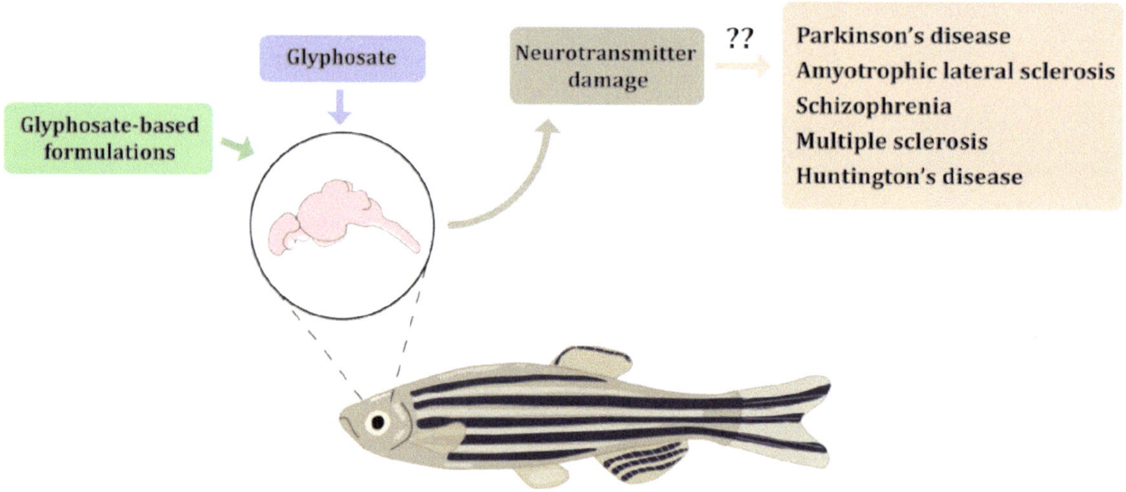


**FIGURE 14.2** Neurotransmitter damage and potential associated neurological diseases due to glyphosate and glyphosate-based commercial formulations.

Parkinson's disease, schizophrenia, and Huntington's disease are associated with the dysregulation of these neurotransmitters in humans (Figure 14.2) (Meisenzahl et al., 2007; Murley and Rowe, 2018; Gasmi et al., 2022). In zebrafish, it is necessary to clarify if neurotransmitter dysregulation is associated with neurological diseases.

Furthermore, (Diaz-Martin et al., 2021b) observed a reduction in acetylated α-tubulin levels and the percentage of polymeric tubulin in zebrafish embryos exposed to glyphosate. The acetylation of α-tubulin on K40 stands out as the main post-translation modification of tubulin, primarily found within the microtubule interior (Eshun-Wilson et al., 2019; Howes et al., 2014). Previous studies have suggested that the inhibition of the levels of K40 acetylation in α-tubulin causes axonal transport defects associated with Huntington's disease, amyotrophic lateral sclerosis, and Parkinson's disease (Eshun-Wilson et al., 2019).

The literature has associated the absolute turn angle impairment with Parkinson-associated symptoms in adult zebrafish exposed to herbicides (Bortolotto et al., 2014). The absolute turn angle is a sensitive measure of motor coordination, and its alteration is associated with neurotransmitter system changes (Irons et al., 2013). The alteration of the absolute turn angle has been widely observed in zebrafish exposed to glyphosate (Bridi et al., 2017; Forner-Piquer et al., 2021; da Costa Chaulet et al., 2019; Pompermaier et al., 2022a).

## Conclusion

The present review shows that exposure to glyphosate or its commercial formulations and AMPA can induce several neurotoxic effects in zebrafish. Glyphosate exposure during the embryonic development of zebrafish can affect normal cell development through the deregulation of different signalling pathways. Moreover, glyphosate exposure impairs behavioural responses in zebrafish. Motor incoordination may be associated with neurotransmitter damage. Glyphosate was found to induce cholinergic system alterations, glutamate and GABA imbalance, and amino neurotransmitter modifications in zebrafish. The damage in the neurotransmitters may be associated with different neurological diseases including multiple sclerosis, Parkinson's disease, schizophrenia, and Huntington's disease. While a direct correlation between glyphosate exposure and the onset of neurological diseases in fish has not been established, there exists a potential association that warrants further investigation into the neurotoxic effects of glyphosate on zebrafish. This research is essential for unravelling the mechanisms through which this herbicide affects organisms.

### Acknowledgement

This work was supported by National Funds by FCT – Portuguese Foundation for Science and Technology, under the projects UID/04033/2023: Centre for the Research and Technology of Agro-Environmental and Biological Sciences and LA/P/0126/2020 (https://doi.org/10.54499/LA/P/0126/2020).

### Conflict of Interest

The authors declare that they have no conflict of interest.

## REFERENCES

Alza-Camacho, W. R., García-Colmenares, J. M. & Chaparro-Acuña, S. P. (2016). Voltammetric quantification of paraquat and glyphosate in surface waters. *Ciencia y Tecnología Agropecuaria*, *17*(3), 331–345.

Ames, J., Miragem, A. A., Cordeiro, M. F., Cerezer, F. O. & Loro, V. L. (2022). Effects of glyphosate on zebrafish: a systematic review and meta-analysis. *Ecotoxicology*, *31*(8), 1189–1204.

Ames, J., Severo, E. S., da Costa-Silva, D. G., Storck, T. R., do Amaral, A. M. B., Miragem, A. A., Rosemberg, D. B. & Loro, V. L. (2023). Glyphosate-based herbicide (GBH) causes damage in embryo-larval stages of zebrafish (Danio rerio). *Neurotoxicol Teratol*, *95*, 107147.

Andre, V. M., Cepeda, C. & Levine, M. S. (2010). Dopamine and glutamate in Huntington's disease: A balancing act. *CNS Neurosci Ther*, *16*(3), 163–178.

Barbosa, E. R., Leiros da Costa, M. D., Bacheschi, L. A., Scaff, M. & Leite, C. C. (2001). Parkinsonism after glycine-derivate exposure. *Mov Disord*, *16*(3), 565–568.

Bellot, M., Carrillo, M. P., Bedrossiantz, J., Zheng, J., Mandal, R., Wishart, D. S., Gomez-Canela, C., Vila-Costa, M., Prats, E., Pina, B. & Raldua, D. (2023). From dysbiosis to neuropathologies: Toxic effects of glyphosate in zebrafish. *Ecotoxicol Environ Saf*, *270*, 115888.

Berman, M. C., Marino, D. J. G., Quiroga, M. V. & Zagarese, H. (2018). Occurrence and levels of glyphosate and AMPA in shallow lakes from the Pampean and Patagonian regions of Argentina. *Chemosphere*, *200*, 513–522.

Bjorklund, A. & Dunnett, S. B. (2007). Dopamine neuron systems in the brain: An update. *Trends Neurosci*, *30*(5), 194–202.

Bonfanti, P., Saibene, M., Bacchetta, R., Mantecca, P. & Colombo, A. (2018). A glyphosate micro-emulsion formulation displays teratogenicity in Xenopus laevis. *Aquat Toxicol*, *195*, 103–113.

Bortolotto, J. W., Cognato, G. P., Christoff, R. R., Roesler, L. N., Leite, C. E., Kist, L. W., Bogo, M. R., Vianna, M. R. & Bonan, C. D. (2014). Long-term exposure to paraquat alters behavioral parameters and dopamine levels in adult zebrafish (Danio rerio). *Zebrafish*, *11*(2), 142–153.

Bridi, D., Altenhofen, S., Gonzalez, J. B., Reolon, G. K. & Bonan, C. D. (2017). Glyphosate and Roundup((R)) alter morphology and behavior in zebrafish. *Toxicology*, *392*, 32–39.

Brovini, E. M., Cardoso, S. J., Quadra, G. R., Vilas-Boas, J. A., Paranaiba, J. R., Pereira, R. O. & Mendonca, R. F. (2021). Glyphosate concentrations in global freshwaters: Are aquatic organisms at risk? *Environ Sci Pollut Res Int*, *28*(43), 60635–60648.

Carretta, L., Cardinali, A., Marotta, E., Zanin, G. & Masin, R. (2019). A new rapid procedure for simultaneous determination of glyphosate and AMPA in water at sub mug/L level. *J Chromatogr A*, *1600*, 65–72.

Cattaneo, R., Clasen, B., Loro, V. L., de Menezes, C. C., Pretto, A., Baldisserotto, B., Santi, A. & de Avila, L. A. (2011). Toxicological responses of Cyprinus carpio exposed to a commercial formulation containing glyphosate. *Bull Environ Contam Toxicol*, *87*(6), 597–602.

Cattani, D., Cesconetto, P. A., Tavares, M. K., Parisotto, E. B., De Oliveira, P. A., Rieg, C. E. H., Leite, M. C., Prediger, R. D. S., Wendt, N. C., Razzera, G., Filho, D. W. & Zamoner, A. (2017). Developmental exposure to glyphosate-based herbicide and depressive-like behavior in adult offspring: Implication of glutamate excitotoxicity and oxidative stress. *Toxicology*, *387*, 67–80.

Cericato, L., Neto, J. G., Fagundes, M., Kreutz, L. C., Quevedo, R. M., Finco, J., da Rosa, J. G., Koakoski, G., Centenaro, L., Pottker, E., Anziliero, D. & Barcellos, L. J. (2008). Cortisol response to acute stress in jundia Rhamdia quelen acutely exposed to sub-lethal concentrations of agrichemicals. *Comp Biochem Physiol C Toxicol Pharmacol*, *148*(3), 281–286.

Choi, T. Y., Choi, T. I., Lee, Y. R., Choe, S. K. & Kim, C. H. (2021). Zebrafish as an animal model for biomedical research. *Experimental and Molecular Medicine*, *53*(3), 310–317.

Costas-Ferreira, C., Duran, R. & Faro, L. R. F. (2022). Toxic effects of glyphosate on the nervous system: A systematic review. *Int J Mol Sci*, *23*(9), 4605.

da Costa Chaulet, F., de Alcantara Barcellos, H. H., Fior, D., Pompermaier, A., Koakoski, G., da Rosa, J. G. S., Fagundes, M. & Barcellos, L. J. G. (2019). Glyphosate- and fipronil-based agrochemicals and their mixtures change zebrafish behavior. *Arch Environ Contam Toxicol*, *77*(3), 443–451.

da Rosa, J. G., de Abreu, M. S., Giacomini, A. C., Koakoski, G., Kalichak, F., Oliveira, T. A., de Alcantara Barcellos, H. H., Barreto, R. E. & Barcellos, L. J. (2016). Fish aversion and attraction to selected agrichemicals. *Arch Environ Contam Toxicol*, *71*(3), 415–422.

de Brito Rodrigues, L., Goncalves Costa, G., Lundgren Tha, E., da Silva, L. R., de Oliveira, R., Morais Leme, D., Cestari, M. M., Koppe Grisolia, C., Campos Valadares, M. & de Oliveira, G. A. R. (2019). Impact of the glyphosate-based commercial herbicide, its components and its metabolite AMPA on non-target aquatic organisms. *Mutat Res Genet Toxicol Environ Mutagen*, *842*, 94–101.

de Esch, C., Slieker, R., Wolterbeek, A., Woutersen, R. & de Groot, D. (2012). Zebrafish as potential model for developmental neurotoxicity testing: a mini review. *Neurotoxicol Teratol*, *34*(6), 545–553.

de Matos, F., de Carvalho, G., Prestes, O., Adaime, M. & Zanella, R. (2023). Simultaneous determination of glyphosate, AMPA and inorganic anions in water samples by gradient capillary ion chromatography. *J Braz Chem Soc*, *34*(11), 1691–1697.

Desmet, N., Touchant, K., Seuntjens, P., Tang, T. & Bronders, J. (2016). A hybrid monitoring and modelling approach to assess the contribution of sources of glyphosate and AMPA in large river catchments. *Sci Total Environ*, *573*, 1580–1588.

Diaz-Martin, R. D., Carvajal-Peraza, A., Yanez-Rivera, B. & Betancourt-Lozano, M. (2021a). Short exposure to glyphosate induces locomotor, craniofacial, and bone disorders in zebrafish (Danio rerio) embryos. *Environ Toxicol Pharmacol*, *87*, 103700.

Diaz-Martin, R. D., Valencia-Hernandez, J. D., Betancourt-Lozano, M. & Yanez-Rivera, B. (2021b). Changes in microtubule stability in zebrafish (Danio rerio) embryos after glyphosate exposure. *Heliyon*, *7*(1), e06027.

Dill, G. M. (2005). Glyphosate-resistant crops: History, status and future. *Pest Manag Sci*, *61*(3), 219–224.

Dill, G. M., Sammons, R. D., Feng, P. C., Kohn, F., Kretzmer, K., Mehrsheikh, A., Bleeke, M., Honegger, J. L., Farmer, D. & Wright, D. (2010). Glyphosate: Discovery, development, applications, and properties. *Glyph Resist Crops Weeds His Dev Manage*, *1*, 344.

Durante, L. D. S., Hollmann, G. & Nazari, E. M. (2023). Impact of exposure to glyphosate-based herbicide on morphological and physiological parameters in embryonic and larval development of zebrafish. *Environ Toxicol*, *39*(3),1822–1835.

EFSA, E. F. S. A. (2023). Glyphosate: No critical areas of concern; data gaps identified.

Eshun-Wilson, L., Zhang, R., Portran, D., Nachury, M. V., Toso, D. B., Lohr, T., Vendruscolo, M., Bonomi, M., Fraser, J. S. & Nogales, E. (2019). Effects of alpha-tubulin acetylation on microtubule structure and stability. *Proc Natl Acad Sci U S A*, *116*(21), 10366–10371.

EUR-Lex (2023). Commission Implementing Regulation (EU) 2023/2660 of 28 November 2023 renewing the approval of the active substance glyphosate in accordance with Regulation (EC) No 1107/2009 of the European Parliament and of the Council and amending Commission Implementing Regulation (EU) No 540/2011.

Falfushynska, H., Khatib, I., Kasianchuk, N., Lushchak, O., Horyn, O. & Sokolova, I. M. (2022). Toxic effects and mechanisms of common pesticides (Roundup and chlorpyrifos) and their mixtures in a zebrafish model (Danio rerio). *Sci Total Environ, 833*, 155236.

Faria, M., Bedrossiantz, J., Ramirez, J. R. R., Mayol, M., Garcia, G. H., Bellot, M., Prats, E., Garcia-Reyero, N., Gomez-Canela, C., Gomez-Olivan, L. M. & Raldua, D. (2021). Glyphosate targets fish monoaminergic systems leading to oxidative stress and anxiety. *Environ Int, 146*, 106253.

Feltracco, M., Barbaro, E., Morabito, E., Zangrando, R., Piazza, R., Barbante, C. & Gambaro, A. (2022). Assessing glyphosate in water, marine particulate matter, and sediments in the Lagoon of Venice. *Environ Sci Pollut Res Int, 29*(11), 16383–16391.

Forner-Piquer, I., Faucherre, A., Byram, J., Blaquiere, M., de Bock, F., Gamet-Payrastre, L., Ellero-Simatos, S., Audinat, E., Jopling, C. & Marchi, N. (2021). Differential impact of dose-range glyphosate on locomotor behavior, neuronal activity, glio-cerebrovascular structures, and transcript regulations in zebrafish larvae. *Chemosphere, 267*, 128986.

Gasmi, A., Nasreen, A., Menzel, A., Gasmi Benahmed, A., Pivina, L., Noor, S., Peana, M., Chirumbolo, S. & Bjorklund, G. (2022). Neurotransmitters regulation and food intake: The role of dietary sources in neurotransmission. *Molecules, 28*(1), 210.

Gaur, H. & Bhargava, A. (2019). Glyphosate induces toxicity and modulates calcium and NO signaling in zebrafish embryos. *Biochem Biophys Res Commun, 513*(4), 1070–1075.

Ge, S., Pradhan, D. A., Ming, G. L. & Song, H. (2007). GABA sets the tempo for activity-dependent adult neurogenesis. *Trends Neurosci, 30*(1), 1–8.

Geng, Y., Jiang, L., Zhang, D., Liu, B., Zhang, J., Cheng, H., Wang, L., Peng, Y., Wang, Y., Zhao, Y., Xu, Y. & Liu, X. (2021). Glyphosate, aminomethylphosphonic acid, and glufosinate ammonium in agricultural groundwater and surface water in China from 2017 to 2018: Occurrence, main drivers, and environmental risk assessment. *Sci Total Environ, 769*, 144396.

Gholami-Seyedkolaei, S. J., Mirvaghefi, A., Farahmand, H. & Kosari, A. A. (2013). Effect of a glyphosate-based herbicide in Cyprinus carpio: assessment of acetylcholinesterase activity, hematological responses and serum biochemical parameters. *Ecotoxicol Environ Saf, 98*, 135–141.

Gholivand, K., Abdollahi, M., Mojahed, F., Alizadehgan, A. M. & Dehghan, G. (2009). Acetylcholinesterase/Butyrylch olinesterase inhibition activity of some new carbacylamidophosphate derivatives. *Journal of Enzyme Inhibition and Medicinal Chemistry, 24*(2), 566–576.

Grandcoin, A., Piel, S. & Baures, E. (2017). AminoMethylPhosphonic acid (AMPA) in natural waters: Its sources, behavior and environmental fate. *Water Res, 117*, 187–197.

Howes, S. C., Alushin, G. M., Shida, T., Nachury, M. V. & Nogales, E. (2014). Effects of tubulin acetylation and tubulin acetyltransferase binding on microtubule structure. *Mol Biol Cell, 25*(2), 257–266.

Irons, T. D., Kelly, P. E., Hunter, D. L., MacPhail, R. C. & Padilla, S. (2013). Acute administration of dopaminergic drugs has differential effects on locomotion in larval zebrafish. *Pharmacol Biochem Behav, 103*(4), 792–813.

Ivantsova, E., Wengrovitz, A. S., Souders, C. L., 2nd & Martyniuk, C. J. (2022). Developmental and behavioral toxicity assessment of glyphosate and its main metabolite aminomethylphosphonic acid (AMPA) in zebrafish embryos/larvae. *Environ Toxicol Pharmacol, 93*, 103873.

Jenkins, T. A., Nguyen, J. C., Polglaze, K. E. & Bertrand, P. P. (2016). Influence of Tryptophan and Serotonin on Mood and Cognition with a Possible Role of the Gut-Brain Axis. *Nutrients, 8*(1), 56.

Jin, M., Zhang, X., Wang, L., Huang, C., Zhang, Y. & Zhao, M. (2009). Developmental toxicity of bifenthrin in embryo-larval stages of zebrafish. *Aquat Toxicol, 95*(4), 347–354.

Koakoski, G., Quevedo, R. M., Ferreira, D., Oliveira, T. A., da Rosa, J. G., de Abreu, M. S., Gusso, D., Marqueze, A., Kreutz, L. C., Giacomini, A. C., Fagundes, M. & Barcellos, L. J. (2014). Agrichemicals chronically inhibit the cortisol response to stress in fish. *Chemosphere, 112*, 85–91.

Kujawski, E. & Nowakowski, A. (2023). Glyphosate is polluting our waters-all across Europe. All across Europe. Pan Europe's water report, September 2023.

Lajmanovich, R. C., Attademo, A. M., Peltzer, P. M., Junges, C. M. & Cabagna, M. C. (2011). Toxicity of four herbicide formulations with glyphosate on Rhinella arenarum (anura: bufonidae) tadpoles: B-esterases and glutathione S-transferase inhibitors. *Arch Environ Contam Toxicol, 60*(4), 681–689.

Lanzarin, G., Venancio, C., Felix, L. M. & Monteiro, S. (2021). Inflammatory, oxidative stress, and apoptosis effects in zebrafish larvae after rapid exposure to a commercial glyphosate formulation. *Biomedicines, 9*(12), 1784.

Lanzarin, G. A. B., Felix, L. M., Santos, D., Venancio, C. A. S. & Monteiro, S. M. (2019). Dose-dependent effects of a glyphosate commercial formulation - Roundup((R)) UltraMax - on the early zebrafish embryogenesis. *Chemosphere, 223*, 514–522.

Lanzarin, G. A. B., Venancio, C. A. S., Monteiro, S. M. & Felix, L. M. (2020). Behavioural toxicity of environmental relevant concentrations of a glyphosate commercial formulation - RoundUp(R) UltraMax - During zebrafish embryogenesis. *Chemosphere, 253*, 126636.

Legradi, J., el Abdellaoui, N., van Pomeren, M. & Legler, J. (2015). Comparability of behavioural assays using zebrafish larvae to assess neurotoxicity. *Environ Sci Pollut Res Int, 22*(21), 16277–16289.

Liu, Z., Shangguan, Y., Zhu, P., Sultan, Y., Feng, Y., Li, X. & Ma, J. (2022). Developmental toxicity of glyphosate on embryo-larval zebrafish (Danio rerio). *Ecotoxicol Environ Saf, 236*, 113493.

Lopes, A. R., Moraes, J. S. & Martins, C. M. G. (2022). Effects of the herbicide glyphosate on fish from embryos to adults: a review addressing behavior patterns and mechanisms behind them. *Aquat Toxicol, 251*, 106281.

Lopes, F. M., Caldas, S. S., Primel, E. G. & da Rosa, C. E. (2017). Glyphosate adversely affects danio rerio males: Acetylcholinesterase modulation and oxidative stress. *Zebrafish, 14*(2), 97–105.

Lutri, V. F., Matteoda, E., Blarasin, M., Aparicio, V., Giacobone, D., Maldonado, L., Becher Quinodoz, F., Cabrera, A. & Giuliano Albo, J. (2020). Hydrogeological features affecting spatial distribution of glyphosate and AMPA in groundwater and surface water in an agroecosystem. Cordoba, Argentina. *Sci Total Environ*, *711*, 134557.

Masiol, M., Gianni, B. & Prete, M. (2018). Herbicides in river water across the northeastern Italy: occurrence and spatial patterns of glyphosate, aminomethylphosphonic acid, and glufosinate ammonium. *Environ Sci Pollut Res Int*, *25*(24), 24368–24378.

McCorry, L. K. (2007). Physiology of the autonomic nervous system. *Am J Pharm Educ*, *71*(4), 78.

Meisenzahl, E. M., Schmitt, G. J., Scheuerecker, J. & Moller, H. J. (2007). The role of dopamine for the pathophysiology of schizophrenia. *Int Rev Psychiatry*, *19*(4), 337–345.

Mills, P. J., Kania-Korwel, I., Fagan, J., McEvoy, L. K., Laughlin, G. A. & Barrett-Connor, E. (2017). Excretion of the herbicide glyphosate in older adults between 1993 and 2016. *Jama-Journal of the American Medical Association*, *318*(16), 1610–1611.

Montiel-Leon, J. M., Munoz, G., Vo Duy, S., Do, D. T., Vaudreuil, M. A., Goeury, K., Guillemette, F., Amyot, M. & Sauve, S. (2019). Widespread occurrence and spatial distribution of glyphosate, atrazine, and neonicotinoids pesticides in the St. Lawrence and tributary rivers. *Environ Pollut*, *250*, 29–39.

Murley, A. G. & Rowe, J. B. (2018). Neurotransmitter deficits from frontotemporal lobar degeneration. *Brain*, *141*(5), 1263–1285.

Nova, P., Calheiros, C. S. C. & Silva, M. (2020). Glyphosate in Portuguese adults - A pilot study. *Environ Toxicol Pharmacol*, *80*, 103462.

Word Health Organization, (WHO) (2022). Chemical fact sheets: Glyphosate and AMPA. https://www.who.int/publications/m/item/chemical-fact-sheets--glyphosate-and-ampa.

Pala, A. (2019). The effect of a glyphosate-based herbicide on acetylcholinesterase (AChE) activity, oxidative stress, and antioxidant status in freshwater amphipod: Gammarus pulex (Crustacean). *Environ Sci Pollut Res Int*, *26*(36), 36869–36877.

Panetto, O. S., Gomes, H. F., Fraga Gomes, D. S., Campos, E., Romeiro, N. C., Costa, E. P., do Carmo, P. R. L., Feitosa, N. M. & Moraes, J. (2019). The effects of Roundup(R) in embryo development and energy metabolism of the zebrafish (Danio rerio). *Comp Biochem Physiol C Toxicol Pharmacol*, *222*, 74–81.

Pelosi, C., Bertrand, C., Bretagnolle, V., Coeurdassier, M., Delhomme, O., Deschamps, M., Gaba, S., Millet, M., Nelieu, S. & Fritsch, C. (2022). Glyphosate, AMPA and glufosinate in soils and earthworms in a French arable landscape. *Chemosphere*, *301*, 134672.

Pereira, A. G., Jaramillo, M. L., Remor, A. P., Latini, A., Davico, C. E., da Silva, M. L., Muller, Y. M. R., Ammar, D. & Nazari, E. M. (2018). Low-concentration exposure to glyphosate-based herbicide modulates the complexes of the mitochondrial respiratory chain and induces mitochondrial hyperpolarization in the Danio rerio brain. *Chemosphere*, *209*, 353–362.

Pompermaier, A., Kirsten, K., Soares, S. M., Fortuna, M., Kalichak, F., Idalencio, R., Koakoski, G., Barreto, R. E. & Barcellos, L. J. G. (2020). Waterborne agrichemicals compromise the anti-predatory behavior of zebrafish. *Environ Sci Pollut Res Int*, *27*(31), 38559–38567.

Pompermaier, A., Tamagno, W. A., Alves, C. & Barcellos, L. J. G. (2022a). Persistent and transgenerational effects of pesticide residues in zebrafish. *Comp Biochem Physiol C Toxicol Pharmacol*, *262*, 109461.

Pompermaier, A., Varela, A. C. C., Mozzato, M. T., Soares, S. M., Fortuna, M., Alves, C., Tamagno, W. A. & Barcellos, L. J. G. (2022b). Impaired initial development and behavior in zebrafish exposed to environmentally relevant concentrations of widely used pesticides. *Comp Biochem Physiol C Toxicol Pharmacol*, *257*, 109328.

Ronco, A. E., Marino, D. J., Abelando, M., Almada, P. & Apartin, C. D. (2016). Water quality of the main tributaries of the Parana Basin: glyphosate and AMPA in surface water and bottom sediments. *Environ Monit Assess*, *188*(8), 458.

Roy, N. M., Carneiro, B. & Ochs, J. (2016). Glyphosate induces neurotoxicity in zebrafish. *Environ Toxicol Pharmacol*, *42*, 45–54.

Scholz, S., Fischer, S., Gundel, U., Kuster, E., Luckenbach, T. & Voelker, D. (2008). The zebrafish embryo model in environmental risk assessment–applications beyond acute toxicity testing. *Environ Sci Pollut Res In*, *15*(5), 394–404.

Serra, L., Estienne, A., Vasseur, C., Froment, P. & Dupont, J. (2021). Review: Mechanisms of glyphosate and glyphosate-based herbicides action in female and male fertility in humans and animal models. *Cells*, *10*(11), 3079.

Silva, V., Montanarella, L., Jones, A., Fernandez-Ugalde, O., Mol, H. G. J., Ritsema, C. J. & Geissen, V. (2018). Distribution of glyphosate and aminomethylphosphonic acid (AMPA) in agricultural topsoils of the European Union. *Sci Total Environ*, *621*, 1352–1359.

Soreq, H. & Seidman, S. (2001). Acetylcholinesterase--new roles for an old actor. *Nat Rev Neurosci*, *2*(4), 294–302.

Sulukan, E., Baran, A., Kankaynar, M., Kiziltan, T., Bolat, I., Yildirim, S., Ceyhun, H. A. & Ceyhun, S. B. (2023a). Global warming and glyphosate toxicity (II): Offspring zebrafish modelling with behavioral, morphological and immunohistochemical approaches. *Sci Total Environ*, *856*(Pt 1), 158903.

Sulukan, E., Baran, A., Senol, O., Kankaynar, M., Yildirim, S., Bolat, I., Ceyhun, H. A., Toraman, E. & Ceyhun, S. B. (2023b). Global warming and glyphosate toxicity (I): Adult zebrafish modelling with behavioural, immunohistochemical and metabolomic approaches. *Sci Total Environ*, *858*(Pt 3), 160086.

Sulukan, E., Kokturk, M., Ceylan, H., Beydemir, S., Isik, M., Atamanalp, M. & Ceyhun, S. B. (2017). An approach to clarify the effect mechanism of glyphosate on body malformations during embryonic development of zebrafish (Daino rerio). *Chemosphere*, *180*, 77–85.

Székács, A. & Darvas, B. (2012). Forty years with glyphosate. *Herb Proper Synt Control Weeds*, *14*, 247–284.

Talyn, B., Muller, K., Mercado, C., Gonzalez, B. & Bartels, K. (2023). The herbicide glyphosate and its formulations impact animal behavior across taxa. *Agrochemicals*, *2*(3), 367–408.

Terrazas-Salgado, L., Yanez-Rivera, B., Llera-Herrera, R., Garcia-Gasca, A., Alvarado-Cruz, I. & Betancourt-Lozano, M. (2022). Transcriptomic signaling in zebrafish (Danio rerio) embryos exposed to environmental concentrations of glyphosate. *J Environ Sci Health B*, *57*(10), 775–785.

Tropepe, V. & Sive, H. L. (2003). Can zebrafish be used as a model to study the neurodevelopmental causes of autism? *Genes, Brain and Behavior*, *2*(5), 268–281.

Tu Vu, C., Binh Chu, D., Mai, H., Herrmann, M., Hoi Bui, V., Thu Le, P., Hoang Anh Chu, N., Tedetti, M. & Behra, P. (2023). First hydrological study on the seasonal occurrence of glyphosate, glufosinate, and their metabolites in the Red River system, North Vietnam. *Environ Nanotechnol Monit Manag*, *20*, 100833.

Uren Webster, T. M., Laing, L. V., Florance, H. & Santos, E. M. (2014). Effects of glyphosate and its formulation, roundup, on reproduction in zebrafish (Danio rerio). *Environ Sci Technol*, *48*(2), 1271–1279.

Van Stempvoort, D. R., Spoelstra, J., Senger, N. D., Brown, S. J., Post, R. & Struger, J. (2016). Glyphosate residues in rural groundwater, Nottawasaga River Watershed, Ontario, Canada. *Pest Manag Sci*, *72*(10), 1862–1872.

Woolf, N. J. & Butcher, L. L. (2011). Cholinergic systems mediate action from movement to higher consciousness. *Behav Brain Res*, *221*(2), 488–498.

Zhang, S., Xu, J., Kuang, X., Li, S., Li, X., Chen, D., Zhao, X. & Feng, X. (2017). Biological impacts of glyphosate on morphology, embryo biomechanics and larval behavior in zebrafish (Danio rerio). *Chemosphere*, *181*, 270–280.

Zhang, Z., Zhang, S., Fu, P., Zhang, Z., Lin, K., Ko, J. K. & Yung, K. K. (2019). Roles of Glutamate Receptors in Parkinson's Disease. *Int J Mol Sci*, *20*(18), 4391.

Zhou, Y. & Danbolt, N. C. (2014). Glutamate as a neurotransmitter in the healthy brain. *Journal of neural transmission*, *121*(8), 799–817.

# 15

# *Zebrafish Model Organism to Investigate Pain in Neurodegenerative Diseases*

**Derya Cansiz**

## Abbreviations

| | |
|---|---|
| **5-HT** | 5-Hydroxytryptamine |
| **AD** | Alzheimer's Disease |
| **AMPA** | A-Amino-3-Hydroxy-5-Methyl-4-İsoxazolepr opionic Acid |
| **AMPA-R** | Amino-3-Hydroxy-5-Methyl-4-İsoxazoleprop ionic Acid Receptors |
| **ASIC** | Acid-Sensing İon Channels |
| **ATP** | Adenosine Triphosphate |
| **BK** | Bradykinin |
| **CGRP** | Calcitonin Gene Related Peptide |
| **CNS** | Central Nervous System |
| **COX** | Cyclooxygenase |
| **DAT** | Dopamine Transporters |
| **DH** | Dorsal Horn |
| **DRG** | Dorsal Root Ganglia |
| **EPSP** | Excitatory Postsynaptic Potentials |
| **GABA** | Gamma-Aminobutyric Acid |
| **GPCR** | G Protein Coupled Receptors |
| **HD** | Huntington's Disease |
| **hpf** | Hours Post-Fertilization |
| **IPSP** | Inhibitory Postsynaptic Potentials |
| **KC** | Potassium Channels |
| **Kv** | Voltage-Activated K+ Channels |
| **MPTP** | 1-Methyl-4-Phenyl-1, 2, 3, 6-Tetrahydropyridine |
| **Nav** | Voltage-Activated Na+ Channel |
| **NE** | Norepinephrine |
| **NGF** | Nerve Growth Factors |
| **NK1** | Neurokinin 1 |
| **NMDA** | N-Methyl-D-Aspartate |
| **NMDA-R** | N-Methyl-D-Aspartate Receptors |
| **NO** | Nitric Oxide |
| **PAG** | Periaqueductal Gray |
| **PD** | Parkinson's Disease |
| **PG** | Prostaglandin |
| **PNS** | Peripheral Nervous System |
| **RVM** | Rostral Ventral Medulla |
| **SP** | Substance P |
| **TAC** | Tachykinin Precursor Gene |
| **TRP** | Transient Receptor Potential |
| **VGCC** | Voltage-Gated Calcium Channels |
| **VGSC** | Voltage-Gated Sodium Channels |
| **VOCC** | Voltage-Operated $Ca^{2+}$ Channels |

## Introduction

In recent years, neurodegenerative diseases have become the subject of more study due to increasing life expectancy and the increasing prevalence of age-related brain disorders. Neurodegenerative diseases encompass a wide range of diseases with diverse etiologies. Common symptoms of these diseases are neurodegeneration, oxidative stress, and mitochondrial dysfunction with aging (Hroudová et al., 2014).

Neurodegenerative diseases have many motor and non-motor symptoms. In addition to important motor symptoms such as bradykinesia, tremor, and rigidity, important non-motor symptoms such as sleep disturbance, constipation, fatigue, and pain are also seen in neurodegenerative diseases.

Movement disorders, coordination difficulties, and cognitive disorders are important chronic conditions seen in neurodegenerative diseases and examined in the clinic. In diseases such as dementia and motoneuronal and extrapyramidal disorders, it is often difficult to establish the connection between pathology and the experience of pain. For this reason, new studies have begun to be included in the evaluation of pain symptoms in neurodegenerative diseases (Tommaso, 2016).

Despite numerous studies on pain, a significant non-motor symptom in neurodegenerative diseases, the mechanism of pain remains incompletely understood. Animal experiments play an important role in research on pain mechanisms and perception. Mammals such as mice and rats, as well as teleosts, are considered important experimental models in the study of pain and neurodegenerative diseases (Manasa et al., 2020; Costa et al., 2019; Deakin et al., 2019; Cansiz et al., 2023; Ünal et al., 2023). Researchers conducted various pain studies on teleosts, testing new analgesic drug formulas and investigating the roles of chemicals on ascending and descending pathways in pain mechanism and pain modulation.

DOI: 10.1201/9781032676555-17

Zebrafish, an important teleost model, are tropical aquatic animals. Zebrafish are considered an important model due to their approximately 70% orthology with humans, ease of growth, transparent appearance during the embryonic period, and the fact that their development can be easily observed under a microscope, enabling the extraction of hundreds of embryos at a time. Additionally, researchers frequently use zebrafish in studies to understand the mechanisms of diseases like cancer, neurodegenerative diseases, pain, diabetes, and obesity.

Due to the presence of specific brain regions and the high compatibility of the dopaminergic, serotonergic, and histaminergic systems, zebrafish are considered a very efficient experimental model in the study of neurodegenerative diseases (Manasa et al., 2020).

## Pain

According to the International Association for the Study of Pain (IASP), pain is defined as "an unpleasant sensory and emotional experience associated with or resembling actual or potential tissue damage" (Raja et al., 2020).

Pain is one of the most common non-motor symptoms seen in neurodegenerative diseases. Non-motor symptoms increasingly outnumber motor symptoms in some neurodegenerative diseases (Moro, 2019). Numerous studies have explored the mechanism of pain in neurodegenerative diseases, revealing a variety of data as the cause (Moro, 2019). However, the mechanism of pain in these diseases has not yet been fully understood.

Motor and cognitive disorders are the primary focus of the management and treatment of neurodegenerative diseases, including Alzheimer's disease and other dementias, Parkinson's disease and its related disorders, motor neuron diseases, and Huntington's disease (Tommaso, 2016).

Despite having different origins, most of these diseases exhibit painful symptoms. However, the treatment process often ignores these symptoms, leaving the clinician solely responsible for managing them. Recent years have recognized pain as a common and unvalidated symptom in some neurodegenerative diseases, including Parkinson's disease (Tommaso, 2016; Ameenudeen et al., 2022).

Pain can be expressed as stimulating the body's defense mechanism to react against a harmful stimulus to prevent further damage. The sensation of pain is associated with the activation of receptors in primary afferent fibers, including the unmyelinated C fiber and the myelinated A$\sigma$ fiber. During homeostasis, in the absence of pain, both nociceptors remain silent but are activated when there is a noxious stimulus potential. In order for the brain to perceive pain and produce a response to harmful stimuli, a series of sensory events must be perceived (Figure 15.1). The "ascending pathway" is made up of the upward nerves that transmit sensory information from the body to the brain through the spinal cord, while the "descending pathway" refers to the nerves that go downward from the brain to the reflex organs through the spinal cord (Yam et al., 2018).

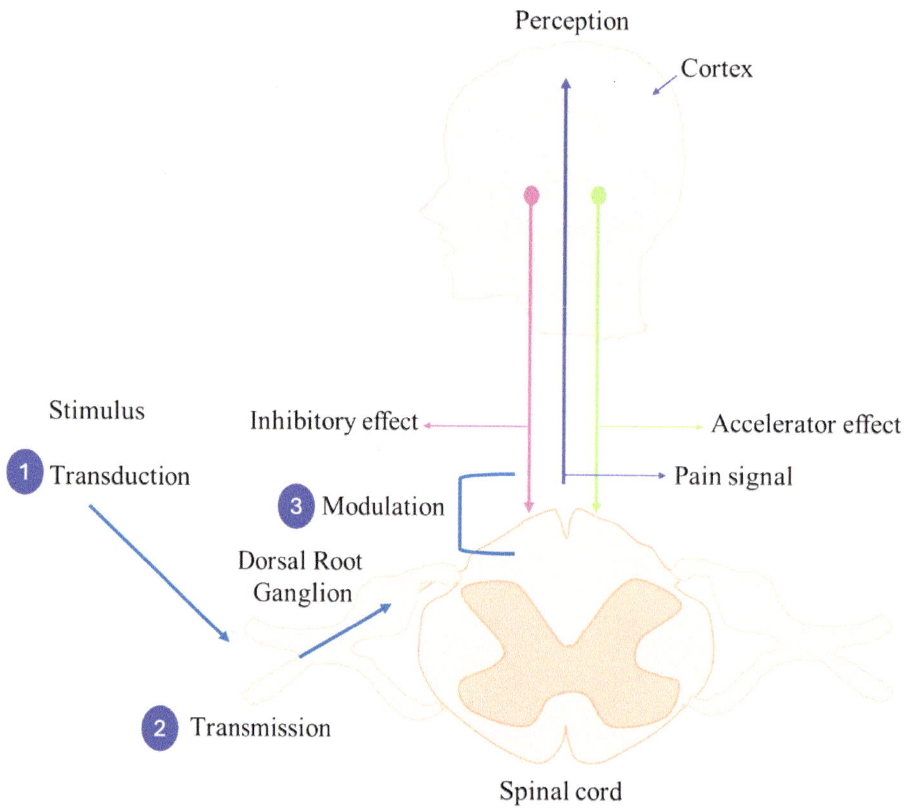


**FIGURE 15.1** Mechanistic action of synapse in pain signals.

The central nervous system (CNS) and peripheral nervous system (PNS) are involved in all mechanisms and pathways of pain perception. The PNS includes nerves and ganglia that are located outside the brain and spinal cord and essentially connect the CNS to the organs and limbs in our body. The CNS consists of the spinal cord and brain. The brain gathers information from the PNS, interprets it, and coordinates all bodily activities before responding to the effector organs (Yam et al., 2018).

## Mechanisms of Pain

Basically, the pain mechanism consists of three stages in the presence of harmful stimuli: transduction, transmission, and modulation. Transduction proceeds in the following order: 1) the stimulus signal is converted into chemical tissue signals; 2) chemical tissue and synaptic cleft signals are then converted into electrical signals in neurons; and 3) electrical signals in neurons are transmitted as chemical signals at synapses. The next stage, after the completion of transduction, is transmission. The transmission process entails transmitting electrical signals along neuronal pathways, whereas neurotransmitters in the synaptic cleft transmit information from one cell's post-synaptic terminal to another's pre-synaptic terminal. The modulation event occurs through up- or downregulation at all levels of nociceptive pathways, including the primary afferent neuron, dorsal horn (DH), and higher brain center. Thus, the stimulus initiates and completes the path of pain, enabling us to experience the pain (Yam et al., 2018).

### Neurons

Neurons are the primary component that connects, receives, and processes all nociceptive information. There are three types of neurons in our body: sensory neurons (afferent neurons), interneurons (functions of transmitting signals between afferent and efferent neurons), and motor neurons (efferent neurons). Soma, axons (myelinated or unmyelinated), and dendrites make up all neurons, which are electrically excitable. Axons transmit signals to neurons after they receive them via dendrites or the soma. The neuron generates short pulses known as action potentials, which propagate through the soma, travel along axons to other neurons to activate synapses, and act as a pathway to carry signals from their source either to the spinal cord or brain. There are two main classifications of specialized neurons: sensory neurons and motor neurons. While sensory neurons located in the dermis and epidermis send these signals when a stimulus is present, the main function of motor neurons is to receive signals from the brain and spinal cord (Yam et al., 2018).

### Axons

Axons are nerve fibers that function to transmit action potentials in one direction, from dendrites to axonal terminals, and from one neuron to another. The presence of the myelin sheath of the axon increases the speed of propagation of impulses because they proceed by jump conduction along the myelinated fiber and act as an insulator during conduction, preventing electrical impulses from leaving the axon.

Afferent or efferent nerve fibers can generally reflect electrical impulses traveling along axons. Afferent nerve fibers, also known as sensory neurons in the PNS, transport sensory information from various body parts to the spinal cord through their axons. Efferent nerve fibers in the PNS are pre- and post-ganglionic motor neurons that carry motor movement signals from the spine to peripheral effector organs, including skeletal and soft muscles. In the PNS, afferent neuron somas reside in the ganglia, and their axons transmit electrical impulses from ganglion to ganglion and eventually to the spine. The dorsal root axons of sensory neurons are primarily responsible for transmitting somatosensory information through interaction with somatosensory receptors (Yam et al., 2018).

When nociceptors are stimulated by a noxious stimulus, two types of potentials are generated and accumulated in the axon hillock: inhibitory postsynaptic potentials (IPSP) and excitatory postsynaptic potentials (EPSP). Upon reaching the trigger threshold, the neurons' axons propagate the action potential. When $Na^+$ enters through a voltage-activated $Na^+$ channel (Nav), the action potential begins. All Nav channels in the axon hillock stimulate and open when the threshold potential reaches, causing a complete depolarization until the neurons reach their peak potential (+40 mV). Nav channels return to their resting state at this point, stimulating and opening voltage-activated $K^+$ channels (Kv) to efflux $K^+$, leading to the repolarization of neurons. During neurons' refractory period, the continuous influx of $K^+$ through Kv and $K^+$ leak channels causes hyperpolarization of the membrane potential. Eventually, K+ channels close, and $Na^+/K^+$ transporters restore the resting potential, allowing three $Na^+$ to enter and two $K^+$ to exit. When the action potential goes to the axon terminal, voltage-operated $Ca^{2+}$ channels (VOCC) let $Ca^{2+}$ into the presynaptic terminal. This makes the synaptic transmission happen (Purves et al., 2018).

Synaptic transmissions are chemical processes that transmit impulses between neurons. There is a synaptic gap between the presynaptic and postsynaptic membranes, and chemical synapses occur there. Synaptic transmission begins when the action potential reaches the presynaptic axon terminal. An action potential at the presynaptic terminal produces membrane depolarization, which causes Nav channels to open in the terminal. The $Na^+$ entry leads to further activation of the VOCC, allowing $Ca^{2+}$ to enter the axon terminal. These calcium ions attach to calcium-sensing proteins at the end of the axon. These proteins then work with soluble N-ethylmaleimide-sensitive factor activating protein receptor (SNARE) proteins. SNARE proteins trigger the fusion of synaptic vesicles (also known as neurotransmitter vesicles) to the presynaptic axon terminal membrane in neurons. This fusion causes neuronal contents, including neurotransmitters and $Ca^{2+}$, to be released into the synaptic cleft by exocytosis. Free neurotransmitters diffuse across the synaptic cleft and bind to ligand-gated ion channels on the membrane of the adjacent postsynaptic neuron, causing an action potential in the axon of the second neuron (Yam et al., 2018).

## Pain Transmission Pathway

The spinal cord, brainstem, and cortical region (cerebral cortex) located in the CNS are important in the transmission of pain. The ascending system's primary afferent nociceptors transmit the noxious stimulus from the projection neurons in the spinal cord's dorsal column. The subunits of these neurons ensure that this harmful stimulus is transmitted to the thalamus and delivered to the somatosensory cortex through the spinothalamic system located in the white matter of the spinal cord. This ascending information reaches neurons in the periaqueductal gray (PAG) and rostral ventral medulla (RVM) in the midbrain to activate descending feedback systems that regulate spinal cord output. The PAG regulates the processing of nociceptive information in the DH of the spinal cord via projection neurons to the RVM and subsequently in the dorsolateral pontine tegmentum (DLPT). Endogenous opioid and cannabinoid systems, neurotransmitters such as 5-hydroxytryptamine (5-HT) and norepinephrine (NE), which are effective in the mechanism of pain, are intensely expressed through PAG/RVM pathways (Basbaum et al., 2009).

## Classification of Pain

Pain is classified according to three characteristics: symptoms, mechanisms, and syndromes. Therefore, pain is divided into three main classes: nociceptive pain, neuropathic pain, and inflammatory pain (Woolf et al., 1998).

### *Nociceptive Pain*

Nociception is the body's response to potentially harmful stimuli. Such stimuli activate nociceptors, triggering the first stage of pain sensation. Aδ- and C-fibers are two types of primary afferent nociceptors that respond to such noxious stimuli. The presence of noxious stimuli activates these two nociceptors, which have specialized free nerve endings. The nociceptive pain class includes visceral somatic pain, which encompasses both deep and superficial somatic sensations. While Aδ and C fibers are located in superficial organs such as the skin, C fibers are located in deep somatic structures such as muscles and joints. Stimulating the receptors in the primary afferent neurons and reaching the pain threshold causes the resulting impulses to spread along the afferent fibers towards the DH (PNS) and medulla (cranial) (Yam et al., 2018).

### *Neuropathic Pain*

Neuropathic pain is often defined as nerve damage and is associated with allodynia (Alles and Smith, 2018). Allodynia is a central pain sensitivity that is the result of repeated painless stimulation of receptors, causing a pain response to a stimulus that is normally considered painless. Neuropathic pain is not actually located in our body's defense system, and it can be a constant sensation or an episodic event. This type of pain may be primarily caused by inflammation, diabetes, trauma, toxins, tumors, or primary neurological diseases. Neuropathic pain can result from nerve damage to the somatosensory nervous system, as well as PNS or CNS disorders (Yam et al., 2018).

### *Inflammatory Pain*

Inflammation is a natural biological process produced by tissues in our body in response to harmful stimuli. The body's first response to inflammation is the accumulation of neutrophils at the site of injury. Then, some mediators, including 5-HT, kinins, histamine, nerve growth factors (NGF), adenosine triphosphate (ATP), PG, glutamate, leukotrienes, nitric oxide (NO), NE, and protons, are produced in the injured tissue area (Franklin and Abbott, 1989) During the inflammation process, these chemical inflammatory mediators are produced from necrotic tissues and interact to activate nociceptors in the inflamed area (Yoo et al., 2011). Pain caused by inflammation can be divided into two classes: acute and chronic. Acute pain is more intense than normal and occurs in a short time, and a healing process mediated by Aδ fibers begins. It is known that chronic pain involves long-term inflammation and a long healing process mediated by C fibers. (Basbaum et al., 2009) Inflammatory pain increases the afferent input to the spinal cord's DC and causes central sensitization.

## Sensitization

Repeated exposure to noxious stimuli causes the action potential to propagate. This allows for $Ca^{2+}$ influx and membrane depolarization through calcium channels. This triggers transmitter release and nociceptors activation in the damaged area. These processes are called sensitization. Sensitization is known as a decrease in the stimulation threshold and an increase in the firing rate as a result of the increased sensitivity of primary afferent nociceptors. The injured area then collects neuroinflammatory mediators such as bradykinin, ATP, serotonin (5-HT), norepinephrine (NE), substance P (SP), PGE2, and NGF. Each inflammatory mediator binds to its specific receptor in postsynaptic neurons, providing activation and interacting with various secondary messengers. They cause the feeling of pain to increase. In addition, the concentration of $H^+$ ions increases in the injured area, thus increasing acidity. Secondary messengers play different roles in pain sensitivity. While cellular cAMP/protein kinase A (PKA) and the protein kinase C/DAG signaling activity are effective in maintaining peripheral hyperalgesia during nociceptor sensitivity, cyclic GMP provides the opposite effect (Purves et al., 2018). The increase in the activities of adenylyl cyclase (AC)/cAMP/PKA, phospholipase c (PLC)/inositol triphosphate (IP3)/$CA^{+2}$, and PLC/DAG/PKC, which are signaling pathways involved in afferent terminal sensitivity, is effective in activating GPCRs and increasing Ca ions (Yam et al., 2018). (Figure 15.2).

Central sensitization is the repetitive stimulation of nociceptors that stimulate the projection neurons in the dorsal horn of the spinal cord. There are two classes of neurotransmitter modulators in the modulation of the pain mechanism: inflammatory and non-inflammatory. Non-inflammatory mediators include CGRP, GABA, and opioid peptides, while inflammatory mediators include PGE2, PGI2, NGF, proton, BK, ATP, SP, NKA, NKB, 5-HT, histamine, glutamate, NE, and NO. Each modulator has specific receptors to which it binds in order to send signals to the cell for pain transmission or inhibition (Figure 15.2). When the action potential reaches the presynaptic terminal as a result of the activation of VOCC,

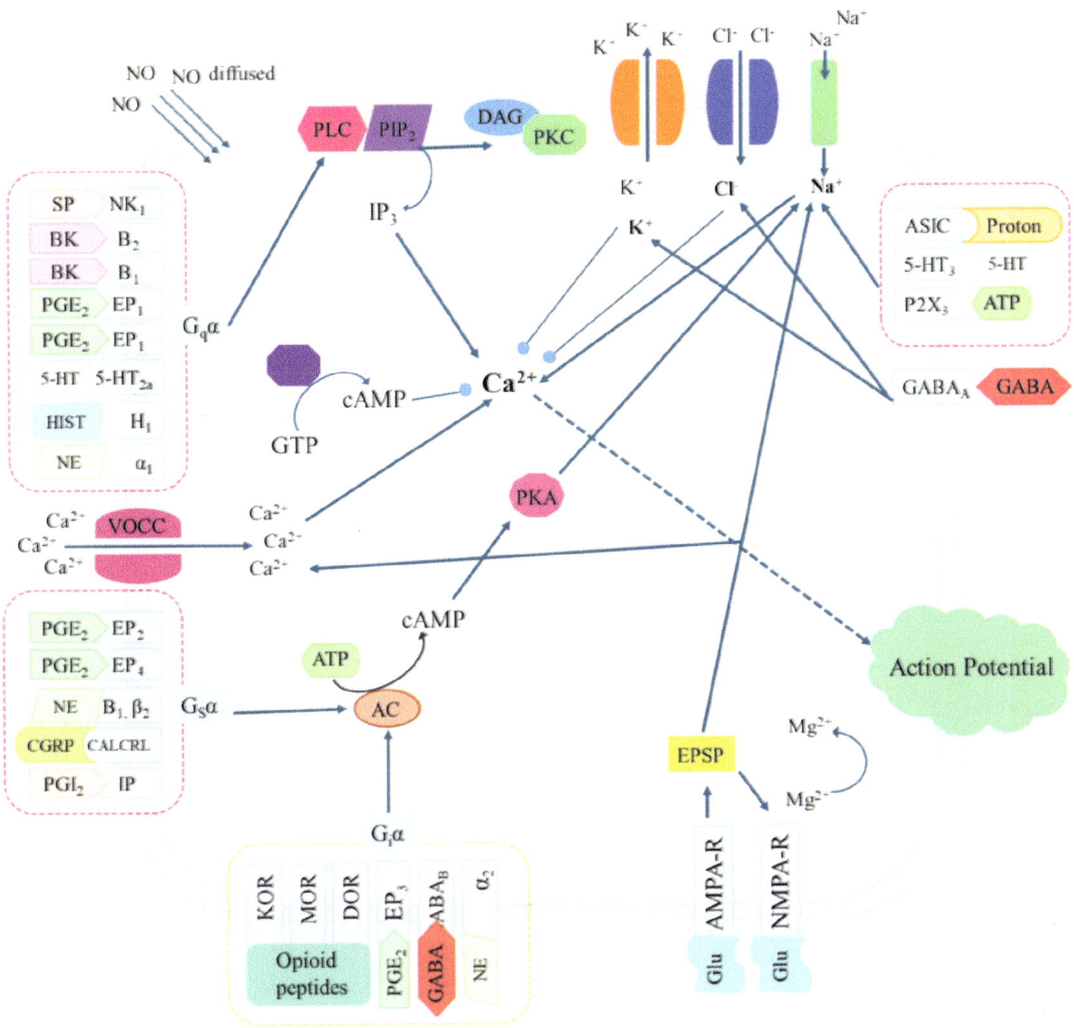


**FIGURE 15.2** Pain-related neurotransmitters signal through specific receptors.

excitatory effects such as SP and glutamate are released from the afferent terminal. Moreover, the modulation also involves the release of chemical modulators like -aminobutyric acid (GABA), NE, and endogenous opioid peptides, which provide an inhibitory effect (Yam et al., 2018).

## The Relationship Between Pain Perception and Mitochondria

Mitochondria, which are important in many biochemical processes, including the production of ATP, play an important role in neuronal function regulation and the inflammatory response (Marchi et al., 2022) Neurons require more energy than other cells. ATP helps maintain the membrane potential in neurons (Pekkurnaz and Wang, 2022). In addition, mitochondria are $Ca^{+2}$ reservoirs that regulate the intracellular $Ca^{+2}$ concentration (Romero-Garcia and Prado-Garcia, 2019; Rueda et al., 2014). Oxygen leakage in the electron transport chain, which enables ATP production in cells, causes free radical formation. Free radical production and oxidative stress play an important role in the pathology of the DRG of the spinal cord, which is effective in pain transmission. Mitochondria produce neurotransmitters that regulate neuronal excitability and signalling (Devine and Kittler, 2018).

The role of mitochondria in neurodegenerative diseases has been the subject of many studies (Wang et al., 2010; Cole et al., 2008). However, studies on the role of mitochondria in regulating pain are very new. Dysfunction in oxidative phosphorylation reduces dendritic synaptic plasticity in neurons and can lead to neuron damage and cell death (Devine and Kittler, 2018; Rangaraju et al., 2019). Exposure of neurons to various inflammatory cytokines causes depolarization of the mitochondrial membrane potential, disrupting oxidative phosphorylation and ATP production (Ferri et al., 2008; Sajic et al., 2018). Mitochondrial dysfunction in neurons also impacts sensory processing and pain. Researchers have shown that complex I and II inhibitors induce mechanical allodynia in mice (Kim et al., 2008). In humans, complex I dysfunction is associated with myalgia (Sánchez-Caballero et al., 2016). A study found that when mitochondrial complex IV activity is disrupted in DRG neurons in mice, ATP production is slowed down and the ADP/ATP ratio goes up. This increase caused hypersensitivity due to the activation of the purinergic receptor P2Y1 in the afferents of sensory neurons (Mitchell et al.,

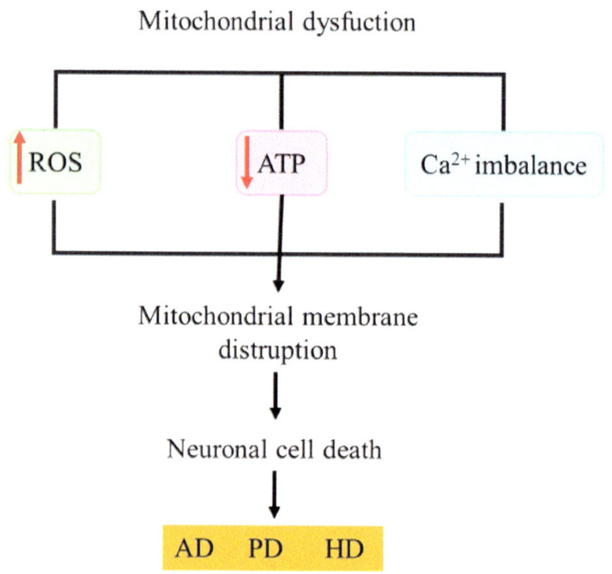


**FIGURE 15.3** Relationship between mitochondrial dysfunction and neurodegenerative diseases.

2019). Mitochondrial dysfunction has been revealed to lead to the pathogenesis of chronic or neuropathic pain (Figure 15.3).

One study revealed that the removal of excess ROS by free radical scavengers reduced both neuropathic and inflammatory pain (Sui et al., 2012). Additionally, increased glutamate receptor activation may be the cause of ROS-induced pain sensitivity (Sui et al., 2012). Consequently, glutamate activates glutamate receptors, including NMDA receptors, resulting in increased $Ca^{2+}$ influx and thus increasing intracellular $Ca^{2+}$ levels (Sui et al., 2012). Increased intracellular $Ca^{2+}$ levels overload the mitochondria and cause a loss of mitochondrial membrane potential stability. Thus, it accelerates the mitochondrial electron transport chain, causing increased mitochondrial ROS production (Sui et al., 2012). The increase in ROS level can then influence the redox-sensitive signaling cascade, increasing neuron pain (Sui et al., 2012). Mitochondrial disorder plays a role in the pathology of neuropathic pain (Flatters, 2015). These elevated ROS can then influence the neuron's redox-sensitive signaling cascade to increase pain (Wu et al., 2019).

## Investigation of Pain Mechanism in Zebrafish

Researchers have conducted numerous studies on advanced organisms like mice and rats to understand the mechanism of pain. In recent years, pain studies have also been conducted on fish (Malafoglia et al., 2013). The most significant reason for this is the perception of fish as an effective model for pain monitoring. Another reason is the close similarity between pain-related peptides and modulators in fish and those in humans. The significant conservation of pain perception mediators from zebrafish to mammals suggests that it may be effective in investigating the perception and mechanism of pain in these fish (Malafoglia, 2013).

Zebrafish provide advantages for experimental studies due to their small size, short reproduction period, large number of offspring, transparent eggs, and external fertilization. Furthermore, zebrafish undergo rapid development; they distinguish structures such as the notochord, main brain parts, heart, and eyes 24 hours after fertilization (hpf1) and complete embryonic development 3–4 days after fertilization. Furthermore, this organism allows for the analysis of numerous systems in vertebrates, such as the excretory system, notochord, and neural tube (Dooley and Zon, 2000).

The ability of zebrafish to sense pain also contributes to their status as an important model system in neurobiological research and the study of pain mechanisms (Gonzalez-Nunez and Rodriguez, 2003a). The zebrafish's nociceptive physiology is well defined and similar to that of mammals. Zebrafish, like humans, exhibit the presence of histamine, glutamate, SP, opioids, transient receptor potential cation channels (TRPs: TRPV1 and TRPA1), acid-sensing ion channels (ASIC), and GABA. Therefore, zebrafish have cellular components to respond to noxious stimuli. These fish exhibit behavioral changes in response to harmful stimuli. Due to these features, it is very easy to conduct studies on pain in zebrafish, which allows efficient and reproducible results to be obtained.

While creating a pain model in zebrafish, chemicals such as acetic acid, formalin, hot or cold stimuli, and electrodes are used as painful agents. Pain studies can be performed efficiently in zebrafish by means of these painful agents, by injection into the tail, lips, or intraperitoneally, or by fin clipping. (Reilly et al., 2008; Maximino, 2011; Correia et al.,2011; Schroeder, 2017; Taylor et al., 2017; Costa et al., 2022; Nascimento et al. 2018). Additionally, our group examined the response to pain using various behavioral analyses (Figure 15.4). We have demonstrated the ease and efficiency of examining pain perception in zebrafish through body curve measurement and behavioral analysis, following an intraperitoneal injection of acetic acid (Cansiz et al., 2023). A pain model was created by injecting various doses of acetic acid into the lip corner of zebrafish; behavioral analyses were performed, and the effects of painkillers were examined (Deakin et al., 2019).

In another study, researchers injected acetic acid intraperitoneally into fish to investigate the role of endogenous opioids in modulating the pain response following pain stimulation in zebrafish. The findings revealed that aquatic animals are a new and promising model for further nociceptive and antinociceptive drug screening (Costa et al., 2019) It has been stated that such pain studies can also be performed in zebrafish embryos (Costa et al., 2019).

## Limitations in Monitoring the Pain Response in Zebrafish

Zebrafish are a very suitable model for pain studies due to their high orthology with humans (70%) and easy monitoring of the response to painful stimuli, but they have some limitations. Genome duplication in teleost fish complicates genetic analyses in pain studies, as several pain-related genes may have two copies or be missing in zebrafish (Costa et al., 2022). Long-term monitoring of endocrine levels in the blood is also difficult due to their small size. This is due to the inability to obtain sufficient blood for analysis over an extended period (Thomson et al., 2019; Costa et al., 2022).

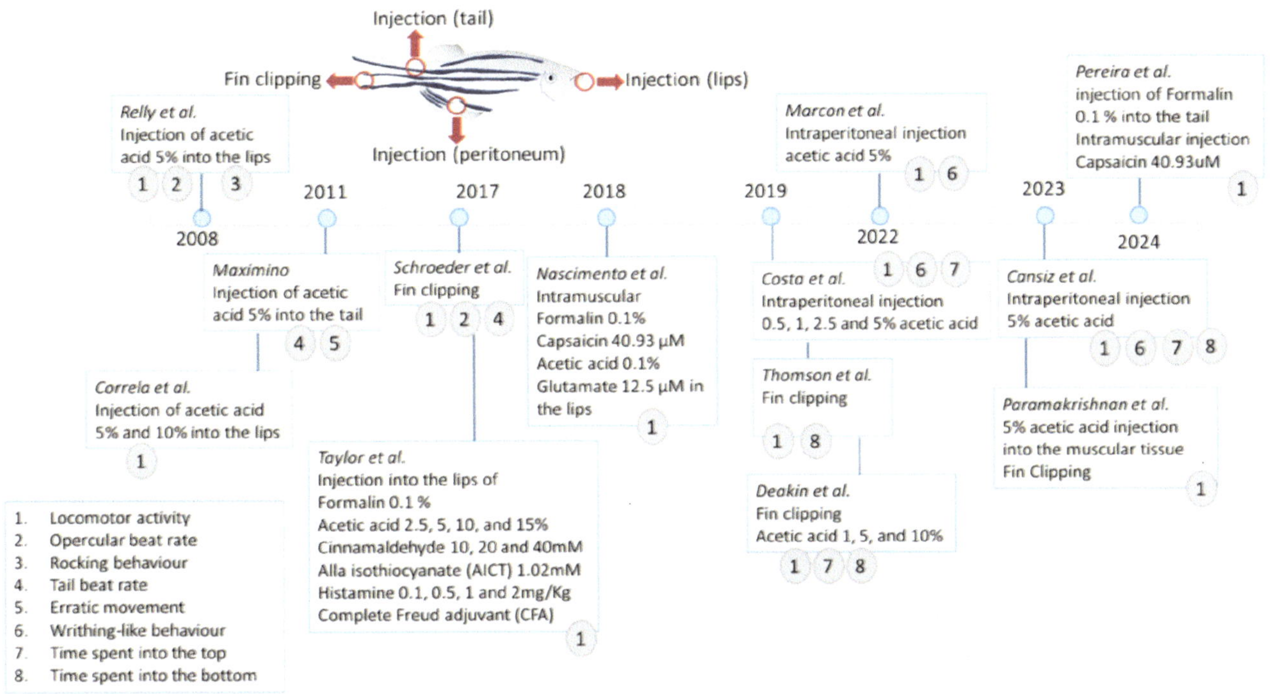


**FIGURE 15.4** Methods used to create pain models in zebrafish and behavioral methods followed to monitor pain.

## Modulators in the Pain Mechanism

### *Calcitonin Gene–Related Peptide (CGRP)*

Both the central and peripheral nervous systems produce calcitonin gene–related peptide (CGRP), which is located in the primary afferent nerves. CGRP, located in the DH, is effective in transmitting noxious stimuli. The CGRP receptor is a calcitonin receptor–like receptor that is Gs-protein-coupled. CGRP can both increase nociceptor sensitivity in the periphery and increase sensory input in the CNS. Thus, it can increase the perception of pain. In a zebrafish pain model, a study that examined the role of neuropeptides against noxious stimuli revealed that conditioned overexpression of CGRP increased the responses to all sensory stimuli (Wood et al., 2014).

### *Bradykinin*

Bradykinin (BK) binds to Gq-protein-coupled bradykinin receptor B in nociceptive afferent nerve fibers and causes sensitization. The injured area releases various chemicals such as histamine, serotonin, norepinephrine, and SP after injury. A study on zebrafish revealed the release of bradykinin following injury. However, in zebrafish, the homolog of this chemical to the Bradykinin1 (B1) and Bradykinin2 (B2) receptors found in humans differs (Dunér et al., 2002; Bromée et al., 2006).

### *Substance P (SP)*

Substance P (SP) is a peptide structure consisting of 11 amino acid residues belonging to the tachykinin small peptide family (Hökfelt et al., 2001). Primary nociceptive sensory neurons (nociceptors) produce SP and release it in response to noxious stimuli (Hökfelt et al., 1975). The release of SP from nociceptive nerve fibers and activation of the neurokinin 1 (NK1) receptor are important factors in the transmission of the pain signal (Lin et al., 2012).

In mammals, there are three genes, TAC1, TAC3, and TAC4, that encode the precursors of TAC (tachykinin precursor gene). The human precursor of the TAC1 gene has four isoforms, and all isoforms contain the region encoding the SP sequence (Pennefather et al., 2004). There are also tacrls, tacr2, and tacr3s genes in zebrafish. These genes code for SP, which includes tachykinin receptors tacl, tac3, and tac4 (Lopez-Bellido, 2013). In light of this information, it can be predicted that it would be fruitful to conduct research on SP and its receptors for pain studies in zebrafish. Research demonstrates that mice lacking the tachykinin-1 gene, responsible for the production of SP and neurokinin A, experience prolonged hyperalgesia (Lin et al., 2012).

### *Cyclooxygenase (COX)*

COX-1 and COX-2 enzymes, known as cyclooxygenases, play a role in the biosynthesis of prostanoids. Nonsteroidal anti-inflammatory drugs (NSAIDs), such as aspirin and ibuprofen, are inhibitors of these enzymes. Prostanoids control many physiological mechanisms, including inflammation. In this regard, the inhibition of COX enzymes causes an anti-inflammatory effect. COX orthologous genes are present in the zebrafish genome, and their similarities with human genes have been revealed by various studies (Grosser et al., 2002; Pini et al., 2005). Receptors for NSAIDs are evolutionarily conserved between mammals and zebrafish. Toxicology studies have been conducted in zebrafish to investigate drug pollution in freshwater due to human use, and these studies have revealed that ibuprofen is rapidly absorbed by embryos and larvae and is metabolized in a similar way to that of mammals (David and Pancharatna, 2009). Additionally, studies have shown that ibuprofen reduces prostaglandin levels in zebrafish (Morthorst et al., 2013). The study supported the formation of pain perception in the acetic acid–induced pain model in adult zebrafish through behavioral analyses and a writhing test, and also demonstrated an increase in the expression of the *cox-1* gene (Cansiz et al., 2023).

### Acid-Sensing Ion Channels (ASICs)

The CNS and PNS of mammals widely express acid-sensitive ion channels (ASICs), which are $Na^+$ channels gated by extracellular $H^+$ (Montalbano et al., 2021). ASICs structurally consist of two hydrophobic transmembrane domains (TMD) consisting of 20 amino acids. Six ASIC proteins (ASIC1a, ASIC1b, ASIC2a, ASIC2b, ASIC3, and ASIC4) have been identified in mammals (Krishtal, 2015). Functionally, ASICs monitor changes in physiological extracellular pH values. ASICs are conserved in zebrafish as well as in mammals. Asic1a, which contributes to synaptic transmission, is widely expressed in the CNS and PNS of zebrafish. (Holzer, 2009; Lingueglia, 2007; Holzer, 2011). The PNS expresses asic1b and asic3, which play a role in pain perception (Montalbano et al., 2021).

Evolutionary preservation of ASICs from fish to mammals is known. While sodium chloride elimination in mammals is achieved through $Na^+$ channels on the mucosal side of the intestinal epithelium and by the combined activity of $Na^+/H^+$ exchangers and $Cl^-/HCO^{3-}$ exchangers, $Na^+$ uptake in teleosts is provided by ASICs (Takei, 2021) It has been proven that ASIC channels are also involved in $Na^+$ uptake in zebrafish (Dymowska et al., 2015). Like mammals, zebrafish have six asic subunits *(zasic1.1*, *zasic1.2*, *zasic1.3*, *zasic2*, *zasic4.1*, and *zasic4.2*). *zasic1.1.*, *zasic2*, and *zasic 4.1* are orthologs of mammalian *ASIC1*, *ASIC2*, and *ASIC4*, while *zasic1.2* and *zasic1.3* are paralogs of *zasic1.1*. However, *zasic4.2* is the paralog of *zasic4.1* (Paukert et al., 2004) (Figure 15.5). Proteins encoded by zasics share 60%–75% of amino acid identity in rats and humans (Chen et al., 2007). ASICs are distributed throughout the zebrafish sensory organs, intestine, gills, and brain.

We have shown that the expression of the *zasic1.3* gene is increased in the acetic acid–induced pain model in zebrafish. In addition, it was revealed that there was a decrease in *zasic1.3* gene expression in the rotenone and acetic acid–induced model. The reason for this was pointed out to be the impairment in pain perception in PD. The same study analyzed locomotor activity, revealing a decrease in distance traveled in the rotenone and acetic acid–induced model compared to the acetic acid only–induced group (Cansiz et al., 2023). The fact that the inhibition of ASICs provides an analgesic effect in animals is thought to be an indication that ASICs play a role in pain. The detection of a decrease in pain perception after tissue inflammation in ASIC3-inhibited mice indicated that ASIC3 has an important role in modulating the perception of stimuli that is effective in pain formation (Montalbano et al., 2021). These animal studies provide important findings demonstrating the role of ASICs in pain.

### Transient Receptor Potential (TRP) Ion Channels

Ion channels, widely distributed in sensory neurons and many pain-related tissues, are considered one of the main factors in regulating the sensation of pain. TRP, $Ca^+$ channels, $Na^+$ channels, and $K^+$ channels were developed in zebrafish as well as in higher mammals.

The mammalian TRP ion channel superfamily consists of 28 cation-permeable channels, grouped into six subfamilies based on sequence homology. TRP channels are located in the dorsal root ganglia (DRG) of the CNS, PNS, and spinal cord. TRP channels, generally expressed in cell membranes, mediate ion homeostasis and are known as important pain transducers (Moore et al., 2018; Lee et al., 2021). Zebrafish, like mammals,

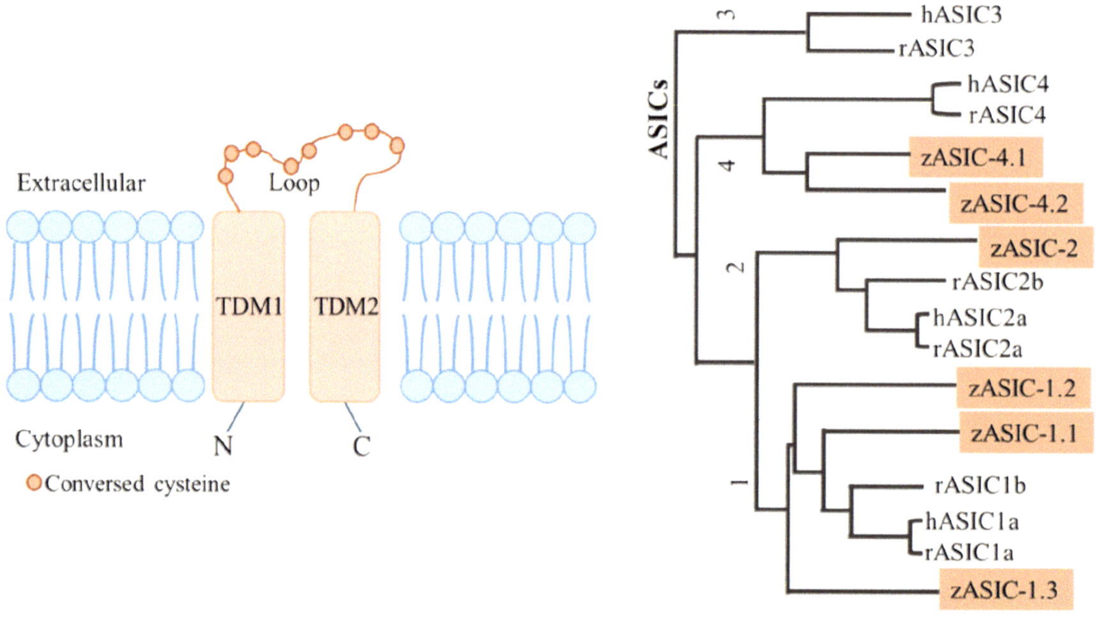


**FIGURE 15.5** Conservation of ASICs across species throughout evolution.

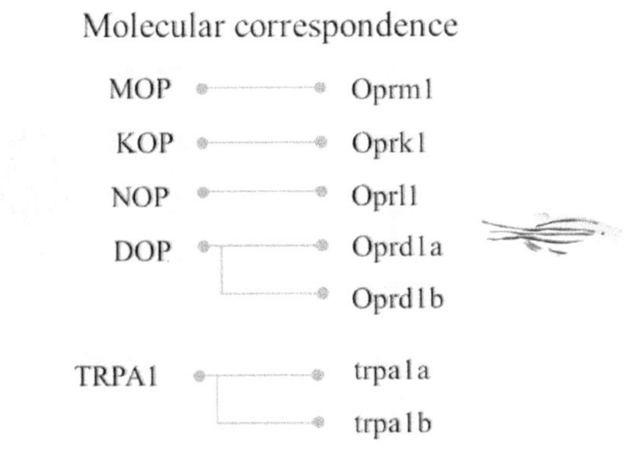


**FIGURE 15.6** Zebrafish counterpart of human opioid receptors and TRPA1 receptor genes.

express TRP channels, which are important in pain modulation (Chen et al., 2016). In zebrafish, many of the molecular receptors that are effective in nociception, such as TRPV1, stimulated by noxious heat, pH, and chemicals; and TRPA1, stimulated by noxious chemosensation, have conserved functions (Curtright et al., 2015). However, zebrafish have two TRPA1 paralogs: TRPA1a and TRPA1b (Figure 15.6).

The *TRPA1a* paralog is highly similar to *TRPA1* in humans. Zebrafish has three TRPM8 paralogues. In zebrafish, *TRPM8* is called *fb95a05*. Similar to *TRPM8* in humans, *fb95a05* is linked to pain signals triggered by cold stimuli (Smith, 2017). The TRPA1 channel, which has a single isoform in humans, is associated with inflammatory pain and hypersensitivity. When TRPA1 receptors are activated by binding to pain fibers, the sensation of pain is triggered, and the brain's sense of escape is activated. TRPA1 agonists may act by increasing cell capacity at the plasma membrane and increasing *TRPA1* expression (Nilius et al., 2012). TRPA1 stimulation can cause spontaneous pain and various hyperalgesias through electrophilic TRPA1 activators. TRPA1 antagonists relieve pain sensation by slowing down or blocking various modulators of pain signaling pathways. In a study investigating possible analgesic compounds, perception and suppression of nociception were observed in zebrafish using TRPA1 agonists and antagonists (Cutright et al., 2015).

### *Voltage-Gated Sodium Calcium Channels (VGCC)*

Voltage-gated $Ca^{2+}$ channels are responsible for regulating neuron excitability and neurotransmitter release in the brain. $Ca^{2+}$ channels include six types: L (long-lived), N (neuronal), P (Purkinje cell), Q (granule cell), R (toxin-resistant), and T (transient). L-, N-, P-, Q-, and R-type $Ca^{2+}$ channels are responsible for regulating neurotransmitter release (Zhou et al., 2008). Defects in P- and Q-type channels cause ataxia and migraine in both animal and human (Kolesnikova, 2022). In pain processing, N- and P/Q-type calcium channels are important voltage-gated $Ca^{2+}$ channels. Mutant mice with deletions of the beta3 subunit (involved in the N-type calcium channel) have demonstrated difficulty in processing pain (Murakami et al., 2002). Zebrafish also exhibit alpha and beta subunits of Ca channels. In zebrafish, the subunit beta2 has been identified as β2.1 and β2.2 (Ebert et al., 2008). Zebrafish have identified VGCC subunits as *cacnb1*, *cacnb2a*, *cacnb2b*, *cacnb3a*, *cacnb3b*, *cacnb4a*, and *cacnb4b*. Various brain regions associated with pain mechanisms express many subunits (Table 15.1).

### *Voltage-gated Sodium Channels (VGSC)*

Voltage-gated sodium channels (VGSC) are responsible for the electrical signaling of cells (Ivo-Ahern et al., 2016). In humans, there are 10 sodium channel alpha genes (see figure scna). All of these, except *SCN7A*, are the *VGSC* genes (Goldin et al., 2000; Catterall et al., 2005). *SCN7A* is the only channel activated depending on extracellular sodium concentration and is sensitive to $Na^+$ concentration (Hiyama et al., 2002; Noda, 2006).

Variants in these genes can cause a variety of diseases related to channelopathies (Kim, 2014). Variants in *SCN1A*, *SCN2A*, and *SCN3A* have been associated with diseases such as various types of epilepsy and migraine (Mulley et al., 2013; Dichgans et al., 2005; Mantegazza et al., 2005; Ogiwara et al., 2009; Vanoye et al., 2014; Zaman et al., 2018). *SCN9A*, *SCN10A*, and *SCN11A*, which are mainly expressed in the sensory nervous system, are considered to be the causes of various pain disorders (Fertleman et al., 2006; Faber et al., 2012).

**TABLE 15.1**

Location and Counterparts of Mammalian CACNB Genes in Zebrafish

| Mammalian Gene | Mammalian expression patterns | Zebrafish homolog | Embryonic expression in zebrafish |
|---|---|---|---|
| **CACNB1** | Skeletal muscle (β1a) neurons (β1b) | *cacnb1* | Skeletal muscle, brain, spinal cord, trigeminal ganglia, olfactory placodes |
| **CACNB2** | Heart, brain, aorta, lung, kidney, pancreas | *cacnb2a* | brain, spinal cord, trigeminal ganglia, optic stalks, retina (inner nuclear layer, photoreceptor) |
| | | *cacnb2b* | brain, spinal cord, retina (ganglion cell layer), olfactory placodes |
| **CACNB3** | Brain, aorta, trachea, lung, heart, pancreas, adrenal gland | *cacnb3a* | brain, spinal interneurons, Rohon-Beard neurons, retina (ganglion cell layer, inner nuclear layer) |
| | | *cacnb3b* | trigeminal ganglia, retina (ganglion cell layer, inner nuclear layer), Rohon-Beard neurons, otic cells |
| **CACNB4** | Brain (predominantly in cerebellum), kidney | *cacnb4a* | brain, spinal cord, heart, retina (inner nuclear layer) |
| | | *cacnb4b* | brain, spinal cord, trigeminal ganglia, retina (ganglion cell layer, inner nuclear layer) |

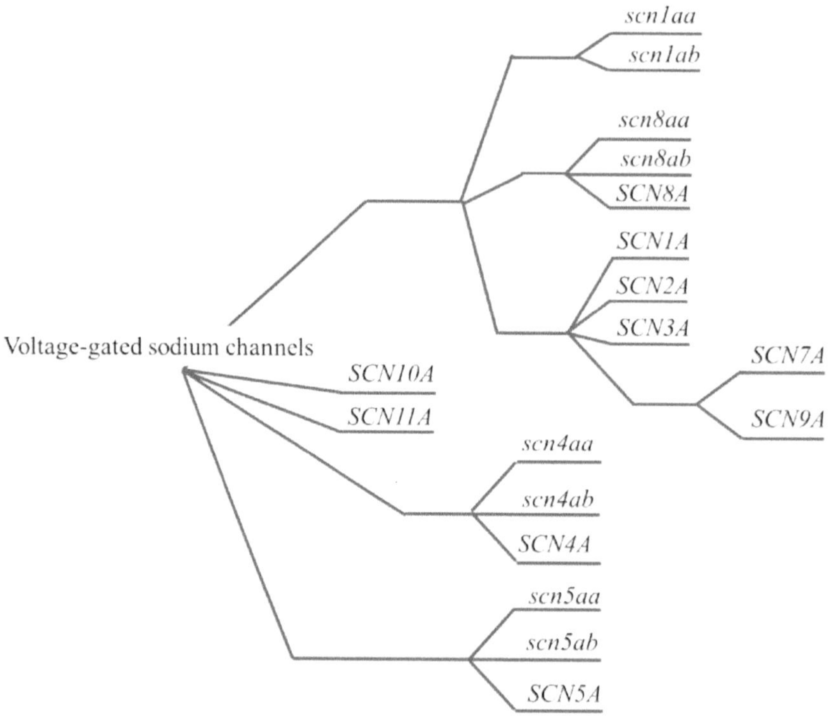


**FIGURE 15.7** Evolutionary tree of zebrafish and human voltage-gated sodium channels.

In addition to indicating that these channels in humans are not fully evolutionarily preserved in zebrafish, the similarity between the two species has also been revealed (Figure 15.7). Different studies have determined that *scn9a* knockout mice have increased *PENK* levels (Minett et al., 2015), which are endogenous opioid precursors, and decreased serotonin levels (Isensee et al., 2017). However, the opioid antagonist naloxone reversed pain insensitivity in *scn9a* knockout mice (Minett et al., 2015). *scn8aa* and *scn8ab* are the zebrafish equivalents of the *SCN8A* gene in humans. These genes in zebrafish are closely related to those in humans (Widmark et al., 2011). In the acetic acid–induced pain model in adult zebrafish, the expression of the *scn8aa* gene increased compared to the control, while it decreased in the rotenone and acetic acid-induced groups. However, researchers have stated that the penk gene exhibits the same change as the *scn8aa* gene (Cansiz et al., 2023). These data indicate that there may be a relationship between VGSCs and the opioid signaling system.

### *Potassium Channels (KC)*

The family of membrane protein potassium channels (KC) is one of the most complex of the ion channels. This channel family contains numerous genes encoding the α subunits in humans (Busserolles et al., 2016). KCs contain transmembrane helices that span the membrane's lipid layer. They are divided into four families according to their structure and functions. The majority of these are expressed in somatosensory neurons and supporting glial cells. KCs generate electrical signals in neurons by aiding the rapid and selective flow of $K^+$ ions across the cell membrane. This function indicates that KCs are important channels in neuronal excitability (Jérôme et al. 2019). Opening potassium channels hyperpolarizes the plasma membrane, preventing other depolarizing channels from stimulating the neuron. KCs are functional for pain processes that regulate pain transmission and modulation, and their dysfunction can cause a variety of pain problems. (Li and Toyoda, 2015; Tsantoulas and McMahon, 2015). Dysregulation of $K^+$ ions has been shown to be the cause of multiple nervous system disorders, such as epilepsy (Bailey, 2019; Kinboshi et al., 2020) and movement disorders (Mochel et al., 2020). Research also reveals that this irregularity contributes to the development of Alzheimer's (Li et al., 2022; Taylor et al., 2022) and Parkinson's disease (Chao et al., 2018).

KCs are important regulators of the electrical activity of sensory neurons and pain perception. This feature has an important role in the development of antinociceptive pharmacology (Busserolles et al., 2019). It is known that potassium channels are expressed in the embryonic period in zebrafish, but when compared to humans, their transcription is more complicated due to the increased number of genes ( Underwood et al., 2023). Researchers have stated that potassium channels play a crucial role in the regeneration of spinal cord injuries in zebrafish (Underwood et al., 2023).

### *Adenosine Triphosphate (ATP)*

Adenosine triphosphate (ATP), the main signaling purinergic messenger, is synthesized in presynaptic terminals, stored, and released into the synaptic cleft via nerve impulses (Gusso, 2022). ATP is an important intracellular messenger released locally from damaged tissues. ATP turns into adenosine and functions by binding to the purine receptor (P1) located in second-order neurons in the DH at the periphery of sensory neurons (Burnstock, 2000; North, 2002). Various studies have shown that the purinergic system modulates pain by activating

P1 adenosine receptors (Gusso, 2022; Liu et al., 2017; Maes et al., 2012). The P1 receptor family includes A1, A2A, A2B, and A3 receptors, which have been identified in zebrafish (Gusso et al., 2022; Boehmler et al., 2009; Bortolotto et al., 2015). It has been observed that AMP hydrolysis increases in acetic acid–induced pain in zebrafish. This result was considered a sign that the purinergic system is involved in pain responses in zebrafish larvae (Gusso et al., 2022).

### Glutamate

Glutamate is an excitatory neurotransmitter released in areas involved in the transmission and sensation of pain in the CNS and PNS following noxious stimulus. This feature makes glutaminergic receptors a target in pain treatments (Wozniak et al., 2012). There are two glutamate receptors: amino-3-hydroxy -5-methyl-4-isoxazolepropionic acid receptors (AMPA-R) and N-methyl-d-aspartate receptors (NMDA-R). NMDA-R activation is an ion channel that causes central sensitivity due to an increase in intracellular $Ca^{+2}$ ions (Yam et al., 2018). Ketamine, an NMDA receptor antagonist, showed an antinociceptive effect on glutamate-induced nociception (Ferreira et al., 2018) Behavioral analyses on the glutamate-induced pain model in zebrafish revealed a decrease in movement, a crucial parameter in pain observation (Sharma et al., 2022; Silva et al., 2020)

### Gamma-aminobutyric Acid (GABA)

GABA, produced by GABAergic neurons, is the most common inhibitory neurotransmitter in the CNS. It can bind to two different receptors in the nervous system: ionotropic GABA (GABAA) and metabotropic GABA (GABAB) receptors. While GABAA receptors are densely found in the CNS, GABAB receptors are also found in presynaptic nerve terminals as well as in the CNS. When GABA binds to the GABAA receptor, extracellular $Cl^-$ ions enter the neurons, thus reducing the membrane potential and creating an inhibitory effect. The GABAB receptor, a Gi protein-coupled receptor, inhibits cAMP when GABA binds to it (Monesson-Olson et al., 2018).

Although the larval zebrafish is a useful model system for neuronal analyses, little information is available regarding the diversity and expression of GABAA receptors in this system. There are 19 different types of GABAA subunits in mammals. Similar to mammalian GABAA receptors, the zebrafish $\alpha$ subfamily is the largest and most diverse of the subfamilies. There are eight $\alpha$-subunits in zebrafish (Monesson-Olson et al., 2018). Studies have demonstrated that activation of Group I metabotropic glutamate receptors (mGlu1) leads to hyperalgesia, while blocking these receptors decreases hyperalgesia in inflammatory pain models (Bhave et al., 2001; Walker et al., 2001). Another study showed that peripheral mGlu1/5 antagonists administered as a dose could reduce neuropathic hypersensitivity in rats (Dogrul et al., 2000).

### Opioids

Opioid peptides synthesized endogenously in vertebrates show their effects in the cell by binding to three different receptors: $\delta$, $\mu$, and $\kappa$. When opioids bind to these Gi protein-dependent receptors, they inhibit AC/cAMP activity. Opioid receptors are densely expressed in both primary afferent neurons and postsynaptic neurons. In addition, the opioid peptides dynorphin and enkephalin are widely expressed in the DC of the spinal cord. These peptides reduce the excitation of neurons by preventing the release of excitatory neurotransmitters in the afferent terminal and thus help reduce the feeling of pain.

There are four opioid receptors in zebrafish. In zebrafish, there are two variants of the $\delta$ opioid receptor, a $\mu$ opioid receptor and a $\kappa$ opioid receptor (Malafoglia et al., 2013; Pinal-Seoane et al., 2006; de Velasco et al., 2009, Alvarez et al., 2006). Furthermore, researchers have identified six opioid genes in zebrafish, which include two proenkephalin genes (Gonzalez-Nunez et al., 2003a), two proopiomelanocortin genes (Gonzalez-Nunez et al., 2003b), one prodynorphin gene (Gonzalez-Nunez et al., 2007), and one pronociceptin gene (Gonzalez-Nunez et al., 2003c). Opioid receptors and peptides are similar to their mammalian homologues in structure, biochemical, and pharmacological properties (Gonzalez-Nunez and Rodríguez, 2009). In zebrafish treated with intraperitoneal acetic acid, naloxone treatment results in a delayed improvement in writhing-like behavior (Fabiano et al., 2019; Costa et al., 2022)

The acetic acid–induced pain model in zebrafish showed an increase in penka and pdyn gene expressions, whereas the rotenone and rotenone-acetic acid–induced groups showed a decrease in these gene expressions. This finding suggests that impairments in the pain mechanism may have occurred due to changes in opioidergic pathways in PD (Cansiz et al., 2023)

### Nitric Oxide (NO)

Nitric oxide (NO) is synthesized from L-arginine via specific neuronal (nNOS) and non-neuronal (endothelial/eNOS or inducible/iNOS) forms of NO synthase (Olson and van der Vliet, 2011). These enzymes and NO play a role in many physiological and pathophysiological processes. NO is a signaling molecule in both the CNS and PNS that plays an important role in pain conditions (Chen et al., 2010). NO is an important mediator of nociception and plays a role in central sensitization (Figure 15.8). However, clinical and experimental data have shown that NO can also induce analgesia (Curry et al., 2011). These data suggest that NO plays a complex and multifaceted role in modulating the nociceptive process.

Depending on the location of NO synthase synthesis, NO synthase inhibitors can have both pro- and antinociceptive effects (Oliveira, et al., 2023). In the spinal cord, NO induces the processing of nociceptive signals, and NO synthase inhibitors generally cause an antinociceptive effect (Schmidtko, 2015). However, in models of peripheral inflammation-induced nociception, NO synthase inhibitors may cause different effects (Gomes et al., 2020). Researchers conducted pain studies in zebrafish, creating pain models using various agents and observing the effect of NO (Oliveira et al., 2023; Cansiz et al., 2023). NO levels were observed to increase in the acetic acid–induced pain model in adult zebrafish (Cansız et al., 2023).

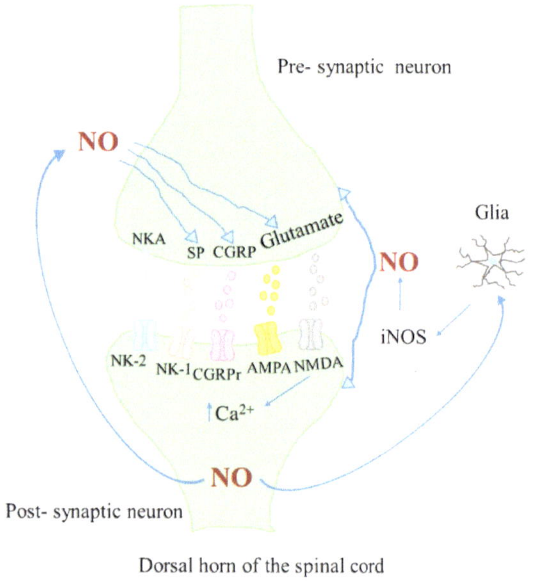


**FIGURE 15.8** Involvement of nitric oxide in central sensitization.

### *Monoaminergic Systems*

Monoaminergic systems consist of a set of evolutionarily conserved chemical structures that play important roles in the formation, function, and plasticity of the nervous system (Azizi, 2022). Dopamine and norepinephrine are synthesized from the amino acid tyrosine, and serotonin from tryptophan. Many areas of the brain contain monoamines. Monoamines are important chemical structures in modulating pain, developing and shaping pain perception at many levels. Pain signals activate ascending and descending monoaninergic pathways, which regulate the transmission of pain (Benarroch, 2016).

One of their common features is that these chemicals in the monoaminergic system act on the cell via G protein coupled receptors (GPCR). There are various types of GPCRs. While the Gs protein structure is a stimulatory receptor, Gi acts as an inhibitory receptor. The Go receptor is responsible for regulating cellular effectors such as ion channels and enzymes to modulate the function of the cell. The Go receptor consists of three subunits: α, β, and γ (Iang and Bajpayee, 2009). Humans and zebrafish share many conserved genes and neurotransmitter signaling pathways (Mathur and Guo, 2010). The relationship of GPCR subtypes with dopaminergic, serotonergic, and noradrenergic pathways in zebrafish is shown in Figure 15.9. The zebrafish genome contains orthologs for all but two human GABAAR subunit isoforms (Cocco et al., 2017). There are also many important differences between species. First, the zebrafish genome encodes a GABAAR β4 subunit with no clear ortholog in mammals. Second, there are six GABAAR α-subunit isoforms in mammals, whereas zebrafish have eight. The α-isoform family is the largest and most diverse family of GABAAR subunits in both humans and zebrafish (Monesson-Olson et al., 2018).

Dopaminergic, noradrenergic, and serotonergic pathways are developed in zebrafish. In zebrafish, dopaminergic neurons are formed at 18 hpf. Homologs of both D1 and D2 receptor subtypes are highly conserved in zebrafish compared to vertebrates (Figure 15.10). Structurally, dopamine neuron groups are commonly found in the telencephalon of zebrafish but not in the midbrain. It is unclear whether these neuron groups are similar to the midbrain population in mammals (Oliveri and Levin, 2019). The D1 receptor family is excitatory, whereas the D2 receptor family is inhibitory with D2 receptors, which are the receptor subtype. The *drd2a* and *drd2b* genes are zebrafish orthologs of human DRD2. *drd2b* is more abundantly expressed in the brain of zebrafish, and *drd2a* is expressed in the spinal cord. Similar to humans, zebrafish have dopaminergic projections descending from the brain to the spinal cord and can thus directly modulate nociception in the spinal cord (van Reij et al., 2021; Reinig et al., 2017; van Reij et al., 2019). A study conducted in zebrafish showed that the drd2a receptor has an antinociceptive effect (van Reij et al., 2021).

NE acts by binding to adrenoreceptors. Adrenoreceptors are also preserved in zebrafish and are similar to those in vertebrates. However, these receptors are diversified in zebrafish. There are five alpha1-adrenergic receptors in zebrafish: *adra1aa*, *adra1ab*, *adra1ba*, *adra1bb*, and *adra1d*. The activation of phospholipase C by alpha1-adrenergic receptors results in an increase in intracellular $Ca^{2+}$ levels. Zebrafish have five alpha2-adrenergic receptors: *adra2a*, *adra2b*, *adra2c*, *adra2aa*, *adra2ab*, and six beta-adrenergic receptors: *adb1*, *adrbk2*, *adrb2a*, *adrb2b*, *adrb3a*, and *adrb3b*. Alpha2 receptors play an inhibitory role, and activation of these inhibitory receptors leads to a decrease in cAMP and $Ca^{2+}$ levels. Activation of beta-adrenergic receptors causes cAMP levels to increase (Singh et al., 2015).

There are two htr1a proteins in zebrafish: htr1aa and htr1ab. These two proteins are 78% similar to each other, and 76% and 60% similar to the human HTR1A protein, respectively. However, the topology of the HTR1B and HTR1D subfamilies has not been fully elucidated. Studies on zebrafish show that *htr1aa* and *htr1bd* are well preserved in zebrafish, whereas *htr1ab* is not (Norton et al., 2008).

## Pain in Neurodegenerative Diseases

### Pain in Alzheimer's Disease

Alzheimer's disease (AD) is the most common form of dementia, accounting for 60%–80% of dementia cases (Zhu et al., 2015). It is characterized by a progressive deterioration in cognitive functions. AD is marked by amyloid β (Aβ) plaques outside of cells and neurofibrillary tangles (NFTs) inside of cells made up of tau protein that is abnormally hyperphosphorylated. AD is linked to a higher incidence of chronic pain (van Kooten et al., 2016). AD shows a variety of responses to experimental pain thresholds, including increased sensitivity and decreased sensitivity, which can vary depending on the type and intensity of the painful stimulus (Morgese and Trabace, 2019).

Studies on pain in AD have revealed that many mechanisms are effective in the formation of pain and the progression of AD. Since increased inflammation and decreased nervous system function are closely related to pain and decreased muscle

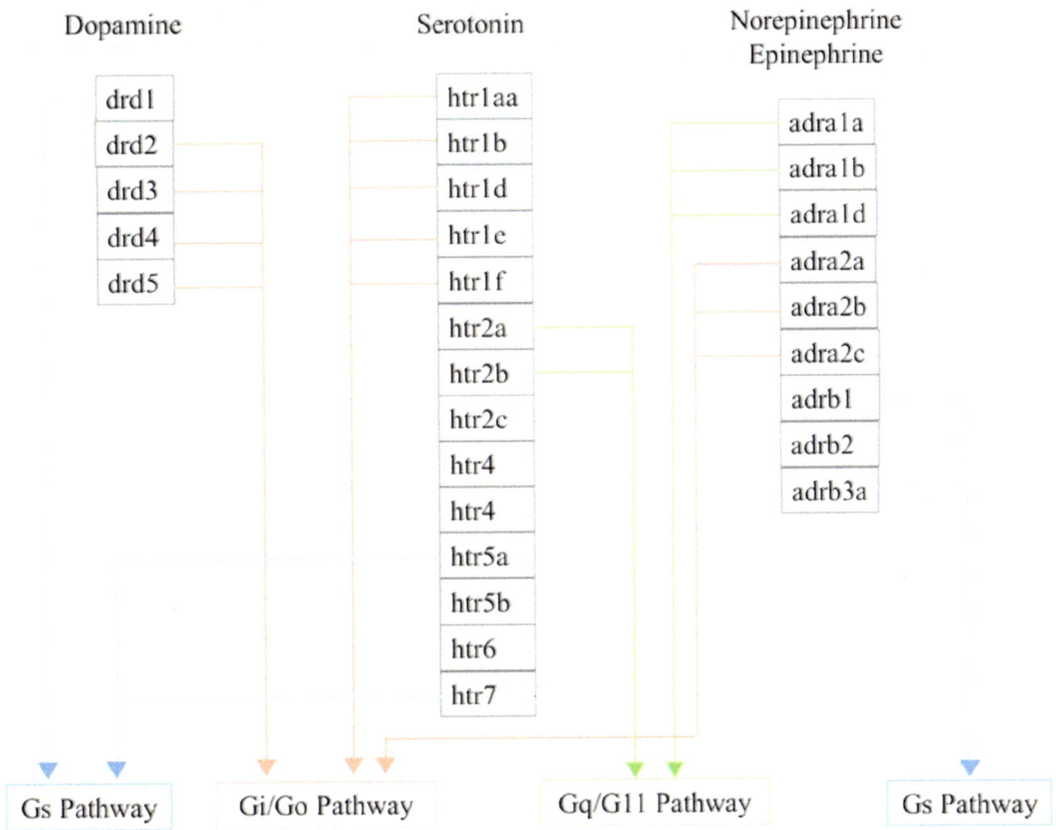


**FIGURE 15.9** Relationship between monoaminergic system and GPCR in the zebrafish.

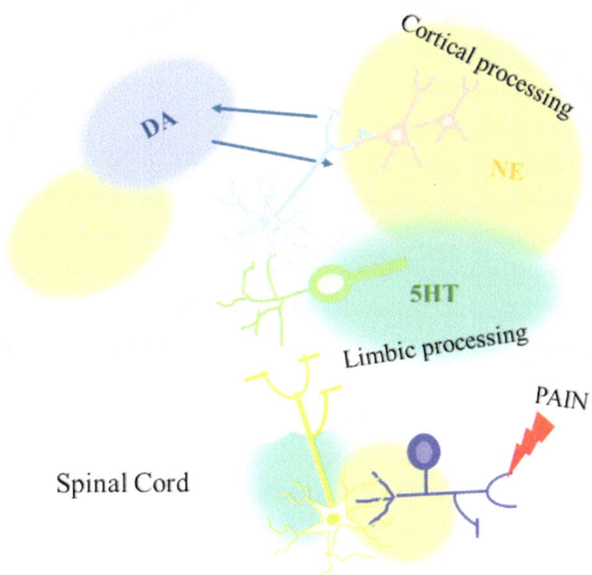


**FIGURE 15.10** Entry of monoamines into the pain system.

strength and balance, it is stated that the increased oxidative stress caused by AD in the central and peripheral nervous systems reduces neurotrophins, range of motion, and balance and causes an increase in chronic and unknown pain (Hosseini et al., 2020).

Numerous systems are involved in pain modulation. Therefore, various pathways have been examined in pain research in AD. Studies have shown that individuals with chronic pain and AD share common pathologies, such as abnormalities in the noradrenergic system in the locus coeruleus (the main center of norepinephrine synthesis and subsequent neurotransmission), activation of microglia, and increased inflammation in certain brain regions (Salter and Stevens, 2017).

AD causes structural and functional changes in the brain, such as changes in gray matter volume and disruptions in functional connectivity between brain networks (Palesi et al., 2016). In addition to changes in brain structure, other common features of Alzheimer's and chronic pain are neuroinflammation and mitochondrial dysfunction. Neuroinflammation plays a key role in the induction and maintenance of chronic pain (Blanton et al., 2023; Ji et al., 2014; Ji et al., 2018).

Neuroinflammation increases vascular permeability, leukocyte infiltration, glial cell activation, and the production of inflammatory mediators such as cytokines (Ji et al., 2014). Cytokines and other glia-produced mediators can induce and maintain central sensitization, which mechanistically causes widespread chronic pain through neuroinflammation. Chronic pain activates microglia, which may play a critical role in pain generation and maintenance (Cao et al., 2019). Microglia-driven neuroinflammation is a characteristic feature of chronic pain (Cao et al., 2019). In fact, microglia are initiators of the post-injury neuroimmune response that contributes to the transition from acute to chronic pain (Cao et al., 2019). In the case of chronic pain, pro-inflammatory microglia secrete inflammation-associated cytokines and chemokines such as IL-6, IL-1β, and TNF-α (Barcelon et al., 2019; Liu et al., 2017). This proinflammatory state leads to changes in synaptic remodeling, brain connectivity, and network functions (Cao et al., 2019). Mitochondria plays a critical role in ATP synthesis, redox balance, and many other biological processes, including ion homeostasis, protein turnover, post-translational modification, and apoptosis (Bozi et al., 2020). Both neuropathic and chronic inflammatory models of pain have been associated with neuronal mitochondrial dysfunction, including impaired bioenergetics, calcium overload, and oxidative stress (Zambelli et al., 2014; Flatters, 2015).

Numerous studies have shown that there is a relationship between chronic pain and mitochondrial dysfunction (Bozi et al., 2020; Doyle and Salvemini, 2021; Lagos-Rodriguez et al., 2020). The hypothalamus is an important brain region involved in pain perception, especially in descending modulation (Bao and Swaab, 2019). It is known that in the early stages of AD, there are changes in the hypothalamic area (Vercruysse et al., 2018). It is known that nociceptive stimuli disrupt dopamine homeostasis in dopaminergic pathways involved in central pain regulation and modulation (Morgese et al., 2022). In addition, it has been reported that soluble Aβ can determine changes in dopamine content in the rodent brain (Morgese and Trabace, 2019).

Soluble A frequently causes depressive-like states, leading to a decrease in central serotonin (5-HT) (Paredes et al., 2019) levels, which play an important role in pain perception (Morgese et al., 2022; Schiavone et al., 2017). In addition, it is known that drugs that increase 5-HT levels are effective for improving cognitive function in rodent models (Ma et al., 2017) and Alzheimer's patients (Xie et al., 2019).

Accumulation of soluble Aβ causes disruption of glutaminergic pathways and synaptic defects (Morgese et al., 2022). It has been reported that soluble Aβ injected into the rodent brain causes an increase in the levels of glutamate (Pereire and Goudet, 2018), which has an important role in pain perception and pain transmission from peripheral regions to central regions (Tucci et al., 2014). In addition, it has been demonstrated that ketamine, an antagonist of the glutaminergic NMDA receptor, reduces pain perception when administered to soluble Aβ-induced rats (Schiavone et al., 2017).

Defects in GABA and its subunits, which are effective in pain modulation, can lead to changes in the perception of pain in AD (Shetty and Bates, 2016) In AD, significant changes may occur in the levels of neurotransmitters such as dopamine, serotonin, acetylcholine, GABA, and glutamate. Zebrafish have evolved to preserve these neurotransmitters, which also play a role in modulating pain. Zebrafish have not yet been the subject of pain studies in the AD model, but future studies could consider zebrafish as a suitable model.

## Pain in Parkinson's Disease

The degeneration of dopaminergic neurons and the accumulation of Lewy bodies in the midbrain cause Parkinson's disease (PD), the second most common neurodegenerative disease (Sprenger et al., 2021). Dopamine is a central neurotransmitter and modulator. This feature leads to the hypothesis that the emergence of various clinical symptoms in brain function can be attributed to the loss of function in dopamine neurons. Its effect is likely to be important in Parkinson's disease due to widespread dopamine projections in the central nervous system (Lu et al., 2021).

Pain is one of the most common non-motor symptoms of PD. However, the incomplete elucidation of the molecular mechanism of pain in PD presents a challenge to the early diagnosis of PD (Cansiz et al., 2023). In addition, pain in PD is highly related to the individual's quality of life, and the prevalence of pain in PD can reach up to 80% (Sprenger et al., 2021).

Pain may be associated with PD itself and occurs as central pain as a result of objective pain processing and pain perception impairment in ascending and descending pathways (Lu et al., 2021). In the clinic, most Parkinson's patients complain of pain. Patients with Parkinson's disease suffer from a number of different pain syndromes that differ in their causes, origins, locations, and chronicities (Blanchet and Brefel-Courbon, 2018) These include musculoskeletal pain, joint/arthritic pain, neuropathic pain, and radicular pain (Lu et al., 2021). Even several years before the onset of motor symptoms, pain can occur (Hawkes et al., 2010). Therefore, it is important to discover the connection between the pathological changes of Parkinson's disease and the basic mechanism of pain. The dopamine system is known to influence pain-related pathways, particularly those related to pain perception such as the anterior cingulate cortex (ACC), insular cortex (IC), and prefrontal cortex (PFC), as well as somatosensory cortices (Lu et al., 2021; Zhao et al., 2005). The substantia nigra pars compacta (SNc) and ventral tegmental area (VTA) send dopaminergic signals up to different parts of the brain, including the striatum, the nucleus accumbens, and the PFC, ACC, and IC. Therefore, the effect of dopamine on these cortical areas may play an important role in Parkinson's disease–related pain (Lu et al., 2021).

The identification of almost every pain localization in PD implies that numerous pain processing structures actively contribute to the progression of PD pathoanatomy in the nervous system at different neuropathological stages (Lu et al., 2021). The basal ganglia ascending and descending centers play an important role in pain processing, and dopamine depletion results in a decrease in the multimodal pain threshold (Lu et al., 2021). In addition, local neurodegenerative changes in spinal excitability or decreased dopaminergic descending inhibition combined with painful hyperactivity in primary afferent nociceptors appear to be pathophysiologically significant in PD (Lu et al., 2021). In PD, the loss of dopamine in the basal ganglia is linked to a lower pain threshold and tolerance. However, the fact that dopaminergic drug treatment does not completely improve symptoms suggests that pain symptoms are not only related to dopamine (Brefel-Courbon et al., 2005; Schapira et al., 2017, Cansiz et al., 2023).

Although loss of DA neurons in the SNpc is a classic pathological feature of PD, studies investigating the effects of DA replacement therapy on pain threshold have yielded conflicting results (Lu et al., 2021; Wang et al., 2017). A clinical study showed that apomorphine had no effect on pain thresholds or pain-related cerebral activities in PD (Dellapina et al., 2011). Another clinical study showed that duloxetine, a serotonin-norepinephrine reuptake inhibitor, relieved pain in Parkinson's patients (Djaldetti et al., 2007). Due to the practical limitations of clinical studies, researchers have conducted various animal experiments to elucidate the pain mechanism in PD. These experiments have revealed that pathways other than dopaminergic pathways affect the pain mechanism. The 6-OHDA model has been an important research model for modeling pain in PD. Numerous studies conducted in rats have reported hypersensitivity to mechanical, chemical, or hot-cold stimuli (Buhidma et al., 2020).

Serotonergic involvement may also contribute to nociceptive hypersensitivity (Wang C-T et al., 2017). In 6-OHDA-lesioned rats with nociceptive hypersensitivity, there was a decrease in the number of 5-HT neurons in the rostral ventromedial medulla (RVM) and dorsal horn of the spinal cord, which are effective in the modulation and transmission of pain (Tong et al., 2015). These results indicate an impaired 5-HT system in PD. Furthermore, in the 6-OHDA-induced PD rat model, administration of citalopram attenuated hyperalgesia. Reports indicate that 5-HT within the RVM regulates the activity of pain-modulating neurons via 5-HT1 and 5-HT2 receptors, resulting in an overall antinociceptive effect (Fields et al., 1991). These results suggest that the mechanisms that modulate nociceptive thresholds in the 6-OHDA lesion rat model of PD involve non-dopaminergic pathways, such as serotonergic pathways, as reflected clinically (Wang C-T et al., 2017).

Reduced GABAergic activity in the contralateral PAG and decreased enkephalin and opioid receptor activities in the bilateral dorsal horns of the spinal cord impaired descending analgesic control in the 6-OHDA-induced PD model in mice (Domenici et al., 2019).

MPTP (1-methyl-4-phenyl-1, 2, 3, 6-tetrahydropyridine) is an important chemical agent to create a PD model in animals (Buhidma et al., 2020). After administration to the animal, MPTP crosses the blood–brain barrier and is converted to 1-methyl-4-phenylpyridinium (MPP+) by astrocyte-derived monoamine oxidase and taken into dopaminergic neurons via dopamine transporters (DAT). MPP+ causes inhibition by binding to complex I in the mitochondrial matrix. This inhibition disrupts oxidative phosphorylation, leading to apoptosis and neurodegeneration (Buhidma et al., 2020). MPTP-induced PD model in mice showed a decrease in heat, chemical, and mechanical nociceptive thresholds. When hypersensitivity was evident, dopamine levels decreased in the spinal cord, but there were no changes in 5-HT or noradrenaline. On the 14th day after MPTP, when the hypersensitivity resolved spontaneously, 5-HT levels were observed to no longer increase (Rosland et al., 1992). These findings suggest that central dopaminergic changes have a knock-on effect at the spinal cord level, leading to hypersensitivity. Increased compensatory activity from intact serotonergic systems in the CNS can reverse this situation.

Rotenone is a pesticide that crosses the blood–brain barrier following systemic injection and inhibits complex I of the mitochondrial respiratory chain, causing oxidative stress and the loss of dopaminergic neurons (Buhidma et al., 2020). In the rotenone-induced PD model in zebrafish, acetic acid was administered intraperitoneally, and the response to pain was examined. This study revealed a decrease in the expression of tyrosine hydroxylase, a key enzyme in dopamine synthesis, in the brain tissues of zebrafish administered rotenone. This result indicates that zebrafish can serve as a suitable model for developing a PD model using rotenone. In addition, it was revealed that there was a significant decrease in the expressions of *penka* and *pdyn*, which are related to the opioid system, in the rotenone-treated groups compared to the acetic acid-induced pain group. In addition, the response to pain was examined with the writhing test, and it was observed that the response to externally administered pain was higher in the PD model compared to the control group. According to the results of this study, it can be stated that pain tolerance decreases in the PD model created with zebrafish, and this may be due to changes in the opioid system as well as dopaminergic pathways (Cansiz et al., 2023).

The involvement of numerous brain and brainstem nuclei in the neurodegenerative process raises the possibility that other neurotransmitters besides dopamine may play a role in the modulation of pain in Parkinson's disease (Obeso et al., 2017). Brainstem monoaminergic projections descending to the dorsal horn of the spinal cord are known to have pain-modulating effects (Mostofi et al., 2021). Their degeneration may increase the transmission of pain signals through the spinal cord, resulting in increased pain sensitivity. In addition, the loss of basal forebrain cholinergic neurons in PD may affect the known role of acetylcholine in modulating pain perception (Mostofi et al., 2021). In the endocannabinoid system, rich expression of receptors in the basal ganglia causes changes in PD, while reductions in endocannabinoid receptor density in pain-modulating regions such as the anterior cingulate cortex, as well as changes in endogenous opioid signaling in the dorsal horn, are associated with increased pain sensitivity following experimental Parkinson's lesions in rodents (Binda et al., 2020; Domenici et al., 2019).

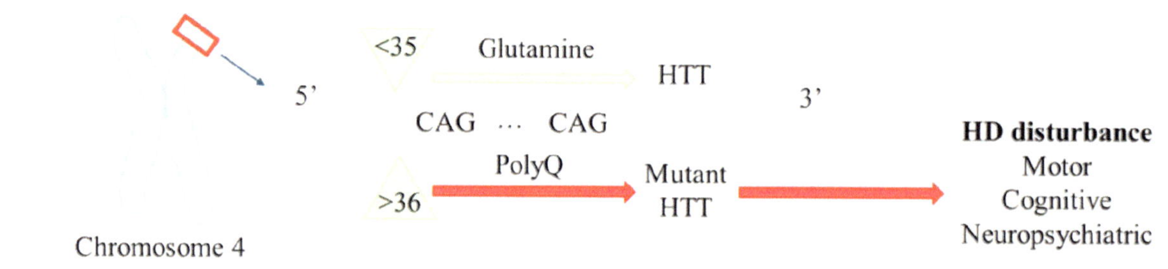


**FIGURE 15.11** Repeated CAG increase on chromosome 4 and formation of HD.

Overall, evidence in humans and animal models has shown that reductions in dopaminergic neurotransmission are associated with increased pain sensitivity. Thus, dopaminergic denervation is a plausible driver for altered central pain processing in PD, but pathological changes in non-dopaminergic neurotransmitter systems also appear to be responsible (Mostofi et al., 2021).

## Pain in Huntington's Disease

Huntington's disease (HD) is a neurodegenerative disorder that develops as a result of excessive expression and mutation of the huntingtin protein in the CNS. Generally, there are approximately 17–20 CAG repeats in the Htt gene located on chromosome 4. The presence of HD is associated with abnormal CAG length (McColgan and Tabrizi, 2018). More than 35 CAG repeats have been reported in HD cases (Figure 15.11). (Rubinsztein et al., 1996).

HD is associated with neuronal loss, especially in the basal ganglia. This condition leads to a variety of symptoms, such as neurocognitive and neurobehavioral disorders (Borsook et al., 2010). The basal ganglia also play a role in acute and chronic pain (Borsook et al., 2010). Dysfunction in the basal ganglia in HD is associated with alterations in the processing and perception of pain. Experimental studies in HD have observed a significant prolongation in the spinal cord processing of painful stimuli in individuals in the premanifest phase compared to healthy controls (De Tommaso et al., 2016; Perrotta et al., 2012). Studies have shown that the pain burden in HD is lower than in PD. However, since there are deficiencies in studies on pain burden in HD, the net value of pain burden is unknown (Sprenger et al., 2021). Although HD is a disease with pain symptoms reported even in the early stages, very few studies have focused on pain in HD (de Tommaso et al., 2016).

The striatum, part of the basal ganglia, regulates pain hypersensitivity at the spinal and trigeminal levels through ipsilateral adrenal dopaminergic descending inhibitory projections and GABAergic descending facilitatory projections (Wei and Pertovaara, 2006).

In HD, dysfunction occurs in many receptors, including GABAA, AMPA, and NMDA receptors, impairing synaptic plasticity and excitability (Zhuo, 2016). Studies have revealed that GABAergic neurons are reduced in neuropathic pain, and stimulation of GABAergic mediators and GABAA receptors can relieve pain (Knabl et al., 2008; St John Smith, 2018). Studies in pain models have reported that stimulation of NMDA receptors plays an important role in reducing central sensitivity and pain hypersensitivity (Kreutzwiser and Tawfic, 2019; Shin et al., 2020). Stimulated NMDA receptors can lead to calcium influx, which is critical for synaptic plasticity. However, the induction of AMPA receptors can regulate synaptic strength and plasticity and alleviate different types of pain hypersensitivity, including neuropathic, neuroinflammatory, and chronic pain (Li J et al., 2023). Some studies have pointed out that the Huntingtin mutation will cause mitochondrial calcium imbalance, disrupt axonal transport of mitochondria, and reduce ATP production (Shirendeb et al., 2011; Reddy and Shirendeb, 2012). There are many ion channels associated with pain in HD, including TRP channels, Ca channels, KC channels, and ASICs. Research has demonstrated the association between HD and the TRP subfamilies TRPC (canonical) and TRPV (vanilloid). Pharmacological experiments demonstrated the potential effects of the TRPV1 channel in HD models, revealing that TRPV1 may contribute to HD dysfunction (Lastres-Becker et al., 2003).

Calcium channels play an important role in neuronal functions and have various subtypes based on their structures and pharmacological properties. Abnormal $Ca^{2+}$ activity in HD may lead to abnormal pain responses (Jimenez-Sanchez et al., 2017). The Cav1.2 subtype is an effective channel in HD (Miranda et al., 2019). Store-operated $Ca^{2+}$ entry (SOCE), an important $Ca^{2+}$ entry pathway, is greatly increased in HD, and this has been associated with $Ca^{2+}$ homeostasis disorder by indirectly activating mHTT (mutant huntingtin) (Czeredys, 2020). Some studies have shown that the interaction between mHTT and inositol-1,4,5-triphosphate receptors (IP3Rs) can cause abnormal $Ca^{2+}$ levels in R6/2 HD mice with high intracellular $Ca^{+}$ and low mitochondrial $Ca^{2+}$ levels (Mackay et al., 2018; Schrank et al., 2020). In addition, subunits of potassium channels, which are effective in pain, are expressed by Glia cells in HD and play a role in the HD process (Zhang et al., 2018).

ASICs, another important channel system, are densely expressed in the nervous system and play an important role in both pain processing and the development of HD (Deval and Linguеglia, 2015; Sluka and Gregory, 2015; Zhu et al., 2022).

## Conclusion

Many studies have been conducted to elucidate the pain mechanism in neurodegenerative diseases, and as a result of these studies, it has been stated that pain, which is a non-motor symptom, may occur before motor symptoms. Therefore, defining pain is crucial in the diagnosis and treatment of neurodegenerative diseases. Researchers not only investigate neurodegenerative diseases through postmortem studies, but also efficiently study them in animal models like rats and mice. Zebrafish, in addition to mammals, serve as an important model for studying neurodegenerative diseases and pain because of their ease of genetic manipulation and the ability to monitor behavioral responses to applied agents. Today, there are many studies conducted to elucidate the mechanism of pain, which is an important symptom in neurodegenerative diseases, but further research is needed to reveal therapeutic approaches to pain.

## REFERENCES

Ahern, C.A., Payandeh, J., Bosmans, F., & Chanda, B. (2016). The hitchhiker's guide to the voltage gated sodium channel galaxy. *J Gen Physiol*, *147*, 1–24.

Alles, S.R.A., Smith, P.A. (2018). Etiology and Pharmacology of Neuropathic Pain. *Pharmacol Rev.* 70(2):315-347. doi: 10.1124/pr.117.014399.

Alvarez, F.A., Rodriguez-Martin, I., Gonzalez-Nuñez, V., de Velasco, E.M., Gonzalez Sarmiento, R., Rodríguez, R.E. (2006). New kappa opioid receptor from zebrafish Danio rerio. *Neurosci Lett*, *405*, 94–99.

Ameenudeen, S., Kashif, M., Banerjee, S., et al. (2022). Cellular and molecular machinery of neuropathic pain: An emerging insight. *Curr Pharmacol Rep*, *8*, 227–235. https://doi.org/10.1007/s40495-022-00294-9.

Azizi, S.A. (2022). Monoamines: Dopamine, norepinephrine, and serotonin, beyond modulation, "switches" that alter the state of target networks. *Neuroscientist*, *28*(2), 121–143. https://doi.org/10.1177/1073858420974336.

Bailey, C.S. (2019). KCNMA1-linked channelopathy. *J Gen Physiol*, *151*(10), 1173–1189.

Bao, A.M., & Swaab, D.F. (2019). The human hypothalamus in mood disorders: The HPA axis in the center. *IBRO Rep*, *6*, 45–53. https://doi.org/10.1016/j.ibror.2018.11.008.

Barcelon, E.E., Cho, W.H., Jun, S.B., & Lee, S.J. (2019). Brain microglial activation in chronic pain-associated affective disorder. *Front Neurosci*, *13*, 213.

Basbaum, A.I., Bautista, D.M., Scherrer, G., & Julius, D. (2009). Cellular and molecular mechanisms of pain. *Cell*, *139*, 267–284. https://doi.org/10.1016/j.cell.2009.09.028.

Benarroch, E.E. (2016). Dorsal horn circuitry: Complexity and implications for mechanisms of neuropathic pain. *Neurology*, *86*(11), 1060–1069.

Bhave, G., Karim, F., Carlton, S.M., & Gereau, R.W. (2001). Peripheral group I metabotropic glutamate receptors modulate nociception in mice. *Nat Neurosci*, *4*, 417–423.

Binda, K.H., Real, C.C., Ferreira, A.F.F., Britto, L.R., & Chacur, M. (2020). Antinociceptive effects of treadmill exercise in a rat model of Parkinson's disease: The role of cannabinoid and opioid receptors. *Brain Res*, *1727*, 146521.

Blanchet, P.J., & Brefel-Courbon, C. (2018). Chronic pain and pain processing in Parkinson's disease. *Prog Neuro-Psychoph*, *87*, 200–206.

Blanton, H., Reddy, P.H., & Benamar, K. (2023). Chronic pain in Alzheimer's disease: Endocannabinoid system. *Exp Neurol*, *360*, 114287. https://doi.org/10.1016/j.expneurol.2022.114287.

Boehmler, W., Petko, J., Woll, M., Frey, C., Thisse, B., Thisse, C., Canfield, V.A., & Levenson, R. (2009). Identification of zebrafish A2 adenosine receptors and expression in developing embryos. *Gene Expr Patterns*, *9*, *3*, 144–151. https://doi.org/10.1016/j.gep.2008.11.006.

Borsook, D., Upadhyay, J., Chudler, E.H., & Becerra, L. (2010). A key role of the basal ganglia in pain and analgesia - insights gained through human functional imaging. *Mol Pain*, *6*, 1–17.

Bortolotto, J.W., de Melo, G.M., de Cognato, G.P., Vianna, M.R.M., & Bonan, C.D. (2015). Modulation of adenosine signaling prevents scopolamine-induced cognitive impairment in zebrafish. *Neurobiol Learn Mem*, *118*, 113–119. https://doi.org/10.1016/j.nlm.2014.11.016.

Bozi, L.H.M., Campos, J.C., Zambelli, V.O., Ferreira, N.D., & Ferreira, J.C.B. (2020). Mitochondrially-targeted treatment strategies. *Mol Asp Med*, *71*, 100836.

Brefel-Courbon, C., Payoux, P., Thalamas, C., et al. (2005). Effects of levodopa on pain threshold in Parkinson's disease: A clinical and positron emission tomography study. *Mov Disord*, *20*(12), 1557–1563.

Brome´e, T., Venkatesh, B., Brenner, S., Postlethwait, J.H., Yan, Y.L., & Larhammar, D. (2006). Uneven evolutionary rates of bradykinin B1 and B2 receptors in vertebrate lineages. *Gene*, *373*, 100–108.

Buhidma, Y., Rukavina, K., Chaudhuri, K.R., & Duty, S. (2020). Potential of animal models for advancing the understanding and treatment of pain in Parkinson's disease. *NPJ Parkinsons Dis*, *6*, 1. https://doi.org/10.1038/s41531-019-0104-6.

Burnstock G. (2000). P2X receptors in sensory neurones. *Br J Anaesth*, *84*, 476–488. https://doi.org/10.1093/oxfordjournals.bja.a013473.

Busserolles, J., Tsantoulas, C., Eschalier, A., López García, J.A. (2016). Potassium channels in neuropathic pain: advances, challenges, and emerging ideas. *Pain.* 157 Suppl 1:S7-S14. doi: 10.1097/j.pain.0000000000000368.

Cansiz, D., Unal I., Beler, M., Ustundag, U.V., Ak, E., Emekli-Alturfan, E., & Alturfan, A.A. (2023). The effect of acetic acid-induced pain in Parkinson's disease model in zebrafish. NeuroToxicology, *99*, 14–23. https://doi.org/10.1016/j.neuro.2023.09.004.

Catterall, W. A., Goldın, A. L., & Waxman, S. G. (2005). International union of pharmacology. XLVII. Nomenclature and structure-function relationships of voltage-gated sodium channels. *Pharmacol Rev*, *57*, 397–409.

Chao, R.Y., Cheng, C.H., Wu, S.N., & Chen, P.C. (2018). Defective trafficking of Kv2.1 channels in MPTP-induced nigrostriatal degeneration. *J Neurochem*, *144*(4), 483–497. https://doi.org/https://doi.org/10.1111/jnc.14282.

Chen, S., Chiu, C.N., McArthur, K.L., Fetcho, J.R., & Prober, D.A. (2016). TRP channel mediated neuronal activation and ablation in freely behaving zebrafish. *Nat Methods*, *13*(2), 147–150. https://doi.org/10.1038/nmeth.3691.

Chen, X., Polleichtner, G., Kadurin, I., & Gründer, S. (2007). Zebrafish Acid-Sensing Ion Channel (ASIC) 4, characterization of homo- and heteromeric channels, and identification of regions important for activation by H+. *J Biol Chem*, *282*, 30406–30413.

Chen, Y., Boettger, M.K., Reif, A., Schmitt, A., Uceyler, N., & Sommer, C. (2010). Nitric oxide synthase modulates CFA-induced thermal hyperalgesia through cytokine regulation in mice *Mol Pain*, *6*, 13. https://doi.org/10.1186/1744-8069-6-13.

Cocco, A., Rönnberg, A.M., Jin, Z., André, G.I., Vossen, L.E., Bhandage, A.K., Thörnqvist, P.O., Birnir, B., & Winberg, S. (2017). Characterization of the γ-aminobutyric acid signaling system in the zebrafish (Danio rerio Hamilton) central nervous system by reverse transcription-quantitative polymerase chain reaction. *Neuroscience*, *20*, 300–321. https://doi.org/10.1016/j.neuroscience.2016.07.018.

Cole, N.B., Dieuliis, D., Leo, P., Mitchell, D.C., & Nussbaum, R.L. (2008). Mitochondrial translocation of alpha-synuclein is promoted by intracellular acidification. *Exp Cell Res*, *314*(10), 2076–2089. https://doi.org/10.1016/j.yexcr.2008.03.012.

Correia, A.D., Cunha, S.R., Scholze, M., & Stevens, E.D. (2011). A novel behavioral fish model of nociception for testing analgesics. *Pharmaceuticals*, *4*(4), 665–680. https://doi.org/10.3390/ph4040665.

Costa, F.V., Rosa, L.V., Quadros, V.A., de Abreu, M.S., Santos, A.R.S., Sneddon, L.U., Kalueff, A.V., & Rosemberg, D.B. (2022). The use of zebrafish as a non-traditional model organism in translational pain research: The knowns and the unknowns. *Curr Neuropharmacol*, *20*(3), 476–493. https://doi.org/10.2174/1570159X19666210311104408.

Costa, F.V., Rosa, L.V., Quadros, V.A., Santos, A.R.S., Kalueff, A.V., & Rosemberg, D.B. (2019). Understanding nociception-related phenotypes in adult zebrafish: Behavioral and pharmacological characterization using a new acetic acid model. *Behav Brain Res*, *359*, 570–578. https://doi.org/10.1016/j.bbr.2018.10.009.

Curtright, A., Rosser, M., Goh, S., Keown, B., Wagner, E., Sharifi, J., Raible, D. W., & Dhaka, A. (2015). Modeling nociception in zebrafish: A way forward for unbiased analgesic discovery. *PLoS One*, *10*(1), e0116766.

Cury, Y., Picolo, G., Gutierrez, V.P., & Ferreira, S.H. (2011). Pain and analgesia: The dual effect of nitric oxide in the nociceptive system. *Nitric Oxide*, *25*(3), 243–254. https://doi.org/10.1016/j.niox.2011.06.004.

Czeredys, M. (2020). Dysregulation of neuronal calcium signaling via store-operated channels in Huntington's disease. *Front Cell Dev Biol*, *8*, 611735. https://doi.org/10.3389/fcell.2020.611735.

David, A., & Pancharatna, K. (2009). Developmental anomalies induced by a non-selective COX inhibitor (ibuprofen) in zebrafish (Danio rerio). *Environ Toxicol Pharmacol*, *27*(3), 390–395. https://doi.org/10.1016/j.etap.2009.01.002.

De Tommaso, M., Franco G., Ricci K., Montemurno A., & Sciruicchio V. (2016). Laser evoked potentials in early and presymptomatic Huntington's disease. *Behav Neurol*, 1–8. https://doi.org/10.1155/2016/8613729.

De Velasco, E.M., Law, P.Y., & Rodríguez, R.E. (2009). Mu opioid receptor from the zebrafish exhibits functional characteristics as those of mammalian mu opioid receptor. *Zebrafish*, *6*, 259–268.

Deakin, A.G., Buckley, J., AlZu'bi, H.S., Cossins, A. R., Spencer, J.W., Al'Nuaimy, W., Young, I. S., Thomson, J.S., & Sneddon, L.U. (2019). Automated monitoring of behaviour in zebrafish after invasive procedures. *Sci Rep*, *9*, 9042. https://doi.org/10.1038/s41598-019-45464-w.

Dellapina, E., Gerdelat-Mas, A., Ory-Magne, F., Pourcel, L., Galitzky, M., Calvas, F., Simonetta-Moreau, M., Thalamas, C., Payoux, P., & Brefel-Courbon, C. (2011). Apomorphine effect on pain threshold in Parkinson's disease: A clinical and positron emission tomography study. *Mov Disord*, *26*, 153–157.

Deval, E., & Lingueglia, E. (2015). Acid-sensing ion channels and nociception in the peripheral and central nervous systems. *Neuropharmacology*, *94*, 49–57.

Devine, M.J., & Kittler, J.T. (2018). Mitochondria at the neuronal presynapse in health and disease. *Nat Rev Neurosci*, *19*(2), 63–80. https://doi.org/10.1038/nrn.2017.170.

Dıchgans, M., Freılınger, T., Ecksteın, G., Babını, E., Lorenz-Depıereux, B., Bıskup, S., Ferrarı, M. D., Herzog, J., Van Den Maagdenberg, A. M., Pusch, M., & Strom, T. M. (2005). Mutation in the neuronal voltage-gated sodium channel SCN1A in familial hemiplegic migraine. *Lancet*, *366*, 371–377.

Djaldetti, R, Yust-Katz, S, Kolianov, V., Melamed, E., & Dabby, Ron. (2007). The effect of duloxetine on primary pain symptoms in Parkinson disease. *Clin Neuropharmacol*, *30*, 201–205.

do Nascimento, J.E.T., de Morais, S.M., de Lisboa, D.S., de Oliveira S.M., Santos, S.A.A.R., Magalhães, F.E.A., & Campos, A.R. (2018). The orofacial antinociceptive effect of Kaempferol-3-O-rutinoside, isolated from the plant Ouratea fieldingiana, on adult zebrafish (Danio rerio). *Biomed Pharmacother*, *107*, 1030–1036. https://doi.org/10.1016/j.biopha.2018.08.089.

Dogrul, A., Ossipov, M.H., Lai, J., Malan, T.P. Jr., & Porreca, F. (2000). Peripheral and spinal antihyperalgesic activity of SIB-1757, a metabotropic glutamate receptor (mGLUR(5)) antagonist, in experimental neuropathic pain in rats. *Neurosci Lett*, *292*, 115–118.

Domenici, R.A., Campos, A.C.P., Maciel, S.T., Berzuino, M.B., Hernandes, M.S., Fonoff, E.T., & Pagano, R.L. (2019). Parkinson's disease and pain: Modulation of nociceptive circuitry in a rat model of nigrostriatal lesion. *Exp Neurol*, *315*, 72–81, https://doi.org/10.1016/j.expneurol.2019.02.007.

Dooley, K., & Zon, L.I. (2000). Zebrafish: a model system for the study of human disease. *Curr Opin Genet Dev*, *10*(3), 252–256. https://doi.org/10.1016/s0959-437x(00)00074-5.

Doyle, T.M., & Salvemini, D. (2021). Mini-review: Mitochondrial dysfunction and chemotherapy-induced neuropathic pain. *Neurosci Lett*, *760*, 136087.

Dune´r, T., Conlon, J.M., Kukkonen, J.P., Akerman, K.E., Yan, Y.L., Postlethwait, J.H., & Larhamma, D. (2002). Cloning, structural characterization and functional expression of a zebrafish bradykinin B2-related receptor. *Biochem J*, *364*, 817–824.

Dymowska, A.K., Boyle, D., Schultz, A.G., Goss, G.G. (2015). The role of acid-sensing ion channels in Epithelial Na+ Uptake in Adult Zebrafish (Danio Rerio). *J Exp Biol*, *218*, 1244–1251.

Ebert, A.M., McAnelly, C.A., Srinivasan, A., Mueller, R.L., Garrity, D. B., & Garrity, D.M. (2008). The calcium channel β2 (CACNB2) subunit repertoire in teleosts. *BMC Molecular Biol*, *9*, 38. https://doi.org/10.1186/1471-2199-9-38

Faber, C.G., Laurıa, G., Merkıes, I.S., Cheng, X., Han, C., Ahn, H. S., Persson, A. K., Hoeıjmakers, J. G., Gerrıts, M. M., Pıerro, T., Lombardı, R., Kapetıs, D., Dıb-Hajj, S. D., & Waxman, S. G. (2012). Gainof-function Nav1.8 mutations in painful neuropathy. *Proc Natl Acad Sci*, *109*, 19444–19449.

Fabiano, V. C., Canzian, J., Stefanello, F.V., Kalueff, A.V., & Rosemberg, D.B. (2019). Naloxone prolongs abdominal constriction writhing-like behavior in a zebrafish-based pain model. *Neurosci Lett*, *708*(24), 134336. https://doi.org/10.1016/j.neulet.2019.134336.

Ferreira, R.C.M., Castor, M.G.M., Piscitelli, F., Di Marzo, V., Duarte, I.D.G., & Romero, T.R.L. (2018). The involvement of the endocannabinoid system in the peripheral antinociceptive action of ketamine. *J Pain*, *19*(5), 487–495.

Ferri, A., Nencini, M., Cozzolino, M., Carrara, P., Moreno, S., & Carrì, M.T. (2008). Inflammatory cytokines increase mitochondrial damage in motoneuronal cells expressing mutant SOD1. *Neurobiol Dis*, *32*(3), 454–460. https://doi.org/10.1016/j.nbd.2008.08.004.

Fertleman, C.R., & Ferrıe, C.D. (2006). What's in a name–familial rectal pain syndrome becomes paroxysmal extreme pain disorder. *J Neurol Neurosurg Psychiatry*, *77*, 1294–1295.

Fields, H.L., Heinricher, M.M., & Mason, P. (1991). Neurotransmitters in nociceptive modulatory circuits. *Annu Rev Neurosci*, *14*, 219–245.

Flatters, S.J.L. (2015). The Contribution of Mitochondria to Sensory Processing and Pain. *Prog Mol Biol Trans Sci*, *131*, 119–146. https://doi.org/10.1016/bs.pmbts.2014.12.004.

Franklin, K.B., & Abbott, F.V. (1989). *Psychopharmacology. Techniques for assessing the effects of drugs on nociceptive responses.* Springer; Berlin, Germany, 145–216.

Goldin, A.L., Barchi, R.L., Caldwell, J.H., Hofmann, F., Howe, J.R., Hunter, J.C., Kallen, R.G., Mandel, G., Meisler, M.H., Netter, Y.B., Noda, M., Tamkun, M.M., Waxman, S.G., Wood, J.N., & Catterall, W.A. (2000). Nomenclature of voltage-gated sodium channels. *Neuron*, *28*, 365–368.

Gomes, F.I.F., Cunha, F.Q., & Cunha, T.M. (2020). Peripheral nitric oxide signaling directly blocks inflammatory pain. *Biochem Pharmacol Junho*, *176*, 113862. https://doi.org/10.1016/j.bcp.2020.113862.

Gonzalez Nunez, V., Gonzalez Sarmiento, R., & Rodríguez, R.E. (2003a). Characterization of zebrafish proenkephalin reveals novel opioid sequences. *Brain Res Mol Brain Res*, *114*, 31–39.

Gonzalez-Nunez, V., Gonzalez-Sarmiento, R., & Rodríguez, R.E. (2003b). Identification of two proopiomelanocortin genes in zebrafish (Danio rerio). *Brain Res Mol Brain Res*, *120*, 1–8.

Gonzalez-Nunez, V., Gonzalez-Sarmiento, R., & Rodríguez, R.E. (2003c). Cloning and characterization of a full-length pronociceptin in zebrafish: Evidence of the existence of two different nociceptin sequences in the same precursor. *Biochim Biophys Acta*, *1629*, 114–118.

Gonzalez-Nuñez, V., Marrón Fernández de Velasco, E., Arsequell, G., Valencia, G., Rodríguez, R.E. (2007) Identification of dynorphin a from zebrafish: a comparative study with mammalian dynorphin A. Neuroscience. 19;144(2):675-84. doi: 10.1016/j.neuroscience.2006.09.028.

Grosser, T., Yusuff, S., Cheskis, E., Pack, M.A., & FitzGerald, G.A. (2002). Developmental expression of functional cyclooxygenases in zebrafish. *Proc Natl Acad Sci*, *99*, 8418–8423.

Gusso, D., Wiprich, M.T., Altenhofen, S., & Bonan, C.D. (2022). Role of the nucleoside-metabolizing enzymes on pain responses in zebrafish larvae. *Neurotoxicol Teratol*, *93*, 107109. https://doi.org/10.1016/j.ntt.2022.107109.

Hawkes, C.H., Del Tredici, K., & Braak, H. (2010). A timeline for Parkinson's disease. *Parkinsonism Relat Disord*, *16*, 79–84.

Hıyama, T. Y., Watanabe, E., Ono, K., Inenaga, K., Tamkun, M. M., Yoshıda, S., & Noda, M. (2002). Na(x) channel involved in CNS sodium-level sensing. *Nat Neurosci*, *5*, 511–512.

Holzer, P. (2009). Acid-sensitive ion channels and receptors. In B.J. Canning & D. Spina (Eds.), *Sensory Nerves*. Berlin/Heidelberg: Springer, 283–332. ISBN 978-3-540-79090-7.

Holzer, P. (2011). Acid sensing by visceral afferent neurones. *Acta Physiol*, *201*, 63–75.

Hosseini, S.A., Salehi, O.R., Farzanegi, P., Farkhaie, F., Darvishpour, A.R., & Roozegar, S. (2020). Interactive effects of endurance training and royal jelly consumption on motor balance and pain threshold in animal model of the alzheimer disease. *Arch Neurosci*, *7*(2), e91857. https://doi.org/10.5812/ans.91857.

Hökfelt, T., Kellerth, J.O., Nilsson, G., & Pernow, B. (1975). Substance P: Localization in the central nervous system and in some primary sensory neurons. *Science*, *190*, 889–890.

Hökfelt, T., Pernow, B., & Wahren, J. (2001). Substance P: A pioneer amongst neuropeptides. *J Intern Med*, *249*, 27–40.

Hroudová, J., Singh, N., & Fišar Z. (2014). Mitochondrial dysfunctions in neurodegenerative diseases: relevance to Alzheimer's disease. *Biomed Res Int*, 175062. https://doi.org/10.1155/2014/175062.

Iang, M., & Bajpayee, N.S. (2009). Molecular mechanisms of go signaling. *Neurosignals*, *17*(1), 23–41. https://doi.org/10.1159/000186688.

Isensee, J., Krahe, L., Moeller, K., Pereıra, V., Sexton, J. E., Sun, X., Emery, E., Wood, J. N., & Hucho, T. (2017). Synergistic regulation of serotonin and opioid signaling contributes to pain insensitivity in Nav1.7 knockout mice. *Sci Signal*, *10*.

Jérôme, B., Gasull, X., & Noël, J. (2019). Potassium channels and pain. In N.W. John (Ed.), *The Oxford handbook of the neurobiology of pain*. Oxford: Oxford University Press.

Ji, R.R., Nackley, A., Huh, Y., Terrando, N., & Maixner, W. (2018). Neuroinflammation and central sensitization in chronic and widespread pain. *Anesthesiology*, *129*(2), 343–366.

Ji, R.R., Xu, Z.Z., & Gao, Y.J. (2014). Emerging targets in neuroinflammation-driven chronic pain. *Nat Rev Drug Discov*, *13*(7), 533–548.

Jimenez-Sanchez, M., Licitra, F., Underwood, B., & Rubinsztein, D. (2017). Huntington's disease: Mechanisms of pathogenesis and therapeutic strategies. *Cold Spring Harbor Perspect Med*, *7*, a024240.

Kim, H.Y., Chung, J.M., & Chung, K. (2008). Increased production of mitochondrial superoxide in the spinal cord induces pain behaviors in mice: the effect of mitochondrial electron transport complex inhibitors. *Neurosci Lett*, *447*(1), 87–91. https://doi.org/10.1016/j.neulet.2008.09.04.

Kim, J.B. (2014). Channelopathies. *Korean J Pediatr*, *57*, 1–18.

Kinboshi, M., Ikeda, A., & Ohno, Y. (2020). Role of Astrocytic Inwardly Rectifying Potassium (Kir) 4.1 Channels in Epileptogenesis. *Front Neurol*, *11*, 626658. https://doi.org/10.3389/fneur.2020.626658.

Knabl, J., Witschi, R., Hosl, K., et al. (2008). Reversal of pathological pain through specific spinal GABAA receptor subtypes. *Nature*, *451*, 330–334.

Kolesnikova, T.O., Demin, K.A., Costa, F.V., Zabegalov, K.N., de Abreu, M.S., Gerasimova, E.V., Kalueff, A.V. (2022) Towards Zebrafish Models of CNS Channelopathies. *Int J Mol Sci*. 12;23(22):13979. doi: 10.3390/ijms232213979.

Kreutzwiser, D., & Tawfic, Q. (2019). Expanding role of NMDA receptor antagonists in the management of pain. *CNS Drugs*, *33*, 347–374. https://doi.org/10.1007/s40263-019-00618-2

Krishtal, O. (2015). Receptor for protons: First observations on acid sensing ion channels. *Neuropharmacology*, *94*, 4–8.

Lagos-Rodriguez, V., Martinez-Palma, L., Marton, S., Miquel, E., Escobar-Pintos, R., Cassina, A., Lago, N., & Cassina, P. (2020). Mitochondrial bioenergetics, glial reactivity, and pain-related behavior can be restored by dichloroacetate treatment in rodent pain models. *Pain*, *161*(12), 2786–2797.

Lastres-Becker, I., de Miguel, R., De Petrocellis, L., Makriyannis, A., Di Marzo, V., & Fernandez-Ruiz, J. (2003). Compounds acting at the endocannabinoid and/or endovanilloid systems reduce hyperkinesia in a rat model of Huntington's disease. *J Neurochem*, *84*, 1097–1109.

Lee, K., Jo, Y., Chung, G., Jung, J., Kim, Y., & Park, C. (2021). Functional importance of transient receptor potential (TRP) channels in neurological disorders. *Front Cell Dev Biol*, *9*, 611773. https://doi.org/10.3389/fcell.2021.611773.

Li, F., Zhou, S. N., Zeng, X., Li, Z., Yang, R., Wang, X. X., Meng, B., Pei, W. L., & Lu, L. (2022). Activation of the TREK-1 potassium channel improved cognitive deficits in a mouse model of Alzheimer's disease by modulating glutamate metabolism. *Mol Neurobiol*, *59*(8), 5193–5206. https://doi.org/10.1007/s12035-022-02776-9.

Li, J., Wang, Y., Yang, R., Ma, W., Yan, J., Li, Y., Chen, G., & Pan, J. (2023). Pain in Huntington's disease and its potential mechanisms. *Front Aging Neurosci*, *6*(15), 1190563. https://doi.org/10.3389/fnagi.2023.1190563.

Li, X.Y., & Toyoda H. (2015). Role of leak potassium channels in pain signalling. *Brain Research Bulletin*, *119*, Part A, 73–79.

Lin, C.C., Chen, W.N., Chen, C.J., Lin, Y.W., Zimmer, A., & Chen, C.C. (2012). An antinociceptive role for substance P in acid-induced chronic muscle pain. *Proc Natl Acad Sci U S A*, *109*(2), E76–E83. https://doi.org/10.1073/pnas.1108903108.

Linguegia, E. (2007). Acid-sensing ion channels in sensory perception. *J Biol Chem*, *282*, 17325–17329.

Liu, X., Ma, L., Zhang, S., Ren, Y., & Dirksen, R.T. (2017). CD73 controls extracellular adenosine generation in the trigeminal nociceptive nerves. *Journal of Dental Research*, *96*(6), 671–677. https://doi.org/10.1177/0022034517692953.

Liu, Y., Zhou, L.J., & Wang, J. (2017). TNF-alpha differentially regulates synaptic plasticity in the hippocampus and spinal cord by microglia-dependent mechanisms after peripheral nerve injury. *J Neurosci*, *37*, 871–81.

López-Bellido, R., Barreto-Valer, K., & Rodríguez, R.E. (2013). Substance P mRNA expression during zebrafish development: influence of mu opioid receptor and cocaine. *Neuroscience*, *242*, 53–68. https://doi.org/10.1016/j.neuroscience.2013.03.022.

Lu, J.S., Chen, Q.Y., Chen, X., Li, X.H., Zhou, Z., Liu, Q., Lin, Y., Zhou, M., Xu, P.Y., & Zhuo, M. (2021). Cellular and synaptic mechanisms for Parkinson's disease-related chronic pain. *Molecular Pain*, *17*. https://doi.org/10.1177/1744806921999025.

Ma, J., Gao, Y., Jiang, L., Chao, F. L., Huang, W., Zhou, C. N., Tang, W., Zhang, L., Huang, C.X., Zhang, Y., Luo, Y.M., Xiao, Q., Yu, H.R., Jiang, R., & Tang, Y. (2017). Fluoxetine attenuates the impairment of spatial learning ability and prevents neuron loss in middle-aged APPswe/PSEN1dE9 double transgenic Alzheimer's disease mice. *Oncotarget*, *8*, 27676–27692. https://doi.org/10.18632/oncotarget.15398.

Mackay, J., Nassrallah, W., & Raymond, L. (2018). Cause or compensation?-Altered neuronal Ca(2+) handling in Huntington's disease. CNS *Neurosci Ther*, *24*, 301–310.

Maes, S. S., Pype, S., Hoffmann, V. L., Biermans, M., & Meert, T. F. (2012). Antihyperalgesic activity of nucleoside transport inhibitors in models of inflammatory pain in guinea pigs. *J Pain Res*, *5*, 391–400. https://doi.org/10.2147/JPR.S35108.

Malafoglia, V., Bryant, B., Raffaeli, W., Giordano, A., & Bellipanni G., (2013). The zebrafish as a model for nociception studies. *J Cell Physiol*, *228*(10). https://doi.org/10.1002/jcp.24379.

Manasa, K., Chitra, V., & Tamilanban, T. (2020). Teleost model as an alternative in Parkinson's disease. *Neurol India*, *68*(5), 979–984. https://doi.org/10.4103/0028-3886.294542.

Mantegazza, M., Gambardella, A., Rusconı, R., Schıavon, E., Annesı, F., Cassulını, R. R., Labate, A., Carrıdeo, S., Chıfarı, R., Canevını, M. P., Canger, R., Franceschettı, S., Annesı, G., Wanke, E., & Quattrone, A. (2005). Identification of an Nav1.1 sodium channel (SCN1A) loss-of-function Mutation associated with familial simple febrile seizures. *Proc Natl Acad Sci U S A*, *102*, 18177–18182.

Marchi, S., Guilbaud, E., Tait, S.W.G., Yamazaki, T., & Galluzzi, L. (2022). Mitochondrial control of inflammation. *Nat Rev Immunol*, 1–15. https://doi.org/10.1038/s41577-022-00760-x.

Mathur, P., & Guo, S. (2010). Use of zebrafish as a model to understand mechanisms of addiction and complex neurobehavioral phenotypes. *Neurobiol Dis*, *40*, 66–72.

Maximino, C. (2011). Modulation of nociceptive-like behavior in zebrafish (Danio rerio) by environmental stressors. *Psychol Neurosci*, *4*(1),149–155. https://doi.org/10.3922/j.psns.2011.1.017.

McColgan, P., & Tabrizi, S. (2018). Huntington's disease: A clinical review. *Eur J Neurol*, *25*, 24–34.

Mınett, M. S., Pereıra, V., Sıkandar, S., Matsuyama, A., Lolıgnıer, S., Kanellopoulos, A. H., Mancını, F., Iannettı, G. D., Bogdanov, Y. D., Santana-Varela, S., Mıllet, Q., Baskozos, G., Macallıster, R., Cox, J. J., Zhao, J., & Wood, J. N. (2015). Endogenous opioids contribute to insensitivity to pain in humans and mice lacking sodium channel Nav1.7. *Nat Commun*, *6*, 8967.

Miranda, A., Cardozo, P., Silva, F., de Souza, J., Olmo, I., Cruz, J., Gomez, M. V., Ribeiro, F.M., & Vieira, L.B. (2019). Alterations of calcium channels in a mouse model of Huntington's disease and neuroprotection by blockage of Ca(V)1 channels. *ASN Neuro*, *11*, 1759091419856811.

Mitchell, R., Campbell, G., Mikolajczak, M., McGill, K., Mahad, D., & Fleetwood-Walker, S.M. (2019). A targeted mutation disrupting mitochondrial complex IV function in primary afferent neurons leads to pain hypersensitivity through P2Y1 receptor activation. *Mol Neurobiol*, *56*(8), 5917–5933. https://doi.org/10.1007/s12035-018-1455-4.

Mochel, F., Rastetter, A., Ceulemans, B., Platzer, K., Yang, S., Shinde, D.N., Helbig, K.L., Lopergolo, D., Mari, F., Renieri, A., Benetti, E., Canitano, R., Waisfisz, Q., Plomp, A.S., Huisman, S.A., Wilson, G.N., Cathey, S.S., Louie, R.J., Gaudio, D.D., Waggoner, D., & Depienne, C. (2020). Variants in the SK2 channel gene (KCNN2) lead to dominant neurodevelopmental movement disorders. *Brain*, *143*(12), 3564–3573. https://doi.org/10.1093/brain/awaa346.

Monesson-Olson, B., McClain, J.J., Case, A.E., Dorman, H.E., Turkewitz, D.R., Steiner, A.B., & Downes, G. B. (2018). Expression of the eight GABAA receptor α subunits in the developing zebrafish central nervous system. *PLoS One*, *13*(4), e0196083. https://doi.org/10.1371/journal.pone.0196083.

Montalbano, G., Levanti, M., Mhalhel, K., Abbate, F., Laurà, R., Guerrera, M.C., Aragona, M., & Germanà, A (2021). Acid-Sensing ion channels in zebrafish. *Animals*, *11*(8), 2471. https://doi.org/10.3390/ani11082471.

Moore, C., Gupta, R., Jordt, S., Chen, Y., & Liedtke W. (2018). Regulation of pain and itch by TRP channels. *Neurosci Bull*, *34*, 120–142.

Morgese, M.G., Bove, M., Di Cesare Mannelli, L., Schiavone, S., Colia, A.L., Dimonte, S., Mhillaj, E., Sikora, V., Tucci, P., Ghelardini, C., & Trabace, L. (2022). Precision medicine in Alzheimer's disease: Investigating comorbid common biological substrates in the rat model of amyloid beta-induced toxicity. *Front Pharmacol*, *12*, 799561. https://doi.org/10.3389/fphar.2021.799561.

Morgese, M.G., Schiavone, S., & Trabace, L. (2017). Emerging role of amyloid beta in stress response: Implication for depression and diabetes. *Eur J Pharmacol*, *817*, 22–29. https://doi.org/10.1016/j.ejphar.2017.08.031.

Morgese, M.G., & Trabace L. (2019). Monoaminergic system modulation in depression and Alzheimer's disease: A new standpoint? *Front Pharmacol*, *10*, 483. https://doi.org/10.3389/fphar.2019.00483.

Moro, A., Moscovich, M., Farah, M., Camargo, C.H.F., Teive, H.A.G., & Munhozet, R.P. (2019). Nonmotor symptoms in spinocerebellar ataxias (SCAs). *Cerebellum Ataxias*, *6*(12). https://doi.org/10.1186/s40673-019-0106-5.

Morthorst, J.E., Lister, A., Bjerregaard, P., & Van Der Kraak, G. (2013). Ibuprofen reduces zebrafish PGE(2) levels but steroid hormone levels and reproductive parameters are not affected. *Comp Biochem Physiol C Toxicol Pharmacol*, *157*(2), 251–257. https://doi.org/10.1016/j.cbpc.2012.12.001.

Mostofi, A., Morgante, F., Edwards, M.J., Brown, P., & Pereira, E.A.C. (2021). Pain in Parkinson's disease and the role of the subthalamic nucleus. *Brain*, *144*(5), 1342–1350. https://doi.org/10.1093/brain/awab001

Mulley, J. C., Hodgson, B., Mcmahon, J. M., Iona, X., Bellows, S., Mullen, S. A., Farrell, K., Mackay, M., Sadleir, L., Bleasel, A., Gill, D., Webster, R., Wirrell, E. C., Harbord, M., Sisodiya, S., Andermann, E., Kivity, S., Berkovic, S. F., Scheffer, I. E., & Dibbens, L. M. (2013). Role of the sodium channel SCN9A in genetic epilepsy with febrile seizures plus and Dravet syndrome. *Epilepsia*, *54*, e122-6.

Murakami, M., Fleischmann, B., De Felipe, C., Freichel, M., Trost, C., Ludwig, A., Wissenbach, U., Schwegler, H., Hofmann, F., Hescheler, J., Flockerzi, V., & Cavalié, A. (2002). Pain perception in mice lacking the beta3 subunit of voltage-activated calcium channels. *J Biol Chem*, *277*(43), 40342–40351. https://doi.org/10.1074/jbc.M203425200.

Nilius, B., Appendino, G., & Owsianik, G. (2012). The Transient receptor potential channel TRPA1: from gene to pathophysiology. *Pflugers Arch- Eur Physiol*, *464*, 425–458.

Noda, M. (2006). The subfornical organ, a specialized sodium channel, and the sensing of sodium levels in the brain. *Neuroscientist*, *12*, 80–91.

North, R.A. (2002). Molecular physiology of p2x receptors. *Physiol Rev*, *82*, 1013–1067. https://doi.org/10.1152/physrev.00015.2002.

Norton, W.H.J., Folchert, A., & Bally-Cuif, L. (2008). Comparative analysis of serotonin receptor (HTR1A/HTR1B families) and transporter (slc6a4a/b) gene expression in the zebrafish brain. *J Comp Neurol*, *511*(4), 521–542. https://doi.org/10.1002/cne.21831.

Obeso, J.A., Stamelou, M., Goetz, C.G., Poewe, W., Lang, A.E., Weintraub, D., Burn, D., Halliday, G.M., Bezard, E., Przedborski, S., Lehericy, S., Brooks, D.J., Rothwell, J.C., Hallett, M., DeLong, M.R., Marras, C., Tanner, C.M., Ross, G.W., Langston, J.W., Klein, C., & Stoessl, A.J. (2017). Past, present, and future of Parkinson's disease: A special essay on the 200th Anniversary of the Shaking Palsy. *Mov Disord*, *32*(9), 1264–1310.

Ogiwara, I., Ito, K., Sawaishi, Y., Osaka, H., Mazaki, E., Inoue, I., Montal, M., Hashikawa, T., Shike, T., Fujiwara, T., Inoue, Y., Kaneda, M., & Yamakawa, K. (2009). De novo mutations of voltage-gated sodium channel alphaII gene SCN2A in intractable epilepsies. *Neurology*, *73*, 1046–1053.

Oliveira, A.S., Biano, L.S., Palmeira, D.N., Almeida, D.R., Lopes-Ferreira, M., Kohlhoff, M., Sousa, J.A.C., Brandão, G.C., Silva, A.M., Grespan, R., & Camargo, E.A. (2023). Antinociceptive effect of Nephelium lappaceum L. fruit peel and the participation of nitric oxide, opioid receptors, and ATP-sensitive potassium channels. *Front Pharmacol*, *14*, 1287580. https://doi.org/10.3389/fphar.2023.1287580.

Oliveri, A.N., & Levin, E.D. (2019). Dopamine D1 and D2 receptor antagonism during development alters later behavior in zebrafish. *Behav Brain Res*, *356*, 250–256. https://doi.org/10.1016/j.bbr.2018.08.028.

Olson, N., & van der Vliet, A. (2011). Interactions between nitric oxide and hypoxia-inducible factor signaling pathways in inflammatory disease. *Nitric Oxide*, *25*(2), 125–137. https://doi.org/10.1016/j.niox.2010.12.010.

Palesi, F., Castellazzi, G., Casiraghi, L., Sinforiani, E., Vitali, P., Gandini Wheeler-Kingshott, C.A.M., & D'Angelo, E. (2016). Exploring patterns of alteration in Alzheimer's disease brain networks: A combined structural and functional connectomics analysis. *Front Neurosci*, *10*, 380. https://doi.org/10.3389/fnins.2016.00380.

Paredes, S., Cantillo, S., Candido, K.D., & Knezevic, N.N. (2019). An association of serotonin with pain disorders and its modulation by estrogens. *Int J Mol Sci*, *20*, 5729. https://doi.org/10.3390/ijms20225729.

Paukert, M., Sidi, S., Russell, C., Siba, M., Wilson, S.W., Nicolson, T., & Gründer, S. A. (2004). Family of acid-sensing ion channels from the zebrafish: Widespread expression in the central nervous system suggests a conserved role in neuronal communication. *J Biol Chem*, *279*, 18783–18791.

Pekkurnaz, G., & Wang, X. (2022). Mitochondrial heterogeneity and homeostasis through the lens of a neuron. *Nat Metab*, *4*(7), 802–812. https://doi.org/10.1038/s42255-022-00594-w

Pennefather, J.N., Lecci, A., Candenas, M.L., Patak, E., Pinto, F.M., & Maggi, C.A. (2004). Tachykinins and tachykinin receptors: A growing family. *Life Sci*, *74*(12), 1445–1463. https://doi.org/10.1016/j.lfs.2003.09.039.

Pereira V., & Goudet C. (2018). Emerging trends in pain modulation by metabotropic glutamate receptors. *Front Mol Neurosci, 11*, 464. https://doi.org/10.3389/fnmol.2018.00464.

Perrotta, A., Serpino, C., Cormio, C., Serrao, M., Sandrini, G., Pierelli, F., & de Tommaso, M. (2012). Abnormal spinal cord pain processing in Huntington's disease. The role of the diffuse noxious inhibitory control. *Clin Neurophysiol, 123*, 1624–1630. https://doi.org/10.1016/j.clinph.2012.01.012.

Pinal-Seoane, N., Martin, I.R., Gonzalez-Nuñez, V., de Velasco, E.M., Alvarez, F.A., Sarmiento, R.G., & Rodriguez, R.E. (2006). Characterization of a new duplicate delta-opioid receptor from zebrafish. *J Mol Endocrinol, 37*, 391–403.

Pini, B., Grosser, T., Lawson, J.A., Price, T.S., Pack, M.A., & FitzGerald, G.A. (2005). Prostaglandin E synthases in zebrafish. *Arterioscler Thromb Vasc Biol, 25*, 315–320.

Purves, D., Augustine, G., Fitzpatrick, D., Hall, W.C., LaMantia, A., Mooney, R., & White, L.E. (2018). *Neuroscience*. Sunderland, MA: Sinauer.

Raja, S.N., Carr, D.B., Cohen, M., Finnerup, N.B., Flor, H., Gibson, S., Keefe, F.J., Mogil, J.S., Ringkamp, M., Sluka, K.A., Song, X.J., Stevens, B., Sullivan, M.D., Tutelman, P.R., Ushida, T., & Vader, K. (2020). The revised International Association for the Study of Pain definition of pain: Concepts, challenges, and compromises. *Pain, 161*(9), 1976–1982. https://doi.org/10.1097/j.pain.0000000000001939.

Rangaraju, V., Lewis, T.L., Hirabayashi, Y., Bergami, M., Motori, E., Cartoni, R., Kwon, S.K., & Courchet, J. (2019). Pleiotropic mitochondria: the influence of mitochondria on neuronal development and disease. *J Neurosci, 39*(42), 8200–8208. https://doi.org/10.1523/JNEUROSCI.1157-19.2019.

Reddy, P., & Shirendeb, U. (2012). Mutant huntingtin, abnormal mitochondrial dynamics, defective axonal transport of mitochondria, and selective synaptic degeneration in Huntington's disease. *Biochim Biophys Acta, 1822*, 101–110.

Reilly, S.C., Quinn, J.P., Cossins, A.R., & Sneddon, L.U. (2008). Behavioural analysis of a nociceptive event in fish: Comparisons between three species demonstrate specific responses. *Appl Anim Behav Sci, 114*, 248–259. https://doi.org/10.1016/j.applanim.2008.01.016.

Reinig, S., Driever, W., & Arrenberg, A.B. (2017). The descending diencephalic dopamine system is tuned to sensory stimuli. *Curr Biol, 27*, 318–333. https://doi.org/10.1016/j.cub.2016.11.059.

Romero-Garcia, S., & Prado-Garcia, H. (2019). Mitochondrial calcium: transport and modulation of cellular processes in homeostasis and cancer (review). *Int J Oncol, 54*(4), 1155–1167. https://doi.org/10.3892/ijo.2019.4696.

Rosland, J.H., Hunskaar, S., Broch, O.J., & Hole, K. (1992). Acute and long term effects of 1-methyl-4-phenyl-1,2,3,6-tetrahydropyridine (MPTP) in tests of nociception in mice. *Pharmacol Toxicol, 70*, 31–37. https://doi.org/10.1111/j.1600-0773.1992.tb00421.x.

Rubinsztein, D., Leggo, J., Coles, R., Almqvist, E., Biancalana, V., Cassiman, J., Chotai, K., Connarty, M., Crauford, D., Curtis, A., Curtis, D., Davidson, M.J., Differ, A.M., Dode, C., Dodge, A., Frontali, M., Ranen, N.G., Stine, O.C., Sherr, M., Abbottet, M.H., & Hayden, M.R. (1996). Phenotypic characterization of individuals with 30–40 CAG repeats in the Huntington disease (HD) gene reveals HD cases with 36 repeats and apparently normal elderly individuals with 36–39 repeats. *Am J Hum Genet, 59*, 16–22.

Rueda, C.B., Llorente-Folch, I., Amigo, I., Contreras, L., González-Sánchez, P., Martínez-Valero, P., Juaristi, I., Pardo, B., del Arco, A., & Satrústegui, J. (2014). Ca2+ regulation of mitochondrial function in neurons. *Biochim Biophys Acta, 1837*(10), 1617–1624. https://doi.org/10.1016/j.bbabio.2014.04.010.

Sajic, M., Ida, K.K., Canning, R., Gregson, N.A., Duchen, M.R., & Smith, K.J. (2018). Mitochondrial damage and "plugging" of transport selectively in myelinated, small-diameter axons are major early events in peripheral neuroinflammation. *J Neuroinflammation, 15*(1), 61. https://doi.org/10.1186/s12974-018-1094-8.

Salter, M.W., & Stevens. B. (2017). Microglia emerge as central players in brain disease. *Nat Med, 23*, 1018–27.

Sánchez-Caballero, L., Ruzzenente, B., Bianchi, L., Assouline, Z., Barcia, G., Metodiev, M.D., Rio, M., Funalot, B., van den Brand, M.A.M., Castillo, S. G., Molenaar, J.P., Koolen, D., Brandt, U., Rodenburg, R.J., Nijtmans, L.G., & Rötig, A. (2016). Mutations in complex I assembly factor TMEM126B result in muscle weakness and isolated complex I deficiency. *Am J Hum Genet, 99*(1), 208–216. https://doi.org/10.1016/j.ajhg.2016.05.022.

Schapira, A.H.V., Chaudhuri, K.R., & Jenner, P. (2017). Non-motor features of Parkinson disease. Vol. 18, Nature Reviews Neuroscience. *Nature Publishing Group, 18*, 435–450.

Schiavone, S., Tucci, P., Mhillaj, E., Bove, M., Trabace, L., & Morgese, M. G. (2017). Antidepressant drugs for beta amyloid-induced depression: A new standpoint? *Prog Neuropsychopharmacol Biol Psychiatry, 78*, 114–122. https://doi.org/10.1016/j.pnpbp.2017.05.00.

Schmidtko, A. (2015). Nitric oxide-mediated pain processing in the spinal cord. *Handb Exp Pharmacol, 227*, 103–117. https://doi.org/10.1007/978-3-662-46450-2_6.

Schrank, S., Barrington, N., & Stutzmann, G. (2020). Calcium-handling defects and neurodegenerative disease. *Cold Spring Harbor Perspect Biol, 12*, a035212.

Schroeder, P.G.S.L.U. (2017). Exploring the efficacy of immersion analgesics in zebrafish using an integrative approach. *Appl Anim Behav Sci, 187*, 93–102. https://doi.org/10.1016/j.applanim.2016.12.003.

Sharma, M., Gupta, P., & Garabadu, D. (2022). Bacopa monnieri attenuates glutamate-induced nociception and brain mitochondrial toxicity in Zebrafish. *Metab Brain Dis, 37*, 383–396. https://doi.org/10.1007/s11011-021-00874-6.

Shetty, A.K., & Bates, A. (2016). Potential of GABA-ergic cell therapy for schizophrenia, neuropathic pain, and Alzheimer's and Parkinson's diseases. *Brain Research, 1638*(Part A), 74–87. https://doi.org/10.1016/j.brainres.2015.09.019.

Shin, H., Na, H., & Do, S. (2020). Magnesium and pain. *Nutrients, 12*, 2184.

Shirendeb, U., Reddy, A., Manczak, M., Calkins, M., Mao, P., Tagle, D., & Reddy, P. H. (2011). Abnormal mitochondrial dynamics, mitochondrial loss and mutant huntingtin oligomers in Huntington's disease: Implications for selective neuronal damage. *Hum Mol Genet, 20*, 1438–1455.

Silva, L.M.R.D., Lima, J.D.S.S., Magalhães, F.E.A., et al. (2020). Graviola fruit bar added acerola by-product extract protects against inflammation and nociception in adult zebrafish (Danio rerio). *J Med Food, 23*, 173–180. https://doi.org/10.1089/jmf.2019.0078.

Singh, C., Oikonomou, G., & Prober, D.A. (2015). Norepinephrine is required to promote wakefulness and for hypocretin-induced arousal in zebrafish. *Elife*, *4*, e07000. https://doi.org/10.7554/eLife.07000.

Sluka, K., & Gregory, N. (2015). The dichotomized role for acid sensing ion channels in musculoskeletal pain and inflammation. *Neuropharmacology*, *94*, 58–63. https://doi.org/10.1016/j.neuropharm.2014.12.013.

Smith, C.J. (2017). Nociception and the experience of pain signaling due to nerve damage. *Master Sci Integr Biol*, *24*. http://digitalcommons.kennesaw.edu/integrbiol_etd/24.

Sprenger, G.P., Roos, R.A.C., van Zwet, E., Reijntjes, R.H., Achterberg, W.P., & de Bot, S.T. (2021). The prevalence of pain in Huntington's disease in a large worldwide cohort. *Parkinsonism and Related Disorders*, *89*, 73–78. https://doi.org/10.1016/j.parkreldis.2021.06.015.

St John Smith, E. (2018). Advances in understanding nociception and neuropathic pain. *J Neurol*, *265*, 231–238.

Sui, B.D., Xu, T.Q., Liu, J.W., Wei, W., Zheng, C.X., Guo, B.L., Wang, Y.Y., & Yang, Y.L. (2012). Understanding the role of mitochondria in the pathogenesis of chronic pain. *Postgrad Med J*, *89*(1058), 709–714. https://doi.org/10.1136/postgradmedj-2012-131068.

Takei, Y. (2021). The digestive tract as an essential organ for water acquisition in marine teleosts: Lessons from euryhaline eels. *Zool Lett*, *7*, 10.

Taylor, J.C., Dewberry, L.S., Totsch, S.K., Yessick, L.R., DeBerry, J.J., Watts, S.A., & Sorge, R.E. (2017). A novel zebrafish-based model of nociception. *Physiology & Behavior*, *174*, 83–88. https://doi.org/10.1016/j.physbeh.2017.03.009.

Taylor, J.L., Pritchard, H.A.T., Walsh, K.R., Strangward, P., White, C., Hill-Eubanks, D., Alakrawi, M., Hennig, G.W., Allan, S.M., Nelson, M.T., & Greenstein, A.S. (2022). Functionally linked potassium channel activity in cerebral endothelial and smooth muscle cells is compromised in Alzheimer's disease. *Proc Natl Acad Sci*, *119*(26), e2204581119. https://doi.org/10.1073/pnas.2204581119.

Thomson, J.S., Al-Temeemy, A.A., Isted, H., Spencer, J.W., & Sneddon, L.U. (2019). Assessment of behaviour in groups of zebrafish (Danio rerio) using an intelligent software monitoring tool, the chromatic fish analyser. *J Neurosci Methods*, *328*, 108433. https://doi.org/10.1016/j.jneumeth.2019.108433.

Tong, Q., Zhang, L., Yuan, Y., Jiang, S., Zhang, R., Xu, Q., Ding, J., Li, D., Zhou, X., & Zhang, K. (2015). Reduced plasma serotonin and 5-hydroxyindoleacetic acid levels in Parkinson's disease are associated with nonmotor symptoms. *Parkinsonism Relat Disord*, *21*, 882–887.

Tsantoulas, C., & McMahon, S.B. (2014). Opening paths to novel analgesics: the role of potassium channels in chronic pain. *Trends Neurosci*, *37*, 146–158.

Tucci, P., Mhillaj, E., Morgese, M. G., Colaianna, M., Zotti, M., Schiavone, S., Cicerale, M., Trezza, V., Campolongo, P., Cuomo, V., & Trabace, L. (2014). Memantine Prevents Memory Consolidation Failure Induced by Soluble Beta Amyloid in Rats. *Front Behav Neurosci*, *8*, 332. https://doi.org/10.3389/fnbeh.2014.00332.

Unal, I., Cansız, C., Beler, M., Sezer, Z., Güzel, E., & Emekli-Alturfan E. (2023). Sodium-dependent glucose co-transporter-2 inhibitor empagliflozin exerts neuroprotective effects in rotenone-induced Parkinson's disease model in zebrafish; mechanism involving ketogenesis and autophagy. *Brain Res*, *1820*, 148536. https://doi.org/10.1016/j.brainres.2023.148536

van Kooten, J., Binnekade, T.T., van der Wouden, J.C., Stek, M.L., Scherder, E.J., Husebø, B.S., Smalbrugge, M., & Hertogh, C.M. (2016). A review of pain prevalence in Alzheimer's, vascular, frontotemporal and Lewy body dementias. *Dement Geriatr Cogn Disord*, *41*(3–4), 220–232. https://doi.org/10.1159/000444791.

van Reij, R.R.I, Salmans, M.M.A., Eijkenboom, I., van den Hoogen, N.J., Joosten, E.A.J., & Vanoevelen, J.M. (2021). Dopamine-neurotransmission and nociception in zebrafish: An anti-nociceptive role of dopamine receptor drd2a. *Eur J Pharmacol*, *912*, 174517. https://doi.org/10.1016/j.ejphar.2021.174517.

van Reij, R.R.I., Joosten, E.A.J., & van den Hoogen, N.J. (2019). Dopaminergic neurotransmission and genetic variation in chronification of post-surgical pain. *Br J Anaesth*, *123*, 853–864. https://doi.org/10.1016/j.bja.2019.07.028.

Vanoye, C.G., Gurnett, C.A., Holland, K.D., George, A.L., Jr., & Kearney, J.A. (2014). Novel SCN3A variants associated with focal epilepsy in children. *Neurobiol Dis*, *62*, 313–322.

Vercruysse, P., Vieau, D., Blum, D., Petersén, Å., & Dupuis L. (2018). Hypothalamic alterations in neurodegenerative diseases and their relation to abnormal energy metabolism. *Front Mol Neurosci*, *11*, 2. https://doi.org/10.3389/fnmol.2018.00002.

Walker, K., Reeve, A., Bowes, M., Winter, J., Wotherspoon, G., Davis, A., Schmid, P., Gasparini F., Kuhn R., & Urban, L. (2001). mGlu5 receptors and nociceptive function II. mGlu5 receptors functionally expressed on peripheral sensory neurones mediate inflammatory hyperalgesia. *Neuropharmacology*, *40*, 10–19.

Underwood, K.L., Walker, W.J., Garrett P,I., Linch, S., Rynes, T.P., & Mruk, K. (2023). Optimizing spinal cord injury in zebrafish larvae: effects of age on the injury response. *bioRxiv*, *18*, 2023.05.18.541337. https://doi.org/10.1101/2023.05.18.541337.

Wang, C.T., Mao, C.J., Zhang, X.Q., Zhang, CY, Lv, D.J., Yang, Y.P., Xia, K.L., Liu, J.Y., Wang, F., Hu, L.F., Xu, G.Y., & Liu, C.F. (2017). Attenuation of hyperalgesia responses via the modulation of 5-hydroxytryptamine signalings in the rostral ventromedial medulla and spinal cord in a 6-hydroxydopamine-induced rat model of Parkinson's disease. *Mol Pain*, *13*, 1744806917691525.

Wang, X., Perry, G., Smith, M.A., & Zhu, X. (2010). Amyloid-beta-derived diffusible ligands cause impaired axonal transport of mitochondria in neurons. *Neurodegener Dis*, *7*(1–3), 56–59. https://doi.org/10.1159/000283484.

Wei, H., & Pertovaara, A. (2006). 5-HT(1A) receptors in endogenous regulation of neuropathic hypersensitivity in the rat. *Eur J Pharmacol*, *535*, 157–165.

Widmark, J., Sundström, G., Daza, D.O., & Larhammar, D. (2011). Differential evolution of voltage-gated sodium channels in tetrapods and teleost fishes. *Mol Biol Evol*, *28*(1), 859–871. https://doi.org/10.1093/molbev/msq257.

Woods, I.G., Schoppik, D., Shi, V.J., Zimmerman, S., Coleman, H. A., Greenwood, J., Soucy, E.R., & Schier, A.F. (2014). Neuropeptidergic signaling partitions arousal behaviors in zebrafish. *J Neurosci*, *34*, 3142–3160.

Woolf, C.J., Bennett G.J., Doherty M., Dubner, R., Kidd, B., Koltzenburg, M., Lipton, R., Loeser, J. D., Payne, R., & Torebjork, E. (1998). Towards a mechanism-based classification of pain? *Pain*, *77*, 227–229. https://doi.org/10.1016/S0304-3959(98)00099-2.

Wozniak, K.M., Rojas, C., Wu, Y., & Slusher, B.S. (2012). The role of glutamate signaling in pain processes and its regulation by GCP II inhibition. *Curr Med Chem*, *19*, 1323–1334. https://doi.org/10.2174/092986712799462630.

Wu, Y., Chen, M., & Jiang, J. (2019). Mitochondrial dysfunction in neurodegenerative diseases and drug targets via apoptotic signaling. *Mitochondrion*, *49*, 35–45. https://doi.org/10.1016/j.mito.2019.07.003.

Xie, Y., Liu, P.P., Lian, Y.J., Liu, H.B., & Kang, J.S. (2019). The effect of selective serotonin reuptake inhibitors on cognitive function in patients with Alzheimer's disease and vascular dementia: Focusing on fluoxetine with long follow-up periods signal. *Transduct Target Ther*, *4*, 30. https://doi.org/10.1038/s41392-019-0064-7.

Yam, M.F., Loh, Y.C., Tan, C.S., Khadijah, A.S., Abdul, M.N., & Basir, R. (2018). General Pathways of Pain Sensation and the Major Neurotransmitters Involved in Pain Regulation. *Int J Mol Sci*, *19*(8), 2164. https://doi.org/10.3390/ijms19082164.

Yoo, S.K., Starnes, T.W., Deng, Q., & Huttenlocher, A. (2011). Lyn is a redox sensor that mediates leukocyte wound attraction in vivo. *Nature*, *480*, 109. https://doi.org/10.1038/nature10632.

Zaman, T., Helbıg, I., Bozovıc, I. B., Debrosse, S.D., Bergqvıst, A.C., Wallıs, K., Medne, L., Maver, A., Peterlın, B., Helbıg, K.L., Zhang, X., & Goldberg, E.M. (2018). Mutations in SCN3A cause early infantile epileptic encephalopathy. *Ann Neurol*, *83*, 703–717.

Zambelli, V.O.,Gross, E.R., Chen, C.H., Gutierrez, V.P., Cury, Y., & Mochly-Rosen D. (2014). Aldehyde dehydrogenase-2 regulates nociception in rodent models of acute inflammatory pain. *Sci Transl Med*, *6*, 251.

Zhang, X., Wan, J., & Tong, X. (2018). Potassium channel dysfunction in neurons and astrocytes in Huntington's disease. *CNS Neurosci Ther*, *24*, 311–318.

Zhao, M.G., Toyoda, H., Lee, Y.S., Wu, L.J., Ko, S.W., Zhang, X.H., Jia, Y.H., Shum, F., Xu, H., Li, B.M., Kaang, B.K., & Zhuo, M. (2005). Roles of NMDA NR2B subtype receptor in prefrontal long-term potentiation and contextual fear memory. *Neuron*, *47*, 859–872.

Zhou, W., Horstick, E.J., Hirata, H., & Kuwada. J.Y. (2008). Identification and expression of voltage-gated calcium channel β subunits in Zebrafish. *Dev Dyn*, *237*(12), 3842–3852. https://doi.org/10.1002/dvdy.21776.

Zhu, X.C., Tan, L., Wang, H.F., Jiang, T., Cao, L., Wang, C., Wang, J., Tan, C.C., Meng, X.F., & Yu, J.T. (2015). Rate of early onset Alzheimer's disease: A systematic review and meta-analysis. *Ann Transl Med*, *3*(3), 38. https://doi.org/10.3978/j.issn.2305-5839.2015.01.19.

Zhu, Y., Hu, X., Wang, L., Zhang, J., Pan, X., Li, Y., Cao, R., Li, B., Lin, H., Wang, Y., Zuo, L., & Huang, Y. (2022). Recent advances in acid-sensitive ion channels in central nervous system diseases. *Curr Pharm Des*, *28*, 1406–1411.

Zhuo, M. (2016). Contribution of synaptic plasticity in the insular cortex to chronic pain. *Neuroscience*, *338*, 220–229.

# 16

## *Neurodegenerative Disease Modeling with CRISPR/Cas Systems in Zebrafish*

**Barış Çerçi, Fulya Yaylacıoğlu Tuncay and Pervin Rukiye Dinçer**

### Abbreviations

| | |
|---|---|
| **AD** | Alzheimer's Disease |
| **Aβ** | Amyloid Beta |
| **APP** | Amyloid Precursor Protein |
| **ALS** | Amyotrophic Lateral Sclerosis |
| **ARSACS** | Autosomal Recessive Spastic Ataxia of Charlevoix-Saguenay |
| **AT** | Ataxia Telangiectasia |
| **CRISPR** | Clustered Regularly Interspaced Palindromic Repeats |
| **Cas** | CRISPR Associated |
| **crRNA** | Crispr RNA |
| **DRD** | Dopa-Responsive Dystonia |
| **ENU** | Ethylnitrosourea |
| **FTD** | Frontotemporal Dementia |
| **GAP** | GTPase Activating Protein |
| **gRNA** | Guide RNA |
| **HSP** | Hereditary Spastic Paraplegia |
| **HDR** | Homology Directed Repair |
| **HD** | Huntington's Disease |
| **Indel** | Insertion and Deletion |
| **MO** | Morpholino |
| **MS** | Multiple Sclerosis |
| **NHEJ** | Non-Homologous End Joining |
| **NMD** | Nonsense-Mediated Decay |
| **PD** | Parkinson's Disease |
| **PTC** | Premature Termination Codon |
| **PAM** | Protospacer Adjacent Motif |
| **RFP** | Red Fluorescent Protein |
| **RNP** | Ribonucleoprotein |
| **RDO** | RNA-DNA Oligonucleotide |
| **SFHR** | Short Fragment Homologous Replacement |
| **SMA** | Spinal Muscular Atrophy |
| **T7EI** | T7 Endonuclease I Digestion Assay |
| **$BH_4$** | Tetrahydrobiopterin |
| **TracrRNA** | Trans-Activating Crispr RNA |
| **TALEN** | Transcription Activator Like Effector Nucleases |
| **TILLING** | Targeting Induced Local Lesions In Genomes |
| **TH** | Tyrosine Hydroxylase |
| **ZFN** | Zinc Finger Nucleases |

### Introduction

Neurodegenerative diseases encompass a wide spectrum of disorders that pathologically share in common the progressive degeneration and loss of neurons in the central and/or peripheral nervous systems. There are many types of rare diseases that clinically present with findings of neurodegeneration. However, the most prevalent neurodegenerative diseases are Alzheimer's disease (AD), Parkinson's Disease (PD), frontotemporal dementia (FTD), amyotrophic lateral sclerosis (ALS), spinal muscular atrophy (SMA), Huntington's disease (HD), and multiple sclerosis (MS). Although these diseases manifest with unique combinations of symptomology, there is a common set of hallmarks that underlies the pathogenesis of these prevalent neurodegenerative diseases. These hallmarks are pathological protein aggregation, neuronal network defects, aberrant proteostasis, cytoskeletal abnormalities, altered energy homeostasis, DNA and RNA defects, inflammation, and neuronal cell death (Wilson et al., 2023). Research has focused on elucidating the genes and molecular pathways participating in these hallmarks' emergence and identifying drugs that can alter these pathological processes. Animal models that can replicate disease phenotype and the underlying hallmarks have been created, and research carried out in these animal models plays crucial roles in uncovering disease pathophysiologies and screening for drugs that can modify disease progression. Although rodent models of neurodegenerative diseases remain the mainstay, zebrafish have emerged as an attractive alternative for neurodegenerative disease modeling in recent years (Chia et al., 2022; Dawson et al., 2018).

Goerge Streisinger pioneered foundational work for establishing zebrafish as a model organism in the 1980s. He was most interested in uncovering the genetic determinants of vertebrate development, especially their nervous system. Due to challenges in tracking chromosome regions of interest in established vertebrate models, he decided to develop new techniques and tools in a new animal model to acquire recessive mutant lines quickly. Zebrafish was his model vertebrate of choice because of three fundamental characteristics of this

DOI: 10.1201/9781032676555-18

species: a vast number of progenies produced in a short time, external fertilization allowing harvest and manipulation of gametes, and external embryonic development that allows visualization of phenotypes during embryogenesis (Grunwald & Eisen, 2002). Moreover, zebrafish is also a useful vertebrate model for phenotype-based drug screening due to its small size, mass production, ease of drug delivery, and phenotype tracking (Patton et al., 2021). In addition to the advantages that make it a suitable model for developmental biology research and drug screening, its optical transparency, the high similarity of its genes with humans, and the ease of genetic manipulation in zebrafish have made it an attractive vertebrate model for a diversity of diseases, including but not limited to bone, cardiovascular, renal, blood, skeletal, and retinal diseases (Adhish & Manjubala, 2023). Zebrafish models of diseases in these organ systems have contributed to our knowledge of disease mechanisms and paved the way for developing new therapeutics (Choi et al., 2021). Especially, the use of zebrafish for phenotype-based drug screening also resulted in the discovery of new molecules for treating several diseases, and some of these have already reached clinical trials (Patton et al., 2021).

Zebrafish have also been used to model most of the common neurodegenerative diseases. The advantages of zebrafish as a model organism have been mentioned above. However, several features make zebrafish especially suitable for neurodegenerative disease modeling. The gross structure and the neuronal circuitry of the central nervous system show remarkable similarities between zebrafish and humans. Moreover, neurotransmitter systems and pathways of humans have corresponding neuronal clusters in zebrafish (Chia et al., 2022). These features are essential for the characterization of disease phenotypes in neurodegenerative disorders. Another critical advantage of zebrafish is its transparency, making imaging cells labeled with fluorescent proteins in live specimens possible. The transparency of zebrafish also allows the optogenetic manipulation of proteins in living specimens (Asakawa et al., 2020). These features of zebrafish open up opportunities for developing new experimental methods to observe and manipulate neuronal activities in living organisms and help us improve our understanding of neurodegenerative diseases.

## Genetic Manipulation Methods in Zebrafish

Since its first use for disease modeling, different methods of genetic manipulation have been successfully implemented in zebrafish. These include ethylnitrosourea (ENU) mutagenesis, targeting induced local lesions in genomes (TILLING) method, viral vector-mediated insertional mutagenesis and gene transduction, and TALEN and ZFN mediated mutagenesis. One of the most widely used methods to knock down genes in zebrafish has been the injection of synthetic antisense oligonucleotide, termed morpholino (MO), that transiently blocks translation of targeted gene transcript. However, this method has significant drawbacks, such as its transient nature, off-target effects, and targeting of all cells of the organism (Collin & Martin, 2017; Kalueff et al., 2014). Subsequently, the advancements in the CRISPR/Cas system for genome editing and its implementation in zebrafish have made these older gene manipulation techniques mostly obsolete for zebrafish disease modeling.

Methods to alter the genetic sequence of an organism have evolved from random mutagenesis with screening, to biased targeting with selection, to true targeting (Segal & Meckler, 2013). Homologous recombination and site-specific recombination have been the mainstay of targeted gene editing for many years since their characterization in the 1970s–1980s (Capecchi et al., 2001; Hinnen et al., 1978; Rothstein, 1983; Sauer, 1994). Several methods have been devised to increase the efficiency and specificity of homologous recombination, such as triplex forming oligonucleotides, RNA-DNA oligonucleotides (RDOs), and short fragment homologous replacement (SFHR) (Kolb et al., 2005). The latest method to increase the effectiveness and site specificity of homologous recombination was the use of natural or engineered endonucleases that can target specific gene sequences to introduce double strand breaks (DBS) (Silva et al., 2011). These engineered endonucleases, zinc finger nucleases (ZFN) and transcription activator–like effector nucleases (TALEN), paved the way for the therapeutic application of genome editing tools in rare genetic diseases. Despite their promises and effectiveness, the major drawback of these engineered endonucleases was the laborious customization for every new target sequence (Gaj et al., 2013). In 2012, a research group led by Jennifer Doudna and Emmanuelle Charpentier demonstrated that the bacterial CRISPR/Cas system, which was previously shown to function as an acquired immunity system of bacteria against bacteriophages (Barrangou et al., 2007), can be used as a genome editing tool that can introduce double strand DNA breaks into specific target sequences, guided by a simple RNA sequence (Jinek et al., 2012). Consequently, the ease of targeting a specific genetic sequence with a short guide RNA (gRNA) has revolutionized the field of genome editing. This technology has since been applied in many fields of biology and medicine, including disease modeling and therapeutics (Çerçi et al., 2023; Wang & Doudna, 2023).

## CRISPR/Cas Systems

The most commonly used CRISPR/Cas9 system comprises two components, a guide RNA and the Cas9 protein. The guide RNA is a synthetic chimera of a crispr RNA (crRNA) that is complementary to the target sequence and a trans-activating crispr RNA (tracrRNA) that activates the Cas9 protein. This protein is composed of two domains, the recognition domain that binds the guide RNA and is required for activation of the protein and target recognition, and the nuclease domain that introduces double strand breaks into target sequence (Wang et al., 2022). A significant feature of the Cas9 protein is that its nuclease domain is not activated unless it recognizes a short 3 bp protospacer adjacent motif (PAM) next to the target sequence, a mechanism evolved to avoid targeting its own DNA (Gleditzsch et al., 2019). When the Cas9 protein introduces a double strand break into the target sequence, the cellular machinery of DNA damage response is activated, and this lesion is repaired by non-homologous end joining (NHEJ) that introduces deletions or insertions (indel) at the breakpoint. If there is a homolog of the targeted sequence in the cell, either supplied exogenously or by the sister chromatid replicated in mitosis, the lesion is repaired by homology directed repair (HDR) that ultimately replaces the target sequence with the homologous sequence, without insertion of indels (Nambiar et al., 2022). Consequently, the canonical CRISPR/Cas system can be used to introduce deleterious mutations into a specific target sequence, or to replace a mutation-carrying target sequence with a wild-type one.

The components of the CRISPR/Cas system can be delivered into cells either as DNA or mRNA to be transcribed and translated within the cell or they can be delivered as preformed ribonucleoprotein (RNP) complexes (Glass et al., 2018). There are unique advantages and disadvantages for each option. Delivery of the CRISPR/Cas system in a plasmid DNA offers stability and an option for viral delivery. However, this comes with an increased potential for off-target effects and development of an immune response. When the components are delivered as mRNA transcripts, genome editing is transient, which can be advantageous for avoiding off-target events and immunity. The major drawback of this option is that mRNAs are very unstable and quickly degraded, making it challenging to deliver them efficiently (Glass et al., 2018). The vectors used to deliver the components of the CRISPR/Cas system can be broadly classified into three major groups: biological vectors, chemical vectors, and physical vectors (Taha et al., 2022). The physical vectors include sonoporation, microinjection, and electroporation. These delivery methods are practical and effective for in vitro transfection of the CRISPR/Cas system components; however, they have limited use in in vivo settings. Biochemical vectors, on the other hand, are primarily used for in vivo delivery. The most studied biochemical vectors are viruses, especially adenoviruses, adeno-associated viruses and lentiviruses. They have the ability to transfect a diversity of cells with high specificity and efficiency (van Haasteren et al., 2020). However, there are important limitations that should be considered when viral vectors are used for in vivo delivery, such as their immunogenicity, limited cargo capacity, and insertional mutagenesis. Chemical vectors, like gold nanoparticles and polymer or lipid-based nanoparticles, have the main advantage of reduced immunogenicity that increases transfection efficiency. In addition, they can deliver the CRISPR/Cas system components in all possible forms, their cargo capacity can be adjusted according to requirements, and they are more cost-effective compared to biochemical vectors. The major consideration for chemical vectors is the small repertoire of cells that can be transfected with these approaches in vivo (Yang et al., 2022). Hence, every delivery method has its own unique limitations, and the choice of any of these vectors for the delivery of the CRISPR/Cas system into cells depends on the unique requirements of each application (Paunovska et al., 2022).

Canonical Cas proteins introduce double strand breaks in the genome where the nucleotide sequence is complementary to a 20-nucleotide segment of the guide RNA. However, the enzyme is not perfect, and Cas proteins can still cut genomic regions with incomplete complementarity to the guide RNA (Fu et al., 2013). This leads to off-target genome editing, not limited to insertions and deletions of a small number of nucleotides in these regions, but also including catastrophic chromosomal events, such as aneuploidy and chromotrypsis (Leibowitz et al., 2021; Nahmad et al., 2022). Prediction and reduction of these off-target effects is one of the most critical challenges in CRISPR/Cas-mediated genome editing. Accordingly, many methods have been developed to predict off-target effects of CRISPR/Cas systems. These can be categorized as in silico modeling, in vitro identification, and in cellulo capturing of off-target events (Tao et al., 2023). In addition to prediction of off-target events, several different ways to reduce these effects have been developed, including, but not limited to, chemically modified guide RNAs, engineering of Cas proteins for higher fidelity, and delivery of the CRISPR/Cas components as ribonucleoprotein complexes for more transient genome editing (Kim et al., 2019; Tao et al., 2023). More recently, modified Cas proteins that introduce only single strand breaks into the target sequence, or the base editing or epigenetic editing CRISPR/Cas systems that do not break DNA strands at all, have been praised for avoiding off-target mutagenesis induced by double strand breaks (Nakamura et al., 2021).

The CRISPR/Cas systems are divided into two classes. Class 1 systems are characterized by a multiprotein nuclease complex composed of a cascade of Cas proteins. Class 2 systems, on the other hand, have a single multidomain Cas protein that performs the target recognition and endonuclease activities (Makarova et al., 2019). Due to their simplicity, class 2 systems have been more widely used for genome editing (Shmakov et al., 2017). However, some unique features of class 1 systems have recently sparked interest in their use for genome editing (Yoshimi & Mashimo, 2022). The class 2 systems are further divided into three types: type II, V, and VI. Each type has its effector Cas protein, Cas9, Cas12, and Cas13, respectively (Shmakov et al., 2017). CRISPR/Cas9 was the first CRISRP/Cas system used for genome editing, and it is by far the most well-characterized system for genome editing (Jinek et al., 2012). Cas9 proteins introduce double strand breaks with blunt ends next to a PAM sequence rich in GC. One of the limitations of Cas9 proteins is the requirement for a stringent PAM sequence next to the 20-nucleotide long target sequence (Mojica et al., 2009). To overcome this barrier, researchers have searched for Cas9 orthologs with different PAM requirements and engineered Cas9 proteins with more flexible PAM sequences (Collias & Beisel, 2021; Gasiunas et al., 2020; Nishimasu et al., 2018). Another limitation of Cas9 proteins is the off-target effects caused by double strand breaks introduced into genomic sequences with near-complete complementarity to guide RNA. Protein engineering efforts have also focused on increasing the fidelity of Cas9 protein to reduce off-target editing without compromising on-target editing efficiency (Pickar-Oliver & Gersbach, 2019). The effector nuclease of the type V system, Cas12, has some features that provide unique advantages over Cas9. It introduces double strand breaks with 5'-overhang that allows DNA integration with a specific orientation and requires only a crispr RNA that it can cleave from a pre-mRNA via its RNAse domain, which facilitates multiplex genome editing, and its PAM sequence is T-rich, hence targeting sequences that cannot be targeted with Cas9 (Swarts & Jinek, 2018). Consequently, the CRISPR/

Cas12 system has been effectively used to edit genomes in various organisms, including zebrafish (Moreno-Mateos et al., 2017). As mentioned above, introducing double strand breaks into DNA is an important vulnerability of canonical CRISPR/Cas systems. Moreover, the knock-in of a desired nucleotide sequence into the genome through HDR is considerably inefficient. To overcome these obstacles, base-editing CRISPR/Cas systems were developed, which can substitute A-T for G-C and vice versa through deamination (Gaudelli et al., 2017; Komor et al., 2016). These base-editing Cas proteins are made nuclease deficient via point mutations, and deaminase proteins are attached to them for catalytic activities. These base editors can efficiently introduce targeted point mutations (Kim et al., 2017). However, undesired base editing, both on-target and off-target, can still occur with low frequency (Anzalone et al., 2020). The two inherent limitations of base editors are that they cannot introduce a desired length of nucleotides into the target region and cannot perform transversion of bases. More recently, a newly engineered CRISPR/Cas system, termed prime editor, was developed to introduce a desired length of deletion, insertion or point mutations into a target sequence without introducing double strand DNA breaks (Anzalone et al., 2019). Prime editing CRISPR/Cas system uses a modified Cas protein with only single functional nuclease domain to cut a single strand of the DNA. The modified Cas protein is equipped with an attached reverse transcriptase domain that reverse transcribes a desired nucleotide sequence into the target region, provided by a single prime editing guide RNA that also guides the Cas protein to the target (Anzalone et al., 2019). Prime editing CRISPR/Cas system is a powerful genome editing tool that can change or insert a short segment of nucleotides within or into a target sequence, respectively. However, new developments in prime editing technology increase the versatility of this powerful genome editing tool and enable manipulation of larger sequences of nucleotides (Chen & Liu, 2022).

## Application of CRISPR/Cas Systems in Zebrafish Genome Editing

Zebrafish was the first vertebrate model organism which was used to demonstrate the applicability of the CRISPR/Cas system for genome editing in vivo (Hwang et al., 2013). Subsequently, the CRISPR/Cas system has been successfully used to create both knock-out and knock-in disease models in zebrafish (Naert & Vleminckx, 2018). The process of generating a disease model in zebrafish using the CRISPR/Cas system can be divided into four main steps: 1) identification of the target gene and generation of a guide RNA; 2) delivery of the components of CRISPR/Cas system into zygote; 3) validation of genome editing, and 4) phenotypic characterization (Figure 16.1). The identification of a target gene is mostly based on findings that arise from human genetic studies. After deciding on the target gene, a guide RNA complementary to a 20 bp sequence of the target gene should be designed. There are several considerations when designing a guide RNA; however, the specificity and efficacy of the guide RNA are the major concerns. In order to satisfy these concerns and design the most efficient guide RNAs, many web-based tools have been developed, each with different purposes (Hanna & Doench, 2020). However, the candidate guide RNAs designed in silico should be validated in vitro before their use in vivo. After a guide RNA is selected, the next step is the delivery of the guide RNA and the Cas protein into the zygote. At this point, a delivery vector and the structure of CRISPR/Cas system components are chosen. As explained above, physical delivery vectors, such as electroporation and microinjection, are suitable for in vitro applications. CRISPR/Cas system components can be delivered as DNA plasmid, mRNA, or RNP complex. However, due to its efficiency and off-target editing superiority, the delivery of RNP complexes is favored over others (Zhang et al., 2021). At the third step, the performed genome editing should be validated. It is a well-known phenomenon that genome editing tools, even when introduced into single-cell embryos, result in genetic mosaicism, most probably due to delayed editing after replication of the genetic material, or even after cell division (P. Singh et al., 2015). In fact, genetic mosaicism can be avoided by implementing several different methods, and non-mosaic knock-out alleles can be produced in founder animals (Burger et al., 2016; Hashimoto et al., 2016; Wu et al., 2018). However, as a more reliable alternative, the founder animals can be bred with wild-types to produce non-mosaic progenies. It was shown that 28 % of F1 progenies carried mutations in the targeted gene when the CRISPR/Cas system was used to introduce indels in single-cell founder zebrafish embryos (Varshney et al., 2015). Therefore, the F1 progenies can be genotyped to detect mutations in the targeted sequence. Several methods can be utilized to detect indels in these progenies cost-effectively, T7 Endonuclease I Digestion Assay (T7EI) being the most widely used one (Carrington et al., 2022). After establishing the mutant line, disease phenotypes of the mutant animals should be characterized using appropriate tests. An important consideration is a phenomenon termed genetic compensation, which may, in some instances, cause subtle phenotypic expression of disease traits in stable knock-out models compared to knock-down models (El-Brolosy & Stainier, 2017). Although there were several hypotheses to explain this discrepancy, the latest research has shown that premature termination codons (PTC) generated by genome editing tools, and the nonsense-mediated decay mechanism (NMD) that is triggered by this, play key roles in the induction of paralogous genes that compensate for the knocked out gene function, leading to milder disease phenotype (El-Brolosy et al., 2019; Ma et al., 2019). Therefore, if a disease phenotype cannot be replicated in a knock-out model that is created by the introduction of indels that form PTCs in the target gene, either the target and its paralogs should be knocked out together to avoid compensation, or the target gene should be knocked out completely to avoid PTC formation or NMD activation. Alternatively, in some instances, mosaic F0 founder mutants can be used to characterize disease phenotypes and potentially overcome genetic compensation that develops in stable mutant lines in F2 generations and beyond. Researchers engineered a *slc25a46* knockout zebrafish using CRISPR/Cas9 mediated mutagenesis. They observed a strong disease phenotype reminiscent of a MO-mediated *slc25a46* knock-down zebrafish model. However, the stable knock-out

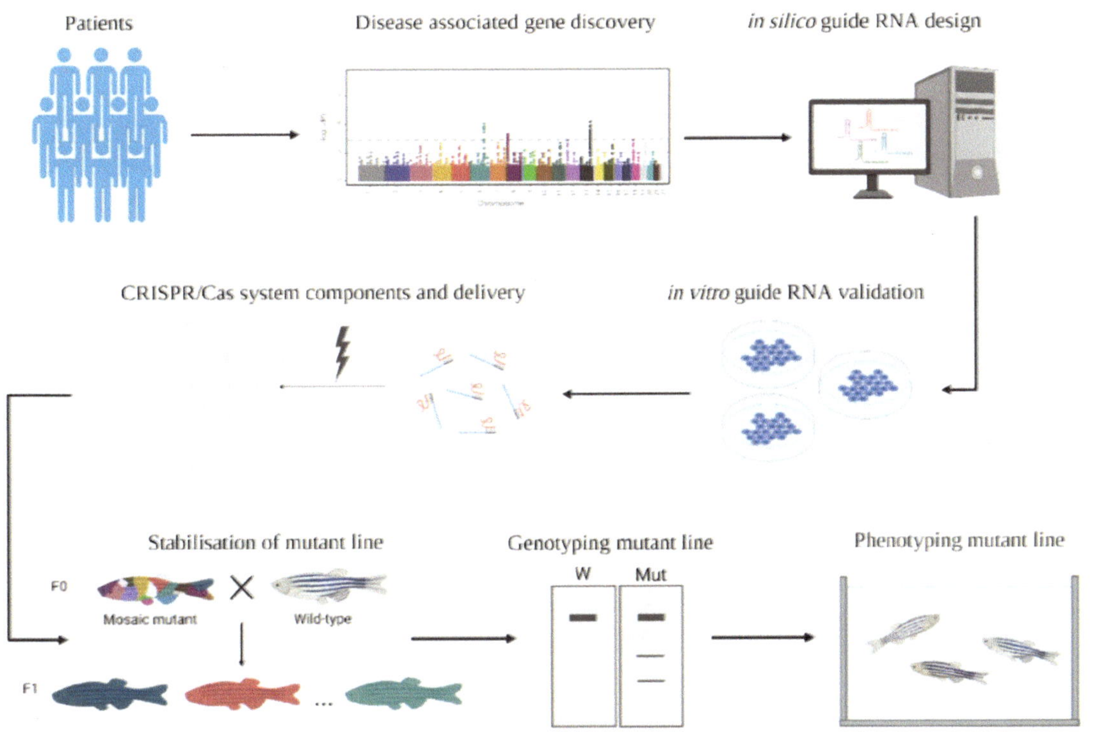


**FIGURE 16.1** The process of generating a disease model in zebrafish using the CRISPR/Cas system.

mutants of F4 generation did not show any disease phenotype. When the transcriptome of mosaic founders and stable mutants were compared, more than 200 genes were differentially expressed; however, none were direct paralogues or associates of the *slc25a46* gene (Buglo et al., 2020). These observations highlight that mosaic founder mutant animals can potentially be used to characterize disease phenotypes when isogenic stable knock-out progenies do not demonstrate disease phenotype due to genetic compensation.

## Zebrafish Models of Neurodegenerative Diseases Created via CRISPR/Cas System

### Alzheimer's Disease

Alzheimer's disease (AD) is the most common neurodegenerative disorder in humans. It affects about 7% of the population aged 80–84 years and 27% of those aged between 90 and 94 years. Although the prevalence of AD is expected to double by 2050, its incidence has been declining in high-income countries since the beginning of the twenty-first century (Knopman et al., 2021). AD is a progressive disorder that starts with a long asymptomatic preclinical phase that progresses, in 15–20 years, to minimal cognitive impairment, primarily in anterograde memory, which in turn evolves to dementia, and death occurs in the following 5–12 years (Long & Holtzman, 2019). The current view on AD pathophysiology is based on the amyloid hypothesis, postulated initially by Hardy & Selkoe (2002) and recently modified by De Strooper and Karran (2016). The modified amyloid hypothesis postulates that AD pathogenesis can be divided into three phases, an initial biochemical phase coinciding with the preclinical phase of the disease, followed by a cellular phase that progresses into the clinical phase of the disease. The biochemical phase is characterized by the aggregation of amyloid beta (Aβ) plaques and neurofibrillary tangles of hyperphosphorylated tau protein that produce a proteopathic stress response in the brain that is tolerated for many years. However, derangements in homeostasis, brought about by advancing age, marks the beginning of a pathologic cellular response to the ongoing proteopathic stress. The main effectors of this cellular response are microglia and astrocytes; however, neurons and cells of neurovascular unit also play important roles in dysfunctional clearance, lipid metabolism, immune response, and amyloid precursor protein (APP) metabolism (De Strooper & Karran, 2016). This pathologic cellular response eventually leads to disruption of synaptic homeostasis and neurodegeneration that manifests as minimal cognitive impairment and dementia in the clinical phase of the disease (De Strooper & Karran, 2016; Knopman et al., 2021). Genetic composition of the patient is responsible for 60%–80% of the AD risk in late onset sporadic cases. However, in very rare autosomal dominant inherited cases of AD, mutations in *APP*, *PSEN1*, and *PSEN2* have full penetrance in the development of AD disease (Scheltens et al., 2021). ε4 and ε3 alleles of *APOE* gene account for much of the genetic risk for sporadic AD; however, dozens of genes in more than 100 loci have also been identified to contribute to AD risk in large genome-wide association studies published in recent years (Andrews et al., 2023). Transgenic mice that express human mutant *APP*,

*PSEN1*, and *PSEN2* genes and *APOE* ε4 allele have been established and examined to dissect the roles of these highly penetrant mutations in AD pathogenesis (Götz & Ittner, 2008). Zebrafish have also been utilized to model AD. Chemicals and heavy metals were used to induce AD-like neurodegeneration in zebrafish. Additionally, $A\beta_{42}$ injection and human *MAPT* transgene overexpression were tried to induce AD-like phenotype (Wang & Cao, 2021). In the last few years, CRISPR/Cas9 genome editing tool was used to knock out *psen1*, *psen2*, and *appa* and *appb*, zebrafish orthologues of human *APP* gene, through introduction of indels that cause frameshift and premature stop codon formation (Banote et al., 2020; Barazzuol et al., 2023; Chebli et al., 2021; Jiang et al., 2018). However, genetic evidence suggests that mutations in *APP, PSEN1*, and *PSEN2* do not mostly act quantitatively but qualitatively and result in increased $A\beta_{42}$ to $A\beta_{40}$ ratio; hence truncating loss-of-function mutations in these genes may not recapitulate AD phenotype precisely (Szaruga et al., 2015). Accordingly, these CRISPR/Cas9-mediated knock-out models were primarily aimed at elucidating physiological roles of the targeted genes in zebrafish. Only one research group used CRISPR/Cas9 genome editing to successfully replicate a known disease causing *PSEN1* mutation in zebrafish *psen1* gene to precisely model AD in zebrafish (Barthelson et al., 2021) (Table 16.1). The authors analyzed subtle transcriptomic changes in brain tissues of relatively young, 6-month-old heterozygous mutants to elucidate potentially novel disease mechanisms that shape the early stages of AD pathology. Transcriptome analysis of these mutants revealed differential expression of genes in oxidative phosphorylation, inflammation, and ribosomal gene sets (Barthelson et al., 2021). Whether characteristic pathologic findings of AD were present in the brains of older mutants was not mentioned by the authors.

## Parkinson's Disease

Parkinson's disease (PD) is the second most common neurodegenerative disorder worldwide. It is characterized clinically by bradykinesia, tremors, rigidity, and postural instability and pathologically by the progressive loss of dopaminergic neurons of nigrostriatal pathway (Poewe et al., 2017). PD is not a single disease entity but a syndrome with heterogenous clinical presentations, pathologies, and pathogenesis (Ye et al., 2023). It is hypothesized that diverse environmental and genetic factors trigger, facilitate, and aggravate the neurodegenerative process which underlies these heterogeneities (Johnson et al., 2019). Genetic comparative studies performed in the last two decades have identified mutations or common variations in more than 100 genetic loci that modify the risk of developing PD and the disease progression. The penetrance of these genetic lesions is variable, and most of the rare mutations cause early-onset disease with high penetrance, while the common variations only slightly increase the risk of disease development (Blauwendraat et al., 2020; Ye et al., 2023). The research on the functions of these genes at cellular level has provided invaluable mechanistic insights about PD pathogenesis. In light of the accumulated research, dysregulations in synaptic function, lysosomal physiology, and mitochondrial homeostasis, as well as oxidative stress and inflammatory response, seem to play important roles in PD initiation and progression. Animal models of PD have been important experimental tools to identify these cellular mechanisms (Ye et al., 2023). In order to produce PD models, genes implicated in autosomal-dominant or recessive familial PD, *lrrk2*, *pink1*, *parkin*, *dj-1*, and *fbxo7* were knocked down in zebrafish by morpholino oligonucleotides. Synucleinopathy observed in Parkinson disease was also modeled in zebrafish by neuron specific overexpression of human α-synuclein or zebrafish γ1-synuclein isoform via plasmid injection into zebrafish embryos (Doyle & Croll, 2022). In the last few years, CRISPR/Cas9 mediated knock-out models of PD have also been established in zebrafish, which increased the number of genes that are targeted (Table 16.2).

Heterozygous mutations in the gene *GBA1*, Glucosylceramidase Beta 1, constitute the most common risk factor for developing PD. Mutations in this gene are detected in about 3% of PD patients worldwide, and 11% of mutation carriers develop PD by age 85 (Rana et al., 2012; Sidransky et al., 2009). Homozygous mutations, on the other hand, are associated with the lysosomal storage disorder, Gaucher disease (Stirnemann et al., 2017). *GBA1* gene encodes the lysosomal enzyme glucocerebrosidase that hydrolyzes glycolipids and sphingolipids. Although the mechanistic relationship between *GBA1* deficiency and development of PD has not been fully understood, numerous observations, based on cellular and animal models and human post-mortem brain tissue analysis, suggest that both loss of enzymatic function as well as gain-of-toxicity of mutant glucocerebrosidase protein seem to predispose to PD via promotion of α-synuclein aggregation, changes in lipid profiles, disruption of lysosomal and autophagic systems, and induction of endoplasmic reticulum stress (Chatterjee & Krainc, 2023). Keatinge et al. (2015) developed *gba1* knock-out zebrafish model using the TALEN genome editing tool. The mutant allele, created by TALEN, had 23 bp deletion in exon 7 that resulted in formation of premature stop codon 66 bp downstream. Homozygous mutant zebrafish had half as much *gba1* mRNA expression as wild-types, indicating nonsense-mediated decay of mutant alleles (Keatinge et al., 2015). These mutants displayed characteristic features of Gaucher disease, such as elevated levels of glucocerebrosidase

**TABLE 16.1**

Zebrafish Models of Alzheimer's Disease Generated via CRISPR/Cas System

| Gene | Genome editing | Isogenicity | Genotype | Pathology | Phenotype | Comments | Reference |
|---|---|---|---|---|---|---|---|
| *psen1* | CRISPR/Cas9 mediated mutagenesis via NHEJ | F3 isogenic stable homozygous mutants | p.T428del in-frame mutation | Not reported | Not reported | Differential expression of genes in oxidative phosphorylation, inflammation and ribosomal gene sets | (Barthelson et al., 2021) |

**TABLE 16.2**

Zebrafish Models of Parkinson's Disease Generated via CRISPR/Cas System

| Reference | (G. Hughes, 2019) | (G. Hughes, 2019) | (G. Hughes, 2019) | (Merhi et al., 2021) | (Edson et al., 2019) | (G. L. Hughes et al., 2020) |
|---|---|---|---|---|---|---|
| **Comments** | Homozygotes display severe disease phenotype reminiscent of Gaucher disease | No NMD of truncated mRNA | Mutant mRNA underwent NMD | Increased expression of *fis1*, associated with mitochondrial fission | Proteomic profiling of young-adult brains revealed changes in mitochondrial electron transport chain, translational control, stress response, glutathione metabolism, and inflammation-related proteins | - |
| **Phenotype** | No disease phenotype | No disease phenotype | No disease phenotype | Increased inactivity periods, reduced responsiveness to repulsive stimulant, cadaverine | Reduced body weight in 16-month-old animals compared to WT | Age-dependent decrease in the duration of movement, distance traveled, velocity, and tail beat frequency at low speed |
| **Pathology** | DA neuron markers not changed | DA neuron markers not changed | DA neuron markers not changed | Reduction in th1 expression, loss of TH+ neurons in olfactory bulb and telencephalon | Reduction in th1 expression | Reduction in th1 and dat expression, reduction in TH+ neurons in posterior tuberculum |
| **Genotype** | c.ex3.615_619del out-of-frame with PTC | c.ex2.224_227del+c224_232ins out-of-frame with PTC | c.ex2.591_691del out-of-frame with PTC | p.M1_R71del | c.ex1.27_31del out-of-frame with PTC | c.ex2.200_201del+c.200_216ins out-of-frame with PTC |
| **Isogenicity** | F2 isogenic stable heterozygous mutants | F2 isogenic stable homozygous mutants | F2 isogenic stable homozygous mutants | F3 isogenic stable homozygous mutants | F3 isogenic stable homozygous mutants | F2 isogenic stable homozygous mutants |
| **Genome editing** | CRISPR/Cas9 mediated mutagenesis via NHEJ | CRISPR/Cas9 mediated mutagenesis via NHEJ | CRISPR/Cas9 mediated mutagenesis via NHEJ | CRISPR/Cas9 mediated mutagenesis via NHEJ | CRISPR/Cas9 mediated mutagenesis via NHEJ | CRISPR/Cas9 mediated mutagenesis via NHEJ |
| **Gene** | *gba1* | *parkin* | *pink1* | *parla* | *park7* | *park7* |
| **Reference** | (Chavali et al., 2023) | (G. Hughes, 2019) | (Nyuzuki et al., 2020) | (Sveinsdóttir et al., 2022) | (Larbalestier et al., 2022) | |
| **Comments** | Upregulation of proteins linked to oxidative stress, mitochondrial homeostasis regulation, and synapse signaling; downregulation of proteins linked to ER-stress response regulation, unfolded protein response, and ubiquitin-proteasome dependent proteolysis | Accumulated lysosome-like vesicles with fingerprint-like structures, severely reduced cathepsin D protein, decrease in levels of proteins annotated with vesicle fusion of Golgi apparatus | L-DOPA normalized distance moved during lights-on period and impaired day-time sleep behavior | Elevation of numerous mRNAs associated with innate immune system activation in brain tissue, activated microglia in brain tissues | | |

*(Continued)*

**TABLE 16.2 (CONTINUED)**
Zebrafish Models of Parkinson's Disease Generated via CRISPR/Cas System

| Reference | (G. Hughes, 2019) | (G. Hughes, 2019) | (G. Hughes, 2019) | (Merhi et al., 2021) | (Edson et al., 2019) | (G. L. Hughes et al., 2020) |
|---|---|---|---|---|---|---|
| **Phenotype** | Loss of weight gain and progressive reduction in total distance travelled beyond 6 month of age, progressive increase in freezing duration and latency to reach upper half of the test tank | No disease phenotype | Not reported | Increased peak velocity in response to lights-off switch, traveled less than wildtype fish in light-on environment, anxiety-like behavior in thigmotaxis assay, abnormalities in day-time sleep behavior | Cachectic, hypomotile, and develop spine curvature, die prematurely at 12 dpf | |
| **Pathology** | Not reported | DA neuron markers not changed | Reduction of $TH^+$ neurons in posterior tuberculum and locus coeruleus | Reduction in $TH^+$ neurons in midline diencephalic neuronal cluster | Dopamine, serotonin, and noradrenaline levels decrease, number of DA neurons in posterior tuberculum not changed | |
| **Genotype** | c.ex1.27_31del out-of-frame with PTC | c.ex2.190_193del out-of-frame with PTC | c.ex2.55_64del out-of-frame with PTC | c.ex3.47_51del out-of-frame with PTC | p.P87_S251del | |
| **Isogenicity** | F3 isogenic stable homozygous mutants | F2 isogenic stable homozygous mutants | F2 or F3 isogenic stable homozygous mutants | F2 or F3 isogenic stable homozygous mutants | F2 or F3 isogenic stable homozygous mutants | |
| **Genome editing** | CRISPR/Cas9 mediated mutagenesis via NHEJ | CRISPR/Cas9 mediated mutagenesis via NHEJ | CRISPR/Cas9 mediated mutagenesis via NHEJ | CRISPR/Cas9 mediated mutagenesis via NHEJ | CRISPR/Cas9 mediated mutagenesis via NHEJ | |
| **Gene** | *park7* | *fbxo7* | *atp13a2* | *slc18a2* | *gch1* | |

substrates and upregulation of upstream enzymes in neurons, Gaucher cell formation and infiltration into solid organs and reduction in mobility, defects of balance, and spinal deformation by age. These mutants also developed dopaminergic neuron loss in posterior tuberculum and widespread neuronal cytoplasmic inclusions, reminiscent of Lewy bodies, at older ages (Keatinge et al., 2015). However, heterozygous mutants did not show any significant phenotypical, pathological, or molecular changes compared to wild-types. Therefore, the mutant zebrafish model created by Keatinge et al. (2015) rather replicates Gaucher disease phenotype but not PD. Recently, Hughes (2019) used CRISPR/Cas9 tool to generate *gba1* deficient zebrafish. The mutant allele carried a 5 bp deletion in exon 3 that caused a premature stop codon formation in exon 4. The resulting protein was severely truncated and lacked majority of enzymatic domain (Hughes, 2019). Although dopaminergic neuron markers were significantly reduced in homozygous mutant zebrafish, the severe motor impairment phenotype was rather similar to that of the Gaucher disease model developed by Keatinge et al. (2015). Similar to findings by Keatinge et al. (2015), heterozygotes did not develop any disease phenotype.

Gain-of-function mutations of *LRRK2* gene are the most common cause of familial Parkinson disease, accounting for 4% of familial PD cases worldwide, but reaching up to 45% in specific populations (Tolosa et al., 2020). *LRRK2* gene consists of two kinase and several protein-protein interaction domains. Although more than 100 mutations in *LRRK2* gene have been discovered, only seven of these are classified as definitely pathogenic, and all of them reside in the two kinase domains. These mutations directly or indirectly increase the enzymatic activity of LRRK2 (Alessi & Sammler, 2018). Both the physiological functions and the roles *LRRK2* mutations play in PD pathogenesis have not been intricately characterized. However, accumulated data indicates that LRRK2 mainly participates in intracellular vesicle and organelle trafficking and implicated in synaptic vesicle cycling, autophagy, mitophagy, and lysosomal damage response. It is thought that hyperactivation of LRRK2 in PD disrupts the fine balance that regulates these processes. To date, *Lrrk2* knock-down, knock-out, and mutant *Lrrk2* overexpression or knock-in mouse models have been established to elucidate the roles of *Lrrk2* in PD pathogenesis, as well as in cellular physiology (Chang et al., 2022). Knock-down and knock-out mouse models, as expected, did not mimic PD but shed light in the physiologic function of *Lrrk2*, especially in peripheral tissues (Seegobin et al., 2020). Mice that overexpress mutant *Lrrk2* gene specifically in dopaminergic neurons developed cardinal features of PD, including progressive dopaminergic neuronal loss, age-dependent L-DOPA responsive locomotor defects, and pathological accumulation of α-synuclein. However, knock-in mouse models of mutant *Lrrk2* that account for physiological expression of the mutant gene did not recapitulate the PD phenotype overtly but manifested only subtle locomotor abnormalities with altered dopamine homeostasis but without neurodegeneration and α-synucleopathy (Seegobin et al., 2020). In zebrafish, *lrrk2* was initially knocked down using MO by three different research groups (Prabhudesai et al., 2016; Ren et al., 2011; Sheng et al., 2010). However, two of the research groups, despite using the identical MO constructs, observed completely contradictory phenotypes; Sheng et al. (2010) observed degeneration of dopaminergic neurons accompanied with L-DOPA responsive locomotor defects, whereas Ren et al. (2011) did not observe any phenotypic or pathologic derangements suggestive of PD in their morphant zebrafish. More recently, Wint and Sirotkin (2020) and Suzzi et al. (2021) used CRISPR/Cas9 gene editing system to knockout *lrrk2* gene function in zebrafish. Wint and Sirotkin (2020) established three mutant lines that had progressively more truncated *lrrk2* gene, lacking either only the serine/threonine kinase function, or both kinase functions or both of the kinase functions and all the protein–protein interaction domains. Suzzi et al. (2021), on the other hand, created zebrafish that lacked the entire *lrrk2* gene. As expected, these zebrafish models did not display any characteristic features of PD phenotype and are not useful animal models of PD.

The most common cause of early-onset PD is the homozygous or compound heterozygous mutations of *PRKN* gene, encoding the RBR E3 ubiquitin protein ligase parkin. Strikingly, in a study of 100 sporadic PD cases, more than two thirds that manifested before the age of 20 harbored *PRKN* gene mutations (Lücking et al., 2000). Parkin is involved in mitophagy, the process of destruction and regeneration of damaged and dysfunctional mitochondria via autophagosome formation. It ubiquitylates with low substrate selectivity proteins on the outer mitochondrial membrane of depolarized mitochondria that initiates mitophagy (Onishi et al., 2021). In order to understand the mechanisms of disease pathogenesis, fruit fly, mouse, and rat models of parkin mutation had previously been created. Although rodent models demonstrated impairments in mitochondrial homeostasis and dopaminergic neuron behaviors, they lacked any disease phenotype and dopaminergic neurodegeneration (Dave et al., 2014; Pickrell & Youle, 2015). Only recently, it was discovered that parkin knock-out mice indeed demonstrate motor impairments compatible with PD and degeneration of dopaminergic neurons, only beyond 120 weeks of age (Noda et al., 2020). In contrast, parkin-deficient fruit flies have severe defects in flight muscles, live shorter, and show dopaminergic neuron shrinkage and loss, as well as susceptibility to oxidative stress and mitochondrial swelling (Pickrell & Youle, 2015). Parkin knock-down zebrafish were also previously created with MO-mediated knock-down method (Fett et al., 2010; Flinn et al., 2009). However, these models had transient knock-down of parkin due to the method used and did not show any overt motor impairments. While one of the studies reported a small reduction in dopaminergic neuron numbers in midbrain and complex 1 deficiency (L. Flinn et al., 2009), the other study did not report any neurodegeneration or mitochondrial pathology (Fett et al., 2010). More recently, Hughes (2019) used CRISPR/Cas9 to knock-out parkin gene in zebrafish embryos. The resultant mutants had 5 bp frameshift at the position of 224. base which resulted in a 97-amino-acids-long truncated protein lacking RING1, RING2, and IBR domains and considered nonfunctional. However, the truncated mRNA product did not undergo nonsense-mediated decay, and the expression of the dopaminergic neuron markers dopamine transporter and tyrosine hydroxylase did not significantly differ between homozygous mutants and wild-type zebrafish (Hughes, 2019).

Moreover, there was no dopaminergic neuron loss in the midbrain of mutants. Hence, to date, parkin knock-down or knockout zebrafish models were not successful in recapitulating the manifestations of human Parkinson's disease.

Another well documented cause of autosomal recessively inherited and early-onset Parkinson's disease is mutations in PTEN-induced kinase 1, encoded by the gene *PINK1* (Poewe et al., 2017). It is a mitochondrial kinase that is responsible for mitochondrial quality control and directly activates and recruits parkin to mitochondria for the initiation of mitophagy (Pickrell & Youle, 2015). Similar to *Prkn* knockout mouse models, *Pink1* knockout models of mice did not have dopaminergic neuronal loss and abnormal motor phenotype but had impairments in dopamine homeostasis, mitochondrial respiratory system, and oxidative stress response (Akundi et al., 2011; Gautier et al., 2008; Kitada et al., 2007). However, knock-out of the homologous gene by ZFN in rats resulted in age-dependent dopaminergic neuronal loss in substantia nigra preceded by motor deficits demonstrated as reduction in rearing frequency, total distance travelled, hindlimb grip strength, and motor incoordination determined by counting the number of forelimb and hindlimb foot slips in the tapered balance beam paradigm (Dave et al., 2014). In zebrafish, three different MO-mediated knock-down models of *Pink1* deficiency have been established (Anichtchik et al., 2008; Sallinen et al., 2010; Xi et al., 2010). The morphants developed by Anichtchik et al. (2008) had severe defects in head and tail development which was only partially rescued by concomitant injection of *Pink1* mRNA with MO, indicating strong off-target effects of MO-mediated knock-down. In other two MO-mediated knock-down models, such severe developmental defects were not observed. Moreover, neurodevelopmental and behavioral defects reported in these three knock-down models were inconsistent with each other. In order to avoid off-target effects of MO-mediated knock-down method, Flinn et al. (2013) screened for *Pink1* mutants in a *N*-ethyl-*N*-nitrosourea (ENU) mutagenized library of zebrafish genomes. They identified a mutant that had a nonsense mutation in the kinase domain of *Pink1* gene. The mutants did not demonstrate any gross developmental abnormalities, as opposed to reports by Anichtchik et al. (2008) and Xi et al. (2010), but had a progressive loss of dopaminergic neurons in diencephalon, accompanied by subtle defects in mitochondrial ultrastructure and complex I and III activities (Flinn et al., 2013). More recently, Hughes (2019) developed a *Pink1* homozygous mutant zebrafish model using CRISPR/Cas9 system that carried a severely truncated *Pink1* gene lacking majority of the kinase domain and expressed only 10 % of *Pink1* mRNA compared to wild-types. However, the resultant mutants did not have any dopaminergic neuron loss, as well as no alterations in molecular markers of dopaminergic neuron function and development (Hughes, 2019).

Mutations in another gene, presenillin-associated rhomboid-like (PARL), that participates in PINK1/PARKIN mediated mitophagy pathway were recently discovered to be associated with two cases of PD (Shi et al., 2011). PARL is a mitochondrial intramembranous serine protease that is responsible for proteolytic cleavage of numerous proteins, including PINK1, and ultimately maintains mitochondrial homeostasis through regulation of several key pathways, including mitophagy (Lysyk et al., 2020). Although the initial report of PD cases with mutations in *PARL* cannot be replicated and its association with the pathogenesis of PD is questioned, its established role in the critical mitophagy pathway robustly associated with PD substantiates the design of PARL deficiency model animals in order to uncover disease mechanisms that contribute to PD pathogenesis (Wüst et al., 2016). Mice lacking PARL protein die prematurely at 8–12 weeks of age due to global cell death associated with increased cytochrome c release from mitochondria (Cipolat et al., 2006). Zebrafish possess two paralogs of PARL, *parla* and *parlb*, and MO-mediated knock-down of both results in larval lethality at 3 days postfertilization (Noble et al., 2012). Knock-down of either *parl* paralog did not cause premature mortality; however, dopaminergic neuron patterning and density was altered in posterior tuberculum, especially in *parla* knock-down morphants. To further characterize *parla* deficiency in zebrafish, Merhi et al. (2021) used the CRISPR/Cas9 system and knocked out *parla* gene. The homozygous mutant fish were morphologically normal and did not die prematurely. Expression of *parla* gene was reduced to 4% and 40% of wild-type levels in homozygous and heterozygous mutants, respectively (Merhi et al., 2021). Expression of *th1* mRNA in the brain of adult and larval fish was significantly decreased, both in heterozygous and homozygous mutants. In accordance with this observation, the number of TH positive dopaminergic neurons was also significantly reduced in the olfactory bulb and telencephalon of homozygous mutant fish. Moreover, the dopaminergic neuron loss in the olfactory bulb manifested as significantly increased time spent in the section of the tank where a repulsive stimulant, cadaverine, was added (Merhi et al., 2021). The homozygous mutants also showed increased inactivity periods that might be explained by dopaminergic neuron loss observed in telencephalic region. Finally, the expression of *fis1*, a gene implicated in mitochondrial fission, was highly increased in the brain of adult and larval homozygous mutants. Collectively, these observations highlight that a zebrafish model of *parla* deficiency could replicate dopaminergic neuronal loss in olfactory and telencephalic regions of brains accompanied by impaired mitochondrial homeostasis and manifest phenotype of impaired olfactory function, all of which are implicated in PD pathogenesis and phenotype.

Loss-of-function mutations in *PARK7* (formerly *DJ-1*), encoding parkinsonism-associated deglycase, is associated with a rare form of familial autosomal recessive and early-onset PD (OMIM: 606324) (Bonifati et al., 2003; Repici & Giorgini, 2019). DJ-1 is a multifunctional protein that is part of oxidative stress response machinery. It has several different enzymatic functions for the detoxification of proteins, scavenges ROS species, transcriptionally regulates key genes in oxidative stress response and mitochondrial homeostasis, and modulates signal transduction pathways implicated in oxidative stress response (Huang & Chen, 2021). Initial DJ-1 gene knock-out models created in mouse and fruit fly demonstrated motor impairments in both and behavioral derangements in the former, compatible with PD. Although loss of dopaminergic neurons and Lewy body aggregation were not observed in these models, dopaminergic outflow impairments and increased susceptibility to oxidative stress were detected in neurons (L. Chen et al., 2005; Goldberg et al., 2005; R. H. Kim et al., 2005; Park et al., 2005). Interestingly, rats with DJ-1 knock-out, in contrast to mice and fruit flies, developed with aging significant neuronal loss in dopaminergic nigrostriatal pathways (Dave et al., 2014). The first attempt to create a *park7* knock-out zebrafish model using MO was not successful, and no disease phenotype was observed (Baulac et al., 2009). More recently, Edson et al. (2019) used a CRISPR/Cas9 gene editing technique to successfully create *park7*-deficient zebrafish. The late adult mutants were smaller and weighed less than wild-type counterparts. The tyrosine hydroxylase levels, an enzyme responsible for the synthesis of dopamine, were significantly reduced in the brains of late adult DJ-1 deficient zebrafish compared to wild-types (Edson et al., 2019). Furthermore, another group observed a reduced number of dopaminergic neurons in the brain region of late adult DJ-1 deficient zebrafish that corresponds to human substantia nigra (Hughes et al., 2020). Age-dependent decreases in the duration of movement, distance traveled, velocity, and tail beat frequency at low speed were detected in mutant zebrafish, reminiscent of bradykinesia, a cardinal feature of PD (Hughes et al., 2020). In addition to motor impairments, anxiety-associated behavioral deficits, like increased time spent in the lower part of a novel tank and increased total time without movement in a novel tank, were also prominent in age-dependent manner in DJ-1 mutant zebrafish (Chavali et al., 2023). These findings suggest that the *park7* knockout zebrafish model recapitulates some features of PD and could be a helpful tool to explore the disease mechanisms and could be used for drug screening. Accordingly, Edson et al. (2019) and Chavali et al. (2023) identified dozens of genes and proteins differentially expressed in brain tissues between wild-type and DJ-1 mutant zebrafish involving many cellular mechanisms that can be implicated in PD pathogenesis.

Loss-of-function mutations of *FBXO7* gene, encoding a F-box protein, is responsible for a very rare familial early-onset parkinsonism syndrome with pyramidal symptoms (PARKIN15 syndrome) (Fonzo et al., 2009). FBXO7 is a multifunctional ubiquitin ligase that participates in SCF ubiquitin E3 ligase complex and specifies targets for proteasomal degradation, independently modulates proteosome function and induces mitophagy (Zhong et al., 2023). It is hypothesized that mutations in *FBXO7* gene contribute to pathogenesis of PARKIN15 syndrome through the disruption of mitochondrial homeostasis and reduction of proteasomal activity which is mediated by loss of function of FBXO7 protein, as well as gain-of-toxicity of the mutant protein that aggregates in mitochondria (Burchell et al., 2013; Joseph et al., 2018; Vingill et al., 2016; Zhou et al., 2015). The first disease model of PARKIN15 syndrome was established in zebrafish by MO-mediated knock-down of *fbxo7* (Zhao et al., 2012). The morphants had a significantly reduced number of tyrosine hydroxylase- or dopamine transporter–positive neurons in the diencephalic region. They also displayed decreased swimming velocity compared to wild-type animals, which was normalized after mutant fish were put into water containing a centrally acting dopamine agonist, apomorphine (Zhao et al., 2012). However, the morphants also displayed developmental defects in several peripheral organs, and the disease phenotype was not rescued with concomitant injection of MO with human *FBXO7* gene. Hence, potential off-target effects of MO-mediated knockdown methodology could not be reliably ruled out. More recently, Hughes (2019) created a *fbxo7* mutant zebrafish line using the CRISPR/Cas9 system which carried a 4 bp deletion within the second exon, resulting in a premature stop codon 57 bp downstream in the same exon. However, the mutant allele did not undergo nonsense-mediated decay, and the expression of dopaminergic neuron markers were not altered in the brain of mutant zebrafish. Consequently, the mutant zebrafish did not reproduce any disease phenotype.

In the last few years, with the advance of CRISPR/Cas9-mediated genome editing technology, new transgenic zebrafish models have been developed that target genes associated with PD but have not previously been targeted in zebrafish by other methods of gene editing. Nyuzuki et al. (2020) created, with CRISPR/Cas9-mediated mutagenesis, *atp13a2* mutant zebrafish that displayed significant reduction of TH positive neurons in the posterior tuberculum and the locus coeruleus. ATP13A2 protein, encoded by *PARK9* gene in humans, is a lysosomal ATPase that maintains functional integrity of lysosomes, and its loss-of-function mutations are associated with a rare form of familial early-onset complex parkinsonism syndrome, termed Kufor-Rakeb syndrome (Fleming et al., 2022). Lysosomal dysfunction, leading to disruption of autophagosome system and α-synuclein aggregation, is observed in patient fibroblasts and primary cortical neurons lacking ATP13A2 function (Dehay et al., 2012; Usenovic et al., 2012). Accordingly, zebrafish with mutant ATP13A2 had accumulated lysosome-like vesicles with fingerprint-like structures and severely reduced cathepsin D protein, a lysosomal protease also implicated in PD pathogenesis, in middle diencephalon (Nyuzuki et al., 2020). Additionally, mass spectrometry analysis of differentially expressed proteins in brains of these mutants revealed a significant decrease in levels of proteins annotated with vesicle fusion of Golgi apparatus. However, whether these mutant zebrafish demonstrate any locomotor phenotype characteristic of PD was not mentioned by the authors.

Several lines of evidence indicate that dysfunction of vesicular monoamine transporter 2 (VMAT2), encoded by *SLC18A2*, which pumps dopamine into synaptic vesicles, may be implicated in PD pathogenesis through disruption of dopamine homeostasis in synaptic terminal (Lohr & Miller, 2014). (Sveinsdóttir et al., 2022) examined whether zebrafish that lacked functional *slc18a2* alleles displayed any phenotype characteristic of PD. Homozygous mutant zebrafish, generated with the CRISPR/Cas9 system, were not viable in adulthood, and mutant larvae had significantly reduced number of TH positive neurons in midline diencephalic neuronal cluster. These mutants also displayed increased peak velocity in response to the lights-off switch but traveled significantly less than wild-type fish in a light-on environment. Furthermore, they had anxiety-like behavior assessed by thigmotaxis assay (Sveinsdóttir et al., 2022). Interestingly, L-DOPA treatment did not normalize the peak velocity phenotype of mutants but normalized the reduction in distance moved during lights-on period. Moreover, mutant larvae showed abnormalities in day-time sleep behavior that was significantly alleviated by L-DOPA treatment (Sveinsdóttir et al., 2022).

Another gene that was recently identified to harbor variants associated with increased risk for PD development is *GCH1*, encoding GTP cyclohydrolyse 1 (GTPCH) (Mencacci et al., 2014; Nalls et al., 2014). GTPCH is responsible for the conversion of GTP to D-erythro-7,8-dihydroneopterin triphosphate, the rate-limiting step of tetrahydrobiopterin ($BH_4$) biosynthesis, which is an essential cofactor of tyrosine hydroxylase (Himmelreich et al., 2021; Kurian et al., 2011). Autosomal dominant loss-of-function mutations in this gene cause dopa-responsive dystonia (DRD) (Kurian et al., 2011). However, several heterozygous variants in *GCH1* have also been shown to underlie the PD diagnosis in relatives of DRD patients, as well as in patients unrelated to any DRD cases (Mencacci et al., 2014). These GCH1-mutant PD patients had reduced DAT tracer uptake in basal ganglia assessed by $^{123}$I-FP-CIT imaging, consistent with dopaminergic neuron denervation. Postmortem analysis of such a patient had also revealed striking gliosis with severe neuronal loss and Lewy body inclusions in surviving neurons in basal ganglia (Gibb et al., 1991). Although these findings undoubtedly underscore a role for GCH1 deficiency in the development of PD, the mechanism underlying GTPCH deficiency and dopaminergic neuronal loss has not been elucidated. Most recently, Larbalestier et al. (2022) used the CRISPR/Cas9 system to knock out *gch1* gene and model *GCH1*-deficient PD disease in zebrafish. Homozygous mutants developed normally until 5 dpf but became cachectic and hypomotile and developed spine curvature by 8 dpf and did not survive beyond 12 dpf. As early as 5dpf, dopamine, serotonin, and noradrenaline levels were significantly decreased in brains of mutant zebrafish; however, dopaminergic neurons in posterior tuberculum that project to striatum were not decreased, even at 8 dpf when the disease phenotype was established (Larbalestier et al., 2022). Interestingly, the hypomotility phenotype could not be rescued by L-DOPA treatment. An unbiased transcriptome analysis of brain tissues of mutant zebrafish revealed significant elevation of numerous genes associated with innate immune system activation. This observation was supported by microglial activation in the brain tissue of homozygous mutants that was detected by morphological and functional analysis (Larbalestier et al., 2022).

## Amyotrophic Lateral Sclerosis (ALS)

Amyotrophic lateral sclerosis (ALS) is the most common neurodegenerative disorder of midlife, with a yearly incidence of 1–3 cases per 100,000. The disease manifests at around 50 years of age and is clinically characterized by insidious onset of muscle weakness with spasticity due to upper motor neuron degeneration, accompanied by development of fasciculations and muscle atrophy due to lower motor neuron degeneration. The pathologic hallmark of ALS is the aggregation of ubiquitylated protein inclusions, mostly composed of TDP43, in degenerated upper and lower motor neurons in cerebral cortex and spinal cord, respectively (Brown & Al-Chalabi, 2017; Hardiman et al., 2017). One in ten cases of ALS is familial, and it is almost always autosomal dominantly inherited. Although dozens of genes have so far been discovered to underlie familial ALS cases, *C9ORF72*, *TARDP*, *SOD1*, and *FUS* gene mutations make up more than two thirds of the cases (G. Kim et al., 2020). Genome-wide association studies of sporadic ALS have also found rare variants with relatively high penetrance in numerous genetic loci (Van Rheenen et al., 2016). The pathogenesis of ALS is poorly understood; however, when these and other genes associated with familial or sporadic ALS are integrated into a scheme of pathogenic pathways, three cellular processes seem to play key roles in ALS pathogenesis: protein homeostasis, RNA homeostasis, and trafficking and cytoskeletal dynamics (Brown & Al-Chalabi, 2017). Numerous disease models of ALS have so far been created in rodents, fruit flies, canines, swine, non-human primates, and zebrafish. Disease models of zebrafish with mutant *fus*, *tardbp*, *sod1*, and *C13H9orf72* manifested many features of ALS, highlighting the utility of zebrafish as a model animal in ALS research (Bonifacino et al., 2021). Subsequently, knock-out models of several causative genes as well as knock-in of point mutations into these genes through the CRISPR/Cas9 system have been established in zebrafish (Table 16.3). C9ORF72 is a DENN domain protein that forms a trimeric complex with SMCR8 and WDR41, which functions as a GTPase activating protein (GAP) complex. This GAP complex participates in the regulation of vesicle trafficking, autophagy, lysosomal homeostasis, actin dynamics, and inflammation (Smeyers et al., 2021; Su et al., 2021). *C9ORF72* gene has a short hexanucleotide repeat motif in the first intron, and very significant expansion of this motif underlies 34% of familial ALS cases, by far the most frequent genetic lesion in this group of patients (Kim et al., 2020). The mechanisms of disease pathogenesis in *C9ORF72* mutant cases of ALS are still debated; however, both gain-of-toxicity due to disrupted RNA and protein homeostasis caused by hexanucleotide repeat expansion and dipeptide repeat proteins, respectively, and loss-of-function of wild-type C9ORF72 protein are suggested to play key roles in this process (Smeyers et al., 2021). A zebrafish model of MO-mediated knock-down of *C13H9orf72*, the zebrafish homologue of *C9ORF72*, displayed reduction in locomotion, demonstrated by reduced touch-evoked escape response, as well as deficits in several parameters of spontaneous swimming (Ciura et al., 2013). However, when *C13H9orf72* was permanently knocked out via CRISPR/Cas9-mediated mutagenesis, homozygous mutants were only mildly disabled in motor activity, and no significant changes in neuromuscular junctions were observed (Rounding, 2018). A similar discordance was also observed in *fus* knock-out zebrafish.

*FUS* encodes a predominantly nuclear RNA-/DNA-binding protein that has a role in RNA processing and maintenance; however, in neurons, FUS protein can also be found in cytoplasm, especially axonal ends and dendrites, where it is associated with local translation (Kim et al., 2020). Mutations in the *FUS* gene are one of the frequent causes of familial ALS disease, and the majority of these mutations disrupt the nuclear localization signal, resulting in cytoplasmic mislocalization. Aggregation of mislocalized FUS proteins in cytoplasm induce formation of stress granules that lead to neurodegeneration and ALS-like phenotype in mouse models (Sharma et al., 2016). On the other hand, depletion of nuclear FUS protein in mouse brain caused significant changes in the expression and splicing patterns of numerous mRNAs, calling into question the contribution of loss-of-function of FUS in ALS

**TABLE 16.3**
Zebrafish Models of Amyotrophic Lateral Sclerosis (ALS) Generated via CRISPR/Cas System

| Reference | (Rounding, 2018) | (Lebedeva et al., 2017) | (Bose et al., 2019) | (Armstrong et al., 2016; Harji, 2019) | (Russell et al., 2021) | (Petel Légaré et al., 2023) |
|---|---|---|---|---|---|---|
| **Comments** | - | - | Abnormalities in fast-twitch muscle electrophysiology, reduced postsynaptic potentials | - | - | Significant reduction of the Ndufa9 subunit (Complex I) in chchd10 KO, several markers of mitochondrial integrated stress response upregulated in double KO |
| **Phenotype** | Subtle motor impairments in spinning task and swim tunnel tests | No distinctive phenotype | Reduction in spontaneous coiling of tail, significant motor impairments in touch-evoked escape response test | Reduced motor activities in free swim assay and swim tunnel tests beyond 2 years of age | Not reported | Reduced mean swim distance and maximum swim velocity in TEER test |
| **Pathology** | No changes in neuromuscular junctions | No motor neuron degeneration | Perturbations in the synaptic connectivity of pre- and postsynaptic components of the NMJ | Not reported | Reduction in SMN, increased apoptosis of SMN, reduced axonal branching of SMN | Significantly more orphaned clusters of presynaptic and postsynaptic markers |
| **Genotype** | p.ex2.166_168del+p. ex2.166_167ins out-of-frame with PTC | p.ex3.15_23del out-of-frame with PTC | p.ex1.1_5del out-of-frame with PTC | c.ex6.409G>A substitution, p.A379T | c.ex4.Δ4del out-of-frame with PTC | *chchd2*: c.ex2.144_150del<br>*chchd10*: c.ex2.141ins |
| **Isogenicity** | F3 isogenic stable homozygous mutants | F3 isogenic stable homozygous mutants | F3 isogenic stable homozygous mutants | F2 or F3 isogenic stable homozygous mutants | F0 isogenic stable homozygous mutants | F2 or F3 isogenic stable homozygous mutants |
| **Genome editing** | CRISPR/Cas9 mediated mutagenesis via NHEJ | CRISPR/Cas9 mediated mutagenesis via NHEJ | CRISPR/Cas9 mediated mutagenesis via NHEJ | CRISPR/Cas9 mediated knock-in via HDR | CRISPR/Cas9 mediated mutagenesis via NHEJ | CRISPR/Cas9 mediated mutagenesis via NHEJ |
| **Gene** | *C13H9orf72* | *Fus* | *tardbpl* | *tardp* | *tp73* | *chchd2*<br>*chchd10* |

pathogenesis (Lagier-Tourenne et al., 2012). In support of this finding, MO-mediated knock-down of *fus* gene in zebrafish resulted in significant impairments in locomotor function, and functional and morphological abnormalities in neuromuscular junctions and motor neurons (Armstrong & Drapeau, 2013). However, more recently, researchers used CRISPR/Cas9 to establish a *fus* knock-out zebrafish line and the mutants did not display any locomotor, neuronal, or neuromuscular junctional abnormalities (Lebedeva et al., 2017). This finding was in parallel with what was observed in *fus* knock-out mice (Kino et al., 2015).

Another nuclear DNA-/RNA-binding protein TDP-43, encoded by the gene *TARDBP*, is an important regulator of alternative splicing and expression levels of thousands of mRNA species. Mutations in this gene can be detected in a small percentage of familial ALS cases; however, almost 90% of ALS cases harbor cytoplasmic inclusions composed of aggregated TDP-43, accompanied by complete depletion of the nuclear TDP-43 pool (Kim et al., 2020). Therefore, it is suggested that both gain-of-toxicity and loss-of-function of TDP-43 may play important roles in the pathogenesis of most ALS cases. In zebrafish, there is a paralogue of *tardbp*, namely *tardbpl*, that compensates for *tardbp* when this is knocked out. Double knock-out of *tardbp* and *tardbpl* in zebrafish via ZFN resulted in early lethality and development of an ALS-like phenotype characterized by impaired axonal outgrowths of spinal motor neurons coupled with muscle degeneration (Schmid et al., 2013). Similar results were obtained when the CRISPR/Cas9 system was used to knock-out *tardbpl* on a *tardbp*$^{-/-}$ background (Bose et al., 2019). Another research group used the CRISPR/Cas9 system to knock-in a homologue of the most common disease-causing mutation in *TARDBP* to zebrafish *tardbp*, to replicate the disease phenotype under physiological expression of *tardbp* (Armstrong et al., 2016). However, they did not knock out the paralogous gene, *tardbpl*, that is known to compensate for lack of *tardbp* function, and no disease phenotype was observed in larvae. However, both 2- and 3-year-old heterozygous or homozygous mutant fish displayed significantly reduced locomotion in a free swim test, compared to wild-types (Harji, 2019). These findings suggest that the mutant protein produced a gain-of-function with time that manifested as disease phenotype in older fish.

Russell et al. (2021) recently reported nonsynonymous deleterious mutations in *TP73* gene among three cohorts of ALS patients. P73, encoded by *TP73*, is a member of the p53-family of proteins and it is evolutionarily closely related to p53, the guardian of the genome. Two isoforms of p73 with opposing functions have been described; one acts in parallel with p53 and promotes apoptosis, and the other promotes survival, a function primarily characterized in neurons (Moll & Slade, 2004). Russell et al. (2021) used the CRISPR/Cas9 system to introduce deleterious mutations in zebrafish *tp73* gene. Homozygous mutant embryos showed reduced numbers of spinal motor neurons, with increased markers of apoptosis in the surviving ones. Moreover, the surviving spinal motor neurons had reduced axonal branching. These findings strongly support previous findings that p73 is required for the development of the nervous system and its absence can cause motor neuron pathologies. However, a more recent analysis of the two largest ALS cohorts revealed that the occurrence of such deleterious mutations in *TP73* were not significantly enriched in ALS patients, as these mutations were also detected in healthy controls (Dilliott et al., 2022). Therefore, despite plausible mechanistic associations, the role of *TP73* in ALS pathogenesis remains questionable.

Disruption in mitochondrial homeostasis is one of the prominent features of ALS pathogenesis, and mutations in *SOD1*, *C9ORF72*, *TARDBP*, and *FUS* genes have been shown to promote mitochondrial pathology in one way or another (Hardiman et al., 2017). More recently, mutations in two mitochondrial genes, *CHCHD2* and *CHCHD10*, were shown to be associated with several cases of familial ALS disease (Johnson et al., 2014). Research suggests that both proteins have partly redundant functions in mitochondrial homeostasis, chiefly in cristae junction integrity and respiratory complex activity, and ALS-associated mutations disrupt this homeostasis (Ikeda et al., 2022). A mouse knock-in model of ALS-associated *CHCHD10* mutation revealed pathologic aggregation of mutant protein in mitochondria that was associated with impaired mitochondrial function and structure (Anderson et al., 2019). In addition, research from mouse neurons indicates that lack of CHCHD10 was associated with cytoplasmic TDP-43 accumulation and synaptic degeneration (Woo et al., 2017). In parallel with these observations, MO-mediated knock-down of *chchd10* in zebrafish resulted in significantly reduced axon length in spinal motor neurons (Brockmann et al., 2018). However, similar to a mouse knock-out model, no mitochondrial pathology was observed in morphants (Anderson et al., 2019; Brockmann et al., 2018). More recently, Petel Légaré et al. (2023) used CRISPR/Cas9 to knock out *chchd10* and *chchd2* genes in zebrafish and investigated the phenotype of homozygous mutants. The mutants had significantly reduced survival at 30 dpf, and double mutants did not survive beyond 15 dpf. Touch-evoked burst swim distance of mutant 2 dpf larvae was significantly reduced compared to wild-type. This impairment in locomotor activity was sustained in 5dpf mutant larvae, confirmed by reduced free swim velocity (Petel Légaré et al., 2023). The researchers also observed disruptions in neuromuscular junctions of mutant larvae, highlighted by an increased number of orphaned presynaptic and postsynaptic markers. Although they did not observe structural pathology in mitochondria of mutant larvae, subtle changes in proteins associated with mitochondrial complex 1 subunit and mitochondrial integrated stress response were detected.

## Spinal Muscular Atrophy (SMA)

Spinal muscular atrophy is characterized by the degeneration of alpha motor neurons, which manifests as progressive flaccid muscle weakness. The global incidence of SMA is about 1 in 10,000 live births. The classical SMA comprises of isolated alpha motor degeneration and is a monogenic disease that is caused by the lack of *SMN1* gene function, mostly due to complete deletion of both alleles but rarely due to biallelic mutations (Alías et al., 2009; Mercuri et al., 2022). In humans, there is a homologous gene, *SMN2*, but due to several nucleotide differences in exon 7 of this gene, only %10 of its transcripts is translated to the full-length SMN protein. The major determinant of disease phenotype in SMA patients is the copy number of *SMN2* gene; those with higher number of copies of this gene manifest with milder phenotype at later ages (Wirth et al., 2020). SMN is crucial for the survival of not only alpha motor neurons during embryonic development but also of the cells in many other organs, demonstrated by embryonic lethality in *smn* knock-out mice (Schrank et al., 1997). The canonical function of SMN is the maintenance of major and minor spliceosomes, and deficiency of SMN results in splicing defects in a set of genes (Pellizzoni et al., 2002). However, many other functions, especially in transcription, translation, cytoskeleton dynamics, cellular trafficking, and mitochondrial homeostasis, have been attributed to SMN (Singh et al., 2017). Many different organisms have so far been used to model SMA to elucidate disease mechanisms and modifiers and to discover and validate disease modifying therapeutics (Wirth et al., 2020). *Smn* gene in zebrafish was knocked down by MO or miRNAs to create SMA disease models. These morphants developed a motor phenotype similar to SMA and showed abnormal branching and truncation of caudal primary motor neurons, compatible with SMA pathology (Singh & Patten, 2022). Recently, Préville et al. (2020) demonstrated that *smn* gene can be successfully knocked-out in zebrafish using CRISPR/Cas9 (Table 16.4). Although the researchers have not yet published full characterization of the SMA disease phenotype in mutant zebrafish, they reported significantly reduced locomotion and SMN levels compared to wild-types in these mutants. Another research group used CRISPR/Cas9-mediated homology directed repair to knock-in the A6T transition into exon 7 of zebrafish *smn*, mimicking the c.840C>T transition of the human *SMN2* gene that is the cause of exon 7 skipping (Tay et al., 2021). They detected a mutant with the desired A6T transition, but the mutant allele also had in-frame 27 bp indel in exon 7, resulting in addition of 9 amino acids within YG box domain. Remarkably, the amount of full-length SMN protein in mutant fish started high at 5 dpf and gradually declined to effectively zero only at 12 dpf. On the other hand, zebrafish with *smn* knock-out had significantly reduced SMN protein levels already at 5 dpf. This was associated with significantly longer survival of homozygous knock-in mutants, on average 39 dpf, compared to only 12 dpf in knock-out mutants. In parallel to this, knock-in mutants displayed delayed onset of motor impairments, which was accompanied by progressive muscle atrophy, motor neuron degeneration, and neuromuscular junction derangements (Tay et al., 2021). These findings support that the knock-in mutants phenocopy the intermediate type SMA that develops at later ages due to higher expression of full-length SMN protein from the *SMN2* gene in some patients.

## Other Rare Neurodegenerative Disease Models

Ataxia telangiectasia (AT) is a disease characterized by progressive degeneration of cerebellar neurons that manifests with a diversity of ataxic symptoms at early ages. Other characteristic features of this disease include premature aging, cancer predisposition, immunodeficiency, and radiation sensitivity (Rothblum-Oviatt et al., 2016). The disease is caused by autosomal recessive mutations of the gene *Ataxia telangiectasia mutated* (*ATM*). ATM is primarily a nuclear kinase that is crucial for the propagation of DNA double-strand break-repair response (Shiloh & Ziv, 2013). The role of cytoplasmic ATM is less well characterized; however, accumulated research indicates that cytoplasmic ATM participates in mitochondrial homeostasis, selective autophagy of peroxisomes, insulin mediated glucose uptake, and hypoxia response (Amirifar et al., 2019). The connection between loss of ATM function and neurodegeneration has not yet been thoroughly elucidated. Several lines of evidence indicate that loss of ATM function can lead to neurodegeneration through more than one mechanism, such as cell cycle re-entry followed by apoptosis, deficiency of oxidative stress response, and increased glial inflammatory response due to cytoplasmic DNA accumulation (Hoche et al., 2012; Quek et al., 2017; Rimkus et al., 2008). Very recently, a new zebrafish model of ATM deficiency was engineered using CRISPR/Cas9 (Chen et al., 2022). The researchers demonstrated that homozygous mutant zebrafish had both reduced average swimming speed and reduced highly active period in 60-minute open field test, highlighting impaired motor function. Furthermore, markers of innate immune system activation and redox homeostasis were elevated in cerebellum and/or hindbrain of homozygous mutants. The mutants also demonstrated other characteristic features of AT disease, such as immunodeficiency and cancer predisposition (Chen et al., 2022).

Autosomal recessive spastic ataxia of Charlevoix-Saguenay (ARSACS) is a progressive neurodegenerative disorder caused by the autosomal recessive mutations of *SACS* gene, encoding the Sacsin protein. It is clinically characterized by lower

**TABLE 16.4**

Zebrafish Models of Spinal Muscular Atrophy (SMA) Generated via CRISPR/Cas System

| Reference | (Préville et al., 2020) | (Tay et al., 2021) |
|---|---|---|
| **Comments** | Reduced SMN protein levels | - |
| **Phenotype** | Significantly reduced locomotion | Slower swimming followed by floating or sinking to the tank bottom and death at 36 dpf |
| **Pathology** | Not reported | Reduced number of motor neurons, retraction of presynaptic nerves, loss of regularly shaped polygonal muscle fibers with increased muscle areas being replaced by fibrotic tissue |
| **Genotype** | Not reported | c.ex8.6A>T + c.ex8.18_25del + c.ex8.18_52ins in-frame mutation |
| **Isogenicity** | Not reported | F2 or F3 isogenic stable homozygous mutants |
| **Genome editing** | CRISPR/Cas9 mediated mutagenesis via NHEJ | CRISPR/Cas9 mediated knock-in via HDR |
| **Gene** | *smn1* | *smn1* |

limb spasticity, cerebellar ataxia, and peripheral neuropathy (Bouhlal et al., 2011). Pathologically, progressive loss of cerebellar neurons is evident in post-mortem analysis of patient tissues (Girard et al., 2012). Experiments performed on Sacsin deficient neurons in vitro and *sacs* knock-out mouse brains in vivo have shown that mitochondrial dysfunction, impaired cytoskeletal organization, and protein trafficking are some of the pathological mechanisms that are associated with neuronal cell death in disease models of ARSACS (Bagaria et al., 2022; Bouhlal et al., 2011; Girard et al., 2012; Romano et al., 2022). Naef et al. (2021) created a *sacs* knock-out zebrafish model using CRISPR/Cas9. The mutant allele had a 10 bp deletion in exon 7 of *sacs* gene that resulted in a PTC formation at residue 495, and these alleles underwent nonsense-mediated decay. The homozygous mutant zebrafish had mild microphthalmia but developed normally otherwise. Concerning motor activity, the mutants had statistically significantly reduced burst activity at 30 hpf and significant reduction in total distance moved and average velocity in 30-minute spontaneous swim test (Naef et al., 2021). Pathological examination revealed absence of gross abnormalities in spinal motor and sensory neurons. To visualize cerebellar area, homozygous mutants were crossed with a transgenic zebrafish line that expressed red fluorescent protein (RFP) and GCaMP5G, a calcium indicator, under a Purkinje cell specific enhancer (Matsui et al., 2014). In fact, these transgenic mutant zebrafish had a significantly reduced RFP signal compared to controls, suggesting cerebellar atrophy. Analysis of mitochondrial respiration markers in mutant zebrafish revealed impairments in baseline respiration, ATP production, maximal respiration, proton leak, and spare respiratory capacity. This was accompanied by increased reactive oxygen species production. Overall, these findings support the validity of this zebrafish model of ARSACS disease.

Hereditary spastic paraplegia (HSP) is a heterogenous group of rare neurodegenerative disorders characterized by progressive lower-limb spasticity and hyperreflexia, either isolated or accompanied by a diversity of neurological findings (Meyyazhagan & Orlacchio, 2022; Shribman et al., 2019). To date, more than 80 disease-causing genetic loci have been identified in patients with HSP, and the mutated genes have been recovered for most of these loci (Murala et al., 2021). Although the genetics of HSP is extremely diverse, post-mortem analysis of CNS tissues of affected individuals show a unifying feature of axonal degeneration, particularly in longer neurons of lateral corticospinal tracts (Murala et al., 2021). The disrupted cellular mechanisms that underlie the neuropathological findings in HSP have not been studied extensively; however, an integrative analysis of known physiological roles of the plethora of genes associated with many types of HSP suggests that impairments in endoplasmic reticulum morphology, axon plotting, lipid metabolism, myelination, motor protein function, endosomal kinetics, mitochondrial homeostasis, and microtubule dynamics play roles in axonal degeneration and several other pathologies observed in HSP (Meyyazhagan et al., 2022). In an effort to determine the genetic mutation in patients with HSP, Fard et al. (2019) identified a novel gene associated with autosomal dominantly inherited HSP in a fairly large group of patients from a diversity of ethnicities. The identified gene, *UBAP1*, is a member of the endosomal sorting complex required for transport-1 (ESCRT-1) and participates in endosome mediated vesicular trafficking. The presence of truncated protein in patient fibroblasts suggested this mutant protein could have dominant negative effects. To test this hypothesis, the researchers knocked out *ubap1* in a transgenic zebrafish with fluorescently labeled motor neurons using CRISPR/Cas9 and injected either wild-type or mutated allele of human *UBAP1* mRNA into genome-edited zygotes. The embryos injected with mutant allele of human *UBAP1* had biologically significantly more truncated and disorganized axonal branching of spinal motor neurons, highlighting the role of mutant UBAP1 protein in axonal degeneration of motor neurons in zebrafish (Table 16.5, Figure 16.2).

**TABLE 16.5**

Zebrafish Models of Ataxia Telangiectasia, Autosomal Recessive Spastic Ataxia of Charlevoix-Saguenay and Hereditary Spastic Paraplegia Generated via CRISPR/Cas System

| Reference | (Chen et al., 2022) | (Naef et al., 2021) | (Farazi Fard et al., 2019) |
|---|---|---|---|
| **Comments** | Cytokines such as *tnfa*, *il-6*, and *il-8* were increased in the cerebellum and hindbrain, expression of genes related to redox homeostasis such as *gpx1b*, *sod1*, and *sod2* was elevated, hindbrains were enriched in GFAP-positive glial cells with a transcriptional elevated level of *tnfa* | Increase of oxidative stress and activation of ROS-mediated apoptosis | Disease pathology was created with supplementation of mutant human *UBAP1* mRNA into homozygous mutant zygotes |
| **Phenotype** | Lower activity levels during the 60-minute open field test, higher prevalence of low-speed tails | Decreased tail-burst activity, reduced locomotor activity with reduced mean velocity and distance travelled | Not reported |
| **Pathology** | Not reported | Significant reduction in cerebellar area | Significantly shorter, more truncated, and misshaped motorneuron axons |
| **Genotype** | c.ex17.101_111ins out-of-frame mutation with PTC | c.ex17.867_877del + c.ex17.867Ains out-of-frame mutation with PTC | Not reported |
| **Isogenicity** | F2 or F3 isogenic stable homozygous mutants | F3 isogenic stable homozygous mutants | F0 isogenic mosaic homozygous mutants |
| **Genome editing** | CRISPR/Cas9 mediated mutagenesis via NHEJ | CRISPR/Cas9 mediated mutagenesis via NHEJ | CRISPR/Cas9 mediated mutagenesis via NHEJ |
| **Gene** | *Atm* | *Sacs* | *ubap1* |

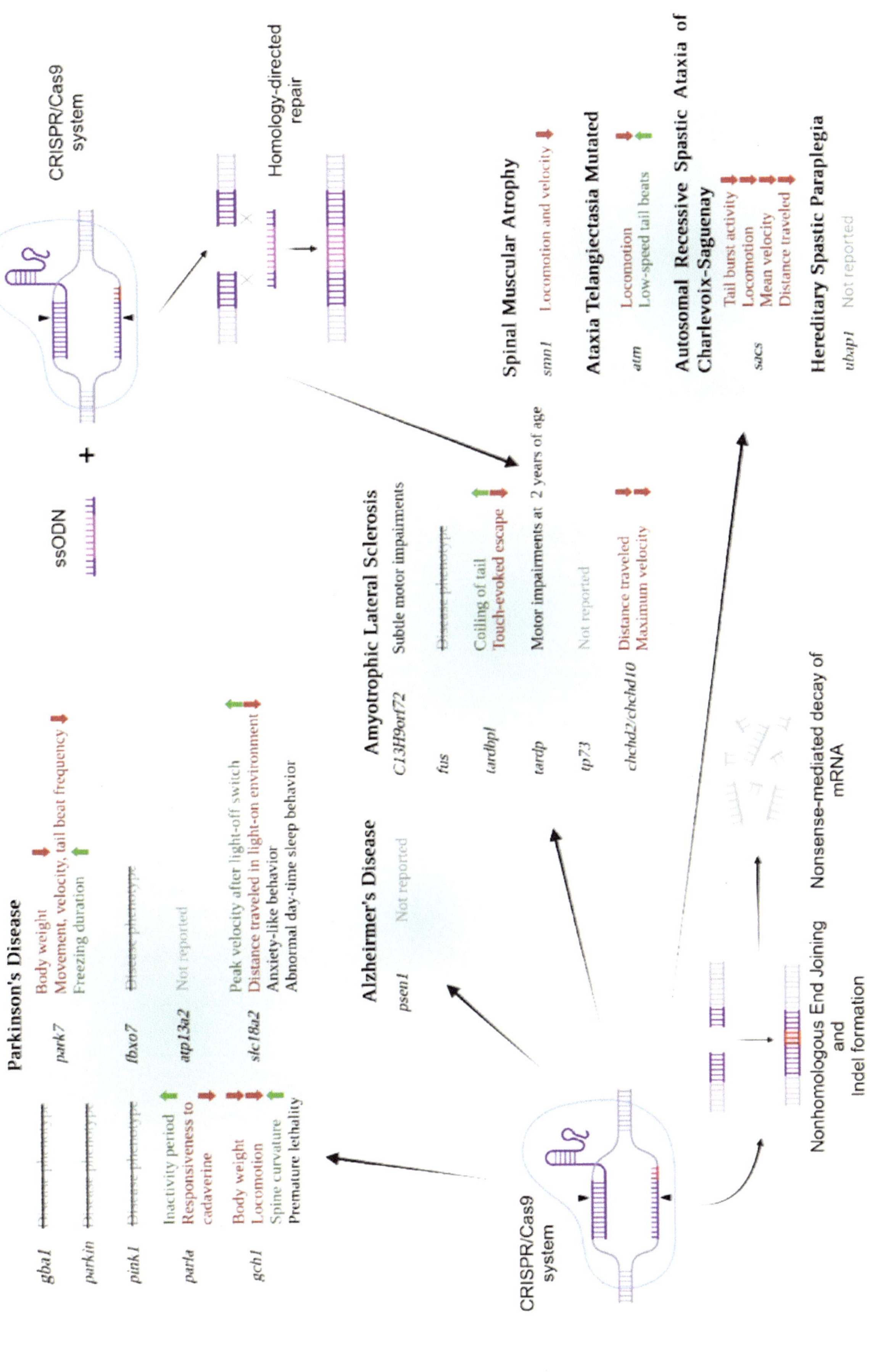


**FIGURE 16.2** Neurodegenerative diseases modeled in zebrafish using CRISPR/Cas systems, the genes targeted, and disease associated phenotypes, when observed.

## Future Directions

At least one zebrafish model for the most common neurodegenerative disorders has so far been established (Singh & Patten, 2022; Wang & Cao, 2021). As discussed above, one of the significant advantages of disease modeling in zebrafish is the ability to quickly generate disease phenotypes and screen for drug candidates to alleviate these phenotypes. Accordingly, rapid, easy, and transient knock-down of genes with morpholino antisense oligomers has been one of the mainstays for disease modeling in zebrafish (Stainier et al., 2017). However, the development of the CRISPR/Cas systems for genome editing has made genetic manipulation in vitro remarkably easy. Subsequently, stable gene knock-out models of neurodegenerative diseases in zebrafish, for which MO-mediated gene knock-down models had been established, were engineered in the last few years. The genetic basis of the common neurodegenerative disorders is very diverse, and the CRISPR/Cas system has only been used to establish a few gene knock-out models (Table 16.1–5) (Figure 16.2). Hence, in the near future, many more neurodegenerative disease models in zebrafish, created with the CRISPR/Cas system by knocking out genes associated with these diseases, are expected to be established. An important consideration that will shape the future of neurodegenerative disease modeling in zebrafish is the phenomenon of genetic compensation. Consequently, the use of mosaic founder animals to characterize disease phenotypes can become a common strategy to overcome genetic compensation (Buglo et al., 2020). Thus far, almost all of the neurodegenerative disease models created via the CRISPR/Cas system in zebrafish have been based on knocking out disease-associated genes through indel formation. However, some genetic mutations associated with the most common neurodegenerative disorders act through gain-of-function mechanisms; hence, modeling the precise mutations is necessary to replicate the disease pathology in model organisms. Efficient and targeted introduction of mutations into the genome of zebrafish with CRISPR/Cas9 mediated HDR, adenine, or cytosine base editors and prime editors have already been demonstrated (Boel et al., 2018; Petri et al., 2021; Rosello et al., 2023; Zhang et al., 2017). In the future, these advanced forms of CRISPR/Cas systems will undoubtedly be utilized to create novel zebrafish models with precise knock-in of disease-causing mutations.

## Conclusion

Zebrafish offer several advantages over other model organisms for neurodegenerative disease modeling. These include potential for mass production, ease of drug delivery, phenotype tracking, and transparency that allows optogenetic manipulations. In recent years, the CRISPR/Cas system has been used to generate zebrafish models of Alzheimer's disease, Parkinson's disease, amyotrophic lateral sclerosis, spinal muscular atrophy, and several rare neurodegenerative diseases. This genome editing tool is ever more commonly preferred over other methods to create disease models in zebrafish. The early research effort has especially concentrated on knocking out genes which were previously transiently knocked down with morpholino oligonucleotides to avoid potential toxic and off-target side effects of morpholinos and generate disease models with permanent genetic perturbations. In most cases, the mutant zebrafish created with the CRISPR/Cas system displayed some features of human disease, supporting their utility as disease models. However, several CRISPR/Cas-mediated gene knock-out models did not manifest disease phenotypes or pathologies, in contrast to their morpholino-mediated knock-down counterparts. Although it can be argued that disease phenotypes observed in morpholino-mediated knock-down models were caused by off-target effects or toxicity, the common occurrence of genetic compensation in zebrafish should be kept in mind. Hence, neurodegenerative disease modeling in zebrafish through CRISPR/Cas mediated genome editing can be complicated with genetic compensation and strategies to avoid this should be implemented. The number of genes and mutations associated with neurodegenerative disorders is extensive and ever growing, and only a tiny minority of these mutations have so far been modeled in zebrafish using the CRISPR/Cas system. The number and diversity of zebrafish models of neurodegenerative diseases will inevitably increase, especially with the widespread use of more advanced CRISPR/Cas systems that can accomplish precise gene editing. This will ultimately help us better understand the disease mechanisms of neurodegenerative disorders, develop new therapeutic interventions, and identify new drug candidates to treat these diseases.

## REFERENCES

Adhish, M., & Manjubala, I. (2023). Effectiveness of zebrafish models in understanding human diseases—A review of models. *Heliyon*, *9*(3), e14557. https://doi.org/10.1016/J.HELIYON.2023.E14557

Akundi, R. S., Huang, Z., Eason, J., Pandya, J. D., Zhi, L., Cass, W. A., Sullivan, P. G., & Büeler, H. (2011). Increased mitochondrial calcium sensitivity and abnormal expression of innate immunity genes precede dopaminergic defects in pink1-deficient mice. *PLoS One*, *6*(1), e16038. https://doi.org/10.1371/JOURNAL.PONE.0016038

Alessi, D. R., & Sammler, E. (2018). LRRK2 kinase in Parkinson's disease. *Science*, *360*(6384), 36–37. https://doi.org/10.1126/SCIENCE.AAR5683

Alías, L., Bernal, S., Fuentes-Prior, P., Barceló, M. J., Also, E., Martínez-Hernández, R., Rodríguez-Alvarez, F. J., Martín, Y., Aller, E., Grau, E., Peciña, A., Antiñolo, G., Galán, E., Rosa, A. L., Fernández-Burriel, M., Borrego, S., Millán, J. M., Hernández-Chico, C., Baiget, M., & Tizzano, E. F. (2009). Mutation update of spinal muscular atrophy in Spain: Molecular characterization of 745 unrelated patients and identification of four novel mutations in the SMN1 gene. *Human Genetics*, *125*(1), 29–39. https://doi.org/10.1007/S00439-008-0598-1/TABLES/3

Amirifar, P., Ranjouri, M. R., Yazdani, R., Abolhassani, H., & Aghamohammadi, A. (2019). Ataxia telangiectasia: A review of clinical features and molecular pathology. *Pediatric Allergy and Immunology*, *30*(3), 277–288. https://doi.org/10.1111/PAI.13020

Anderson, C. J., Bredvik, K., Burstein, S. R., Davis, C., Meadows, S. M., Dash, J., Case, L., Milner, T. A., Kawamata, H., Zuberi, A., Piersigilli, A., Lutz, C., & Manfredi, G. (2019). ALS/FTD mutant CHCHD10 mice reveal a tissue-specific toxic gain-of-function and mitochondrial stress response. *Acta Neuropathologica*, *138*(1), 103–121. https://doi.org/10.1007/S00401-019-01989-Y/FIGURES/10

Andrews, S. J., Renton, A. E., Fulton-Howard, B., Podlesny-Drabiniok, A., Marcora, E., & Goate, A. M. (2023). The complex genetic architecture of Alzheimer's disease: novel insights and future directions. *EBioMedicine*, *90*, 104511. https://doi.org/10.1016/J.EBIOM.2023.104511

Anichtchik, O., Diekmann, H., Fleming, A., Roach, A., Goldsmith, P., & Rubinsztein, D. C. (2008). Loss of PINK1 function affects development and results in neurodegeneration in zebrafish. *Journal of Neuroscience*, *28*(33), 8199–8207. https://doi.org/10.1523/JNEUROSCI.0979-08.2008

Anzalone, A. V., Koblan, L. W., & Liu, D. R. (2020). Genome editing with CRISPR–Cas nucleases, base editors, transposases and prime editors. *Nature Biotechnology, 38*(7), 824–844. https://doi.org/10.1038/s41587-020-0561-9

Anzalone, A. V., Randolph, P. B., Davis, J. R., Sousa, A. A., Koblan, L. W., Levy, J. M., Chen, P. J., Wilson, C., Newby, G. A., Raguram, A., & Liu, D. R. (2019). Search-and-replace genome editing without double-strand breaks or donor DNA. *Nature*, *576*(7785), 149–157. https://doi.org/10.1038/s41586-019-1711-4

Armstrong, G. A. B., & Drapeau, P. (2013). Loss and gain of FUS function impair neuromuscular synaptic transmission in a genetic model of ALS. *Human Molecular Genetics*, *22*(21), 4282–4292. https://doi.org/10.1093/HMG/DDT278

Armstrong, G. A. B., Liao, M., You, Z., Lissouba, A., Chen, B. E., & Drapeau, P. (2016). Homology directed knockin of point mutations in the zebrafish tardbp and fus genes in ALS using the CRISPR/Cas9 system. *PLoS One*, *11*(3), e0150188. https://doi.org/10.1371/JOURNAL.PONE.0150188

Asakawa, K., Handa, H., & Kawakami, K. (2020). Optogenetic modulation of TDP-43 oligomerization accelerates ALS-related pathologies in the spinal motor neurons. *Nature Communications*, *11*(1), 1–16. https://doi.org/10.1038/s41467-020-14815-x

Bagaria, J., Bagyinszky, E., & An, S. S. A. (2022). Genetics of autosomal recessive spastic ataxia of charlevoix-saguenay (ARSACS) and role of sacsin in neurodegeneration. *International Journal of Molecular Sciences, 23*(1), 552. https://doi.org/10.3390/IJMS23010552

Banote, R. K., Chebli, J., Şatır, T. M., Varshney, G. K., Camacho, R., Ledin, J., Burgess, S. M., Abramsson, A., & Zetterberg, H. (2020). Amyloid precursor protein-b facilitates cell adhesion during early development in zebrafish. *Scientific Reports, 10*(1), 1–16. https://doi.org/10.1038/s41598-020-66584-8

Barazzuol, L., Cieri, D., Facchinello, N., Calì, T., Washbourne, P., Argenton, F., & Pizzo, P. (2023). Unraveling presenilin 2 functions in a knockout zebrafish line to shed light into Alzheimer's disease pathogenesis. *Cells*, *12*(3), 376. https://doi.org/10.3390/CELLS12030376/S1

Barrangou, R., Fremaux, C., Deveau, H., Richards, M., Boyaval, P., Moineau, S., Romero, D. A., & Horvath, P. (2007). CRISPR provides acquired resistance against viruses in prokaryotes. *Science*, *315*(5819), 1709–1712. https://doi.org/10.1126/SCIENCE.1138140/SUPPL_FILE/BARRANGOU.SOM.PDF

Barthelson, K., Dong, Y., Newman, M., & Lardelli, M. (2021). PRESENILIN 1 mutations causing early-onset familial Alzheimer's disease or familial acne inversa differ in their effects on genes facilitating energy metabolism and signal transduction. *Journal of Alzheimer's Disease*, *82*(1), 327–347. https://doi.org/10.3233/JAD-210128

Baulac, S., Lu, H., Strahle, J., Yang, T., Goldberg, M. S., Shen, J., Schlossmacher, M. G., Lemere, C. A., Lu, Q., & Xia, W. (2009). Increased DJ-1 expression under oxidative stress and in Alzheimer's disease brains. *Molecular Neurodegeneration*, *4*(1), 1–14. https://doi.org/10.1186/1750-1326-4-12/FIGURES/7

Blauwendraat, C., Nalls, M. A., & Singleton, A. B. (2020). The genetic architecture of Parkinson's disease. *The Lancet Neurology*, *19*(2), 170–178. https://doi.org/10.1016/S1474-4422(19)30287-X

Boel, A., De Saffel, H., Steyaert, W., Callewaert, B., De Paepe, A., Coucke, P. J., & Willaert, A. (2018). CRISPR/Cas9-mediated homology-directed repair by ssODNs in zebrafish induces complex mutational patterns resulting from genomic integration of repair-template fragments. *DMM Disease Models and Mechanisms*, *11*(10). https://doi.org/10.1242/DMM.035352/2796

Bonifacino, T., Zerbo, R. A., Balbi, M., Torazza, C., Frumento, G., Fedele, E., Bonanno, G., & Milanese, M. (2021). Nearly 30 years of animal models to study amyotrophic lateral sclerosis: A historical overview and future perspectives. *International Journal of Molecular Sciences*, *22*(22). https://doi.org/10.3390/IJMS222212236

Bonifati, V., Rizzu, P., Van Baren, M. J., Schaap, O., Breedveld, G. J., Krieger, E., Dekker, M. C. J., Squitieri, F., Ibanez, P., Joosse, M., Van Dongen, J. W., Vanacore, N., Van Swieten, J. C., Brice, A., Meco, G., Van Duijn, C. M., Oostra, B. A., & Heutink, P. (2003). Mutations in the DJ-1 gene associated with autosomal recessive early-onset parkinsonism. *Science*, *299*(5604), 256–259. https://doi.org/10.1126/SCIENCE.1077209/SUPPL_FILE/1077209S.PDF

Bose, P., Armstrong, G. A. B., & Drapeau, P. (2019). Neuromuscular junction abnormalities in a zebrafish loss-of-function model of TDP-43. *Journal of Neurophysiology*, *121*(1), 285–297. https://doi.org/10.1152/JN.00265.2018/ASSET/IMAGES/LARGE/Z9K0011949100006.JPEG

Bouhlal, Y., Amouri, R., El Euch-Fayeche, G., & Hentati, F. (2011). Autosomal recessive spastic ataxia of charlevoix–saguenay: An overview. *Parkinsonism & Related Disorders*, *17*(6), 418–422. https://doi.org/10.1016/J.PARKRELDIS.2011.03.005

Brockmann, S. J., Freischmidt, A., Oeckl, P., Müller, K., Ponna, S. K., Helferich, A. M., Paone, C., Reinders, J., Kojer, K., Orth, M., Jokela, M., Auranen, M., Udd, B., Hermann, A., Danzer, K. M., Lichtner, P., Walther, P., Ludolph, A. C., Andersen, P. M., … Weishaupt, J. H. (2018). CHCHD10 mutations p.R15L and p.G66V cause motoneuron disease by haploinsufficiency. *Human Molecular Genetics*, *27*(4), 706–715. https://doi.org/10.1093/HMG/DDX436

Brown, R. H., & Al-Chalabi, A. (2017). Amyotrophic lateral sclerosis. *New England Journal of Medicine*, *377*(2), 162–172. https://doi.org/10.1056/NEJMRA1603471

Buglo, E., Sarmiento, E., Martuscelli, N. B., Sant, D. W., Danzi, M. C., Abrams, A. J., Dallman, J. E., & Züchner, S. (2020). Genetic compensation in a stable slc25a46 mutant zebrafish: A case for using F0 CRISPR mutagenesis to study phenotypes caused by inherited disease. *PLoS One*, *15*(3), e0230566. https://doi.org/10.1371/JOURNAL.PONE.0230566

Burchell, V. S., Nelson, D. E., Sanchez-Martinez, A., Delgado-Camprubi, M., Ivatt, R. M., Pogson, J. H., Randle, S. J., Wray, S., Lewis, P. A., Houlden, H., Abramov, A. Y., Hardy, J., Wood, N. W., Whitworth, A. J., Laman, H., & Plun-Favreau, H. (2013). The Parkinson's disease–linked proteins Fbxo7 and Parkin interact to mediate mitophagy. *Nature Neuroscience, 16*(9), 1257–1265. https://doi.org/10.1038/nn.3489

Burger, A., Lindsay, H., Felker, A., Hess, C., Anders, C., Chiavacci, E., Zaugg, J., Weber, L. M., Catena, R., Jinek, M., Robinson, M. D., & Mosimann, C. (2016). Maximizing mutagenesis with solubilized CRISPR-Cas9 ribonucleoprotein complexes. *Development (Cambridge)*, *143*(11), 2025–2037. https://doi.org/10.1242/DEV.134809/VIDEO-1

Capecchi, M. R., Evans, M. J., & Smithies, O. (2001). Generating mice with targeted mutations. *Nature Medicine*, *7*(10), 1086–1090. https://doi.org/10.1038/nm1001-1086

Carrington, B., Bishop, K., & Sood, R. (2022). A comprehensive review of indel detection methods for identification of zebrafish knockout mutants generated by genome-editing nucleases. *Genes*, *13*(5), 857. https://doi.org/10.3390/GENES13050857

Çerçi, B., Uzay, I. A., Kara, M. K., & Dinçer, P. (2023). Clinical trials and promising preclinical applications of CRISPR/Cas gene editing. *Life Sciences*, *312*, 121204. https://doi.org/10.1016/J.LFS.2022.121204

Chang, E. E. S., Ho, P. W. L., Liu, H. F., Pang, S. Y. Y., Leung, C. T., Malki, Y., Choi, Z. Y. K., Ramsden, D. B., & Ho, S. L. (2022). LRRK2 mutant knock-in mouse models: therapeutic relevance in Parkinson's disease. *Translational Neurodegeneration*, *11*(1), 1–19. https://doi.org/10.1186/S40035-022-00285-2

Chatterjee, D., & Krainc, D. (2023). Mechanisms of glucocerebrosidase dysfunction in Parkinson's disease. *Journal of Molecular Biology*, *435*(12), 168023. https://doi.org/10.1016/J.JMB.2023.168023

Chavali, L. N. M., Yddal, I., Bifulco, E., Mannsåker, S., Røise, D., Law, J. O., Frøyset, A. K., Grellscheid, S. N., & Fladmark, K. E. (2023). Progressive motor and non-motor symptoms in park7 knockout zebrafish. *International Journal of Molecular Sciences*, *24*(7), 6456. https://doi.org/10.3390/IJMS24076456/S1

Chebli, J., Rahmati, M., Lashley, T., Edeman, B., Oldfors, A., Zetterberg, H., & Abramsson, A. (2021). The localization of amyloid precursor protein to ependymal cilia in vertebrates and its role in ciliogenesis and brain development in zebrafish. *Scientific Reports, 11*(1), 1–19. https://doi.org/10.1038/s41598-021-98487-7

Chen, K., Wang, P., Chen, J., Ying, Y., Chen, Y., Gilson, E., Lu, Y., & Ye, J. (2022). Loss of atm in zebrafish as a model of ataxia–telangiectasia syndrome. *Biomedicines*, *10*(2), 392. https://doi.org/10.3390/BIOMEDICINES10020392/S1

Chen, L., Cagniard, B., Mathews, T., Jones, S., Koh, H. C., Ding, Y., Carvey, P. M., Ling, Z., Kang, U. J., & Zhuang, X. (2005). Age-dependent motor deficits and dopaminergic dysfunction in DJ-1 null mice. *Journal of Biological Chemistry*, *280*(22), 21418–21426. https://doi.org/10.1074/JBC.M413955200

Chen, P. J., & Liu, D. R. (2022). Prime editing for precise and highly versatile genome manipulation. *Nature Reviews Genetics*, *24*(3), 161–177. https://doi.org/10.1038/s41576-022-00541-1

Chia, K., Klingseisen, A., Sieger, D., & Priller, J. (2022). Zebrafish as a model organism for neurodegenerative disease. *Frontiers in Molecular Neuroscience*, *15*, 940484. https://doi.org/10.3389/FNMOL.2022.940484/BIBTEX

Choi, T. Y., Choi, T. I., Lee, Y. R., Choe, S. K., & Kim, C. H. (2021). Zebrafish as an animal model for biomedical research. *Experimental & Molecular Medicine*, *53*(3), 310–317. https://doi.org/10.1038/s12276-021-00571-5

Cipolat, S., Rudka, T., Hartmann, D., Costa, V., Serneels, L., Craessaerts, K., Metzger, K., Frezza, C., Annaert, W., D'Adamio, L., Derks, C., Dejaegere, T., Pellegrini, L., D'Hooge, R., Scorrano, L., & De Strooper, B. (2006). Mitochondrial rhomboid PARL regulates cytochrome C release during apoptosis via OPA1-dependent cristae remodeling. *Cell*, *126*(1), 163–175. https://doi.org/10.1016/j.cell.2006.06.021

Ciura, S., Lattante, S., Le Ber, I., Latouche, M., Tostivint, H., Brice, A., & Kabashi, E. (2013). Loss of function of C9orf72 causes motor deficits in a zebrafish model of amyotrophic lateral sclerosis. *Annals of Neurology*, *74*(2), 180–187. https://doi.org/10.1002/ANA.23946

Collias, D., & Beisel, C. L. (2021). CRISPR technologies and the search for the PAM-free nuclease. *Nature Communications*, *12*(1), 1–12. https://doi.org/10.1038/s41467-020-20633-y

Collin, J., & Martin, P. (2017). Zebrafish as a research organism: Danio rerio in biomedical research. *Basic Science Methods for Clinical Researchers*, 235–261. https://doi.org/10.1016/B978-0-12-803077-6.00013-8

Dave, K. D., De Silva, S., Sheth, N. P., Ramboz, S., Beck, M. J., Quang, C., Switzer, R. C., Ahmad, S. O., Sunkin, S. M., Walker, D., Cui, X., Fisher, D. A., McCoy, A. M., Gamber, K., Ding, X., Goldberg, M. S., Benkovic, S. A., Haupt, M., Baptista, M. A. S., … Frasier, M. A. (2014). Phenotypic characterization of recessive gene knockout rat models of Parkinson's disease. *Neurobiology of Disease*, *70*, 190–203. https://doi.org/10.1016/J.NBD.2014.06.009

Dawson, T. M., Golde, T. E., & Lagier-Tourenne, C. (2018). Animal models of neurodegenerative diseases. *Nature Neuroscience*, *21*(10), 1370–1379. https://doi.org/10.1038/s41593-018-0236-8

De Strooper, B., & Karran, E. (2016). The cellular phase of Alzheimer's disease. *Cell*, *164*(4), 603–615. https://doi.org/10.1016/J.CELL.2015.12.056

Dehay, B., Ramirez, A., Martinez-Vicente, M., Perier, C., Canron, M. H., Doudnikoff, E., Vital, A., Vila, M., Klein, C., & Bezard, E. (2012). Loss of P-type ATPase ATP13A2/PARK9 function induces general lysosomal deficiency and leads to Parkinson disease neurodegeneration. *Proceedings of the National Academy of Sciences of the United States of America*, *109*(24), 9611–9616. https://doi.org/10.1073/PNAS.1112368109/SUPPL_FILE/PNAS.201112368SI.PDF

Dilliott, A. A., Rouleau, G. A., & Farhan, S. M. K. (2022). Lack of association of TP73 with amyotrophic lateral sclerosis in a large cohort of cases. *Neurobiology of Aging*, *115*, 109–111. https://doi.org/10.1016/J.NEUROBIOLAGING.2022.04.004

Doyle, J. M., & Croll, R. P. (2022). A critical review of zebrafish models of Parkinson's disease. *Frontiers in Pharmacology*, *13*, 835827. https://doi.org/10.3389/FPHAR.2022.835827/BIBTEX

Edson, A. J., Hushagen, H. A., Frøyset, A. K., Elda, I., Khan, E. A., Di Stefano, A., & Fladmark, K. E. (2019). Dysregulation in the brain protein profile of zebrafish lacking the Parkinson's disease-related protein DJ-1. *Molecular Neurobiology*, *56*(12), 8306–8322. https://doi.org/10.1007/S12035-019-01667-W/FIGURES/6

El-Brolosy, M. A., Kontarakis, Z., Rossi, A., Kuenne, C., Günther, S., Fukuda, N., Kikhi, K., Boezio, G. L. M., Takacs, C. M., Lai, S. L., Fukuda, R., Gerri, C., Giraldez, A. J., & Stainier, D. Y. R. (2019). Genetic compensation triggered by mutant mRNA degradation. *Nature*, *568*(7751), 193–197. https://doi.org/10.1038/s41586-019-1064-z

El-Brolosy, M. A., & Stainier, D. Y. R. (2017). Genetic compensation: A phenomenon in search of mechanisms. *PLOS Genetics*, *13*(7), e1006780. https://doi.org/10.1371/JOURNAL.PGEN.1006780

Farazi Fard, M. A., Rebelo, A. P., Buglo, E., Nemati, H., Dastsooz, H., Gehweiler, I., Reich, S., Reichbauer, J., Quintáns, B., Ordóñez-Ugalde, A., Cortese, A., Courel, S., Abreu, L., Powell, E., Danzi, M., Martuscelli, N. B., Bis-Brewer, D. M., Tao, F., Zarei, F., … Faghihi, M. A. (2019). Truncating mutations in UBAP1 cause hereditary spastic paraplegia. *The American Journal of Human Genetics*, *104*(4), 767–773. https://doi.org/10.1016/J.AJHG.2019.03.001

Fett, M. E., Pilsl, A., Paquet, D., van Bebber, F., Haass, C., Tatzelt, J., Schmid, B., & Winklhofer, K. F. (2010). Parkin is protective against proteotoxic stress in a transgenic zebrafish model. *PLoS One*, *5*(7), e11783. https://doi.org/10.1371/JOURNAL.PONE.0011783

Fleming, A., Bourdenx, M., Fujimaki, M., Karabiyik, C., Krause, G. J., Lopez, A., Martín-Segura, A., Puri, C., Scrivo, A., Skidmore, J., Son, S. M., Stamatakou, E., Wrobel, L., Zhu, Y., Cuervo, A. M., & Rubinsztein, D. C. (2022). The different autophagy degradation pathways and neurodegeneration. *Neuron*, *110*(6), 935–966. https://doi.org/10.1016/J.NEURON.2022.01.017

Flinn, L. J., Keatinge, M., Bretaud, S., Mortiboys, H., Matsui, H., De Felice, E., Woodroof, H. I., Brown, L., McTighe, A., Soellner, R., Allen, C. E., Heath, P. R., Milo, M., Muqit, M. M. K., Reichert, A. S., Köster, R. W., Ingham, P. W., & Bandmann, O. (2013). TigarB causes mitochondrial dysfunction and neuronal loss in PINK1 deficiency. *Annals of Neurology*, *74*(6), 837–847. https://doi.org/10.1002/ANA.23999

Flinn, L., Mortiboys, H., Volkmann, K., Kster, R. W., Ingham, P. W., & Bandmann, O. (2009). Complex I deficiency and dopaminergic neuronal cell loss in parkin-deficient zebrafish (Danio rerio). *Brain*, *132*(6), 1613–1623. https://doi.org/10.1093/BRAIN/AWP108

Fonzo, A. D., Dekker, M. C. J., Montagna, P., Baruzzi, A., Yonova, E. H., Guedes, L. C., Szczerbinska, A., Zhao, T., Dubbel-Hulsman, L. O. M., Wouters, C. H., De Graaff, E., Oyen, W. J. G., Simons, E. J., Breedveld, G. J., Oostra, B. A., Horstink, M. W., & Bonifati, V. (2009). FBXO7 mutations cause autosomal recessive, early-onset parkinsonian-pyramidal syndrome. *Neurology*, *72*(3), 240–245. https://doi.org/10.1212/01.WNL.0000338144.10967.2B/ASSET/43B01A16-24A8-45CC-B883-376A466C1ECA/ASSETS/GRAPHIC/9FSM1.GIF

Fu, Y., Foden, J. A., Khayter, C., Maeder, M. L., Reyon, D., Joung, J. K., & Sander, J. D. (2013). High-frequency off-target mutagenesis induced by CRISPR-Cas nucleases in human cells. *Nature Biotechnology*, *31*(9), 822–826. https://doi.org/10.1038/nbt.2623

Gaj, T., Gersbach, C. A., & Barbas, C. F. (2013). ZFN, TALEN, and CRISPR/Cas-based methods for genome engineering. *Trends in Biotechnology*, *31*(7), 397–405. https://doi.org/10.1016/J.TIBTECH.2013.04.004

Gasiunas, G., Young, J. K., Karvelis, T., Kazlauskas, D., Urbaitis, T., Jasnauskaite, M., Grusyte, M. M., Paulraj, S., Wang, P. H., Hou, Z., Dooley, S. K., Cigan, M., Alarcon, C., Chilcoat, N. D., Bigelyte, G., Curcuru, J. L., Mabuchi, M., Sun, Z., Fuchs, R. T., … Siksnys, V. (2020). A catalogue of biochemically diverse CRISPR-Cas9 orthologs. *Nature Communications 2020 11:1*, *11*(1), 1–10. https://doi.org/10.1038/s41467-020-19344-1

Gaudelli, N. M., Komor, A. C., Rees, H. A., Packer, M. S., Badran, A. H., Bryson, D. I., & Liu, D. R. (2017). Programmable base editing of A•T to G•C in genomic DNA without DNA cleavage. *Nature*, *551*(7681), 464–471. https://doi.org/10.1038/nature24644

Gautier, C. A., Kitada, T., & Shen, J. (2008). Loss of PINK1 causes mitochondrial functional defects and increased sensitivity to oxidative stress. *Proceedings of the National Academy of Sciences of the United States of America*, *105*(32), 11364–11369. https://doi.org/10.1073/PNAS.0802076105/SUPPL_FILE/0802076105SI.PDF

Gibb, W. R. G., Narabayashi, H., Yokochi, M., Iizuka, R., & Lees, A. J. (1991). New pathologic observations in juvenile onset parkinsonism with dystonia. *Neurology*, *41*(6), 820–822. https://doi.org/10.1212/WNL.41.6.820

Girard, M., Larivieṙe, R., Parfitt, D. A., Deane, E. C., Gaudet, R., Nossova, N., Blondeau, F., Prenosil, G., Vermeulen, E. G. M., Duchen, M. R., Richter, A., Shoubridge, E. A., Gehring, K., McKinney, R. A., Brais, B., Chapple, J. P., & McPherson, P. S. (2012). Mitochondrial dysfunction and Purkinje cell loss in autosomal recessive spastic ataxia of Charlevoix-Saguenay (ARSACS). *Proceedings of the National Academy of Sciences of the United States of America*, *109*(5), 1661–1666. https://doi.org/10.1073/PNAS.1113166109/SUPPL_FILE/PNAS.201113166SI.PDF

Glass, Z., Lee, M., Li, Y., & Xu, Q. (2018). Engineering the delivery system for crispr-based genome editing. *Trends in Biotechnology*, *36*(2), 173–185. https://doi.org/10.1016/J.TIBTECH.2017.11.006

Gleditzsch, D., Pausch, P., Müller-Esparza, H., Özcan, A., Guo, X., Bange, G., & Randau, L. (2019). PAM identification by CRISPR-Cas effector complexes: Diversified mechanisms and structures. *RNA Biology*, *16*(4), 504–517. https://doi.org/10.1080/15476286.2018.1504546

Goldberg, M. S., Pisani, A., Haburcak, M., Vortherms, T. A., Kitada, T., Costa, C., Tong, Y., Martella, G., Tscherter, A., Martins, A., Bernardi, G., Roth, B. L., Pothos, E. N., Calabresi, P., & Shen, J. (2005). Nigrostriatal dopaminergic deficits and hypokinesia caused by inactivation of the familial parkinsonism-linked gene DJ-1. *Neuron*, *45*(4), 489–496. https://doi.org/10.1016/J.NEURON.2005.01.041

Götz, J., & Ittner, L. M. (2008). Animal models of Alzheimer's disease and frontotemporal dementia. *Nature Reviews Neuroscience*, *9*(7), 532–544. https://doi.org/10.1038/nrn2420

Grunwald, D. J., & Eisen, J. S. (2002). Headwaters of the zebrafish — emergence of a new model vertebrate. *Nature Reviews Genetics*, *3*(9), 717–724. https://doi.org/10.1038/nrg892

Hanna, R. E., & Doench, J. G. (2020). Design and analysis of CRISPR–Cas experiments. *Nature Biotechnology*, *38*(7), 813–823. https://doi.org/10.1038/s41587-020-0490-7

Hardiman, O., Al-Chalabi, A., Chio, A., Corr, E. M., Logroscino, G., Robberecht, W., Shaw, P. J., Simmons, Z., & Van Den Berg, L. H. (2017). Amyotrophic lateral sclerosis. *Nature Reviews Disease Primers*, *3*(1), 1–19. https://doi.org/10.1038/nrdp.2017.71

Hardy, J., & Selkoe, D. J. (2002). The amyloid hypothesis of Alzheimer's disease: Progress and problems on the road to therapeutics. *Science*, *297*(5580), 353–356. https://doi.org/10.1126/SCIENCE.1072994

Harji, Z. (2019). *Preliminary characterization of a novel tardbp knock-in zebrafish model of familial amyotrophic lateral sclerosis*. McGill University Libraries.

Hashimoto, M., Yamashita, Y., & Takemoto, T. (2016). Electroporation of Cas9 protein/sgRNA into early pronuclear zygotes generates non-mosaic mutants in the mouse. *Developmental Biology*, *418*(1), 1–9. https://doi.org/10.1016/J.YDBIO.2016.07.017

Himmelreich, N., Blau, N., & Thöny, B. (2021). Molecular and metabolic bases of tetrahydrobiopterin (BH4) deficiencies. *Molecular Genetics and Metabolism*, *133*(2), 123–136. https://doi.org/10.1016/J.YMGME.2021.04.003

Hinnen, A., Hicks, J. B., & Fink, G. R. (1978). Transformation of yeast. *Proceedings of the National Academy of Sciences of the United States of America*, *75*(4), 1929–1933. https://doi.org/10.1073/PNAS.75.4.1929

Hoche, F., Seidel, K., Theis, M., Vlaho, S., Schubert, R., Zielen, S., & Kieslich, M. (2012). Neurodegeneration in ataxia telangiectasia: What is new? What is evident? *Neuropediatrics*, *43*(3), 119–129. https://doi.org/10.1055/S-0032-1313915/ID/JR110766RA-14/BIB

Huang, M., & Chen, S. (2021). DJ-1 in neurodegenerative diseases: Pathogenesis and clinical application. *Progress in Neurobiology*, *204*, 102114. https://doi.org/10.1016/J.PNEUROBIO.2021.102114

Hughes, G. (2019). Machine learning discriminates parkinsonian movement disorders in zebrafish. *Disease Models & Mechanisms*, *13*(10), dmm045815.

Hughes, G. L., Lones, M. A., Bedder, M., Currie, P. D., Smith, S. L., & Pownall, M. E. (2020). Machine learning discriminates a movement disorder in a zebrafish model of Parkinson's disease. *DMM Disease Models and Mechanisms*, *13*(10). https://doi.org/10.1242/DMM.045815/267205/AM/MACHINE-LEARNING-DISCRIMINATES-A-MOVEMENT-DISORDER

Hwang, W. Y., Fu, Y., Reyon, D., Maeder, M. L., Tsai, S. Q., Sander, J. D., Peterson, R. T., Yeh, J. R. J., & Joung, J. K. (2013). Efficient genome editing in zebrafish using a CRISPR-Cas system. *Nature Biotechnology*, *31*(3), 227–229. https://doi.org/10.1038/nbt.2501

Ikeda, A., Imai, Y., & Hattori, N. (2022). Neurodegeneration-associated mitochondrial proteins, CHCHD2 and CHCHD10–what distinguishes the two? *Frontiers in Cell and Developmental Biology*, *10*, 996061. https://doi.org/10.3389/FCELL.2022.996061/BIBTEX

Jiang, H., Newman, M., & Lardelli, M. (2018). The zebrafish orthologue of familial Alzheimer's disease gene PRESENILIN 2 is required for normal adult melanotic skin pigmentation. *PLoS One*, *13*(10), e0206155. https://doi.org/10.1371/JOURNAL.PONE.0206155

Jinek, M., Chylinski, K., Fonfara, I., Hauer, M., Doudna, J. A., & Charpentier, E. (2012). A programmable dual-RNA-guided DNA endonuclease in adaptive bacterial immunity. *Science*, *337*(6096), 816–821. https://doi.org/10.1126/SCIENCE.1225829/SUPPL_FILE/JINEK.SM.PDF

Johnson, J. O., Glynn, S. M., Raphael Gibbs, J., Nalls, M. A., Sabatelli, M., Restagno, G., Drory, V. E., Chiò, A., Rogaeva, E., & Traynor, B. J. (2014). Mutations in the CHCHD10 gene are a common cause of familial amyotrophic lateral sclerosis. *Brain*, *137*(12), e311–e311. https://doi.org/10.1093/BRAIN/AWU265

Johnson, M. E., Stecher, B., Labrie, V., Brundin, L., & Brundin, P. (2019). Triggers, facilitators, and aggravators: Redefining Parkinson's disease pathogenesis. *Trends in Neurosciences*, *42*(1), 4–13. https://doi.org/10.1016/j.tins.2018.09.007

Joseph, S., Schulz, J. B., & Stegmüller, J. (2018). Mechanistic contributions of FBXO7 to Parkinson disease. *Journal of Neurochemistry*, *144*(2), 118–127. https://doi.org/10.1111/JNC.14253

Kalueff, A. V., Stewart, A. M., & Gerlai, R. (2014). Zebrafish as an emerging model for studying complex brain disorders. *Trends in Pharmacological Sciences*, *35*(2), 63–75. https://doi.org/10.1016/J.TIPS.2013.12.002

Keatinge, M., Bui, H., Menke, A., Chen, Y. C., Sokol, A. M., Bai, Q., Ellett, F., Da Costa, M., Burke, D., Gegg, M., Trollope, L., Payne, T., McTighe, A., Mortiboys, H., De Jager, S., Nuthall, H., Kuo, M. S., Fleming, A., Schapira, A. H. V., … Bandmann, O. (2015). Glucocerebrosidase 1 deficient Danio rerio mirror key pathological aspects of human Gaucher disease and provide evidence of early microglial activation preceding alpha-synuclein-independent neuronal cell death. *Human Molecular Genetics*, *24*(23), 6640–6652. https://doi.org/10.1093/HMG/DDV369

Kim, D., Luk, K., Wolfe, S. A., & Kim, J. S. (2019). Evaluating and enhancing target specificity of gene-editing nucleases and deaminases. *Annual Review of Biochemistry*, *88*, 191–220. https://doi.org/10.1146/ANNUREV-BIOCHEM-013118-111730

Kim, G., Gautier, O., Tassoni-Tsuchida, E., Ma, X. R., & Gitler, A. D. (2020). ALS genetics: Gains, losses, and implications for future therapies. *Neuron*, *108*(5), 822–842. https://doi.org/10.1016/J.NEURON.2020.08.022

Kim, K., Ryu, S. M., Kim, S. T., Baek, G., Kim, D., Lim, K., Chung, E., Kim, S., & Kim, J. S. (2017). Highly efficient RNA-guided base editing in mouse embryos. *Nature Biotechnology*, *35*(5), 435–437. https://doi.org/10.1038/nbt.3816

Kim, R. H., Smith, P. D., Aleyasin, H., Hayley, S., Mount, M. P., Pownall, S., Wakeham, A., You-Ten, A. J., Kalia, S. K., Horne, P., Westaway, D., Lozano, A. M., Anisman, H., Park, D. S., & Mak, T. W. (2005). Hypersensitivity of DJ-1-deficient mice to 1-methyl-4-phenyl-1,2,3,6- tetrahydropyrindine (MPTP) and oxidative stress. *Proceedings of the National Academy of Sciences of the United States of America*, *102*(14), 5215–5220. https://doi.org/10.1073/PNAS.0501282102/SUPPL_FILE/01282FIG6.JPG

Kino, Y., Washizu, C., Kurosawa, M., Yamada, M., Miyazaki, H., Akagi, T., Hashikawa, T., Doi, H., Takumi, T., Hicks, G. G., Hattori, N., Shimogori, T., & Nukina, N. (2015). FUS/TLS deficiency causes behavioral and pathological abnormalities distinct from amyotrophic lateral sclerosis. *Acta Neuropathologica Communications*, *3*(1), 24. https://doi.org/10.1186/S40478-015-0202-6/FIGURES/5

Kitada, T., Pisani, A., Porter, D. R., Yamaguchi, H., Tscherter, A., Martella, G., Bonsi, P., Zhang, C., Pothos, E. N., & Shen, J. (2007). Impaired dopamine release and synaptic plasticity in the striatum of PINK1-deficient mice. *Proceedings of the National Academy of Sciences of the United States of America*, *104*(27), 11441–11446. https://doi.org/10.1073/PNAS.0702717104/SUPPL_FILE/02717FIG6.PDF

Knopman, D. S., Amieva, H., Petersen, R. C., Chételat, G., Holtzman, D. M., Hyman, B. T., Nixon, R. A., & Jones, D. T. (2021). Alzheimer disease. *Nature Reviews Disease Primers*, *7*(1), 1–21. https://doi.org/10.1038/s41572-021-00269-y

Kolb, A. F., Coates, C. J., Kaminski, J. M., Summers, J. B., Miller, A. D., & Segal, D. J. (2005). Site-directed genome modification: nucleic acid and protein modules for targeted integration and gene correction. *Trends in Biotechnology*, *23*(8), 399–406. https://doi.org/10.1016/J.TIBTECH.2005.06.005

Komor, A. C., Kim, Y. B., Packer, M. S., Zuris, J. A., & Liu, D. R. (2016). Programmable editing of a target base in genomic DNA without double-stranded DNA cleavage. *Nature*, *533*(7603), 420–424. https://doi.org/10.1038/nature17946

Kurian, M. A., Gissen, P., Smith, M., Heales, S. J. R., & Clayton, P. T. (2011). The monoamine neurotransmitter disorders: An expanding range of neurological syndromes. *The Lancet Neurology*, *10*(8), 721–733. https://doi.org/10.1016/S1474-4422(11)70141-7

Lagier-Tourenne, C., Polymenidou, M., Hutt, K. R., Vu, A. Q., Baughn, M., Huelga, S. C., Clutario, K. M., Ling, S. C., Liang, T. Y., Mazur, C., Wancewicz, E., Kim, A. S., Watt, A., Freier, S., Hicks, G. G., Donohue, J. P., Shiue, L., Bennett, C. F., Ravits, J., … Yeo, G. W. (2012). Divergent roles of ALS-linked proteins FUS/TLS and TDP-43 intersect in processing long pre-mRNAs. *Nature Neuroscience*, *15*(11), 1488–1497. https://doi.org/10.1038/nn.3230

Larbalestier, H., Keatinge, M., Watson, L., White, E., Gowda, S., Wei, W., Koler, K., Semenova, S. A., Elkin, A. M., Rimmer, N., Sweeney, S. T., Mazzolini, J., Sieger, D., Hide, W., McDearmid, J., Panula, P., MacDonald, R. B., & Bandmann, O. (2022). GCH1 deficiency activates brain innate immune response and impairs tyrosine hydroxylase homeostasis. *Journal of Neuroscience*, *42*(4), 702–716. https://doi.org/10.1523/JNEUROSCI.0653-21.2021

Lebedeva, S., de Jesus Domingues, A. M., Butter, F., & Ketting, R. F. (2017). Characterization of genetic loss-of-function of Fus in zebrafish. *RNA Biology*, *14*(1), 29–35. https://doi.org/10.1080/15476286.2016.1256532

Leibowitz, M. L., Papathanasiou, S., Doerfler, P. A., Blaine, L. J., Sun, L., Yao, Y., Zhang, C. Z., Weiss, M. J., & Pellman, D. (2021). Chromothripsis as an on-target consequence of CRISPR–Cas9 genome editing. *Nature Genetics*, *53*(6), 895–905. https://doi.org/10.1038/s41588-021-00838-7

Lohr, K. M., & Miller, G. W. (2014). VMAT2 and Parkinson's disease: Harnessing the dopamine vesicle. *Expert Review of Neurotherapeutics*, *14*(10), 1115–1117. https://doi.org/10.1586/14737175.2014.960399

Long, J. M., & Holtzman, D. M. (2019). Alzheimer disease: An update on pathobiology and treatment strategies. *Cell*, *179*(2), 312–339. https://doi.org/10.1016/J.CELL.2019.09.001

Lücking, C. B., Dürr, A., Bonifati, V., Vaughan, J., De Michele, G., Gasser, T., Harhangi, B. S., Meco, G., Denèfle, P., Wood, N. W., Agid, Y., Nicholl, D., Breteler, M. M. B., Oostra, B. A., De Mari, M., Marconi, R., Filla, A., Bonnet, A.-M., Broussolle, E., … Brice, A. (2000). Association between early-onset Parkinson's disease and mutations in the Parkin gene. *New England Journal of Medicine*, *342*(21), 1560–1567. https://doi.org/10.1056/NEJM200005253422103

Lysyk, L., Brassard, R., Touret, N., & Lemieux, M. J. (2020). PARL protease: A glimpse at intramembrane proteolysis in the inner mitochondrial membrane. *Journal of Molecular Biology*, *432*(18), 5052–5062. https://doi.org/10.1016/J.JMB.2020.04.006

Ma, Z., Zhu, P., Shi, H., Guo, L., Zhang, Q., Chen, Y., Chen, S., Zhang, Z., Peng, J., & Chen, J. (2019). PTC-bearing mRNA elicits a genetic compensation response via Upf3a and COMPASS components. *Nature*, *568*(7751), 259–263. https://doi.org/10.1038/s41586-019-1057-y

Makarova, K. S., Wolf, Y. I., Iranzo, J., Shmakov, S. A., Alkhnbashi, O. S., Brouns, S. J. J., Charpentier, E., Cheng, D., Haft, D. H., Horvath, P., Moineau, S., Mojica, F. J. M., Scott, D., Shah, S. A., Siksnys, V., Terns, M. P., Venclovas, Č., White, M. F., Yakunin, A. F., … Koonin, E. V. (2019). Evolutionary classification of CRISPR–Cas systems: a burst of class 2 and derived variants. *Nature Reviews Microbiology*, *18*(2), 67–83. https://doi.org/10.1038/s41579-019-0299-x

Matsui, H., Namikawa, K., Babaryka, A., & Köster, R. W. (2014). Functional regionalization of the teleost cerebellum analyzed in vivo. *Proceedings of the National Academy of Sciences of the United States of America*, *111*(32), 11846–11851. https://doi.org/10.1073/PNAS.1403105111/SUPPL_FILE/PNAS.1403105111.SM05.MOV

Mencacci, N. E., consortium, on behalf of the I. P. D. G. C. and U., Isaias, I. U., consortium, on behalf of the I. P. D. G. C. and U., Reich, M. M., consortium, on behalf of the I. P. D. G. C. and U., Ganos, C., consortium, on behalf of the I. P. D. G. C. and U., Plagnol, V., consortium, on behalf of the I. P. D. G. C. and U., Polke, J. M., consortium, on behalf of the I. P. D. G. C. and U., Bras, J., consortium, on behalf of the I. P. D. G. C. and U., Hersheson, J., consortium, on behalf of the I. P. D. G. C. and U., Stamelou, M., consortium, on behalf of the I. P. D. G. C. and U., Pittman, A. M., … consortium, on behalf of the I. P. D. G. C. and U. (2014). Parkinson's disease in GTP cyclohydrolase 1 mutation carriers. *Brain*, *137*(9), 2480–2492. https://doi.org/10.1093/BRAIN/AWU179

Mercuri, E., Sumner, C. J., Muntoni, F., Darras, B. T., & Finkel, R. S. (2022). Spinal muscular atrophy. *Nature Reviews Disease Primers*, *8*(1), 1–16. https://doi.org/10.1038/s41572-022-00380-8

Merhi, R., Kalyn, M., Zhu-Pawlowsky, A., & Ekker, M. (2021). Loss of parla function results in inactivity, olfactory impairment, and dopamine neuron loss in zebrafish. *Biomedicines*, *9*(2), 205. https://doi.org/10.3390/BIOMEDICINES9020205

Meyyazhagan, A., Kuchi Bhotla, H., Pappuswamy, M., & Orlacchio, A. (2022). The puzzle of hereditary spastic paraplegia: From epidemiology to treatment. *International Journal of Molecular Sciences*, *23*(14), 7665. https://doi.org/10.3390/IJMS23147665

Meyyazhagan, A., & Orlacchio, A. (2022). Hereditary spastic paraplegia: An update. *International Journal of Molecular Sciences*, *23*(3), 1697. https://doi.org/10.3390/IJMS23031697

Mojica, F. J. M., Díez-Villaseñor, C., García-Martínez, J., & Almendros, C. (2009). Short motif sequences determine the targets of the prokaryotic CRISPR defence system. *Microbiology*, *155*(3), 733–740. https://doi.org/10.1099/MIC.0.023960-0/CITE/REFWORKS

Moll, U. M., & Slade, N. (2004). p63 and p73: Roles in development and tumor formation. *Molecular Cancer Research*, *2*(7), 371–386. https://doi.org/10.1158/1541-7786.371.2.7

Moreno-Mateos, M. A., Fernandez, J. P., Rouet, R., Vejnar, C. E., Lane, M. A., Mis, E., Khokha, M. K., Doudna, J. A., & Giraldez, A. J. (2017). CRISPR-Cpf1 mediates efficient homology-directed repair and temperature-controlled genome editing. *Nature Communications*, *8*(1), 1–9. https://doi.org/10.1038/s41467-017-01836-2

Murala, S., Nagarajan, E., & Bollu, P. C. (2021). Hereditary spastic paraplegia. *Neurological Sciences*, *42*(3), 883–894. https://doi.org/10.1007/S10072-020-04981-7/TABLES/4

Naef, V., Marchese, M., Ogi, A., Fichi, G., Galatolo, D., Licitra, R., Doccini, S., Verri, T., Argenton, F., Morani, F., & Santorelli, F. M. (2021). Efficient neuroprotective rescue of sacsin-related disease phenotypes in zebrafish. *International Journal of Molecular Sciences*, *22*(16), 8401. https://doi.org/10.3390/IJMS22168401/S1

Naert, T., & Vleminckx, K. (2018). CRISPR/Cas9 disease models in zebrafish and Xenopus: The genetic renaissance of fish and frogs. *Drug Discovery Today: Technologies*, *28*, 41–52. https://doi.org/10.1016/J.DDTEC.2018.07.001

Nahmad, A. D., Reuveni, E., Goldschmidt, E., Tenne, T., Liberman, M., Horovitz-Fried, M., Khosravi, R., Kobo, H., Reinstein, E., Madi, A., Ben-David, U., & Barzel, A. (2022). Frequent aneuploidy in primary human T cells after CRISPR–Cas9 cleavage. *Nature Biotechnology*, *40*(12), 1807–1813. https://doi.org/10.1038/s41587-022-01377-0

Nakamura, M., Gao, Y., Dominguez, A. A., & Qi, L. S. (2021). CRISPR technologies for precise epigenome editing. *Nature Cell Biology*, *23*(1), 11–22. https://doi.org/10.1038/s41556-020-00620-7

Nalls, M. A., Pankratz, N., Lill, C. M., Do, C. B., Hernandez, D. G., Saad, M., Destefano, A. L., Kara, E., Bras, J., Sharma, M., Schulte, C., Keller, M. F., Arepalli, S., Letson, C., Edsall, C., Stefansson, H., Liu, X., Pliner, H., Lee, J. H., … Ansorge, O. (2014). Large-scale meta-analysis of genome-wide association data identifies six new risk loci for Parkinson's disease. *Nature Genetics*, *46*(9), 989–993. https://doi.org/10.1038/ng.3043

Nambiar, T. S., Baudrier, L., Billon, P., & Ciccia, A. (2022). CRISPR-based genome editing through the lens of DNA repair. *Molecular Cell*, *82*(2), 348–388. https://doi.org/10.1016/J.MOLCEL.2021.12.026

Nishimasu, H., Shi, X., Ishiguro, S., Gao, L., Hirano, S., Okazaki, S., Noda, T., Abudayyeh, O. O., Gootenberg, J. S., Mori, H., Oura, S., Holmes, B., Tanaka, M., Seki, M., Hirano, H., Aburatani, H., Ishitani, R., Ikawa, M., Yachie, N., … Nureki, O. (2018). Engineered CRISPR-Cas9 nuclease with expanded targeting space. *Science*, *361*(6408), 1259–1262. https://doi.org/10.1126/SCIENCE.AAS9129/SUPPL_FILE/PAPV2.PDF

Noble, S., Ismail, A., Godoy, R., Xi, Y., & Ekker, M. (2012). Zebrafish Parla- and Parlb-deficiency affects dopaminergic neuron patterning and embryonic survival. *Journal of Neurochemistry*, *122*(1), 196–207. https://doi.org/10.1111/J.1471-4159.2012.07758.X

Noda, S., Sato, S., Fukuda, T., Tada, N., Uchiyama, Y., Tanaka, K., & Hattori, N. (2020). Loss of Parkin contributes to mitochondrial turnover and dopaminergic neuronal loss in aged mice. *Neurobiology of Disease*, *136*, 104717. https://doi.org/10.1016/J.NBD.2019.104717

Nyuzuki, H., Ito, S., Nagasaki, K., Nitta, Y., Matsui, N., Saitoh, A., & Matsui, H. (2020). Degeneration of dopaminergic neurons and impaired intracellular trafficking in Atp13a2 deficient zebrafish. *IBRO Reports*, *9*, 1–8. https://doi.org/10.1016/J.IBROR.2020.05.002

Onishi, M., Yamano, K., Sato, M., Matsuda, N., & Okamoto, K. (2021). Molecular mechanisms and physiological functions of mitophagy. *The EMBO Journal*, *40*(3), e104705. https://doi.org/10.15252/EMBJ.2020104705

Park, J., Sung, Y. K., Cha, G. H., Sung, B. L., Kim, S., & Chung, J. (2005). Drosophila DJ-1 mutants show oxidative stress-sensitive locomotive dysfunction. *Gene*, *361*(1–2), 133–139. https://doi.org/10.1016/J.GENE.2005.06.040

Patton, E. E., Zon, L. I., & Langenau, D. M. (2021). Zebrafish disease models in drug discovery: From preclinical modelling to clinical trials. *Nature Reviews Drug Discovery*, *20*(8), 611–628. https://doi.org/10.1038/s41573-021-00210-8

Paunovska, K., Loughrey, D., & Dahlman, J. E. (2022). Drug delivery systems for RNA therapeutics. *Nature Reviews Genetics*, *23*(5), 265–280. https://doi.org/10.1038/s41576-021-00439-4

Pellizzoni, L., Yong, J., & Dreyfuss, G. (2002). Essential role for the SMN complex in the specificity of snRNP assembly. *Science (New York, N.Y.)*, *298*(5599), 1775–1779. https://doi.org/10.1126/SCIENCE.1074962

Petel Légaré, V., Rampal, C. J., Gurberg, T. J. N., Aaltonen, M. J., Janer, A., Zinman, L., Shoubridge, E. A., & Armstrong, G. A. B. (2023). Loss of mitochondrial Chchd10 or Chchd2 in zebrafish leads to an ALS-like phenotype and complex I deficiency independent of the mitochondrial integrated stress response. *Developmental Neurobiology*, *83*(1–2), 54–69. https://doi.org/10.1002/DNEU.22909

Petri, K., Zhang, W., Ma, J., Schmidts, A., Lee, H., Horng, J. E., Kim, D. Y., Kurt, I. C., Clement, K., Hsu, J. Y., Pinello, L., Maus, M. V., Joung, J. K., & Yeh, J. R. J. (2021). CRISPR prime editing with ribonucleoprotein complexes in zebrafish and primary human cells. *Nature Biotechnology*, *40*(2), 189–193. https://doi.org/10.1038/s41587-021-00901-y

Pickar-Oliver, A., & Gersbach, C. A. (2019). The next generation of CRISPR–Cas technologies and applications. *Nature Reviews Molecular Cell Biology*, *20*(8), 490–507. https://doi.org/10.1038/s41580-019-0131-5

Pickrell, A. M., & Youle, R. J. (2015). The roles of PINK1, parkin, and mitochondrial fidelity in Parkinson's disease. *Neuron*, *85*(2), 257–273. https://doi.org/10.1016/J.NEURON.2014.12.007

Poewe, W., Seppi, K., Tanner, C. M., Halliday, G. M., Brundin, P., Volkmann, J., Schrag, A. E., & Lang, A. E. (2017). Parkinson disease. *Nature Reviews Disease Primers*, *3*(1), 1–21. https://doi.org/10.1038/nrdp.2017.13

Prabhudesai, S., Bensabeur, F. Z., Abdullah, R., Basak, I., Baez, S., Alves, G., Holtzman, N. G., Larsen, J. P., & Møller, S. G. (2016). LRRK2 knockdown in zebrafish causes developmental defects, neuronal loss, and synuclein aggregation. *Journal of Neuroscience Research*, *94*(8), 717–735. https://doi.org/10.1002/JNR.23754

Préville, M., Zaouter, C., & Patten, S. A. (2020). Generating a CRISPR/Cas9 zebrafish model to unravel SMA pathogenesis and for drug discovery. *11e European Zebrafish Meeting*. https://www.zebrafish2020.org/files/Zebrafish2020_Program_and_Abstract_Book.pdf

Quek, H., Luff, J., Cheung, K., Kozlov, S., Gatei, M., Lee, C. S., Bellingham, M. C., Noakes, P. G., Lim, Y. C., Barnett, N. L., Dingwall, S., Wolvetang, E., Mashimo, T., Roberts, T. L., & Lavin, M. F. (2017). A rat model of ataxia-telangiectasia: Evidence for a neurodegenerative phenotype. *Human Molecular Genetics*, *26*(1), 109–123. https://doi.org/10.1093/HMG/DDW371

Rana, H. Q., Balwani, M., Bier, L., & Alcalay, R. N. (2012). Age-specific Parkinson disease risk in GBA mutation carriers: information for genetic counseling. *Genetics in Medicine*, *15*(2), 146–149. https://doi.org/10.1038/gim.2012.107

Ren, G., Xin, S., Li, S., Zhong, H., & Lin, S. (2011). Disruption of LRRK2 does not cause specific loss of dopaminergic neurons in zebrafish. *PLoS One*, *6*(6), e20630. https://doi.org/10.1371/JOURNAL.PONE.0020630

Repici, M., & Giorgini, F. (2019). DJ-1 in Parkinson's disease: Clinical insights and therapeutic perspectives. *Journal of Clinical Medicine*, *8*(9), 1377. https://doi.org/10.3390/JCM8091377

Rimkus, S. A., Katzenberger, R. J., Trinh, A. T., Dodson, G. E., Tibbetts, R. S., & Wassarman, D. A. (2008). Mutations in String/CDC25 inhibit cell cycle re-entry and neurodegeneration in a Drosophila model of Ataxia telangiectasia. *Genes & Development*, *22*(9), 1205–1220. https://doi.org/10.1101/GAD.1639608

Romano, L. E. L., Aw, W. Y., Hixson, K. M., Novoselova, T. V., Havener, T. M., Howell, S., Taylor-Blake, B., Hall, C. L., Xing, L., Beri, J., Nethisinghe, S., Perna, L., Hatimy, A., Altadonna, G. C., Graves, L. M., Herring, L. E., Hickey, A. J., Thalassinos, K., Chapple, J. P., & Wolter, J. M. (2022). Multi-omic profiling reveals the ataxia protein sacsin is required for integrin trafficking and synaptic organization. *Cell Reports*, *41*(5), 111580. https://doi.org/10.1016/J.CELREP.2022.111580

Rosello, M., Serafini, M., Concordet, J.-P., Filippo, &, & Bene, D. (2023). Precise mutagenesis in zebrafish using cytosine base editors. *Nature Protocols*, *18*(9), 2794–2813. https://doi.org/10.1038/s41596-023-00854-3

Rothblum-Oviatt, C., Wright, J., Lefton-Greif, M. A., McGrath-Morrow, S. A., Crawford, T. O., & Lederman, H. M. (2016). Ataxia telangiectasia: a review. *Orphanet Journal of Rare Diseases*, *11*(1), 1–21. https://doi.org/10.1186/S13023-016-0543-7

Rothstein, R. J. (1983). [12] One-step gene disruption in yeast. *Methods in Enzymology*, *101*(C), 202–211. https://doi.org/10.1016/0076-6879(83)01015-0

Rounding, N.P. (2018). *Zebrafish C9orf72 loss-of-function models of amyotrophic lateral sclerosis and frontotemporal dementia*. University of Sheffield.

Russell, K. L., Downie, J. M., Gibson, S. B., Tsetsou, S., Keefe, M. D., Duran, J. A., Figueroa, K. P., Bromberg, M. B., Murtaugh, L. C., Bonkowsky, J. L., Pulst, S. M., & Jorde, L. B. (2021). Pathogenic effect of TP73 gene variants in people with amyotrophic lateral sclerosis. *Neurology*, *97*(3), e225–e235. https://doi.org/10.1212/WNL.0000000000012285

Sallinen, V., Kolehmainen, J., Priyadarshini, M., Toleikyte, G., Chen, Y. C., & Panula, P. (2010). Dopaminergic cell damage and vulnerability to MPTP in Pink1 knockdown zebrafish. *Neurobiology of Disease*, *40*(1), 93–101. https://doi.org/10.1016/J.NBD.2010.06.001

Sauer, B. (1994). Site-specific recombination: Developments and applications. *Current Opinion in Biotechnology*, *5*(5), 521–527. https://doi.org/10.1016/0958-1669(94)90068-X

Scheltens, P., De Strooper, B., Kivipelto, M., Holstege, H., Chételat, G., Teunissen, C. E., Cummings, J., & van der Flier, W. M. (2021). Alzheimer's disease. *The Lancet*, *397*(10284), 1577–1590. https://doi.org/10.1016/S0140-6736(20)32205-4

Schmid, B., Hruscha, A., Hogl, S., Banzhaf-Strathmann, J., Strecker, K., Van Der Zee, J., Teucke, M., Eimer, S., Hegermann, J., Kittelmann, M., Kremmer, E., Cruts, M., Solchenberger, B., Hasenkamp, L., Van Bebber, F., Van Broeckhoven, C., Edbauer, D., Lichtenthaler, S. F., & Haass, C. (2013). Loss of ALS-associated TDP-43 in zebrafish causes muscle degeneration, vascular dysfunction, and reduced motor neuron axon outgrowth. *Proceedings of the National Academy of Sciences of the United States of America*, *110*(13), 4986–4991. https://doi.org/10.1073/PNAS.1218311110/SUPPL_FILE/SM05.MOV

Schrank, B., Götz, R., Gunnersen, J. M., Ure, J. M., Toyka, K. V., Smith, A. G., & Sendtner, M. (1997). Inactivation of the survival motor neuron gene, a candidate gene for human spinal muscular atrophy, leads to massive cell death in early mouse embryos. *Proceedings of the National Academy of Sciences*, *94*(18), 9920–9925. https://doi.org/10.1073/PNAS.94.18.9920

Seegobin, S. P., Heaton, G. R., Liang, D., Choi, I., Blanca Ramirez, M., Tang, B., & Yue, Z. (2020). Progress in LRRK2-associated Parkinson's disease animal models. *Frontiers in Neuroscience*, *14*, 539964. https://doi.org/10.3389/FNINS.2020.00674/BIBTEX

Segal, D. J., & Meckler, J. F. (2013). Genome engineering at the dawn of the golden age. *Annual Review of Genomics and Human Genetics*, *14*, 135–158. https://doi.org/10.1146/ANNUREV-GENOM-091212-153435

Sharma, A., Lyashchenko, A. K., Lu, L., Nasrabady, S. E., Elmaleh, M., Mendelsohn, M., Nemes, A., Tapia, J. C., Mentis, G. Z., & Shneider, N. A. (2016). ALS-associated mutant FUS induces selective motor neuron degeneration through toxic gain of function. *Nature Communications*, *7*(1), 1–14. https://doi.org/10.1038/ncomms10465

Sheng, D., Qu, D., Kwok, K. H. H., Ng, S. S., Lim, A. Y. M., Aw, S. S., Lee, C. W. H., Sung, W. K., Tan, E. K., Lufkin, T., Jesuthasan, S., Sinnakaruppan, M., & Liu, J. (2010). Deletion of the WD40 domain of LRRK2 in zebrafish causes Parkinsonism-like loss of neurons and locomotive defect. *PLOS Genetics*, *6*(4), e1000914. https://doi.org/10.1371/JOURNAL.PGEN.1000914

Shi, G., Lee, J. R., Grimes, D. A., Racacho, L., Ye, D., Yang, H., Ross, O. A., Farrer, M., McQuibban, G. A., & Bulman, D. E. (2011). Functional alteration of PARL contributes to mitochondrial dysregulation in Parkinson's disease. *Human Molecular Genetics*, *20*(10), 1966–1974. https://doi.org/10.1093/HMG/DDR077

Shiloh, Y., & Ziv, Y. (2013). The ATM protein kinase: Regulating the cellular response to genotoxic stress, and more. *Nature Reviews Molecular Cell Biology*, *14*(4), 197–210. https://doi.org/10.1038/nrm3546

Shmakov, S., Smargon, A., Scott, D., Cox, D., Pyzocha, N., Yan, W., Abudayyeh, O. O., Gootenberg, J. S., Makarova, K. S., Wolf, Y. I., Severinov, K., Zhang, F., & Koonin, E. V. (2017). Diversity and evolution of class 2 CRISPR–Cas systems. *Nature Reviews Microbiology*, *15*(3), 169–182. https://doi.org/10.1038/nrmicro.2016.184

Shribman, S., Reid, E., Crosby, A. H., Houlden, H., & Warner, T. T. (2019). Hereditary spastic paraplegia: from diagnosis to emerging therapeutic approaches. *The Lancet Neurology*, *18*(12), 1136–1146. https://doi.org/10.1016/S1474-4422(19)30235-2

Sidransky, E., Nalls, M. A., Aasly, J. O., Aharon-Peretz, J., Annesi, G., Barbosa, E. R., Bar-Shira, A., Berg, D., Bras, J., Brice, A., Chen, C.-M., Clark, L. N., Condroyer, C., De Marco, E. V., Dürr, A., Eblan, M. J., Fahn, S., Farrer, M. J., Fung, H.-C., … Ziegler, S. G. (2009). Multicenter analysis of glucocerebrosidase mutations in Parkinson's disease. *New England Journal of Medicine*, *361*(17), 1651–1661. https://doi.org/10.1056/NEJMOA0901281/SUPPL_FILE/NEJM_SIDRANSKY_1651SA1.PDF

Silva, G., Poirot, L., Galetto, R., Smith, J., Montoya, G., Duchateau, P., & Paques, F. (2011). Meganucleases and other tools for targeted genome engineering: Perspectives and challenges for gene therapy. *Current Gene Therapy*, *11*(1), 11–27. https://doi.org/10.2174/156652311794520111

Singh, J., & Patten, S. A. (2022). Modeling neuromuscular diseases in zebrafish. *Frontiers in Molecular Neuroscience*, *15*, 1054573. https://doi.org/10.3389/FNMOL.2022.1054573/BIBTEX

Singh, P., Schimenti, J. C., & Bolcun-Filas, E. (2015). A mouse geneticist's practical guide to CRISPR applications. *Genetics*, *199*(1), 1–15. https://doi.org/10.1534/GENETICS.114.169771/-/DC1

Singh, R. N., Howell, M. D., Ottesen, E. W., & Singh, N. N. (2017). Diverse role of survival motor neuron protein. *Biochimica et Biophysica Acta (BBA) - Gene Regulatory Mechanisms*, *1860*(3), 299–315. https://doi.org/10.1016/J.BBAGRM.2016.12.008

Smeyers, J., Banchi, E. G., & Latouche, M. (2021). C9ORF72: What it is, what it does, and why it matters. *Frontiers in Cellular Neuroscience*, *15*, 661447. https://doi.org/10.3389/FNCEL.2021.661447/BIBTEX

Stainier, D. Y. R., Raz, E., Lawson, N. D., Ekker, S. C., Burdine, R. D., Eisen, J. S., Ingham, P. W., Schulte-Merker, S., Yelon, D., Weinstein, B. M., Mullins, M. C., Wilson, S. W., Ramakrishnan, L., Amacher, S. L., Neuhauss, S. C. F., Meng, A., Mochizuki, N., Panula, P., & Moens, C. B. (2017). Guidelines for morpholino use in zebrafish. *PLoS Genetics*, *13*(10), e1007000. https://doi.org/10.1371/JOURNAL.PGEN.1007000

Stirnemann, J. Ô., Belmatoug, N., Camou, F., Serratrice, C., Froissart, R., Caillaud, C., Levade, T., Astudillo, L., Serratrice, J., Brassier, A., Rose, C., De Villemeur, T. B., & Berger, M. G. (2017). A review of gaucher disease pathophysiology, clinical presentation and treatments. *International Journal of Molecular Sciences*, *18*(2), 441. https://doi.org/10.3390/IJMS18020441

Su, M. Y., Fromm, S. A., Remis, J., Toso, D. B., & Hurley, J. H. (2021). Structural basis for the ARF GAP activity and specificity of the C9orf72 complex. *Nature Communications*, *12*(1), 1–8. https://doi.org/10.1038/s41467-021-24081-0

Suzzi, S., Ahrendt, R., Hans, S., Semenova, S. A., Chekuru, A., Wirsching, P., Kroehne, V., Bilican, S., Sayed, S., Winkler, S., Spieß, S., Machate, A., Kaslin, J., Panula, P., & Brand, M. (2021). Deletion of lrrk2 causes early developmental abnormalities and age-dependent increase of monoamine catabolism in the zebrafish brain. *PLoS Genetics*, *17*(9), e1009794. https://doi.org/10.1371/JOURNAL.PGEN.1009794

Sveinsdóttir, H. S., Decker, A., Christensen, C., Lucena, P. B., Þorsteinsson, H., Richert, E., Maier, V. H., Cornell, R., & Karlsson, K. Æ. (2022). Motility phenotype in a zebrafish vmat2 mutant. *PLoS One*, *17*(1), e0259753. https://doi.org/10.1371/JOURNAL.PONE.0259753

Swarts, D. C., & Jinek, M. (2018). Cas9 versus Cas12a/Cpf1: Structure–function comparisons and implications for genome editing. *Wiley Interdisciplinary Reviews: RNA*, *9*(5), e1481. https://doi.org/10.1002/WRNA.1481

Szaruga, M., Veugelen, S., Benurwar, M., Lismont, S., Sepulveda-Falla, D., Lleo, A., Ryan, N. S., Lashley, T., Fox, N. C., Murayama, S., Gijsen, H., De Strooper, B., & Chávez-Gutiérrez, L. (2015). Qualitative changes in human γ-secretase underlie familial Alzheimer's disease. *Journal of Experimental Medicine*, *212*(12), 2003–2013. https://doi.org/10.1084/JEM.20150892

Taha, E. A., Lee, J., & Hotta, A. (2022). Delivery of CRISPR-Cas tools for in vivo genome editing therapy: Trends and challenges. *Journal of Controlled Release*, *342*, 345–361. https://doi.org/10.1016/J.JCONREL.2022.01.013

Tao, J., Bauer, D. E., & Chiarle, R. (2023). Assessing and advancing the safety of CRISPR-Cas tools: From DNA to RNA editing. *Nature Communications*, *14*(1), 1–16. https://doi.org/10.1038/s41467-023-35886-6

Tay, S. H., Ellieyana, E. N., Le, Y., Sarusie, M. V., Grimm, C., Ohmer, J., Mathuru, A. S., Fischer, U., & Winkler, C. (2021). A novel zebrafish model for intermediate type spinal muscular atrophy demonstrates importance of Smn for maintenance of mature motor neurons. *Human Molecular Genetics*, *30*(24), 2488–2502. https://doi.org/10.1093/HMG/DDAB212

Tolosa, E., Vila, M., Klein, C., & Rascol, O. (2020). LRRK2 in Parkinson disease: Challenges of clinical trials. *Nature Reviews Neurology*, *16*(2), 97–107. https://doi.org/10.1038/s41582-019-0301-2

Usenovic, M., Tresse, E., Mazzulli, J. R., Taylor, J. P., & Krainc, D. (2012). Deficiency of ATP13A2 leads to lysosomal dysfunction, α-synuclein accumulation, and neurotoxicity. *Journal of Neuroscience*, *32*(12), 4240–4246. https://doi.org/10.1523/JNEUROSCI.5575-11.2012

van Haasteren, J., Li, J., Scheideler, O. J., Murthy, N., & Schaffer, D. V. (2020). The delivery challenge: Fulfilling the promise of therapeutic genome editing. *Nature Biotechnology*, *38*(7), 845–855. https://doi.org/10.1038/s41587-020-0565-5

Van Rheenen, W., Shatunov, A., Dekker, A. M., McLaughlin, R. L., Diekstra, F. P., Pulit, S. L., Van Der Spek, R. A. A., Võsa, U., De Jong, S., Robinson, M. R., Yang, J., Fogh, I., Van Doormaal, P. T. C., Tazelaar, G. H. P., Koppers, M., Blokhuis, A. M., Sproviero, W., Jones, A. R., Kenna, K. P., … Veldink, J. H. (2016). Genome-wide association analyses identify new risk variants and the genetic architecture of amyotrophic lateral sclerosis. *Nature Genetics*, *48*(9), 1043–1048. https://doi.org/10.1038/ng.3622

Varshney, G. K., Pei, W., Lafave, M. C., Idol, J., Xu, L., Gallardo, V., Carrington, B., Bishop, K., Jones, M., Li, M., Harper, U., Huang, S. C., Prakash, A., Chen, W., Sood, R., Ledin, J., & Burgess, S. M. (2015). High-throughput gene targeting and phenotyping in zebrafish using CRISPR/Cas9. *Genome Research*, *25*(7), 1030–1042. https://doi.org/10.1101/GR.186379.114

Vingill, S., Brockelt, D., Lancelin, C., Tatenhorst, L., Dontcheva, G., Preisinger, C., Schwedhelm-Domeyer, N., Joseph, S., Mitkovski, M., Goebbels, S., Nave, K., Schulz, J. B., Marquardt, T., Lingor, P., & Stegmüller, J. (2016). Loss of FBXO 7 ( PARK 15) results in reduced proteasome activity and models a parkinsonism-like phenotype in mice . *The EMBO Journal*, *35*(18), 2008–2025. https://doi.org/10.15252/EMBJ.201593585/SUPPL_FILE/EMBJ201593585-SUP-0007-SDATAFIG6.PDF

Wang, J., & Cao, H. (2021). Zebrafish and medaka: Important animal models for human neurodegenerative diseases. *International Journal of Molecular Sciences*, *22*(19), 10766. https://doi.org/10.3390/IJMS221910766

Wang, J. Y., & Doudna, J. A. (2023). CRISPR technology: A decade of genome editing is only the beginning. *Science*, *379*(6629). https://doi.org/10.1126/SCIENCE.ADD8643/ASSET/61B6D348-788F-4A7A-9EFA-AADEF0FF5842/ASSETS/IMAGES/LARGE/SCIENCE.ADD8643-FA.JPG

Wang, J. Y., Pausch, P., & Doudna, J. A. (2022). Structural biology of CRISPR–Cas immunity and genome editing enzymes. *Nature Reviews Microbiology*, *20*(11), 641–656. https://doi.org/10.1038/s41579-022-00739-4

Wilson, D. M., Cookson, M. R., Van Den Bosch, L., Zetterberg, H., Holtzman, D. M., & Dewachter, I. (2023). Hallmarks of neurodegenerative diseases. *Cell*, *186*(4), 693–714. https://doi.org/10.1016/J.CELL.2022.12.032

Wint, J. M., & Sirotkin, H. I. (2020). Lrrk2 modulation of Wnt signaling during zebrafish development. *Journal of Neuroscience Research*, *98*(10), 1831–1842. https://doi.org/10.1002/JNR.24687

Wirth, B., Karakaya, M., Kye, M. J., & Mendoza-Ferreira, N. (2020). Twenty-five years of spinal muscular atrophy research: From phenotype to genotype to therapy, and what comes next. *Annual Review of Genomics and Human Genetics*, *21*, 231–261. https://doi.org/10.1146/ANNUREV-GENOM-102319-103602

Woo, J. A. A., Liu, T., Trotter, C., Fang, C. C., De Narvaez, E., Lepochat, P., Maslar, D., Bukhari, A., Zhao, X., Deonarine, A., Westerheide, S. D., & Kang, D. E. (2017). Loss of function CHCHD10 mutations in cytoplasmic TDP-43 accumulation and synaptic integrity. *Nature Communications*, *8*(1), 1–15. https://doi.org/10.1038/ncomms15558

Wu, R. S., Lam, I. I., Clay, H., Duong, D. N., Deo, R. C., & Coughlin, S. R. (2018). A rapid method for directed gene knockout for screening in G0 zebrafish. *Developmental Cell*, *46*(1), 112-125.e4. https://doi.org/10.1016/J.DEVCEL.2018.06.003

Wüst, R., Maurer, B., Hauser, K., Woitalla, D., Sharma, M., & Krüger, R. (2016). Mutation analyses and association studies to assess the role of the presenilin-associated rhomboid-like gene in Parkinson's disease. *Neurobiology of Aging*, *39*, 217.e13-217.e15. https://doi.org/10.1016/J.NEUROBIOLAGING.2015.11.025

Xi, Y., Ryan, J., Noble, S., Yu, M., Yilbas, A. E., & Ekker, M. (2010). Impaired dopaminergic neuron development and locomotor function in zebrafish with loss of pink1 function. *European Journal of Neuroscience*, *31*(4), 623–633. https://doi.org/10.1111/J.1460-9568.2010.07091.X

Yang, P., Lee, A. Y. T., Xue, J., Chou, S. J., Lee, C., Tseng, P., Zhang, T. X., Zhu, Y., Lee, J., Chiou, S. H., & Tseng, H. R. (2022). Nano-vectors for CRISPR/Cas9-mediated genome editing. *Nano Today*, *44*, 101482. https://doi.org/10.1016/J.NANTOD.2022.101482

Ye, H., Robak, L. A., Yu, M., Cykowski, M., & Shulman, J. M. (2023). Genetics and pathogenesis of Parkinson's syndrome. *Annual Review of Pathology: Mechanisms of Disease*, *18*, 95–121. https://doi.org/10.1146/ANNUREV-PATHMECHDIS-031521-034145

Yoshimi, K., & Mashimo, T. (2022). Genome editing technology and applications with the type I CRISPR system. *Gene and Genome Editing*, *3–4*, 100013. https://doi.org/10.1016/J.GGEDIT.2022.100013

Zhang, S., Shen, J., Li, D., & Cheng, Y. (2021). Strategies in the delivery of Cas9 ribonucleoprotein for CRISPR/Cas9 genome editing. *Theranostics*, *11*(2), 614. https://doi.org/10.7150/THNO.47007

Zhang, Y., Qin, W., Lu, X., Xu, J., Huang, H., Bai, H., Li, S., & Lin, S. (2017). Programmable base editing of zebrafish genome using a modified CRISPR-Cas9 system. *Nature Communications*, *8*(1), 1–5. https://doi.org/10.1038/s41467-017-00175-6

Zhao, T., Zondervan-van der Linde, H., Severijnen, L. A., Oostra, B. A., Willemsen, R., & Bonifati, V. (2012). Dopaminergic neuronal loss and dopamine-dependent locomotor defects in Fbxo7-deficient zebrafish. *PLoS One*, *7*(11), e48911. https://doi.org/10.1371/JOURNAL.PONE.0048911

Zhong, Y., Li, J., Ye, M., & Jin, X. (2023). The characteristics of FBXO7 and its role in human diseases. *Gene*, *851*, 146972. https://doi.org/10.1016/J.GENE.2022.146972

Zhou, Z. D., Xie, S. P., Sathiyamoorthy, S., Saw, W. T., Sing, T. Y., Ng, S. H., Chua, H. P. H., Tang, A. M. Y., Shaffra, F., Li, Z., Wang, H., Ho, P. G. H., Lai, M. K. P., Angeles, D. C., Lim, T. M., & Tan, E. K. (2015). F-box protein 7 mutations promote protein aggregation in mitochondria and inhibit mitophagy. *Human Molecular Genetics*, *24*(22), 6314–6330. https://doi.org/10.1093/HMG/DDV340

# 17

## *Zebrafish Genetic Models of Alzheimer's Disease*

**Goksu Kasarci Kavsara, Baris Ertugrul, Sinem Bireller and Bedia Cakmakoglu**

### Abbreviations

| | |
|---|---|
| **AD** | Alzheimer's Disease |
| **APP** | Amyloid Precursor Protein |
| **APOE** | Human Apolipoprotein E |
| **apoE** | Zebrafish Apolipoprotein E |
| **ApoE** | Mice Apolipoprotein E |
| **BBB** | Blood–Brain Barrier |
| **CNS** | Central Nervous System |
| **FAD** | Familial AD |
| **MAPT** | Microtubule-Associated Protein Tau |
| **NFT** | Neurofibrillary Tangles |
| **PSEN1** | Presenilin 1 |
| **PSEN2** | Presenilin 2 |
| **PEN2** | Presenilin Enhancer 2 |
| **Psen1** | Zebrafish Homolog |
| **Psen2** | Zebrafish Homolog |
| **N140fs** | Psen2N140fs |
| **T141_L142delinsMISLISV** | Psen2T141_L142delinsMISLISV |
| **SAD** | Sporadic AD |
| **Aβ** | β-Amyloid |

### Introduction

Aging is a multifactorial process with chronological, social, biological, and psychological dimensions that cannot be prevented. Due to the gradual increase in the world's population, especially over the age 65, old age and the difficulties of old age are becoming one of the important issues in all societies.

According to the World Health Organization, since 2019, 9.1% of the world's population has consisted of individuals aged 65 and over, and the old age population is expected to increase to 15% by 2050. Dementias, which constitute the most common disease group in old age, occur due to many different factors, and approximately 60%–70% of them are Alzheimer-type senile dementias (Alzheimer's disease facts and figures, 2023).

Alzheimer's disease (AD) was first described in 1907 by Alois Alzheimer, who demonstrated the neurofibrillary bundles and senile plaques that characterized the disease in a woman who died of dementia at the age of 51. Clinical genetic, pathological, and neurochemical studies pointed out that AD has a multifactorial underlying scale and is a common neurodegenerative disease. Moreover, it is a progressive disease that causes a decrease in cognitive abilities, memory loss, and deterioration in thinking and language skills. Age is a risk factor for AD, which is why the incidence of the disease increases significantly in individuals over the age of 65. Moreover, genetic factors (especially the *APOE* gene), familial history, general health status, and lifestyle might affect and contribute to the risk of AD.

The main pathological features of AD are known as brain atrophy caused by neuron loss, intracellular neurofibrillary tangles (NFT), extracellular senile plaques, and cerebrovascular amyloidosis (Sehar et al., 2022; Sehar et al., 2023). In addition to these features, the disease is characterized by abnormal protein accumulations in the brain in the form of beta amyloid plaques and helical fibrils formed by tau protein in neurons.

The diagnosis relies on cognitive tests, neurological examination, imaging techniques (MRI, PET scans), and cerebrospinal fluid analysis. Early diagnosis is very important not only for the management and the treatment of the disease but also for minimizing its effects on the quality of life of individuals. Even though the techniques are increasingly improved day by day, there is still no actual and punctual treatment for AD yet that eliminates all the symptoms. However, some medications might relieve the symptoms and make the progression of the disease slower.

AD research is still an ongoing and popular subject not only because the treatment options offer limited therapeutic approaches, but also because the advancing technology provides the researchers several opportunities with new and different perspectives such as using zebrafish *(Danio rerio)* as an animal model. Animal models used in research on AD are important for understanding the mechanisms of the disease and for developing potential treatment strategies. Some of the animal models frequently used for this purpose are mouse models (APP/PS1 mouse, 3XTg-AD mouse), rat models (Tg2576 rat), transgenic mouse (rTg4510 mouse) model, *Drosophila melanogaster* model, *C. Elegans* model, and primate models.

Zebrafish have been shown to be a good candidate for biomedical research, including AD research, because of its numerous advantages that have been mentioned before (Giong et al., 2021). In addition to these, as a vertebrate, zebrafish are closer to humans than nematodes and drosophilas, which are other popular model organisms used in neurodegenerative disorder studies (Giong et al., 2021). Even though zebrafish do



DOI: 10.1201/9781032676555-19

not have a cortex and hippocampus, which are defined in the mammalian brain, they possess learning and memory capabilities supported by anatomical structures similar to those in humans (Calvo and Schluessel, 2021).

Most importantly, as an animal model, zebrafish are easy to monitor phenotypically, even at the genetic level, and allow very fast follow-up of experiments due to the success of genetic and chemical manipulations on laboratory conditions (Shenoy et al., 2022; Sassen and Koster, 2015). Zebrafish are a valuable model for AD research, not only because they possess several orthologous genes and share neuroanatomical similarities with humans, but also because their embryos, larvae, and adults offer unique advantages, such as rapid growth mechanisms and transparency for monitoring (Shenoy et al., 2022; Gupta et al., 2018).

Beta amyloid injection and transgenic models are among the most common zebrafish models, and related coherent research for AD includes neurological assessment, behavioral tests, optogenetics and imaging approaches, drug screening tests, and nerve regeneration research. The reasons for choosing zebrafish models include rapid development and proliferation, regeneration capacity, optical transparency of anatomic appearance, similarities of nervous system and amyloid beta deposits. Moreover, zebrafish offer easy genetic manipulation, are compatible with behavioral tests, and are advantageous due to their low economic cost and ethical feasibility. On the other hand, there are a few disadvantages, such as being an invertebrate model, which may prevent it from reflecting some of the complex and detailed features found in higher organisms. Therefore, complementary findings can be obtained by using different model organisms.

AD is described by the presence of pathophysiological features, including the accumulation of amyloid beta plaques and NFT as described previously (Shenoy et al., 2022; Hampel et al., 2021). However, AD is divided into two groups: familial AD (FAD) (<1%) and sporadic AD (SAD) (>99%) (Piaceri et al., 2013). FAD is related to the increased production of β-amyloid (Aβ) peptide of 42 amino acids caused by the mutations in three genes: amyloid-β precursor protein (APP), presenilin 1 (PSEN1), and, to a lesser extent, presenilin 2 (PSEN2). In contrast, SAD is linked to multifactorial susceptibility ranging from genetic to environmental risks, with the Apolipoprotein E (*APOE*) gene, particularly allele ε4 (*APOE*ε4) (Corder et al.,1993; Santana et al., 2012). This chapter discusses the amyloid precursor protein (APP) cascade and Aβ peptides, NFT, tau protein functions, presenilins, and *APOE* in AD, with a focus on the use of zebrafish as an animal model.

## Amyloid Precursor Protein (APP) Cascade and β-Amyloid (Aβ) Peptides

The production and aggregation of Aβ peptides have a critical role in the pathogenesis of AD. Aβ peptides are peptide sequences consisting of 36–43 aa residues derived from the APP. APP is a type I transmembrane glycoprotein expressed in many tissues, especially the brain (Abyadeh et al., 2024). APP undergoes posttranslational proteolytic processing by α-, β-, and γ-secretases. As a result of proteolytic processes, α-secretase produces soluble amyloid protein, while β- and γ-secretases produce APP components with amyloidogenic properties (Figure 17.1) (Sery et al., 2013; Maisam et al., 2023; Gholami, 2023).

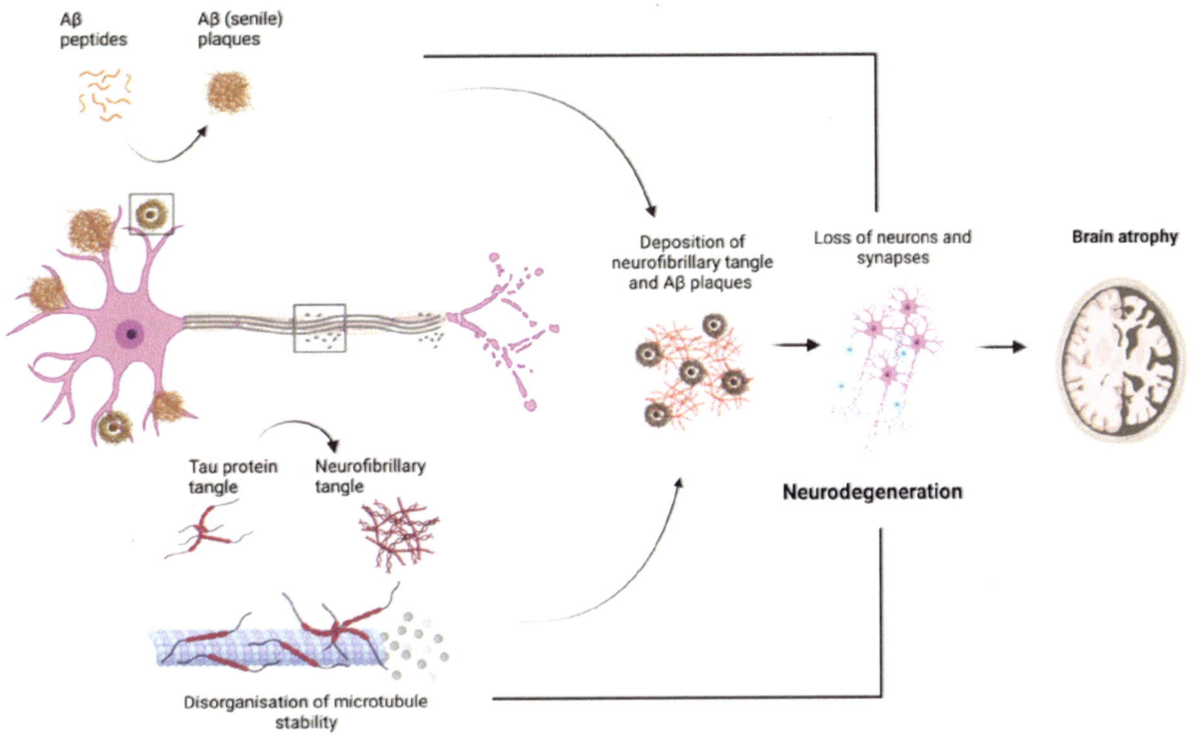


**FIGURE 17.1** The diagram highlights the two main hallmarks in AD pathology. Loss of synapses and neurons due to progressive accumulation of extracellular amyloid-β plaques and intraneuronal neurofibrillary tangles eventually causes neurodegeneration (created with BioRender).

Even though Aβ peptides play a critical role in the pathophysiology of AD, research shows that this is not the only function of these proteins (Chia et al., 2022). In a study using human endothelial cells and zebrafish as a model organism, it has been reported that Aβ peptides induce blood-vessel formation and play a role in maintaining healthy cerebrovascular functioning (Cameron et al., 2012). Similar to these findings, in a different study, it was suggested that reducing Aβ levels by knocking down APP in zebrafish larvae was associated with serious cerebrovascular defects, and a direct correlation between vascular density and Aβ levels in the development of the hindbrain was demonstrated (Luna et al., 2013).

Moreover, there are many studies highlighting the importance of App in zebrafish development (Liu, 2023). Zebrafish have two *app* genes (*appa* and *appb*) that are homologous to the human APP gene (Musa et al., 2001; Kiper and Freeman, 2022). According to molecular genetic studies of these genes, it was observed that *appa* was expressed mostly in the lens and otic vesicles, while *appb* was expressed mostly in the hindbrain (Musa et al., 2001). In another study, Banote et al. examined the biological functions of APP in embryonic development on the zebrafish model organism, where cellular processes can be analyzed and observed more easily than in many other vertebrates, in order to better understand the role of APP, which is constantly and highly expressed in the central nervous system (CNS). According to this study, a zebrafish model carrying a homozygous mutation in the *appb* gene was created using CRISPR-Cas9 technology, and it was shown that *appb* plays a role in cell adhesion in the early stages of zebrafish development (Banote et al., 2020). Furthermore, different studies have also shown that *appa* and *appb* mutations in the zebrafish model cause defects in the development of the brain and ependymal cilia (Chebli et al., 2021).

Although the etiology of the disease is not fully clear, extracellular amyloid plaque accumulation and the formation of intracellular NFT from hyperphosphorylated tau proteins are two typical features encountered in the brains of AD patients (Blennow et al., 2006). Abnormal accumulation of Aβ peptides may be due to uncontrolled Aβ peptide production owing to a mutational reason occurring in the APP pathway, or it may result from a defect in the clearance mechanisms of accumulated Aβ peptides (Maisam et al., 2023). Disruption of the balance between the production and clearance processes of Aβ peptides causes accumulation of Aβ peptides as amyloid or senile plaques in extracellular compartment between neurons, and eventually resulting in a decrease in neuronal integrity and synaptic plasticity (Tarasoff-Conway et al., 2015; Maisam et al., 2023). Therefore, in order to understand the pathophysiology of AD and establish a therapeutic approach that eliminates Aβ toxicities, it is crucial to also clarify how various forms of Aβ are cleared in both healthy and pathological states.

The mechanism of clearance of Aβ peptides from the brain is carried out through two main pathways: 1) through blood circulatory clearance, in which Aβ peptides enter the peripheral circulation and are eliminated by proteases; or 2) through the intracellular clearance in the brain, in which Aβ peptides are eliminated directly by intracellular degradation in glial cells (Jeong et al., 2021). Both mechanisms are important in maintaining homeostasis in the organism by removing waste materials and/or abnormal protein aggregates.

When it comes to the development of model organisms for neurodegenerative diseases that increase by age, it is essential to evaluate if humans and the model organisms have similar aging processes and clearance systems in the brain, as this could have a major impact on how adaptable the findings are to human systems (Kanoh et al., 2024; Jeong et al., 2021). Recent studies point out that precisely for this reason, research conducted between human and zebrafish models are more suitable for these comparisons than mice. Along with lifespan, rapid growth, and ease of reproduction, heart rate is one of the key differences between zebrafish and mice and plays an interesting role in AD pathology. Aβ peptides that enter the peripheral circulation are removed from the brain and are ultimately eliminated by proteases produced by macrophages and red blood cells in the bloodstream or by peripheral organs such as the kidney or liver (Xin et al., 2018). A high heart rate accelerates the blood circulation in the organism and thus serves to clear waste substances or aberrant deposition of Aβ peptides more quickly before they accumulate too much (Kanoh et al., 2024). While the resting heart rate of mice is around 500 bpm, it is around 60–70 bpm in humans and around 120 bpm in zebrafish (Wang et al., 2017; Mousavi and Patil, 2020). From this perspective, the rate of clearance and elimination of waste substances in mice is higher than in humans (Qosa et al., 2014). Accordingly, Aβ peptide accumulation occurs over a longer period or through different mechanisms. On the other hand, the idea that zebrafish may show a course more similar to human AD pathogenesis in terms of their heart rate being closer to that of humans and their ability to clear Aβ peptide accumulation in the brain stands out (Kanoh et al., 2024).

## Neurofibrillary Tangles and Tau Protein Function

NFT are large filaments, located inside the cell, formed by hyperphosphorylated tau proteins, which contribute to synaptic and neuronal loss. This process involves molecular events such as the hyperphosphorylation of tau and its subsequent oligomerization (Shenoy et al., 2022; Kiper and Freeman, 2022; Moloney et al., 2021). Continuously, any dysfunction at the synaptic level and neurotoxic events invite the inflammation process and finally lead to changes in neuronal ionic forces, as well as alterations in kinase and phosphatase activities. These changes eventually contribute to the abnormal phosphorylation of tau, leading to its hyperphosphorylation and oligomerization and the formation of NFT (Figure 17.1) (Kiper and Freeman, 2022; Hardy and Selkoe, 2002; Newman et al., 2011; Maccioni et al., 2010).

Tau protein had not been a popular subject for scientists who were studying the brain until the findings showed abnormal tangles in AD patients' brains (Giong et al., 2021). Tau protein plays a key role in the stabilizing the microtubules network in CNS through various post-translational modifications, including aggregation, glycosylation, glycation, nitration, polyamination, polyphoshorilation, and oxidation (Giong et al., 2021).

When tau undergoes pathological changes through a series of phosphorylation events, the accumulation of the aggregates causes the formation of tangles. In addition to this, tau protein has been shown to localize to the plasma membrane and neuron's nucleus (Barbereau et al., 2021; Brandt et al., 1995; Loomis et al., 1990).

The dysfunction of tau protein has garnered significant attention due to its involvement in several neurological disorders, including frontotemporal dementia, corticobasal degeneration, and progressive supranuclear palsy (Morgan and Carlyle, 2024; Santana et al., 2012; Pittman et al., 2006). The mutual mechanism underlying these diseases, including AD, involves abnormal functioning or microtubule-associated protein tau (MAPT), though the severity and specific manifestations may differ (Giong et al., 2021).

Tauopathy is the term for these groups of neurodegenerative disorders with NFT that lead to the intracellular aggregate accumulation in the brain including AD. Given the increasing prevalence of the neurodegenerative disorders related to dementia every year, targeted MAPT therapy for each tauopathy is still needed and will remain needed. That is why animal models are very important as they mimic the disease, helping to understand its mechanism and develop treatment options. AD animal models basically involve having NFT and plaques in the brain which are clinical features. NFT, along with hyperphosphorylated tau, are found in the spinal cord of AD patients, extending to the CNS, including the brainstem and brain (Kiper and Freeman, 2022; Dugger et al., 2013; Saito and Murayama, 2000).

One of the orthologous genes in the zebrafish genome is MAPT, represented by MAPTA and MAPTB that are the focus points for AD pathogenesis (Kiper and Freeman, 2022). MAPT-related tauopathies in humans show similar biochemical outcomes in transgenic zebrafish that express human MAPT, highlighting the genetic mechanisms' similarities and convenience as a study model (Lim et al., 2015). It has been reported that the zebrafish MAPTA and MAPTB genes along with tau isoforms are conserved similar to human tau, though their expression patterns in embryos are not identical (Morgan and Carlyle 2024; Chen et al., 2009). Common to both, these genes are expressed during the CNS development, although MAPTB shows higher expression in some part of the ganglion and dorsal sensory neurons of the spinal cord (Morgan and Carlyle 2024).

Tau protein is encoded by the MAPT gene, which produces 6 isoforms that result in 3R or 4R forms due to alternative splicing (Barbereau et al., 2021). The isoforms of 3R/4R ratio for tau aggregates have been identified in equimolar ratio in adult neurons and they are involved in some tauopathies, including AD. Besides, from 24 h post fertilization, both MAPTA and MAPTB, which correspond to human MAPT in zebrafish, are expressed in mRNA level in both isoforms within the CNS of the zebrafish. Mapta might have 4–6 microtubule binding domains while maptb has 3 domains. All these isoforms correspond to the human 3R and 4R isoforms.

If pathologies related to cell degeneration occur, tau proteins adopt different forms that contribute to the formation of oligomers, aggregates, or NFT. These post-translational modifications alter tau protein function, making it toxic and promoting aggregations (Barbereau et al., 2021). Another important modification in NFT formation is the hyperphosphorylation of tau, where normally 2–3 phosphates are attached under normal physiological conditions, while in pathological conditions, this increases to 7–8 phosphates per tau protein (Barbereau et al., 2021; Köpke et al., 1993).

## Presenilins in Alzheimer's Disease

There is a robust genetic background, with many genes associated with the pathogenesis of AD. One key factor believed to contribute to this forceful infrastructure and linked to pivotal mechanisms is presenilins. Presenilins were first identified in 1995 during studies investigating the genes responsible for early-onset, autosomal-dominant forms of FAD (Sherrington et al., 1995). APP forms the catalytic core structure of the γ-secretase complex, which supports the intramembranous proteolysis of various type I membrane proteins such as Notch1, Wnt, Delta1, E- and N-cadherins, CD44, LRP, Syndecan, Jagged, ErbB4, and Nectin1a (Hernandez-Sapiens et al., 2022).

Presenilins are a family of 9-helix multipass transmembrane proteins with two conserved aspartate residues on transmembrane domains 6 and 7. Their main localization sites include different cellular structures such as endosomes, lysosomes, nuclear envelope, mitochondria, trans-Golgi network, and endoplasmic reticulum (Escamilla-Ayala et al., 2020). After folding, the hydrophilic, flexible N-terminus resides in the cytosol, while the C-terminus is either embedded in the lumenal face of the lipid bilayer or protrudes into the lumen or extracellular space. During assembly and maturation of presenilin complexes, PRESENILIN1 (PSEN1) or PRESENILIN2 (PSEN2) subunits undergo endoproteolytic cleavage into N- and C-terminal fragments (Zheng and Koo, 2011) (Figure 17.2).

Presenilins, which are γ-secretase functional subunit genes, contain two known presenilin genes: PSEN1, located on chromosome 14q24.2; and PSEN2, located on chromosome 1q42.13. Both PSEN1 and PSEN2 are ~50 kDa polytopic transmembrane proteins that interact with nicastrin, Presenilin Enhancer 2 (PEN2), and APH1 to form the active four-membered γ-secretase structure. Mutations in the γ-secretase structural subunits may contribute to FAD, principally by disrupting the complex balance and changing the amyloid beta peptide accumulation with an increase in the Aβ42/Aβ40 ratio. However, it should be underlined that there is a need for more studies that will point to the precise contributions of presenilins at these points. Although PSEN2 is 70% similar to PSEN1, the presence of FAD-related autosomal dominant mutations detected in PS1 is considerably higher than those detected in PS2. Similarly, PSEN1 stands out as the subunit primarily responsible for the early onset of FAD (Tang and Kepp, 2018; Kabir et al., 2020).

Considering their active regulatory roles in processes such as neurogenesis and neuronal function, mutations and signaling defects occurring on PSEN1, PSEN2, and therefore γ-secretase create a favorable environment for the development of AD pathogenesis (Hernandez-Sapiens et al., 2022; Feng et

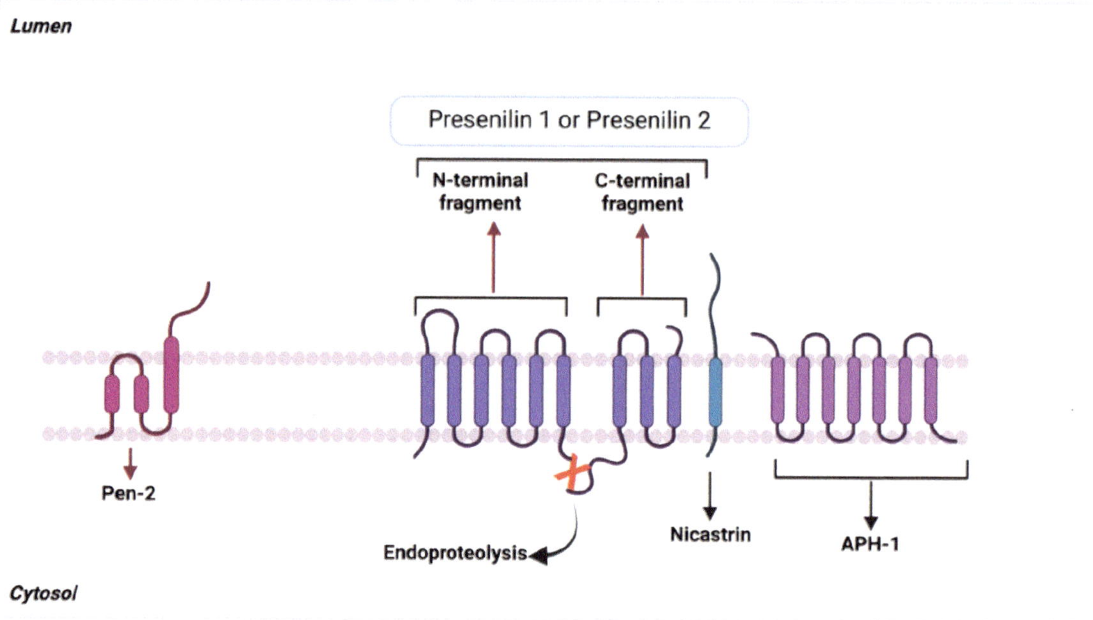


**FIGURE 17.2** The diagram illustrates the subunits of the γ-secretase complex. Presenilins (PSEN1 and PSEN2) function as the catalytic elements of the γ-secretase complexes (created with BioRender).

al., 2004). Disruptions in Notch signaling, chiefly involving APP, which is responsible for amyloid beta peptide accumulation, have been highlighted in previous studies as contributing to the Alzheimer's infrastructure and are noteworthy. Studies using zebrafish models to evaluate the relationships between these substrates with presenilins may assist in devising practical outcomes regarding AD (Lok et al., 2013; Newman et al., 2014; Sundvik et al., 2013). On one hand, the role of $Ca^{2+}$ homeostasis in different processes such as neurotransmitter release, memory, and learning is necessary for normal brain function. On the other hand, disruptions in $Ca^{2+}$ homeostasis are implicated in harmful mechanisms such as cell death and degeneration, making its role indisputable. In particular, studies suggesting that mitochondrial $Ca^{2+}$ dysregulation may be a pathological pathway linking defective $Ca^{2+}$ signals, organelle dysfunction, and neurodegeneration for PSEN1 and PSEN2-dependent AD (Sarasija et al., 2018a, 2018b).

When direct or indirect effects of presenilins are evaluated, it is extra critical to conduct pathway analyses at these points. Those can contribute to the clinical outcomes and help investigate new treatment candidates in more detail using human-like models, especially to elucidate complex and unknown interactions (Lok et al., 2013). In the literature, there are detailed analyses of presenilin in neurodegenerative diseases and AD in particular, and different animal models, including transgenic mice and drosophila, are used to elucidate these complex and still unanswered molecular mechanisms. These studies highlight several factors, such as the fact that commonly used models are not closely related to humans, their long life cycles, and the low homology of the presenilin gene in humans. These issues underscore the need to use alternative models in studies (Toda et al., 2011; Folwell et al., 2010).

In this context, compared to other animal models, zebrafish offer significant advantages for studies, including their rapid reproductive cycle and transparent developmental processes, and the fact that they allow simultaneously observation of behavioral and cerebral functions together. Studies investigating presenilins in zebrafish highlight the advantages of the zebrafish model in Alzheimer's diagnosis and treatment, considering the high homology of the PSEN1 zebrafish homolog (Psen1) and PSEN2 zebrafish homolog (Psen2), for their human counterparts (Santana et al., 2012).

Zebrafish modeling studies aimed at explaining the early changes in the brains of human FAD mutation carriers and other individuals prone to AD are gradually gaining momentum. The data obtained in this way may enable us to establish a relationship with genes that play a role in susceptibility to the disease, and also to make predictions about the course of the disease by monitoring instant or long-term markers such as presenilin expressions, which can change during definite periods of the disease. In a zebrafish modeling study investigating the relationship of the K115fs mutation in Psen2 with Alzheimer's, an increase in brain aging rates and differences in glucocorticoid signaling were detected in young fish with heterozygous mutants. Researchers argue that when forced expression of a PS2V-like isoform is stimulated, there is an increase in the immune and stress responses necessary to promote AD pathogenesis. This study is just one of the fundamental pieces of evidence that stress the importance of researchers using zebrafish genetic models to investigate the molecular mechanisms involved in AD pathogenesis (Hin et al., 2020).

Compared to similar studies in humans, the existence of alternative mutations presented by this point represents a substantial point for FAD. Researchers have shown that individuals with the PSEN2 K115Efs*11 mutation secrete less Aβ 1–40. They pointed out that this situation may play a crucial role in the pathogenesis of Alzheimer's by causing abnormal γ-secretase activity (Braggin et al., 2019). In the study by Bi et al., it has been reported that loss of PEN2 causes neuronal loss with astrogliosis followed by age-related cortical atrophy, indicating that the survival of neurons in the adult cerebral cortex relies on a PEN2-dependent state (Bi et al., 2021). On the other hand, based on the results of their study, Escamilla-Ayala et al. suggested that Notch signaling may not be active in the progression of PEN2-induced neurodegeneration. While PEN2 is essential for presenilin endoproteolysis, presenilins can differentially modulate gamma-secretase integrity through their reciprocal regulatory relationship. In this regard, it would not be wrong to assume that it is essential to focus on these points in studies to be implemented with zebrafish models in the evaluation of the FAD mechanism (Escamilla-Ayala et al., 2020). In their study investigating the effects of presenlins on the zebrafish embryogenesis process, Nornes et al. showed that blocking protein translation with the morpholino injection method could cause a decrease in Psen1 or Psen2 expression. The authors also found that the expression loss led to a decrease in the number of melanocytes and Notch signaling, causing impairments in neurogenesis and trunk and tail neural crest development. The study results indicate that an opposite relationship may exist between Psen1 and Psen2, although it has not been clearly confirmed (Nornes et al., 2009). In a different study in which Psen1 was inhibited, again using morpholino, Newman et al. showed that Psen1 inhibition plays an active role in the balance of activation and inhibition of Notch and APP signaling. It is anticipated that detailed investigation of these molecular interactions could provide valuable information regarding the role of presenilins in vital cellular pathways, particularly those associated with Alzheimer's (Newman et al., 2014). In their study investigating the relationship between Psen1 and histaminergic system plasticity, Sundvik et al. compared Psen1 mutant zebrafish with wild-type zebrafish regarding the number of histamine neurons at different times after fertilization. Compared to those in the wild type, Psen1 levels in mutant fish were lower at the end of the seventh day, at equal levels in the second month, and at higher levels than in the wild type at the end of one year. Due to these changes, differences in histamine-induced behaviors have been observed. Their results suggest that Psen1 mutation and Notch1a activity may affect brain plasticity throughout life (Sundvik et al., 2013). In a transcriptome analysis study using zebrafish brain samples, they determined that the expression of 105 genes for the mutant allele Q96_K97del of Psen1 was increased in heterozygous mutant brains compared to wild-type sibling brains. At the end of the study, researchers reported that the expression of 146 genes decreased. The researchers also point out that this early onset FAD–like mutation of zebrafish Psen1 is located at a pivot point in the functioning of cellular energy metabolism and transmembrane transport processes (Newman et al., 2019). Newman et al. emphasized in their study that Psen1 mutations accelerate the hypoxia response in zebrafish with aging-related alterations. Moreover, researchers asserted that these alterations occur due to aging and constitute a suitable basis for AD but are not a sufficient element on their own (Newman et al., 2020). In their subsequent studies, the same researchers aimed to observe the expression of wild-type and mutant Psen2S4Ter allele transcripts under acute hypoxia conditions. Although they could not obtain significant differences in the expression of mutant allele transcripts under normoxia or hypoxia, it was observed that 6-month-old zebrafish exhibited a behavior to increase their expression among individuals carrying both mutant and wild type heterozygosity (Jiang et al., 2020).

A different work, using CRISPR-Cas9, reported that protein deficiency in Psen2 knockout zebrafish did not directly affect Notch signaling at early developmental stages. On the other hand, when the larvae were stimulated, excessive reactivity in locomotor responses, decreased ER–mitochondria connection in their neurons, and increases in basal autophagy were observed. In addition, researchers have emphasized the potent role of Psen2 in mitochondrial axonal transport and point out that the presence of accompanying mutations may further develop AD (Barazzuol et al., 2023). At this point, to comment on the course of Notch signaling of presenilins and their contribution to the pathogenesis of AD, it seems that studies on the Notch relationship of presenilins are generally limited. However, according to the available data in the literature, we can conclude that the control of Psen1 on the Notch mechanism is more robust than Psen2 and that more comprehensive research is needed to associate presenilin with the pathogenesis of AD. A study investigating the effects of frameshift versus frame-conserving mutations in zebrafish Psen2 transcripts revealed opposing phenotypic outcomes in AD pathogenesis. Notably, only frameshift mutations—not frame-conserving variants—were associated with early-onset FAD and disruptions in oxidative phosphorylation-related substructures (Barthelson et al., 2021). Nery et al. found an increase in Aβ1-42 levels in zebrafish larvae after inhibition of Psen1 with morpholino, while decreases in cognitive abilities and the synaptic marker PSD-95 in the brain were found. They showed that these findings are prominent in relation to AD and draw attention to the potential role of Psen1 in neurodegenerative processes (Nery et al., 2017).

In 2010, Jayadev et al. examined individuals with the same N141I mutations in Psen2 in a specific family group and performed detailed neuropathological analyses of 18 autopsies in these families. Findings indicate that 30% of the individuals experienced seizures, while Lewy bodies were reported in the amygdala of more than 75% of the individuals. Although the N141I mutation parallels both early- and late-onset AD, it has been reported to produce an AD phenotype with onset across a wide age range, frequently associated with seizures, high penetrance, and typical AD neuropathology (Jayadev et al., 2010). Studies suggest that the N141I mutation increases the ratio of Aβ42 to Aβ40 through abnormal γ-secretase activity (Raska et al., 2021). N141I, which stands out as the most common and best-characterized FAD mutation in PSEN2, is represented by the N140 equivalent codon in zebrafish Psen2. Using genome editing technology in zebrafish, researchers discovered two mutations: Psen2N140fs (N140fs), which causes a truncation of the coding sequence; and Psen2T141_L142delinsMISLISV (T141_L142delinsMISLISV), which deletes two codons immediately downstream of N140 and replaces them with seven codons. Like almost every FAD mutation in the presenilin genes, this last mutation did not interrupt the open reading frame of the gene. It has been noted that individuals homozygously harboring both mutations are viable, but N140fs transcripts are associated with nonsense-mediated degradation and do not have the coding required for γ-secretase enzyme activation. Furthermore, although N140fs homozygous larvae initially showed mostly normal melanotic skin pigmentation, they lost this ability as they grew older. With these studies, researchers pointed out that null mutations in human presenilin genes do not cause AD, but that it is important to focus on the effects of null and FAD-like presenilin mutations on brain aging (Jiang et al., 2018).

In summary, to elucidate the mechanisms of AD pathogenesis, research on Psen1 and Psen2, which have a very high percentage of homology to the human PSEN1 and PSEN2 genes, should increasingly focus on zebrafish.

## Apolipoprotein E (*APOE*)

APOE is highly expressed in multiple tissues, with predominant levels observed in the liver, brain astrocytes, and a subset of microglia. This glycoprotein works as a receptor-mediated endocytosis ligand for lipoproteins such as cholesterol, which are essential for myelin production and normal brain functions (Breijyeh and Karaman, 2020). The *APOE* gene is located on chromosome 19 with three isoforms: *APOE2*, *APOE3*, and *APOE4*. The *APOE4* allele is an important risk factor for both early onset AD and late onset AD, in contrast to the *APOE2* and *APOE3* alleles (Kim et al., 2009; Isbir et al., 2001). *APOE4* is also involved in senile plaque accumulation and is known to be a biomarker for AD in cerebral amyloid angiopathy (Liu et al., 2013). In addition to these, *APOE4* has been shown to interact with cerebrovascular dysfunction, which contributes to the pathogenesis of Alzheimer's disease (Giau et al., 2015; Breijyeh and Karaman, 2020).

*APOE* is known as one of the major genetic risk factors for SAD. Further studies are needed to elucidate issues such as the precise role of *APOE4* in mediating interactions between the peripheral circulatory system and the CNS and how this may relate to brain and cognitive aging (Santana et al., 2012). When *APOE4* is associated with AD, it has been shown that single allele carriage is associated with a 3-fold increased risk, and two allele copy carriage is associated with an approximately 12-fold increased risk of the disease (Saunders et al., 1993). However, the allele *APOE4* could not be held solely responsible for the disease (Verghese et al., 2011). Although there are many studies examining the contribution of *APOE* to AD, the molecular mechanisms remain incompletely understood (Kim et al., 2009). FAD cases, on the other hand, are shown to be mainly associated with APP, PSEN1, and PSEN2 mutations. These mutations increase the Aβ42/Aβ40 ratio, causing the formation of amyloid plaques (Koehler and Williams, 2018; Hamley, 2012). The biggest risk factor in the development of SAD, besides aging, is the decrease in the ε4 allele levels of *APOE*, which plays a role in the clearance of Aβ fragments (Wildsmith et al., 2013; Koehler and Williams, 2018). Studies have yielded important findings that APOE4 proteolysis is one of the disease risk factors. The presence of 17–20 kDa APOE4 fragments located within NFT in the prefrontal cortex tissues of individuals with AD is an important finding supporting this idea (McCarthy et al., 2022). Despite meticulous studies in the literature, no results have been obtained to support or clarify the pathophysiological relationship regarding the contribution of the APOE4 allele to the development of late-onset AD (Harris et al., 2003; Rohn et al., 2012; Harris et al., 2004; Brecht et al., 2004; Muñoz et al., 2019; McCarthy et al., 2022). The fact that the APOE4 isoform is highly susceptible to proteolysis compared to E3 and E2 may be a finding that may help explain the association with dementia risk (Rohn, 2013; McCarthy et al., 2022). Some of the studies support the hypothesis that APOE4 fragmentation via the metalloproteinase-9 may contribute to AD pathology and inflammation (Love et al., 2017).

Since *APOE4* carriers account for 80% of all cases, it is cited as the greatest risk factor for late-onset AD (McCarthy et al., 2022). Since there is no evidence yet on how *APOE4* contributes to the pathogenesis of AD, studies on animal models genetically similar to humans are still needed (Koehler et al., 2018).

The nucleotide sequence of zebrafish genes has significant homology to that of human genes, and 84% of genes known to be associated with human diseases have a zebrafish gene counterpart (Kalueff et al., 2014). The basic zebrafish brain structure is highly protected compared to the mammalian brain. The organization of mammalian and zebrafish brains looks quite similar. Beyond this similarity, brain organization in zebrafish continues into telencephalon, mesencephalon, metencephalon, and myelencephalon divisions (Santana et al., 2012; Kalueff et al., 2014). The similarities between mammals and zebrafish are not limited to these. In terms of the blood–brain barrier (BBB), mammals and zebrafish BBB have a wonderful similarity. Thanks to this similarity, the BBB permeability of new neurodrugs can be easily investigated using zebrafish (Vitek et al., 2021; Chen et al., 2023). Following the identification of the gene homologous to mammalian *APOE* in zebrafish, studies in the literature began to gain momentum accordingly (Durliat et al., 2000). This similarity might be interpreted as indicating a different genomic organization, as the similarity rate is reflected at a low rate of 28% (Babin et al., 1997). apoE is highly expressed in the egg syncytial layer, suggesting that *apoE* represents a paramount point in zebrafish nutrition. Since apoE is intensely expressed in zebrafish eyes and especially in brain regions at stages of embryo development, it is thought that it might have an important function in the CNS (Babin et al., 1997). Zebrafish might be used as a simple and convenient model to examine the relationship between *apoE*, which might have nutritional importance, and zebrafish brain development. However, as far as we know, *apoE* zebrafish transgenic models have not been developed to date, and the number of studies in the literature in which *APOE* has been genetically examined in zebrafish models is insufficient.

With the dominating strength of the genetic association between *APOE* genotypes and risk of AD, inclusion of human *APOE* isoforms in all animal AD models is a critical requirement to improve translatability between these models and humans (Vitek et al., 2021). Unlike animals, humans are the only species to express multiple isoforms of APOE protein, which derives from a single polymorphic *APOE* gene. Despite the *APOE4* allele being carried by only 14% of the human population, roughly half of all AD patients are *APOE4* carriers, whereas *APOE2* carriers (8% prevalence) are relatively protected from the disease and tend to live to 100 years or more (Reiman et al., 2020). Non-human primates, dogs, and rodents all express single versions of *APOE* that are similar, but non-identical, to the human epsilon-4 APOE protein (Chambers et al., 2012).

Studies in *ApoE* knockout mice provided evidence that ApoE proteins are the cause of the accumulation of amyloid plaques and NFT. The fact that such pathologies do not occur in the absence of *ApoE* provides prime evidence for the relationship of these pathologies to the presence of *ApoE*. Researchers have shown that transgenic mice with a human *APOE4* background exhibit human-like behavioral and pathological deficits (Holtzman et al., 2000; Shi et al., 2017; Tai et al., 2017). The more advanced the genetic infrastructure in transgenic animals becomes, resembling that of humans with AD, the more effectively these models can provide intelligible ideas about AD in studies on *APOE* genotypes.

Lewkowicz et al. investigated how APOE interacts with Aβ fibrils associated with AD. The NMR structure of APOE was determined using molecular modeling together with Aβ1-40 and Aβ1-42 fibril structures obtained from patients. APOE was shown to undergo conformational changes to bind to the hydrophobic surfaces of amyloid fibrils, and the mechanism may inhibit secondary nucleation by fibril disintegration (Lewkowicz et al., 2023).

Guan et al. aimed to investigate how the *APOE4* allele affects the risk of AD and how this genetic factor is associated with diet, blood markers, brain structure, and cognitive functions (Guan et al., 2023). In older adults, demographic, genetic, dietary, blood, and, brain MRI data were collected at baseline and 12 years later to compare *APOE4* carriers with non-carriers. While *APOE4* carriers showed lower white matter integrity and cognitive scores than non-carriers, diet and blood markers showed enormous relationships with white matter structure in carriers. As a result, it has been suggested that *APOE4* carrier state may affect the risk of Alzheimer's by increasing the communication between central and peripheral systems (Guan et al., 2023).

Although there is evidence that inheritance of the *APOE4* allele increases the risk of developing SAD, the molecular basis underlying this increased risk is still unclear (Vitek et al., 2021). At this point, studies that will investigate how the *APOE4* mechanism contributes to the risk of AD using zebrafish modeling continue to gain importance.

## Conclusion

Zebrafish offer numerous advantages for study, such as their suitable gene capacity of biochemical pathways and anatomical features. Studying AD through gene interactions responsible for the disease in zebrafish models could significantly contribute to the literature.

With genome editing, knock-in approaches, zebrafish embryos, larvae, and even adults can be easily monitored and studied in a short time and at a lower cost compared to other animal models. Thanks to these advantages, zebrafish genetic models continue to provide valuable perspectives for researchers who continue their studies to find answers to unresolved questions regarding the disease mechanism.

## REFERENCES


Abyadeh, M., Gupta, V., Paulo, J. A., Mahmoudabad, A. G., Shadfar, S., Mirshahvaladi, S., Gupta, V., Nguyen, C. T. O., Finkelstein, D. I., You, Y., Haynes, P. A., Salekdeh, G. H., Graham, S. L., & Mirzaei, M. (2024). Amyloid-beta and tau protein beyond Alzheimer's disease. *Neural Regeneration Research*, *19*(6), 1262–1276. https://doi.org/10.4103/1673-5374.386406

Alzheimer's disease facts and figures. (2023). *Alzheimer's Dement*, *19*, 1598–1695. https://doi.org/10.1002/alz.13016

Babin, P. J., Thisse, C., Durliat, M., Andre, M., Akimenko, M. A., & Thisse, B. (1997). Both apolipoprotein E and A-I genes are present in a nonmammalian vertebrate and are highly expressed during embryonic development. *Proceedings of the National Academy of Sciences of the United States of America*, *94*(16), 8622–8627. https://doi.org/10.1073/pnas.94.16.8622

Banote, R. K., Chebli, J., Şatır, T. M., Varshney, G. K., Camacho, R., Ledin, J., Burgess, S. M., Abramsson, A., & Zetterberg, H. (2020). Amyloid precursor protein-b facilitates cell adhesion during early development in zebrafish. *Scientific Reports*, *10*(1), 10127. https://doi.org/10.1038/s41598-020-66584-8

Barazzuol, L., Cieri, D., Facchinello, N., Calì, T., Washbourne, P., Argenton, F., & Pizzo, P. (2023). Unraveling presenilin 2 functions in a knockout zebrafish line to shed light into Alzheimer's disease pathogenesis. *Cells*, *12*(3), 376. https://doi.org/10.3390/cells12030376

Barbereau, C., Cubedo, N., Maurice, T., & Rossel, M. (2021). Zebrafish models to study new pathways in tauopathies. *International Journal of Molecular Sciences*, *22*(9), 4626. https://doi.org/10.3390/ijms22094626

Barthelson, K., Pederson, S. M., Newman, M., Jiang, H., & Lardelli, M. (2021). In-frame and frameshift mutations in zebrafish presenilin 2 affect different cellular functions in young adult brains. *Journal of Alzheimer's Disease Reports*, *5*(1), 395–404. https://doi.org/10.3233/ADR-200279

Bi, H. R., Zhou, C. H., Zhang, Y. Z., Cai, X. D., Ji, M. H., Yang, J. J., Chen, G. Q., & Hu, Y. M. (2021). Neuron-specific deletion of presenilin enhancer2 causes progressive astrogliosis and age-related neurodegeneration in the cortex independent of the Notch signaling. *CNS Neuroscience & Therapeutics*, *27*(2), 174–185. https://doi.org/10.1111/cns.13454

Blennow, K., de Leon, M. J., & Zetterberg, H. (2006). Alzheimer's disease. *Lancet*, *368*(9533), 387–403. https://doi.org/10.1016/S0140-6736(06)69113-7

Braggin, J. E., Bucks, S. A., Course, M. M., Smith, C. L., Sopher, B., Osnis, L., Shuey, K. D., Domoto-Reilly, K., Caso, C., Kinoshita, C., Scherpelz, K. P., Cross, C., Grabowski, T., Nik, S. H. M., Newman, M., Garden, G. A., Leverenz, J. B., Tsuang, D., Latimer, C., Gonzalez-Cuyar, L. F., … Jayadev, S. (2019). Alternative splicing in a presenilin 2 variant associated with Alzheimer disease. *Annals of Clinical and Translational Neurology*, *6*(4), 762–777. https://doi.org/10.1002/acn3.755

Brandt, R., Léger, J., & Lee, G. (1995). Interaction of tau with the neural plasma membrane mediated by tau's amino-terminal projection domain. *The Journal of Cell Biology, 131*(5), 1327–1340. https://doi.org/10.1083/jcb.131.5.1327

Brecht, W. J., Harris, F. M., Chang, S., Tesseur, I., Yu, G. Q., Xu, Q., Dee Fish, J., Wyss-Coray, T., Buttini, M., Mucke, L., Mahley, R. W., & Huang, Y. (2004). Neuron-specific apolipoprotein e4 proteolysis is associated with increased tau phosphorylation in brains of transgenic mice. *The Journal of Neuroscience:* The Official Journal *of the Society for Neuroscience, 24*(10), 2527–2534. https://doi.org/10.1523/JNEUROSCI.4315-03.2004

Breijyeh, Z., & Karaman, R. (2020). Comprehensive review on Alzheimer's disease: Causes and treatment. *Molecules, 25*(24), 5789. https://doi.org/10.3390/molecules25245789

Calvo, R., & Schluessel, V. (2021). Neural substrates involved in the cognitive information processing in teleost fish. *Animal Cognition, 24*(5), 923–946. https://doi.org/10.1007/s10071-021-01514-3

Cameron, D. J., Galvin, C., Alkam, T., Sidhu, H., Ellison, J., Luna, S., & Ethell, D. W. (2012). Alzheimer's-related peptide amyloid-β plays a conserved role in angiogenesis. *PLoS One, 7*(7), e39598. https://doi.org/10.1371/journal.pone.0039598

Chambers, J. K., Uchida, K., & Nakayama, H. (2012). White matter myelin loss in the brains of aged dogs. *Experimental Gerontology, 47*(3), 263–269. https://doi.org/10.1016/j.exger.2011.12.003

Chebli, J., Rahmati, M., Lashley, T., Edeman, B., Oldfors, A., Zetterberg, H., & Abramsson, A. (2021). The localization of amyloid precursor protein to ependymal cilia in vertebrates and its role in ciliogenesis and brain development in zebrafish. *Scientific Reports, 11*(1), 19115. https://doi.org/10.1038/s41598-021-98487-7

Chen, M., Martins, R. N., & Lardelli, M. (2009). Complex splicing and neural expression of duplicated tau genes in zebrafish embryos. *Journal of Alzheimer's Disease: JAD, 18*(2), 305–317. https://doi.org/10.3233/JAD-2009-1145

Chen, Y., He, Y., Han, J., Wei, W., & Chen, F. (2023). Blood-brain barrier dysfunction and Alzheimer's disease: Associations, pathogenic mechanisms, and therapeutic potential. *Frontiers in Aging Neuroscience, 15*, 1258640. https://doi.org/10.3389/fnagi.2023.1258640.

Chia, K., Klingseisen, A., Sieger, D., & Priller, J. (2022). Zebrafish as a model organism for neurodegenerative disease. *Frontiers in Molecular Neuroscience, 15*, 940484. https://doi.org/10.3389/fnmol.2022.940484.

Corder, E. H., Saunders, A. M., Strittmatter, W. J., Schmechel, D. E., Gaskell, P. C., Small, G. W., Roses, A. D., Haines, J. L., & Pericak-Vance, M. A. (1993). Gene dose of apolipoprotein E type 4 allele and the risk of Alzheimer's disease in late onset families. *Science, 261*(5123), 921–923. https://doi.org/10.1126/science.8346443

Dugger, B. N., Hidalgo, J. A., Chiarolanza, G., Mariner, M., Henry-Watson, J., Sue, L. I., & Beach, T. G. (2013). The distribution of phosphorylated tau in spinal cords of Alzheimer's disease and non-demented individuals. *Journal of Alzheimer's Disease: JAD, 34*(2), 529–536. https://doi.org/10.3233/JAD-121864

Durliat, M., André, M., & Babin, P. J. (2000). Conserved protein motifs and structural organization of a fish gene homologous to mammalian apolipoprotein E. *European Journal of Biochemistry, 267*(2), 549–559. https://doi.org/10.1046/j.1432-1327.2000.01033.x

Escamilla-Ayala, A., Wouters, R., Sannerud, R., & Annaert, W. (2020). Contribution of the Presenilins in the cell biology, structure and function of γ-secretase. *Seminars in Cell & Developmental Biology, 105*, 12–26. https://doi.org/10.1016/j.semcdb.2020.02.005

Feng, R., Wang, H., Wang, J., Shrom, D., Zeng, X., & Tsien, J. Z. (2004). Forebrain degeneration and ventricle enlargement caused by double knockout of Alzheimer's presenilin-1 and presenilin-2. *Proceedings of the National Academy of Sciences of the United States of America, 101*(21), 8162–8167. https://doi.org/10.1073/pnas.0402733101

Folwell, J., Cowan, C. M., Ubhi, K. K., Shiabh, H., Newman, T. A., Shepherd, D., & Mudher, A. (2010). Abeta exacerbates the neuronal dysfunction caused by human tau expression in a Drosophila model of Alzheimer's disease. *Experimental Neurology, 223*(2), 401–409. https://doi.org/10.1016/j.expneurol.2009.09.014

Gholami A. (2023). Alzheimer's disease: The role of proteins in formation, mechanisms, and new therapeutic approaches. *Neuroscience Letters, 817*, 137532. https://doi.org/10.1016/j.neulet.2023.137532

Giau, V. V., Bagyinszky, E., An, S. S., & Kim, S. Y. (2015). Role of apolipoprotein E in neurodegenerative diseases. *Neuropsychiatric Disease and Treatment, 11*, 1723–1737. https://doi.org/10.2147/NDT.S84266

Giong, H. K., Subramanian, M., Yu, K., & Lee, J. S. (2021). Non-rodent genetic animal models for studying tauopathy: Review of drosophila, zebrafish, and C. elegans models. *International Journal of Molecular Sciences, 22*(16), 8465. https://doi.org/10.3390/ijms22168465

Guan, Y., Cheng, C. H., Bellomo, L. I., Narain, S., Bigornia, S. J., Garelnabi, M. O., Scott, T., Ordovás, J. M., Tucker, K. L., Bhadelia, R., & Koo, B. B. (2023). APOE4 allele-specific associations between diet, multimodal biomarkers, and cognition among Puerto Rican adults in Massachusetts. *Frontiers in aging neuroscience, 15*, 1285333. https://doi.org/10.3389/fnagi.2023.1285333

Gupta, T., Marquart, G. D., Horstick, E. J., Tabor, K. M., Pajevic, S., & Burgess, H. A. (2018). Morphometric analysis and neuroanatomical mapping of the zebrafish brain. *Methods, 150*, 49–62. https://doi.org/10.1016/j.ymeth.2018.06.008

Hamley I. W. (2012). The amyloid beta peptide: A chemist's perspective. Role in Alzheimer's and fibrillization. *Chemical Reviews, 112*(10), 5147–5192. https://doi.org/10.1021/cr3000994

Hampel, H., Hardy, J., Blennow, K., Chen, C., Perry, G., Kim, S. H., Villemagne, V. L., Aisen, P., Vendruscolo, M., Iwatsubo, T., Masters, C. L., Cho, M., Lannfelt, L., Cummings, J. L., & Vergallo, A. (2021). The Amyloid-β pathway in Alzheimer's disease. *Molecular Psychiatry, 26*(10), 5481–5503. https://doi.org/10.1038/s41380-021-01249-0

Hardy, J., & Selkoe, D. J. (2002). The amyloid hypothesis of Alzheimer's disease: Progress and problems on the road to therapeutics. *Science, 297*(5580), 353–356. https://doi.org/10.1126/science.1072994

Harris, F. M., Brecht, W. J., Xu, Q., Mahley, R. W., & Huang, Y. (2004). Increased tau phosphorylation in apolipoprotein E4 transgenic mice is associated with activation of extracellular signal-regulated kinase: Modulation by zinc. *The Journal of Biological Chemistry*, *279*(43), 44795–44801. https://doi.org/10.1074/jbc.M408127200

Harris, F. M., Brecht, W. J., Xu, Q., Tesseur, I., Kekonius, L., Wyss-Coray, T., Fish, J. D., Masliah, E., Hopkins, P. C., Scearce-Levie, K., Weisgraber, K. H., Mucke, L., Mahley, R. W., & Huang, Y. (2003). Carboxyl-terminal-truncated apolipoprotein E4 causes Alzheimer's disease-like neurodegeneration and behavioral deficits in transgenic mice. *Proceedings of the National Academy of Sciences of the United States of America*, *100*(19), 10966–10971. https://doi.org/10.1073/pnas.1434398100

Hernandez-Sapiens, M. A., Reza-Zaldívar, E. E., Márquez-Aguirre, A. L., Gómez-Pinedo, U., Matias-Guiu, J., Cevallos, R. R., Mateos-Díaz, J. C., Sánchez-González, V. J., & Canales-Aguirre, A. A. (2022). Presenilin mutations and their impact on neuronal differentiation in Alzheimer's disease. *Neural Regeneration Research*, *17*(1), 31–37. https://doi.org/10.4103/1673-5374.313016

Hin, N., Newman, M., Kaslin, J., Douek, A. M., Lumsden, A., Nik, S. H. M., Dong, Y., Zhou, X. F., Mañucat-Tan, N. B., Ludington, A., Adelson, D. L., Pederson, S., & Lardelli, M. (2020). Accelerated brain aging towards transcriptional inversion in a zebrafish model of the K115fs mutation of human PSEN2. *PLoS One*, *15*(1), e0227258. https://doi.org/10.1371/journal.pone.0227258

Holtzman, D. M., Bales, K. R., Tenkova, T., Fagan, A. M., Parsadanian, M., Sartorius, L. J., Mackey, B., Olney, J., McKeel, D., Wozniak, D., & Paul, S. M. (2000). Apolipoprotein E isoform-dependent amyloid deposition and neuritic degeneration in a mouse model of Alzheimer's disease. *Proceedings of the National Academy of Sciences of the United States of America*, *97*(6), 2892–2897. https://doi.org/10.1073/pnas.050004797

Isbir, T., Agaçhan, B., Yilmaz, H., Aydin, M., Kara, I., Eker, D., & Eker, E. (2001). Interaction between apolipoprotein-E and angiotensin-converting enzyme genotype in Alzheimer's disease. *American Journal of Alzheimer's Disease and Other Dementias*, *16*(4), 205–210. https://doi.org/10.1177/153331750101600407

Jayadev, S., Leverenz, J. B., Steinbart, E., Stahl, J., Klunk, W., Yu, C. E., & Bird, T. D. (2010). Alzheimer's disease phenotypes and genotypes associated with mutations in presenilin 2. *Brain: A Journal of Neurology*, *133*(Pt 4), 1143–1154. https://doi.org/10.1093/brain/awq033

Jeong, Y. M., Lee, J. G., Cho, H. J., Lee, W. S., Jeong, J., & Lee, J. S. (2021). Differential clearance of Aβ species from the brain by brain lymphatic endothelial cells in zebrafish. *International Journal of Molecular Sciences*, *22*(21), 11883. https://doi.org/10.3390/ijms222111883

Jiang, H., Newman, M., & Lardelli, M. (2018). The zebrafish orthologue of familial Alzheimer's disease gene PRESENILIN 2 is required for normal adult melanotic skin pigmentation. *PLoS One*, *13*(10), e0206155. https://doi.org/10.1371/journal.pone.0206155

Jiang, H., Pederson, S. M., Newman, M., Dong, Y., Barthelson, K., & Lardelli, M. (2020). Transcriptome analysis indicates dominant effects on ribosome and mitochondrial function of a premature termination codon mutation in the zebrafish gene psen2. *PLoS One*, *15*(7), e0232559. https://doi.org/10.1371/journal.pone.0232559

Kabir, M. T., Uddin, M. S., Setu, J. R., Ashraf, G. M., Bin-Jumah, M. N., & Abdel-Daim, M. M. (2020). Exploring the Role of PSEN Mutations in the Pathogenesis of Alzheimer's Disease. *Neurotoxicity Research*, *38*(4), 833–849. https://doi.org/10.1007/s12640-020-00232-x

Kalueff, A. V., Stewart, A. M., & Gerlai, R. (2014). Zebrafish as an emerging model for studying complex brain disorders. *Trends in Pharmacological Sciences*, *35*(2), 63–75. https://doi.org/10.1016/j.tips.2013.12.002

Kanoh, T., Mizoguchi, T., Tonoki, A., & Itoh, M. (2024). Modeling of age-related neurological disease: Utility of zebrafish. *Frontiers in Aging Neuroscience*, *16*, 1399098. https://doi.org/10.3389/fnagi.2024.1399098

Kim, J., Basak, J. M., & Holtzman, D. M. (2009). The role of apolipoprotein E in Alzheimer's disease. *Neuron*, *63*(3), 287–303. https://doi.org/10.1016/j.neuron.2009.06.026

Kiper, K., & Freeman, J. L. (2022). Use of zebrafish genetic models to study etiology of the amyloid-beta and neurofibrillary tangle pathways in Alzheimer's disease. *Current Neuropharmacology*, *20*(3), 524–539. https://doi.org/10.2174/1570159X19666210524155944

Koehler, D., & Williams, F. E. (2018). Utilizing zebrafish and okadaic acid to study Alzheimer's disease. *Neural Regeneration Research*, *13*(9), 1538–1541. https://doi.org/10.4103/1673-5374.237111

Köpke, E., Tung, Y. C., Shaikh, S., Alonso, A. C., Iqbal, K., & Grundke-Iqbal, I. (1993). Microtubule-associated protein tau. Abnormal phosphorylation of a non-paired helical filament pool in Alzheimer disease. *The Journal of Biological Chemistry*, *268*(32), 24374–24384.

Lewkowicz, E., Nakamura, M. N., Rynkiewicz, M. J., & Gursky, O. (2023). Molecular modeling of apoE in complexes with Alzheimer's amyloid-β fibrils from human brain suggests a structural basis for apolipoprotein co-deposition with amyloids. *bioRxiv: The Preprint Server for Biology*, 2023.08.04.551703. https://doi.org/10.1101/2023.08.04.551703

Lim, A., Moussavi Nik, S. H., Ebrahimie, E., & Lardelli, M. (2015). Analysis of nicastrin gene phylogeny and expression in zebrafish. *Development Genes and Evolution*, *225*(3), 171–178. https://doi.org/10.1007/s00427-015-0500-9

Liu Y. (2023). Zebrafish as a model organism for studying pathologic mechanisms of neurodegenerative diseases and other neural disorders. *Cellular and Molecular Neurobiology*, *43*(6), 2603–2620. https://doi.org/10.1007/s10571-023-01340-w

Liu, C. C., Liu, C. C., Kanekiyo, T., Xu, H., & Bu, G. (2013). Apolipoprotein E and Alzheimer disease: Risk, mechanisms and therapy. Nature reviews. *Neurology*, *9*(2), 106–118. https://doi.org/10.1038/nrneurol.2012.263

Lok, K., Zhao, H., Shen, H., Wang, Z., Gao, X., Zhao, W., & Yin, M. (2013). Characterization of the APP/PS1 mouse model of Alzheimer's disease in senescence accelerated background. *Neuroscience Letters*, *557 Pt B*, 84–89. https://doi.org/10.1016/j.neulet.2013.10.051

Loomis, P. A., Howard, T. H., Castleberry, R. P., & Binder, L. I. (1990). Identification of nuclear tau isoforms in human neuroblastoma cells. *Proceedings of the National Academy of Sciences of the United States of America*, *87*(21), 8422–8426. https://doi.org/10.1073/pnas.87.21.8422

Love, J. E., Day, R. J., Gause, J. W., Brown, R. J., Pu, X., Theis, D. I., Caraway, C. A., Poon, W. W., Rahman, A. A., Morrison, B. E., & Rohn, T. T. (2017). Nuclear uptake of an amino-terminal fragment of apolipoprotein E4 promotes cell death and localizes within microglia of the Alzheimer's disease brain. *International Journal of Physiology, Pathophysiology and Pharmacology*, *9*(2), 40–57.

Luna, S., Cameron, D. J., & Ethell, D. W. (2013). Amyloid-β and APP deficiencies cause severe cerebrovascular defects: Important work for an old villain. *PLoS One*, *8*(9), e75052. https://doi.org/10.1371/journal.pone.0075052

Maccioni, R. B., Farías, G., Morales, I., & Navarrete, L. (2010). The revitalized tau hypothesis on Alzheimer's disease. *Archives of Medical Research*, *41*(3), 226–231. https://doi.org/10.1016/j.arcmed.2010.03.007

Maisam, M., Khan, M. T., Lodhi, M. S., Mou, K., Liu, Z., & Wei, D. (2023). Alzheimer's disease; Mechanism, mutations, and applications of nano-medicine. *Frontiers in Bioscience (Landmark Edition)*, *28*(10), 258. https://doi.org/10.31083/j.fbl2810258

McCarthy, M. M., Hardy, M. J., Leising, S. E., LaFollette, A. M., Stewart, E. S., Cogan, A. S., Sanghal, T., Matteo, K., Reeck, J. C., Oxford, J. T., & Rohn, T. T. (2022). An amino-terminal fragment of apolipoprotein E4 leads to behavioral deficits, increased PHF-1 immunoreactivity, and mortality in zebrafish. *PLoS One*, *17*(12), e0271707. https://doi.org/10.1371/journal.pone.0271707

Moloney, C. M., Lowe, V. J., & Murray, M. E. (2021). Visualization of neurofibrillary tangle maturity in Alzheimer's disease: A clinicopathologic perspective for biomarker research. *Alzheimer's & Dementia: The Journal of the Alzheimer's Association*, *17*(9), 1554–1574. https://doi.org/10.1002/alz.12321

Morgan, G.R., Carlyle, B.C. (2024). Interrogation of the human cortical peptidome uncovers cell-type specific signatures of cognitive resilience against Alzheimer's disease. *Sci Rep.* 26;14(1):7161.

Mousavi, S. E., & Patil, J. G. (2020). Light-cardiogram, a simple technique for heart rate determination in adult zebrafish, Danio rerio. Comparative biochemistry and physiology. *Part A, Molecular & Integrative Physiology*, *246*, 110705. https://doi.org/10.1016/j.cbpa.2020.110705

Muñoz, S. S., Garner, B., & Ooi, L. (2019). Understanding the role of ApoE fragments in Alzheimer's Disease. *Neurochemical Research*, *44*(6), 1297–1305. https://doi.org/10.1007/s11064-018-2629-1

Musa, A., Lehrach, H., & Russo, V. A. (2001). Distinct expression patterns of two zebrafish homologues of the human APP gene during embryonic development. *Development Genes and Evolution*, *211*(11), 563–567. https://doi.org/10.1007/s00427-001-0189-9

Nery, L. R., Silva, N. E., Fonseca, R., & Vianna, M. R. M. (2017). Presenilin-1 targeted morpholino induces cognitive deficits, increased brain Aβ1-42 and decreased synaptic marker PSD-95 in zebrafish larvae. *Neurochemical Research*, *42*(10), 2959–2967. https://doi.org/10.1007/s11064-017-2327-4

Newman, M., Ebrahimie, E., & Lardelli, M. (2014). Using the zebrafish model for Alzheimer's disease research. *Frontiers in Genetics*, *5*, 189. https://doi.org/10.3389/fgene.2014.00189

Newman, M., Hin, N., Pederson, S., & Lardelli, M. (2019). Brain transcriptome analysis of a familial Alzheimer's disease-like mutation in the zebrafish presenilin 1 gene implies effects on energy production. *Molecular Brain*, *12*(1), 43. https://doi.org/10.1186/s13041-019-0467-y

Newman, M., Nik, H. M., Sutherland, G. T., Hin, N., Kim, W. S., Halliday, G. M., Jayadev, S., Smith, C., Laird, A. S., Lucas, C. W., Kittipassorn, T., Peet, D. J., & Lardelli, M. (2020). Accelerated loss of hypoxia response in zebrafish with familial Alzheimer's disease-like mutation of presenilin 1. *Human Molecular Genetics*, *29*(14), 2379–2394. https://doi.org/10.1093/hmg/ddaa119

Newman, M., Verdile, G., Martins, R. N., & Lardelli, M. (2011). Zebrafish as a tool in Alzheimer's disease research. *Biochimica et biophysica acta*, *1812*(3), 346–352. https://doi.org/10.1016/j.bbadis.2010.09.012

Newman, M., Wilson, L., Verdile, G., Lim, A., Khan, I., Moussavi Nik, S. H., Pursglove, S., Chapman, G., Martins, R. N., & Lardelli, M. (2014). Differential, dominant activation and inhibition of Notch signalling and APP cleavage by truncations of PSEN1 in human disease. *Human Molecular Genetics*, *23*(3), 602–617. https://doi.org/10.1093/hmg/ddt448

Nornes, S., Newman, M., Wells, S., Verdile, G., Martins, R. N., & Lardelli, M. (2009). Independent and cooperative action of Psen2 with Psen1 in zebrafish embryos. *Experimental Cell Research*, *315*(16), 2791–2801. https://doi.org/10.1016/j.yexcr.2009.06.023

Piaceri, I., Nacmias, B., & Sorbi, S. (2013). Genetics of familial and sporadic Alzheimer's disease. *Frontiers in Bioscience (Elite Edition)*, *5*(1), 167–177. https://doi.org/10.2741/e605

Pittman, A. M., Fung, H. C., & de Silva, R. (2006). Untangling the tau gene association with neurodegenerative disorders. *Human Molecular Genetics*, *15*(2), R188–R195. https://doi.org/10.1093/hmg/ddl190

Qosa, H., Abuasal, B. S., Romero, I. A., Weksler, B., Couraud, P. O., Keller, J. N., & Kaddoumi, A. (2014). Differences in amyloid-β clearance across mouse and human blood-brain barrier models: Kinetic analysis and mechanistic modeling. *Neuropharmacology*, *79*, 668–678. https://doi.org/10.1016/j.neuropharm.2014.01.023

Raska, J., Hribkova, H., Klimova, H., Fedorova, V., Barak, M., Barta, T., Pospisilova, V., Vochyanova, S., Vanova, T., & Bohaciakova, D. (2021). Generation of six human iPSC lines from patients with a familial Alzheimer's disease (n = 3) and sex- and age-matched healthy controls (n = 3). *Stem Cell Research*, *53*, 102379. https://doi.org/10.1016/j.scr.2021.102379

Reiman, E. M., Arboleda-Velasquez, J. F., Quiroz, Y. T., Huentelman, M. J., Beach, T. G., Caselli, R. J., Chen, Y., Su, Y., Myers, A. J., Hardy, J., Paul Vonsattel, J., Younkin, S. G., Bennett, D. A., De Jager, P. L., Larson, E. B., Crane, P. K., Keene, C. D., Kamboh, M. I., Kofler, J. K., Duque, L., … Alzheimer's Disease Genetics Consortium. (2020). Exceptionally low likelihood of Alzheimer's dementia in APOE2 homozygotes from a 5,000-person neuropathological study. *Nature Communications*, *11*(1), 667. https://doi.org/10.1038/s41467-019-14279-8

Rohn T. T. (2013). Proteolytic cleavage of apolipoprotein E4 as the keystone for the heightened risk associated with Alzheimer's disease. *International Journal of Molecular Sciences*, *14*(7), 14908–14922. https://doi.org/10.3390/ijms140714908

Rohn, T. T., Catlin, L. W., Coonse, K. G., & Habig, J. W. (2012). Identification of an amino-terminal fragment of apolipoprotein E4 that localizes to neurofibrillary tangles of the Alzheimer's disease brain. *Brain Research*, *1475*, 106–115. https://doi.org/10.1016/j.brainres.2012.08.003

Saito, Y., & Murayama, S. (2000). Expression of tau immunoreactivity in the spinal motor neurons of Alzheimer's disease. *Neurology*, *55*(11), 1727–1729. https://doi.org/10.1212/wnl.55.11.1727

Santana, S., Rico, E. P., & Burgos, J. S. (2012). Can zebrafish be used as animal model to study Alzheimer's disease?. *American Journal of Neurodegenerative Disease*, *1*(1), 32–48.

Sarasija, S., & Norman, K. R. (2018a). Role of presenilin in mitochondrial oxidative stress and neurodegeneration in caenorhabditis elegans. *Antioxidants (Basel, Switzerland)*, *7*(9), 111. https://doi.org/10.3390/antiox7090111

Sarasija, S., Laboy, J. T., Ashkavand, Z., Bonner, J., Tang, Y., & Norman, K. R. (2018b). Presenilin mutations deregulate mitochondrial Ca2+ homeostasis and metabolic activity causing neurodegeneration in Caenorhabditis elegans. *eLife*, *7*, e33052. https://doi.org/10.7554/eLife.33052

Sassen, W. A., & Köster, R. W. (2015). A molecular toolbox for genetic manipulation of zebrafish. *Advances in Genomics and Genetics*, *5*, 151–163. https://doi.org/10.2147/AGG.S57585

Saunders, A. M., Strittmatter, W. J., Schmechel, D., George-Hyslop, P. H., Pericak-Vance, M. A., Joo, S. H., Rosi, B. L., Gusella, J. F., Crapper-MacLachlan, D. R., & Alberts, M. J. (1993). Association of apolipoprotein E allele epsilon 4 with late-onset familial and sporadic Alzheimer's disease. *Neurology*, *43*(8), 1467–1472. https://doi.org/10.1212/wnl.43.8.1467

Sehar, U., Kopel, J., & Reddy, P. H. (2023). Alzheimer's disease and its related dementias in US Native Americans: A major public health concern. *Ageing Research Reviews*, *90*, 102027. https://doi.org/10.1016/j.arr.2023.102027

Sehar, U., Rawat, P., Reddy, A. P., Kopel, J., & Reddy, P. H. (2022). Amyloid beta in aging and Alzheimer's disease. *International Journal of Molecular Sciences*, *23*(21), 12924. https://doi.org/10.3390/ijms232112924

Serý, O., Povová, J., Míšek, I., Pešák, L., & Janout, V. (2013). Molecular mechanisms of neuropathological changes in Alzheimer's disease: A review. *Folia Neuropathologica*, *51*(1), 1–9. https://doi.org/10.5114/fn.2013.34190

Shenoy, A., Banerjee, M., Upadhya, A., Bagwe-Parab, S., & Kaur, G. (2022). The brilliance of the zebrafish model: Perception on behavior and Alzheimer's disease. *Frontiers in Behavioral Neuroscience*, *16*, 861155. https://doi.org/10.3389/fnbeh.2022.861155

Sherrington, R., Rogaev, E. I., Liang, Y., Rogaeva, E. A., Levesque, G., Ikeda, M., Chi, H., Lin, C., Li, G., Holman, K., Tsuda, T., Mar, L., Foncin, J. F., Bruni, A. C., Montesi, M. P., Sorbi, S., Rainero, I., Pinessi, L., Nee, L., Chumakov, I., … St George-Hyslop, P. H. (1995). Cloning of a gene bearing missense mutations in early-onset familial Alzheimer's disease. *Nature*, *375*(6534), 754–760. https://doi.org/10.1038/375754a0

Shi, Y., Yamada, K., Liddelow, S. A., Smith, S. T., Zhao, L., Luo, W., Tsai, R. M., Spina, S., Grinberg, L. T., Rojas, J. C., Gallardo, G., Wang, K., Roh, J., Robinson, G., Finn, M. B., Jiang, H., Sullivan, P. M., Baufeld, C., Wood, M. W., Sutphen, C., … Holtzman, D. M. (2017). ApoE4 markedly exacerbates tau-mediated neurodegeneration in a mouse model of tauopathy. *Nature*, *549*(7673), 523–527. https://doi.org/10.1038/nature24016

Sundvik, M., Chen, Y. C., & Panula, P. (2013). Presenilin1 regulates histamine neuron development and behavior in zebrafish, danio rerio. *The Journal of Neuroscience: The Official Journal of the Society for Neuroscience*, *33*(4), 1589–1597. https://doi.org/10.1523/JNEUROSCI.1802-12.2013

Tai, L. M., Balu, D., Avila-Muñoz, E., Abdullah, L., Thomas, R., Collins, N., Valencia-Olvera, A. C., & LaDu, M. J. (2017). EFAD transgenic mice as a human APOE relevant preclinical model of Alzheimer's disease. *Journal of Lipid Research*, *58*(9), 1733–1755. https://doi.org/10.1194/jlr.R076315

Tang, N., & Kepp, K. P. (2018). Aβ42/Aβ40 ratios of presenilin 1 mutations correlate with clinical onset of Alzheimer's disease. *Journal of Alzheimer's Disease: JAD*, *66*(3), 939–945. https://doi.org/10.3233/JAD-180829

Tarasoff-Conway, J. M., Carare, R. O., Osorio, R. S., Glodzik, L., Butler, T., Fieremans, E., Axel, L., Rusinek, H., Nicholson, C., Zlokovic, B. V., Frangione, B., Blennow, K., Ménard, J., Zetterberg, H., Wisniewski, T., & de Leon, M. J. (2015). Clearance systems in the brain-implications for Alzheimer disease. Nature reviews. *Neurology*, *11*(8), 457–470. https://doi.org/10.1038/nrneurol.2015.119

Toda, T., Noda, Y., Ito, G., Maeda, M., & Shimizu, T. (2011). Presenilin-2 mutation causes early amyloid accumulation and memory impairment in a transgenic mouse model of Alzheimer's disease. *Journal of Biomedicine & Biotechnology*, *2011*, 617974. https://doi.org/10.1155/2011/617974

Verghese, P. B., Castellano, J. M., & Holtzman, D. M. (2011). Apolipoprotein E in Alzheimer's disease and other neurological disorders. *The Lancet. Neurology*, *10*(3), 241–252. https://doi.org/10.1016/S1474-4422(10)70325-2

Vitek, M. P., Araujo, J. A., Fossel, M., Greenberg, B. D., Howell, G. R., Rizzo, S. J. S., Seyfried, N. T., Tenner, A. J., Territo, P. R., Windisch, M., Bain, L. J., Ross, A., Carrillo, M. C., Lamb, B. T., & Edelmayer, R. M. (2021). Translational animal models for Alzheimer's disease: An Alzheimer's association business consortium think tank. *Alzheimer's & Dementia (New York, N. Y.)*, *6*(1), e12114. https://doi.org/10.1002/trc2.12114

Wang, L. W., Huttner, I. G., Santiago, C. F., Kesteven, S. H., Yu, Z. Y., Feneley, M. P., & Fatkin, D. (2017). Standardized echocardiographic assessment of cardiac function in normal adult zebrafish and heart disease models. *Disease Models & Mechanisms*, *10*(1), 63–76. https://doi.org/10.1242/dmm.026989

Wildsmith, K. R., Holley, M., Savage, J. C., Skerrett, R., & Landreth, G. E. (2013). Evidence for impaired amyloid β clearance in Alzheimer's disease. *Alzheimer's Research & Therapy*, *5*(4), 33. https://doi.org/10.1186/alzrt187

Xin, S. H., Tan, L., Cao, X., Yu, J. T., & Tan, L. (2018). Clearance of amyloid beta and tau in Alzheimer's disease: From mechanisms to therapy. *Neurotoxicity Research*, *34*(3), 733–748. https://doi.org/10.1007/s12640-018-9895-1

Zheng, H., & Koo, E. H. (2011). Biology and pathophysiology of the amyloid precursor protein. *Molecular Neurodegeneration*, *6*(1), 27. https://doi.org/10.1186/1750-1326-6-27

# 18

# *Neurodegenerative Effects of Boron and Other Elements on Zebrafish and Zebrafish Embryos*

**A. Begüm Özel-Korlu and Ayşen Yarat**

## Abbreviations

| | |
|---|---|
| **AChE** | Acetylcholinesterase |
| **Akt** | Protein Kinase B |
| **Al** | Aluminum |
| **ALS** | Amyotrophic Lateral Sclerosis |
| **As** | Arsenic |
| **B** | Boron |
| **CAT** | Catalase |
| **Cd** | Cadmium |
| **CNS** | Central Nervous System |
| **Cr** | Chromium |
| **Co-NMS** | Cobalt Complexes Bearing NMS Ligand |
| **Co** | Cobalt |
| **Cu** | Copper |
| **EAAT4** | Excitatory Amino Acid Transporter 4 |
| **GPX** | Glutathione Peroxidase |
| **Hg** | Mercury |
| **Hpf** | Hours Post-Fertilization |
| **MDA** | Malondialdehyde |
| **MeHg** | Methylmercury |
| **Mg** | Magnesium |
| **MgT** | Magnesium-L-Threonate |
| **Mn** | Manganese |
| **mTOR** | Rapamycin |
| **NAD** | Nicotinamide Adenine Dinucleotide |
| **ND** | Neurodegenerative Disease |
| **Ni** | Nickel |
| **NMDA** | N-Methyl-D-Aspartate |
| **NRGN** | Neurogranin |
| **Pb** | Lead |
| **PI3K** | Phosphatidylinositol 3-Kinase |
| **PPAR** | Peroxisome Proliferator-Activated Receptors |
| **PQ** | Paraquat |
| **ROS** | Reactive Oxygen Species |
| **SAM** | S-Adenosylmethionine |
| **Se** | Selenium |
| **SeNs** | Selenium Nanoparticles |
| **Si** | Silicon |
| **SOD** | Superoxide Dismutase |
| **SodSe** | Sodium Selenite |
| **Ti** | Titanium |
| **Sn** | Tin |
| **TiO2 NPs** | Titanium Dioxide Nanoparticles |
| **TiO2** | Titanium Dioxide |
| **WHO** | World Health Organization |
| **Zn** | Zinc |
| **ZnONPs** | Zinc Oxide Nanoparticles |

## Introduction

Neurodegeneration involves the progressive loss of neurons, their structure, and their functions. Diseases characterized by neurodegeneration are collectively referred to as neurodegenerative diseases (NDs). Alzheimer's disease, Parkinson's disease, amyotrophic lateral sclerosis (ALS), Huntington's disease, prion disease, spinocerebellar ataxia, motor neuron disease, and spinal muscular atrophy are common NDs (Lamptey et al., 2022). Millions of people are affected by NDs globally. According to the World Health Organization (WHO), after cardiovascular disease, NDs are projected to become the second leading cause of death (Wang et al., 2021).

Zebrafish (*Danio rerio*) are tropical freshwater fish originating from southern Asia. George Streisinger introduced zebrafish into biological research in the 1970s, and since then, they have become one of the primary preclinical animal models (Tan et al., 2022).

Zebrafish are considered a suitable model for conducting experimental research in mammalian neurobiology, especially for understanding the functions of the human brain (Paduraru et al., 2023).

The increasing interest in zebrafish as a model species is due to their numerous advantageous features (Figure 18.1). They have a rapid developmental cycle and high spawning rates, reaching sexual maturity by 12 weeks and producing hundreds of embryos weekly. This efficiency reduces time and cost while minimizing individual variability and increasing the study's throughput. Zebrafish embryos develop outside the body, which permits experimental manipulation and


DOI: 10.1201/9781032676555-20

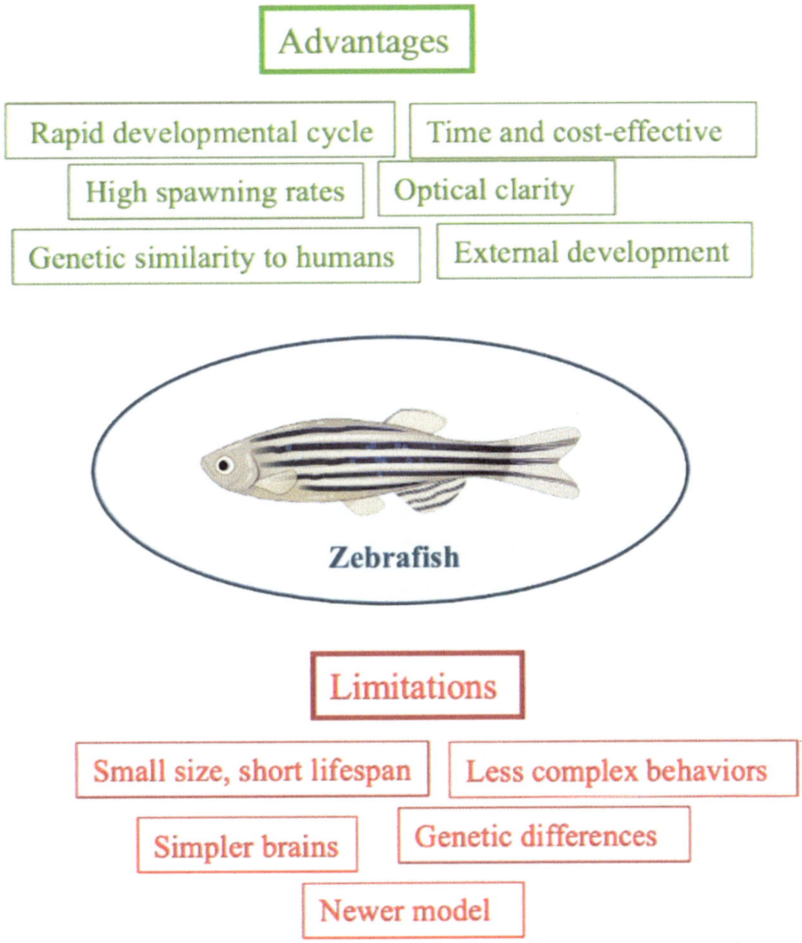


**FIGURE 18.1** Advantages and limitations of zebrafish as a model organism in biomedical research compared to mammalian models.

observation starting from the one-cell stage. Within 72 hours post-fertilization (hpf), all major organs and the central nervous system (CNS) fully develop. Their transparency makes them ideal for optical manipulation and visualization of the entire CNS (Adhish & Manjubala, 2023; Chia et al., 2022; Veldman & Lin, 2008).

Zebrafish exhibit high genomic similarity to humans, sharing at least one ortholog with over 70% of human genes. This genetic overlap encompasses numerous genes linked to NDs, such as Parkinson's, Alzheimer's, ALS, and Huntington's. The zebrafish brain shares both structural and functional similarities with the mammalian brain. The neuroendocrine functions and neurotransmitter systems of zebrafish are closely parallel to those of mammals (Chia et al., 2022).

The zebrafish model also has some limitations and disadvantages (Figure 18.1). Their brains are simpler than those of mammals, with some regions either missing or simplified, making it challenging to apply results from zebrafish studies to humans directly. Furthermore, modeling complex behaviors and cognitive functions in zebrafish can be difficult due to the less complex nature of their behaviors than humans. Despite genetic similarities between zebrafish and humans, notable genetic differences limit the generalizability of zebrafish findings to mammals. While zebrafish and humans share genetic homology, there are significant genetic differences. Additionally, their small size and short lifespan, and the fact they are a relatively new model compared to mice and rats, add to the challenges in conducting detailed and long-term research (Doszyn et al., 2024; Paduraru et al., 2023; d'Amora & Giordani, 2018).

Several elements such as boron (B), zinc (Zn), magnesium (Mg), selenium (Se), iron, manganese (Mn), copper (Cu), lead (Pb), mercury (Hg), arsenic (As), cadmium (Cd), chromium (Cr), aluminum (Al), silicon (Si), cobalt (Co), nickel (Ni), titanium (Ti) and tin (Sn) have been studied for their neurodegenerative or potential neuroprotective effects on zebrafish. These elements interact with various cellular pathways and mechanisms to exert their effects. Among these, arsenic, mercury, and cadmium are considered to be among the most toxic metals. They are listed in the substance priority list of the Agency for Toxic Substances and Disease Registry (ATSDR) (ATSDR, 2022). Although the precise mechanisms have not been identified, heavy metals can cause neurotoxicity through molecular pathways such as oxidative stress, neuroinflammation, and mitochondrial dysfunction (Figure 18.2). Research in this area continues to uncover the specific mechanisms and potential therapeutic applications of these elements in protecting neural tissues and promoting healthy neurodevelopment in zebrafish.

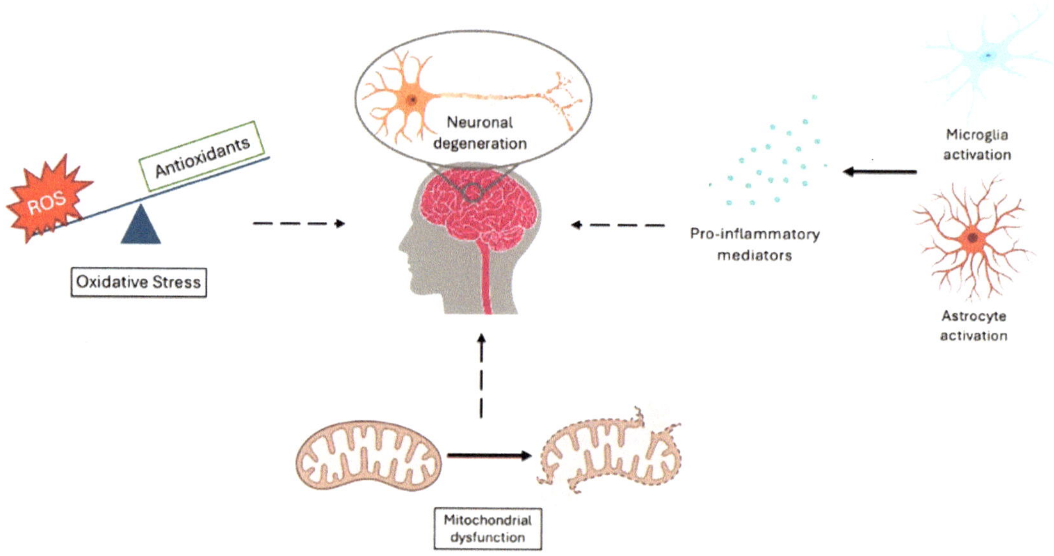


**FIGURE 18.2** Key mechanisms of neurodegeneration include oxidative stress, neuroinflammation, and mitochondrial dysfunction.

## Major Elements That Have Neurodegenerative Effects on Zebrafish and Zebrafish Embryos

### Boron

B is a trace mineral that is considered a probably essential element by the WHO. It supports bone growth and brain functions and regulates hormone activity, while also alleviating arthritic symptoms (Turkez et al., 2022). B is important for growth in plants, animals, and humans. Recent studies revealed that B supplementation strengthens antioxidant defenses and provides significant protection against oxidative damage (Turkez et al., 2022; Kan & Kucukkurt, 2023). It reduces heavy-metal toxicity and boosts antioxidant enzymes such as superoxide dismutase (SOD), glutathione peroxidase (GPX), and catalase (CAT), providing cellular and genetic protection (Aliem et al., 2023). B-based compounds, such as boric acid ($H_3BO_3$), borax ($Na_2B_4O_7{\cdot}10H_2O$), ulexite ($NaCaB_5O_9{\cdot}8H_2O$), and colemanite ($Ca_2B_6O_{11}{\cdot}5H_2O$), also known as natural borates, possess anti-inflammatory, antibacterial, antifungal, antiviral, antimutagenic, and anticancer properties (Turkez et al., 2022).

Determining the primary mechanisms behind B's bioactivity is challenging. However, its biochemistry may reveal potential explanations for its effects. B forms compounds with elements such as sodium, calcium, and Mg (Turkez et al., 2022). Many of these are electron-deficient, allowing them to accept protons ($H^+$ ions), leading to the creation of complexes with biologically essential sugars like ribose, a component of adenosine (Alien et al., 2023). Diverse actions of B may result from its interactions with biomolecules containing adenosine, such as S-adenosylmethionine (SAM), which has a high affinity for B. SAM is a key enzyme used in methylation reactions, influencing the activity of DNA, RNA, proteins, phospholipids, hormones, and transmitters (Perlinska et al., 2020). It is suggested that B plays a role in synthesizing auto-inducer-2 (AI-2), a bacterial quorum-sensing molecule that is a furanosyl borate ester (Majdura et al., 2023). Quorum-sensing involves bacterial communication via extracellular signaling molecules (Waters & Bassler, 2005).

Additionally, studies have shown that B deprivation increases plasma homocysteine levels and reduces liver SAM in rats (Nielsen, 2017). Moreover, B can bind to nicotinamide adenine dinucleotide (NAD) and influence reactions involving NAD. It can also form diester borate complexes with glycoproteins, glycolipids, and phosphoinositides in cellular membranes. These interactions could affect the integrity and functionality of the membranes, potentially impacting hormonal responses. The borate transporter NaBC1 regulates boron homeostasis by facilitating the transport of sodium and hydroxide ions across cell membranes, potentially affecting ion transduction and hormonal responses such as those involving vitamin D, insulin, and thyroid hormones (Nielsen, 2017).

B-deficient zebrafish exhibited photophobia, likely due to photoreceptor dystrophy. B is crucial for zebrafish embryogenesis and early development (Nielsen, 2017; Rowe & Eckhert, 1999). Embryonic development halts in zebrafish raised in low B conditions and resumes with B supplementation (Rowe et al., 1998). Rowe and Eckher observed that zygotes from B-deficient parents started to die at the one-cell stage, with early signs of death marked by blebbing of the cell and yolk-sac membranes. However, B supplementation halted this early mortality in the low-B zygotes (Rowe & Eckhert, 1999).

Alak et al. investigated the effects of ulexite, a B mineral containing calcium and sodium, on the zebrafish brain by evaluating oxidative parameters. Adult zebrafish were exposed to four different ulexite concentrations (5, 10, 20, and 40 mg/L) for 96 hours. Biochemical analyses of brain tissues revealed that low doses of ulexite did not induce oxidative stress, DNA damage, or apoptosis in the zebrafish brain. The findings suggest that ulexite can be safely used in aquatic environments (Alak et al., 2021).

The effects of colemanite on oxidative parameters in the brain, liver, and blood tissues of zebrafish were also investigated. Zebrafish were exposed to colemanite at different concentrations (5, 10, and 20 mg/L) for 96 hours. Various biochemical analyses were conducted on brain and liver tissues including measuring DNA damage (8-OHdG), apoptosis (Caspase-3), and antioxidant and lipid peroxidation levels. The results showed that low doses of colemanite enhanced the antioxidant system without causing oxidative stress, DNA damage, or apoptosis in zebrafish, and it was concluded that colemanite can be used safely and provide positive impacts on aquatic environments (Alak et al., 2020).

Santos et al. evaluated the effects of boron nanoparticles (BNPs) and vanadium nanoparticles (VNPs) on zebrafish embryos and larvae. BNPs and VNPs showed different impacts in various assays. BNPs caused no mortality or hatching delays but induced malformations at 10 mg/L, while VNPs led to hatching delays and malformations at the same concentration. Locomotor behavior tests revealed increased swimming distance with VNPs and changes in path angle frequencies with both NPs. Biochemically, BNPs had minimal effects, only reducing triglyceride content and GST activity at both concentrations, with no impact on lipid, carbohydrate, protein contents, or cholinesterase activity. In contrast, VNPs significantly altered biochemical markers by increasing lipid content, reducing carbohydrate and triglyceride levels, and decreasing cholinesterase and GST activities, except for protein content (Santos et al., 2021).

Like any element, B can be toxic in high concentrations. Excessive amounts have been shown to cause adverse effects, including impaired growth rates, DNA damage, oxidative stress, histopathological changes, and reduced feed intake in various fish species, such as Oncorhynchus mykiss (Alak et al., 2019; Oz et al., 2020) and Cirrhinus mrigala (Adhikari & Mohanty, 2012).

The acute genotoxic effects of boric acid (BA) and borax (BX) on zebrafish were examined, revealing that both compounds cause significant DNA damage in fish erythrocytes. Zebrafish were exposed to BA and BX concentrations ranging from 1 to 64 mg/L, with daily media changes to ensure stable exposure. DNA damage, measured using the comet assay, increased significantly with BA and BX treatments compared to controls. In the BA group, the highest DNA damage occurred after 24 hours at 64 mg/L but did not consistently rise over time, while in the BX group, DNA damage was dose- and time-dependent, peaking after 96 hours. The findings suggest that BA and BX have different genotoxic effects, with BX causing more consistent DNA damage (Gülsoy et al., 2015). Although this study was conducted using blood samples, DNA damage observed in blood can sometimes reflect broader systemic effects, including brain damage. Exposure to neurotoxic substances can result in systemic oxidative stress, and DNA damage can be an essential marker of this stress, offering valuable insights into neurodegenerative processes (Cao et al., 2021; Martínez & Segura, 2019; Ohayon et al., 2012).

## Zinc

Zn is an essential trace element for animal and human health, playing crucial catalytic, structural, and signaling roles in the body (Takesono et al., 2023). It regulates synaptic transmission, reduces the harmful effects of glutamate, and boosts antioxidant enzyme activity, thus protecting neurons from oxidative stress (Sensi et al., 2011).

In the brain, ionic Zn ($Zn^{2+}$) ions are found in high concentrations at many glutamatergic nerve terminals, where they are released during neuronal activity and play a vital role in brain functions. However, they can also be toxic, contributing to neuronal and glial death in conditions like epilepsy, ischemia, and brain trauma. Due to their presence in amyloid plaques and cerebral amyloid angiopathy, they are also linked to Alzheimer's disease (Sensi et al., 2009).

Zn deficiency is a major global health issue linked to diseases such as depression, growth failure, and endocrine disorders, according to the WHO (Tuerk & Fazel, 2009).

Conversely, excessive Zn can also be harmful, causing lethargy, neuronal deficits, and immune dysfunction; disrupting brain excitability; and leading to memory issues and neurological problems (Plum et al., 2010; Sarasamma et al., 2018).

However, the chronic toxic effects of Zn on aquatic vertebrates are still ambiguous. A study investigated Zn accumulation in zebrafish after chronic Zn exposure and toxic effects on the intestines, muscles, and gills. The results showed that after 25 days of exposure to Zn, there was a marked accumulation in the intestine, muscle tissue, and gills of zebrafish. The primary toxic effects observed included oxidative stress, possible neurotoxic impacts, and changes in the intestinal microbiome. Significant alterations were noted in the levels of SOD, CAT, metallothionein, glutathione, and malondialdehyde (MDA), suggesting that Zn compromised the antioxidant defenses of adult zebrafish. Moreover, Zn exposure significantly reduced acetylcholinesterase (AChE) activity, which is linked to abnormal neural signaling. The exposure also led to an increase in the diversity of intestinal microbes while decreasing the Simpson index in these fish. Additionally, the predominant microbial communities at both the phylum and genus levels in the intestines were altered due to Zn exposure (Yuana et al., 2023).

Zn oxide nanoparticles (ZnONPs) are widely produced and used in various products (e.g., industrial products, pharmaceuticals, food additives, cosmetics, and sunscreens) due to their beneficial properties, such as UV absorption, antibacterial actions, and catalytic functions. (Sirelkhatim et al., 2015; Vimercati et al., 2020). They are also used as micronutrients in fertilizers and animal feed (Singh et al., 2022). Despite their widespread use, concerns have emerged regarding the potential health and environmental risks of ZnONPs. Research by Takesonoa et al. revealed that ZnONPs negatively impact the olfactory sensory system in zebrafish, leading to impairments in behavior related to olfaction (Takesono et al., 2023).

The study exposed a 6-month-old female zebrafish to environmentally relevant Zn concentrations for 6 weeks to assess Zn accumulation in the brain and intestine, behavioral changes, and underlying mechanisms. Zn significantly accumulated, leading to anxiety-like and altered social behaviors. It affected neurotransmitter levels, which were directly associated with changes in behavior, caused oxidative damage and mitochondrial dysfunction, and disrupted energy metabolism in the brain. Zn exposure also led to disruption of DNA replication and cell cycle, affecting intestinal cell renewal and impaired carbohydrate and peptide metabolism (Yu et al., 2023).

## Magnesium

Mg is one of the most abundant ions in the body. Beyond acting as a cofactor for over 300 enzymatic reactions and participating in numerous biochemical pathways, it plays a vital role in various physiological processes. It is essential for muscle contraction, nerve signal transmission, maintaining the integrity of the blood–brain barrier, blood pressure regulation, insulin metabolism, and DNA synthesis (Kirkland et al., 2018; Maier et al., 2022).

One of the primary neurological functions of Mg is its role in modulating the N-methyl-D-aspartate (NMDA) receptor. By inhibiting NMDA receptor activity, Mg prevents prolonged stimulation that can lead to neuronal death. It also prevents apoptosis by blocking the opening of the mitochondrial permeability transition pore (Maier et al., 2022; Slutsky et al., 2010).

Dietary and environmental factors significantly influence learning and memory. Research has shown that elevating brain Mg levels by administering Mg-L-threonate (MgT) enhances both short-term and long-term memory, learning abilities, and working memory in rats (Slutsky et al., 2010).

Additionally, Mg deficiency is associated with chronic low-grade brain inflammation, which can lead to neurodegenerative disorders. Research indicates a connection between Mg imbalance and diseases such as multiple sclerosis, Alzheimer's disease, and Parkinson's disease. Therefore, Mg supplementation might offer therapeutic advantages in the management of neurodegeneration (Maier et al., 2022). Kim et al. used a hypoxic zebrafish model to investigate the impact of Mg on cognition. Their findings demonstrated that MgT treatment improved neuronal viability under hypoxic conditions, reduced cerebral infarction, and maintained learning and memory in zebrafish post-hypoxia. The study also focused on excitatory amino acid transporter 4 (EAAT4), a glutamate receptor responsible for clearing glutamate. Dysfunction in EAAT4, leading to excessive extracellular glutamate, is associated with various NDs. MgT treatment significantly increased EAAT4 levels, whereas hypoxia led to a reduction in EAAT4 expression. This upregulation of EAAT4 by MgT appeared to safeguard the brain from hypoxic damage (Kim et al., 2020).

## Selenium

Se is an essential micronutrient. It is a cofactor for antioxidant enzymes like GPX, which helps diminish oxidative stress and protects cells from oxidative damage (Cardoso et al., 2015). It is essential for brain health, contributing to redox balance, mitochondrial function, calcium channel regulation, and neurogenesis. Additionally, Se is necessary for the normal growth and development of vertebrates (Cardoso et al., 2015; Lin et al., 2024). Recent studies indicate that Se may play a vital role in the development of NDs due to its impact on oxidative damage and cellular homeostasis (Zhou et al., 2023).

In vertebrate proteins, Se exists in the form of selenocysteine. Eighteen zebrafish genes encoding selenoproteins were identified, including orthologs of human SelT, GPX 1, and GPX4, along with single orthologs of several other selenoproteins. Notably, three of these zebrafish selenoproteins do not have known mammalian counterparts. The zebrafish genome includes a selenoprotein that has the highest number of selenocysteine residues identified so far (Kryukov & Gladyshev, 2000).

The impact of Se on organisms is often described as a "double-edged sword," as both deficiency and excess can pose health risks to humans and animals. To better understand Se's nutritional benefits and potential toxicity, extensive research has been conducted using zebrafish as a model organism. Despite this, comprehensive research on how Se affects zebrafish is lacking. Research highlights the benefits of Se for zebrafish, including providing anti-tumor effects, promoting growth, boosting immune function, and counteracting pollutants. Conversely, Se toxicity in zebrafish can affect the nervous system and behavior, reproductivity, and growth and cause damage to various organs (Lin et al., 2024).

The neuroprotective effects of sodium selenite against paraquat (PQ)-induced neurotoxicity was explored in zebrafish, focusing on behavioral and biochemical parameters. PQ is a hazardous herbicide commonly employed in agricultural practices due to its cost-effectiveness and swift efficacy. PQ treatment led to reduced distance traveled, increased freezing episodes, increased time spent in the top area of the tank, and heightened aggressive behavior, indicating anxiety-like responses. The Se diet effectively mitigated these adverse behavioral effects. Biochemically, PQ treatment significantly increased oxidative and nitrosative damage markers, such as TBARS, carbonylated protein levels, and reactive oxygen species (ROS) levels, while the Se diet reduced this oxidative damage. Antioxidant enzyme activities (SOD, CAT, GPx, GR, GST) were altered by PQ, with Se pretreatment enhancing the activity of several antioxidant enzymes, counteracting PQ's effects. PQ treatment also decreased NPSH levels, a marker of antioxidant capacity, which was prevented by Se pretreatment (Müller et al., 2018).

On the other hand, it has been suggested that an excess of Se can be toxic. Zhao et al. investigated how excessive Se affects zebrafish embryos, focusing on its impact on the nervous system and locomotor behavior. Exposure to 0.5 μM Se reduced movement speed, distance, and touch response, causing impairments such as increased cell apoptosis, proliferation, neuroinflammation, and decreased neuron count. Gene expression related to various neurons and neurotransmitter transport was significantly reduced. Se also downregulated genes related to peroxisome proliferator-activated receptors (PPAR) signaling. PPARs are non-steroid nuclear receptors that play a key role in nervous system development and function and can mitigate neurological dysfunctions and improve locomotor activity (Zhao et al., 2022).

Another study investigated the effects of Se nanoparticles (SeNs) and sodium selenite (SodSe) on zebrafish embryos and larvae. SeNs exhibited lower toxicity compared to SodSe, with higher survival rates and less pronounced developmental defects. SodSe caused more severe deformities at lower concentrations, indicating higher toxicity. Both SeNs and SodSe induced apoptosis in zebrafish larvae, with SodSe having a more pronounced effect. Behavioral assays revealed that increasing concentrations of SeNs and SodSe led to anxiety-like symptoms, irregular swimming patterns, and increased cortisol levels, with SodSe having a more deleterious impact. Additionally, both SeNs and SodSe inhibited AChE activity, disrupting cholinergic signaling and indicating potential neurological toxicity (Hariharan et al., 2024).

Larval zebrafish was used to compare the toxicity and tissue distribution of different Se compounds. Organic Se forms, such as Se-methyl-L-selenocysteine and L-selenomethionine, were significantly more toxic than inorganic forms like selenite and selenate, with toxicity levels linked to bioaccumulation. Se was found to accumulate notably in melanin-rich tissues, indicating that these pigmented tissues, rather than the brain, might function as reservoirs for Se (Dolgova et al., 2016).

## Iron

Iron is vital for numerous cellular functions, such as oxygen transport, energy production, and the synthesis of DNA. In the brain, it supports neuronal development, myelination, neurotransmitter synthesis and breakdown, electron transport, and respiration (Rouault, 2013, McCann et al., 2020). Although altered brain iron homeostasis is a characteristic feature of many NDs, it remains uncertain whether iron plays a primary role in the neurodegenerative process or if its alteration is a result of the degenerative process (Levi et al., 2024).

Both iron deficiency and iron overload significantly impact cellular health. Iron deficiency hinders the development of neurons and glial cells, causing cognitive impairments and neuropsychiatric conditions like depression and anxiety (Levi et al., 2024; Gao et al., 2023). In contrast, excess iron can be harmful to the brain. It leads to oxidative stress due to its electron exchange capability, producing free radicals that can damage tissue components, resulting in neurotoxicity and exacerbating neurological diseases (Levi et al., 2024; Salvador et al., 2010).

Various theories propose that brain iron contributes to cognitive decline. One key mechanism involves Fenton's reaction, where excess iron causes the production of highly reactive hydroxyl radicals. These radicals can further trigger the release of iron from mitochondrial iron-sulphur cluster proteins and storage proteins, leading to a cycle of increased ROS generation. When the ROS and free radicals surpass the brain cells' antioxidant defenses, oxidative stress ensues, causing damage to DNA, lipids, and proteins; disrupting mitochondria; and eventually leading to neuronal death. Additionally, ferroptosis, an iron-dependent form of necrosis characterized by shrunken mitochondria and outer membrane rupture, may also play a role in neurodegeneration (Wu et al., 2023).

Zebrafish absorb iron from the water through their gastrointestinal tracts and gills. The uptake of iron in the gills has both high- and low-affinity components. At low iron concentrations (<40 nmol/L), the Km is 5.9 nmol/l Fe and the Vmax is 2.1 pmol/g·h. At higher iron concentrations (40–200 nmol/L), the uptake rate increases linearly. High levels of cadmium (Cd) can block iron transport in zebrafish gills. Additionally, zebrafish that were fed a low-iron diet displayed a notable rise in the accumulation of cadmium in their tissues. This indicates a possible relationship between the assimilation of iron and cadmium in fish (Zhao et al., 2014).

Elevated brain iron levels have been shown to adversely affect cognitive function. It is implicated in the process of aging and NDs. The distribution of iron in different brain regions is uneven, with high concentrations in certain areas impairing their specific functions. For example, excess iron in the caudate nuclei, hippocampus, and thalamus is associated with poorer memory performance, whereas increased iron in the putamen is linked to reduced overall cognitive abilities. Understanding whether iron buildup is a direct cause of brain atrophy or a consequence of it is essential (Spence et al., 2020).

The effects of iron exposure on larval zebrafish were evaluated, and significant physiological and behavioral disruptions were reported. Iron exposure notably increased whole-body iron content in 3 dpf and 5 dpf larvae, with elevated iron levels also detected in the head region of 5 dpf larvae. Dysregulation of Mn, Zn, Ni, and calcium homeostasis was also observed. The increased generation of ROS indicated oxidative stress–induced damage in specific larval regions. Droplet digital PCR analysis revealed temporal changes in the expression of oxidative stress-responsive genes due to iron exposure. Furthermore, iron-exposed larvae displayed impaired escape responses, reduced swimming activity, and diminished anxiety-like behavior. The results suggested that iron exposure leads to a disruption in metal homeostasis and adversely affects neurophysiological performance in developing zebrafish (Hassan et al., 2020).

Another study examined how iron exposure affects AChE activity, a neurotransmitter crucial for learning, memory, motor control, and autonomic functions, in the brain and liver of zebrafish (Behra et al., 2002). Various iron concentrations were analyzed both in vitro and in vivo. In vitro experiments revealed that high iron levels significantly boosted AChE activity by 52% in the brain and 53% in the liver. Likewise, in vivo experiments demonstrated that exposure to 15 mg/L of iron markedly enhanced AChE activity by 62% in the brain and 70% in the liver. Semiquantitative RT-PCR analysis indicated that there were no significant changes in the mRNA levels of AChE. These findings suggest that iron exposure leads to significant changes in AChE activity (Sant'Anna et al., 2011).

Additionally, another study highlights the necessity of huntingtin (Htt) protein in iron metabolism using a zebrafish model to understand its role in Huntington's disease, a neurodegenerative disorder. It reveals that Htt deficiency leads to signs of cellular iron deficiency despite the presence of iron within the cells, resulting in a temporary reduction in hemoglobin production. The experiments show increased levels of transferrin receptor mRNA levels in Htt-deficient embryos, providing strong support for cellular iron starvation, thus highlighting its critical role in iron homeostasis (Lumsden et al., 2007).

## Manganese

Mn is an essential trace element involved in numerous cellular functions. +2 and +3 are the most biologically relevant oxidation states (Neal et al., 2013). Mn acts as a cofactor for various enzymes such as glutamine synthetase and MnSOD and is crucial for mitochondrial function, antioxidant defense mechanisms, and neurotransmitter synthesis. However, excessive Mn exposure can lead to neurotoxicity and NDs like manganism, characterized by motor and cognitive impairments, since it can accumulate in the brain (Altenhofen et al., 2017).

Mn exposure can occur from various sources and pathways, including mines, industrial particle emissions, and contaminated water, air, food, and soil (Hassani et al., 2012; Okereafor et al., 2020; Wang et al., 2021). Given its neurotoxicity and widespread presence in the environment, increasing attention has been paid to its effects on human health.

Mn-induced neurotoxicity involves several mechanisms, including oxidative stress, inflammation, apoptosis, mitochondrial dysfunction, glutamate excitotoxicity, protein misfolding, endoplasmic reticulum stress, autophagy, and mitophagy (Pajarillo et al., 2022). Different studies use zebrafish models to elucidate the impact of Mn on neuronal health and the underlying mechanisms of its neurotoxic effects (Zhang et al., 2023).

Zebrafish were exposed to varying concentrations of Mn for 30 days, resulting in decreased behavioral abilities and memory capacity, and neural damage. Exposure to low concentrations of Mn promoted neurogenesis, while high concentrations suppressed neurogenesis and disrupted neural function. Differentially expressed genes analysis highlighted genes associated with oxidative stress, inflammation, and neural damage. Gene set variation analysis suggested that pathways related to oxidative damage, inflammation, and apoptosis were activated by Mn exposure. Additionally, it was found that Mn exposure, especially at higher doses, reduced intercellular communication among various cell types (Wang et al., 2024).

In a study, zebrafish embryos were subjected to various concentrations of Mn chloride (25, 50, and 100 μM). It is observed that Mn exposure did not impact the hatching rate of zebrafish larvae. However, it resulted in increased mortality and malformation rates and adversely affected the motor behavior of the zebrafish. Additionally, Mn exposure led to a higher proportion of apoptotic dopaminergic neurons and a notable decrease in tyrosine hydroxylase expression. Moreover, exposure to Mn promoted the expression of genes related to apoptosis and DNA damage, and it may trigger the activation of the MAPK signaling pathway (Xu et al., 2022).

Neurodegeneration and neuronal injuries can lead to variations in the expression of synaptic proteins such as neurogranin (NRGN), which is a calmoduin-binding protein. In humans, conditions such as Alzheimer's disease correspond with decreased brain NRGN expression and increased NRGN levels in cerebrospinal fluid, which are linked to cognitive decline. Alba-González et al. used larval zebrafish and investigated whether Mn overexposure influences NRGN expression and results in behavioral deficits. Their findings indicated that excessive Mn exposure causes significant morphological changes, modifies NRGN expression, and leads to behavioral impairments in zebrafish larvae (Alba-González et al., 2024).

Nadig et al. developed an economical and dependable Parkinson's disease model in zebrafish by chronically exposing them to Mn chloride for 21 days. The Mn chloride–treated fish exhibited Parkinson's disease–like symptoms, including reduced locomotory activity, anxiety, depression-like behavior, and olfactory dysfunction. These behavioral changes were linked to decreased dopamine, increased 3,4-dihydroxyphenylacetic acid levels, and activation of the oxidative stress–mediated apoptotic pathway. Additionally, chronic Mn chloride exposure resulted in decreased brain-derived neurotrophic factor and elevated pro-inflammatory cytokine levels in the zebrafish brain (Nadig et al., 2022).

## Copper

Cu is an important trace element that commonly exists in two oxidation states, $Cu^{+}$ and $Cu^{2+}$, in the body. It is abundant in the brain due to its high metabolic and signaling needs. Redox properties of Cu enable it to be both beneficial and harmful to cells (An et al., 2022). Cu plays a crucial role in various important physiological processes in the brain, including energy production, iron metabolism, antioxidant defense, cellular signaling, catecholamine homeostasis, neuronal myelin formation, synaptic transmission, and gene transcription (Scheiber et al., 2014). It acts as a cofactor for several key enzymes such as Cu/Zn SOD, ceruloplasmin, cytochrome c oxidase, tyrosinase, dopamine-β-hydroxylase, and peptidylglycine α-amidating mono-oxygenase (Desai & Kaler, 2008). However, the same beneficial redox properties can lead to oxidative stress and neurodegeneration if Cu levels are not properly regulated. Excessive intake or dysregulation of its homeostasis can result in its accumulation in the brain, leading to cytotoxicity or cell death (Gromadzka et al., 2020). Therefore, the absorption, distribution, and excretion of Cu are tightly regulated to maintain appropriate levels and prevent toxicity. Research suggests that neurodegeneration may be connected to both increased and decreased Cu bioavailability (Scheiber et al., 2013).

It has been demonstrated that Cu deficiency harms both the embryonic growth and survival of zebrafish. Research using drugs that bind to Cu has revealed a high sensitivity of the notochord to Cu deficiency. This sensitivity has been attributed to the inactivation of the lysyl oxidase, which is a cuproenzyme (Gansner et al., 2007). The importance of Cu was also highlighted genetically. A mutation in the atp7a transporter gene, known for its association with Menkes disease, results in a similar phenotype (Mendelsohn et al., 2006). Additionally, the knockdown of high-affinity Cu transporter 1, a transmembrane protein that facilitates the uptake of Cu ions into cells, results in the death of the larvae approximately 4 days post-fertilization (Mackenzie et al., 2004).

On the other hand, another study examined the effects of Cu exposure on the early life stages of zebrafish, specifically looking at mortality, growth, neurogenesis, and epigenetic changes. Results showed increased mortality and decreased body length in Cu-exposed zebrafish compared to controls. Neurotoxic effects were observed, with significant downregulation of key neurogenesis-related genes such as neuroD and olig2, suggesting disturbances in neural development and differentiation. There was also a notable decrease in the expression of DNA methyltransferase genes, indicating potential epigenetic modifications due to Cu exposure (Santos et al., 2022).

Cu contamination in surface water often arises from activities such as mining, agriculture, and industry. Although acceptable levels of Cu in water vary between countries, they are typically in the range of μg/L. Acosta et al. investigated the potential impact of environmentally relevant Cu concentrations (0, 5, 9, 20, and 60 μg/L) on both larval and adult zebrafish, focusing on assessing morphological and behavioral changes in larval zebrafish, as well as examining the influence on biochemical markers and behavior in adult zebrafish. Cu exposure reduced larval body length at a concentration of 60 μg/L. Larvae exposed to 9 μg/L Cu showed a significant increase in swimming performance, with an increase in distance traveled and an increase in mean velocity compared to the control group. No significant changes in swimming distance or velocity were observed at Cu concentrations of 5 and 60 μg/L. Cu exposure did not affect locomotor activity, as measured by distance traveled and mean velocity, nor did it alter spatial memory or response to novelty in adult zebrafish, except at 60 μg/L. However, fish exposed to 9 μg/L Cu failed to retain memory in the inhibitory avoidance task, showing a significant decrease in latency time compared to the control group. Aggressive behavior and social interaction were not impacted by any Cu concentration tested. Overall, memory retention was disrupted at certain Cu levels, while other behaviors remained unaffected. GST activity reduced significantly in the gills at 9 and 60 μg/L, but not in the brain or muscle. GPx activity increased in the muscle at 20 μg/L but decreased in the gills, while GR activity remained unchanged across all tissues and concentration (Acosta et al., 2016).

Takemura et al. investigated the toxicity of $CuSO_4$ on larval zebrafish at concentrations relevant to their natural environments, identifying a median lethal concentration of 8.4 mg/L at 72 hpf. Lower, non-lethal concentrations of 0.16, 0.32, and 1.6 mg/L were tested, showing significant adverse effects on morphological features. Behavioral assays revealed impairments in the larvae's escape response and swimming abilities, along with elevated levels of ROS and lipid peroxidation. The study also noted decreases in non-protein thiols and overall cellular viability, suggesting toxicity at these concentrations (Takemura et al., 2024).

Pilehvar et al. evaluated how exposure to copper impacted the behavior of zebrafish by using the novel tank test and a visual discrimination learning test in a T-maze. Thirty zebrafish were divided into three groups (Control, 0.77 μM Cu, and 1.52 μM Cu) for 10 days, with Cu administered as $CuSO_4.5H_2O$. In the novel tank test, Cu exposure minimally impacted behavioral responses, with only a temporary increase in freezing bouts in the 1.52 μM Cu group on day 4. In the T-maze test, zebrafish showed associative learning abilities, but Cu exposure at both concentrations significantly impaired their learning performance. Overall, Cu exposure did not affect exploratory behavior in the novel tank test but did impede learning capabilities in the T-maze (Pilehvar et al., 2020).

Another study investigated the effects of Cu overload on apoptotic responses by exposing zebrafish to Cu over periods of 6, 12, 24, and 48 hours. The TUNEL assay revealed increased DNA fragmentation, indicating apoptosis, with significant differences in the relative volume of apoptotic cells at various times compared to controls. mRNA expression analysis demonstrated time-dependent changes in apoptotic genes, including p53, caspase-8, caspase-9, and AIF. These changes varied according to Cu concentration and exposure duration (Luzio et al., 2013).

## Lead

Pb is classified as a post-transition metal. Its atomic number is 82. It can be found in both organic and inorganic forms and typically exhibits oxidation states of +2 or +4 (Lee & Freeman, 2014).

Pb does not serve any physiological role, and even small levels can cause toxicity. It can pass through the blood–brain barrier, accumulating in astrocytic cells and other cell types in the brain. This can result in neurodevelopmental disorders, cognitive impairment, and later neurodegenerative disorders (Shvachiy et al., 2022; Wani et al., 2015).

It interferes with enzymes involved in vitamin D synthesis and maintaining cell membrane integrity and also interferes with DNA transcription (Wani et al., 2015). It can form complexes with glutathione, inhibit GPX, and impair antioxidant mechanisms (Li et al., 2022). Additionally, Pb acts as a potent inhibitor of the NMDA receptor, which is essential for brain development and cognitive functions (Sanders et al., 2009). Pb also imitates calcium and other divalent cations, disrupting various cellular processes that depend on these cations (Komoike et al., 2022).

Several studies have demonstrated the neurotoxic effects of Pb exposure on zebrafish. Dou and Zhang conducted a study to examine the impact of Pb acetate exposure on the early embryonic stages of zebrafish. They observed neurotoxic symptoms, such as sluggish behavior, slow swimming movements, and delayed escape responses, which were found to be dose-dependent. In Pb acetate–treated groups, the expression of the "gfap" and "huC" genes in the brain was diminished. The TUNEL assay additionally showed that the reduction in nerve cells resulted from increased apoptosis of both neurons and glial cells (Dou & Zhang, 2011).

In another study, fertilized zebrafish eggs were exposed to 100 ppb of Pb acetate, a concentration below occupational safety standards, from 6 to 72 hpf. Pb-exposed embryos exhibited early hatching and reduced body length compared to the controls at 48 and 72 hpf. The ROS levels in Pb-exposed embryos were observed to be similar at 24 hpf, increased at 48 hpf, and decreased at 72 hpf compared to the control. The study also noted an upregulation in genes related to oxidative stress response at 48 and 72 hpf, and endoplasmic reticulum stress at 48 hpf in Pb-exposed embryos. These findings suggest that even low-level Pb exposure can cause oxidative and endoplasmic reticulum stresses, leading to developmental disruptions (Komoike et al., 2022).

Xu et al. investigated the impact of Pb nitrate exposure during embryonic stages on the learning abilities of adult zebrafish and their third-generation (F3) descendants. Adult zebrafish exposed to lower concentrations of Pb (0.0 or 0.1 μM) during embryonic stages demonstrated significant learning capabilities during the training and testing phases. Conversely, zebrafish exposed to higher concentrations (1.0 or 10.0 μM) showed no significant improvement in avoidance responses, indicating learning impairments. Similarly, the F3 generation zebrafish exposed to the same Pb concentrations mirrored the learning patterns of their ancestors, with only those exposed to lower concentrations showing enhanced learning abilities. These results suggest that developmental exposure to Pb can induce learning deficits that persist across generations, highlighting the trans-generational impacts of embryonic Pb exposure (Xu et al., 2016).

Additionally, a study revealed that 10 μM Pb acetate exposure led to decreased expression of the Nrxn2a protein in zebrafish. Nrxn2 is one of the genes that encode neurexins, synaptic proteins involved in various neural activities and associated with neurological disorders. This exposure also led to increased apoptosis and changes in locomotor behavior in the zebrafish (Tu et al., 2018).

## Mercury

Hg is classified as a transition metal with an atomic number of 80. Hg can exist in various chemical forms and can be categorized into inorganic Hg ($Hg^0$, $Hg^{2+}$) and organic Hg compounds. Generally, inorganic Hg has low bioavailability, whereas organic Hg compounds, such as methylmercury (MeHg), are more toxic and can accumulate in the nervous system, leading to neurotoxic effects (Wu et al., 2024).

The electrophilic nature of MeHg allows it to readily react with various thiol groups, resulting in the S-mercuration of several proteins. These interactions can change the structure and biological function of the proteins. Also, it leads to oxidative stress because MeHg can covalently bind to antioxidant enzymes and glutathione, disrupting their normal function. This interference compromises the body's antioxidant defenses, leading to increased ROS levels (Novo et al., 2021).

One of the primary mechanisms of MeHg neurotoxicity involves the disruption of glutamate transport. Glutamate, the main excitatory neurotransmitter in the brain, is essential for normal neuronal activity. MeHg also disrupts intracellular calcium ion homeostasis, triggering a cascade of harmful events. These events include the activation of calcium-dependent enzymes that can degrade cellular structures, the production of ROS, and the initiation of apoptotic pathways. Ultimately, these disruptions result in neuronal damage (Abbott & Nigussie, 2021; Branco et al., 2021; Antunes et al., 2018; Farina et al., 2011).

Ho et al. used zebrafish embryos to investigate the neurotoxic effects of MeHg exposure. Their study, which involved whole mount in situ hybridization with 88 genes, confirmed alterations in the expression of 60 genes. Notably, 24 of these genes related to processes such as oxidative stress, apoptosis, acute inflammatory responses, cell redox homeostasis, transcriptional elongation, fatty acid metabolism, and DNA repair exhibited specific changes, indicating potential disruptions in various biological activities in the zebrafish brain due to developmental MeHg exposure (Ho et al., 2013).

Hassan et al. exposed zebrafish embryos to various concentrations of MeHg starting at 6 hpf. Low concentrations (5, 10, and 50 μg/L) did not significantly affect mortality, but mortality increased at 80 μg/L and sharply rose at 200 μg/L, with 100% lethality at 500 μg/L. Hatching rates were significantly delayed at 80 and 100 μg/L. Exposure to 200 μg/L resulted in extremely low hatching rates. A substantial decrease in cellular proliferation was found within the neural tube of embryos exposed to 10, 50, and 80 μg/L of m of MeHg (Hassan et al., 2012).

In another study, zebrafish embryos were exposed to two different Hg compounds, MeHg and $HgCl_2$. An increase in the number of apoptotic brain cells in the zebrafish larvae and downregulation of genes related to neural development were observed. Additionally, Hg exposure led to a decline in tyrosine, dopamine, and tryptophan levels (Zhu et al., 2019).

Assessment of the developmental neurotoxicity of compounds by measuring locomotor activity in zebrafish is an alternative to developmental neurotoxicity testing (Hil et al., 2021). Abu Bakar et al. exposed zebrafish embryos to different levels of $HgCl_2$ from 5 hpf to 72 hpf. They observed that exposure to 100 nM $HgCl_2$ led to locomotor deficits and disrupted anxiety-like behavior in the zebrafish embryos Also, it leads to biochemical alterations in proteins, lipids, carbohydrates, and nucleic acids structures (Abu Bakar et al., 2017).

## Arsenic

As has three ionized states, arsines ($As^{3-}$), trivalent arsenite ($As^{3+}$), and pentavalent arsenate ($As^{5+}$). However, the trivalent or pentavalent states are the most common. All states can exist in either inorganic or organic forms. In general, inorganic forms of As are considered to be more toxic. Among these, trivalent arsenite is known to be the most toxic form (Lee & Freeman, 2014). Arsenate, owing to its similar chemical structure to phosphate, mainly enters the cell through phosphate transporters whereas arsenite ions mainly use aquaglyceroporins (Muzaffar et al., 2023; Sadiku & Rodríguez-Seijo, 2022).

As is a toxic element that has no physiological role in the human body. It can inhibit several biochemical processes (Sadiku & Rodríguez-Seijo, 2022). It can penetrate the blood–brain barrier and accumulate in different regions of the brain. It can alter the levels of neurotransmitters such as serotonin, dopamine, and norepinephrine. It targets specific enzymes involved in energy metabolism. For example, it can inhibit the pyruvate dehydrogenase complex and α-ketoglutarate dehydrogenase complex, thereby blocking the Krebs cycle and oxidative phosphorylation. As also induces neuronal apoptosis through the activation of p38 mitogen-activated protein kinase and c-Jun N-terminal kinase 3 pathways. Furthermore, As inhibits mitochondrial complexes, promoting the uncoupling of oxidative phosphorylation and generating superoxide radicals, which leads to oxidative damage in the brain (Medda et al., 2020; Mochizuki, 2019; Thakur et al., 2021).

Liu et al. highlight the neurotoxic effects of sodium arsenite exposure in zebrafish. It is observed that exposure to As resulted in significant brain damage, including glial scar formation and ventricular enlargement. Swimming ability was significantly reduced compared to the control group. As exposure resulted in a dose-dependent decrease in brain AChE enzyme activity and dopamine and pentraxin levels, an increase in cortisol levels, and a decrease in epinephrine levels. It also led to changes in gene expression associated with neurotransmitter pathways, neuronal development, and the hypothalamic–pituitary–interrenal axis (Liu, et al., 2024).

Another study demonstrated that exposure to sodium arsenite resulted in a significant reduction in social ability, social cognition ability, learning, memory capacity, aggression, and cohesion of zebrafish. Ultrastructural analysis of the zebrafish telencephalon and midbrain revealed brain damage. Furthermore, the study found that As exposure reduced the expression of genes related to cognition, social behavior, and neurodevelopment in the zebrafish brain (Ma et al., 2024).

Chronic exposure to dietary sodium arsenite in zebrafish resulted in impaired latent learning abilities, elevated brain dopamine levels, and increased oxidative stress, characterized by a decreased GSH:GSSG ratio and elevated MDA levels. These effects occurred in a dose-dependent manner. Additionally, significant alterations were observed in the expression of genes involved in dopamine signaling, antioxidant enzyme activity, and neurotrophic factor (Rachamalla et al., 2023).

In a study on the effects of sodium arsenite exposure on juvenile zebrafish, those exposed to 500 ppb had significantly shorter body lengths than the control and 100 ppb groups. Furthermore, exposure to sodium arsenite resulted in a notable rise in ROS levels in the zebrafish brains. This was accompanied by reduced activities of antioxidant enzymes like SOD and CAT, as well as decreased levels of GSH. Behavioral tests showed decreased locomotor activity and social withdrawal, with erratic interactions indicating reduced synchronization (Wang et al., 2023)

## Cadmium

Cd is a toxic post-transition metal with no physiological function. Its atomic number is 48. Almost all Cd compounds exhibit an oxidation state of +2 (Genchi et al., 2020).

The exact mechanisms through which Cd elicits its neurotoxic effects remain unresolved. However, mitochondria have emerged as primary targets. Cd affects both the mitochondrial electron transport chain and membrane permeability, halting ATP synthesis and signaling apoptosis. Cd exposure also promotes apoptosis by upregulating proapoptotic proteins Bcl-2–interacting mediator of cell death (Bim) and Bcl-2–associated X protein (Bax) (Ji et al., 2011). Cd can interfere with the biological functions of calcium and Zn. It can block the entry of $Ca^{2+}$ into the nerve terminal via membrane channels after an action potential. Cd can also hinder the uptake of Zn in various systems due to their similar size and charge. It disrupts the packaging of neurotransmitters, which reduces the amount available for release. Additionally, it diminishes the activity of antioxidant enzymes, including CAT, Cu-Zn SOD, and MnSOD (Arruebarrena et al., 2023; Genchi et al., 2020; Wang & Du, 2013).

Cd exposure in zebrafish leads to significant developmental and neurotoxic effects. Embryos exposed to $CdCl_2$ exhibited developmental delays, impaired locomotor activity, and disrupted reactivity to environmental signals. Neurogenesis and the expression of neurodevelopment-associated genes were adversely affected. The exposure caused cell cycle arrest and extensive apoptosis, along with upregulation of proapoptotic genes. Additionally, $Cd^{2+}$ exposure triggered neuroinflammation, resulting in significant morphological changes in microglia (Xu et al., 2022a).

Similarly, Xu et al. stated that $CdCl_2$ exposure led to developmental delays, impaired locomotion, disrupted reactivity to environmental signals, and neurodevelopmental disorders in zebrafish larvae. They also linked the detrimental effects of Cd to the Wnt/β-catenin signaling pathway, observing that Cd exposure reduced β-catenin levels and downregulated the expression of Wnt-related genes. Activation of this pathway using a pharmacological activator (TWS119) significantly alleviated Cd-induced developmental and neurotoxic effects, while inhibition exacerbated these effects (Xu et al., 2022b).

Additionally, different studies have shown that $CdCl_2$ exposure leads to apoptosis; marked pathological lesions; increased levels of ROS, nitric oxide, and MDA; inflammation; and decreased levels of glial fibrillary acidic protein (an astroglial cytoskeleton protein) in zebrafish (Monaco et al., 2017; Monaco et al., 2016; Zheng et al., 2016; Favorito et al., 2011). Therefore, exposure to Cd can trigger various pathological processes and adversely affect normal brain functioning.

## Chromium

Cr is a heavy metal that occurs naturally and is recognized as a group 1 carcinogen by the International Agency for Research on Cancer (IARC). It has seven oxidation states (0–VI), with the trivalent form, Cr (III), and hexavalent form, Cr (VI), being the most prevalent and stable. The chemical properties of Cr play a crucial role in determining how it enters cells and exerts its toxic effects. The neurotoxic effects of Cr (VI) and its underlying mechanism remain insufficiently explored (Yan et al., 2023; Wise et al., 2022).

Cr (VI) is more toxic than Cr (III) because it is more soluble and reactive. This enables it to cross the cell membrane easily and enter cells through different transporters on the cell surface. Exposure of cells to Cr (VI) leads to diverse toxic effects impacting multiple cellular pathways. When Cr (VI) enters the body, it exerts its toxic effects by being reduced to Cr (III). This reduction reaction produces ROS-like hydroxyl radicals, hydrogen peroxide, and superoxide anions. These ROS cause considerable oxidative stress, leading to extensive damage to cell components such as lipids, proteins, mitochondria, DNA, and RNA (DesMarais & Costa, 2019; Mishra & Bharagava, 2015). Moreover, Cr (VI) stimulates increased transcription of genes associated with apoptosis, such as p53 and caspase (3 and 9), and reduces the expression of genes involved in the antioxidant pathway. It also causes epigenetic alterations, including alterations in histone modifications, DNA methylation, and miRNA expression, which exacerbate cellular damage and inflammatory responses. Ultimately, contaminating cells with Cr (VI) leads to various toxic effects across multiple cellular pathways (Singh et al., 2022; DesMarais & Costa, 2019).

Xu et al. exposed zebrafish embryos to different valence states of Cr, Cr (III), and Cr (VI). The study showed that Cr exposure led to malformations, with Cr (VI) causing more severe effects than Cr (III). Both Cr-exposed groups exhibited reduced body length, eye diameter, and inhibited locomotor behavior, though Cr (III) and Cr (VI) impacted behavior differently across various developmental stages. Exposure to Cr (VI) significantly increased 8OH-dG levels, indicating DNA oxidative damage. Specifically, Cr (VI) exposure led to elevated ROS levels. Additionally, Cr exposure reduced AchE activity, suggesting potential acetylcholine accumulation in the larvae. Cr (III) exposure upregulated neurodevelopmental genes like *α1-tubulin*, *elavl3*, *sox19b*, and *neurogenin1*, while Cr (VI) also upregulated *gap43* and *neuroD1*. Exposure to Cr (III) and Cr (VI) reduced the expression of DNA methyltransferase (dnmt1 and dnmt3) and led to hypomethylation of neurogenin1 and neuroD1, especially in Cr (VI) exposure. Additionally, Cr exposure decreased H3K27me3 levels, a marker of gene silencing. Cr (III) reduced H3K27me3 binding on neurogenin1 and neuroD1, while Cr (VI) upregulated neuroD1 expression without affecting H3K27me3 binding, further indicating epigenetic regulation (Xu et al., 2021).

Another study investigated the impact of varying concentrations of Cr (VI) (0.1 μg/L–100 μg/L) on the development of zebrafish embryos. The findings revealed that elevated levels of Cr (≥3.125 μg/L) resulted in decreased survival rates and delayed hatching. Exposure to increasing concentrations of Cr led to reduced body length in zebrafish larvae compared to the control group. Real-time RT-PCR analysis showed that Cr exposure (≥1 μg/L) reduced the expression of cell-cycle-related genes, such as *cdk4* and *cdk6*, in zebrafish embryos. However, there was a partial recovery of *cdk6* expression by day 3. Antioxidant-related genes showed mixed responses: *sod1* was upregulated on day 3, while *sod2* expression remained low throughout. Apoptosis-related genes *caspase 3* and *bax* were upregulated following exposure, while *bcl2* expression showed no significant difference on day 1 but was downregulated by day 3, indicating increased apoptosis in Cr-treated embryos. Western blot analysis showed that Cr exposure increased the expression of Caspase 3 and Bax proteins after 1 day, particularly at higher Cr concentrations. After 3 days of exposure, Caspase 3 and Bax protein expression remained elevated, while Bcl2 expression was reduced (Dang et al., 2023).

Shaw et al. found that chronic exposure to Cr (VI) led to neurotoxic stress and significant changes in the brain structure of zebrafish. These changes included detachment of the stratum marginale from the stratum opticum, vacuolation in the stratum opticum, and an increase in mononuclear cells with clear spaces in the stratum griseum centrale. Exposure to Cr (VI) in zebrafish brains resulted in stable GSH levels on day 1, followed by a decrease from day 7 to day 30, and then an increase by day 60. MDA levels remained elevated throughout, reaching a peak on day 15. CAT activity significantly increased after day 15, reaching its highest level on day 30, with an initial drop on day 7. Furthermore, the exposure to Cr (VI) significantly altered the expression levels of genes responsive to stress in zebrafish brains. Until day 30, there was no change in the expression of *nrf2*, but by day 60, it had increased significantly. *nqo1* expression showed minimal change on day 15, yet it increased sharply by day 30 and continued to increase by day 60. The pattern of *ho1*'s expression was similar to that of *nrf2*, increasing from day 30 and reaching its highest at day 60. On the other hand, *ucp2* decreased on day 15 but increased by day 60. AChE levels remained almost unchanged until day 30, then upregulated by day 60. Cr (VI) exposure led to significant alterations in apoptotic gene expression in the zebrafish brain. The expression of the *p53* gene remained unchanged until the 30th day but increased markedly on the 60th day. The *bax* gene exhibited a similar pattern on day 60. *bcl2* expression was downregulated on day 60. *caspase 9* expression decreased on days 15 and 30 but increased on the 60th day. *caspase 3* expression showed a similar pattern to *caspase 9* (Shaw et al., 2020).

## Aluminum

Aluminum (Al) is classified as a post-transition metal with an atomic number of 13. It is combined with elements such as oxygen, silicon, and fluorine. In compounds, Al predominantly exists in its +3 oxidation state (Rahimzadeh et al., 2022).

Although It is known that Al accumulates in the brain, its exact mechanism is not fully understood. Oxidative stress and mitochondrial dysfunction are suggested to be among the earliest pathological effects of Al neurotoxicity. Increased production of ROS, triggered by the activation of prooxidant systems and a reduction in antioxidant capacity, leads to Al-induced oxidative stress. This oxidative stress disrupts cellular signaling and neurotransmission, resulting in cytotoxicity, neural injury, and neurodegeneration. Additionally, Al forms complexes with superoxide anions, creating potent oxidants that exacerbate oxidative damage. Mitochondrial dysfunction is also an early response to Al exposure, causing altered energy metabolism, oxidative stress, and apoptosis (Rahimzadeh et al., 2022; Skalny et al., 2021; Andrade et al., 2017).

Capriello et al. examined the time-dependent neurotoxic effects of $AlCl_3$ on zebrafish. Al accumulation in the brain peaked at 15 days, then decreased by 20 days. Histological analysis revealed severe cellular disorganization and edema at 10 and 15 days, with partial recovery at 20 days. The behavioral and locomotory responses indicate elevated anxiety levels. Al exposure increased ROS production and affected enzyme activities, with a peak in oxidative stress markers at 15 days. Glutathione levels and oxidative damage biomarkers (thiobarbituric acid reactive substances, carbonyls, and metallothioneins) fluctuated, showing significant changes in oxidative stress and lipid peroxidation (Capriello et al., 2021).

Another study investigated the effects of different doses of $AlCl_3$ (50, 100, and 200 μg/L) on zebrafish. $AlCl_3$ exposure led to memory deficits, elevated anxiety, and spatial memory impairments. Biochemical analyses indicated elevated AChE levels; reduced neurotransmitters like GABA, serotonin, noradrenaline, and dopamine; and increased glutamate. Oxidative stress markers showed increased ROS and MDA levels, with decreased GSH and SOD. Histological examination revealed neuronal degeneration, with higher doses of $AlCl_3$ leading to increased pyknotic neurons. The nuclear factor erythroid 2–related factor 2 (Nrf-2) immunohistochemistry showed reduced expression of the Nrf2 protein, correlating with increased oxidative stress (Kaur, et al., 2022).

Another study investigated the effects of Al exposure on cognitive function, revealing significant impairments in learning and memory performance. The findings also indicated a dose-dependent downregulation in the expression of genes associated with the mammalian target of rapamycin (mTOR), protein kinase B (Akt), and phosphatidylinositol 3-kinase (PI3K) signaling pathway compared to controls. It is claimed that the PI3K/Akt/mTOR pathway may be involved in Al-induced cognitive impairment (Shang et al., 2020).

## Silicon

Silicon, symbolized as Si with an atomic weight of 28, is a chemical element. Silicon is widely distributed in the Earth's crust and is present in the human body. In nature, it is rarely found as a pure element, as it usually forms strong bonds with oxygen. Compounds that consist of silicon and oxygen are known as silica. The primary water-soluble form of silicon is orthosilicic acid, which is recognized for its biological availability and is often used as a liquid nutritional supplement (Ferreira et al.,2018; Price et al., 2013).

Although its exact mechanism is not understood yet, silicon is believed to be involved in bone mineralization and collagen synthesis. Also, it has been suggested that silicon exhibits a protective effect against NDs by acting as a natural antagonist to Al, decreasing its absorption and enhancing its excretion (Götz, et al., 2019, Garcimartín, et al., 2015).

On the other hand, it has been shown that silicon, in the form of silica nanoparticles, exhibits deleterious effects. These particles can accumulate in vital organs such as lungs, spleen, liver, and brain, leading to adverse effects (Jarrar et al., 2021). A study found that exposure to silica nanoparticles resulted in neurotoxicity by significantly reducing axon length and causing damage to axon structure in transgenic zebrafish lines. Genes involved in the neuroactive ligand-receptor interaction pathway, crucial for neurofunction and memory, demonstrated significantly reduced expression levels in silicon-treated zebrafish. However, brain of zebrafish embryos was not significantly affected by apoptosis (Wei et al., 2020).

Another study demonstrated that treating adult zebrafish with silica nanoparticles resulted in suppressed locomotion, increased anxiety, reduced exploratory behavior, and impaired learning and memory abilities. Exposure to nano silica also affected the activity of certain genes related to autophagy and Parkinson's disease in the zebrafish brain. Autophagy-related genes showed upregulation. On the other hand, low doses (300 μg/mL) of silica nanoparticles downregulated the expression of Parkinson-related genes (β-synuclein, pink1, and uch-l1), whereas high doses (1000 μg/mL) upregulated the expression of these genes. It is suggested that high doses of silica nanoparticles could potentially increase the risk of NDs (Li et al., 2020).

## Cobalt

Co is a hard, silvery-gray metallic element that primarily exists in two oxidation states: +2 and +3. The most biologically active Co compound is vitamin B12 (cyanocobalamin), where Co is bound to a ring structure composed of four pyrrole groups (Genchi et al., 2023; Catalani et al., 2011).

Only minimal amounts of Co are needed by humans, and excessive exposure can cause toxicity. Acute toxicity is uncommon and usually linked to over-supplementation, while chronic toxicity is associated with occupational exposure or malfunctioning metal-on-metal hip implants. Co exerts toxic effects in biological tissues through increasing hypoxia-inducible factor HIF1a, leading to hypoxia and the generation of ROS, contributing to oxidative stress (Tang et al., 2023; Oria et al., 2022).

Chai et al. examined the impact of Co exposure on the early developmental stages of zebrafish embryos. Embryo survival and development remained largely unaffected at Co concentrations below 100 μg/L. However, at concentrations exceeding 100 μg/L, there was a noticeable decline in survival and the onset of developmental abnormalities. The levels of lipid peroxidation were found to be elevated in embryos that were exposed to 100 and 500 μg/L of Co. The mRNA levels of CAT and SOD showed a dose-dependent increase, while GPX 1a expression remained relatively unchanged. Additionally, genes like *p53*, *caspase-3*, and *caspase-9* were also upregulated in a concentration-dependent manner. TUNEL assays detected abnormal apoptotic activity in the brain, trunk, and tail of embryos treated with 500 μg/L of Co (Chai et al., 2012).

In another study, Co complexes bearing NMS ligand (Co-NMS) were synthesized and evaluated for their toxicity in normal brain tissue, radiosensitizing effects on glioma cells, and potential therapeutic benefits. Zebrafish model tests revealed that higher doses of Co-NMS led to neurotoxicity, increased HSP70 expression, and behavioral impairments, while lower doses showed no significant brain damage. In glioma cells, Co-NMS enhanced radiosensitivity, inhibited cell growth, induced G2/M cell cycle arrest, and triggered apoptosis via mitochondrial dysfunction. The combination of Co-NMS with irradiation significantly reduced mitochondrial membrane potential, increased oxidative stress, decreased ATP production, and altered cellular metabolism, demonstrating the compound's potential to enhance the efficacy of radiation therapy in glioma treatment (Liu et al., 2020).

## Nickel

Ni is a durable, silvery-white metal commonly found in its +2 oxidation state. Although Ni-containing enzymes are not known in higher organisms, they are present in various simpler organisms such as bacteria, archaea, and plants (Genchi et al., 2020).

Ni exposure, based on the time and amount, can result in a range of health problems, particularly affecting the nervous system due to its accumulation in the brain. Although the precise mechanisms are not fully understood, oxidative stress and mitochondrial dysfunction are thought to be key factors behind Ni-induced neurotoxicity. Ni can impair mitochondrial functions by disrupting membrane potential, depleting ATP levels, and leading to the breakdown of mitochondrial DNA. These impairments can hinder the electron transport chain and intensify oxidative stress. It is also suggested that Ni may disrupt the release of neurotransmitters, thereby impeding the proper synaptic connections (Anyachor et al.,2022; Wei et al., 2022; Lamtai et al., 2018). Research involving zebrafish is relatively limited.

Nabinger et al. explored the effects of Ni chloride exposure on zebrafish larvae and adults, focusing on metal accumulation, physiological responses, and behavioral changes. Zebrafish exposed to higher concentrations of Ni chloride (5.0 mg/L and 15.0 mg/L) showed increased Ni levels in their bodies, particularly in the brain, without impacting survival rates. However, subchronic exposure delayed larval hatching, reduced body length, and decreased eye size at early stages. Behavioral tests revealed that higher Ni chloride doses affected heart rates, exploratory behavior, and escape responses in larvae, while adult zebrafish displayed reduced aggression and impaired memory after prolonged exposure (Nabinger et al., 2018).

In another study, Ni exposure caused dose-dependent developmental delays in zebrafish embryos, affecting survival rates, hatching, heart rate, and body length. Spontaneous tail coiling also decreased with higher Ni levels. Ni exposure in zebrafish larvae led to significant reductions in swimming distance, velocity, and mobility in a dose-dependent manner, indicating neurobehavioral impairments. Additionally, their response to vibration stimuli was notably inhibited at higher Ni levels. Transcriptome analysis of zebrafish larvae exposed to Ni showed significant changes in gene expression, particularly affecting neurodevelopment, iron homeostasis, and ferroptosis pathways. Ni exposure in zebrafish led to reduced neurogenesis, impaired vascular development, and activated immune responses, as seen through in vivo imaging and gene expression analysis. Neuronal loss, microglia activation, and constriction of blood vessels were observed, highlighting significant disruptions in neurodevelopment and angiogenesis. Ni exposure in zebrafish larvae activated ferroptosis, with increased oxidative stress and disrupted iron balance. This was confirmed by altered gene expression profiles and damaged mitochondria observed through electron microscopy (Wang et al., 2023).

## Titanium

Titanium, symbolized by Ti, is used in a variety of industries. Ti dioxide ($TiO_2$), a natural oxide of Ti in the form of nanoparticles ($TiO_2$ NPs), is one of the most commonly produced nanoparticles, typically ranging from 1 to 100 nm in size. TiO2 NPs exhibit different crystalline forms, such as anatase, rutile, and brookite. Among these, anatase and rutile are the most commonly employed in various personal care products, photocatalysts, paints, and cosmetics. These nanoparticles are also utilized as a food additive owing to their low toxicity, known as E171 (Aschner et al., 2023; Marin & Lanzutti, 2023; Musial et al., 2020).

Although the exact mechanisms remain unclear, $TiO_2$ NPs can accumulate in the brain. Their accumulation is associated with neurotoxic effects such as behavioral changes, nervous system impairment, and oxidative stress, which leads to structural damage, autophagy, and inflammation. These potential health risks are not yet fully understood, and more comprehensive studies are required to better assess the neurotoxicity of TiO2 NPs and their long-term impact on human health (Kaur et al., 2024; Zhang et al., 2023; Song et al., 2015).

The effects of TiO2 particles, both micro and nano-sized, were investigated in zebrafish embryos and larvae. Toxicity tests showed no significant differences in embryo hatching or mortality rates, but larvae exposed to nano-TiO2 displayed reduced body weight, length, and swimming activity in a dose-dependent manner. Nano-TiO2 exposure reduced zebrafish larvae's swimming speed and clockwise rotation in a dose-dependent manner, with higher concentrations (0.10 and 1.00 mg/L) causing significant declines in activity. The study also investigated genes associated with oxidative stress, neurogenesis, and neural maturation at 72 hpf. The genes *nrd* and *elavl3* involved in early neurogenesis were significantly downregulated in the group treated with 1.00 mg/L nano-TiO2. The expression of *α1-tubulin* for neural maturation was significantly upregulated at all nano-TiO2 concentrations (0.01, 0.1, and 1.0 mg/L). Additionally, the expression of *mbp* and *gap43*, which are associated with neural maturation, were significantly increased in the 0.1 or 1.0 mg/L nano-TiO2 groups. No significant changes were observed in the genes associated with oxidative stress (Gu et al., 2021).

Zebrafish embryos were exposed to TiO2 NPs at concentrations of 0, 0.1, 1, and 10 μg/mL from 2 to 96 hpf. Exposure to TiO2 nanoparticles (NPs) in zebrafish embryos did not impact mortality but accelerated hatching and caused malformations, indicating developmental risks. TiO2 NPs exposure also impaired locomotion in larvae, suggesting Parkinson's disease–like symptoms. Transmission electron microscopy confirmed that TiO2 NPs crossed the blood–brain barrier and accumulated in brain tissue, increasing neurodegenerative risk. Additionally, TiO2 NPs elevated ROS levels, induced cell death in the hypothalamus, and significantly reduced dopamine neurons by 60%, further supporting Parkinson's disease–like neurotoxicity. TiO2 NPs also significantly increased the mRNA levels of *pink1*, *parkin*, and *α-syn* in a dose-dependent manner, suggesting dysfunction of the ubiquitin-proteasome system (UPS). This dysfunction may contribute to the formation of Lewy bodies, a key feature in Parkinson's disease pathology (Hu et al., 2017).

## Tin

Sn is a natural element present in the earth's crust, existing in organic and inorganic forms. It is introduced into the environment from both natural and human-made sources. Although inorganic Sn compounds generally exhibit low toxicity due to their limited solubility and absorption, they can be converted into organotin compounds through methylation processes, which are highly soluble and toxic to various organisms. Inorganic Sn can react with elements such as chlorine or oxygen, creating inorganic Sn compounds used in products such as toothpaste, soaps, and food additives. Additionally, Sn can bond with carbon to produce organotins, commonly used in plastics, pesticides, and pest repellents. Although Sn has no known essential role in human health, excessive intake has been linked to reduced copper levels, alterations in cholesterol metabolism, and increased risk of heart disease (Ostrakhovitch, 2015; Gad & Pham, 2014; Ostrakhovitch & Cherıan, 2007).

There are very few studies on the neurotoxicity of Sn in zebrafish. Sisman highlighted the teratogenic and genotoxic impacts of $SnCl_2$ exposure on adult zebrafish and their embryos. Adult fish exposed to concentrations of up to 250 μM $SnCl_2$ did not experience mortality; however, their body weight decreased, and their feeding behavior changed. The frequency of micronuclei in adult fish increased only at higher doses of $SnCl_2$ (≥50 μM). In embryos, exposure to a concentration of 250 μM resulted in complete mortality within 24 hpf, whereas lower concentrations ranging from 10 to 25 μM showed no signs of toxicity. However, concentrations of 50 μM and above led to increased mortality and developmental defects, including tail deformities and smaller heads and eyes. Embryos with intact chorions were more susceptible to the toxicity of $SnCl_2$ compared to dechorionated embryos, as the chorions appeared to retain toxic debris. Although dechorionated embryos initially exhibited better survival rates, the LC50 values for both groups converged by 120 hpf, indicating that the absence of the chorion provided only temporary protective effects (Sisman, 2011).

A different study examined how trimethyltin exposure impacts motor behavior, tactile response, morphology, photomotor behavior, and apoptosis in zebrafish embryos and larvae. Exposure to trimethyltin at 10 µM significantly raised spontaneous tail bending at 20 hpf, while the 5 and 10 µM doses later diminished the frequency of tail bending and the response to touch stimuli at 25–36 hpf. At 48–72 hpf, trimethyltin exposure led to malformations, including yolk sac and pericardial edemas, and resulted in hypoactivity during light phases at 4–5 dpf. Furthermore, embryos exposed to trimethyltin exhibited localized apoptosis, particularly in the brain, eyes, ears, and tail, with significant cell death noted specifically in the tail region (Chen et al., 2011).

## Conclusion

Several elements play a crucial role in human health and brain function, acting as cofactors in enzymatic reactions and participating in various biochemical pathways. However, disturbances in their homeostasis can lead to neurotoxicity, contributing to NDs such as Alzheimer's, Parkinson's disease, and ALS. Both excessive essential elements and exposure to toxic elements can hasten neurodegeneration. Studies on zebrafish have shown that such elements can trigger neurodegenerative effects through oxidative stress, inflammation, and mitochondrial dysfunction, although the exact mechanism remains unclear. Therefore, further research is necessary to understand these mechanisms and develop therapeutic approaches to alleviate metal-induced neurotoxicity and promote healthy neurodevelopment.

## REFERENCES


Abbott, L. C., & Nigussie, F. (2021). Mercury toxicity and neurogenesis in the mammalian brain. *International Journal of Molecular Sciences*, *22*(14), 7520.

Abu Bakar, N., Mohd Sata, N. S., Ramlan, N. F., Wan Ibrahim, W. N., Zulkifli, S. Z., Che Abdullah, C. A., Ahmad, S., & Amal, M. N. (2017). Evaluation of the neurotoxic effects of chronic embryonic exposure with inorganic mercury on motor and anxiety-like responses in zebrafish (Danio rerio) larvae. *Neurotoxicology and Teratology*, *59*, 53–61.

Acosta, D. D. S., Danielle, N. M., Altenhofen, S., Luzardo, M. D., Costa, P. G., Bianchini, A., Bonan, C. D., da Silva, R. S., & Dafre, A. L. (2016). Copper at low levels impairs memory of adult zebrafish (Danio rerio) and affects swimming performance of larvae. *Comparative Biochemistry and Physiology. Toxicology & Pharmacology: CBP*, *185–186*, 122–130.

Adhikari, S., & Mohanty, M. (2012). Effect of waterborne boron and molybdenum on survival, growth and feed intake of Indian major carp, Cirrhinus mrigala (Hamilton). *Chemistry and Ecology*, 28(2), 113–121.

Adhish, M., & Manjubala, I. (2023). Effectiveness of zebrafish models in understanding human diseases: A review of models. *Heliyon*, *9*(3), e14557.

Agency for Toxic Substances and Disease Registry (ATSDR). (2022). *Substance priority list*. https://www.atsdr.cdc.gov/spl/index.html

Alak, G., Parlak, V., Aslan, M. E., Ucar, A., Atamanalp, M., & Turkez, H. (2019). Borax supplementation alleviates hematotoxicity and DNA damage in rainbow trout (oncorhynchus mykiss) exposed to copper. *Biological trace element research*, *187*(2), 536–542.

Alak, G., Parlak, V., Ucar, A., Yeltekin, A. Ç., Ozgeris, F. B., Caglar, O., Atamanalp, M., Turkez, H. (2020). Oxidative and DNA damage potential of colemanite on zebrafish: Brain, liver and blood. *Turkish Journal of Fisheries and Aquatic Sciences*, *20*(8), 593–602.

Alak, G., Ucar, A., Parlak, V., Yeltekin, A. Ç., Özgeriş, F. B., Atamanalp, M., & Türkez, H. (2021). Antioxidant potential of ulexite in zebrafish brain: Assessment of oxidative DNA damage, apoptosis, and response of antioxidant defense system. *Biological Trace Element Research*, *199*(3), 1092–1099.

Alba-González, A., Dragomir, E. I., Haghdousti, G., Yáñez, J., Dadswell, C., González-Méndez, R., Wilson, S. W., Tuschl, K., & Folgueira, M. (2024). Manganese overexposure alters neurogranin expression and causes behavioral deficits in larval zebrafish. *International Journal of Molecular Sciences*, *25*(9), 4933.

Aliem, R. A., Khaled, H. E., Soliman, B. A., Mourad, M., Dighies, H. S. (2023) Boron, the forgotten element. *Frontiers in Scientific Research and Technology*, *6*(1), 20–28.

Altenhofen, S., Wiprich, M. T., Nery, L. R., Leite, C. E., Vianna, M. R. M. R., & Bonan, C. D. (2017). Manganese(II) chloride alters behavioral and neurochemical parameters in larvae and adult zebrafish. *Aquatic Toxicology (Amsterdam, Netherlands)*, *182*, 172–183.

An, Y., Li, S., Huang, X., Chen, X., Shan, H., & Zhang, M. (2022). The role of copper homeostasis in brain disease. *International Journal of Molecular Sciences*, *23*(22), 13850.

Andrade, V. M., Aschner, M., & Marreilha dos Santos, A. P. (2017). Neurotoxicity of metal mixtures. *Neurotoxicity of Metals*, *18*, 227–265.

Antunes Dos Santos, A., Ferrer, B., Marques Gonçalves, F., Tsatsakis, A. M., Renieri, E. A., Skalny, A. V., Farina, M., Rocha, J. B. T., & Aschner, M. (2018). Oxidative stress in methylmercury-induced cell toxicity. *Toxics*, *6*(3), 47.

Anyachor, C. P., Dooka, D. B., Orish, C. N., Amadi, C. N., Bocca, B., Ruggieri, F., Senofonte, M., Frazzoli, C., & Orisakwe, O. E. (2022). Mechanistic considerations and biomarkers level in nickel-induced neurodegenerative diseases: An updated systematic review. *IBRO Neuroscience Reports*, *13*, 136–146.

Arruebarrena, M. A., Hawe, C. T., Lee, Y. M., & Branco, R. C. (2023). Mechanisms of cadmium neurotoxicity. *International Journal of Molecular Sciences*, *24*(23), 16558.

Aschner, M., Skalny, A. V., Santamaria, A., Buha Djordjevic, A., Tizabi, Y., Jiang, Y., Lu, R., Virgolini, M. B., & Tinkov, A. A. (2023). From mechanisms to implications: Understanding the molecular neurotoxicity of titanium dioxide nanoparticles. *Frontiers in Bioscience (Landmark edition)*, *28*(9), 204.

Behra, M., Cousin, X., Bertrand, C., Vonesch, J. L., Biellmann, D., Chatonnet, A., & Strähle, U. (2002). Acetylcholinesterase is required for neuronal and muscular development in the zebrafish embryo. *Nature Neuroscience*, *5*(2), 111–118.

Branco, V., Aschner, M., & Carvalho, C. (2021). Neurotoxicity of mercury: An old issue with contemporary significance. *Advances in Neurotoxicology*, *5*, 239–262.

Cai, G., Zhu, J., Shen, C., Cui, Y., Du, J., & Chen, X. (2012). The effects of cobalt on the development, oxidative stress, and apoptosis in zebrafish embryos. *Biological Trace Element Research*, *150*(1–3), 200–207.

Cao, Y., Li, B., Ismail, N., Smith, K., Li, T., Dai, R., & Deng, Y. (2021). Neurotoxicity and underlying mechanisms of endogenous neurotoxins. *International Journal of Molecular Sciences*, *22*(23), 12805.

Capriello, T., Félix, L. M., Monteiro, S. M., Santos, D., Cofone, R., & Ferrandino, I. (2021). Exposure to aluminium causes behavioural alterations and oxidative stress in the brain of adult zebrafish. *Environmental Toxicology and Pharmacology*, *85*, 103636.

Cardoso, B. R., Roberts, B. R., Bush, A. I., & Hare, D.J. (2015). Selenium, selenoproteins and neurodegenerative diseases. *Metallomics*, *7*(8), 1213–1228.

Catalani, S., Rizzetti, M., Padovani, A., & Apostoli, P. (2011). Neurotoxicity of cobalt. *Human & Experimental Toxicology*, *31*(5), 421–437.

Chen, J., Huang, C., Zheng, L., Simonich, M., Bai, C., Tanguay, R., & Dong, Q. (2011). Trimethyltin chloride (TMT) neurobehavioral toxicity in embryonic zebrafish. *Neurotoxicology and Teratology*, *33*(6), 721–726.

Chia, K., Klingseisen, A., Sieger, D., & Priller, J. (2022). Zebrafish as a model organism for neurodegenerative disease. *Frontiers in Molecular Neuroscience*, *15*, 940484.

d'Amora, M., & Giordani, S. (2018). The utility of zebrafish as a model for screening developmental neurotoxicity. *Frontiers in neuroscience*, *12*, 976.

Dang, K. D., Ho, C. N. Q., Van, H. D., Dinh, S. T., Nguyen, Q. T. T., Nguyen, T. T. T., Kien, X. T. N., Dao, T. V., Nong, H. V., Nguyen, M. T., Doan, C. C., Hoang, S. N., Nguyen, T. T. P., & Le, L. T. (2023). Hexavalent chromium inhibited zebrafish embryo development by altering apoptosis- and antioxidant-related genes. *Current Issues in Molecular Biology*, *45*(8), 6916–6926.

Desai, V., & Kaler, S. G. (2008). Role of copper in human neurological disorders. *The American Journal of Clinical Nutrition*, *88*(3), 855S–858S.

DesMarais, T. L., & Costa, M. (2019). Mechanisms of chromium-induced toxicity. *Current Opinion in Toxicology*, *14*, 1–7.

Dolgova, N. V., Hackett, M. J., MacDonald, T. C., Nehzati, S., James, A. K., Krone, P. H., George, G. N., & Pickering, I. J. (2016) Distribution of selenium in zebrafish larvae after exposure to organic and inorganic selenium form. *Metallomics*, *8*, 305–312.

Doszyn, O., Dulski, T., & Zmorzynska, J. (2024). Diving into the zebrafish brain: Exploring neuroscience frontiers with genetic tools, imaging techniques, and behavioral insights. *Frontiers in Molecular Neuroscience*, *17*, 1358844.

Dou, C., & Zhang, J. (2011). Effects of lead on neurogenesis during zebrafish embryonic brain development. *Journal of Hazardous Materials, 194, 277–282.*

Farina, M., Rocha, J. B., & Aschner, M. (2011). Mechanisms of methylmercury-induced neurotoxicity: evidence from experimental studies. *Life sciences*, *89*(15–16), 555–563.

Favorito, R., Chiarelli, G., Grimaldi, M. C., De Bonis, S., Lancieri, M., & Ferrandino, I. (2011). Bioaccumulation of cadmium and its cytotoxic effect on zebrafish brain. *Chemistry and Ecology*, 27(sup2), 39–46.

Ferreira, A. O., Freire, É. S., Polonini, H. C., Da Silva, P.J.L.C., Brandão, M.A.F., & Raposo, N.R.B. (2018). Anti-aging effects of monomethylsilanetriol and maltodextrin-stabilized orthosilicic acid on nails, skin and hair. *Cosmetics*, *5*, 41.

Gad, S. C., & Pham, T. (2014). Tin. In B. Anderson, A. de Peyster, S. C. Gad, P. J. Bert Hakkinen, M. Kamrin, B. Locey, H. M. Mehendale, C. Pope, & L. Shugart (Eds.), *Encyclopedia of toxicology* (pp. 574–576). Elsevier.

Gansner, J. M., Mendelsohn, B. A., Hultman, K. A., Johnson, S. L., & Gitlin, J. D. (2007). Essential role of lysyl oxidases in notochord development. *Developmental Biology*, *307*(2), 202–213.

Gao, G., You, L., Zhang, J., Chang, Y. Z., & Yu, P. (2023). Brain iron metabolism, redox balance and neurological diseases. *Antioxidants (Basel, Switzerland)*, *12*(6), 1289.

Garcimartín, A., Merino, J. J., Santos-López, J. A., López-Oliva, M. E., González, M. P., Sánchez-Muniz, F. J., & Benedí, J. (2015). Silicon as neuroprotector or neurotoxic in the human neuroblastoma SH-SY5Y cell line. *Chemosphere*, *135*, 217–224.

Genchi, G., Carocci, A., Lauria, G., Sinicropi, M. S., & Catalano, A. (2020). Nickel: Human health and environmental toxicology. *International Journal of Environmental Research and Public Health*, *17*(3), 679.

Genchi, G., Lauria, G., Catalano, A., Carocci, A., & Sinicropi, M. S. (2023). Prevalence of cobalt in the environment and its role in biological processes. *Biology*, *12*(10), 1335.

Genchi, G., Sinicropi, M. S., Lauria, G., Carocci, A., & Catalano, A. (2020). The effects of cadmium toxicity. *International Journal of Environmental Research and Public Health*, *17*(11), 3782.

Götz, W., Tobiasch, E., Witzleben, S., & Schulze, M. (2019). Effects of silicon compounds on biomineralization, osteogenesis, and hard tissue formation. *Pharmaceutics*, *11*(3), 117.

Gromadzka, G., Tarnacka, B., Flaga, A., & Adamczyk, A. (2020). Copper dyshomeostasis in neurodegenerative diseases-therapeutic implications. *International Journal of Molecular Sciences*, *21*(23), 9259.

Gu, J., Guo, M., Huang, C., Wang, X., Zhu, Y., Wang, L., Wang, Z., Zhou, L., Fan, D., Shi, L., & Ji, G. (2021). Titanium dioxide nanoparticle affects motor behavior, neurodevelopment and axonal growth in zebrafish (Danio rerio) larvae. *The Science of the Total Environment*, *754*, 142315.

Gülsoy, N., Yavas, C., & Mutlu, Ö. (2015). Genotoxic effects of boric acid and borax in zebrafish, Danio rerio using alkaline comet assay. *EXCLI Journal*, *14*, 890–899.

Hariharan, S., Chauhan, S., Marcharla, E., Alphonse, C. R. W., Rajaretinam, R. K., & Ganesan, S. (2024). Developmental toxicity and neurobehavioral effects of sodium selenite and selenium nanoparticles on zebrafish embryos. *Aquatic toxicology (Amsterdam, Netherlands)*, *266*, 106791.

Hassan, A. T., & Kwong, R. W. M. (2020). The neurophysiological effects of iron in early life stages of zebrafish. *Environmental Pollution (Barking, Essex: 1987)*, *267*, 115625.

Hassan, S. A., Moussa, E. A., & Abbott, L. C. (2012). The effect of methylmercury exposure on early central nervous system development in the zebrafish (Danio rerio) embryo. *Journal of Applied Toxicology: JAT*, *32*(9), 707–713.

Hill, B. N, Coldsnow, K. D., Hunter, D. L., Hedge, J. M., Korest, D., Jarema, K. A., & Padilla, S. (2021). Assessment of larval zebrafish locomotor activity for developmental neurotoxicity screening. In J. Llorens & M. Barenys (Eds.), *Experimental neurotoxicology methods. Neuromethods* (Vol. 172). Humana.

Ho, N. Y., Yang, L., Legradi, J., Armant, O., Takamiya, M., Rastegar, S., & Strähle, U. (2013). Gene responses in the central nervous system of zebrafish embryos exposed to the neurotoxicant methyl mercury. *Environmental Science & Technology*, *47*(7), 3316–3325.

Hu, Q., Guo, F., Zhao, F., & Fu, Z. (2017). Effects of titanium dioxide nanoparticles exposure on parkinsonism in zebrafish larvae and PC12. *Chemosphere*, *173*, 373–379.

Jarrar, B., Al-Doaiss, A., Shati, A., Al-Kahtani, M., & Jarrar, Q. (2021). Behavioural alterations induced by chronic exposure to 10 nm silicon dioxide nanoparticles. *IET Nanobiotechnology*, *15*(2), 221–235.

Ji, Y. L., Wang, H., Zhao, X. F., Wang, Q., Zhang, C., Zhang, Y., Zhao, M., Chen, Y. H., Meng, X. H., & Xu, D. X. (2011). Crosstalk between endoplasmic reticulum stress and mitochondrial pathway mediates cadmium-induced germ cell apoptosis in testes. *Toxicological Sciences: An Official Journal of the Society of Toxicology*, *124*(2), 446–459.

Kan, F., & Kucukkurt, I. (2023). The effects of boron on some biochemical parameters: A review. *Journal of Trace Elements in Medicine and Biology: Organ of the Society for Minerals and Trace Elements (GMS)*, *79*, 127249.

Kaur, K., Narang, R. K., & Singh, S. (2022). $AlCl_3$ induced learning and memory deficit in zebrafish. *Neurotoxicology*, *92*, 67–76.

Kaur, K., Narang, R.K., & Singh, S. (2024). Neuroprotective potential of Betulinic acid against $TIO_2NP$ induced neurotoxicity in zebrafish. *International Immunopharmacology*, *138*, 112604.

Kim, Y. S., Won, Y. J., Lim, B. G., Min, T. J., Kim, Y. H., & Lee, I. O. (2020). Neuroprotective effects of magnesium L-threonate in a hypoxic zebrafish model. *BMC Neuroscience*, *21*(1), 29.

Kirkland, A. E., Sarlo, G. L., & Holton, K. F. (2018). The role of magnesium in neurological disorders. *Nutrients*, *10*(6), 730.

Komoike, Y., & Matsuoka, M. (2022). Developmental adverse effects of trace amounts of lead: Evaluation using zebrafish model. *Frontiers in Pharmacology*, *13*, 1014912.

Kryukov, G. V., & Gladyshev, V. N. (2000). Selenium metabolism in zebrafish: Multiplicity of selenoprotein genes and expression of a protein containing 17 selenocysteine residues. *Genes to Cells*, *5*(12), 1049–1060.

Lamptey, R. N. L., Chaulagain, B., Trivedi, R., Gothwal, A., Layek, B., & Singh, J. (2022). A review of the common neurodegenerative disorders: Current therapeutic approaches and the potential role of nanotherapeutics. *International Journal of Molecular Sciences*, *23*(3), 1851.

Lamtai, M., Chaibat, J., Ouakki, S., Zghari, O., Mesfioui, A., El Hessni, A., Rifi, E. H., Marmouzi, I., Essamri, A., & Ouichou, A. (2018). Effect of chronic administration of nickel on affective and cognitive behavior in male and female rats: Possible implication of oxidative stress pathway. *Brain Sciences*, *8*(8), 141.

Lee, J., & Freeman, J. (2014). Zebrafish as a model for developmental neurotoxicity assessment: The application of the zebrafish in defining the effects of arsenic, methylmercury, or lead on early neurodevelopment. *Toxics*, *2*(3), 464–495.

Levi, S., Ripamonti, M., Moro, A. S., & Cozzi, A. (2024). Iron imbalance in neurodegeneration. *Molecular Psychiatry*, *29*(4), 1139–1152.

Li, X., Chen, C., He, M., Yu, L., Liu, R., Ma, C., Zhang, Y., Jia, J., Li, B., & Li, L. (2022). Lead exposure causes spinal curvature during embryonic development in zebrafish. *International Journal of Molecular Sciences*, *23*(17), 9571.

Li, X., Ji, X., Wang, R., Zhao, J., Dang, J., Gao, Y., & Jin, M. (2020). Zebrafish behavioral phenomics employed for characterizing behavioral neurotoxicity caused by silica nanoparticles. *Chemosphere*, *240*, 124937.

Lin, Y., Hu, L., Li, X., Ma, J., Li, Q., Yuan, X., & Zhang, Y. (2024). The beneficial and toxic effects of selenium on zebrafish. A systematic review of the literature. *Toxicology Research*, *13*(2), tfae062.

Liu, M., Deng, P., Li, G., Liu, H., Zuo, J., Cui, W., Zhang, H., Chen, X., Yao, J., Peng, X., Peng, L., Liu, J., Zheng, W., Yan, W., & Luan, N. (2024). Neurotoxicity of combined exposure to the heavy metals (Pb and As) in zebrafish (*Danio rerio*). *Toxics*, *12*(4), 282.

Liu, Y., Zhang, T., Li, G., Li, S., Li, J., Zhao, Q., Wu, Q., Xu, D., Hu, X., Zhang, L., Li, Q., Zhang, H., & Liu, B. (2020). Radiosensitivity enhancement by Co-NMS-mediated mitochondrial impairment in glioblastoma. *Journal of Cellular Physiology*, *235*(12), 9623–9634.

Lumsden, A. L., Henshall, T. L., Dayan, S., Lardelli, M. T., & Richards, R. I. (2007). Huntingtin-deficient zebrafish exhibit defects in iron utilization and development. *Human Molecular Genetics*, *16*(16), 1905–1920.

Luzio, A., Monteiro, S. M., Fontaínhas-Fernandes, A. A., Pinto-Carnide, O., Matos, M., & Coimbra, A. M. (2013). Copper induced upregulation of apoptosis related genes in zebrafish (Danio rerio) gill. *Aquatic Toxicology (Amsterdam, Netherlands)*, *128–129*, 183–189.

Ma, H., Yang, W., Li, Y., Li, J., Yang, X., Chen, Y., Ma, Y., Sun, D., & Sun, H. (2024). Effects of sodium arsenite exposure on behavior, ultrastructure and gene expression of brain in adult zebrafish (Danio rerio). *Ecotoxicology and Environmental Safety*, *273*, 116107.

Mackenzie, N. C., Brito, M., Reyes, A. E., & Allende, M. L. (2004). Cloning, expression pattern and essentiality of the high-affinity copper transporter 1 (ctrl) gene in zebrafish. *Gene*, *328*, 113–120.

Maier, J. A. M., Locatelli, L., Fedele, G., Cazzaniga, A., & Mazur, A. (2022). Magnesium and the brain: A focus on neuroinflammation and neurodegeneration. *International Journal of Molecular Sciences*, *24*(1), 223.

Majdura, J., Jankiewicz, U., Gałązka, A., Orzechowski, S. (2023). The Role of Quorum Sensing Molecules in Bacterial-Plant Interactions. *Metabolites*. 10;13(1):114.

Marin, E., & Lanzutti, A. (2023). Biomedical applications of titanium alloys: A comprehensive review. *Materials (Basel, Switzerland)*, *17*(1), 114.

Martínez Leo, E. E., & Segura Campos, M. R. (2019). Systemic oxidative stress: A key point in neurodegeneration - a review. *The Journal of Nutrition, Health & Aging*, *23*(8), 694–699.

McCann, S., Perapoch Amadó, M., & Moore, S. E. (2020). The role of iron in brain development: A systematic review. *Nutrients*, *12*(7), 2001.

Medda, N., Patra, R., Ghosh, T. K., & Maiti, S. (2020). Neurotoxic mechanism of arsenic: Synergistic effect of mitochondrial instability, oxidative stress, and hormonal-neurotransmitter impairment. *Biological Trace Element Research*, *198*(1), 8–15.

Mendelsohn, B. A., Yin, C., Johnson, S. L., Wilm, T. P., Solnica-Krezel, L., & Gitlin, J. D. (2006). Atp7a determines a hierarchy of copper metabolism essential for notochord development. *Cell Metabolism*, *4*(2), 155–162.

Mishra, S., & Bharagava, R. N. (2015). Toxic and genotoxic effects of hexavalent chromium in environment and its bioremediation strategies. *Journal of Environmental Science and Health, Part C*, *34*(1), 1–32.

Mochizuki, H. (2019). Arsenic neurotoxicity in humans. *International Journal of Molecular Sciences*, *20*(14), 3418.

Monaco, A., Capriello, T., Grimaldi, M. C., Schiano, V., & Ferrandino, I. (2017). Neurodegeneration in zebrafish embryos and adults after cadmium exposure. *European Journal of Histochemistry: EJH*, *61*(4), 2833.

Monaco, A., Grimaldi, M. C., & Ferrandino, I. (2016). Neuroglial alterations in the zebrafish brain exposed to cadmium chloride. *Journal of Applied Toxicology: JAT*, *36*(12).

Musial, J., Krakowiak, R., Mlynarczyk, D. T., Goslinski, T., & Stanisz, B. J. (2020). Titanium dioxide nanoparticles in food and personal care products-what do we know about their safety? *Nanomaterials (Basel, Switzerland)*, *10*(6), 1110.

Muzaffar, S., Khan, J., Srivastava, R., Gorbatyuk, M. S., & Athar, M. (2023). Mechanistic understanding of the toxic effects of arsenic and warfare arsenicals on human health and environment. *Cell Biology and Toxicology*, *39*(1), 85–110.

Müller, T. E., Nunes, M. E., Menezes, C. C., Marins, A. T., Leitemperger, J., Gressler, A. C. L., Carvalho, F. B., de Freitas, C. M., Quadros, V. A., Fachinetto, R., Rosemberg, D. B., & Loro, V. L. (2018). Sodium selenite prevents paraquat-induced neurotoxicity in zebrafish. *Molecular Neurobiology*, *55*(3), 1928–1941.

Nabinger, D. D., Altenhofen, S., Bitencourt, P. E. R., Nery, L. R., Leite, C. E., Vianna, M. R. M. R., & Bonan, C. D. (2018). Nickel exposure alters behavioral parameters in larval and adult zebrafish. *The Science of the Total Environment*, *624*, 1623–1633.

Nadig, A. P. R., Huwaimel, B., Alobaida, A., Khafagy, E. S., Alotaibi, H. F., Moin, A., Lila, A. S. A., Suman, M, S., & Krishna, K. L. (2022). Manganese chloride ($MnCl_2$) induced novel model of Parkinson's disease in adult Zebrafish; Involvement of oxidative stress, neuroinflammation and apoptosis pathway. *Biomedicine & Pharmacotherapy = Biomedecine & Pharmacotherapie*, *155*, 113697.

Neal, A. P., & Guilarte, T. R. (2013). Mechanisms of lead and manganese neurotoxicity. *Toxicology Research*, *2*(2), 99–114.

Nielsen, F. H. (2017). Historical and recent aspects of boron in human and animal health. *Journal of Boron*, 2(3), 153 – 160.

Novo, J. P., Martins, B., Raposo, R. S., Pereira, F. C., Oriá, R. B., Malva, J. O., & Fontes-Ribeiro, C. (2021). Cellular and molecular mechanisms mediating methylmercury neurotoxicity and neuroinflammation. *International Journal of Molecular Sciences*, *22*(6), 3101.

Ohayon, S., Boyko, M., Saad, A., Douvdevani, A., Gruenbaum, B. F., Melamed, I., Shapira, Y., Teichberg, V. I., & Zlotnik, A. (2012). Cell-free DNA as a marker for prediction of brain damage in traumatic brain injury in rats. *Journal of Neurotrauma*, *29*(2), 261–267.

Okereafor, U., Makhatha, M., Mekuto, L., Uche-Okereafor, N., Sebola, T., Mavumengwana, V. (2020). Toxic Metal Implications on Agricultural Soils, Plants, Animals, Aquatic life and Human Health. *Int J Environ Res Public Health*. 25;17(7):2204. doi: 10.3390/ijerph17072204.

Oria, R. S., Ben, R. B., Esomonu, U. G., Essien, P. I., Odinaka, L. E., Ettah, G. E., Eyong, O. O., & Ijomone, O. M. (2022). Cobalt exposure triggers impairments in cognitive and anxiety-like behaviors, brain oxidative stress and inflammation, and hippocampo-amygdala histomorphological alterations: Protective role of aqueous *Prosopis africana* seed extract. *Iranian Journal of Basic Medical Sciences*, *25*(12), 1528–1536.

Ostrakhovitch, E. A. (2015). Tin. In G. F. Nordberg, B. A. Fowler, M. Nordberg, & L. T. Friberg (Eds.), *Handbook on the toxicology of metals* (pp. 1241–1285). Elseiver.

Ostrakhovitch, E. A., & Cherıan, M. G. (2007). Tin. In G. F. Nordberg, B. A. Fowler, M. Nordberg, & L. T. Friberg (Eds.), *Handbook on the toxicology of metals* (pp. 839–859). Elseiver.

Oz, M., Yavuz, O., & Bolukbas, F. (2020). Histopathology changes in the rainbow trout (Onchorhyncus mykiss) consuming boric acid supplemented fish fodder. *Journal of Trace Elements in Medicine and Biology: Organ of the Society for Minerals and Trace Elements (GMS)*, *62*, 126581.

Paduraru, E., Iacob, D., Rarinca, V., Plavan, G., Ureche, D., Jijie, R., & Nicoara, M. (2023). Zebrafish as a potential model for neurodegenerative diseases: A focus on toxic metals implications. *International Journal of Molecular Sciences*, *24*(4), 3428.

Pajarillo, E., Nyarko-Danquah, I., Digman, A., Multani, H. K., Kim, S., Gaspard, P., Aschner, M., & Lee, E. (2022). Mechanisms of manganese-induced neurotoxicity and the pursuit of neurotherapeutic strategies. *Frontiers in Pharmacology*, *13*, 1011947.

Perlinska, A.P., Stasiulewicz, A., Nawrocka, E.K., Kazimierczuk, K., Setny, P., Sulkowska, J.I. (2020). Restriction of S-adenosylmethionine conformational freedom by knotted protein binding sites. *PLoS Comput Biol*. 26;16(5):e1007904.

Pilehvar, A., Town, R. M., & Blust, R. (2020). The effect of copper on behaviour, memory, and associative learning ability of zebrafish (Danio rerio). *Ecotoxicology and Environmental Safety*, *188*, 109900.

Plum, L. M., Rink, L., & Haase, H. (2010). The essential toxin: Impact of zinc on human health. *ınternational Journal of Environmental Research and Public Health*, *7*(4), 1342–1365.

Price, C. T., Koval, K. J., & Langford, J. R. (2013). Silicon: A review of its potential role in the prevention and treatment of postmenopausal osteoporosis. *International Journal of Endocrinology*, *2013*, 316783.

Rachamalla, M., Salahinejad, A., Khan, M., Datusalia, A. K., & Niyogi, S. (2023). Chronic dietary exposure to arsenic at environmentally relevant concentrations impairs cognitive performance in adult zebrafish (Danio rerio) via oxidative stress and dopaminergic dysfunction. *The Science of the Total Environment*, *886*, 163771.

Rahimzadeh, M. R., Rahimzadeh, M. R., Kazemi, S., Amiri, R. J., Pirzadeh, M., & Moghadamnia, A. A. (2022). Aluminum poisoning with emphasis on its mechanism and treatment of intoxication. *Emergency Medicine International*, *2022*, 1480553.

Rouault T. A. (2013). Iron metabolism in the CNS: Implications for neurodegenerative diseases. *Nature Reviews. Neuroscience*, *14*(8), 551–564.

Rowe, R. I., & Eckhert, C. D. (1999). Boron is required for zebrafish embryogenesis. *The Journal of Experimental Biology*, *202*, 1649–1654.

Rowe, R. I., Bouzan, C., Nabiu, S., Eckhert, C.D. (1998). The response of trout and zebrafish embryos to low and high boron concentrations ıs u-shaped. *Biological Trace Element Research*, *66*, 261–270.

Sadiku, O. O., & Rodríguez-Seijo, A. (2022). Metabolic and genetic derangement: A review of mechanisms involved in arsenic and lead toxicity and genotoxicity. *Arhives of Industrial Hygiene and Toxicology*, *73*(4), 244–255.

Salvador, G. A., Uranga, R. M., & Giusto, N. M. (2010). Iron and mechanisms of neurotoxicity. *International Journal of Alzheimer's Disease*, *2011*, 720658.

Sanders, T., Liu, Y., Buchner, V., & Tchounwou, P. B. (2009). Neurotoxic effects and biomarkers of lead exposure: A review. *Reviews on Environmental Health*, *24*(1), 15–45.

Sant'Anna, M. C., Soares V.deM., Seibt, K. J., Ghisleni, G., Rico, E. P., Rosemberg, D. B., de Oliveira, J. R., Schröder, N., Bonan, C. D., & Bogo, M. R. (2011). Iron exposure modifies acetylcholinesterase activity in zebrafish (Danio rerio) tissues: Distinct susceptibility of tissues to iron overload. *Fish Physiology and Biochemistry*, *37*(3), 573–581.

Santos, D., Luzio, A., Bellas, J., & Monteiro, S. M. (2022). Microplastics- and copper-induced changes in neurogenesis and DNA methyltransferases in the early life stages of zebrafish. *Chemico-Biological Interactions*, *363*, 110021.

Santos, J., Barreto, Â., Almeida, C., Azevedo, C., Domingues, I., Amorim, M. J. B., & Maria, V. L. (2021). Toxicity of boron and vanadium nanoparticles on Danio rerio embryos - Phenotypical, biochemical, and behavioral alterations. *Aquatic Toxicology (Amsterdam, Netherlands)*, *238*, 105930.

Sarasamma, S., Audira, G., Juniardi, S., Sampurna, B. P., Liang, S. T., Hao, E., Lai, Y. H., & Hsiao, C. D. (2018). Zinc Chloride Exposure Inhibits Brain Acetylcholine Levels, Produces Neurotoxic Signatures, and Diminishes Memory and Motor Activities in Adult Zebrafish. *International Journal of Molecular Sciences*, *19*(10), 3195.

Scheiber, I., Dringen, R., & Mercer, J. F. (2013). Copper: effects of deficiency and overload. *Metal ions in Life Sciences*, *13*, 359–387.

Scheiber, I. F., Mercer, J. F., & Dringen, R. (2014). Metabolism and functions of copper in brain. *Progress in Neurobiology*, *116*, 33–57.

Sensi, S. L., Paoletti, P., Bush, A. I., & Sekler, I. (2009). Zinc in the physiology and pathology of the CNS. *Nature Reviews. Neuroscience*, *10*(11), 780–791.

Sensi, S. L., Paoletti, P., Koh, J. Y., Aizenman, E., Bush, A. I., & Hershfinkel, M. (2011). The neurophysiology and pathology of brain zinc. *The Journal of Neuroscience: The Official Journal of the Society for Neuroscience*, *31*(45), 16076–16085.

Shang, N., Zhang, P., Wang, S., Chen, J., Fan, R., Chen, J., Huang, T., Wang, Y., Duncan, J., Zhang, L., Niu, Q., & Zhang, Q. (2020). Aluminum-induced cognitive impairment and PI3K/Akt/mTOR signaling pathway involvement in occupational aluminum workers. *Neurotoxicity Research*, *38*(2), 344–358.

Shaw, P., Mondal, P., Bandyopadhyay, A., & Chattopadhyay, A. (2020). Environmentally relevant concentration of chromium induces nuclear deformities in erythrocytes and alters the expression of stress-responsive and apoptotic genes in brain of adult zebrafish. *The Science of the Total Environment*, *703*, 135622.

Shvachiy, L., Amaro-Leal, Â., Outeiro, T. F., Rocha, I., & Geraldes, V. (2022). From molecular to functional effects of different environmental lead exposure paradigms. *Biology*, *11*(8), 1164.

Singh, V., Singh, N., Verma, M., Kamal, R., Tiwari, R., Sanjay Chivate, M., Rai, S. N., Kumar, A., Singh, A., Singh, M. P., Vamanu, E., & Mishra, V. (2022). Hexavalent-chromium-induced oxidative stress and the protective role of antioxidants against cellular toxicity. *Antioxidants (Basel, Switzerland)*, *11*(12), 2375.

Sirelkhatim, A., Mahmud, S., Seeni, A., Kaus, N.H.M, Ann, L.C., Bakhori, S.K.M., Hasan, H., Mohamad, D. (2015) Review on Zinc Oxide Nanoparticles: Antibacterial Activity and Toxicity Mechanism. *Nanomicro Lett*. 7(3), 219–242.

Şişman T. (2011). Early life stage and genetic toxicity of stannous chloride on zebrafish embryos and adults: toxic effects of tin on zebrafish. *Environmental Toxicology*, *26*(3), 240–249.

Skalny, A. V., Aschner, M., Jiang, Y., Gluhcheva, Y. G., Tizabi, Y., Lobinski, R., & Tinkov, A. A. (2021). Molecular mechanisms of aluminum neurotoxicity: Update on adverse effects and therapeutic strategies. *Advances in Neurotoxicology*, *5*, 1–34.

Slutsky, I., Abumaria, N., Wu, L. J., Huang, C., Zhang, L., Li, B., Zhao, X., Govindarajan, A., Zhao, M. G., Zhuo, M., Tonegawa, S., & Liu, G. (2010). Enhancement of learning and memory by elevating brain magnesium. *Neuron*, *65*(2), 165–177.

Song, B., Liu, J., Feng, X., Wei, L., & Shao, L. (2015). A review on potential neurotoxicity of titanium dioxide nanoparticles. *Nanoscale Research Letters*, *10*(1), 1042.

Spence H, McNeilCJ, Waiter GD (2020) The impact of brain iron accumulation on cognition: A systematic review. *PLoS One*, *15*(10), e0240697.

Takemura Mariano, M. V., Paganotto Leandro, L., Gomes, K. K., Dos Santos, A. B., de Rosso, V. O., Dafre, A. L., Farina, M., Posser, T., & Franco, J. L. (2024). Assessing the disparity: Comparative toxicity of Copper in zebrafish larvae exposes alarming consequences of permissible concentrations in Brazil. *Journal of Toxicology and Environmental Health. Part A*, *87*(4), 166–184.

Takesono, A., Dimitriadou, S., Clark, N. J., Handy, R. D., Mourabit, S., Winter, M. J., Kudoh, T., & Tyler, C. R. (2023). Zinc oxide nanoparticles disrupt development and function of the olfactory sensory system impairing olfaction-mediated behaviour in zebrafish. *Environment International*, *180*, 108227.

Tan, J. K., Nazar, F. H., Makpol, S., & Teoh, S. L. (2022). Zebrafish: A pharmacological model for learning and memory research. *Molecules (Basel, Switzerland)*, *27*(21), 7374.

Tang, J., Li, Y., Liu, X., Yu, G., Zheng, F., Guo, Z., Zhang, Y., Shao, W., Wu, S., & Li, H. (2023). Cobalt induces neurodegenerative damages through impairing autophagic flux by activating hypoxia-inducible factor-1α triggered ROS overproduction. *The Science of the Total Environment*, *857*(Pt 2), 159432.

Thakur, M., Rachamalla, M., Niyogi, S., Datusalia, A. K., & Flora, S. J. S. (2021). Molecular mechanism of arsenic-induced neurotoxicity including neuronal dysfunctions. *International Journal of Molecular Sciences*, *22*(18), 10077.

Tu, H., Peng, T., Liu, J., Chen, X., Fan, C., Huang, Z., Zhang, Y., Zou, F., & Meng, X. (2018). Role of neurexin2a in lead-induced locomotor defect in developing zebrafish. *Aquatic Toxicology (Amsterdam, Netherlands)*, *194*, 167–175.

Tuerk, M. J., & Fazel, N. (2009). Zinc deficiency. *Current Opinion in Gastroenterology*, *25*(2), 136–143.

Turkez, H., Yıldırım, S., Sahin, E., Arslan, M. E., Emsen, B., Tozlu, O. O., Alak, G., Ucar, A., Tatar, A., Hacimuftuoglu, A., Keles, M. S., Geyikoglu, F., Atamanalp, M., Saruhan, F., & Mardinoglu, A. (2022). Boron compounds exhibit protective effects against aluminum-induced neurotoxicity and genotoxicity: In vitro and in vivo study. *Toxics*, *10*(8), 428.

Veldman, M. B., & Lin, S. (2008). Zebrafish as a developmental model organism for pediatric research. *Pediatric Research*, *64*(5), 470–476.

Vimercati, L., Cavone, D., Caputi, A., De Maria, L., Tria, M., Prato, E., Ferri, G.M. (2020) Nanoparticles: An Experimental Study of Zinc Nanoparticles Toxicity on Marine Crustaceans. General Overview on the Health Implications in Humans. *Front Public Health*. 21;8:192. doi: 10.3389/fpubh.2020.00192.

Wang, B., & Du, Y. (2013). Cadmium and its neurotoxic effects. *Oxidative Medicine and Cellular Longevity*, *2013*, 898034.

Wang, X., Yang, X., He, W., Zhang, S., Song, X., Zhang, J., Ma, J., Chen, L., Niu, P., & Chen, T. (2024). Single-cell transcriptomics analysis of zebrafish brain reveals adverse effects of manganese on neurogenesis. *Environmental Pollution*, *341*, 122908.

Wang, X., Zhang, J. B., He, K. J., Wang, F., & Liu, C. F. (2021). Advances of zebrafish in neurodegenerative disease: From models to drug discovery. *Frontiers in Pharmacology*, *12*, 713963.

Wang, Y. H., Wang, Y. Q., Yu, X. G., Lin, Y., Liu, J. X., Wang, W. Y., & Yan, C. H. (2023). Chronic environmental inorganic arsenic exposure causes social behavioral changes in juvenile zebrafish (Danio rerio). *The Science of the Total Environment*, *867*, 161296.

Wang, Z., Li, K., Xu, Y., Song, Z., Lan, X., Pan, C., Zhang, S., Foulkes, N. S., & Zhao, H. (2023). Ferroptosis contributes to nickel-induced developmental neurotoxicity in zebrafish. *The Science of the Total Environment*, *858*(Pt 3), 160078.

Wani, A. L., Ara, A., & Usmani, J. A. (2015). Lead toxicity: a review. *Interdisciplinary Toxicology*, *8*(2), 55–64.

Waters, C.M., Bassler, B.L. (2005). Quorum sensing: cell-to-cell communication in bacteria. *Annu Rev Cell Dev Biol.* 21:319–46.

Wei, J., Liu, J., Liang, S., Sun, M., & Duan, J. (2020). Low-dose exposure of silica nanoparticles induces neurotoxicity via neuroactive ligand-receptor interaction signaling pathway in zebrafish embryos. *International Journal of Nanomedicine*, *15*, 4407–4415.

Wei, L., Zuo, Z., Yang, Z., Yin, H., Yang, Y., Fang, J., Cui, H., Du, Z., Ouyang, P., Chen, X., Chen, J., Geng, Y., Zhu, Y., Chen, Z., Huang, C., Wang, F., & Guo, H. (2022). Mitochondria damage and ferroptosis involved in Ni-induced hepatotoxicity in mice. *Toxicology*, *466*, 153068.

Wise, J. P., Jr, Young, J. L., Cai, J., & Cai, L. (2022). Current understanding of hexavalent chromium [Cr(VI)] neurotoxicity and new perspectives. *Environment International*, *158*, 106877.

Wu, Q., Ren, Q., Meng, J., Gao, W. J., & Chang, Y. Z. (2023). Brain iron homeostasis and mental disorders. *Antioxidants (Basel, Switzerland)*, *12*(11), 1997.

Wu, Y. S., Osman, A. I., Hosny, M., Elgarahy, A. M., Eltaweil, A. S., Rooney, D. W., Chen, Z., Rahim, N. S., Sekar, M., Gopinath, S. C. B., Mat Rani, N. N. I., Batumalaie, K., & Yap, P. S. (2024). The toxicity of mercury and its chemical compounds: Molecular mechanisms and environmental and human health implications: A comprehensive review. *ACS Omega*, *9*(5), 5100–5126.

Xu, X., Weber, D., Burge, R., & VanAmberg, K. (2016). Neurobehavioral impairments produced by developmental lead exposure persisted for generations in zebrafish (Danio rerio). *Neurotoxicology*, *52*, 176–185.

Xu, Y., Liu, J., Tian, Y., Wang, Z., Song, Z., Li, K., Zhang, S., & Zhao, H. (2022b). Wnt/β-catenin signaling pathway is strongly implicated in cadmium-induced developmental neurotoxicity and neuroinflammation: Clues from zebrafish neurobehavior and in vivo neuroimaging. *International Journal of Molecular Sciences*, *23*(19), 11434.

Xu, Y., Peng, T., Xiang, Y., Liao, G., Zou, F., & Meng, X. (2022). Neurotoxicity and gene expression alterations in zebrafish larvae in response to manganese exposure. *The Science of the Total Environment*, *825*, 153778.

Xu, Y., Wang, L., Zhu, J., Jiang, P., Zhang, Z., Li, L., & Wu, Q. (2021). Chromium induced neurotoxicity by altering metabolism in zebrafish larvae. *Ecotoxicology and Environmental Safety*, *228*, 112983. Advance online publication.

Xu, Y., Zhao, H., Wang, Z., Gao, H., Liu, J., Li, K., Song, Z., Yuan, C., Lan, X., Pan, C., & Zhang, S. (2022a). Developmental exposure to environmental levels of cadmium induces neurotoxicity and activates microglia in zebrafish larvae: From the perspectives of neurobehavior and neuroimaging. *Chemosphere*, *291*(Pt 1), 132802.

Yan, G., Gao, Y., Xue, K., Qi, Y., Fan, Y., Tian, X., Wang, J., Zhao, R., Zhang, P., Liu, Y., & Liu, J. (2023). Toxicity mechanisms and remediation strategies for chromium exposure in the environment. *Frontiers in Environmental Science*, 11. https://doi.org/10.3389/fenvs.2023.1131204

Yu, F., Hou, Z. S., Luo, H. R., Li, H. X., Cui, X. F., Li, J. L., Feng, W. R., Tang, Y. K., Su, S. Y., Gao, Q. F., Xiao, J., & Xu, P. (2023). Neurobehavioral disorders induced by environmental zinc in female zebrafish (Danio rerio): Insights from brain and intestine transcriptional and metabolic signatures. *Chemosphere*, *335*, 138962.

Yuana, Z., Lia, R., Lib, S., Qiuc, D., Lia, G., Wang, C., Nid, J., Sun, Y., & Hue, H. (2023). Oxidative stress, neurotoxicity, and intestinal microbial regulation after a chronic zinc exposure: An experimental study on adult zebrafish (Danio rerio). *Water Reuse*, 13 (1), 82–96.

Zhang, S., Wu, L., Zhang, J., Wang, X., Yang, X., Xin, Y., Chen, L., Li, J., & Niu, P. (2023). Multi-omics analysis reveals Mn exposure affects ferroptosis pathway in zebrafish brain. *Ecotoxicology and Environmental Safety*, *253*, 114616.

Zhang, X., Song, Y., Gong, H., Wu, C., Wang, B., Chen, W., Hu, J., Xiang, H., Zhang, K., & Sun, M. (2023). Neurotoxicity of titanium dioxide nanoparticles: A comprehensive review. *International Journal of Nanomedicine*, *18*, 7183–7204.

Zhao, G., Hu, J., Gao, M., Zhu, Y., & Hong, Y. (2022). Excessive selenium affects neural development and locomotor behavior of zebrafish embryos. *Ecotoxicology and Environmental Safety*, *238*, 113611.

Zhao, L., Xia, Z., & Wang, F. (2014). Zebrafish in the sea of mineral (iron, zinc, and copper) metabolism. *Frontiers in Pharmacology*, *5*, 33.

Zheng, J.L., Yuan, S.S., Wu, C.W., & Lv, Z.M. (2016). Acute exposure to waterborne cadmium induced oxidative stress and immunotoxicity in the brain, ovary and liver of zebrafish (Danio rerio). *Aquatic Toxicology*, *180*, 36–44.

Zhou, J., Zhang, W., Cao, Z., Lian, S., Li, J., Nie, J., Huang, Y., Zhao, K., He, J., & Liu, C. (2023). Association of selenium levels with neurodegenerative disease: A systemic review and meta-analysis. *Nutrients*, *15*(17), 3706.

Zhu, J., Wang, C., Gao, X., Zhu, J., Wang, L., Cao, S., Wu, Q., Qiao, S., Zhang, Z., & Li, L. (2019). Comparative effects of mercury chloride and methylmercury exposure on early neurodevelopment in zebrafish larvae. *RSC Advances*, *9*(19), 10766–10775.

# 19

# *Effects of Microplastics and Nanoplastics on Neurodevelopment and Neurodegeneration in Zebrafish*

**Zülal Mızrak, Derya Cansiz and Ebru Emekli-Alturfan**

## Abbreviations

| | |
|---|---|
| ***α1-tubulin*** | Microtubule Formation |
| **ACh** | Acetylcholine |
| **AChE** | Acetylcholinesterase |
| **ALT** | Alanine Aminotransferase |
| **AST** | Aspartate Aminotransferase |
| **ATP** | Adenosine Triphosphate |
| ***bdnf*** | Brain-Derived Neurotrophic Factor |
| **BPA** | Bisphenol A |
| **CAP** | Caudal Primary Motoneuron |
| **CAT** | Catalase |
| **CNS** | Central Nervous System |
| **DNA** | Deoxyribonucleic Acid |
| **DOPAC** | 3,4-Dihydroxyphenylacetic Acid |
| **EDC** | Endocrine Disrupting Chemical |
| **EEG** | Electroencephalography |
| ***elavl3*** | ELAV Like Neuron-Specific RNA Binding Protein 3 |
| **GABA** | Gamma-Aminobutyric Acid |
| ***gap43*** | Growth Associated Protein 43 |
| ***gfap*** | Glial Fibrillary Acidic Protein |
| **Gln** | L-Glutamine |
| **GSH** | Glutathione |
| **GST** | Glutathione S-Transferase |
| **hiPSCs** | Human Induced Pluripotent Stem Cells |
| **HPF** | Hours Post Fertilization |
| ***isl2a*** | *ISL LIM Homeobox 2a* |
| **LDH** | Lactate Dehydrogenase |
| ***mbp*** | Myelin Basic Protein |
| **MP** | Microplastic |
| **MNP** | Microplastic and Nanoplastic |
| ***neurod1*** | Neuronal Differentiation 1 |
| **NP** | Nanoplastic |
| ***nrf2*** | Nuclear Factor Erythroid 2–Related Factor 2 |
| **OCP** | Organochlorine Pesticide |
| **P53** | Tumor Protein |
| **PAH** | Polycyclic Aromatic Hydrocarbon |
| **PCB** | Polychlorinated Biphenyl |
| **PE** | Polyethylene |
| **PET** | Polyethylene Terephthalate |
| **PP** | Polypropylene |
| **PS** | Polystyrene |
| **PS-MP** | Polystyrene Microplastic |
| **PS-NP** | Polystyrene Nanoplastic |
| **PVC** | Polyvinyl Chloride |
| **ROS** | Reactive Oxygen Species |
| ***shha*** | Sonic Hedgehog Signaling Molecule |
| **SOD** | Superoxide Dismutase |
| ***syn2α*** | Synaptophysin 2α |
| **TBBPA** | Tetrabromobisphenol A |
| ***th1*** | Tyrosine Hydroxylase |
| ***zfblue*** | Opsin 1, Short-Wave-Sensitive 2 |
| ***zfrho*** | Rhodopsin |

## Introduction

### Plastics

Plastics were developed after the Industrial Revolution in the nineteenth century. Plastic materials are important in today's industrial economy due to their light weight, durability, low cost, and easy formability (De-la-Torre, 2020; Lehner et al.,2019). Plastics are very suitable materials for daily use because they have structures that can withstand different temperatures, light, and chemical conditions. In addition, they are frequently preferred materials by people due to their ease of use (Andrady and Neal, 2009). Single-use products such as bottles, straws, and bags are important causes of plastic pollution in the environment, and the large number of single-use plastic products used during the COVID-19 pandemic made a significant contribution to plastic waste (Adyel, 2020). Consequently, world plastic production reached 400 million tons in 2022, and China leads the world in total global plastic production with a rate of 32% (PlasticsEurope, 2024). Türkiye, with a daily output of 144 tons, is one of the countries in the Mediterranean region with the highest plastic waste rate (Osman et al., 2023).

Plastic waste can undergo various physical, chemical, or biological degradation processes in the environment for many



DOI: 10.1201/9781032676555-21

years. The main processes are exposure to UV light from the sun, thermal reactions, hydrolysis, and microbial decomposition, leading to degradation of larger plastic materials into smaller pieces (De-la-Torre, 2020; Lehner et al., 2019). However, the widespread use of plastics and their low degradation rates have led to the accumulation of plastic particles in a range of different environments, including seawater, rivers, lakes, and sediments, as well as soil and the atmosphere (Lai et al., 2021).

## Submillimeter-sized Plastic Particles: Microplastics and Nanoplastics

The term "microplastic" (MP) was first used by Thompson (Thompson et al., 2004) to describe particles with a diameter between 1 μm and 5 mm. Particles with a diameter below 1 μm are classified as nanoplastics (NP). There are five main types of microplastics: fragments, fibers, foam, pellets, and films. The most common types of microplastics and nanoplastics (MNPs) with chemical properties include polyethylene (PE), polypropylene (PP), polystyrene (PS), polyvinyl chloride (PVC), and polyethylene terephthalate (PET). MNPs are divided into two categories; primary MNPs are small particles produced industrially for use in products such as clothing, cosmetics, and cleaning. Resin pellets, microbeads found in cosmetic products and toothpaste, and powder from textile coatings are examples. Secondary ones such as microfiber and debris are mainly formed when large disposable plastics such as water bottles and bags degrade over time through various processes such as mechanical abrasion, photolysis, hydrolysis, heat, and biological degradation (De-la-Torre, 2020; Lehner et al., 2019).

## *Accumulation of Plastic (Micro-Nanoparticles) Waste in the Environment*

Microplastic pollution has attracted the attention of scientists and the public in the last decade. The use of plastic fibers and beads in micro- and nanosized forms is prevalent in the production of consumer products such as cosmetics and textiles. As a consequence of the manufacturing, application, and disposal processes, these materials inevitably become discharged into the environment (Figure 19.1) (Lai et al., 2021).

Due to their very small size, MNPs cannot be adequately treated from wastewater and can easily mix into fresh waters and the sea, causing pollution. As a result of their use for many years, they have become pervasive in every layer of the seas (De-la-Torre, 2020; Lai et al., 2021; Rezania et al, 2018). They are not just confined to populated areas, but are found in many parts of the world, from deep sea to the poles (Van Cauwenberghe et al., 2013). Land-based waste such as plastic bottles and bags, cosmetics, textiles, medical supplies, and construction materials account for 80%–90% of microplastic pollution in aquatic areas. Ocean-sourced waste such as commercial fishing, marine vessels, and coastal industries account for 10%–20% (Lehner et al., 2019; Osman et al., 2023).

MNPs can absorb various chemicals. Pollutants released from electronic devices and dyes include heavy metals, polycyclic aromatic hydrocarbons (PAHs), polychlorinated biphenyls (PCBs), organochlorine pesticides (OCPs), and pharmaceuticals at trace concentrations (De-la-Torre, 2020; Lehner et al., 2019). Endocrine disrupting chemicals (EDCs) such as bisphenol A (BPA), nonylphenol, phthalate esters, and octylphenol are frequently used in plastics production and are found in microplastics produced as additives or reaction markers. Therefore,

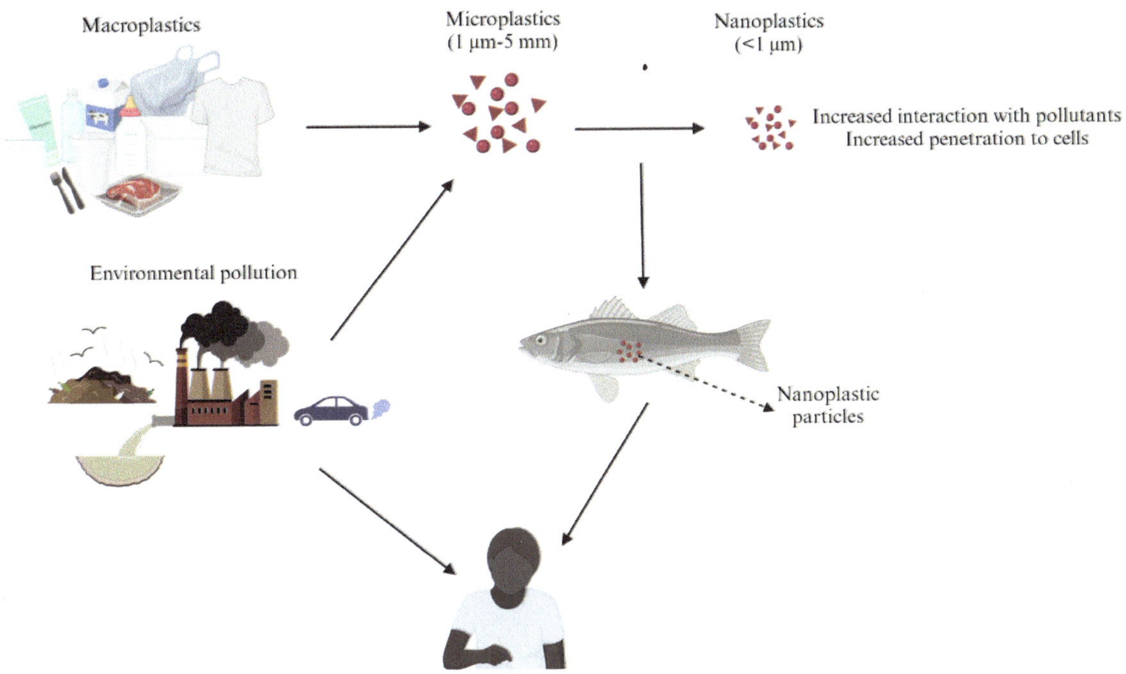


**FIGURE 19.1** The spread of micro and nanoplastics.

damage to the endocrine system is one of the potential effects of MNPs. EDCs can be released from MNPs when they are ingested or come into contact with the organism. This can lead to negative effects on hormonal balance, reproductive function, and development. Indeed, exposure to MNPs and related chemicals led to altered gene expression and abnormal germ cell proliferation in Japanese medaka (Emenike et al., 2023).

MPs can also affect bacteria and other organisms, leading to the formation of complex biofilms. The term "plastisphere" was first used by Zettler to describe unique microbial communities living on microplastic surfaces in the marine environment (Zettler et al., 2013). By acting as carriers for pathogenic bacteria, MNPs can contaminate water supplies and the food chain and lead to the spread of disease. Some pathogenic bacteria found in MPs have been associated with gastrointestinal diseases, respiratory infections, and skin diseases (Emenike et al., 2023).

## Effects of Microplastics and Nanoplastics on Organisms

Plastic particles were first found in the gastrointestinal tract of seabirds in 1960 (Ugwu et al., 2021) and since then, their presence has been detected in a variety of marine organisms such as zooplankton, crustaceans, bivalves, fish species, and seabirds, both in their natural environment and in controlled laboratory studies (De-la-Torre, 2020; Emenike et al., 2023; Ugwu et al., 2021). Field investigations have demonstrated the presence of a variety of fibrous, spherical, and fragmentary MPs within the gastrointestinal tract and gills of fish, particularly within muscle and liver tissues (Ma et al., 2021). Their ingestion can lead to a number of problems such as blockage of the digestive tract, entry into the circulatory system, and impairment of body functions (Lai et al., 2021).

Humans are exposed to MNPs mainly through three routes: skin contact, inhalation (synthetic textiles and urban dust), and diet (contaminated seafood and other packaged foods) (Figure 19.2) (De-la-Torre, 2020; Lehner et al., 2019; Wang et al., 2021)

Microbeads in personal care products and synthetic fibers in textile products are examples of MP that we are exposed to through skin contact. Synthetic textiles, construction materials, landfills, sewage, abrasion of plastic materials, and waste incineration are assumed to be exposure sources of MPs through inhalation. MPs have also been detected in indoor air and in lung tissue and sputum. Some small fibers can accumulate in the alveoli and bronchioles and cause inflammation (Lehner et al., 2019). MPs have been detected in products consumed as food or used for cooking (drinking water, tap water, table salt, canned food, honey, and sugar). It is inevitable that

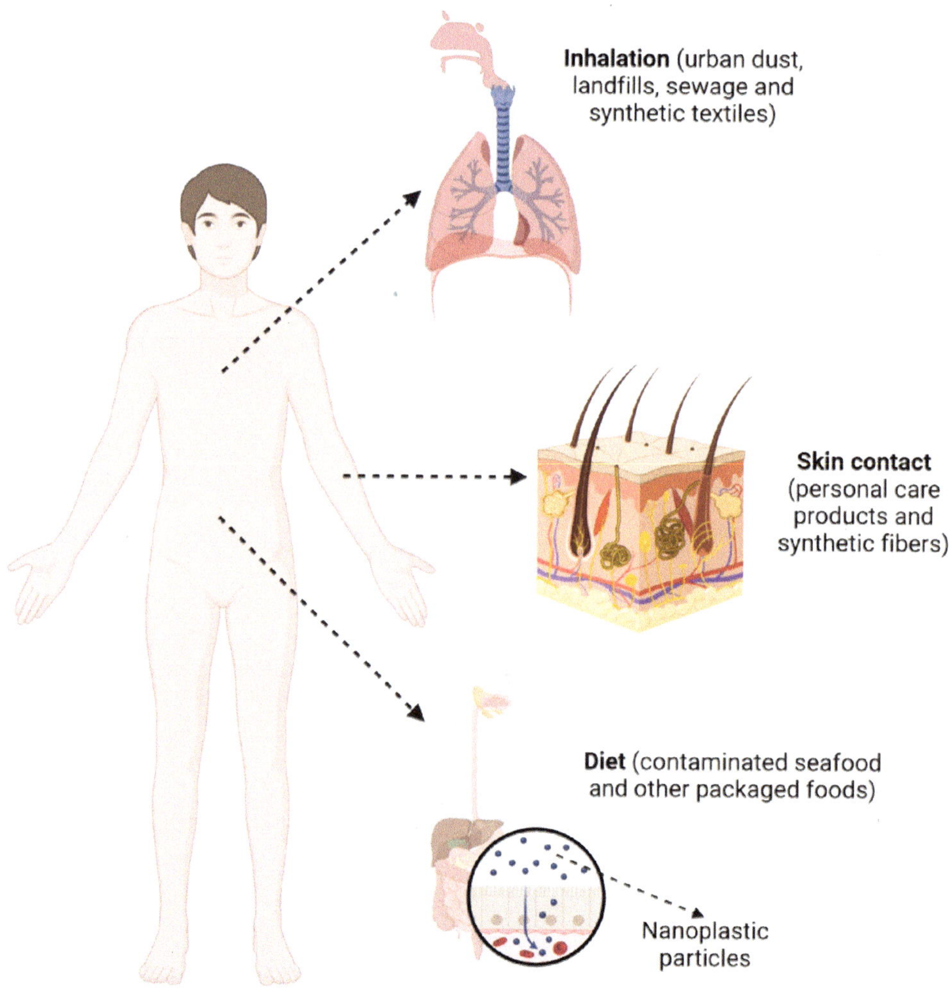


FIGURE 19.2 Exposure ways to micro and nanoplastics of humans.

people are chronically exposed to MPs even at low concentrations (De-la-Torre, 2020).

The size of MNPs facilitates their easier penetration into living organisms. The ingestion of MNPs by organisms and their subsequent accumulation in biological tissues may result in the transfer of these particles through food chains, thereby increasing the associated health and environmental risks (Lai et al., 2021; Wang et al., 2021).

Polymeric substances such as polystyrene, which are commonly utilized in various consumer products, have the capacity to gain access to cells through a number of different pathways. Endocytosis is the primary route of this uptake, which includes clathrin- and caveolae-mediated endocytosis, macropinocytosis, and phagocytosis. Nanoparticles have been shown to interact with a number of different biomolecules as a result of their expansive surface area and relatively small size. The attachment of these particles to proteins results in the formation of a polystyrene-protein complex, which in turn creates a protein coat, known as the corona. This formation allows MNPs to gain a biological identity, thereby enabling them to evade the immune system, remain in circulation, and disrupt a range of cellular and molecular processes (Gopinath et al., 2019; Yee et al., 2021).

Several studies have shown that MPs of 0.1–10 μm in size can cross cell membrane and the blood–brain barrier and accumulate in various biological samples and organs such as blood, heart, saliva, sputum, feces, placenta, liver, lung, and brain (Wang et al., 2021). It has been demonstrated that both HeLa cells, which are of mammalian origin, and coelomocytes, which are derived from the nematode worm Caenorhabditis elegans, are capable of endocytic uptake of latex beads with a diameter of less than 200 nm (Ali et al., 2024). Since polystyrene particles are not biodegradable and can remain intracellular, in addition to accumulation, cellular stress and cytotoxicity may occur in long-term chronic exposures. Adverse effects in pregnancy and fetal exposure may also occur. Changes in gene expression also can be seen in long-term exposures (Wang et al., 2021). The effects of MP exposure on oxidative stress, cytotoxicity, genotoxicity, neurotoxicity, apoptosis, inflammatory responses, immune dysfunction, and altered biochemical and energy metabolism have been demonstrated in both model organisms and human cells (Ali et al., 2024; Wang et al., 2021).

MNPs encounter biological barriers following their entry into the body, specifically in the lungs and digestive tract. The beads have been observed to accumulate in a number of different organs and tissues, where they have been shown to exert a range of effects, including biochemical, metabolic and genotoxic impacts, as well as potential carcinogenic effects. The presence of cell damage, inflammation, and apoptosis was identified in adenocarcinoma human alveolar basal epithelial cell lines exposed to NPs, accompanied by an increase in pro-inflammatory cytokines and pro-apoptotic proteins. In human bronchial epithelial cell lines, the induction of inflammation and cytotoxicity was observed, resulting in a reduction in lung cell viability and an imbalance in the redox levels (Ali et al., 2024).

The ingestion of MPs has been demonstrated to result in damage and inflammation of the gut, leading to the disruption of the intestinal homeostasis and consequently affecting the energy homeostasis. Co-administration of MPs and tetrabromobisphenol A (TBBPA) results in the enhancement of TBBPA-induced cytotoxic effects and alteration of gut microbiota and metabolic pathways. The gut microbiota has evolved mechanisms to cope with adverse circumstances, as evidenced by the presence of enzymes capable of degrading plastic in the human gut. A mouse model revealed that co-exposure to polystyrene nanoplastics (PS-NPs) and polystyrene microplastics (PS-MPs) resulted in apoptosis of epithelial cells and impairment of the intestinal barrier function. The toll-like receptor 4 (TLR4) signaling pathway has been proposed as a potential mechanism for the disruption of the microbiota and inflammatory response (Ali et al., 2024).

MP exposure has been demonstrated to impact energy metabolism and liver function. An increase in apoptotic cells and alterations in molecular markers such as alanine aminotransferase (ALT), aspartate aminotransferase (AST), and lactate dehydrogenase (LDH) can lead to hepatotoxicity and lipotoxicity. Liver organoids have been observed to exhibit a number of metabolic abnormalities, including lipid accumulation, diminished adenosine triphosphate (ATP) production, elevated reactive oxygen species (ROS) production, and the release of pro-inflammatory markers. In addition, mitochondrial cristae are damaged by MPs, resulting in liver damage (Ali et al., 2024).

A 24-hour exposure to PS-NPs resulted in a notable impact on the development of the atrioventricular valve in a human-induced pluripotent stem cells (hiPSCs) model. MNPs were observed to cause hemolysis, a reduction in thrombin production, and coagulation. In vivo exposure of rats to PS-MPs resulted in cardiac damage, mitochondrial dysfunction, and elevated levels of creatine kinase-MB and cardiac troponin I. Oral exposure of chickens to PS-MPs led to myocardial fibre damage and mitochondrial lesions (Ali et al., 2024).

## Effects of Microplastics and Nanoplastics on the Central Nervous System

The human central nervous system is vulnerable to the effects of environmental pollutants, particularly during embryonic development. MNPs have been demonstrated to cause cellular damage by triggering oxidative stress and thereby alter the function of the central nervous system (CNS). A reduction in acetylcholinesterase (AChE) activity results in an accumulation of acetylcholine (ACh), which can lead to the onset of neurological disorders. A decrease in the amount of phenylalanine, a precursor of neurotransmitters, has been observed in mice. Prolonged exposure in a human brain model resulted in cell death (Ali et al., 2024).

The effects of exposure to MNPs with different sizes and concentrations on neural stem cells and neural stem cell–derived astrocytes, oligodendrocytes, and neurons have been examined. The results demonstrated that astrocytes exhibited the most pronounced effects, with a notable decline in cell viability and the induction of an inflammatory response indicative of astrogliosis. Other cell types demonstrated minimal or negligible cytotoxicity. Tumor protein (P53) signaling was stimulated, resulting in increased activity of pathways associated with cell senescence and apoptosis. Upregulated pathways were found to be associated with neuroinflammation, cell migration, proliferation, extracellular matrix remodeling, and cytoskeleton structures. The downregulated pathways were linked to lipid metabolism. Smaller particles activated a greater number of pathways in comparison with bigger ones. The link between MNPs and neurobehavioral and neurodevelopmental effects remains a topic of ongoing research (Marcellus et al., 2024).

## Accumulation of Microplastics and Nanoplastics on Zebrafish

Zebrafish represent a promising model organism for human diseases, drug development, and toxicology. In recent years, they have been used frequently for the study of diverse organ systems and biological pathways. Zebrafish share many similar physiological characteristics with land organisms, including an endocrine system, hormones, receptors, and signal transduction pathways. In comparison to other model organisms, it offers technical advantages such as easy maintenance, low cost, high fertility, and short experimental periods. Moreover, rapid embryonic development, suitability for phenotypic imaging with transparent eggs and embryos, relatively shorter generation time (3–4 months), and genetic similarity to humans are among the advantages of zebrafish. Zebrafish are capable of producing around 200 embryos per spawning. Because of the yolk sac, the embryos can survive for up to five days without external feeding. Fish can be exposed to chemicals by injection or diffusion. These attributes collectively make zebrafish a suitable vertebrate model for toxicological studies (Bhagat et al., 2020; Segner H., 2009).

A number of studies have been conducted to investigate the potential toxic effects of MNPs in zebrafish. PSNP was observed to accumulate in the yolk and in various organs, including the digestive tract, liver, brain (Hwang et al., 2022; Pitt et al., 2018a; Sarasamma et al., 2020; Torres-Ruiz et al., 2023; Zhang et al., 2020), pancreas, gallbladder, heart (Hwang et al., 2022; Pitt et al., 2018a; Torres-Ruiz et al., 2023; Zhang et al., 2020), eyes (Hwang et al., 2022; Torres-Ruiz et al., 2023, Zhang et al., 2020), neuromasts (Torres-Ruiz et al., 2023), and gonads (Sarasamma et al., 2020). Microplastics can gain access to the internal organs of fish at the early developmental stage by adherence, and in adult fish, this occurs through intestinal absorption or epidermal penetration. At the 0–72 hpf stage, the primary routes of penetration for NPs are through the chorion and epidermis. In general, NPs with a diameter of less than 500 nm will penetrate the chorionic membrane, given that the diameter of the blastopore on the surface of a zebrafish embryo is approximately 500 nm. From 72 hpf, direct ingestion becomes the main route of uptake for NPs. The process of larval morphogenesis is typically complete by 120 hpf (Ma et al., 2021). The results of the electron microscopy analysis revealed the presence of PSNP aggregation, thereby indicating that the chorion serves as a defensive barrier against PSNP penetration. The embryos exhibited PSNP fluorescence predominantly in the yolk during the initial 48 hours post fertilization (hpf), followed by an increase in other organs after 72 hours (Hwang et al., 2022; Pitt et al., 2018a; Santos et al., 2020; Torres-Ruiz et al., 2023; Zhang et al., 2020).

The use of confocal imaging has enabled the clear visualization of NPs in the gills of the zebrafish, demonstrating a close proximity to the blood vessels. Furthermore, imaging has revealed the presence of NPs within the gut, as well as their migration outside the intestinal epithelium (Parenti et al., 2019). The reduction in particle size results in a significant increase in surface area, which facilitates the adsorption of PS-NPs to cells, including but not limited to blood cells, hepatic cells, and epithelial cells, leading to the destruction of the cell membrane (Umamaheswari et al., 2021). A study revealed that 50 nm PS-NPs exhibited a marked accumulation in the brain region, with a notable prevalence in the forebrain, optic tectum, cerebellum, and hindbrain. 50 nm PS-NPs demonstrated higher blood–brain barrier penetration compared to 100 nm PS-NPs (Hwang et al., 2022). Due to the distribution of NPs in gonad tissues, the next generation can be affected by these particles. As a matter of fact, it has been demonstrated that NPs of approximately 42 nm in size can transfer from the maternal parent to the yolk region of embryos (Pitt et al., 2018b).

In a study, larval and juvenile zebrafish were exposed to microplastics with a particle size range of 1–5 μm and 40–47 μm, incorporated into their diet. It was observed that particles measuring 40–47 μm were not absorbed by the intestine and passed into the gut lumen. There was a reduction in mucosal folds and a corresponding increase in mucous cells. In contrast, particles measuring between 1 and 5 μm can successfully traverse the intestinal barrier and be transported to other organs, such as the liver and muscle. The accumulation of particles in the liver led to an increase in oxidative stress markers in the juvenile specimens. Conversely, a smaller number of particles were able to reach the muscle, suggesting that the liver plays a significant role in the retention of MPs (Cattaneo et al., 2023).

The common results of numerous studies on the effects of exposure to MNPs of varying sizes and concentrations on zebrafish are as follows: an increase in oxidative stress (Chen et al., 2017; Sökmen et al., 2020; Martin-Folgar et al., 2023), disruption of neurotransmitter metabolism (inhibition of acetylcholine and dopamine activity) (Chen et al, 2017; Hwang et al, 2022; Teng et al, 2022; Martin-Folgar et al, 2023; Torres Ruiz et al, 2023; Suman et al, 2023), induction of apoptosis (Teng et al, 2022; Martin-Folgar et al, 2023; Suman et al, 2023), and decreased locomotor activity (Chen et al, 2017; Pitt et al, 2018a) (Figure 19.3).

### Effects of Microplastics and Nanoplastics on Zebrafish Neurodevelopment and Neurodegeneration

Transcriptional and molecular studies have demonstrated that exposure to MNPs significantly suppresses gene expression. The genetic factors that exhibited the most significant alterations were those related to apoptosis, the endocrine system, and nervous system development. These included the differentiation and function of nerve cells, synapse formation and function, axon formation, myelination, motor neuron development, and the retinal system (LeMoine et al., 2018; Martin-Folgar et

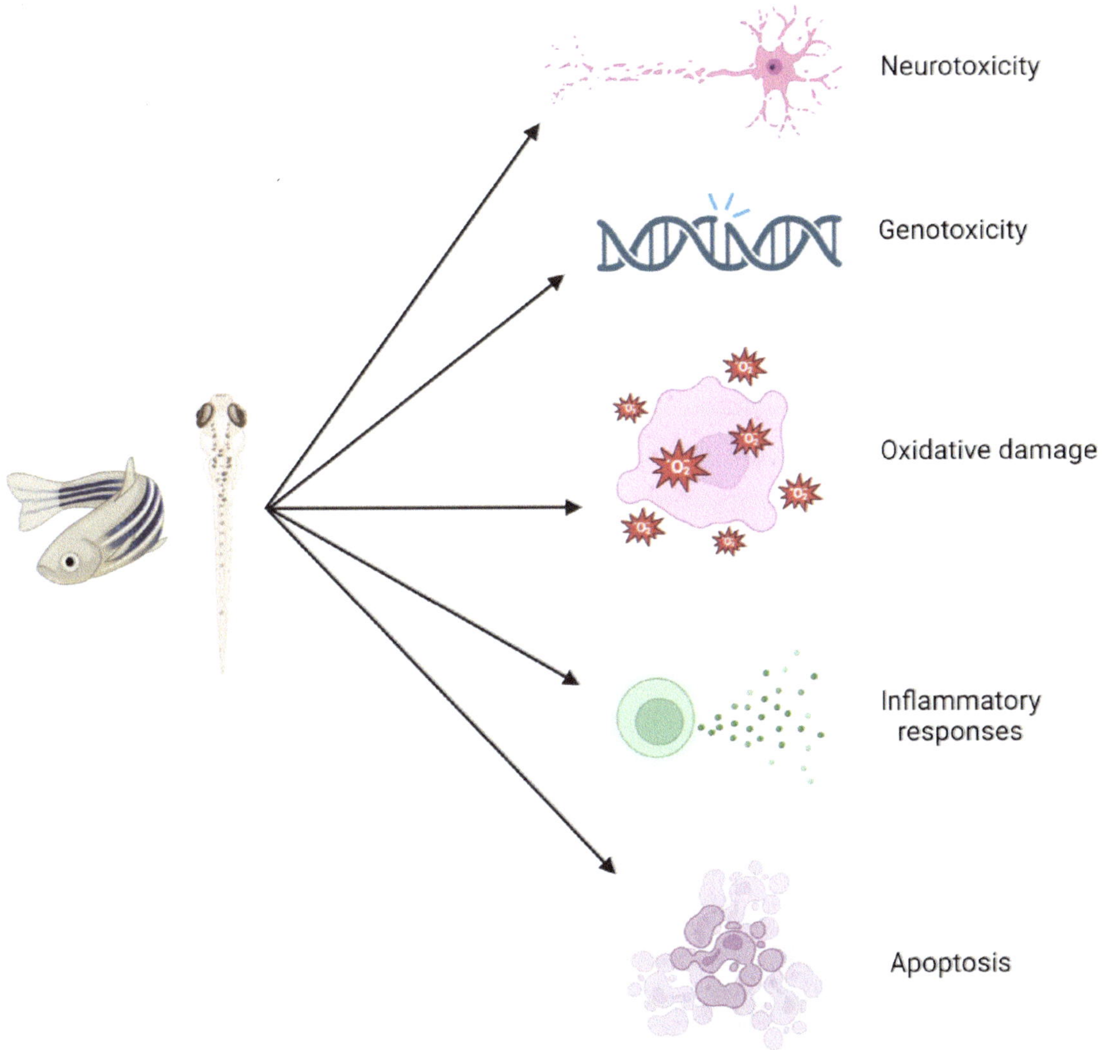


**FIGURE 19.3** The effects of micro and nanoplastics on zebrafish.

al., 2023; Pedersen et al., 2020; Suman et al., 2023). Exposure to 42 nm and 100 nm PSNPs in zebrafish caused a reduction in the expressions of neurodevelopment-related genes that are involved in axonal myelination (myelin basic protein (mbp)), synaptogenesis and neurotransmitter release (synaptophysin 2α (syn2α)), microtubule formation (α1-tubulin), sonic hedgehog signaling molecule (shha), growth associated protein 43 (gap43) and early neurogenesis (neuronal differentiation 1 (nrd, neurod1) and ELAV like neuron-specific RNA binding protein 3 (elavl3)) (Zhang et al., 2020; Zhang et al., 2022).

These genes are expressed in the central nervous system during early developmental stages and can indicate developmental neurotoxicity. Moreover, exposure has been found to downregulate opsin genes, which may lead to a decrease in opsin synthesis in photoreceptors, potentially affecting retinal development and function (Zhang et al., 2020). Also, decrease of zfrho (rhodopsin), and zfblue (opsin 1, short-wave-sensitive 2) genes and upregulation of the gene that regulates the astrocyte cytoskeleton glial fibrillary acidic protein (gfap) was observed in another study (Chen et al., 2017).

The neurotransmission processes were also found to be altered (Chen et al., 2017; Hwang et al., 2022; Teng et al., 2022; Umamaheswari et al., 2021). In particular, an inhibition in acetylcholinesterase (AChE) activity was observed (Chen et al., 2017; Suman et al., 2023; Torres-Ruiz et al., 2023). AChE, a cholinergic enzyme in postsynaptic neuromuscular junctions, terminates neuronal transmission and inhibits the release of acetylcholine, playing a crucial role in the nervous system's functionality (Trang and Khandhar, 2023). Gene transcripts associated with dopamine metabolism were also affected (Hwang et al., 2022), displaying a notable upregulation of th1 (tyrosine hydroxylase) in a concentration-dependent manner (Torres-Ruiz et al., 2023). In another study, a significant reduction was evident in the levels of 3,4-dihydroxyphenylacetic acid, or DOPAC (a metabolite of dopamine), and acetylcholine chloride. Acetylcholine chloride functions as a modulator of dopaminergic neuronal activity, whereas DOPAC plays a crucial role as an indicator of neuronal function, and its decline may indicate underlying alterations in neural activity (Teng et al., 2022).

A downregulation of the brain-derived neurotrophic factor (bdnf), which plays a vital role in the survival, development, and functioning of neurons (Bathina and Das, 2015), gene expression and the activation of downstream apoptotic signaling pathways upon exposure to 0.5 µm PS-MPs were noted. These results indicate the potential for neurotoxicity (Suman et al., 2023). The electroencephalographic signals indicated a significant increase in the number and total duration of seizure-like behaviors in the presence of 10 µm PS-MPs corroborated by the altered expression of genes associated with seizure-related pathways. Moreover, an increase in neurochemical levels, such as choline, betaine, dopamine, 3-methoxytyramine, and gamma-aminobutyric acid, indicated a potential association between the seizure-like behavior and the dysregulation of the cholinergic, dopaminergic, and GABAergic systems (Jeong et al., 2022).

The exposure to 100 nm PS-NP resulted in the shortening of Cap motoneuron axons and reduced expression of isl2a (ISL LIM homeobox 2a) gene, thereby impairing locomotor behavior (Zhang et al., 2022). A notable reduction in larval locomotor activity was observed in other studies as well (Chen et al., 2017; Luan et al., 2023; Pitt et al., 2018a). The presence of microplastics was found to have a significant negative impact on the swimming competence of zebrafish larvae, as evidenced by both the free-swimming test and light-to-dark alternation stimulation (Qiang and Cheng, 2019). In addition, 120 hpf zebrafish larvae demonstrated elevated thigmotaxis behavior, a tendency to remain at the edge of the arena, under both light and dark conditions, which may be an indicative of underlying anxiety-like responses (Luan et al., 2023). In another study, the electroencephalography (EEG) data demonstrated that the delta waves of zebrafish larvae exposed to 50 nm PS-NPs exhibited a reduction, while the beta waves exhibited an increase, indicating the presence of anxious behaviors (Hwang et al., 2022).

MPs and NPs have been demonstrated to induce the generation of ROS at varying concentrations, thereby inducing oxidative stress and affecting the immune system (Martin-Folgar et al., 2023; Umamaheswari et al., 2021; Zhang et al., 2022). In a study, the activities of catalase (CAT) and superoxide dismutase (SOD) exhibited a decline, and there was an induction of lipid peroxidation (LPO) levels in the 0.10–0.12 µm PS-MPs group. The elevated transcription of inflammatory markers indicates an enhanced immune response resulting in the production of inflammatory mediators in the zebrafish. The induction of ROS has been shown to stimulate the activation of the p53 apoptotic cascade (Umamaheswari et al., 2021). PSMP exposure also has been demonstrated to cause extreme production of ROS with the inhibition of nuclear factor erythroid 2–related factor 2 (nrf2) signaling (Zhang et al., 2022). Nrf2 has recently emerged as a regulator of cellular resistance to oxidants (Ma, 2013).

Both 5 µm and 70 nm PS-MPs and PS-NPs have been shown to alter lipid metabolism and induce hepatic lipid accumulation in zebrafish, which could indicate that MPs and NPs have the potential to induce metabolic disorders (Lu et al., 2016). The presence of 5-µm and 50-µm (PS-MPs) has been demonstrated to result in gut microbiota dysbiosis in larval zebrafish, accompanied by an alteration in gene transcription associated with glucose and lipid metabolism (Wan et al., 2019). In a study, the brain–gut axis was observed for a period of 30 days. The findings revealed that intestinal permeability was adversely affected by 44 nm PS-NPs exposure. Alteration of microbial populations decreased serotonin and L-glutamine (Gln) levels were noted. Gln plays a role in maintaining gut health by strengthening the cell barrier and preventing damage (Teng et al., 2022).

## Conclusion

MNPs have been demonstrated to accumulate in zebrafish in a concentration-dependent manner. It has been reported that strong fluorescence is visible around the embryonic chorion prior to hatching, which suggests that the yolk sac, pericardium, and gastrointestinal tract are the main sites of MNP accumulation in zebrafish larvae. Subsequently, the particles can even reach other organs, including the brain. In adults, the liver and muscle were identified as the primary organs for accumulation. A number of adverse effects, including oxidative stress, neurotoxicity, apoptosis, and genotoxicity, have been documented in zebrafish. The evidence of MNPs accumulating in early life stages in zebrafish, along with the toxic effects observed, highlights the urgent need for a more comprehensive understanding of the impact of MNPs on organisms.

MNPs have been detected across a range of ecosystems. These include the atmosphere, soil, and marine and aquatic environments. Due to their high resilience to degradation, MNPs remain in the environment for extended periods. Thus far, no solution has been identified for the removal of MNPs from the environment. Furthermore, the global overconsumption of plastics is exacerbating the accumulation of these particles in natural environments.

It is inevitable that both human and aquatic animals will come into contact with MNPs. The extant evidence indicates that MNPs can be ingested by aquatic organisms and mammals. Once ingested, MNPs have the potential to reach other organs. It has been demonstrated that MNP particles are absorbed from the gastrointestinal system into the lymph and circulatory system. The toxicity of MNPs is closely correlated with particle size and exposure duration. Nevertheless, the number of particles that reach the organs, as well as the potential toxicity of these tiny plastic particles, is a field of study that has yet to reach a sufficient level of understanding.

## REFERENCES

Adyel, T.M. (2020). Accumulation of plastic waste during COVID-19. *Science, 369*(6509), 1314–1315.

Ali, N., Katsouli, J., Marczylo, E. L., Gant, T. W., Wright, S., & Bernardino de la Serna, J. (2024). The potential impacts of micro-and-nano plastics on various organ systems in humans. *EBioMedicine, 99*, Article 104901. https://doi.org/10.1016/j.ebiom.2023.104901

Andrady, A. L., & Neal, M. A. (2009). Applications and societal benefits of plastics. Philosophical transactions of the Royal Society of London. *Series B, Biological Sciences, 364*(1526), 1977–1984. https://doi.org/10.1098/rstb.2008.0304

Bathina, S., & Das, U. N. (2015). Brain-derived neurotrophic factor and its clinical implications. *Archives of Medical Science: AMS*, *11*(6), 1164–1178. https://doi.org/10.5114/aoms.2015.56342

Bhagat, J., Zang, L., Nishimura, N., & Shimada, Y. (2020). Zebrafish: An emerging model to study microplastic and nanoplastic toxicity. *The Science of the Total Environment*, *728*, 138707. https://doi.org/10.1016/j.scitotenv.2020.138707

Cattaneo, N., Zarantoniello, M., Conti, F., Frontini, A., Chemello, G., Dimichino, B., Marongiu, F., Cardinaletti, G., Gioacchini, G., & Olivotto, I. (2023). Dietary Microplastic Administration during Zebrafish (*Danio rerio*) Development: A Comprehensive and Comparative Study between Larval and Juvenile Stages. *Animals: An Open Access Journal from MDPI*, *13*(14), 2256. https://doi.org/10.3390/ani13142256

Chen, Q., Gundlach, M., Yang, S., Jiang, J., Velki, M., Yin, D., & Hollert, H. (2017). Quantitative investigation of the mechanisms of microplastics and nanoplastics toward zebrafish larvae locomotor activity. *The Science of the total environment*, *584–585*, 1022–1031. https://doi.org/10.1016/j.scitotenv.2017.01.156

De-la-Torre G. E. (2020). Microplastics: Ān emerging threat to food security and human health. *Journal of food science and technology, 57*(5), 1601–1608. https://doi.org/10.1007/s13197-019-04138-1

Emenike, E. C., Okorie, C. J., Ojeyemi, T., Egbemhenghe, A., Iwuozor, K. O., Saliu, O. D., Okoro, H. K., & Adeniyi, A. G. (2023). From oceans to dinner plates: The impact of microplastics on human health. *Heliyon*, *9*(10), e20440. https://doi.org/10.1016/j.heliyon.2023.e20440

Gopinath, P. M., Saranya, V., Vijayakumar, S., Mythili Meera, M., Ruprekha, S., Kunal, R., Pranay, A., Thomas, J., Mukherjee, A., & Chandrasekaran, N. (2019). Assessment on interactive prospectives of nanoplastics with plasma proteins and the toxicological impacts of virgin, coronated and environmentally released-nanoplastics. *Scientific Reports*, *9*(1), 8860. https://doi.org/10.1038/s41598-019-45139-6

Hwang, K. S., Son, Y., Kim, S. S., Shin, D. S., Lim, S. H., Yang, J. Y., Jeong, H. N., Lee, B. H., & Bae, M. A. (2022). Size-dependent effects of Polystyrene Nanoparticles (PS-NPs) on behaviors and endogenous neurochemicals in zebrafish larvae. *International Journal of Molecular Sciences*, *23*(18), 10682. https://doi.org/10.3390/ijms231810682

Jeong, S., Jang, S., Kim, S. S., Bae, M. A., Shin, J., Lee, K. B., & Kim, K. T. (2022). Size-dependent seizurogenic effect of polystyrene microplastics in zebrafish embryos. *Journal of Hazardous Materials*, *439*, 129616. https://doi.org/10.1016/j.jhazmat.2022.129616

Lai, Y., Dong, L., Li, Q., Li, P., & Liu, J. (2021). Sampling of micro-and nano-plastics in environmental matrixes. *TrAC Trends in Analytical Chemistry*, *145*, 116461.

Lehner, R., Weder, C., Petri-Fink, A., & Rothen-Rutishauser, B. (2019). Emergence of nanoplastic in the environment and possible impact on human health. *Environmental Science & Technology*, *53*(4), 1748–1765. https://doi.org/10.1021/acs.est.8b05512

LeMoine, C. M. R., Kelleher, B. M., Lagarde, R., Northam, C., Elebute, O. O., & Cassone, B. J. (2018). Transcriptional effects of polyethylene microplastics ingestion in developing zebrafish (Danio rerio). *Environmental pollution*, *243*(Pt A), 591–600. https://doi.org/10.1016/j.envpol.2018.08.084

Lu, Y., Zhang, Y., Deng, Y., Jiang, W., Zhao, Y., Geng, J., Ding, L., & Ren, H. (2016). Uptake and accumulation of polystyrene microplastics in zebrafish (danio rerio) and toxic effects in liver. *Environmental Science & Technology*, *50*(7), 4054–4060. https://doi.org/10.1021/acs.est.6b00183

Luan, J., Zhang, S., Xu, Y., Wen, L., & Feng, X. (2023). Effects of microplastic exposure on the early developmental period and circadian rhythm of zebrafish (Danio rerio): A comparative study of polylactic acid and polyglycolic acid. *Ecotoxicology and Environmental Safety*, *258*, 114994. https://doi.org/10.1016/j.ecoenv.2023.114994

Ma, C., Chen, Q., Li, J., Li, B., Liang, W., Su, L., & Shi, H. (2021). Distribution and translocation of micro- and nanoplastics in fish. *Critical Reviews in Toxicology*, *51*(9), 740–753. https://doi.org/10.1080/10408444.2021.2024495

Ma, Q. (2013). Role of nrf2 in oxidative stress and toxicity. *Annual Review of Pharmacology and Toxicology*, *53*, 401–426. https://doi.org/10.1146/annurev-pharmtox-011112-140320

Marcellus, K. A., Bugiel, S., Nunnikhoven, A., Curran, I., & Gill, S. S. (2024). Polystyrene nano- and microplastic particles induce an inflammatory gene expression profile in rat neural stem cell-derived astrocytes in vitro. *Nanomaterials*, *14*(5), 429. https://doi.org/10.3390/nano14050429

Martin-Folgar, R., Torres-Ruiz, M., de Alba, M., Cañas-Portilla, A. I., González, M. C., & Morales, M. (2023). Molecular effects of polystyrene nanoplastics toxicity in zebrafish embryos (Daniorerio). *Chemosphere*, *312*(Pt 1), 137077. https://doi.org/10.1016/j.chemosphere.2022.137077

Osman, A. I., Hosny, M., Eltaweil, A. S., Omar, S., Elgarahy, A. M., Farghali, M., Yap, P. S., Wu, Y. S., Nagandran, S., Batumalaie, K., Gopinath, S. C. B., John, O. D., Sekar, M., Saikia, T., Karunanithi, P., Hatta, M. H. M., & Akinyede, K. A. (2023). Microplastic sources, formation, toxicity and remediation: A review. *Environmental Chemistry Letters*, 1–41. Advance online publication. https://doi.org/10.1007/s10311-023-01593-3

Parenti, C. C., Ghilardi, A., Della Torre, C., Magni, S., Del Giacco, L., & Binelli, A. (2019). Evaluation of the infiltration of polystyrene nanobeads in zebrafish embryo tissues after short-term exposure and the related biochemical and behavioural effects. *Environmental Pollution*, *254*(Pt A), 112947. https://doi.org/10.1016/j.envpol.2019.07.115

Pedersen, A. F., Meyer, D. N., Petriv, A. V., Soto, A. L., Shields, J. N., Akemann, C., Baker, B. B., Tsou, W. L., Zhang, Y., & Baker, T. R. (2020). Nanoplastics impact the zebrafish (Danio rerio) transcriptome: Associated developmental and neurobehavioral consequences. *Environmental Pollution*, *266*(Pt 2), 115090. https://doi.org/10.1016/j.envpol.2020.115090

Pitt, J. A., Kozal, J. S., Jayasundara, N., Massarsky, A., Trevisan, R., Geitner, N., Wiesner, M., Levin, E. D., & Di Giulio, R. T. (2018a). Uptake, tissue distribution, and toxicity of polystyrene nanoparticles in developing zebrafish (Danio rerio). *Aquatic Toxicology*, *194*, 185–194. https://doi.org/10.1016/j.aquatox.2017.11.017

Pitt, J. A., Trevisan, R., Massarsky, A., Kozal, J. S., Levin, E. D., & Di Giulio, R. T. (2018b). Maternal transfer of nanoplastics to offspring in zebrafish (Danio rerio): A case study with nanopolystyrene. *The Science of the total environment*, *643*, 324–334. https://doi.org/10.1016/j.scitotenv.2018.06.186

Qiang, L., & Cheng, J. (2019). Exposure to microplastics decreases swimming competence in larval zebrafish (Danio rerio). *Ecotoxicology and Environmental Safety*, *176*, 226–233. https://doi.org/10.1016/j.ecoenv.2019.03.088

PlasticsEurope. (2024). *The circular economy for plastics – a European analysis.* https://plasticseurope.org/wp-content/uploads/2024/05/Circular_Economy_report_Digital_light_FINAL.pdf

Rezania, S., Park, J., Md Din, M. F., Mat Taib, S., Talaiekhozani, A., Kumar Yadav, K., & Kamyab, H. (2018). Microplastics pollution in different aquatic environments and biota: A review of recent studies. *Marine Pollution Bulletin, 133*, 191–208. https://doi.org/10.1016/j.marpolbul.2018.05.022

Santos, D., Félix, L., Luzio, A., Parra, S., Cabecinha, E., Bellas, J., & Monteiro, S. M. (2020). Toxicological effects induced on early life stages of zebrafish (Danio rerio) after an acute exposure to microplastics alone or co-exposed with copper. *Chemosphere*, *261*, 127748. https://doi.org/10.1016/j.chemosphere.2020.127748

Sarasamma, S., Audira, G., Siregar, P., Malhotra, N., Lai, Y. H., Liang, S. T., Chen, J. R., Chen, K. H., & Hsiao, C. D. (2020). Nanoplastics cause neurobehavioral impairments, reproductive and oxidative damages, and biomarker responses in zebrafish: Throwing up alarms of widespread health risk of exposure. *International Journal of Molecular Sciences*, *21*(4), 1410. https://doi.org/10.3390/ijms21041410

Segner H. (2009). Zebrafish (Danio rerio) as a model organism for investigating endocrine disruption. Comparative biochemistry and physiology. *Toxicology & Pharmacology: CBP, 149*(2), 187–195. https://doi.org/10.1016/j.cbpc.2008.10.099

Sökmen, T. Ö., Sulukan, E., Tu̇rkoğlu, M., Baran, A., Özkaraca, M., & Ceyhun, S. B. (2020). Polystyrene nanoplastics (20 nm) are able to bioaccumulate and cause oxidative DNA damage in the brain tissue of zebrafish embryo (Danio rerio). *Neurotoxicology, 77*, 51–59. https://doi.org/10.1016/j.neuro.2019.12.010

Suman, A., Mahapatra, A., Gupta, P., Ray, S. S., & Singh, R. K. (2023). Polystyrene microplastics modulated bdnf expression triggering neurotoxicity via apoptotic pathway in zebrafish embryos. *Comparative Biochemistry and Physiology. Toxicology & Pharmacology: CBP, 271*, 109699. https://doi.org/10.1016/j.cbpc.2023.109699

Teng, M., Zhao, X., Wang, C., Wang, C., White, J. C., Zhao, W., Zhou, L., Duan, M., & Wu, F. (2022). Polystyrene nanoplastics toxicity to zebrafish: Dysregulation of the brain-intestine-microbiota axis. *ACS Nano*, *16*(5), 8190–8204. https://doi.org/10.1021/acsnano.2c01872

Thompson, R. C., Olsen, Y., Mitchell, R. P., Davis, A., Rowland, S. J., John, A. W., McGonigle, D., & Russell, A. E. (2004). Lost at sea: where is all the plastic? *Science, 304*(5672), 838. https://doi.org/10.1126/science.1094559

Torres-Ruiz, M., de Alba González, M., Morales, M., Martin-Folgar, R., González, M. C., Cañas-Portilla, A. I., & De la Vieja, A. (2023). Neurotoxicity and endocrine disruption caused by polystyrene nanoparticles in zebrafish embryo. *The Science of the Total Environment*, *874*, 162406. https://doi.org/10.1016/j.scitotenv.2023.162406

Trang, A., Khandhar, P.B. (2023). Physiology, Acetylcholinesterase. In: StatPearls [Internet]. Treasure Island (FL): StatPearls Publishing; PMID: 30969557.

Ugwu, K., Herrera, A., & Gómez, M. (2021). Microplastics in marine biota: A review. *Marine Pollution Bulletin, 169*, 112540. https://doi:10.1016/j.marpolbul.2021.112540

Umamaheswari, S., Priyadarshinee, S., Kadirvelu, K., & Ramesh, M. (2021). Polystyrene microplastics induce apoptosis via ROS-mediated p53 signaling pathway in zebrafish. *Chemico-Biological Interactions*, *345*, 109550. https://doi.org/10.1016/j.cbi.2021.109550

Van Cauwenberghe, L., Vanreusel, A., Mees, J., & Janssen, C. R. (2013). Microplastic pollution in deep-sea sediments. *Environmental Pollution, 182*, 495–499. https://doi.org/10.1016/j.envpol.2013.08.013

Yee, M. S., Hii, L. W., Looi, C. K., Lim, W. M., Wong, S. F., Kok, Y. Y., Tan, B. K., Wong, C. Y., & Leong, C. O. (2021). Impact of microplastics and nanoplastics on human health. *Nanomaterials*, *11*(2), 496. https://doi.org/10.3390/nano11020496

Wan, Z., Wang, C., Zhou, J., Shen, M., Wang, X., Fu, Z., & Jin, Y. (2019). Effects of polystyrene microplastics on the composition of the microbiome and metabolism in larval zebrafish. *Chemosphere*, *217*, 646–658. https://doi.org/10.1016/j.chemosphere.2018.11.070

Wang, J., Zhao, X., Wu, F.,Niu, L., Tang, Z., Liang,W., Zhao, T., Fang, M.,Wang, H., & Wang, X. (2021). Characterization, occurrence, environmental behaviors, and risks of nanoplastics in the aquatic environment: Current status and future perspectives. *Fundamental Research, 1*(3), 317–328. https://doi.org/10.1016/j.fmre.2021.05.00

Zettler, E. R., Mincer, T. J., & Amaral-Zettler, L. A. (2013). Life in the "plastisphere": Microbial communities on plastic marine debris. *Environmental Science & Technology*, *47*(13), 7137–7146. https://doi.org/10.1021/es401288x

Zhang, R., Silic, M. R., Schaber, A., Wasel, O., Freeman, J. L., & Sepúlveda, M. S. (2020). Exposure route affects the distribution and toxicity of polystyrene nanoplastics in zebrafish. *The Science of the Total Environment*, *724*, 138065. https://doi.org/10.1016/j.scitotenv.2020.138065

Zhang, Y., Zhang, X., Yan, Q., Xu, C., Liu, Q., Shen, Y., Xu, J., Wang, G., & Zhao, P. (2022). Melatonin attenuates polystyrene microplastics induced motor neurodevelopmental defect in zebrafish (Danio rerio) by activating nrf2- isl2a Axis. *Ecotoxicology and Environmental Safety, 241*, 113754. https://doi.org/10.1016/j.ecoenv.2022.113754

# 20

# *Investigation of the Effects of Vitamins on Neurodegenerative Diseases Using Zebrafish as a Model Organism*

**Semanur Işıkoğlu, İsmail Ünal and Ebru Emekli-Alturfan**

## Abbreviations

| | |
|---|---|
| **5-HT** | 5-Hydroxytryptamine (Serotonin) Receptor |
| **5-HTP** | 5-Hydroxytryptophan |
| **AD** | Alzheimer's Disease |
| **ALS** | Amyotrophic Lateral Sclerosis |
| **ATP** | Adenosine Triphosphate |
| **CNS** | Central Nervous System |
| **Coa** | Coenzyme A |
| **FAD** | Flavin Adenine Dinucleotide |
| **FMN** | Flavin Mononucleotide |
| **GABA** | Gamma-Aminobutyric Acid |
| **Hcy** | Homocysteine |
| **MCI** | Mild Cognitive Impairment |
| **MCM** | Methylmalonyl Coenzyme A Mutase |
| **Mecbl** | Methylcobalamin |
| **MS** | Multiple Sclerosis |
| **NAD** | Nicotinamide Adenine Dinucleotide |
| **NADP** | Nicotinamide Adenine Dinucleotide Phosphate |
| **PD** | Parkinson's Disease |
| **PL** | Pyridoxal |
| **PLP** | Pyridoxal 5-Phosphate |
| **PM** | Pyridoxamine |
| **PN** | Pyridoxine |
| **PNS** | Peripheral Nervous System |
| **SAM** | S-Adenosylmethionine |
| **SNc** | Substantia Nigra Pars Compacta |
| **α-Syn** | α-Synuclein |

## Introduction

Neurodegenerative diseases are critical conditions affecting millions of people worldwide, characterized by the progressive degeneration of brain cells. Neurodegenerative disorders such as Huntington's disease, Alzheimer's disease, Parkinson's disease, and multiple sclerosis manifest with severe symptoms, including cognitive decline, memory loss, and motor control dysfunction (Gruber-Bzura et al., 2017; Dayalu and Albin, 2015; Masters et al., 2015). To date, there is no definitive treatment for these diseases, so research into their pathogenesis and the development of new therapeutic strategies are highly significant (Poewe et al., 2017).

The zebrafish (*Danio rerio*) has emerged as an important model organism for genetic and neurological research. Due to its transparent embryos, short reproductive cycle, and ease of genetic manipulation, this small fish is preferred in biomedical research (Kalueff et al., 2014). Additionally, the zebrafish central nervous system is organized similarly to the human brain, allowing it to serve as an effective model for understanding the etiology of neurodegenerative diseases (Guo, 2004).

Vitamins are essential micronutrients for the healthy functioning of the organism, and deficiencies can lead to serious neurological disorders. For instance, vitamin $B_{12}$ deficiency is associated with neurological dysfunctions and demyelination (Hogenkamp et al., 2017), while vitamin D deficiency can increase the risk of neuroinflammation and cognitive decline (Mao et al., 2018). Similarly, vitamin E, as a potent antioxidant, can play a protective role against neurodegenerative processes (Kang et al., 2021), and the effects of vitamin K on brain health have recently garnered more attention (Hsu et al., 2019).

Studies examining the effects of vitamin deficiencies on neurodegenerative disease models in zebrafish provide valuable insights into the pathophysiology of these diseases and suggest potential therapeutic pathways (Kalueff et al., 2014). Understanding the roles of vitamins such as B, D, E, and K, which are directly related to nervous system health, in neurodegenerative processes is crucial for developing prevention and treatment strategies (Hogenkamp et al., 2017; Mao et al., 2018; Kang et al., 2021; Hsu et al., 2019; Nair and Prakash, 2017).

This chapter will comprehensively address the effects of various vitamin deficiencies on neurodegenerative diseases, based on research utilizing the zebrafish model. In this context, the protective effects of vitamins on nervous system health and the consequences of their deficiencies will be explored through zebrafish models.

DOI: 10.1201/9781032676555-22

## Water-Soluble Vitamins and Their Functions

### Vitamin $B_1$ (Thiamine)

Vitamin $B_1$ is named as such because it was the first B vitamin to be identified (Strandler and Strand, 2023). In the human body, thiamine exists in various phosphorylated forms, including free thiamine, thiamine monophosphate, thiamine triphosphate, and thiamine diphosphate (thiamine pyrophosphate). The active form of thiamine is thiamine pyrophosphate, which serves as a cofactor for certain enzymes involved in energy metabolism. These enzymes are typically involved in carbohydrate catabolism, and a deficiency in thiamine results in reduced activity of these enzymes. This leads to impaired ATP synthesis, which causes ATP deficiency in the brain and consequently cell death due to the lack of ATP (Bunik and Aleshin, 2017).

Vitamin $B_1$ is essential for carbohydrate metabolism and nerve function and is naturally found in grains and whole grain foods, milk, legumes, meat, vegetables, fruits, rice bran, and yeast. Besides being used as a dietary supplement in many foods and enriched flours, it is also used as a flavoring agent in some foods (Manzetti et al., 2014 ).

### Vitamin $B_2$ (Riboflavin)

Vitamin $B_2$ (riboflavin) was named lactoflavin because it was first obtained from milk (Yarat A.,2023). Its chemical structure consists of a flavin isoalloxazine ring attached to a sugar side chain called ribitol (Pinto and Zempleni, 2016). Riboflavin is a component of the coenzymes flavin adenine dinucleotide (FAD) and flavin mononucleotide (FMN) and therefore plays a role in energy metabolism. It is sensitive to light. This vitamin can be synthesized by plants and microorganisms but not by mammals (Yarat, 2023)

Riboflavin is found naturally in a wide variety of foods. In addition to being present in significant amounts in dairy products, it is also present in eggs, liver, yeast, vegetables, and fruits (Suwannasom et al., 2020).

### Vitamin $B_3$ (Niacin)

Vitamin $B_3$, also known as niacin, is a water-soluble vitamin and one of the B vitamins. It is found in various foods such as eggs, peanuts, bran, yeast, poultry, red meat, fish, whole grain cereals, seeds, and legumes. Nicotinamide adenine dinucleotide (NAD or coenzyme I) and nicotinamide adenine dinucleotide phosphate (NADP or coenzyme II) actively participate in fundamental oxidation-reduction reactions in the body, including glycolysis, protein and amino acid metabolism, pyruvate metabolism, glycerol metabolism, pentose biosynthesis, high-energy phosphate bond synthesis, and fatty acid metabolism. Vitamin $B_3$ is considered a vital vitamin because NAD and NADP are essential in these metabolic processes. Vitamin $B_3$ also plays a role in managing hyperlipidemia and is used in the treatment of pellagra, as an adjunct in treating discoid lupus erythematosus, and to alleviate neurological and dermatological symptoms associated with NAD and NADP deficiencies (Peechakara and Gupta, 2025).

### Vitamin $B_6$ (Pyridoxine)

Vitamin $B_6$ is a water-soluble vitamin that can be converted into coenzymes required for over 100 enzymes in the human body. The three natural forms of this vitamin – pyridoxine (PN), pyridoxal (PL), and pyridoxamine (PM) – are converted into pyridoxal 5-phosphate (PLP or P5P), which is the active form in the body. P5P functions as a coenzyme primarily in amino acids, protein, carbohydrate, and lipid metabolism and also plays a role in neurotransmitter synthesis. Additionally, Vitamin $B_6$ is involved in gluconeogenesis and glycogenolysis (Abosamak and Gupta, 2023).

### Vitamin $B_9$ (Folate)

Folate is a general term for a group of compounds important in the biosynthesis of deoxyribonucleic acid (DNA) and is also known as vitamin $B_9$. Folinic acid is a synthetic derivative of folate. Folate is converted into tetrahydrofolate (THF), which is crucial for the synthesis of nitrogenous bases in both DNA and ribonucleic acid (RNA) and for the maturation of red blood cells (RBCs).

Folic acid deficiency can result in malabsorption disorders, chronic alcoholism, hemolytic anemia, or macrocytic megaloblastic anemia due to increased demands during pregnancy. Folate is naturally present in some food sources and must be obtained regularly through diet since the body cannot synthesize it. Dietary sources of folate include green leafy vegetables such as broccoli, spinach, and lettuce, as well as meats (especially liver), eggs, and milk (Merrell and McMurry, 2023).

### Vitamin $B_{12}$ (Cobalamin)

Vitamin $B_{12}$, also known as cobalamin, is a water-soluble vitamin found in animal-derived foods such as eggs, red meat, and dairy products (Ankar and Kumar, 2022). Vitamin $B_{12}$ is essential for DNA stability (Halczuk et al., 2023). It acts as a cofactor for enzymes involved in DNA, fatty acid, and myelin synthesis. A deficiency in Vitamin $B_{12}$ can lead to hematological and neurological symptoms. Additionally, Vitamin $B_{12}$ has antioxidant properties, and excess amounts are stored in the liver, which helps reduce the likelihood of deficiency. However, deficiencies can arise from dietary insufficiency, malabsorption, or intrinsic factor deficiency, leading to depleted hepatic stores (Ankar and Kumar, 2022).

### Vitamin C (Ascorbic Acid)

Vitamin C is an antioxidant vitamin involved in the synthesis of carnitine, collagen, and norepinephrine (Karaağaç and Pınarlı, 2023). Discovered in 1920, Vitamin C refers to compounds demonstrating the biological activity of ascorbic acid. Although many animals and plants produce it, humans cannot synthesize vitamin C due to the absence of the enzyme L-gulonolactone oxidase and must obtain it through their diet. It is biologically important due to its low redox potential and functions as an antioxidant in many enzymatic reactions.

The chemical structure of vitamin C includes two enantiomeric pairs: L and D ascorbic acid, and L and D isoascorbic acid. L-ascorbic acid is the natural and essential form for vitamin C activity. Vitamin C also plays a significant role in the production of various neurotransmitters. It is involved in the synthesis of collagen, an important connective tissue component for wound healing, and helps regenerate other antioxidants in the body, such as α-tocopherol. Vitamin C may also help protect against certain cancers and cardiovascular diseases due to its antioxidant effects. Epidemiological studies have shown that diets rich in vitamin C from fruits and vegetables are associated with a lower risk of cardiovascular diseases and cancer. Additionally, vitamin C plays a role in preventing and treating visual disorders such as oxidative stress, macular degeneration, and cataracts (Emekli-Alturfan, 2023).

## Effects of Water-Soluble Vitamins on the Nervous System

### *Vitamin $B_1$*

The effects of vitamin $B_1$ on the nervous system have been well-known for a long time. Deficiency of this vitamin can lead to conditions such as polyneuropathy and/or cardiovascular symptoms and can result in a syndrome known as beriberi that endangers the peripheral nervous system (PNS). It can also cause neuropsychiatric disorders such as Wernicke-Korsakoff syndrome, characterized by encephalopathy and psychosis. Vitamin $B_1$ plays numerous physiological roles in the body. Among these, it is especially crucial for glucose metabolism, maintaining nerve membrane function, and synthesizing myelin and various neurotransmitters (such as acetylcholine, serotonin, and amino acids). Vitamin $B_1$'s most significant role in the nervous system is contributing to cellular energy metabolism and acting as a vital cofactor in the conversion of carbohydrates, thus helping to supply energy to nerve cells. Nerve cells, especially in the brain, require large amounts of energy to function and cannot store high-energy compounds themselves. Therefore, supplying energy is critical. Vitamin $B_1$ is essential in this regard, as one of its key functions is to activate biochemical steps in energy production processes such as the pentose phosphate pathway, glycolysis, and the Krebs cycle (citric acid cycle) (Calderón-Ospina et al., 2023).

In these processes, energy is supplied to nerves in the form of adenosine triphosphate (ATP) or NADPH. ATP and NADPH are also necessary for other cellular activities in nerves (Isenberg-Grzeda et al., 2012). Therefore, thiamine is also required for the energy-dependent synthesis of nucleic acids, neurotransmitters, and myelin. In other words, thiamine helps protect myelin sheaths and, consequently, affects nerve conduction speed. Additionally, thiamine has an antioxidative effect on nerve cells. Due to its antioxidative properties, adequate thiamine can even prevent cell damage caused by hyperglycemia (Page et al., 2011).

Since vitamin $B_1$ is necessary for energy production and the proper functioning of nerve cells, a deficiency can lead to damage or even death of neurons. Consequently, it affects both the central nervous system (CNS) and the PNS and can present with various clinical symptoms. General neurological symptoms of thiamine deficiency include confusion, impaired memory retention and cognitive function, visual deficiencies, psychomotor retardation, confabulation, tremors, ataxia, and loss of positional sense. If the body does not have adequate levels of thiamine for the CNS, sensitive areas of the brain, such as the mammillary bodies and the thalamus, can be damaged. Wernicke's encephalopathy and Wernicke-Korsakoff syndrome are considered the most severe CNS manifestations of thiamine deficiency. In Wernicke's encephalopathy, it is believed that thiamine deficiency triggers apoptotic cell death due to N-methyl-D-aspartate toxicity, leading to neurological symptoms. Typical signs of thiamine deficiency in the PNS include polyneuropathy and paralysis (e.g., beriberi disease). Thiamine deficiency in the sensory system can cause pain, affect the sense of touch, alter temperature sensitivity, and result in the loss of vibration sense. Paralysis, affecting the motor system, typically begins in the distal regions of the lower extremities and spreads. In this condition, tendon reflexes are impaired, muscle weakness increases, and atrophy develops in the leg muscles. Overall, these symptoms clearly demonstrate the critical importance of thiamine for the nervous system, given its antioxidative effects and its activating role in neuronal excitability and metabolism (Calderón-Ospina et al., 2023).

### *Vitamin $B_2$*

Vitamin $B_2$, also known as riboflavin, is associated with antioxidant properties and includes two flavoprotein coenzymes, FMN (flavin mononucleotide) and FAD (flavin adenine dinucleotide). These coenzymes are crucial limiting factors in many cellular enzymatic processes. FMN and FAD are important for the synthesis, conversion, and recycling of folate, niacin, and vitamin B6. They are also essential for the production of hemoproteins, including nitric oxide synthases, hemoglobin, P450 enzymes, and proteins involved in oxygen transport and storage. Additionally, flavoproteins play a role in iron absorption and utilization, metabolism of essential fatty acids in brain lipids, and regulation of thyroid hormones. Deficiency of riboflavin can lead to negative effects on brain function due to the disruption of these processes (Kennedy, 2016).

### *Vitamin $B_3$*

Niacin derivatives, such as NAD and NADP, are responsible for most processes and enzymes involved in peripheral and brain cell functions. Examples of these processes include energy production, DNA metabolism and repair, oxidative reactions, cellular signaling events (through intracellular calcium), antioxidant protection, and the conversion of folate to tetrahydrofolate (Wuerch et al., 2023). Furthermore, niacin receptors are distributed both peripherally in adipose tissue and immune cells, and throughout the brain (Kennedy, 2016).

### *Vitamin $B_5$*

Vitamin $B_5$, also known as pantothenic acid, is used as a substrate in the synthesis of coenzyme A (CoA), which is involved in many processes in the body. CoA plays a crucial role not only in oxidative metabolism but also in the synthesis of amino acids, cholesterol, phospholipids, and fatty acids, thus affecting the structure and functionality of brain cells. Pantothenic acid, through CoA, also contributes to the formation of steroid hormones and various neurotransmitters (Bauerly and Rucker, 2007).

### Vitamin $B_6$

Vitamin $B_6$ is associated with over 140 enzymatic functions (Parra et al., 2018). Besides its various bodily functions, it is particularly known for its role in synthesizing important neurotransmitters such as dopamine from L-DOPA, gamma-aminobutyric acid (GABA) from glutamate, and 5- HTfrom 5-HTP (Sigvard et al., 2022 ). It affects the adrenergic, glutamatergic, and serotonergic systems. Vitamin $B_6$ also has a neuroprotective role, related to its ability to regulate GABA and glutamate levels. Since GABA serves as a fundamental inhibitory neurotransmitter, its deficiency can lead to severe consequences such as seizures. In such cases, GABA or pyridoxine administration can terminate seizures. Moreover, pyridoxine administration can even alleviate the excitatory effects of the neurotoxin domoic acid. Vitamin $B_6$ is also essential for brain development during pregnancy and after birth, as its deficiency in children during these periods can lead to reduced GABA levels and permanent brain damage.

Only non-phosphorylated forms of vitamin $B_6$ can cross cell membranes, but they can then be phosphorylated inside the cell to active forms (5′-phosphate esters pyridoxine 5′-phosphate (PNP), pyridoxal 5′-phosphate (PLP; the most important coenzyme variant), and pyridoxamine 5′-phosphate (PMP)). PLP is primarily involved in neurotransmitter production but also acts as a coenzyme in homocysteine metabolism, supports carbohydrate/fat synthesis and degradation, and contributes to the release of necessary energy in amino acid and protein metabolism. Additionally, it plays a role as a cofactor in sphingolipid synthesis, making it necessary for myelin formation.

The role of pyridoxine in the nervous system is notably demonstrated by its use in treating pyridoxine-dependent seizures (a congenital abnormality) in infants with seizures that do not respond to common anticonvulsants. Vitamin $B_6$ deficiency can significantly impair the CNS and PNS due to its crucial roles in neurotransmitter and myelin synthesis. In cases of partial vitamin $B_6$ deficiency, some enzymes may be more affected than others biochemically, leading to greater depletion of certain neurotransmitters and imbalances in neurotransmitter levels. Neurological symptoms of pyridoxine deficiency generally include cognitive function impairment, convulsive seizures, depression, and even early neuron aging, burning sensations, paresthesia, painful dysesthesias, and thermal sensations, such as those seen in carpal tunnel syndrome. All these conditions can be treated.with pyridoxine, though it is important to note that long-term and high doses of pyridoxine can trigger sensory neuropathy. Even in such cases, symptoms resolve after discontinuing pyridoxine, and no permanent nerve damage has been observed. Additionally, high amounts of pyridoxine are needed to support fetal brain development *p*articularly during the preconception period, and pyridoxine supplementation can even reduce nausea in the early stages of pregnancy. Thus, pyridoxine significantly contributes to the proper functioning of the nervous system by facilitating neurotransmitter and myelin synthesis and controlling glutamate excitability and neuro-renal metabolism (Calderón-Ospina et al., 2023).

### Vitamin $B_9$

In genomic or non-genomic methylation reactions and purine/pyrimidine synthesis in brain tissue, a real or functional folate deficiency can occur due to decreased activity. This situation can lead to decreased DNA stability and repair and reduced gene expression/transcription, which may compromise membrane phospholipid integrity, promote hippocampal atrophy and demyelination, and hinder neuronal differentiation and repair (Moll and Graber, 2022). Effective folate cycling is also crucial for the synthesis and renewal of tetrahydrobiopterin, an essential cofactor for enzymes that convert amino acids into monoamine neurotransmitters (melatonin, serotonin, dopamine, adrenaline, noradrenaline) and nitric oxide (Merrell and McMurry, 2023).

### Vitamin $B_{12}$

Vitamin $B_{12}$ was initially recognized for its role in hematopoiesis but also functions as a coenzyme in many biochemical steps essential for maintaining nervous system health. Vitamin $B_{12}$ plays a significant role in the DNA synthesis of oligodendrocytes, which produce myelin, and thus in myelin synthesis (Calderón-Ospina et al., 2023).

Vitamin $B_{12}$ contributes to the synthesis of the myelin sheath, which surrounds nerve axons and facilitates nerve conduction speed by providing electrical insulation. It supports nerve regeneration in the event of injury (Geller M et al., 2017). In addition to this vital role in the nervous system, it is involved in energy production, fatty acid and nucleic acid synthesis, homocysteine metabolism, nerve metabolism (transmethylation processes), and cell maturation processes. It also plays a role in maintaining healthy gastrointestinal mucosa. Vitamin $B_{12}$ levels affect the antioxidant activity of reduced glutathione in blood cells and the liver. Consequently, a deficiency of cobalamin results in lower levels of reduced glutathione in cells, making them more susceptible to oxidative stress (Moshfegh and Goldman, 2019).

In the mitochondria, methylmalonyl CoA is converted to succinyl CoA, an important intermediate in the Krebs cycle, requiring MCM (methylmalonyl coenzyme A mutase) and AdoCbl (adenosylcobalamin). In the cytosol, methionine synthase (MS) requires MeCbl (methylcobalamin) as a cofactor to methylate homocysteine to methionine, which is necessary for the adequate synthesis of DNA, proteins, and neurotransmitters (Obeid et al., 2015).

In cells, if there is a deficiency of vitamin $B_{12}$, plasma concentrations of methylmalonic acid and homocysteine (Hcy), which are functional markers of cobalamin, will increase. Additionally, disturbances in myelin synthesis and the incorporation of abnormal fatty acids into neurons may be observed.

Neurological disorders associated with vitamin $B_{12}$ deficiency include myelopathy, neuropathy, subacute combined sclerosis of the spinal cord, optic nerve atrophy, polyneuropathy, and cognitive dysfunctions, though these are not exhaustive. Fundamentally, elevated levels of Hcy and methylmalonic acid are related to impaired neurotransmitter production and myelin lesions. The synthesis of S-adenosylmethionine (SAM) is critically dependent on vitamin $B_{12}$, and SAM is a universal methyl donor in neuronal demyelination. Demyelination generally affects both central and peripheral nerves, particularly impacting long white-matter tracts in the spinal cord's posterior and lateral columns, which contain sensory fibers related to vibration and positional sense. Motor fibers may also be demyelinated. Affected individuals may exhibit symptoms such as impaired positional sense, numbness in limbs, symmetric dysesthesia, paresthesias, spastic paraparesis or tetraparesis, and difficulties with daily activities such as writing or buttoning (Liao and Zhang, 2021).

Based on these points, it can be concluded that vitamin $B_{12}$ is essential for myelin synthesis, nerve metabolism, and neuronal regeneration, and is crucial for the nervous system (Calderón-Ospina et al., 2023).

## Fat-Soluble Vitamins and Their Functions

### Vitamin A (Retinoids)

Vitamin A is a general term used to describe several compounds with functions similar to those of retinol. These include retinal, retinyl esters, and all-trans-retinoic acid (ATRA). These molecules, along with synthetic analogs of vitamin A and other molecules with similar structures or functions, are collectively referred to as retinoids, a group of fat-soluble vitamins. The retinoid group plays roles in vision, tissue growth and cell differentiation, epithelial integrity, reproduction, immune response, central nervous system formation, and the development of the cornea, bones, and fetus. Additionally, it is important for fertility, iron utilization, glycoprotein synthesis, gene expression, and phagocytosis. With its antioxidant properties, vitamin A is essential throughout human life, from embryogenesis to adulthood (Menezes and Almeida, 2024).

### Vitamin D (Calciferol)

Vitamin D is a sterol derivative primarily obtained in humans through exposure to ultraviolet radiation (UVR) and certain dietary sources (such as fish or mushrooms) (Liu & Chen, 2022). Vitamin D has critical roles in protecting humans from invasive pathogens, reducing autoimmune risks, and maintaining health. By facilitating calcium and phosphorus absorption, vitamin D maintains calcium homeostasis in bone mineralization. It also plays roles in the proliferation and differentiation of neurons and microglia, and in dopamine signaling. Vitamin D regulates synaptic plasticity and molecular transport in cell organelles, and maintains the cytoskeleton, contributing significantly to brain development (Khan and Tiwari 2019). Its regulatory effects extend to controlling the expression of antioxidants. Besides its crucial role in the immune system, Vitamin D also controls the synthesis of essential neurotransmitters such as acetylcholine (ACh), serotonin (5-HT), dopamine (DA), and gamma-aminobutyric acid (GABA). As a fat-soluble vitamin essential for the human body, adequate intake and absorption of vitamin D are necessary for a healthy life (Wang et al., 2024; Kim et al., 2019).

### Vitamin E (Tocopherol)

Vitamin E is a type of fat-soluble vitamin that acts as an antioxidant in the body (Medina and Gupta, 2023). Due to its antioxidant effects, vitamin E protects the nervous system in the brain during aging and prevents cognitive decline (Kluge et al., 2016). Besides serving as a protective agent in lipid peroxidation, vitamin E also regulates humoral and cell-mediated immunity. Therefore, vitamin E is essential for the body, and its deficiency can lead to various health problems (Altıner et al., 2017). However, the human body cannot produce vitamin E and must obtain it from external sources (Medina and Gupta, 2023). Rich sources of vitamin E include vegetable oils, nuts and seeds, cheese, soybeans, wheat germ, egg yolk, avocado, olives, margarine, oatmeal, and green leafy vegetables (Colombo, 2010).

### Vitamin K

Vitamin K consists of a group of substances with chlorophyll quinone bioactivity and it is a fat-soluble vitamin. This vitamin exists in nature in the forms of K1 and K2, while K3 is its synthetic form. The first discovery related to vitamin K was its promotion of blood clotting, and hence it is also known as the clotting vitamin. In addition, vitamin K has roles in immunity, antioxidant capacity, regulation of gut microbiota, epithelial development, bone protection, and anti-inflammatory and immunomodulatory effects (Xie et al., 2024).

### Effects of Fat-Soluble Vitamins on the Nervous System

#### *Vitamin A*

Derivatives of vitamin A, known as retinoids, play a role in a complex signaling pathway that controls neuronal differentiation and neural tube patterning in the central nervous system while regulating gene expression. Retinoic signaling has been proposed to have significant functional effects on synaptic plasticity, sleep, schizophrenia, learning and memory, depression, Alzheimer's disease, and Parkinson's disease; however, the underlying mechanisms in the adult brain continue to be investigated.

Numerous experimental and clinical findings have demonstrated that vitamin A and its active derivatives, retinoids, play a crucial role in cell proliferation and differentiation, embryogenesis, and apoptosis. In addition to vitamin A deficiency, high doses of retinoids can also lead to developmental abnormalities (Tafti and Ghyselinck, 2007).

#### *Vitamin D*

Vitamin D and the enzymes that metabolize the vitamin D receptor are present in various regions of the brain, including the hippocampus, prefrontal cortex, hypothalamus, cingulate gyrus, and substantia nigra. The anti-inflammatory, neuroprotective, and antioxidant effects of vitamin D on neurons also play a role in maintaining brain health (Memiç, 2021; Stanger and Seidman, 2016).

Additionally, vitamin D is involved in regulating the gene expression of several neurotransmitters, such as dopamine, serotonin, acetylcholine, and gamma-aminobutyric acid (Schellenberg and Hulette, 2023; Sultan et al., 2020).

### *Vitamin E*

As the plasma level of α-tocopherol, the most abundant and bioavailable form of Vitamin E in human tissues, decreases, the transport of apolipoprotein E4 (APOE4) in the brain increases, leading to moderate cognitive impairment (Grimm et al., 2016). It has been shown that individuals with less than half of the recommended daily intake of vitamin E perform cognitively worse compared to those with higher intake levels (Memiç, 2021).

### *Vitamin K*

The forms of vitamin K, including phylloquinone (K1), menaquinone (K2), and synthetic menadione (K3), play roles in mechanisms that affect cognitive performance. Additionally, the menaquinone-4 (MK-4) form has been found in brain regions with high myelination, suggesting that this could contribute to the pathogenesis of cognitive performance impairment associated with vitamin deficiency. It is proposed that the potential effects of vitamin K on cognitive performance occur through protein S and Gas-6 molecules, which exert a protective effect by preventing oxidative damage to neurons (Ejder and Ekici, 2024).

## The Effects of Vitamins on the Development of Neurodegenerative Diseases

### Alzheimer's Disease

Alzheimer's disease (AD), an important type of neurodegenerative disease that ranks among the top in terms of prevalence worldwide, has a growing patient population. AD is the main cause of dementia. The initial symptoms of AD include short-term memory loss and cognitive impairment, and as the disease progresses, other symptoms such as disorientation, language problems, and mood changes also occur (Wang et al., 2024). AD neuropathology includes the formation of amyloid (Aβ) plaques, progressive degeneration of neurons, and development of neurofibrillary tangles of hyperphosphorylated tau protein. Thus, post-mortem analysis of AD patient brains reveals abnormal deposits within and around neurons and a reduction in the size of the cerebral cortex. As with other neurodegenerative diseases, one of the major causes of progressive neurodegeneration in AD is oxidative stress. Evidence from numerous studies suggests that vitamins may be effective in treating AD due to their antioxidant activity. Therefore, vitamins are used in AD treatment to help boost the immune response (Rai et al., 2021).

#### *Vitamins in Alzheimer's Disease*

When looking at clinical studies based on vitamins in AD, there is considerable evidence that levels of vitamins A, B, and C are reduced in the blood serum of AD patients compared to healthy individuals without neurocognitive impairment (De Wilde et al., 2017). Meta-analytic studies have also revealed decreases in serum levels of vitamin A, vitamin B, vitamin C, vitamin D, vitamin E, and vitamin K in AD patients. Cross-sectional and prospective studies have shown that supplementation with a combination of vitamins C and E reduces the incidence and prevalence of AD.

Studies have shown that some American women who received long-term treatment with retinoids exhibited significant improvements in cognitive function. The gut microbiota in AD can also influence retinoid signaling. The dynamic gut microbiota can directly establish a connection with retinoic acid to mediate therapeutic responses in AD (Endres, 2019). There is no correlation between retinol-binding protein 4 and AD in the early stages. Therefore, RBP4 is not used as a clinical biomarker in early AD. Disruptions in retinoic acid signaling and the resulting retinoic acid deficiency are associated with age-related cognitive decline in AD. Consequently, the therapeutic activity of retinoic acid is directly related to AD and AD-related symptoms.

Types of vitamin B, particularly vitamins $B_2$, $B_6$, $B_9$, and $B_{12}$, are involved in homocysteine metabolism. Elevated total plasma levels of homocysteine can lead to cognitive impairments, potentially resulting in dementia (Price et al., 2018; Tiffany and Norman, 2020). Numerous studies have shown that B-vitamin supplementation reduces homocysteine levels in the treatment of cognitive decline. However, a meta-analysis of randomized controlled trials indicated that although B-vitamin supplementation reduces total homocysteine, there was no improvement in cognitive impairment. Nevertheless, a randomized placebo-controlled trial demonstrated that B-vitamin supplementation prevented cognitive decline in elderly mild cognitive impairment (MCI) patients (Kwok et al., 2019). Another cross-sectional study on AD showed that higher levels of $B_{12}$ and folate exhibited strong therapeutic activity and improved cognitive performance. All these studies suggest that the role of B vitamins and folic acid in MCI and AD is speculative and requires comprehensive research.

Significant cognitive impairment seen in progressive dementia is also associated with thiamine deficiency. This result was observed based on the improvement of symptoms in dementia patients with thiamine supplementation. Furthermore, reversible dementia in elderly patients has been associated with cobalamin deficiency.

When looking at AD in terms of folate, some studies have shown that cognitive impairment improves for a short period with folate supplementation in patients with folate deficiency (Hama et al., 2020). Inflammation is one of the leading factors causing progressive neurodegeneration in AD. A clinical study suggested that folic acid exhibits anti-inflammatory activity and prevents neurodegeneration in AD.

Many studies related to AD indicate that vitamin D deficiency may also be a risk factor for AD. A meta-analysis conducted in 2019 showed a positive correlation between vitamin D deficiency (<10 ng/mL) and AD risk, suggesting that every 10 ng/mL increase in vitamin D supplementation could reduce AD risk by approximately 17%. An experiment on mice found that vitamin D deficiency in the early stages of AD could increase amyloid load in the hippocampus and cortex and strongly inhibit neurogenesis, cell proliferation, and neuron differentiation in AD-like mice in the later stages. All these studies demonstrate that AD development may increase in cases of vitamin D deficiency. Additionally, many studies and experiments on this topic suggest that vitamin D supplementation can prevent AD development.

Vitamin D also benefits AD by reducing oxidative stress and exerting anti-inflammatory effects. $1,25(OH)_2D_3$ acts as an anti-inflammatory agent in brain pericytes by affecting transcription events, suggesting that vitamin D may have a protective effect on the brain's capillaries during chronic inflammation associated with AD (Wang et al., 2024). Additionally, vitamin D exhibits anti-aging effects by increasing the expression of Klotho, an anti-aging gene. It also affects AD by reducing the expression and inhibiting the activity of mTOR, a protein associated with aging (Forster et al., 2011). Furthermore, calcitriol can influence memory processes and learning through the septohippocampal pathway (Wang et al., 2024).

Vitamin E plays an effective role in both memory deficits and the improvement of cognitive function. To significantly enhance these effects, vitamin E can be combined with other antioxidants (Browne et al., 2019). Clinical studies on the activity of vitamin E in AD have yielded conflicting findings (Gugliandolo et al., 2017). One study on patients with mild cognitive impairment (MCI), an early stage of AD, found no additional benefit from vitamin E supplementation. However, another study where MCI patients were treated with 2000 IU of vitamin E daily and 10 mg of donepezil or a placebo for three years found that while vitamin E had no effect on MCI, donepezil reduced the rate of progression from MCI to AD, particularly in apolipoprotein E4 carriers, during the first 12 months and for a longer period (Baroni et al., 2019).

In conclusion, the long-term and synergistic benefits of vitamin E in the management of AD are uncertain, indicating a need for larger population studies to better understand the impact of Vitamin E on AD (Browne et al., 2019).

## Parkinson's Disease

Parkinson's disease (PD) is a common neurodegenerative disorder characterized by motor system problems, such as tremor, rigidity, bradykinesia, and postural instability, as well as non-motor symptoms, including anosmia, sleep disturbances, depression, and constipation. In PD, there is a loss of dopaminergic neurons in the substantia nigra pars compacta (SNc), an accumulation of insoluble cytoplasmic protein inclusions known as Lewy bodies, and the presence of Lewy neurites. The specific pathogenesis of PD is not fully understood, but α-synuclein (α-Syn) protein aggregation is widely accepted as a key factor in the disease's pathogenesis. The aggregation of α-Syn is cytotoxic, leading to dysfunction and cellular damage (Wang et al., 2024).

### *Vitamins in Parkinson's Disease*

In a clinical study on PD, high concentrations of Vitamin A and its derivatives were found in the post-mortem frontal lobe cortex of human brains (Rai et al., 2021). This suggests a potential therapeutic effect of Vitamin A in the frontal lobe cortex. However, an age-related decline in retinol and its derivatives was observed in the frontal lobe cortex compared to the occipital cortex, but no definitive conclusion was reached from this study. Additionally, a cohort study conducted on Chinese individuals in Singapore suggested no linear relationship between the intake of carotenoids and other vitamins and the risk of developing PD (Ying et al., 2020). Another comparative study involving both male and female PD patients found a relationship between the intake of carotenoids, vitamin E, and vitamin C and the risk of PD. In this study, patients were followed up for 12–14 years, with their dietary frequency noted. None of the vitamins or even multivitamins taken by the patients were scientifically linked to the risk of PD. However, the results showed that high dietary intake of vitamin E, rather than vitamins A and C, was associated with a reduced risk of PD in both male and female patients. Another study showed that the intake of carotenoids and vitamins E and C did not significantly reduce the risk of PD, but a multivitamin approach rich in vitamin E provided good therapeutic effects. Therefore, a diet rich in vitamin E compared to carotenoids or vitamin C was suggested to reduce the risk of PD. As a result, vitamin E supplementation is protective against PD, but more extensive studies are needed to understand the therapeutic effects of vitamins A, C, and E and their derivatives on PD (Rai et al., 2021).

In most neurodegenerative diseases, including PD, elevated levels of homocysteine have been found to have neurotoxic effects (Moretti and Caruso, 2019). Therefore, there is a strong linear relationship between homocysteine levels and PD pathogenesis in PD patients with high homocysteine levels (Saadat et al., 2018). High homocysteine levels in PD patients lead to the death of nigrostriatal dopaminergic neurons. Thus, regulating homocysteine levels may prevent the progression of PD (Rai et al., 2021). At this point, vitamin B, which serves as a cofactor in the synthesis of methionine from homocysteine, comes to mind, and in one study, researchers suggested a strong relationship between vitamin B levels and homocysteine levels (Markišić et al., 2017). It has been shown that supplementation of vitamin B can reduce homocysteine levels in the blood plasma. Studies on the roles of different vitamin B types in PD pathology have highlighted that vitamin $B_{12}$ levels are lower in PD patients compared to healthy controls. Additionally, sufficient intake of vitamin $B_6$ in the diet has been shown to reduce the risk of PD. However, no significant changes were observed in vitamin $B_9$ concentrations in PD patients compared to healthy controls (Dietiker et al., 2019).

Research on the relationship between homocysteine, vitamin B, and brain function has mainly focused on three of the eight B-vitamins (vitamin $B_9$, vitamin $B_{12}$, vitamin $B_6$), with equivalent results obtained so far. However, it has been suggested that combined treatments involving all B vitamins could yield better results due to their interconnected cellular functions (Rai et al., 2021). In PD, dopaminergic neuron death occurs in the presence of thiamine deficiency. Thiamine supplementation in PD delays the death of dopaminergic neurons. Another study showed that high doses of thiamine administered to Parkinson's patients without side effects significantly reversed Parkinson's symptoms (Costantini and Fancellu, 2016).

In addition to these studies, olfactory dysfunction, observed 2–8 years before the onset of PD symptoms, has been linked to low folate (Vitamin $B_9$) and thiamine (Vitamin $B_1$) concentrations in the diet (Håglin et al., 2016). Therefore, in the early stages of PD, folate and thiamine levels effectively regulate the olfactory system and could be used as an important screening tool to detect the risk of PD. Another study on vitamins in PD showed that niacin treatment significantly improved rigidity and bradykinesia. In summary, due to the limited and sometimes contradictory results of studies on various types of vitamin B for PD treatment and/or diagnosis, more research is needed before advocating for the use of vitamin B types in PD treatment (Rai et al., 2021).

Vitamin C has effects such as reducing oxidative stress, which is one of the main causes of neurodegenerative diseases, but to achieve this, vitamin C levels need to be maintained at homeostatic concentrations. While vitamin C could be effective in PD due to its antioxidant properties, it has also been suggested that vitamin C significantly enhances the absorption of levodopa (10 mg carbidopa and 100 mg levodopa) in PD patients. Therefore, combination therapy with levodopa and ascorbic acid may yield better results than monotherapy in PD patients. However, the use of vitamin C in PD treatment is still debated, and more studies are needed to reach a definitive conclusion (Zhao et al., 2020).

Vitamin D deficiency is common in PD patients, and insufficient serum vitamin D levels increase the risk of PD. On the other hand, it has been shown that exposure to sunlight for at least 15 minutes a week and vitamin D supplementation have a preventive effect against PD (Zhou et al., 2019). A review showed that high doses of vitamin D can improve balance and postural stability and reduce the risk of falls, especially in younger PD patients, though not in elderly PD patients. Additionally, some studies have shown that higher serum vitamin D levels are associated with better cognition in PD patients without dementia. Other studies have linked vitamin D to olfactory function and mood in PD (Fullard and Duda, 2020). However, there are also studies with contradictory results.

Vitamin E is known to be a good antioxidant with numerous therapeutic uses. Vitamin E has been shown to play a role in many neurological diseases, including PD. A recent study found an inverse relationship between vitamin E intake and the development of PD, independent of gender and age (Schirinzi et al., 2019). Therefore, diets rich in vitamin E may minimize the risk of PD (Rai et al., 2021).

## Multiple Sclerosis

Multiple sclerosis (MS) is a chronic inflammatory and neurodegenerative CNS disorder (Jagannath et al., 2018). In MS, the immune system attacks the myelin sheaths in the CNS, which serve to cover and protect nerve fibers (Correale et al., 2019). Visual disturbances, movement abnormalities, fatigue, and pain are common symptoms of MS. The etiology of this disease is not fully understood, but it is thought to be caused by both genetic and environmental factors (Dobson and Giovannoni, 2019). There is no definitive treatment for MS; however, the usefulness of vitamins due to their antioxidant activities in MS treatment has been investigated in many studies (Rai et al., 2021).

### *Vitamins in Multiple Sclerosis*

Vitamin D has a protective activity in MS, which is often dependent on the stage of the disease. Vitamin D supplementation has been found to be more effective in the juvenile/early-stage rat models of MS compared to older animals. Many observational studies and clinical trials show that low levels of vitamin D in the blood are a risk factor for the development of MS. Vitamin D supplementation has a strong anti-inflammatory effect on the body. However, excessive intake of vitamin D can lead to hypercalcemia (Sandberg et al., 2016). Studies have shown that vitamin D supplementation increases tissue oxygenation by enhancing blood perfusion, thereby reducing neurodegeneration and neuroinflammation in MS patients (Muller et al., 2019). Some studies suggest that the levels of apolipoprotein E and vitamin D-binding protein (DBP) in cerebrospinal fluid could be used as biomarkers for MS diagnosis. However, another study found no connection between MS, DBP, and vitamin D in blood samples from MS patients. Therefore, broader clinical trials are needed to assess the effectiveness of vitamin D supplementation in MS treatments (Rai et al., 2021).

There are many in vitro and in vivo studies demonstrating the benefits of vitamin A and retinoic acid in MS treatment (Khosravi-Largani et al., 2018). These studies indicate that vitamin A levels are reduced in MS patients and that vitamin A exhibits antioxidant and anti-inflammatory activities in the brain. Vitamin A has been shown to promote remyelination and improve astrocyte function in MS patients. However, some studies have found no correlation between vitamin A serum concentration and the progression of MS (Rai et al., 2021).

Few studies have investigated the potential therapeutic effects of vitamin E or vitamin C in MS due to their antioxidant properties (Evans et al., 2018). However, several studies show that MS patients have lower levels of vitamin C compared to healthy individuals (Polachini et al., 2016). A study in an MS mouse model demonstrated that direct intrahippocampal injections of vitamin C improved memory function impairments in passive avoidance learning (Babri et al., 2015). Studies have shown that vitamin E inhibits factors related to necrosis and improves oligodendrocyte function in MS. Nevertheless, due to limited information on vitamin C and vitamin E in MS, more research is needed (Rai et al., 2021).

The role of B vitamins in MS is highly debated. Some studies show no relationship between vitamin B levels and MS, while many others have observed reduced vitamin B levels in MS patients. Vitamins $B_{12}$ and $B_9$ effectively improve homocysteine levels, which are responsible for myelin sheath synthesis, thereby regulating immune function in MS. Research on the roles of vitamin $B_1$, vitamin $B_3$, and vitamin $B_6$ in MS is also present. Studies suggest that vitamin $B_3$ may be effective in MS treatment by promoting remyelination (Lempriere, 2020). High-dose vitamin $B_1$ has been observed to improve MS-related fatigue. However, more extensive studies are needed to confirm the role of vitamin B in MS treatments.

In summary, many studies have associated vitamin D levels with MS risk, and significant results have been obtained validating the relationship due to vitamin D's anti-inflammatory effects. Additionally, other vitamins like vitamin A and vitamin E, due to their antioxidant and anti-inflammatory effects, may be beneficial in MS treatments (Liu and Hu, 2016). The role of vitamin B in MS remains highly debated, and more clinical research is needed to validate these findings (Rai et al., 2021).

## Amyotrophic Lateral Sclerosis (ALS)

Amyotrophic lateral sclerosis (ALS) is a fatal neurodegenerative disease characterized by the permanent loss of motor neurons in the brain and spinal cord, with an unknown etiology and no known cure (Rojas et al., 2020 , ).

### *Vitamins in Amyotrophic Lateral Sclerosis*

Similar to MS, there is limited research on the role of vitamins in ALS. However, as with MS, a strong correlation between ALS and vitamin D supplementation has been observed. Studies have shown that the active form of vitamin D is reduced in ALS patients and ALS animal models (Libonati et al., 2017). Some studies have also demonstrated that vitamin D plays a protective role in vitro in motor neurons, and that plasma levels of vitamin D are directly linked to the severity of ALS in patients (Camu et al., 2014). Genetic studies have indicated that vitamin D is associated with ALS pathology through the regulation of various immune components. Additionally, vitamin D affects ALS pathology through various cellular signaling mechanisms. Many studies on ALS patients have shown that vitamin D supplementation reduces muscle weakness symptoms and improves motor functional capacity in transgenic mouse models of ALS, although it does not prevent the progression of the disease (Rai et al., 2021).

In contrast, a recent study found no reduction in vitamin D levels in ALS patients and no benefit of vitamin D supplementation in preventing the disease (Libonati et al., 2017). Therefore, new studies are needed to confirm the role of vitamin D in ALS pathophysiology (Rai et al., 2021).

A clinical study conducted on ALS patients observed the effects of vitamin E supplementation. The results of this study showed that individuals who did not receive regular vitamin E doses experienced early death compared to those who received regular vitamin E supplementation. The same study also concluded that vitamin E significantly improved motor functions in ALS patients (Ascherio et al., 2010).

In summary, vitamins D and E may be beneficial in ALS by improving motor symptoms and protecting motor neurons. However, further research is needed to fully understand the roles of these vitamins in ALS and to explore the potential use of other vitamins in treatment (Rai et al., 2021).

## Studies on Vitamins in Zebrafish Models of Neurodegenerative Diseases

Zebrafish are frequently preferred as model organisms in scientific research and experiments due to their resistance to environmental conditions, small size, short reproductive period, rapid development, high reproductive ability throughout the year under laboratory conditions, low cost, ability of females to produce several hundred eggs every 2–3 days and at one time, external fertilisation, high fertility, optical transparency during development and openness to manipulation during all developmental stages of living embryos, ease of traceability in genetic screening, and genetic similarity with humans as one of the most important reasons (Yipel and İlhan, 2021).

There are studies determining the toxicity of valproic acid (VPA) in zebrafish. These studies have shown that VPA causes autism-like behaviors in zebrafish (Camussi et al., 2025). A study using zebrafish models with VPA has demonstrated that vitamin A improves VPA-induced autism-like behaviors in zebrafish larvae by mitigating oxidative stress and apoptosis (Wang et al., 2024).

In studies where zebrafish, used as a model organism, were exposed to toxic metals (such as lead and mercury) known to be associated with neurodegenerative diseases, it was observed that zebrafish exposed to neurotoxins mimicking the effects of PD showed improvements in motor functions and reductions in neuronal death when treated with B vitamins. This suggests that B vitamins might play a role in slowing the progression of neurodegenerative diseases (Paduraru et al., 2023).

There are also studies examining the relationship between vitamin $B_{12}$ and sociability/asociability in zebrafish. One such study found that high-dose acute exposure to pesticides, followed by intervention with vitamin $B_{12}$, resulted in significant improvements in locomotor activity parameters and positive effects on sociability (Robea et al., 2024).

A study investigating the effects of vitamin C on zebrafish exposed to butachlor-induced developmental toxicity showed that vitamin C could ameliorate butachlor-induced toxicity. This study also demonstrated the antioxidant properties of vitamin C in zebrafish (Xiang et al., 2018).

In a study examining the effects of vitamin $D_3$ on PD using zebrafish models, it was found that vitamin $D_3$ supplementation reduced neuroinflammation and oxidative stress in zebrafish. The results suggest that vitamin $D_3$ has the potential to alleviate the pathological symptoms of PD (Mao et al., 2018).

A study on the effects of vitamin E in zebrafish models of AD investigated how vitamin E treatment could alleviate cognitive impairments and neuroinflammation. The results indicated that vitamin E could be effective in reducing the symptoms of AD (Kang et al., 2021).

Another study assessed the protective effects of vitamin E in PD models in zebrafish and examined how vitamin E affects neurodegeneration, oxidative stress, and motor function loss. The results indicated that vitamin E has potential in slowing the progression of PD (Zhao et al., 2020).

A study investigating the effects of vitamin $K_2$ in zebrafish models of PD found that vitamin $K_2$ reduced neurodegeneration and improved motor functions. The results suggest that vitamin $K_2$ could be effective in alleviating symptoms of PD (Eckhardt et al., 2020).

In zebrafish models related to AD, a study on vitamin $K_1$ demonstrated that vitamin $K_1$ treatment improved cognitive functions. Notably, performance improvements were observed in learning and memory tests. Additionally, vitamin $K_1$ was effective in reducing neuroinflammation caused by AD (Kang et al., 2021).

A study investigating the protective effects of vitamin $K_3$ in zebrafish models of neurodegenerative diseases found that vitamin $K_3$ effectively reduced oxidative stress and neuronal damage. Zebrafish treated with vitamin $K_3$ showed significant reductions in oxidative stress markers compared to untreated groups. Vitamin $K_3$ was also effective in reducing neuronal damage, with substantial reductions in damage to neuronal structures and functions (Hsu et al., 2019).

## Conclusion

In conclusion, the effect of vitamins on neurodegeneration is a topic that will always maintain its relevance. Studies to be conducted on this subject will shed light on the mechanisms related to the subject. Furthermore, studies conducted on the effect of vitamins on neurodegeneration in zebrafish to date have shown that zebrafish, as a model organism, are a very useful model for examining the effects of different nutritional components on neurodegeneration.

## REFERENCES


Abosamak, N. R., & Gupta, V. (2023). Vitamin B6 (Pyridoxine). In *StatPearls* [Internet]. Treasure Island (FL): StatPearls Publishing. https://pubmed.ncbi.nlm.nih.gov/32491368/

Altıner, A., Atalay, H., & Bilal, T. (2017). Bir antioksidan olarak E vitamini. *Balıkesir Sağlık Bilimleri Dergisi, 6*(3), 149–157.

Ankar, A., & Kumar, A. (2022). Vitamin B12 Deficiency. In *StatPearls* [Internet]. Treasure Island (FL): StatPearls Publishing. https://www.ncbi.nlm.nih.gov/books/NBK441923/

Ascherio, A., Munger, K. L., & Simon, K. C. (2010). Vitamin D and Multiple Sclerosis. *Lancet Neurology, 9*(6), 599–612. https://doi.org/10.1016/S1474-4422(10)70086-7

Babri, S., Mehrvash, F., Mohaddes, G., Hatami, H., & Mirzaie, F. (2015). Effect of intrahippocampal administration of vitamin C and progesterone on learning in a model of multiple sclerosis in rats. *Advanced Pharmaceutical Bulletin, 5*(1), 83–89. https://doi.org/10.5681/apb.2015.011

Baroni, L., Bonetto, C., Rizzo, G., Bertola, C., Caberlotto, L., & Bazzerla, G. (2019). Association between cognitive impairment and vitamin B12, folate, and homocysteine status in elderly adults: A retrospective study. *Journal of Alzheimer's Disease, 70*(2), 443–453. https://doi.org/10.3233/JAD-190249

Bauerly, K., Rucker, R.B. (2007) Pantothenic acid. In: Zempleni J, Rucker RB, McCormick DB, Suttie JW, eds. Handbook of vitamins. 4th ed. Boca Raton, Fl: CRC Press; 289-314.

Browne, D., McGuinness, B., Woodside, J. V., & McKay, G. J. (2019). VITAMIN E and Alzheimer's disease: What do we know so far? *Clinical Interventions in Aging, 14*, 1303–1313. https://doi.org/10.2147/CIA.S186760

Bunik, V.I., Aleshin, V.A. (2017) Analysis of the Protein Binding Sites for Thiamin and Its Derivatives to Elucidate the Molecular Mechanisms of the Noncoenzyme Action of Thiamin (Vitamin B1), Editor(s): Atta-ur-Rahman, Studies in Natural Products Chemistry, Elsevier, Volume 53, 2017, 375-429,

Calderón-Ospina, C. A., & Nava-Mesa, M. O. (2023). B vitamins in the nervous system: Current knowledge of the biochemical modes of action and synergies of thiamine, pyridoxine, and cobalamin. *CNS Neuroscience & Therapeutics, 26*(1), 5–13. https://doi.org/10.1111/cns.13207

Camu, W., Tremblier, B., Plassot, C., Alphandery, S., Salsac, C., Pageot, N., Juntas-Morales, R., Scamps, F., Daures, J. P. (2014). VITAMIN D confers protection to motoneurons and is a prognostic factor of amyotrophic lateral sclerosis. *Neurobiology of Aging, 35*(5), 1198–1205. https://doi.org/10.1016/j.neurobiolaging.2013.11.005

Camussi, D., Marchese, M., Nicoletti, F., Santorelli, F. M., & Ogi, A. (2025). Valproate-Induced Model of Autism in Adult Zebrafish: A Systematic Review. Cells, 14(2), 109. https://doi.org/10.3390/cells14020109

Colombo, M. L. (2010). An update on VITAMIN E, tocopherol and tocotrienol-perspectives. *Molecules, 15*(4), 2103–2113. https://doi.org/10.3390/molecules15042103

Correale, J., Marrodan, M., & Ysrraelit, M. C. (2019). Mechanisms of neurodegeneration and axonal dysfunction in progressive multiple sclerosis. *Biomedicines, 7*(1), 14. https://doi.org/10.3390/biomedicines7010014

Costantini, A., & Fancellu, R. (2016). An open-label pilot study with high-dose thiamine in Parkinson's disease. *Neural Regeneration Research, 11*(3), 406–407. https://doi.org/10.4103/1673-5374.179047

Dayalu, P., & Albin, R. L. (2015). Huntington disease: Pathogenesis and treatment. *Neurology Clinics, 33*(1), 101–114. https://doi.org/10.1016/j.ncl.2014.09.003

De Wilde, M. C., Vellas, B., Girault, E., Yavuz, A. C., & Sijben, J. W. (2017). Lower brain and blood nutrient status in Alzheimer's disease: Results from meta-analyses. *Alzheimer's & Dementia: Translational Research & Clinical Interventions, 3*, 416–431. https://doi.org/10.1016/j.trci.2017.06.002

Dietiker, C., Kim, S., Zhang, Y., & Christine, C. W. (2019). Characterization of vitamin B12 supplementation and correlation with clinical outcomes in a large longitudinal study of early Parkinson's disease. *Journal of Movement Disorders, 12*(2), 91–101. https://doi.org/10.14802/jmd.18049

Dobson, R., & Giovannoni, G. (2019). Multiple sclerosis—A review. *European Journal of Neurology, 26*(1), 27–40. https://doi.org/10.1111/ene.13819

Ebru Emekli-Alturfan. (2023). Vitamins and Cardiovascular Diseases. In A Guide to Vitamins and Their Effects on Diseases. Ed: Alturfan AA, Emekli-Alturfan E. 371-418, Cambridge Scholars Publishing, UK.

Eckhardt, S., Aro, K., Burch, M., & Kringel, H. (2020). Vitamin K2 supplementation reduces neurodegeneration and improves motor functions in zebrafish models of Parkinson's disease. *Frontiers in Neuroscience, 14*, 586752. https://doi.org/10.3389/fnins.2020.586752

Ejder, Z. B., & Ekici, E. M. (2024). K Vitamini ve Bilişsel Performans: Mini Bir Derleme. *Avrasya Sağlık Bilimleri Dergisi, 7*(1), 61–67. https://doi.org/10.53493/avrasyasbd.1339322

Endres, K. (2019). Retinoic acid and the gut microbiota in Alzheimer's disease: Fighting back-to-back? *Current Alzheimer Research, 16*(5), 405–417. https://doi.org/10.2174/1567205016666190321163705

Evans, E., Piccio, L., & Cross, A. H. (2018). Use of vitamins and dietary supplements by patients with multiple sclerosis: A review. *JAMA Neurology, 75*(8), 1013–1021. https://doi.org/10.1001/jamaneurol.2018.0611

Forster, R. E., Jurutka, P. W., Hsieh, J. C., et al. (2011). Vitamin D receptor controls expression of the anti-aging klotho gene in mouse and human renal cells. *Biochemical and Biophysical Research Communications, 414*(3), 557–562. https://doi.org/10.1016/j.bbrc.2011.09.035

Fullard, M.E., Duda, J.E. (2020). A Review of the Relationship Between Vitamin D and Parkinson Disease Symptoms. Front Neurol. 27;11:454. doi: 10.3389/fneur.2020.00454.

Geller, M., Oliveira, L., Nigri, R., Mezitis, S.G., Ribeiro, M.G., et al. (2017) B Vitamins for Neuropathy and Neuropathic Pain. Vitam Miner 6:161.

Grimm, M.O., Mett, J., Hartmann, T. (2016) The Impact of Vitamin E and Other Fat-Soluble Vitamins on Alzheimer´s Disease. *Int J Mol Sci.* 26;17(11):1785. doi: 10.3390/ijms17111785.

Gruber-Bzura, B. M., Krzysztoń-Russjan, J., Bubko, I., Syska, J., Jaworska, M., Zmysłowski, A., Rosłon, M., Drozd, J., & Drozd, E. (2017). Role of thiamine in Huntington's disease pathogenesis: In vitro studies. *Advances in Clinical and Experimental Medicine, 26*(5), 751–760. https://doi.org/10.17219/acem/63091

Gugliandolo, A., Bramanti, P., & Mazzon, E. (2017). Role of Vitamin E in the treatment of Alzheimer's disease: Evidence from animal models. *International Journal of Molecular Sciences, 18*(12), 2504. https://doi.org/10.3390/ijms18122504

Guo, S. (2004). Zebrafish as a model system for studying neurodegenerative diseases. *Frontiers in Bioscience, 9*, 190–196. https://doi.org/10.2741/1242

Håglin, L., Johansson, I., Forsgren, L., Bäckman, L. (2017). Intake of vitamin B before onset of Parkinson's disease and atypical parkinsonism and olfactory function at the time of diagnosis. *Eur J Clin Nutr.* 71(1):97-102. doi: 10.1038/ejcn.2016.181.

Halczuk, K., Kaźmierczak-Barańska, J., Karwowski, B. T., Karmańska, A., & Cieślak, M. (2023). Vitamin B12-Multifaceted in vivo functions and in vitro applications. *Nutrients, 15*(12), 2734. https://doi.org/10.3390/nu15122734

Hama, Y., Hamano, T., Shirafuji, N., Hayashi, K., Ueno, A., Enomoto, S., Nagata, M., Kimura, H., Matsunaga, A., Ikawa, M., et al. (2020). Influences of folate supplementation on homocysteine and cognition in patients with folate deficiency and cognitive impairment. *Nutrients, 12*(10), 3138. https://doi.org/10.3390/nu12103138

Hogenkamp, P. S., MacDonald, P. M., Van Riper, J., & Bartels, L. (2017). Vitamin B12 deficiency and neurological dysfunction. *Nutritional Neuroscience, 20*(6), 282–290. https://doi.org/10.1080/1028415X.2016.1180850

Hsu, T. H., Chen, Y. H., & Yu, W. L. (2019). Vitamin K3 administration reduces oxidative stress and neuronal damage in zebrafish models of neurodegenerative diseases. *Neurotoxicity Research, 36*(3), 415–425. https://doi.org/10.1007/s12640-019-00076-1

Isenberg-Grzeda, E., Kutner, H. E., & Nicolson, S. E. (2012). Wernicke–Korsakoff syndrome: Under-recognized and under-treated. *Psychosomatics, 53*(6), 507–516. https://doi.org/10.1016/j.psym.2012.06.001

Jagannath, V.A., Filippini, G., Di, Pietrantonj, C., Asokan, G.V., Robak, E.W., Whamond, L., Robinson, S.A. (2018) Vitamin D for the management of multiple sclerosis. Cochrane Database Syst Rev. 24;9(9):CD008422. doi: 10.1002/14651858.CD008422.pub3.

Kalueff, A. V., Stewart, A. M., & Gerlai, R. (2014). Zebrafish as an emerging model for studying complex brain disorders. *Trends in Neurosciences, 37*(2), 71–84. https://doi.org/10.1016/j.tins.2013.10.005

Kang, J., Zhang, L., & Zhao, S. (2021). The effects of vitamin K1 on cognitive function and neuroinflammation in zebrafish models of Alzheimer's disease. *Journal of Alzheimer's Disease, 82*(4), 1425–1436. https://doi.org/10.3233/JAD-201326

Kang, K. H., Kim, Y. S., Kim, K. J., & Park, S. K. (2021). Vitamin E ameliorates cognitive deficits and neuroinflammation in zebrafish models of Alzheimer's disease. *Journal of Neuroinflammation, 18*(1), 150. https://doi.org/10.1186/s12974-021-02269-5

Karaağaç, M., & Pınarlı, Ç. (2023). Suda Çözünen Vitaminlerin Bağırsak Mikrobiyotası Üzerine Etkileri. *Hitit Sağlık Dergisi,* 1, 32-44.

Kennedy, D. O. (2016). B vitamins and the brain: Mechanisms, dose and efficacy—a review. *Nutrients, 8*(2), 68. https://doi.org/10.3390/nu8020068

Khan, S., & Tiwari, M. (2019). VITAMIN D deficiency and neurodegenerative diseases: A review. *Journal of Neurology and Neuroscience, 10*(5), 204–211. https://doi.org/10.21767/2171-6625.1000303

Khan, U., & Lippman, R. J. (2020). Nutritional strategies to mitigate neurodegeneration: The role of vitamins and minerals. *Nutritional Neuroscience, 23*(9), 715–732. https://doi.org/10.1080/1028415X.2019.1645680

Khosravi-Largani, M., Pourvali-Talatappeh, P., Rousta, A.M., Karimi-Kivi, M., Noroozi, E., Mahjoob, A., Asaadi, Y., Shahmohammadi, A., Sadeghi, S., Shakeri, S., Ghiyasvand, K., Tavakoli-Yaraki, M. (2018). A review on potential roles of vitamins in incidence, progression, and improvement of multiple sclerosis. eNeurologicalSci. 28;10:37-44. doi: 10.1016/j.ensci.2018.01.007.

Kim, J. Y., Ko, K. H., & Kim, H. J. (2019). Vitamin D and neurodegenerative diseases: A review. *Brain Sciences, 9*(12), 339. https://doi.org/10.3390/brainsci9120339

Kluge, M. A., Furr, H. C., & Davis, J. T. (2016). Vitamin E and its effects on the nervous system in aging and disease. *Neurochemical Research, 41*(4), 632–642. https://doi.org/10.1007/s11064-016-1911-4

Kwok, T., Wu, Y., Lee, J., Lee, R., Yung, C.Y., Choi, G., Lee, V., Harrison, J., Lam, L., Mok, V. A.(2020). Randomized placebo-controlled trial of using B vitamins to prevent cognitive decline in older mild cognitive impairment patients. Clin Nutr. 39(8):2399-2405. doi: 10.1016/j.clnu.2019.11.005

Lemprière, S. (2020). NLRP3 inflammasome activity as biomarker for primary progressive multiple sclerosis. Nat Rev Neurol. 16(7):350. doi: 10.1038/s41582-020-0366-y.

Liao, Y., & Zhang, C. (2021). Vitamin B12 and cognition: A review. *Journal of Alzheimer's Disease, 83*(1), 159–176. https://doi.org/10.3233/JAD-201097

Libonati, L., Onesti, E., Gori, M.C., Ceccanti, M., Cambieri, C., Fabbri, A., Frasca, V., Inghilleri, M. (2017). Vitamin D in amyotrophic lateral sclerosis. *Funct Neurol.* 32(1):35-40. doi: 10.11138/fneur/2017.32.1.035.

Liu, J., & Hu, Y. (2016). Role of VİTAMİN A in the treatment of multiple sclerosis: A meta-analysis. *Multiple Sclerosis and Related Disorders, 9*, 36–42. https://doi.org/10.1016/j.msard.2016.07.015

Lu, K., & Chen, H. (2022). Vitamin B6 and its role in neurological disorders: A review. *Nutrients, 14*(2), 284. https://doi.org/10.3390/nu14020284

Manzetti, S., Zhang, J., Van der Spoel D. (2014). Thiamin function, metabolism, uptake, and transport. *Biochemistry.* 2014;53:821–835. doi: 10.1021/bi401618y.

Mao, J., Yang, Z., Liu, Y., & Zhang, C. (2018). VITAMIN D3 supplementation alleviates neuroinflammation and oxidative stress in zebrafish models of Parkinson's disease. *Neurobiology of Disease, 119*, 12–23. https://doi.org/10.1016/j.nbd.2018.07.004

Markišić, M., Pavlović, A.M., Pavlović, D.M. (2017). The Impact of Homocysteine, Vitamin B12, and Vitamin D Levels on Functional Outcome after First-Ever Ischaemic Stroke. *Biomed Res Int.* 2017:5489057.

Masters, C. L., Bateman, R. J., Blennow, K., Rowe, C. C., & Sperling, R. A. (2015). Alzheimer's disease. *Nature Reviews Disease Primers, 1*, 15056. https://doi.org/10.1038/nrdp.2015.56

Medina, J., Gupta, V. (2023) Vitamin E. In: StatPearls [Internet]. Treasure Island (FL): StatPearls Publishing; 2025 Jan-. Available from: https://www.ncbi.nlm.nih.gov/books/NBK557737/

Memiç İnan, C. (2021). Bazı Besin Ögelerinin Sinir Sistemi Üzerine Etkileri. *Cumhuriyet Üniversitesi Sağlık Bilimleri Enstitüsü Dergisi, 6*(1), 49–58. https://doi.org/10.51754/cusbed.807327

Menezes, M. S. S., & Almeida, C. M. M. (2024). Structural, functional, nutritional and clinical aspects of vitamin A: A review. PharmaNutrition, 27, Article 100383. https://doi.org/10.1016/j.phanu.2024.100383

Merrell, B.J., McMurry, J.P. (2023) Folic Acid. In: StatPearls [Internet]. Treasure Island (FL): StatPearls Publishing; 2025 Jan. PMID: 32119374.

Moll, T., & Graber, J. J. (2022). The role of folate in neurological health. *Journal of Neurology and Neuroscience, 13*(3), 242–249. https://doi.org/10.1016/j.jnn.2022.07.002

Moretti, R., Caruso, P. (2019). The Controversial Role of Homocysteine in Neurology: From Labs to Clinical Practice. *Int J Mol Sci.* 8;20(1):231. doi: 10.3390/ijms20010231.

Moshfegh, A. J., & Goldman, J. D. (2019). Dietary vitamin B12 intake and absorption. *Journal of Nutrition, 149*(12), 2042–2054. https://doi.org/10.1093/jn/nxz186

Müller, T., Lohse, L., Blodau, A., Frommholz, K. (2019). Vitamin D rise enhances blood perfusion in patients with multiple sclerosis. J Neural Transm (Vienna). 126(12):1631-1636. doi: 10.1007/s00702-019-02093-x.

Nair, V., & Prakash, S. (2017). The role of vitamins in neurodegenerative diseases: A review. *Journal of Neurodegenerative Disorders, 14*(2), 120–135. https://doi.org/10.1186/s12883-017-0704-6

Obeid, R., Fedosov, S.N., Nexo, E. (2015). Cobalamin coenzyme forms are not likely to be superior to cyano- and hydroxylcobalamin in prevention or treatment of cobalamin deficiency. Mol Nutr Food Res. 59(7):1364-72. doi: 10.1002/mnfr.201500019.

Paduraru, E., Iacob, D., Rarinca, V., Plavan, G., Ureche, D., Jijie, R., & Nicoara, M. (2023). Zebrafish as a potential model for neurodegenerative diseases: A focus on toxic metals implications. *International Journal of Molecular Sciences, 24*(4), 3428. https://doi.org/10.3390/ijms24043428

Page, G.L, Laight, D., Cummings, M.H. (2011). Thiamine deficiency in diabetes mellitus and the impact of thiamine replacement on glucose metabolism and vascular disease. *Int J Clin Pract.* 65(6):684-90. doi: 10.1111/j.1742-1241.2011.02680.x.

Parra, M., Stahl, S., Hellmann, H. (2018). Vitamin $B_6$ and Its Role in Cell Metabolism and Physiology. Cells. 22;7(7):84. doi: 10.3390/cells7070084.

Peechakara, B.V., Gupta, M. (2025). Vitamin B3. 2024 Feb 29. In: StatPearls [Internet]. Treasure Island (FL): StatPearls Publishing; PMID: 30252363.

Pinto JT, Zempleni J. (2016) Riboflavin. *Adv Nutr.* 2016 Sep 15;7(5):973-5. doi: 10.3945/an.116.012716.

Poewe, W., Seppi, K., Tanner, C. M., & Weintraub, D. (2017). Parkinson disease. *Nature Reviews Disease Primers, 3*, 17013. https://doi.org/10.1038/nrdp.2017.13

Polachini, C.R., Spanevello, R.M., Zanini, D., Baldissarelli, J., Pereira, L.B., Schetinger, M.R., (2016). Evaluation of Delta-Aminolevulinic Dehydratase Activity, Oxidative Stress Biomarkers, and Vitamin D Levels in Patients with Multiple Sclerosis. *Neurotox Res.* 29(2):230-42. doi: 10.1007/s12640-015-9584-2

Price, B.R., Wilcock, D.M., Weekman, E.M. (2018). Hyperhomocysteinemia as a Risk Factor for Vascular Contributions to Cognitive Impairment and Dementia. *Front Aging Neurosci.* 31;10:350. doi: 10.3389/fnagi.2018.00350.

Rai, S.N., Singh, P., Steinbusch, H.W.M., Vamanu, E., Ashraf, G., Singh, M.P. (2021) The Role of Vitamins in Neurodegenerative Disease: An Update. *Biomedicines.* 22;9(10):1284

Robea, M. A., Ilie, O. D., Nicoara, M. N., Solcan, G., Romila, L. E., Ureche, D., & Ciobica, A. (2024). Vitamin B12 ameliorates pesticide-induced sociability impairment in zebrafish (Danio rerio): A prospective controlled intervention study. *Animals, 14*(3), 405. https://doi.org/10.3390/ani14030405

Rojas, P., Ramírez, A.I., Fernández-Albarral, J.A., López-Cuenca, I., Salobrar-García, E., Cadena, M., Elvira-Hurtado, L., Salazar, J.J., de Hoz, R., Ramírez, J.M. (2020). Amyotrophic Lateral Sclerosis: A Neurodegenerative Motor Neuron Disease With Ocular Involvement. Front Neurosci. 25;14:566858. doi: 10.3389/fnins.2020.566858.

Saadat, P., Ahmadi Ahangar, A., Samaei, S.E., Firozjaie, A., Abbaspour, F., Khafri, S., Khoddami, A. (2018). Serum Homocysteine Level in Parkinson's Disease and Its Association with Duration, Cardinal Manifestation, and Severity of Disease. Parkinsons Dis. 2;2018:5813084. doi: 10.1155/2018/5813084.

Sandberg, L., Biström, M., Salzer, J., Vågberg, M., Svenningsson, A., Sundström, P. (2016) Vitamin D and axonal injury in multiple sclerosis. *Mult Scler.* 22(8):1027-31. doi: 10.1177/1352458515606986.

Schellenberg, G. D., & Hulette, C. M. (2023). Vitamin D and cognitive decline: Evidence from population-based studies. *Neurobiology of Aging, 102*, 198–206. https://doi.org/10.1016/j.neurobiolaging.2021.07.010

Schirinzi, T., Martella, G., Imbriani, P., Di Lazzaro, G., Franco, D., Colona, V.L., et al., (2019). Dietary Vitamin E as a Protective Factor for Parkinson's Disease: Clinical and Experimental Evidence. *Front Neurol.* 26;10:148. doi: 10.3389/fneur.2019.00148.

Sigvard, A.K., Bojesen, K.B., Ambrosen, K.S., Nielsen, M.Ø., Gjedde, A., Tangmose, K., et al. (2022). Dopamine Synthesis Capacity and GABA and Glutamate Levels Separate Antipsychotic-Naïve Patients With First-Episode Psychosis From Healthy Control Subjects in a Multimodal Prediction Model. *Biol Psychiatry Glob Open Sci.* 30;3(3):500-509. doi: 10.1016/j.bpsgos.2022.05.004.

Stanger, M., & Seidman, M. (2016). Vitamin D and neurological health: A review. *Journal of Clinical Neuroscience, 24*, 15–19. https://doi.org/10.1016/j.jocn.2015.10.006

Strandler, H.S., Strand, T.A. (2023) Thiamin (Vitamin $B_1$) - A scoping review for Nordic Nutrition Recommendations. *Food Nutr Res.* 13;67. doi: 10.29219/fnr.v67.10290

Sultan, S., Taimuri, U., Basnan, S.A., Ai-Orabi, W.K., Awadallah, A, Almowald, F., Hazazi, A. (2020). Low Vitamin D and Its Association with Cognitive Impairment and Dementia. *J Aging Res.* 30;2020:6097820. doi: 10.1155/2020/6097820.

Suwannasom, N., Kao, I., Pruß, A., Georgieva, R., Bäumler, H. (2020). Riboflavin: The Health Benefits of a Forgotten Natural Vitamin. *Int J Mol Sci.* 31;21(3):950. doi: 10.3390/ijms21030950.

Tafti M, Ghyselinck NB. (2007). Functional implication of the vitamin A signaling pathway in the brain. *Archives of Neurology, 64*(12), 1706–1711.

Tiffany, L. M., & Norman, H. (2020). Folate, homocysteine, and cognitive function: An integrative review. *Journal of Clinical Nutrition, 24*(1), 65–74. https://doi.org/10.1016/j.jcln.2019.09.003

Wang, J., Zou, L., Jiang, P., Yao, M., Xu, Q., Hong, Q., Zhu, J., & Chi, X. (2024). Vitamin A ameliorates valproic acid-induced autism-like symptoms in developing zebrafish larvae by attenuating oxidative stress and apoptosis. *Journal of Experimental Biology, 227*(12), jeb242423. https://doi.org/10.1242/jeb.242423

Wuerch, E, Urgoiti, GR, Yong, V.W. (2023) The Promise of Niacin in Neurology. *Neurotherapeutics.* Jul;20(4):1037-1054. doi: 10.1007/s13311-023-01376-2.

Xiang, Q., Xu, B., Ding, Y., Zhang, Y., & Wang, J. (2018). Oxidative stress response induced by butachlor in zebrafish embryo/larvae: The protective effect of vitamin C. *Bulletin of Environmental Contamination and Toxicology, 100*(2), 208–215. https://doi.org/10.1007/s00128-017-2245-9

Xie, C., Gong, J., Zheng, C., Zhang, J., Gao, J., Tian, C., Guo, X., Dai, S., Gao, T. (2024) Effects of vitamin K supplementation on bone mineral density at different sites and bone metabolism in the middle-aged and elderly population. *Bone Joint Res.* 11;13(12):750-763. doi: 10.1302/2046-3758.1312.BJR-2024-0053.R1.

Yarat, A. (2023). Vitamins in neurological disorders. In A. Alturfan & E. Emekli-Alturfan (Ed.), *A guide to vitamins and their effects on disease* (Vol. 1, pp. 453–494). Cambridge Scholars Publishing.

Ying, A.F., Khan, S., Wu, Y., Jin, A., Wong, A.S.Y., Tan, E.K., Yuan, J.M., Koh, W.P., Tan, L.C.S. (2020). Dietary Antioxidants and Risk of Parkinson's Disease in the Singapore Chinese Health Study. *Mov Disord.* 35(10):1765-1773. doi: 10.1002/mds.28173.

Yipel, M., & İlhan, A. (2021). Zebra Balığı (Danio rerio): Toksikolojik Çalışmalar İçin Uygun Bir Model Organizma. *Veteriner Farmakoloji ve Toksikoloji Derneği Bülteni, 12*(3), 161–167. https://doi.org/10.38137/vftd.1007548

Zhao, L., Zheng, X., Zhang, M., & Wu, S. (2020). The protective effects of Vitamin E on neurodegeneration in zebrafish models of Parkinson's disease. *Frontiers in Neuroscience, 14*, 556567. https://doi.org/10.3389/fnins.2020.556567

Zhou, Z., Zhou, R., Zhang, Z., Li, K. (2019). The Association Between Vitamin D Status, Vitamin D Supplementation, Sunlight Exposure, and Parkinson's Disease: A Systematic Review and Meta-Analysis. *Med Sci Monit.* 2019 Jan 23;25:666-674. doi: 10.12659/MSM.912840.

# 21

## *Usability of Zebrafish as a Model Organism to Investigate the Relationship between Breast Milk and Neurodegeneration*

**Tuğba Tunalı-Akbay**

### Abbreviations

| | |
|---|---|
| **AD** | Alzheimer's disease |
| **ADHD** | Attention-Deficit Hyperactivity Disorder |
| **BCM** | β- Casomorphine |
| **BDE-47** | 2,2'4,4'-Brominated Diphenyl Ether |
| **BFR** | Brominated Flame Retardants |
| **BPDP** | Tert-Butylated Phenyl Diphenyl Phosphate |
| **CNS** | Central Nervous System |
| **EHDP** | 2-Ethylhexyl Diphenyl Phosphate |
| **FXS** | Fragile X Syndrome |
| **GABA** | Gamma Amino Butyric Acid |
| **GABAA PAM** | Positive Allosteric Modulator of Γ-Aminobutyric Acid Subtype A Receptors |
| **hBCM** | Human β- Casomorphine |
| **HMO** | Human Milk Oligosaccharides |
| **5-HT1A** | 5-Hydroxytryptamine 1A |
| **IDDP** | Isodecyl Diphenyl Phosphate |
| **IPP** | Isopropylated Phenyl Phosphate |
| **IQ** | Intelligence Quotient |
| **LSTb** | Sialyllacto-N-Tetraose b |
| **MFG-E8** | Milk Fat Globule-Epithelial Growth Factor 8 Protein |
| **MFGM** | Milk Fat Globule Membrane |
| **MiRNA** | Microsomal RNA |
| **mGluR5** | Metabotropic Glutamate Receptor 5 |
| **OCP** | Organochlorine Pesticides |
| **PBDE** | Polybrominated Diphenyl Ethers |
| **PCB** | Polychlorinated Biphenyls |
| **PFAS** | Perfluoroalkyl Substances |
| **Shh** | Sonic Hedgehog |
| **TBBPA** | 3,3',5,5'-Tetrabromobisphenol A |
| **TCEP** | Tris(2-chloroethyl) Phosphate |
| **TDCIPP** | Tris(1,3dichloroisopropyl) Phosphate |
| **TMPP** | Trimethyl Phenyl Phosphate |
| **TPHP** | Triphenyl Phosphate |

### Breast Milk and Neurological Health

Breast milk has been considered the gold standard when it comes to infant nutrition. It is a food that alone meets the nutritional needs of babies in the first 6 months of their lives, providing them with essential nutrients and antibodies that promote optimal growth and development. Its multifaceted effects on infant health, in addition to its nutritional properties, are facts unanimously accepted by all health organizations (Newton, 2004). One of the most significant benefits of breastfeeding is its potential to reduce the incidence of sudden infant death syndrome. Studies have consistently shown that breastfed infants exhibit a lower risk of sudden infant death syndrome compared to those who are formula-fed. Additionally, the bioactive components in breast milk, such as antibodies, lactoferrin, and cytokines, contribute to rapid recovery and overall better health outcomes in breastfed infants, underscoring the critical role of breastfeeding in early childhood development. Research has indicated that breastfed infants are less likely to develop dental malocclusions and have a lower risk of obesity and increased intelligence (Picó et a.,, 2021).

When evaluating the relationship between breast milk and the brain, breast milk promotes early brain development and offers long-term neurological health advantages via epigenetic modifications, microbiome modulation, and neuroprotective effects. These effects can benefit an individual's cognitive and neurological health, even in older people (Erick, 2018). According to the European Consensus on the Scientific Concepts of Functional Foods, "a portion of food can be considered functional if, beyond adequate nutritional effects, it is demonstrated to beneficially affect one or more target functions in the body in a way that is related to improved health and well-being and/or reduced risk of disease" ("Scientific Concepts of Functional Foods in Europe Consensus Document," 1999). Breast milk is the most complete functional food, aligning perfectly with the definition. Belfort et al. determined that breastfeeding 180 premature infants in the first 28 days of life had a positive effect on their neurodevelopment at age 7. It was reported that breastfeeding in the first 28 days of life resulted in more deep nuclear gray matter and hippocampal volume at age 7 in premature infants. This finding was also associated with higher IQ, academic achievement, working memory, and motor scores (Belfort et al., 2016).



DOI: 10.1201/9781032676555-23

There is evidence that psychological and physical contact between mother and child during breastfeeding plays a role in the development of cortical connections and the limbic system in the baby's brain. Breastfeeding has positive effects on both mother and child. Research has generally focused on the health benefits for the baby, but breastfeeding also has important implications for maternal health (Chowdhury et al., 2015). Some studies have emphasized that there may be a link between the physiological effects of breastfeeding and the etiology of Alzheimer's disease and that the relationship between breastfeeding history and Alzheimer's disease should be investigated (Fox et al., 2013). Brain development is at its maximum level in the first 2 years of life. It is certain that during this period, the baby's brain makes new connections as it communicates with the environment, and this will have positive effects on normal growth and brain development. The results and evidence obtained from the studies prove that breast milk and breastfeeding are necessary for the optimal neuropsychological, cognitive, social, and behavioral development of the baby. The mechanisms by which breastfeeding affects infant neurodevelopment are not yet fully understood. However, the potential epigenetic effects of breast milk through its various bioactive components, including non-coding RNAs, stem cells, exosomes, bioactive peptides, and microbiomes, help us to provide information about this relationship (Brown Belfort, 2017; Di et al., 2024; Hatmal et al., 2022).

The breast milk microbiota contributes to the creation of the baby's gut microbiome, which plays a role in brain development through epigenetic modifications and the regulation of key molecules (Notarbartolo et al., 2022). Our knowledge of the effects of breast milk on infant brain development is increasing each day, but the important thing is to determine whether some components of breast milk can also be beneficial in neurodegenerative diseases in adults. The beneficial effects of breastfeeding on neurological development continue to be investigated in the treatment of neurological disorders ranging from anxiety and depression in adults to brain damage in premature babies. In these studies, alternative models are being developed to provide more accurate information in a shorter time. Therefore, zebrafish come together with breast milk in this niche point.

Malyshev et al. measured the binding potential of peptides obtained from *Bos Taurus* milk to metabotropic glutamate receptor 5 (mGluR5) in the central nervous system (CNS). GluR5 is expressed at high levels in various regions of the brain (Malyshev et al., 2021). It is implicated in numerous brain-related diseases, including depression, Fragile X syndrome (FXS), Parkinson's disease, Alzheimer's disease, attention deficit hyperactivity disorder (ADHD), and addictions (Malyshev et al., 2024). Inactivation of this receptor using negative allosteric modulators ensures that they are unaffected by the high glutamate concentrations that may be present in neurological disease states (e.g., stroke, epilepsy, neuropathic pain, etc.) (Pin & Acher, 2002).

Zebrafish are a valuable screening tool for pharmacological studies because they respond to various classes of psychoactive compounds, such as diazepam, a positive allosteric modulator of γ-aminobutyric acid subtype A receptors (GABAA PAM); buspirone, a 5-HT1A receptor agonist; desipramine, a serotonin-norepinephrine reuptake inhibitor; and the selective serotonin reuptake inhibitors fluoxetine, caffeine, and ethanol (Cueto-Escobedo et al., 2022; Maximino et al., 2013). Peptides obtained from *Bos Taurus* milk have been found to exhibit a specific activity profile similar to that observed after diazepam treatment in zebrafish (Malyshev et al., 2021). It has also been reported that these milk-derived peptides affect similar motor stimulant effects of caffeine (10 mg/kg) and bupropion (30 mg/kg) on locomotion.

When milk-derived peptides were given to zebrafish in stressful situations, it was determined that their anxiety behaviors were reduced. For example, zebrafish preferred dark areas, while when milk-derived peptides were given, zebrafish spent more time in the anxiety-inducing light tank (Maximino et al., 2014).

Cheng et al. have revealed that human milk oligosaccharides (HMOs) help build the infant immune system and gut microbiome and prevent infections. It is stated that human milk–derived oligosaccharides (OM001 and OM003), which can alleviate social stress associated with autism, may act by affecting gut bacteria that communicate with the brain through gut–nervous system connections. It has been stated that these safe and shelf-stable compounds are also suitable for use in adults, especially in the treatment of complex neurogenerative diseases (Cheng et al., 2021).

Brandt et al. investigated the potential therapeutic effects of breast milk components in white-matter damage, a brain injury that causes loss of myelination in the brain and is the most common in premature infants. Because there is no cure for this damage, and it can only be diagnosed with magnetic resonance imaging, the earliest diagnosis of white-matter damage is at or after 37 weeks gestation, when the baby is approaching term (Brandt et al., 2023).

Micro and long non-coding RNAs found in breast milk–derived exosomes and breast milk–derived stem cells are not digested and can cross the blood–brain barrier into the circulation. Some non-coding RNAs potentially regulate genes related to brain development and function, while nestin-positive stem cells may differentiate into neurons and/or act as epigenetic regulators in the brain (Di et al., 2024; Mourtzi et al., 2021). Breast milk is the biological fluid with the highest exosome amount and is rich in microRNAs (miRNAs). MiRNAs are key regulators of gene expression networks in normal physiological and disease contexts. Human breast tissue has a specific miRNA expression profile, and breast milk–derived miRNAs originate primarily from the mammary gland epithelium rather than the maternal circulation (Alsaweed et al., 2015). They are transferred to the baby through breast milk and regulate various functions, including the strengthening of the immune system and the development of the baby, in a tissue-specific manner. miRNAs also perform various developmental functions, including the formation of specific organelles in neuronal and other differentiated cells (Denzler & Stoffel, 2015). miRNAs are regulated during the maturation of the CNS and have active roles in adult neural functions and in coordinating gene functions related to learning and memory. In addition to their versatile roles, miRNAs have been associated with various neurological diseases. Exosomes can transport miRNA long-distance due to their lipid bilayer membrane enclosure and adhesion molecules, thus enabling regulated endosomal transfer through intestinal epithelial cells (Melnik et al., 2021).

When the effects of the fats present in milk are examined, their vital roles in providing energy, supporting cellular structures, and facilitating the absorption of fat-soluble vitamins become clear. Milk fat is found in fat globule form in breast milk. It is one of the main components, accounting for about half of the total calories of milk (Raza et al., 2021). Additionally, certain bioactive lipids–based components in milk contribute to immune modulation and may have anti-inflammatory properties. The structure of fat globules consists of the core, triglycerides, covered by a membrane composed of three layers of phospholipids and proteins derived from the mammary cell membrane and other membrane-associated substances (Singh & Gallier, 2017).

The membrane of milk fat globules (MFGM) contains some proteins that are embedded in the phospholipid bilayer or attached to the membrane surface (Arranz & Corredig, 2017). The composition of fat globule membrane proteins varies in different mammalian species. MFGM contains important milk-derived functional components associated with neurodevelopment (Breij et al., 2019). MFGM supports the development of the nervous system and cognitive functions in infants and children (Deoni et al., 2024). MFGM proteins are beneficial in various neurological disorders in children and adults over 65 years old. Milk fat globule-EGF factor 8 protein (MFG-E8), also known as lactadherin, is a protein encoded by the MFGE8 gene in humans. MFG-E8 plays a role in the attenuation of neurodegenerative diseases such as Alzheimer's disease, stroke, and traumatic brain injury in older adults (Cheyuo et al., 2019). The main mechanisms include recognition of phosphatidylserine signaling in cells, recognition and binding to cell surface integrin αvβ5, initiation of the intracellular CrkII–DOCK180–Rac1 signaling pathway, induction of cytoskeletal rearrangement in phagocytes, participation in phagocytosis of apoptotic cells and potentially toxic substances, reduction of neuroinflammation, and maintenance of normal physiological functions in the brain (Cheyuo et al., 2019).

Studies have shown that cholesterol derivatives called oxysterols, natural components in breast milk, activate and display the Sonic Hedgehog signaling pathway. This pathway is the main regulator of vertebrate embryonic development and has functions such as homeostasis, regeneration, and maintenance of stem cells in adults. Therefore, it is thought that the Sonic Hedgehog (Shh) pathway may support the increase in the regenerative capacity of the central nervous system because it protects neural stem cells and supports oligodendrogenesis (Meda et al., 2018).

## Investigating Breast Milk and Neurodegeneration Using Zebrafish

Chemicals and drugs that pass into breast milk are transferred from the mother's bloodstream to the mammary glands and mix into the milk. The time it takes for these chemicals to pass from the blood into the milk can vary depending on the chemical structure of the drug, the mother's metabolism, and the frequency of breastfeeding. The rate at which drugs pass into breast milk is generally determined by molecular weight, lipid solubility, and ionization state. Larger molecules are less likely to pass, but fat-soluble drugs are more likely to pass (Anadón et al., 2022).

Zebrafish embryos are a viable vertebrate model organism with many advantages for high-throughput testing of these chemicals that have toxicological consequences in humans (Bauer et al., 2021). Therefore, studies on zebrafish embryos to investigate the biological effects of chemicals and drugs excreted in breast milk provide information about the possible effects of these substances (Table 21.1).

### Modeling the Effects of Harmful Chemicals in Breast Milk: Exposure Studies with Identified Toxicants

Exposure to chemicals that disrupt the endocrine system may negatively affect neuronal connectivity (Stamou, Streifel, Goines, & Lein, 2013), possibly altering GABAergic, glutamatergic, serotonergic, and dopaminergic systems via oxidative stress and causing epigenetic changes (Quaak et al., 2013). The effects of the polychlorinated biphenyls (PCBs), organochlorine pesticides (OCPs), perfluoroalkyl substances (PFASs), and polybrominated diphenyl ethers (PBDEs) that pass into breast milk were investigated using zebrafish. The risk of these chemicals causing autism was investigated in zebrafish based

**TABLE 21.1**

Some Neurotoxic Agents Excreted into Breast Milk

| Agent | Effects |
|---|---|
| Heavy metals | Exposure to high levels of heavy metals such as mercury, lead, and cadmium, which pass into breast milk, has harmful effects on the nervous system of babies. |
| Pesticides | Some pesticides used in agriculture can pass into breast milk and cause neurotoxic effects, affecting the brain development of babies. |
| Polychlorinated biphenyls | These chemicals, used in electrical equipment and industrial applications, can pass into breast milk and cause neurotoxic effects, causing developmental problems in babies. |
| Dioxins | Dioxins are industrial byproducts that pass into breast milk and can have adverse effects on the nervous system. |
| Bromine flame retardants | Flame retardants used in electronics and furniture manufacturing have the potential to pass into breast milk and cause neurotoxic effects. |
| Alcohol and nicotine | They cause adverse effects on the nervous system in babies. |
| Some medicines | Some anticonvulsants and antidepressants may pass into breast milk and cause neurotoxic effects. |

on their concentrations in breast milk. These chemicals were added to the zebrafish development medium at concentrations 10 times higher than those in breast milk. Brain development was assessed by proliferation assay and whole in situ hybridization, and effects on tactile response, locomotion, and social behavior in larvae were also measured (Desalegn et al., 2023).

Polystyrene and nanoplastics (NPs) can pass into breast milk, potentially leading to neurodegenerative disorders in infants. Researchers often use the zebrafish model to investigate these effects, as they may be related to neurodegenerative diseases. Studies have shown that polystyrene and NPs cause developmental and behavioral abnormalities in zebrafish larvae, as well as neuronal loss and motor neuron issues in transgenic larvae. Additionally, these substances induce neurodevelopmental and apoptotic genetic changes in the larvae. NPs enter the organism through ingestion, where they can accumulate and trigger various toxic responses, including neurotoxicity (Santos et al., 2024). Such research may offer valuable insights into the long-term effects of microplastics that infants are exposed to via breast milk, contributing to neuronal loss and motor neuron abnormalities in transgenic larvae (Santos et al., 2024).

A study investigated the effects of various organophosphorus flame retardants – namely, triphenyl phosphate (TPHP), isopropylated phenyl phosphate (IPP), 2-ethylhexyl diphenyl phosphate (EHDP), tert-butylated phenyl diphenyl phosphate (BPDP), trimethyl phenyl phosphate (TMPP), isodecyl diphenyl phosphate (IDDP), tris(1,3-dichloroisopropyl) phosphate (TDCIPP), and tris(2-chloroethyl) phosphate (TCEP) – as well as two brominated flame retardants (BFRs): 3,3',5,5'-tetrabromobisphenol A (TBBPA) and 2,2',4,4'-brominated diphenyl ether (BDE-47). The study focused on their developmental toxicity, neurotoxicity, cardiotoxicity, and hepatotoxicity in zebrafish embryos. The concentrations of these chemicals in breast milk were used to determine the doses given to the zebrafish. This approach serves as a valuable tool for screening the effects of a group of chemical compounds that may be linked to human exposure. The use of zebrafish models as a tool to screen for the effects of chemical compounds excreted in breast milk has yielded results that correlate with human exposure (Alzualde et al., 2018).

## Beneficial Effects of Breast Milk on Neurodegeneration

Breast milk contains biologically active substances, including β-casomorphin (BCM) peptides. Research has shown that β-casomorphins (hBCM) derived from human milk possess opioid-like properties in the brain (Brantl & Teschemacher, 1983). These bioactive peptides, which have the sequences Tyr-Pro-Phe-Val-Glu and Tyr-Pro-Phe-Val-Glu-Pro-Ile, are generated from β-casein protein through proteolytic degradation in the intestine (Teschemacher, Koch, & Brantl, 1997). Studies using zebrafish have indicated that these bioactive peptides in breast milk may play a significant role in the early development of the pancreas in infants. In one study, bioactive peptides were purchased and added to the E3 zebrafish development medium at specific concentrations, allowing researchers to observe the development of the fish from 0 to 7 days post-fertilization (dpf) (Singh et al., 2020). Research into the effects of beta-casomorphin on neurodegeneration in zebrafish is an emerging area.

Studying the processes of nerve damage and myelin regeneration in zebrafish may provide important information in understanding the neuroprotective effects of breast milk–derived components. Such studies may further elucidate the link between breast milk and multiple sclerosis and possible protective mechanisms. Breast milk contains approximately 100 times more vitamin $B_{12}$-binding protein, haptocorrin, than serum. Most of this haptocorrin is in the unsaturated apo-haptocorrin form. Haptocorrin is a glycoprotein that plays a role in binding and stabilizing vitamin $B_{12}$ (cobalamin). Therefore, a decrease or increase in the amount of haptocorrin will also affect vitamin $B_{12}$ metabolism. Vitamin $B_{12}$ deficiency in children and infants, resulting from malnutrition in breastfeeding mothers or a strict vegan diet, can affect their development, causing developmental delays and various cognitive and movement disorders (Lönnerdal, 2013). Vitamin $B_{12}$ deficiency in adults can lead to some neurological disorders, including numbness and tingling in the limbs, depression, confusion, dementia, and optic neuropathy. Additionally, Vitamin $B_{12}$ deficiency is associated with cognitive functions, and, although not fully confirmed, some researchers suggest that it may be a risk factor for Alzheimer's disease (AD) (McCaddon et al., 2002). It is also associated with many other neurological disorders, such as Wernicke encephalopathy and subacute combined degeneration of the spinal cord and peripheral neuropathy. Vitamin $B_{12}$ deficiencies, potentially linked to altered haptocorrin function, have been associated with elevated homocysteine levels, a risk factor for neurodegeneration and AD. Elevated homocysteine levels may cause oxidative stress, inflammation, and vascular damage and may exacerbate pathological hallmarks of AD, such as amyloid-beta accumulation and tau hyperphosphorylation. Studies have shown that cobalamin transport proteins (Tcn2, Tcn-beta-a (Tcnba), and Tcn-beta-b (Tcnbb)) concentrations were high (Mathew et al., 2024). This suggests that zebrafish use an alternative receptor or pathway for cobalamin uptake into cells.

Oligosaccharides in breast milk are carbohydrates that serve as bioactive factors in various physiological functions, including brain development. Although infants cannot digest these oligosaccharides, they reach the large intestine, where they promote the growth of beneficial microbiota. Additionally, they act as protective agents against toxins, pathogenic bacteria, and viruses, helping to prevent intestinal infections. These molecules are synthesized in the mammary gland, and over 200 different types of milk oligosaccharides have been identified through extensive research (Oliveira et al., 2015; Smilowitz et al., 2014).

Neutral oligosaccharides are the most abundant oligosaccharides found in human breast milk. The total sialylated oligosaccharide content in animal milk is 10–100 times lower than that of human breast milk. Sialylated milk oligosaccharides appear to play an important role in brain development because they are the main suppliers of sialic acid, an important component of basic brain proteins such as gangliosides and myelin-associated glycoprotein, which play a role in the formation of brain structures that sustain cognitive functions. In a study, sialylated oligosaccharide was administered to zebrafish embryos using a microinjection technique to increase zebrafish yolk reserves in the early stages of embryogenesis, and the effects of this procedure on mortality, locomotor behavior, and gene expression were examined. The results of the study showed that larvae given sialylated oligosaccharides were more adventurous and/or less fearful compared to control individuals, showing an improved locomotor behavior in the dark phase and an increased expression of genes involved in cell cycle control, chromosomal replication, synaptogenesis, and neuronal function (Licitra et al., 2023).

Sialyllacto-N-tetraose b (LSTb) is a type of sialylated oligosaccharide found in breast milk. There are individual differences in the concentrations and distributions of galactose and sialyllacto-N-tetraose b, and research on its effects is limited due to the low levels of sialyllacto-N-tetraose b in milk. Similar to other oligosaccharides derived from breast milk, it is not well understood whether sialyllacto-N-tetraose b affects the development of the central nervous system. To investigate this effect, a transgenic zebrafish line (Hb9:GFP), commonly used to evaluate the motor neurons of the central nervous system, was incubated with sialyllacto-N-tetraose b to gain insight into its effect on nervous system development (Li et al., 2024; Lis-Kuberka & Orczyk-Pawiłowicz, 2019). The study provided evidence that sialyllacto-N-tetraose b has a significant effect on the development and function of motor neurons in zebrafish despite its low concentration in human milk. Sialyllacto-N-tetraose b has been shown to regulate neurodevelopment by regulating Slit2 and Slit3 genes, which are critical for axonal growth and guidance. These findings contribute to our understanding of the role of breast milk–derived oligosaccharides in neurodevelopment and provide new avenues for further investigation of the potential of sialyllacto-N-tetraose b as a therapeutic agent for neurodevelopmental disorders (Li et al., 2024).

Carboxyl ester lipase is a lipolytic enzyme found in breast milk and secreted by the pancreas that hydrolyzes triglycerides, diglycerides, monoglycerides, phospholipids, cholesteryl esters, and ceramides. This enzyme aids in the digestion of milk triglycerides and cholesteryl/retinyl esters in breastfed infants (Brejchova et al., 2022). Lipids are essential components of the Wnt and Sonic Hedgehog signaling pathways. Since the yolk is the sole source of lipids required for zebrafish embryonic development, any defective lipid transport can lead to growth-retardation phenotypes and multiple embryonic abnormalities. A study found that carboxyl ester lipase knockdown disrupted brain development and the Sonic Hedgehog signaling pathway in zebrafish embryos. Morphological analysis of embryos showed that carboxyl ester lipase morphants, unlike control embryos, exhibited shrunken midbrain tectum, reduced diameter eyes (microphthalmia), and delayed pigmentation of the eyes. It was stated in the study that there is a limitation in producing morphant embryos by injecting morpholino oligos, which are synthetic molecules that are the product of redesigning the natural nucleic acid structure, into zebrafish embryos and that it would be better to use stable zebrafish knockout lines produced by CRISPR/Cas9 in future studies to further investigate the role of carboxyl ester lipase 1 and 2 genes (Qiu et al., 2020).

## Conclusion

Although there are few studies in the literature examining the impact of toxic chemical substances that can pass into breast milk and the effects of breast milk components on various systems in zebrafish, direct research linking breast milk to neurodegeneration is limited. Future studies should focus on understanding the potential new effects of bioactive molecules found in breast milk by incorporating them into zebrafish models. Additionally, investigating the long-term effects of breast milk–derived molecules and cells on neurodegenerative markers, particularly Alzheimer's and Parkinson's disease models in zebrafish, could reveal new insights and add to the known remarkable benefits of breast milk. The effects of breast milk on neurodevelopment, as described in this section, are suitable for experimental studies with zebrafish to explore treatments for neurodegenerative diseases and remain an open area of research.

## REFERENCES

Alsaweed, M., Hartmann, P. E., Geddes, D. T., & Kakulas, F. (2015). MicroRNAs in breastmilk and the lactating breast: potential immunoprotectors and developmental regulators for the infant and the mother. *International Journal of Environmental Research and Public Health, 12*(11), 13981–14020.

Alzualde, A., Behl, M., Sipes, N. S., Hsieh, J.-H., Alday, A., Tice, R. R., ... Quevedo, C. (2018). Toxicity profiling of flame retardants in zebrafish embryos using a battery of assays for developmental toxicity, neurotoxicity, cardiotoxicity and hepatotoxicity toward human relevance. *Neurotoxicology and Teratology, 70*, 40–50.

Anadón, A., Martínez-Larrañaga, M. R., Ares, I., & Martínez, M. A. (2022). Drugs and chemical contaminants in human breast milk. In *Reproductive and developmental toxicology* (pp. 1019–1052). Academic Press.

Arranz, E., & Corredig, M. (2017). Invited review: Milk phospholipid vesicles, their colloidal properties, and potential as delivery vehicles for bioactive molecules. *Journal of Dairy Science, 100*(6), 4213–4222.

Bauer, B., Mally, A., & Liedtke, D. (2021). Zebrafish embryos and larvae as alternative animal models for toxicity testing. *International Journal of Molecular Sciences, 22*(24), 13417.

Belfort, M. B., Anderson, P. J., Nowak, V. A., Lee, K. J., Molesworth, C., Thompson, D. K., ... Inder, T. E. (2016). Breast milk feeding, brain development, and neurocognitive outcomes: A 7-year longitudinal study in infants born at less than 30 weeks' gestation. *The Journal of Pediatrics, 177*, 133–139.e131. https://doi.org/10.1016/j.jpeds.2016.06.045

Brandt, M. J., Nijboer, C. H., van Elburg, R. M., Benders, M. J., & de Theije, C. G. (2023). Nutritional supplementation for the prevention and treatment of neonatal brain injury: Breast milk and beyond. In *Treatments, nutraceuticals, supplements, and herbal medicine in neurological disorders* (pp. 833–853). Academic Press. https://doi.org/10.1016/B978-0-323-90052-2.00032-9

Brantl, V., & Teschemacher, H. (1983). Opioids in milk. *Trends in Pharmacological Sciences, 4*(5), 193–193.

Breij, L. M., Abrahamse-Berkeveld, M., Vandenplas, Y., Jespers, S. N., de Mol, A. C., Khoo, P. C., ... Norbruis, O. F. (2019). An infant formula with large, milk phospholipid–coated lipid droplets containing a mixture of dairy and vegetable lipids supports adequate growth and is well tolerated in healthy, term infants. *The American Journal of Clinical Nutrition, 109*(3), 586–596.

Brejchova, K., Paluchova, V., Brezinova, M., Cajka, T., Balas, L., Durand, T., ... Kuda, O. (2022). Triacylglycerols containing branched palmitic acid ester of hydroxystearic acid (PAHSA) are present in the breast milk and hydrolyzed by carboxyl ester lipase. *Food Chemistry, 388*, 132983.

Brown Belfort, M. (2017). The science of breastfeeding and brain development. *Breastfeeding Medicine, 12*(8), 459–461. https://doi.org/10.1089/bfm.2017.0122

Cheng, L., Akkerman, R., Kong, C., Walvoort, M. T., & de Vos, P. (2021). More than sugar in the milk: human milk oligosaccharides as essential bioactive molecules in breast milk and current insight in beneficial effects. *Critical Reviews in Food Science and Nutrition, 61*(7), 1184–1200.

Cheyuo, C., Aziz, M., & Wang, P. (2019). Neurogenesis in Neurodegenerative Diseases: Role of MFG-E8. *Frontiers in Neuroscience, 13*, 569. https://doi.org/10.3389/fnins.2019.00569

Chowdhury, R., Sinha, B., Sankar, M. J., Taneja, S., Bhandari, N., Rollins, N., ... Martines, J. (2015). Breastfeeding and maternal health outcomes: a systematic review and meta-analysis. *Acta Paediatrica, 104*(467), 96–113. https://doi.org/10.1111/apa.13102

Cueto-Escobedo, J., German-Ponciano, L. J., Guillén-Ruiz, G., Soria-Fregozo, C., & Herrera-Huerta, E. V. (2022). Zebrafish as a useful tool in the research of natural products with potential anxiolytic effects. *Frontiers in Behavioral Neuroscience, 15*, 795285.

Denzler, R., & Stoffel, M. (2015). Uptake and function studies of maternal milk-derived microRNAs. *Journal of Biological Chemistry, 290*(39), 23680–23691.

Deoni, S. C., Beauchemin, J., D'Sa, V., Bonham, K., & Klepac-Ceraj, V. (2024). Enhanced Brain Myelination and Cognitive Development in Young Children Associated with Milk Fat Globule Membrane (MFGM) Intake: A Temporal Cohort Study. *Research Sq*uare, 3.rs, 4999582. https://doi.org/10.21203/rs.3.rs-4999582/v1

Desalegn, A. A., van der Ent, W., Lenters, V., Iszatt, N., Stigum, H., Lyche, J. L., ... Eggesbø, M. (2023). Perinatal exposure to potential endocrine disrupting chemicals and autism spectrum disorder: From Norwegian birth cohort to zebrafish studies. *Environment International, 181*, 108271.

Di, S.-J., Cui, X.-W., Liu, T.-J., & Shi, Y.-Y. (2024). Therapeutic potential of human breast milk-derived exosomes in necrotizing enterocolitis. *Molecular Medicine, 30*(1), 243. https://doi.org/10.1186/s10020-024-01010-7

Erick, M. (2018). Breast milk is conditionally perfect. *Medical Hypotheses, 111*, 82–89.

Fox, M., Berzuini, C., & Knapp, L. A. (2013). Maternal breastfeeding history and Alzheimer's disease risk. *Journal of Alzheimers Disease, 37*(4), 809–821. https://doi.org/10.3233/jad-130152

Hatmal, M. m. M., Al-Hatamleh, M. A. I., Olaimat, A. N., Alshaer, W., Hasan, H., Albakri, K. A., ... Mohamud, R. (2022). Immunomodulatory properties of human breast milk: MicroRNA contents and potential epigenetic effects. *Biomedicines, 10*(6), 1219. https://www.mdpi.com/2227-9059/10/6/1219

Li, P., Chen, P., Zheng, Y., Suo, G., Shen, F., Li, H., ... Wu, Y. (2024). Enhancement of motor neuron development and function in zebrafish by sialyllacto-N-tetraose b. *Translational Pediatrics, 13*(7), 1201.

Licitra, R., Naef, V., Marchese, M., Damiani, D., Ogi, A., Doccini, S., ... Santorelli, F. M. (2023). Short-term effects of human versus bovine sialylated milk oligosaccharide microinjection on zebrafish larvae survival, locomotor behavior and gene expression. *International Journal of Molecular Sciences, 24*(6), 5456.

Lis-Kuberka, J., & Orczyk-Pawiłowicz, M. (2019). Sialylated oligosaccharides and glycoconjugates of human milk. The impact on infant and newborn protection, development and well-being. *Nutrients, 11*(2), 306.

Lönnerdal, B. (2013). Bioactive proteins in breast milk. *Journal of Paediatrics and Child Health, 49*, 1–7.

Malyshev, A. V., Pavshintcev, V. V., Mitkin, N. A., Sukhanova, I. A., Gedzun, V. R., Zlobin, A. S., ... Sawyer, T. K. (2024). The novel peptide LCGM-10 attenuates metabotropic glutamate receptor 5 activity and demonstrates behavioral effects in animal models. *Frontiers in Behavioral Neuroscience, 18*, 1333258.

Malyshev, A. V., Sukhanova, I. A., Zlobin, A. S., Gedzun, V. R., Pavshintsev, V. V., Vasileva, E. V., ... Golovin, A. V. (2021). In silico screening and behavioral validation of a novel peptide, LCGA-17, with anxiolytic-like properties. *Frontiers in Neuroscience, 15*, 705590.

Mathew, A. R., Di Matteo, G., La Rosa, P., Barbati, S. A., Mannina, L., Moreno, S., ... Fidaleo, M. (2024). Vitamin B12 deficiency and the nervous system: Beyond metabolic decompensation—comparing biological models and gaining new insights into molecular and cellular mechanisms. *International Journal of Molecular Sciences, 25*(1), 590.

Maximino, C., da Silva, A. W. B., Araújo, J., Lima, M. G., Miranda, V., Puty, B., ... Oliveira, K. R. M. (2014). Fingerprinting of psychoactive drugs in zebrafish anxiety-like behaviors. *Plos One, 9*(7), e103943.

Maximino, C., Puty, B., Benzecry, R., Araújo, J., Lima, M. G., Batista, E. d. J. O., ... Herculano, A. M. (2013). Role of serotonin in zebrafish (Danio rerio) anxiety: relationship with serotonin levels and effect of buspirone, WAY 100635, SB 224289, fluoxetine and para-chlorophenylalanine (pCPA) in two behavioral models. *Neuropharmacology, 71*, 83–97.

McCaddon, A., Regland, B., Hudson, P., & Davies, G. (2002). Functional vitamin B12 deficiency and Alzheimer disease. *Neurology, 58*(9), 1395–1399.

Meda, F., Rampon, C., Dupont, E., Gauron, C., Mourton, A., Queguiner, I., ... Vriz, S. (2018). *Nerves, $H_2O_2$ and Shh: Three players in the game of regeneration.* Paper presented at the Seminars in cell & developmental biology.

Melnik, B. C., Stremmel, W., Weiskirchen, R., John, S. M., & Schmitz, G. (2021). Exosome-derived microRNAs of human milk and their effects on infant health and development. *Biomolecules, 11*(6), 851.

Mourtzi, N., Siahanidou, T., Tsifintaris, M., Karamichali, E., Tasiopoulou, A., Sertedaki, A., ... Giannakakis, A. (2021). lncRNA NORAD is consistently detected in breastmilk exosomes and its expression is downregulated in mothers of preterm infants. *International Journal of Molecular Medicine, 48*(6). https://doi.org/10.3892/ijmm.2021.5049

Newton, E. R. (2004). Breastmilk: The gold standard. *Clinical Obstetrics and Gynecology, 47*(3), 632–642.

Notarbartolo, V., Giuffrè, M., Montante, C., Corsello, G., & Carta, M. (2022). Composition of Human Breast Milk Microbiota and Its Role in Children's Health. *Pediatric Gastroenterology Hepatology and Nutrition, 25*(3), 194–210. https://doi.org/10.5223/pghn.2022.25.3.194.

Oliveira, D. L., Wilbey, R. A., Grandison, A. S., & Roseiro, L. B. (2015). Milk oligosaccharides: A review. *International Journal of Dairy Technology, 68*(3), 305–321.

Picó, C., Palou, M., Pomar, C. A., & Palou, A. (2021). Benefits of breastfeeding in infant health: A role for milk signaling peptides. In *Molecular nutrition: Mother and infant* (pp. 29–56).

Pin, J.-P., & Acher, F. (2002). The metabotropic glutamate receptors: Structure, activation mechanism and pharmacology. *Current Drug Targets-CNS & Neurological Disorders, 1*(3), 297–317.

Qiu, Y., Sun, S., Yu, X., Zhou, J., Cai, W., & Qian, L. (2020). Carboxyl ester lipase is highly conserved in utilizing maternal supplied lipids during early development of zebrafish and human. *Biochimica et Biophysica Acta (BBA)-Molecular and Cell Biology of Lipids, 1865*(6), 158663.

Quaak, I., Brouns, M. R., & de Bor, M. V. (2013). The dynamics of autism spectrum disorders: how neurotoxic compounds and neurotransmitters interact. *International Journal of Environmental Research and Public Health, 10*(8), 3384–3408.

Raza, G. S., Herzig, K.-H., & Leppäluoto, J. (2021). Invited review: Milk fat globule membrane—A possible panacea for neurodevelopment, infections, cardiometabolic diseases, and frailty. *Journal of Dairy Science, 104*(7), 7345–7363.

Santos, A. L., Rodrigues, L. C., Rodrigues, C. C., Cirqueira, F., Malafaia, G., & Rocha, T. L. (2024). Polystyrene nanoplastics induce developmental impairments and vasotoxicity in zebrafish (Danio rerio). *Journal of Hazardous Materials, 464*, 132880.

Scientific Concepts of Functional Foods in Europe Consensus Document. (1999). *British Journal of Nutrition, 81*(4), S1–S27. https://doi.org/10.1017/S0007114599000471

Singh, A., Enjapoori, A. K., Gibert, Y., & Dwyer, K. M. (2020). The protective effects of human milk-derived peptides on the pancreatic islet biology. *Biology open, 9*(8), bio049304.

Singh, H., & Gallier, S. (2017). Nature's complex emulsion: The fat globules of milk. *Food Hydrocolloids, 68*, 81–89.

Smilowitz, J. T., Lebrilla, C. B., Mills, D. A., German, J. B., & Freeman, S. L. (2014). Breast milk oligosaccharides: structure-function relationships in the neonate. *Annual Review of Nutrition, 34*(1), 143–169.

Stamou, M., Streifel, K. M., Goines, P. E., & Lein, P. J. (2013). Neuronal connectivity as a convergent target of gene× environment interactions that confer risk for Autism Spectrum Disorders. *Neurotoxicology and Teratology, 36*, 3–16.

Teschemacher, H., Koch, G., & Brantl, V. (1997). Milk protein-derived opioid receptor ligands. *Peptide Science, 43*(2), 99–117.

# 22

# *Antimicrobial Agents in Mouthwashes: Unveiling Neurotoxic and Neurodegenerative Impacts through Zebrafish Models*

**Atakan Karagöz and Ebru Emekli-Alturfan**

## Abbreviations

| | |
|---|---|
| **AChE** | Acetylcholinesterase |
| **CELF2** | CUGBP Elav-Like Family Member 2 |
| **CHG** | Chlorhexidine Digluconate |
| **CNS** | Central Nervous System |
| **$H_2O_2$** | Hydrogen Peroxide |
| ***mbpa*** | Myelin Basic Protein a |
| **QAC** | Quaternary Ammonium Compound |
| ***plp1b*** | Proteolipid Protein 1b |
| **ROS** | Reactive Oxygen Species |
| ***sema7a*** | Semaphorin 7a |
| ***scrt2*** | Scratch Family Transcriptional Repressor 2 |

## Introduction

The zebrafish has emerged as a powerful model organism for studying human disease processes, including neurodegenerative disorders (Xi et al., 2011). They share many genetic and biological similarities with humans, allowing researchers to investigate conditions like Alzheimer's, Parkinson's, and Huntington's diseases (Chia et al., 2022). Additionally, zebrafish have practical advantages, such as fast growth, transparent bodies, and high reproduction rates, which aid the study of disease processes (Hsu et al., 2007).

Recent evidence suggests that exposure to certain antimicrobial agents found in common consumer products, such as mouthwashes, may have detrimental neurological effects. This review examines the use of the zebrafish model to investigate the neurotoxic and neurodegenerative impacts of antimicrobial agents, with a focus on substances like triclosan, chlorhexidine digluconate, cetylpyridinium chloride, and hydrogen peroxide.

## Zebrafish as a Model for Developmental Neurotoxicity Assessment

The zebrafish has been widely recognized as a valuable model organism for studying developmental biology and neuroscience. Several key characteristics make zebrafish an attractive complementary model for assessing developmental neurotoxicity, including their high genetic homology to humans, rapid development, and transparency during early life stages (Lee & Freeman, 2014a).

Zebrafish share a high degree of genetic similarity with humans, with over 70% of human genes having a zebrafish ortholog. This genetic conservation extends to genes associated with neurodegenerative diseases, such as those linked to Alzheimer's and Parkinson's diseases. The rapid development of zebrafish, with the entire nervous system forming within the first 5 days of embryogenesis, allows for efficient screening and monitoring of neurodevelopmental processes and potential disruptions (Kalueff et al., 2014; Lee & Freeman, 2014a).

Compared to other vertebrate models, the zebrafish model provides a number of advantages. These consist of outward development, translucent embryos, and high fertility. It is simple to visualize and manipulate zebrafish in experiments (Xi et al., 2011) These unique features make zebrafish an ideal model for investigating the developmental neurotoxic effects of environmental exposures, including those from common consumer products containing antimicrobial agents.

## Antimicrobial Agents and Neurodegenerative Effects in Zebrafish

### Triclosan

Triclosan is a widely used antimicrobial agent found in various consumer products, including toothpaste and mouthwashes (Dann & Hontela, 2010). Due to its lipophilic nature, triclosan readily accumulates in the brain, liver, and other fat-rich tissues, potentially leading to adverse effects on various organs (Wang et al., 2020; Wang et al., 2021). Despite the US Food and Drug Administration's ban on triclosan in certain consumer products between 2016 and 2019, the substance remains incorporated into a variety of household and workplace items sold in Canada and Asian countries (Wang et al., 2019).

Studies have found the presence of triclosan in the central nervous systems of humans and other organisms, indicating that it may contribute to neurotoxicity (Pullaguri et al., 2022). Specifically, zebrafish exposed to triclosan have been observed to exhibit impaired neuronal development, characterized by disruptions in axon formation, reduced synapse density, and

DOI: 10.1201/9781032676555-24

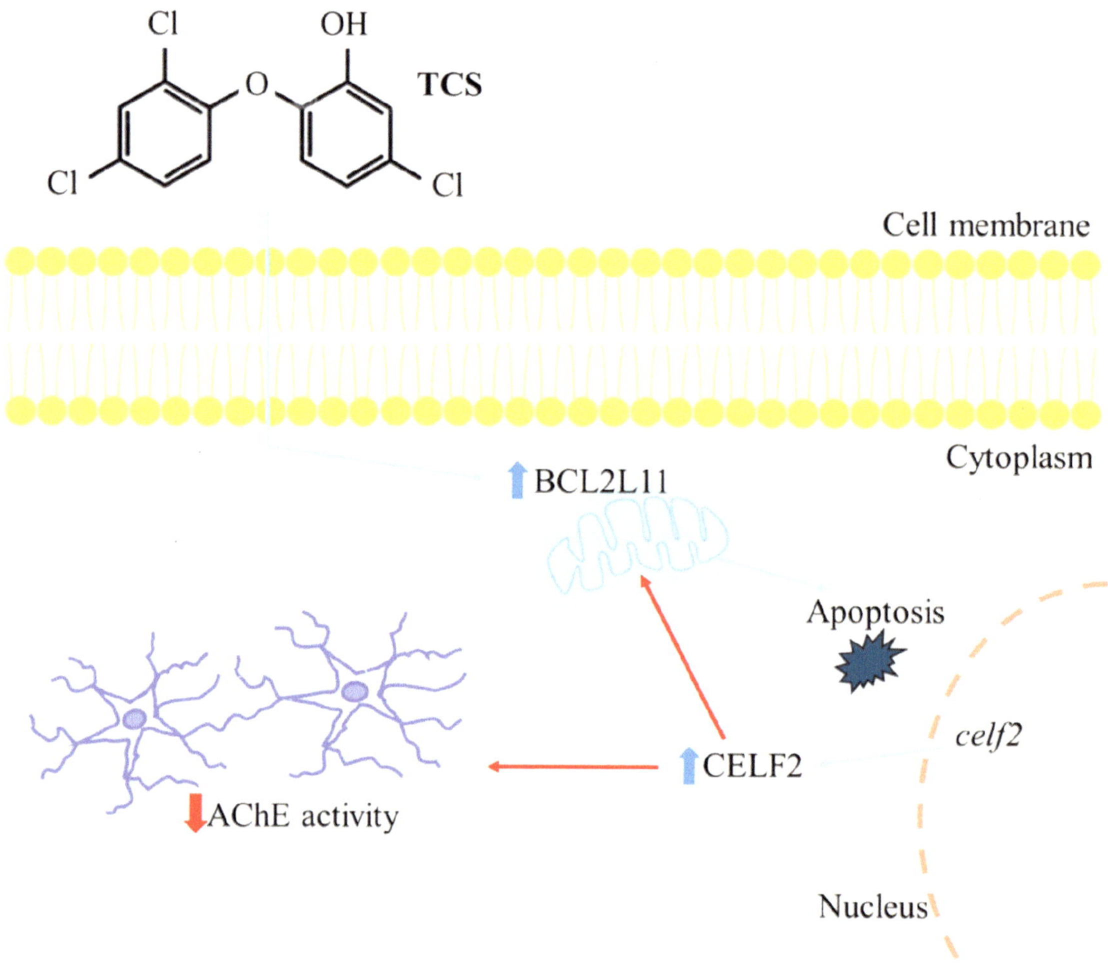


**FIGURE 22.1** Effect of TCS-induced neurotoxicity on zebrafish neuronal system.

structural abnormalities in neuronal processes (Kim et al., 2018). These developmental neurotoxic impacts are accompanied by altered gene expression patterns, including upregulation of genes like alkbh5, miR-137, and CELF2, which play crucial roles in neurogenesis, neuronal connectivity, and myelination (Figure 22.1) (Liu et al., 2018; Wang et al., 2023; Kim et al., 2018).

Furthermore, triclosan exposure has been linked to neurobehavioral changes, such as disruptions in the gut–brain axis, including gut microbiota dysbiosis and pathological damage to both the brain and intestine. These findings collectively suggest that triclosan, a commonly used antimicrobial agent, has the potential to exert detrimental effects on the developing nervous system and overall neurological function (Kalueff et al., 2014; Lee & Freeman, 2014b).

Triclosan's influence on alternative splicing of nerve-related genes through the RNA-binding protein CELF2, further contributing to neurotoxicity, has been demonstrated. As the concentration of triclosan increased, a significant rise in the number of apoptotic (programmed cell death) cells was observed in the head region. Elevated triclosan concentrations were associated with a reduction in primitive cells within the neural crest, as well as a loosening of their spatial arrangement (Wang et al., 2021).

Triclosan exposure was found to induce developmental neurotoxicity, particularly by impacting the maturation and differentiation of oligodendrocyte precursor cells. High-throughput sequencing revealed that miR-219, a key regulator of oligodendrocyte precursor cell differentiation, was dysregulated by triclosan exposure, and its depletion led to phenotypes like those observed with triclosan exposure (Ling et al., 2020).

Triclosan exposure disrupted normal neurodevelopment by triggering apoptotic pathways in neuronal cells. Furthermore, triclosan exposure significantly activated acetylcholinesterase (AChE) activity, while dopamine levels were moderately suppressed. These findings suggest that triclosan exposure resulted in abnormal alterations in marker genes associated with neurodevelopment and apoptosis, as well as dysregulation of key neurotransmitters. TrkB signaling pathway was also found to be affected by triclosan exposure. TrkB inhibitors and agonists had a rescue effect on the neurotoxicity caused by triclosan exposure. It has been suggested that triclosan exposure may induce neurotoxicity by inhibiting the BDNF/TrkB signaling pathway (Wang et al., 2023).

Exposure to 0.6 mg/L of triclosan during zebrafish embryonic development resulted in impaired axonal growth, decreased synaptic density, and a loss of axonal continuity. Additionally, apoptotic neurons and vacuoles were observed in

**TABLE 22.1**

Examples of Neurological Consequences Associated with Antimicrobials Utilized in Mouthwashes

| Classes of Drugs | Representative Drugs | Neurotoxic Effects | Reference |
|---|---|---|---|
| **Antibacterials** | | | |
| Phenols | Triclosan | Showed impaired neuronal development, marked by disruptions in axon formation, diminished synapse density, and structural abnormalities in neuronal processes, alongside the upregulation of genes such as alkbh5, miR-137, and CELF2, which are essential for neurogenesis, neuronal connectivity, and myelination. | Liu et al., 2018; Wang et al., 2023; Kim et al., 2018 |
| Bisbiguanides | Chlorhexidine Digluconate | Destroys the cell wall and membrane structure, resulting in the depletion of lipid constituents; disturbs redox equilibrium via free radicals and induces cytotoxicity by infiltrating periradicular tissue; induces abnormalities in oligodendrocytes within the CNS, impeding myelin synthesis and eliciting structural modifications in myelin. | Jesus et al., 2013; Choi et al., 2022 |
| Quaternary ammonium salts/ Cationic surfactants | Cetylpyridinium Chloride | Alter neurotransmitter levels in the zebrafish brain; induce behavioral and developmental adverse effects in zebrafish, associated with oxidative stress. | Dong et al., 2022; Qiu et al., 2022a |
| **Others** | | | |
| Topical anti-infectives | Hydrogen Peroxide | Several physical abnormalities (e.g., behavioral alterations); initiates pro-apoptotic pathways within the neurological system, resulting in neurodegeneration; linked to oxidative stress, which disrupts critical cellular processes, including apoptosis, antioxidant defenses, and estrogen synthesis; investigated its potential effects for cerebral and neural connections. | Saputra et al., 2024; Rieger & Sagasti, 2011; |

the brain and spinal cord, indicating that triclosan exposure led to developmental abnormalities in the central nervous system. Utilization of HuC-GFP and Olig2-dsRED transgenic zebrafish has been suggested to investigate the impact of triclosan on the development of sensory and motor neurons within CNS morphology. High amounts of triclosan exposure decreased axon length in both transgenic zebrafish types, resulting in a loss of axonal continuity. This indicates that triclosan exposure may disrupt several phases or cell types (Kim et al., 2018).

Triclosan exposure during zebrafish embryonic development resulted in impaired maturation and connectivity of motor neurons. Specifically, the antimicrobial agent induced a delay in the formation and differentiation of secondary motor neurons in the developing zebrafish embryos (Muth-Köhne et al., 2012) (Table 22.1).

Triclosan exposure in zebrafish induced neurotoxicity by upregulating miR-137, which inhibits neurodevelopmental target genes and metabolic processes. This upregulation influences phenotypic deformities, neuromast maturation, and behavioral assessment. MiR-137 is one of the most significantly associated genes with schizophrenia and has been connected to spinal cord injury and neurological disorders such as schizophrenia, post-stroke Parkinson's disease, and Alzheimer's disease (Liu et al., 2018).

Triclosan affects the alternative splicing of genes related to nerve function, impacting nerve cell migration, axon guidance, and myelination. Specifically, triclosan increased the expression of the RNA-binding protein CELF2, which plays a crucial role in these alternative splicing events (Wang et al., 2021).

Chronic triclosan exposure leads to observable neurobehavioral abnormalities in zebrafish, as indicated by (Wang et al., 2022). Triclosan triggered aberrant development of nervous system, caused neurobehavioral abnormalities including hypoactivity in locomotor behavior, light–dark arrhythmia, and turning behaviors (Wang et al., 2021).

Triclosan exposure was also found to alter swimming behavior, impair motility, and precipitate autism and depression in adult zebrafish. The research additionally examined the atypical regulation of marker genes inside the gut–brain axis. RT-qPCR was utilized to analyze the mRNA expression of BDNF/NTRK2 signaling in the brain and the gut of zebrafish (Wang et al., 2022).

Gyimah et al. (2020) suggested that triclosan-induced excess reactive oxygen species could lead to reduced activity of the antioxidant enzyme superoxide dismutase, which in turn may result in inhibition of AChE activity in the brain. This indicates that the brain AChE may be a potential target for triclosan's neurotoxic effects. The inhibition in superoxide dismutase and catalase activities indicated that triclosan could alter the antioxidant system of the brain of zebrafish.

Triclosan caused significantly enriched immune system development processes, including immune system process, innate immune response, and immune response of zebrafish, and downregulated the expressions of Cxcl12, Tnfsf10, and Csf1ra in the brain (Wang et al., 2019).

## The Dual Nature of Hydrogen Peroxide: Friend or Foe?

Hydrogen peroxide ($H_2O_2$) is effective as a mouthwash to minimize plaque and gingival bleeding, exhibiting no adverse effects. Some studies indicate that concentrations exceeding 5% may result in soft tissue damage; therefore, lower concentrations ranging from 1% to 3% are typically used in mouthwashes for their antibacterial properties (Brookes, McGrath and McCullough, 2023). The concentration of hydrogen peroxide in natural aquatic environments varies between 0.004 and 199 micro molars. When high concentrations are reached, it may cause some physical abnormalities (e.g. behavioral changes) that pose a potential threat to aquatic creatures (Saputra et al., 2024).

The connection between reactive oxygen species (ROS) and the progression towards oxidative damage, ultimately leading to neurodegeneration, is well established. Hydrogen peroxide, a prominent ROS molecule, is metabolized through its interaction with catalase. Although the concentration of hydrogen peroxide in mouthwashes is low enough to avoid toxicity and primarily facilitates disinfection through local metabolism, studies on exogenous hydrogen peroxide exposure have explored its potential implications for brain and neural connections (Engur et al., 2023).

Hydrogen peroxide activates pro-apoptotic pathways in the neurological system, leading to neurodegeneration (Rieger and Sagasti, 2011). Research demonstrates that hydrogen peroxide exposure can significantly impair visual development in zebrafish. Exposure to concentrations of 0.1 mM or higher has been shown to disrupt normal eye growth and development, as indicated by alterations in the eye-to-body length ratio. Additionally, hydrogen peroxide exposure increases retinal apoptosis, a process of programmed cell death essential for development but detrimental when occurring excessively, as it can lead to vision impairment. The optomotor response, a reflexive behavior enabling zebrafish to maintain their position in a moving environment, was also significantly impaired, suggesting that the visual processing pathways in the brain are adversely affected. These developmental disruptions are attributed to oxidative stress, which interferes with critical cellular processes, including apoptosis, antioxidant defenses, and estrogen production (Saputra et al., 2024) (Table 22.1).

Hydrogen peroxide has been shown to play a critical role in peripheral nerve regeneration and developmental processes in zebrafish. Following injuries such as tail fin amputation or keratinocyte ablation, hydrogen peroxide production is localized near the injury site, promoting the regeneration of damaged sensory axons. Exogenous application of hydrogen peroxide further stimulates axon growth without causing additional damage to surrounding cells, while inhibiting its production hinders the regenerative process. These findings suggest that hydrogen peroxide signaling is essential for coordinating wound healing with the reinnervation of sensory neurons in the skin (Rieger & Sagasti, 2011).

In addition to its role in regeneration, hydrogen peroxide is crucial for axon pathfinding during zebrafish brain development. Its levels are dynamic during early developmental stages, peaking during somitogenesis and organogenesis before declining in mature tissues. This spatio-temporal distribution is regulated primarily through enzymatic degradation. Hydrogen peroxide is enriched in specific regions of the developing brain, and its depletion has been shown to impair retinal ganglion cell axonal projections. These defects can be rescued by either the exogenous addition of hydrogen peroxide or by ectopically activating the Hedgehog signaling pathway, suggesting a critical interplay between hydrogen peroxide and Hedgehog signaling in guiding axon growth (Gauron et al., 2016).

Overall, research in zebrafish highlights the multifaceted roles of hydrogen peroxide in developmental processes, particularly within the visual and nervous systems.

## Cetylpyridinium Chloride

Cetylpyridinium chloride, a quaternary ammonium compound (QAC), is utilized in mouthwashes at concentrations ranging from 0.045% to 0.1%, exhibiting low toxicity to humans (Brookes et al., 2023). Cetylpyridinium chloride is commonly encountered as a surfactant in aquatic systems; however, information regarding its toxic effects remains limited. Its presence in aquatic environments raises concerns about its potential adverse effects on aquatic organisms. Zebrafish, a popular model organism in toxicological studies, has been used to investigate the impacts of cetylpyridinium chloride exposure. In general, the impact of surfactants on aquatic organisms is primarily observed in fundamental aspects such as swimming behavior, growth, and development. Studies conducted on zebrafish at varying doses further confirm these effects of cetylpyridinium chloride (Qiu et al., 2022a).

The effects of cetylpyridinium chloride on locomotor activity and social interaction have also been investigated in both juvenile and adult zebrafish, revealing that juveniles are more sensitive to cetylpyridinium chloride exposure than adults (Figure 22.2). This heightened sensitivity in younger zebrafish underscores the potential for developmental neurotoxicity (Dong et al., 2022). Time-series responses in the behavior of male zebrafish exposed to short-term cetylpyridinium chloride exposure provide a more nuanced understanding of the temporal dynamics of cetylpyridinium chloride's behavioral effects (Liu et al., 2022) (Table 22.1).

Cetylpyridinium chloride exposure can disrupt neurotransmitter levels in the zebrafish brain. It has been found to differentially affect neurotransmitter levels in juvenile and adult zebrafish, suggesting that the specific neurochemical changes induced by cetylpyridinium chloride may vary depending on the developmental stage. This differential effect could contribute to the observed life-stage-specific sensitivities to cetylpyridinium chloride exposure. Further research is needed to fully characterize the specific neurotransmitter systems affected by cetylpyridinium chloride and the functional consequences of these changes (Dong et al., 2022).

On the other hand, recent studies indicate that sublethal exposure to cetylpyridinium chloride can negatively impact the antioxidant enzymes in aquatic worms and common carp (Bhattacharya et al., 2022). Exposure to cetylpyridinium chloride, either alone or in combination with other substances, has been shown to induce behavioral and developmental toxic effects in zebrafish, which have been linked to oxidative stress (Qiu et al., 2022a, 2022b). The long-term impacts of oxidative stress are known to contribute to the development of neurodegenerative diseases. Oxidative stress occurs through the accumulation of ROS, leading to attacks on macromolecules and metabolic damage within the organism. This process sets the stage for the progression of neurodegenerative diseases through various mechanisms, such as neuronal cell death pathways (Kabir et al., 2022) (Table 22.1).

## Chlorhexidine Digluconate

Chlorhexidine digluconate, a bis-biguanide compound, is a widely utilized antimicrobial agent incorporated into various

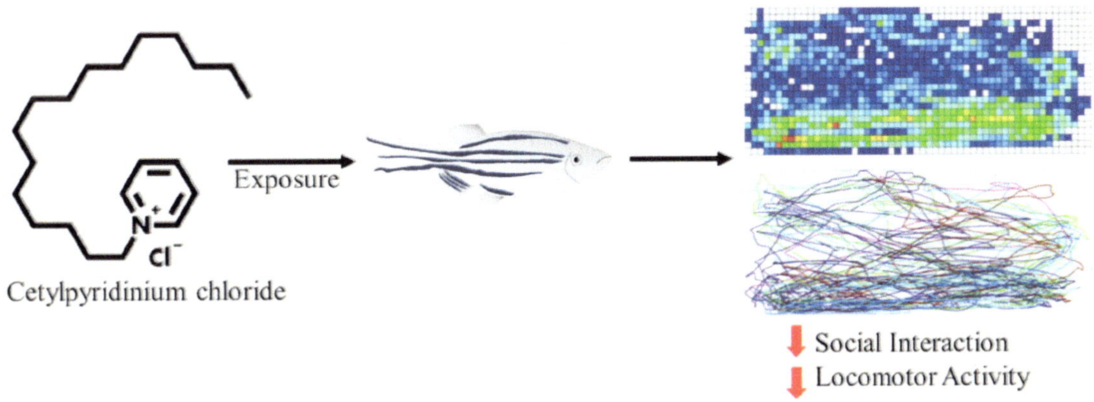


**FIGURE 22.2** Effect of cetylpyridinium chloride on behavior of zebrafish.

personal hygiene goods, particularly mouthwashes and disinfectants. This agent exhibits broad-spectrum antimicrobial efficacy against both Gram-positive and Gram-negative bacteria, along with certain fungi and viruses (Brookes et al., 2023). Chlorhexidine digluconate disrupts the cell wall and membrane structure, leading to the depletion of lipid components (Jesus et al., 2013).

Currently, limited research has been conducted on the potential neurodegenerative effects of chlorhexidine digluconate. Chlorhexidine is a substance whose toxic effects should not be overlooked, as it disrupts redox balance through free radicals and causes cytotoxicity by spreading into periradicular tissue (Valan et al., 2024). Excessive production of cytotoxicity and ROS, in turn, can be associated with neurodegenerative diseases such as Alzheimer's and Parkinson's (Chong et al., 2013).

Hexachlorophene, a phenol derivative that is comparable to chlorhexidine digluconate and was removed in the 1970s due to occurrences of vacuolar encephalopathy in children, notably preterm infants, raises questions regarding chlorhexidine digluconate–related neurodevelopmental toxicity. In both the central and peripheral nervous systems, hexachlorophene causes myelin vacuolation, which is equivalent to intramyelinic edema of the white matter. At different ages, the distribution of the reticular formation's vacuolation changes, indicating that neuronal development determines myelin's susceptibility (Shuman et al., 1975; Kimbrough, 1973).

In a study that examines the impact of chlorhexidine on myelin in zebrafish, the researchers discovered that exposure to chlorhexidine digluconate during early embryonic development causes abnormalities in oligodendrocytes within the central nervous system (CNS). Chlorhexidine digluconate exposure decreased the quantity of dorsally migrating Olig2+ cells and MBP+ cells in the spinal cord. The myelin sheath of dorsally situated oligodendrocytes in the spinal cord was examined, uncovering uncompacted, aberrant myelin structures in chlorhexidine digluconate–exposed nerves. The neurobiological locomotor response was assessed, revealing diminished locomotion alongside heightened total distance traveled and mean velocity in chlorhexidine digluconate–treated larvae.

Chlorhexidine digluconate exposure dramatically downregulated the expression of myelin basic protein a (mbpa) and proteolipid protein 1b (plp1b), which encode key myelin proteins. The scratch family transcriptional repressor 2 (scrt2) gene was downregulated in larvae exposed to chlorhexidine digluconate. Chlorhexidine digluconate exposure downregulated the expression of semaphorin 7a (sema7a), which has a role in oligodendrocyte guidance and neuroinflammation, albeit the change was not statistically significant. In summary, chlorhexidine digluconate exposure can cause abnormalities in oligodendrocytes inside the CNS, hinder myelin production, and provoke structural alterations in myelin (Choi et al., 2022).

The cytotoxic effects of chlorhexidine digluconate, such as apoptosis, have been observed in different cell types, suggesting that the observed induction of ChE activity in zebrafish embryos is linked to the cytotoxic effects of chlorhexidine digluconate (Jesus et al., 2013) (Table 22.1).

## Conclusion

Neurology has proven to be an invaluable model organism for investigating the complex biological effects of environmental chemicals, offering insights into their behavioral, neurological, and developmental impacts. The research reviewed here highlights the utility of the zebrafish system in elucidating the mechanisms underlying xenobiotic toxicity, particularly during critical stages of development. Notably, studies on the antimicrobial agent triclosan have revealed its potential to disrupt fundamental neural processes, leading to impaired neuronal connectivity, altered gene expression, and neurobehavioral changes. Furthermore, the association between triclosan-induced gut dysbiosis and neurotoxicity emphasizes the pivotal role of the gut–brain axis in mediating systemic responses to environmental exposures. Building on these findings, future research utilizing zebrafish models will be essential for advancing our understanding of developmental neurotoxicity and informing strategies to mitigate the detrimental effects of environmental pollutants on both ecosystem and human health.

## REFERENCES

Bhattacharya, R., Chatterjee, A., Chatterjee, S., Saha, N.C. (2022). Commonly used surfactants sodium dodecyl sulphate, cetylpyridinium chloride and sodium laureth sulphate and their effects on antioxidant defence system and oxidative stress indices in Cyprinus carpio L.: An integrated in silico and in vivo approach. *Environ Sci Pollut Res, 29*(20), 30622–30637.

Brookes, Z. L. S., McGrath, C., & McCullough, M. (2023). Antimicrobial mouthwashes: An overview of mechanisms—what do we still need to know?. *Int Dent J, 73 Suppl 2*(Suppl 2), S64–S68.

Chia, K., Klingseisen, A., Sieger, D., & Priller, J. (2022). Zebrafish as a model organism for neurodegenerative disease. *Front Mol Neurosci, 13*(15), 940484. https://doi.org/10.3389/fnmol.2022.940484.

Choi, E. K., Choi, B. M., Cho, Y., & Kim, S. (2022). Myelin toxicity of chlorhexidine in zebrafish larvae. *Pediatr Res, 93*(4), 845–851.

Chong, C. M., Zhou, Z. Y., Razmovski-Naumovski, V., Cui, G. Z., Zhang, L. Q., Sa, F., Hoi, P. M., Chan, K., & Lee, S. M. (2013). Danshensu protects against 6-hydroxydopamine-induced damage of PC12 cells in vitro and dopaminergic neurons in zebrafish. *Neurosci Lett, 543*, 121–125.

Dann, A. B., & Hontela, A. (2010). Triclosan: Environmental exposure, toxicity and mechanisms of action. *J Appl Toxicol, 31*(4), 285–311.

Dong, K., Li, L., Chen, C., Tengbe, M. S., Chen, K., Shi, Y., Wu, X., & Qiu, X. (2022). Impacts of cetylpyridinium chloride on the behavior and brain neurotransmitter levels of juvenile and adult zebrafish (Danio rerio). *Comp Biochem Physiol C Toxicol Pharmacol, 259*, 109393.

Engur, D., Ercan, I., Kiser, C., Tufekci, K. U., Soy, S., Micili, S. C., Ozhan, G., Guven, S., Kumral, A., & Genc, S. (2023). Hydrogen peroxide signaling modulates neuronal differentiation via microglial polarization and Wnt/beta-catenin pathway. *Eur Rev Med Pharmacol Sci, 27*(11) 5083–5096.

Gauron, C., Meda, F., Dupont, E., Albadri, S., Quenech'Du, N., Ipendey, E., Volovitch, M., Bene, F. D., Joliot, A., Rampon, C., & Vriz, S. (2016). Hydrogen peroxide (H2O2) controls axon pathfinding during zebrafish development. *Dev Biol, 414*(2), 133–141.

Gyimah, E., Dong, X., Qiu, W., Zhang, Z., & Xu, H. (2020). Sublethal concentrations of triclosan elicited oxidative stress, DNA damage, and histological alterations in the liver and brain of adult zebrafish. *Environ Sci Pollut Res Int, 27*(14), 17329–17338. https://doi.org/10.1007/s11356-020-08232-2.

Hsu, C.H., Wen, Z.H., Lin, C.S., Chakraborty, C. (2007). The zebrafish model: Use in studying cellular mechanisms for a spectrum of clinical disease entities. *Curr Neurovasc Res, 4*(2), 111–120. https://doi.org/10.2174/156720207780637234.

Jesus, F., Oliveira, R., Silva, A., Catarino, A. L., Soares, A. M. V. M., Nogueira, A. J. A., & Domingues, I. (2013). Lethal and sub lethal effects of the biocide chlorhexidine on aquatic organisms. *Ecotoxicology, 22*(9), 1348–1358. https://doi.org/10.1007/s10646-013-1121-6.

Kabir, M. T., Rahman, M. H., Shah, M., Jamiruddin, M. R., Basak, D., Al-Harrasi, A., Bhatia, S., Ashraf, G. M., Najda, A., El-Kott, A. F., Mohamed, H. R. H., Al-Malky, H. S., Germoush, M. O., Altyar, A. E., Alwafai, E. B., Ghaboura, N., & Abdel-Daim, M. M. (2022). Therapeutic promise of carotenoids as antioxidants and anti-inflammatory agents in neurodegenerative disorders. *Biomed Pharmacother, 146*, 112610.

Kalueff, A. V., Stewart, A., & Gerlai, R. (2014). Zebrafish as an emerging model for studying complex brain disorders. *Trends Pharmacol Sci, 35*(2), 63–75. https://doi.org/10.1016/j.tips.2013.12.002.

Kim, J., Oh, H., Ryu, B., Kim, U., Lee, J. M., Jung, C., Kim, C., & Park, J. (2018). Triclosan affects axon formation in the neural development stages of zebrafish embryos (Danio rerio). *Environ Pollut, 236*, 304–312.

Kimbrough, R. D. (1973). Review of the toxicity of hexachlorophene, including its neurotoxicity. *J Clin Pharmacol, 13*(11), 439–444.

Lee, J., & Freeman, J. L. (2014a). Zebrafish as a model for developmental neurotoxicity assessment: The application of the zebrafish in defining the effects of arsenic, methylmercury, or lead on early neurodevelopment. *Toxics, 2*(3), 464–495. https://doi.org/10.3390/toxics2030464.

Lee, J., & Freeman, J. L. (2014b). Zebrafish as a model for investigating developmental lead (Pb) neurotoxicity as a risk factor in adult neurodegenerative disease: a mini-review. *Neurotoxicology, 43*, 57–64. https://doi.org/10.1016/j.neuro.2014.03.008.

Ling, Y., Sun, L., Wang, D., Jiang, J., Sun, W., Ai, W., Wang, X., & Wang, H. (2020). Triclosan induces zebrafish neurotoxicity by abnormal expression of miR-219 targeting oligodendrocyte differentiation of central nervous system. *Arch Toxicol, 94*(3), 857–871.

Liu, J., Xiang, C., Huang, W., Mei, J., Sun, L., Ling, Y., Wang, C., Wang, X., Dahlgren, R. A., & Wang, H. (2018). Neurotoxicological effects induced by up-regulation of miR-137 following triclosan exposure to zebrafish (Danio rerio). *Aquat Toxicol, 206*, 176–185.

Liu, L., Chen, C., Chen, K., Shi, Y., Qiu, X., Shimasaki, Y., & Oshima, Y. (2022). Time–series responses in behaviors of male zebrafish to short–term cetylpyridinium chloride exposure. *Journal of the Faculty of Agriculture Kyushu University, 67*(2), 189–195.

Muth-Köhne, E., Wichmann, A., Delov, V., & Fenske, M. (2012). The classification of motor neuron defects in the zebrafish embryo toxicity test (ZFET) as an animal alternative approach to assess developmental neurotoxicity. *Neurotoxicol Teratol, 34*(4), 413–424. https://doi.org/10.1016/j.ntt.2012.04.006.

Pullaguri, N., Umale, A., & Bhargava, A. (2022). Neurotoxic mechanisms of triclosan: The antimicrobial agent emerging as a toxicant. *J Biochem Mol Toxicol, 37*(2), e23244. https://doi.org/10.1002/jbt.23244.

Qiu, X., Liu, L., Xu, W., Chen, C., Li, M., Shi, Y., Wu, X., Chen, K., Wang, C. (2022b). Zeolitic imidazolate framework-8 nanoparticles exhibit more severe toxicity to the embryo/larvae of zebrafish (danio rerio) when co-exposed with cetylpyridinium chloride. *Antioxidants, 11*(5), 945. https://doi.org/10.3390/antiox11050945.

Qiu, X., Tengbe, M. S., Xia, X., Dong, K., Chen, C., Shi, Y., Li, M., Xu, H., Wu, X., & Chen, K. (2022a). Impacts of cetylpyridinium chloride on the survival, development, behavior, and oxidative stress of early-life-stage zebrafish (Danio rerio). *Antioxidants (Basel), 11*(4), 676.

Rieger, S., & Sagasti, A. (2011). Hydrogen peroxide promotes injury-induced peripheral sensory axon regeneration in the zebrafish skin. *PLoS Biol, 9*(5), e1000621.

Saputra, F., Kishida, M., & Hu, S. (2024). Oxidative stress induced by hydrogen peroxide disrupts zebrafish visual development by altering apoptosis, antioxidant and estrogen related genes. *Sci Rep*, *14*(1), 22020.

Shuman, R. M., Leech, R. W. & Alvord, E. C. Jr (1975). Neurotoxicity of topically applied hexachlorophene in the young rat. *Arch Neurol*, *32*(5), 315–319.

Valan, A., Krithikadatta, J., Guru, A. (2024). Exploring periostracum as an alternative root canal irrigant: Insights from zebrafish embryo experiments. *Cureus*, *16*(3), e56638. https://doi.org/10.7759/cureus.56638.

Wang, C., Huang, W., Lin, J., Fang, F., Wang, X., & Wang, H. (2020). Triclosan-induced liver and brain injury in zebrafish (Danio rerio) via abnormal expression of miR-125 regulated by PKCa/Nrf2/p53 signaling pathways. *Chemosphere, 241*, 125086.

Wang, D., Wang, X., Huang, H., & Wang, H. (2021). Triclosan regulates alternative splicing events of nerve-related genes through RNA-binding protein CELF2 to induce zebrafish neurotoxicity. *J Hazard Mater*, *413*, 125414.

Wang, F., Wang, R., Liu, F., & Wan-guang, C. (2019). Gene expression profiles in brain of male juvenile zebrafish (Danio rerio) treated with triclosan. *Toxicol Appl Pharmacol*, *362*, 35–42. https://doi.org/10.1016/j.taap.2018.10.014.

Wang, W., Li, X., Qian, Q., Yan, J., Huang, H., Wang, X., & Wang, H. (2023). Mechanistic exploration on neurodevelopmental toxicity induced by upregulation of ALKBH5 targeted by triclosan exposure to larval zebrafish. *J Hazard Mater, 457*, 131831.

Wang, Y., Song, J., Wang, X., Qian, Q., & Wang, H. (2022). Study on the toxic-mechanism of triclosan chronic exposure to zebrafish (Danio rerio) based on gut-brain axis. *Sci Total Environ*, *844*, 156936.

Xi, Y., Noble, S., & Ekker, M. (2011). Modeling neurodegeneration in zebrafish. *Curr Neurol Neurosci Rep*, *11*(3), 274–282. https://doi.org/10.1007/s11910-011-0182-2.

# Index